Bioenergetic Processes of Cyanobacteria

Günter A. Peschek • Christian Obinger • Gernot Renger
Editors

Bioenergetic Processes of Cyanobacteria

From Evolutionary Singularity to Ecological Diversity

Springer

Editors
Prof. Dr. Günter A. Peschek
Molecular Bioenergetics Group
Inst. Physikalische Chemie
University of Vienna
Währingerstr. 42, 1090 Wien
Austria
guenter.peschek@univie.ac.at
cyano1@aon.at

Prof. Dr. Gernot Renger
Fak. II, Mathematik/Naturwissenschaften,
Institut für Chemie
Max-Volmer-Laboratorium
TU Berlin
Straße des 17. Juni 135, 10587 Berlin
Germany
gernot.renger@mailbox.tu-berlin.de

Prof. Dr. Christian Obinger
Department für Chemie
Abteilung für Biochemie
Universität für Bodenkultur Wien
Muthgasse 18, 1190 Wien
Austria
christian.obinger@boku.ac.at

ISBN 978-94-024-0488-3 ISBN 978-94-007-0388-9 (eBook)
DOI 10.1007/978-94-007-0388-9
Springer Dordrecht Heidelberg London New York

Printed on acid-free paper

Springer is part of Springer Science+Business Media (www.springer.com)

Dedication: Jack Edgar Myers (1913–2006)

We dedicate this book on **Cyanobacteria** to Jack Edgar Myers (1913–2006), the most influential algal and cyanobacterial physiologist of the twentieth century. I have known Jack's work since I became a graduate student of Robert Emerson and Eugene Rabinowitch in 1956.

Jack was born in a farmhouse in eastern Pennsylvania to Dr. Gary Cleveland Myers and Mrs. Caroline Clark Myers. Jack was one of the most wonderful and jovial persons I have met in my life. He was highly inquisitive and always asked questions very slowly and clearly, and answered questions in a thoughtful and a detailed manner. He also had a great humor, and was always good natured. He was a social drinker and one enjoyed his company and stories. I can count several such occasions at Conferences that we attended together. Jack was a remarkable family man as he not only took care of his family, but that of his brother when he passed away. He was a great Science educator for children all over the World through his simple descriptions of many aspects of Science through his down-to-earth clear articles in "*Highlights for Children*", a magazine that had been started by his parents.

Jack received his BS in 1934 in Chemistry from Juanita College, Huntington, Pennsylvania, his MS from Montana State (Bozeman), and his PhD (in Botany), in 1939, from the University of Minnesota under George Burr. Jack did his postdoctoral work with E.D. MacAlister before his appointment on the faculty of University of Texas in Austin, in 1941. Jack had been recognized with the Charles F. Kettering Award for Excellence in Photosynthesis Research, a Guggenheim Fellowship and an honorary Doctor of Science degree from Juniata College. Jack was a member of the U.S. National Academy of Sciences.

The best way to understand Jack's thinking about *Photosynthesis Research* is to read his thoughtful essays on the conceptual developments in photosynthesis (Myers 1974). His insight into the work and understanding of the *Photosynthetic Unit* concept is seen in his article on the 1932 experiments of Robert Emerson and William Arnold (Myers 1994). In his autobiographical article (Myers 1996), we obtain a deep understanding of Jack as a young man and his scientific pursuits leading to his work up to 1949, some of which I will highlight below. Jack's last article on the evolution of his thoughts in photosynthesis research titled '*In one era and out the other*' (Myers 2002) contains his views on the work of his favorite scientists (William Arnold, Bessel Kok and C. Stacy French).

A detailed tribute of Jack by Brand et al. (2008) must be consulted for understanding the full impact of Jack on the field of photosynthesis research. Here, I focus on some aspects of Jack's work, rather chronologically. Jack's PhD thesis had dealt with effects of high light on the green alga Chlorella (Myers and Burr 1940) and on pigments produced by some green algae grown in darkness (Myers 1940), both were unique discoveries. During his postdoctoral work, he obtained data on wheat showing antiparallel relationship, up to 4-minute illumination, between chlorophyll *a* fluorescence intensity and CO_2 fixation rate (McAlister and Myers 1940). In the 1940s, when Jack was in Texas, detailed and thorough work on the growth and cultivation of algae began: an apparatus for the continuous culture of algae was invented (see Myers and Clark 1944); and carbon and nitrogen balance of algae was investigated (Myers and Johnston 1949). Research on cyanobacteria was in full swing in the 1950s. The paper of Kratz and Myers (1955) on nutrition and growth of cyanobacteria was used by all who were working on cyanobacteria at that time. Myers and Kratz (1955) described thoroughly the pigments and photosynthetic characteristics of cyanobacteria. When I started working on cyanobacteria in late 1950s and early 1960s, these were the first papers I read before growing what was then known as *Anacystis nidulans*. Jack's work on the mass cultivation of algae and cyanobacteria was done with his long-time associate J.R. Graham (see e.g., Myers and Graham 1959).

After the discovery of the Emerson Enhancement Effect in photosynthesis at the University of Illinois at Urbana (in 1957), Jack went to Stacy French's laboratory and, using an oxygen electrode, confirmed the work of Emerson and showed equivalence of the action spectra of the enhancement effect with the chromatic transients discovered by Larry Blinks (Myers and French 1960). This was at the time I was finishing my PhD and had shown that in addition to chlorophyll b (that Emerson and Myers had shown), a short-wavelength form of chlorophyll a was present in the chlorophyll b-containing system (Govindjee and Rabinowitch 1960). Jack then focused more on cyanobacteria, the topic of this book. Fujita and Myers (1965) studied hydrogenase and NADP reduction in the cyanobacterium *Anabaena cylindrica*. Holten and Myers (1967a, b) provided the most thorough research on the many cytochromes in the cyanobacterium *Anacystis nidulans*. I will end describing Jack's research by mentioning his remarkable work with C. Bonaventura, where he thoroughly investigated oxygen evolution and chlorophyll a fluorescence (remember his 1940 work with Macalister, cited above) and discovered the phenomena of opposite effects in the photochemical efficiencies of the two light reactions as caused by changes due to light absorbed by *pigment systems I* or *II*; they had called it *light state 1* or *light state 2* (Bonaventura and Myers 1969); most now call this regulation phenomenon *"state transitions"*, but some call it *"state shifts"* (see Papageorgiou and Govindjee 2011, for a full history).

Jack was a humble man. I was particularly humbled when once he wrote: "other more patient and perceptive (as Govindjee, see Govindjee 1995) would later find gold in the mine". Jack, you have always been much more patient and perceptive that I could ever hope to be, and you certainly have discovered more gold than I could ever hope to see.

References

Bonaventura C, Myers J (1969) Fluorescence and oxygen evolution from *Chlorella pyrenodoisa.* Biochim Biophys Acta 189: 366–383

Brand JJ, Krogmann DW, Patterson CO (2008) Jack Edgar Myers (1913–2006), an algal physiologist par excellence. Photosynth Res 96: 9–14

Fujita Y, Myers J (1965) Hydrogenase and NADP-reduction in a cell-free preparation of *Anabaena cylindrica.* Arch Biochem Biophys 111: 619–625

Govindjee (1995) Sixty-three years since Kautsky: Chlorophyll a fluorescence. Aust J Plant Physiol (*now Functional Plant Biology*) 22: 131–160

Govindjee and Rabinowitch E (1960) Two forms of chlorophyll *a in vivo* with distinct photochemical functions. Science 132: 355–356

Holton RW, Myers J (1967a) Water-soluble cytochromes from a bluegreen alga. I. Extraction, purification and spectral properties of cytochromes c (549, 552 and 554, *Anacystis nidulans*). Biochim Biophys Acta 131: 362–374

Holton RW, Myers J (1967b) Water-soluble cytochromes from a blue-green alga. II. Physiochemical properties and quantitative relationships of cytochromes c (549, 552, and 554, *Anacystis nidulans*). Biochim Biophys Acta 131: 375–384

Kratz WA, Myers J (1955) Nutrition and growth of several blue-green algae. Am J Bot 42: 282–287

McAlister ED, Myers J (1940) Time course of photosynthesis and fluorescence. Science 92: 241–243

Myers J (1940) A study of the pigments produced in darkness by certain green algae. Plant Physiol 15:575–588

Myers J (1974) Conceptual developments in photosynthesis, 1924–1974. Plant Physiol 54: 420–426

Myers J (1994) The 1932 experiments. Photosynth Res 40: 303–310

Myers J (1996) Country boy to a scientist. Photosynth Res 50: 195–208

Myers J (2002) In one era and out the other. Photosynth Res 73: 21–28

Myers J, Burr GO (1940) Some effects of high light intensity on *Chlorella.* J Gen Physiol 24: 45–67

Myers J, Clark LB (1944) An apparatus for the continuous culture of *Chlorella.* J Gen Physiol 28: 103–112

Myers J, French CS (1960) Evidences from action spectra for a specific participation of chlorophyll *b* in photosynthesis. J Gen Physiol 43: 723–736

Myers J, Graham J-R (1959) On the mass culture of algae II. Yield as a function of cell concentration under continuous sunlight irradiance. Plant Physiol 34: 345–352

Myers J, Johnston JA (1949) Carbon and nitrogen balance of *Chlorella* during growth. Plant Physiol 24: 111–119

Myers J, Kratz WA (1955) Relations between pigment content and photosynthetic characteristics in a blue-green alga. J Gen Physiol 39: 11–22

Papageorgiou GC, Govindjee (2011) Photosystem II fluorescence: Slow changes—Scaling from the Past. J Photochem Photobiol B. Biology doi: 10.1016/j.jphotobiol2011.03.008

University of Illinois at Urbana—Champaign
Urbana, Illinois, USA

Govindjee
(gov@life.illinois.edu)

A tribute to Mamoru Mimuro (1949–2011)

Professor Mamoru Mimuro, the principal author of Chapter 9 **"Bioenergetics in a Primordial Cyanobacterium *Gloeobacter violaceus* PCC 7421"**, passed away on February 8, 2011, only few days after he sent back the corrected author's proof.

Three years ago Mamo Mimuro was diagnosed with colon cancer. In spite of the after effects of surgery and chemotherapy, he continued his active research and teaching at Kyoto University. After three months of intravenous feeding at home, he even attended the Master's thesis presentations of his students at the university on February 2, 2011. His untimely death at the age of 61 has left us with a feeling of deep sadness about the loss of a remarkable scientist and a wonderful friend.

Mamo Mimuro started his research on excitation energy transfer in cyanobacteria in 1973, while he was a Master's student in the laboratory of Dr. Yoshihiko Fujita at the Ocean Research Institute of the University of Tokyo. After he moved to the National Institute of Basic Biology at Okazaki in 1980 as an assistant professor, he started a series of studies on time-resolved fluorescence spectral analysis of excitation energy transfer in photosynthesis. In collaboration with the physical chemist, Dr. Iwao Yamazaki, and his coworkers, Mamo made beautiful and fantastic discoveries on the excited singlet state energy transfer between different spectral components of the antenna complexes, using a single photon counting method with very fast time (less than 10 ps) and high spectral resolution (better than 1.5 nm). Kinetic analysis of spectral components among phycobilin pigments in red algae and cyanobacteria revealed that the rise and decay kinetics of spectral components were proportional to the square root of time. This method was applied by Mamo and collaborators to analyze antenna systems of plants and bacteria, especially chlorosomes of the green bacteria, and the presence and relevance of various minor spectral components were reported. His pioneering work on excitation energy transfer in red algae has been recognized by Govindjee in his 2004 review on the basics and history of chlorophyll fluorescence.

After Mamo moved to Yamaguchi University as a professor in 1997, he started the study of a primordial cyanobacterium, *Gloeobacter violaceus,* with interests in early evolution of oxygenic photosynthesis. Details of the *Gloeobacter* studies from his group are reviewed in Chapter 9 of this book. His earlier research on **"Photon Capture, Exciton Migration and Trapping and Fluorescence Emission in Cyanobacteria and Red Algae"** has been beautifully reviewed by Mamo himself in Chapter 7 of the 2004 book "Chlorophyll a Fluorescence: A Signature of Photosynthesis", volume 19 of Advances in Photosynthesis and Respiration, edited by G.C. Papageorgiou and Govindjee, and published by Springer.

Mamo was enthusiastic in stimulating international exchange of information and ideas in the field of photosynthesis. He organized many meetings and symposia. Among these, he organized the "International Symposium on Molecular Structure and Regulation of Photosynthetic Pigment Systems" in 1992 at Sanda, Japan, as a satellite meeting of the IXth International Congress on Photosynthesis. He also took an initiative to invite 11th International Symposium on Photosynthetic Prokaryote in 2003 in Tokyo. In 2007, he organized "7th International conference on the tetrapyrrole photoreceptors in photosynthetic organisms" in Kyoto. In those symposia, Mamo was the key person for the relaxed atmosphere and fruitful discussions.

In the words of Govindjee (University of Illinois at Urbana-Champaign, USA), who speaks on behalf of a very large community of friends and well-wishers of Mamo: "The entire photosynthesis community has lost a great scientist, a great friend, a great human being, and the major pioneer of this decade in the field of excitation energy transfer in photosynthesis. He will be sorely missed by all of us; no one can come close to his breadth of knowledge and dedication."

Tokyo Metropolitan University
Japan

Katsumi Matsuura
(matsuura-katsumi@tmu.ac.jp)

One of the editors of this book (Günter Peschek) has highly enjoyed the privilege to stay in Mamoru Mimuro's laboratory at the University of Kyoto as a visiting coworker of "Mamo" several times during almost a decade from 2000 and 2009, when being on sabbatical leaves from the Institute of Physical Chemistry at the University of Vienna, Austria. Günter not only performed experiments on cyanobacteria in Mamo's group but also joined him in teaching and instructing his graduate students. During these visits Günter had the great chance to enjoy Mamo's extremely hospitable, helpful and friendly personality and, in particular, he will never forget the great welcome and farewell parties that Mamo had arranged each time, together with his whole research group. Everybody who ever got the opportunity to know him personally will keep an everlasting brilliant memory of Professor Mamoru Mimuro. Another editor of this book (Gernot Renger) gratefully remembers enjoyable talks and discussions at international meetings and in particular he was deeply impressed by the extraordinary constructive and exceptionally friendly cooperation when he invited Mamo as the author of a chapter for the book on the *"Primary Processes of Photosynthesis"*.

We all – the colleagues and friends in Japan and around the world – will very much miss the big smile and friendly words of this outstanding person.

University of Vienna
Austria

Günter Peschek
(cyano1@aon.at)

Technical University
Berlin, Germany

Gernot Renger
(gernot.renger@mailbox.tu-berlin.de)

Preface

The evolutionary "invention" of a system that enabled water-splitting into molecular dioxygen and metabolically bound hydrogen was the "Big-Bang" in the development of the biosphere. This event occurred about three billion years ago at the level of cyanobacteria and enriched the hitherto essentially anoxic terrestrial atmosphere with O_2 which had dramatic global consequences: (1) the efficiency of free energy exploitation from food increased by more than a factor of ten through the possibility of aerobic respiration thus opening the "thermodynamic door" for the development and sustenance of all higher forms of life, (2) the ozone layer was formed and provided the essential protective umbrella to UV radiation as prerequisite for populating the land with plants and animals, and (3) the oxidation of minerals gave rise to large-scale changes of the earth crust including geological phenomena. At the same time most of the existing organisms were killed due to oxidative degradation (in geological terms called "the O_2 cataclysm or O_2 catastrophe") unless they were able to develop suitable protective mechanisms or to find anaerobic ecological niches for survival.

Cyanobacteria have effectively colonized our whole planet. They are rather flexible in adaptation to a great variety of different environments and even capable of occupying arctic water bodies, or hot (alkaline) springs by forming the significant family of thermophilic organisms. A unique and common characteristics of all species is their "dual nature" of a prokaryotic (bacterial) cell structure and an O_2-evolving photosynthesis which is typical for (green) plants. Accordingly, the cyanobacteria can be classified in two ways, i.e. either as blue-green or *cyano*-bacteria or as (prokaryotic) blue-green algae (plants). In the past, both terms have been used but nowadays the term cyanobacteria—introduced about 50 years ago by the leading French-Canadian microbiologist Roger Y. Stanier—is universally accepted in the biological literature of nomenclature and systematics.

Due to the paramount importance for our entire biosphere it is not surprising and highly justified that several books and monographs on cyanobacteria have been published in recent years, and specialized scientific meetings and congresses exclusively or predominantly devoted to cyanobacteria are being regularly organized. Therefore the question arises: Why another book on this subject? Mainly two reasons should be mentioned: (a) A book is missing which focuses on *bioenergetics* of cyanobacteria primarily addressing questions of energy conversion by the fundamental bioenergetic processes: (Oxygenic) photosynthesis, (aerobic) respiration, and (anaerobic) fermentation which uniquely occur together in these prokaryotic cells, and (b) thermophilic cyanobacteria offer the most suitable material for high resolution structure analyses of Photosystem I and II and other electron transport complexes by X-ray crystallography (for example, at present the structure of Photosystem II at atomic resolution is only known for these organisms). These achievements during the last decade represent a milestone in our understanding of the complexes which are crucial for solar energy exploitation through photosynthetic water splitting.

Based on these considerations we feel confident that it is worth publishing this book and sincerely hope that it will find a positive resonance. It represents an ambitious attempt to achieve the goal of a synoptic state-of-the-art picture on the bioenergetics of cyanobacteria by casting together the mosaics of detailed knowledge described by leading experts in the field. It contains 24 chapters written by 51 authors from Austria, Finland, France, Germany, India, Israel, Japan, Netherlands, Portugal, Spain, UK and USA. The book is aimed at reaching a broad audience ranging from students to experienced scientists in chemistry, physics, biology and physiology. It is devided into seven parts: **Part I** offers in seven chapters a general description of cyanobacteria and their environment, with occasional historic and philosophic excursions into the phenomenon of life as such, the living cell and our universe. **Part II** contains three chapters on the history and function of electron transport and ATP-synthesis including brand-new results on *Gloeobacter*. Details on the oxygenic photosynthesis and aerobic respiration are outlined in **Part III** by presenting five chapters. **Part IV** focuses in two chapters on electron entry systems. **Part V** addresses in five chapters the connection between the various electron transfer complexes, and **Part VI** in three chapters reactions of terminal oxidation. The book finishes in **Part VII** by a single chapter which describes most recent progress with tools for genetic manipulation of cyanobacteria.

We have many people to thank. First of all, the authors for their efforts to offer the readers excellent chapters and for their positive response to our suggestions. Without their invaluable cooperation there would be no book. The book has grown out from a pertinent discussion of Govindjee with one of us (G.A.P.) in March 2007 and the editors want to express their particular thanks to Govindjee for his untiring encouragement during the years. We also gratefully acknowledge the editorial staff of Springer, especially Dr. Jacco Flipsen and Mrs. Ineke Ravesloot, for their generous help in all respects. Last, but not the least, we also thank Martin Pairer, BOKU—University of Natural Resources and Life Sciences, Vienna, who has taken the utmost care in achieving a uniform style of all manuscripts and to satisfy all editorial requirements set by Springer, Dordrecht. We made all efforts to reach the goal of presenting a book which may significantly contribute to the deepening and broadening of scientific knowledge and interest in cyanobacteria.

We wish all the readers a pleasant and stimulating journey through the fascinating "world" of the bioenergetics of cyanobacteria. It is our sincere hope that this book will entice young people into this exciting research area with the aim to address successfully the challenging problems of high relevance that are still waiting for satisfactory answers.

The Editors:

- Günter A. Peschek (address: Molecular Bioenergetics Group, Institute of Physical Chemistry, University of Vienna, Währingerstrasse 42, A-1090 Wien, Austria, Phone: +43-1-4277-52430; E-mail: guenter.peschek@univie.ac.at and cyano1@aon.at; homepage: http://homepage.univie.ac.at/Guenter.Peschek)
- Christian Obinger (address: Vienna Institute of BioTechnology, Department of Chemistry and Biochemistry, BOKU—University of Natural Resources and Life Sciences, Muthgasse 18, A-1190 Vienna, Austria, Phone: +43-1-47654-6073; Fax: +43-1-47654-6059; E-mail: christian.obinger@boku.ac.at) and
- Gernot Renger (address: Technische Universität Berlin, Institut für Chemie, Max-Volmer-Laboratorium für Biophysikalische Chemie PC14, Strasse des 17. Juni 135, D-10623 Berlin, Germany, Fax: +49-(0)30 314 211 22; E-mail: gernot.renger@mailbox.tu-berlin.de)

Austria
Germany

Günter A. Peschek, Christian Obinger and Gernot Renger

Contents

Dedication: Jack Edgar Myers (1913–2006) v

A tribute to Mamoru Mimuro (1949–2011) ix

Preface xiii

Contributors xxi

About the Editors xxv

Abbreviations xxix

Part I Cyanobacteria and Their Environment

1 **Life Implies Work: A Holistic Account of Our Microbial Biosphere Focussing on the Bioenergetic Processes of Cyanobacteria, the Ecologically Most Successful Organisms on Our Earth** 3
Günter A. Peschek, Margit Bernroitner, Samira Sari, Martin Pairer and Christian Obinger

2 **The Photosynthetic Apparatus of the Living Fossil, *Cyanophora paradoxa*** 71
Jürgen M. Steiner and Wolfgang Löffelhardt

3 **An Alternate Hypothesis for the Origin of Mitochondria** 89
Roschen Sasikumar, Jijoy Joseph and Günter A. Peschek

4 **The Complex Regulation of the Phosphate Uptake System of Cyanobacteria** 109
Gernot Falkner and Renate Falkner

5 **The Site of Respiratory Electron Transport in Cyanobacteria and Its Implication for the Photoinhibition of Respiration** 131
Ioan I. Ardelean and Günter A. Peschek

6 **Nitrogenases and Hydrogenases in Cyanobacteria** 137
Hermann Bothe, Oliver Schmitz, M. Geoffrey Yates and William E. Newton

7 **Hydrogen Peroxide Degradation in Cyanobacteria** 159
Marcel Zamocky, Margit Bernroitner, Günter A. Peschek and Christian Obinger

Part II Cyanobacteria—The First Electric Power Plants ("Discovery" of Biological Electron Transport, the Clue to Most Efficient Bioenergetic Processes)

8 **History and Function: The Respiratory and Photosynthetic Electron Transport Chains** 189
Peter Nicholls

9 **Bioenergetics in a Primordial Cyanobacterium *Gloeobacter violaceus* PCC 7421** 211
Mamoru Mimuro, Tohru Tsuchiya, Kohei Koyama and Günter A. Peschek

10 **ATP Synthase: Structure, Function and Regulation of a Complex Machine** 239
Dirk Bald

Part III Cyanobacteria and Light: Oxygenic Photosynthesis (and Other Uses of Light)

11 **The Evolution of Cyanobacteria and Photosynthesis** 265
Gerhart Drews

12 **Structure of Cyanobacterial Photosystems I and II** 285
Petra Fromme and Ingo Grotjohann

13 **Mechanism of Photosynthetic Production and Respiratory Reduction of Molecular Dioxygen: A Biophysical and Biochemical Comparison** 337
Gernot Renger and Bernd Ludwig

14 Photoprotection in Cyanobacteria: The Orange Carotenoid Protein and Energy Dissipation 395
Cheryl A. Kerfeld and Diana Kirilovsky

15 Mn Transport and the Assembly of Photosystem II 423
Eitan Salomon, Gernot Renger and Nir Keren

Part IV Electron Entry (Dehydrogenation)

16 Structure and Physiological Function of NDH-1 Complexes in Cyanobacteria 445
Natalia Battchikova, Eva-Mari Aro and Peter J. Nixon

17 The Superfamily of Succinate:Quinone Oxidoreductases and its Implications for the Cyanobacterial Enzymes 469
C. Roy D. Lancaster

Part V Connecting (−) and (+): Electron and Protein Circuits Between Dehydrogenases and Oxygen Reductases in Cyanobacteria

18 The Water-Soluble Cytochromes of Cyanobacteria 515
Kwok Ki Ho, Cheryl A. Kerfeld and David W. Krogmann

19 Transient Interactions Between Soluble Electron Transfer Proteins. The Case of Plastocyanin and Cytochrome *f* 541
Derek S. Bendall, Beatrix G. Schlarb-Ridley and Christopher J. Howe

20 Center of the Cyanobacterial Electron Transport Network: The Cytochrome b_6f Complex 573
Gábor Bernát and Matthias Rögner

21 The Convergent Evolution of Cytochrome c_6 and Plastocyanin Has Been Driven by Geochemical Changes 607
Miguel A. De la Rosa, José A. Navarro and Manuel Hervás

22 Flavodiiron Proteins and Their Role in Cyanobacteria 631
Vera L. Gonçalves, João B. Vicente, Lígia M. Saraiva and Miguel Teixeira

Part VI Electron Exit (Terminal Oxidation)

23 Cyanobacterial Respiratory Electron Transport: Heme-Copper Oxidases and Their Electron Donors 657
Margit Bernroitner, Marcel Zamocky, Martin Pairer, Günter A. Peschek and Christian Obinger

Part VII Progress in the Genetic Manipulation of Cyanobacteria

24 Tools for Genetic Manipulation of Cyanobacteria 685
Annegret Wilde and Dennis Dienst

Index 705

Contributors

Ioan I. Ardelean Molecular Bioenergetics Group, Institute of Physical Chemistry, University of Vienna, UZA2 Althanstrasse 14, 1090 Vienna, Austria
e-mail: ioan.ardelean@ibiol.ro

Eva-Mari Aro Department of Biochemistry and Food Chemistry, Molecular Plant Biology, University of Turku, 20520 Turku, Finland
e-mail: evaaro@utu.fi

Dirk Bald Department of Molecular Cell Biology, Faculty of Earth- and Life Science, VU University Amsterdam, De Boelelaan 1085, 1081 HV Amsterdam, The Netherlands
e-mail: dirk.bald@falw.vu.nl

Natalia Battchikova Department of Biochemistry and Food Chemistry, Molecular Plant Biology, University of Turku, 20520 Turku, Finland

Derek S. Bendall Department of Biochemistry, University of Cambridge, Downing Site, Cambridge, CB2 1QW, UK
e-mail: dsb4@mole.bio.cam.ac.uk

Gábor Bernát Plant Biochemistry, Ruhr-University Bochum, 44780 Bochum, Germany
e-mail: gabor.bernat@rub.de

Margit Bernroitner Metalloprotein Research Group, Vienna Institute of Bio-Technology, Department of Chemistry, Division of Biochemistry, BOKU-University of Natural Resources and Life Sciences, Muthgasse 18, 1190 Vienna, Austria

Hermann Bothe Botanical Institute, The University of Cologne, Zuelpicherstr. 47b, 50923 Köln, Germany
e-mail: hermann.bothe@uni-koeln.de

Miguel A. De la Rosa Instituto de Bioquímica Vegetal y Fotosíntesis, Universidad de Sevilla & CSIC, Américo Vespucio 49, 41092 Sevilla, Spain
e-mail: marosa@us.es

Dennis Dienst Institute of Biology, Humboldt-University Berlin, Chausseestr. 117, 10115 Berlin, Germany

Gerhart Drews Institute of Biology 2, Microbiology, Albert-Ludwigs-Universität, Schänzlestrasse 1, 79104 Freiburg, Germany
e-mail: gerhart.drews@biologie.uni-freiburg.de

Gernot Falkner Cell Biology Department, Plant Physiology Division, Neurodynamics and Signaling Group, University of Salzburg, Hellbrunnerstr. 34, 5020 Salzburg, Austria
e-mail: gernot.falkner@sbg.ac.at

Renate Falkner Plant Physiology Division, Cell Biology Department, Neurodynamics and Signaling Group, University of Salzburg, Hellbrunnerstr. 34, 5020 Salzburg, Austria

Petra Fromme Department of Chemistry and Biochemistry, Arizona State University, PO Box 871604, Tempe, AZ 85287-1604, USA
e-mail: pfromme@asu.edu

Vera L. Gonçalves Instituto de Tecnologia Química e Biológica, Universidade Nova de Lisboa, Av. da República, 2780-157 Oeiras, Portugal

Ingo Grotjohann Department of Chemistry and Biochemistry, Arizona State University, PO Box 871604, Tempe, AZ 85287-1604, USA
e-mail: ingo.grotjohann@asu.edu

Manuel Hervás Instituto de Bioquímica Vegetal y Fotosíntesis, Universidad de Sevilla & CSIC, Américo Vespucio 49, 41092 Sevilla, Spain

Kwok Ki Ho Department of Biochemistry, Purdue University, West Lafayette, IN 47907-2063, USA

Christopher J. Howe Department of Biochemistry, University of Cambridge, Downing Site, Cambridge CB2 1QW, UK

Jijoy Joseph CSIR, Thiruvananthapuram, Kerala, India
e-mail: jijoyjos@yahoo.com

Nir Keren Department of Plant and Environmental Sciences, The Silberman Institute of Life Sciences, The Hebrew University of Jerusalem, Edmond Safra Campus, Givat-Ram, Israel
e-mail: nirkeren@vms.huji.ac.il

Cheryl A. Kerfeld United States Department of Energy, Joint Genome Institute, 2800 Mitchell Drive, Walnut Creek, CA 94598, USA
e-mail: ckerfeld@lbl.gov
Department of Plant and Microbial Biology, University of California, Berkeley, CA 94720, USA
e-mail: kerfeld@mbi.ucla.edu

Diana Kirilovsky Institut de Biologie et Technologies de Saclay (iBiTec-S), Commissariat à l'Energie Atomique (CEA), CEA Saclay, 91191 Gif sur Yvette, France

URA 2096, Center National de la Recherche Scientifique (CNRS), CEA Saclay, 91191 Gif sur Yvette, France

Kohei Koyama Graduate School of Human and Environmental Studies, Kyoto University, Kyoto 606-8501, Japan

David W. Krogmann Department of Biochemistry, Purdue University, 175 S. University Street, West Lafayette, IN 47907-2063, USA
e-mail: krogmann@purdue.edu

C. Roy D. Lancaster Faculty of Medicine, Department of Structural Biology, Institute of Biophysics, Saarland University, Building 60, 66421 Homburg, Germany
e-mail: roy.lancaster@structural-biology.eu

Wolfgang Löffelhardt Max F. Perutz Laboratories, Department of Biochemistry, University of Vienna, Dr. Bohrgasse 9, 1030 Vienna, Austria
e-mail: wolfgang.loeffelhardt@univie.ac.at

Bernd Ludwig Institute Biochemistry, Biocenter of Goethe University Inst, Max-von-Laue-Str. 9, 60438 Frankfurt, Germany

Mamoru Mimuro Graduate School of Human and Environmental Studies, Kyoto University, Kyoto 606-8501, Japan
e-mail: mamo-mi@mm1.mbox.media.kyoto-u.ac.jp

José A. Navarro Instituto de Bioquímica Vegetal y Fotosíntesis, Universidad de Sevilla & CSIC, Américo Vespucio 49, 41092 Sevilla, Spain

William E. Newton Department of Biochemistry, Virginia Polytechnic Institute & State University, Blacksburg, VA 24061, USA

Peter Nicholls Department of Biological Sciences, University of Essex, Colchester, Essex, CO4 3SQ, UK
e-mail: pnicholl@essex.ac.uk

Peter J. Nixon Department of Life Sciences, Biochemistry Building, Imperial College London, S. Kensington Campus, London, SW7 2AZ, UK
e-mail: p.nixon@imperial.ac.uk

Christian Obinger Vienna Institute of BioTechnology, Department of Chemistry, Division of Biochemistry, BOKU—University of Natural Resources and Life Sciences, Muthgasse 18, 1190 Vienna, Austria
e-mail: christian.obinger@boku.ac.at

Martin Pairer Metalloprotein Research Group, Department of Chemistry, Division of Biochemistry, BOKU-University of Natural Resources and Life Sciences, Muthgasse 18, 1190 Vienna, Austria

Günter A. Peschek Molecular Bioenergetics Group, Institute of Physical Chemistry, University of Vienna, UZA2 Althanstrasse 14, 1090 Vienna, Austria
e-mail: guenter.peschek@univie.ac.at, cyano1@aon.at

Gernot Renger Max-Volmer-Laboratorium für Biophysikalische Chemie, Technische Universität Berlin, Berlin, Germany
e-mail: gernot.renger@mailbox.tu-berlin.de

Matthias Rögner Plant Biochemistry, Ruhr-University Bochum, 44780 Bochum, Germany

Eitan Salomon Department of Plant and Environmental Sciences, The Silberman Institute of Life Sciences, The Hebrew University of Jerusalem, Edmond Safra Campus, Givat-Ram, Israel

Beatrix G. Schlarb-Ridley Department of Biochemistry, University of Cambridge, Downing Site, Cambridge CB2 1QW, UK

Lígia M. Saraiva Instituto de Tecnologia Química e Biológica, Universidade Nova de Lisboa, Av. da República, Oeiras 2780-157, Portugal

Samira Sari Molecular Bioenergetics Group, Institute of Physical Chemistry, University of Vienna, UZA2 Althanstrasse 14, 1090 Vienna, Austria

Roschen Sasikumar CSIR, Thiruvananthapuram, India
e-mail: roschen.csir@gmail.com

Oliver Schmitz Johannes-Brahms-Str.16, 14624 Dalgow-Döberitz, Germany

Jürgen M. Steiner Department of Plant Physiology, Martin-Luther-University Halle-Wittenberg, FRG, 06120 Halle (Saale), Germany

Miguel Teixeira Instituto de Tecnologia Química e Biológica, Universidade Nova de Lisboa, Av. da República, 2780-157 Oeiras, Portugal
e-mail: miguel@itqb.unl.pt

Tohru Tsuchiya Graduate School of Human and Environmental Studies, Kyoto University, 606-8501 Kyoto, Japan

João B. Vicente Instituto de Tecnologia Química e Biológica, Universidade Nova de Lisboa, Av. da República, 2780-157 Oeiras, Portugal

Annegret Wilde Institute of Microbiology and Molecular Biology, Justus-Liebig-University Giessen, Heinrich-Buff-Ring 26-32, 35392 Giessen, Germany
e-mail: annegret.wilde@mikro.bio.uni-giessen.de

M. Geoffrey Yates Kingston Ridge, Fir Trees, Kingston, Lewes, Sussex BN7 3JU, UK

Marcel Zamocky Vienna Institute of BioTechnology, Department of Chemistry, Division of Biochemistry, BOKU-University of Natural Resources and Life Sciences, Muthgasse 18, 1190 Vienna, Austria

Institute of Molecular Biology, Slovak Academy of Sciences, Dubravska cesta 21, 84551 Bratislava, Slovakia

About the Editors

Günter A. Peschek was born in Linz on Danube, Austria, in January 1944, five months after his father had died in action (World War II) aboard a grounded submarine. Since 1952, the second husband of his mother had proved to be his true and loving father up to his sudden death in January 2008. Having passed all examinations in Elementary School (1950–1954 in Linz), High School (1954–1963) and at the University of Vienna with the best marks possible, and having finished his doctoral thesis ("The bioenergetic processes of the blue-green alga *Anacystis nidulans*") under the supervision of the renowned physicochemist and evolutionary biologist Professor E. Broda at the University of Vienna (1971–1976), Günter graduated with Ph.D. (in Chemistry) in 1976 "*Sub auspiciis praesidentis rei publicae Austriae*" (known to be still a little bit above *summa cum*) followed by a career of Broda's *ad personam* University Assistant until 1980, habilitation ("*venia legendi*") in 1980 based on his thesis on "The role of hydrogenases in *Anacystis nidulans*". Günter was appointed a Professor of Biophysical Chemistry in 1987. In October 2009, he retired from official University duties. His academic awards include: special Ph.D. graduation by the University of Vienna in the presence of, and "decorated" by, the President of the Federal Republic of Austria, Dr. Rudolf Kirchschläger, special science award by the county of Upper Austria ("Talentförderungsprämie") in 1975, four research awards by the City of Vienna between 1976 and 1983, and a major Austrian Science Award ("Kulturpreis") in 1983. Günter has worked all his life with blue-green algae (cyanobacteria); he believes that these intriguing organisms have a unique position in our world, and they merit a much better and more firmly accepted position both in academic teaching as well as in research. In his attempt to reach this goal, he has organized seven major international scientific meetings (among them the 7th International Symposium on Phototrophic Prokaryotes in Vienna, 1997, and the European Research Conferences on the Molecular Bioenergetics of Cyanobacteria in Gmunden, Austria, 1999, a series of highly successful conferences which, since then are still taking place, on a bi- to tri-ennial basis up to the present day). Günter has also worked in several renowned research groups overseas for several months, each, e.g., the Biochemistry Department of the Weizmann Institute of Science, Rehovot, Israel (with the late Mordhay Avron, in 1982), Institute of Molecular Biology, Berkeley, California (with Lester Packer, 1984, 1989), National Institute of Basic Biology, Okazaki, Japan (with Norio Murata, 1987), Biological Laboratories, Harvard University, USA (with the late Lawrence Bogorad, 1989), Department of Biology, Brock University, Canada (with Peter Nicholls, 1991), Banaras Hindu University, India (with L. C. Rai and J. P. Gaur, 2007), and University of Kyoto (with Mamoru Mimuro, 2000, 2007, 2008 and 2009). He has hosted, in his laboratory, many leading cyanobacteriologists from all over the world. More than 90 graduate students have worked with him at the University of Vienna; he has published 350 refereed papers in international scientific journals and has delivered more than 150 invited lectures worldwide. (See http://homepage.univie.ac.at/Guenter.Peschek).

Christian Obinger was born in Salzburg, Austria, in 1962. He obtained his degree in Chemistry and Biochemistry in 1988 and completed his Ph.D. in Günter Peschek's lab at the Institute of Physical Chemistry at the University of Vienna, Austria (1992). In the same year he obtained a position as research and teaching assistent at the Department of Chemistry at BOKU—University of Natural Resources and Life Sciences in Vienna. In 1995 he did postdoc training with Peter Nicholls at the Brock University in Canada and got his habilitation in biochemistry in 1999. His awards include the Novartis award for Biochemistry (2000). At the moment he is Associate Professor in Biochemistry at the Department of Chemistry at BOKU—University of Natural Resources and Life Sciences and heads the Metalloprotein Research Group. He is author of more than 120 science publications in international journals in the field of bioinorganic chemistry and leads the International PhD-program on Biomolecular Technology of Proteins—BioToP.

The main objective of his research is to understand the structural basis of metalloprotein functions, with particular interest in iron and copper-containing oxidoreductases. He uses the tools of recombinant protein expression and mutagenesis in combination with biophysical methods (electron paramagnetic resonance spectroscopy, resonance Raman spectroscopy, UV-Vis- and fluorescence spectroscopy, circular dichroism spectroscopy, differential scanning calorimetry, multi-mixing stopped-flow spectroscopy, spectroelectrochemistry and X-ray crystallography) for elucidation of structure and function of heme peroxidases and catalases as well as of components of the cyanobacterial respiratory chain downstream of plastquinol. His research includes detailed characterization of redox intermediates as well as elucidation of the role of post-translational modifications in prosthetic groups and in the protein matrix that are relevant for catalysis of metalloproteins.

Gernot Renger was Professor of Physical Chemistry at the Technical University in Berlin for 22 years. After his retirement in 2003, he has continued to do research on the *Primary Processes of Photosynthesis*.

Gernot was born in 1937 in Rumburg (North Bohemia). As a consequence of World War II, the family had to leave the home town and moved to a small village near Görlitz in Eastern Germany (former German Democratic Republic). He received the degree of an engineer in Leipzig (1959) and worked in a company of chemical industry. In 1960 he left East Germany, studied at the Technical University of Berlin and graduated with a diploma in Chemistry (1966). In 1965 he joined the group of Horst Tobias Witt, an international authority in the field of photosynthesis The atmosphere in Witt's group that had a number of young ambitious and brilliant scientists was outstanding and highly stimulating. In 1969, Gernot finished his PhD thesis "*Investigations on the system of photosynthetic water splitting in photosynthesis*". This topic turned out to be so fascinating that Gernot remained "engaged" with the challenging secrets of this unique machinery for the rest of his scientific life. He found that lipophilic uncouplers selectively destabilize intermediates of the redox sequence in the water oxidizing complex (WOC), designated as ADRY (Acceleration of the Deactivation of the Reactions of the water splitting enzyme system Y) effect (1972) and later proposed a model of oxidative water splitting where the essential O–O bond is formed at the level of a complexed peroxide (1978). In cooperation with late Christoph Wolff the recombination reaction between $P680^{+\bullet}$ and $Q_A^{-\bullet}$ was discovered (1976). Gernot's finding that DCMU type herbicides bind to a protein, which can be degraded by mild trypsin treatment (1976) triggered the discovery of the "herbicide-binding protein" D1 by Marvin Edelman (1978) and numerous studies on both herbicide binding and the origin of a great variety of herbicide resistant mutants by Charles Arntzen and co-workers.

In 1980, Gernot was appointed as Professor at the Technical University in Berlin; his research group tackled different facets of Photosystem II (PSII): discovery of UV absorption changes reflecting the turnover of the WOC (1982; this work was done independently of the 1981 research of Bruno Velthuys); coupling of electron and proton transport in $P680^{+\bullet}$ reduction by tyrosine Y_Z (1988); first detailed analyses of the reaction coordinates (activation energies, kinetic isotope exchange effects) of the WOC. Further, Gernot's group worked on excitation energy transfer, exciton trapping and light stress (the long lasting fruitful and friendly cooperation with laser specialist Hans Joachim Eichler and his group from the Institute of Optics and Atomic Physics of Technical University was key to the success of this research). Neutron scattering is now used in combination with fluorescence studies to unravel relations between protein dynamics and PS II electron transport.

Gernot's scientific work is documented in almost 400 papers that include many invited reviews and book chapters. He is the editor of a two volume book set on the primary processes of photosynthesis (2008) and a co-editor of two books on photosynthesis. He organized several meetings on photosynthesis. In 1996, he was chairman and organizer of a European Research Conference on Biophysics of Photosynthesis and in 1997 he had the honor of becoming *Eminent Scientist* of RIKEN (Rikagaku Kenkyushu, Japan).

Gernot has always been very appreciative of all of his students, colleagues and friends from all over the world who joined the interesting journey through the wonderful "nanoscale world" of the water splitting machinery of photosynthesis. Last but not the least, he has been most grateful to his wife Eva for her constant support over more than four decades.

Abbreviations

ΔpH_T	pH gradient across the thylakoid membrane
Δp	Electrochemical proton potential (proton motive force) across a membrane
$\Delta\psi$	Electrical proton potential across a membrane
μ	Growth rate
μ_{max}	Maximal growth rate
$^1(RC\text{-}PC)^*$	^{1}P680*, excited singlet state of RC pigments
ADP	Adenosine diphosphate
APC	Allophycocyanin
ARTO	Alternative terminal oxidase
ATP	Adenosine triphosphate
BChl	Bacteriochlorophyll
BN/SDS-PAGE	2-Dimensional electrophoresis consisting of BN-PAGE in the first dimension followed by SDS-PAGE in the second
BNC	Binuclear center in terminal oxidases
BN-PAGE	1-Dimensional blue-native polyacrylamide gel electrophoresis
bp	Base pair
Bphe	Bacteriopheophytin
Brs	Complex of barnase and barstar
BYA	Billion years ago
Car	Carotenoid
Cc_6	Cytochrome c_6
CCCP	Carbonyl cyanide m-chloro-phenylhydrazone
C*c*O	Cytochrome *c* oxidase
ccpc	Complex of yeast cytochrome *c* peroxidase and *iso*-1-cytochrome *c*
C*f*	Cytochrome *f*
CGR	Chaos game representation
Chl	Chlorophyll
Chl a	Chlorophyll a
Chl_{D1}, Chl_{D2}	"Accessory" chlorophylls of the D1- and D2-branches, respectively of the RC

Chls	Chlorophylls
C_i	Inorganic carbon
cIP	Calculated isoelectric point
CM	Cytoplasmic *or* plasma *or* cell membrane(s)
COX	Cyt *c* oxidase
coxBAC	Operon encoding COX subunits II, I and III;
CP43, CP47	Core antenna subunits of PS II
CUP-A module	Portion of NDH-1MS containing NdhD3, NdhF3, CupA and CupS
CUP-B module	Portion of NDH-1MS′ containing NdhD4, NdhF4 and CupB
cyt	Cytochrome
Cyt *b*559	Cytochrome *b* 559
Cyt b_6f	Cytochrome b_6f complex
Cyt c_6 (CYT c_6)	Cytochrome c_6
Cyt c_M (CYT c_M)	Cytochrome c_M
Da	Dalton
DBMIB	2,5-Dibromo-3-methyl-6-isopropyl-*p*-benzoquinone
DCMU	3-(3,4-Dichlorophenyl)-1,1-dimethylurea
DFT	Density functional theory
DiHCCP	di-Heme cytochrome *c* peroxidase
DM	n-Dodecyl β-D-maltoside
EC	Energy charge
E_m^{work}	"Working" midpoint potential
EDTA	Ethylene diamine tetraacetic acid
EET	Excitation energy transfer
E_m	Midpoint redox potential
$E_{M6.5}$	Oxidation–reduction midpoint potential at pH 6.5
E_{M7}	Oxidation–reduction midpoint potential at pH 7
ENDOR	Electron-nuclear double resonance
EPR	Electron paramagnetic resonance
ER	Endoplasmic reticulum
ESA	Excited-state-absorption band
ET	Electron transfer *or* Electron transport
ETC	Electron transport chain
ETP	Electron transport phosphorylation
EXAFS	Extended X-ray absorption fine structure
FAD	Flavin adenine dinucleotide
Fd	Ferredoxin
FDP	Flavodiiron protein
Fe-S (FeS)	Iron–sulfur center
FlRd	Flavorubredoxin
FMN	Flavin mononucleotide
FNR	Ferredoxin-$NADP^+$ reductase
FQR	Fd:PQ oxidoreductase

FRET	Förster Resonance Energy Transfer
FTIR	Fourier transform infrared
Gfp	Green fluorescent protein
GOE	Great oxidation event
GOI	Gene of interest
H^+_T, H^+_C	Protons in the thylakoid and cytoplasmic space, resp.
heme b_H and b_L	*b*-Type hemes with high and low midpoint redox potential
heme c_i	*c*-Type heme located between heme b_L and Q_i in the Cyt b_6f complex
HL	High light
HMWC	High molecular weight compounds
ICM	Intracytoplasmic *or* thylakoid membrane(s)
IP	isoelectric point
J_p	Phosphate uptake rate
K'	Equilibrium constant of polyphosphate formation from phosphate under the prevailing external conditions
KatG	Catalase-peroxidase
katG	Gene encoding catalase-peroxidase
K_M	Michaelis constant
K_s	Concentration of the growth limiting substrate at which the growth rate is $\mu_{max}/2$
K_{sp}	Solubility-product constant
L. P.	Low potential
L'	Proportionality factor
LHC	Light harvesting complex
LHCs	Light-harvesting-complexes
LMWC	Low molecular weight compounds
L_p, L	Phenomenological coefficient of the linear and non linear flow-force relationship
m	Exponential number characterizing the non linear dependence of the uptake rate on the driving force of this process
Ma	million years before present
Mbp	Million base pair
$M_jL_kW_l$	Detailed symbol for redox states S_i of the WOC
MK and MQ	Menaquinone
MKH_2	Menaquinol
MnCat	Manganese catalase-peroxidase
mRNA	Messenger RNA *or* transcript(s)
MS-EPT	Multiple site electron and proton transfer
NAD(P)H	Nicotinamide adenine dinucleotide (phosphate), reduced form
NCBI	National Center for Biotechnology Information
ncRNA	Non-coding RNA
NDH	NADH-dehydrogenase
NDH/Ndh	NDH-1 complex found in chloroplasts

NDH-1	Bacterial NADH:quinone oxidoreductase or type I NADH dehydrogenase
NDH-1L	Large NDH-1 complex
NDH-1M	Medium NDH-1 complex
NDH-1MS	Protein complex composed of NDH-1M and NDH-1S
NDH-1MS′	Protein complex composed of NDH-1M and NDH-1S’
NDH-1S	Small NDH-1 complex containing NdhD3, NdhF3, CupA and CupS
NDH-1S′	Small NDH-1 complex containing NdhD4, NdhF4 and CupB
NET	Nonadiabatic electron transfer
n_P	Number of protons passing the thylakoid membrane per incorporated phosphate molecule
NPQ	Nonphotochemical quenching
NQNO	2-*n*-nonyl-4-hydroxyquinoline-*N*-oxide
NRO	NADH:rubredoxin oxidoreductase
N-side	Negatively charged membrane side.
OCM	Cyanobacterial outer membrane
OCP	Orange carotenoid protein
OEC	Oxygen-evolving complex
OM	Outer membrane
ORF	Open reading frame
OSCP	Oligomycin sensitivity conferring protein
P	Product
$P680^{+\bullet}$	Oxidized state of the RC pigments of PS II
P700	The primary electron donor of PS I
PAL	Present Atmospheric Level
PB	Phycobilisomes
PBRC	RC of purple bacteria
PBS	Phycobilisome
Pc	Plastocyanin
PC	phycocyanin
PCC	Pasteur culture collection
pcf	Complex of cytochrome *f* and plastocyanin from higher plants
P_{D1}, P_{D2}	“Special pair” chlorophylls of the D1- and D2-branches, respectively, of the RC of PS II
PDB	Protein data bank
P_e	External phosphate
PE	Phycoerythrin
PEC	Phycoerythrocyanin
pers. comm.	Personal communication
PET	Photosynthetic electron transport
petE	Gene encoding plastocyanin
petJ	Gene encoding cytochrome c_6;
$Pheo_{D1}$, $Pheo_{D2}$	Pheophytins of the D1- and D2-branches, respectively, of the RC of PS II

pI	Isoelectric point
P_i	Cytoplasmic phosphate
plpcf	Complex of cytochrome *f* and plastocyanin from the cyanobacterium *Phormidium laminosum*
PLS	Proton loading site
PM	Plasma membrane
P_M	Mixed valence state of COX
PMF	Proton motive force
P_n, P_{n+1}	Polyphosphates
PQ	Plastoquinone
PQH_2	Plastoquinol
Prx	Peroxiredoxin
PS	Photosytem
PS I	Photosystem I
PS II	Photosystem II
PSA	Photosynthetic apparatus
P-side	Positively charged membrane side
PT	Proton transfer
$[P_e]_A$	Threshold value, i.e. external phosphate concentration at which net uptake ceases
PTOX	Plastid terminal oxidase
Q	Ubiquinone
Q_A, Q_B	Plastoquinones of the RC of PS II
qE	High-energy quenching
QFR	Quinol:fumarate reductase
QH_2	Ubiquinol
Q_i and Q_o site	Plastoquinone/quinol binding sites in the Cyt b_6f complex
QOX	Quinol oxidase
qoxBAC	operon encoding QOX subunits II, I and III
RC	Reaction center
RET	Respiratory electron transport
ROO	Rubredoxin:oxygen oxidoreductase
ROS	Reactive oxygen species
RRL	Remaining respiration in light (% of dark respiration)
S	Substrate
SDH	Succinate dehydrogenase
SLP	Substrat level phosphorylation
Sq	Semiquinone
SQDG	Sulfoquinovosyl diacylglycerol
SQOR	Succinate:quinone oxidoreductase
SQR	Succinate:quinone reductase
SU	Subunit
SU-I/II	Subunit I/II
TDS	Tridecylstigmatellin
Tfp	Type IV pili

TM	Thylakoid membrane
TMH	Transmembrane helix
TRO	Terminal respiratory oxidase
WOC	Water oxidizing complex
WT	Wild type
X	Driving force for the incorporation of phosphate into the polyphosphate pool
XANES	X-ray absorption near edge spectroscopy
XRDC	X-ray diffraction crystallography
Y_Z, Y_D	Redox active tyrosines of the D1- and D2-branches, respectively

Part I
Cyanobacteria and Their Environment

Chapter 1
Life Implies Work: A Holistic Account of Our Microbial Biosphere Focussing on the Bioenergetic Processes of Cyanobacteria, the Ecologically Most Successful Organisms on Our Earth

Günter A. Peschek, Margit Bernroitner, Samira Sari, Martin Pairer and Christian Obinger

1.1 Introduction

The central purpose of this review article is to recall the pivotal importance of bioenergetics for the whole living world or, in other words, for the entirety of all living matter on our earth. Thus the first half of this review will extensively discuss all fundamental aspects of bioenergetics—from general philosophical aspects ("what is life?") [90, 76–81] to historical and evolutionary as well as physical aspects of energy, work, and force. Included in this option is a detailed discussion of the physical, biological, biochemical, energetic, and genetic aspects of "*biological energy*" together with minor philosophical excursions on life in general. At the very beginning of our reflections may stand the absolutely fundamental and ever-lasting dogma valid for all living cells ("living matter"): *Life implies work* (pointing out in this context already the physical equivalence or identity of the conceptions of *work* and *energy*!). Heuristically, for a more rational and holistic discussion of "life" and its components or constituents it seems wise to follow the well-known classical *atomistic approach*, dating back to the Abderitic philosophers Leukippos and Demokritos, approx. 300 B.C. (see [76–80]). Originally, the most fundamental and irreducible enigma was *being per se* (or, in the original language of Parmenides: το ειναι, or: το ον) and the very first and constitutive question of all philosophy which latter is rightly called "the mother of *all* sciences" in general, coined in increasingly precise form from Herakleitos and Parmenides (approx. 500 B.C., see [76–80]) to Leibniz [81] and Heidegger [82], is: "Why is there being (or: why is there *anything* at all), and not simply *nothing*?",—immediately followed by the second-fundamental constituent of all philosophy: *Astonishment* and *amazement*, according to Aristotle (384–322 B.C.)!

G. A. Peschek (✉)
Molecular Bioenergetics Group, Institute of Physical Chemistry, University of Vienna, Währingerstr. 42, Rm. 2119, 1090 Vienna, Austria
e-mail: guenter.peschek@univie.ac.at, cyano1@aon.at

G. A. Peschek et al. (eds.), *Bioenergetic Processes of Cyanobacteria*,
DOI 10.1007/978-94-007-0388-9_1, © Springer Science+Business Media B.V. 2011

Following further up the *atomistic approach*, what we need in this review, is: The *atom* of life (the living cell!), the *atom* of light (the light quantum or photon, hν, for photosynthesis), the *atom* of biological energy (ATP), and the *atoms* that make up living matter: C, O, H, N, S, and P. Our discussion will have to span from the origin of our universe, the origin of our blue planet and why it is "*blue*", the absolutely unique and incredibly improbable phenomenon and origin of life (of living matter) on it, the origin and role of ATP (and other "freely convertible" forms of biologically utilizable energy) in living cells for what is called "*energy conversion*" or its synonym: *phosphorylation* (and de-phosphorylation, the reverse process). Thus FUNDAMENTAL BIOENERGETICS will be dealt with in the first part of this chapter (Part I). In this context we will have to discuss the physical chemistry of ATP and its chemical interconversions (*Gibbs-Helmholtz equation*, enthalpy, entropy, etc.), the *phenomenologically* different (photo- and chemo-trophic) and *mechanistically* different (substrate-level) (SLP) [111, 112] and electron transport (ETP) [105–110] or chemiosmotic types of phosphorylation or energy conversion, the basic environmental "*ingredients*", *constituents*, or *sources of life* (energy, electrons, and carbon), and their comprehensive mutual combination or interaction as most clearly shown by the *ergotrophic hexagon* yielding the systematically consistent unitary concept of *all* living cells as such (also see [161, 430]). It seems correct, therefore, to start the discussion with *the atom of life* (of living matter), viz., the living cell.

The concluding chapter (Part II) of this comprehensive review article will be made up by a more specialized survey of *cyanobacterial electron transport* (in particular: *respiration*, the still most neglected aspect of cyanobacterial biology, biochemistry, genetics and bioenergetics), and finally highlighting the cyanobacteria as the nonplus-ultra of microbial bioenergetics and physiology, thereby also touching the most modern area of biological "*in-silico-research*". Previous reviews of cyanobacterial electron transport and respiration, usually together with photosynthesis, may be found in [32, 33, 34, 36–38, 86, 88, 104, 132, 134, 142, 232, 265–267, 268, 275–278, 366, 368]. However, even before discussing the ever-significant energetic performances of living cells as best exemplified by the cyanobacteria, the nonplus-ultra of microbial bioenergetics and physiology, it may be permitted, somehow in the style of an ancient Greek Satyr Play, which usually was an integral part of the classical tragedies of Aischylos, Sophokles, Euripides, Agathon, Aristodemos, etc., to start our *strictly scientific reflections on the tragedy of life* with a similarly satyric prelude:

THE LIVING CELL—A very special *punchinello* in evolution

Well, there are two extreme ways of looking at the *phenomenon of life*: Either, life as a divine miracle—as it might seem to the naïve minds of common, usually less-educated people overlooking the truly inconceivably "polychromatic" multiplicity of life on earth, or, otherwise: Life as a ridiculous Punch and Judy Show of evolution where, during the aeons, every individual has been nothing but foodstuff to the other, somehow following the philosophical maxim of Thomas Hobbes (1588–1679): "Homo homini lupus" or, more generally: "Alius alii praeda"…

Yet, in all events, life's occurrence on earth may well be regarded kind of "mystery", at least an extremely improbable and irreproducible incident—or better: accident?—though, actually, life is nothing but the inevitable consequence of an extremely long, more or less undisturbed, sequence of "simple" (though each of them: highly improbable!) evolutionary processes all governed by the "simple" laws of nature that we all have known for long already (as will be evident from this essay, too). And since we just celebrated the "Darwin Year" in 2009 (Charles R. Darwin: 1809–1882)—hoping thereby that the well-proven concept of mutation and selection, even if basically synonymous with rather cruel-sounding notions such as "struggle for existence" or "survival of the fittest"—has found firm hold even in the brains of the general public due to countless pertinent newspaper and street-journal articles and low-level lectures in schools and pubs around the corner all over 2009. In the sense of Darwin, therefore, it seems all the more justified to devote a major biological review article to the most important "creation" (or better: result) of this Darwinian evolution, viz., to the cyanobacteria.

1.2 Fundamental Bioenergetics

1.2.1 The Structural Basis of Life

Recalling the basic structural differences of prokaryotic and eukaryotic cells (Fig. 1.1a, b), it is barely appreciated, unfortunately, that these two types of cell architecture (structure) are the only really different types on earth. There is not a single transitional form ("missing link") of whatever kind between them, and also Woese's "archaebacteria" [1–3] are in essence nothing but prokaryotes. It is worth noting that, in spite of profound and unbridgeable structural and genetic differences, and despite the evolutionary distance between prokaryotes and eukaryotes as well as between eubacteria and archaebacteria, the *all-decisive biochemical or functional features* that make a living cell what it is, are strikingly uniform in all existing types of living cells, just think of Nirenberg's four-letter *genetic code* or of Watson's and Crick's *double helix*. This *universal dogma of life* is already implicit in the famous unitary principle of biochemistry first coined by Kluyver and Donker in 1926 [4] (also see [5–7]). In particular, there is not a single biochemical trait in any eukaryote that would not also be found in one or the other of the prokaryotes. Rather, in the opposite direction: there are a few specifically prokaryotic functions that had apparently been lost in *all* eukaryotes, e.g., nonmitotic proliferation [8] by, e.g., binary fisson, the "key to immortality", so to speak, or N_2 fixation [9], the renouncement of which is obviously due to an irreversible loss of the *nif genes* coding for nitrogenase enzymes. Thus it first seems appropriate to spend a few words on "the structural home" or "the atom" of life, the living cell, proper. In doing so, the relation between pro-karyotes and eu-karyotes deserves most attention.

There is little doubt that the "atom of" life, the minimum structural entity or unity of *living matter*, to which the albeit rather ill-defined phenomenon of life does apply in all its entirety, is *the living cell* sensu Th. Schwann (1810–1882) and M. J. Schleiden (1804–1881). Yet, still the primary origin of (the first) living cell is absolutely enigmatic. After any belief in a *generatio spontanea* or *generatio aequivoca* or *abiogenesis* or *archaeobiosis* or *necrogenesis* had been entirely dropped by 1862, even if no minor than Louis Pasteur (1822–1895) had still believed in a *generatio spontanea* up to the mid-eighteens before, Pasteur himself became outrider of the

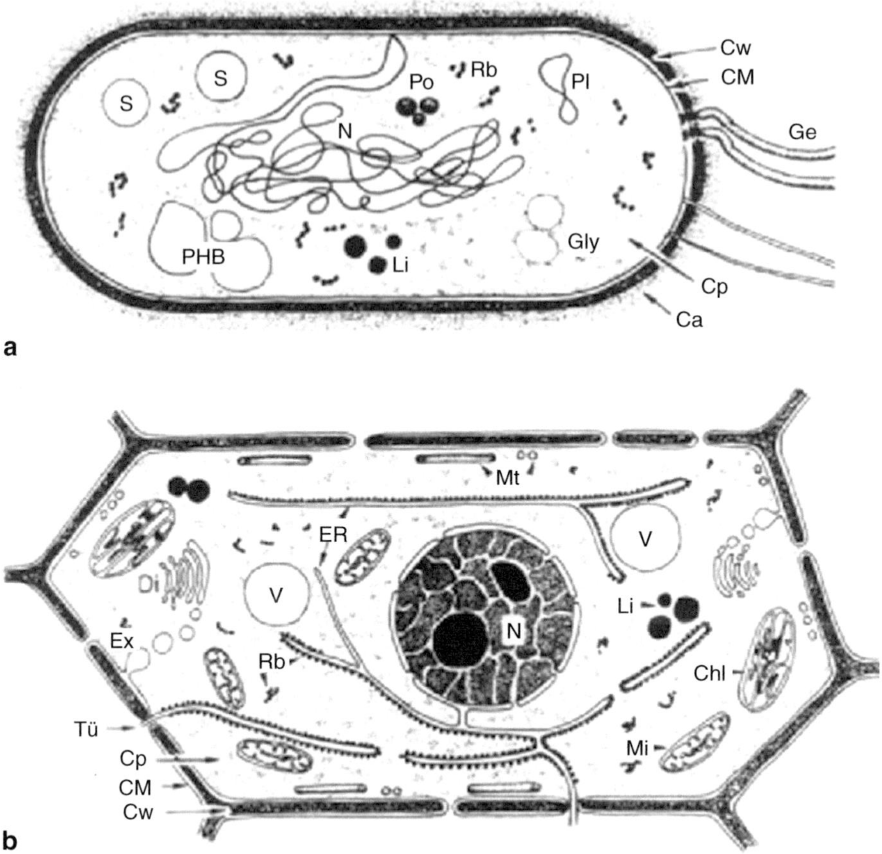

Fig. 1.1 Structural sketches of a typical prokaryotic cell (**a**) and a typical eukaryotic plant cell (**b**). (Abbreviations: *Ca* capsule (glycocalyx), *CM* cytoplasmic membrane, *Cp* cytoplasma, *Cw* cell wall, *Ge* flagellae, *Gly* glycogen granules, *Li* droplets of lipids, *N* nuclear region (DNA freely suspended in cytoplasma in prokaryotes but membrane-enclosed "true" nucleus in eukaryotes), *Pl* plasmid, *PHB* poly-β-hydroxybutyric acid, *Po* polyphosphate granules, *ER* endoplasmatic reticulum (an extension of the nuclear membrane in eukaryotes, absent from prokaryotes), *S* inclusion of elemental sulfur, *Chl* chloroplasts, *Mi* mitochondria, *Mt* microtubuli, *Di* dictyosome, *Ex* exocytotic vesicles, *Tü* plasamodesmata or "Tüpfel", i.e. permeable connections between two adjacent eukaryotic cells, *V* vacuole, *Rb* 70S (30S+50S subunits) ribosomes in prokaryotes inhibited by the typically prokaryotic translational inhibitor chloroamphenicol but NOT by cycloheximide, and 80S (60S+40S subunits) ribosomes in eukaryotes attached to the ("rough") ER and inhibited by e.g. the typically eukaryotic translational inhibitor cycloheximide but NOT by chloroamphenicol. Note that exactly the same ("prokaryotic") inhibition profile is typical of the (genetically semi-autonomous) mitochondria and chloroplasts (see also Fig. 1.2). Also note that the prokaryotic cell not only lacks an ER but also, in general, any type of cytoskeleton [49]. This could mean that a prokaryote had already been somehow "predestined" to form a self-supporting cell ("organelle") within another cell without much need for any further internal structural communication)

group of scientists definitely rejecting any such *generatio spontanea* (for fascinating historical reviews see [10, 11]), and the dogma of *omne vivum e vivo* or *omnis cellula e cellula* (R. Virchov, 1821–1902) has long been firmly established by now.

Clearly, the undeniable and uninterrupted continuity of living cells *per se* for almost four billion years [12, 13] poses the severest problem notably since not even most gifted and clever scientists or molecular geneticists have ever been able to synthesize a living cell in the laboratory. No wonder, therefore, that rational human spirit has designed several fictive scenarioes how the origin of a living cell could be conceived of in strict correspondence with all existing scientific reality. Also, beyond any science fiction, it seems clear that in evolution *sensu* Charles R. Darwin (1809–1882) [14, 15]—if not to be replaced by a doubtful creationism (see, e.g., [16–18])—the origin of the delicate and vulnerable structure of the first living cell must have taken place in the absence of any high-energy impact such as ionizing radiation or molecular oxygen, and that the structurally simpler prokaryotic cell must have preceded eukaryotic cells (Fig. 1.1) [19–23]. Most significantly, as it cannot be emphasized too often, there are no transitional forms ("missing links") between a prokaryotic and a eukaryotic cell. Thus, a eukaryotic cell does not simply and gradually evolve from a prokaryotic cell. Rather, at some stage in evolution a "suitable" prokaryote, most probably some Archaeon, became host for another prokaryotic cell by way of establishing a stable endosymbiosis [22–28].

Originally, the well-known (classical) endosymbiont hypothesis [29, 30] had been given birth to by the botanists Schmitz, Meyer and Schimper around 1880 already but has first become more widely known through the publication of Constantin Mereschkowsky in 1905 [29]. Such primary endosymbiont event (between two prokaryotes) has never been observed experimentally nor could it ever be studied in nature. Again, therefore, it must have been an extremely improbable event. Nevertheless, the classical endosymbiont hypothesis which describes the endosymbiotic origin of a green plant's or algal chloroplast from an engulfed, formerly free-living cyanobacterium [29] is almost unequivocally accepted by now [30] as it is based on fairly sound and uncontradictory comparative biological and biochemical evidence. Modern molecular biology has added much to this striking relatedness, e.g., with respect to the inhibition profiles of protein synthesis in free-living prokaryotes (viz., *cyanobacteria*, but also, e.g., α-proteobacteria and others) and in *chloroplasts* (and mitochondria). Clearly, the improbability of the (primary endosymbiont) event would not be alleviated by the assumption that the engulfment of the cyanobacterium was achieved by an already "mitochondriate prokaryote" [19] which anyway would have had to originate by an endosymbiosis between a (fermenting?) Archaeon and another (respiring?) prokaryote of the α-proteobacterial type (e.g., *Paracoccus denitrificans*, according to current hypotheses [19, 22–31]). Thus, the orthodox endosymbiont hypothesis assumes two separate and consecutive events: First, the "uptake" of a respiring purple bacterium [19, 31] (e.g., *Paracoccus denitrificans* or another, *Rickettsia*-like, bacterium), and secondly: Endosymbiosis between this proto-mitochondriate "eu-oid-karyote" and an ordinary (unicellular) cyanobacterium leading to a "full-fledged eukaryote" with mitochondrion *and* chloroplast (Fig. 1.2b). In order to avoid such cumulative improbabilities, and also in view of the striking *cyanobacterial* feature of being already kind of a free-living proto-

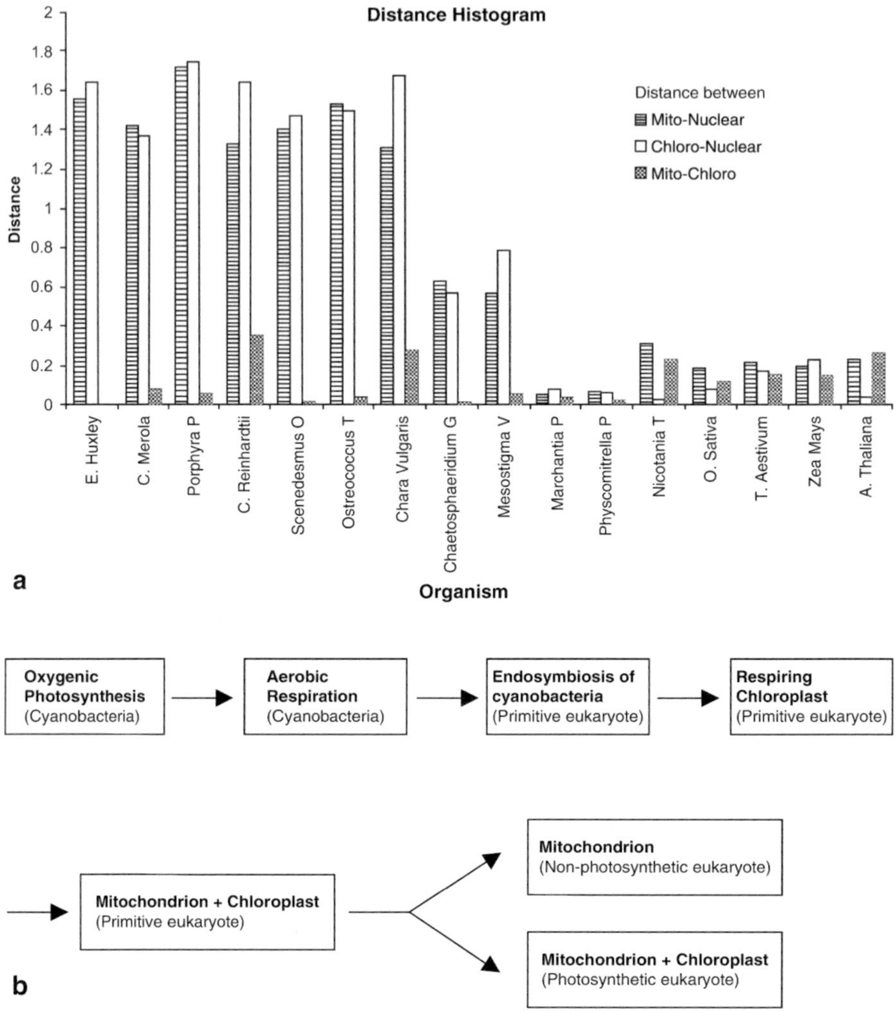

Fig. 1.2 **a** Genomic signature-[41–43]-based evolutionary distances [44] between the chloroplast and mitochondrion (*grey*), between the chloroplast and the nuclear genome (*white*) and between mitochondrion and the nuclear genome (*striped*) for a number of photosynthetic eukaryotes. Chloroplasts and mitochondria (of the same cell) are very similar, especially in lower organisms, where the organelle genomes are very distant from their nuclear genomes proving that the organelle genomes have been introduced through endosymbiosis from a different organism. **b** Schematic representation of the path of endosymbiosis. Based on the assumption that genome signatures reflect evolutionary closeness, it is proposed that a cyanobacterium could have given rise to mitochondria and chloroplasts in a single endosymbiont event

chloromitochondrion [39, 40], capable of both oxygenic photosynthesis and aerobic respiration [32–39], it was recently proposed that both chloroplast and mitochondrion could well have originated from a single (thus less improbable) endosymbiont event between an Archaeon and a cyanobacterium [25, 39, 40]. A recent comparison of genome signatures [41–43] and proteomes of Rickettsiae, mitochondria, chloroplasts and cyanobacteria [44] has shed much probability on this new and unifying or generalized endosymbiont hypothesis (Fig. 1.2a). This approach opens a free and easy path to the conclusion that, during the aeons of evolution, not only *chloroplasts* but also *mitochondria* might have evolved from an endosymbiotic cyanobacterium, very much in accordance with the conversion hypothesis (Fig. 1.2b).

1.2.2 The "Conversion Hypothesis"

Figure 1.3 compares the intracellular arrangement of membrane *functions*, viz., electron transfer, proton movement, and ATP synthesis (thereby showing the "sidedness" or "vectoriality" of the membranes), of a cyanobacterium (top), a mitochondrion (middle), and a chloroplast (below). The comparison suggests not only accidental *analogy* but also evolutionary *homology*, i.e., a common monophyletic evolutionary origin of all three "*bioenergetic machines*". Such monophyletic origin of respiration and photosynthesis is being amply discussed by the so-called *conversion hypothesis* [45–47] which now receives much support also from the analysis of genome signatures [44] (also see chap. 3 in this book).

It seems to be a general strategy of biological evolution to adhere to a significant functional or structural "achievement" or progress once attained, viz., sorted out from the countless variants or, e.g., "metabolic and structural combinations" offered by the typical "trial-and-error movement" characteristic of Darwinian evolution which is known to proceed by mutation and selection. Therefore, it would be surprising if a cellular (prokaryotic) functional structure would not have been handed over from one to the next generation in the form of a free living cell, of a mitochondrion, and of a chloroplast at the same time, so to speak, always, of course, in response to prevailing environmental constraints such as light or dark, and anoxic or oxic, conditions, etc. Reasons for the strikingly greater evolutionary success of eukaryotes compared with prokaryotes doubtless have to do with a peculiar, viz., *mitotic* type of (*sexual*) reproduction [48], with improved intracellular structural and communication networks (due to ER and cytoskeleton [49]) with improved cell-to-cell communication [50] for the building up of multicellular organisms as has been amply discussed in the literature [23, 28, 51–58]. Alas, life thereby unfortunately and irreversibly seems to obey another rule: The farther away from simplicity, the farther away from immortality….

1.2.3 Origin and Evolution of Life and the Role of Molecular Oxygen

Admittedly, yet, even when understanding the "formation" of eukaryotes from prokaryotes, we are still very far from understanding the formation of the first living cell as such., i.e., we have not yet resolved the question of *the origin of life* [59, 60]. The best-known and at the same time also the most plausible and comprehensible

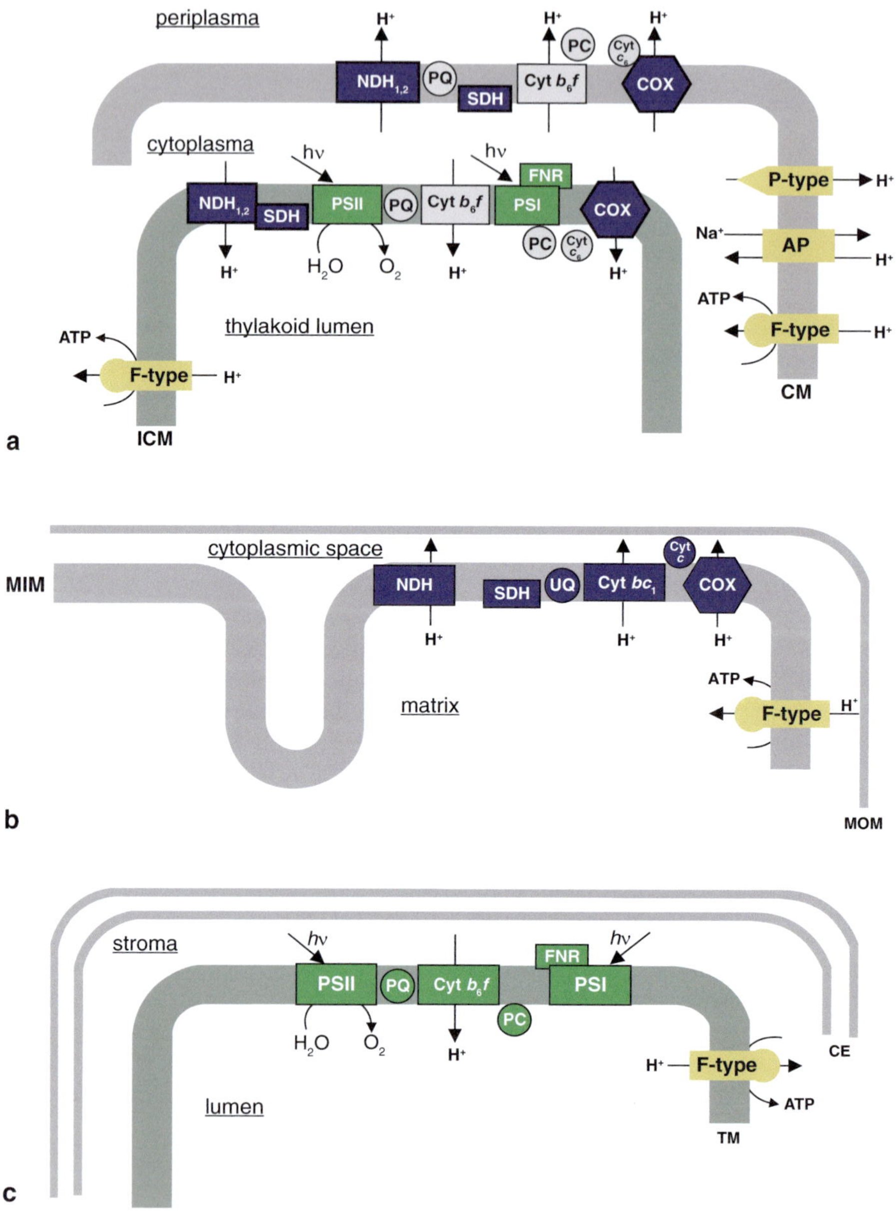

Fig. 1.3 Comparison of bioenergetic membrane functions in a cyanobacterium **a**, a mitochondrion **b**, and a chloroplast **c**. (Abbreviations: *CM* plasma or cytoplasmic membrane, *ICM* or *TM* intracytoplasmic or thylakoid membrane, *CE* chloroplast envelope, *MOM* mitochondrial outer membrane, *MIM* mitochondrial innermembrane.) Anastomoses between CM and ICM which, though clearly demonstrated in anoxyphototrophs [302, 303] and also implicated for cyanobacteria [37, 278] and chloroplasts [417] by both phylogenetic and functional considerations, could so far not be *physically* detected in cyanobacteria by electron microscopy or similar techniques [278, 413–416].

scenario for the origin of life doubtless is Oparin's [61] and Haldane's [62] (heterotrophic) primeval soup scenario (both original articles, in the form of English reprints, may be found in [63, 64]): for discusson see, e.g., [45, 48, 59, 60, 63–65]: As Urey and Miller [66, 67] have convincingly demonstrated, abiogenesis of practically all the constituents and chemical components (*monomers*) of a living cell—amino acids, carbohydrates, lipids of whatever kind, and structures (up to porphyrins!) could have taken place on the still life-less earth, in the absence of any living cell, under the high-energy impact of ionizing radiation from the outer space and/or electrical discharges in the primeval atmosphere, as well as through the then fervid volcanism and intense UV radiation, from (then ubiquitous) reactants such as H_2O, CO_2, CO, HCN, $(CN)_2$, H_2S, NH_3, N_2, etc. which must have been abundant in the reducing to neutral conditions of the atmosphere of the early earth shortly after a solid crust had been formed [68] (also see internet: http://planetquest.jpl.nasa.gov) (Fig. 1.4a). The (heterotrophic) primeval soup scenario of Oparin and Haldane thus presupposes the abiogenesis of all the constituents of (not yet) living matter as was experimentally demonstrated by Urey and Miller [66, 67] in their famous so-called Urey-Miller-type experiments which were consistently repeated many times afterwards (see, e.g., [69]).

However, regarding scenarios for *the origin of life*, at present, in the form of the autotrophic surface-FeS-redox concept of Wächtershäuser and associates [70–72] we do have a seriously competing hypothesis which, even more, on the basis of primordial *inorganic* surface-catalysis, claims to be able to explain the *structural formation* of the first living cell [73]. At any rate, and without going into too many details, the present authors want to stress that, owing to the highly specialized, exclusive, and selective physicochemical prerequisites for the existence of life in a form conceivable to us—and what other form should it finally be?—"in-silico-life", instead of protein-based life? (just to mention the obligate availability of *liquid water* within a rather narrow range of temperatures usually not encountered in any region of the universe about which we know), the still extremely mysterious and unexplained (perhaps even unexplainable?) *phenomenon of life and its origin* [59, 60], i.e., *the existence of living matter* of whatever form, is an unbelievably improbable accident or hazard since literally countless physicochemical conditions had to be *the right ones*. Yet, whenever these conditions, viz., C, O, H, N, S, and P in the appropriate mixture, at the appropriate temperature, and under the appropriate energy input, indeed were the right ones [59, 60], life simply *had to happen* (see "La chance et la necessite" by Monod [74]; also see [426])—and thus it did happen on our earth about four billion years ago [12, 13]. Since, however, nobody of us was present dur-

(Further abbreviations: *AP* Na^+/H^+-antiporter, *COX* aa_3-type cytochrome *c* oxidase, Cyt *c*, soluble *c*-type cytochrome, Cyt c_6, cytochrome c_6, *PC* plastocyanin, Cyt b_6f, cytochrome b_6f, *PQ* plastoquinol, *UQ* ubiquinol, *PSI* photosystem I, *PSII* photosystem II, *FNR* ferredoxin $NADP^+$ oxidoreductase, *NDH* NAD(P)H dehydrogenase, *SDH* succinate dehydrogenase, P-type, proton-pumping P-type ATPase, F-type ATPase, i.e. ATP synthase.) Note the mutually opposite polarity (sidedness) of thylakoid and inner mitochondrial membranes with an otherwise functionally and structurally quite similar electron transport assembly. For more details about cyanobacterial structure-function relationships see [32, 37, 278]. For a cyanobacterium, also see Fig. 1.11

בְּרֵאשִׁית
בָּרָא אֱלֹהִים
אֵת הַשָּׁמַיִם
וְאֵת הָאָרֶץ׃
וְהָאָרֶץ הָיְתָה
תֹהוּ וָבֹהוּ
וְחֹשֶׁךְ עַל־פְּנֵי
תְהוֹם
וְרוּחַ אֱלֹהִים
מְרַחֶפֶת

Origin or appearance of	Years ago ($\times 10^9$)
Universe	13.7
Sun	10
Earth (solid crust)	4.6
Life (prokaryotes)	3.5
CYANOBACTERIA (O_2!)	3.2
Atmospheric O_2	2
Eukaryotes	1.6
Terrestrial biosphere (O_3-shield)	0.4
Homo erectus	0.0007-4
Homo sapiens	0.00005

a

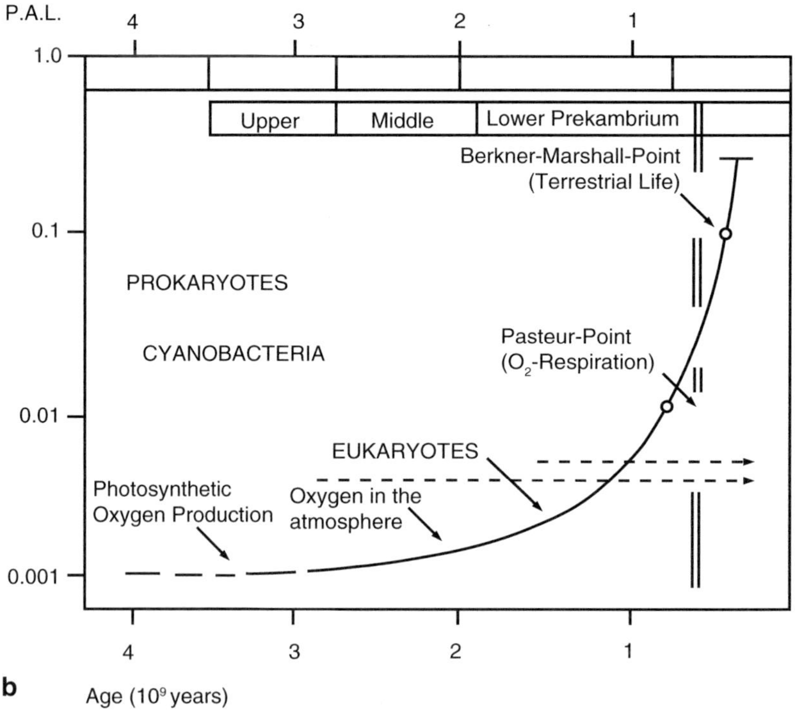

b

Fig. 1.4 **a** Crucial steps in the evolution of the universe, of our Earth, and of life on this Earth. The Wilkinson Microwave Anisotropy Probe Satellite, launched by NASA in 2003/2004 and having measured very accurately possible anisotropies in the 3-K-background radiation which is considered a remnant of the (cosmological) Big-Bang, has determined the age of the universe T_o as $13.7 \pm 1\%$ (!) billion years. The unique advent of the cyanobacteria and of O_2 at around 3.2 billion ing the formation of the first living cell and since this "achievement" cannot be experimentally re-accomplished in our laboratories either, not even by most gifted scientists or sorcerers, most evolutionary speculations are nothing but educated guess. Yet, what a serious (natural) scientist *can* do nowadays is to carefully reconstruct the time course of evolution during the millennia, using the enormous wealth of data from astronomy, astrophysics, geology, geophysics, geochemistry, climatology, paleontology, comparative biochemistry and genetics, etc. Figure 1.4a, b underline that molecular oxygen is the driving force or the "guiding angel" of, both geological and biological, terrestrial evolution as such. At any rate, O_2 or what is originally "synonymous" with it: the cyanobacteria [48], are "the things" which have shaped our earth to what it is: "The Blue Planet". Due to this decisive and extraordinary role of O_2 for "life on earth", which originally and uniquely stems from the cyanobacteria and which appears to be an absolutely exceptional situation in our whole universe, later in this review a special section will be devoted to the photosynthetic production of O_2. In this context we also want to draw your attention very briefly to the following (more philosophical and epistemological) aspect:

years ago is often called "*the evolutionary* or *biological Big-Bang*" [180, 181]. By the way, the total number of stars in our universe is about 10^{22} (!) so that even a most pessimistic estimate "confronts" us with at least several thousands of "human-like" races in our universe from which, however and fortunately, we are and will forever be separated by unbridgeable gaps in both space and time. Thus, remembering that the *cosmological Big-Bang* represents a so-called *singularity* [418] (something that exists in pure mathematics only, as we were taught in school… However, if the gentle reader wants to learn more about the conflict between *ordinary reality* and *cosmologic singularity*, particularly in the context of the origin of our Universe, in the context of Space and Time, and in the context of "infinity" and "eternity" he may be advised to try also references [419–426] and he will see how difficult it is to reconcile even most learned *common sense* with the suspected attributes of the *Big Bang*. However, for the time being, in the absence of any more comprehensible and plausible proof from astrophysics, the *left-hand side* of the Figure gives the original Hebrew sentence (Mose 1, 1: "In the beginning God made heaven and earth") introducing the creation myth in the "Holy Bible"—and who claims to know how it really happened? Note that in view of the tremendous humanitarian achievements of man (war, murder, crime, capitalism, poverty, global pollution, etc.) *homo*'s second "*sapiens*" might better be replaced by something like "destructive", "evolutionary dead-end" or "cul de sac". Well, be all that as it may be, conceived as undeniable "products of a cosmic singularity", the degrees of freedom in our Western mind [131] are high enough at least to appreciate a good joke as a reason to lough about and to enjoy the sunset on a clear day as a reason to be happy… unless there is another unexpected "singularity" [418–426] somewhere nearby…. **b** Oxygen content of the Earth's biosphere and atmosphere during the previous four billion years. P.A.L., present atmospheric level, i.e., fractions of 21% ($^{v}/_{v}$). Compare this with the present 0.03% ($^{v}/_{v}$) CO_2 in our atmosphere! The Pasteur point marks the time when the global efficiency of respiration (RET), in terms of ATP production, because of increasing availability of O_2, exceeded that of fermentation (SLP, substrate level phosphorylation). The Berkner-Marshall point marks the time when the ozone layer in the upper atmosphere, a result of the photochemical reactions of O_2, started to prevent deleterious ionizing radiation from the outer space from reaching the surface of the earth. The transient overshoot of atmospheric O_2 concentrations peaking at 35% ($^{v}/_{v}$) during the carboniferous period about 300 million years ago was due to the massive development of terrestrial (in particular: fern) plants under the shielding ozone layer in the upper atmosphere. It is the time to which all our (diminishing!) fossil fuel deposits date back. And the overshooting O_2 at first could not be compensated for by a likewise massive development of animals. The figure is compiled from data in [45, 48, 203, 427–429]

Looking back to the history of mankind, all science had started from *philosophy* which, rightly therefore, is called *the mother of all sciences*. In a still somewhat broader context, epistemologically speaking, someone still wondering about *life on earth* should rather ask the question: Why is there "existence" of whatever kind at all, why is there "being and not simply nothing?". Exactly this is the constitutive question of all philosophy, which has been being asked ever since Herakleitos and Parmenides, 500 B.C. [76–80], and repeatedly re-coined in increasingly precise form up to Leibniz [81] and Heidegger [82]. Yet, even if we accept—without asking too many questions—"existence" as such (="being", the famous Parmenideic "ειναι"/einai), and above all: our own existence,—as trained "rational scientists" we are nevertheless allowed to ask how this all, viz., how "life in its present form", might have come about. Thus it is nice to see that the joint effort of so many "exact" sciences such as astronomy, astrophysics, mathematics, geology, geochemistry, paleontology, comparative biochemistry, etc. (see above), has indeed been able to depict a consistent sequence and time-table of events that have eventually brought about our present universe and our present terrestrial biosphere, not to forget about us ourselves, but we should also remember that it has always been, is, and will ever be, absolutely inevitable to be mentally confronted, at one or the other point of our lives, with the question "where do we come from?" and "where do we go to?". However, since all phantastic metaphysical constructions are nothing but cowardly attempts to hide against the undeniable scientific facts of the all-embracing contingency that makes up, yet: that *is*, our world and our lives, a clear-headed study of Fig. 1.4a, b is highly recommended: It is important to realize that, as there *was nothing* (we could remember of) *before us*, there *will be nothing behind us* as well (thank heavens!), and "redemption", following the archaic but invincibly wise pessimism of Homer (see, e.g., Hom. Il. 6, *146*; 17, *446*; Od. 180, *130*), might best be envisaged as the *ontological nothingness* of Heidegger [82], Sartre [83], and Camus [16, 84].

Returning to the origin of life, irrespective of Oparin and Haldane or of Wächtershäuser, it is sufficiently clear that the extremely sensitive structures of a living cell *in statu nascendi* (protocell, coacervate, etc.) could never have been formed under any high-energy impact such as ionizing (UV) radiation, or free O_2 with its extremely positive standard reduction potential of +820 mV [85, 86]. Molecular O_2, e.g., from purely thermodynamic reasoning, could instantaneously oxidize (i.e., literally: *devour*) every constituent of the biosphere including N_2 and, *a fortiori*, all carbon-based organic matter—provided there were no *kinetic constraints* in the form of quantum mechanical laws (Hund's rule, Pauli principle, etc.), or in the form of lacking catalysts (enzymes), as is a bit more extensively discussed in [48, 85, 86]. Quite understandably, for this reason, the uniquely *paramagnetic* diatomic gas "free molecular oxygen" (or dioxygen or simply O_2), a "*bi-radical*" with two "three-electron bonds" in one molecule [87]—on one hand the real driving force of both biological and geological evolution (Fig. 1.4a, b) and the "biophysical synonym", so to speak, for the cyanobacteria [46–48, 86, 88] but, on the other, the reason for combustion and destruction, of ageing and death (of multicellular organisms) [89]—is frequently called a *devil in disguise* [86] and the cyanobacteria, having most sucessfully "learnt" to both evolve O_2 (in the light) and consume O_2 (in the dark) and use it for most efficient bioenergetic processes such as respiration [32–38, 88], present

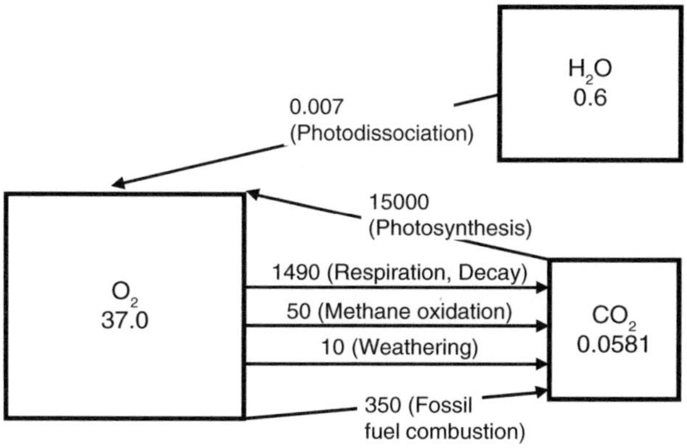

Fig. 1.5 Global oxygen exchange pattern in the atmosphere and biosphere of our earth. The rate of O_2 production from photosynthesis equals the sum of the rates of O_2 removal by respiration, decay, methane oxidation, and weathering. This is the entire global O_2 cycle. In the Figure, reservoirs (*squares*) are in Emoles O_2 and rates are in Emoles O_2/Myr. 1 Emole (erda mole) = 10^{18} moles, 1 Myr (mega year) = 10^6 years. The mean residence or turnover time of oxygen is 6,000 years in (free) O_2, 500 years in CO_2, and 20 million years in H_2O. [45, 203]

themselves as *the tamers of* O_2 [86]. Being already aware of the absolutely exceptional and outstanding role that O_2 plays in and for our terrestrial biosphere, Fig. 1.5 describes quantitative aspects of the global oxygen cycle, showing exchange rates and respective pool sizes (O_2, H_2O, and CO_2) from which corresponding residence times can be easily calculated. The main *biological part* of this cycle, reflecting the *syntrophic community* of oxygenic photosynthesizers ("plants") and aerobic respirers ("animals"), will be discussed in detail below.

1.3 Energy and Life

"Life implies work", the basic axiom of all life on earth without exception, means: There is no life without constant input of environmental energy (energy from outside) which, in turn, according to fundamental thermodynamical principles (see below) is used to maintain, at least for a short while ("life time"), the characteristic low-entropy ("high-order") state synonymous with a living cell. This review article about "the life of cyanobacteria" as our paradigms of the bioenergetic nonplus-ultra among living cells will therefore predominantly deal with *energy*, its "production" (viz. "conversion" by the *bioenergetic processes* fermentation, photosynthesis, and respiration), and its productive utilization. Note that all three bioenergetic processes can uniquely be found in a single *cyanobacterial cell* (see below). Life, this ill-defined *state of living matter*, something that *happens* rather than *is* [90–92], this *doubtful happening*, so to speak, irrevocably depends on the continuous input of *environmental energy* which, in the living cell, is immediately and permanently transformed into biologically useful forms of energy, viz. ATP (and other nucleoside triphosphates [93, 94] (Fig. 1.6a and Table 1.1). In equilibrium with the latter, in more advanced (viz. "*electron-transporting*") living cells, we have the ion electrochemical gradient ($\Delta\mu_H{}^+$) conserved

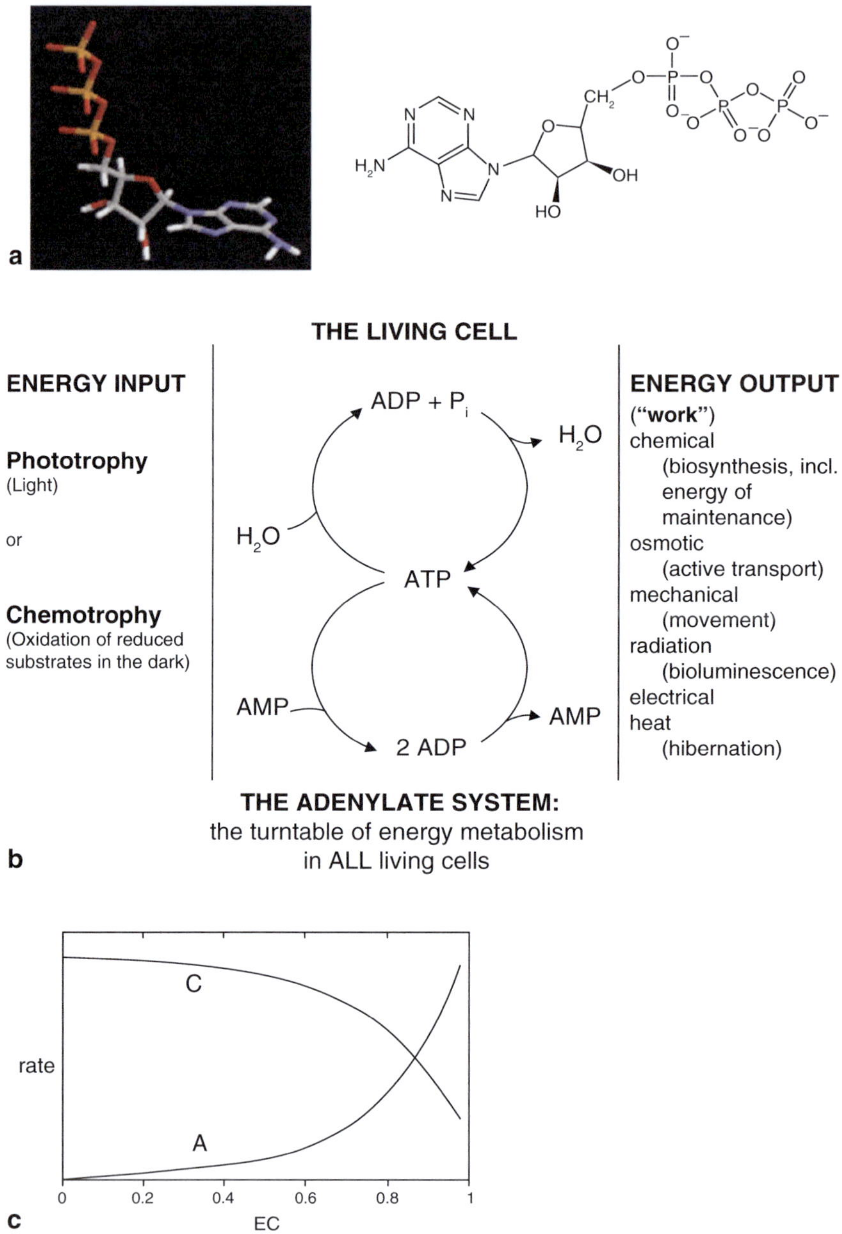

Fig. 1.6 ATP, energy flow, and energy charge. **a** Structural formula of ATP (adenosin 5′-triphosphate). **b** The flow of energy from the environment through the adenylate system in a living cell (see Eqs. 1.1 and 1.2) and, in the form of work, back to the environment again. During this flow, the energy is "degraded" (it becomes less and less reversible, thermodynamically speaking), i.e., its entropy content increases steadily. The entropy is taken up from the living cell which thereby is left behind in the typical low-entropy (high-order) state characteristic of a living cell [45, 81, 90–92]. **c** Generalized representation of enzyme activity responses to energy charge (EC). C, enzymes involved in ATP regeneration (*catabolism*); A, biosynthetic enzymes that use ATP (*anabolism*). After Atkinson [115–117]

Table 1.1 Phosphate group transfer potentials ($-\Delta G°'$) of various metabolically important phosphorylated high-energy intermediates in physicochemical standard conditions (1 M concentration, 25°C, and at pH 7). Water is taken as the standard phosphate acceptor in ordinary hydrolysis reactions. The compounds shown are of particular relevance to substrate-level phosphorylations. $\Delta G°'$-values mean $\Sigma Gi°'$ of products minus $\Sigma Gi°'$ of reactants, terminal state minus initial state. See Eq. (1.1) and (1.4)

Phosphorylated compound	$-\Delta G°'$ (kJ/mol)
Phosphoenole pyruvate	62.2
1,3-Bisphosphoglycerate	49.6
Phosphocreatin	43.2
Acetylphosphate	42.4
Phosphoarginine	32.2
ATP (γ-phosphate)	28.5
GTP (γ-phosphate)	28.5
UTP (γ-phosphate)	28.5
Glucose 1-phosphate	21.0
Glucose 6-phosphate	13.9
Fructose 6-phosphate	16.0
Glycerol 1-phosphate	9.2
Carbamyl phosphate (at pH 9.5)	52

in an *energized* biomembrane [93–96] which is "heaved up" into an "excited (=high-energy) state" according to Peter Mitchell's chemiosmotic hypothesis [97–100] (also see Fig. 1.8). In a few specialized organisms the $\Delta\mu_H{}^+$ is replaced by a sodium-motive force or $\Delta\mu_{Na}{}^+$ [100–102]. This mutual, reversible, and *vectorial* transformation of energy is the essence of so-called *biological energy conversion*: Energy is taken up from the environment by the living cell and given off as work (Fig. 1.6b). Energy conversion in both photosysnthesis and respiration is impossible without *chemiosmosis*, i.e. without the *membrane principle* (a term which may be taken synonymous with *chemiosmosis*). In most cases—except the retinal-based halobacterial photosynthesis which is based on a purely conformational chemiosmotic proton pump (see below)—chemiosmosis implies membrane-bound ion-translocating electron transport [93–110]. The theoretical foundations of biological electron transport, ET, both inter- and intra-protein ET, were laid by Marcus already [105, 106] and more recently refined by Dutton and colleagues, quantifying and rationalizing the basic proportionality between Log (reaction rate) vs. Distance (between electron-exchangeing reactants, in Angströms) which has become colloquially famous among bioenergeticists as "Dutton's ruler" [107–109]. Only in fermentative energy conversions through substrate level phosphorylation (SLP) [111, 112], which can proceed also anaerobically in darkness, membranes are *mechanistically* not involved.

1.3.1 ATP, Adenylate Kinase and Energy Charge

Comparable with a living cell which is the "atom" or "quantum" of life, and similar to *hν* which is known as a *light quantum*, ATP (Fig. 1.6a) may be called the freely convertible *biological energy quantum* (biological energy currency). This biological or biochemical energy quantum must be synthesized, without exception, in an endergonic reaction, from ADP and inorganic phosphate, P_i: Eq. (1.1), with K being the equilibrium constant, R the universal gas constant, T the absolute temperature (K) and $\Delta G^{o\prime}$ the difference of Gibbs energies (free enthalpies) between *after* and *before* the reaction, in physiological standard conditions [T = 298 K, pressure = 1 atm = 10^5 Pa, concentration of each reactant, except H^+, = 1 M (1 mol/L), but $[H^+] = 10^{-7}$ M (pH = 7)].

$$\mathrm{ADP} + \mathrm{P_i} \leftrightarrows \mathrm{ATP} + \mathrm{H_2O} \quad K = 10^{-5}; \quad \Delta G^{o\prime} = -R.T.\ln K = +28.5\ \mathrm{kJ/mol} \tag{1.1}$$

Being aware that the *free enthalpy* of a living cell, viz. the energy that is capable of performing useful work (Fig. 1.6b) in biological terms is ATP (or another equivalent energy currency), it would be desirable for a biochemist (bioenergeticist) to have an unequivocal, experimentally amenable, easily quantifiable, and biologically meaningful and comprehensible measure or "coefficient" of the free energy content of a living cell under given conditions. One might assume that the experimental determination of the intracellular ATP concentration could be such measure. However, because of the absolutely ubiquitous occurrence in each and all living cells of the enzyme *adenylate kinase* (formerly: myokinase), this is not the case. Adenylate kinase catalyses the extremely fast (k_{cat} of several millions per min) and energetically neutral *equilibrium* between the three adenylates according to Eq. (1.2).

$$\mathrm{ADP} + \mathrm{ADP} \leftrightarrows \mathrm{ATP} + \mathrm{AMP} \tag{1.2}$$

Thus the "*ATP cycle*" that describes the chemistry of "phosphorylation" is represented by Eqs. (1.1) and (1.2). Together with other nucleoside triphosphate interconversions (Table 1.1) and the energy stored in an "energized biomembrane", it serves as *energy buffer* in living cells [93–110, 119]. Equation (1.2) demonstrates, that 50% of intracellular ADP, the primary reaction product after ATP has fulfilled its P_i-donating, monomer-activating role, are still energetically equivalent to ATP. Exactly for this reason, around 1966 already, the Angloamerican biochemist Daniel E. Atkinson has proposed the so-called *energy charge* (EC) as a complete and sufficient measure of the *entire cellular free energy content or working capacity* of an integral living cell (Eq. 1.3, Fig. 1.6c):

$$\mathrm{EC} = \{[\mathrm{ATP}] + 0.5 \times [\mathrm{ADP}]\}/\{[\mathrm{ATP}] + [\mathrm{ADP}] + [\mathrm{AMP}]\} \tag{1.3}$$

Equation (1.3) immediately shows that the (intra)cellular EC only can assume values between 0.0 (totally de-energized cell: all adenylates present as AMP) and 1.0 (fully or over-energized cell, all adenylates present as ATP). Both states are unrealistic, of course. But most interestingly, due to the many and multifarious allosterically regulated enzymes in a living cell [113, 114], the overall metabolic activity crucially and sensibly responds to the EC value of the cell. Mere common sense together with basic laws of biochemical kinetics (see below), demand that, simply speaking, the rate of ATP production be the slower the higher the intracellular ATP concentration (i.e., the EC), and *vice versa*. Since, however, the "overall rate of metabolism" is a damned complex thing which cannot easily be expressed in a simple equation or coefficient, Atkinson already divided metabolism into a *catabolic* (dissimilatory and energy-producing) half and into an *anabolic* (assimilatory and biomass-producing) half and he indicated (rate-determining) so-called *key enzymes* for each metabolic half, e.g. fructose 6-phosphate kinase or glucose 6-phosphate dehydrogenase as catabolic key enzymes and e.g. ribulose bisphosphate carboxylase-oxygenase (Rubisco) or fatty acid synthetase as anabolic key enzymes. Then he drew diagrams plotting the enzyme activity *in vitro* against EC values artificially adjusted in the reaction mixture, *in vitro*, too [115–117] (Fig. 1.6c). The result of literally innumerable such experiments performed on any thinkable type of living cell invariably yielded a distinguished capital point (cross-over point between C- and A-curves in Fig. 1.6c) at EC values between 0.8 and 0.85. This point marks identical rates of catabolism and anabolism, in other words: a steady-state or "static head" of "cellular equilibrium", better: *Flow-equilibrium* [91, 92] which latter term was coined by the Austrian biologist and systems theoreticist Ludwig von Bertalanffy (1901–1972). This uniquely distinguished EC-value is amazingly invariant in each and all living cells investigated thus far. Metaphorically, it may be said that this point truly signifies the mysterious, *inherently dynamic* state of life itself, the point where the living cell is in the metastable condition of *being alive*, degrading and rebuilding itself at equal rates [90–92].

ATP, through a multitude of various (soluble) transphosphorylases (phosphotransferases) is in rapid (isoenergetic) metabolic equilibrium with other "energy-rich" [118] nucleoside (tri-) phosphates such as GTP, CTP, UTP, etc. (Table 1.1) as well as with an energizable biomembrane [93–100, 103–110, 119] (also see Fig. 1.8). It is biochemically significant that *all enzymatic* reactions of nucleoside phosphates (particularly adenylates) are absolutely dependent on Mg^{2+} which is complexed by the nucleoside and nucleotide phosphates [114]. However, it is *always* and *irrevocably* **ATP** which forms the primary and exclusively *first* biochemically utilizable energy pool in all living cells. And the reason for this unique role of **ATP** (in contrast to GTP, CTP, or UTP, etc., which do store practically the same amount of chemical energy in their anhydride γ- and β-phosphate bonds), must be sought, not in a possibly peculiar constitution of ATP itself but in *evolving life* as such, viz. in the entirety of a living cell which has evolved according to the Darwinian principles of *struggle for life* and *survival of the fittest*. And the fittest cell that "had finally overcome" and evolved further on, i.e. handed over its possibly even improved (genetic) identity to the next generation, for whatever—be it even a more or less statistical, population-dynamical—reason in detail, *by chance* was the one that used **ATP** as the metabolic energy source. This "decision- and pace-making power" of evolving life itself had already been pointed out by Wald many decades ago [120–121] and similar reasoning must apply, by the way, to other *unexplained biological dichotomies* ("one-sidednesses") such as the preponderance of L-amino acids (over D-amino acids) in natural proteins, and of D-sugars (over L-sugars) in natural polymeric carbohydrates. This type of reasoning—which might almost be seen to ascribe some "*intelligent*" or "*teleological*" capacity to evolving life—does, of course, imply a strictly *monophyletic origin of life* in general. Irrespective of *lateral gene transfer*

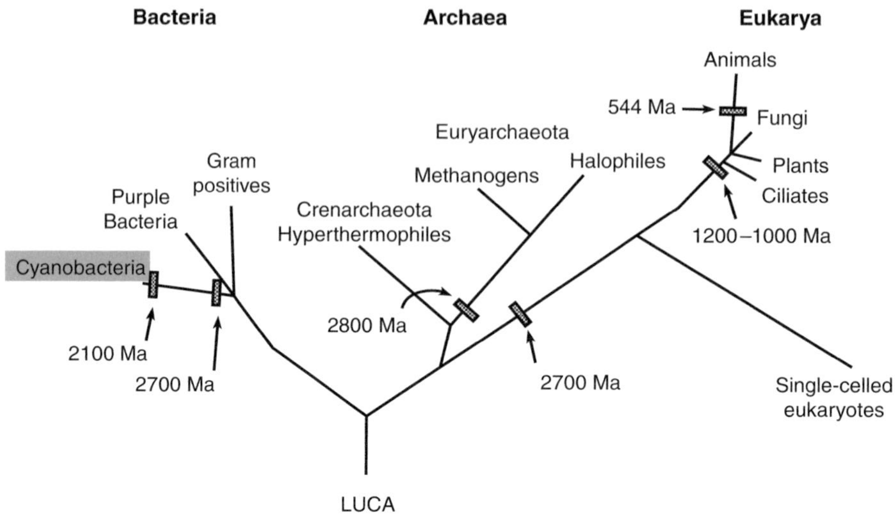

Fig. 1.7 The monophyletic origin of life on Earth is supported by this "rooted" universal tree of life. The tree shows Woese's three domains of life [124–129]. It is based on sequence comparisons of ribosomal RNA, analysed by Carl Woese and his colleagues [124–126]. The order and length of branches are proportional to the sequence similarities within and between the domains and the three kingdoms of life, i.e. they are directly proportional to the genetic similarities between species. In no way does this tree conflict with a generalized endosymbiont hypothesis (Fig. 1.2) or with an evolutionary path as discussed here (Fig.1.4). The common root represents LUCA, the last universal common ancestor [123]. Ma, 10^6 years

etc., hardly any biologist nowadays would like to question a monophyletic origin of life which, by the way, is also elegantly supported by the *LUCA* theorem of Carl Woese and asssociates ("LUCA"=last universal common ancestor (Fig. 1.7), see [123–129]; also see [48] for a comprehensive discussion). At any rate, the decision-making "switch-points" [Fig. 1.4] in the evolution of living cells, leading to the major unexplained *dichotomies* of Nature, most evidently must have been "governed" by the famous "chance" of Monod [74]. They bear convincing witness of the undeniable and all-embracing *contingency of our world and our lives* [130, 131].

1.3.2 Biological Energy Conversion

In order to understand the basics of *biological energy conversion* (Figs. 1.6 and 1.8) we must first deal with the *driving force* of a (chemical or physical) reaction, i.e. with the fundamental *Gibbs-Helmholtz equation* (Eq. 1.4). The total or *free* environmental enthalpy difference, in physicochemical terms: the maximum reversible reaction work either performed by (–), or needed to effect (+), the reaction, is commonly symbolized by $\pm\Delta G$ and called the *Gibbs energy* or free enthalpy difference between the two states in question, viz. *after* and *before* the reaction. The ΔG tapped by living cells in order to sustain their life processes is determined by the Gibbs-Helmholtz equation which comprises two terms: The difference of *heat contents* ΔH

and the difference of the states of *disorder* = *entropy*, Δ*S*, *with* Δ being "*after* minus *before*", and it reads as shown below:

$$\Delta G = \Delta H - \mathrm{T}\Delta S \tag{1.4}$$

At thermodynamic equilibrium (no reaction) Δ*G* = 0. In case of ΔG < 0, the process (reaction) is *exergonic*, i.e. energy-yielding, and in case of ΔG > 0, it is *endergonic*, i.e. energy-requiring.

Δ*S* represents the difference of degree of disorder, i.e. the thermodynamical probability according to Boltzmann's equation $S = k \times \ln W$, with W being the thermodynamical probability or number of possibilities for the verification of the macro-state in question from a given number of micro-states. Conceptually, W is the number of the degrees of freedom in the system [T is the absolute temperature (K), *k*, Boltzmann's constant ($k = R/N_L = 1.38 \times 10^{-16}$ erg/K $= 1.38 \times 10^{-23}$ J/K)]. Schrödinger [90] has named (–S) "*negentropy*" to have a *direct* measure for the *degree of order*.

The flow of energy through a living cell, from an environmental energy *source* to the *work* given off or better: *performed* (terminal *consumer*), can best be seen from Fig. 1.6b. Only two forms of environmental energy can be directly used by living cells, either radiation energy (*hv*) from (sun)light (*phototrophy*) or the energy released in the oxidation or "biological combustion" of (reduced) food-stuff, either organic (ordinary carbon nutrition) or inorganic (lithotrophic "nutrition"), in the dark (*chemotrophy*).

The primary energy from the environment, i.e., energy of chemical oxidations in darkness (*chemotrophy*) or the energy of absorbed light (*phototrophy*), is inevitably and universally first converted into, and temporarily stored in, ATP and the adenylate system (Fig. 1.6). In each and all living cells the biologically utilizable energy from the environment is always *first* captured in ATP through "*phosphorylation*", i.e., the endergonic synthesis of ATP from ADP and P_i (Eq. 1.1).

Systematically, this "*phosphorylation*" ("chemical slang" for: ATP biosynthesis) can be classified using the criteria of biochemical *mechanism* and energetic *efficiency*: (1) Poorly efficient substrate-level phosphorylation (SLP) [111, 112] in the completely soluble and isotropic cytosolic system of a cell, taking advantage of a wide variety of different transphosphorylase or phosphotransferase enzymes (see Table 1.1); (2) The other, and *mechanistically* entirely distinct, type of "*phosphorylation*" is the energetically much more highly efficient chemiosmotic electron transport phosphorylation, ETP [104, 110].

1.3.2.1 Substrate-Level Phosphorylation (SLP)

The essence of SLP is *intramolecular redox reactions*. In SLP, the "missing link" between metabolic redox energy and ATP is various (soluble) phosphorylated "high-energy" intermediates [118] from which the P_i finally disembarks *exergonically* (Eq. 1.1) on ADP to yield ATP (Table 1.1). Thus, membrane-bound processes are not needed for SLP. Fermentations are usually catalysed by a completely soluble cytosolic enzyme assembly. The many different types of fermentations, energetically all based on SLP, may be looked up in every textbook on microbiology or biochemistry [113, 114].

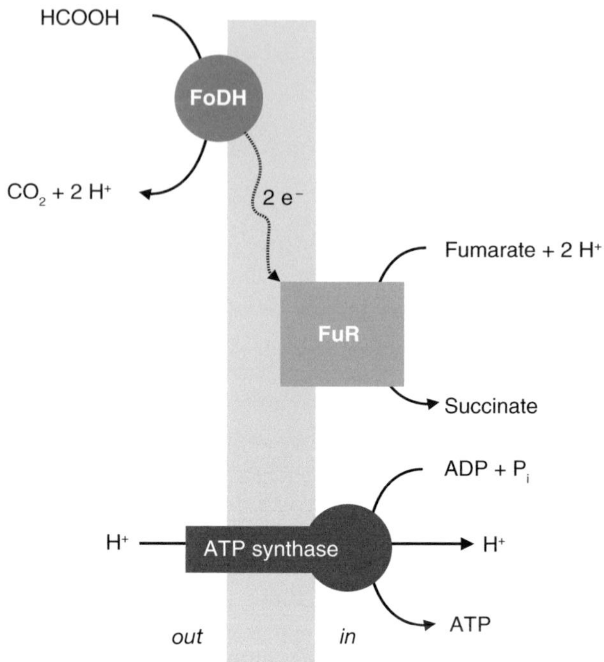

Fig. 1.8 Energy-conserving pre-respiratory transmembrane electron transfer. Electron transfer from formate dehydrogenase (*FoDH*) to fumarate reductase (*FuR*), *chemically* coupled to the formation of a p.m.f. as "fuel" for the ATP synthase: Model of the evolutionary origin of chemiosmotic energy conversion in the absence of a more advanced membrane-bound electron transport chain [99, 100, 104, 158]. The scheme strongly resembles the extant fumarate respiration (Table 1.3), i.e. the transmembrane formate-fumarate reductase complex [342–347]

The well-established and almost colloquial term "high-energy or energy-rich compound" which dates back to the German biochemist and bioenergeticist Fritz A. Lipmann (1899–1986) [118] may be somehow confusing in terms of physical chemistry but is anyhow usually and implicitly always applied to organic high-energy *phosphates* for which the term "high-energy" or "energy-rich" just means a strongly negative value for the Gibbs energy difference ($\Delta G^{\circ\prime}$) during their hydrolysis (Table 1.1). The $-\Delta G^{\circ\prime}$ values play an important role for the multifarious transphosphorylations in SLP as well as within the network of the whole intermediary energy metabolism of a living cell. Therefore, some more prominent and important representatives of such "*Lipmann-compounds*" are compiled in Table 1.1, with special reference to substrate-level phosphorylation [111, 112].

According to most researchers in the field, SLP (biochemically synonymous with *fermentation*) is the evolutionarily most ancient mode of biological energy conversion as it functions, without participation of "advanced" membrane-bound ("vectorial") enzymes, in the soluble ("isotropic") system of the cytosol, in the dark and without obligate participation of electron transport processes [104, 110]. Note, however, that, as no energy-conversion is possible without *oxidation* and as no oxidation is possible without conjugated *reduction*, also SLP does imply *redox reactions* though in this case, in the absence of light or an external oxidant (viz., e.g., O_2!), these redox reactions are *intramolecular* as most clearly seen in glycolysis [112] where, anaerobically and in the dark, without involving membrane-bound processes, half of an, e.g., glucose molecule is, as a net result, *oxidized* and the other half is *reduced*. In the absence of an external oxidant (more precisely: in a primordial O_2-free biosphere, see Fig. 1.4) the reduced half (e.g., lactate or alcohol) must be discarded (=excreted) in order to be removed from the intracellular metabolic equilibrium network while the other half (e.g., glyceraldehyde phosphate, still a carbohydrate, in the form of a triose) is retained and used as a building block for cell material. The intramolecular redox reactions during energy conversion in SLP nicely illustrate the "two halves" of cellular metabolism, the (oxidative, energy-converting and ATP synthesizing) *catabolism* or dissimilation, and the (reductive, cell material-increasing) *anabolism* or assimilation.

The knowledge of the "free-enthalpy changes" [$\Delta G^{\circ\prime}$] during the mutual transformations of the many metabolic intermediates in a living cell are of pivotal importance for an understanding of the actual biochemical driving forces within the metabolic network in a given organism, particularly in fermentatively active organisms (Table 1.1). But what about *cyanobacteria* in this context? True, cyanobacteria, combining all of the three bioenergetic processes—fermentation, respiration, and (oxygenic) photosynthesis—together in a single prokaryotic (bacterial) cell [32, 36, 37, 104, 132], are absolutely unique among living cells. But ecologically speaking, just substrate-level phosphorylation, though potentially active in cyanobacteria in the form of, e.g., lactic acid fermentation [132–135], plays a very marginal role only: Not a single cyanobacterium is known that is able to grow in the absence of both light and oxygen. The overwhelming physiological and ecological domain of cyanobacteria is *oxygenic photosynthesis* (see below) which was "invented" and brought to perfection by themselves [48, 86, 88, 136]. Also the previously much-discussed *anoxygenic photosynthesis* of certain cyanobacteria [137–139], let alone the rather doubtful "light-activated heterotrophic growth" of *Synechocystis* PCC6803 [140], turned out to be more or less limited bioenergetic short-term emergency valves rather than long-term alternative growth physiologies. And as concerns generally cyanobacterial energy metabolism in darkness [86, 88, 132, 141, 142], even effective growth at the expense of dark *respiration* on whatever (reduced) carbon substrates is all else but the rule among cyanobacteria. On the contrary, more than half of all cyanobacteria that have been successfully cultured so far [143] are obligate *photo*trophs, mostly even photo*auto*trophs, though the term *obligate autotroph* sometimes has been discredited [144, 147]. At any rate, however, as a matter of fact, a large number of cyanobacterial species, without palliation, do exclusively need light as an external (*environmental*) energy source for growth and proliferation [143, 146] while compound organic matter, primarily carbohydrates, may at least partly serve as carbon source [143, 148], as is best shown by the "artificial" *photo(organo)heterotrophic* growth on glucose of many cyanobacteria in the presence of light and DCMU [143]. This *obligate phototrophy* [145–147] is to be compared with obligate lithotrophs and methylotrophs which need specific inorganic (or 1-carbon-) reductants to be taken up from the environment [145–149].

Chemo- and photo-litho*auto*trophs depending on CO_2 for carbon supply [145–150] may face the problem that the metabolic reductants for CO_2 needed, e.g., by the Calvin cycle, i.e. ferredoxin ($E^{\circ\prime}$=−420 mV) and NAD(P)H ($E^{\circ\prime}$=−320 mV), cannot be "reached" by their natural electron donors (including light! See [146, 149, 150], also see Tables 1.2 and 1.3): Available redox potentials may be too less nega-

Table 1.2 Energetics of (mostly aerobic) chemolithoautotrophic bacteria. Molecular oxygen was taken as the standard acceptor throughout [$E^{o\prime}(O_2/H_2O)=+820$ mV]. $E^{o\prime}$, standard reduction potential; $\Delta G^{o\prime}$, change in free enthalpy Eq. (1.4)

Redox couple	$E^{o\prime}$ (mV)	Chemical reaction	$\Delta G^{o\prime}$ (kJ)	Bacteria
CO_2/CO	–513	$CO+0.5\ O_2 \rightarrow CO_2$	–257.2	Carboxydo-
$2H^+/H_2$	–420	$H_2+0.5\ O_2 \rightarrow H_2O$	–239.3	Knallgas-
CO_2/CH_4	–244	$CH_4+2\ O_2 \rightarrow CO_2+2\ H_2O$	–818.0	Methylotrophs
SO_4^{2-}/H_2S	–223	$H_2S+2\ O_2 \rightarrow H_2SO_4$	–805.3	Thiobacilli
$SO_4^{2-}/S°$	–205	$S+1.5\ O_2+H_2O \rightarrow H_2SO_4$	–593.6	Thiobacilli
NO_2^-/NH_4^+	+334	$NH_3+1.5\ O_2 \rightarrow HNO_2+H_2O$	–281.5	Nitroso-
NO_3^-/NO_2^-	+425	$NO_2^-+0.5\ O_2 \rightarrow NO_3^-$	–76.1	Nitro-
Fe^{3+}/Fe^{2+}	+762	$Fe^{2+}+0.25\ O_2+H^+ \rightarrow Fe^{3+}+0.5\ H_2O$	–4.35 (–44.3 at pH 0)	Thio-(ferro-)
NO_3^-/N_2	+752	$N_2+2.5\ O_2+H_2O \rightarrow 2\ NO_3^-+2H^+$	–65.6	Hypothetical

For calculations: $\Delta G = -n.F.\Delta E$, ΔE = standard reduction potential of acceptor minus donor; n = number of transferred electrons; $F = 96.5$ kJ.mol^{-1} .V^{-1}

Table 1.3 Energetics of several types of (aerobic and anaerobic) respiration with H_2 taken as the standard donor throughout [$E^{o\prime}(2H^+/H_2)=-420$ mV]

Redox couple	$E^{o\prime}$ (mV)	Chemical reaction	$\Delta G^{o\prime}$ (kJ)	Bacteria
O_2/H_2O	+820	$O_2+2H_2 \rightarrow 2H_2O$	–478.6	Knallgas
NO_3^-/N_2	+740	$NO_3^-+2.5\ H_2+H^+ \rightarrow 0.5\ N_2+3H_2O$	–560.0	Nitrate resp.
NO_3^-/NO_2^-	+425	$NO_3^-+H_2 \rightarrow NO_2+H_2O$	–163.2	Denitrification
Fumarate/ succinate	+33	Fumarate + H_2 → succinate	–87.4	Fumarate resp.
SO_4^{2-}/H_2S	–223	$SO_4^{2-}+4H_2 \rightarrow S^{2-}+4H_2O$	–152.2	Sulphate resp.
$S°/H_2S$	–271	$S°+H_2 \rightarrow H_2S$	–28.8	Sulphur resp.
CO_2/CH_4	–244	$CO_2+4H_2 \rightarrow CH_4+2H_2O$	–135.0	Carbonate resp.
CO_2/ CH_3COO^-	–285	$CO_2+2H_2 \rightarrow 0.5$ AcOH$+H_2O$	–52.3	*Clostridium aceticum*
N_2/NH_4^+	–284	$N_2+3H_2+2H^+ \rightarrow 2\ NH_4^+$	–78.7	Hypothetical

For calculations: $\Delta G = -n.F.\Delta E$, ΔE = standard reduction potential of acceptor minus donor; n = number of transferred electrons; $F = 96.5$ kJ.mol^{-1} .V^{-1}

tive to perform the task (of CO_2 reduction). For these cases, Nature has "invented" the so-called "reverse electron transport" ingeniously put forward and discussed by Gest [117, 150]. Reverse electron transport which had originally been discovered in artificially energized mitochondria by Chance and Klingenberg in the late 1960s, demands that primarily conserved metabolic energy (ATP etc.) is used to *invert* electron transfer reactions from the *exergonic* (–)→(+) to the *endergonic*, energy-dependent (+)→(–)-direction [86, 99, 100, 104]. Note that, thermodynamically speaking, the possibility of such inversion of the usual sequence: (exergonic or "downhill") electron transport → energized membrane (p.m.f.) → ATP synthesis into the endergonic sequence: ATP hydrolysis → p.m.f. → reverse electron transport is due to the fact that, in physiological conditions in a living cell, the enzyme responsible, viz. the ARMA or ATP synthase (see Fig. 1.8), is working very close to thermodynamical equilibrium. For a comprehensive overview on all possible growth physiologies ("ergotrophic combinations" reflecting the biophysical interplay between energy and nutrients) see Fig. 1.9.

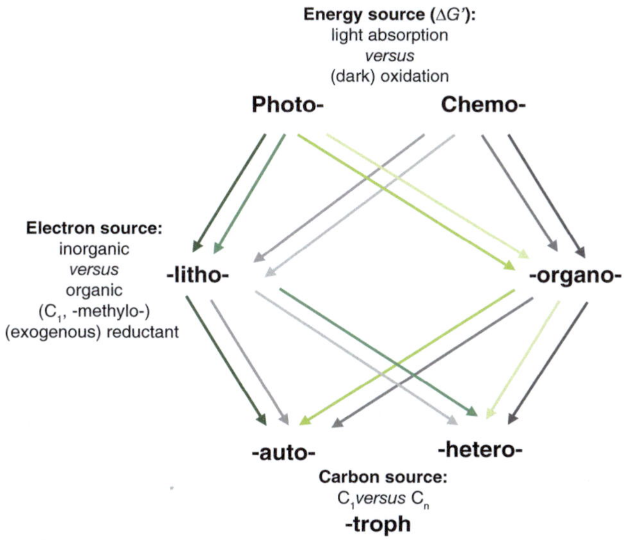

Fig. 1.9 The ergotrophic hexagon. The "magic" ergotrophic hexagon, which classifies all extant living cells according to the three "sources" of life, viz., energy source (for catabolism), electron source (for anabolism), and carbon source (for nutrition). Note that even the *eight principal ergotrophic modes* are usually *facultative* in one or the other respect but sometimes also obligate [143–148]: Typical (*facultative*) photoautotrophs, e.g., (eukaryotic) *green plants* and *algae*, may also grow (though with *etiolation*=loss of chlorophyll) as chemo-organo-hetero-trophs on, e.g., glucose in darkness, while most *cyanobacteria*, e.g., are obligate photoautotrophs unable to grow in the dark at all but perhaps capable of growing photoorganoheterotrophically, i.e., in the light (enery source) with organic substrates, primarily carbohydrates [143], as electron (-organo-) *and* carbon sources (-hetero-). Facultative photo- and chemo-lithotrophs may also grow on organic substrates [148, 261, 430–432]. Quite easily and unconstrainedly, the "*ergotrophic hexagon*" also accommodates large but rather "exotic" groups of bacteria such as (chemotrophic) hydrogen bacteria [431], CO-oxidizers [432], and other chemolitho(auto)trophs ([261, 430]; also see Table 1.2). The scheme does not distinguish between the two different fundamental biochemical *mechanisms* of energy conversion ("*phosphorylation*"), viz., SLP [111, 112] and ETP [105–110]. However, without any doubt, during the millennia of biological evolution the bioenergetically much more efficient and mechanistically much more versatile *chemiosmotic membrane principle* ("chemiosmosis") has by far won the race [104]. Thereby, particularly with respect to *energetic efficiency*, it must be considered that, e.g., during the glycolytic degradation of glucose [111, 112] along the well-known Embden-Meyerhof-Parnas pathway, 2 *mol ATP* are produced from 1 mol glucose in net effect. By contrast, aerobic respiration (see Eq. 1.5) produces, also from 1 mol glucose, 36–38 *mol ATP* (depending on the coenzymes used by the primary dehydrogenases). This huge discrepancy reflects the difference between 24% and 42% energetic net efficiency (ATP produced vs. glucose degraded) in fermentation (SLP) and respiration (ETP), respectively (see sect. 1.3.5, p. 33)

1.3.2.2 Chemiosmotic ("Electron Transport") Phosphorylation (ETP)

Compared to SLP [111, 112] chemiosmotic electron transport phosphorylation, ETP, is energetically much more efficient [104, 110]. Its essence is *membrane energization* (Figs. 1.8 and 1.10) by an electrochemical ion potential gradient across the membrane, either $\Delta\mu_H{}^+$/protonmotive force (p.m.f.) [97–100], or occasionally, in certain other living cells, alternatively a sodium-motive force ($\Delta\mu_{Na}{}^+$) [100–102]. Electron transport phosphorylation can be classified either as *photophosphorylation* (photosynthesis) or *oxidative phosphorylation* (respiration). Both types of phosphorylation are in essence ETPs [104]. However, in the former the primary oxidative step, proper, e.g. oxidation of H_2S or H_2O etc., is *endergonic*, viz. light-dependent (*photooxidations*), while in the latter the oxidation of the primary electron donor, e.g., of NADH, is *exergonic* and occurs also in darkness [104]. There is a single type of photophosphorylation that occurs without (conventional) electron transport in the membrane: This is the retinal-based photophosporylation in the purple membrane of halobacteria [151–154]. Strangely, in evolution, this absolutely deviating type of *halobacterial (retinal-based) photosynthesis*—as elegant and simple and promising as it might appear at first glance—has not been further followed up and brought to perfection in the biological world. Rather note that halobacteria (strictly aerobic respirers!), though photo*synthetic*ally capable of making some ATP in the light, are *non-phototrophic* as they are devoid of an autotrophic CO_2-reducing mechanism [154].

Clearly, both chlorophyll-based and halobacterial (retinal-based) photosynthesis depend on a *chemiosmotic* type of ATP synthesis (see [97–100, 104], i.e. they need an "*ARMA*", an anisotropic, reversible, membrane-bound ATPase (ATP synthase or reversible F_oF_1-ATPase) [119, 155, 156]. The latter enzyme is by far the most ancient and most widely spread bioenergetic enzyme in the whole biosphere. Figure 1.8 shows how this enzyme might have been crucially involved in the evolution of membrane-bound electron transport as such, already at the stage of *pre-*

Fig. 1.10 Evolutionary transition from anoxygenic photosynthesis to oxygenic photosynthesis and/or to aerobic respiration. *Top*: Simplified linear scheme for the evolutionary transition from anoxygenic ("bacterial") photosynthesis as found in purple and green sulfur bacteria to oxygenic (cyanobacterial or plant-type) photosynthesis and/or to aerobic respiration as envisaged by the generalized endosymbiont hypothesis. *Bottom*: Quantitative illustration of the redox potential's "split-and-shift" which occurred upon transition from the primordially separated two bacterial photosystems to the two extant photosystems I (low potential, FeS-type RC) and II (high-potential, pheophytin-quinone type RC), hand in hand with an independent (convergent?) evolution of a WOC (OEC) from a di-Mn-containing catalase. Finally this water-photolysing, oxygenic enzyme complex, instead of more or less specified dehydrogenases, brought about the globally unlimited bioenergetic electron supply from water (see text and [48, 104]). PSI, photosystem I; PSII, photosystem II; BChl, bacteriochlorophyll; Phe, pheophytine; BPhe, bacterial pheophytine; Q, quinol; Cyt, cytochrome; Cyt b_6f, cytochrome b_6f complex; PC, plastocyanin; A_0 and A_1, electron acceptors of P700*; FeS, iron-sulfur cluster; F_x, F_A, F_B, types of 4Fe-4S clusters; Fd, ferredoxin; $NADP^+$, nicotinamide adenine dinucleotide phosphate; DH, dehydrogenase; COX, cytochrome *c* oxidase; Cyt *c*, cytochrome *c* *respiration* [45–47, 119, 157], and it has, therefore, been extremely well studied during the decades (see [88, 155, 156]) (also see chap. 10 in this book). The simple vectorial transmembrane electron transfer e.g. from a membrane-bound formate dehydrogenase facing the outside to a fumarate reductase facing the inside of the cell (Fig. 1.8), i.e. a *vectorial* ("Mitchellian") reaction, is automatically (simply "chemically") coupled to outside acidification and inside alkalinization, creating a p.m.f. across the membrane which serves as the driving force of a membrane-bound

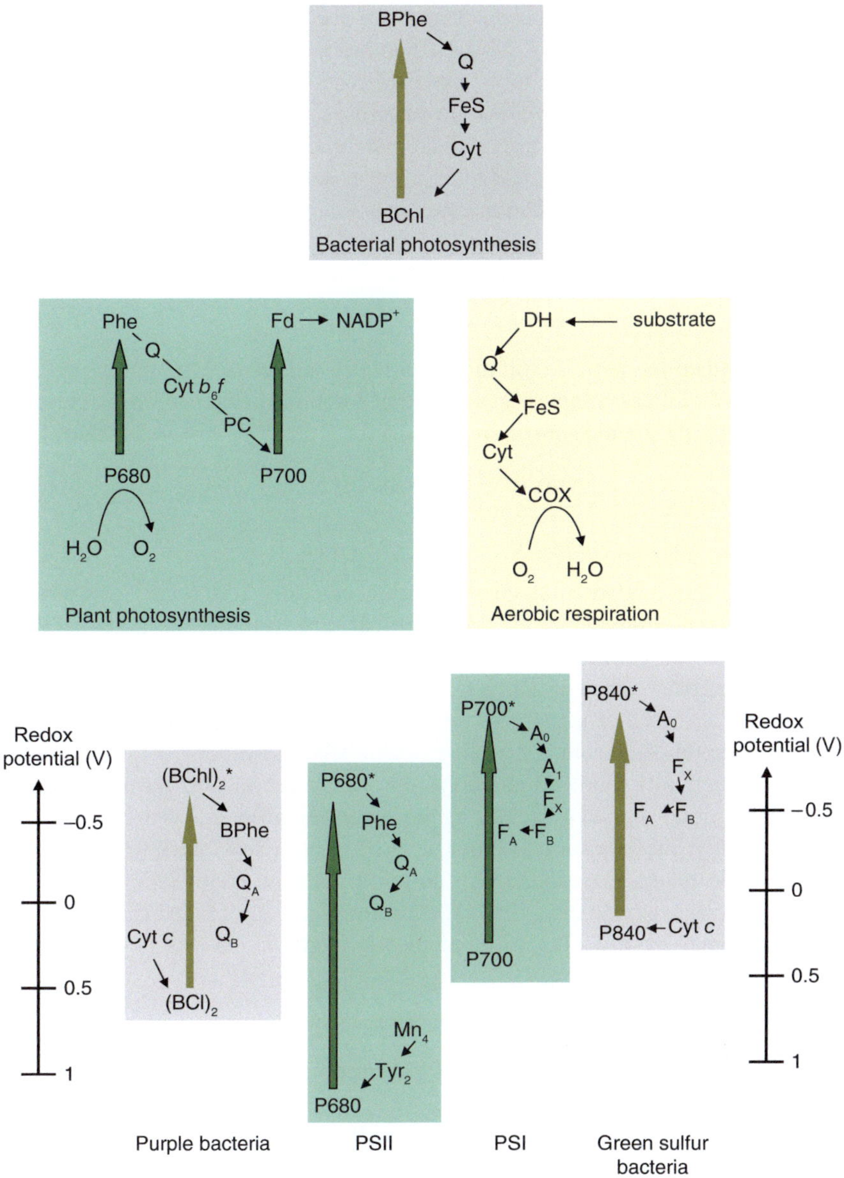

ATP synthase of proper orientation. Such mechanism represents a feasible primordial chemiosmotic system for energy conversion (also see [46, 47, 119, 157]). The energy-conserving (=ATP-synthesizing) pre-*respiratory* transmembrane electron transfer in the dark probably preceded any more elaborate and streamlined intramembrane electron transport systems employing FeS proteins, quinones and cytochromes as we know them today [158], and in evolution it might thus have preceded even photosynthesis as a whole [46, 47, 157].

Besides a membrane-bound ATP synthase of proper orientation (see Figs. 1.3, 1.8 and [99] for a more extensive discussion), a further inevitable *chemiosmotic precondition* is the occurrence of closed (membrane-enclosed) cellular compartments with defined "in's" and "out's" since Mitchell's "*vectoriality*" necessarily implies (non-equivalent) *directionality*. In general, the *free enthalpy* ΔG (kJ/mol) stored in a transmembrane electrochemical potential gradient is represented by (Eq. 1.5), with $\Delta\psi$ being the electrical membrane potential (defined as *p*-side minus *n*-side) and ΔpH being the pH-difference (again *p*-side minus *n*-side). In the specific case of proton translocation, ΔG is usually expressed as the proton electrochemical gradient $\Delta\mu_H{}^+$ (with $\Delta\mu_H{}^+ = -F\Delta\psi + 2.3RT\Delta\text{pH}$).

$$\Delta G = -nF\Delta\psi + 2.3RT\,\Delta\text{pH} \tag{1.5}$$

Mitchell has defined the term proton motive force (p.m.f. or Δp) in units of voltage (Eq. 1.6), which facilitates comparison with redox potential differences in the electron transfer chains. Using Δp and substituting values for R and T at 25°C gives (Eq. 1.7).

$$\Delta p(\text{mV}) = -(\Delta\mu_H{}^+)/F \tag{1.6}$$

$$\Delta p(\text{mV}) = \Delta\psi - 59\Delta\text{pH} \tag{1.7}$$

Equation (1.7), the fundamental chemiosmotic equation ("*Mitchell equation*"), shows the energetic equivalence of the electrical term (membrane potential $\Delta\psi$) and the chemical (concentration) term ΔpH (pH gradient). The total proton electrochemical gradient is always composed, to varying degrees, of both ΔpH and $\Delta\psi$. However, e.g., in chloroplasts the former prevails, while in mitochondria it is the latter. A comprehensive survey of all existent "chemiosmotic", i.e. membrane-bound and energetically relevant, chemotrophic electron transport reactions which actually can be found in extant bacteria is presented in Tables 1.2 and 1.3 (also see [104]). Apparently due to a subconscious shift of interest from dark to light much more, viz. whole books, have been written and published about *light-dependent*, photosynthetic *electron transport and* O_2 evolution (see, e.g., [159–162], also see [104] and the somewhat older but still matchlessly instructive booklet by Rabinowitch and Govindjee [163]).

1.3.2.3 Oxidative Phosphorylation and Photophosphorylation

Quantitatively, the largest amount of biologically synthesized ATP in our contemporary biosphere is the result of respiration and photosynthesis (see Eq. 1.8).

A 75-kg human being makes a total 70 kg of albeit short-lived ATP per day (turnover rate of an individual ATP molecule in mammals: 10^3/day). For comparison, the N_2-fixing soil bacterium *Azotobacter vinelandii* (the organism that holds the world record of respiratory oxygen uptake with a $Q_{O2} = 8$ ml O_2/h.mg d.w.!) synthesizes 300 μmol ATP$\times s^{-1}\times$(mg d.w.)$^{-1}$ corresponding to roughly 10^8 erg=10 J (turnover rate of bacterial ATP is about 3 s^{-1}). Similarly, the complete protein inventory of a living cell is in a constant state of *dissipative* degradation by thermodynamically favored (exergonic!) hydrolysis, which processes must be poised against the energy-requiring (endergonic) *rebuilding* or *reconstruction* through biosynthesis, with an overall mean turnover number of 1 per day for the entire protein inventory of a living cell, on an average. This is only another "definition" of the inherent flow-equilibrium (*dynamic* or *steady* state) of living matter as is amply and instructively discussed in the work of Ludwig von Bertalanffy (1901–1972) [91, 92]. To say it again, in a more straight-forward language: Once per day the whole protein inventory of an average living cell is replaced by newly synthesized, but otherwise identical, material. In all events, thus, life most conspicuously illustrates the classical philosophic axiom of the pre-socratic philosopher Herakleitos of Ephesos (approx. 500 B.C.): "All is flowing and nothing rests", or in his own language: παντα χωρει και ουδεν μενει [76–80], colloquially better known as "παντα ρει" ("panta rhei").

Equation 1.8 shows the chemical reaction schemes of (aerobic) respiration and (oxygenic) photosynthesis revealing that both processes which are by far the most important bioenergetic processes on earth are represented by one and the same chemical equation, just read in opposite directions:

$$12H_2O + 6CO_2 \leftrightarrows C_6H_{12}O_6 + 6O_2 + 6H_2O \qquad \Delta G^{\circ\prime} = \pm 2821.5\,\text{kJ/mol} \qquad (1.8)$$

The equation describes the present steady-state of our terrestrial biosphere and atmosphere as is *biologically* established by the concurrent and equally efficient actions of oxygenic (plant-type) photosynthesis and aerobic respiration. Equation (1.8) is nothing but the famous equation of C. B. Van Niel (1897–1985) (see R. Y. Stanier (1916–1982) [164, 165]) for *autotrophic* CO_2 *fixation* as applied to oxygenic photosynthesis, if read in the endergonic direction (from left to right: +2821.5 kJ/mol). When reversed, the same—now exergonic—reaction *releases* an equal amount of energy through respiration ("biological combustion"): –2821.5 kJ/mol. Equation (1.8), which is kind of "*life's token*" on earth, is also sometimes called the water-water cycle [166], but it could as well also be called the (*aerobic*) "CO_2–CO_2 cycle" or even, in a fit of reductionism, "H–H cycle", formally similar to the sulfur-sulfur cycle ("sulphuretum") [167], which will be discussed together with other *syntrophic communities* later in this chapter (sect. 1.3.5), but also see Fig. 1.9. Here we will just very briefly point out that, based on (Eq. 1.8), the whole biosphere on earth seems to form such a *syntrophic community*. A closer biochemical look on the reactions of (Eq. 1.8) shows that, even mechanistically, the oxidation of H_2O to O_2 shares important steps in common with the reduction of O_2 to H_2O [168, 169] (also see chap. 13 in this book). It is doubtless the reaction that also keeps our own (human) lives going, based on the biosynthesis of ATP (energy) and all other essential and nonessential ("secondary") biomaterials, i.e. biomass.

Ultimately, all primary production on earth is energetically driven by (sun) light (start of terrestrial food chains). *Chemo*synthetic primary production, as is associated with the chemoautotrophic microbial communities around so-called black smokers [170] is without any significant quantitative ecological impact on earth. However, *photosynthetically* an estimated 10^{11} t of carbon (in the form of CO_2) per year is converted into biomass by plant-type photosynthesis (to which the primordial cyanobacteria have given birth to, approx. 3.2 billion years ago [171–174]; also see Fig. 1.4) and the equivalent amount of O_2 is thereby released from water according to (Eq. 1.8). Recent estimates assign between 20 and 30% of this worldwide primary production to cyanobacteria, in particular to small unicellular marine *Synechococcus* species [175] and likewise planktonic Prochlorophytes [176] which, though not especially concentrated anywhere in the euphotic zone of the oceans, are nevertheless extremely widespread in *all* oceans. Bewilderment must be caused by the alleged finding of vestiges of oxygenic photosynthetic organelles ("thylakosomes") in several chemoheterotrophic protists such as *Psalteriomonas lanterna* and members of the parasitic genus *Apicomplexa* [177] though such findings might well go with the unifying or generalized endosymbiont hypothesis as outlined above [39, 40, 44].

On a more global scale the total biological O_2-cycle on earth, together with respective pool sizes (H_2O, CO_2, and O_2) is described in Fig. 1.5. Ecologically speaking it describes the fundamental and life-supporting *steady-state of our present-day aerobic biosphere* and atmosphere with 0.03% $^v/_v$ CO_2 and 21% $^v/_v$ O_2 which (presently!) are balanced and "stabilized" by about equal rates of O_2 production and uptake. Also compare Fig. 1.5 with (Eq. 1.8) with respect to the quantitative global oxygen exchange pattern and particularly note the dimensions of pool sizes and exchange rates.

1.3.3 Photosynthetic Oxygen Evolution

Without any doubt *cyanobacteria* are born "as creatures of light", they are "born to photosynthesize", so to speak: It had been the cyanobacteria or their immediate ancestors that had "invented" the *oxygenic photosynthesis*. More or less simultaneously, the water-oxidizing or oxygen-evolving complex (WOC or OEC) (see below) together with a "high-potential" photosystem II (pheophytin-quinone type reaction center (RC, see Fig. 1.10) must have evolved (also see Sect. 1.2.4.). Figure 1.4 shows how they provided for the so-called *biological Big-Bang* in evolution [178–181] and thus paved the path for all other "higher forms" of life, from simple eukaryotes (protozoa) up to "*Homo sapiens sapiens*". At the same time, covering all their needs for growth and proliferation with ubiquitous water, light, and CO_2—together with a few dissolved minerals—*cyanobacteria* are also rightly called the "*nonplus-ultra of bioenergetics*".

To the terrestrial biosphere as a whole the advent of the cyanobacteria and, together with them, of bulk amounts of O_2, i.e. the *Promethean Fire of Evolution* [104] is, indeed, comparable to the "discovery (and controlled use) of the *fire* or the *wheel*" by human beings. The cyanobacteria's "invention" of the water-splitting and O_2-evolving photosynthetic mechanism, doubtess deserves a few more words: Quite a number of scientific facts have now accumulated which point to an unexpectedly early origin of oxygenic photosynthesis [178–192]. Oxygen-radical damage had been proven to provoke similar biological effects as does "radiation damage" [193]: There must have been kind of an oxidative stress already before the advent of O_2 (see the broad and penetrating discussion in [48]). A substantial level of photochemically formed H_2O_2, a mildly *reactive oxygen species* (ROS), must have been washed out from the primeval, still anoxic but H_2O_2-containing atmosphere [186–188, 191, 192] leading to an enrichment of up to [μM] and [mM] H_2O_2 in localized water bodies. In defence against this H_2O_2, and supported by relatively high solubility of Mn^{2+} (compared, e.g., to Fe^{2+} [194] (also see chap. 21 in this book)), certain still anoxygenic (cyano?)-bacteria in these habitats could have started to synthesize a *di-manganese catalase* as a forerunner of likewise Mn-containing photosystem II (PSII) [48, 191, 192]. Equation (1.9) and (1.10) are (endergonic photochemical) reactions that might have been responsible for the photochemical formation of H_2O_2 (and a negligible amount of abiogenic O_2 [195]) on the early earth.

$$\mathrm{H_2O + H_2O} + UV \leftrightarrows \mathrm{H_2O_2 + 2[H]} \qquad \Delta G^{\circ\prime} = +340.2\ \text{kJ, equivalent to } 1\,\textit{Einstein}\,(\text{mol}) \text{ of 350-nm-UV} \tag{1.9}$$

$$\mathrm{H_2O + H_2O} + UV \leftrightarrows \mathrm{O_2 + 4[H]} \qquad \Delta G^{\circ\prime} = +478.6\ \text{kJ, equivalent to } 1\,\textit{Einstein}\,(\text{mol}) \text{ of 250-nm-UV} \tag{1.10}$$

Such photochemical redox reactions (Eqs. 1.9 and 1.10) could have been greatly facilitated by the hydrogen-accepting capability of compounds like CO and $(CN)_2$ which certainly were abundant, too, in the primeval (secondary) terrestrial atmosphere.

The structure of the active center of di-manganese catalase ("pseudocatalase") from *Lactobacillus plantarum* and *Thermus thermophilus* [196, 197], when compared with presently available WOC structures as discussed in [198–200], also see [104], could have evolved in a primordial (still "anaerobic"?) cyanobacterium in response to the "pre-O_2 oxidative stress" [48, 193] resulting from the photochemically formed H_2O_2. It could have been a starting point, in turn, for the evolution of the tetranuclear Mn_4O_xCa cluster of the WOC from the former, by gene-doubling [201, 202] and incorporation of the catalytic Ca^{2+} (for schematic illustration see [104]). For more comprehensive information on the topic of "water-splitting and electron transport in oxygenic photosynthesis" see [159–162, 198–200]. Summing up, the present state of our knowledge about the geochemistry of terrestrial O_2 [203] and the biochemistry of oxygenic photosynthesis (see, in particular, [48]), together with the evidence for a surprisingly early origin of oxygenic photosynthesis [178–

192] and the almost identical biophysical impacts of short-wave UV radiation and oxygen radicals [193], and also taking into account recent findings from comparative structural biochemistry [196–200, 207–214], seem to corroborate the evolution of the water-splitting complex from a primordial di-manganese ("pseudo-") catalase, i.e., a *biological or evolutionary Big-Bang* [178–183] in the infancy of our mother earth already [Fig. 1.4].

1.3.4 The Evolution of Photosynthetic Reaction Centers

Roughly in parallel to the evolution of the WOC or OEC must have proceeded, probably also by some gene-doubling events [201, 202], the evolution of a two-photosystem- (2-RC-) photosynthetic apparatus (plant-type photosynthesis or *photophytotrophy* [45]), as shown in Fig. 1.10: Under the impact of an increasing oxygen partial pressure in the atmosphere (Fig. 1.4) the primordial single photosynthetic system of an anoxygenic purple or green photosynthetic bacterium underwent a redox potential "shift-and-split" (or "*redox disproportionation*", in chemical language) into a *high-potential* and into a *low-potential* photosystem, both of which first must have separately co-existed, side by side, in still unconnected form, in a single (prokaryotic) cell.

Unfortunately, an anoxyphototrophic bacterium which houses *both* bacterial RCs, in still unconnected form, in one and the same cell, has not yet been found in Nature. Rather, in the course of further evolution and economization, the two individual photosystems, let us say "pre-PSII" and "pre-PSI", must have merged into a full-fledged cyanobacterial (plant-type) Z-scheme [204–207]. At the same time, by the same path, a "*respiratory chain*", instead of a high-potential PSII, might have "split off" the common branch between ("pre-") PSII and ("pre-") PSI, now with an "alternative" (aerobic) cytochrome *c* oxidase (COX) as a terminal electron acceptor (schematically described in Fig. 1.10). Be it noted that the recently achieved X-ray structural analyses of PSI [208–210], PSII [197–200, 211, 212], and the cytochrome b_6f [213, 214] complex, together with sequence comparisons of the respective proteins (see [199, 200, 215], also see respective chapters in [162]), are in good agreement with an evolutionary scenario as put forth here. A similarly convergent evolution possibly also involving several gene-doubling events as for the WOC or OEC (see above) and the RC's in photosynthetic electron transport (PET) (see Fig. 1.10) is being discussed for the "parallel" formation of plant-type *ferredoxin* from bacterial *ferredoxin* in general [45] as well as for the evolution of cyanobacterial cytochrome c_6 and plastocyanin [194] (also see chap. 21 in this book), the cyanobacterial C-phycocyanin family [201], and several other soluble cyanobacterial electron transport catalysts [202].

Obviously, however, on the other hand, ecologically speaking, some other organisms "refused to adapt" (to O_2) but "decided" to entirely hide against O_2 and retreat into anoxic ecological niches such as the bottom of eutrophic lakes or the paunch of ruminants. Thereby, again, another *biological dichotomy* was created which now led to the still prevailing and rather significant group of "obligate anaerobes" in our extant, essentially oxic biosphere [216–218].

1.3.5 The Ergotrophic Hexagon

The "magic" *ergotrophic hexagon* is shown in Fig. 1.9 (sect. 1.3.2.1., p. 25). It includes and consistently classifies ***all*** living cells according to the three fundamental *life requirements* or "ingredients": Energy, electrons (reducing power), and carbon. The three fundamental and absolutely indispensable bioenergetic constituents or *sources*, viz., *energy* source (light, or dark oxidation), *electron* source = reducing power (organic or inorganic electron donors), and *carbon* source (C_1 compounds, or pre-synthesized C_n compounds) may thus be derived each from two exclusively different "origins". The eight different combinations yield eight distinctly different ergotrophic types of living cells displaying very peculiar metabolic types in each case. For each of the ergotrophic modes a living representative can be found among extant organisms. Unfortunately, it is not the place here to discuss in detail all the different and partly exotic ergotrophic cell types, e.g., from chemo-litho-heterotrophic to photo-organo-autotrophic organisms, together with all the needed *accessory characterizations* such as obligate (strict)—facultative, aerobic—anaerobic, thermophilic—psychrophilic, acidophilic—alkalo(i?)philic (see, e.g., textbooks of microbiology [113, 114, 261]).

In the context of Fig. 1.9 it is noteworthy that, of the three fundamental "caterers to living cells": energy, reducing power ("electrons"), and carbon, each of the three "sources" occurs in Nature in two strictly different forms, i.e. they represent a *triple dichotomy*: The only environmental energy sources (*fuel of life*) that can be converted into biologically useful and freely convertible (intracellular) energy (ATP and/or an energized biomembrane) are (sun) light for *phototrophs* and the oxidation of reduced substrates (for SLP or ETP) in *chemotrophs* (see Fig. 1.6b). Yet, the *ergotrophic hexagon* (Fig. 1.9), which focuses on the energy *source* does not discriminate between different biochemical mechanisms of energy conversion: A *chemotrophic* organism, e.g., can recruit its metabolic energy either through substrate-level phosphorylation (SLP) or through a chemiosmotic (usually electron-transporting) mechanism (ETP).

The next ergotrophic dichotomy is presented by the utilizable environmental *reducing power* (electron source) for inevitably reductive, energy-requiring cellular *anabolism* (biosynthesis of cell material) which is indispensable for growth and proliferation but also for the permanent *anaplerotic* reactions necessary for maintaining the steady-state ("flow-equilibrium") of living matter in general [90–92]. External electron sources can be complex organic C_n compounds (*organotrophy*), as might be provided by usual carbon nutrition which, in ordinary *chemo-organo-heterotrophs*, is energy-, electron-, and carbon-source at the same time (usually abbreviated as "*heterotrophy*" in short, the normal growth physiology of *animals* and "colorless microorganisms"). Electron source, however, may also be an inorganic compound (*lithotrophy*) including, by definition, also reduced C_1 fragments ("methylotrophy" [219, 220]; see Fig. 1.9; also see Table 1.2). In case of lithotrophy, carbon supply may be separate from reductant supply: For example, desulfuricants (*Desulfovibrio* sp., *Desulfotomaculum* sp., etc.) "produce (biological) energy" by the oxidation of H_2 with SO_4^{2-} (i.e. chemolithotrophically) but must assimilate, for carbon supply, amino acids or other organic acids, resulting in chemo-litho-heterotrophy (Fig. 1.9). Photo-organo-heterotrophs, e.g. Pringsheim's well-known (eukaryotic) *acetate algae*, assimilate acetate at the expense of absorbed light energy (acetate is electron and carbon source, energy source is light). The *chemo-organo-autotrophic Pseudomonas ocalaticus* splits oxalate to formate, decarboxylates formate, and assimilates CO_2 by way of a conventional Calvin cycle, i.e. autotrophically.

The third ergotrophic dichotomy is formed by *the compositional building blocks of living cells*, viz. *carbon nutrition*. By far the most abundant atom, together with H, in living matter is C. The carbon sources of living cells may be either complicated organic molecules (C_n-compounds) pre-formed in other (originally: phototrophic) organisms (*heterotrophy*), or C_1 compounds (*auto*trophy). Autotrophs, therefore, are capable of performing an *organic total synthesis* of the whole biomass from C_1 fragments, an unprecedented biosynthetic task. By far the most common C_1 fragment assimilated (usually through the Calvin cycle) is CO_2 but also CH_4 and other reduced C_1-compounds (*assimilated*, as HCHO, by methanogens [5, 221] and methylotrophs [210, 211]) do constitute an autotrophic life style.

It seems important to note, however, that *methylotrophs* (strict aerobes) [219, 220] and *methanogens* (strict anaerobes) [5, 221] (also see Tables 1.2 and 1.3) naturally are phylogenetically farthest apart and absolutely unrelated. Nevertheless, both assimilate carbon in the form of metabolic formaldehyde, HCHO, through identical biochemical pathways indiscriminately, viz., the serine or the ribulose monophosphate pathways [5, 220]. Look at the following metabolic parallelism (better: *anti*-parallelism) between from CH_4 to CO_2 (oxidation) and from CO_2 to CH_4 (reduction). In either case, the four electrons will pass HCHO, i.e. formaldehyde (C with oxidation level *zero*) as a common intermediate: ***Methylo***trophs: $CH_4+O_2 \leftrightarrows HCHO+H_2O$ and $HCHO+O_2 \leftrightarrows CO_2+H_2O$ (lithotrophic, aerobic oxidation of CH_4). ***Methano***gens: $2H_2+CO_2 \leftrightarrows HCHO+H_2O$ and $HCHO+2H_2 \leftrightarrows CH_4+H_2O$ (anaerobic carbonate respiration). The C_1 fragment autotrophically assimilated is formaldehyde in either case (see Tables 1.2 and 1.3). Since, for energy conversion in darkness, (one of) the first catabolic step(s) *must be* an oxidation, even a carbon substrate on the carbon-oxidation level of biomass already (viz. ±0, such as, e.g., acetate) cannot be assimilated without reduction. In many respects, therefore, methanogens and methylotrophs, though physiologically absolutely unrelated, offer a nice paradigm of convergent evolution! (See above).

1.4 Cyanobacterial Photosynthetic and Respiratory Electron Transport

The fundamental difference in biochemical mechanism and energetic efficiency between SLP and ETP (chemiosmosis) has already been pointed out. The much higher energetic efficiency (in terms of ATP production) and mechanistic versatility (as a *membrane process*, independent of special chemical "high-energy" intermediates) of ETP have by far "*won the race*" [104]. In the words of J. B. Hall [222] versatile membrane-bound electron transport assemblies have "freed the organisms from the slavish dependence on particular phosphorylatable substrates" (Table 1.1). *Cyanobacteria*, as the electron-transporting photosynthesizers *par excellence* [104], certainly were among the first that have supplemented and streamlined the primordial chemiosmotic transmembrane electron transfer system as shown in Figs. 1.10 and 1.11 with additional membrane-bound electron carriers such as, first of all, an electron/proton-transporting quinone [46, 88, 99, 100, 104]. As kind of "free-living proto-chloromitochondria" [39, 40] the *cyanobacteria* thus may also be regarded as *electron transport protagonists*. Moreover, *cyanobacteria* were the first living cells that produced bulk amounts of free, molecular O_2 in the light (see sect. 2.3.). In doing so, they have successfully and persistently learnt to cope with it. But even before the functioning of a full-fledged aerobic respiration (most probably from foregoing photosynthesis [45–47, 99]; see Fig. 1.10), naturally, there must have been a rather long (geologic) period of mere detoxification of O_2 and other ROSs.

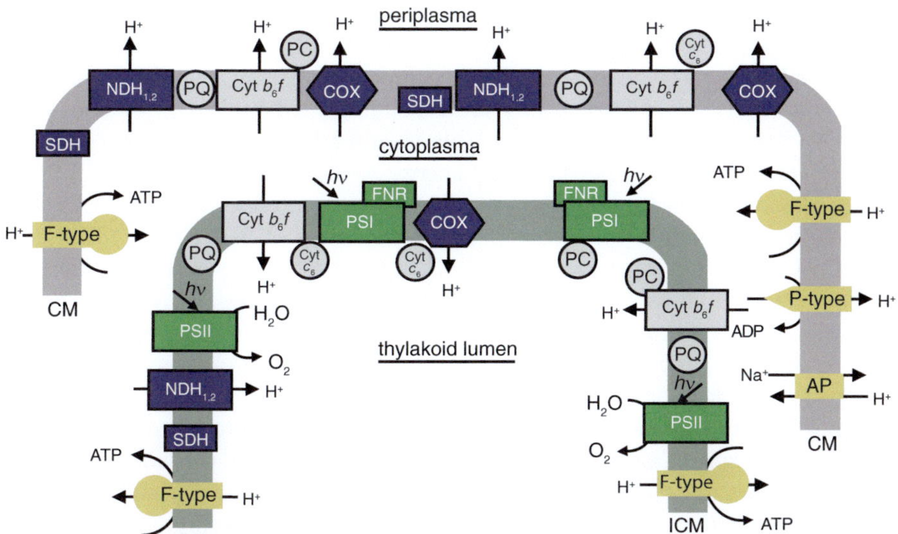

Fig. 1.11 General structural and functional organization of electron transport carriers and ATP synthases in the membrane systems of a model cyanobacterium. Electron donor and acceptor, as well as protein complexes involved exclusively in PET, are depicted in *green*, those found only in RET in *blue color*. Electron components used in both PET and RET are shown in *grey*. CM, cytoplasmic (plasma) membrane(s); ICM, intracytoplasmic (thylakoid) membrane(s); PSII and PSI, photosystems II and I; PQ, plastoquinone; Cyt b_6f, cytochrome b_6f; PC, plastocyanin; Cyt c_6, cytochrome c_6; FNR, ferredoxin-NADP$^+$ reductase [254]; NDH_1, (mitochondrial-like) NDH-1 enzyme [321, 322, 325] and NDH_2, (nonproton-pumping) NDH-2 enzyme [330, 331, 335]; SDH, succinate dehydrogenase [333, 334, 340–347]; COX, cytochrome *c* oxidase; proton-pumping P-type ATPase (ATP hydrolase [240, 411, 412, 433], and F-type ATPase (ATP synthase [155, 156, 240]), and AP, the Na^+/H^+ antiporter [397, 398], are also shown. Note that neither in cyanobacteria [413, 416] nor in mature chloroplasts [417] any conspicuous structural continuity between CM and ICM could so far be discovered

For this rather long transition period of *adaptation to* O_2 the *cyanobacteria*, owing to the well-documented occurrence of a variety of O_2- and ROS-detoxifying oxidoreductases (superoxide dismutases, catalase-peroxidases, heme catalases, manganese catalases, peroxiredoxins, etc.) [223–232], were perfectly well equipped, too. In all, therefore, they are also sometimes called the *tamers of* O_2 [86].

1.4.1 Photosynthetic Electron Transport (PET)

With a single exception (i.e. *Gloeobacter violaceus*, also see chap. 9 in this book) all cyanobacteria contain two bioenergetically competent but physically separated intracellular membrane systems, the (yellow) chlorophyll-free cytoplasmatic or plasma membrane (CM) and the green chlorophyll-containing thylakoid or intracytoplasmic membrane (ICM) (Fig. 1.11). Either of the two encloses an osmotically autonomous compartment, the cytosol (surrounded by the CM) and the intrathylakoid or lumenal space (surrounded by the ICM). Only in *Gloeobacter violaceus*, which does not at all form a thylakoid membrane [233], does the CM also contain chlorophyll *a* and all other photosynthetic components. Being nevertheless an obligate phototroph [143], *Gloeobacter* is correspondingly tedious and difficult to grow.

As has been outlined above, oxygenic photosynthesis is the main process on Earth that converts solar energy into chemical energy. All higher life on Earth depends on this process, since all the oxygen in the atmosphere is produced by oxygenic photosynthesis (Eq. 1.8). In cyanobacteria, all components of PET (besides RET, see below) are located at, or associated with, ICM (Fig. 1.11). Photosynthetic electron transport starts with capturing of the light by antenna systems, which transfer the excitation energy to reaction centers that, in turn, catalyse light-induced electron transfer across the thylakoid membrane. The highly efficient job of capturing light energy, for which chlorophyll *a* molecules (absorbance maxima at 430–440 nm and 670 nm) contained in the photosynthetic reaction centers are responsible, is in fact not enough to supply all the needed light. Cyanobacteria have partially solved this problem by utilizing the phycobilisome (PBS) located on the stromal side of ICM, a water-soluble antenna complex of fluorescent, pigmented proteins associated with PSII. PBS is responsible for as much as 50% of protein mass within a cyanobacterial cell. Depending on the light intensity and wavelength, cyanobacteria can also undergo state transitions and transfer the PBS to PSI.

The phycobilisome consists of a basis consisting of three cylinders attached to ICM, and several rods that protrude from it in a fan-like arrangement. The rods are trimers or hexamers of phycoerythrin, phycocyanin and allophycocyanin. Although each protein directly absorbs light (within 450 nm and 660 nm, with each protein having its own spectral features), but the absorbed light energy is also passed over from one to the other protein with a fixed order of energy transfer: phycoerythrin → phycocyanin → allophycocyanin → photosystem chlorophyll. Phycoerythrocyanin may replace phycoerythrin in certain organisms [201]. In addition to the membrane extrinsic phycobilisomes, cyanobacteria also have membrane intrinsic antenna complexes, which all belong to the Pcb family. The group of Pcb proteins, that serve as antenna for both PSII & PSI, is diverse as they show much variation in their pigment composition [234, 235].

Photosystem II (PSII) catalyzes light-driven electron transfer from water to plastoquinone (PQ) and has the same dimeric oligomeric structure in plants and cyanobacteria [161]. It consists of 17 protein subunits, which bind 35 to 40 chlorophylls and 8 to 12 carotinoids [198–200, 211, 244–247]. The subunits are named according to their genes PsbA to PsbO, PsbU, PsbV and PsbX. This protein complex is unique and provides the whole photosynthetic process with electrons by extracting them from water [159–162]: $H_2O \rightarrow O_2 + 4H^+ + 4e^-$. Light for the photoreactions in PSII is captured by the PBS and the core antenna of PSII [237]. Excitation energy is transferred from the antenna into the center of the complex. Each PSII monomer harbors the electron transfer chain consisting of PsbA (D1) and PsbD (D2). The structural similarity between the central core of PSI and PSII is remarkable, as the sequence identity is less than 15% [215]. The structural comparison clearly suggests that all photo-reaction centers have evolved from a common ancestor as previously proposed [236, 238]. The cofactors of the ETC are four chlorophyll *a* molecules, two pheophytins, two plastoquinones, two tyrosines (TyrZ and TyrD) and the manganese cluster.

P680 has the highest redox potential reported for any cofactor in biological systems and this is the thermodynamic prerequisite for water oxidation. The four chlorophylls are arranged in two symmetrically related pairs and all may be able to perform the initial charge separation across ICM. In each charge separation event, $P680^+$ is re-reduced by extracting one electron from the oxygen evolving complex (Mn cluster, WOC or OEC) consisting of 4 Manganese ions and a Calcium ion. The structure of the Mn cluster is still unsolved. A redox active tyrosine functions as the intermediate electron carrier between $P680^+$ and the Mn cluster. One molecule of O_2 is evolved after 4 positive charges have been accumulated. During this process, 4 protons are released into the lumen of the thylakoids. From P680, the electron is transferred to pheophytin and via tightly bound plastoquinone PQ_A to mobile PQ_B at the Q_B-side that serves as the terminal electron acceptor in PSII. After 2 charge separation events, PQ is doubly reduced, takes up two protons from the stroma, leaves PSII as PQH_2 and diffuses through the membrane to the b_6f complex. A PQ from the PQ-pool re-fills the Q_B-site (Fig. 1.11).

Electrons from mobile Q_B are transferred to the cytochrome b_6f complex that is also located in ICM within proximity to PSII and PSI. Electrons move internally through the cofactors of the cytochrome b_6f complex via the cytochrome *f* subunit directly to one of the lumenal mobile electron carriers, such as plastocyanin (PC) or cytochrome c_6 (Cyt c_6) (Figs. 1.10 and 1.11). In 2003 the structure of the cytochrome b_6f complex (Cyt b_6f) from the cyanobacterium *Mastigocladus laminosus* was solved [213] (also see [214]) comprising a dimeric multiprotein complex with each monomer consisting of eight subunits and several cofactors ligated by the protein. The cofactors are cytochrome *f*, a 2Fe-2S-cluster (Rieske iron-sulfur protein) and cytochrome b_6. Electrons are transferred from mobile PQ having two protons added in its doubly reduced form. Differently from the photosystem proteins, in complex III the two electrons of the reduced quinone do not take the same path: The first electron moves through the high-potential transfer chain including the Rieske protein and cytochrome *f* both located at the lumenal side of the ICM. The second electron is transferred by two *b*-type hemes (referred to by their absorbance maxima as b_{562} and b_{565}, or as b_n and b_p, the latter subscripts referring to whether the particular heme is closer to the *n*-side (stromal) or *p*-side (lumenal) of ICM) to the stromal side of the membrane. Alternatively, it may also be moved to the anionic plastosemiquinol of the original donor quinone if cyclic electron transfer is taking place. In either case, an extra proton is taken up from the stromal side, and either released to the lumen or taken up by the deprotonized plastoquinol. This "Q-cycle" may be compared to the electron transfer cycle in the cytochrome bc_1-complex [337–339] however, in spite of the functional similarity, Cyt b_6f and cytochrome bc_1 are only distantly related in structural terms [239]. In any case, cyanobacterial membranes seem to possess kind of "chimeric *b.c*-complex" that shares immunological as well as inhibition characteristics with both mitochondrial bc_1- and chloroplast b_6f-complexes [240–242], somehow reminiscent of the analogous complexes in green sulfur bacteria [353] and *Heliobacteriaceae* [243].

In connection with Cyt b_6f the surprising discovery of a novel semi-*c*-type cytochrome, termed cytochrome *x* or c_n, deserves special attention. Its heme group is covalently attached to the protein via only one thioether bond and the fifth and sixth coordination sites of its iron ion are H_2O only and free, respectively. Cytochrome c_x resides between heme b_n and PQ on the stromal side of ICM in close vicinity to a conspicuous positively charged amino acid region of the b_6f-complex. Cytochrome c_x is a highly plausible candidate for the long-sought-for ferredoxin:PQ oxidoreductase necessary for cyclic endogenous PET around PSI [213].

Cytochrome *f* (a purely historical name: "f" from *folium*, the Latin word for *leaf*) is the largest subunit of Cyt b_6f and has the largest membrane-extrinsic domain projecting into the lumen. This domain contains the *c*-type heme ligated by a histidine and a tyrosine and covalently bound with the protein. The positive midpoint potential of cytochrome *f* (290–360 mV) is thought to be a result of the heme's protein environment. Cytochrome *f* donates electrons to either plastocyanin (PC) or cytochrome c_6 (Cyt c_6), both being small soluble electron carriers present on the lumenal side of ICM (Fig. 1.11).

While higher plants and many cyanobacteria utilize solely PC, in some green algae and cyanobacteria PC may be functionally substituted by cytochrome c_6 ([194, 376–379]; also see [380–383]). Actually, despite having different structures, the two proteins can replace each other and play the same physiological role in cyanobacterial PET, but the synthesis of either one is controlled by copper availability within each organism. Plastocyanin (encoded by the *petE* gene) consists of a single polypeptide chain forming a β-barrel with eight β-strands and a small α-helix, along with a type-1 blue copper center (235–238). On the other hand, Cyt c_6 (encoded by the *petJ* gene) is a typical class I *c*-type cytochrome containing four α-helices and a covalently-linked heme group in which the iron atom is axially coordinated by a histidine and a methionine [244–247]. This peculiar case of 'convergent evolution' is extensively dealt with in chapter 21 of this book.

Despite their very different structures, both proteins share a number of physico-chemical properties: their molecular masses are about 10 kDa and their midpoint redox potential values are around 350 mV at pH 7, which is in accordance with their role as electron shuttle between cytochrome b_6f and PSI [194] (also see chap. 19 in this book). Moreover, their isoelectric points are often similar within the same organism but can vary in parallel from one organism to another [250]. Beside *Synechococcus* sp RCC307 and *Thermosynechococcus el.* BP-1, all cyanobacteria contain one *petE* gene (exception: *Gloeobacter violaceus* with two paralogs). Roughly, in filamentous cyanobacteria PC is usually positively charged at pH 7, whereas PC from unicellular species is mostly acidic (exception: *Gloeobacter*) (Table 1.5) (also see [250]). Especially in cyanobacteria that have several, viz. [2–4] *petJ* genes, the isoelectric points of PC and Cyt c_6 can also be dissimilar [250]. Nevertheless, regarding PET the kinetic information, along with structural data, indicate that PC and Cyt c_6 possess similar surface regions to interact with their redox partners in PET, but also that they are responsible for determining the reaction mechanism and the spatial conformation of the transient reaction complex (for a review of the alternative, yet functionally identical roles of PC and Cyt c_6 in PET see [194]; further details of the molecular interaction of soluble (cyanobacterial) redox proteins are discussed in [251–253]) (See also chap. 19 in this book).

Both PC and Cyt c_6 approach PSI through either electrostatic or hydrophobic interaction with the nature of interaction varying between species. In any case the carrier docks in a groove on the lumenal side of PSI and donates its electron to the photo-oxidized chlorophyll dimer, P700$^+$ [213]. PSI is a light-driven oxidoreductase that actually transfers the electron from the lumenal electron carrier to ferredoxin on the stromal side. That particular process supplies electrons for subsequent reduction of NADP$^+$ to NADPH by the ferredoxin-NADP$^+$ reductase (FNR) ([254]; also see Fig. 1.11). It also serves generation of a proton electrochemical gradient across the thylakoid membrane. The first crystal structures of PSI were obtained from *Thermosynechococcus elongatus* [208, 209]. PSI belongs to the largest membrane proteins ever crystallized in its entirety with 12 subunits (PsaA-PsaF, PsaI-PsaM, PsaX) and 127 cofactors (chlorophylls *a*, phylloquinones, iron-sulfur clusters, carotinoids). PsaA and PsaB are the largest subunits making up the core of the (heterodimeric) RC and provide the binding side for PC or Cyt c_6 as well as two conserved tryptophans important for electron transfer to P700$^{+\bullet}$. The entire functional PSI is usually found in the ICM as a trimer of these individual heterodimeric RCs. In total, each of the PSI trimer's monomers contains 95 Chl *a* molecules, with six of them actually forming part of the ETC (see below).

The reaction center core harbors the ETC and a core antenna system. The ETC is located at the very center of the reaction center core, at the "interlocked" region of PsaA and PsaB. It is bifurcated and consists of two diametrically opposed branches. The electron proceeds up one of the branches, from P700 to the uppermost Fe-S cluster. Regarding residues ligating the ETC's cofactors, all residues involved in coordinating the Mg^{2+} of ETC chlorophylls or in H-bonding or otherwise ligating other co-factors are very strictly conserved from cyanobacteria to higher plants. At the basis of the ETC P700 is found which is a heterodimer of overlapping and parallel chlorophylls *a*. Here charge separation takes place. Above the P700 dimer a second and a third pair of Chl *a* molecules constitute the pathway of electrons followed by a pair of firmly bound phylloquinone cofactors and, finally, three 4Fe-4S clusters (F_X, F_A, F_B).

The final cofactor in PSI that actually transfers electrons to the electron acceptor proteins, ferredoxin or flavodoxin, is F_B. Ferredoxin is a small (10.6 kDa) soluble, globular and acidic protein that approaches PSI from the stromal side of the ICM. In cyanobacteria ferredoxin (encoded by light-induced *fedI* gene) is characterized by a 2Fe-2S cluster of −420 mV, making reduced ferredoxin one of the strongest soluble natural reducing agents, though still not as strong as PSI. In conditions of iron deprivation, ferredoxin is replaced by its functional but iron-less analogue, flavodoxin, which is also a small, highly acidic protein. Compared to ferredoxin its overall structure is completely different, whereas charge and hydrophobic surface patterns are comparable, very similar to interchangeable PC and Cyt c_6 (see above). The low-potential cofactor in flavodoxin is flavin mononucleotide (FMN). Ferredoxin or flavodoxin finally transfers the electron to the ferredoxin:$NADP^+$ oxidoreductase (FNR), a low-potential flavoprotein, which reduces $NADP^+$ to NADPH [253]. NADPH is the biochemical reductant of 3-phosphoglycerate forming glyceraldehyde phosphate, a triose phosphate and the first carbohydrate as the primary product of CO_2 fixation through the Calvin cycle, also see chap. 12 in this book.

Figure 1.11 clearly demonstrates that ICM harbors both PET and RET. Before describing the interplay between both ET pathways (see sect. 1.4.2.3) the respiratory electron transport will be briefly highlighted.

1.4.2 Aerobic Respiration and Respiratory Electron Transport in General

1.4.2.1 The Origin of Aerobic Respiration

The evolutionarily laborious and time-consuming way of establishing or adapting a *full-fledged and sustained mechanism* of coupled *aerobic respiration in an evolving living cell* (viz., in a *cyanobacterium*) certainly does require the availability of O_2, i.e., an *oxic biosphere* (However, see [304, 305]. Also see below). Nevertheless, based on phylogenetic gene analysis (sequence comparisons) the so-called "respiration-early" hypothesis has been proposed [255–259]. This hypothesis, the essence of which had been published years before already [260], rests on the striking molecular sequence similarity between aerobic COX and certain electron transport components of anaerobic nitrate respiration (see Table 1.3), including, e.g., the presence of Cu_A in *Pseudomonas stutzeri* N_2O reductase. However, molecular similarities ("homologies") of aerobic and anaerobic respiratory electron transport components do not necessarily mean that the former have evolved from the latter. The same similarities would be found if the latter had evolved from the former [48, 86] in the sense of some long-lasting adaptation to anaerobic habitats. Such "reverse evolution" ("re-volution"?) also seems to be valid for many representatives of the so-called Archaea [1–3, 5, 7]. The problem thus may not have been so much one of the living cell itself (whose versatility and adaptability in the course of aeons has turned out to be almost unlimited) but rather one of environmental (geological) constraints [45, 48, 203, 261].

Geological evidence [48, 203] lets it appear highly unlikely that, before the advent of bulk O_2 (set free by *cyanobacteria* and/or their immediate ancestors) in the biosphere, substantial quantities of potential (oxygenated) electron acceptors such as N-O compounds could have been available at all [188–190, 203]. In contrast to oxygenated N-compounds, in an anoxic biosphere, fumarate, CO_2, and SO_4^{2-}, rather than NO_3^-, would have been readily available for "tinkering up" some anaerobic RET system: Fumarate, as a frequent chemical intermediate of many (anaerobic) fermentative pathways [113, 114], and SO_4^{2-} as a component of the *ecological sulfur cycle* [167] where, in the light, anoxyphototrophic representatives of e.g., *Chromatium* spp. or *Chlorobium* spp. would anaerobically photooxidize sulfide to sulfate while desulfuricants (likewise anaerobically and, if necessary, in the dark) would reduce SO_4^{2-} back to S^{2-} (see Tables 1.2 and 1.3). Both reactions are independent of O_2!. The respective *syntrophic community* of sulfide-oxidizers and sulfate-reducers is responsible for the bacterial *sulfur cycle* [167]: $H_2S+2CO_2+2H_2O \leftrightarrows H_2SO_4+2[CH_2O]$ (reductive *anabolism* for the synthesis of biomass), and $H_2SO_4+8[H] \leftrightarrows H_2S+4H_2O$ ("reductively dissimilatory", yet overall oxidative!, *catabolism*, since the metabolic hydrogen [H] for sulphate reduction comes from the oxidation of organic substrates or H_2) for ATP synthesis. A similar "*anaerobic CO_2 cycle*" powered by a (fictitious) syntrophic community of (lithotrophic) *methane oxidizers* (Table 1.2) and *methanogens* (carbonate respirers, Table 1.3) exists on paper only since all naturally occurring methanogens have been found to be (very) strict anaerobes [5, 221] while methane oxidizers were generally characterized as strict aerobes [219, 220]. Yet, nevertheless, there is indeed also an "anaerobic CO_2-cycle" in our biosphere: The strictly anaerobic syntrophic community of *Methanobacterium* spp. and *Ruminococcus* spp. in the paunch of ruminants (together representing kind of "anaerobic CO_2-cycle" as contrasted with the "aerobic CO_2-cycle" of Eq. (1.8), see also [166]). In this highly specialized ecological niche *Ruminococcus* just replaces the inexistent "anaerobic methane oxidizer" by an anaerobic bacterium that ferments glucose to H_2 (and CO_2). In *Ruminococcus* spp., an obligatory inhabitant of the paunch of Ruminants, glucose is the hydrolytic end-product of the cellulose present in *grass*, the usual food-stuff of a cow—which immediately reminds us of the famous saying of the prophet Jesaia: "*All flesh is grass*!"…). The H_2 and CO_2 produced in the "hydrogen fermentation" of *Ruminococcus* spp. according to: $C_6H_{12}O_6+6H_2O \leftrightarrows 6CO_2+12H_2$ are the reactants for CH_4 production by *Methanobacterium* spp. Note that the methanogenic bacterium's affinity for H_2 is so high that in a "normal paunch in its steady-state" the partial pressure of H_2 is measured to be as low as 10^{-6} atm (10^{-1} Pa)! According to the Norwegian physicochemists C. M. Guldberg (1836–1902) and P. Waage (1833–1900), this has a tremendous impact on, e.g., the chemical equilibrium between NADH and H_2 ($NADH+H^+ \leftrightarrows NAD^++H_2$): Equilibrium constant K of this reaction is 6.7×10^{-4} at 1 atm H_2, but 6.7×10^2 at 10^{-6} atm H_2. This means that, in standard conditions, the production of

H_2 ($E^{\circ\prime}$=−420 mV) from NADH ($E^{\circ\prime}$=−320 mV)—the essential overall reaction of Howard Gest's famous "hydrogen relief valve" [111, 117], which many strict anaerobes critically depend on (in order to escape from dying on over-reduction in an anoxic ecological habitat)—could never work. Proper "poising of the reactants' concentrations", however, may change the situation. The "hydrogen relief valve", e.g., becomes an *exergonic* (thermodynamically favored) reaction under exceedingly low hydrogen partial pressures. In the paunch of ruminants at least, where—biochemically speaking—the ultimate biochemical source of H_2 naturally is also NADH (the "metabolic hydrogen" in the hydrogen fermentation of *Ruminococcus* spp.), this intriguing *syntrophic community* works so efficient that a standard adult ruminant (cattle or cow) releases up to 20 L (converted to standard conditions) of CH_4 per day! Unfortunately, this has most deleterious consequences for the much-discussed "greenhouse effect" in the earth's atmosphere notably since the photochemical (photo-physical) "heating effect" of CH_4 (per mol) is about twenty-times more pronounced than that of CO_2.

Strictly speaking, the whole biosphere on our Earth may be regarded as a single huge *syntrophic community* (between plants and animals, to say it in short). This is shown by the so-called water-water cycle [166] or "aerobic CO_2-" or "H-H-cycle" (Eq. 1.8), which looks quite alike to, e.g., the sulfur-sulfur cycle ("sulphuretum") [167] or the anaerobic "CO_2-cyle" in the paunch of ruminants discussed here shortly before. Yet, in contrast to the ecological insignificance of the latter, Eq. (1.8) [166] is the fundamental and irrevocable biological basis of *all life on earth*.

Resuming the question of *nitrate respiration* again, in all events we are wrongly advised searching the evolutionary origin of aerobic respiration in some sort of nitrate respiration. Just oxygenated N-compounds, even if formed on the early Earth in anoxic Urey-Miller conditions, show an extremely low stability (negligibly short life time) in an artificial primordial atmosphere so that it must have been hardly feasible for them to trigger complicated evolutionary processes over aeons ([69]; also see [48, 203]). On the other hand, regarding more readily available oxidants in an anoxic biosphere, aerobic respiration has never been even suggested to have evolved from either carbonate or sulfate respiration (see our *syntrophic discussions* above; also see Table 1.3) surely because respective electron transport components are molecularly much too different from each other. Thus, however, if (practically) no N-O compound was available on Earth before the advent of bulk O_2 [203] (i.e., before the advent of cyanobacteria, see Fig. 1.4a, b), how could nitrate respiration (in whatever form) have preceded oxygen respiration? Or to say it *parabolically*: How could the encyclopedia Britannica have been written and printed without letters? Or how could human language have evolved in a throat without vocal cords?

As has been outlined above, comparisons between signature-based whole genome sequences of mitochondria, chloroplasts and cyanobacteria strongly support the so-called conversion hypothesis (Fig. 1.2). The similarity of the photosynthetic apparatus of a free living cyanobacterium and a chloroplast is more than striking in both functional and structural respects. Additionally, there is close mechanistic and structural similarity between respiration and photosynthesis in general (cf. Eq. 1.8), as also analysed in detail by the conversion hypothesis [45–47]. This has led to the conclusion [39, 40] that both respiration and photosynthesis might share a common evolutionary (viz., a *monophyletic*) origin. The *unifying or generalized endosymbiont hypothesis* [39, 40] is corroborated by recent findings of unexpected genetic similarities (similar "gene signatures") in (prokaryotic) Rickettsiales, mitochondria and cyanobacteria [41–44]. Also, the discovery of "thylakosomes", i.e. vestiges of oxygenic photosynthetic organelles, in several (chemoheterotrophic!) protists such as *Psalteriomonas lanterna* and representatives of the parasitic genus *Apicomplexa* [177] would be in line with a "generalized endosymbiont hypothesis" according to which both chloroplasts and mitochondria might have started as endosymbiotically engulfed cyanobacteria. Intracellularly, then, the "enslaved" cyanobacterium would have had more than enough time to gradually drop its genetic independence, exchange genes with a possibly second (the "primary"?) endosymbiont or with the host itself, and to develop into either a truely obligate endosymbiont (see e.g., *Cyanophora paradoxa* [262–264]), into a mitochondrion, or into a chloroplast (Fig. 2b), or into still other true organelles, according to physiological needs of specialization or intracellular ("eukaryotic") constraints of e.g. *intracellular communication*. Evolutionary questions similar to those touched in this section are discussed more fully in chap. 2 and 3 of this book.

1.4.2.2 Cyanobacterial Respiratory Electron Transport (RET)

Cyanobacterial respiration has been a traditionally much-neglected bioenergetic process in cyanobacteria [86, 88, 104, 132, 250]. No longer than 40 years ago, in a major review article, the bioenergetic significance of respiration in cyanobacteria was denied at all [136] though it should have been clear from the beginning that an (obligately) photo(auto)trophic organism devoid of sufficient fermentative ATP production [132–135] but nevertheless facing only 50% availability of light, worldwide and over the year, simply *must*, in an oxic biosphere, rely on more or less efficient aerobic respiration for the "generation of metabolic energy" (ATP!), be it only of *maintenance energy* and for *stress defence* [265] in situations when photosynthesis is not possible [86, 88, 104, 132, 250].

The first serious and modern research article on cyanobacterial respiration, dealing with isolated membranes, did not appear before 1969 [266], even if several important and significant whole-cell studies on cyanobacterial photosynthesis *and* respiration had been published already before that time [36, 37, 132, 136, 267, 268, 390]. By all means, however, it must be remarked here that the first published results on a typically bacterial respiratory chain with an aa_3-type cytochrome *c* oxidase (COX) in both CM and ICM preparations from *Anacystis nidulans* (*Synechococcus* PCC6301) and a few other cyanobacteria [269–278], the pronounced increase of respiratory activities in stress conditions [275, 279–288] as well as the unexpected heme promiscuity in the cyanobacterial COX [289–292] and indications of adenylate regulation in this *prokaryotic* (!) COX ([293], also see [294]), had initially met a lot of suspicion in the established scientific community. Yet, most of the discrepancies in the original experimental results published by the Japanese ([295, 296];

also see [297]) and the Viennese ([271–278]; also see [298]) research groups could later be reconciled with each other [299] when it became clear that (less of a surprise to microbiologists than to plant physiologists) growth conditions are of crucial importance for the degrees of "expression" of RET in CM and/or ICM [265, 276]. At any rate, and nicely corresponding with common sense, *light-limited growth* in general has always given *more respiration* (particularly in CM [276, 300, 301]) than light-saturated growth [276]. Just at this point now it seems important to remind the microbiological scientific community of the fact that respiration in cyanobacteria is not only an "accidental appendage to photosynthesis" [38] but an important and self-sustaining bioenergetic process indispensable at least for the generation of maintenance energy in darkness as well as in stress conditions [265].

Table 1.4 presents a survey of cytochrome *c* oxidase (COX) and reductase activities of cytoplasmic and intracytoplasmic membranes. It seems appropriate here to compare these rates with whole cell *photosynthetic* rates (O_2 release = CO_2 fixation). The physiological significance of electron transport rates measured in isolated membranes compared with intact cells may be assessed from the following typical rates of *Anacystis nidulans* (*Synechococcus* sp. PCC6301), the model cyanobacterium in earlier investigations: 15–35 nmol NAD(P)H or succinate oxidized per min/mg membrane protein (i.e. 2×nmol (horse heart) ferricytochrome *c* reduced per min/mg membrane protein). Concomitant O_2 uptake by these membranes measured polarographically in the dark was determined to be 10–20 nmol O_2 per min/mg membrane protein, which compares with an O_2 uptake by intact cells of 130–150 nmol O_2/hr.mg d.w. under identical conditions (approximate correlations: 1 mg d.w. of cells=0.5 mg protein=0.25 mg (crude) membrane protein=0.02 mg chlorophyll). These data clearly show that rates of RET in isolated cyanobacterial membranes (measured spectrophotometrically or polarographically) are fully compatible with rates of *in vivo* dark oxygen uptake by intact cells (30–35°C). However, compare with the 6.5–7.5 µmol O_2/mg chl. per hour taken up in dark respiration the rates of photosynthetic O_2 evolution of up to 250 µmol O_2/mg chl. per hour under saturating light [132, 133]!

As in most other phototrophic prokaryotes, in cyanobacteria the ICM contains a dual-function PET-RET system [302, 303]. Yet, somewhat different from anoxyphototrophs, CM contains a pure-bred respiratory chain without photosynthetic reaction centers (Fig. 1.11). The distinct physical partition of the two membrane types (CM and ICM) in cyanobacteria is more similar to a chloroplast (chlorophyll-free boundary membrane and thylakoid membranes without obvious anastomoses, than to an anoxyphototrophic bacterium with its multifarious invaginations of the CM forming various ICM structures, which latter, when isolated, are called chromatophores) [302, 303].

Cyanobacterial RET comprises, as any other biological RET systems, the following five invariant components (from low (–) to high (+) potential): One or more dehydrogenases, an electron transport quinone (lipid-soluble mobile carrier), a Cyt (*bc*-type) b_6f complex, PC and/or Cyt c_6 (water-soluble mobile carrier), and a terminal respiratory oxidase (TRO) as the final electron acceptor ("electron sink"). Significant chemical deviations from the common theme of electron transport com-

Table 1.4 Survey of cytochrome *c* oxidase (COX) and reductase activities of cytoplasmic (plasma, CM) and intracytoplasmic membrane (ICM) separated and purified from nineteen cyanobacterial species followed spectrophotometrically. Oxidase and reductase activities are expressed as nmol horse heart ferro(ferri)cytochrome *c* oxidized (reduced) per min per mg membrane protein. Rates of O_2-reduction were measured also polarographically being exactly four-times the corresponding rates of spectrophotometrically determined rates. All cytochrome *c* reductase activity was abolished after extraction (by *n*-pentane or *n*-octane) or destruction (by appropriate UV irradiation) of the electron transport quinone, viz. of the plastoquinone-9 [see 348, 349]. In addition, apparent molar masses (kDa) of subunits I and II of COX in isolated and purified CM and ICM are shown. Apparent molar masses were obtained from SDS-PAGE/Western Blot analysis and identification of the proteins by means of polyclonal antisera raised against subunits I and II (and the holoenzyme) of COX from *Paracoccus denitrificans* and beef heart mitochondria [300, 301]. Moreover, the Table depicts molar masses calculated from *cox* genes (Table 1.6), number of amino acids and calculated isoelectric points, cIPs

Cyanobacteria		Cytochrome *c* oxidase		Cytochrome *c* reductase				Apparent molar mass from SDS-PAGE (kDa)		Molar mass calculated from genes (kDa)		Number of amino acids/ cIP	
		CM	ICM	CM		ICM							
				+NADH	+NADPH	+NADH	+NADPH	SUI	SUII	SUI	SUII	SUI	SUII
Synechococcus sp. PCC7003 *)		30	16	110	38	5	0.1	40	25	-	-	-	-
Synechococcus sp. PCC6715 **)		32	35	-	5	1	10	34	14-19	-	-	-	-
Synechococcus elongatus PCC6301 (*Anacystis nidulans*)		60	3	20	3	5	3	46-55	29-32	60	34	541/8.6	315/5.6
Synechococcus elongatus PCC7942 (*Anacystis* R2)		52	8	18	10	3	-	45-52	30-35	-	-	-	-
Synechocystis sp. PCC6714		2	81	-	-	20	23	34-36	21-23	-	-	-	-
Synechocystis sp. PCC6803		3	90	55	10	20	16	34-36	21-23	61	36	551/5.8	332/4.6
Mircrocystis sp. PCC7005 (*Synechocystis sp.* PCC7005)		5	85	43	28	22	18	34-38	22-25	-	-	-	-
Gloeobacter violaceus PCC7421		20	-	0.3	0.1	-	-	52	25	62	33	560/6.6	308/6.4
Spirulina sp. PCC6313		7	78	93	34	17	13	46-50	44	-	-	-	-
Oscillatoria sancta PCC7515		66	9	23	9	32	11	49	35	-	-	-	-
Prochlorothrix		145	20	23	0.1	0.1	0.1	50-51	26	-	-	-	-
Chlorogloeopsis fritschii PCC6718		48	12	33	13	9	19	40	33	-	-	-	-
Leptolyngbya PCC73110 (*Plectonema boryanum* IAM M-129)		72	5	19	2	2	4	48-50	42-43	-	-	-	-
Calothrix desertica PCC7102		58	13	33	13	44	14	48-51	19-25	-	-	-	-
Fischerella muscicola PCC7414		83	34	22	6	10	6	43-47	14-18	-	-	-	-
Anabaena variabilis PCC7937/ATCC29413	Veg	35	46	430	17	5	0.1	48-49	36	64	39	575/6.1	355/4.8
	Het	350	2,400	170	26	20	0.1			62	35	559/6.8	324/4.9
Nostoc punctiforme PCC73102	Veg	35	42	380	68	4	0.1	48-51	36	64	39	580/6.0	361/7.8
	Het	530	4,200	28	16	13	5			62	35	562/7.3	329/5.0
Nostoc sp. PCC7120	Veg	38	50	450	20	8	0.1	50	36	64	39	575/6.0	355/4.8
	Het	420	3,500	195	48	17	-			62	35	559/6.8	327/5.0
Nostoc sp. PCC8009	Veg	5	33	370	42	-	3	50	36	-	-	-	-
	Het	62	320	12	53	39	-			-	-	-	-

ponents are found in the so-called *archaea* only [1–3, 5, 304, 305]. In this context the interesting note may be appropriate that up to the present day not a single *phototrophic* archaeon has been discovered and that even among Gram positive eubacteria phototrophy is a rare exception, whatever this striking correlation might mean in systematic terms. More prominent, yet still pace-keeping variations of the common *monophyletic* scheme (between chloroplasts and mitochondria, in particular) are the following: Photosystem II instead of dehydrogenases, plastoquinone instead of ubiquinone, a Cyt b_6f instead of a bc_1 complex, and photosystem I instead of a TRO (see below).

For chemiosmotic energy conversion [96–100] it is obligatory that the membranes (i.e. CM and ICM) form a closed, osmotically autonomous compartment and that the same membrane also possess a reversible F_0F_1-ATPase or ATP synthase (F-type ATPase in Fig. 1.11) of appropriate orientation to catalyse the endergonic synthesis of ATP from ADP and P_i ("phosphorylation"). This enzyme, in functional terms, is the most strictly conserved biochemical device in the whole biosphere and certainly was present even before any sort of electron transport was "invented" [45–47, 119, 157]. Figure 1.11 depicts the topology of ATP synthase in ICM and CM underlying its important role of energy transduction in both PET and RET (also see chap. 10 in this book). Its architecture is very similar in all organisms though eukaryotic mitochondrial ATP synthases are much more complex than the prokaryotic (including cyanobacteria) and chloroplast enzymes [306–309] comprising two structurally and functionally distinct domains: A membrane-intrinsic proton translocation system (the F_0 part) and a membrane-extrinsic catalytic domain (the F_1 part) that contains six nucleotide-binding sites. Unfortunately, no high-resolution X-ray structure is yet available of an intact integral F_0F_1 enzyme but only of the F_1 part [155]. In detail, F_1 consists of a hexamer of three α- (with three regulatory nucleotide-binding sites) and three β-subunits (with three catalytic binding sites of different nucleotide occupancy), as well as a γ-subunit located asymmetrically in the middle of the hexamer. The structure of F_1 strongly supports the binding change mechanism proposed by Boyer [309]. F_1 and F_0 are structurally and functionally coupled by at least two stalks. The F_0 part consists of three different proteins: one subunit *a*, two subunits *b* and 10–15 subunits *c* (with the nomenclature in cyanobacteria being often different). Only marginal variations of the basic architecture are found in the so-called Archaea [5] but in all organisms the enzyme works as a molecular motor: Ion flux through the F_0 part drives rotation of the rings of subunit *c*, coupled by the central stalk to the rotation of the γ-subunit, which drives "step-wise" ATP synthesis at the catalytic sites of the β-subunits [309].

As to the low-potential or electron-entry end or "start" of cyanobacterial RET, the physiological functions of both a photosynthetic ("reversible") and a respiratory ("unidirectional" or *uptake*) hydrogenase had been described, obviously as an evolutionary relic, in the obligately phototrophic, unicellular, and non-nitrogen-fixing cyanobacterium *Anacystis nidulans* (*Synechococcus* sp. PCC6301) many years ago already [138, 142, 310–312]. These results were later confirmed by genome analysis [313]. Generally, hydrogenases are extremely wide-spread and quite randomly distributed among all bacteria including cyanobacteria (for reviews see [314–316]). In dinitrogen-fixing (cyano)bacteria *uptake* hydrogenases are useful for the recycling of the 25% electrons inevitably going to H^+ instead of to N_2 [317]. In obligately aerobic hydrogen bacteria *uptake* hydrogenases act as ultimate electron donors to the TRO of ordinary respiratory chains while in obligately anaerobic methanogens *uptake* hydrogenases reduce the alternative terminal electron acceptor CO_2 in a membrane-bound *respiratory* and chemiosmotic electron transport reaction giving rise for the production of CH_4 [5]. In many other anaerobic bacteria "reversible" hydrogenases help to prevent metabolic over-reduction in the absence of suitable external oxidants ("hydrogen relief valve" [117]). It should be noted in this context that hydrogenases *sensu stricto* metabolize H_2 (either consuming or producing it) *without* hydrolysis of ATP while H_2 production by nitrogenase which, in usual physiological conditions, mediates the hydrogen relief valve (see above, [117]) needs, as for typical triple bond reductions (reduction of N_2, CN^-, HN_3, N_2O, C_2H_2), on an average 2 ATP/e^- transferred from the Fe- to the MoFe-protein. This net energy requirement (ATP hydrolysis) for thermodynamically *exergonic* reductions such as N_2 fixation (with H_2 via ferredoxin, e.g.) remains enigmatic (see Table 1.3). However, a primordial and (in a reducing environment) energy-requiring hydrogenase might, in the course of evolution, with acquirement of Mo but basically retained FeS-scaffold, have evolved into a nitrogenase as we know it today [318–320]. Note that, essentially, both hydrogenase and nitrogenase are phylogenetically old iron-sulfur enzymes operating at a very low (= negative) redox level of around $E^{\circ\prime}=-420$ mV and that both are severely inactivated or even damaged in the presence of free O_2, the latter usually even more than the former.

The occurrence of a "mitochondrial" energy-transducing, multi-subunit NADH dehydrogenase in both CM and ICM of cyanobacteria as well as in chloroplasts (i.e., in *oxyphototrophs*) was first described by Steinmüller and associates [321, 322]. Competent reviews on mitochondrial and bacterial energy-transducing (i.e., proton-pumping) NADH dehydrogenases, so-called NDH-1 enzymes (NDH_1 in Fig. 1.11) may be found in [323–327] (also see chap. 16 in this book). However, in all oxyphototrophs three of the 14 NDH-1 subunits that form the minimal functioning complex I [325], viz., NuoE, F and G which make up the dehydrogenase module, proper [322], are not at all coded for. Also, in contrast to previous claims, hydrogenase genes *hoxE*, *F*, and *U* do not substitute for *nuoE*, *F*, and *G* genes [328]. Thus, the question remains unresolved: How can the observed oxidation of NAD(P)H by cyanobacterial membranes (CM and/or ICM) be explained at all with an enzyme lacking the dehydrogenase module? For NADPH oxidation in cyanobacteria, (soluble, photosynthetic) FNR once was invoked [329] which, however, cannot be reconciled with solid experimental facts (see [88] for detailed discussion). Secondly, cyanobacteria synthesize two quite different types of NADH dehydrogenases: The classical multi-subunit NDH-1 enzyme pumps protons, uses FMN and several FeS clusters as redox co-factors, oxidizes both deamino-NADH and NAD(P)H, and is strongly inhibited by rotenone or piericidin A (this enzyme is marked NDH_1 in Fig. 1.11). The alternative NDH-2 enzyme [330–332] (see NDH_2 in Fig. 1.11) usually consists of one subunit only, does not pump protons, utilizes FAD instead of FMN, and is devoid of FeS clusters. The (non-energy-transducing) NDH-2 enzyme does not oxidize NADPH [330] nor deamino-NADH while the cyanobacterial NDH-1 enzyme is assumed to oxidize *both* NADH *and* NADPH (as well as deamino-NADH) [333, 334] most probably just *because* it is lacking the advanced 'mitochondrial' NADH dehydrogenase module. A preliminary survey of the occurrence and activity of NDH-1 and NDH-2 enzymes in isolated and purified CM and ICM from a number of cyanobacteria has been presented [335] (also see Table 1.4). In several protists, NDH-2 functions as an energy-overflow valve removing excess reducing equivalents without at the same time giving rise for ATP synthesis [322].

Another potential confusion surrounding cyanobacterial NAD(P)H dehydrogenation is the following: At least two of the *ndh* (= *nuo*) genes are multicopy genes, each copy with a distinct primary structure, leading to six chemically (and possibly also physiologically) different proteins for subunit 4 and three different proteins for subunit 6 of the cyanobacterial NDH-1 complex. This way, in principle, each of the different holoenzymes could catalyse for a distinct physiological function, e.g. in photosynthesis [322], respiration [241] and possibly even C_i transport [336] as has, indeed, been envisaged (see Ref. 88 for discussion). The preliminary kinetic characterization of NAD(P)H dehydrogenation by CM and ICM preparations of cyanobacteria (see also Table 1.4) has yielded the following three distinct K_M values: 5 and 25 µM toward NADH in both CM and ICM (to be attributed to NDH-1 and NDH-2, respectively), only one K_M toward NADPH in CM (due to the primordial NDH-1 enzyme of oxyphototrophs) but again two distinct K_M's, of 5 and 2 µM, toward NADPH in ICM due to the bifunctional 'oxyphototrophic' NDH-1 and the photosynthetic FNR, respectively. These results were obtained on isolated and purified CM and ICM from *Synechococcus* 6301, *Synechocystis* 6803, *Anabaena* 7120, and *Nostoc* 8009 ('Mac'), they were strikingly uniform (i.e. each characteristic of a distinct and individual *enzyme*) and reproducible within 10–20% throughout [333, 334]. Exclusive oxidation of NADPH via FNR as had previously been envisaged [329] could be ruled out by the observation of NADPH oxidation in CM preparations with which a strictly monospecific antibody raised against cyanobacterial FNR did not react at all while it did react with ICM [88, 333, 334]. Succinate oxidation by the same membrane preparations (see SDH, succinate dehydrogenase, in Fig. 1.11) using either ambient O_2 (in the presence of catalytic concentrations of ferrocyanide) or [µM] ferricyanide as oxidants, gave K_M values (toward succinate) of between 600 and 800 µM [333, 334] (also see chap. 17 in this book). Further (including molecular) details of the dehydrogenation of NAD(P)H and succinate in (cyano)bacteria (see SDH, succinate dehydrogenase, in Fig. 1.11) are discussed in [337–347].

A further remarkable peculiarity of cyanobacterial RET is the practically exclusive role of *plasto*quinone as the lipid-soluble mobile carrier in both PET and RET (and thus in CM and ICM, see Fig. 1.11). No ubiquinone or menaquinone has ever been detected. And if the strikingly differential effects of plastoquinone and phylloquinone for the reconstitution of PET and RET in quinone depleted (pentane extracted) membranes [348, 349] might explain the tremendously different rates of

RET and PET (in the order of up to 1:100) in cyanobacterial CM and ICM, respectively, still remains to be seen.

The next speciality of cyanobacterial RET is the (apparently chimeric) cytochrome b_6f complex. As has been mentioned already, b_6f complexes from the cyanobacterium *Mastigocladus laminosus* [197] (as well as from *Chlamydomonas rheinhardtii* chloroplasts [198]) have been structurally resolved in all details quite recently, and the occurrence of "split" b_6f complexes in chloroplasts and cyanobacteria (and in Gram-positive bacilli [350]) had been known for decades. However, it sprang a surprise when it was shown that the cyanobacterial b_6f complex occurs in both ICM and (chlorophyll-free!) CM [240, 351, 352] (Fig. 1.11), that it is immunologically cross-reactive with monoclonal and strictly monospecific antibodies raised against cytochrome c_1 from beef heart mitochondria and *P. denitrificans* [240, 241], though cytochromes f and c_1 do not share sufficient amino acid homology [239], and that it is strongly inhibited by the classical complex III inhibitor antimycin *A* which normally does not at all affect canonical b_6f complexes [240, 241]. A similarly chimeric nature has been suggested for the *bc* complexes from green sulphur bacteria [353] and heliobacteriaceae [243]. Extensive reviews on structure-function relationships of Cyt bc_1 and b_6f complexes are found in [242, 337–339].

In RET (similar to PET), downstream of Cyt b_6f ("complex III"), extrinsic, water-soluble mobile carriers, either plastocyanin or cytochrome c_6 transport electrons to TRO(s), i.e. complex IV. Certain (respiring) bacteria, usually Gram positive bacteria and archaea, short-circuit RET between the quinone and complex IV by using (a) quinol oxidase(s) instead of COX(s). In doing so, they not only by-pass both complex III and the soluble mobile carrier but also renounce an important coupling site for ATP synthesis. Putative genes for quinol oxidases are also found in cyanobacteria, however, their physiological role is still under discussion (see below).

Contrary to previous claims [354], at least either Cyt c_6 or PC is absolutely indispensable for integral PET *and* RET [248, 249] and in cyanobacteria *both* proteins, which can substitute for each other, are capable of reducing P700 of photosystem I *and* the COX (see below) (also see [250]), with reaction rates primarily depending on IEPs of the redox proteins involved [249, 250, 355, 356] as also shown by corresponding Brönsted plots [251–253].

The many completely sequenced cyanobacterial genomes have given astonishing evidence for a variety of several different terminal respiratory oxidases (TROs) that *might be* coded for in many cyanobacterial species, sometimes even up to three different TROs in one and the same species (Table 1.5) [86, 250]. The situation is all the more embarrassing as most of those species are *obligate photoautotrophs* incapable of making major ecological use of respiration for growth and proliferation anyway, and the species known to express those different TRO genes were shown to synthesize very small amounts of transcripts only which do not make it likely that it will ever be feasible to characterize *the enzymes* on a protein level [250, 357].

Generally, heme-copper oxidases (or oxygen reductases) play a key role in aerobic RET. They are redox driven proton pumps that couple the four-electron reduction of molecular oxygen to water to the vectorial translocation of protons across

Table 1.5 ORFs and genes of cyanobacterial terminal respiratory heme-copper oxidases [i.e. cytochrome *c* oxidases (COX) and *bo*-quinol oxidases (QOX)] and non-heme copper quinol oxidases, *bd*-QOX. In addition the (putative) electron donors for COX are given, i.e., plastocyanin (PC), cytochrome c_6 (Cyt c_6) and cytochrome c_M (C cyt c_M). 44 completely or partially (*) sequenced strains were analyzed. The genome size is included. Nitrogen-fixing cyanobacteria are highlighted in grey, heterocyst-forming species are highlighted in dark-grey

Heme-copper oxidases COX	*bo*-QOX	*bd*-QOX	Cyanobacteria	Classification	Genome size	N_2-fixation	Heterocysts	PC	Cytc_6	Cytc_M
1	0	0	***Prochlorococcus marinus* str. MIT9301**	*Pleurocapsales*	1.6 Mb	no	no	1	1	1
1	1	0	***Prochlorococcus marinus* str. MIT9515**	*Pleurocapsales*	1.7 Mb	no	no	1	0	0
1 1 1	0 90% 1 10%	0	**10 x *Prochlorococcus marinus*** [a]	*Pleurocapsales*	1.7 - 2.7 Mb	no	no	1 1 1	0 40% 1 40% 2 20%	1 1 1
1[#]	0	1	***Gloeobacter violaceus* PCC7421**	*Chroococcales*	4.6 Mb	no	no	2	2	1
1	0	0	***Microcystis aeruginos* NIES-843**	*Chroococcales*	5.8 Mb	no	no	1	1	1
1	0	1	***Synechococcus elongatus* PCC6301**	*Chroococcales*	2.7 Mb	no	no	1	3	1
1	0	1	***Synechococcus elongatus* PCC7942**	*Chroococcales*	2.7 Mb	no	no	1	3	1
1 1 1 1	0 30% 1 70%	0 85% 1 15%	**14 x *Synechococcus sp.*** [b]	*Chroococcales*	2.2 - 3.0 Mb	no	no	1 0 1 1	1 10% 2 50% 3 10% 4 30%	1 1 1 1
1	1	1	***Synechocystis sp.* PCC6803**	*Chroococcales*	3.6 Mb	no	no	1	1	1
1	0	1	***Thermosynechococcus elongatus* BP-1**	*Chroococcales*	2.6 Mb	no	no	0	1	1
1	0	1	***Acaryochloris marina* MBIC11017**	unclassified	6.5 Mb	no	no	1	2	1
1	1	0	***Crocosphaera watsonii* WH8501***	*Chroococcales*	6.2 Mb	yes	no	1	3	1
1	1	1	***Cyanothece sp.* ATCC51142**	*Chroococcales*	4.9 Mb	yes	no	1	2	1
1	1	1	***Cyanothece sp.* CCY0110***	*Chroococcales*	5.9 Mb	yes	no	1	2	1
1	1	1	***Cyanothece sp.* PCC7424***	*Chroococcales*	6.4 Mb	yes	no	2	2	1
1	0	0	***Cyanothece sp.* PCC8801***	*Chroococcales*	4.6 Mb	yes	no	1	2	1
1	1	1	***Lyngbya sp.* PCC8106***	*Oscillatoriales*	7.0 Mb	yes	no	1	2	1
1	1	0	***Trichodesmium erythraeum* IMS101**	*Oscillatoriales*	7.7 Mb	yes	no	1	2	1
2	2[#]	4	***Anabaena variabilis* ATCC29413**	*Nostocales*	6.3 Mb	yes	yes	1	3	1
2	1	1	***Nodularia spumigena* CCY9414***	*Nostocales*	5.3 Mb	yes	yes	1	4	1
2	1[#]	0	***Nostoc punctiforme* PCC73102***	*Nostocales*	9.0 Mb	yes	yes	1	3	1
2	1	1	***Nostoc (Anabaena) sp.* PCC7120**	*Nostocales*	6.4 Mb	yes	yes	1	2	1

a) The *Prochlorococcus marinus* genus includes following strains: Pro9601, Pro9211, Pro9215, Pro9303, Pro9312, Pro9313, ProNATL1A, ProNATL2A, Pro1375, Pro1986.
b) Following strains are included in the *Synechococcus* group: Syn107*, Syn9311, Syn9605, Syn9902, SynJA23, SynJA33, Syn7002, Syn307, Syn9916*, Syn9917*, Syn5701*, Syn7803, Syn7805*, Syn8102.
Genome analysis shows the presence of additional but incomplete operons for COX or QOX (absence of subunit III).

the membrane. They are characterized for having a heme-copper binuclear reaction center comprising a Cu_B electronically coupled with a high-spin heme. Additionally, all heme-copper oxidases contain a low-spin heme in subunit I [360]. Depending on the nature of electron donors, the superfamily of (cyanide sensitive) heme-copper oxidases is divided into two branches, namely cytochrome *c* oxidases (COX, *aa*$_3$-type cytochrome oxidase) or quinol oxidases (QOX, *bo*-type quinol oxidase) [357, 361, 362]. In contrast, the cytochrome *bd*-quinol oxidases do not belong to this protein superfamily, since they lack the binuclear reaction center and instead contain two heme groups. Additionally, they do not pump protons.

Cytochrome *c* oxidase (COX) has been found in all cyanobacteria both at genomic and protein levels [250]. Thus, eventually the occurrence of orthodox RET (albeit united with PET in ICM) and of a proton-translocating [363–367] *aa*$_3$-type cytochrome *c* oxidase in both CM and ICM of cyanobacteria had been accepted by the scientific community [299]. Cytochrome *c* oxidases have been discovered spectrophotometrically and characterized biochemically and immunologically in crude membrane preparations of *Anabaena* sp. [271], *Nostoc* MAC [270] and *Anacystis nidulans* [269, 278] as well as in all cyanobacteria tested so far [38, 368, 369] and their bioenergetic capabilities are absolutely sufficient to meet the modest respiratory maintenance requirements of cyanobacteria in general. A gene cluster encoding subunits II, I and III of COX from *Synechocystis* PCC6803 has been cloned [370–372]. Operons encoding subunits I-III in the same order have also been cloned from other species (for review see [250]). The presently available genome data unequivocally demonstrate that all cyanobacteria contain at least one complete COX-operon, whereas *Nostocales* (heterocyst-forming N_2-fixing cyanobacteria) have two such operons (Table 1.5; see also [250]).

So far the 3D structure of a cyanobacterial COX could not be elucidated. However, sequence alignment clearly demonstrates [250] that subunits I and II contain all redox cofactors and residues involved in O_2 reduction and proton pumping. Several high-resolution 3D structures of heme-copper oxidases are known including COX from *Paracoccus denitrificans* [373] and *Rhodobacter sphaeroides* [374]. Similar to *Paracoccus* COX, all cyanobacterial COX subunits I have 12 (predicted) transmembrane helices with the core (most probably) forming three pores ("proton channels") within and across the membrane, with each pore being shaped by four transmembrane helices [357, 351, 373, 375]. The first pore is filled with aromatic residues, the second holds the binuclear reaction center, and the third one the low-spin heme. Amino acid residues coordinating the metal centers are strictly conserved in all cyanobacteria sequenced so far [250]. Additionally, the first and second pores are known to form two uptake pathways for protons from the *n*-side, viz., the so-called D- and K-channels [357–360].

Cytochrome *c* oxidase subunit II includes a typical hairpin-like structure comprising two transmembrane helices and a peripheral domain on the *p*-side of the membrane, which latter can be either CM or ICM. In case of cyanobacterial COXs, this extrinsic domain (the donor binding and electron entry site), that contains the Cu_A center, has been shown to be uniquely located in two entirely distinct environments, viz., the periplasmic space and the thylakoid lumen (Fig. 1.11). Subunit II of COX participates in both electron transfer and proton pumping. Electrons from a mobile donor (i.e., PC or Cyt c_6) are transferred to the Cu_A center, which is composed of two electronically coupled, mixed-valence copper ions [373, 374]. In all (including cyanobacterial) COXs this center is binuclear and the strictly conserved ligands are located near the C-terminus of subunit II.

The role of both PC and cytochrome c_6 as mobile electron donors for COX is still under discussion. A recent investigation of two deletion mutants of *Synechocystis* PCC6803 (each mutant lacking either *petE* or *petJ*) and their photoautotrophic and heterotrophic growth rate in copper-free and copper-supplemented medium clearly demonstrated that efficient function of respiration requires the presence of either Cyt c_6 or PC [248, 249]. These data confirm an early suggestion of Lockau [376] that cyanobacterial RET can use both PC and Cyt c_6 as electron donor. (Also see [377–379]). This is also underlined by investigations of the electron transfer kinetics between PC or Cyt c_6 and the recombinant Cu_A domain of COX from *Synechocystis* PCC6803 [380–382]. Recently, it could be demonstrated that both mobile electron carriers can efficiently be oxidized by COX in *Nostoc* PCC7119 [383]. Both Cyt c_6 and PC are located inside the thylakoid lumen, and also a periplasmic Cyt c_6 was identified as an electron donor to CM-bound COX (Figs. 1.11 and 1.12) [384, 385]. The occurrence of PC in the periplasmic space has not yet been proven.

Deletion of the operon that encodes COX in *Synechocystis* PCC6803 resulted in a mutant strain that was still viable under both photoautotrophic and photomixotrophic conditions. The strain respired at near wild-type rates, and this respiration was cyanide-sensitive [386]. Isolated membranes from the mutant were unable to oxidize reduced horse heart cytochrome *c*. This was the first evidence for a second cyanide-sensitive terminal respiratory oxidase in cyanobacteria. Although COX appears to be the main terminal oxidase, as is reflected by its occurrence in all cyanobacteria (Table 1.5) there is also evidence of the presence and function of the other branch of the heme-copper oxidase superfamily [i.e. the quinol oxidase (QOX) or *bo*-type quinol oxidase] in many cyanobacteria. QOX is encoded by a set of genes very similar in sequence to the operon that encodes COX. Similarities between the two branches of heme-copper oxidases include in addition the presence of one low-spin (six-coordinated heme *b* in QOX) and one high-spin (five-coordinated heme o_3 in QOX) heme, exchange coupling between high-spin heme and Cu_B in the binuclear reaction site in subunit I, heme-heme interaction as well as alignment of these hemes with respect to the membrane bilayer [387]. QOX subunit I contains all redox cofactors and is homologous to the corresponding subunit in COX. Sequence analysis of cyanobacterial genomes demonstrated that all ligands of the two heme groups and of Cu_B are strictly conserved [250]. In contrast to cytochrome *c* oxidases, QOX subunit II has neither a binuclear Cu_A center nor a cytochrome *c* binding site [250, 387]. Instead, heme *b* receives electrons directly from a quinol molecule dissolved in the membrane core matrix (ubiquinol in case of *E. coli* QOX, plastoquinol in case of cyanobacteria?). The protons produced upon quinol oxidation ("scalar protons") are released on the *p*-side of the membrane. Table 1.5 shows that many unicellular cyanobacteria lack the genes encoding QOX, whereas all nitrogen

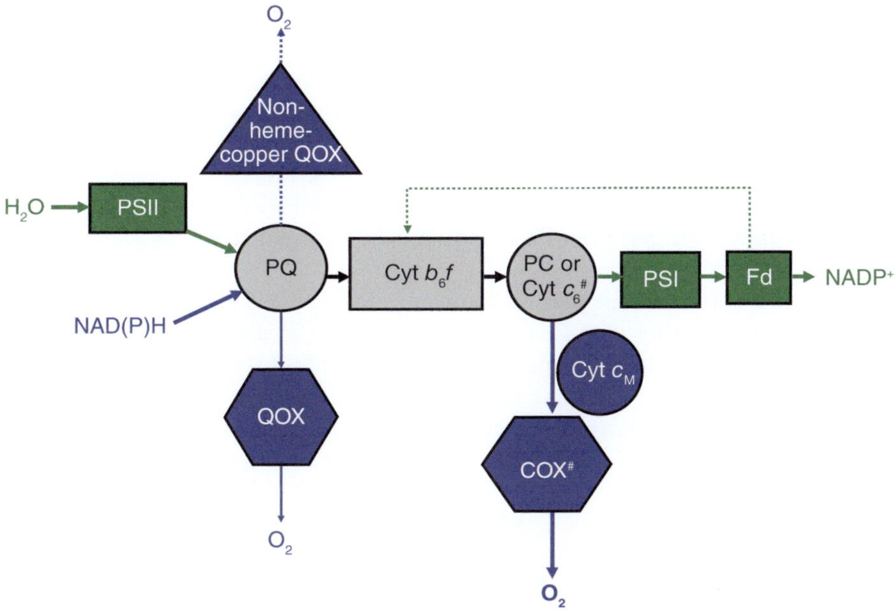

Fig. 1.12 Scheme of the branched dual-function photosynthetic (PET)/respiratory (RET) electron transport chain(s) in cyanobacteria. Note that, in all cyanobacteria except *Gloeobacter violaceus*, PET is only localized in intracytoplasmic or thylakoid membranes (ICM), whereas RET is active in both ICM and cytoplasmic membrane (CM). Electron donor and acceptor as well as protein complexes involved exclusively in PET are depicted in *green*, those found only in RET in *blue* color. Electron components used in both PET and RET are shown in *grey*. In contrast to cytochrome *c* (aa_3-type) oxidase (COX, hexagon), which has been detected on a protein level in both CM and ICM, there is no unequivocal information about the physiological function and localization of heme-copper (QOX, hexagon) and non-heme-copper quinol oxidases (triangle). Depending on copper-availability plastocyanin (PC) and/or cytochrome c_6 (Cyt c_6) alternatively can functions as electron donor for both, photosystem I (PSI) or COX. Another alternative (respiratory) electron carrier exclusively found in cyanobacteria (cytochrome c_M, Cyt c_M) most probably is involved in electron shuttling between cytochrome b_6f (Cyt b_6f) and COX during environmental stress [265, 399, 400]. COX is the main terminal oxidase in RET, but—depending on growth conditions and nitrogen supply—may partially be substituted by QOX or non-heme-copper oxidases which both would have to use, then, plastoquinol-9 (PQ) as electron donor. In contrast to PC and Cyt c_M, up to four paralogs for Cyt c_6 are found in cyanobacterial genomes (#). In *Nostocales* two paralogs for COX are found (#). Cyclic electron flow around PSI is shown as a *green dashed line* [250]

fixing species usually have one QOX-operon. (Marginal) transcription of operons encoding QOX has been detected in several species, and studies using deletion mutants and inhibitor studies (though unfortunately performed on whole cells only) have suggested QOX to be active and to contribute to energy metabobism (for an extensive review see [250]). One major obstacle in understanding the role of QOX in cyanobacterial RET is the absence of its characterization on the protein level. The operon is transcribed but the exceedingly low levels of mRNAs synthesized indicate that characterization at the protein level may remain problematic [357].

Additionally, genome analysis suggests the presence of genes or ORFs for another terminal oxidase, namely the *bd*-quinol oxidase [250], which is not a member of the heme-copper protein superfamily, since it does not contain a binuclear reaction site and is incapable of proton pumping. The role of this enzyme in cyanobacterial electron transport chains is even more speculative.

1.4.2.3 Interaction Between PET and RET in Cyanobacteria

Figure 1.12 summarizes the present knowledge of branched RET in cyanobacteria and its interrelationship with PET. As has been outlined above, the dual-function PET-RET pathway in ICM downstream of plastquinol (PQ) shares common membrane protein complexes and mobile electron carriers involved in both pathways: PQ, cytochrome b_6f, plastocyanin and cytochrome c_6. Nowadays, the respective identities and locations of the various electron transport components are fairly well documented and proven, amongst others, by immunogold labeling of thin-sectioned whole cells [271, 300, 301, 352, 385, 388]. An immediate interaction ("*competition*") between PET and RET can only be expected for the ICM-contained electron transport system. Figure 1.12 shows this intramembrane "electron transport bottleneck" which, for understandable reasons, must hold for all "uncompartmentalized" dual-function membranes of *all* prokaryotes capable of *both* respiration *and* photosynthesis, irrespective of the peculiar electron transport components involved. Since, in cyanobacteria, the "production of electrons" in the light from water-oxidation is much more rapid than from substrate dehydrogenation, the "electronic bottle-neck" is practically "saturated" and therefore, in the light, electrons from substrate dehydrogenation (which can basically, of course, proceed in the light as well) cannot be 100% accommodated any more "in the bottle-neck" but "have to wait", i.e., are excluded from the bottle-neck. This results in the well-known "light-inhibition of respiration" in illuminated prokaryotes. Since the competition between PET and RET in a dual-function biomembrane (of a respiring phototrophic prokaryote) naturally is mutual, an inhibition of RET by PET (discernible in the form of an inhibitory effect of O_2 on photosynthesis) can also be measured in special conditions. However, as the inhibition of photosynthesis by oxygen (known as "Warburg effect" in the early to mid-20th century [389]) may result, in an intact cell, from widely different molecular effects (e.g., "photorespiration", ribulose-bisphosphate carboxylase/oxygenase competition, glycolate oxidase action, other more or less unspecific oxygenases, etc.), the competition between PET and RET, proper, can only be recognized, as such, in experiments on isolated membranes. In case of isolated cyanobacterial ICM such studies, confirming much earlier propositions for a competition between photosynthesis and respiration in cyanobacteria [390], had been presented some time ago already [391, 392].

Interaction between RET and PET as well as the role of alternative TROs is obvious under (environmental) stress conditions as well as nitrogen fixation (which, regarding the high need for energy and micro-anaerobic conditions, is a stress, too, for the living cell). With respect to stress and adaptation, it is the CM and not the ICM that is closest to the environment and, thus to practically any environmental stress. It will therefore also be primarily affected. Respiratory membrane energization and ATP supply are crucial to stressed cyanobacteria, which usually shut down PET first (viz., at PS II). It has been demonstrated that reversible H^+-translocating ATPases (or ATP synthases) are located also in the CM [265, 271, 300], for review also see [36, 37]) and that energization of CM is a KCN-sensitive process [393]. Altogether, it has taken some time, until it became clear that cyanobacteria are uniquely capable of recruiting biological energy from both photosynthesis *and* respiration, and that their CM can be energized directly by RET *but*, more secondarily, also by proton-translocating ATPases which may utilize ATP from whatever bioenergetic process (see respective sections in this article; also see [287, 288, 394, 395]).

In *A. nidulans* (*Synechococcus* PCC6301) exposed to a variety of seemingly unrelated stress conditions (e.g., high-salt or light limitation) RET in CM was shown to be significantly enhanced while ICM-bound respiratory chain was much less affected ('salt respiration') [265]. Additionally, the amount of COX in CM (but not in ICM) was increased [300, 301]. The increased respiration was proposed to generate a proton gradient required to drive H^+/Na^+ exchange [265, 397]. H^+/Na^+ antiporters and proton-pumping P-type ATPases have been identified in *Synechocystis* PCC6803 (Fig. 1.11) [396]. COX was shown to be necessary and sufficient for 'salt respiration' [265, 276, 393].

Recently, the interrelation between RET and PET under conditions of photoinhibition and high light tolerance has been demonstrated in *Synechococcus* PCC7002 [398]. Both COX and QOX were reported to be dispensable for growth under normal to moderately high light conditions, whereas at more stressful conditions (higher light intensities) COX seems to be the primary oxidase in RET. Usually, photoinhibition occurs in conditions of excessive light intensities or electron flow limitations due to environmental stresses. Changes in the electron flow around the photosystems in a COX-lacking strain were demonstrated to be very similar to those under photoinhibitory conditions [398]. Additionally, this mutant exhibited also a decrease in the total amount of chlorophyll as well as in the number of active PSII complexes. This underlines the important role of COX (and RET) for the protection against excessive reductive stress through diverting of electrons from the reduced PQ pool to TROs (Fig. 1.12).

Recently, it has been demonstrated that an additional small *c*-type membrane-associated cytochrome c_M (Cyt c_M in Fig. 1.12), encoded by all cyanobacteria so far sequenced [250, 399, 400], can reduce Cu_A in COX [399]: The determined rates of RET were comparable with those effected by Cyt c_6 and PC as inferred from (1) thermodynamic considerations (Cyt c_M has a midpoint potential of 150 mV that is to be compared to +320 mV of cytochrome *f*) and (2) laser-induced kinetic analysis of PSI reduction by recombinant Cyt c_M [383]. Thus it has been proposed that Cyt c_M exclusively participates in RET [399]. This is plausible since under stress conditions cyanobacteria not only shut down PET [265] but also suppress the expression of both Cyt c_6 and PC, whereas expression of *cyt* c_M (that is scarcely expressed under normal growth conditions) is enhanced [400]. This gives further evidence that cytochrome c_M may also in vivo participate in COX reduction. Its function could be to act as intermediary electron sink in peculiar conditions such as light, salt or temperature stress, thereby diverting electrons (via plastoquinol?) towards COX (Fig. 1.12).

Additionally, the interrelation between PET and RET is evident during nitrogen-fixation: Cyanobacteria are the only diazotrophs that produce oxygen as a by-product of PET, and therefore must reconcile the inevitable (intracellular!) presence of molecular oxygen with an essentially anaerobic enzyme. The biological reduction of N_2 is absolutely restricted to prokaryotes and is catalysed by the multimeric metalloenzyme nitrogenase, which is irreversibly inhibited by molecular oxygen and reactive oxygen species [401] (see Sect. 3.2.2). Quite in accordance with what has been found about the role of dinitrogen fixation as a bioenergetic stress-response (see above), rates of RET go steeply up when cyanobacteria are grown with atmospheric N_2 as the sole N-source [265, 402]. Moreover, RET efficiently lowers the concentration of dissolved O_2 (primarily within the O_2-producing cell itself) and may thus help to maintain nitrogenase in a virtually oxygen-free cellular environment ("respiratory protection" [388, 401]) with simultaneous satisfaction of the severe energy (ATP) requirement [265, 317, 388, 402–404].

During the course of planetary evolution, cyanobacteria have co-evolved with the changing oxidation state of the ocean and atmosphere to accommodate the machinery of PET and RET and oxygen-sensitive N_2-fixation within the same cell and/or colony of cells. These strategies have been generally described as temporal and/or spatial separation of photosynthesis and N_2-fixation (for review see [250, 401]). The simplest adaptation is seen for the genus *Trichodesmium* (order *Oscillatoriales*) that fixes nitrogen during the day [405]. Here, oxygen protection is a complex interaction between spatial and temporal segregation of PET, RET and N_2-fixation within the photoperiod. In *Trichodesmium*, nitrogenase is compartmentalized in a fraction of the cells (10–20%) that are often arranged consecutively along the trichome. However, active photosynthetic components are found in all cells, even those harboring nitrogenase [406]. Light initiates PET, providing energy and reductants for carbohydrate synthesis and storage, stimulating cyclic and pseudocyclic (Mehler) electron flow through PSI, and poising the PQ pool at reduced levels. Interestingly, RET has been shown to be active also early in the photoperiod at high rates offering respiratory protection [388, 402] and supplying carbon skeletons for amino acid synthesis (the primary sink for fixed nitrogen) [402, 406]. RET reduces the PQ pool further, sending negative feedback to linear PET. As a consequence, PSII becomes down-regulated and oxygen consumption exceeds oxygen production providing a "near-anaerobic" intracellular environment. This opens a window of opportunity for N_2-fixation during the photoperiod. As the carbohydrate pool is consumed, PET through the PQ pool slows down. The PQ pool becomes increasingly oxidized, net O_2-production exceeds O_2-consumption and nitrogenase activity is lost during the following day.

With the evolution of cyanobacteria and the subsequent generation of a gradually more and more oxic atmosphere, more sophisticated oxygen protective mechanisms were needed. A full temporal separation, in which nitrogen is fixed only at night, then developed in other filamentous genera of the order *Oscillatoriales* as well as

Table 1.6 List of all mentioned cyanobacterial strains, abbreviations and access codes of all genomes

Abbreviations	Sequenced cyanobacteria	Accession numbers
Aca.mar	*Acaryochloris marina* MBIC11017	NC_009925
Ana.var	*Anabaena variabilis* ATCC29413	NC_007413
Cro.wat	*Crocosphaera watsonii* WH8501	NZ_AADV00000000
Cya51142	*Cyanothece sp.* ATCC51142	NC_010546
Cya0110	*Cyanothece sp.* CCY0110	NZ_AAXW00000000
Cya7424	*Cyanothece sp.* PCC7424	NC_011729
Cya8801	*Cyanothece sp.* PCC8801	NC_011726
Glo.vio	*Gloeobacter violaceus* PCC7421	NC_005125
Lyn8106	*Lyngbya sp.* PCC8106	NZ_AAVU00000000
Mic.aer	*Microcystis aeruginosa* NIES-843	NC_010296
Nod.spu	*Nodularia spumigena* CCY9414	NZ_AAVW00000000
Nos.pun	*Nostoc punctiforme* PCC73102	NC_010628
Nos7120	*Nostoc* (*Anabaena*) *sp.* PCC7120	NC_003272
Pro9601	*Prochlorococcus marinus* str. AS9601	NC_008816
Pro9211	*Prochlorococcus marinus* str. MIT9211	NC_009976
Pro9215	*Prochlorococcus marinus* str. MIT9215	NC_009840
Pro9301	*Prochlorococcus marinus* str. MIT9301	NC_009091
Pro9303	*Prochlorococcus marinus* str. MIT9303	NC_008820
Pro9312	*Prochlorococcus marinus* str. MIT9312	NC_007577
Pro9313	*Prochlorococcus marinus* str. MIT9313	NC_005071
Pro9515	*Prochlorococcus marinus* str. MIT9515	NC_008817
ProNATL1A	*Prochlorococcus marinus* str. NATL1A	NC_008819
ProNATL2A	*Prochlorococcus marinus* str. NATL2A	NC_007335
Pro1375	*Prochlorococcus marinus subsp. marinus* str. CCMP1375	NC_005042
Pro1986	*Prochlorococcus marinus subsp. marinus* str. CCMP1986	NC_005072
Syn6301	*Synechococcus elongatus* PCC6301	NC_006576
Syn7942	*Synechococcus elongatus* PCC7942	NC_007604
Syn107	*Synechococcus sp.* BL107	NZ_AATZ00000000
Syn9311	*Synechococcus sp.* CC9311	NC_008319
Syn9605	*Synechococcus sp.* CC9605	NC_007516
Syn9902	*Synechococcus sp.* CC9902	NC_007513
SynJA23	*Synechococcus sp.* JA-2-3B′a(2–13)	NC_007776
SynJA33	*Synechococcus sp.* JA-3-3Ab	NC_007775
Syn7002	*Synechococcus sp.* PCC7002	NC_010475
Syn307	*Synechococcus sp.* RCC307	NC_009482
Syn9916	*Synechococcus sp.* RS9916	NZ_AAUA00000000
Syn9917	*Synechococcus sp.* RS9917	NZ_AANP00000000
Syn5701	*Synechococcus sp.* WH5701	NZ_AANO00000000
Syn7803	*Synechococcus sp.* WH7803	NC_009481
Syn7805	*Synechococcus sp.* WH7805	NZ_AAOK00000000
Syn8102	*Synechococcus sp.* WH8102	NC_005070
Syc6803	*Synechocystis sp.* PCC6803	NC_000911
The.elo	*Thermosynechococcus elongatus* BP-1	NC_004113
Tri.ery	*Trichodesmium erythraeum* IMS101	NC_008312

in unicellular members of the order *Chroococcales*. These organisms fix nitrogen at night and nitrogenase is typically found in all cells. They also show extensive metabolic periodicities ("circadian rhythms") of PET, RET and nitrogen fixation when grown under N_2-fixing conditions. Nitrogen fixation and respiration peak at 24 h intervals early in the dark or in a subsequent dark period, whereas photosynthesis is approximately 12 h out of phase and peaks toward the end of the light phase.

Finally, a highly refined specialization is found in heterocystous cyanobacteria *Nostocales* and also reflected by the increased number of genes encoding components of RET including TROs (terminal heme-copper oxidases and their putative electron donors) (Table 1.5). In these organisms, nitrogenase is confined to a micro-anaerobic cell, the heterocysts [407], which differentiates completely and irreversibly 12–20 h after combined nitrogen sources are removed from the medium. Heterocysts are enveloped by two superimposed layers, one consisting of polysaccharides and the other of glycolipids, the latter slowing down the diffusion of O_2. Heterocysts have lost the capacity of cell division, exhibit high PSI activity (but lack PSII) and increased RET. They rely on vegetative cells as source of carbon and reductant and, in return, they supply the surrounding vegetative cells with fixed nitrogen. Cyclic electron flow around PSI and RET supplies ATP. Interestingly, heterocysts prefer ICM as the site of RET, while unicellular N_2-fixing species prefer CM [402, 403]. For a comprehensive list of all cyanobacterial species mentioned in this chapter (together with abbreviations and accession numbers) see Table 1.6.

1.4.2.4 Conclusion and Outlook

In conclusion, cyanobacteria are *the* paradigmatic organisms of oxygenic (plant-type) photosynthesis and aerobic (mitochondrial-like) respiration in bioenergetic, evolutionary, and ecological respects. They have uniquely accomodated both a photosynthetic and a respiratory electron transport chain within a single prokaryotic cell. Since many cyanobacteria in addition are also capable of N_2-fixation, they are often called the bioenergetic "nonplus-ultra" among living beings. They represent ideal objects to study the evolution and molecular physiology of resourcing energy for metabolic needs under various environmental conditions (and restraints). Cyanobacteria not only were the first oxygenic (plant-type) photosynthesizers but also the first aerobic respirers and the cyanobacterial COX (most probably) the first functional aerobic terminal oxidase [408].

Our research attempts focusing on *cyanobacterial respiration* often confronted us with the question: "Why devote research efforts to a cyanobacterial bioenergetic process that, by itself, is generally insufficient to sustain growth and proliferation?" However, since everything that happens must have a reason for why it happens, and a mechanism for how it happens, we were nevertheless able in the course of decades to contribute significantly to an unraveling of the salient features of cyanobacterial respiration, the most important of which features might be summarized as the following: (1) Functional oxidative phosphorylation in intact cyanobacteria with a quite normal P/O value of 3 [93–96, 409, 410], (2) function of the (NAD(P)H-)-plastoquinol-cytochrome c_6/PC-electron transport segment as a common component in both photosynthesis and respiration [391, 392], (3) occurrence of both F-type and P-type ATPases in cyanobacterial CM [240, 411, 412], (4) detection and characterization of a fully functional aa_3-type cytochrome *c* oxidase (COX) in both CM and ICM isolated and purified from a wide variety of cyanobacteria subjected to very different growth conditions (contents of the overwhelming majority of the sections in this article), and (5) unraveling of the interplay between RET and PET under various stress conditions including nitrogen fixation ([265] and this article).

Finally, looking back to our scientific lives under the auspices of the cyanobacteria, we can only hope that *Homo sapiens sapiens* with his inborn destructive attitude (see Fig. 1.4a) will not be able to destroy within a few decades what cyanobacteria have built up during millennia.

Acknowledgments This work was supported by the Austrian Science Funds (FWF-project P17928 at present). Devoted and invaluable technical assistance has always been provided by Mr. Otto Kuntner.

References

1. Danson MJ, Hough DW and Lunt GG (1992) The archaebacteria: biochemistry and biotechnology. Portland Press, London, UK
2. Woese CR and Fox GE (1977) Proc Natl Acad Sci U S A 74: 5088
3. Woese CR, Magrum LJ and Fox GE (1978) J Mol Evol 11: 245
4. Donker HJL and Kluyver AJ (1926) Die Einheit in der Biochemie. Chem Zelle Gewebe 13: 134
5. Schäfer G, Engelhard M and Müller V (1999) Microbiol Mol Biol Rev 63: 570
6. Starr C, Taggart R and Star L (2001) The unity and diversity of life. Brook and Cole Publishers, USA
7. Zillig W (1991) Curr Opin Genet Dev 1: 544
8. Sagan L (1967) J Theoret Biol 14: 125
9. Gallon JR (1992) New Phytol 122: 571
10. Schlegel HG (1999) Geschichte der Mikrobiologie. Acta Historica Leopoldina 28, Halle (Saale), Germany
11. Drews G (2010) Mikrobiologie: Die Entdeckung der unsichtbaren Welt. Springer, Berlin
12. Margulis L and Sagan D (1986) Microcosmos four billion years of microbial evolution. University of California Press, Berkeley, USA
13. Mojzsis SJ, Arrhenius G, McKeegan KD, Harrison TM, Nutman A and Friend CRL (1996) Nature 384: 5
14. Darwin CR (1859, 1872) The origin of species, 1st & 6th ed., John Murray, London, UK
15. Denton M (1985) Evolution: a theory in crisis. Adler & Adler, London, UK
16. Illies J (1980) Schöpfung oder Evolution? Interform, Zürich, Switzerland
17. Junker R and Scherer S (1988) Entstehung und Entwicklung der Lebewesen, 2nd ed., Weyel Biologie Verlag, Gießen, Germany
18. Atkins PW (1992) Creation revisited. WH Freeman & Co., Oxford, UK
19. Margulis L (1970) Origin of eukaryotic cells. Yale Univ Press, New Haven, USA
20. Cavalier-Smith T (2006) Philos Trans R Soc Lond B Biol Sci 361: 969
21. Knoll AH (1992) Science 256: 622
22. Stanier RY (1970) Symp Soc Gen Microbiol 20: 1
23. Dyall SD, Brown MT and Johnson PJ (2003) Science 304: 253

24. Poole A, Jeffares D and Penny D (1999) Bioessays 21: 880
25. Margulis L (1996) Proc Natl Acad Sci U S A 93: 1071
26. Van Valen LM and Maiorana VC (1980) Nature 287: 248
27. Moreira D and Lopez-Garcia P (1998) J Mol Evol 47: 517
28. Hackstein JHP and Yarlett N (2005) In: Overmann J (Ed), Molecular Basis of Symbiosis, Springer Verlag, Berlin, Germany, 117–142
29. Mereschkowsky C (1905) Biol Zentralbl 25: 593
30. Gray MW and Doolittle WF (1982) Microbiol Rev 46: 1
31. Cavalier-Smith T (2006) Proc R Soc Lond B Biol Sci 273: 1943
32. Peschek GA (1996) Biochem Soc Trans 24: 729
33. Peschek GA, Wastyn M, Trnka M, Molitor V, Fry IV and Packer L (1989) Biochemistry 28: 3057
34. Peschek GA, Hinterstoisser B, Wastyn M, Pineau B, Missbichler A and Lang J (1989) J Biol Chem 264: 11827
35. Peschek GA, Hinterstoisser B, Riedler M, Muchl R and Nitschmann WH (1986) Arch Biochem Biophys 247: 40
36. Peschek GA (1996) Biochim Biophys Acta 1275: 27
37. Peschek GA (1999) In: Peschek GA, Löffelhardt W and Schmetterer G (Eds), The phototrophic prokaryotes, Kluwer Academic/Plenum Publishers, New York, 201–209
38. Peschek GA, Wastyn M, Molitor V, Kraushaar H, Obinger C and Matthijs HCP (1989) In: Kotyk A, Skoda J, Paces V and Kosta V (Eds), Highlights of modern biochemistry, 1st vol. VSP Publishers, Zeist, The Netherlands, 893–902
39. Peschek GA (2000) Plant Physiol Biochem 38: 266
40. Peschek GA (2005) In: van der Est A and Bruce D (Eds), Photosynthesis: fundamental aspects to global perspectives, The international Society of Photosynthesis, Toronto, CDN, 746–749
41. Wang Y, Hill K, Singh S and Kari L (2005) Gene 346: 173
42. Karlin S and Burge C (1995) Trends Genet 11: 283
43. Gupta RS (1998) Microbiol Mol Biol Rev 62: 1435
44. Sasikumar R, Jijoy J and Peschek GA (2007) In: Galambos C (Ed), Abstr 10th international colloquium on endocytobiology and symbiosis, Gmunden, Austria, International Society of Endocytobiology (ISE), 51
45. Broda E (1975) The evolution of the bioenergetic processes. Pergamon Press, Oxford, UK
46. Broda E and Peschek GA (1979) J Theor Biol 81: 201
47. Peschek GA (1981) Photosynthetica 15: 543
48. Lane N (2002) Oxygen: the molecule that made the world. Oxford University Press, Oxford, UK
49. Wiche G (1998) J Cell Sci 111: 2477
50. Mikulecky DC (1996) Acta Biotheoretica 44: 179
51. Brocks JJ, Logan GA, Buick R and Summons RE (1999) Science 285: 1033
52. Hartman H and Fedorov A (2002) Proc Natl Acad Sci U S A 99: 1420
53. Kurland CG, Collins LJ and Penny D (2006) Science 312: 1011
54. Horiike T, Hamada K, Kanaya S and Shinozawa T (2001) Nature Cell Biol 3: 210
55. Rivera MC and Lake JA (2004) Nature 431: 152
56. Vellai T, Takacs K and Vida G (1998) J Mol Evol 46: 499
57. Vellai T and Vida G (1999) Proc R Soc Lond B Biol Sci 266: 1571
58. Embley TM and Martin W (2006) Nature 440: 623
59. Davies P (1998) The fifth miracle. The search for the origin of life. Penguin Books, London, UK
60. Cairns-Smith G (1985) Seven clues to the origin of life. Cambridge University Press, Cambridge, UK
61. Oparin AI (1924) The origin of life (translated in Bernal JD 1967). Plenum Press, London, UK
62. Haldane JBS (1954) New Biol 16: 12

63. Bernal JD (1967) The origin of life. Weidenfel Nicholson, London, UK
64. Haldane JBS (1970) In: Cloud P (Ed), Adventures in earth history. Freeman WH, San Francisco, USA, 377–384
65. Albrecht P (1971) Angew Chemie 83: 221
66. Miller SL and Urey HC (1959) Science 130: 245
67. Miller SL and Orgel LE (1974) The origins of life on Earth. Prentice Hall, Englewood Cliffs, USA
68. Urey HC (1952) The planets: their origin and development. New Haven. USA
69. Zohner A and Broda E (1979) Origins Life 9: 291
70. Wächtershäuser G (1988) Microbiol Rev 52: 452
71. Wächtershäuser G (1998) Syst Appl Microbiol 21: 473
72. Huber C and Wächtershäuser G (1997) Science 276: 245
73. Martin W and Russell MJ (2003) Phil Trans R Soc Lond B Biol Sci 358: 59
74. Monod J (1970) Le hasard et la necessite. Editions du Seuil, Paris, France
75. Hartman H (1998) Orig Life Evol Biosph 28: 515
76. Schrödinger E (1954) Nature and the greeks. Cambridge University Press, Cambridge, UK
77. Diels H and Kranz W (1903) Die Fragmente der Vorsokratiker, 1st ed., 1st–3rd vol. Weidemann Verlag, Zürich, Switzerland
78. Mansfeld J (1983) Die Vorsokratiker (two volumes, Greek/German). Philipp Reclam, Stuttgart, Germany
79. Gemelli Marciano ML (2007–2010) Die Vorsokratiker (three volumes, Greek/German, Sammlung Tusculum). Artemis und Winkler, Düsseldorf, Germany
80. Kirk GS, Raven JE and Schofield M (1983) The presocratic philosophers. A critical history with a selection of texts. Cambridge University Press, Cambridge, UK
81. Eigen M (1994) Ber Bunsenges Phys Chem 98: 135
82. Heidegger M (1929) Was ist Metaphysik? Inaugural Lecture, University of Freiburg. University Press, Freiburg, Germany
83. Sartre J-P (1956) Being and nothingness: a phenomenological essay on ontology. Citadel Press, New York (English Translation of "L'Etre et le Neant", Paris, France 1943)
84. Camus A (1942) Le Mythe de Sisyphe. Librairie Gallimard, Paris, France
85. Djerassi C, and Hoffman R (2001) Oxygen. Wiley-VCH, Weinheim, Germany
86. Paumann M, Regelsberger G, Obinger C and Peschek GA (2005) Biochim Biophys Acta 1707: 231
87. Pauling L (1931) J Amer Chem Soc 53: 3225
88. Peschek GA, Obinger C and Paumann M (2004) Physiol Plant 120: 358
89. Skulachev VP (1996) Q Rev Biophys 29: 169
90. Schrödinger E (1951) What is Life? Cambridge University Press, Cambridge, UK
91. von Bertalanffy KL (1972) Theoretische Biologie, Stoffwechsel und Wachstum, 2nd vol. A. Francke Verlag, Bern, Switzerland
92. von Bertalanffy KL, Beier W and Laue R (1977) Biophysik des Fließgleichgewichts. Zsolnay, Vienna, Austria
93. Nitschmann WH (1986) J Theor Biol 122: 409
94. Nitschmann WH and Peschek GA (1986) J Bacteriol 168: 1285
95. Nitschmann WH and Peschek GA (1985) Arch Microbiol 141: 330
96. Peschek GA, Nitschmann WH and Czerny T (1988) Methods Enzymol 167: 361
97. Mitchell P (1961) Nature 191: 144
98. Mitchell P (1966) Biol Rev 41: 455
99. Nicholls DG and Ferguson SJ (1992) Bioenergetics, 2nd ed., Academic Press, London, UK
100. Skulachev VP (1992) In: L Ernster (Ed), Molecular mechanisms in bioenergetics. Elsevier Science Publishers, Amsterdam, The Netherlands, 37–73
101. Dimroth P (1987) Microbiol Rev 51: 3287
102. Müller V and Gottschalk G (1994) In: HL Drake (Ed), Acetogenesis. Chapman and Hall, New York, USA, 127–156
103. Nitschmann WH and Packer L (1996) Biochem Mol Biol Intern 40: 1201

104. Peschek GA (2008) In: G Renger (Ed), Primary processes of photosynthesis: principles and applications, 2nd vol. European Society of Photobiology, RSC Publishing, Cambridge, UK, 383–415
105. Marcus RA (1956) J Chem Phys 24: 966
106. Marcus RA and Sutin N (1985) Biochim Biophys Acta 811: 265
107. Moser CC, Keske JM, Warncke K, Farid RS and Dutton PL (1992) Nature 355: 796
108. Moser CC and Dutton PL (1992) Biochim Biophys Acta 1101: 171
109. Moser CC, Page CC, Chen X and Dutton PL (1997) J Bioinorg Chem 2: 393
110. Baltscheffsky H and Baltscheffsky M (1974) Annu Rev Biochem 43: 871
111. Thauer RK, Jungermann K and Decker K (1977) Bacteriol Rev 41: 100
112. Racker E (1965) Mechanisms in bioenergetics. Academic Press, New York, USA
113. Gross T, Faull J, Ketteridge S and Springham D (1995) Introductory microbiology. Chapman & Hall, London, UK
114. Stryer H (1981) Biochemistry. Freeman & Company, San Francisco, USA
115. Atkinson DE (1969) Annu Rev Microbiol 23: 47
116. Atkinson DE (1977) Cellular energy metabolism and its regulation. Academic Press, New York, USA
117. Gest H (1999) In: Peschek GA, Löffelhardt W and Schmetterer G (Eds), The phototrophic prokaryotes, Kluwer Academic/Plenum Publishers, New York, USA, 11–19
118. Lipmann FA (1941) Advan Enzymol 1: 99
119. Raven JA and Smith FA (1976) J Theor Biol 57: 301
120. Wald G (1957) Ann N Y Acad Sci 69: 352
121. Wald G (1964) Proc Natl Acad Sci U S A 52: 595
122. Wald G (1966) In: Kaplan NO and Kennedy EP (Eds), Current Aspects of Biochemical Energetics. New York, USA, 119–133
123. Forterre P (1999) Bioessays 21: 871
124. Woese CR, Kandler O and Wheelis ML (1990) Proc Natl Acad Sci U S A 87: 4576
125. Woese CR (1998) Proc Natl Acad Sci U S A 95: 6854
126. Woese CR (2000) Proc Natl Acad Sci U S A 97: 8392
127. Doolittle WF and Brown JR (1994) Proc Natl Acad Sci U S A 91: 6728
128. Doolittle WF (1999) Science 284: 2124
129. Doolittle WF (2000) Curr Opin Struct Biol 10: 355
130. Dawkins R (1989) The selfish gene. Oxford University Press. Oxford, UK
131. Tarnas R (1991) The passion of western mind. Random House, Inc., New York, USA
132. Peschek GA (1975) Doctoral Thesis, University of Vienna, Austria (in German)
133. Peschek GA, and Broda E (1973) Naturwissenschaften 60: 479
134. Peschek GA (1974) In: Avron M (Ed), Proceedings of the 3rd International Congress on Photosynthesis. Elsevier Science Publishers, Amsterdam, The Netherlands, 921
135. Stal LJ and Moezelaar R (1997) FEMS Microbiol Rev 21: 179
136. Holm-Hansen O (1968) Annu Rev Microbiol 21: 47
137. Padan E (1979) Annu Rev Plant Physiol 30: 27
138. Peschek GA (1978) Arch Mikrobiol 119: 313
139. Peschek GA (1979) FEBS Lett 106: 34
140. Anderson S and McIntosh L (1991) J Bacteriol 173: 2761
141. Peschek GA (1976) In: Codd GA and Stewart WDP (Eds), Phototrophic prokaryotes. Proceedings of the 2nd International Symposium. 209
142. Peschek GA (1980) Arch Microbiol 125: 123
143. Rippka R, Deruelles J, Waterbury JB, Herdman M and Stanier RY (1979) J Gen Microbiol 111: 1–120
144. Rittenberg SC (1972) Antonie van Leeuwenhoek. J Microbiol Serol 38: 457
145. Smith AJ and Hoare DS (1977) Bact Rev 41: 419
146. Benedict CR (1978) Annu Rev Plant Physiol 29: 67
147. Whittenbury R and Kelly SP (1977) In: Hoddock BA and Hamilton WA (Eds), Microbial Energetics, Symposium 27, Soc Gen Microbiol 121–129

148. Kelly DP (1971) Annu Rev Microbiol 25: 177
149. Suzuki I (1974) Annu Rev Microbiol 28: 85
150. Gest H (1972) Adv Microb Physiol 7: 243
151. Renthal R (1992) In: Ernster L (Ed), Molecular mechanisms in bioenergetics. Elsevier Science Publishers, Amsterdam, The Netherlands, 75–101
152. Racker E and Stoeckenius W (1974) J Biol Chem 249: 662
153. Oesterhelt D and Stoeckenius W (1971) Nat New Biol 233: 149
154. Ovchinikov YA, Abdulaev NG, Feigina MY, Kiselev AV and Lobanov NA (1979) FEBS Lett 100: 219
155. Abrahams JP, Leslie AGW, Lutter R and Walker JE (1994) Nature 370: 621
156. Pedersen PL and Amzel M (1993) J Biol Chem 268: 9937
157. Peschek GA and Broda E (1980) In: Schopf JW (Ed), An interdisciplinary study of the origin and evolution of earth's earliest biosphere. UCLA, Los Angeles, USA, 23
158. Nicholls P (1999) In: S Papa (Ed), Frontiers of cellular bioenergetics. Kluwer Academic/Plenum Publishers, New York, 1–22
159. Renger G (2001) Biochim Biophys Acta 1503: 210
160. Renger G (2004) Biochim Biophys Acta 1655: 195
161. Renger G and Holzwarth AR (2005) In: Wydrzynski T and Satoh K (Eds), Photosystem II: the light-driven water: plastoquinone oxidoreductase. Springer Verlag, The Netherlands, 139–175
162. Renger G (Ed) (2008) Primary processes of photosynthesis. Principles and applications, 2 vols. European Society of Photobiology, RSC Publishing, Cambridge, UK
163. Rabinowitch E and Govindjee (1969) Photosynthesis. John Wiley & Sons, New York, USA
164. Stanier RY (1961) Bacteriol Rev 25: 1
165. Stanier RY and Van Niel CB (1962) Arch Mikrobiol 42: 17
166. Asada K (1999) Annu Rev Plant Physiol Plant Mol Biol 50: 601
167. Van Gemerden H (1967) Ph. D. Thesis, University of Leiden, The Netherlands. (Printed by JH Pasmans, S'Gravenhage)
168. Babcock GT and Wikström M (1992) Nature 356: 301
169. Hoganson CW, Pressler MA, Proshlyakov DA and Babcock GT (1998) Biochim Biophys Acta 1365: 170
170. Crabtree RH (1997) Science 276: 222
171. Schopf JW (1970) Biol Rev 45: 319
172. Schopf JW (1999) Cradle of life. Princeton University Press, Princeton, USA
173. Barghoorn ES and Schopf JW (1965) Science 150: 337
174. Barghoorn ES and Schopf JW (1966) Science 152: 758
175. Waterbury JB, Watson SW, Guillard RRL and Brand LE (1979) Nature 277: 293
176. Chisholm SW, Olson RJ, Zettler FR, Goericke R, Waterbury JB and Welschmeyer NA (1988) Nature 340: 340
177. Hackstein JHP, Schubert H, Rosenberg J, Mackenstedt U, van den Berg M, Brul S, Derksen J and Matthijs HCP (1997) In: Schenk HEA, Herrmann RG, Jeon KW, Müller NE and Schwemmler W (Eds), Eukaryotism and symbiosis. Intertaxonic combination versus symbiotic adaptation. Endocytobiology VI. Springer Verlag, Berlin, Germany, 49–55
178. Des Marais DJ (2000) Science 289: 1703
179. Canfield DE and A Teske (1996) Nature 382: 127
180. Barber J (2004) Biochim Biophys Acta 1655: 132
181. Barber J (2008) Philos Trans R Soc Lond B Biol Sci 363: 2665
182. Dismukes GC, Klimov VV, Baranov SV, Das Gupta J and Tyryshkin A (2001) Proc Natl Acad Sci U S A 98: 2170
183. Canfield DE (1999) Nature 400: 503
184. Berner RA (1988) Paleogeogr Paleoclimatol Paleoecol 75: 97
185. Cloud P Jr (1968) Science 160: 729
186. Blankenship RE and Hartman H (1998) Trends Biochem Sci 23: 94
187. Ioannidis N, Schansker G, Barynin VV and Petrouleas V (2000) J Bioinorg Chem 5: 354

188. Kasting JF, Holland HD and Pinto JP (1985) J Geophys Res 90: 10497
189. Kasting JF (1993) Science 259: 920
190. Kasting JF (2006) Nature 443: 643
191. McKay CP and Hartman H (1991) Orig Life Evol Biosph 21: 157
192. Rye R and Holland HD (1998) Am J Sci 298: 621
193. Gerschman R, Gilbert DL, Nye SW, Dwyer P and Fenn WO (1954) Science 119: 623
194. De la Rosa MA, Molina-Heredia FP, Hervas M and Navarro JA (2006) In: JH Golbeck (Ed), The Light-Driven Plastocyanin: Ferredoxin Oxidoreductase. Springer Verlag, The Netherlands, 683–696
195. Towe KM (1978) Nature 274: 657
196. Barynin VV, Whittaker MM, Antonyuk SV, Lamzin VS, Harrison PM, Artymiuk PJ and Whittaker JW (2001) Structure 9: 725
197. Kono A and Fridovich I (1983) J Biol Chem 258: 6015
198. Ferreira KN, Iverson TM, Maghlaoui K, Barber J and Iwata S (2004) Science 303: 1831
199. Biesiadka J, Loll B, Kern J, Irrgang K-D and Zouni A (2004) Phys Chem Chem Phys 6: 4733
200. Loll B, Kern J, Saenger W, Zouni A and Biesiadka J (2005) Nature 438: 1040
201. Sidler WA (1994) In: Bryant DA (Ed), The molecular biology of cyanobacteria. Kluwer Academic Publishers, Dordrecht, The Netherlands, 139–216
202. Morand LZ, Cheng RH, Krogmann DW and Ho KK (1994) In: Bryant DA (Ed), The molecular biology of cyanobacteria. Kluwer Academic, Dordrecht, The Netherlands, 381–407
203. Gilbert DL (Ed) (1981) Oxygen and living processes. An interdisciplinary approach. Springer Verlag, New York, USA
204. Hill R (1937) Nature 139: 281
205. Hill R and Bendall F (1960) Nature 186: 136
206. Nitschke W and Rutherford AW (1991) Trends Biochem Sci 16: 241
207. Xiong J, Fischer WM, Inoue K, Nakahara M and Bauer CE (2000) Science 289: 1724
208. Jordan T, Fromme P, Witt HT, Klukas O, Saenger W and Krauß N (2001) Nature 411: 909
209. Krauß N, Schubert WD, Klukas O, Fromme P, Witt HT and Saenger W (1996) Nat Struct Biol 3: 965
210. Ben-Shem A, Frolow F and Nelson N (2003) Nature 426: 630
211. Zouni A, Witt HT, Kern J, Fromme P, Krauß N, Saenger W and Orth P (2001) Nature 409: 739
212. Kamiya N and Shen JR (2003) Proc Natl Acad Sci U S A 100: 98
213. Kurisu G, Zhang H, Smith JL and Cramer WA (2003) Science 302: 1009
214. Stroebel D, Choquet Y, Popot J-L and Picot D (2003) Nature 426: 413
215. Schubert WD, Klukas O, Saenger HT, Witt HT, Fromme P and Krauß N (1998) J Mol Biol 280: 297
216. Morris JG (1975) Adv Microb Physiol 12: 169
217. Cole JA (1976) Adv Microb Physiol 14: 1
218. Cadenas E (1989) Annu Rev Biochem 58: 79
219. Higgins IJ, Best DJ, Hammond RC and Scott D (1981) Microbiol Rev 45: 556
220. Haber CL, Allen LN, Zhao S and Hanson RS (1983) Science 221: 1147
221. Balch WE, Fox GE, Magrum LJ, Woese CR and Wolfe RS (1979) Microbiol Rev 43: 260.
222. Hall JB (1971) J Theor Biol 30: 429
223. Obinger C, Regelsberger G, Strasser G, Burner U and Peschek GA (1997) Biochem Biophys Res Commun 235: 545
224. Obinger C, Regelsberger G, Pircher A, Sevcik-Klöckler A, Strasser G and Peschek GA (1999) In: Peschek GA, Löffelhardt W, Schmetterer G (Eds) The phototrophic prokaryotes. Kluwer Academic/Plenum Publishers, New York, USA, 719–731
225. Regelsberger G, Obinger C, Zoder R, Altmann F and Peschek GA (1999) FEMS Microbiol Lett 170: 1
226. Regelsberger G, Jacopitsch C, Rüker F, Krois D, Peschek GA and Obinger C (2000) J Biol Chem 275: 22854

227. Regelsberger G, Jakopitsch C, Plasser L, Schwaiger H, Furtmüller PG, Peschek GA, Zamocky M and Obinger C (2002) Plant Physiol Biochem 40: 479
228. Regelsberger G, Atzenhofer W, Rüker F, Peschek GA, Jacopitsch C, Paumann M, Furtmüller PG and Obinger CJ (2002) Biol Chem 277: 43615
229. Regelsberger G, Laaha U, Ditmann D, Rüker F, Canini A, Grilli-Caiola M, Furtmüller PG, Jakopitsch C, Peschek GA and Obinger C (2004) J Biol Chem 279: 44384
230. Jakopitsch C, Regelsberger G, Furtmüller PG, Rüker F, Peschek GA and Obinger C (2002) J Inorg Biochem 91: 78
231. Atzenhofer W, Regelsberger G, Jacob U, Peschek GA, Furtmüller PG, Huber R and Obinger C (2002) J Mol Biol 321: 479
232. Bernroitner M, Zamocky M, Furtmüller PG, Peschek GA and Obinger C (2009) J Exp Bot 60: 423
233. Rippka R, Waterbury JB, Cohen-Bazire G (1974) Arch Mikrobiol 100: 419
234. Grossman AR (2003) Photosynth Res 76: 207
235. Bibby TS, Mary I, Nield J, Partensky F and Barber J (2003) Nature 424: 1051
236. Hillier W and Babcock GT (2001) Plant Physiol 125: 33
237. DeRuyter YS and Fromme P (2008) In: Herrero A and Flores E (Eds), The cyanobacteria: molecular biology and evolution. Caister Academic Press, Norfolk
238. Blankenship RE (1992) Photosynth Res 33: 91
239. Hauska G, Nitschke W and Herrmann RG (1988) J Bioenerg Biomemb 20: 211
240. Dworsky A, Mayer B, Regelsberger G, Fromwald S and Peschek GA (1995) Bioelectrochem Bioenerg 38: 35
241. Dzelzkalns VA, Obinger C, Regelsberger G, Niederhauser H, Kamensek M, Peschek GA and Bogorad L (1994) Plant Physiol 105: 1435
242. Berry EA, Guergova-Kuras M, Huang L and Crofts AR (2000) Annu Rev Biochem 69: 1005
243. Nitschke W, Schoepp B, Floss B, Schricker A, Rutherford AW Liebl U (1996) Eur J Biochem 242: 695
244. Bertini I, Ciurli S, Dikiy A, Fernandez CO, Luchinat C, Safarov N, Shumilin S and Vila AJ (2001) J Am Chem Soc 123: 2405
245. Frazao C, Soares CM, Carrondo MA, Pohl E, Dauter Z, Wilson KS, Hervas M, Navarro MA, De la Rosa MA and Sheldrick GM (1995) Structure 3: 1159
246. Kerfeld CA, Anwar HP, Interrante R, Merchant S and Yeates TO (1995) J Mol Biol 250: 627
247. Ubbink M, Ejdebäck M, Karlsson BG and Bendall DS (1998) Structure 6: 323
248. Duran V, Hervas MA, De la Rosa MA and Navarro JA (2004) J Biol Chem 279: 7229
249. Moser D, Nicholls P, Wastyn M and Peschek GA (1991) Biochem Int 24: 757
250. Bernroitner M, Zamocky M, Pairer M, Furtmüller PG, Peschek GA and Obinger C (2008) Chem Biodivers 5: 1927
251. Schlarb-Ridley BG, Bendall DS and Howe CJ (2002) Biochemistry 41: 3279
252. Schlarb-Ridley BG, Navarro JA, Spencer M, Bendall DS, Hervas M, Howe CJ and De la Rosa MA (2002) Eur J Biochem 269: 5893
253. Schlarb-Ridley BG, Bendall DS and Howe CJ (2003) Biochemistry 42: 4057
254. Karplus PA, Daniels MJ and Herriott JR (1991) Science 251: 60
255. Saraste M and Castresana J (1994) FEBS Lett 341: 1
256. Castresana J and Saraste M (1995) Trends Biochem Sci 20: 443
257. Castresana J, Lübben M and Saraste M (1995) J Mol Biol 250: 202
258. Viebrock A and Zumft W (1988) J Bacteriol 170: 4658
259. Scott RA, Zumft WG, Coyle CL and Dooley DM (1989) Proc Natl Acad Sci U S A 86: 4082
260. Egami F (1974) Orig Life 5: 405
261. Fenchel T and Blackburn TH (1979) Bacteria and mineral cycling. Academic Press, London, UK

262. Jakowitsch J, Neumann-Spallart C, Ma Y, Steiner J, Schenk HEA, Bohnert HJ and Löffelhardt W (1996) FEBS Lett 381: 153
263. Löffelhardt W and Bohnert HJ (1994) In: Bryant DA (Ed), The molecular biology of cyanobacteria. Kluwer Academic Publishers, Dordrecht, The Netherlands, 65–89
264. Pfanzagl B, Zenker A, Pittenauer E, Allmaier G, Martinez-Torrecuadrada J, Schmid ER, De Pedro MA and Löffelhardt W (1996) J Bacteriol 178: 332
265. Peschek GA and Zoder R (2001) In: Rai LC and Gaur JP (Eds), Algal adaptation to environmental stresses. Physiological, biochemical, and molecular mechanisms. Springer Verlag, Berlin, Germany, 203–258
266. Biggins J (1969) J Bacteriol 99: 570
267. Wolk CP (1973) Bacteriol Rev 37: 32
268. Binder A (1982) J Bioenerg Biomemb 14: 271
269. Peschek GA (1981) Biochim Biophys Acta 635: 470
270. Peschek GA (1981) Biochem Biophys Res Commun 98: 72
271. Peschek GA, Schmetterer G, Lockau W and Sleytr UB (1981) In: Akoyunoglu G (Ed), Photosynthesis V. chloroplast development. Balaban International Science Services, Philadelphia, USA, 707–719
272. Molitor V and Peschek GA (1986) FEBS Lett 195: 145
273. Trnka M and Peschek GA (1986) Biochem Biophys Res Commun 136: 235
274. Molitor V, Trnka M and Peschek GA (1987) Curr Microbiol 14: 263
275. Peschek GA, Molitor V, Trnka M, Wastyn M and Erber W (1988) In: Rogers LJ and Gallon JR (Eds), Biochemistry of the algae and cyanobacteria. Clarendon Press, Oxford, UK, 178
276. Peschek GA, Molitor V, Trnka M, Wastyn M and Erber W (1988) Methods enzymol. 167: 437
277. Peschek GA, Schmetterer G, Muchl R, Nitschmann WH and Riedler M (1984) In: Sybesma C (Ed), Advances in Photosynthesis Research, 3rd vol. Martinus Nijhoff/Dr. W. Junk Publishers, The Hague, The Netherlands, 335–339
278. Peschek GA (1987) In: Fay P and van Baalen C (Eds), The cyanobacteria. Elsevier Science Publishers, Amsterdam, The Netherlands, 119–161
279. Fry IV, Peschek GA, Huflejt M and Packer L (1985) Biochem Biophys Res Commun 129: 109
280. Fry IV and Peschek GA (1988) Methods Enzymol 167: 450
281. Molitor V, Erber W and Peschek GA (1986) FEBS Lett 204: 251
282. Rivière M-E, Arrio B, Steffan I, Molitor V, Kuntner O and Peschek GA (1990) Arch Biochem Biophys 280: 159
283. Molitor V, Trnka M, Erber W, Steffan I, Riviere M-E, Arrio B, Springer-Lederer H and Peschek GA (1990) Arch Microbiol 154: 112
284. Molitor V, Kuntner O, Sleytr UB and Peschek GA (1990) Protoplasma 157: 112
285. Fry IV, Huflejt M, Erber WWA, Peschek GA and Packer L (1986) Arch Biochem Biophys 244: 686
286. Erber WWA, Nitschmann WH, Muchl R and Peschek GA (1986) Arch Biochem Biophys 247: 28
287. Peschek GA, Hinterstoisser B, Riedler M, Muchl R and Nitschmann WH (1986) Arch Biochem Biophys 247: 40
288. Peschek GA, Czerny T, Schmetterer G and Nitschmann WH (1985) Plant Physiol 79: 278
289. Peschek GA, Wastyn M, Fromwald S and Mayer B (1995) FEBS Lett 371: 89
290. Peschek GA, Alge D, Fromwald S and Mayer B (1995) J Biol Chem 270: 27937
291. Auer G, Mayer B, Wastyn M, Fromwald S, Eghbalzad K, Alge D and Peschek GA (1995) Biochem Mol Biol Int 37: 1173
292. Fromwald S, Zoder R, Wastyn M, Lübben M and Peschek GA (1999) Arch Biochem Biophys 367: 122
293. Alge D, Wastyn M, Mayer C, Jungwirth C, Zimmermann U, Zoder R, Fromwald S and Peschek GA (1999) IUBMB Life 48: 187
294. Arnold S and Kadenbach B (1997) Eur J Biochem 249: 350

295. Omata T and Murata N (1984) Biochim Biophys Acta 766: 395
296. Omata T and Murata N (1985) Biochim Biophys Acta 810: 354
297. Bisalputra T, Brown DL and Weier TE (1969) J Ultrastruct Res 27: 182
298. Peschek GA, Schmetterer G and Sleytr UB (1981) FEMS Microbiol Lett 11: 121
299. Scherer S, Almon H and Böger P (1988) Photosynth Res 15: 95
300. Peschek GA, Obinger C, Fromwald S and Bergman B (1994) FEMS Microbiol Lett 124: 431
301. Peschek GA, Obinger C, Sherman DM and Sherman LA (1994) Biochim Biophys Acta 1187: 369
302. Oelze J and Drews G (1972) Biochim Biophys Acta 265: 209
303. Oelze J (1981) Subcell Biochem 8: 1
304. Lübben M (1994) Biochim Biophys Acta 1229: 1
305. Lübben M and Morand J (1994) J Biol Chem 269: 21473
306. Walker JE and Dickson VK (2006) Biochim Biophys Acta 1757: 286
307. Hausrath AC, Capaldi RA and Matthews BW (2001) J Biol Chem 276: 47227
308. Rubinstein JL, Walker JE and Henderson R (2003) EMBO J 22: 6182
309. Kohlbrenner WE and Boyer PD (1983) J Biol Chem 258: 10881
310. Peschek GA (1979) Biochim Biophys Acta 548: 203
311. Peschek GA (1979) Biochim Biophys Acta 548: 187
312. Peschek GA (1979) Arch Microbiol 123: 81
313. Schmitz O, Boison G, Hilscher R, Hundeshagen B, Zimmer W, Lottspeich F and Bothe H (1995) Eur J Biochem 233: 266
314. Bothe H, Boison G and Schmitz O (1999) In: Peschek GA, Löffelhardt W and Schmetter G (Eds), The phototrophic prokaryotes. Kluwer Academic/Plenum Publishers, New York, USA, 589–601
315. Houchins P (1984) Biochim Biophys Acta 768: 227
316. Tamagnini P, Axelsson R, Lindberg P, Oxelfelt F, Wünschiers R and Lindblad P (2002) Microbiol Mol Biol Rev 66: 1
317. Jensen BJ (1983) J Gen Microbiol 129: 2633
318. Broda E and Peschek GA (1980) Biosystems 13: 47
319. Broda E and Peschek GA (1983) Biosystems 16: 1
320. Broda E and Peschek GA (1984) Orig Life 14: 653
321. Berger S, Ellersiek U and Steinmüller K (1991) FEBS Lett 286: 129
322. Friedrich T, Steinmüller K and Weiss H (1995) FEBS Lett 367: 107
323. Friedrich T and Weiss H (1997) J Theor Biol 187: 529
324. Friedrich T and Böttcher B (2004) Biochim Biophys Acta 1608: 1
325. Leif H, Sled VD, Ohnishi T, Weiss H and Friedrich T (1995) Eur J Biochem 230: 538
326. Weiss H, Friedrich T, Hofhaus G and Preis D (1991) Eur J Biochem 197: 563
327. Yagi T (1993) Biochim Biophys Acta 1141: 1
328. Boison G, Bothe H, Hansel A and Lindblad P (1999) FEMS Microbiol Lett 174: 159
329. Scherer S, Alpes I, Sadowski H and Böger P (1988) Arch Biochem Biophys 267: 228
330. Alpes I, Scherer S and Böger P (1989) Biochim Biophys Acta 973: 41
331. Melo AMP, Bandeiras TM and Teixeira M (2004) Microbiol Mol Biol Rev 68: 603
332. Ohkawa H, Pakrasi HB and Ogawa T (2000) J Biol Chem 275: 31630
333. Festetics T (2004) Diploma Thesis, University of Vienna, Austria
334. Flasch H (1997) Diploma Thesis, University of Vienna, Austria
335. Peschek GA, Karner A-H, Zeitler G, Festetics T, Flasch H, Özcelik M and Ruppert J (2006) In: Teixeira M (Ed), Abstr. EUROBIC 8, Aveiro, Portugal, PS 5.49.
336. Ohkawa H, Price GD, Badger MR and Ogawa T (2000) J Bacteriol 182: 2591
337. Malkin R (1992) Photosynth Res 33: 121
338. Trumpower BL (1990) Microbiol Rev 54: 101
339. Trumpower BL (1990) J Biol Chem 265: 11409
340. Cooley JW, Howitt CA and Vermaas WFJ (2000) J Bacteriol 182: 714
341. Cooley W and Vermaas WFJ (2001) J Bacteriol 183: 4251

342. Lancaster CR, Kröger A, Auer M and Michel H (1999) Nature 402: 377
343. Lancaster CR (2001) In: Messerschmidt A, Huber R, Poulos T and Wieghart K (Eds), Handbook of metalloproteins. John Wiley & Sons Ltd, Chichester, UK, 379–388
344. Lancaster CR (2002) Biochim Biophys Acta 1565: 215
345. Lancaster CR (2003) FEBS Lett 555: 21
346. Hägerhäll C (1997) Biochim Biophys Acta 1320: 107
347. Yankovskaya V, Horsefield R, Törnroth S, Lunba-Chavez C, Miyoshi H, Leger C, Byrne B, Cecchini G and Iwata S (2003) Science 299: 700
348. Peschek GA (1980) Biochem J 186: 515
349. Peschek GA and Kuntner O (1987) In: Favre A, Tyrrell R, Cadet J (Eds), From Photophysics to Photobiology. Elsevier, Amsterdam, 157–166
350. Sone N, Sawa G, Sone N and Noguchi S (1995) J Biol Chem 270: 10612
351. Kraushaar H, Hager S, Wastyn M and Peschek GA (1990) FEBS Lett 273: 227
352. Sherman DM, Troyan TA and Sherman LA (1994) Plant Physiol 106: 251
353. Klughammer C, Hager V, Padan E, Schütz W, Schreiber U, Shahak Y and Hauska G (1995) Photosynth Res 43: 27
354. Zhang L, Pakrasi HB and Whitmarsh J (1994) J Biol Chem 269: 5036
355. Nicholls P, Obinger C, Niederhauser H and Peschek GA (1991) Biochem Soc Trans 19: 252S
356. Nicholls P, Obinger C, Niederhauser H and Peschek GA (1992) Biochim Biophys Acta 1098: 184
357. Hart SEE, Schlarb-Ridley BG, Bendall DS and Howe CJ (2005) Biochem Soc Trans 33: 832
358. Pereira MM, Santana M and Teixeira M (2001) Biochim Biophys Acta 1505: 185
359. Pereira MM, Gomes CM and Teixeira M (2002) FEBS Lett 522: 14
360. Pereira MM and Teixeira M (2004) Biochim Biophys Acta 1665: 340
361. Musser SM, Stowell HB and Chan SI (1993) FEBS Lett 327: 131
362. Howitt CA and Vermaas WFJ (1998) Biochemistry 37: 17944
363. Kienzl PF and Peschek GA (1982) Plant Physiol 69: 580
364. Peschek GA, Schmetterer G and Kienzl PF (1981) FEBS Lett 131: 11
365. Peschek GA (1983) J Bacteriol 153: 539
366. Peschek GA, Schmetterer G, Lauritsch G, Muchl R, Kienzl PF and Nitschmann WH (1983) In: Papageorgiou GC and Packer L (Eds), Photosynthetic prokaryotes: cell differentiation and Function. Elsevier Science Publishers, Amsterdam, The Netherlands, 147–162
367. Peschek GA, Czerny T, Schmetterer G and Nitschmann WH (1985) Plant Physiol 79: 278
368. Peschek GA (2000) Plant Cell Physiol 41 (Suppl): 14
369. Peschek GA, Schmetterer G, Lauritsch G, Nitschmann WH, Kienzl PF and Muchl R (1982) Arch Microbiol 131: 261
370. Alge D and Peschek GA (1993) Biochem Mol Biol Int 29: 511
371. Alge D and Peschek GA (1993) Biochem Biophys Res Commun 191: 9
372. Alge D, Schmetterer G and Peschek GA (1994) Gene 138: 127
373. Iwata S, Ostermeier C, Ludwig B and Michel H (1995) Nature 376: 660
374. Malatesta F, Nicoletti F, Zickermann V, Ludwig B and Brunori M (1998) FEBS Lett 434: 322
375. Svensson-Ek M, Abramson J, Larrson G, Törnroth S, Brzezinski P and Iwata S (2002) J Mol Biol 321: 329
376. Lockau W (1981) Arch Microbiol 128: 336
377. Sandmann G and Böger P (1980) Plant Sci Lett 17: 417
378. Siegelman H, Rasched IR, Kunert K-J, Kroneck P and Böger P (1976) Eur J Biochem 64: 131
379. Peikert R (1995) Diploma Thesis, University of Vienna, Austria
380. Paumann M, Lubura B, Regelsberger G, Feichtinger M, Köllensberger G, Jakopitsch C, Furtmüller PG, Obinger C and Peschek GA (2004) J Biol Chem 27: 10293
381. Paumann M, Feichtinger M, Bernroitner M, Goldfuhs J, Jakopitsch C, Furtmüller PG, Regelsberger G, Peschek GA and Obinger C (2004) FEBS Lett 576: 101

382. Paumann M, Bernroitner M, Lubura B, Peer M, Jakopitsch C, Furtmüller PGG, Regelsberger G, Peschek GA and Obinger C (2004) FEMS Microbiol Lett 239: 301
383. Navarro JA, Duran RV, De la Rosa MA and Hervas M (2005) FEBS Lett 579: 3565
384. Obinger C, Knepper J-C, Zimmermann U and Peschek GA (1990) Biochim Biophys Res Commun 169: 492
385. Serrano A, Gimenez P, Scherer S and Böger P (1990) Arch Microbiol 154: 614
386. Schmetterer G, Alge D and Gregor W (1994) Photosynth Res 42: 43
387. Abramson J, Riistama S, Larrson G, Jasaitis A, Svensson-Ek M, Laakkonen L, Puustinen A, Iwata S and Wikström M (2000) Nat Struct Biol 7: 910
388. Bergman B, Siddiqui PJA, Carpenter EJ and Peschek GA (1993) Appl Environ Microbiol 59: 3239
389. Turner JS and Brittain EG (1962) Biol Rev 37: 170
390. Jones LW and Myers J (1963) Nature 199: 670
391. Peschek GA and Schmetterer G (1982) Biochem Biophys Res Commun 108: 1188
392. Peschek GA (1983) Biochem J 210: 269
393. Pils D and Schmetterer G (2001) FEMS Microbiol Lett 203: 217
394. Peschek GA (1984) Plant Physiol 75: 968
395. Erber W, Nitschmann WH, Muchl R and Peschek GA (1986) Arch Biochem Biophys 247: 28
396. Inaba M, Sakamoto A and Murata N (2001) J Bacteriol 183: 1376
397. Padan E and Schuldiner S (1996) In: Konings WN, Kaback HR and Lolkema JS (Eds), Handbook of biological physics, 2nd vol. Elsevier Science Publishers, Amsterdam, The Netherlands, 501–531
398. Nomura CT, Persson S, Shen G, Inoue-Sakamoto K and Bryant DA (2006) Photosynth Res 87: 215
399. Bernroitner M, Tangl D, Lucini C, Furtmüller PG, Peschek GA and Obinger C (2009) Biochim Biophys Acta 1787: 135
400. Malakhov MP, Malakhova OA and Murata N (1999) FEBS Lett 444: 281
401. Fay P (1992) Microbiol Rev 56: 340
402. Peschek GA, Villgrater K and Wastyn M (1991) Plant Soil 127: 17
403. Wastyn M, Achatz A, Molitor V and Peschek GA (1988) Biochim Biophys Acta 935: 217
404. Wastyn M, Achatz A, Trnka M and Peschek GA (1987) Biochem Biophys Res Commun 149: 102
405. Zehr JP, Mellon MT and Zani S (1998) Appl Environ Microbiol 64: 3444
406. Rai AN, Borthakur M and Bergman B (1992) J Gen Microbiolol 138: 481
407. Jones KM, Buikema WJ and Haselkorn R (2003) J Bacteriol 185: 2306
408. Peschek GA, Niederhauser H and Obinger C (1992) EBEC Short Reports, Elsevier Science Publishers, Amsterdam, The Netherlands, 7: 48
409. Nitschmann WH and Peschek GA (1982) FEBS Lett 139: 77
410. Nitschmann WH, Schmetterer G and Peschek GA (1982) Biochim Biophys Acta 682: 293
411. Fromwald S, Dworsky A and Peschek GA (1995) In: Mathis P (Ed) Photosynthesis: from light to biosphere, 3rd vol. Kluwer Academic Publishers, Dordrecht, The Netherlands, 47–50
412. Neisser A, Fromwald S, Schmatzberger A and Peschek GA (1994) Biochem Biophys Res Commun 200: 884
413. Nierzwicki-Bauer SA, Balkwill DL and Stevens SE (1983) J Cell Biol 97: 713
414. Liberton M, Berg H, Heuser J, Roth R and Pakrasi HB (2006) Protoplasma 227: 129
415. Peschek GA and Sleytr UB (1983) J Ultrastruct Res 82: 233
416. Schmetterer G, Peschek GA and Sleytr UB (1983) Protoplasma 115: 202
417. Douce R, Block A, Dorne AJ and Joyard J (1984) Subcell Biochem 10: 1
418. de Duve C (2005) Singularities. Cambridge University Press, New York
419. Born M (1964) Die Relativitätstheorie Einsteins, 4th ed., (328 pp.), Heidelberger Taschenbücher. Springer Verlag, Berlin etc., Germany
420. Einstein A (1965) Vier Vorlesungen über Relativitätstheorie, 6th ed., Fr. Vieweg & Sohn, Braunschweig, Germany

421. Einstein A (1963) Über die spezielle und die allgemeine Relativitätstheorie, 19th ed., Fr. Vieweg & Sohn, Braunschweig, Germany
422. Fraser JT (1987) Time—The familiar stranger. The University of Massachusetts Press, Amherst, USA
423. Davies P (1987) About time. Einstein's Unfinished Revolution, 3rd ed., Simon and Schuster, New York
424. Hawking S (1988) A brief history of time: From the big bang to black holes. Bantham Books, New York
425. Hawking S (1993) Black holes and baby universes and other essays. Bantham Books, New York
426. Hawking S and Mlodinov L (2010) The grand design. Bantham Books, New York
427. Rutten GM (1966) Paleogeogr Paleoclimatol Paleooecol 2: 47
428. Schidlowski M (1971) Geol Rundsch 60: 1351
429. Rhoads DC and Morse JW (1970) Lethaia 4: 413
430. Schlegel HG (1976) In: Kinne O (Ed), Marine ecology, 2nd vol. part 1, Wiley and Sons, London, UK, 9–60
431. Schlegel HG (1976) Antonie van Leeuwenhoek 42: 181
432. Meyer O and Schlegel HG (1983) Annu Rev Microbiol 37: 277
433. Geisler M, Richter J and Schumann J (1993) J Mol Biol 234: 1284

Chapter 2
The Photosynthetic Apparatus of the Living Fossil, *Cyanophora paradoxa*

Jürgen M. Steiner and Wolfgang Löffelhardt

Muroplasts, the peculiar plastids (previously called cyanelles) of glaucocystophyte algae did retain—with some modifications—the peptidoglycan wall of their cyanobacterial ancestor. This is not only a convincing proof of the endosymbiotic theory, but earns glaucocystophytes the status of living fossils, as peptidoglycan is found nowhere else among eukaryotes. Muroplasts show even more cyanobacterial features than other primitive plastids, e.g., rhodoplasts. Almost all data available at present come from one species, *Cyanophora paradoxa*. The plastome, containing a surplus of 50 protein genes compared to chloroplast genomes (Table 2.1) and about 30% of the transcriptome (ESTs) of this organism are sequenced and a genome project is in progress. While *Cyanophora* does not offer such possibilities for genetic analysis as the *Chlamydomonas* system, due to its reasonable growth rate it is amenable to biochemical investigations.

The light-harvesting antennae, the pathways of protein sorting into and within plastids, the genetic compartmentation of photosynthesis genes, the composition of the oxygen-evolving complex, and the nature of the carbon-concentrating mechanism will be compared for muroplasts, rhodoplasts and (green algal) chloroplasts. Preliminary data on the biosynthesis of starch in *C. paradoxa* will be put in the context of reserve carbohydrate nature and localization in various oxygenic autotrophs. Based on sequence alignments and phylogenetic analyses, *Cyanophora* occupies a bridge position between cyanobacteria, red algae and green algae, in some cases taking sides with the ancestors, in other cases with the green lineage or the red lineage, respectively.

W. Löffelhardt (✉)
Max F. Perutz Laboratories, Department of Biochemistry, University of Vienna, Dr. Bohrgasse 9, 1030 Vienna, Austria
e-mail: wolfgang.loeffelhardt@univie.ac.at

G. A. Peschek et al. (eds.), *Bioenergetic Processes of Cyanobacteria*,
DOI 10.1007/978-94-007-0388-9_2, © Springer Science+Business Media B.V. 2011

Table 2.1 Muroplast genes from *Cyanophora paradoxa* specifying proteins of the photosynthetic apparatus

Phycobiliproteins (7): *apcA, apcB, apcD, apcE, apcF, cpcA, cpcB*
Photosystem I proteins: *psaA, psaB, psaC, psaE, psaF, psaI, psaJ, psaM*
Photosystem II proteins: *psbA, psbB, psbC, psbD, psbE, psbF, psbH, psbI, psbJ, psbK, psbL, psbM, psbN, psbT, psbV, psbX, psbY, psbZ, psb30*
ATP synthase subunits: *atpA, atpB, atpD, atpE, atpF, atpG, atpH*
Subunits of the cytochrome b_6/f complex & ferredoxin: *petA, petB, petD, petG, petL, petN, petX; petF*
Calvin cycle enzymes: *rbcL, rbcS*
Chlorophyll biosynthesis: *hemA**, *chlB, chlI, chlL, chlN*
Carotenoid & prenylquinone biosynthesis: *crtE**, *preA*
ORFs with putative function in photosynthesis: *ycf3*[a], *ycf4*[a], *ycf5*[b], *ycf16*[c], *ycf24*[d], *ycf27*[e], *ycf33*[f], *orf333*[g]

Gene nomenclature according to the guidelines for chloroplast genes (Stoebe et al. 1998). Genes marked with an asterisk are not found on any other plastid genome. Genes underlined are absent from the chloroplast genomes of higher plants

[a] Role in PS I assembly

[b] Role in PS I function

[c] ABC transporter subunit, ortholog to bacterial *sufC*, involved in [Fe-S] cluster biogenesis

[d] ABC transporter subunit, ortholog to bacterial *sufB*, involved in [Fe-S] cluster biogenesis

[e] Response regulator of PS I genes (*rpaB*)

[f] Role in cyclic electron transport

[g] Role in assembly/stability of PS II (*hcf136*)

2.1 Introduction

Phototrophic eukaryotes emerged from the so-called primary endosymbiotic event, about 1.5 billion years ago (Fig. 2.1). It is generally accepted that cyanobacteria gave rise to plastids after massive gene transfer from the endosymbiont genome to the nucleus of the heterotrophic host cell. Most, but not all researchers in the field consider the primary endosymbiotic event a single one, i.e., the immensely complicated establishment of a hereditary symbiosis leading to a photosynthetic organelle happened only once (Martin et al. 1998; Rodríguez-Ezpeleta et al. 2005; Reyes-Prieto et al. 2007a), between a suitable host cell and a suitable—likely filamentous and nitrogen-fixing—cyanobacterium (Deusch et al. 2008). This view implies monophyly of the kingdom "Plantae": all plastid types, regardless of their morphology and pigmentation are derived from an ancestral organelle, the "protoplastid". This primordial organelle had already performed 90% of the total gene transfer events to the nucleus, was still armored by a peptidoglycan wall, and had developed a specific protein import apparatus at its envelope. The Archaeplastida (glaucocystophytes, rhodophytes, chlorophytes and streptophytes) are phototrophic eukaryotes containing primary plastids, surrounded by two envelope membranes (Adl et al. 2005). The phagosomal membrane was lost early in plastid evolution as evidenced by two much more recent endosymbioses: the photosynthetic chromatophores (cyanelles) from the thecamoeba *Paulinella chromatophora* (Nowack et al.

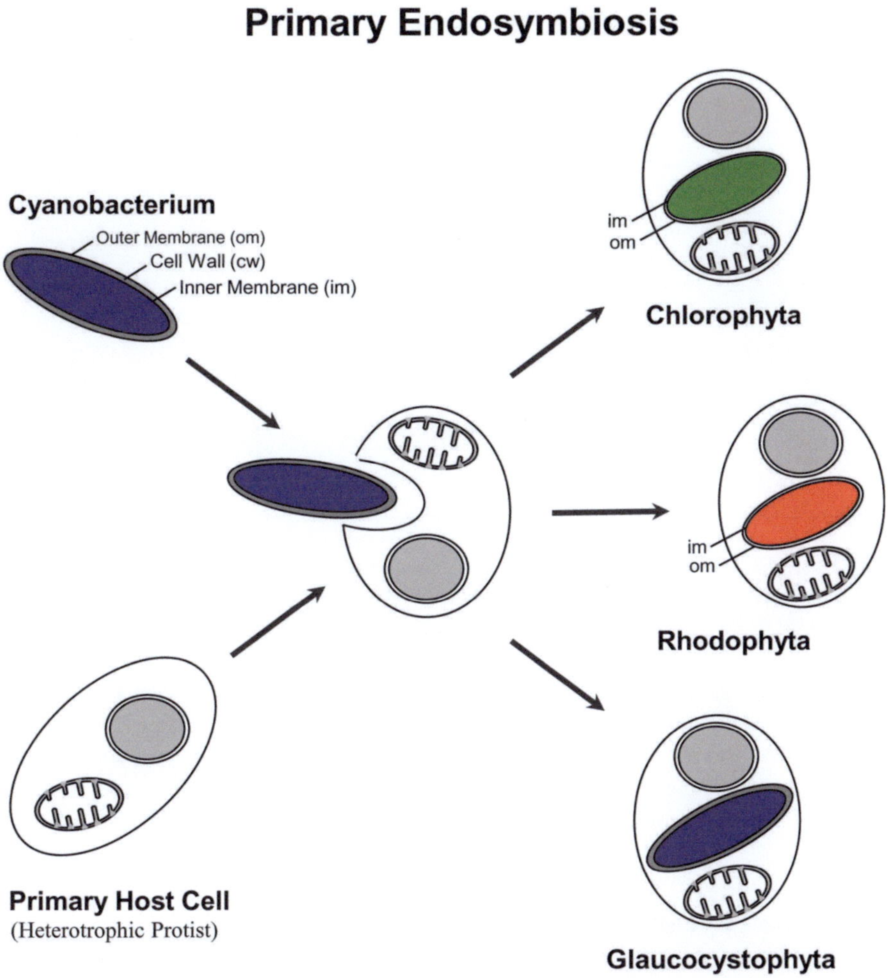

Fig. 2.1 The single primary endosymbiotic event. (Modified from Stoebe and Maier 2002)

2008; derived from a *Synechococcus*-like ancestor, estimated age 60 My) are surrounded by two membranes only, whereas the nitrogen-fixing spheroid bodies from the diatom *Rhopalodia gibba* (Kneip et al. 2008; derived from a *Cyanothece*-like cyanobacterium, estimated age 25–35 My) are still surrounded by three membranes (Kovacevic et al. 2009). The inner (IEM) and outer (OEM) envelope membranes of the Archaeplastida clearly correspond to the respective membranes of the ancestral cyanobacterium. The majority of algal phyla arose from superimposed (multiple and polyphyletic) secondary endosymbiotic events where hetrotrophic protists engulfed red algae or green algae, i.e. phototrophic eukaryotes. The resulting secondary or complex plastids are surrounded by three or four membranes (Stoebe and Maier 2002).

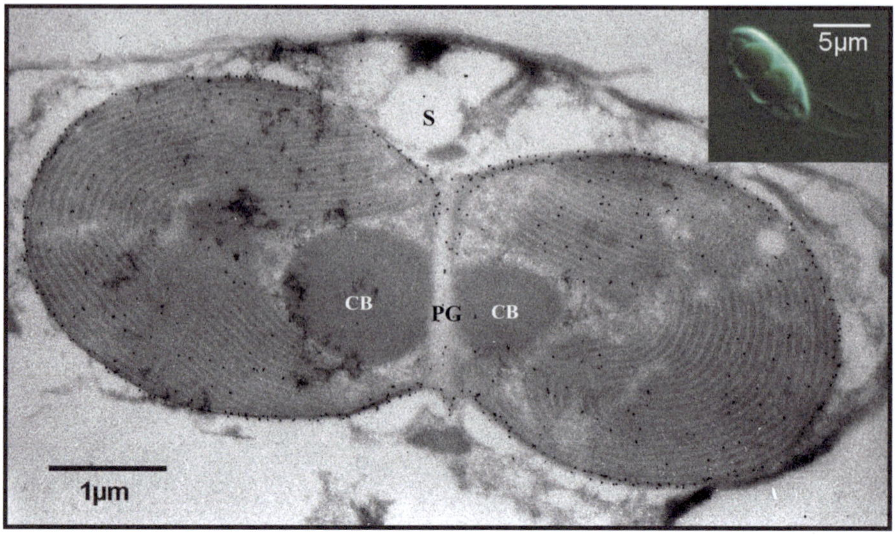

Fig. 2.2 Eukaryotic peptidoglycan in the muroplast envelope of *Cyanophora paradoxa* visualized via immunogold electron microscopy. Primary antibodies are directed against peptidoglycan from *Escherichia coli*. *PG*, peptidoglycan septum; *CS*, putative carboxysome

Most phylogenetic analyses hold the glaucocystophytes as the most ancient phototrophic eukaryotes known to date (Rodríguez-Ezpeleta et al. 2005; Reyes-Prieto et al. 2007b) and their muroplasts (Fig. 2.2) occupy a bridge position between plastids and free-living cyanobacteria. It is the purpose of this review to elaborate these special features mimicking early events in plastid evolution.

2.2 The Phycobilisome Antenna System

This major light harvesting system of cyanobacteria (MacColl 1998), rhodoplasts, and glaucocystophyte muroplasts was studied in some detail in *C. paradoxa* (Table 2.2). While its cyanobacterial origin is obvious, modifications typical for eukaryotic phycobilisomes (PBS) are encountered: some of the putative rod linker polypeptides are of unusual size, i.e., 50–60 kDa, as also found in rhodoplasts (Egelhoff and Grossman 1983; Liu et al. 2008a). In contrast to cyanobacterial PBS, no binding of FNR to the outer segments of rods (Jakowitsch et al. 1993; Steiner et al. 2003) was observed. Unlike red algae, there is no phycoerythrin (PE) and the genetic compartmentalization is more rigorous: the chromophorylated phycobiliproteins are all muroplast-encoded whereas all colorless linker polypeptides are products of nuclear genes and thus major import products into muroplasts (Table 2.2). However, there are PBS components with features from both groups as the chromophore-bearing, linker-like γ–phycoerythrin from the red alga *Aglaothamnion neglectum* (Apt et al. 1993), which is nucleus-encoded. On the other hand, the "genuine" rod-core linker

Table 2.2 Components of purified, intact muroplast PBS (Steiner et al. 2003)

Apparent MW (kDa)	Abundance	Phycobiliprotein	Correlated gene	Assignment
98	Medium	Yes	*apcE*[a]	Core-membrane linker
55	Medium	No		Rod linker?
53	Medium	No		Rod linker?
38	Low	No		Rod linker[b]?
31	Low	No	*cpcG1, cpcG2*[b]	Rod-core linker
19–21	High	Yes	*apcA*[a],*B*[a],*D*[a],*F*[a]	Allophycocyanin subunits
16–18	High	Yes	*cpcA*[a],*B*[a]	Phycocyanin subunits[b]
9	Low	No		Terminal rod linker?
8	Low	No	*apcC*	Core linker

[a] Muroplast-encoded
[b] Potential components of a PS I associated PBS subcomplex

CpcG is encoded by the plastid genome in *A. neglectum* (Apt and Grossman 1993). The accessory allophycocyanins ApcD and ApcF have functions in energy transfer from the PBS to PS I and PS II, respectively, in cyanobacterial (Dong et al. 2009) and, likely, also in glaucocystophyte PBS. Recently, evidence was obtained that PBS serve both photosystems in rhodoplasts (Liu et al. 2008a; Busch et al. 2010).

In analogy to cyanobacteria (Wang et al. 2004), the expression of the rod-core linker gene *cpcG1* is downregulated upon shift of the cultures from high to low [CO_2] indicating a reduction of antenna size during CO_2-limitation (Burey et al. 2007). The 7.8 kDa core linker ApcC (45–51% sequence identity to cyanobacterial counterparts of almost the same size) was expressed through *in vitro* transcription/translation of the cloned gene and imported into isolated muroplasts where its fast assembly into PBS could be demonstrated (Steiner et al. 2003). This is convincing evidence for the dynamic behavior of PBS. ApcC from *C. paradoxa* shows an insertion of four amino acids (AKKT) between the highly conserved positions P-11 and S-12 of cyanobacterial homologs which is not found in the corresponding ESTs from red algae. Partial cDNA sequence as well as peptide sequence information is available for the other linker polypeptides of Table 2.2 with the exception of the 9 kDa protein (J. Steiner, unpublished). The small linker polypeptides described for the unicellular red alga, *Porphyridium cruentum*, are of higher MW (13 kDa and 14 kDa; Liu et al. 2008a). Interestingly, the large linkers (Table 2.2) seem to have arisen from a duplication of the *cpcG* gene, which itself exists in two variants, *cpcG1* and *cpcG2* (J. Steiner, unpublished). A similar gene fusion event has been reported for an unusual, chromophore-bearing phycoerythrin linker protein from a marine *Synechococcus* sp. (Six et al. 2005). Recently, CpcG2 was shown to be part of a novel photosystem I antenna, consisting of a rod only, in *Synechocystis* sp. PCC 6803 (Kondo et al. 2007). It remains to be seen if this can be generalized to muroplast CpcG2, the C-terminus of which is less hydrophobic than that of the cyanobacterial counterpart. However, there is convincing evidence for a dual PSI antenna in the red alga *C. merolae* consisting of crescent-shaped Lhcr proteins and a PBS subcomplex of the *Synechocystis* type (Busch et al. 2010). There is only one gene for CpcG in the (streamlined) genome of the extremophile, that also lacks a hydrophobic stretch at the C-terminus. Taken together, a bridge position seems to emerge for PSI of glaucocystophytes: no extrinsic chlorophyll antenna (unlike red algae) but a (reduced) special PBS antenna. CpcG2, a fraction of the alpha- and beta-phycocyanins, and the putative 38 kDa rod linker (substoichiometric in Table 2.2) could function as subunits thereof.

The prevalent muroplast pigments are chlorophyll *a*, β-carotene, zeaxanthin, β-cryptoxanthin, allophycocyanin, and C-phycocyanin. Carotenoids typical for cyanobacteria such as echinenone and myxoxanthophyll are absent. A carotenoid–rich protein (CRP) was isolated from *C. paradoxa* through its cross-reactivity with LHCP (Rissler and Durnford 2005). However, the 28 kDa protein appeared to be a peripheric rather than an integral thylakoid membrane protein with a preference for zeaxanthin binding. Elucidation of its function, likely photoprotective as in cyanobacteria (Liu et al. 2008b), must await further studies.

2.3 Photosystem II

The photosynthetic apparatus of cyanobacteria (including primitive plastids) and chloroplasts (from green algae and higher plants) show some differences in spite of their high overall similarity. The oxygen evolving complex (OEC) in cyanobacteria (Shen et al. 1998) consists of three peripheral membrane proteins: a 33 kDa protein (OE33, the product of the *psbO* gene), cytochrome c_{550} (*psbV*), and a 9–12 kDa protein (*psbU*). Chloroplasts possess only one of these components: the 33 kDa protein. An unrelated 23 kDa protein, OE23 (*psbP*), fulfills the function of cytochrome c_{550}, whereas PsbU is replaced by an unrelated 16 kDa protein, OE16 (*psbQ*). Primitive plastids comprise the muroplasts of *Cyanophora paradoxa*, the rhodoplasts of e.g., *Porphyra purpurea* and *Cyanidium caldarium* and, likely, those complex plastids (e.g., of the cryptomonad *Guillardia theta* and the diatom *Odontella sinensis*) that arose from a red algal secondary endosymbiont. Evidence for their cyanobacterial type OEC is the *psbV* gene which resides on the plastid genomes in all known cases (Kowallik et al. 1995; Reith and Munholland 1995; Löffelhardt et al. 1997; Douglas and Penny 1999; Glöckner et al. 2000) and the nuclear *psbU* genes of e.g., *C. caldarium* (Ohta et al. 1999) and *C. paradoxa* (Steiner et al. 2005b) which show bipartite presequences. In addition, the photosystem II and OEC complex was characterized at the protein level for *C. caldarium* (Enami et al. 1995) and *C. paradoxa* (Shibata et al. 2001). The nucleus-encoded *psbP* and *psbQ* genes of higher plants and green algae display "twin-arginine" motifs in the thylakoid transfer domains of their bipartite presequences, which is an indication for the usage of the ΔpH or Tat pathway for thylakoid translocation. This pathway is employed by lumenal proteins unable to translocate via the alternative SecA/Y/E pathway because of folding in the stroma with or without cofactors (Dalbey and Robinson 1999). At present, there is no evidence for any cofactors of OE23 and OE16. On the other hand, the twin-arginine motif is missing from the presequences of *psbU* genes from cyanobacteria.

Muroplast SecA appeared to be quite susceptible to inhibition by sodium azide, but not to the Tat pathway inhibitor Nigericin, during import experiments with homologous precursors: thylakoid translocation of the larger intermediate form of PsbO was completely abolished (Steiner et al. 2005b). Cyanobacterial thylakoids do not form tight vesicles upon isolation and thus are not suitable to demonstrate protease protection of internalized, processed lumenal proteins. With improved muroplast fractionation methods it was possible, at least for PsbO, to show Sec-dependent translocation *in organello* and, after muroplast lysis and thylakoid isolation, for the first time protease protection of the mature protein inside of phycobilisome-bearing thylakoids. Recently, a psbQ-like protein corresponding to ORF *sll*1638 has been identified by proteomic analysis of a highly active PSII preparation from *Synechocystis* sp. PCC6803 (Kashino et al. 2002). In the red alga *Cyanidium caldarium* the OEC is of "primitive" type consisting of PsbO, PsbU and PsbV (cyt c_{550}), but additionally contains a 20 kDa protein with significant sequence similarity to PsbQ, now named PsbQ′, which has been shown to be able to functionally replace the green algal 17 kDa protein on binding to green algal, but not higher plant PSII and restore oxygen evolution (Ohta et al. 2003). Later on, PsbQ′ was established as an essential component of the cyanobacterial OEC (Roose et al. 2007). A crystalline preparation of *C. caldarium* PS II dimers with excellent resolution of protein bands (Adachi et al. 2009) also revealed the additional extrinsic (lumenal) protein of 20 kDa (PsbQ′). In the dBEST data base, a *Cyanophora* EST with sequence similarity (http://amoebidia.bch.umontreal.ca/pepdb/pepdb.html) to PsbQ′ is contained, pointing towards even greater analogies between the OECs from cyanobacteria and primitive algae than previously expected.

PSII reaction center core and intrinsic subunits are very similar for all oxygenic phototrophs. The muroplast genome encodes in total 19 subunits of PS II (Table 2.1), the same applies for rhodoplasts (Reith and Munholland 1995). Shibata et al. (2001) investigated the subunit composition of purified oxygen-evolving PS II particles of *C. paradoxa*: more than 30 protein bands were resolved including PsbO and cytochrome c_{550} whereas PsbU obviously was lost during the preparation.

2.4 Cytochrome b_6/f Complex and Electron Donors to Photosystem I

As in the case of PSII, the muroplast genome is a treasure trove containing seven genes for membrane-bound subunits (Table 2.1). Pre-cytochrome *f*(*petA*) contains a signal sequence with a highly hydrophobic core domain. The only nuclear-encoded subunit known to date is Rieske iron sulfur protein (PetC). About five aa of its N-terminus with a noncleavable signal-anchor domain protrude into the stroma, followed by one trans-membrane helix and the lumenally exposed bulk of the protein containing the Fe_2S_2 cluster. The "twin arginine" motiv of cyanobacterial PetC is replaced by RK: nevertheless there is good evidence that the Tat translocase assists in thylakoid membrane translocation/insertion of this folded holoprotein that has to attain its prosthetic group in the stroma (Steiner et al. 2005b). There are no data at present on purification of the cytochrome b_6/f complex and subunit characterization at the protein level for *C. paradoxa* or red algae.

Cytochrome c_6 (PetJ) of *C. paradoxa* was characterized at the protein and cDNA level (Steiner et al. 2000). The bipartite presequence of the nuclear-encoded precursor shows a signal peptide with a core domain of moderate hydrophobicity, unlike the muroplast-encoded pre-cytochrome *f* and pre-cytochrome c_{550}. The intermediate form, i-cytochrome c_6 accumulated in the stroma after import into isolated cyanelles and transit peptide cleavage when sodium azide was added to the incubation buffer (Steiner et al. 2005b). As expected c-type cytochromes attain their heme in the thylakoid lumen and are translocated across the thylakoid membrane as unfolded Sec passengers. PetJ is solely responsible for electron transport between the cytochrome b_6/f complex and PS II as there is no plastocyanin in *C. paradoxa.* PetJ is present in the periplasmic space as well as in the thylakoid lumen of muroplasts and might play a role as electron donor to yet unknown terminal oxidases in the inner envelope membrane (G. Peschek, personal communication). There is no indication for a continuity of IEM and TM in muroplasts, a question which is also answered in the negative for cyanobacteria (Liberton et al. 2006). Nevertheless, under freeze-thaw conditions muroplasts are completely drained of cytochrome c_6 whereas no loss of phycobiliproteins was observed (Steiner et al. 2000).

2.5 Photosystem I

At the present state of knowledge the very well investigated PS I of cyanobacteria is trimeric, whereas PS I of algae is a monomer (good evidence for green and red algae, by analogy for glaucocystophytes). The contribution of the muroplast genome—eight genes –is less pronounced than for PS II (Table 2.1). PS I preparations from *C. paradoxa* revealed, in addition to the major polypeptides, a number of smaller subunits. PsaD is nuclear-encoded in contrast to red algae and its N-terminus resembled the counterparts from the “green lineage”. On the other hand, the N-terminal sequence of PsaL was closer to the cyanobacterial homologs. (Koike et al. 2000). In this study, immunodetection of Lhca proteins was unsuccessful. Thus, in contrast to other algal phyla, the lack of an extrinsic chlorophyll antenna in PS I must be assumed for *C. paradoxa*, a parallel to cyanobacteria. It remains to be seen upon the completion of the *C. paradoxa* genome sequence if PsaGHK (involved in Lhca and Lhcr binding) are absent.

Muroplast ferredoxin was the first plastid-encoded ferredoxin reported (Table 2.1, Löffelhardt et al. 1997). Ferredoxin-$NADP^+$ oxidoreductase (FNR) of *Cyanophora paradoxa* was characterized at the protein and cDNA level (Gebhart et al. 1992; Jakowitsch et al. 1993). The 34 kDa protein shows high amino acid sequence similarity to higher plant counterparts and lacks the C-terminal extension of cyanobacterial enzymes (around 45 kDa) responsible for binding to phycobilisomes. The availability of the ^{35}S-labeled precursor was important for the establishment of a muroplast *in vitro* import system (Jakowitsch et al. 1996). Muroplasts and likely rhodoplasts appear to import precursors displaying a phenylalanine residue in the N-terminal domain of the transit sequences via a primordial Toc translocon with postulated receptor and pore functions. The bulk import products of chloroplasts, Rubisco SSU and Lhcb, which necessitated the development of specific receptors, are not relevant here (Steiner et al. 2005a; Steiner and Löffelhardt 2005).

2.6 Calvin Cycle Enzymes

A NAD(P)-dependent glyceraldehyde-3-phosphate dehydrogenase was purified from a muroplast extract of *C. paradoxa* as a 142 kDa homotetramer with features similar to the cyanobacterial counterpart (Serrano and Löffelhardt 1994). This is in agreement with the postulated duplication of the *gapA* gene early in streptophyte evolution (Petersen et al. 2006). In addition, the NAD-dependent cytosolic enzyme involved in glycolysis was purified (Serrano and Löffelhardt 1994).

The gene for CP12 protein involved in the formation of inactive complexes of Calvin cycle enzymes during night was also characterized (Petersen et al. 2006). Cytosolic and muroplast-localized fructose-1,6-bisphosphate aldolases of class II were fractionated from *C. paradoxa* extracts as 90 and 85 kDa proteins, respectively (Gross et al. 1994). The latter was shown to be bifunctional for fructose-1,6-bisphosphate and sedoheptulose-1,7-bisphosphate cleavage (Flechner et al. 1999). The gene for the cyanelle enzyme was subject to phylogenetic analysis. It proved to belong to type B, hitherto not found in plastids, with *Synechocystis* sp. as the nearest neighbor (Nickol et al. 2000). The cDNA of pre-transketolase from *C. paradoxa* was sequenced. In a neighbor-net graph, the *Cyanophora* enzyme occupied a position intermediate to the plastid and cyanobacterial homologs. *In vitro* import of the precursor into muroplasts and pea chloroplasts was equally efficient (Ma et al. 2009). The single copy gene was downregulated upon shift to low CO_2 conditions (Burey et al. 2007)

2.7 Starch Metabolism

Early diverging phototrophic eukaryotes seem to play an important role in the conversion of cyanobacterial glycogen into the starch of green algae and higher plants during evolution (Deschamps et al. 2008). *C. paradoxa* starch showed a (high) amylose and amylopectin content with chain length distributions and crystalline organization similar to green algae and land plants. However, several starch synthase activities were found utilizing UDP-glucose, this time in analogy to rhodophytes that also synthesize starch in the cytosol. In addition, a multimeric isoamylase complex (in support of the proposed correlation between the presence of starch and of isoamylase) and multiple starch phosphorylases were demonstrated. These results were obtained at the zymogram level and in some cases also at the gene level (Plancke et al. 2008). Transcription of a granule-bound starch synthase (responsible for amylose formation) was shown to be upregulated upon shift to low [CO_2] (Burey et al. 2007). Recently, the transglucosidase DPE2 (disproportionating enzyme 2, transferring one glucose moiety from maltose to a cytosolic heteroglucan) from *C. paradoxa* could be demonstrated on zymograms. Despite the differential location of starch in the two cases, DPE2 is a cytosolic enzyme as in *Arabidopsis* (Fettke et al. 2009).

The properties of a number of transporters were studied in the muroplast inner envelope membrane: a phosphate translocator (TPT) similar to the chloroplast homolog and a glucose transporter (Schlichting and Bothe 1993) as well as a glutamine carrier with unusual features (Kloos et al. 1993) whereas a malate/oxaloacetate shuttle and an ATP/ADP translocator could not be demonstrated (Schlichting et al. 1990). The efficient triose phosphate (TP)/phosphate exchange resembles that observed for rhodophytes: in both cases starch is synthesized and stored in the cytosol, necessitating rapid export of photosynthate from the plastids, whereas the low-affinity plant TPTs allow for sufficient TP levels for starch biosynthesis inside of the chloroplasts (Linka et al. 2008). An interesting hypothesis postulates the insertion of host cell transporters into the endosymbiont plasma membrane as a primordial step in the establishment of endosymbioses: candidates are the TPT (Weber et al. 2006) and a transporter for ADP-glucose required for relocation of starch synthesis from the endosymbiont into the host cytosol (Deschamps et al. 2008).

2.8 The Case of the Eukaryotic Carboxysome

All cyanobacteria (Badger and Price 2003) and most algae (Giordano et al. 2005) possess an inorganic carbon-concentrating mechanism (CCM) that involves a microcompartment—carboxysomes in prokaryotes, pyrenoids in eukaryotes—harboring the bulk of cellular (plastidic) Rubisco. Inorganic carbon (bicarbonate) is concentrated at the very site where the bulk of Rubisco is involved in efficient CO_2 fixation, i.e., superior to the rate that soluble Rubisco acting below substrate saturation could achieve. Carbonic anhydrase (CA) is important for converting the "storage form" bicarbonate into the Rubisco substrate CO_2. CA is co-packaged with Rubisco in cyanobacterial carboxysomes whereas it localizes to the lumen of thylakoids traversing the pyrenoid in *C. reinhardtii*. The two microcompartment types also differ in size, number and morphology. Typically, one pyrenoid (diameter 500 nm–1 μm) of rounded shape without distinct contours is seen per plastid. In cyanobacteria, several smaller (diameter 100–200 nm) polyhedral carboxysomes with a hexagonal cross-section are found in the centroplasm, i.e., distant from the thylakoid membranes. Carboxysomes are never penetrated by thylakoids. At higher magnification, the quasicrystalline arrangement of the Rubisco matrix and the surrounding proteinaceous layer consisting of the shell proteins CcmK-N (Kerfeld et al. 2005) become apparent.

For *C. paradoxa*, the operation of a CCM was a matter of debate. Recently, microarray data revealing 142 CO_2-responsive genes (induced or repressed through a shift from high to low CO_2 conditions), gas exchange measurements, and measurements of photosynthetic affinity provided strong support for a CCM in the muroplasts (Burey et al. 2007). The microarray results compared well with corresponding data on *Synechocystis* PCC6803 (Wang et al. 2004) and *C. reinhardtii* (Miura et al. 2004). Photosynthesis genes specifying subunits of photosystem I and II and Calvin cycle enzymes were downregulated, whereas genes for proteins expected to be involved in the CCM as carbonic anhydrases, the putative bicarbonate transporter LciA, Rubisco activase, and granule-bound starch synthase were upregulated (Table 2.3; Burey et al. 2007).

A recent hypothesis claims that glaucocystophyte muroplasts as the closest cousins to cyanobacteria among plastids contain "eukaryotic carboxysomes": bicarbonate enrichment within muroplasts should be considerably higher than in chloroplasts with their pyrenoid-based CCM. Thus, the stress-bearing function of the peptido-

Table 2.3 A selection of CO_2-responsive genes in *Cyanophora paradoxa* (From Burey et al. 2007)

Function	Gene	Protein
1. Up-regulated upon shift to low [CO_2]		
Photosynthesis (CCM)	*rca*	Rubisco activase[a]
		Granule-bound starch synthase[a]
CCM	*CAH3-1?*	beta-CA (mitochondrial)[a]
	CAH3-2?	beta-CA (mitochondrial)[a]
		CA (cytosolic?)
	LciA	Bicarbonate transporter[a]
ROS inactivating enzymes	Prdx1	Peroxiredoxin1
	cat1	Catalase
		Glutaredoxin
Protein degradation	UBC4	Ubiquitin-conjugating enzyme E2
		Ubiquitin/ribosomal protein S27a
Chaperones		Protein disulfide-isomerase
		Peptidyl-propyl cis-trans-isomerase
2. Down-regulated upon shift to low [CO_2]		
Phycobilisome antenna	*cpcG*	Phycobilisome rod-core linker[b]
Calvin cycle		Sedoheptulose-1,7-bisphosphatase[a,b]
	pgk	Phosphoglycerate kinase[ab]
	tktC	Transketolase[a,b]
Photosynthetic electron transport	psaC	PS I reaction center subunit II[a,b]
	psaK	PS I reaction centre subunit X[a,b]
		PS I reaction center subunit XI[a,b]
	petJ	Cytochrome c_6
Protein synthesis		Translation elongation factor 1 beta-2[a]
		Translation elongation factor 1 alpha[a]
Cytoskeleton		Tubulin alpha-2[a]
		Tubulin beta-1[a]
		Tubulin gamma[a]

[a] Paralleled in *C. reinhardtii* (Miura et al. 2004)
[b] Paralleled in *Synechocystis* sp. PCC6803 (Wang et al. 2004)

glycan layer (Pfanzagl et al. 1996), the other unique cyanobacterial heritage, would be indispensable (Raven 2003). An isolation method for muroplast "carboxysomes" was developed and the protein components other than Rubisco analysed by mass spectrometry. Rubisco activase was identified for the first time in a carboxysome-like microcompartment and corroborated by western blotting. Rubisco activase incorporation into isolated muroplast carboxysomes after *in vitro* import followed a time course and increased from approximately 15% at 7 min to 25% at 25 min (Burey et al. 2005; Fathinejad et al. 2008). Other protein bands yielded good mass spectra but no match to existing databases was found—the incomplete knowledge on *C. paradoxa* nuclear genes certainly is an obstacle.

The geological record indicates a decrease in [CO_2] and a concomitant increase in [O_2] during the Phanerozoic period about 400 million years ago, which certainly posed a problem for Rubisco-dependent CO_2 fixation. This led Badger and Price (2003) to propose a scenario different from Raven's: independent origin of the carboxysomal CCM in cyanobacteria and the pyrenoidal CCM in algae during that period.

Glaucocystophyte muroplasts contain a conspicuous, electron-dense aggregate of Rubisco (Mangeney and Gibbs 1987) of pyrenoid size in the center (only in *G. nostochinearum* at a polar position) of the coccoid organelles surrounded by multiple layers of concentric thylakoids (Kies 1992). For *C. paradoxa*, the muroplast nucleoid was shown to occupy the narrow space between the central body and the innermost thylakoids (Löffelhardt and Bohnert 2001). In no known case are these microcompartments traversed or penetrated by thylakoid membranes (Figs. 2.2, 2.3 and 2.4). They remain stable during muroplast division and are neatly bisected

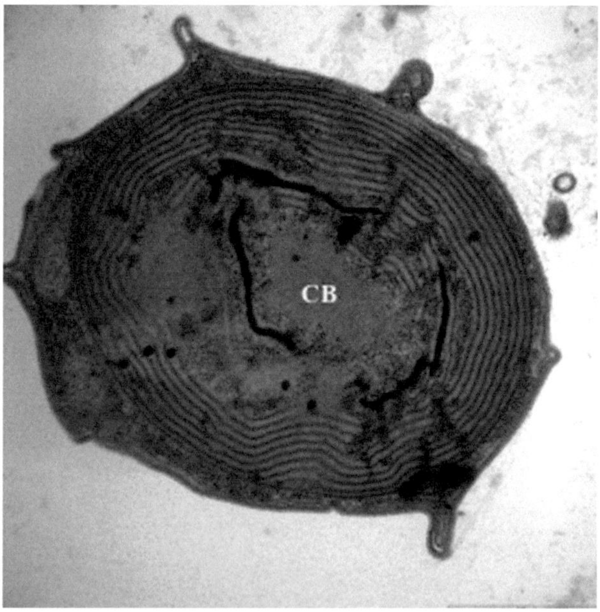

Fig. 2.3 Transmission electron micrograph of a *Cyanophora* cell grown under high [CO_2] Electron-dense regions around the innermost thylakoids are notable. (scale bar, 1 μM). Chemically fixed and epoxy resin embedding. *CB*, central bodies; *Env*, envelope. Under high [CO_2] the contours of the CBs are not clearly defined. Electron-dense material is abundant among thylakoids proximal to the center. Cyanelle ribosomes appear around the central body (for Rubisco synthesis) and between the thylakoids (for phycobiliprotein synthesis)

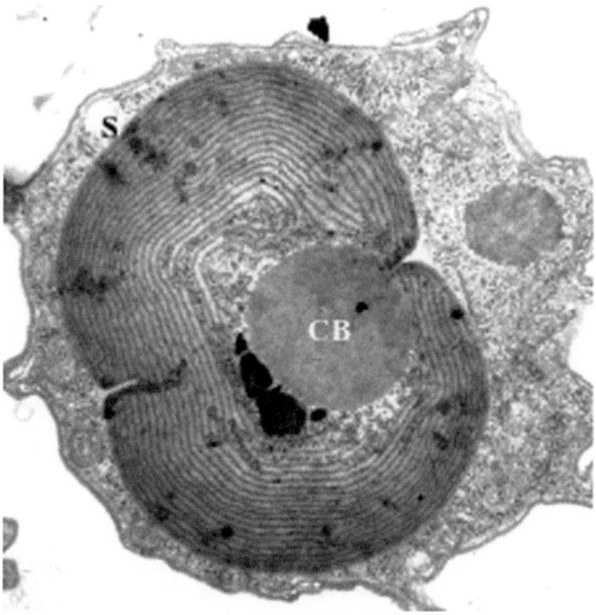

Fig. 2.4 Transmission electron micrograph of a *Cyanophora* cell grown under low [CO_2] (scale bar, 1 μM). Chemically fixed and epoxy resin embedding. *CB*, central body; *S*, starch granule; The onset of septum growth is observed. Indication for a polyhedral shape of CB (*arrows*)

by the growing peptidoglycan septum (Fig. 2.2). In *C. paradoxa* and *G. nostochinearum* the microcompartment has a rounded shape though some polyhedral elements become apparent, especially under low CO_2 conditions. Then, the central bodies also appear to be enlarged (Fig. 2.4; Fathinejad et al. 2008). Ribosomes are now concentrated around the central body (involved in Rubisco synthesis, Fig. 2.4) whereas under high CO_2 conditions they are also abundant in the thylakoid area (involved in phycobiliprotein synthesis, Fig. 2.3). This may be consistent with a dynamic structure for the microcompartment which expands and shrinks according to CO_2 availability. In *Gloeochaete wittrockiana* and *Cyanoptyche gloeocystis* the single, large central bodies clearly are polyhedral (with a pentagonal cross-section) and are confined by a distinct, electron-dense layer (Kies 1992). This structure is suggestive of a proteinaceaous shell as found in cyanobacterial carboxysomes. According to Raven (2003), earlier low CO_2 episodes could have occurred beyond the 600 million years covered by the geological record. An acquisition of the CCM by cyanobacteria prior to the primary endosymbiotic event could explain the retainment of carboxysomes solely in the muroplasts of the most ancient glaucocystophytes whereas all other algae shifted towards a pyrenoidal CCM and lost the plastid peptidoglycan wall. A final decision between carboxysome or pyrenoid must await the identification of muroplast CA and, especially, the demonstration of shell proteins. Also, the internal accumulation of Ci within muroplasts under low CO_2 conditions has to be determined: is the concentration factor in the range of >1,000 (carboxysomal CCM) or just around 70 (pyrenoidal CCM)? Intermediate stages between carboxysomes and pyrenoids can also be envisaged for the muroplast microcompartments.

Acknowledgements The authors are indebted to the Austrian "Fonds zur Förderung der wissenschaftlichen Forschung" for continuous support of their research on glaucocystophytes.

References

Adachi H, Umena Y, Enami I, Henmi T, Kamiya N and Shen J-R (2009) Towards structural elucidation of eukaryotic photosystem II: purification, crystallization and preliminary X-ray diffraction analysis of photosystem II from a red alga. Biochim Biophys Acta 1787: 121–128

Adl SM, Simpson AGB, Farmer MA, Andersen R, Anderson OR, Barta JA, Bowser SS, Bragerolle G, Fensome RA, Fredericq S, James TY, Karpov S, Kugrens P, Krug J, Lane CE, Lewis LA, Lodge J, Lynn DH, Mann DG, McCourt RM, Mendoza L, Moestrup Ø, Mozley-Standridge SE, Nerad TA, Shearer CA, Smirnov AV, Spiegel FW and Taylor MFJR (2005) The new higher level classification of eukaryotes with emphasis on the taxonomy of protists. J Eukaryot Microbiol 52: 399–451

Apt KE and Grossman AR (1993) Characterization and transcript analysis of the major phycobiliprotein subunit genes from *Aglaothamnion neglectum* (Rhodophyta). Plant Mol Biol 21: 27–38

Apt KE, Hoffman NE and Grossman AR (1993) The gamma subunit of R-phycoerythrin and its possible mode of transport into the plastid of red algae. J Biol Chem 268: 16208–16215

Badger MR and Price GD (2003) CO_2 concentrating mechanisms in cyanobacteria: molecular components, their diversity and evolution. J Exp Bot 54: 609–622

Burey SC, Fathi-Nejad S, Poroyko V, Steiner JM, Löffelhardt W and Bohnert HJ (2005) The central body of the cyanelles of *Cyanophora paradoxa*: a eukaryotic carboxysome? Can J Bot 83: 758–764

Burey SC, Poroyko V, Ozturk ZN, Fathi-Nejad S, Schüller C, Ohnishi N, Fukuzawa H, Bohnert HJ and Löffelhardt W (2007) Acclimation to low CO_2 by an inorganic carbon concentrating mechanism in *Cyanophora paradoxa*. Plant Cell Environ 30: 1422–1435

Busch A, Nield J and Hippler M (2010) The composition and structure of photosystem I associated antenna from *Cyanidioschyzon merolae.* Plant J 62: 886–897

Dalbey RE and Robinson C (1999) Protein translocation into and across the bacterial plasma membrane and the plant thylakoid membrane. Trends Biochem Sci 24: 17–22

Deschamps P, Colleoni C, Nakamura Y, Suzuki E, Putaux J-L, Buléon A, Haebel S, Ritte G, Steup M, Falcón L, Moreira D, Löffelhardt W, Raj JN, Plancke C, d'Hulst C, Dauvillée D and Ball S (2008) Metabolic symbiosis and the birth of the plant kingdom. Mol Biol Evol 25: 536–548

Deusch O, Landan G, Roettger M, Gruenheit N, Kowallik KV, Allen JF, Martin W and Dagan T (2008) Genes of cyanobacterial origin in plant nuclear genomes point to a heterocyst-forming plastid ancestor. Mol Biol Evol 25: 748–761

Dong CX, Tang AH, Zhao JD, Mullineaux CW and Bryant DA (2009) ApcD is necessary for efficient energy transfer from phycobilisomes to photosystem I and helps to prevent photoinhibition in the cyanobacterium *Synechococcus* sp. PCC 7002. Biochim Biophys Acta 1787: 1122–1128

Douglas SE and Penny SL (1999) The plastid genome of the cryptophyte alga, *Guillardia theta*: complete sequence and conserved synteny groups confirm its common ancestry with red algae. J Mol Evol 48: 236–244

Egelhoff T and Grossman AR (1983) Cytoplasmic and chloroplast synthesis of phycobilisome polypeptides. Proc Natl Acad Sci U S A 80: 3339–3343

Enami I, Murayama H, Ohta H, Kamo M, Nakazato K and Shen JR (1995) Isolation and characterization of a Photosystem II complex from the red alga *Cyanidium caldarium*: association of cytochrome c-550 and a 12 kDa protein with the complex. Biochim Biophys Acta 1232: 208–216

Fathinejad S, Steiner JM, Reipert S, Marchetti M, Allmaier G, Burey SC, Ohnishi N, Fukuzawa H, Löffelhardt W and Bohnert HJ (2008) A carboxysomal CCM in the cyanelles of the "coelacanth" of the algal world, *Cyanophora paradoxa*? Physiol Plant 133: 27–32

Fettke J, Hejazi M, Smirnowa J, Höchel E, Stage M and Steup M (2009) Eukaryotic starch degradation: integration of plastidial and cytosolic pathways. J Exp Bot 60: 2907–2922

Flechner A, Gross W, Martin WF and Schnarrenberger C (1999) Chloroplast class I and II aldolases are bifunctional for fructose-1,6-bisphosphate and sedoheptulose-1,7-bisphosphate cleavage in the Calvin cycle. FEBS Lett 447: 200–202

Gebhart UB, Maier TL, Stefanovic S, Bayer MG and Schenk HEA (1992) Ferredoxin-NADP oxidoreductase of *Cyanophora paradoxa*: purification, partial characterization, and N-terminal amino acid sequence. Protein Expr Purif 3: 228–235

Giordano M, Beardall J and Raven JA (2005) CO_2 concentrating mechanisms in algae: mechanisms, environmental modulation, and evolution. Annu Rev Plant Biol 56: 99–131

Glöckner G, Rosenthal A and Valentin K (2000) The structure and gene repertoire of an ancient red algal plastid genome. J Mol Evol 51: 382–390

Gross W, Bayer MG, Schnarrenberger C, Gebhart UB, Maier TL and Schenk HEA (1994) Two distinct aldolases of class II type in the cyanoplasts and in the cytosol of the alga *Cyanophora paradoxa*. Plant Physiol 105: 1393–1398

Jakowitsch J, Bayer MG, Maier TL, Lüttke A, Gebhart UB, Brandtner M, Hamilton B, Neumann-Spallart C, Michalowski CB, Bohnert HJ, Schenk HEA and Löffelhardt W (1993) Sequence analysis of pre-ferredoxin-$NADP^+$-reductase cDNA from *Cyanophora paradoxa* specifying a precursor for a nucleus-encoded cyanelle polypeptide. Plant Mol Biol 21: 1023–1033

Jakowitsch J, Neumann-Spallart C, Ma Y, Steiner JM, Schenk HEA, Bohnert HJ and Löffelhardt W (1996) *In vitro* import of pre-ferredoxin-$NADP^+$-oxidoreductase from *Cyanophora paradoxa* into cyanelles and into pea chloroplasts. FEBS Lett 381: 153–155

Kashino Y, Lauber WM, Carroll JA, Wang Q, Whitmarsh J, Satoh K and Pakrasi HB (2002) Proteomic analysis of a highly active photosystem II preparation from the cyanobacterium *Synechocystis* sp. PCC 6803 reveals the presence of novel polypeptides. Biochemistry 41: 8004–8012

Kerfeld CA, Sawaya MR, Tanaka S, Nguyen CV, Phillips M, Beeby M and Yeates TO (2005) Protein structures forming the shell of primitive bacterial organelles. Science 309: 936–938

Kies L (1992) Glaucocystophyceae and other protists harboring prokaryotic endosymbionts. In: W Reisser (ed) Algae and symbioses, pp. 353–377. Biopress, Bristol

Kloos K, Schlichting R, Zimmer W and Bothe H (1993) Glutamine and glutamate transport in *Cyanophora paradoxa*. Bot Acta 106: 435–440

Kneip C, Voss C, Lockhart PJ and Maier UG (2008) The cyanobacterial endosymbiont of the unicellular alga *Rhopalodia gibba* shows reductive genome evolution. BMC Evol Biol 8: 30

Koike H, Shibata M, Yasutomi K, Kashino Y and Sato K. (2000) Identification of photosystem I components from a glaucocystophyte, *Cyanophora paradoxa*: the PsaD protein has an N-terminal stretch homologous to higher plants. Photosynth Res 65: 207–217

Kondo K, Ochiai Y, Katayama M and Ikeuchi M (2007) The membrane-associated CpcG2-phycobilisome in *Synechocystis*: a new photosystem I antenna. Plant Physiol 144: 1200–1210

Kovacevic G, Steiner JM and Löffelhardt W (2009) Cyanobacterial and algal symbioses. In: UE Schaible and A Haas (eds) Intracellular niches of microbes, pp. 527–545, Wiley-VCH, Weinheim

Kowallik KV, Stoebe B, Schaffran I and Freier U (1995) The chloroplast genome of a chlorophyll a+c containing alga, *Odontella sinensis*. Plant Mol Biol Rep 13: 336–342

Liberton M, Howard Berg R, Heuser J, Roth R and Pakrasi H (2006) Ultrastructure of the membrane systems in the unicellular cyanobacterium *Synechocystis* sp. strain PCC 6803. Protoplasma 227: 129–138

Linka M, Jamai A and Weber APM (2008) Single ancient origin of a plastidic phosphate translocator gene family from the thermo-acidophilic alga *Galdieria sulfuraria* reveals specific adaptations of primary carbon partitioning in green plants and red algae. Plant Physiol 148: 1487–1496

Liu L-N, Aartsma TJ, Thomas J-C, Lamers GEM, Zhou B-C and Zhang Y-Z (2008a) Watching the native supramolecular architecture of photosynthetic membrane in red algae. J Biol Chem 283: 34946–34953

Liu L-N, Elmalk AT, Aartsma TJ, Thomas J-C, Lamers GEM, Zhou BC and Zhang YZ (2008b) Light-induced energetic decoupling as a mechanism for phycobilisome-related energy dissipation in red algae: a single molecule study. PLoS ONE 3: e3134

Löffelhardt W and Bohnert HJ (2001) The cyanelle (muroplast) of *Cyanophora paradoxa*: a paradigm for endosymbiotic organelle evolution. In: J Seckbach (ed) Symbiosis, pp. 111–130. Kluwer Academic, Dordrecht

Löffelhardt W, Bohnert HJ and Bryant DA (1997) The complete sequence of the *Cyanophora paradoxa* cyanelle genome (*Glaucocystophyceae*). Plant Syst Evol (Suppl) 11: 149–162

Ma Y, Jakowitsch J, Deusch O, Henze K, Martin W and Löffelhardt W (2009) Transketolase from *Cyanophora paradoxa*: *in vitro* import into cyanelles and pea chloroplasts and a complex history of a gene often, but not always, transferred in the context of secondary endosymbiosis. J Eukaryot Microbiol 56: 568–576

MacColl R (1998) Cyanobacterial phycobilisomes. J Struct Biol 124: 311–334

Mangeney E and Gibbs SP (1987) Immunocytochemical localization of ribulose-1,5-bisphosphate carboxylase/oxygenase in the cyanelles of *Cyanophora paradoxa* and *Glaucocystis nostochinearum*. Eur J Cell Biol 43: 65–70

Martin W, Stoebe B, Goremykin V, Hansmann S, Hasegawa M and Kowallik K (1998) Gene transfer to the nucleus and the evolution of chloroplasts. Nature 393: 162–165

Miura K, Yamano T, Yoshioka S, Kohinata T, Inoue Y, Taniguchi F, Asamizu E, Nakamura Y, Tabata S, Yamato KT, Ohyama K and Fukuzawa H (2004) Expression profiling-based identification of CO_2-responsive genes regulated by CCM1 controlling a carbon-concentrating mechanism in *Chlamydomonas reinhardtii*. Plant Physiol 135: 1–13

Nickol AA, Müller NE, Bausenwein U, Bayer MG, Maier TL and Schenk HEA (2000) Nucleotide sequence and phylogeny of the nucleus-encoded muroplast fructose-1,6-bisphosphate aldolase. Z Naturforsch 55c: 991–1003

Nowack ECM, Melkonian M and Glöckner G (2008) Chromatophore genome sequence of *Paulinella* sheds light on acquisition of photosynthesis by eukaryotes. Curr Biol 18: 410–418

Ohta H, Okumura A, Okuyama S, Akiyama A, Iwai M, Yoshihara S, Shen JR, Kamo M and Enami I (1999) Cloning and expression of the psbU gene and functional studies of the recombinant 12 kDa protein of photosystem II from the red alga *Cyanidium caldarium*. Biochem Biophys Res Commun 260: 245–250

Ohta H, Suzuki T, Ueno M, Okumura A, Yoshihara S, Shen JR and Enami I (2003) Extrinsic proteins of photosystem II: an intermediate member of PsbQ protein family in red algal PS II. Biochem Biophys Res Commun 270: 4156–4163

Petersen J, Teich R, Becker B, Cerff R and Brinkmann H (2006) The GapA/B gene duplication marks the origin of the streptophyta (Charophytes and land plants). Mol Biol Evol 23: 1109–1118

Pfanzagl B, Allmaier G, Schmid ER, de Pedro MA and Löffelhardt W (1996) N-Acetylputrescine as a characteristic constituent of cyanelle peptidoglycan in glaucocystophyte algae. J Bacteriol 179: 6994–6997

Plancke C, Colleoni C, Deschamps P, Dauvillée D, Nakamura Y, Haebel S, Steup M, Buléon A, Putaux J-L, Dupeyre D, d'Hulst C, Ral J-P, Löffelhardt W, Maes E and Ball SG (2008) The pathway of starch synthesis in the model glaucophyte *Cyanophora paradoxa*. Eukaryot Cell 7: 247–257

Raven JA (2003) Carboxysomes and peptidoglycan walls of cyanelles: possible physiological functions. Eur J Phycol 38: 47–53

Reith M and Munholland J (1995) Complete nucleotide sequence of the *Porphyra purpurea* chloroplast genome. Plant Mol Biol Rep 13: 333–335

Reyes-Prieto A and Bhattacharya D (2007a) Phylogeny of Calvin cycle enzymes supports plantae monophyly. Mol Phylogenet Evol 45: 384–391

Reyes-Prieto A and Bhattacharya D (2007b) Phylogeny of nuclear-encoded plastid-targeted proteins supports an early divergence of glaucophytes within plantae. Mol Biol Evol 24: 2358–2361

Rissler HM and Durnford DG (2005) Isolation of a novel carotenoid-rich protein in *Cyanophora paradoxa* that is immunologically related to the light-harvesting complexes of photosynthetic eukaryotes. Plant Cell Physiol 46: 416–424

Rodríguez-Ezpeleta N, Brinkmann H,Burey SC, Roure B, Burger G, Löffelhardt W, Bohnert HJ, Philippe H and Lang BF (2005) Monophyly of primary photosynthetic eukaryotes: green plants, red algae and glaucophytes. Curr Biol 15: 1325–1330

Roose JL, Kashino Y and Pakrasi H (2007) The PsbQ protein defines cyanobacterial photosystem II complexes with highest activity and stability. Proc Natl Acad Sci U S A 104: 2548–2553

Schlichting R and Bothe H (1993) The cyanelles (organelles of a low evolutionary scale) possess a phosphate-translocator and a glucose-carrier in *Cyanophora paradoxa*. Bot Acta 106: 428–434

Schlichting R, Zimmer W and Bothe H (1990) Exchange of metabolites in *Cyanophora paradoxa* and its cyanelles. Bot Acta 103: 392–398

Serrano A and Löffelhardt W (1994) Identification of 2 different glyceraldehyde-3-phosphate dehydrogenases (phosphorylating) in the photosynthetic protist *Cyanophora paradoxa*. Arch Microbiol 162: 14–19

Shen J-R, Ming Q, Inoue Y and Burnap RL (1998) Functional characterization of *Synechocystis* sp. PCC 6803 ΔpsbU and ΔpsbV mutants reveal important roles of cytochrome c-550 in cyanobacterial oxygen evolution. Biochemistry 37: 1551–1558

Shibata M, Kashino Y, Sato K and Koike H (2001) Isolation and characterization of oxygen-evolving thylakoid membranes and photosystem II particles from a glaucocystophyte, *Cyanophora paradoxa*. Plant Cell Physiol 42: 733–741

Six C, Thomas J-C, Thion L, Lemoine Y, Zal F and Partensky F (2005) Two novel phycoerythrin-associated linker proteins in the marine cyanobacterium *Synechococcus* sp. strain WH8102. J Bacteriol 187: 1685–1694

Steiner JM and Löffelhardt W (2005) Protein translocation into and within cyanelles. Mol Membr Biol 22: 123–132

Steiner JM, Serrano A, Allmaier G, Jakowitsch J and Löffelhardt W (2000) Cytochrome c_6 from *Cyanophora paradoxa*: characterization of the protein and the cDNA of the precursor and import into isolated cyanelles. Eur J Biochem 267: 4232–4241

Steiner JM, Pompe JA and Löffelhardt W (2003) Characterization of *apcC*, the nuclear gene for the phycobilisome core linker polypeptide $L_c^{7.8}$ from the glaucocystophyte alga *Cyanophora paradoxa*. Import of the precursor into cyanelles and integration of the mature protein into intact phycobilisomes. Curr Genet 44: 132–137

Steiner JM, Yusa F, Pompe JA and Löffelhardt W (2005a) Homologous protein import machineries in chloroplasts and cyanelles. Plant J 44: 646–652

Steiner JM, Berghöfer J, Yusa F, Pompe JA, Klösgen RB and Löffelhardt W (2005b) Conservative sorting in a primitive plastid: the cyanelle of *Cyanophora paradoxa*. FEBS J 272: 987–998

Stoebe B and Maier UG (2002) One, two, three: nature's tool box for building plastids. Mol Biol Evol 25: 748–761

Stoebe B, Martin W and Kowallik KV (1998) Distribution and nomenclature of protein–coding genes in 12 sequenced chloroplast genomes. Plant Mol Biol Rep 16: 243–255

Wang HL, Postier BL and Burnap RL (2004) Alterations in global patterns of gene expression in *Synechocystis* sp. PCC 6803 in response to inorganic carbon limitation and the inactivation of *ndhR*, a LysR family regulator. J Biol Chem 279: 5739–5751

Weber APM, Linka M and Bhattacharya D (2006) Single ancient origin of a plastid metabolite translocator family in Plantae from an endomembrane-derived ancestor. Eukaryot Cell 5: 609–612

Chapter 3
An Alternate Hypothesis for the Origin of Mitochondria

Roschen Sasikumar, Jijoy Joseph and Günter A. Peschek

3.1 Introduction

The evolutionary origin of the eukaryotic cell is an intensely debated topic. Intimately tied up with this is the origin of the eukaryotic organelles, mitochondria and chloroplasts, which perform the functions of aerobic respiration and oxygenic photosynthesis respectively. The well-known endosymbiont hypothesis proposes that "symbiotic consortiums" of prokaryotic cells were the ancestors of the present-day eukaryotic cells. According to the theory, these organelles were once free-living bacteria which got engulfed in a host cell, survived endocytosis, and entered into a symbiotic relationship with the host cell. The currently accepted view (Yang et al. 1985; Andersson et al. 1998; Gray and Doolittle 1982; Gray et al. 2001) is that mitochondria and chloroplasts originated from two separate endosymbiont events—the engulfment of an aerobic respirer belonging to the α-subdivision of proteobacteria leading to formation of mitochondria, and the subsequent engulfment of a photosynthetic cyanobacterium leading to formation of chloroplasts. In this chapter we put forth a more parsimonious hypothesis, namely that both mitochondria and chloroplasts originated through the single endosymbiont event of a host cell engulfing a cyanobacterium. We propose that a cyanobacterium could have been engulfed by a proto-eukaryote in an extremely unlikely rare-chance event very early in evolution (~2.4 Ga ago). Possessing the unique capability of both aerobic respiration and oxygenic photosynthesis (Peschek 1996a, b, 2005, 2008), the endosymbiotic cyanobacterium—in the sense of a formerly free-living chloromitochondrion—could have conferred such a selective advantage to the host that this "lucky host" out-competed all other proto-eukaryotes.

The current view of an α-proteobacterial origin of mitochondria rests on the remarkable similarity of the amino acid sequences of mitochondrial proteins to corresponding α-proteobacterial proteins. We argue that phylogenetic evidence based solely on aminoacid sequence similarity must be treated with caution. More and

R. Sasikumar (✉)
CSIR, Thiruvananthapuram, Kerala, India
e-mail: roschen.csir@gmail.com

G. A. Peschek et al. (eds.), *Bioenergetic Processes of Cyanobacteria*,
DOI 10.1007/978-94-007-0388-9_3, © Springer Science+Business Media B.V. 2011

more evidence is turning up that selectional constraints work not only at the amino acid level but also at the nucleotide level itself. The so-called "silent mutations" producing synonymous codons are proving to be *not* really *silent*. The non-coding regions also face evolutionary constraints because of their involvement in the regulation of gene expression. This point to the necessity to support the amino-acid sequence-based evidence with nucleotide sequence-based evidence at the level of individual genes as well as at the level of the whole genome. We show here that whole genome signatures of mitochondria as well as the nucleotide sequences of a number of mitochondrial genes show more similarity to those of cyanobacteria than of most α-proteobacteria. We argue that, instead of mitochondria originating from α-proteobacteria, the two evolved in parallel, adapting to an environment getting progressively richer in oxygen. Therefore, the similarity of their proteins could be partly due to convergent evolution.

3.2 A Plausible Alternate Hypothesis

3.2.1 The Sequence of Events

The fossil record of cyanobacteria has been argued to extend to 3.5 Ga backwards but reliable biomarker evidence at 2.7–3.2 Ga ago is usually considered to be the earliest undisputable micro-fossil record of cyanobacteria (Barghoorn and Schopf 1965, 1966; Schopf 1970). And if, as it seems well justified, we may equal cyanobacteria=O_2 the novel data that have accumulated concerning an unexpectedly early origin of oxygenic photosynthesis (Hartmann 1998; Blankenship 1992; Blankenship and Hartmann 1998; Des Marais 2000; Xiong et al. 2000; Kasting 2006) the origin of cyanobacteria would just have to correspond to these data.

The date of emergence of eukaryotes is still debated and estimates vary widely, from as early as 3.5 billion years ago to no more than 0.9–1.3 billion years ago. The latter estimates (<1.3 Ga) of the origin of eukaryotes rest mostly on the lack of undisputed fossil evidence, which is not a strong argument by itself (de Duve 2007), since it is possible that the earliest eukaryotes did not leave fossils at all, or that they are yet to be found. On the other hand there are a large number of arguments that strongly favor an early origin of eukaryotes. There are a large number of apparently ancient eukaryotic innovations that do not have a prokaryotic counterpart (de Duve 2007). The characteristics that are unique to eukaryotes seem to show that the proto-eukaryotes were essentially anaerobic organisms that had developed long before the atmosphere became oxygenated (~2.4 Ga ago). The earliest reliable biomarker evidence of eukaryotes is reported at 2.7 Ga ago (Brocks et al. 1999) and the earliest fossils are reported to be 2.1 Ga old (Han and Runnegar 1992). A genomic timescale fixes the origin of eukaryotes around 2.6 Ga (Hedges et al. 2001). All these findings have strengthened the view that modern eukaryotic and prokaryotic cells had long followed separate evolutionary trajectories (Kurland et al. 2006). A number of studies have suggested that the divergence of the archaeal, bacterial, and eukaryotic lineages is ancient (Forterre 2001; Sicheritz-Ponten and Andersson 2001) and, what seems most important in the present context, that the divergence of the eukaryotic lineage predates the divergence of α-proteobacteria (Canback et al. 2002).

Alpha-proteobacteria have been shown to be a rather late diverging group within bacteria by evidence provided by signature sequences in a number of proteins (Fig. 3.1) (Gupta 1998, 2000, 2003). The α-proteobacterial endosymbiosis is considered to have taken place around 1.8 Ga ago (Farquhar et al. 2007), 600 million years after the atmosphere became oxygen-rich. This late acquisition of mitochondria raises questions on how the anaerobic proto-eukaryotes survived until they met their α-proteobacterial rescuers and became aero-tolerant and even aerobic (de Duve 2007). It seems more plausible that eukaryotic aero-tolerance was accomplished much earlier than 1.8 Ga ago. Cyanobacteria possessing aerobic respiratory systems were already around from the early days of atmospheric oxygenation. Endosymbiosis of a cyanobacterium by an anaerobic proto-eukaryote before 2.4 Ga ago and subsequent evolution of the respiratory system under increasing oxygen is more consistent with an ancient origin of eukaryotes.

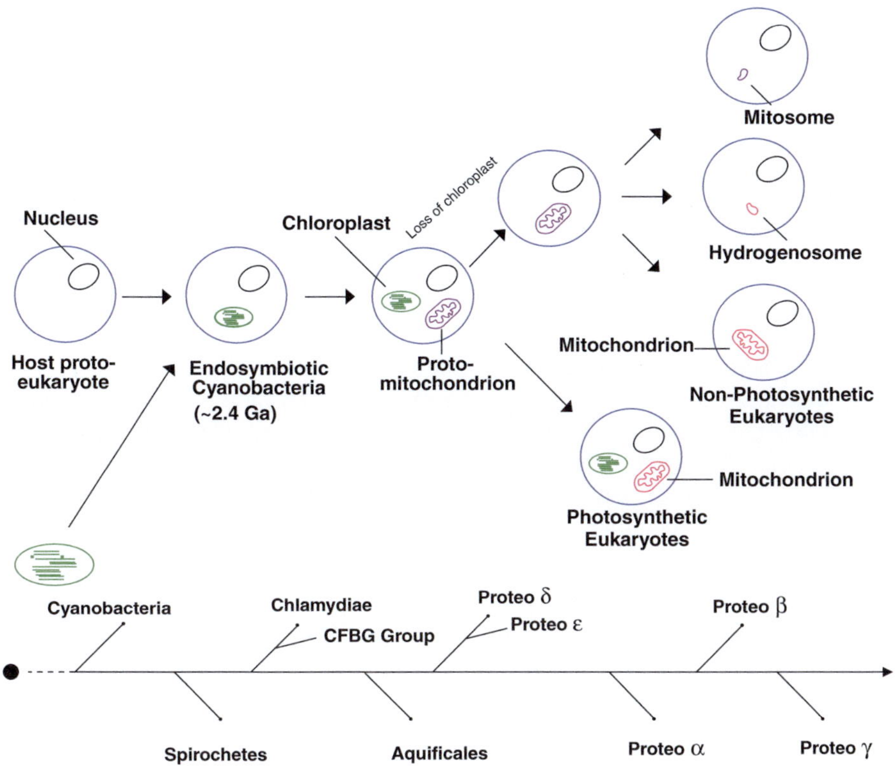

Fig. 3.1 Proposed sequence of events leading to eukaryotic organelles. (Parallel evolution of free-living bacteria is also shown schematically)

It is generally believed that mitochondria were acquired before chloroplasts because most extant eukaryotes possess some form of mitochondria while only a subset of them possess chloroplasts. However, recent evidence suggests that a number of today's non-photosynthetic eukaryotic microbes have evolved from a photosynthetic ancestor by loss of pigments and/or chloroplasts (a "laboratory example" would be *Euglena gracilis* and *Astasia longa*; also compare the (pigmented) cyanobacteria *Oscillatoria* sp., *Calothrix* sp, or Spirulina sp., now: *Arthrospira* sp., with their colorless counterparts or "derivatives" *Vitreoscilla* sp., *Leucothrix* and *Thiothrix* sp., or *Saprospira* sp., and the many others of Pringsheim's *colorless algae* (Pringsheim 1949, 1963; Pringsheim and Wiessner 1960). At any rate, in the light of all newly accumulated knowledge, the primary endosymbiont event leading to chloroplasts occurred much earlier in eukaryotic evolution than currently envisaged (Andersson and Roger 2002; Purton 2002). Phylogenetic data on the origin of chloroplast Hsp90 suggest that the common ancestor of animals and plants once harbored chloroplasts (Emelyanov 2002), and the recent discovery of vestiges of photosynthetic structures ("thylakosomes") in (chemoheterotrophic!) protists such as *Psalteriomonas lanterna* and some representatives of the parasitic genus *Apicomplexa* (Hackstein et al. 1997) would rather conform to this type of reasoning. Even some peroxisomal enzymes from animals have been shown to have originated from cyanobacteria (Gabaldon et al. 2006; Tabak et al. 2006). Therefore, it is well possible that the ancestors of all the current non-photosynthetic organisms had once harbored chloroplasts in the form of cyanobacteria.

Stanier (1970) proposed that chloroplast endosymbiosis took place first because oxygenic photosynthesis must have preceded aerobic respiration. Cavalier Smith (1987) proposed a simultaneous origin of chloroplasts and mitochondria by (nearly) simultaneous endosymbiont events around 0.9 Ga ago, that should have led to uptake of both the progenitors (of chloroplasts and mitochondria) by a proto-eukaryotic host possessing phagocytic capability. He argued that, as soon as eukaryotes acquired the phagocytic machinery, all kinds of symbionts could be taken up and it is unlikely that photosynthetic ones would be taken up appreciably earlier or later than respiratory ones. In a more recent paper (Cavalier Smith 2006) he reported that TOM70, a protein of crucial importance in the import of proteins into mitochondria, shows a clear cyanobacterial origin thereby supporting his hypothesis that the primordial host also harbored symbiotic cyanobacteria along with the proto-mitochondrial endosymbiont.

3.2.2 The Victory of Parsimony and the Selective Advantage of a Single Primordial Cyanobacterial Endosymbiosis

Cavalier Smith's hypothesis of late endosymbiosis between a proto-eukaryote with well-developed phagocytic capability and some α-proteobacterium does not go well with the monophyletic nature of mitochondria. If "all kinds of symbionts could be taken up" (Cavalier Smith 1987), it is difficult to see why one and only one combination could out-compete all the others. The monophyletic nature of mitochondria suggests that the mitochondrial endosymbiosis was a highly unusual and unlikely event. Evidence that other organelles like mitosomes and hydrogenosomes are also derived from mitochondria is another indication of the extreme parsimony involved in the primary endosymbiont event. Therefore, the event must have occurred very early in evolution at a time when the mechanism for attaining a stable endosymbiotic system was not yet well established.

In view of the physiological unlikeliness of the so-called respiration-early hypothesis (see Paumann et al. 2005) it is logical to hypothesize that the trigger for the emergence of aerobic respiration came from the molecular oxygen first released by the cyanobacterial oxygenic photosynthesis and that the aerobic respiratory chain was derived from the photosynthetic electron transfer chain, as extensively discussed by the so-called conversion hypothesis (Broda 1975; Broda and Peschek 1979; Peschek 1996a, b, 2004, 2005, 2008; Paumann et al. 2005). Initially the function of the respiratory chain must have been the detoxification of O_2 and there must have been a long intermediate phase before it developed into the full-fledged respiratory chain currently found in mitochondria and aerobic bacteria (Paumann et al. 2005). As the inventors of oxygenic photosynthesis cyanobacteria must naturally have been the first organisms to face the challenge of molecular oxygen which they generated within themselves. They are thus likely to have been the first organisms to have developed oxygen-detoxifying systems—finally most efficiently and even bioenergetically useful—via aerobic respiration. A wide variety of detoxifying enzymes for O_2 and its even more dangerous partially reduced intermediates have been identified in cyanobacteria (Regelsberger et al. 2002; Paumann et al. 2005; Bernroitner et al. 2009).

We propose that, when primitive organisms faced the challenge of toxic oxygen, one of them, a proto-eukaryote *sensu* Woese et al. (1978), made a lucky breakthrough by managing to enslave a cyanobacterium and availed the enormous selective advantage conferred by simultaneous acquisition of oxygenic photosynthesis and aerobic respiration, thus taking the famous quantum leap in evolution. A dramatic increase in atmospheric oxygen levels known as the Great Oxidation Event took place around 2.4 Ga ago though the cyanobacteria had emerged at least 300 million years before already (Kasting 2006). This time lag has not yet received complete explanation, though several hypotheses have been offered (Lenton et al. 2004; Goldblatt et al. 2006; Kasting 2006). Be all that as it might be, it is more than plausible that the efficient transfer of photosynthetic capability to eukaryotes made photosynthesis extremely wide-spread and powerful, and thus contributed significantly to the dramatic rise in oxygen. This would place the timing of cyanobacterial endosymbiosis around 2.4 Ga ago.

3.2.3 Separation of the Organelles

During replication of the cyanobacterial endosymbiont within the host, some of the progeny could have lost the thylakoid membrane possibly by mutation of a critical gene involved in biosynthesis of the membranes. This part of the off-spring was transformed into proto-mitochondria specializing in the function of aerobic respiration using the respiratory electron transfer chain that had been situated in the cyanobacterial plasma membrane from the very beginning (Peschek et al. 2004; Paumann et al. 2005). The main function of the proto-mitochondrion must have been oxygen detoxification by aerobic respiration. As the atmosphere became progressively richer in oxygen, the proto-mitochondria developed more and more efficient and refined mechanisms for aerobic respiration as nowadays found in full-fledged mitochondria. In some of the eukaryotes occupying anaerobic niches, the proto-mitochondria degenerated into hydrogenosomes and mitosomes. It might also be speculated that peroxisomes are vestiges of the proto-mitochondria still contributing to the scavenging of oxygen free radicals.

The original cyanobacterium probably survived longer in its endosymbiont form and was converted to a chloroplast at a later stage (Cavalier Smith 2006). Somewhere along the way a lineage of non-photosynthetic eukaryotes emerged by secondary loss of pigments and/or the photosynthetic function (see Fig 3.1).

3.3 Structural and Functional Characteristics of Cyanobacterial and Mitochondrial Membranes

Mitochondria, in their inner membrane, contain the highly sophisticated system of chemiosmotic oxidative phosphorylation which, inherently dependent on membrane-bound electron transport, must be the result of a long process of evolution under aerobic conditions. The cyanobacterial respiratory system, though basically similar to the mitochondrial one, is still more primitive ("simpler") showing signs of having evolved under an atmosphere poorer in oxygen (Peschek et al. 2004).

Like other Gram-negative bacteria, cyanobacteria have a cell envelope consisting of an outer membrane, a peptidoglycan layer, and a plasma membrane. In addition, these organisms possess an elaborate internal system of intracellular (thylakoid) membranes that host a dual-function photosynthetic–respiratory electron transport chain (Peschek 1996a, b). Their plasma membrane carries a pure respiratory chain without the photosynthetic reaction centers (Peschek et al. 2004). Perhaps one of the critical evolutionary innovations of the cell membrane in cyanobacteria was its ability to invaginate to create a space between the cell membrane and the cell wall (Koning 1994). The "invaginations", which form contact points between plasma and thylakoid membranes, are sometimes called "mesosomes" or thylakoid centers (Hinterstoisser et al. 1993). These areas of the cell membrane are rich in respiratory electron transport proteins. It has been debated whether the thylakoid membranes themselves are invaginations of the plasma membrane, or if they form separate compartments within the interior of the cyanobacterial cell. Recent electron microscopic studies indicate that the thylakoid membranes are physically discontinuous from the plasma membrane (Liberton et al. 2006). They may, however, be physically connected with plasma membrane through thylakoid centers (Hinterstoisser et al. 1993).

Mitochondria and chloroplasts are both surrounded by double membranes. Chloroplasts contain, in addition, an internal (thylakoid) membrane system which, in cyanobacteria, carries both respiratory and photosynthetic electron transport chains. The mitochondrial respiratory chain is contained in the inner mitochondrial membrane which would thus be analogous (and homologous?) to the cyanobacterial plasma membrane, and which exhibits numerous invaginations called cristae. Both the mitochondrial inner membrane and the chloroplast thylakoid membrane form closed, osmotically autonomous (thus chemiosmotically competent) compartments and both membranes are stuffed with a reversible ATP synthase of appropriate orientation to catalyze the phosphorylation of ADP to ATP. Thus physiologically the chloroplast structure is virtually identical to that of cyanobacteria whereas mitochondria could be considered to resemble cyanobacteria without thylakoid membranes.

The respiratory electron transport chain carried by the cyanobacterial membranes (Fig. 3.2) has seen shown to be quite similar to that carried by the mitochondrial inner membrane, and in general—as is one of the main conclusions of the conversion hypothesis (see before)—functionally speaking the electron transport components of both respiratory and photosynthetic chains are strikingly similar to each other (see Peschek 2008).

The electron transport sequence (Peschek et al. 2004; Paumann et al. 2005; Peschek 2008) will be briefly discussed in the following in order to realize the re-

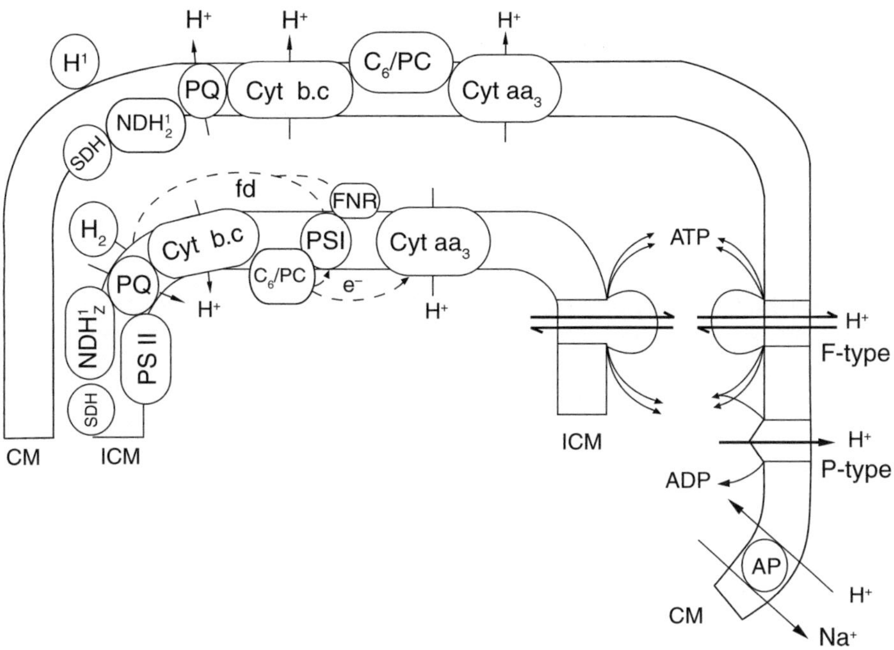

Fig. 3.2 Details of the photosynthetic and respiratory belectron transport systems in a cyanobacterium

markable similarity of electron transport systems in cyanobacteria, chloroplasts, and mitochondria:

Complex I—NAD(P)H Dehydrogenase Cyanobacteria possess either a multi-subunit "mitochondrial" energy-transducing NDH-1 enzyme or a 1-subunit non-mitochondrial, non-energy transducing NDH-2 enzyme.

Complex II—(SDH) Succinate-Dehydrogenase The occurrence of succinate dehydrogenase has been firmly established in both cyanobacterial membranes. Interestingly, just like the mitochondrial SDH, the cyanobacterial enzyme is inhibited by thenoyltrifluoro acetone (TTFA), and it cross-reacts with monospecific antibody against mitochondrial SDH.

Lipid-Soluble Mobile Carrier This pool component is ubiquinone in mitochondria but plastoquinone inevitably in all cyanobacteria. Redox potentials of ubi- and plasto-quinone (both benzoquinones) are almost the same and they can functionally substitute for each other. Side chains in positions 3 and 4 are methoxy- in ubiquinone, but methyl- in plastoquinone which may indicate that ubiquinone replaced plastoquinone as the atmosphere became richer in oxygen.

Complex III: The Cytochrome b.c Complex Functionally speaking this complex is basically the same in cyanobacteria, chloroplasts, and mitochondria. The cytochrome $\boldsymbol{b_6 f}$ complex in chloroplasts and cyanobacteria, and the $\boldsymbol{b.c_1}$ complex in mitochondria, have exactly the same function, viz. as a quinol:cytochrome *c*/PC oxidoreductase.

Water-Soluble Mobile Carrier In cyanobacteria this can be cytochrome c_6 or PC or, as an electron donor to COX, also cytochrome c_M (Bernroitner et al. 2008). In chloroplasts of higher plants this carrier is the blue copper protein PC while in mitochondria it invariably is cytochrome *c*, functionally and topologically very similar to cyanobacterial cytochrome c_6. In prokaryotic and eukaryotic algae cytochrome c_6 and PC may be physiologically interchangeable according to availability of Cu in the medium

The Terminal Respiratory Oxidase (TRO) While there may be several TRO-types in cyanobacteria, e.g. aa_3-type, bo_3-type, *bd*-type, etc., the only TRO unequivocally characterized as a functional protein in up to 35 different cyanobacteria so far is a canonical aa_3-type cytochrome *c* oxidase. The mitochondrial TRO also is an aa_3-type cytochrome *c* oxidase. O_2-affinity, redox properties and inhibition profiles of mitochondrial and cyanobacterial TRO are identical. Like other bacterial TROs its main redox-group-carrying subunit I shows the well-known property of the promiscuity of heme groups. The characteristic difference is that the enzyme is composed of 13 protein subunits in mitochondria but only 3–4 in cyanobacteria (and other bacteria), and that the TON (turnover number) of the cyanobacterial enzyme is lower by a factor of almost 100. Therefore, in spite of its high O_2-affinity it is obviously not as well adapted to aerobic respiration in fully oxic conditions as is the mitochondrial TRO. Summing up, therefore, following all the arguments discussed above, it is more than plausible that a cyanobacteria-derived proto-mitochondrion in eukaryotes could have evolved into a full fledged mitochondrion as the level of oxygen in the atmosphere gradually, yet incessantly increased.

3.4 A Reappraisal of the Molecular Phylogenetic Evidence in Favor of α-Proteobacteria

3.4.1 A Case for Looking at Nucleotide Sequences (Rather than Amino Acid Sequences)

Molecular sequencing of macromolecules, such as genes and proteins, have surpassed morphological and other organismal characters as the most popular forms of data for phylogenetic analyses. Molecular phylogenetic analysis based on amino acid sequence comparisons of mitochondrial proteins with corresponding prokaryotic proteins doubtless shows α-proteobacteria as the closest prokaryotic relatives of mitochondria.

The literature contains different opinions about whether nucleotide sequences or amino acid sequences should be used to infer ancient phylogenetic relationships. Slowly evolving sequence characters have generally been favored over faster evolving characters and this has led to the preferential use of amino acid characters over nucleotide characters. However, Simmons et al. (2004) point out that there are at least six factors that need to be considered in making this choice. Three of these factors, favor the use of nucleotide sequences and the other three favor amino acid sequences. Using amino acid character results in loss of more hierarchical information, loss of phylogenetic information, and increased chances of convergence while they have the advantages of increased character space, lower sensitivity to base composition, and lower probability of saturation. Fairly long amino acid sequences have always been preferred for the inference of ancient phylogenies, however, unless the phylogenies supported by amino acid sequence variations are supported by the underlying nucleotide sequences, there is room for legitimate doubt, especially when chances of convergent evolution are high.

A basic assumption underlying the use of the amino acid sequence variation in deducing evolutionary relationships is that selection takes place predominantly at the level of protein *function* which is, of course, decided by the amino acid sequence. This implies that nucleotide character changes that lead to synonymous codons can take place without selection constraints. However, recent studies have shown that these so-called "silent" substitutions are *not* functionally neutral (Sarfaty et al. 2007) and, therefore, they are not evolutionarily neutral. This implies that selection can constrain changes at the nucleotide sequence level itself. Another important factor to consider is that it is not only individual protein function but also the regulation of their expression that contributes to the making of an organism which is more than just the sum of several (be it even millions of) amino acid sequences. Therefore "tinkering" with gene regulation has been recognized as a particularly powerful mode of evolution (Jacob 1977), and gene regulation involves non-coding nucleotide sequences as well. Therefore it is obvious that we lose important evolutionary information by looking only at amino acid sequence changes without considering the nucleotide sequence changes.

Over the last decade considerable evidence has come up that DNA sequences evolve under the constraint of a "genome signature" which is related to the frequency of occurrence of short oligonucleotides in the sequence (Karlin and Ladunga 1994; Edwards et al. 2002; Qi et al. 2004; Dehnert et al. 2005; Wang et al. 2005; Chapus et al. 2005). Oligonucleotide frequency differences between species are seen to change too slowly to be purely the result of random mutational drift. This slow pattern of change reflects the direct or indirect action of purifying selection and the presence of functional constraints. Bush and Lahn (2006) in a recent paper show that the distances based on di-nucleotide and tetra-nucleotide frequencies from any part of the genome, and octa-nucleotide frequencies from the promoter, regions are correlated to evolutionary time. They suggest that, while genome-wide processes like DNA replication and repair constrain the divergence of the shorter DNA words, the octa-nucleotide frequencies may be getting constrained by the slow evolution of the gene regulatory machinery. The remarkable agreement between phylogenetic relationships based on genomic signature distances and some of the well-established phylogenetic relationships (Qi et al. 2004; Dehnert et al. 2005; Wang et al. 2005) is a strong indication that sequence diversification does take place under a strong constraint to conserve the *genomic signature*. These observations open to question the tacit assumption in most phylogenetic analyses that selection acts at the level of protein function only, and therefore amino acid sequence variations hold the key to evolutionary relationships.

In the following we take a look at what nucleotide sequences have to say about the relationship between chloroplasts, mitochondria, cyanobacteria, and α-proteobacteria.

3.4.2 Phylogenetic Evidence of Nucleotide Sequences of Genes Based on Multiple Sequence Alignment

To see whether the nucleotide-based results are consistent with the amino acid sequence based evidence we performed phylogenetic analyses with the same genes that are used in one of the classical works that showed the closeness of mitochondrial proteins to α-proteobacterial proteins (Andersson et al. 1998). Multiple sequence alignment was performed on the nucleotide sequences using ClustalW (Thompson et al. 1994). The alignment thus obtained was given as input to the logdet program of PHYLIP package to compute the logdet distance matrix. Logarithmic distances were computed since sequences were taken from organisms with widely varying mutation rates. The tree was drawn from the distance matrix using the Kitsch program of PHYLIP (Felsenstein 1989). The unrooted trees are visualized using the Phylodraw (Choi et al. 2000) program.

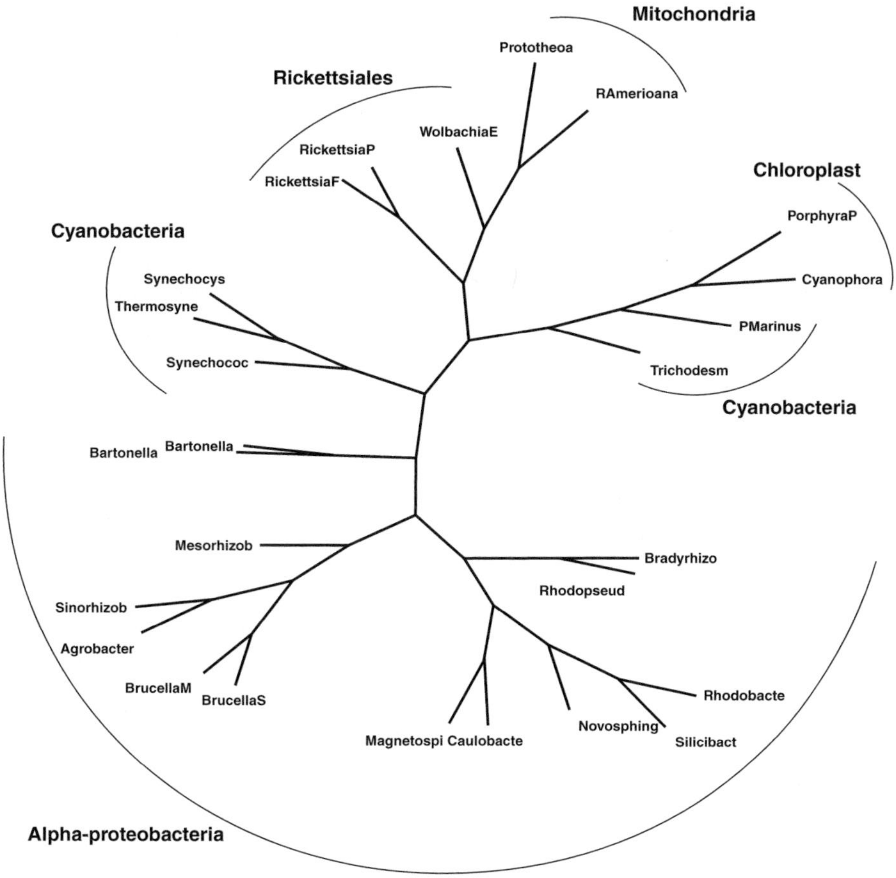

Fig. 3.3 Tree of concatenated ribosomal genes S2, S3, S7, S10, S11, S12, S13, S14, S19, L5, L6 and L16 (Unrooted)

Figures 3.3 and 3.4 show unrooted phylogenetic trees using concatenated ribosomal genes and concatenated respiratory genes, respectively, of mitochondria, chloroplasts, α-proteobacteria, and cyanobacteria. Both trees clearly show Rickettsiales clustering with mitochondria. However, the cluster containing cyanobacteria and chloroplasts is significantly closer to the mitochondria-Rickettsiales cluster than the rest of the α-proteobacterial cluster.

The assembly of iron–sulfur clusters is considered one of the most essential functions of the mitochondrial compartments and Iron-Sulphur Cluster Assembly gene IscS is found in all eukaryotic lineages including the a-mitochondrial ones containing hydrogenosomes and mitosomes.

Figure 3.5 shows a phylogenetic tree based on the nucleotide sequences of IscS genes of eukaryotes, α-proteobacteria, and cyanobacteria. We find that the eukaryotic genes cluster together and that this cluster is closest to a cluster containing both

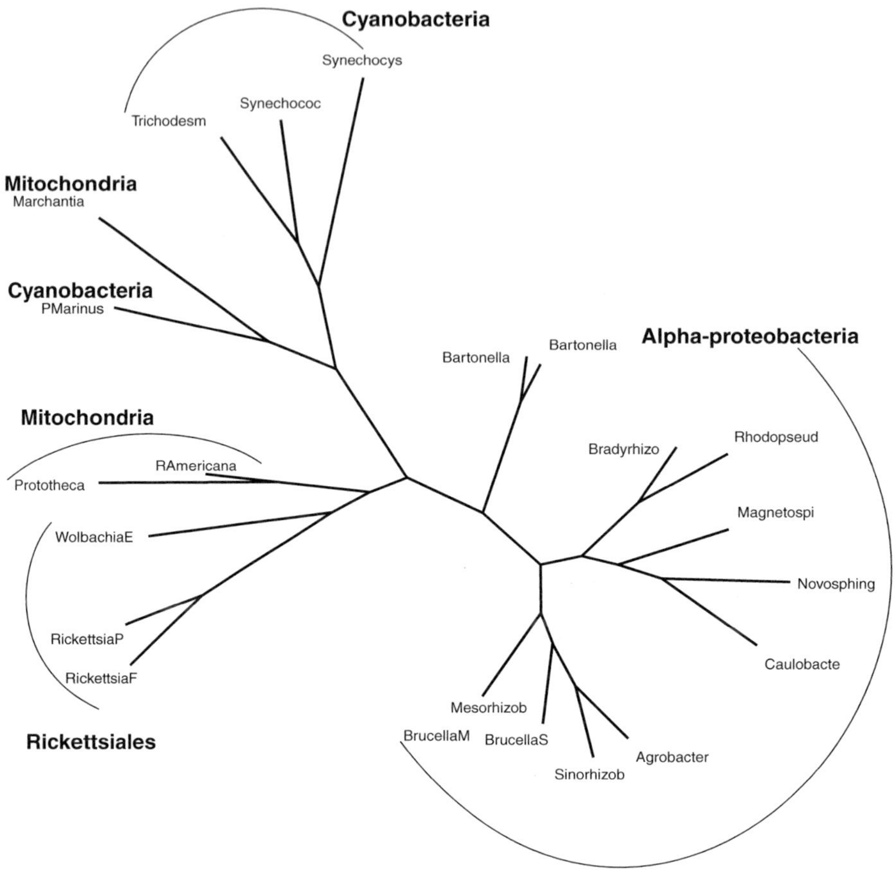

Fig. 3.4 Tree of concatenated NADH genes nuoA, J, K, L, M, N

cyanobacteria and Rickettsiales while the rest of the α-proteobacteria are considerably farther apart.

3.4.3 Phylogenetic Evidence Derived from Whole Genome Signatures

Inference of genomic distances based on differences in sequence signature is based on simple techniques free from the ambiguities involved in aligning multiple sequences. Comparisons can be made at the whole genome level or at the level of individual genes. Several measures for the genomic signature and signature-based distance measures have been proposed in the literature. These include Dinucleotide Relative Abundance Profile (DRAP) (Karlin and Burge 1995), Frequency Chaos

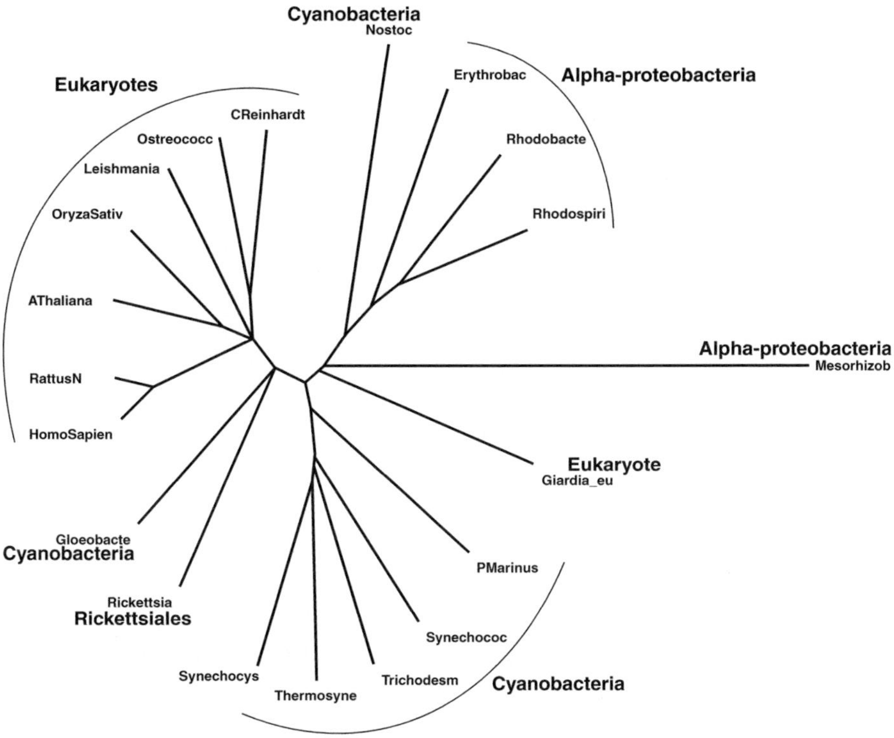

Fig. 3.5 Tree of iron-sulphur cluster assembly genes, IscS

Game Representation (FCGR) matrices (Almeida et al. 2001) and relative Frequency Chaos Game Representation (Wang et al. 2005). Chaos Game Representation (CGR) is an iterative mapping technique that processes nucleotide sequences (Jeffrey 1990). Every nucleotide of the sequence corresponds to a point in the CGR and this point contains information on the sequence up to this nucleotide.

A kth order FCGR of a sequence is a $2^k \times 2^k$ matrix in which each term represents the number of times a particular k-mer occurs in the sequence. When the CGR is resolved using a $2^k \times 2^k$ grid, the number of points inside each cell can be shown to be the frequency of occurrence of a k-mer. In this paper we use FCGR of resolution three as our genomic signature. We find hardly any difference between the results for $k=2$, 3 or 4. For long sequences like the plant nuclear genomes we use two randomly chosen segments of approximately 5 kb to produce the FCGR signature.

Various measures for the distances between FCGR based genome signatures have been proposed (Wang et al. 2005). These include Euclidean distance, Hamming distance and image distance, all of which are geometric distances. We use here a statistical distance based on weighted Pearson correlation proposed by Almeida et al. (2001). An advantage of the Pearson distance over other distances is that we can immediately compare sequences of very different sizes using this distance type.

The weighted Pearson correlation distance between two FCGRs, X and Y, can be calculated as follows. Consider X and Y expressed as arrays x_i and y_j respectively:

$$nw = \sum_{i=1}^{n} x_i y_i$$

$$\bar{x}w = \frac{\sum_{i=1}^{N} x_i^2 y_i}{nw} \qquad \bar{y}w = \frac{\sum_{i=1}^{N} y_i^2 x_i}{nw}$$

$$sx = \frac{\sum_{i=1}^{N} (x_i - \bar{x}w)^2 x_i y_i}{nw} \qquad sy = \frac{\sum_{i=1}^{N} (y_i - \bar{y}w)^2 x_i y_i}{nw}$$

$$rw_{x,y} = \frac{\sum_{i=1}^{N} \frac{x_i - \bar{x}w}{\sqrt{sx}} \frac{y_i - \bar{y}w}{\sqrt{sy}} x_i y_i}{nw}$$

The weighted Pearson correlation distance is defined as

$$\mathrm{d}(\mathrm{X}, \mathrm{Y}) = 1 - rw_{x,y}$$

In the results reported here we compute the weighted Pearson correlation distances between the sequences, and input the distance matrices to the Kitsch program of the PHYLIP software for constructing phylogenetic trees.

3.4.4 Genome signature Relationships Between Cyanobacteria, α-Proteobacteria, and the Eukaryotic Organelles

Figure 3.6 shows the genome signature-based distance tree of the whole genomes of mitochondria, cyanobacteria, and α-proteobacteria. One interesting result is that the eukaryotic mitochondria separate into clusters of protists, lower animals, plants, fungi and vertebrates. The cyanobacteria and Rickettsiales cluster close to the protist mitochondria while the other α-proteobacteria form a far more distant cluster.

In Fig. 3.7 we examine the genome signature distance between chloroplasts and mitochondria as well as the distance of the organellar genomes from the respective nuclear genomes of lower and higher photosynthetic eukaryotes. We find that, in the lower eukaryotes the organellar signatures are very distinct from the nuclear signatures while the two organellar signatures are strikingly close to each other. This is indicative of the organelles having originated from a single endosymbiont engulfed by a host with a nucleus of distinctly different signature. In higher plants the nuclear signature is seen to come closer to the organellar signature probably because of massive gene transfer between the organelles and the nucleus.

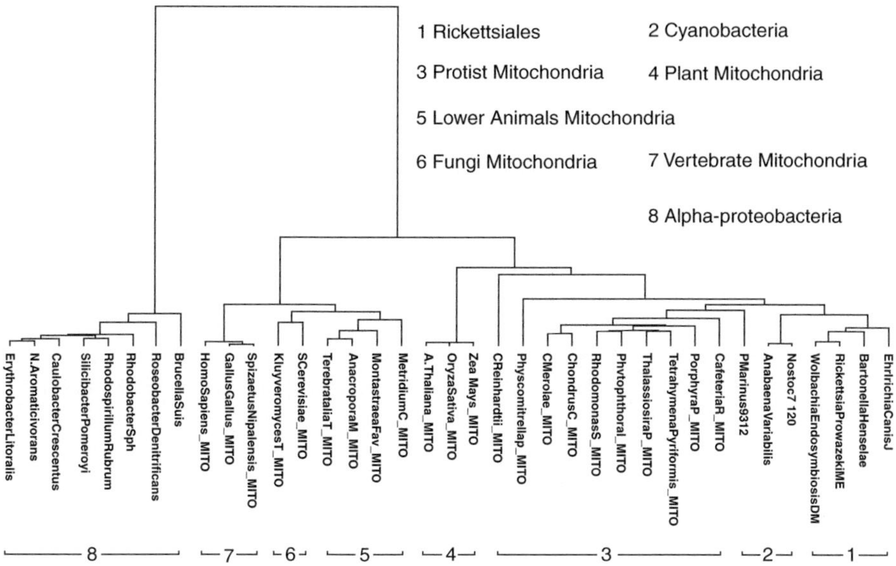

Fig. 3.6 Trinucleotide relative frequency tree with cyanobacteria, α-proteobacteria and mitochondria

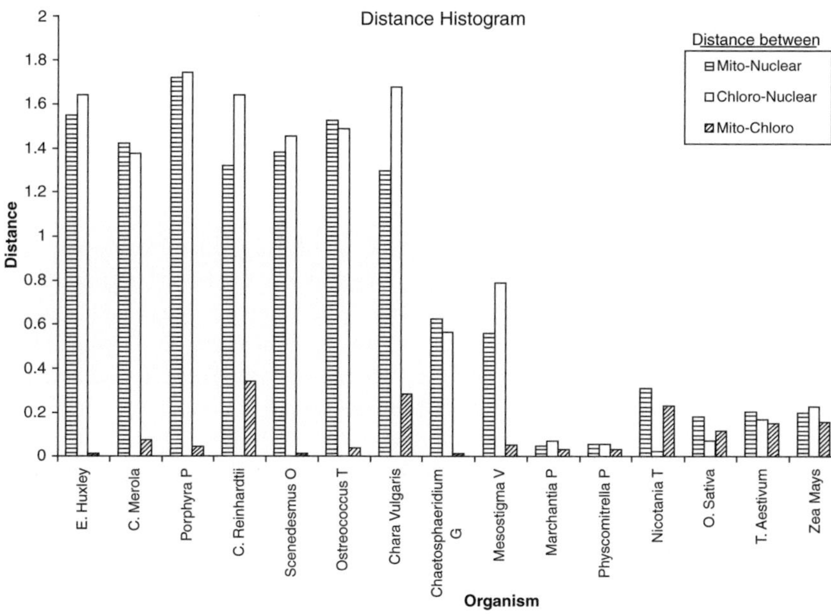

Fig. 3.7 The CGR distance histogram showing mito-nuclear and chloro-nuclear distances decreasing from lower to higher organisms

3.5 Alternate Explanation for the Similarity of α-Proteobacterial Proteins to Mitochondrial Proteins

The search for the bacterial endosymbiont of mitochondria pointed to α-proteobacteria because, among the bacterial clades, the α-proteobacteria were found to be the closest identified relatives of mitochondria on the basis of sequence similarities of several protein-encoding genes on the mitochondrial genomes. However, the inference that the most closely related bacterial clade contains the mitochondrial ancestor is true only if we are sure that this bacterial clade existed before the emergence of mitochondria. According to our hypothesis mitochondrial (and chloroplast) endosymbiosis took place around 2.4 Ga, viz., long before the emergence of α-proteobacteria. If we consider that the proto-mitochondria and the ancestors of α-proteobacteria evolved together in an oxygen rich environment, it is quite possible that they developed functionally similar proteins by a process of convergent evolution as a result of having to adapt to similar environments. The predominantly parasitic nature of α-proteobacteria adds to the similarity of the environments in which mitochondria and α-proteobacteria evolved. It has been suggested that there could also have been some "lateral gene transfer" from the parasitic α-proteobacteria to mitochondria (Brindefalk et al. 2007). Symbionts belonging to the Rickettsiales have been found in the mitochondria of animal cells (Beninati et al. 2004) and some species of Rickettsiales have been observed to transfer their genes into the nuclear genomes of their hosts (Kondo et al. 2002). Therefore, some of the "well-conserved" mitochondrial genes, including the rRNA-genes, could also have been acquired much later in evolution from parasitic α-proteobacteria.

3.6 Conclusion

Our objective in this article has been to build a case for a plausible alternate hypothesis for the origin of mitochondria. We put together a number of arguments to establish the plausibility that a single ancient endosymbiosis of a cyanobacterium in a proto-eukaryote could have resulted in the origin of all the energy-related eukaryotic organelles such as mitochondria *and* chloroplasts. As with most hypotheses regarding ancient evolutionary events it is difficult to offer a conclusive proof of this hypothesis. We have, however, extensively shown that the "proof" based on amino acid similarity of the proteins of α-proteobacteria and mitochondria becomes open to doubt when we consider the underlying nucleotide sequences which appear to clearly demonstrate that the cyanobacteria are closer to the mitochondria than most α-proteobacteria with the exception of the Rickettsiales.

Acknowledgements One of the authors Jijoy Joseph wishes to acknowledge the financial support received from the Council for Scientific and Industrial Research, India.

References

Almeida JS, Carrico JA, Maretzek A, Noble PA and Fletcher M (2001) Analysis of genomic sequences by Chaos Game Representation, Bioinformatics 17(5): 429–437
Andersson JO and Roger AJ (2002) A cyanobacterial gene in nonphotosynthetic protists-An early chloroplast acquisition in Eukaryotes?, Curr Biol 12: 115–119
Andersson SGE, Zomorodipour A, Andersson JO, Sicheritz-Ponten T, Alsmark UCM, Podowski RM, Naslund AK, Eriksson AS, Winkler HH and Kurland CG (1998) The genome sequence of Rickettsia prowazekii and the origin of mitochondria, Nature 396 (6707): 133–140
Barghoorn ES and Schopf JW (1965) Microorganisms from the late Precambrian of Central Australia, Science 150: 337–339
Barghoorn ES and Schopf JW (1966) Microorganisms three billion years old from the precambrian of South Africa, Science 152: 758–763
Beninati T, Lo N, Sacchi L, Genchi L, Noda H and Bandi C (2004) A novel alpha-proteobacterium resides in the mitochondria of ovarian cells of the tick Ixodes ricinus, Appl Environ Microbiol 70: 2596–2602
Bernroitner M, Tangl D, Lucini C, Furtmüller PG, Peschek GA and Obinger C (2008) Cyanobacterial cytochrome c_M: Probing its role as electron donor for Cu_A of cytochrome *c* oxidase, Biochim Biophys Acta 1787(3): 135–143
Bernroitner M, Zamocky M, Furtmüller PG, Peschek GA and Obinger C (2009) Occurrence, phylogeny, structure and function of catalases and peroxidases in Cyanobacteria, J Exp Bot 60(2): 423–440
Blankenship RE (1992) Origin and early evolution of photosynthesis, Photosynth Res 33: 91–111
Blankenship RE and Hartman H (1998) The origin and evolution of oxygenic Photosynthesis, Trends Biol Sci 23: 94–97
Brindefalk B, Viklund J, Larsson D, Thollesson M and Andersson SGE (2007) Origin and Evolution of the Mitochondrial Aminoacyl-tRNA Synthetases, Mol Biol Evol 24(3): 743–756
Brocks JJ, Logan GA, Buick R and Summons RE (1999) Archean molecular fossils and the early rise of eukaryotes, Science 285: 1033–1036
Broda E (1975) The evolution of the bioenergetic processes, Pergamon Press, Oxford
Broda E and Peschek GA (1979) Did respiration or photosynthesis come first?, J Theor Biol 81: 201–212
Bush EC and Lahn BT (2006) The evolution of word composition in metazoan promoter sequence, PLoS Comput Biol 2(11): e150
Canback B, Andersson SGE and Kurland CG (2002) The global phylogeny of glycolytic enzymes, Proc Natl Acad Sci U S A 99(9): 6097–6102
Cavalier-Smith T (1987) The simultaneous symbiotic origin of mitochondria, chloroplasts, and microbodies, Ann N Y Acad Sci 503: 55–71
Cavalier-Smith T (2006) Origin of mitochondria by intracellular enslavement of a photosynthetic purple bacterium, Proc R Soc B 273: 1943–1952
Chapus C, Dufraigne C, Edwards S, Giron A, Fertil B and Deschavanne P (2005) Exploration of phylogenetic data using a global sequence analysis method, BMC Evol Biol 5: 63
Choi JH, Jung HY, Kim HS and Cho HG (2000) PhyloDraw: a phylogenetic tree drawing system, Bioinformatics 16(11): 1056–1058
Christian de Duve (2007) The origin of eukaryotes: a reappraisal, Nat Rev Genet 8: 395–403
Dehnert M, Plaumann R, Helm WE and Hutt MT (2005) Genome phylogeny based on short-range correlations in dna sequences, J Comput Biol 12(5): 545–553
Des Marais DJ (2000) Science 289: 1703
Edwards SV, Fertil B, Giron A and Deschavanne PJ (2002) A genomic schism in birds revealed by phylogenetic analysis of dna strings, Syst Biol 51: 599–613
Emelyanov VV (2002) Phylogenetic relationships of organellar Hsp90 homologs reveal fundamental differences to organellar Hsp70 and Hsp60 evolution, Gene 299(1, 2): 125–133

Farquhar J, Peters M, Johnston DT, Strauss H, Masterson A, Wiechert U and Kaufman AJ (2007) Isotopic evidence for Mesoarchaean anoxia and changing atmospheric sulphur chemistry, Nature 449: 706–709
Felsenstein J (1989) PHYLIP—Phylogeny Inference Package (Version 3.2), Cladistics 5: 164–166
Forterre P (2001) Genomics and early cellular evolution. The origin of the DNA world, C R Acad Sci Ser III 324: 1067–1076
Gabaldón T, Snel B, Frank van Zimmeren, Hemrika W, Tabak H and Huynen MA (2006) Origin and evolution of the peroxisomal proteome, Biol Direct 1: 8
Goldblatt C, Lenton TM and Watson AJ (2006) The Great Oxidation at ~2.4 Ga as a bistability in atmospheric oxygen due to UV shielding by ozone, Geophys Res Abstr 8: 00770
Gray MW and Doolittle WF (1982) Has the endosymbiont hypothesis been proven? Microbiol Rev 46: 1–42
Gray MW, Burger G and Lang BF (2001) The origin and early evolution of mitochondria, Genome Biol Rev 2(6): 1018.1–1018.5
Gupta RS (1998) Protein phylogenies and signature sequences: a reappraisal of evolutionary relationships among Archaebacteria, Eubacteria, and Eukaryotes, Microbiol Mol Biol Rev 62: 1435–1491
Gupta RS (2000) The phylogeny of Proteobacteria: relationships to other Eubacterial phyla and eukaryotes, FEMS Microbiol Rev 24: 367–402
Gupta RS (2003) Evolutionary Relationships among Photosynthetic Bacteria, Photosynth Res 76: 173–183
Hackstein JHP, Schubert H, Rosenberg J, Mackenstedt M, Berg Mvd, Brul S, Derksen J and Matthijs HCP (1997) Plastid-like organelles in anaerobic mastigotes and parasitic Apicomplexans, In: Schenk HEA, Herrmann RG, Jeon KW, Müller NE and Schwemmler W (eds) Eukaryotism and Symbiosis. Intertaxonic Combination versus Symbiotic Adaptation, pp 49–55. Springer Verlag, Berlin
Han TM and Runnegar B (1992) Megascopic eukaryotic algae from the 2.1 billion-year-old negaunee iron-formation, Michigan, Science 257: 232–235
Hartmann H (1998) Photosynthesis and the origin of life, Orig Life Evol Biosph (OLEB) 28: 515–521
Hedges SB, Chen H, Kumar S, Wang DYC, Thompson AS and Watanabe H (2001) A genomic timescale for the origin of eukaryotes, BMC Evol Biol 1: 4
Hinterstoisser B, Cichna M, Kuntner O and Peschek GA (1993) Cooperation of plasma and thylakoid membranes for the biosynthesis of chlorophyll in cyanobacteria: The role of the thylakoid centers, J Plant Physiol 142: 407–413
Jacob F (1977) Evolution and tinkering, Science 196(4295): 1161–1166
Jeffrey HJ (1990) Chaos game representation of gene structure, Nucleic Acids Res 18: 2163–2170
Karlin S and Burge C (1995) Dinucleotide relative abundance extremes: a genomic signature, Trends Genet 11: 283–290
Karlin S and Ladunga I (1994) Comparisons of eukaryotic genomic sequences, Proc Natl Acad Sci U S A 91: 12832–12836
Kasting JF (2006) Ups and downs of ancient oxygen, Nature 443: 643–645
Kondo N, Nikoh N, Ijichi N, Shimada M and Fukatsu T (2002) Genome fragment of Wolbachia endosymbiont transferred to X chromosome of host insect, Proc Natl Acad Sci USA 99: 14280–14285
Koning and Ross E (1994) Cyanophyta. Plant Physiology Information http://plantphys.info/plant_biology/cyanophyta.shtml (June 22, 2006)
Kurland CG, Collins LJ and Penny D (2006) Genomics and the irreducible nature of Eukaryote cells, Science 312: 1011–1014
Lenton TM, Schellnhuber HJ and Szathmáry E (2004) Climbing the co-evolutionary ladder, Nature 431: 913
Liberton M, Berg HR, Heuser J, Roth R and Pakrasi HB (2006) Ultrastructure of the membrane systems in the unicellular cyanobacterium Synechocystis sp. strain PCC 6803, Protoplasma 227: 129–138

Paumann M, Regelsberger G, Obinger C and Peschek GA (2005) The bioenergetic role of dioxygen and the terminal oxidase(s) in cyanobacteria, Biochim Biophys Acta (BBA)—Bioenergetics 1707(2–3): 231–253
Peschek GA (1996a) Cytochrome *c* oxidase and the *cta* operon of cyanobacteria, Biochim Biophys Acta 1275: 27–32
Peschek GA (1996b) Structure-function relationships in the dual-function photosynthetic-respiratory electron transport assembly of cyanobacteria, Biochem Soc Trans 24: 729–733
Peschek GA (2005) Cyanobacteria viewed as free-living chloromitochondrtia,In: Est Avd And Bruce D (eds) Photosynthesis: Fundamental Aspects to Global Perspectives, pp 746—749, The International Society of Photosynthesis, Toronto
Peschek GA (2008) Electron transport chains in oxygenic cyanobacteria, In: Renger G (ed) Primary Processes of Photosynthesis: Principles and Applications, 2: 383–415, European Society of Photobiology, The Royal Society of Chemistry, Great Britain
Peschek GA, Obinger C and Paumann M (2004) The respiratory chain of blue-green algae (cyanobacteria), Physiologia Plantarum 120(3): 358–369
Pringsheim EG (1949) Colorless algae, Bact Rev 13: 47–56
Pringsheim EG (1963) Farblose Algen, pp 471 Gustav Fischer Verlag, Stuttgart
Pringsheim EG and Wiessner W (1960) Colorless phototrophs? Nature 188: 919–920
Purton S (2002) Going green: the evolution of photosynthetic eukaryotes, Microbiol Today 29: 126–128
Qi J, Wang B and Hao BI (2004) Whole proteome prokaryote phylogeny without sequence alignment: A k-string composition approach, J Mol Evol 58: 1–11
Regelsberger G, Jakopitsch C, Plasser L, Schwaiger HJ, Furtmüller PG,Peschek GA, Zamocky M and Obinger C (2002) Occurrence and biochemistry of hydroperoxidases in oxygenic phototrophic prokaryotes (cyanobacteria), Plant Physiol Biochem 40: 479–490
Sarfaty CK, Mi Oh J, Kim IW, Sauna ZE, Calcagno AM, Ambudkar SV, Gottesman MM (2007) A "silent" polymorphism in the MDR1 gene changes substrate specificity, Science 315: 525–528
Schopf JW (1970) Precambrian microorganisms and evolutionary events prior to the origin of vascular plants, Biol Rev 45: 319–352
Sicheritz-Ponten T and Andersson SG (2001) A phylogenomic approach to microbial evolution, Nucleic Acids Res 29: 545–552
Simmons MP, Carr TG and Neill KO (2004) Relative character-state space, amount of potential phylogenetic information and heterogeneity of nucleotide and amino acid characters, Mol Phylogenet Evol 32: 913–926
Stanier RY (1970) Some aspects of the biology of cells and their possible evolutionary significance, Symp Soc Gen Microbiol 20: 1–38
Tabak HF, Hoepfner D, Zand Avd, Geuze HJ, Braakman I and Huynen MA (2006) Formation of peroxisomes: Present and past, Biochim Biophys Acta (BBA)—Molecular Cell Research 1763(12): 1647–1654
Thompson JD, Higgins DG and Gibson TJ (1994) CLUSTAL W: improving the sensitivity of progressive multiple sequence alignment through sequence weighting, position specific gap penalties and weight matrix choice, Nucleic Acids Res 22: 4673–4680
Wang Y, Hill K, Singh S and Kari L (2005) The spectrum of genomic signatures: from dinucleotides to chaos game representation, Gene 346: 173–185
Woese CR, Magrum LJ and Fox GE (1978) Archaebacteria, J Mol Evol 11: 245–252
Xiong J, Fischer WM, Inoue K, Nakahara M and Bauer CE (2000) Molecular evidence for the early evolution of photosynthesis, Science 289: 1724–1730
Yang D, Oyaizu Y, Oyaizu H, Olsen GJ and Woese CR (1985) Mitochondrial origins, Proc Natl Acad Sci U S A 82(13): 4443–4447

Chapter 4
The Complex Regulation of the Phosphate Uptake System of Cyanobacteria

Gernot Falkner and Renate Falkner

4.1 Introduction

Cyanobacteria usually grow under environmental conditions, under which the supply of the essential nutrient phosphate varies to a great extent. Periods of excessive phosphate supply are interrupted by periods, in which no phosphate is available. Nevertheless cyanobacteria—as well as many eukaryotic algae—are capable of performing in such a scenario a continuous growth process in that they utilize a polyphosphate store, established during occasional rises of the external phosphate concentration (Fig. 4.1). This remarkable ability is a precondition for survival in oligotrophic lakes, where the concentration of external phosphate can become so low that the uptake of this nutrient is impaired for energetic reasons (Hudson et al. 2000; Falkner et al. 1989). In such a situation phosphate can only be re-incorporated and stored in polyphosphate granules, when the external concentration exceeds a certain *threshold value*, above which the available energy suffices to drive the transport through the cell membrane (for example after excretion of faeces by aquatic animals). Simultaneously and because of the uptake activity of the whole community the external concentration decreases after a transient phosphate supply more or less rapidly to the threshold value again. Due to these energetic and kinetic constraints, fluctuations of the external concentration are experienced by the cells as a pattern of *phosphate pulses*, in which periods without phosphate uptake are interrupted by transient rises in the external concentration. The growth rate is then somehow conformed to the amount of polyphosphates; in continuous cultures the amount of polyphosphates increases with an increase in the growth rate (Droop 1973).

Since under growth conditions described above phosphate incorporation is a discontinuous event that occurs occasionally during periods of continuous growth, phosphate uptake and growth are not directly coupled. The need to conform the growth rate to the amount of stored polyphosphate gives rise to a regulatory prob-

G. Falkner (✉)
Neurodynamics and Signaling Group, Plant Physiology Division, Cell Biology Department, University of Salzburg, Hellbrunnerstr. 34, 5020 Salzburg, Austria
e-mail: gernot.falkner@sbg.ac.at

G. A. Peschek et al. (eds.), *Bioenergetic Processes of Cyanobacteria,*
DOI 10.1007/978-94-007-0388-9_4, © Springer Science+Business Media B.V. 2011

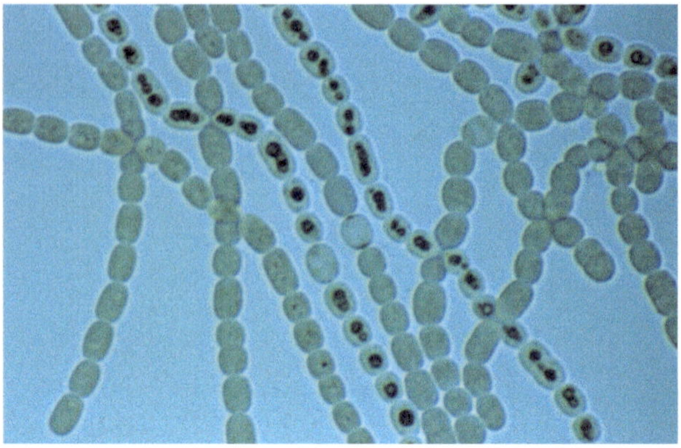

Fig. 4.1 *Anabaena variabilis* with and without polyphosphates. The polyphosphate granules have been stained with lead sulphide. (Ebel et al. 1958)

lem: on the one hand cyanobacteria must incorporate, during short-term rises of the external concentration, sufficient phosphate for maintaining a growth process in periods without phosphate supply. On the other hand excessive storage of phosphate must be avoided, since otherwise the polyphosphate granules become so large that cellular structures are disrupted. A direct effect of the size of the granules on the activity of the uptake system is not conceivable, because the granules are segregated in the cytoplasm as osmotically inert structures. To solve this regulatory problem, the cells have to employ a mechanism that can guide the activity of the phosphate uptake by some sort of "*memory*" or "learning" based on previous phosphate accumulations. This memory is revealed in an interdependence of adaptive events, in which information about former adaptations to phosphate pulses regulates subsequent adaptations, such that the amount of accumulated phosphate meets the demand of the anticipated growth rate, which—in turn—depends on the amount of stored phosphate (Falkner et al. 1995, 1996; Falkner and Falkner 2003; Wagner et al. 1995, 2000).

An investigation of the adaptive features of the uptake system confronts an experimentalist with a primary problem: namely, the cells adapt to the experimental conditions already during the course of an investigation, making futile any analysis in objectivistic terms. In this case a mechanistic description must be replaced by a more refined analysis, which accounts for the fact that the experimentalist has become part of the investigated system and of the organismic response to the experimentally imposed conditions (Falkner and Falkner 2000).

In the following we present the results of such an analysis, using a relationship between the flow of external phosphate into the polyphosphate pool and the driving force of this process. This relationship offers a possibility of studying on a phenomenological basis the energetic constraints that determine adaptive responses of phosphate deficient cells to an abrupt rise of the external phosphate concentration. In this case the uptake system, previously conformed to operation close to equilibrium at low external concentrations, is suddenly forced to operate in a region far from equilibrium. This provokes an intracellular self-organization process, in which the kinetic and energetic properties of the phosphate uptake system are altered in a complex manner that reflects intracellular information processing about alterations of phosphate supply. Thereby, cyanobacteria are able of 'interpreting' changes of the external phosphate concentrations in the light of environmental phosphate fluctuations, experienced by the cells in the past, and to reconstruct the energy dependent phosphate transport system in the plasmalemma and the ATP-synthase in the thylakoids according to this interpretation. The outcome of this reconstruction has a distinct and lasting influence on cellular energy conversion, contributing to transfer of information from one adaptive event to the next over a prolonged period of time. This allows cyanobacteria to discriminate between different patterns of phosphate fluctuations, enables information about phosphate pulses to be transferred to daughter generations and explains a persisting effect of transient increases in the external phosphate concentration on CO_2 fixation and on the flow of carbon in the glycogen pool.

4.2 The Adaptive Properties of the Phosphate Uptake System

4.2.1 Energetic Properties of the Constituents of the Phosphate Uptake System

Incorporation of external phosphate into the polyphosphate pool proceeds in three steps: (1) Transport of external phosphate P_e into the cell. (2) Conversion of the incorporated internal phosphate P_i to ATP *via* photophosphorylation and respiration. (3) Formation of polyphosphates from ATP.

A carrier protein catalyzes the transport of phosphate into the cell (reaction 1 in Fig. 4.2). At low external concentrations an energy source is needed for the translocation against the existing electrochemical gradient at the cell membrane. In the cyanobacterium *Anacystis nidulans* the necessary amount of energy is provided by an ATPase (reaction 2 in Fig. 4.2), which can be coupled to the phosphate carrier to a variable degree (Wagner and Falkner 1992). The degree of coupling is increased, when the concentration in the growth medium is diminished. This affects the threshold value, which becomes lower with an increase in the degree of coupling (Falkner et al. 1993, 1994). A decrease of the threshold value is paralleled by a dramatic activation of the transport system (Falkner et al. 1989, 1993, 1994).

Also the next step, the conversion of internal phosphate into ATP (reaction 3 in Fig. 4.2), is energy dependent and potentially coupled to a variable degree to the driving process, i.e. the flow of protons from the thylakoid space into the cytoplasmic space. Under phosphate deficient growth conditions about four protons are required for the formation of one ATP, during growth on higher external concentrations this stoichiometry decreases (to three or less, Wagner and Falkner 1992). The variable H^+/ATP stoichiometry affects the stationary phosphate concentration in the cell in that the cytoplasmic concentration becomes higher with a decrease in the stoichiometry.

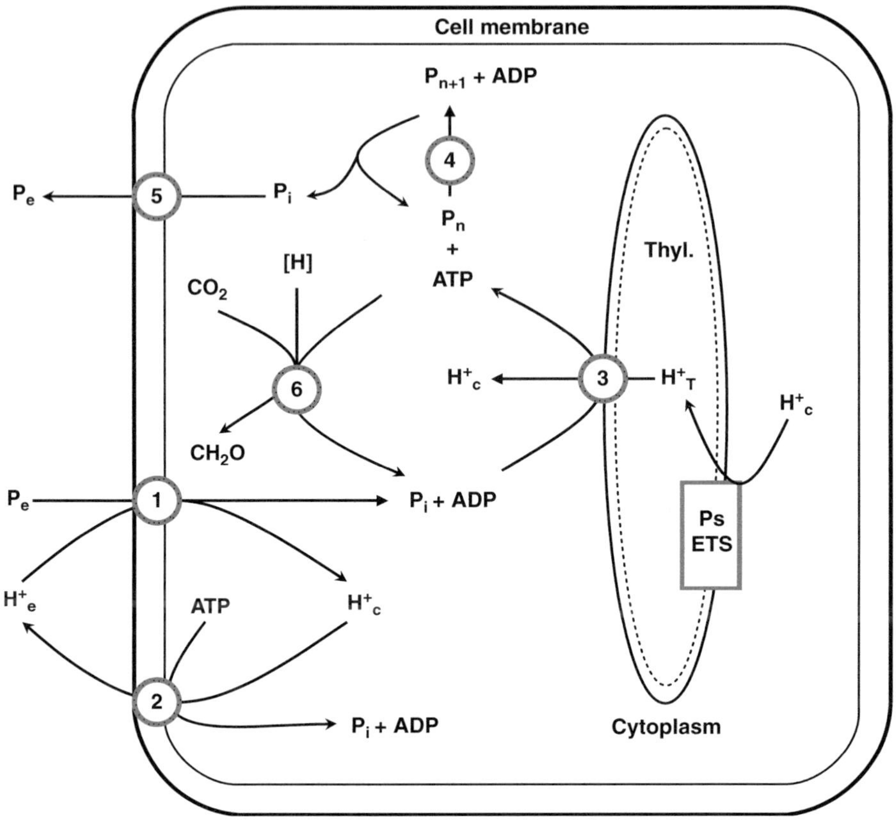

Fig. 4.2 Schematic presentation of phosphate utilization by cyanobacteria. The correct stoichiometries are not indicated in the figure. P_e: external phosphate; P_i: internal phosphate; P_n, P_{n+1}: polyphosphates; [CH_2O]: products of CO_2-fixation. H^+_C and H^+_T are the proton concentration in the cytoplasmic and thylakoid space. Ps ETS: photosynthetic electron transport system. For further explanations, see text

From this and the corresponding pH-gradient under the prevailing growth conditions, a stationary phosphate concentration in the cytoplasm can be estimated: it is around 10 μM for phosphate deficient cyanobacteria and increases into the millimolar range for non-phosphate deficient bacteria (Wagner and Falkner 1992). In adapted states the Michaelis constant of the ATP-synthase for phosphate is conformed to the steady state concentration, resulting from the corresponding H^+/ATP-stoichiometry (in the phosphate deficient state the K_M is around 30 μM, whereas in a non-phosphate deficient state K_M-values are higher than 200 μM; Wagner and Falkner 1992). These energetic and kinetic adjustments lead to an extended range of validity over which there is a proportional relation between the rate of ATP formation and the driving force for this process, a precondition for efficient operation of the ATP-synthase during fluctuations of the cytoplasmic phosphate concentration within one order of magnitude above the stationary level (Van der Meer et al. 1980). Since this enzyme also provides the substrate for ATPases in the cell membrane, the proton flux across the thylakoid membrane can be coupled with the transport process in a great many ways. In adapted states the degree of coupling between the proton flux and the phosphate flow into the polyphosphate pool is conformed to the external concentration such that energy dissipation is minimal (Falkner et al. 1994).

Only the last step, the transphosphorylation of the terminal phosphate group from ATP to polyphosphates (reaction 4 in Fig. 4.2) does not require an energy source.

Reaction 5 refers to a dissipative leakage of phosphate at the cell membrane and reaction 6 to photosynthesis.

The adaptive properties of the transport system and the ATP-synthase, described above, determine how the threshold value and the kinetic properties of the uptake system are conformed to transient increases in the external concentration. We will first consider the dependence of the threshold value on the energetic properties of the two subsystems and then present the time course of the uptake process, revealed during increases in the external concentration above the threshold value.

4.2.2 The Threshold Value

The overall process of polyphosphate (P_n, P_{n+1}) formation from external *via* internal phosphate (P_e and P_i, resp.), driven by the flow of n_P protons from the thylakoid into the cytoplasmic space, can be described by the following reactions:

1. Translocation through the cell membrane: $P_e \leftrightarrow P_i$
2. ATP formation: $P_i + ADP + n_P H^+{}_T \leftrightarrow ATP + n_P H^+{}_C$
3. Transphosphorylation to polyphosphates: $ATP + P_n \leftrightarrow P_{n+1} + ADP$

Sum reaction: $P_e + P_n + n_P H^+{}_T \leftrightarrow P_{n+1} + n_P H^+{}_C$

In this scheme reaction 2 also provides the ATP molecules for the ATPase that drives the transport of phosphate through the cell membrane. For this reason the sum reaction not only contains the number of protons, necessary for the conversion of internal phosphate to polyphosphates, but also the number of protons required for the formation of ATP that is consumed by the ATPase, coupled with the transport process.

Using the equation for the driving force of chemical reactions, $X=-\Delta G=RT \ln K'[S]/[P]$, where [S] and [P] refer to the corresponding substrate and product concentrations, the sum reaction allows calculating the driving force for polyphosphate formation from external phosphate:

$$
\begin{aligned}
X &= 2.3\ RT\ \log\left(K'[P_e]([H^+]_T/[H^+]_C)\right)^{np} \\
&= 2.3\ RT\ (\log[P_e] + \log K' + n_P \Delta pH_T).
\end{aligned}
\tag{4.1}
$$

P_n and P_{n+1} have the same activity and can be cancelled out. ΔpH_T is the pH-gradient across the thylakoid membrane, depending on the photosynthetic or respiratory electron transport chain. K′ is the equilibrium constant of the conversion of external phosphate to polyphosphate under the prevailing external conditions. Since the driving force X is zero at the threshold value, Eq. (4.1) allows estimating with $K' = 10^{-5}\ M^{-1}$ (Yoshida 1955) the logarithmic threshold value $\log[P_e]_A$: From $0 = 2.3\ RT\ (\log[P_e]_A + \log K' + n_P\ \Delta pH_T)$, one obtains for the logarithmic threshold value: $\log[P_e]_A = -\log K' - n_P \Delta pH_T$. Thus, for a pH-gradient of about 1.75 under phosphate deficient conditions and a H^+/ATP stoichiometry of 4 (Wagner and Falkner 1992), the threshold value decreases to nanomolar levels, presupposing that eight protons are required for the conversion of external phosphate to polyphosphates (four protons for the formation of ATP from cytoplasmic phosphate and four more protons for one ATP, involved in the transport of phosphate into the cell). However, one must be aware that this calculation has been performed for the unrealistic situation of complete coupling between the ATPase reaction and the transport process on the one hand and the proton flux and the ATP-formation on the other hand. Concomitantly occurring ATP-consuming and other dissipative processes, lead to a decrease in the apparent stoichiometric factor n_P and, in consequence, to a higher threshold value (Falkner et al. 1994). Also a lower H^+/ATP stoichiometry, established by cyanobacteria that grow on elevated phosphate concentrations at a higher rate, results in higher threshold values, although the pH-gradient in a non-phosphate deficient state is augmented (to about three, Falkner et al. 1976). Furthermore, the degree of coupling between the (driven) phosphate flow through the cell membrane and the activity of the ATPase that drives this process can rapidly be altered during an adaptive response to changes in the external concentration. By this mechanism cyanobacteria are able to regulate the phosphate flow into the cell by increasing or decreasing the threshold level.

In principle, the threshold value can be determined in net uptake experiments by measuring the concentration of phosphate that remains in the incubation medium after the uptake process has come to an end. Since the threshold value of phosphate deficient cyanobacteria is usually in the nanomolar range and thus below the limit of detection by conventional analytical methods, this measurement should be performed with radioactive phosphate. A possible isotopic exchange with non radioactive intracellular phosphate, leading to an alteration of the specific activity of the labelled phosphate, can be avoided using cyanobacteria that have been grown on radioactive phosphate of the same specific activity (non radioactive algae are only appropriate for this kind of investigation, if the amount of radioactive phosphate employed in the uptake experiment is much greater than the exchangeable intracellular phosphate pool).

A determination of the threshold value by uptake experiments using phosphate deficient *Anacystis nidulans* confronted with the peculiarities of adaptive phenomena in that the threshold value was influenced by the measurement conditions, in particular by the exposure time to elevated phosphate concentrations during the experiment. In consequence, different values were obtained, when the uptake was followed using different dilutions of the same algal suspension, even if the same amount of phosphate had been accumulated per average cell (Wagner et al. 1995; Falkner et al. 1995); this indicates that in this physiological state the measured threshold value is potentially influenced by concomitantly occurring adaptive responses of the uptake system to the experimentally employed phosphate concentrations, affecting the degree of coupling between the phosphate transport into the cell and the ATPase activity in the cell membrane. The lowest threshold values, determined with phosphate deficient *Anacystis n.* under conditions, under which the uptake process was so fast that the cells did not have sufficient time to adapt to the experimental phosphate concentrations, averaged around 1 nM, in accordance with the estimation given above.

4.2.3 The Adaptive Response of the Phosphate Uptake System to Alterations of the External Phosphate Concentration

Adaptive processes also affect the time course of ^{32}P-phosphate influx into unlabelled and phosphate deficient cyanobacteria that were grown under conditions, under which the external phosphate concentration remains close to the threshold value for a prolonged period of time (as in a continuous or discontinuous cultivation mode). Since isotopic exchange during ^{32}P-phosphate influx is insignificant in *Anacystis nidulans* (Falkner et al. 1974), the tracer influx practically represents net uptake. For these measurements the cultures have to be sufficiently diluted so that the uptake process does not significantly alter the external concentration. Under this condition the time course is linear, when the external concentrations are of the same order of magnitude as the threshold value (a typical example is given in Fig. 4.3, curve C). In contrast, at higher external concentrations the time course

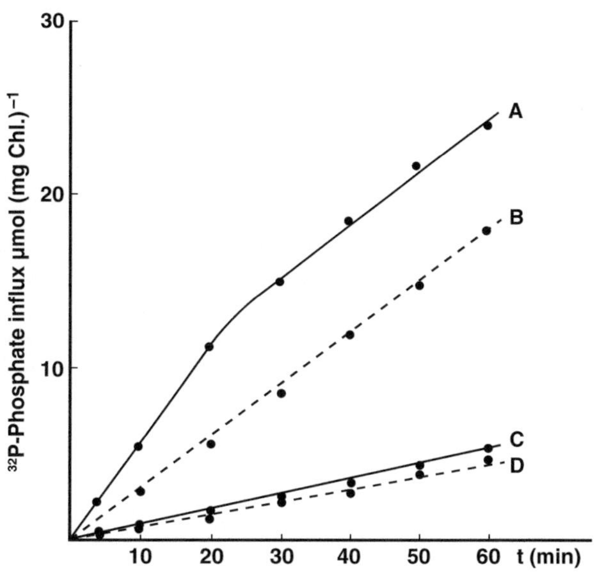

Fig. 4.3 Time course of ^{32}P-phosphate influx in *Anacystis nidulans* at two external concentrations before (*A* and *C*) and after (*B* and *D*) an incubation with 10 µM phosphate for 15 min. *A* and *B*: 20 nM, *C* and *D*: 3 nM. (For experimental details see Falkner et al. (1995))

shows a biphasic pattern. In this case a higher initial uptake rate is followed by a transition to a substantially lower rate that usually remains constant for at least one hour (curve A), indicating that exposure to higher phosphate concentrations results in a conversion from a more-active to a less-active state. When the population is exposed prior to the uptake experiment to 10 μM phosphate for 15 min and then washed phosphate free (by centrifugation and re-suspension in phosphate free medium), a biphasic shape does not appear, apparently because the adaptive self-organization has already taken place during the pre-incubation period (curve B). As will be shown below, microbial information processing about alterations of the external phosphate concentration is essentially based on these two manifestations of the self-organization process, namely its adaptive potentiality, which depends on the growth history, and on adapted states, finally attained after a response to a rise in the external concentration. An energetic analysis of the two manifestations in respect to an adaptive self-organization is given in the subsequent section.

4.2.4 Energetic Characterization of the Adaptive Self-Organization of the Uptake System

According to general principles of non equilibrium thermodynamics self-organization occurs in regions far from thermodynamic equilibrium, where the flows into the system depend on the driving forces in a non linear manner. In contrast, stationary situations in vicinity to the equilibrium, in which no self-organization takes place, are characterized by a linear dependence of flows on the driving forces. The effect of the driving force on the phosphate flow into the cell before and after adaptation allows investigating to what extent the adaptive behaviour of the phosphate uptake system can be explained along this line. Phosphate deficient cells are adapted to growth on low phosphate concentration, close to the threshold value, where the uptake system operates near the equilibrium. When these cells are abruptly exposed to higher phosphate concentrations, the uptake system is forced to function far from the equilibrium. The characteristic features of this adaptive self-organization are revealed in a 'Thellier-plot' of uptake rates at different external concentrations against the logarithm of these concentrations (Thellier 1970; Thellier et al. 1993). This plot allows distinguishing the non adaptive, linear region from the concentration range, in which adaptive processes take place, for the following reason: in the linear range, i.e. at concentrations close to the threshold value, the uptake rate J_P can be expected to be proportional to the affinity for this process. Hence, with $J_P = L'X$ and the expression for the affinity from Eq. (4.1) one obtains the flow-force relationship:

$$J_P = L_P(\log[P_e] + \log K' + n_P \Delta pH_T). \quad (4.2)$$

The proportionality factor L_P, comprising the term $2.3RTL'$, represents a conductivity coefficient which is proportional to the maximal velocity of the uptake system

(Wagner et al. 1995; Falkner et al. 1995). Inserting into this flow-force relationship the logarithmic expression for the threshold value $\log[P_e]_A = -\log K - n_P \Delta pH_T$ (see above) leads to the simple function:

$$J_P = L_P(\log[P_e] - \log[P_e]_A). \tag{4.3}$$

Equation (4.3) shows that a Thellier plot of the uptake rate against the logarithm of the external concentration follows a straight line within the validity range of this linear flow-force relation. This line intercepts the $\log[P_e]$-axis at the logarithmic threshold value and reflects in its slope the conductivity coefficient L_P. The onset of adaptive self organization processes should then be characterized by deviations from the linear dependence of J_P on $\log[P_e]$. In such a case the concentration dependence of uptake rates obeys a flow-force relation when Eq. (4.3) is supplemented with non linear terms (Thellier 1970):

$$J_P = L_P(\log[P_e] - \log[P_e]_A) + L(\log[P_e] - \log[P_e]_A)^m$$

where $m > 1$. An appropriate value for the exponential number m that accounts for the non linear dependence of the uptake rate on the driving force must be found empirically.

4.2.5 *Analysis of the Concentration Dependence of Uptake Rates by the Flow Force Relationship*

Thellier plots of the time course of phosphate uptake before and after the observed rate transition, described above, led to an interesting discovery. In the case of the initial velocities the resulting function always revealed an upward curved shape (Fig. 4.4, curve A), in accordance with the principles of non equilibrium thermodynamics. A plot of the second steady state incorporation rates, however, indicated that adaptation to the experimentally employed phosphate concentrations resulted in the development of an extended range of validity, over which there is a linear dependence of J_P on $\log[P_e]$ (curve B). In this case the obtained straight line intercepts the $\log[P_e]$-axis at a logarithmic concentration that corresponded to the lowest threshold value observed directly in net uptake experiments, using a parallel culture that had been simultaneously grown on radioactive phosphate (1.4 nM, arrow below the $\log[P_e]$-axis). This 'evolution to linearity' resulted in divergence of the two curves A and B at higher concentrations. The difference between the two curves shows to what extent the transition from a stationary state close to equilibrium to an operation in regions far from equilibrium provoked an adaptive self-organization of the uptake system. Thus, in the region where curve A and B coincide, no self-organization took place during the experiment. This explains the lack of transition in the time course at 3 nM and its occurrence at 20 nM in Fig. 4.3. Pre-incubation of cyanobacteria with micromolar phosphate concentrations resulted in the development of new energetic properties, characterized by a substantial extension of the linear region (curve C).

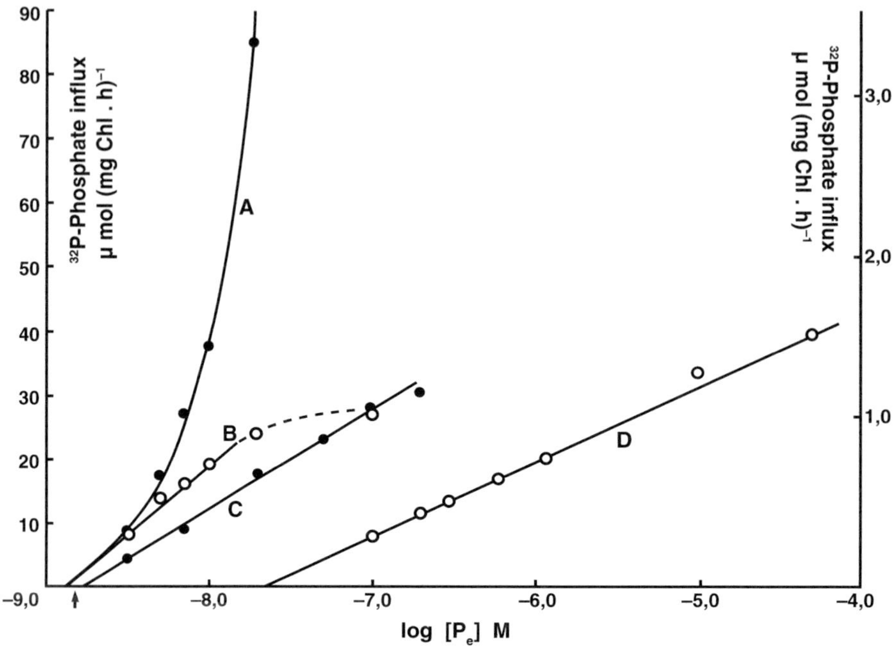

Fig. 4.4 Thellier plot of the concentration dependence of initial uptake rates of ^{32}P-phosphate by *Anacystis nidulans* before (*A*) and after (*C*) an incubation with 10 μM phosphate for 15 min. Graph *B* presents the concentration dependence of steady state uptake rates observed after the rate transition, caused by adaptation to the phosphate concentration employed in the experiment. Graph *D* represents a plot of steady state uptake rates after an increase of the phosphate concentration in the reservoir to 40 μM, leading to a growth concentration of 40.4 in the chemostat (determined photometrically with an ammonium molybdate method). Graphs *A*, *B* and *C* refer to the left scale, graph *D* to the right scale on the ordinate. (For experimental details see Falkner et al. (1995))

If the external concentration was raised to 40 μM and maintained at that level under continuous culture conditions, a new steady state of growth developed. A Thellier plot of the concentration dependence of uptake rates (Fig. 4.4, D) again showed the existence of a wide linear region, extending from (a higher) threshold value to the logarithmic growth concentration ($\log[P_e]=-4.4$). We therefore may conclude that the development to a 'linear operation mode' is a general feature of adaptation to elevated phosphate concentrations. The extended range of validity of the linear logarithmic function has been explained by the simultaneous function of several uptake systems with different, but coordinated kinetic parameters (Falkner et al. 1995; Wagner et al. 1995). Since the efficiency of an energy converter is much higher in the linear than in the non linear operation mode (Stucki et al. 1983), the observed evolution to linearity reflects an inherent tendency of the uptake system to operate with optimal efficiency. Model calculations about the efficiency of phosphate incorporation at different external concentrations were in accordance with this conclusion (Falkner et al. 1994). As will be shown in the following section, this observation provides an important criterion for distinguishing adapted from non adapted states of the uptake system, necessary for analyzing microbial information processing about experienced phosphate fluctuations. It is notable that the linear flow-force relationship $J_P = L_P(\log[P_e] - \log[P_e]_A)$ has the same structure as Weber-Fechner's law, when the uptake rate J_P is interpreted as the response to the stimulus $[P_e]$. Apparently adaptation of cyanobacteria to elevated phosphate concentrations leads to a physiological state, in which the relation between stimulus and response follows a similar logic structure as the sensory perception of higher organisms.

Both the time required for an adaptive reconstruction and the critical concentration, above which this reconstruction occurs, depend on the growth conditions and are different in different organisms under the same culture conditions. In order to illustrate this we present an example, in which *Anabaena variabilis* had been grown in two different discontinuous cultivation modes (in this mode every 24 h half of the culture was discarded and replaced by fresh medium containing either 0.1 or 5 μmol l^{-1} phosphate. The remaining cyanobacteria in the growth medium incorporated the freshly added phosphate within less than half an hour and grew then at the expense of stored phosphate). A comparison of the time course of ^{32}P-influx at different external concentrations showed significant differences (Figs. 4.5 and 4.6), indicating that cells originating from the lower growth concentrations adapted faster and at lower concentrations.

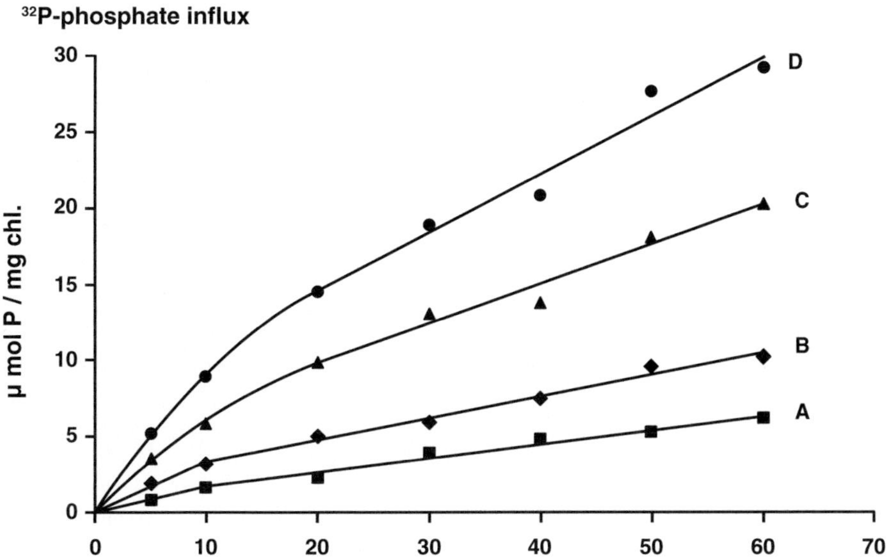

Fig. 4.5 Time course of ^{32}P-phosphate influx in *Anabaena v.* (cultivated on a total phosphorus content of 0.1 μmol l^{-1}) at different external concentrations. *A*: 20 nM; *B*: 50 nM; *C*: 200 nM; *D*: 500 nM. The experimental conditions are described in Falkner et al. (2006). For the measurement the original suspension was diluted 500-fold to ensure that the uptake process did not alter the external concentration. (Reproduced with permission from Falkner et al. (2006). Copyright Landes Bioscience)

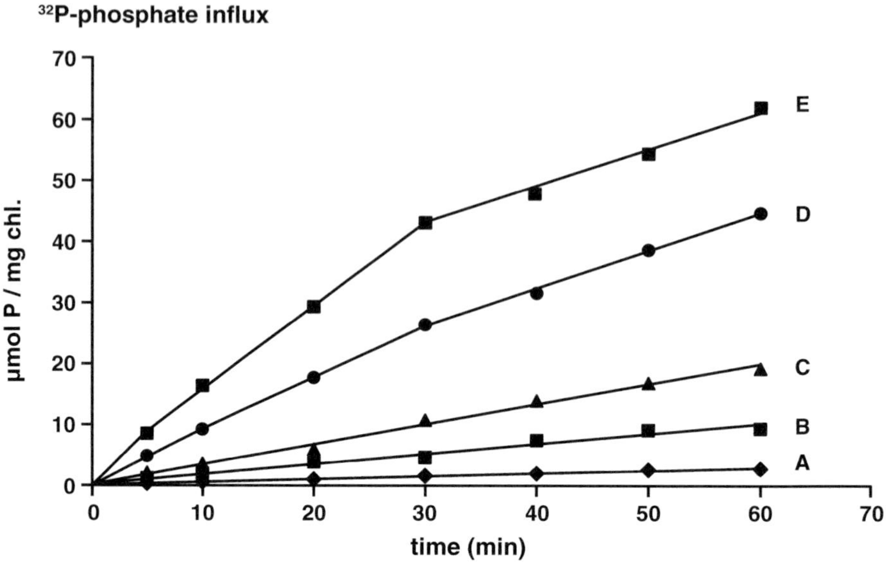

Fig. 4.6 Time course of ^{32}P-phosphate influx in *Anabaena v.* (cultivated on a total phosphorus content of 5 µmol l^{-1}) at different external concentrations. *A*: 20 nM; *B*: 100 nM; *C*: 200 nM; *D*: 500 nM; *E*: 1 µM. Experimental conditions as in Fig. 4.5. (Reproduced with permission from Falkner et al. (2006). Copyright Landes Bioscience)

4.3 Microbial Information Processing About Changes in the External Phosphate Concentration

4.3.1 The Dependence of the Adaptive Response on the Experimentally Employed Conditions

Owing to the energetic and kinetic attributes, described above, an uptake system displays an extremely complex behaviour, when the population is exposed to a series of phosphate pulses, in which the external concentration is first abruptly raised and then diminished by the uptake activity of the whole population. In this situation the change of the ambient phosphate concentration and the uptake system depend on each other, so that neither the change in the concentration nor in the uptake system can be analyzed in terms of a simple cause-effect relationship: each transiently occupied adapted state is cause and effect simultaneously and the change of the concentration and of the transport system reflect each other. Under this condition short-term rises of the external phosphate concentration above a threshold value provoke a variety of adaptive responses that reflect complex information processing about preceding phosphate supply. In the following we give several characteristic examples.

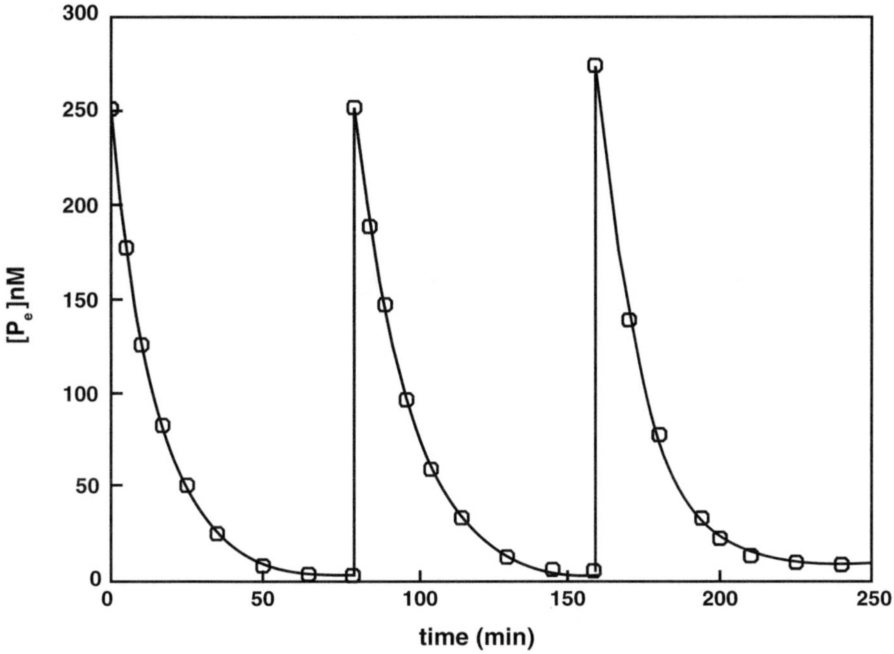

Fig. 4.7 Time course of ^{32}P-phosphate removal from the external medium during three consecutive pulses of 250 nM phosphate. The content of bacterial phosphorus of the suspension of *Anabaena v.* was likewise 250 nmol l^{-1}. The experimental conditions are described in Falkner and Falkner (2003) and Falkner et al. (2006). The curves represent the best computer fit of the MLAB program using the non linear equation: $d[P_e]/dt = -0.645\ (\ln[P_e] - \ln[P_e]_A) - L\cdot(\ln[P_e] - \ln[P_e]_A)^m$, with the parameters: $L = 9.83\cdot10^{-3}$ nM min^{-1}, $[P_e]_A = 3.43$ nM, m = 5 (first pulse); $L = 7.21\cdot10^{-3}$ nM min^{-1}, $[P_e]_A = 3.24$ nM, m = 5 (second pulse); $L = 0.318$ nM min^{-1}, $[P_e]_A = 9.59$ nM, m = 3 (third pulse)

In the first example two differently diluted suspensions of the filamentous cyanobacterium *Anabaena variabilis* (containing 250 nmol l^{-1} and 5 μmol l^{-1} resp.) were exposed to three phosphate pulses and the time course of the decrease of the external phosphate concentration was recorded. In the presented experiment the phosphorus content of the two cell suspensions and of each of the three respective pulses was of the same magnitude, so that both suspensions had stored the same amount of phosphate per cell after each pulse (Figs. 4.7 and 4.8). The suspension with the lower cell number per unit volume incorporated the supplied amounts of phosphate without persisting adaptation, so that the incorporation kinetics during the three subsequent pulses was practically identical. In contrast, the suspension with the twenty times higher number of cells revealed a totally different behaviour. Here the kinetics of incorporation varied from pulse to pulse, indicating that an *adapted state*, attained by the uptake system during the foregoing pulse, influenced the *adaptive operation mode* during the subsequent pulse. Apparently in this case information about exposures to external phosphate was transferred from one adaptive event to the next.

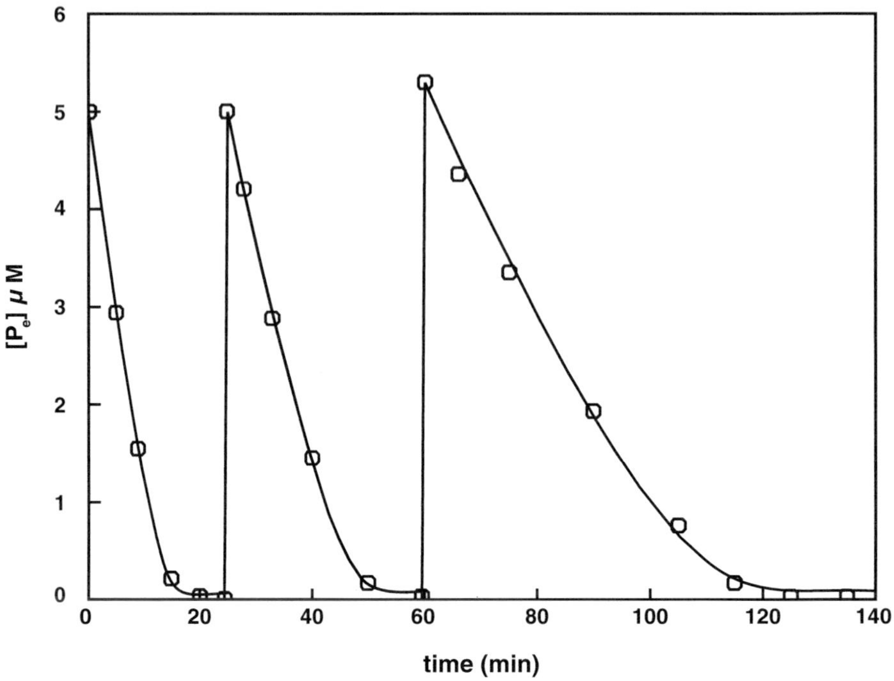

Fig. 4.8 Time course of ^{32}P-phosphate removal from the external medium during three consecutive pulses of 5 μM. The content of bacterial phosphorus of the suspension of *Anabaena v.* was likewise 5 μmol l^{-1}. Experimental conditions as in Fig. 4.7. The curves represent the best computer fit of the MLAB program using the equation: $d[P_e]/dt=-L_p\,(\ln[P_e]-\ln[P_e]_A)$ with the parameters: $L_p=0.100$ μM min^{-1}, $[P_e]_A=0.066$ μM, (first pulse); $L_p=0.067$ μM min^{-1}, $[P_e]_A=0.085$ μM, (second pulse); $L_p=0.032$ μM min^{-1}, $[P_e]_A=0.090$ μM, (third pulse)

The next example demonstrates the discriminatory potential of such information transfer, revealed when two identical samples of a population of *Anabaena v.* are exposed to five small pulses of 1 μM each and one greater pulse with 5 μM, but in reverse order: one sample experienced first the five smaller pulses and then the great pulse, the other sample *vice versa*. A comparison of the uptake kinetics in a subsequent pulse of identical maximum concentration revealed significant differences (Fig. 4.9), indicating a more pronounced response of the sample that experienced

Fig. 4.9 Time course of ^{32}P-phosphate removal from the external medium by two identical populations of *A. variabilis* (*upper graph*): 5 pulses of 1 μM followed by 1 pulse of 5 μM; (*lower graph*): 1 pulse of 5 μM followed by 5 pulses of 1 μM. For the small pulses only the concentration before and after addition of phosphate are given in the graph. The two different patterns initiate a different adaptive behaviour during the final pulse. The curves of the final pulse represent the best computer fit of the MLAB program using the equation: $d[P_e]/dt=-L_p(\ln[P_e]-\ln[P_e]_A)$. The following parameters were obtained: upper graph: $L_p=0.026$ μM min^{-1}; $[P_e]_A=0.13$ μM; lower graph: $L_p=0.018$ μM min^{-1}; $[P_e]_A=0.42$ μM. (Reproduced with permission from Falkner et al. (2006). Copyright Landes Bioscience)

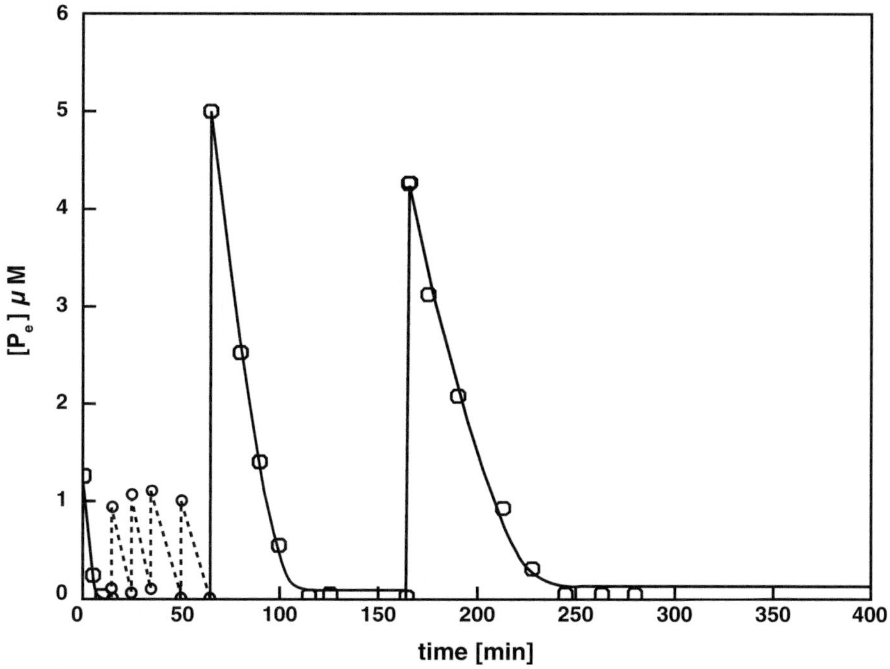

[P_e] μM
0
1
2
3
4
5
6
0
50
100
150
200
250
300
350
400
time [min]

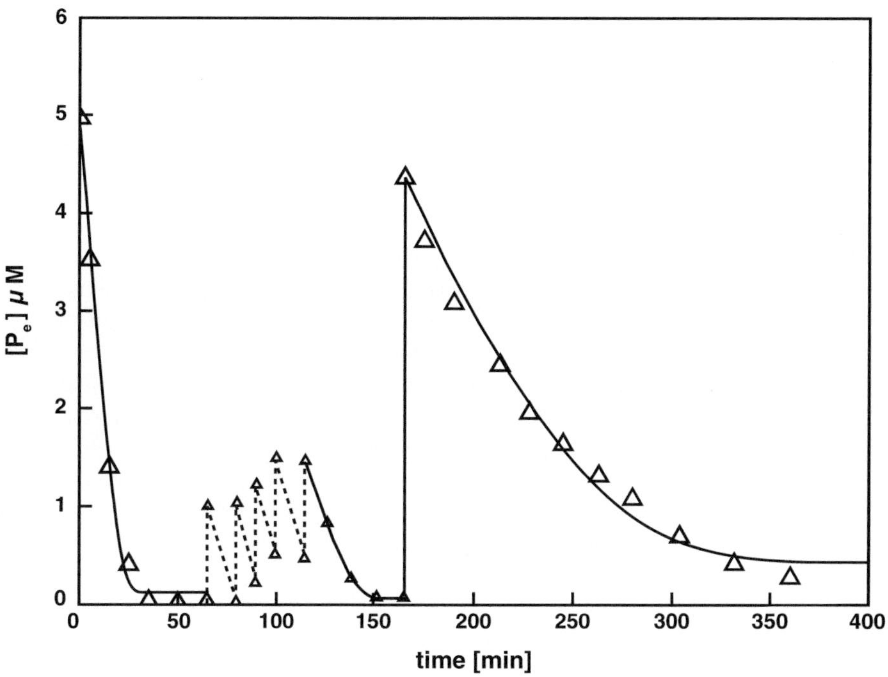

[P_e] μM
0
1
2
3
4
5
6
0
50
100
150
200
250
300
350
400
time [min]

the smaller pulses *after* the great pulse. When this experiment was performed with a lower number of small pulses, essentially the same result was obtained; however, the difference in the resulting adaptive behaviour was smaller (data not shown). Since under this condition a difference in the pattern of phosphate fluctuations leads to a corresponding difference in the subsequent adaptive behaviour, we may conclude that already a simple prokaryotic organism is capable of some sort of pattern recognition and to transcribe information about former environmental alterations into a distinct subsequent adaptive response. In this conclusion we employed a notion of information, given originally by Bateson (2000); accordingly an elementary unit of information is definable as a difference (in the environment) that makes a difference (in the adaptive self-constitution of the uptake system).

4.3.2 Inheritance of Acquired Traits

Distinct information about the pattern of experienced pulses is potentially transferred to daughter generations. When the two patterns of phosphate pulses, to which two different samples of the same population have been exposed, do differ significantly, a distinct adaptive behaviour can even be observed after subsequent cell division. In the following example two identical suspensions of *Anabaena v.* were challenged with two different supply modes, in which one suspension experienced one pulse of 10 μM ('high pulse culture'), the other ten pulses of 1 μM ('low pulse culture'). After this treatment the two suspensions were cultivated for 24 h at the expense of the stored phosphorus, which was the same in both cultures. Thereby, the chlorophyll content of the two cultures increased from 0.13 to 0.33 and 0.31 mgl^{-1} for the high and low pulse culture respectively, indicating that the number of cells had tripled during that time. The two suspensions were appropriately diluted by growth medium and then exposed to three identical test pulses for a comparison of their adaptive behaviour. Figure 4.10 shows that the culture that had been subjected to one great pulse during its previous growth deactivated the uptake system more rapidly than the reference low pulse culture, in which the external concentration never attained high values. Hence, it appears as if cells that had experienced a high phosphate concentration in the past, 'expect' a continuation of exuberant phosphate supply, when again challenged by elevated phosphate levels. In contrast, the same challenge led to a different adaptive *interpretation* in cells of the 'low pulse culture'; the data show that an adaptive event has a temporal vectorial character, connecting antecedent experiences with anticipated environmental changes.

The experiments, presented above, can be explained as follows: no information transfer takes place, when either the decrease in the external concentration is too rapid for adaptive processes to occur or when adaptive processes after the abrupt increase in the external phosphate concentration at the beginning of a pulse are confined to the transport system. But in such a case a possibly acquired adapted state would not be stable, since the concentration is further decreased by the newly adapted uptake properties. Therefore the uptake system had to readapt to an even lower

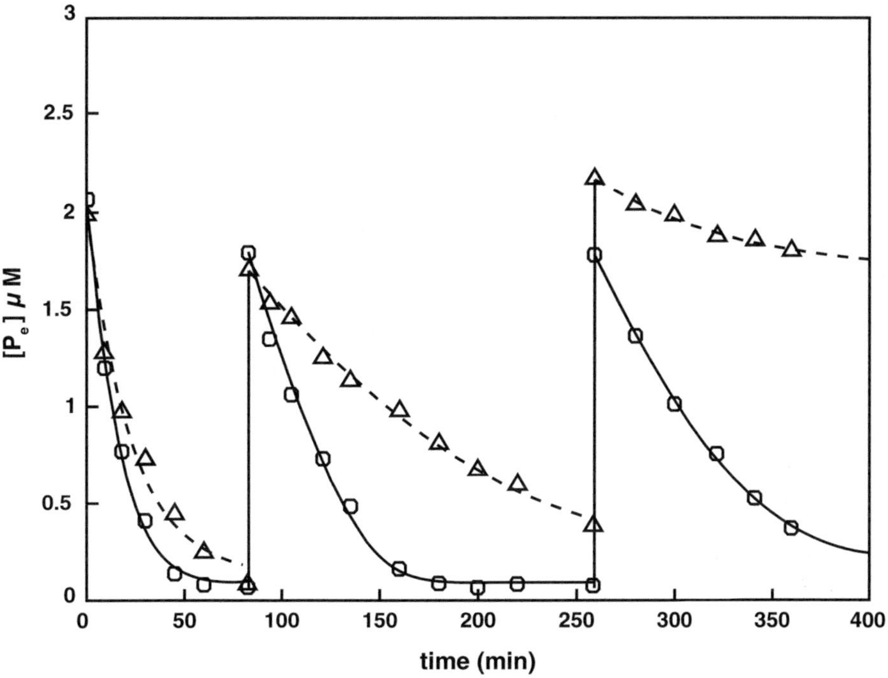

Fig. 4.10 Time course of ^{32}P-phosphate removal from the external medium by two populations of *Anabaena var*. produced by different pre-treatments of the same mother culture. *Circles*: low pulse culture, *triangles*: high pulse culture. The curves represent the best computer fit of the MLAB program using the equations: First pulse: $d[P_e]/dt = -0.01 \cdot (\ln[P_e] - \ln[P_e]_A) - (L \cdot (\ln[P_e] - \ln[P_e]_A))^3$ for the low pulse culture and $d[P_e]/dt = -0.004 \cdot (\ln[P_e] - \ln[P_e]_A) - (L \cdot (\ln[P_e] - \ln[P_e]_A))^3$ for the high pulse culture. All other pulses: $d[P_e]/dt = -Lp \cdot (\ln[P_e] - \ln[P_e]_A)$. The obtained parameters were for the low pulse culture: first pulse: $L = 0.135\ \mu M\ min^{-1}$; $[P_e]_A = 0.089\ \mu M$; second pulse: $L_p = 0.011\ \mu M\ min^{-1}$; $[P_e]_A = 0.093\ \mu M$; third pulse: $L_p = 0.0096\ \mu M\ min^{-1}$; $[P_e]_A = 0.205\ \mu M$; For the high pulse culture: first pulse: $L = 0.137\ \mu M\ min^{-1}$; $[P_e]_A = 0.105\ \mu M$; second pulse: $L_p = 0.0062\ \mu M\ min^{-1}$; $[P_e]_A = 0.261\ \mu M$; third pulse: $L_p = 0.0256\ \mu M\ min^{-1}$; $[P_e]_A = 1.69\ \mu M$. (Reproduced with permission from Falkner et al. (2006). Copyright Landes Bioscience)

level and so on. This process, in which no stable state is produced, would continue until the lowest possible threshold level is attained (corresponding to the highest possible degree of coupling) so that the system goes back to its original properties at the end of the foregoing pulse. In the subsequent pulse the same sequence of adaptive events is repeated, so that all pulses reveal similar kinetics, in which the dependence of uptake flows on the driving force is non-linear and Weber-Fechner's law is therefore not valid over a wide concentration range (Fig. 4.7).

A different situation arises, when during a pulse the external concentration is so high that also the ATP-synthase adapts to the transient increase of the cytoplasmic phosphate concentration, resulting in a persistent alteration of the stationary cytoplasmic phosphate concentration (Fig. 4.8, see Sect. 4.2.1). Naturally in this case a potential adaptive response of the ATP-synthase is favoured by a primary adaptive deactivation of the transport system, because this leads to a prolonged exposure of the cells to the elevated external concentration. If, however, the stationary cytoplasmic phosphate concentration has been changed, the phosphate transport system is no longer conformed to the new energetic situation, which therefore becomes a constraint for a new adaptive reconstruction. This may be followed by another adaptive response of the ATP-synthase, when the cytoplasmic concentration is further affected by alterations of the external concentration. Hence, the adaptive interplay between the transport system and ATP-synthase is potentially governed by some sort of 'communication', in which adaptive changes of the two systems influence each other, such that one system *interprets* in its adaptive operation mode the metabolic manifestation of an adapted state of the other and *vice versa*. The 'communication' between the two subsystems potentially leads to distinct sequences of adaptive events and—as a result—to adapted states at the finally attained threshold value that reflect the preceding adaptive events. These stable states are then reflected in the subsequent pulse in an energetically favourable linear dependence of the phosphate flow into the cell on the driving force over a wide concentration range, affecting a potential adaptive response to this pulse in a characteristic way. A computer simulation of such an adaptive interplay was in accordance with this explanation (Plaetzer et al. 2005). The preservation of a certain adapted state at the threshold value over a prolonged period of time can be explained by attainment of a stable self-referential state, in which the kinetic properties of the transport system are conformed to the existing gradient between the threshold value and the stationary cytoplasmic phosphate concentration, determined by the energetic properties of the ATP-synthase. Stability is also provided by an accommodation of the kinetic properties of the ATP-synthase to the stationary cytoplasmic phosphate concentration, resulting from the H^+/ATP-stoichiometry, such that fluctuations of the cytoplasmic phosphate level are buffered by a rapid phosphate flow into and from the polyphosphate pool. Due to the mutual adjustment the characteristic attributes of two subsystems are arrested at the corresponding threshold level and influence then the adaptive operation mode in the subsequent pulse in a distinct way. Transfer of information from one pulse to the next, exhibited in Figs. 4.8, 4.9 and 4.10, can be explained along that line. We also may conclude that a given self-referential state is not affected by a possible turnover of proteins. When one of the proteins is occasionally replaced during a subsequent growth period without phosphate supply, the new protein re-adapts to the pre-existing condition and, by this mechanism, "inherits" the properties of the protein it had substituted. This could explain the prolonged influence of antecedent adaptive events on the subsequent uptake behaviour after cell division.

4.3.3 The Influence of Pulse Patterns on Photosynthetic Activity

A persistent increase in the cytoplasmic phosphate level during transient exposure to elevated phosphate concentrations can be expected to affect CO_2 fixation. An alteration of the cytoplasmic phosphate level can be indirectly estimated via an

Table 4.1 CO_2 fixation in the phosphate-deficient cyanobacterium *Anacystis nidulans* before and after phosphate pulses. (Adopted from Wagner et al. (2000))

P_e incubation prior to CO_2 fixation experiment	Total CO_2 fixation nmol(10^9 cells min)$^{-1}$	LMWC nmol(10^9 cells min)$^{-1}$	HMWC nmol(10^9 cells min)$^{-1}$
Control, without P_e addition	2.6	1.1	1.5
Ten pulses of 0.1 µM P_e	2.7	1.9	0.8
One pulse of 1.0 µM P_e	12.5	10.4	2.1

Two bacterial suspensions (40 µg Chl a l^{-1}) were incubated with the same amount of phosphate (P_e) per cell but supplied by either ten pulses of 0.1 µM or one pulse of 1.0 µM phosphate respectively. After the threshold was attained, CO_2 fixation of the two suspensions was determined under identical conditions, and the flow of fixed carbon into the individual pools of low molecular weight compounds (LMWC) and high molecular weight compounds (HMWC) was analysed. The values for total CO_2 fixation, synthesis of LMWC and HMWC were determined from linear regression of the respective time courses (R^2 >0.98)

investigation of the flux of assimilated CO_2 into the pools of low- and high-molecular weight compounds (LMWC and HMWC resp., the latter mainly representing intracellular glycogen (Heldt at al. 1977), for the following reason: ADP-glucose-pyrophosphorylase, an allosteric enzyme involved in the regulation of glycogen formation, is inhibited by orthophosphate (Levi and Preiss 1976). This enzyme catalyses the biosynthesis of ADP-glucose, a precursor for glycogen formation, and thus regulates the flow of photosynthetically fixed carbon into either the pool of low molecular weight carbon compounds or into glycogen as a carbon storage of high molecular weight.

Table 4.1 presents an experiment, in which the CO_2 fixation and the resulting flow of fixed carbon before and after phosphate pulses has been followed with phosphate-deficient *Anacystis nidulans*. In this experiment two identical cyanobacterial suspensions were exposed to two different pulse patterns, containing *in toto* the same amount of phosphate per average cell: one suspension was challenged by one pulse of 1 µmol l^{-1}, the other by 10 pulses of 0,1 µmol l^{-1}. After the threshold had been attained, CO_2 fixation of the two suspensions was determined under identical conditions and compared with a reference culture without phosphate addition. In this experiment the suspension that had been exposed to 10 small pulses showed essentially the same photosynthetic activity as the control suspension. In contrast, the suspension that had been exposed to one high pulse, revealed a much higher photosynthetic activity. A separation of the LMWC and the HMWC showed that the increase could be attributed essentially to an enhanced biosynthesis of low-molecular weight compounds, whereas starch formation remains relatively uneffected.

It is obvious that the complex dependence of photosynthetic activity on the pattern of phosphate pulses renders difficult predictions of the growth behaviour of cyanobacteria under natural conditions. Since phosphate uptake and growth are not directly coupled, it is not possible to establish a simple relationship that describes the dependence of the growth rate on the external concentration. An application of the Monod equation $\mu=\mu_{max}$ $[P_e]/([P_e]+K_s)$, frequently employed for investigation of the growth behaviour of continuous cultures, is not useful for two reasons: first, stationary uptake rates at constant external concentrations usually do not exist in an environment, in which the amount of biomass formation is determined by phosphate supply. Second, Monod's equation does not contain a threshold value. The same refers to the Michaelis equation, from which the Monod equation originates. The Michealis equation is valid only for initial velocities of enzyme reactions, characterized by a product concentration of zero and therefore of infinite affinity. Replacing in the Michaelis equation the phosphate concentration $[P_e]$ by $([P_e]-[P_e]_A)$ is not justified, because $[P_e]_A$ itself is a function of $[P_e]$. Doubts about this replacement have been already expressed by Droop in 1974: "To put $(S-S_0)$ for S into the Michaelis equation, though from a practical point of view the obvious thing to do, knocks away any theoretical foundation the equation may have had."

4.4 Conclusion

A generalization of the regulatory principles, discussed above, to all other cellular energy converting subsystems, by which an energy source is coupled to a variable degree with a driven flow of substrates (or information, as in signal transduction pathways), leads to a dynamic view of the relation between microorganisms and their environment. In this view, an environmental alteration is experienced, when it initiates an intracellular communication, in which an excitation of cellular energy converter propagates in a wave pattern through the organism. Such an excitation would then incite to all kinds of molecular changes that continue, until an ordered state of least energy dissipation emerges. Consistent with this view is the postulate that an organismic ensemble of subsystems *experiences* a state of tension, when adaptive subsystems are not conformed to each other. Tension would then function as a regulatory constraint that determines how fast and to what extent an adaptive reconstruction takes place, aimed at maintenance of the organismic identity in a constant flux of structural rearrangements under ever changing environmental conditions. In this way the capacity of microorganisms to recreate themselves *autopoietically* (Maturana and Varela 1980) in every new adaptive event would become a characteristic feature of the intra-cellular energy flow, striving for specific states of intra-mundane stabilities that correspond to the nature of the particular species. The postulated tension as a physiological foundation for the ability of processing information about environmental phosphate fluctuations provides an internal relatedness among energy converting subsystems in that it reflects any deviation of an idealized manifestation of the organism under the prevailing ambient situation. By this, energy converting subsystems establish an 'adaptive representational network' (Trewavas 2005), by which a microorganism attains a 'model of itself' (Rosen 1985), representing a 'target value' that is anticipated in every adaptive event. In this way intracellular tension explains the observed connectivity between adaptive events, because a potential deviation from an idealized manifestation (for example due to an inappropriate anticipation of an environmental change during an antecedent event, see above), is corrected in a subsequent event. The resulting historical order of events is the precondition for the occurrence of memory phenomena at different levels of organization of living systems. This memory refers to a(n) (micro)organismic *self*, which apparently accounts for the difference between adapted states and adaptive operation modes; in the latter the organism seems to decide 'subjectively', to what extent an external influence on an objectively given adapted state 'has significance for itself' (Whitehead 1929/1978, p. 25).

Acknowledgements The work has been supported by the Austrian Science Fund.

References

Bateson G (2000): Steps to an ecology of mind. Chicago: The University of Chicago Press

Droop MR (1973): Some thoughts on nutrient limitations in algae. *J Phycol* 9: 264–272

Droop MR (1974): The nutrient status of algal cells in continuous culture. *J Mar Biol Assoc UK* 54: 825–855

Ebel JB, Colas J and Muller S (1958): Recherches cytochimiques sur les polyphosphates inorganiques contenus dans les organismes vivants. II. Mise au point des méthodes de détection cytochimiques spécifiques des polyphosphates. *Exp Cell Res* 15: 29–36

Falkner G, Werdan K, Horner F and Heldt HW (1974): Energieabhängige Phosphataufnahme der Blaualge *Anacystis nidulans*. *Ber Dtsch Bot Ges* 87: 263–266

Falkner G, Horner F, Werdan K and Heldt HW (1976): pH-changes in the cytoplasm of the blue-green alga *Anacystis nidulans* caused by light-dependent proton flux into the thylakoid space. *Plant Physiol* 58: 717–718

Falkner G, Falkner R and Schwab AJ (1989): Bioenergetic characterization of transient state phosphate uptake by the cyanobacterium *Anacystis nidulans*. Theoretical and experimental basis for a sensory mechanism adapting to varying environmental phosphate levels. *Arch Microbiol* 152: 353–361

Falkner G, Falkner R and Wagner F (1993): Adaptive phosphate uptake behaviour of the cyanobacterium *Anacystis nidulans*: analysis by a proportional flow-force relation. *C R Acad Sci Paris, Sciences de la vie/Life sciences* 316: 784–787

Falkner G, Wagner F and Falkner R (1994): On the relation between phosphate uptake and growth of the cyanobacterium *Anacystis nidulans*. *C R Acad Sci Paris, Sciences de la vie/Life sciences* 317: 535–541

Falkner G, Wagner F, Small JV and Falkner R (1995): Influence of fluctuating phosphate supply on the regulation of phosphate uptake by the blue-green alga *Anacystis nidulans*. *J Phycol* 31: 745–753

Falkner G, Wagner F and Falkner R (1996): The bioenergetic coordination of a complex biological system is revealed by its adaptation to changing environmental conditions. *Acta Biotheor* 44: 283–299

Falkner G and Falkner R (2000): Objectivistic views in biology: an obstacle to our understanding of self-organization processes in aquatic ecosystems. *Freshw Biol* 44: 553–559

Falkner R and Falkner G (2003): Distinct adaptivity during phosphate uptake by the cyanobacterium *Anabaena variabilis* reflects information processing about preceding phosphate supply. *J Trace Microprobe Tech* 21: 363–375

Falkner R, Priewasser M and Falkner G (2006): Information processing by Cyanobacteria during adaptation to environmental phosphate fluctuations. *Plant Signal Behav* 1: 212–220

Heldt HW, Chon CJ, Maronde D, Herold A, Stankoviv ZS, Walker DA, Kraminer A, Kirk MR and Heber U (1977): Role of orthophosphate and other factors in the regulation of starch formation in leaves and isolated chloroplasts. *Plant Physiol* 59: 1146–1155

Hudson JJ, Taylor WD and Schindler DW (2000): Phosphate concentrations in lakes. *Nature* 406: 54–56
Levi C and Preiss J (1976): Regulatory properties of the ADP-glucose pyrophosphorylase of the blue green bacterium Synechococcus 6301. *Plant Physiol* 58: 753–756
Maturana H and Varela F (1980): Autopoiesis and cognition: the realization of the living. Boston: Reidel
Plaetzer K, Thomas SR, Falkner R and Falkner G (2005): The microbial experience of environmental phosphate fluctuations. An essay on the possibility of putting intentions into cell biochemistry. *J Theor Biol* 235: 540–554
Rosen R (1985): Anticipatory Systems. Philosophical, mathematical and methodological foundations. Oxford: Pergamon Press
Stucki J, Compiani M and Caplan S (1983): Efficiency of energy conversion in model biological pumps: optimization by linear equilibrium thermodynamic relations. *Biophys Chem* 18: 101–109
Thellier M (1970): An electrokinetic interpretation of the functioning of biological systems and its application to the study of mineral salts absorption. *Ann Bot* 34: 983–1009
Thellier M, Ripoll C, Vincent JC and Mikulecki D (1993): Interpretation of the fluxes of substances exchanged by cellular systems with their external medium. In: Greppin H, Bonzon M and Degli Agosti R (eds) Some physicochemical and mathematical tools for understanding of living systems. Greppin: University of Geneva, pp. 221–277
Trewavas A (2005): Green plants as intelligent organisms. *Trends in Plant Science* 10: 413–419
Wagner F and Falkner G (1992): Concomitant changes in phosphate uptake and photophosphorylation in the blue-green alga *Anacystis nidulans* during adaptation to phosphate deficiency. *J Plant Physiol* 140: 163–167
Wagner F, Falkner R and Falkner G (1995): Information about previous phosphate fluctuations is stored via an adaptive response of the high-affinity phosphate uptake system of the cyanobacterium *Anacystis nidulans*. *Planta* 197: 147–155
Wagner F, Sahan E and Falkner G (2000): The establishment of coherent phosphate uptake behaviour by the cyanobacterium *Anacystis nidulans*. *Eur J Phycol* 35: 243–253
Van der Meer R, Westerhoff HV and Van Dam K (1980): Linear relation between rate and thermodynamic force in enzyme-catalyzed reactions. *Biochim Biophys Acta* 591: 488–493
Whitehead AN (1929/1978): Chapter II: The categoreal scheme. In: Griffin DR and Sherburne DW (eds) Process and reality (Corrected edition). An essay in cosmology. London: The Free Press, Collier Macmillan
Yoshida A (1955): Studies on metaphosphate. II. Heat of hydrolysis of metaphosphate extracted from yeast cells. *J Biochem (Tokyo)* 42: 163–168

Chapter 5
The Site of Respiratory Electron Transport in Cyanobacteria and Its Implication for the Photoinhibition of Respiration

Ioan I. Ardelean and Günter A. Peschek

5.1 Introduction

Cyanobacteria are the most diversified, ecologically most successful and evolutionary most important group of prokaryotes (Peschek et al. 1994b) clearly defined by their ability to carry out oxygenic photosynthesis in the thylakoid membranes and respiration in both plasma (CM) and thylakoid membranes (ICM) (Peschek 1996; Peschek and Zoder 2001). The only exception is *Gloeobacter violaceus* which does not contain thylakoids at all and, hence, must contain its photosynthetic pigments in/at the CM (Rippka et al. 1974).

Light inhibition of respiration in cyanobacteria was first reported in *Anabaena* sp. (Brown 1953; Brown and Webster 1953). In *Anacystis nidulans* (Hoch 1963) it was first demonstrated "that the inhibition of oxygen uptake is mainly sensitized by chlorophyll *and* 'light' as many times confirmed later (e.g., Katoh and Ohki (1975))". Jones and Myers (1963) suggested that "the electron acceptor for the chlorophyll *a* reaction, photosystem I (P700), is at least partially a competitor for oxygen as an electron acceptor". This explanation was substantiated by Imafuku and Kazoh (1976) who suggested "that PSI, upon illumination, absorbs electrons otherwise funneled to oxygen".

One basic idea behind these considerations is the occurrence of electron carriers active in both photosynthesis and respiration which has indeed been proved for plastoquinone, cytocrome b_6f, cytochrome c_6 and plastocyanin (Peschek 1987; Koike and Satoh 1996).

The first *in vitro* studies (Peschek and Schmetterer 1982) demonstrated that cytochrome oxidase mediates the inhibition of photosynthetic electron transport by oxygen, and P700 mediates the inhibition of respiratory electron transport by light in membrane preparations of *Anacystis nidulans*. The original proposal of Hoch (1963) was further corroborated by Scherer and Boger (1982) and Scherer et al.

I. I. Ardelean (✉)
Institute of Biology, Center of Microbiology, Splaiul Independentei 296, Bucharest 79651, Romania
e-mail: ioan.ardelean@ibiol.ro

G. A. Peschek et al. (eds.), *Bioenergetic Processes of Cyanobacteria,* 131
DOI 10.1007/978-94-007-0388-9_5, © Springer Science+Business Media B.V. 2011

(1982) who sustained that the inhibition of respiration was due to competition between electrons from the water splitting system and electrons from carbohydrate breakdown. The light inhibition of respiration in *Aphanocapsa* (*Synechocystis*) was interpreted as a competition between cytochrome oxidase and photosystem I (PSI) for reduced plastocyanin or cytochrome c_{553} (the actual cytochrome c_6) in the light (Sandmann and Malkin 1984) clearly stressing that one requirement for the above interpretation is the localisation of all electron transfer components in the same membrane, in agreement with (Scherer and Boger 1982).

Thus, further challenges and improvements to explain this inhibition came from the work devoted to cellular localisation of electron carriers involved in (both) respiration and photosynthesis, with special emphasis from results concerning the occurrence and cellular localisation of cytochrome *c* oxidase as *the* terminal oxidase in cyanobacteria (Peschek 1996; Peschek and Zoder 2001).

Within the framework of knowledge concerning the competition for electrons between cytochrome *c* oxidase and PSI, and taking into account all the progress with respect to cellular localisation of cytocrome *c* oxidase (Peschek 1996; Peschek and Zoder 2001), we put forward a hypothesis (Ardelean and Peschek 2002) concerning the role of cellular localisation of cytochrome *c* oxidase for controlling the degree of inhibition of respiration by light as follows: the inhibition is expected to be higher in strains where the predominant localization of cytochrome *c* oxidase is at intracellular (=thylakoid) membranes (e.g., in *Synechocystis* PCC 6803) (Peschek et al. 1989, 1994) compared to strains housing the major level of cytochrome *c* oxidase activity in the cytoplasmic (plasma) membrane (e.g., in *Anacystis nidulans*) (Trnka and Peschek 1986; Peschek et al. 1989, 1994).

The aim of this contribution is to develop our original hypothesis showing that the inhibition of respiration by light is stronger in *Synechocystis* PCC 6803 most of whose main cytochrome *c* oxidase activity occurs at the thylakoid membranes as opposed to *Anacystis nidulans* (=*Synechococcus* PCC 6301) whose main cytochrome *c* oxidase activity occurs at the plasma membrane.

5.2 Materials and Methods

Strains *Synechocystis* PCC 6803 and *Synechococcus* PCC 6301 were grown in BG11 and Kratz & Myers mediums, respectively, in the absence or presence of 0.4 M additional NaCl (Peschek et al. 1994). The cultures were bubbled with CO_2-enriched air. Batch cultures used for measurements were either early log cells (OD of the culture at 750 nm in between 0.1 and 0.2, 1 cm light pass cuvettes) or (light limited!) late log cells (OD of the culture at 750 nm in between 2.9 and 3.3, in 1 cm light pass cuvettes). For oxygen uptake measurements the cells were collected by centrifugation at 4000×*g* for 5 minutes and the pellet was suspended in 0.67 M phosphate buffer (pH=7.5), supplemented or not supplemented with 0.4 M NaCl for salt adapted cells and control cultures, respectively. In all cases the cell density of the suspension to be used for measurements was adjusted to an OD_{750} of 3 units (corresponding to 22–24 μL packed cell volume (pcv)×mL^{-1} for *Synechocystis* PCC 6803 and 4.0–4.5 μL pcv mL^{-1} for *Synechococcus* PCC 6301).

Oxygen uptake by illuminated cells was measured in the presence of 10 μM DCMU in case of control cells and 25 μM DCMU in case of salt adapted cells. DCMU had been added in darkness and oxygen consumption in dark in the absence of DCMU was the same as in its presence, in agreement with previous measurements (Hoch 1963; Valiente et al. 1992; Ardelean et al. 1998, 1999). After 5–10 minutes of DCMU addition the light (white light from an Orion Xenon Lamp) was switched on and oxygen consumption was monitored at different white light intensities. The intensity of light was changed using neutral filters and the light intensity was controlled with a YSI radiobolometer. The different light intensities used were: 1.3, 5, 6.7 and 17.8 μW s^{-1} m^{-2}. The measurement of oxygen consumption either in darkness or at different light intensities was performed at 32°C, taking for calculations the decrease in oxygen concentration from 100 and 50% saturation. The remaining respiration in the light (RRL) was calculated as percentage of the respiration occurring at the lowest light intensity used in this study (1.3 μW s^{-1} m^{-2}) from the dark respiratory. The measurements were repeated in 3 independent experiments, standard deviation being within 10%.

5.3 Results and Discussion

Composite Table 5.1 presents the results concerning the inhibition of respiration by white light of different intensities in intact cells of late log, light limited *Synechocystis* PCC 6803 and *Anacystis nidulans* (*Synechococcus* PCC 6301). As one can see, even at low light intensity there is a clear inhibition of respiration (Table 5.1) by light which is stronger in *Synechocystis* PCC 6803 than in *Anacystis nidulans*, thus the remaining respiration in light (RRL) (% of the dark respiration) is higher in *Anacystis nidulans* (52%) than in *Synechocystis* PCC 6803 (26%). These results are in agreement with our hypothesis which sustains that the inhibition of respiration by light must be stronger in *Synechocystis* PCC 6803 whose main cytochrome *c* oxidase occurs at thylakoidal membranes (Peschek et al. 1989, 1994) than in *Anacystis nidulans* (=*Synechococcus* PCC 6301) whose main cytochrome *c* oxidase occurs at plasma membrane (Trnka and Peschek 1986; Peschek et al. 1989). These results that are in agreement with those published about *Anacystis nidulans* (Brown 1953; Scherer et al. 1982) with the remaining respiration in light being at about 50% of the rate in darkness.

In order to further check our assumption we took advantage of the fact that in early logarithmic, low density *Anacystis nidulans* culture, the main cytochrome *c* oxidase activity is housed at the intracellular membrane (Peschek et al. 1988). Thus, in these cells, if our hypothesis is correct, the inhibition of respiration by light must be more pronounced than in late log (light limited!) high density cells. As one can see from Table 5.1, the inhibition is as strong as in early logarithmic, low density *Synechocystis* PCC 6803 whose main cytochrome *c* oxidase activity is at the ICM *a priori*, and the remaining respiration in light is only 35%.

Table 5.1 (Combining results from 10 to 20 experiments). The remaining respiration in light (RRL) at 1.3 W s^{-1} m^{-2} given as percent of the corresponding dark respiration. pcv, packed cell volume

Cells from	*Synechocystis* PCC 6803 RRL	*Synechococcus* PCC 6301 (*Anacystis nidulans*) RRL
Batch culture, light limited, late log	26%	52%
Batch culture, light limited, late log, salt adapted	53%	73%
Batch culture, low density culture, early log	35%	30%

Dark- and remaining respiration in light (RRL). *Synechocystis* PCC 6803 and *Synechococcus* sp. 6301, fully adapted to NaCl (0.4 M) (μmol O_2 min^{-1}/μL pcv). Intensity of white light is in μW s^{-1} m^{-2}

			Light intensity			
Strain	Dark	Dark (+DCMU)	1.3	5.0	6.7	17.8
Synechocystis PCC 6803	0.34	0.34	0.18	0.14	0.13	0.13
Anacystis nidulans	0.80	0.80	0.59	0.49	0.42	0.43

Dark- and remaining respiration in light (RRL). *Synechocystis* PCC 6803 and *Anacystis nidulans*, control cells (μmol O_2 min^{-1}/μL pcv). Intensity of white light is in μW s^{-1} m^{-2}

			Light intensity			
Strain	Dark	Dark (+DCMU)	1.3	5.0	6.7	17.8
Synechocystis PCC 6803	0.13	0.13	0.03	0.03	0.03	0.03
Anacystis nidulans	0.66	0.66	0.35	0.33	0.33	0.33

Dark- and remaining respiration in light (RRL). *Synechocystis* PCC 6803 and *Synechococcus* sp. 6301 control cells, early log cells (μmols O_2 min^{-1}/μL pcv). The intensity of the white light is in μW s^{-1} m^{-2}

			Light intensity			
Strain	Dark	Dark (+DCMU)	1.3	5.0	6.7	17.8
Synechocystis PCC 6803	0.40	0.40	0.14			
Synechococcus PCC 6301	1.88	1.88	0.57			

The dramatic change in the remaining respiration in light in *Anacystis nidulans* following the cellular localisation of the main cytochrome *c* oxidase activity (52% for cells with the main activity at the plasma membrane and 30% for the cells with the main activity at intracellular membranes) further sustains our hypothesis that the cellular localisation of cytochrome *c* oxidase is a major factor controlling the degree of inhibition of respiration by light in cyanobacteria.

Theses results are important because the kinetic *in vivo* measurements (oxidation of horse heart cytochrome *c* or oxygen consumption) correspond with *in vitro* measurements of horse heart cytochrome *c* oxidation detected by immunoblotting using antibodies against *P. denitrificans* COX (Peschek 1987). Thus, we claim that, by determining the inhibition of respiration by light in whole cells of cyanobacteria grown in normal media and suspended in 0.67 M phosphate buffer (pH=7.5), one could predict if the main cytochrome *c* oxidase activity is associated either with CM (remaining respiration in light, RRL, above 50%) or with ICM (RRL below 35%).

Furthermore we used salt adapted cells (batch cultures, late log cells, light limited) of *Anacystis nidulans* and *Synechocystis* PCC 6803 whose main cytochrome *c* oxidase activity is associated with CM or ICM just as in non salt adapted cells (Peschek et al. 1989). We expected that in salt adapted cells the remaining respiration in light must be higher in *Anacystis nidulans* than in *Synechocystis* PCC 6803. Indeed this is the case (Table 5.1), the obtained results showing that RRL is higher in *Anacystis nidulans* (73%) than in *Synechocystis* PCC 6803 (53%).

These findings are in agreement with the previous prediction that under salt conditions the inhibition of respiration by light should be lower than under low salt conditions (Peschek et al. 1989), as we had already demonstrated for *Synechocystis* PCC 6803 (Ardelean et al. 1998, 1999) (using a strain physiologically identical with that used in this work). When it comes to the molecular mechanism(s) by which such a decrease in the inhibition occurs in salt adapted cells, it could be speculated that for the increase of the COX activity at CM, both in *Anacystis nidulans* and in *Synechocystis* PCC 6803 (Peschek et al. 1989) for salt adapted cells as compared with control cells, identical mechanisms could be involved. These data (Peschek et al. 1989) show that there is an absolute increase of activity from 55 to 155 (nmol cyt *c*/min per mg protein) in COX activity at CM in *Anacystis nidulans* in cells fully adapted to high salinity, while in *Synechocystis* PCC 6803 the increase in activity, also at the CM, only is from 4 to 20 nmol cyt c/min per mg protein.

Furthermore, the full adaptation to growth at high salinity involves extensive modifications in cell structure and functions (Joset et al. 1996; Peschek and Zoder 2001) apart from content and activity of cytochrome *c* oxidase alone: some of these modifications may affect the competition for electrons between cytochrome *c* oxidase and P700, as for example the increase in the activity of PSI (Jeanjean et al. 1993; Ardelean et al. 2002).

All these results corroborate our original hypothesis concerning the role of the intracellular topography of cytochrome *c* oxidase for controlling the degree of inhibition of respiration by light. The main conclusion is that the intracellular topography of cytochrome *c* oxidase is a major factor controlling the inhibition of respiration by light in cyanobacteria.

These results refer to the problem concerning the RRL at ICM and CM. The degree of light inhibition of respiration at either CM or ICM in *Synechocystis* PCC 6803 and *Anacystis nidulans* is yet unknown. Pioneering work done more than twenty years ago on the mutual inhibition of respiration and photosynthesis in isolated membranes of *Anacystis nidulans* and *Nostoc* sp. strain Mac (Peschek and Schmetterer 1982) in the context of our hypothesis together with the tools and concepts of molecular biology not only would improve our knowledge on basic mechanisms such as the detailed molecular interplay between (dark) catabolic and (light) anabolic processes in cyanobacteria as well as in other phototrophic procaryotes in general but also could help the scientist to control this interplay, practical aspects of which could eventually be useful for certain technical applications such as the large scale photoproduction of O_2 with concomitant removal of CO_2, e.g., in manned spacecraft technology and thus could be interesting even for NASA.

Acknowledgements Thanks are due to FEBS and UNESCO for short term fellowships to I.I.A, and to Mr. Otto Kuntner for most skilful technical assistance.

References

Ardelean I and Peschek GA (2002) 28th FEBS Meeting, Abstract 258

Ardelean I, Enache M, Tunaru S, Flonta ML, Dumitru L and Zarnea G (1998) In: Garab G (ed) Photosynthesis: Mechanisms and Effects, vol IV, pp 2609–2612, Kluwer Academic Publishers, Dordrecht

Ardelean I, Tunaru S, Flonta ML, Teodosiu G, Mădălin E, Dumitru L and Zarnea G (1999) In: Peschek GA, Löffelhardt W, Schmetterer G (eds) Phototrophic Prokaryotes, pp 403–409, Plenum Publisher, New York

Ardelean I, Matthjis HCP, Havaux M, Joset F and Jeanjean R (2002) FEMS Microbiol Lett 213: 113–119

Brown AH (1953) Am J Bot 40: 719–729

Brown AH and Webster GC (1953) Am J Bot 40: 753–758

Hoch G, Owens OH and Kok B (1963) Arch Biochem Biophys 101: 171–180

Imafuku H and Katoh T (1976) Plant Cell Physiol 17: 515–524

Jeanjean R, Matthjis HCP, Onana B, Havaux M and Joset F (1993) 34: 1073–1079

Jones LW and Myers J (1963) Nature 199: 670–672

Joset F, Jeanjean R and Hagemann M (1996) Physiol Planta, 96: 738–744

Katoh T and Ohki K (1975) Plant Cell Physiol 16: 815–828

Koike H and Satoh K (1996) J Sci Ind Res 55: 564–582

Peschek GA (1987) In: Fay P and Van Baalen C (eds) Respiration, pp 119–161, Elsevier Science Publishers, Amsterdam

Peschek GA (1996) Biochem Biophys Acta 1275: 27–32

Peschek GA and Schmetterer G (1982) Biochem Biophys Res Commun 108: 1188–1195

Peschek GA and Zoder R (2001) In: Rai LC and Gaur JP (eds) Algal Adaptation to Environmental Stress Temperature Stress and Basic Bioenergetic Strategies for Stress Defence, pp 203–258, Springer, Berlin

Peschek GA, Molitor V, Trnka M, Wastyn M and Erber M (1988) In: Packer L and Glazer N (eds) Methods in Enzymology: Characterization of Cytochrome *c* Oxidase in Isolated and Purified Plasma and Thylakoid Membranes from Cyanobacteria, vol 167, pp 437–449, Academic Press, Inc, San Diego

Peschek GA, Wastyn M, Molitor V, Kraushaar H, Obinger C and Matthjis HCPJ (1989) In: Kotik A, Skoda J, Paces V and Kosta V (eds) Highlights Modern Biochem Self-Contained or Accessory Respiration in the Phototrophic Cyanobacteria (Blue-green Algae)?, vol 1, pp 893–902, VSP Publishers, Zeist

Peschek GA, Obinger C, Fromwald S and Bergman B (1994a) FEMS Microbiol Lett 124: 431–438

Peschek GA, Obinger C, Sherman DM, and Sherman LA (1994b) Biochem Biophys Acta, 1187: 369–372

Rippka R, Waterbury JB and Cohen Bazire G (1974) Arch Microbiol 100: 419–436

Sandmann G and Malkin R (1984) Arch Biochem Biophys 243: 105–111

Scherer S and Boger P (1982) Arch Microbiol 132: 329–332

Scherer S, Sturzl E and Boger P (1982) Arch Microbiol 132: 333–337

Trnka M and Peschek GA (1986) Biochem Biophys Rres Commun 136: 235–241

Valiente EF, Nieva M, Avendano MC and Maeso ES (1992) Plant Cell Physiol 33: 307–313

Chapter 6
Nitrogenases and Hydrogenases in Cyanobacteria

Hermann Bothe, Oliver Schmitz, M. Geoffrey Yates and William E. Newton

6.1 Introduction

Both enzyme complexes discussed in this chapter are functionally linked. Nitrogenase catalyzes the reaction:

$$N_2 + 8H^+ + 8e^- \rightarrow 2NH_3 + H_2 \quad (6.1)$$

Hydrogenase consumes hydrogen gas by the equation:

$$H_2 \leftrightarrow 2e^- + 2H^+ \quad (6.2)$$

In most (di)nitrogen fixing organisms, the H_2 gas produced by nitrogenase is immediately recycled by hydrogenase so that no net H_2 gas production is measurable in many intact cells. Most N_2-fixing organisms possess a hydrogenase but there are many examples where hydrogenase occurs in organisms that do not contain the gene sets coding for nitrogenase. Several enzyme types exist both for nitrogenase and hydrogenase. Whereas nitrogenases are restricted to prokaryotes, hydrogenases can also occur in eukaryotic cells, but not in higher plants. The distribution of nitrogenase and hydrogenase in cyanobacteria is haphazard and does not follow the taxonomic criteria based, for example, on a 16S r-DNA phylogeny. N_2-fixation is extremely energy demanding: the transfer of one e^- in nitrogenase requires the expenditure of 2 ATP hydrolysed. Formation or uptake of H_2 by hydrogenase is ATP independent and is not affected by CO in contrast to N_2-fixation. Recent reviews on nitrogenases and hydrogenases are available (Vignais and Colbeau 2004; Newton 2007).

H. Bothe (✉)
Botanical Institute, The University of Cologne, Zuelpicherstr. 47b, 50923 Köln, Germany
e-mail: hermann.bothe@uni-koeln.de

G. A. Peschek et al. (eds.), *Bioenergetic Processes of Cyanobacteria*,
DOI 10.1007/978-94-007-0388-9_6, © Springer Science+Business Media B.V. 2011

6.2 General Properties of Mo-Nitrogenases

All nitrogenases are composed of two component proteins neither of which shows catalytic activity alone. The smaller subunit ((di)nitrogenase reductase, nitrogenase protein-2 or Fe-protein) has a molecular mass of ~64 kDa and is a homodimer. Its prosthetic group is a single [4Fe-4S] cluster that bridges the subunit interface and is attached to each subunit by two cysteinyl residues. This cluster accepts reductants from the electron carrier which is either ferredoxin or flavodoxin, depending on the organism. In cyanobacteria, a ferredoxin-type FdxH (or FdxN) is the electron carrier in Fe-replete media. It is substituted by flavodoxin (originally termed phytoflavin, Trebst and Bothe 1966) under Fe-deficiency in cyanobacterial cultures. Nitrogenase reductase is encoded by the *nifH* gene which is most conserved among all nitrogenase genes (Ruvkun and Ausubel 1980). Therefore probes developed from *nifH* are commonly used to screen for nitrogenase in organisms or environmental samples.

The larger component protein (nitrogenase protein-1 or MoFe-protein) of about 240 kDa and of an $\alpha_2\beta_2$ subunit composition contains two unique prosthetic groups, the P-cluster and the MoFe-cofactor. The P-cluster consists of a [4Fe-4S] and a [4Fe-3S] that are bridged by a common S^{2-}. One copy sits at each α/β subunit interface. The P-cluster was proposed to serve in transferring electrons from the Fe-protein to the MoFe-cofactor. This might be the case under high but not low electron flux through nitrogenase proteins (Fisher et al. 2007). The MoFe cofactor contains 1Mo, 7Fe and 9S and a light atom (presumably N, C or O) at its centre. Each α subunit binds one MoFe-cofactor. This can be subdivided into a [Mo-3Fe-3S] and a [4Fe-3S] subcluster and is unusually bound to the protein by homocitrate. The substrate N_2 may be reduced at a 4Fe-4S face with the participation of the light atom. This face forms a surface where other substrates such as nitriles, isonitriles, nitrous oxide or acetylene might also be reduced. These substrates all have in common a partial triple bond. However, the substrate binding and reduction site has not been identified unambiguously as yet. Substrate reduction may alternatively proceed at the Mo-homocitrate part of the MoFe-cofactor.

The formation of NH_3 by nitrogenase is paralleled by the production of H_2-gas presumably with a stoichiometry of 2:1 (Eq. 6.1). Hydrogen is produced with the formation of NH_3 but not with the reduction of other nitrogenase substrates. In the absence of any other substrate, nitrogenase catalyzes an ATP-dependent formation of H_2. In intact cells, however, no or little net H_2 production by nitrogenase is observed, since the gas is immediately recycled by the uptake hydrogenase of most organisms (see below).

Nitrogenases are encoded by the three structural genes *H* (for the nitrogenase reductase) and *DK* (for the $\alpha_2\beta_2$-subunits of the larger protein). In addition, further 17 genes, all located contiguously on the chromosome in *Klebsiella pneumoniae*, are required for nitrogenase cofactor biosynthesis, maturation or electron transport to nitrogenase (Arnold et al. 1988). In other organisms, they may be scattered throughout the chromosome, and additional fix genes may be required to express an active nitrogenase.

The complex prosthetic groups of nitrogenase are irreversibly damaged by O_2; therefore extreme care against exposure to air is a prerequisite during all nitrogenase purification steps. Intact cells of different organisms have developed versatile devices to protect their nitrogenase against O_2. One example of this is the deposition of the enzyme into specialized cells, the heterocysts of cyanobacteria.

The production of ammonia by nitrogenase is accompanied by an extensive hydrolysis of ATP. Together with reductant, ATP is required to activate each of the two subunits of the Fe-protein which then undergoes a conformational change accompanied by a decrease of the redox potential of some 200 mV. Docking of this activated Fe protein to the MoFe-protein causes a further reduction of the redox potential to a final value of −600 mV. Altogether 16 ATP are required per N_2-molecule reduced, thus the equivalent of 2 ATP per e^- transferred.

Gene probing with *nifH* to search for the occurrence of nitrogenase in environmental samples surprisingly showed that most nitrogenase sequences retrieved were new or belonged to uncultured, non-characterized bacteria that have only sequence deposits in the databanks as yet (for example see Eilmus et al. 2007; Rösch and Bothe 2009).

6.3 Alternative Nitrogenases

Somewhat surprisingly, the Mo atom in the cofactor can be substituted by V (or by Fe) in alternative nitrogenases (reviewed by Bishop and Joerger 1990; Pau 1991; Eady 1996; Zhao et al. 2006), However, this change is not simply a switch of atoms, because the protein environment is also altered. In fact, the V-nitrogenase and the Fe-nitrogenase proteins are encoded by a different set of genes, *vnfHDGK* and *anfHDGK*, respectively. Their *DGK* gene products are $\alpha_2\beta_2\gamma_2$ heterohexamers that constitute the VFe- and FeFe-proteins, respectively. The function of the additional γ-subunit has not been finally resolved. It might serve in the processing of the apoproteins of the alternative nitrogenases to the functional enzyme complex by assisting in the insertion of the VFe- or FeFe-cofactor, respectively.

Alternative nitrogenases reduce N_2 and C_2H_2 with significantly lower rates but produce more H_2 than Mo-nitrogenase. The alternative nitrogenases reduce C_2H_2 partly beyond C_2H_4 to C_2H_6. Although this ethane-formation amounts to only ~3% of the total C_2H_2-reduction activity, it can easily be measured by gas chromatography. The C_2H_6-formation is thus indicative of the expression of an alternative nitrogenase in an organism.

Alternative nitrogenase occurs only occasionally in organisms, apparently without any correlation to taxonomic affiliation. *Azotobacter vinelandii, A. paspali, Rhodopseudomonas palustris* and *Methanosarcina acetivorans* possess the gene sets of all three nitrogenases. The Mo- and V-nitrogenases occur in *Azotobacter chroococcum, A. salinestris, Methanosarcina barkeri* 27 and in some cyanobacteria (see below). The Mo- and Fe-nitrogenases, but not the V-enzyme occur in *Clostridium pasteurianum, Azomonas macrocytogenes, Azospirillum brasilense* Cd,

Rhodospirillum rubrum, Rhodobacter capsulatus and *Heliobacterium gestii* (Loveless et al. 1999; Betancourt et al. 2008). Sequences coding for alternative nitrogenases were also detected in samples from aquatic environments (Loveless et al. 1999; Betancourt et al. 2008). They mainly belonged to fluorescent pseudomonads and azotobacteria. All the other N_2-fixing organisms investigated possess only the Mo-nitrogenase.

Under laboratory culture conditions, alternative nitrogenases are only expressed under Mo-deficiency in the medium. V-nitrogenase synthesis requires the presence of V in the medium, and Fe-nitrogenase expression occurs only in the absence of both Mo and V in the medium. In natural environments, Mo is generally not growth-limiting. Therefore organisms possessing an alternative nitrogenase might not have a competitive advantage, and it appears somewhat surprising that these alternative nitrogenase genes were retained in the organisms during evolution.

The rather close sequence similarities in the azotobacter-fluorescent pseudomonad group suggest that alternative nitrogenase may have arisen by gene duplication from Mo-nitrogenase. Outside this group, there is fairly little sequence similarity between the *nifH*, *vnfH* and *anfH* genes and the inferred phylogeny from the 16S rRNA gene sequences. Thus alternative nitrogenase may have been developed by lateral gene transfer among non-members of the azotobacter-fluorescent pseudomonad group. This seems to be the case in *Methanobacterium barkeri* 27 where the *DG* genes show close sequence similarities to other *vnfDG* genes, particularly of the cyanobacterium *Anabaena variabilis*. In contrast, the *H* gene is a separate entity clustering with *anfH* genes from other organisms (Chien et al. 2000). In *Methanococcus maripaludis*, *nifDK* clusters with these genes of other organisms, whereas *nifH* is an amalgam of both Mo- and V-nitrogenase *H* genes, indicating that nitrogenase genes of this organism were acquired by two independent transfers from other organisms (Kessler et al. 1997).

6.4 Nitrogenases in Cyanobacteria

Many cyanobacteria have developed differentiated cells, the heterocysts, to accommodate their nitrogenase. Heterocysts possess a thick cell wall composed of long chain, densely packed glycolipids that prevent the entry of air. Nitrogen and oxygen gas might therefore penetrate into heterocysts mainly by the plasma channels that connect vegetative cells with heterocysts. Protein might narrow these pores so that they are just permeable enough to supply sufficient N_2 (Walsby 2007). The high respiratory activity in heterocysts may consume all the O_2 coming in and so provide an anaerobic environment for nitrogenase catalysis.

The occurrence of non-specific intercellular channels between vegetative cells and heterocysts ("microplasmodesmata", Giddings and Staehelin 1978) has recently been confirmed (Mullineaux et al. 2008). The exchange of metabolites between both cell types might follow the source-sink gradient along these plasma bridges. Alternatively, the periplasmic space between the peptidoglycan layer and the outer membrane of these Gram-negative bacteria may serve as a continuous communication conduit between heterocysts and vegetative cells (Flores et al. 2006).

Differentiation of vegetative cells to heterocysts demands N-deprivation and occurs over a period of approximately 24 h in *Anabaena* sp. Differentiation is under control of the master genes *hetR* and *ntcA* and requires the degradation and synthesis of about 50% of all proteins. Most proteins of photosystem II and of the Calvin cycle are degraded. Thus heterocysts do not evolve O_2 photosynthetically and must be supplied with organic carbon from vegetative cells. They have a very active ferredoxin-dependent cyclic photophosphorylation (Bothe 1969) which might supply most of the ATP required for N_2-fixation. The reader is referred to details of heterocyst biochemistry and molecular biology in review articles on heterocyst differentiation (Golden and Yoon 2003; Haselkorn 2005).

Nitrogenase synthesis in *Anabaena* sp. requires a peculiar gene rearrangement prior to its start. The specific excisase (XisA) mediates the excision of an 11 kb element within *nifD* and the ligation of the resulting two fragments. The excisase gene *xisA* is located on the excised DNA. It may have been derived from an ancient virus that has come under control of the host. Similar gene rearrangements as a prerequisite to protein synthesis are found in a special ferredoxin (FdxN) and in HupL, encoding the larger subunit of the uptake hydrogenase of *Anabaena* 7120. Newer publications on the subject are available (Carrasco et al. 2005; Shah et al. 2007; Henson et al. 2008).

As regard to alternative nitrogenases (Fig. 6.1), the occurrence of the V-enzyme complex in *Anabaena variabilis* was first inferred from physiological experiments (Kentemich et al. 1988). The enzyme was then characterized in much detail (Thiel

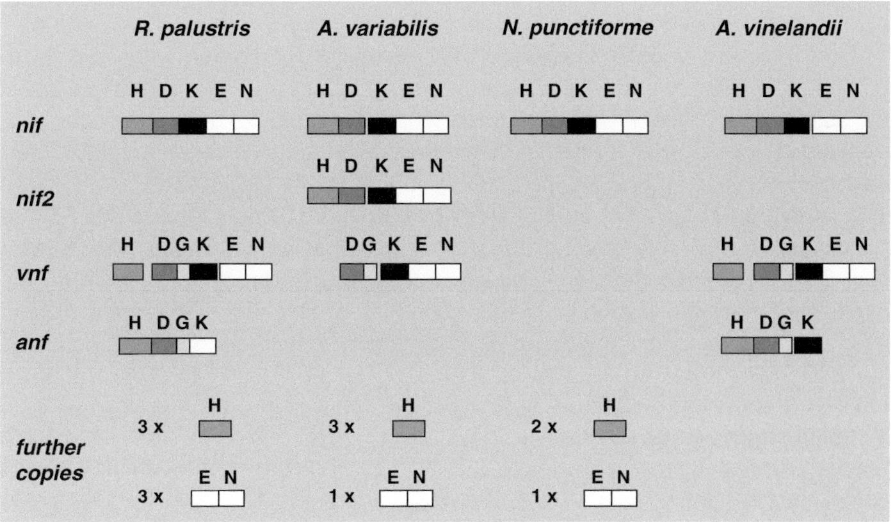

Fig. 6.1 Multiple nitrogenase gene sets in four bacteria. The structural genes *nif* genes *HDK* are in grey, the additional *G* gene of alternative nitrogenases in black. (From Thiel, see Thiel et al. (1998))

1993). The genes *vnfDGKEN* cluster whereas two other *H* genes, in addition to *nifH* are located elsewhere on the chromosome of *A. variabilis*. Thus it is not yet clear which one is *vnfH*. Two copies of *nifH* also occur in *Nostoc punctiforme* which has only the Mo-nitrogenase. V-nitrogenase has been found as yet only in *A. variabilis*, in an *Anabaena* isolate from the fern *Azolla* and recently in *Anabaena azotica* and *Anabaena* CH1, both from Southern Chinese rice fields (Boison et al. 2006). The ecological role of the cyanobacterial V-nitrogenase is not known. Both Mo- and V-nitrogenase are expressed to the same extent throughout various light and temperature regimes (Boison et al. 2006). The third nitrogenase, the Fe-only enzyme complex has not yet been detected in cyanobacteria. There was some physiological evidence for its occurrence in *A. variabilis* (Kentemich et al. 1991a). However, this observation has not been corroborated by any sequence data from cyanobacteria.

A peculiarity of cyanobacteria, though as yet only of *A. variabilis*, is the possession of a second Mo-nitrogenase (Schrautemeier et al. 1995; Thiel et al. 1995). One enzyme occurs in heterocysts, while a distinctly different nitrogenase is expressed in vegetative cells under anaerobic conditions or, better to say, under microaerobic conditions, since these cells produce O_2 photosynthetically. The nitrogenase in vegetative cells resembles the enzyme from non-heterocystous, filamentous cyanobacteria such as *Plectonema boryanum* (Stewart and Lex 1970).

Non-heterocystous species that have the capability to perform N_2-fxation are also known. Most of them do so under anaerobic or microaerobic conditions and separate in time the incompatible reactions of N_2-fixation during the night and photosynthetic O_2-evolution in light (Gallon 2001). However, *Goeothece* and *Synechococcus* perform both photosynthesis and N_2-fixation during the day and, in cultures under continuous illumination, during the whole day, albeit with lower rates than under the light-dark rhythm. The filamentous *Trichodesmium* appears to solve the issue by performing photosynthesis in some cells and N_2-fixation in others.

Until recently, major contributors of the oceanic N_2-fixation were thought to be *Trichodesmium* sp. and the heterocystous *Richelia intracellularis* living inside the diatom *Rhizoselenia*. However, microscopically small N_2-fixing cyanobacteria, such as *Crocosphaera watsonii*, and also other, uncultured and as yet non-characterized cells occur in abundance, particularly in the Pacific Ocean (Zehr et al. 2008). The nitrogenase sequences of the latter resemble those of the "spheroid bodies" of the diatoms *Rhopalodia gibba* and *Epithemia* sp. (Floener and Bothe 1980; Prechtl et al. 2004). The spheroid bodies and uncultured cyanobacteria could either perform cyclic photophosphorylation or could be completely dependent on a supply of both ATP and reductant from the ocean or host, respectively. Such N_2-fixing inclusion bodies inside eukaryotic cells might attract special attention in the near future for obvious reasons.

In all cyanobacteria, N_2-fxation is largely stimulated by light, and reductant to nitrogenase mainly comes from reduced ferredoxin (FdxH, Masepohl et al. 1997) generated via photosystem I (Fig. 6.2). Reducing equivalents for this may come from NAD(P)H and an NAD(P)H dehydrogenase or H_2 and uptake hydrogenase (see below) feeding in electrons at or close to the plastoquinone site. Alternatively, ferredoxin may be reduced in darkness, either by NAD(P)H or pyruvate (Fig. 6.2).

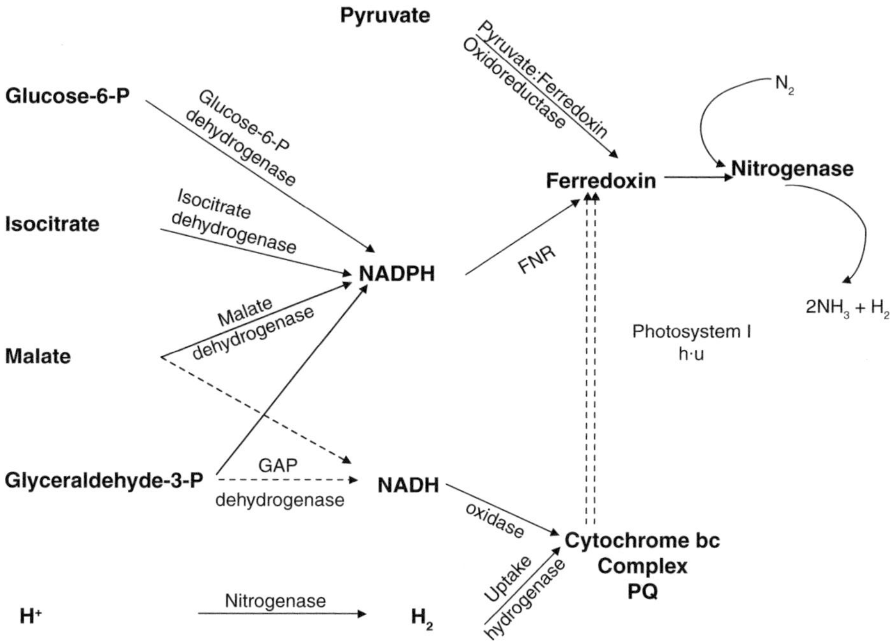

Fig. 6.2 Electron donation to nitrogenase in cyanobacteria

Cyanobacteria possess two isoforms of NAD(P)H:ferredoxin oxidoreductase (FNR, Thomas et al. 2006). One of them may be specifically involved in the reduction of ferredoxin by NAD(P)H. However this would require about a 1,000 fold excess of NAD(P)H over $NAD(P)^+$ since a reduction of ferredoxin ($E_0' \sim -420$ mV) is thermodynamically unfavourable (E_0' of $NAD(P)H/NAD(P)^+ = -320$ mV). In cyanobacteria, several catabolic enzymes such as isocitrate dehydrogenase, glyceraldehyde-3-phosphate dehydrogenase or malate dehydrogenase preferentially utilize $NADP^+$ as electron acceptor. Therefore NADPH but not NADH might serve to reduce ferredoxin by FNR.

Cyanobacteria can perform a pyruvate phosphoroclastic reaction where pyruvate, coenzyme A and oxidized ferredoxin are cleaved to acetyl-coenzyme A, CO_2 and reduced ferredoxin (Leach and Carr 1971; Neuer and Bothe 1982). The genome sequencing projects of *Anabaena* 7120 and other cyanobacteria revealed that heterocystous cyanobacteria contain two different pyruvate:ferredoxin oxidoreductases (PFOs) catalyzing this pyruvate cleavage. One PFO, in *Anabaena* 7120, is expressed only under Fe-deficiency (Bauer et al. 1993), whereas the other, as shown in *A. variabilis* (Schmitz et al. 1993), is constitutively expressed in Fe-replete media. Both PFOs show only a low sequence similarity of some 75%, whereas the sequence divergence of other genes from the two organisms is generally less than 5%. Remarkably, PFO from *A. variabilis* is constitutively expressed even in air grown cells but its sequence clusters with those from strict anaerobes such as *Clostridium pasteurianum* and *Desulfovibrio* sp. (Schmitz et al. 1993). The genome sequencing projects revealed that cyanobacteria also possess genes coding for phosphotransacetylase and acetate kinase. Thus acetyl-coenzyme A formed in the pyruvate cleavage could be converted to acetyl-phosphate and ATP. This has, however, never been demonstrated in cyanobacteria. Similarly, the fate of the acetate formed, if any, is unknown. On the other hand, the cleavage of pyruvate in darkness would provide ATP and reduced ferredoxin and would thus ideally meet the demands for nitrogenase function.

6.5 General Properties of Hydrogenases

Recent general reviews on hydrogenases are available (Vignais and Colbeau 2004; Vignais and Billoud 2007; Ghirardi et al. 2007). An [FeFe]-hydrogenase occurs in anaerobic bacteria and in green algae, but has not yet been detected in cyanobacteria. The two hydrogenases of cyanobacteria are Ni-containing enzymes which are differentiated and defined by their physiological roles. They function either in the consumption of H_2 ("uptake hydrogenase") or in both evolution and uptake of the gas ("bidirectional" or "reversible" hydrogenase). A regulatory hydrogenase encoded by *hupUV* and found in *Ralstonia eutropha* has not been described for cyanobacteria. Cyanobacterial hydrogenases also do not contain Se as in some [FeFe] hydrogenases of anaerobic bacteria. Detailed accounts of cyanobacterial hydrogenases are available (Houchins 1984; Appel and Schulz 1998; Madamwar et al. 2000; Tamagnini et al. 2002, 2007; Schütz et al. 2004; Tsygankov 2007; Ghirardi et al. 2007).

The Ni-hydrogenases are composed of two different subunits. The larger protein of about 60 kDa possesses the deeply buried binuclear NiFe active site of remarkable composition. The Fe in this centre binds two CN^- and one CO. This active centre is ligated to the protein by thiolate groups of four cysteines. The smaller subunit of some 30 kDa possesses up to three FeS centres which mediate the transfer of electrons to and from the active centre. Hydrophobic channels from the surface of the protein to the active site provide access for H^+ and egress for H_2. Ni-hydrogenases are often membrane-bound. They have a low K_M (high affinity) for H_2 indicating that they mainly function in H_2-uptake. Indeed, they are often functionally linked to nitrogenase and utilize the H_2-gas formed by N_2-fixation (Eqs. 6.1 and 6.2).

Ni-hydrogenases are often synthesized as longer peptides with some 30–50 amino acids that are excised upon incorporation into membranes. They feed electrons into the respiratory chain to ubi-(mena-)quinone or a cytochrome b at the respiratory complex III. They can be divided into several classes according to their function (Vignais et al. 2001; Vignais and Colbeau 2004). In the oxidized form, [NiFe]-hydrogenases are inactive due to a bridging hydroxo-ligand between the Ni- and Fe-atoms. This ligand is removed by reduction to water and the simultaneous reduction of Ni^{3+} to Ni^{2+}. Then the enzyme can bind H_2 likely at the Fe-atom and can heterolytically cleave it to $2H^+$ and $2e^-$. The details of the reaction mechanism have been described (Vignais and Colbeau 2004). The Ni-hydrogenases differ in their sensitivity to O_2. None of them couple with ferredoxin or with any other electron carrier with a midpoint potential of less than −300 mV. The biosynthesis of the NiFe cluster including the incorporation of the ligands CO and CN^- has been elucidated by A. Böck and coworkers in Munich and has been reviewed (Vignais and Colbeau 2004; Böck et al. 2006; Vignais and Billoud 2007).

6.6 Uptake Hydrogenase of Cyanobacteria

This enzyme is encoded by the cotranscribed genes *hupSL*. Transcription starts before *hupS* and terminates behind *hupL*. Thus the electron acceptor is independently transcribed and is not known. The enzyme has never been purified to homogeneity or biochemically characterized in detail. Cyanobacterial uptake hydrogenase is bound to the thylakoids, to the cytoplasmic membrane or both. H_2-uptake by this enzyme proceeds either by photosystem I or the respiratory pathway (Bothe and Neuer 1988; Fig. 6.2). The cytochrome bc complex in cyanobacteria is shared both by the photosynthetic and respiratory electron transport. After entry of the electrons from H_2 to the electron transport chain at the plastoquinone pool or close to the plastoquinone site and passage through the cytochrome bc complex, the electrons are allocated either to photosystem I or to respiration under O_2-uptake. Factors that regulate the electron transfer to either photosystem I or respiration in light have not yet been elucidated.

H_2-utilization by uptake hydrogenase may have several functions as first stated for *Rhizobium* (Dixon 1972): (a) it minimizes the loss of ATP in H_2-formation by nitrogenase, since ATP is regained in respiration with H_2 as electron donor; (b) it removes O_2 by the respiratory "Knallgas" reaction and thereby protects nitrogenase against damage (H_2-dependent O_2-uptake might significantly contribute to provide an anaerobic environment within heterocysts); (c) it prevents a deleterious build-up of a high concentration of H_2 in cells such as heterocysts (H_2 in high concentrations affects N_2-fixation); (d) it provides additional reductant to photosynthetic and respiratory processes.

Uptake hydrogenase-deficient mutants produce about three times more H_2 than wild-type cells (see Tamagnini et al. 2007; Weyman et al. 2008). Growth rates of the mutants and the wild type are, however, essentially the same (Lindblad et al. 2002). The effect of H_2 on N_2-fixation can easily be seen in C_2H_2-reduction experiments with intact *A. variabilis* (Bothe et al. 1978, 2008). The reaction is distinctly less sensitive to O_2 when the vessels with this cyanobacterium are supplemented with H_2.

Thus H_2-uptake is closely linked with nitrogenase functioning, and uptake hydrogenase appears to be present in all N_2-fixing cyanobacteria (Tamagnini et al. 2000). However this hydrogenase is not detected in the N_2-fixing unicellular *Synechococcus* sp. BG 043511 (Ludwig et al. 2006) nor in some *Chroococcidiopsis* isolates (see below). The enzyme has also been reported in ammonia-grown *A. variabilis* which does not perform N_2-fixation under these conditions (Weyman et al. 2008). It may also be expressed in parallel with the second Mo-nitrogenase in vegetative cells of *A. variabilis*; a possibility which still needs to be examined.

Uptake hydrogenase transcripts do not show typical motifs for membrane insertion. However, HupL of cyanobacteria as that of other organisms contains a C-terminal extension. At the last step of maturation this extension is cleaved by a specific endopeptidase encoded by *hupW*. The *hupL* gene of *Anabaena* 7120 and of approximately half of the other heterocystous strains undergoes a rearrangement during the late state of heterocyst differentiation which is a prerequisite for its expression. A 9.5 kb element is excised in *hupL* by a specific recombinase XisC with its gene located on the excised DNA (Carrasco et al. 2005). The physiological role of this excision is not obvious.

Uptake hydrogenase synthesis seems to be regulated at the transcriptional and translational levels. General factors are Ni-availability, anaerobiosis, H_2-presence, absence of combined nitrogen and thus differentiation to heterocysts. However, the factors have not always been unambiguously identified and seem to vary from one cyanobacterium to the next. The general transcriptional regulator of nitrogen metabolism NtcA also regulates *hupSL* expression (Weyman et al. 2008; Holmqvist et al. 2009). However, the NtcA binding site upstream of the transcriptional start site has not been unambiguously identified as yet. A promoter fragment, covering 57 bp upstream and 258 bp downstream of the transcriptional start site, suffices for a high and heterocyst-specific expression of *hupSL* independently of NtcA (Holmqvist et al. 2009). At the translational level, uptake hydrogenase is activated by reduced thioredoxin (Papen et al. 1986). Thioredoxin is reduced in light, and H_2-production by nitrogenase is also largely light-stimulated. Thus it makes sense to activate uptake hydrogenase under these conditions.

6.7 Bidirectional Hydrogenase of Cyanobacteria

Cyanobacteria may contain a second hydrogenase, the bidirectional or reversible enzyme, which is constitutively expressed, even under aerobic growth conditions and independently of N_2-fixation (Houchins 1984). Molecular analysis provided the unexpected information that it is a pentameric enzyme encoded by *hoxEFUYH* (Schmitz et al. 1995). It is related to the soluble, NAD^+-reducing hydrogenase of *Ralstonia eutropha* (Friedrich and Schwartz 1993). Extracts from cyanobacteria catalyze both a NAD(P)H-dependent H_2-evolution and a reduction of $NAD(P)^+$ with H_2 as electron donor (Schmitz and Bothe 1996). HoxYH constitutes the hydrogenase proper part, which contains the motifs for binding the Ni-Fe-S and Fe-S centres. HoxFU comprises the diaphorase part which serves in transfer electrons from H_2 to $NAD(P)^+$ and contains the binding sites for $NAD(P)^+$, FMN and Fe-S centres. The enzyme contains another subunit, HoxE, which copurifies in the other four HoxFUYH proteins (Schmitz et al. 2002) but which is not present with the *Ralstonia eutropha* enzyme. The *hoxE* gene product possesses the motif for binding a [4Fe-4S] cluster and may therefore be involved in binding the enzyme to thylakoid or cytoplasmic membranes, a possibility which remains to be proven. The function of HoxE, present in organisms other than cyanobacteria, such as *Thiocapsa roseopersicina* and *Allochromatium vinosum*, is also not yet resolved.

The bidirectional hydrogenase occurs in unicellular, filamentous and heterocystous cyanobacteria and is thus widespread. It is always encoded by the five structural genes *hoxEFUYH* which, however, may be contiguous, interspersed with ORFs or partly separated from each other by several kb, depending on the cyanobacterium (Boison et al. 1998; Tamagnini et al. 2007). The enzyme is apparently not present in some marine cyanobacteria of the open ocean (Ludwig et al. 2006). Bidirectional hydrogenase seems to be more O_2-sensitive than the uptake enzyme (Cournac et al. 2004). When reduced, it can, however, be purified without major problems. It might function as a dimeric assembly complex $Hox(EFUYH)_2$ (Schmitz et al. 2002).

The expression of the bidirectional hydrogenase genes is under control of the circadian clock, as shown for two promoters of the gene cluster in *Synechoccocus* sp. PCC 7942 (Schmitz et al. 2001a). HoxH undergoes maturation at the C-terminus catalyzed by a specific endopeptidase encoded by *hoxW* (Tamagnini et al. 2007). The regulation of the expression of bidirectional hydrogenase is different from one cyanobacterium to the next and is dependent on the location of the *hox* genes on the chromosome. In *Synechococcus* sp. PCC 7942 the genes are separated into the two clusters hoxEF and *hoxUYHWhypAB* and three promoters, located before *hoxE*, *hoxU* and *hoxW*, were identified (Boison et al. 2000; Schmitz et al. 2001a). The genes *hoxEFUYH* of *Synechocystis* PCC 6803 are cotranscribed with the transcription start point at 168 bp upstream from the start codon (Oliveira and Lindblad 2005; Gutekunst et al. 2005). These two investigations also showed that expression is under control of the activator protein LexA. The binding site of LexA has not been finally resolved; it may bind either to a region from −198 to −338 bp from the translational start point (Oliveira and Lindblad 2005) or may interact in the region −592 to −690 from the *hoxE* start codon (Gutekunst et al. 2005).

The location of this soluble enzyme in the cyanobacterial cells is not known. Immunological studies were performed with antibodies before its complex structure as pentameric enzyme was recognized. New investigations with freshly prepared antibodies are timely. Solubilization studies (Kentemich et al. 1991b) seemed to indicate an association of the bidirectional hydrogenase to the cytoplasmic membrane (Fig. 6.3).

The hyp genes required for the biosynthesis of hydrogenase 3 of *E. coli* have been elucidated (Böck et al. 2006). Since homologous hyp genes occur in cyanobacteria, biosynthesis of both enzymes in these organisms might be similar to *E. coli.*

The physiological function of bidirectional hydrogenase has been much debated in the past. The high affinity (low K_M) of the enzyme for H_2 tends to suggest that the enzyme functions in H_2-consumption. H_2 and hydrogenase can support reductive processes such as CO_2-fixation, nitrate- or sulphate-reduction, albeit with marginal rates compared with the activities when H_2O is the photosynthetic electron donor (see Bothe and Neuer 1988). Also when does the aerobically expressed bidirectional hydrogenase come into contact with H_2 in non-N_2-fixing cyanobacteria? Some authors consider the enzyme to function as a valve to remove excess reductant

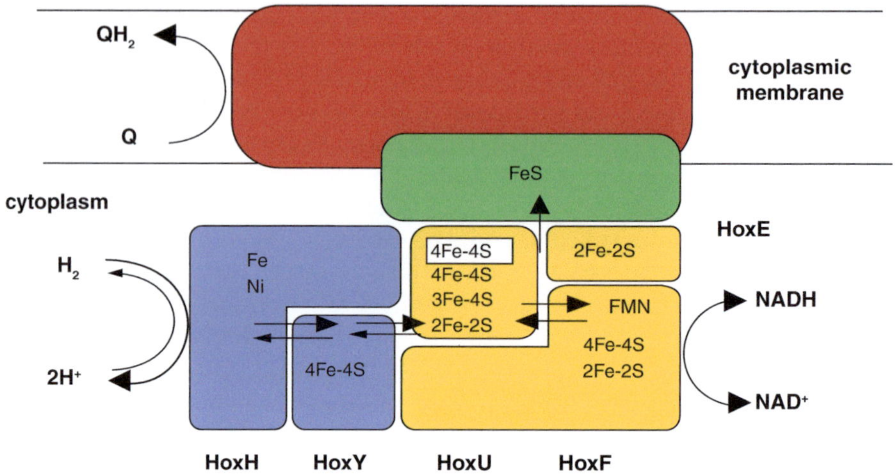

Fig. 6.3 Composition and possible coupling of the bidirectional hydrogenase to the cytoplasmic membrane. Solubilization studies suggested that the enzyme is bound to the cytoplasmic membrane. (Kentemich et al. 1991b)

(Appel et al. 2000; Cournac et al. 2004; Ludwig et al. 2006). Several authors noted a burst of H_2 lasting at best a few minutes when cyanobacteria are suddenly exposed to high light intensities. However, this might be a physiological artefact of laboratory experiments, since the sun does not rise so suddenly in the morning that soil cyanobacteria are suddenly faced with the problem of too much light, particularly since most of them are protected by a slime layer. In water, turbulences are hardly so violent that cyanobacteria have to face sudden exposure to toxic light intensities. Cyanobacteria may, however, be forced to continuously produce H_2 on a very bright summer's day when exposed to high light intensities over longer periods, as first experimentally documented by Laczkó (1986). Then H_2 may be produced via photosynthetic electron transport, reduced ferredoxin, NADPH:ferredoxin oxidoreductase (FNR), NADPH and bidirectional hydrogenase.

More likely, the bidirectional hydrogenase functions in darkness and under anaerobic conditions (Fig. 6.4). Several authors suggested a function for the bidirectional hydrogenase in cyanobacterial fermentation (Van der Oost et al. 1989; Stal and Mozelaar 1997). However, as stated in an extensive review (Stal and Mozelaar 1997), most cyanobacteria are obligate autotrophs and have only a limited capability to consume organic carbon. Cyanobacteria accumulate glycogen in the light which is degraded via the oxidative pentose phosphate pathway to pyruvate (Ernst et al. 1990). Their tricarboxylic acid cycle is incomplete, since an enzyme catalyzing the degradation of oxoglutarate (either oxoglutarate dehydrogenase or an oxoglutarate:ferredoxin oxidoreductase) is missing. Several catabolic enzymes, such as isocitrate dehydrogenase, glyceraldehyde-3-phosphate-dehydrogenase or malate dehydrogenase, are mainly $NADP^+$-dependent. Many cyanobacteria form dense cultures in mats or biofilms where they may rapidly become O_2-limited. In

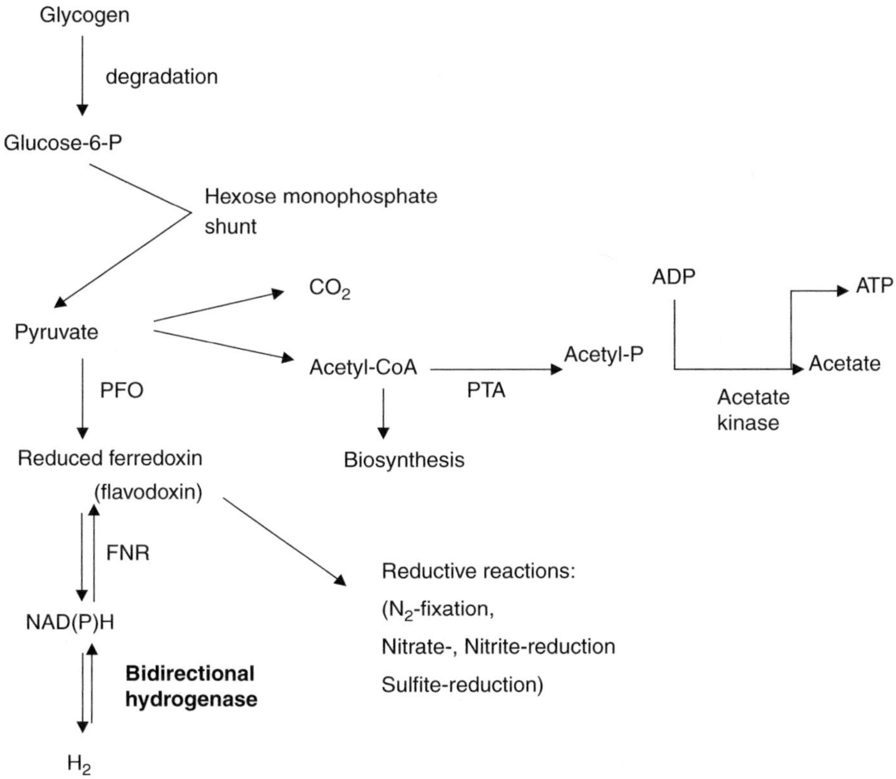

Fig. 6.4 The degradation of carbohydrate reserves under anaerobic conditions and in darkness. The bidirectional hydrogenase then functions to dispose the reductants generated. (see text)

darkness, they may be forced to degrade the pyruvate formed via pyruvate:ferredoxin oxidoreductase, reduced ferredoxin, FNR, NADPH and bidirectional hydrogenase. H_2 must then be excreted, since O_2 is not available for the respiratory utilization of NADPH. In line with such suggestions is the observation that both bidirectional hydrogenase and pyruvate:ferredoxin oxidoreductase are constitutively expressed under all conditions (Schmitz et al. 2001b) and that two isoforms of FNR occur in cyanobacteria (Thomas et al. 2006). When cyanobacteria, such as *Anabaena variabilis*, *Synechocystis* PCC 6803 or *Synechoccocus* sp. PCC 7942, are transferred to darkness and anaerobic conditions, a distinct H_2-production commences without any lag-phase, and this H_2-formation stops when the assays are supplemented with O_2 (early and ongoing observations from the Cologne and other laboratories, for example, Benemann and Weare 1974). Thus the bidirectional hydrogenase must have an essential function in darkness and under O_2-limitation to avoid over-reduction of the cells. Then a more than 100-fold excess of NAD(P)H over $NAD(P)^+$ may rapidly be achieved to allow a reduction of ferredoxin by NAD(P)H and production of H_2 to proceed.

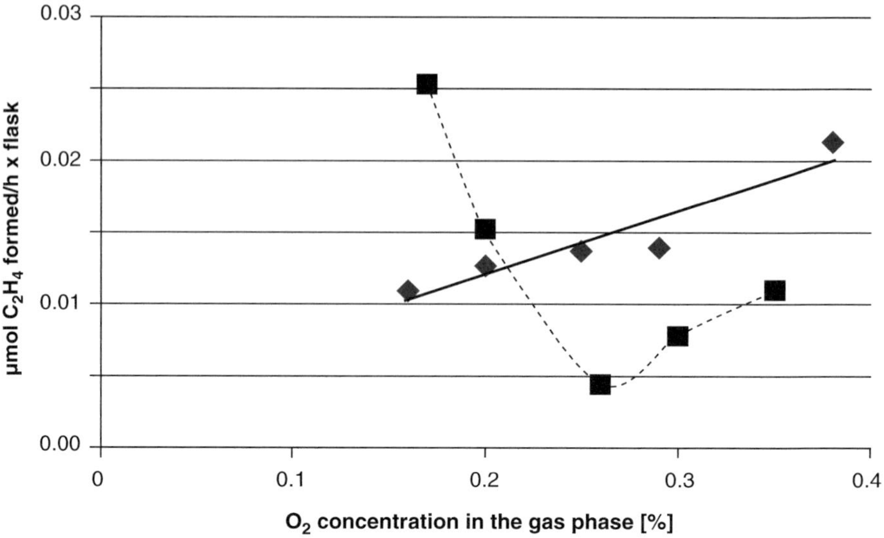

Fig. 6.5 N_2-fixation (C_2H_2-reduction) by *Chroococcidiopsis thermalis* strain PCC 7203 in dependence of the O_2-concentration in the gas phase. A slight positive effect of H_2 on C_2H_2-reduction at a concentration of 0.2–0.5% O_2 in the gas phase was detected in many, but in less than 50% of the 130 different experiments performed on the subject. O_2 was always completely inhibitory above 0.5% in the headspace of the vessels. The positive effect was dependent on the age of the cultures, their photosynthetic activity, their density in the assay vessels and the light intensity applied to the cells. The experiments were performed in 7 ml Fernbach flasks on a horizontal rotary shaker for 5 h with *C. thermalis* cells (about 0.02 mg chlorophyll/vessel) grown in BG11 medium and preincubated under H_2 overnight before the start of the experiment. The low light intensity at the vessels of 30 µE/m^2 s was optimal for C_2H_2-reduction by *C. thermalis*. *Straight line*: presence of 15% H_2 in the vessels; *dashed line* without H_2

In ancient times, the bidirectional hydrogenase may have had an additional role. Such speculation is taken from observations of the current cyanobacterium *Chroococcidiopsis* which is regarded as a fossil record (Fewer et al. 2002). *Chroococcidiopsis* is discussed as the best suited organism to go on exploratory missions to Mars (Cockell et al. 2005). Nowadays it only thrives in extreme habitats where competition between organisms is low (Boison et al. 2004). Some *Chroococcidiopsis* strains, such as *C. thermalis* PCC 7203, do not possess an uptake hydrogenase, only the bidirectional enzyme. Unpublished experiments from the Cologne laboratory showed that H_2 and bidirectional hydrogenase can support light-dependent C_2H_2-reduction, but only at the very low O_2-concentration of 0.1–0.3% in the gas phase (Fig. 6.5). Above that concentration, O_2 blocks this activity, presumably by oxidizing hydrogenase to the inactive form and/or by inhibiting an extremely sensitive nitrogenase. Since the cells evolve O_2 photosynthetically and since the range of the O_2-concentration is so narrow, problems were encountered in getting reproducible positive effects in the many experiments performed. Anyway, in early earth conditions of low O_2, H_2 and bidirectional hydrogenase may have served to protect nitrogenase against O_2-damage. Later, at higher O_2-concentrations, more effective protective devices such heterocysts may have developed. Indeed, *Chroococcidiopsis* is regarded as a progenitor of heterocystous cyanobacteria (Fewer et al. 2002).

6.8 The Use of Cyanobacterial Cells for the Generation of Molecular Hydrogen as a Source for New and Clean Energy

This topic has been of continuous interest for years, at least from the first energy crisis in 1973, though its support by the funding agencies has been cyclic. Currently there is a renewed interest, since photosynthetic H_2-production is not accompanied by the formation of greenhouse gases as by-products. Cyanobacteria have the simplest nutrient demands of all organisms in nature. They thrive on inorganic media photosynthetically and many of them can even meet their N-demands from N_2. After ceasing to produce H_2, the cells may be utilized either to supplement cattle food, though only at 5% maximally, or as biofertilizers.

Many attempts to exploit cyanobacteria for the generation of H_2 have been made and, mainly depending on their own research interests, authors suggest using either hydrogenases or nitrogenases. With regard to hydrogenases, both cyanobacterial enzymes do not couple with ferredoxin, as said above. Therefore the clostridial ferredoxin-dependent hydrogenase was heterologously expressed in the unicellular cyanobacterium *Synechoccocus* PCC 7942 which resulted in a three-fold enhancement of hydrogenase activity compared to the wild-type situation (Asada et al. 2000). Another attempt was to genetically engineer photosystem I PsaE subunit so that it could couple with the O_2-insensitive, membrane-bound hydrogenase of *Ralstonia eutropha* (Ihara et al. 2006a, b). The published data seemed to indicate that the increase in photosystem I-dependent H_2-production was quite low in this approach.

Upon sudden exposure to high light intensities, hydrogenase catalyzes an outburst of H_2 (Cournac et al. 2004), but this burst can hardly be expected to be sustained. Alternatively, cyanobacteria were suggested to form photosynthate in light and this could subsequently be degraded in darkness by the use of hydrogenase (Ananyev et al. 2008). An alternative approach could be to degrade cyanobacterial photosynthate by a mixed culture with fermentative bacteria, since H_2-formation rates by clostridia, for example, are at least three orders of magnitude higher than in cyanobacteria.

Cyanobacterial nitrogenases are ferredoxin-dependent and thus couple directly with photosystem I. Recent experiments indicate that all photosynthetically generated reductant can be exploited for H_2-production by nitrogenase in *Anabaena variabilis* or *A. azotica* (Bothe et al. 2008). In the presence of high concentrations of C_2H_2 and H_2 these cyanobacteria produce large amounts of H_2 anaerobically on top

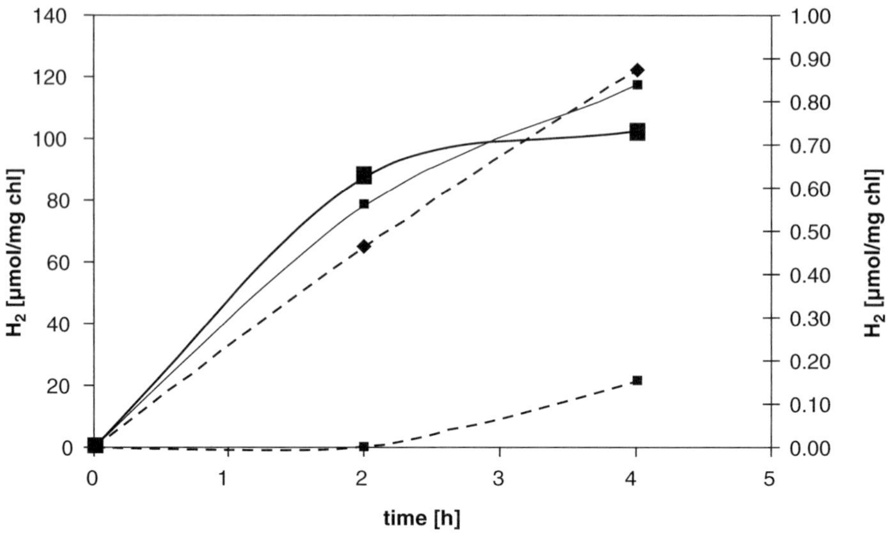

Fig. 6.6 Kinetics of H_2-formation by V-grown *Anabaena azotica*. The upper two curves (*full lines, thick (upper) line* Mo-grown cells, *thin (lower) line* V-grown cells) represent the data obtained in the vessels supplemented with 15% H_2, the lower two, dashed curves are obtained from the vessels without H_2 (*upper dashed line*: V-grown cells, *lower dashed line*: Mo-grown cells). Note the difference in scale. The experiments were performed in 7 ml Fernbach flasks with 3 ml *A. azotica* cells (total chlorophyll content 0.035 ml at 20°C and a light intensity at the vessels of about 300 µE/m^2 s. The gas phase was argon supplemented with about 15% C_2H_2 and about 15% H_2. (For further details see Bothe et al. 2008)

of the H_2 added to the vessels (Fig. 6.6). Under the same conditions, C_2H_2-reduction is almost totally blocked so that all electrons reaching nitrogenase are utilized to produce H_2. The effect cannot be mechanistically explained as yet. *Nostoc muscorum* incubated under argon and H_2 was also described to produce H_2, although not with such high activities (Scherer et al. 1980). The rates of H_2-production are higher in V- than in Mo-grown cultures (Fig. 6.6) and amount to ~40 µmol H_2 formed/per h×mg chlorophyll (Bothe et al. 2008). This is about the maximal rate of N_2-fixation that can be achieved with filamentous, heterocystous cyanobacteria. Their CO_2-fixation activities amount to roughly 100 µmol/h×mg chlorophyll. Since the C/N ratio in cells is about 6, maximal N_2-fixation is unlikely to exceed 20 µmol/h×mg chlorophyll. If all electrons transferred to nitrogenase were re-allocated to reduce H^+, H_2-production by cyanobacteria would be around 40 µmol H_2 produced/h×mg chlorophyll based on the fact that $4e^-$ are needed for NH_4^+-production but only $2e^-$ for H_2-formation.

Published data indicate that maximal rates of H_2-production of ~40 µmol/h×mg chlorophyll are already achieved with both nitrogenase (Markov et al. 1995; Bothe et al. 2008) and hydrogenase (Asada et al. 2000; Laczkó 1986). Thus it does not appear to be a realistic goal to enhance cyanobacterial H_2-production significantly by genetically engineering a more effective hydrogenase or nitrogenase into cyanobacteria. A better chance to improve H_2-production is to enhance the photosynthetic electron flow for the generation of reductant as outlined years ago (Hall et al. 1995). The exploitation of the incident light by cyanobacteria is only 1–2% mainly due to self-shadowing, slime formations and other factors in the cells. Immobilization of the cyanobacterial filaments by absorption on solid matrices or by entrapment in gels may result in a better exploitation of light, may enhance the life time of the cells and may also increase the number of heterocysts in the filaments (Hall et al. 1995). Such approaches may result in long lasting H_2-production. Heterocyst frequencies in the filaments may also be enhanced by mutagenesis (Meeks and Elhai 2002) or treatment with chemicals such as 7-azatryptophan (Bothe and Eisbrenner 1977).

In intact cells, the H_2 produced by nitrogenase is immediately recycled—more or less completely—by hydrogenase so that no net H_2-production is observable. Uptake hydrogenase but not the bidirectional enzyme is effective in re-cycling the gas (Masukawa et al. 2002). Hydrogenase-negative mutants obtained by classical NTG-treatment (Mikheeva et al. 1995) or by site directed insertions (Masukawa et al. 2002; Happe et al. 2000b) show significantly higher H_2-formation than the wild-type.

In the long run, with the exhaustion of fossil energy, biological systems such as cyanobacterial H_2-production may offer realistic prospects as an alternative and environmentally friendly source of energy production.

References

Ananyev G, Carrieri D and Dismukes CG (2008) Optimization of metabolic capacity and flux through environmental cues to maximize hydrogen production by the cyanobacterium *Arthospira (Spirulina) maxima*. Appl Environ Microbiol 74: 6102–6113

Appel J and Schulz R (1998) Hydrogen metabolism in organisms with oxygenic photosynthesis: Hydrogenase as important regulatory devices for a proper redox poising? J Photochem Photobiol Biol 47: 1–11

Appel JSP, Steinmüller K and Schulz R (2000) The bidirectional hydrogenase of *Synechocystis* PCC 6803 works as an electron valve during photosynthesis. Arch Microbiol 173: 333–338

Arnold W, Rump A, Klipp W, Priefer UB and Pühler A (1988) Nucleotide sequence of a 24,206-base-pair DNA fragment carrying the entire nitrogen fixation gene cluster of *Klebsiella pneumoniae*. J Mol Biol 203: 715–738

Asada Y, Koike Y, Schnackenberg J, Miyake M, Uemura I and Miyake J (2000) Heterologous expression of clostridial hydrogenase in the cyanobacterium *Synechococcus* PCC7942. Bioch Biophys Acta 1490: 269–278

Bauer CC, Scappino L and Haselkorn R (1993) Growth of the cyanobacterium *Anabaena* on molecular nitrogen: NifJ is required when iron is limited. Proc Natl Acad Sci U S A 90: 8812–8816

Benemann JR and Weare NM (1974) Hydrogen evolution by nitrogen fixing *Anabaena cylindrical* cultures. Science 129: 174–175

Betancourt DA, Loveless TM, Brown J and Bishop PE (2008) Characterization of diazotrophs containing Mo-independent nitrogenases, isolated from diverse natural environments. Appl Environ Microbiol 74: 3471–3480

Bishop P and Joerger RD (1990) Genetics and molecular biology of alternative nitrogen fixing systems. Ann Rev Plant Physiol Plant Mol Biol 41: 109–125

Böck A, King PW, Blokesch M and Posewitz MC (2006) Maturation of hydrogenases. Adv Microbiol Physiol 51: 1–71

Boison G, Schmitz O, Schmitz B and Bothe H (1998) Unusual gene arrangement of the bidirectional hydrogenase and functional analysis of its diaphorase subunit HoxU in respiration of the unicellular cyanobacterium *Anacystis nidulans*. Curr Microbiol 36: 253–258

Boison G, Bothe H and Schmitz O (2000) Transcriptional analysis of hydrogenase genes in the cyanobacteria *Anacystis nidulans* and *Anabaena variabilis* monitored by RT-PCR. Curr Microbiol 40: 315–321

Boison G, Mergel A, Jolkver H and Bothe H (2004) Bacterial life and dinitrogen fixation at a gypsum rock. Appl Environ Microbiol 70: 7070–7077

Boison G, Steingen C, Stal LJ and Bothe H (2006) The rice field cyanobacteria *Anabaena azotica* and *Anabaena* sp.CH1 express vanadium-dependent nitrogenase. Archiv Microbiol 186: 367–376

Bothe H (1969) Ferredoxin als Kofaktor der cylischen Photophosphorylierung in einem zellfreien System aus der Blaualge *Anacystis nidulans*. Z Naturforsch 24b: 1574–1582

Bothe H and Eisbrenner G (1977) Effect of 7-azatryptophan on nitrogen fixation and heterocyst formation in the blue-green alga *Anabaena cylindrica*. Biochem Physiol Pflanz 133: 323–332

Bothe H and Neuer G (1988) Electron donation to nitrogenase in heterocysts. Methods Enzym 167: 496–501

Bothe H, Distler E and Eisbrenner G (1978) Hydrogen metabolism in blue-green algae. Biochimie 60: 277–289

Bothe H, Winkelmann S and Boison G (2008) Maximizing hydrogen production by cyanobacteria. Z Naturforsch 63c: 226–232

Carrasco CD, Holliday SD, Hansel A, Lindblad P and Golden JW (2005) Heterocyst-specific excision of the *Anabaena* strain PCC 7120 *hupL* element requires *xisC*. J Bacteriol 187: 6031–6038

Chien YT, Auerbruch V, Brabban AD and Zinder SH (2000) Analysis of genes encoding an alternative nitrogenase in the archaeon *Methanosarcina barkeri* 227. J Bacteriol 182: 3247–3253

Cockell CS, Schuerger AC, Billi D, Friedmann EI and Panitz C (2005) Effects of a simulated martian UV flux on the cyanobacterium, *Chroococcidiopsis* sp. 029. Astrobiol 5: 127–140

Cournac L, Guedeney G, Peltier G and Vignais PM (2004) Sustained photoevolution of molecular hydrogen in a mutant of *Synechocystis* sp strain PCC 6803 deficient in the type I NADH-dehydrogenase complex. J Bacteriol 186: 1737–1746

Dixon ROD (1972) Hydrogenase in legume root nodule bacteroids, occurrence and properties. Archiv Microbiol 85: 193–201

Eady RR (1996) Structure-function relationship of alternative nitrogenase. Chem Rev 96: 3013–3030

Eilmus S, Rösch C and Bothe H (2007) Prokaryotic life in a potash-polluted marsh with emphasis on N-metabolizing microorganisms. Environ Poll 146: 478–491

Ernst A, Reich S and Böger P (1990) Modification of dinitrogen reductase in the cyanobacterium *Anabaena variabilis* due to C-starvation and ammonia. J Bacteriol 172: 748–755

Fewer D, Friedl T and Büdel B (2002) *Chroococcidiopsis* and heterocysts-differentiating cyanobacteria are each other's closest living relatives. Mol Phylogenet Evol 23: 82–90

Fisher K, Lowe, DJ, Tavares P, Pereira AS, Hynh BH, Edmondson D and Newton WE (2007) Conformations generated during turnover of the *Azotobacter vinelandii* MoFe protein and their relationship to physiological function. J Inorg Biochem 101: 1649–1656

Floener L and Bothe H (1980) Nitrogen fixation in *Rhopalodia gibba*, a diatom containing blue-greenish inclusions symbiotically. In: Schwemmler W, Schenk HEA (eds) Endocytobiology, Endosymbiosis and Cell Biology, pp 541–552, Walter de Gruyter & Co, Berlin

Flores E, Herrero A, Wolk CP and Maldener I (2006) Is the periplasm continuous in filamentous cyanobacteria? Trends Microbiol 14: 439–443

Friedrich B and Schwartz E (1993) Molecular biology of hydrogen utilization in aerobic chemolithotrophs. Annu Rev Microbiol 47: 351–383

Gallon JR (2001) N_2 fixation in phototrophs. Adaptation to a specialized way of life. Plant Soil 239: 39–48

Ghirardi ML, Posewitz MC, Maness PC, Dubini A, Yu J and Seibert M (2007) Hydrogenases and hydrogen photoproduction in oxygenic photosynthetic organisms. Annu Rev Plant Biol 58: 71–91

Giddings JW and Staehelin LA (1978) Plasma membrane architecture of *Anabaena cylindrica*: occurrence of microplasmodesmata and changes associated with heterocyst development and the cell cycle. Eur J Cell Biol 16: 235–249

Golden JW and Yoon HS (2003) Heterocyst development in *Anabaena*. Curr Opin Microbiol 6: 557–563

Gutekunst K, Phunpruch S, Schwarz C, Schuchardt S, Schulz-Friedrich R and Appel J (2005) LexA regulates the bidirectional hydrogenase in the cyanobacterium *Synechocystis* sp. PCC 6803. Mol Microbiol 58: 810–823

Hall DO, Markov SA, Watanabe Y and Rao KK (1995) The potential applications of cyanobacterial photosysthesis for clean technologies. Photosynth Res 46: 159–167

Happe T, Schütz K and Böhme H (2000b) Transcriptional and mutational analysis of the uptake hydrogenase of the filamentous cyanobacterium *Anabaena variabilis* ATCC 29413. J Bacteriol 182: 1624–1631

Haselkorn R (2005) Heterocyst differentiation and nitrogen fixation in *Anabaena*. In: Wang YP, Lin M, Tian ZX, Elmerich C and Newton WE (eds) Biological nitrogen fixation, sustainable agriculture and the environment, Proceedings of the fourteenth international nitrogen fixation congress, pp 65–68, Springer, Dordrecht

Henson BJ, Pennington LE, Watson LE and Barnum SR (2008) Excision of the *nifD* element in heterocystous cyanobacteria. Arch Microbiol 189: 357–366

Holmqvist M, Stensjö K, Oliveira P, Lindberg P and Lindblad P (2009) Characterization of the *hupLS* promoter activity in *Nostoc punctiforme* ATCC 29133. BMC Microbiol 9:54

Houchins JP (1984) The physiology and biochemistry of hydrogen metabolism in cyanobacteria. Biochim Biophys Acta 768: 227–255

Ihara M (2006a) Light-driven production by a hybrid complex of a [NiFe]-Hydrogenase and the cyanobacterial photosystem I. Photochem Photobiol 82: 676–682

Ihara M, Nakamoto H, Kamachi T, Okura I and Maeda M (2006b) Light-driven hydrogen production by a hybrid complex of a [NiFe]-hydrogenase and the cyanobacterial photosystem I. Photochem Photobiol 82: 1677–1685

Kentemich T, Danneberg G, Hundeshagen B and Bothe H (1988) Evidence for the occurrence of the alternative, vanadium-containing nitrogenase in the cyanobacterium *Anabaena variabilis*. FEMS Microbiol Lett 51: 19–24

Kentemich T, Haverkamp G and Bothe H (1991a) The expression of a third nitrogenase in the cyanobacterium *Anabaena variabilis*. Z Naturforsch 46c: 217–222

Kentemich T, Casper M and Bothe H (1991b) The reversible hydrogenase in Anacystis nidulans is a component of the cytoplasmic membrane. Naturwissenschaften 78: 559–560

Kessler PS, McLarnan J and Leigh JA (1997) Nitrogenase phylogeny and the molybdenum dependence of nitrogen fixation in *Methanococcus maripaludis*. J Bacteriol 179: 541–543

Laczkó I (1986) Appearance of a reversible hydrogenase activity in *Anabaena cylindrica* grown in high light. Physiol Plant 67: 634–637

Leach CK and Carr NG (1971) Pyruvate:ferredoxin oxidoreductase and its activation by ATP in the blue-green alga *Anabaena variabilis*. Biochim Biophys Acta 245: 165–174

Lindblad P, Christensson K, Lindberg P, Federov A, Pinto G and Tsygankov A (2002) Photoproduction of H_2 by wildtype *Anabaena* PCC 7120 and a hydrogenase deficient mutant: from laboratory experiments to outdoor culture. Int J Hydrogen Energy 27: 1271–1281

Loveless TM, Saah JR and Bishop PE (1999) Isolation of nitrogen-fixing bacteria containing molybdenum—independent nitrogenases from natural environments. Appl Environ Microbiol 65: 4223–4225

Ludwig M, Schulz-Friedrich R and Appel J (2006) Occurrence of hydrogenases in cyanobacteria and anoxygenic photosynthetic bacteria: implications for the phylogenetic origin of cyanobacterial and algal hydrogenases. J Mol Evol 63: 758–768

Madamwar D, Garg N and Shah V (2000) Cyanobacterial hydrogen production. World J Microbiol Biotechnol 16: 757–767
Markov SA, Lichtl R, Bazin MJ and Hall DO (1995) Hydrogen production and carbon dioxide uptake by immobilised *Anabaena variabilis* in a hollow fibre photobioreactor. Enzyme Microb Biotechnol 17: 306–310
Masepohl B, Scholisch K, Görlitz K, Kiutski C and Böhme H (1997) The heterocyst-specific *fdxH* gene product of the cyanobacterium *Anabaena* sp. PCC 7120 is important but not essential for nitrogen fixation. Mol Gen Genet 253: 770–776
Masukawa H, Mochimaru M and Sakurai H (2002) Disruption of the uptake hydrogenase gene, but not of the bidirectional hydrogenase gene, leads to enhanced photobiological hydrogen production by the nitrogen-fixing cyanobacterium *Anabaena* sp. PCC 7120. Appl Microbiol Biotech 58: 618–624
Meeks JC and Elhai J (2002) Regulation of cellular differentiation in filamentous cyanobacteria in free-living and plant-associated symbiotic growth states. Microbiol Mol Biol Rev 66: 94–121
Mikheeva LE, Schmitz O, Shestakov SV and Bothe H (1995) Mutants of the cyanobacterium *Anabaena variabilis* altered in hydrogenase activities. Z Naturforsch 50c: 505–510
Mullineaux CW (2008) Mechanisms of intercellular molecule exchange in heterocyst-forming cyanobacteria. Embo J 27: 1299–1308
Neuer G and Bothe H (1982) The pyruvate: ferredoxin oxidoreductase in heterocyts of the cyanobacterium *Anabaena cylindrica*. Biochim Biophys Acta 716: 358–365
Newton WE (2007) Physiology, biochemistry, and molecular biology of nitrogen fixation. In: Bothe H, Ferguson SJ and Newton WE (eds), Biology of the nitrogen cycle, pp 109–129. Elsevier, Amsterdam
Oliveira P and Lindblad P (2005) LexA. A transcriptional regulator binding in the promoter region of the bidirectional hydrogenase in the cyanobacterium *Synechocystis* sp. strain PCC6803. FEMS Microbiol Lett 251: 59–66
Papen H, Kentemich T, Schmülling T and Bothe H (1986) Hydrogenase activities in cyanobacteria. Biochimie 68: 121–132
Pau RN (1991) The alternative nitrogenases. In: Dilworth MJ, Glenn AR (eds) Biology and biochemistry of nitrogen fixation, pp 37–57. Elsevier, Amsterdam
Prechtl J, Kneip C, Lockhart P, Wenderoth K and Maier UG (2004) Intracellular spheroid bodies of *Rhopalodia gibba* have nitrogen-fixing apparatus of cyanobacterial origin. Mol Biol Evol 21: 1477–1481
Rösch C and Bothe H (2009) Diversity of total nitrogen-fixing and denitrifying bacteria in an acid forest soil. Eur J Soil Sci 60: 883–894
Ruvkun GB and Ausubel FM (1980) Interspecies homology of nitrogenase genes. Proc Natl Acad Sci U S A 77: 191–195
Scherer S, Kerfin W and Böger P (1980) Increase of nitrogenase activity in the blue-green alga *Nostoc muscorum* (cyanobacterium). J Bacteriol 144: 1017–1023
Schmitz O, Boison G, Hilscher R, Hundeshagen B, Zimmer W, Lottspeich F and Bothe H (1995) Molecular biological analysis of a directional hydrogenase from cyanobacteria. Eur J Biochem 233: 266–276
Schmitz O, Boison G, Salzmann H, Bothe H, Schütz K, Wang S and Happe T (2002) HoxE—a subunit specific for the pentameric bidirectional hydrogenase complex (HoxEFUYH) of cyanobacteria. Biochim Biophys Acta 1554: 66–74
Schmitz O, Bothe H (1996) $NAD(P)^+$-dependent hydrogenase activity in extracts from the cyanobacterium *Anacystis nidulans*. FEMS Microbiol Lett 69: 176–182
Schmitz O, Kentemich T, Zimmer W, Hundeshagen B and Bothe H (1993) Identification of the *nifJ* gene coding for pyruvate: ferredoxin oxidoreductase in dinitrogen-fixing cyanobacteria. Arch Microbiol 160: 62–67
Schmitz O, Boison G and Bothe H (2001a) Quantitative analysis of two circadian clock-controlled gene clusters coding for the birectional hydrogenase in the cyanobacterium *Synechoccus* sp. PCC7942. Mol Microbiol 41: 1409–1417

Schmitz O, Gurke J and Bothe H (2001b) Molecular evidence for the aerobic expression of *nifJ*, encoding pyruvate: ferredoxin oxidoreductase, in cyanobacteria. FEMS Microbiol Lett 195: 97–102

Schrautemeier B, Neveling U and Schmitz S (1995) Distinct and differentially regulated Mo-dependent nitrogen-fixing systems evolved for heterocysts and vegetative cells of *Anabaena variabilis* ATCC 29413: characterization of the *fdX1/2* gene regions as part of the *nif1/2* gene clusters. Mol Microbiol 18: 357–359

Schütz K (2004) Cyanobacterial H_2-production—a comparative analysis. Planta 218: 350–359

Shah GR, Karunakaran R and Kumar GN (2007) In vivo restriction endonuclease activity of the *Anabaena* PCC 7120 XisA protein in *Escherichia coli*. Res Microbiol 158: 679–684

Stal LJ and Mozelaar R (1997) Fermentation in cyanobacteria. FEMS Microb Rev 21: 179–211

Stewart WDP and Lex M (1970) Nitrogenase activity in the blue-green alga *Plectonema boryanum*. Arch Mikrobiol 73: 250–260

Tamagnini P (2007) Cyanobacterial hydrogenases. Diversity, regulation and application. FEMS Microb Rev 31: 692–720

Tamagnini P, Costa J-L, Almeida L, Oliveira M-J, Salema R and Lindblad P (2000) Diversity of cyanobacterial hydrogenases, a molecular approach. Curr Microbiol 40: 356–361

Tamagnini P, Axelsson R, Lindberg P, Oxelfelt F, Wünschiers R and Lindblad P (2002) Hydrogenases and hydrogen metabolism of cyanobacteria. Microbiol Mol Biol Rev 66: 1–20

Thiel T (1993) Characterization of genes for an alternative nitrogenase in the cyanobacterium *Anabaena variabilis*. J Bacteriol 175: 6276–6286

Thiel T, Lyons EM, Erker J and Ernst A (1995) A second nitrogenase in vegetative cells of a heterocyst-forming cyanobacterium. Proc Natl Acad Sci U S A 92: 9358–9362

Thiel T, Lyons EM and Thielemeyer J (1998) Organization and regulation of two clusters of *nif* genes in the cyanobacterium *Anabaena variabilis*. In: Peschek GA, Loeffelhardt W, Schmetterer G (eds) Phototrophic Prokaryotes, pp 517–521, Plenum Press, New York

Thomas JC, Ughy B, Lagoutte B and Ajlani G (2006) A second isoform of ferredoxin:NADP oxidoreductase generated by an in-frame initiation of translation. Proc Natl Acad Sci U S A 103: 18368–18373

Trebst A and Bothe H (1966) Zur Rolle des Phytoflavins im photosynthetischen Elektronentransport. Ber Dtsch Bot Ges 79: 44–47

Tsygankov A (2007) Nitrogen fixing cyanobacteria: a review. Appl Biochem Microbiol 43: 250–259

Van der Oost J, Builthuis BA, Feitz S, Krab K and Kraayenhof R (1989) Fermentation metabolism of the unicellular cyanobacterium *Cyanothece* PCC 7822. Arch Microbiol 151: 415–419

Vignais PM and Billoud B (2007) Occurrence, classification and biological function of hydrogenases: an overview. Chem Rev 107: 4206–4272

Vignais PM and Colbeau A (2004) Molecular biology of microbial hydrogenases. Curr Issues Mol Biol 6: 159–188

Vignais PM, Billoud B and Meyer J (2001) Classification and phylogeny of hydrogenases. FEMS Microbiol Rev 25: 455–501

Walsby AE (2007) Cyanobacterial heterocysts. Terminal pores proposed as sites of gas exchange. Trends Microbiol 15: 340–349

Weyman PD, Pratte B and Thiel T (2008) Transcription of *hupSL* in *Anabaena variabilis* ATTC 29143 is regulated by NtcA and not by hydrogen. Appl Environ Microbiol 74: 2103–2110

Zehr JP (2008) Globally distributed uncultivated oceanic N_2-fxing cyanobacteria lack oxygenic photosystem II. Science 322: 1110–1112

Zhao Y, Bian SM, Zhou HN and Huang JF (2006) Diversity of nitrogenase systems in diazotrophs. J Integrative Plant Biol 48: 745–755

Chapter 7
Hydrogen Peroxide Degradation in Cyanobacteria

Marcel Zamocky, Margit Bernroitner, Günter A. Peschek and Christian Obinger

7.1 Introduction

Our aerobic biosphere ultimately dates back to the advent of cyanobacteria. Primordial blue-green algae (cyanobacteria) have evolved ca. 3×10^9 years ago from ancient phototrophic organisms that already lived on our planet Earth (Barghoorn 1971; Rasmussen et al. 2008). They had succeeded to link photosynthetic electron flow from water as the (energy-requiring!) photoreductant through an oxygen-evolving complex of a newly elaborated photosystem II, which is thought to have originated from a uniform primordial photosystem by gene duplication. The resulting tandem operation of two photosystems (namely a high-potential water-oxidizing photosystem II and a low-potential ferredoxin-reducing photosystem I) is now known as oxygenic or plant-type photosynthesis. This most decisive evolutionary step marked a turning point in evolution on our Earth, opening up the era of an aerobic, oxygen-containing biosphere and atmosphere. The impact of this free dioxygen on the further evolution of life on Earth can hardly be over-emphasized, and cyanobacteria thus introduce themselves as true pace-makers of (terrestrial) evolution, in both a geological and biological sense.

Clearly, the first evolution of molecular O_2 must have been a deadly threat to all extant organisms in those anoxic days and it is logical to assume that cyanobacteria must have been among the first organisms to elaborate mechanisms for the detoxification of partially reduced, reactive oxygen species (ROS) originating from one- (i.e. $O_2^{\bullet-}$ or $HO_2^{\bullet}$) and two-electron reduction of O_2 (i.e. H_2O_2) (Regelsberger et al. 2002; Bernroitner et al. 2009).

In the 1980s and even still in the beginning of the 1990s only crude extracts of cyanobacteria were tested for peroxide-degrading enzymes. Catalase activity (Tel-Or et al. 1985), ascorbate peroxidase activity (Miyake et al. 1991), and glutathi-

C. Obinger (✉)
Department of Chemistry, Division of Biochemistry, Vienna Institute of BioTechnology, BOKU-University of Natural Resources and Life Sciences, Muthgasse 18, 1190 Vienna, Austria
e-mail: christian.obinger@boku.ac.at

G. A. Peschek et al. (eds.), *Bioenergetic Processes of Cyanobacteria*,
DOI 10.1007/978-94-007-0388-9_7, © Springer Science+Business Media B.V. 2011

one peroxidase activity (Tözum and Gallon 1979) have been reported. Originally, cyanobacteria were divided into two groups, in those that have and those that lack ascorbate peroxidase (Miyake et al. 1991). This was based on the observation that the first group scavenges hydrogen peroxide with a peroxidase using a photoreductant as electron donor. Thus, it was thought that—similar to chloroplasts—the oxidation products of ascorbate peroxidase are reduced by dehydroascorbate reductase (Hossain and Asada 1984) and monodehydroascorbate reductase (Hossain and Asada 1985) with glutathione and NADPH as electron donors (the production of both depending on photosynthetic activity). However, the concentration of ascorbate in the cyanobacterial cytosol (20–100 μM) is about 250-fold lower than in the stroma of chloroplasts (Tel-Or et al. 1985) and analysis of cyanobacterial genomes (Passardi et al. 2007) unequivocally has demonstrated that cyanobacteria do not contain ascorbate peroxidases.

Bifunctional catalase-peroxidases (KatGs) were the first H_2O_2 dismutating cyanobacterial enzymes which were characterized by both biochemical and genetic methods. KatGs were originally isolated from the unicellular species *Synechococcus* PCC7942 (Mutsuda et al. 1996; Perelman et al. 2003), *Synechococcus* PCC6301 (Engleder et al. 2000) and *Synechocystis* PCC6803 (Jakopitsch et al. 1999; Tichy and Vermaas 1999). Later on, the occurrence of a gene in *Nostoc punctiforme* that encodes a typical (monofunctional) catalase has been reported and also genes for putative manganese catalases were ascribed to *Nostoc punctiforme* and *Nostoc* sp. PCC7120 (Regelsberger et al. 2002). Furthermore, it has been demonstrated that (multiple) genes for peroxiredoxins (Yamamoto et al. 1999; Perelman et al. 2003) and glutathione peroxidases (Gaber et al. 2001) exist in cyanobacteria and that functional enzymes are expressed. Apparently, these enzymes (and not ascorbate peroxidase) were responsible for the 'photoreductant peroxidase' described earlier in the literature. These hydroperoxidases will be described in more detail in the following chapters.

7.2 Peroxidase *Versus* Catalase Activity

Peroxidases and catalases are ubiquitous oxidoreductases capable of the reductive heterolytic cleavage of the peroxidic bond, predominantly in hydrogen peroxide (H-O-O-H), but also in organic peroxides (R-O-O-H). Peroxidases reduce peroxides by means of two one-electron donors (Reaction 7.1) or one two-electron donor (Reaction 7.2). One-electron donors (AH) can be aromatic (e.g. phenols) or aliphatic (e.g. glutathione, GSH) molecules, small inorganic anions (e.g. NO_2^-) or even metal cations (e.g. Mn^{2+}) and the corresponding oxidation product (e.g. a radical, $^{\bullet}A$) eventually dimerizes (e.g. forming glutathione disulfide, GSSG). Two-electron donors (X^-) can be halides (chloride, bromide, iodide or thiocyanate) with the corresponding oxidation product (HOX) being hypohalous acid or hypothiocyanate. In this respect, all peroxidases that catalyze Reaction 7.2 are often denominated haloperoxidases (irrespective of their evolutionary origin).

$$HOOH(ROOH) + 2AH \rightarrow 2\,^{\bullet}A + 2\,H_2O(H_2O + ROH) \quad (7.1)$$

$$HOOH + X^- + H^+ \rightarrow HOX + H_2O \quad (7.2)$$

Catalases have the unique catalytic capacity to dismutate hydrogen peroxide (Reaction 7.3) by their striking ability to evolve molecular oxygen (O_2) by oxidation of hydrogen peroxide.

$$H_2O_2 + H_2O_2 \rightarrow O_2 + 2H_2O \quad (7.3)$$

Several gene families evolved in the ancestral genomes capable of reduction of H_2O_2 or organic peroxides. Among them are heme and non-heme oxidoreductases. Heme containing enzymes can be divided in three main (super)families, namely (1) typical or "monofunctional" catalases found in all domains of life (Zamocky et al. 2008a), (2) the peroxidase-catalase superfamily with members in plants, fungi, protists, and (archae)bacteria (Welinder 1992; Passardi et al. 2007) and (3) the peroxidase-cyclooxygenase superfamily also distributed in all domains of life (Zamocky et al. 2008b). In addition, some other minor groups of heme-containing peroxidases are found including the novel dye-decolorizing peroxidase family found in archaea, bacteria and fungi (Zubieta et al. 2007) or bacterial di-heme peroxidases (Echalier et al. 2006).

Non-heme peroxide utilizing enzymes are manganase catalases (Zamocky et al. 2008a), vanadium peroxidases (Littlechild 1999) and ubiquitous thiol peroxidases (peroxiredoxins and glutathione peroxidases), which catalyze the reduction of peroxides by catalytic cysteine residues and thiol-containing proteins as reductants (Rouhier and Jacquot 2005).

7.3 Catalase

Several gene families evolved in the ancestral genomes capable of H_2O_2 dismutation according to Reaction 7.3. The most abundant are heme-containing enzymes that are spread among Bacteria, Archaea and Eukarya. They are divided in two main groups, **typical** or "monofunctional" **catalases** (EC 1.11.1.6) and **catalase-peroxidases** (KatGs). Both types of heme enzymes exhibit high catalase activities, but have significant differences, including absence of any sequence similarity and very different active-site, tertiary, and quaternary structures. Enzymatic classification of bifunctional catalase-peroxidase is not clear because, besides their catalase activity (EC 1.11.1.6, hydrogen peroxide, hydrogen peroxide oxidoreductase), they exhibit a peroxidase activity similar to that of conventional peroxidases (EC 1.11.1.7, hydrogen peroxide, donor oxidoreductase). Nonheme **manganese-containing catalases** constitute a third (minor) group of enzymes with catalase activity. Mn-catalases (EC 1.11.1.6), initially referred to as pseudo-catalases, are present only in bacteria.

7.3.1 Typical (Monofunctional) Catalase

Though being the largest group of H_2O_2 dismutating enzymes that segregated rather early in the evolution into three main clades through at least two gene-duplicating events (Zamocky et al. 2008a) and even occurs in anaerobic bacteria, it was interesting to see that genes for typical catalases are very unusual in cyanobacteria (Table 7.1). At the moment, the only complete and nonfused gene in which all essential amino acids of typical catalases are conserved (i.e. proximal heme ligand Tyr and the conserved distal residues His, Arg and Ser) (Zamocky et al. 2008a) is found in *Nostoc punctiforme* PCC73102. The genome of this species in addition shows two ORFs that encode manganese catalases (see below). *Nostoc punctiforme* catalase belongs to clade 3 of small-subunit catalases that contain heme *b* at the active site and use NADPH as a second redox-active cofactor. A recently performed phylogenetic analysis (Zamocky et al. 2008a) suggests a lateral gene transfer (LGT) from an ancestral proteobacterium to *Nostoc punctiforme*. Incomplete (C-terminal truncated) catalase genes are found in *Synechococcus elongatus* PCC7942 and *Cyanothece* sp. ATCC 51142. The putative proteins would contain all essential amino acids but lack the C-terminal α-helical domain that is known to be essential in folding of typical catalases (Chelikani et al. 2004). Another protein of this family in *Nostoc* sp. PCC7120 is part of a fusion protein between a typical catalase-related domain and a putative lipoxygenase domain, with significant similarity to the well-investigated fusion protein of allene oxide synthase from *Plexaura homomalla* and related corals (Oldham et al. 2005).

The physiological role of the typical catalase in *Nostoc punctiforme* is unknown. From the published apparent kinetic parameters of this protein family [k_{cat} within 54,000 s^{-1} to 833,000 s^{-1}, K_M within 38–600 mM and activities that are essentially pH-independent from pH 5 to pH 10] (Chelikani et al. 2004), it is reasonble to assume that it has a protective role against environmental H_2O_2 generated in the ecosystem similar to the proposed role of bifunctional catalase-peroxidase (see below).

7.3.2 Bifunctional Catalase-Peroxidase

In contrast to typical catalases, about 30% of all known cyanobacterial genomes contain one gene (*katG*) encoding a bifunctional catalase-peroxidase (KatG). KatG represents the only peroxidase of both heme peroxidase superfamilies with a reasonable high catalase activity (Reaction 7.3), besides a usual peroxidase activity (Reaction 7.1). Together with ascorbate peroxidases, present mainly in chloroplastic organisms, and cytochrome *c* peroxidase, found mainly in the cytosol an/or mitochondria of fungi, KatGs constitute Class I of the peroxidase-catalase superfamily (Welinder 1992; Passardi et al. 2007). The predominant form of heme *b* containing KatGs in solution is a dimer or tetramer (Welinder 1992; Chelikani et al. 2004) and each subunit is composed of two distinct sequence-related N- and C-terminal do-

Table 7.1 ORFs and genes of cyanobacterial catalases and peroxidases in 44 completely or partially (*) sequenced strains (including genome size) Nitrogen-fixing cyanobacteria are highlighted in *grey*, heterocyst-forming species are highlighted in *dark-grey*. a) The *Prochlorococcus marinus* genus includes following strains: Pro9601, Pro9211, Pro9215, Pro9301, Pro9303, Pro9312, Pro9313, Pro9515, ProNATL1A, ProNATL2A, Pro1375, Pro1986. b) The following strains are included in the *Synechococcus* group: Syn107*, Syn9311, Syn9605, Syn9902, SynJA23, SynJA33, Syn7002, Syn307, Syn9916*, Syn9917*, Syn5701*, Syn7803, Syn7805*. Classification of cyanobacteria: I, Pleurocapsales; II, Chroococcales; III, Oscillatoriales; IV, Nostocales; V, unclassified. Mb, Megabase; #, pseudo-gene (incomplete or fusion gene); %, quantifies differences within the species of the *Prochlorococcus* and the *Synechococcus* group

Cyanobacteria	Genome size (Mb)	Catalases			Peroxidases							
		(Mono-functional) Heme catalase	(Class I) Catalase-peroxidase	(Binuclear) Manganese catalase	Peroxi-dockerin-related peroxidase	Diheme peroxi-dase	Vanadium peroxidase	Peroxiredoxin				Glutathione peroxidase-related peroxidase
								1-cys Prx	2-cys Prx	Type II Prx	Prx Q	
12 x *Prochlorococcus marinus* [a] (I)	1.6 - 2.7								1	15% 1 85% 0	85% 2 15% 1	1
Gloeobacter violaceus PCC7421 (II)	4.6		1	1				1	1		4	1
Microcystis aeruginosa NIES-843 (II)	5.8							1	1	1	3	
Synechococcus elongatus PCC6301 (II)	2.7		1					1	1		4	1
Synechococcus elongatus PCC7942 (II)	2.7	1#	1					1	1		4	1
14 x *Synechococcus sp.* [b] (II)	2.2 - 3.0		60% 1 40% 0				5% 1 95% 0	40% 1 60% 0	1	15% 1 85% 0	5% 4 45% 3 45% 2 5% 1	5% 2 80% 1 15% 0
Synechocystis sp. PCC 6803 (II)	3.6		1					1	1	1	2	2
Thermosynechococcus el. BP-1 (II)	2.6							1	1		3	
Acaryochloris marina (V)	6.5		1		1	(2?)	2 (1)	2	1	1	4	
Crocosphaera watsonii WH8501* (II)	6.2				1		1	1	1	1	2	1
Cyanothece sp. ATCC51142* (II)	4.9	1#			1		(1)	1	1	1	2	
Cyanothece sp. CCY0110* (II)	5.9		1	1	1			1	1	1	3	
Cyanothece sp. PCC7424* (II)	5.9			3				1	1	1	4	
Cyanothece sp. PCC8801* (II)	5.9							1	1	1	3	1
Lyngbya sp. PCC8106*(III)	7.0			1	1			1	1	1	3	
Trichodesmium erythraeum IMS101 (III)	7.7							1	1	1	3	
Anabaena variabilis ATCC29413 (IV)	6.3			1		(1?)		1	1	1	4	
Nodularia spumigena CCY9414* (IV)	5.3			1			1	1	1	1	4	
Nostoc punctiforme PCC73102* (IV)	9.0	1		2				1	1	1	4	1
Nostoc (Anabaena) sp. PCC7120 (IV)	6.4	1#		2				1	1	1	4	

mains, which led to the proposal that the large gene size of KatG had arisen through a gene duplication and fusion event of a primordial peroxidase gene after which the C-terminal domain likely lost its functionality (Welinder 1991).

The evolution of catalase-peroxidases was analysed recently (Passardi et al. 2007). The most important output of this investigation is that *katG* genes are distributed in ~40% of bacterial genomes and sometimes even closely-related species differ in possessing *katG* genes of different origin or even do not possess any *katG* genes. It has to be mentioned that KatGs are also found in eukaryotes including some algae, protists and mainly fungi. Phylogenetic analysis (Fig. 7.1) reveals the presence of a distinct and well-segregated clade of cyanobacterial catalase-peroxidases suggesting early segregation in evolution. In addition at least two independent lateral gene transfers occurred during the evolution of this protein family between various bacterial taxa and ancestors of (1) a group formed by *Synechococcus elongatus* PCC6301, *Synechococcus* PCC7002 and *Synechocystis* PCC6803, (2) *Gloeobacter violaceus,* and (3) most probably *Acaryochloris marina* (Fig. 7.1). *Cyanothece* sp. CCY0110 is the only diazotrophic cyanobacterium with a *katG* gene, all other species that contain this heme protein are unable to fix nitrogen (Table 7.1). No paralogues are found within one cyanobacterial species.

The four available crystal structures of KatGs from *Haloarcula marismortui* (1ITK) (Yamada et al. 2002), *Burkholderia pseudomallei* (1MWV) (Carpena et al. 2003), *Mycobacterium tuberculosis* (1SJ2) (Bertrand et al. 2004) and of the cyanobacterium *Synechococcus* PCC7942 (1UB2) (Wada et al. 2002) revealed that KatGs have proximal and distal conserved amino acids at almost identical positions as in other Class I peroxidases (Smulevich et al. 2006). In particular both the triads His/Trp/Asp (His259, Trp341 and Asp402; *Synechocystis* PCC6803 numbering) and His/Arg/Trp (His123, Arg119, Trp122) are conserved (Fig. 7.2a). Moreover, the distal His is hydrogen-bonded to a conserved asparagine (Asn153) and the proximal His is H-bonded to the carboxylate side chain of the nearby Asp residue which, in turn, is H-bonded to the nitrogen atom of the indole group of the nearby Trp residue (Fig. 7.2a). However, the X-ray structures also revealed features unique to KatG. In the vicinity of the active site, novel covalent bonds are formed among the side chains of three distal residues including the conserved Trp122 (Fig. 7.2a). In particular, both X-ray crystallization data (Yamada et al. 2002; Carpena et al. 2003; Bertrand et al. 2004) and mass spectrometric analysis (Donald et al. 2003; Jakopitsch et al. 2003) have confirmed the existence of a covalent adduct between Trp122, Tyr249 and Met275. Sequence alignment suggests that all these structural features are fully conserved in cyanobacterial KatGs (Fig. 7.2a).

Other KatG-typical features also found in cyanobacterial proteins are three large loops, two of them show highly conserved sequence patterns (Zamocky et al. 2001) and constrict the access channel of H_2O_2 to the prosthetic heme *b* group at the distal side. The channel is characterized by a pronounced funnel shape and a continuum of water molecules. At the narrowest part of the channel, which is similar but longer and more restricted than that in typical (monofunctional) peroxidases, two highly conserved residues, namely Asp152 and Ser335 control the access to distal heme side. Together with a conserved glutamate residue at the entrance both

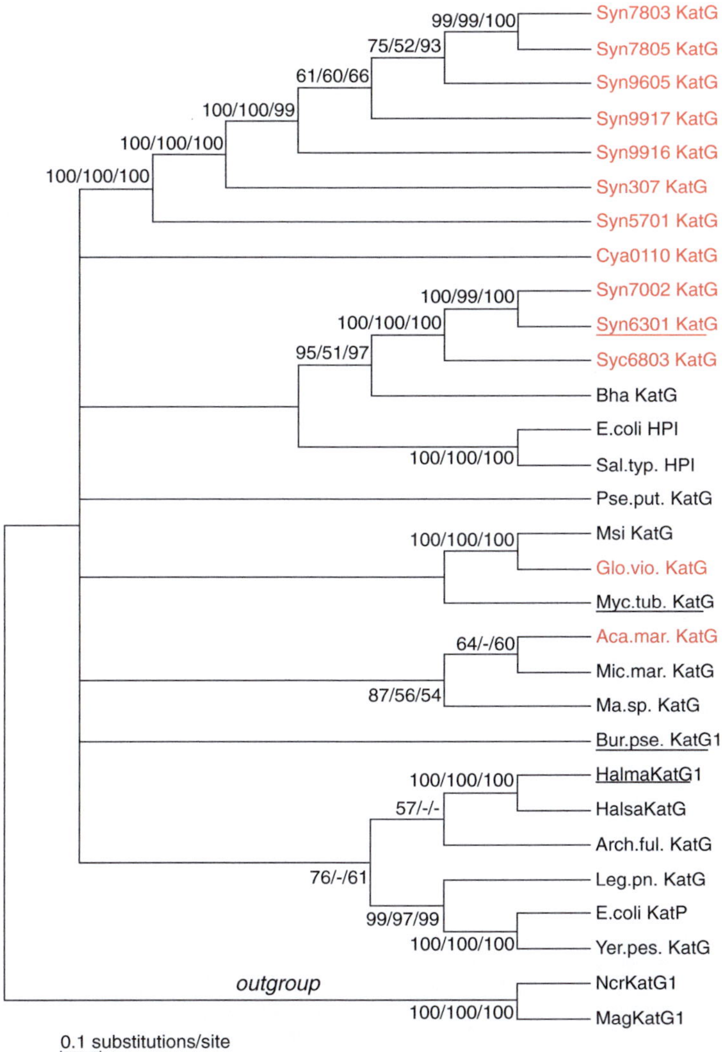

Fig. 7.1 Reconstructed phylogenetic tree of 30 catalase-peroxidases (KatGs) rooted with an out-group, i.e. fungal KatGs (originally published in Bernroitner et al. 2009). Presented is the tree obtained by the MEGA4 package (Tamura et al. 2007). Neighbor-joining method was applied with 1000 bootstrap replications and Jones-Taylor-Thornton model of amino acid substitutions. Further, complete deletion of gaps, homogenous pattern among lineages and uniform rates among sites were used. Nearly identical trees were obtained also with maximum parsimony method (within MEGA4 package, 1000 bootstraps) and with maximum likelihood method (from Phylip package 3.68 http://evolution.gs.washington.edu/phylip.html 100 bootstraps). The reconstructed consensus tree is presented using the Tree Explorer included in the MEGA package. Numbers on the branches represent bootstrap values as obtained from NJ/MP/ML, respectively (only values above 50% are presented, lower values are denoted "–"). Cyanobacterial KatGs are highlighted in red. For abbreviations, gene assignments and accession numbers of cyanobacterial KatGs see Table 7.2. Sequences with known 3D structures are underlined. The scale bar indicates the frequency of substitutions per site

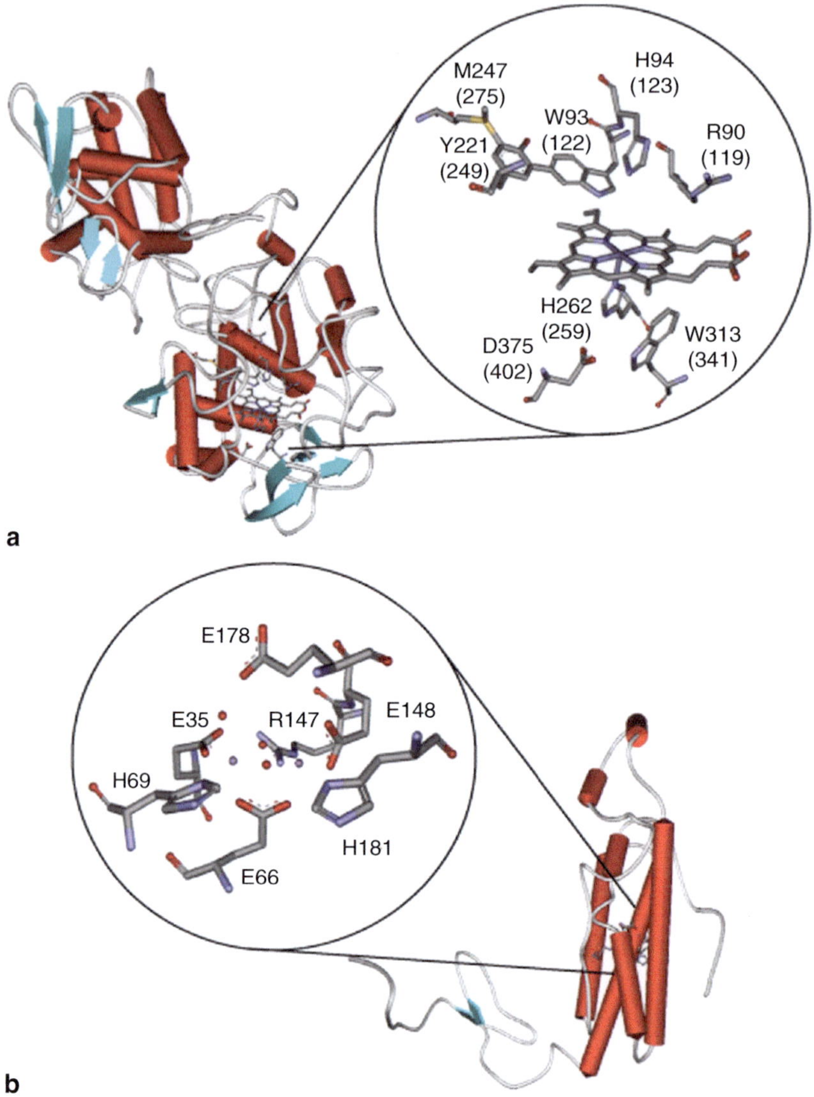

Fig. 7.2 **a** KatG monomer and assignment of secondary structure elements as well as active site of dimeric catalase-peroxidase from the cyanobacterium *Synechococcus* PCC7942 (accession code 1UB2) (Wada et al. 2002) (*Synechocystis* numbering in parenthesis). Each subunit is composed of two distinct sequence-related N- and C-terminal domains. Only the N-terminal domain contains heme *b*. Active site residues include the proximal triad His-Trp-Asp and the distal triad His-Arg-Trp as well as the unique covalent adduct Trp-Tyr-Met. **b** Monomeric structure and view of the dinuclear manganese complex at the active site of dimeric manganese catalase from *Lactobacillus plantarum* (accession code 1JKU) (Barynin et al. 2001). In the monomeric structure all secondary structure elements and the prosthetic group are assigned. In the active site structure all coordinating ligands are depicted. Manganese ions are represented by violet spheres, and coordinated solvent is represented in red spheres. The figures were built using SWISS-PDB Viewer (www.expasy.ch/spdbv)

acidic residues seem to be critical for stabilizing the solute matrix and orienting the water dipoles in the channel. Exchange of both residues affect the catalase but not the peroxidase activity (Jakopitsch et al. 2005).

This extensive distal site hydrogen-bonding network causes KatGs to differ from typical peroxidases. Typically, the catalase but not the peroxidatic activity is very sensitive to mutations that disrupt this network (Heering et al. 2002; Jakopitsch et al. 2004). Moreover, the integrity of this network is crucial for the formation of distinct protein radicals that are formed upon incubation of KatG with peroxides (Ivancich et al. 2003). Mutational analysis clearly underlined the importance of the KatG-typical covalent adduct Trp-Tyr-Met for H_2O_2 dismutation. Its disruption significantly decreased the catalase but not the peroxidase activity. Similarly all other mutations so far performed in the heme cavity or substrate channel of a KatG affected the oxidation but not the reduction reaction of hydrogen peroxide (Smulevich et al. 2006), which is the initial reaction in all heme catalases and peroxidases.

Dismutation of hydrogen peroxide according to Reaction 7.3 follows a two-step mechanism. Heterolytic cleavage of hydrogen peroxide (i.e. compound I formation) follows Reaction 7.4. It is important to note that the exact electronic structure of formed compound I, that actually oxidizes the second H_2O_2 according to Reaction 7.5 (i.e. compound I reduction), is still speculative (Jakopitsch et al. 2007). Conventional compound I is an oxoiron(IV) porphyrin π-cation radical species [$^{\bullet+}$Por Fe(IV)=O] that in KatG could be in equilibrium with an alternative (H_2O_2 oxidising?) compound I [$^{\bullet+}$AA Por Fe(IV)=O] with AA being a so far not localized amino acid radical (Jakopitsch et al. 2007). One role of the KatG-specific Trp-Tyr-Met adduct might be to provide the (at least transient) radical site ($^{\bullet+}$MYW Fe(IV)=O) that quenches the porphyryl radical according to Reaction 7.4 and participates in H_2O_2 oxidation (Reaction 7.5). Similar to typical peroxidases, compound I formation is clearly assisted by the conserved distal residues His123 and Arg119 (Smulevich et al. 2006) but the exact role of specific distal amino acids in H_2O_2 oxidation is still speculative.

$$\text{Por Fe(III)} + H_2O_2 \rightarrow [^{\bullet+}\text{Por Fe(IV)} = O \rightleftarrows {}^{\bullet+}\text{AA Por Fe(IV)} = O] + H_2O \tag{7.4}$$

$$\begin{aligned}&[^{\bullet+}\text{Por Fe(IV)} = O \rightleftarrows {}^{\bullet+}\text{AA Por Fe(IV)} = O] + H_2O_2 \\ &\quad\rightarrow \text{Por Fe(III)} + H_2O + O_2\end{aligned} \tag{7.5}$$

Both k_{cat} and K_M values of KatGs are significant lower compared to typical catalases. Apparent values range from 3500 to 7500 s^{-1} and 3.7–8 mM (Smulevich et al. 2006), respectively, and the two so far kinetically investigated cyanobacterial representatives fall within this range (Mutsuda et al. 1996; Jakopitsch et al. 1999). In contrast to typical catalases and Mn-catalases, the catalase activity of KatG has a sharp maximum activity around pH 6.5.

KatG is a bifunctional enzyme. It oxidizes typical artificial peroxidase substrates like *o*-dianisidine, guaiacol or ABTS. The pH profile of the peroxidase activity of KatG has a maximum within pH 5–6 depending on the nature of the one-electron donor. Additionally, KatGs have been reported to have also halogenation (Jakopitsch et al. 2001) and NADH oxidase (Singh et al. 2004) activity.

The naturally occurring peroxidase substrate is unknown. In physiological conditions it is well known that KatG from *Mycobacterium tuberculosis* can activate the anti-tuberculosis drug isoniazid (Zhao et al. 2006). Due to the restricted access, only small peroxidase substrates can enter the main entrance channel. A second access route, found in monofunctional peroxidases, approximately in the plane of the heme, is blocked by the KatG-typical loops. However, another potential route that provides access to the core of the protein, between the two domains of the subunit, has been described (Carpena et al. 2003). It has been speculated that this could be the binding site for substrates with extended, possibly even polymeric character. In any case neither from prokaryotic nor from eukaryotic KatGs the naturally occurring one-electron donor(s) is known nor the function of a high peroxidase activity in an enzyme with catalase activity. A reasonable role at low peroxide concentration could be the reduction of inactive compound II species to ferric KatG.

It has been proposed that KatG in *Synechocystis* PCC6803 has a protective role against environmental H_2O_2 generated in the ecosystem and that this protective role is most apparent at a high cell density of the cyanobacterium (Tichy and Vermaas 1999). Deletion of the *katG* gene in *Synechocystis* PCC6803 resulted in a mutant strain with normal phenotype and resistance to H_2O_2 and methyl viologen indistinguishable from those of the wild-type. Nevertheless the rate of H_2O_2 decomposition in the Δ*katG* mutant was about 30 times lower than in the wild-type (Tichy and Vermaas 1999). Apparently the residual H_2O_2-scavenging capacity was more than sufficient to deal with the rate of H_2O_2 production by the cell. In the Δ*katG* mutant hydrogen peroxide degradation was light-dependent and could be stimulated by addition of thiol such as dithiothreitol. The authors concluded that this activity might derive from a thiol-specific peroxidase, for which thioredoxin is the physiological electron donor. This has been underlined by a bioinformatic analysis of the genome of *Synechocystis* PCC6803 (Stork et al. 2005) that revealed the presence of five open reading frames with similarity to peroxiredoxins (see below and also Table 7.1).

Similarly, investigations of *Synechococcus* sp. PCC7942 (Miller et al. 2000; Perelman et al. 2003) has also demonstrated the existence of two classes of H_2O_2 degrading enzymes, namely a (cytosolic) KatG, that is essential for survival and elimination of relatively high concentrations of externally added H_2O_2, and light-dependent peroxidase(s) that are essential during excessive radiation (see also Table 7.1). Despite addition of ^{18}O-labeled H_2O_2 in the light only $^{16}O_2$ production was observed, indicative of $H_2{}^{16}O$ oxidation by photosystem II and formation of a photoreductant. However, in the dark added $H_2{}^{18}O_2$ led to release of $^{18}O_2$ that could be completely blocked by NH_2OH, an inhibitor of KatG (Miller et al. 2000). This suggests also a temporal difference in activity of peroxiredoxins and KatG.

7.3.3 Manganese Catalase

Manganese catalases (non-heme or di-manganese catalases, MnCats; EC 1.11.1.6) represent only a minor gene family with catalase activity that is also found in cyanobacteria. With the exception of *Gloebacter violaceus* PCC7421, only diazotrophic species contain at least one open reading frame with similarity to MnCat. Two paralogs are found in heterocyst-forming *Nostoc punctiforme* (that also contains a typical catalase) and *Nostoc* PCC7120 (Table 7.1) and three paralogs exist in the genome of *Cyanothece* sp. PCC7424. Interestingly, with the exception of *Gloeobacter violaceus* and *Cyanothece* sp. CCY0110, cyanobacteria contain either a *katG* gene or gene(s) encoding manganese catalase(s) and about 50% of the investigated species contain neither a typical catalase, nor a KatG nor a MnCat (e.g. all *Prochlorococcus marinus* strains, *Microcystis aeruginosa*, *Synechococcus* el. BP-1 etc., compare with Table 7.1). Only in all so far known genomes of heterocyst-forming diazotrophic cyanobacteria at least one gene encoding MnCat was dedected (Table 7.1).

A recently performed phylogenetic analysis of bacterial Mn-catalases (Zamocky et al. 2008a) reveals the presence of five distinct and well segregated clades. Figure 7.3, that includes all available cyanobacterial genes, demonstrates that cyanobacterial MnCats are found in clade 3 (with Firmicutes and proteobacteria) as well as in clade 4 (with the genus *Bacteroides*) that was separated rather early from the common ancestor. It is interesting to note that some authors argued for a role of di-manganese catalase in still anoxygenic (cyano?)bacteria as a forerunner of Mn-containing photosystem II (McKay and Hartman 1991; Lane 2002). It was noted that the active site structure of manganese catalase reveals similarity to the tetranuclear manganese complex of the oxgygen evolving center of photosystem II (Blankenship et al. 1998).

So far neither an expression study nor a functional or structural analysis of a cyanobacterial MnCat is found in literature. However, multiple sequence alignment (Bernroitner et al. 2009) revealed a rather high identity mainly in the active site residues with the two enzymes of known crystal structure, namely MnCat from *Thermus thermophilus* (Antonyuk et al. 2000) and *Lactobacillus plantarum* (Barynin et al. 2001). These proteins contain six identical subunits (~30 kDa) each with a bridged binuclear Mn-center located within a conserved closely packed four-helix bundle domain (Fig. 7.2b). The ligands for the di-manganese center are almost invariantly conserved among all known sequences (Zamocky et al. 2008a). A bridging glutamate (Glu 66 in *Lactobacillus plantarum* numbering) anchors the two ions in the binuclear cluster. Each Mn ion is further coordinated by one histidine (His69 to Mn1 and His181 to Mn2) and one glutamate (Glu35 to Mn1 and Glu148 to Mn2) bound to opposite faces of the cluster (Whittaker et al. 2003) (Fig. 7.2b). These essential ligands are highly conserved forming typical signatures of MnCat sequence, whereas the environments of these ligands differ slightly. The manganese core is completed by two solvent-derived oxygen atom bridges.

Access to the binuclear center is via a central channel that extends the full width of the hexamer, with branches into each subunit. Each di-manganese center is embedded in a network of hydrogen bonds that radiate from the metal center towards the outer sphere environment. The length (≈15 Å) and narrowness of these channels provide a restricted access to the only substrate hydrogen peroxide. Mutational analysis of an essential outer sphere tyrosine (Tyr 42), which is conserved in all so far available sequences clearly demonstrated the importance of an intact hydro-

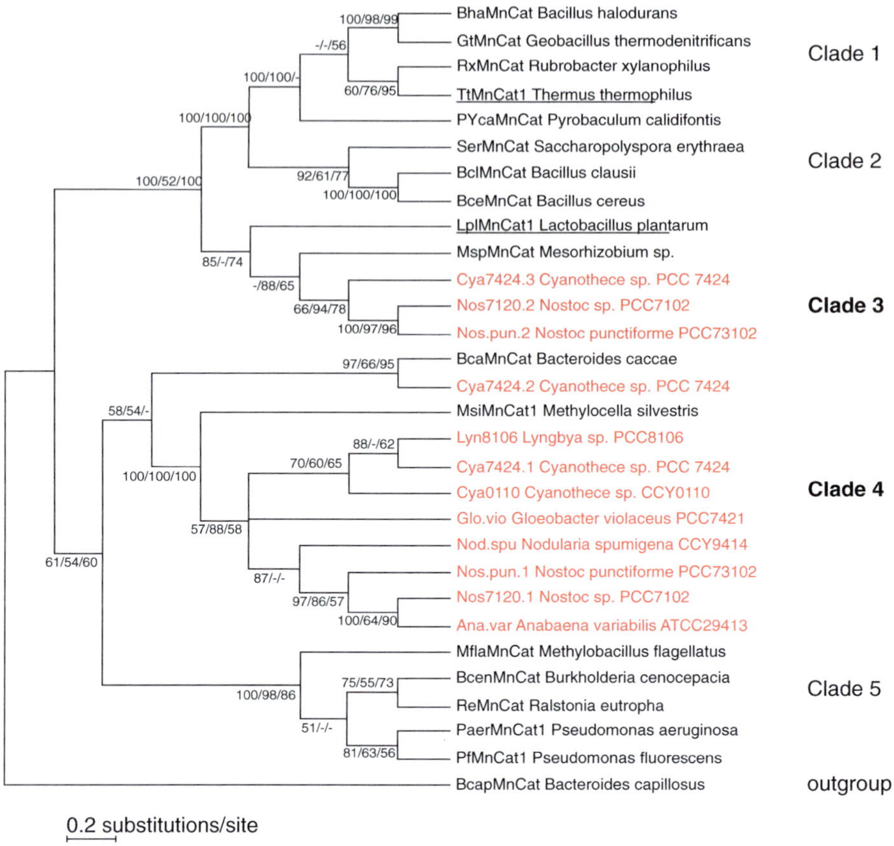

Fig. 7.3 Reconstructed phylogenetic tree of 30 manganese catalases rooted with an outgroup (originally published in Bernroitner et al. 2009). Presented is the tree obtained with the MEGA4 package (Tamura et al. 2007). Neigbor-joining method was applied with 1000 bootstrap replications and Jones-Taylor-Thornton model of amino acid substitution. Very similar trees were obtained also with maximum parsimony method (within MEGA4 package, 1000 bootstraps) and with maximum likelihood method (from Phylip package 3.68 http://evolution.gs.washington.edu/phylip.html using 100 bootstraps). Numbers on the branches represent bootstrap values as obtained from NJ/MP/ML, respectively (only values above 50% are presented, lower values are denoted "–"). Cyanobacterial manganese catalases are highlighted in red. For abbreviations, gene assignments and accession numbers of cyanobacterial sequences see Table 7.2. Sequences with known 3D structures [*Thermus thermophilus* (TtMnCat1) and *Lactobacillus plantarum* (LplMnCat1)] are underlined. The scale bar indicates the frequency of substitutions per site

gen bonding network also in Mn-catalases (Whittaker et al. 2003). In the variant Tyr42Phe a solvent bridge was broken and an "open" form of the di-manganese cluster was generated, thereby considerably influencing the catalytic turnover, pH optimum and interaction with H_2O_2.

Though following the same overall reaction (Reaction 7.3) the mechanism of the catalase reaction in MnCat differs significantly from catalase-peroxidases.

The di-manganese cluster is equally stable in either Mn^{2+}-Mn^{2+} or Mn^{3+}-Mn^{3+} oxidation states. Upon isolation the protein is frequently in a mixture of both oxidation states. Fully reduction to Mn^{2+}-Mn^{2+} is mediated by hydroxylamine, whereas re-oxidation to homogenous Mn^{3+}-Mn^{3+} state can be achieved by molecular oxygen at pH > 7.0 (Whittaker et al. 2003). The catalase reaction can be written as a two stage process. During catalytic turnover, the active site has to accommodate both the reduced and the oxidized state of the di-manganese core. The Mn^{2+}-Mn^{2+} cluster is expected to polarize the O-O bond favouring a heterolytic cleavage of the peroxidic bond and release of water (Reaction 7.6) (Whittaker et al. 2003). Hydrogen peroxide oxidation and dioxygen release occurs by a simple electron transfer (Reaction 7.7). Principally there is no temporal order to the reduction and oxidation stages.

$$H_2O_2 + Mn^{2+}\text{-}Mn^{2+}\ (2H^+) \rightarrow Mn^{3+}\text{-}Mn^{3+} + 2H_2O \quad (7.6)$$

$$H_2O_2 + Mn^{3+}\text{-}Mn^{3+} \rightarrow Mn^{2+}\text{-}Mn^{2+}\ (2H^+) + O_2 \quad (7.7)$$

It is assumed that the change in the oxidation status is responsible for protonation or deprotonation of both bridging oxygens. Protonation is proposed for the reduced Mn^{2+}-Mn^{2+} state (see Reaction 7.6). In contrast to heme-containing catalases no reactive intermediate is formed and both product waters are formed in one reaction (Reaction 7.6). There is no evidence for involvement of free radicals in manganese catalase turnover and nothing is known about a peroxidatic reactivity of these oxidoreductases.

The H_2O_2 dismutation rates are lower compared to typical catalases and catalase-peroxidases, which could explain that Mn-catalases may not have become as widespread in nature. The activity varies only slightly over the pH range pH 5–10, very similar to typical catalases. The apparent K_m values are reported to be around 220 mM (Whittaker et al. 2003) suggesting a low catalytic efficiency at low H_2O_2 concentrations. Their exact physiological role(s) and expression pattern are more or less unknown.

7.4 Peroxidase

Here, it is reasonable to distinguish between heme and non-heme enzymes. From the two heme peroxidase superfamilies, i.e. the peroxidase-catalase superfamily (Welinder 1992) and the peroxidase-cyclooxygenase superfamily (Zamocky et al. 2008b) two types of heme peroxidases are found. Bifunctional (Class I) catalase-peroxidase is the only cyanobacterial member of the first superfamily and—due to its overwhelming catalase activity—has been discussed in the chapter before. Most interestingly, cyanobacterial peroxidases might be at the origin of the second (i.e. peroxidase-cyclooxygenase) superfamily since ORFs with similarity to animal peroxidases where found within the subfamily of peroxidockerins (Zamocky et al. 2008b). Whereas genes encoding typical catalases, KatGs and MnCat appear to have been acquired by lateral gene transfer in cyanobacteria, there is phylogentic evidence that proteins from the peroxidase-cyclooxygenase superfamily originate from ancient cyanobacteria (see below). Furthermore, two cyanobacterial genomes show the occurrence of ORFs with similarity to putative di-heme peroxidase.

The most important group of non-heme peroxide degrading enzymes is represented by thiol-specific peroxidases including peroxiredoxins and glutathione peroxidases. All four subclasses of peroxiredoxins found in plants occur also in cyanobacteria (Table 7.1) (Stork et al. 2005). ORFs with similarity to glutathione peroxidases are found in the majority of the cyanobacterial genomes but so far no glutathione peroxidase activity could be measured in these organisms (Tichy and Vermaas 1999). Furthermore few cyanobacterial genomes show the occurrence of ORFs with similarity to putative vanadium peroxidase.

7.4.1 Peroxidockerin-Related (Primordial) Peroxidase

It was interesting to see that some cyanobacterial genomes contain genes that encode a putative protein with similarity to mammalian peroxidases like lactoperoxidase (LPO) or eosinophil peroxidase (EPO, EC 1.11.1.7). So far in six cyanobacteria the corresponding ORF of this putative peroxidase is found that—together with peroxidockerins—constitute an old subfamily at the root of the peroxidase-cyclooxygene superfamily of heme peroxidases (Zamocky et al. 2008b). Peroxidockerins are multidomain proteins of unknown physiological role composed of a transmembrane domain, Dockerin type 1 repeats and the catalytic peroxidase domain. The homologous cyanobacterial proteins lack the non-enzymatic domains and cluster separately within this subfamily (Zamocky et al. 2008b).

Recent analysis demonstrated that all catalytic residues known to be essential in the catalysis of mammalian peroxidases (Furtmüller et al. 2006) are also found in these primordial heme peroxidases: (1) The essential motif—WGQXV**DHD**- with the distal histidine in immediate neighbourhood of two aspartate residues that are involved in heme to protein linkage and Ca^{2+}-binding. Moreover this region also contains a highly conserved glutamine that is essential in halide binding (Furtmüller et al. 2006), (2) a second motif with two essential residues is -**R**XX**E**- that contains the catalytic distal arginine and a glutamate that is also involved in the second heme to protein ester bond in LPO, (3) the proximal ligand histidine and its H-bonding partner asparagine.

It is completely unknown whether these proteins are expressed in cyanobacteria. However, inspection of the sequence (Bernroitner et al. 2009) suggests a functionality very similar to the corresponding mammalian peroxidases that use small anionic substrates like halides (bromide, iodide) and thiocyanate as substrates to fulfill their physiological role, i.e. to produce hypohalous acids and hypothiocyanate according to Reaction 7.2. These oxidizing and halogenating agents participate in unspecific immune defence of mammals against pathogens. It is tempting to speculate whether some cyanobacterial species have also acquired the ability to produce antimicrobial compounds possibly providing selective advantage.

7.4.2 *Di-heme Peroxidase and Vanadium Peroxidase*

Bacterial di-heme cytochrome *c* peroxidases (DiHCCPs) catalyze the two-electron reduction of H_2O_2 to water according to Reaction 7.1 by soluble one-electron donors such as cytochrome *c* and cupredoxins (Echalier et al. 2006). However, unlike their eukaryotic counterparts, they possess two heme *c* prosthetic groups, that are covalently attached to a single polypeptide chain via two cysteines (-**C**XX**C**H-), and allow reduction of H_2O_2 without the need to generate (semi-stable) radicals. The two heme groups are designated (low potential, N-terminal) heme P (peroxidatic), that actually reacts with peroxides and ligands, and (high potential, C-terminal) heme E (electron-transferring). In the resting (oxidized) state heme P is coordinated by two histidine ligands and the E heme is coordinated by a histidine and a methionine (Fülöp et al. 1995). Activation of the peroxidase needs reduction of heme E by a small redox donor protein in order to activate heme P that includes dissociation of the coordinating distal histidine in order to allow access of the substrate(s) (Echalier et al. 2006).

Three ORFs with similarity to bacterial DiHCCP were found in the genomes, one in *Anabaena variabilis* ATCC29413 and two in *Acaryochloris marina* (Table 7.1). Analyis of the three sequences indicated the presence of two heme *c* prosthetic groups (Bernroitner et al. 2009). However, N-terminal heme P seems to lack the distal histidine and at least in two sequences the distal coordination of heme E is unclear. Thus, the functionality of these putative proteins is fully unclear and needs further investigation. No information about (bacterial) DiHCCP-related proteins in cyanobacteria was found in the literature.

Some cyanobacterial strains in addition contain ORFs with similarity to vanadium-containing peroxidases. Vanadium-dependent enzymes are often designated as haloperoxidases (bromoperoxidase, iodoperoxidase) since they exclusively oxidize halides according to Reaction 7.2 and are usually found in brown and red algae, fungi and lichens (Weyland et al. 1999). They are nearly all-helical homodimeric proteins that contain trigonal-bipyramidal coordinated vanadium atoms at their two active centres. The catalytically active histidine is described to be in a different environment in bromoperoxidases (-SG**H**-) and in iodoperoxidases (-AG**H**-) (Weyland et al. 1999; Colin et al. 2005). The present analysis demonstrates the occurrence of vanadium-dependent iodoperoxidases in at least five cyanobacterial species (Table 7.2). All essential structural motifs seem to be conserved (not shown). In addition ORFs were found with the catalytic histidine located within the sequence -GG**H**-. Whether this has an impact on the functionality of these putative haloperoxidases is unknown (thus these proteins were set in brackets in Table 7.1) as is the physiological role of vanadium peroxidases in cyanobacteria.

Table 7.2 List of abbreviations and access codes of all cyanobacterial genomes or catalase and peroxidase genes of all mentioned sequences

Abbreviations	Sequenced cyanobacteria	Accession numbers
Aca.mar	*Acaryochloris marina* MBIC11017	YP_001516080
Ana.var	*Anabaena variabilis* ATCC29413	YP_3214520.1
Cro.wat	*Crocosphaera watsonii* WH8501	ZP_00514853
Cya51142	*Cyanothece sp.* ATCC51142	ZP_001804678
Cya0110	*Cyanothece sp.* CCY0110	ZP_01731923
Cya7424	*Cyanothece sp.* PCC7424	ZP_02974490
Cya8801	*Cyanothece sp.* PCC8801	ZP_02940003
Glo.vio	*Gloeobacter violaceus* PCC7421	NP_925728
Lyn8106	*Lyngbya sp.* PCC8106	ZP_01623400
Mic.aer	*Microcystis aeruginosa* NIES-843	YP_001655793
Nod.spu	*Nodularia spumigena* CCY9414	ZP_01628632
Nos.pun	*Nostoc punctiforme* PCC73102	YP_001868279
Nos7120	*Nostoc* (*Anabaena*) *sp.* PCC7120	NP_485408
Pro9601	*Prochlorococcus marinus* str. AS9601	YP_001009553
Pro9211	*Prochlorococcus marinus* str. MIT9211	YP_001551018
Pro9215	*Prochlorococcus marinus* str. MIT9215	YP_0014844393
Pro9301	*Prochlorococcus marinus* str. MIT9301	YP_001091387
Pro9303	*Prochlorococcus marinus* str. MIT9303	YP_001016925
Pro9312	*Prochlorococcus marinus* str. MIT9312	YP_397563
Pro9313	*Prochlorococcus marinus* str. MIT9313	NP_894962
Pro9515	*Prochlorococcus marinus* str. MIT9515	NC_008817
ProNATL1A	*Prochlorococcus marinus* str. NATL1A	YP_001015351
ProNATL2A	*Prochlorococcus marinus* str. NATL2A	YP_291889
Pro1375	*Prochlorococcus marinus subsp. marinus* str. CCMP1375	NP_875535
Pro1986	*Prochlorococcus marinus subsp. marinus* str. CCMP1986	NP_893174
Syn6301	*Synechococcus elongatus* PCC6301	YP_173174
Syn7942	*Synechococcus elongatus* PCC7942	YP_400495
Syn107	*Synechococcus sp.* BL107	ZP_01468329
Syn9311	*Synechococcus sp.* CC9311	YP_730179
Syn9605	*Synechococcus sp.* CC9605	YP_382244
Syn9902	*Synechococcus sp.* CC9902	YP_376729
SynJA23	*Synechococcus sp.* JA-2-3B'a(2-13)	YP_476915
SynJA33	*Synechococcus sp.* JA-3-3Ab	YP_47007
Syn7002	*Synechococcus sp.* PCC7002	YP_001733641
Syn307	*Synechococcus sp.* RCC307	YP_001226989
Syn9916	*Synechococcus sp.* RS9916	ZP_01470366
Syn9917	*Synechococcus sp.* RS9917	ZP_01080660
Syn5701	*Synechococcus sp.* WH5701	ZP_01084959
Syn7803	*Synechococcus sp.* WH7803	YP_001225319
Syn7805	*Synechococcus sp.* WH7805	ZP_01124489
Syn8102	*Synechococcus sp.* WH8102	NP_896813
Syc6803	*Synechocystis sp.* PCC6803	NP_441492
The.elo	*Thermosynechococcus elongatus* BP-1	NP_683219
Tri.ery	*Trichodesmium erythraeum* IMS101	YP_721847

Table 7.2 (continued)

Abbreviations	Non cyanobacterial sequences	Accession numbers
Arch.ful.KatG	*Archaeoglobus fulgidus* DSM 4304	NP_071058
BcaMnCat	*Bacteroides caccae* ATCC 43185	ZP_01960863
BcapMnCat	*Bacteroides capillosus* ATCC 29799	ZP_02036380
BceMnCat	*Bacillus cereus* E33L	YP_084426
BcenMnCat	*Burkholderia cenocepacia* AU 1054	YP_621456
BclMnCat	*Bacillus clausii* KSM-K16	YP_175570
BhaKatG	*Bacillus halodurans* C-125	NP_241772
BhaMnCat	*Bacillus halodurans* C-125	NP_241935
BmaPxDo1	*Blastopirellula marina* DSM 3645	ZP_01093887
B.t._LPO	*Bos taurus*	NP_776358
Bur.pse. KatG1	*Burkholderia pseudomallei* K96243	YP_109459
DdPxDo1	*Dictyostelium discoideum* AX4	XP_642775
E.coli HPI	*Escherichia coli* str. K12 substr. MG1655	NP_418377
E.coli KatP	*Escherichia coli* O157:H7 EDL933	YP_325577
GtMnCat	*Geobacillus thermodenitrificans* NG80-2	YP_001127084
HalmaKatG1	*Haloarcula marismortui* ATCC 43049	YP_135827
HalsaKatG	*Halobacterium sp.* NRC-1	NP_395796
Leg.pn. KatG	*Legionella pneumophila subsp. pneumophila* str. Philadelphia 1	YP_094248
LplMnCat1	*Lactobacillus plantarum*	P60355
Ma. sp. KatG	*Marinomonas sp.* MWYL1	YP_001340448
MagKatG1	*Magnaporthe grisea* 70-15	XP_361863
MflaMnCat	*Methylobacillus flagellatus* KT	YP_545545
Mic.mar. KatG	*Microscilla marina* ATCC 23134	ZP_01693331
Msi KatG	*Methylocella silvestris* BL2	ZP_02946304
MsiMnCat1	*Methylocella silvestris* BL2	ZP_02947034
MspMnCat	*Mesorhizobium sp.* BNC1	YP_674434
Myc.tub. KatG	*Mycobacterium tuberculosis* H37Rv	NP_216424
NcrKatG1	*Neurospora crassa* OR74A	XP_959745
NmoPxDo1	*Nitrococcus mobilis* Nb-231	ZP_01126301
OsiPxDo1	*Oryza sativa* Indica Group	CT836978
PaerMnCat1	*Pseudomonas aeruginosa* PAO1	NP_250875
PfMnCat1	*Pseudomonas fluorescens* Pf-5	YP_261802
PlmaPxDo1	*Planctomyces maris* DSM 8797	ZP_01857045
Pse.put. KatG	*Pseudomonas putida* KT2440	NP_745804
PYcaMnCat	*Pyrobaculum calidifontis* JCM 11548	YP_001055621
RbaPxDo1	*Rhodopirellula baltica* SH 1	NP_869799
RbaPxDo2	*Rhodopirellula baltica* SH 1	NP_864006
ReMnCat	*Ralstonia eutropha* JMP134	YP_298605
RpPxDo2	*Rhodopseudomonas palustris* BisA53	YP_783237
RxMnCat	*Rubrobacter xylanophilus* DSM 9941	YP_643463
Sal.typ. HPI	*Salmonella typhimurium*	CAA37187
SarPxDo1	*Salinispora arenicola* CNS-205	YP_001537753
SerMnCat	*Saccharopolyspora erythraea* NRRL 2338	YP_001105249
TtMnCat	*Thermus thermophilus* HB27	YP_005841
Yer.pes. KatG	*Yersinia pestis* CO92	NP_406785

7.4.3 *Peroxiredoxins*

Peroxiredoxins (Prx-s, EC 1.11.1.15) are a ubiquitous family of antioxidant enzymes which have been identified in all domains of life (Dietz 2003; Wood et al. 2003). *Synechocystis* PCC6803 has been the first cyanobacterium where genes encoding Prx-s as well as the expression of the corresponding proteins have been demonstrated (Yamamoto et al. 1999; Kobayasgi et al. 2004; Dietz et al. 2005; Hosaya-Matsuda et al. 2005). A bioinformatic analysis of the genomes of *Synechocystis* PCC6803 and *Synechococcus elongatus* PCC7942 (Stork et al. 2005) revealed the existence of five and six ORFs, respectively, with similarity to peroxide-detoxifying Prx-s in plants.

All peroxiredoxins share the same basic catalytic mechanism composed of two steps. All Prx-s appear to have the first step in common (Reaction 7.8), in which an activated cysteine (Cys-SH, the peroxidative cysteine) is oxidized to a sulphenic acid (Cys-SOH) by the peroxide substrate. The peroxide decomposition probably requires a base to deprotonate the peroxidatic cysteine as well as an acid to protonate the poor RO^- leaving group (Wood et al. 2003). All Prx-s conserve an active-site Arg, which lowers the pK_a of the peroxidatic cysteine somewhat by stabilizing its thiolate form. Genenerally, Prx-s have a rather low activity, but possess a broad substrate specificity. Possible substrates are H_2O_2, alkyl hydroperoxides, and peroxynitrite. Regarding detoxification of H_2O_2 it has been shown that Prx-s have a K_M in the low µM range and thus are more efficient scavengers of trace amounts of hydrogen peroxide compared to enzymes with catalase activity (see above). Reductive regeneration of the oxidized catalytic thiol (Reaction 7.9) depends on glutathione (R′SH) or thioredoxin, glutaredoxin, cyclophilin, and tryparedoxin (Wood et al. 2003; Stork et al. 2005). Mechanism of Reaction 7.9 distinguishes the Prx subclasses.

$$\text{Cys-SH} + \text{ROOH} \rightarrow \text{Cys-SOH} + \text{ROH} \tag{7.8}$$

$$\text{Cys-SOH} + 2\text{R}'\text{SH} \rightarrow \text{Cys-SH} + \text{R}'\text{SSR}' + \text{H}_2\text{O} \tag{7.9}$$

Peroxiredoxin activity can be regulated *in vivo* by cysteine oxidation, aggregation state, phosphorylation, or limited proteolysis (Wood et al. 2003; Stork et al. 2005; Low et al. 2008). These regulatory mechanisms have, so far, mainly been investigated in eukaryotes. It is important to mention that—besides peroxide detoxification—Prx-s have also a function as regulators of redox-mediated signal transduction at least in some eukaryotes (Dietz et al. 2003; Veal et al. 2004). Thus Prx-s are important components of both the cellular antioxidant defense system and the redox homeostasis.

In plants, Prx proteins are categorized into four subclasses based on subunit composition, number and location of conserved cysteine residues as well as the sequence environment of the catalytic centre. They also show differences with respect to the reductant which is predominantly used. According to the work of Stork and colleagues (2005) we have used this categorization for the analysis of all so far available cyanobacterial genomes. Table 7.1 demonstrates that with a few exceptions all cyanobacteria have at least one representative Prx for each of the four subclasses.

7.4.3.1 1-Cys Peroxiredoxin

All investigated cyanobacterial genomes contain an ORF with similarity to 1-Cys peroxiredoxin (Table 7.1). Only *Acaryochloris marina* contains two paralogues. 1-Cys Prx contains a single conserved catalytic cysteine with the N-terminal highly conserved sequence -VLFSHPXDYTPV<u>C</u>TTE-. Its catalytic cycle is not fully understood. Reduction of the oxidized peroxidatic cysteine (Reaction 7.9) is directly by the reductant molecule. In plants it has a dual location in the cytosol and nucleus, and is preferentially expressed in the embryo and aleurone (Stacy et al. 1996). Similarly to its suggested function in plants, 1-Cys Prx in cyanobacteria may be involved in protecting nucleic acids from oxidative damage, particularly under stress. With the exception of H_2O_2- and methylviologen-induced oxidative stress, in *Synechocystis* PCC6803 *1-cys prx* mRNA increased in response to all kinds of metabolic imbalances, including irradiation, salinity, and iron deficiency (Stork et al. 2005). Heterologously expressed 1-Cys Prx from *Synechocystis* had only a low peroxidase activity but disruption of the gene significantly reduced the growth rate of *Synechocystis* PCC6803 cells (Hosaya-Matsuda et al. 2005).

7.4.3.2 2-Cys Peroxiredoxin

All investigated species have one ORF with similarity to 2-Cys peroxiredoxin (Table 7.1). Generally, 2-Cys Prx-s, including the cyanobacterial enzymes, contain two conserved cysteins far away in the sequence within fully conserved motifs (N-terminal <u>Cys</u>: -FFYPLDFTFTFV<u>C</u>PTE- and C-terminal "resolving" <u>Cys</u>: -V<u>C</u>P-). 2-Cys-Prx-s are homodimeric enzymes where the two subunits interact in the catalytic cycle and are linked via a disulfide bond in the oxidized form. In detail, regeneration of the oxidized peroxidatic cysteine from one subunit occurs via attack of the C-terminal "resolving" cysteine of the second subunit to form an intersubunit disulfide bond, which is then reduced by one of several reductant molecules. In atypical Prx (see below) the peroxidatic cysteine and its resolving cysteine are in the same polypeptide, so their reaction forms an intrachain disulfide bond.

Cyanobacterial 2-Cys Prx-s are very similar to the corresponding 2-Cys Prx-s in plants, which are localized exclusively in the chloroplast. The plant protein has a defined function in photosynthesis. Depending on its redox related oligomeric state, 2-Cys Prx cycles between a thylakoid-bound and stromal state (Konig et al. 2002). In addition to ascorbate peroxidase-mediated detoxification of Mehler reaction-driven hydrogen peroxide, a redox signaling function of 2-Cys Prx has been proposed in the context of photosynthesis (Dietz et al. 2003, 2005). Sequence comparisons of 2-Cys Prx from *Synechocystis* PCC6803 and the red alga *Porphyra*, where 2-Cys Prx is still encoded in the chloroplast genome, and the higher plant 2-Cys Prx (encoded in the nucleus) has led to the hypothesis that 2-Cys Prx originated from a former cyanobacterial endosymbiont (Baier and Dietz 1997).

Generally, in bacteria 2 Cys-Prx are suggested to detoxify peroxides generated at low rates (Wood et al. 2003). In cyanobacteria, *2-cys prx* mRNA levels were highly responsive to any type of metabolic disturbance, i.e. light, methylviologen, H_2O_2, NaCl etc. (Stork et al. 2005) which contrasts the constitutively high level expression in photosynthesizing plant cells. In *Synechocystis* PCC6803 the essential role of 2-Cys Prx in photosynthetic adaptation has been established in a deletion mutant that showed increased stress sensitivity and a disturbed peroxide detoxification capacity (Klughammer et al. 1998; Yamamoto et al. 1999). It uses thioredoxin as reductant, thus its activity is coupled to the photosynthetic electron transport system. Moreover, from gene disruption analysis in *Synechococcus elongatus* PCC7942, it was concluded that 2-Cys Prx is also essential for growth during excessive radiation and that under such conditions the mutant strain could not compete with the wild-type (Perelman et al. 2003).

7.4.3.3 Type II Peroxiredoxin

About 40% of cyanobacterial genomes contain one ORF with similarity to type II peroxiredoxin (atypical 2-Cys Prx) which can use thioredoxin and glutaredoxin as the reductant and can exist in multiple isoforms localized in plants in many subcellular compartments, including one in plastids (Horling et al. 2002). Cyanobacterial type II Prx-s contain a fully conserved N-terminal region with one cystein (-LP-GAFTPT**C**SS-), however the second (C-terminal) cysteine is not conserved in all species. It is absent in the putative type II Prx-s from all *Prochlorococcus* species, in *Microcysts aeruginosa*, *Trichodesmium erythraeum* and all *Nostocales*. However, it has been demonstrated recently (Hong et al. 2008) that all the putative type II Prx-s of these mentioned species (with exception of *Prochlorococcus* species and *Trichodesmium erythraeum*) are hybrids with a fused glutaredoxin domain at the C-terminus that containing the conserved -CXXC- domain. The corresponding fusion protein from *Nostoc* PCC7120 has been expressed heterologously in *E. coli* and was shown to have the highest peroxidase activity toward H_2O_2 using glutathione as electron donor. The calculated k_{cat} and k_{cat}/K_M values for H_2O_2 were reported to be 48 s^{-1} and $3.3 \times 10^6\ M^{-1}\ s^{-1}$, respectively (Hong et al. 2008). Immunoblot analysis revealed its occurrence in both vegetative cells and heterocysts. Expression of this hybrid protein is enhanced during late phase of vegetative and heterocyst growth (Hong et al. 2008).

(Non-hybrid) *Synechocystis* type II Prx has been demonstrated to be strongly upregulated upon treatment with methylviologen, hydrogen peroxide and, to a lesser extent, in response to light, salt, iron and nitrogen deprivation (Li et al. 2004; Kobayasgi et al. 2004; Stork et al. 2005). Moreover, it was shown that a Fur-type transcription factor plays a regulatory role in the induction of the gene in response to oxidative stress (Kobayasgi et al. 2004). Genetic disruption of the gene indicated that the gene product is essential for aerobic phototrophic growth, essential in high light (Stork et al. 2005).

7.4.3.4 Peroxiredoxin Q

Peroxiredoxin Q (Prx Q, atypical 2-Cys Prx-s) is found in all cyanobacteria. Interestingly, in contrast to *Arabidopsis thaliana* that has only a single gene, up to four paralogues are found in cyanobacterial genomes (Table 7.1). Thus, this subfamily constitutes the largest group of peroxidases in cyanobacteria. Prx-s Q are homologues of the *E. coli* bacterioferritin co-migrating protein and function as monomer (Kong et al. 2000). In typical eukaryotic Prx-s Q two cysteins are spaced by only a few amino acids (-**C**TXXX**C**-) and this pattern is also followed by about 48% of cyanobacterial Prx Q, whereas the other putative proteins lack the second cystein. For example both putative Prx Q from *Synechocystis* PCC6803 and one (out of 4) proteins from *Synechococcus elongatus* PCC7942 lack this second cysteine and thus represent atypical Prx Q-type enzymes. One of the *Synechocystis* proteins was hardly expressed under any condition tested, whereas the four Prx-s Q from *Synechococcus elongatus* PCC7942 genes showed a time dependency and stress-specific pattern of expression (Stork et al. 2005). Prx-s Q of higher plants function in the context of photosynthesis. They are localized in the chloroplast (Lamkemeyer et al. 2003), upregulated upon oxidative stress (Horling et al. 2003) and preferentially use thioredoxins as most efficient electron donors. All these data support the conclusion that the plant Prx Q functions in the context of antioxidant defence and in the redox homeostasis of photosynthesis. A similar role in cyanobacterial metabolism is plausible.

7.4.4 NADPH-Dependent Glutathione Peroxidase-Like Proteins

Glutathione peroxidase (GPx, EC 1.11.1.9 and EC 1.11.1.12) is the general name for a family of multiple isozymes that also catalyse the reduction of H_2O_2 or organic hydroperoxides to water or corresponding alcohols (Reaction 7.1) using reduced glutathione (GSH) as an electron donor ($H_2O_2 + 2\ GSH \rightarrow GSSG + 2\ H_2O$). In plants GPx-s, that are distributed in different subcellular compartments, have been often classified as the fifth group of peroxiredoxins because they can reduce peroxide with higher efficiency (sometimes exclusively) by the thioredoxin system rather than by using GSH as a reducing agent. A recent performed phylogenetic analysis demonstrates the occurrence of GPx-s in cyanobacteria as well as their scattered distribution inside the group of proteobacteria (Margis et al. 2008). The present analysis (Table 7.1) demonstrates that about 70% of all cyanobacterial genomes have an ORF with homology to plant GPx-s, with *Synechocystis* PCC6803 and two *Synechococcus* species having two paralogues.

At the molecular level plant GPx-s are closely related to animal phospholipid hydroperoxide glutathione peroxidase (PHGPx) and the corresponding proteins have three widely conserved cysteine residues, which are assumed to be essential for catalysis (Margis et al. 2008). In cyanobacterial putative GPx-s these three cysteines are found within the following conserved surroundings (from N- to C-terminus): -X**C**G-, -P**C**N-, and –F**C**X-, respectively, with the latter being absent in the *Synechocystis* PCC6803 and *Synechococcus* PCC7002 protein. It has to be mentioned that the animal enzymes may contain selenocysteine, whereas plant proteins exclusively contain cysteine.

The two GPx-s from *Synechocystis* PCC6803 were overexpressed in *E. coli* and showed no peroxide-reducing activity with glutathione but were able to use NADPH as electron donor (Tichy and Vermaas 1999; Gaber et al. 2001). This fits with the fact that Arg and Lys residues conserved in GPx-s are replaced by other amino acids and that both proteins contain an NADPH-binding domain. Only unsaturated fatty acid hydroperoxides and alkyl peroxides in combination with NADPH could be used as substrates but not H_2O_2, NADH, glutathione, ascorbate and cytochrome *c* (Gaber et al. 2001). This fits with the data reported by Tichy and Vermaas (1999) demonstrating that in a *Synechocystis* Δ*katG* mutant in the presence of 100 μM mercaptosuccinate, which is a potent and specific inhibitor of glutathione peroxidase, no changes in the peroxidase activity with H_2O_2 as oxidant and in the growth rate were observed.

In crude extracts of *Synechocystis*, NADPH-dependent GPx-like activity was measured using NADPH and α-linolenic acid hydroperide as substrates. Immunoblotting showed that both *Synechocystis* proteins are constitutively expressed. These data suggest that GPx-s in cyanobacteria seem to play a role in protection of cellular membranes from oxidative damage by reducing unsaturated fatty acids with NADPH as electron donor.

7.5 Conclusion

Peroxiredoxins are the only peroxide degrading enzymes found in all so far sequenced cyanobacterial genomes. All four classes of plant-type Prx-s are found. Apparently these enzymes have acquired specific and indispensible functions in these oxygenic phototrophic organisms and are responsible for the detoxification of hydrogen peroxide and alkyl peroxides usually produced by the cell under normal growth conditions. Additionally, they seem to have a role in stress adaptation. It has been demonstrated that their expression is upregulated in response to all kinds of metabolic imbalances and that their activity is coupled to the photosynthetic electron transport (photosynthetic adaptation).

The majority of cyanobacterial species contain also a putative NADPH-dependent glutathione peroxidase-like protein that—in addition to H_2O_2 and alkyl peroxides—can also reduce fatty acid hydroperoxides produced in membranes during oxidative stress. Whether the three other (minor) groups of peroxidases (i.e. primordial heme peroxidase, diheme cytochrome *c* peroxidase and vanadium-dependent iodoperoxidase) found in some species are expressed and play a specific physiological role in metabolism is completely unknown at present.

Genes encoding typical heme catalase are very unusual in cyanobacteria and many species (~50%) even lack a gene encoding also alternative H_2O_2 dismutating enzymes. These include bifunctional catalase-peroxidase found primarily in non-diazotrophic species and manganese catalase found mainly in nitrogen-fixing organisms. As has been demonstrated at least for KatG, these H_2O_2 dismutating enzymes might have a protective role against higher concentrations of (environmental) H_2O_2 and/or when the light-dependent peroxidases have reduced activity due to limited availability of electron donor(s).

References

Antonyuk SV, Melik-Adamyan VR, Popov AN, Lamzin VS, Hempstead PD, Harrison PM, Artymiuk PJ and Barynin VV (2000) Three-dimensional structure of the enzyme dimanganese catalase from *Thermus thermophilus* at 1 Å resolution. Crystallogr Rep 45: 105–116

Baier M and Dietz KJ (1997) The plant 2-Cys peroxiredoxin BAS1 is a nuclear-encoded chloroplast protein: its expression regulation, phylogenetic origin, and implications for its specific physiological function in plants. Plant J 12: 179–190

Barghoorn ES (1971) The oldest fossils. Sci Am 224: 30–42

Barynin VV, Whittaker MM, Antonyuk SV, Lamzin VS, Harrison PM, Artymiuk PJ and Whittaker JW (2001) Crystal structure of manganese catalase from *Lactobacillus plantarum*. Structure 9: 725–738

Bernroitner M, Zamocky M, Furtmüller PG, Peschek GA and Obinger C (2009) J Exp Bot 60: 423–440

Bertrand T, Eady NAJ, Jones JN, Jesmin JM, Nagy JM, Jamart-Gregoire B, Raven EL and Brown KA (2004) Crystal structure of *Mycobacterium tuberculosis* catalase-peroxidase. J Biol Chem 279: 38991–38999

Blankenship RE and Harman H (1998) The origin and evolution of oxygenic photosynthesis. Trends Biochem Sci 23: 94–97

Carpena X, Loprasert S, Mongkolsuk S, Switala J, Loewen PC and Fita I (2003) Catalase-peroxidase KatG of *Burkholderia pseudomallei* at 1.7 Å resolution. J Mol Biol 327: 475–489

Chelikani P, Fita I and Loewen PC (2004) Diversity of structures and properties among catalases. Cell Mol Life Sci 61: 192–208

Colin C, Leblanc C, Michel G, Wagner E, Leize-Wagner E, Van Dorsselaer A and Potin P (2005) Vanadium-dependent iodoperoxidases in *Laminaria digitata*, a novel biochemical function diverging from brown algal bromoperoxidases. J Biol Inorg Chem 10: 156–166

Dietz KJ (2003) Plant peroxiredoxins. Annu Rev Plant Biol 54: 93–107

Dietz KJ, Stork T, Finkemeier I, Lamkemeyer P, Li WX, El-Tayeb MA, Michel KP, Pistorius EK and Baier M (2005) The role of peroxiredoxins in oxygenic photosynthesis of cyanobacteria and higher plants: peroxide detoxification or redox sensing? In: Demmig-Adams B, Adams W and Mattoo A (eds) Photoprotection, photoinhibition, and environment. Dordrecht: Kluwer

Donald LJ, Krokhin OV, Duckworth HW, Wiseman B, Deemagarn T, Singh R, Switala J, Carpena X, Fita I and Loewen PC (2003) Characterization of the catalase-peroxidase KatG from *Burkholderia pseudomallei* by mass spectrometry. J Biol Chem 278: 35687–35692

Echalier A, Goodhew CF, Pettigrew GW and Fülöp V (2006) Activation and catalysis of the diheme cytochrome *c* peroxidase from *Paracoccus pantotrophus*. Structure 14: 107–117

Engleder M, Regelsberger G, Jakopitsch C, Furtmüller PG, Rüker F, Peschek GA and Obinger C (2000) Nucleotide sequence analysis, overexpression in *Escherichia coli* and kinetic characterization of *Anacystis nidulans* catalase-peroxidase. Biochimie 82: 211–219

Fülöp V, Ridout CJ, Greenwood C and Hajdu J (1995) Crystal structure of the di-haem cytochrome *c* peroxidase from *Pseudomonas aeruginosa*. Structure 3: 1225–1233

Furtmüller PG, Zederbauer M, Jantschko W, Helm J, Bogner M, Jakopitsch C and Obinger C (2006) Active site structure and catalytic mechanisms of human peroxidases. Arch Biochem Biophys 445: 199–213

Gaber A, Yoshimura K, Tamoi T, Takeda T, Nakano Y and Shigeoka S (2001) Induction and functional analysis of two reduced nicotinamid adenine dinucleotide phosphate-dependent glutathione peroxidase-like proteins in *Synechocystis* PCC6803 during the progression of oxidative stress. Plant Physiol 136: 2855–2861

Heering HA, Indiani C, Regelsberger G, Jakopitsch C, Obinger C and Smulevich G (2002) New insights into the heme cavity structure of catalase-peroxidase: a spectroscopic approach to the recombinant *Synechocystis* enzyme and selected distal cavity mutants. Biochemistry 41: 9237–9247

Hong S-K, Cha M-K and Kim I-H (2008) A glutaredoxin-fused thiol peroxidase acts as an important player in hydrogen peroxide detoxification in late-phased growth of *Anabaena* sp. PCC7120. Arch Biochem Biophys 475: 42–49

Horling F, König J and Dietz KJ (2002) Type II peroxiredoxin C, a member of the peroxiredoxin family in *Arabidopsis thaliana*: its expression and activity in comparison with other peroxiredoxins. Plant Physiol Biochem 40: 491–499

Horling F, Lamkemeyer P, Konig J, Finkemeier I, Kandlbinder A, Baier M and Dietz KJ (2003) Divergent light-, ascorbate-, and oxidative stress-dependent regulation of expression of the peroxiredoxin gene family in *Arabidopsis*. Plant Physiol 131: 317–325

Hossain MA and Asada K (1984) Purification of dehydroascorbate reductase from spinach and its characterization as a thiol enzyme. Plant Cell Physiol 25: 85–92

Hossain MA and Asada K (1985) Monodehydroascorbate reductase from cucumber is a flavin adenine dinucleotide enzyme. J Biol Chem 260: 12920–12926

Hosaya-Matsuda B, Motohashi K, Yoshimura H, Nozaki A, Inoue K, Ohmuri M and Hisabori T (2005) Anti-oxidative stress system in cyanobacteria. Significance of type II peroxiredoxin and the role of 1-Cys peroxiredoxin in *Synechocystis* sp. strain PCC 6803. J Biol Chem 280: 840–846

Ivancich A, Jakopitsch C, Auer M, Un S and Obinger C (2003) Protein-based radicals in the catalase-peroxidase of *Synechocystis* PCC6803: a multifrequency EPR investigation of wild-type and variants. J Am Chem Soc 125: 14093–14102

Jakopitsch C, Rüker F, Regelsberger G, Dockal F, Peschek GA and Obinger C (1999) Catalase-peroxidase from the cyanobacterium *Synechocystis* PCC 6803: cloning, overexpression in *Escherichia coli*, and kinetic characterization. Biol Chem 380: 1087–1096

Jakopitsch C, Regelsberger G, Furtmüller PG, Rüker F, Peschek GA and Obinger C (2001) Catalase-peroxidase from *Synechocystis* is capable of chlorination and bromination reactions. Biochem Biophys Res Commun 287: 682–687

Jakopitsch C, Kolarich D, Petutschnig G, Furtmüller PG and Obinger C (2003) Distal side tryptophan, tyrosine and methionine in catalase-peroxidases are covalently linked in solution. FEBS Lett 552: 135–140

Jakopitsch C, Ivancich A, Schmuckenschlager F, Wanasinghe A, Pöltl G, Furtmüller PG, Rüker F and Obinger C (2004) Influence of the unusual covalent adduct on the kinetics and formation of radical intermediates in *Synechocystis* catalase-peroxidase. J Biol Chem 279: 46082–46095

Jakopitsch C, Droghetti E, Schmuckenschlager F, Furtmüller PG, Smulevich G and Obinger C (2005) The role of the main access channel of catalase-peroxidase in catalysis. J Biol Chem 280: 42411–42422

Jakopitsch C, Vlasits J, Wiseman B, Loewen PC and Obinger C (2007) Redox intermediates in the catalase cycle of catalase-peroxidases from *Synechocystis* PCC 6803, *Burkholderia pseudomallei*, and *Mycobacterium tuberculosis*. Biochemistry 46: 1183–1193

Klughammer B, Baier M and Dietz KJ (1998) Inactivation by gene disruption of 2-cysteine-peroxiredoxin in *Synechocystis* sp. PCC6803 leads to increased stress sensitivity. Physiol Plant 104: 699–706

Kobayasgi M, Ishizuka T, Katayama M, Kanehisa M, Bhattacharyya-Pakrasi M, Pakrasi HB and Ikeuchi M (2004) Response to oxidative stress involves a novel peroxiredoxin gene in the unicellular cyanobacterium *Synechocystis* sp. PCC 6803. Plant Cell Physiol 45: 290–299

Kong W, Shiota S, Shi Y, Nakayama H and Nakayama K (2000) A novel peroxiredoxin of the plant Sedum Lineare is a homologue of *Escherichia coli* bacterioferritin co-migratory protein (Bcp) Biochem J 351: 107–114

Konig J, Naier M, Horling F, Kahmann U, Harris G, Schurmann P, Dietz KJ (2002) The plant-specific function of 2-Cys peroxiredoxin-mediated detoxification of peroxides in the redox-hierarchy of photosynthetic electron flux. Proc Natl Acad Sci USA 99: 5738–5743

Lamkemeyer P, Finkemeier I, Kandlbinder A, Baier M and Dietz KJ (2003) The role of peroxiredoxin Q in the antioxidant defence system of the chloroplast. Free Radic Res 37(Suppl): 40

Lane N (2002) Oxygen. The Molecule that made the World. Oxford: Oxford University Press

Li H, Singh AK, McIntyre LM and Sherman LA (2004) Differential gene expression in response to hydrogen peroxide and the putative PerR regulon of *Synechocysti*s sp. strain PCC6803. J Bacteriol 186: 3331–3345

Littlechild J (1999) Haloperoxidases and their role in biotransformation reactions. Curr Opin Chem Biol 3: 28–34

Low FM, Hampton MB and Winterbourn CC (2008) Peroxiredoxin 2 and peroxide metabolism in the erythrocyte. Antioxid Redox Signal 10: 1–9

Margis R, Dunand C, Teixeira FK and Margis-Pinheiro M (2008) Glutathion peroxidase family—an evolutionary overview. FEBS J 275: 3959–3970

McKay CP and Hartman H (1991) Hydrogen peroxide and the evolution of photosynthesis. Orig Life Evol Biosph 21: 157–163

Miller AG, Hunter KJ, O'Leary SJB and Hart LJ (2000) The photoreduction of H_2O_2 by *Synechococcus* sp. PCC7942 and UTEX 625. Plant Physiol 123: 625–635

Miyake C, Michihata F and Asada K (1991) Scavenging of hydrogen peroxide in prokaryotic and eukaryotic algae: acquisition of ascorbate peroxidase during the evolution of cyanobacteria. Plant Cell Physiol 32: 33–43

Mutsuda M, Ishikawa T, Takeda T and Shigeoka S (1996) The catalase-peroxidase of *Synechococcus* PCC 7942: purification, nucleotide sequence analysis and expression in *Escherichia coli*. Biochem J 316: 251–257

Oldham ML, Brash AR and Newcomer ME (2005) The structure of coral allene oxide synthase reveals a catalase adapted for metabolism of a fatty acid hydroperoxide. Proc Natl Acad Sci USA 102: 297–302

Passardi F, Zamocky M, Favet J, Jakopitsch C, Penel C, Obinger C and Dunand C (2007) Phylogenetic distribution of catalase-peroxidases: are there patches of order in chaos? Gene 397: 101–113

Perelman A, Avraham U, Hacohen D and Schwarz R (2003) Oxidative stress in *Synechococcus* sp. strain PCC7942: various mechanisms for H_2O_2 detoxification with different physiological roles. J Bacteriol 185: 3654–3660

Rasmussen B, Fletcher IR, Brocks JJ and Kilburn MR (2008) Reassessing the first appearance of eukaryotes and cyanobacteria. Nature 455: 1101–1104

Regelsberger G, Jakopitsch C, Plasser L, Schwaiger H, Furtmüller PG, Zamocky M, Peschek GA and Obinger C (2002) Occurrence and biochemistry of hydroperoxidases in oxygenic phototrophic prokaryotes (cyanobacteria). Plant Physiol Biochem 40: 479–490

Rouhier N and Jacquot J-P (2005) The plant multigenic family of thiol peroxidases. Free Radic Biol Med 38: 1413–1421

Singh R, Wiseman B, Deemagarn T, Donald LJ, Duckworth HW, Carpena X, Fita I and Loewen PC (2004) Catalase-peroxidases (KatG) exhibit NADH oxidase activity. J Biol Chem 279: 43098–43106

Smulevich G, Jakopitsch G, Droghetti E and Obinger C (2006) Probing the bifunctionality of catalase-peroxidase (KatG). J Inorg Biochem 100: 568–585

Stacy RA, Munthe E, Steinum T, Sharma B and Aaalen RB (1996) A peroxiredoxin antioxidant is encoded by a dormancy-related gene, Per1, expressed during late development in the aleurone and embryo of barley grains. Plant Mol Biol 31: 1205–1216

Stork T, Michel K-P, Pistorius EK and Dietz K-J (2005) Bioinformatic analysis of the genomes of the cyanobacteria *Synechocystis* sp. PCC 6803 and *Synechococcus elongatus* PCC 7942 for the presence of peroxiredoxins and their transcript regulation under stress. J Exp Bot 56: 3193–3206

Tamura K, Dudley J, Nei M and Kumar S (2007) MEGA4: molecular evolutionary genetics analysis (MEGA) software version 4.0. Mol Biol Evol 24: 1596–1599

Tel-Or E, Huflejt M and Packer L (1985) The role of glutathione and ascorbate in hydroperoxide removal in cyanobacteria. Biochem Biophys Res Commun 132: 533–539

Tichy M and Vermaas W (1999) In vivo role of catalase-peroxidase in *Synechocystis* sp. strain PCC6803. J Bacteriol 181: 1875–1882

Tözum SRD and Gallon JR (1979) The role of glutathione and ascorbate in hydroperoxide removal in cyanobacteria. J Gen Microbiol 111: 313–326

Veal EA, Findlay VJ, Day AM, Bozonet SM, Evans JM, Quinn J and Morgan BA (2004) A 2-Cys peroxiredoxin regulates peroxide-induced oxidation and activation of a stress-activated MAP kinase. Mol Cell 15: 129–139

Wada K, Tada T, Nakamura Y, Kinoshita T, Tamoi M, Shigeoka S and Nishimura K (2002) Crystallization and preliminary X-ray diffraction studies of catalase-peroxidase from *Synechococcus* PCC7942. Acta Crystallogr D 58: 157–159

Welinder KG (1991) Bacterial catalase-peroxidases are gene duplicated members of the plant peroxidase superfamily. Biochim Biophys Acta 1080: 215–20

Welinder KG (1992) Superfamily of plant, fungal and bacterial peroxidases. Curr Opin Struct Biol 2: 388–393

Weyland M, Hecht H-J, Kieß M, Liaud M-F, Vilter H and Schomburg D (1999) X-ray structure determination of a vanadium-dependent haloperoxidase from *Ascophyllum nodosum* at 2.0 Å resolution. J Mol Biol 293: 595–611

Whittaker MM, Barynin VV, Igarashi T and Whittaker JW (2003) Outer sphere mutagenesis of *Lactobacillus plantarum* manganese catalase disrupts the cluster core. Eur J Biochem 270: 1102–1116

Wood ZA, Schroder E, Robin Harris J and Poole LB (2003) Structure, mechanism and regulation of peroxiredoxins. Trends Biochem Sci 28: 32–40

Yamamoto H, Miyake C, Dietz K-J, Tomizawa K-I, Murata N and Yokota A (1999) Thioredoxin peroxidase in the cyanobacterium *Synechocystis* sp. PCC6803. FEBS Lett 447: 269–273

Yamada Y, Fujiwara T, Sato T, Igarashi N and Tanaka N (2002) The 2.0 Å crystal structure of catalase-peroxidase from *Haloarcula marismortui*. Nat Struct Biol 9: 691–695

Zamocky M, Regelsberger G, Jakopitsch C and Obinger C (2001) The molecular peculiarities of catalase-peroxidases. FEBS Lett 492: 177–182

Zamocky M, Furtmüller PG and Obinger C (2008a) Evolution of catalases from bacteria to humans. Antioxid Redox Signal 10: 1527–1547

Zamocky M, Jakopitsch C, Furtmüller PG, Dunand C and Obinger C (2008b) The peroxidase-cylooxygenase superfamily: reconstructed evolution of critical enzymes of the innate immune system. Proteins 71: 589–605

Zhao X, Yu H, Yu S, Wang F, Sacchettini JC and Magliozzo RS (2006) Hydrogen peroxide-mediated isoniazid activation catalyzed by *Mycobacterium tuberculosis* catalase-peroxidase (KatG) and its S315T mutant. Biochemistry 45: 4131–4140

Zubieta C, Krishna SS, Kapoor M, Kozbial P, McMullan D, Axelrod HL, Miller MD, Abdubek P, Ambing E, Astakhova T, Carlton D, Chiu H-J, Clayton T, Deller MC, Duan L, Elsliger, M-A, Feuerhelm J, Grzechnik SK, Hale J, Hampton E, Han GW, Jaroszewski L, Jin KK, Klock HE, Knuth MW, Kumar A, Marciano D, Morse AT, Nigoghossian E, Okach L, Oomachem S, Reyes R, Rife CL, Schimmel P, van den Bedem H, Weekes D, White A, Xu Q, Hodgson KO, Wooley J, Deacon AM, Godzik A, Leslrey SA and Wilson IA (2007) Crystal structures of two novel dye-decolorizing peroxidases reveal a β-barrel fold with a conserved heme binding motif. Proteins 70: 223–233

Part II
Cyanobacteria—The First Electric Power Plants ("Discovery" of Biological Electron Transport, the Clue to Most Efficient Bioenergetic Processes)

Chapter 8
History and Function: The Respiratory and Photosynthetic Electron Transport Chains

Peter Nicholls

8.1 Introduction

Grass and 'go' signals are green while blood and 'stop' signals are red. This statement not only tells us colour properties of grass, of blood, and of signals; it also reminds us of the psychological salience of colour. Grass and green soothe and reassure, while blood and red alarm and provoke action. Red is also psychologically ambivalent. Our primate ancestors' evolution of trichromatic vision (from the dichromatic type in most other mammals), by developing a red-green cone cell pigment, may reflect the usefulness of identifying ripe fruit among the leaves. So red can be both exciting and dangerous. But that there were specific pigments (Fig. 8.1) involved, in blood and in leaves, and close chemical similarities between these pigments, haems and chlorophylls, was an idea that did not develop until the latter part of the nineteenth century. It required the invention of the spectroscope—the high dispersion spectroscope for solar and stellar studies (Schellen 1872; Lockyer 1873) and the low dispersion microspectroscope for biological studies (Browning 1870; Valentin 1863; MacMunn 1880).

Haem and chlorophyll complexes are typically associated with specialised cells, organelles or with microorganisms. Chloroplasts, which contain both chlorophyll and haem systems, and mitochondria, which use only haems, are both found in green plants. Chlorophyll and haem systems, respiratory and photosynthetic are also found together in the membranes of several bacterial groups, purple sulphur bacteria, purple nonsulphur bacteria, green bacteria and cyanobacteria. Mitochondria are present in all eukaryote groups, were identified as organelles fairly late—perhaps not until the end of the nineteenth century and even then were not functionally characterised as respiratory organelles until the mid-twentieth century. Chloroplasts have been known for much longer—perhaps even having been identified by Leeuwenhoek—but functional understanding came much later.

P. Nicholls (✉)
Department of Biological Sciences, University of Essex, Colchester, Essex, CO4 3SQ, UK
e-mail: pnicholl@essex.ac.uk

G. A. Peschek et al. (eds.), *Bioenergetic Processes of Cyanobacteria,*
DOI 10.1007/978-94-007-0388-9_8,

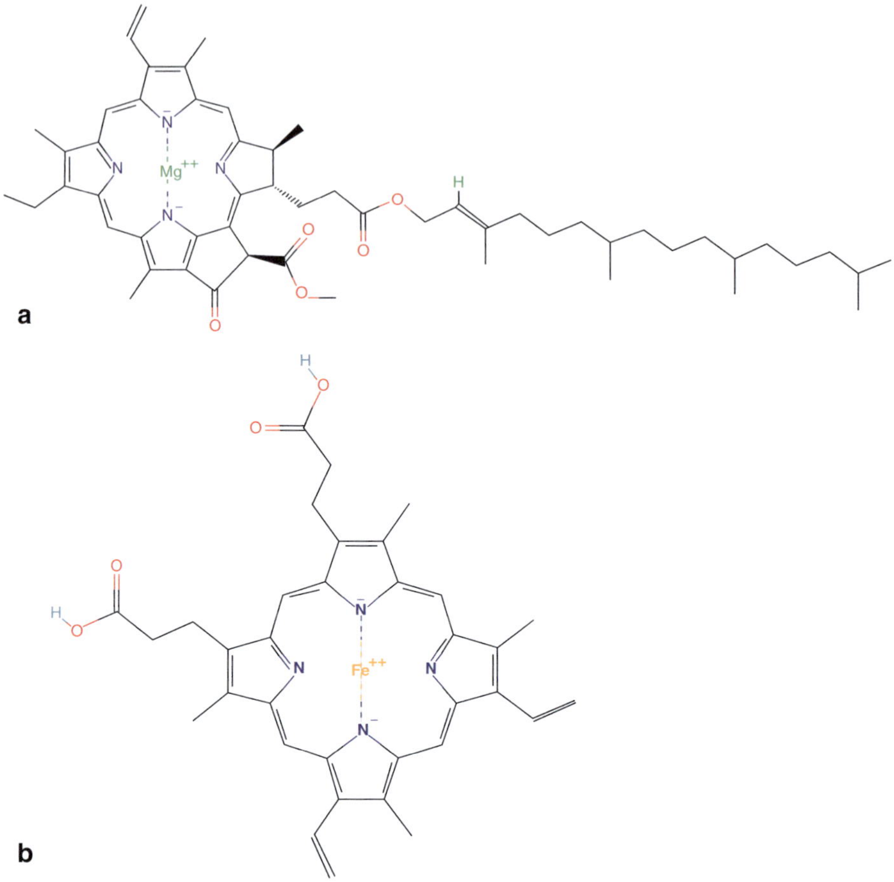

Fig. 8.1 Chlorophyll and haem structure. **a** Chlorophyll a. **b** Protohaem (haem B). Haem structure http://pubchem.ncbi.nlm.nih.gov/summary/summary.cgi?cid=444098 Chlorophyll a structure http://pubchem.ncbi.nlm.nih.gov/summary/summary.cgi?cid=6433192. For other haems and chlorophylls see the indicated pubchem web site

8.2 The Ubiquity of Chlorophyll

Chlorophyll, both in plant material and as extracted by organic solvents, was a major focus of interest in the latter half of the nineteenth century. Extensive studies were undertaken by a number of workers, among whom we may note the name of Gregor Kraus, whose extensive monograph with numerous spectral plates is amazingly now completely and freely web accessible (Kraus 1872). MacMunn, the original discoverer of the cytochromes (histohaematins), together with his nemesis in the haem area, Felix Hoppe-Seyler, were students of many natural pigments including chlorophyll derivatives. MacMunn's major 'green' focus was upon chlorophylls in animal material (MacMunn 1886) the so-called 'enterochlorophylls'. At that time, chlorophylls as well as haems could have been present in a variety of organisms other than green plants. The development of chemical understanding of the pigments came more slowly.

The towering and tragic figure who laid the basis for our current knowledge of chlorophyll chemistry (Fig. 8.1) was Richard Willstätter (Willstätter and Stoll 1913, 1928). Willstätter (1872–1942) resigned his university position presciently (in 1925) as a protest against antisemitism in German universities long before the Nazis' rise to power. He and Stoll had been able to identify and characterise several types of chlorophyll (Fig. 8.1) and examine their chemical behaviour. Unlike the haems, free chlorophylls engage in a variety of photochemical processes, some irreversible but some catalytic and reversible, which seemed at one time to be the clue to their involvement in the fixation of CO_2 in photosynthesis. Willstätter later examined the catalytic action of enzymes and developed the idea of 'trager' and 'ferment'—the enzyme being the low molecular weight chemically specific 'ferment' associated with a less specific carrier (trager) protein. In that earlier era chlorophyll seemed to be a 'ferment' with no need for 'trager'. Chlorophyll photochemistry as an *in vitro* analogue of natural photosynthesis was further developed in elaborate but ultimately heuristically barren ways by Krasnovsky and coworkers (Krasnovsky 1972). Liposomal chlorophyll photocatalysis of cytochrome *c* reduction by ascorbate (Nicholls et al. 1974) could still be used as a model system long after the distinction between 'light' and 'dark' processes had eliminated it from chemical responsibility for CO_2 fixation.

The location of chlorophyll within the chloroplast, and its relationship to proteins, also remained a puzzle for many years after the corresponding properties of haems had been clarified. Willstätter and Stoll (1928) thought of leaf chlorophyll as a colloidal aggregate, based upon a comparison between the spectral properties of chlorophyll in vivo and those of isolated preparations. Later considerations had chlorophylls defined as lipids, on the basis of solubility properties and the phytyl side chains, and hence located dissolved in the chloroplast membranes (Nicholls et al. 1974). Because magnesium was the central metal in chlorophyll, rather than iron, and as magnesium's fifth and sixth coordination positions provide rather weak ligand binding, the concept of chlorophyll as prosthetic group was not given much attention. Chlorophyll as a clear prosthetic group was first identified in photosynthetic reaction centers (where the actual photochemical species is a dimer) and secondarily in the multiple light harvesting complexes that occur in green plants and some bacteria. Cyanobacteria, along with the related anomalous eukaryotic red algae, are unique in using phycobilisomes containing the phycocyanin pigments as light harvesting systems instead of the peripheral LHCII complexes of green plants. Nevertheless the photosynthetic reaction centers, even in cyanobacteria, are always fed excitation energy by chlorophyll-containing 'core' antenna complexes.

Has opinion moved towards all chlorophyll being in complex with proteins of some kind? It certainly seems rather undiscriminating in its associations. Both cytochrome b_6f (Wenk et al. 2005; Cramer et al. 2006) and chloroplast ATPase (Varco-

Merth et al. 2008) contain bound chlorophyll of uncertain function if any. In the specialised chlorosome structures of green photosynthetic bacteria there seem to be stacked arrays of chlorophylls without nearby proteins (Kakitani et al. 2009; Jochum et al. 2008).

Whether other 'free' chlorophylls, like low concentrations of free haems, exist in plant membranes, seems uncertain. But the amount must be small compared with the total bound and free; this change in concept (the return of the 'trager') has changed the analysis of the photochemical process (below).

8.3 The Specificity of Haems

Willstätter's monograph (Willstätter and Stoll 1913) also treats the chemistry of haematin as an analogue of chlorophyll, while pointing out the differences due to the different central metal atom. Similar chemical studies on haem and haematin were carried out by Hoppe-Seyler (1865) and others (e.g. Kuster 1896a, b) during the same critical period at the end of the nineteenth century. The debate between Hoppe-Seyler and MacMunn (cf. Nicholls 1999) was in part driven by Hoppe-Seyler's appreciation of the complexity of haem chemistry contrasted with MacMunn's 'biological' intuitive assurance of the new haem species as native to the organisms under study.

That haems are usually permanent prosthetic groups of a number of oxidative enzymes as well as of oxygen carriers took a long time to be established. Willstätter as Hoppe-Seyler's follow-on twentieth century substitute and world authority on haem and chlorin chemistry, failed to show the haematin nature of horse radish peroxidase, a recognition left to Keilin and Mann (1937). Haems, before chlorophylls, were slowly recognised as essentially always protein associated. Free iron and free haem are dangerous in isolation—reacting with molecular oxygen to produce reactive oxygen intermediates including superoxide, hydrogen peroxide and possibly hydroxyl radicals (Oliveira and Oliveira 2002). 'Free' ferrous and ferric haems in the blood plasma are sequestered by serum albumin (Rosenfeld and Surgenor 1950). Intracellular haems released from (degraded) haem proteins are normally metabolized by haem oxygenase, recycling the iron, producing possibly reactive bile pigments and possible messenger carbon monoxide (Quinlan et al. 2008).

Other mechanisms may be needed if large amounts of haem are released during hemoprotein degradation. Thus the red cell occupying malarial parasite uses haemoglobin as its amino acid source for growth and multiplication (Egan et al. 2002). Its intracellular viability is threatened by the consequent haem accumulation. Its defence against haem-induced toxicity is an enzymatic process that aggregates the free haems into large crystalline arrays termed haemozoin (Egan et al. 2002; Slater and Cerami 1992). The protective (for the parasite) formation of haemozoin may provide one of the targets for antimalarial drugs such as chloroquine (Slater and Cerami 1992).

8.4 The Respiratory Chain

That the mammalian or other (yeast) cellular respiratory system involves a ‘chain’ process had its probable origin in Keilin’s wish to reconcile the oxygen activation theory of Warburg and the hydrogen activation theory of Wieland. It must be remembered that electrons played no part in biological/biochemical mechanistic modelling in the first few decades of the twentieth century. The electron had been discovered in the 1897–1899 era, primarily by J. J. Thomson (cf. the analysis by Pais 1986), but only as part of physics (to explain beta particles and cathode rays). Even the physical existence of atoms and molecules was uncertain during the first couple of decades of the twentieth century. Establishing the reality of atoms depended upon the work of Einstein and others, especially Perrin (1913, 1923) and Boltzmann (cf. Broda 1983).

Warburg’s interpretation (1949) of the relationship between Keilin’s cytochromes (Keilin 1966) and his own Atmungsferment specifically conflated electron transfer (in terms of the valency change of Fe) and oxygen carriage. And neither he nor Keilin seem to have had a respiratory chain concept as we understood it during the second half of the twentieth century—as a sequence of specific proteins transferring electrons or reducing equivalents ‘down’ a chain from low to higher redox potentials. Such a formulation required the insights of Chance and Williams (1956) and Slater (1958) whose single chain models dominated the field for several decades (see Nicholls 1963, 1978, 1999). These thermodynamically ‘downhill’ processes were not transferred conceptually to an analysis of photosynthetic electron transfer. The latter concept is quite recent. The conceptual linkage required the ‘straight’ biochemistry of mitochondrial membrane fractionation and the identification of distinct but multihaem and multicenter respiratory complexes (Green and Hatefi 1961; Tzagoloff et al. 1967) and the linking low molecular mass intermediary carriers coenzyme Q and cytochrome *c*.

8.5 The Photosynthetic Chain

Before the Z scheme there had been very little focus on possible electron carrier components in green plants and especially in or associated with chloroplasts. Hill came to the subject from his initial work with haemoglobin and in consequence of his awareness of Keilin (Hill and Scarisbrick 1950). Very few others involved in photosynthesis research were interested in electron carriers or chloroplast haem proteins. Why was this the case? Firstly the comparative side of photosynthesis, especially the behaviour of prokaryotes, including both eubacteria and cyanobacteria, was neglected for many years. Secondly there was a conceptual difficulty in integrating ideas about energy transfer with ideas about electron transfer.

Warburg, who may have been the first to propose something like a redox respiratory chain for cell respiration, never considered adding this to his ideas about photosynthesis. This was seen entirely as a matter of energy and his major effort was to establish the value of the quantum yield for CO_2 fixation—seen as involving a direct reaction between chlorophyll and carbon dioxide (Warburg 1951), with either a 'chemical' stoichiometry of four or a transient stoichiometry as low as one. Others with higher measured quantum requirements for CO_2 fixation did not link this to a mechanism. The discovery of the need for two different lights (Emerson and Lewis 1941, 1943), which reemphasised the mechanistic complexity of photosynthesis and justified the quantum requirement calculations of those led by Emerson and his team who obtained the higher (8–10) values, did not immediately lead to an electron transfer model.

There could be other interpretations. Franck's analysis of the energy requirements and two light phenomenon retained CO_2 fixation events that were driven by a single chlorophyll species (Franck 1941, 1949; Franck and Rosenberg 1964; Rosenberg 2004). Franck's photosynthetic model (see Fig. 8.2) is symptomatic. At first sight it is a double Z-scheme with four ('Warburgian') steps needed for a com-

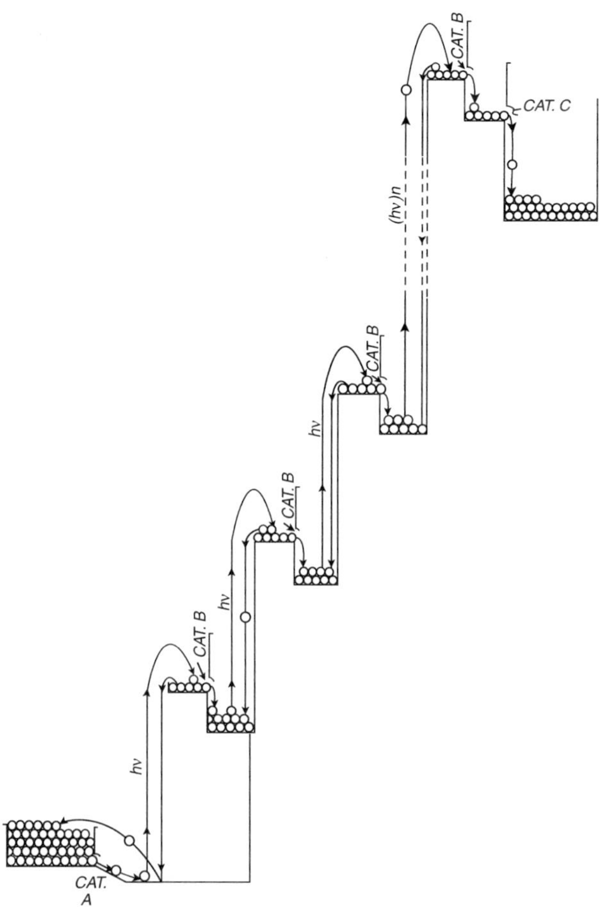

Fig. 8.2 Franck's energetic carbon dioxide fixation ladder. (From Fig. 1.5 of Franck (1949))

plete reaction. But these are not electron transfer events but energy transfer events, a sequence of energy inputs by light and intervening thermal losses of the trapped energy. The final chemical species has retained the energy of four quantum events less the thermal costs. But Franck gives no indication as to what the chemical species might be or what its relationship to the reduction of CO_2. The Nobel prize-winning physicist did not see the need for chemistry even though the distinctions between light and dark processes were concurrently being studied by Calvin and Massini (1952). The atomic bomb programme to which both Franck (whose name is remembered as the initiating author of the never delivered letter to Truman) and Kamen (1985) contributed was responsible for both the focus upon physics as a basis for understanding photosynthesis and the 'invention' of carbon-14 that permitted an analysis of the dark chemistry (Kamen 1947). Kamen later took up an interest in bacterial cytochromes that helped establish a basis for our current understanding of both bacterial respiration and photosynthesis. But the major conceptual contribution was that of Van Niel (1949) whose water-splitting model of bacterial photosynthesis broke the old photosynthetic mould in two ways—by emphasising the separation of photosynthetic biochemistry from the biochemistry of CO_2 fixation and by separating the energetics (light absorption) from the redox (H/oxygen) events. This was easier with photosynthetic eubacteria than with chloroplasts because the bacteria only catalyse a single photoreaction.

But once the theoretical conceptual apparatus was assembled the two photoreactions of green plants could be placed in their relative chemical positions. The Z scheme was the result (Hill 1954, 1965; Hill and Whittingham 1955). Figure 8.3 summarises the two steps needed to achieve this. Remarkably enough the literature tracks the changes in the thinking of Hill and Bendall (1960). Too often the process whereby ideas are generated is hidden within the final polished account. Here we are able to see the initial representation of two cytochromes dividing the redox gap between hydrogen and oxygen (Fig. 8.3a)—a thermodynamic pattern—changed into a kinetic pattern of three steps (Fig. 8.3b and 8.3c). The three steps are initially all endergonic and thus each must correspond to a photochemical event (Fig. 8.3b). But finally the central step is exergonic and thus a thermal process flanked by two photochemical processes (Fig. 8.3c). After the Z scheme was proposed a rearguard action continued for a while in which other versions were canvassed—James Franck still espoused his single site double hit model (Franck and Rosenberg 1964) and Dan Arnon and colleagues (who had earlier elaborated an even more complex three-light model) remained concerned about various inconsistent pieces of evidence that needed assimilating to the Z scheme (Arnon 1995).

The Z scheme appeared to validate theoretically the Emersonian higher values (8–10) for the quantum requirement of green plant or algal photosynthesis. Each reducing equivalent (electron) requires two quanta and therefore the reduction of CO_2 (via two NADPH molecules) needs a minimum of eight quanta. As the fixation steady state also requires small amounts of ATP (at least one ATP/CO_2 fixed) a little extra light may even be needed in excess of 8. However this assumes that no reducing equivalents can find their way to the level of NAD(P) by a single photo-

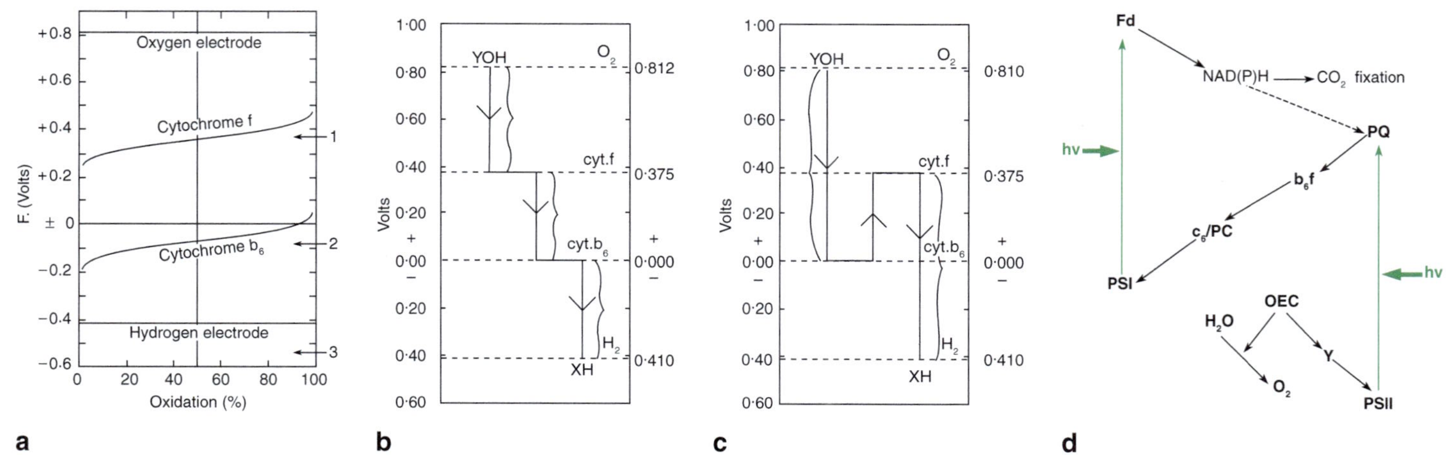

Fig. 8.3 The redox relationship between cytochrome b_6 and cytochrome *f*. **a** The original redox pattern: % oxidation of cytochromes b_6 and *f* plotted against the redox potential and compared with the hydrogen and oxygen electrode potentials (From Hill 1954). **b** A possible 3-light scheme (From Hill and Bendall 1960). **c** The original 2-light Z scheme (From Hill and Bendall 1960). **d** A more current Z scheme, as may be functioning in cyanobacterial photosynthesis. PSI: photosystem 1; PSII: photosystem 2; c_6: cyanobacterial cytochrome c_6; PC: plastocyanin; Fd: ferredoxin; b_6f: complex III (the *bc* complex); OEC: oxygen-evolving complex; Y: primary electron donor to PSII; PQ: plastoquinone

event. Osborne and Geider (1987), in a careful analysis of published data, find that the minimum quantum requirement may be as low as 6 rather than 8. How could this arise? Figure 8.3d outlines the possible pathways within an updated Z scheme (cf. Fig. 8.5b below). If the chloroplast (or cyanobacterial thylakoid membrane) contains a reversible NADH or NADPH dehydrogenase system (the equivalent of the respiratory chain's complex I) then the membrane energisation (reversing the 'cyclic' electron flow pathway in Fig. 8.3d indicated by the dashed arrow) could conceivably drive electron transfer uphill from the PQ to the pyridine nucleotide level and hence permit the lower quantum requirements calculated by Osborne and Geider (1987).

Does such a dehydrogenase exist? The evidence is equivocal. Complex I is a widespread enzyme of some antiquity (Friedrich and Scheide 2000) and some chloroplasts show NAD(P)H-PQ reductase activity both in higher plants (Endo et al. 1997) and green algae (Jans et al. 2008). There are at least three types of NAD(P)H dehydrogenase (NDH-1) in cyanobacterial (*Synechocystis*) membranes, NDH-1L, large and reportedly definitely involved in both respiration and cyclic electron flow around PSI (Battchikova and Aro 2007) and NDH-1 M and NDH-1S (smaller, thylakoid-located, and involved in CO_2 fixation) (Zhang et al. 2004). Related complexes occur in higher plant chloroplasts (Rumeau et al. 2005; Sirpio et al. 2009). Both types of NDH can be visualised *in situ* by electron microscopy as membrane-associated particles (Fole et al. 2008). The smaller particles(1 M,1S) markedly increase in number and membrane density at low CO_2 levels. The chloroplast b_6f complex, unlike the bc_1 complex, may contain NDH-active components such as ferredoxin-NAD reductase (Zhang et al. 2001). But there is no evidence for proton pumping or proton-motive force utilising mechanisms at this level in green plants. If however the cyanobacterial respiratory chain is a complete one (with three energy conserving sites between NADH and oxygen) and if it is directly linked to the photosynthetic chain, then reversed electron transfer and the concomitant reduction in quantum requirement must be possible for any such organism. Experimental confirmation or refutation is awaited.

8.6 Bacterial Photosynthesis

The first membrane-bound electron transfer complex to be structurally analysed using X-ray crystallography was the *Rhodobacter sphaeroides* photoreaction center (Deisenhofer et al. 1984), a success that brought a Nobel prize to the three major participants. *Rhodobacter* also expresses an aa_3-like cytochrome *c* oxidase whose X-ray crystal structure is also known (Svensson-Ek et al. 2002). Electron transfer from the quinol level to the cytochrome *c* level (the thermodynamic reverse of the light reaction) is catalysed by a bc_1 complex. The overall pathways are illustrated in Fig. 8.4. Figure 8.4a shows the electron transfer pathways in a purple non-sulfur bacterium such as *Rhodobacter spheroides*. Sites of energy conservation, both respiratory and photosynthetic, are shown by the dashed red

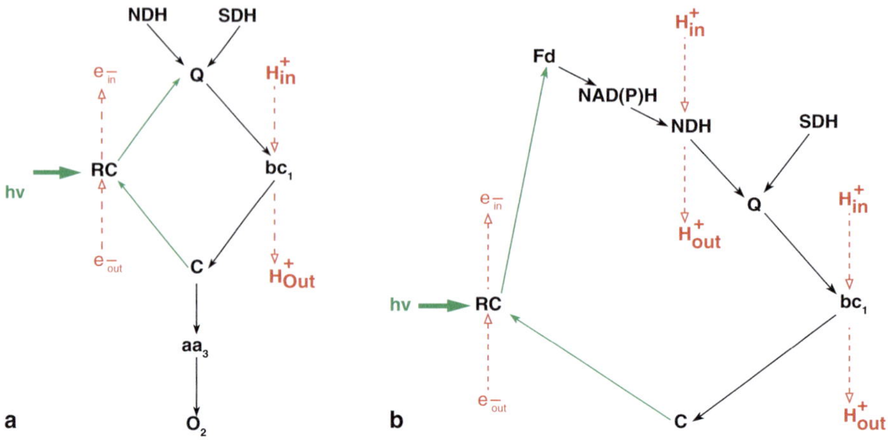

Fig. 8.4 **a** The relationships between the *Rhodobacter* electron transfer complexes: purple non-sulphur bacterial photosynthetic and respiratory pathways. Postulated energy conserving (trans-membrane charge or proton transfer) sites indicated by the arrows linking H^+_{in} to H^+_{out} or $e-_{out}$ to $e-_{in}$. **b** The relationships between the *Chlorobium* electron transfer complexes: green sulphur bacterial photosynthetic and possible respiratory(?) pathways. Postulated energy conserving (transmembrane charge or proton transfer) sites indicated by the arrows linking H^+_{in} to H^+_{out} or $e-_{out}$ to $e-_{in}$

arrows (proton transfer from inside to outside or electron transfer from outside to inside). The quinol pool can be reduced either chemically or photochemically. But as there is no way of reducing the *c* cytochromes other than via the bc_1 complex the action of the photosynthetic system is pure energisation, formation of a proton motive force and thence ATP. If the stoichiometric requirement for proton movement through complex I to reverse its redox direction is 2/electron this can in principle be generated by a single excitation event followed by the return of the electron via the bc_1 complex. If all that were required to fix carbon dioxide was two molecules of NAD(P)H the consequence would be a Warburgian quantum requirement of four. However about three moles of ATP are also needed to fix one CO_2. The stoichiometry of the *Rhodobacter* ATPase is uncertain but probably lies between 3 and 4 translocated protons per mole ATP requiring an additional 9–12 charge translocations or 4.5 to 6 excitation events per CO_2 fixed. The theoretical quantum requirement then becomes 8.5 to 10, close to some classical values (Larsen et al. 1952).

8.7 Other Photosynthetic Bacteria

Photosynthetic green bacteria such as *Chlorobium* (Ke 2001) are presumed to be directly capable of reducing NAD(P) via FeS centers rather than quinols. The delivery of reducing equivalents to the P-840 photocenter is via the bc_1 complex. Little is known of the transmembrane charge movements involved but analogy would suggest one charge per electron traversing the bc_1 complex and one more following the excitation process (Fig. 8.4b). Under these conditions all the electrons from the donor system will reach NAD(P) each with an associated pair of charge separations ('pumped' protons). A role for cyclic electron transfer in the green bacteria is uncertain. Figures 8.4a and b illustrate the possible (still putative) differences between the green and purple non-S electron transfer pathways.

8.8 Cyanobacterial Membranes

The only major organisms containing a membrane system with both complete (2 light reactions) photosynthetic (oxygen-evolving) and respiratory (oxygen-absorbing) systems are the cyanobacteria. The cyanobacterial membranes typically occur in two forms (Peschek et al. 2004)—the inner or thylakoid membrane (ICM) and the outer or plasma membrane (OCM). The thylakoid membrane typically contains both photosynthetic and respiratory complexes. The plasma membrane, like that of eubacteria and the plant chloroplast outer membrane (Jager-Vottero et al. 1997), contains only respiratory complexes with some extra cytochrome *c* components and carotenoids.

Electron transport involves equivalents of all four of the eukaryotic respiratory complexes (Paumann et al. 2005; Peschek 1987) but with the normally photosynthetic cytochrome b_6f complex and plastoquinone (Peschek 1983) replacing the more usual cytochrome bc_1 and ubiquinone (coenzyme Q)—and therefore forming a unique respiratory chain (Fig. 8.5a). How much electron transfer takes place from the NDH and SDH complexes to oxygen under conditions where respiration predominates over photosynthesis? Vermaas (2001) estimates that respiratory capacity in *Synechocystis* PCC 6803 is only about 10% that of photosynthesis. Both SDH and cytochrome oxidase have lower capacities and of the four respiratory complexes only the b_6f complex can transfer electrons and keep pace with the two photosystems in the light. NADPH dehydrogenase works even more slowly (about 1–2% of the photosynthetic rate). The succinate pool will therefore be the major system keeping ATP levels up in the dark. If there is no means of sustaining succinate levels outside the operation of the Krebs cycle then this activity will slow as cycle intermediates deplete. The other role for b_6f reductants and oxidants is to keep a balance between the two photochemical processes. If excitation and charge separation involving PSI predominates then the dehydrogenase complexes can supply the needed electrons. If excitation and charge separation involving PSII predominates then the oxidase complexes can remove the extra electrons. Figure 8.5b illustrates the possible patterns.

To achieve this the dehydrogenases have to be adapted to reduce plastoquinone. More unusually the aa_3-type oxidases need to be able to react with the PSI reductant cytochrome c_6 or plastocyanin (see below). These linking carriers are discussed elsewhere in this volume (Bendall et al. 2011; Krogmann 2011). The probable over-

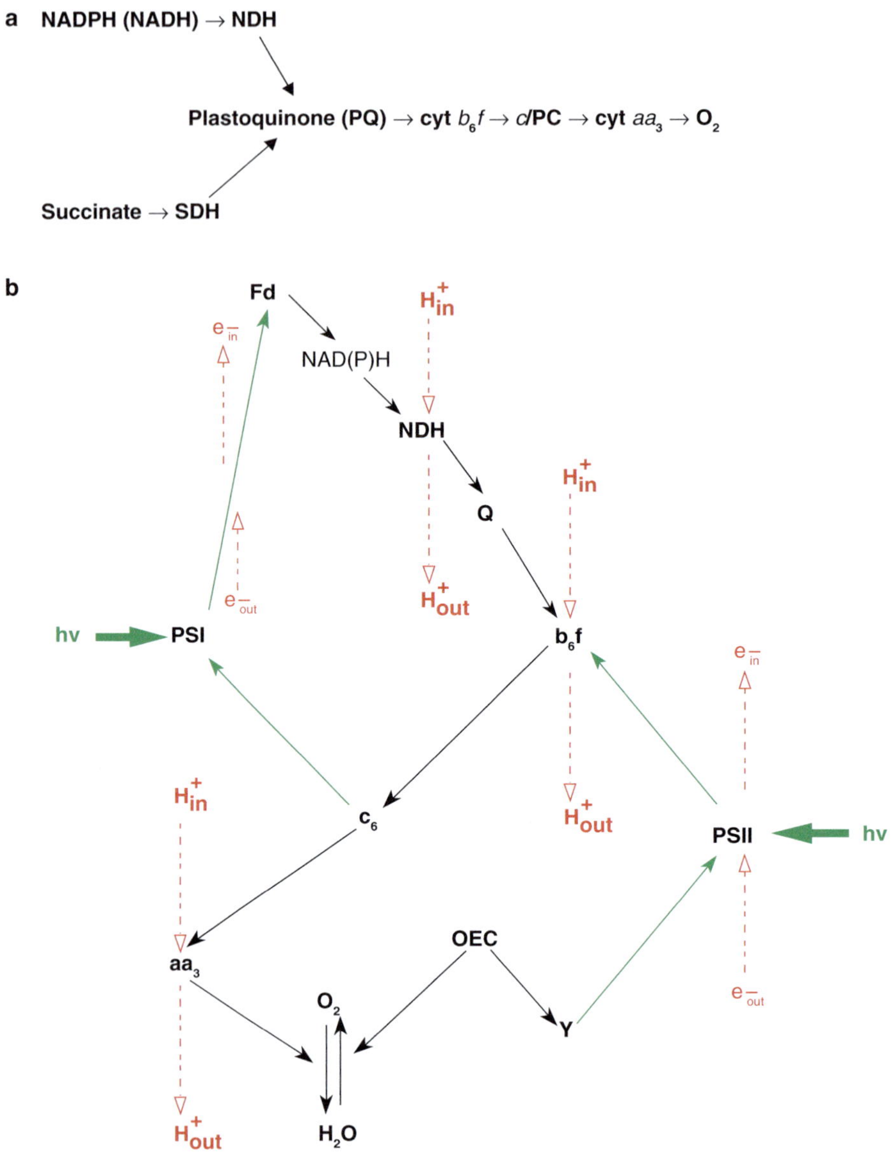

Fig. 8.5 Cyanobacterial chains. **a** The unique respiratory chain complex sequence. **b** The combined respiratory and photosynthetic chain complexes. Components as in Fig. 8.3d with the addition of the respiratory chain components of Fig. 8.5a and the putative energy conserving (transmembrane charge or proton transfer) sites as indicated by the arrows linking H^+_{in} to H^+_{out} or $e-_{out}$ to $e-_{in}$

all respiratory and photosynthetic redox chain components in the two membranes are shown in Fig. 8.5b, which combines the classical if modified respiratory chain of Fig. 8.5a with the cyanobacterial Z scheme of Fig. 8.3a, and includes the possible sites of energy conservation. The cyanobacterial inner membrane may contain at least five complexes capable of transmembrane charge (electron or proton) transfer—the two photosystems and the three energy-conserving respiratory complexes (complexes I, III and IV). But we still need to prove energy conservation at some of these proposed sites.

8.9 Cyanobacterial Oxidases

The presence and functioning of an aa_3-type cytochrome *c* oxidase in both cyanobacterial outer membranes and inner thylakoid membranes was demonstrated conclusively in a substantial series of papers by Peschek et al. (1989). Under the usual growth conditions expression levels are however low (suggesting a redox controlling rather than major bioenergetic role) but appreciable levels of aa_3 are seen in the specialised heterocyst cells (Wastyn et al. 1988) and the levels in vegetative cells are increased under salt stress (Molitor et al. 1986). Spectroscopy *in situ* is difficult due to low expression levels and the presence of perturbing pigments including chlorophylls, phycobilins and carotenoids, some of which mimic responses to reductants and oxidants. The kinetic behaviour towards ascorbate-cytochrome *c* is however unmistakable (Nicholls et al. 1991; Moser et al. 1991). Molecular biology subsequently showed the presence of at least two sets of cytochrome or ubiquinol oxidase genes in all typical cyanobacteria (Peschek 1996; Hart et al. 2005). *Synechocystis* species may also express a *bd*-type of terminal oxidase (Hart et al. 2005). The three classical subunits I, II and III are all present in a set of genes representing the *c* oxidase while a similar set of an analogous trio represent a ubiquinol or plastoquinol oxidase analogous to the *bo* cytochrome of *E. coli*. Although the cytochrome c_6/plastocyanin oxidase has been demonstrated as expressed and its functional properties studied the conditions for expression of the quinol oxidase have apparently not yet been discovered.

Cyanobacterial subunit II—containing the major reductant binding sites and the purple binuclear copper center (CuA)—has been expressed separately in *E. coli* (Paumann et al. 2004). It is unique among CuA subunits in containing a lengthy extra insert whose function may be to provide the extra binding needed to accommodate plastocyanin as well as cytochrome c_6. The extra 'loop' in *Synechocystis* is rich in methionine and histidine residues which could suggest another metal-binding site. However there is little sequence homology amongst the cyanobacterial family in this part of the subunit II gene (Paumann et al. 2004). The polypeptide 'turn' which seems to interact most closely with bound cytochrome *c* in eubacteria including *Paracoccus* and *Rhodobacter* is conserved in the cyanobacterial CuA subunits (the WIFTY sequence in Fig. 8.6). The NIR spectrum of the copper dimer in the expressed subunit II most closely resembles that of the other thermophile

Fig. 8.6 Two key amino acid sequences in cytochrome *c* oxidases (From Paumann et al. 2005; Bernroitner et al. 2008) Cytochrome *c* oxidase subunit II: the cytochrome *c* electron transfer site sequences. *Phormidium* CtaC (subunit II) courtesy of Chris Howe. Cytochrome *c* oxidase subunit I: the haem a_3 and proton transfer site sequences

Cytochrome *c* oxidase Subunit I –
the haem–linked proton transfer sequence

Beef heart residues 236–246	WFFGHPEVYIL
Paracoccus residues 272–282	WFFGHPEVYII
Synechocystis residues gene a 247–257	WFYSHPAVYIM
Synechocystis residues gene b 251–261	WFYSHPAVYVM

Cytochrome *c* oxidase Subunit II –
the cytochrome *c* interacting sequence

Beef heart residues 103–114	QWYWSYEYTDYE
Paracoccus residues 148–159	QWYWSYEYPNDA
Synechocystis residues gene a 178–189	QYAWIFTYPETG
Synechocystis residues gene b 162–173	QWLWTFTYPNGG
*Phormidium** residues 205–216	QYAWLFNYPNSG

Thermus thermophilus (Fig. 8.7)—the *Paracoccus* and *Rhodobacter* spectra are red-shifted (Paumann et al. 2004; Slutter et al. 1996; Lappalainen et al. 1993; Zhen et al. 1999).

The oxidase as a whole, and the major subunit I which carries the two haems *a* and a_3 and the CuB center, have not been expressed independently. The only reliable spectra available for the holoenzyme are those from the heterocysts (Wastyn et al. 1988) but the derived amino acid sequences show the usual conserved histidines responsible for ligating the three metal centers. However again, like *Thermus thermophilus*, the cyanobacterial genes show a modification of the putative proton-transferring group close to the cytochrome a_3-CuB binuclear center (sequences in Fig. 8.6). The sequence YSHPAVY replaces the highly conserved GSHPEVY sequence seen elsewhere, in which the glutamate residue seems to be closely associated with the oxidase proton pump. In the helix involved the replacement tyrosine lies almost exactly above the position occupied by glutamate in the other oxidases (now replaced by an alanine); it can therefore act as an analogous proton carrier from the so-called 'D' channel to the intrathylakoid space but at pH values more alkaline than those encountered by the eukaryotic mitochondrion and prokaryotes such as *Paracoccus* and *Rhodobacter*. Despite these differences cyanobacterial, as well as thermophilic eubacteria such as *Thermus*, pump protons like more conventional oxidase systems, as demonstrated using spheroplast preparations (Peschek 1984). The plasma membrane oxidase appears to be as effective a proton pump as the GHPEVY family, and as the thylakoid enzyme seems to be identical the latter will be expected to act in the same way (Peschek 1987). We do not however have full respiratory chain proton pumping stoichiometries for such spheroplasts (proton pumping in complexes I and III). The efficiency of overall cyanobacterial respiration and the possibility of reverse electron flow between plastoquinol and $NAD(P)^+$ are thus currently undecided questions.

In *Anacystis nidulans* (now *Synechococcus sp.*) and some other cyanobacteria (Moser et al. 1991) the expressed cytochrome *c* oxidase is present in both plasma membrane (OCM) and thylakoid membrane (ICM). It is uncertain whether both membrane types contain full respiratory chains of the type discussed above

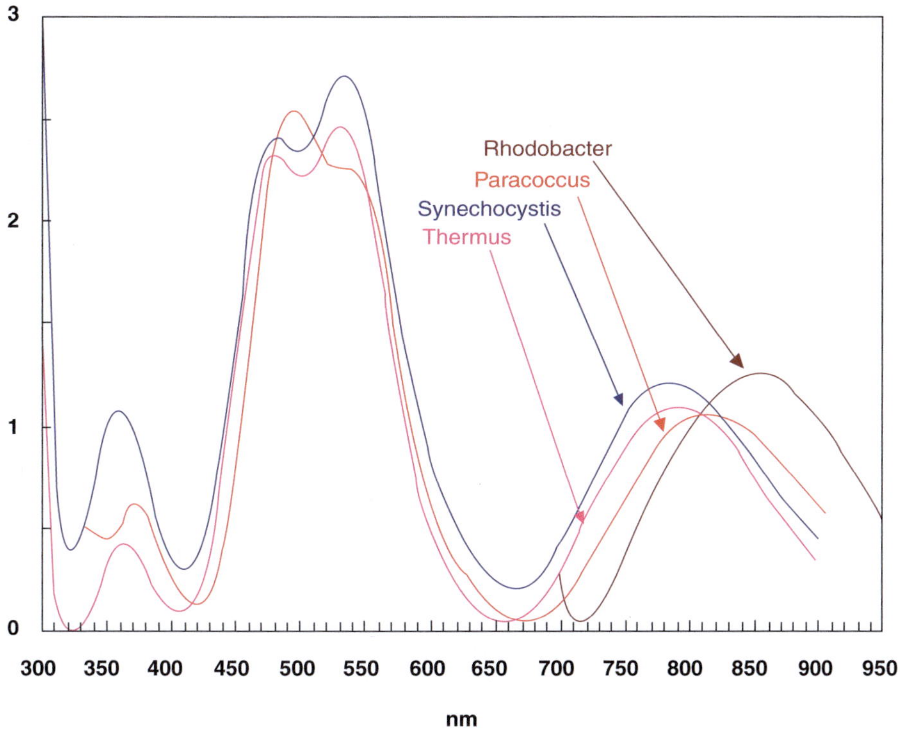

Fig. 8.7 Near infrared and visible spectra of subunit II containing the binuclear 'purple copper' center: difference spectra of three prokaryotic CuA-containing subunit II samples (*Synechocystis*, *Paracoccus* and *Thermus*) and the NIR spectral region of *Rhodobacter sphaeroides* holoenzyme, digitised and replotted from published spectra in Paumann et al. (2004), Slutter et al. (1996), Lappalainen et al. (1993), and Zhen et al. (1999). Neutral pH buffers (pH 7–7.5). Arbitrary approximate extinction coefficient values adjusted to facilitate comparisons

(Fig. 8.5). The highest activities are often seen in the thylakoid membrane. Remarkably the *Synechococcus* enzyme and some other cyanobacterial oxidases can react rapidly with both the strongly basic eukaryotic cytochromes *c* and with the equally strongly acidic cyanobacterial c_6 species. In the former case the apparent affinity of cytochrome *c* measured kinetically decreases with increasing ionic strength; in the latter case the apparent affinity of cytochrome c_6 measured kinetically increases with increasing ionic strength (Nicholls et al. 1991; Moser et al. 1991). Figure 8.8 illustrates the two patterns of ionic strength effect on the apparent (kinetically determined) affinity (as a K_m) for the two types of cytochrome *c* and the *Synechococcus* oxidase-containing thylakoid membranes. The c_6 cytochrome must engage in hydrophobic interactions at high ionic strengths when the electrostatic interactions have been neutralised. But this interaction is of similar magnitude to the attractive effect of the electrostatic interactions with cationic (horse heart) cytochrome

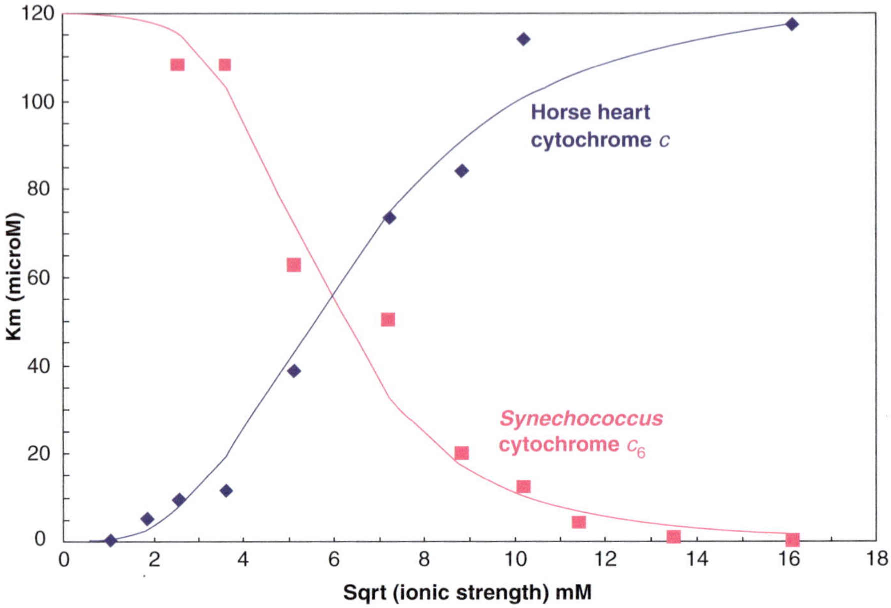

Fig. 8.8 Cyanobacterial cytochrome *c* oxidase or plastocyanin oxidase: effect of ionic strength on cytochrome c_6 binding. Data as from Nicholls et al. (1991) replotted; pH 7.8 phosphate buffers at 35°C. Apparent 'high affinity' K_m is plotted as a function of the square root of ionic strength

c. And the residual (non-electrostatic) cationic cytochrome *c* affinity is of similar magnitude to the residual c_6 affinity at low ionic strengths. The two types of interaction must represent evolutionarily adaptive responses to the *c*-oxidase interaction requirements. The seemingly anomalous behaviour of the anionic cytochrome c_6 binding also permits ionic strength control of its distribution between photosystem I and the oxidase; at low ionic strengths the photosystem with a cationic binding site can sequester most of the cytochrome; at high ionic strength the cytochrome will be largely oxidase-associated. Whether this can be of physiological significance we do not know.

Hart, Howe and coworkers (Hart et al. 2008; Bendall et al. 2011) have been able to model the docking of both cytochrome c_6 and plastocyanin to the *Phormidium luminosum* oxidase, using the sequence information for the latter and the Xray crystal structure information obtained for *Paracocccus* cytochrome *c* oxidase, plus the structural information available for plastocyanin (Paumann et al. 2005) and cytochrome c_6 (Worrall et al. 2007). At least two possible docking sites have been identified, one of which may be specialised for interaction with plastocyanin, which does not react with oxidases other than the cyanobacterial (Bendall et al. 2010; Hart et al. 2008). The need for this interaction site may account for the complexity of and inserts into the amino acid sequence of the cyanobacterial subunit II (above). A structural analysis of the latter is sorely needed.

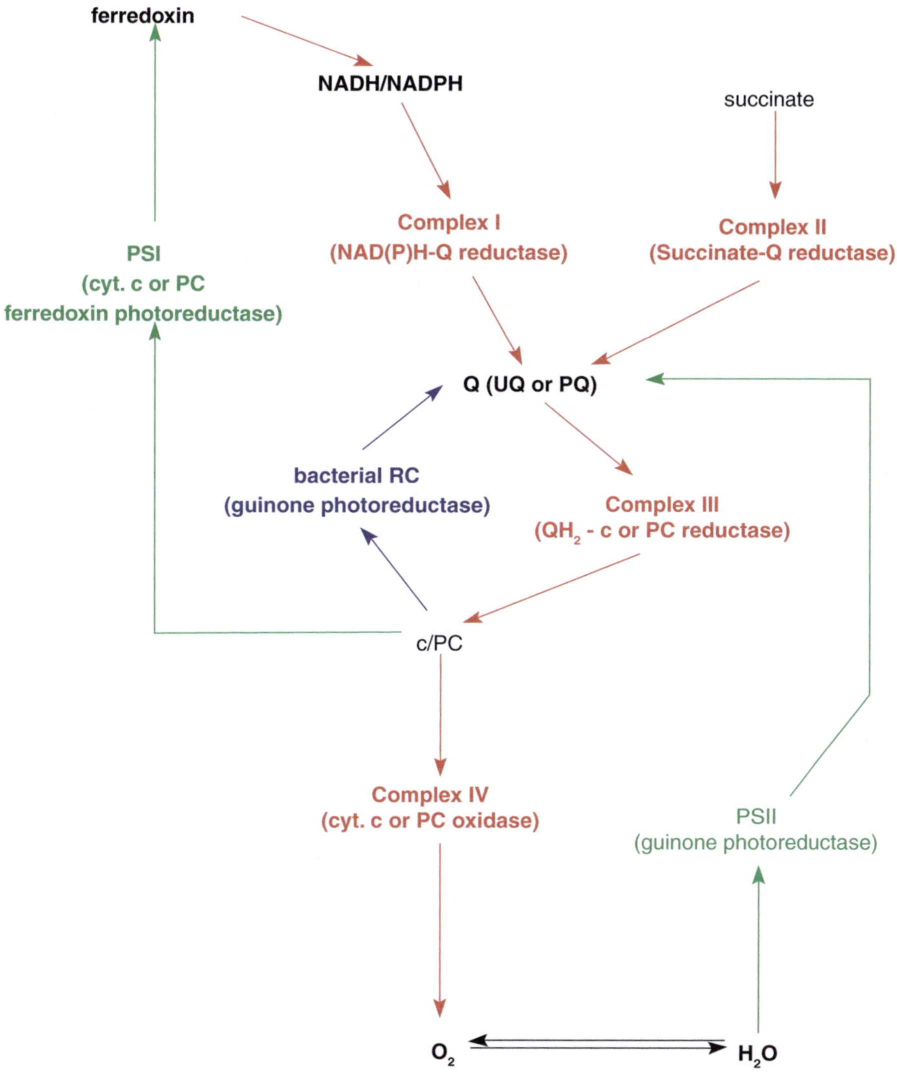

Fig. 8.9 The possible relationships between the respiratory and photosynthetic electron transport chains in cyanobacteria and photosynthetic eubacteria. *Rhodobacter* spp. (non-sulphur purple bacteria) have the blue and red systems in the same membrane. Cyanobacteria may have the green and red systems in the same membrane. Green plants have the green and red systems in different membranes (chloroplast and mitochondria) but some analogous common complexes are present in both, including: cyt. bc_1 and b_6f (quinol-c/PC reductases), including possible complex I equivalents but not complex II or complex IV equivalents

8.10 The Overall Pattern

Figure 8.9 combines the photosynthetic bacterial, cyanobacterial and green plant electron transfer systems (cf. Figs. 8.4a, b and 8.5b) into a complete scheme for a virtual organism capable of 'everything'. No existing organism maintains all these pathways. Cyanobacteria are the presumed ancestors of green plant chloroplasts and the latter, which also contain mitochondria, have lost some of the capabilities of the former, especially the cytochrome *c*/plastocyanin oxidase capability. If the latter provide a 'release valve' for excess of reductant at the c level the green plant has learned other ways to balance the activities of the two photosystems. The precursor of the dark respiration capability is an organism similar to the present day *Paracoccus*. But the precursors of the two photosystems are the ancestors of the non-sulphur purple bacteria and the green bacteria. The typical non-sulphur purple bacterium however contains a bc_1 rather than a b_6f quinol-*c* reductase. The two types are analogous and the *b* haem component may be evolutionarily homologous but this does not apply to the *f* haem component, which is not homologous with cytochrome c_1 and thus must have arisen independently. The evolutionary as well as the functional puzzles remain to be solved.

Acknowledgments The author wishes to thank Dr. Alison Telfer (Imperial College, London), Dr. Chris Howe (Cambridge), Dr. Derek Bendall (Cambridge) and Dr. Richard Geider (Essex) for helpful discussions and for providing material before publication. I also thank Ms. Julie Double (Essex) for the careful digitizing of the journal figures used to create Fig. 8.7 and Profs. Mike Wilson and Chris Cooper (Essex) for the hospitality of their laboratories which permitted the preparation of this review.

References

Arnon DI (1995) Divergent pathways of photosynthetic electron transfer: the autonomous oxygenic and anoxygenic photosystems. Photosyn Res 46: 47–71

Battchikova N and Aro EM (2007) Cyanobacterial NDH-1 complexes: multiplicity in function and subunit composition. Physiol Plant 131: 22–32

Bendall DS, Schlarb-Ridley BG and Howe CJ (2011) Transient interactions between soluble electron transfer proteins—the case of plastocyanin and cytochrome *f*. In: G. A. Peschek et al (eds) Bioenergetic Processes of Cyanobacteria, chapter 19 (this volume). Springer Science-Business Media

Bernroitner M, Zamocky M, Pairer M, Furtmueller PG, Peschek GA and Obinger C (2008) Heme-copper oxidases and their electron donors in cyanobacterial respiratory electron transport. Chem Biodivers 5: 1927–1961

Broda E (1983) Ludwig Boltzmann: man, physicist, philosopher. Ox Bow Press, Woodbridge

Browning J (1870) On a method of measuring the position of absorption bands with a microspectroscope. Monthly Microsc J 2: 68

Calvin M and Massini P (1952) The path of carbon in photosynthesis XX. Experientia 8: 445

Chance B and Williams GR (1956) The respiratory chain and oxidative phosphorylation. Adv Enzymol 17: 65–134

Cramer WA, Zhang H, Yan J, Kurisu G and Smith JL (2006) Transmembrane traffic in the cytochrome $b_6 f$ Complex. Annu Rev Biochem 75: 769–790

Deisenhofer J, Epp O, Miki K, Huber R and Michel H (1984) X-ray structure analysis of a membrane protein complex: electron density map at 3Å resolution and a model of the chromophores of the photosynthetic reaction center from *Rhodopseudomonas viridis*. J Mol Biol 180: 385–398

Egan TJ, Combrinck JM, Egan J, Hearne GR, Marques HM, Ntenteni S, Sewell BT, Smith PJ, Taylor D, Van Schalkwyk DA and Walden JC (2002) Fate of haem iron in the malaria parasite Plasmodium falciparum. Biochem J 365: 343–347

Emerson R and Lewis CM (1941) Carbon dioxide exchange and the measurement of the quantum yield of photosynthesis. Am J Bot 28: 789–804

Emerson R and Lewis CM (1943) The dependence of the quantum yield of *Chlorella* photosynthesis on wavelength of light. Am J Bot 30: 165–178

Endo T, Mi H, Shikanai T and Asada K (1997) Donation of electrons to plastoquinone by NAD(P)H dehydrogenase and by ferredoxin-quinone reductase in spinach chloroplasts. Plant Cell Physiol 38: 1272–1277

Fole M, Zhang P, Nowaczyk MM, Ogawa T, Aro EM and Boekema EJ (2008) Single particle analysis of thylakoid proteins from *Thermosynechococcus elongatus* and *Synechocystis* 6803: localization of the CupA subunit of NDH-1. FEBS Lett 582: 249–254

Franck J (1941) Some fundamental aspects of photosynthesis. Sigma Xi Quart 29: 81–105

Franck J (1949) Cited by Loomis WE. In: Franck J, Loomis WE (eds) Photosynthesis in plants, pp. 1–17, Lowa State College Press, Ames

Franck J and Rosenberg JL (1964) A theory of light utilization in photosynthesis. J Theor Biol 7: 276–301

Friedrich T and Scheide D (2000) The respiratory complex I of bacteria, archaea and eukarya and its module common with membrane-bound multisubunit hydrogenases. FEBS Lett 479: 1–5

Green DE and Hatefi Y (1961) The mitochondrion and biochemical machines. Science 133: 13–19

Hart SE, Schlarb-Ridley BG, Bendall DS and Howe CJ (2005) Terminal oxidases of cyanobacteria. Biochem Soc Trans 33: 832–835

Hart SE, Howe CJ, Mizuguchi K and Fernandez-Recio J (2008) Docking of cytochrome c_6 and plastocyanin to the aa_3-type cytochrome *c* oxidase in the cyanobacterium *Phormidium laminosum*. Protein Eng Des Sel—21: 689–698

Hill R (1954) The cytochrome *b* component of chloroplasts. Nature 174: 501–502

Hill R (1965) The biochemists' green mansions: the photosynthetic electron-transport chain in plants. Essay Biochem 1: 121–151

Hill R and Bendall F (1960) Function of the two cytochrome components in chloroplasts: a working hypothesis. Nature 186: 136–137

Hill R and Scarisbrick R (1950) The haematin compounds of leaves. New Phytol 50: 98–111

Hill R and Whittingham CP (1955) Photosynthesis. Methuen & Co., London

Hoppe-Seyler F (1865) Handbuch der physiologisch- und pathologisch- Chemischen Analyse für Ärzte und Studierende. Verlag von August Hirschwald, Berlin

Jager-Vottero P, Dorne A J, Jordanov J, Douce R and Joyard J (1997) Redox chains in chloroplast envelope membranes: spectroscopic evidence for the presence of electron carriers including iron-sulfur centers. PNASUS 94: 1597–1602

Jans F, Mignolet E, Houyoux P A, Cardol P, Ghysels B, Cuiné C, Cournac L, Peltier G, Remacle C and Franck F (2008) A type II NAD(P)H dehydrogenase mediates light-independent plastoquinone reduction in the chloroplast of *Chlamydomonas*. Proc Natl Acad Sci USA 105: 20546–20551

Jochum T, Reddy CM, Eichhofer A, Buth G, Szmytkowski J, Kalt H, Moss D and Balaban TS (2008) The supramolecular organization of self-assembling chlorosomal bacteriochlorophyll c, d, or e mimics. Proc Natl Acad Sci USA 105: 12736–12741

Kakitani Y, Koyama Y, Shimoikeda Y, Nakai T, Utsumi H, Shimizu T and Nagae H (2009) Stacking of bacteriochlorophyll c macrocycles in chlorosome from *Chlorobium limicola* as revealed by intermolecular 13C magnetic-dipole correlation, X-ray diffraction, and quadrupole coupling in 25 Mg NMR. Biochemistry 48: 74–86

Kamen MD (1947) Radioactive tracers in biology. Academic, New York

Kamen MD (1985) Radiant science, dark politics: a memoir of the nuclear age. University of California Press, Berkeley

Ke B (2001) The green bacteria. II. The reaction center—photochemistry and electron transport. In: Photosynthesis: photobiochemistry and photobiophysics. Advances in photosynthesis, vol 10, pp 159–178, Kluwer Academic, Dordrecht

Keilin D (1966) The history of cell respiration and cytochrome (Keilin J ed). Cambridge University Press, Cambridge

Keilin D and Mann T (1937) On the haematin compound of peroxidase. Proc Roy Soc B 122: 119–133

Krasnovsky AA (1972) The fragments of the photosynthetic electron transfer chain in model systems. Biophys J 12: 749–763

Kraus G (1872) Zur Kenntniss der Chlorophyllfarbstoffe und ihrer Verwandten; spectralanalytische Untersuchungen. E. Schweizerbart's Verlagshandlung (E. Koch), Stuttgart

Krogmann D (2011) The water-soluble cytochromes of cyanobacteria. In: G. A. Peschek et al (eds) Bioenergetic processes of cyanobacteria, chapter 18 (this volume). Springer Science-Business Media

Kuster W (1896a) Beiträge zur Kenntniss des Hämatins (Excerpt from the author's Habilitationsschrift). Berichte der Deutschen Chemischen Gesellschaft 29, Issue 1, 821–824

Kuster W (1896b) Beiträge zur Kenntniss des Hämatins (Habilitationsschrift). Verlag von Franz Pietzker (Tubingen)

Lappalainen P, Aasa R, Malmstrom BG and Saraste M (1993) Soluble CuA-binding domain from the *Paracoccus* cytochrome *c* oxidase. J Biol Chem 268: 26416–26421

Larsen H, Yocum CS and van Niel CB (1952) On the energetics of the photosynthesis in green sulphur bacteria. J Gen Physiol 36: 161

Lockyer JN (1873) The spectroscope and its applications. MacMillan, London

MacMunn CA (1880) The spectroscope in medicine. Lindsay & Blakiston, Philadelphia

MacMunn CA (1886) Further observations on enterochlorophyll and allied pigments. Phil Trans Roy Soc 177: 235–266

Molitor V, Erber W and Peschek GA (1986) Increased levels of cytochrome *c* oxidase and sodium-proton antiporter in the plasma membrane of *Anacystis nidulans* after growth in sodium-enriched media. FEBS Lett 204: 251–256

Moser D, Nicholls P, Wastyn M and Peschek GA (1991) Acidic cytochrome c_6 of unicellular bacteria is an indispensable and kinetically competent electron donor to cytochrome oxidase in plasma and thylakoid membranes. Biochem Int 24: 757–768

Nicholls P (1963) Cytochromes; a survey. In: Boyer P, Lardy H, Myrback K (eds) The enzymes. Academic, New York, pp 3–40

Nicholls P (1978) Cytochromes and biological oxidation, 2nd ed. Carolina Biology Reader, vol 66. Carolina Biological Supply Co., Burlington, Vermont USA

Nicholls P (1999) The mitochondrial and bacterial respiratory chains: from MacMunn and Keilin to current concepts. Chap. 1, pp 1–22 in Frontiers of cellular bioenergetics. In Papa S, Tager J, Guerriero F (eds) Kluwer Academic, New York

Nicholls P, West J and Bangham AD (1974). Chlorophyll b-containing liposomes: effect of temperature on spectrum and catalytic activity. Biochim Biophys Acta 363: 190–201

Nicholls P, Obinger C, Niederhauser H and Peschek GA (1991). Cytochrome oxidase in *Anacystis nidulans*: stoichiometries and possible functions in the cytoplasmic and thylakoid membranes. Biochim Biophys Acta 1098: 184–190

Oliveira PL and Oliveira MF (2002) Vampires, pasteur and reactive oxygen species: is the switch from aerobic to anaerobic metabolism a preventive antioxidant defence in blood-feeding parasites? FEBS Lett 525: 3–6

Osborne BA and Geider RJ (1987) The minimum photon requirement for photosynthesis: an analysis of the data of Warburg and Burk (1950) and Yuan, Evans and Daniels (1955). New Phytol 106: 631–644

Pais A (1986) Inward bound: of matter and forces in the physical world. Clarendon Press, Oxford
Paumann M, Lubura B, Regelsberger G, Feichtinger M, Kollensberger G, Jakopitsch C, Furtmueller PG, Peschek GA and Obinger C (2004) Soluble CuA domain of cyanobacterial cytochrome *c* oxidase. J Biol Chem 279: 10293–10303
Paumann M, Regelsberger G, Obinger C and Peschek GA (2005) The bioenergetic role of dioxygen and the terminal oxidase(s) in cyanobacteria. Biochim Biophys Acta 1707: 231–253
Perrin J (1913, 1923) Atoms (trans DL Hammick). Constable & Co., London
Peschek GA (1983) The cytochrome *f-b* electron-transport complex: a common link between photosynthesis and respiration in the cyanobacterium *Anacystis nidulans*. Biochem J 210: 269–272
Peschek GA (1984) Characterization of the proton-translocating cytochrome *c* oxidase activity in the plasma membranes of intact *Anacystis nidulans* spheroplasts. Plant Physiol 75: 968–973
Peschek GA (1987) Respiratory electron transport. In: Fay P, van Baalen C (eds) The cyanobacteria. Elsevier, Amsterdam
Peschek GA (1996) Cytochrome oxidase and the cta operon of cyanobacteria. Biochim Biophys Acta 1275: 27–32
Peschek GA, Wastyn M, Trnka M, Molitor V, Fry IV and Packer L (1989) Characterization of the cytochrome *c* oxidase in isolated and purified plasma membranes from the cyanobacterium *Anacystis nidulans*. Biochemistry 28: 3057–3063
Peschek GA, Obinger C and Paumann M (2004) The respiratory chain of blue-green algae (cyanobacteria). Physiol Plantarum 120: 358–369
Quinlan GJ, Lagan AL and Evans TW (2008) Haem oxygenase: a model for therapeutic intervention. Intensive Care Med 34: 595–597
Rosenberg JL (2004) The contributions of James Franck to photosynthesis research: a tribute. Photosynth Res 80: 71–76
Rosenfeld M and Surgenor DM (1950) Methemalbumin: interaction between human serum albumin and ferriprotoporphyrin IX. J Biol Chem 183: 663–677
Rumeau D, Bécuwe-Linka N, Beyly A, Louwagie M, Garin J and Peltier G (2005) New subunits NDH-M, -N, and -O, encoded by nuclear genes, are essential for plastid Ndh complex functioning in higher plants. Plant Cell 17: 219–232
Schellen H (1872) Spectrum analysis. Longman's Green & Co., London
Sirpio S, Allahverdiyeva Y, Holmstrom M, Khrouchtchova A, Haldrup A, Battchikova N and Aro EM (2009) Novel nuclear-encoded subunits of the chloroplast NAD(P)H dehydrogenase complex. J Biol Chem 284: 905–912
Slater EC (1958) The constitution of the respiratory chain in animal tissues. Adv Enzymol 20: 147–199
Slater AFG and Cerami A (1992) Inhibition by chloroquine of a novel haem polymerase enzyme activity in malaria trophozoites. Nature 355: 167–169
Slutter CE, Sanders D, Wittung P, Malmstrom BG, Aasa R, Richards JH, Gray HB and Fee JA (1996) Water-soluble, recombinant CuA-domain of the cytochrome ba_3 subunit II from *Thermus thermophilus*. Biochemistry 35: 3387–3395
Svensson-Ek M, Abramson J, Larsson G, Törnroth S, Brzezinski P and Iwata S (2002) The X-ray crystal structures of wild-type and EQ(I-286) mutant cytochrome *c* oxidases from *Rhodobacter sphaeroides*. J Mol Biol 321: 329–339
Tzagoloff A, MacLennan D, McConnell D and Green DE (1967) Studies on the electron transfer system LXVIII. Formation of membranes as the basis of the reconstitution of the mitochondrial electron transfer system. J Biol Chem 242: 2051–2061
Valentin G (1863) Der gebrauch des spectroskopes zu physiologischen und artzlichen zwecken. CF Winter's Verlagshandlung, Leipzig
Van Niel CB (1949) The comparative biochemistry of photosynthesis. In: Franck J, Loomis VE (eds) Photosynthesis in plants, Chap. 22. Lowa State College Press, Ames, pp 437–495
Varco-Merth B, Fromme R, Wang M and Fromme P (2008) Crystallization of the c14-rotor of the chloroplast ATP synthase reveals that it contains pigments. Biochim Biophys Acta 1777: 605–612

Vermaas WJ (2001) Photosynthesis and respiration in cyanobacteria. In: Encyclopedia of life sciences, Wiley, Chichester

Warburg O (1949) Heavy metal prosthetic groups and enzyme action. Clarendon Press, Oxford

Warburg O (1951) 1-quanten-mechanismus der photosynthese. Z Elektrochem Angew Phys Chem 55: 447

Wastyn M, Achatz A, Molitor V and Peschek GA (1988) Respiratory activities and the aa_3-type cytochrome oxidase in plasma and thylakoid membranes from vegetative cells and heterocysts of the cyanobacterium *Anabaena* ATCC 29413. Biochim Biophys Acta 935: 217–224

Wenk SO, Schneider D, Boronowsky U, Jager C, Klughammer C, de Weerd FL, van Roon H, Vermaas WFJ, Dekker JP and Rogner M (2005) Functional implications of pigments bound to a cyanobacterial cytochrome b_6f complex. FEBS J 272: 582–592

Willstätter R and Stoll A (1913, 1928) Investigations on chlorophyll (trans. Schertz FM, Merz AR). Science Press Printing, Lancaster

Worrall JA, Schlarb-Ridley BG, Reda T, Marcaida MJ, Moorlen RJ, Wastl J, Hirst J, Bendall DS, Luisi BF and Howe CJ (2007) Modulation of heme redox potential in the cytochrome c_6 family. J Am Chem Soc 129: 9468–9475

Zhang H, Whitelegge JP and Cramer WA (2001) Ferredoxin: NADP oxidoreductase is a subunit of the chloroplast cytochrome b_6f complex. J Biol Chem 276: 38159–38165

Zhang P, Battchikova N, Jansen T, Appel J, Ogawa T and Aro EM (2004) Expression and functional roles of the two distinct NDH-1 complexes and the carbon acquisition complex NdhD3/NdhF3/CupA/Sll1735 in synechocystis sp PCC 6803. Plant Cell 16: 3326–3340

Zhen Y, Hoganson CW, Babcock GT and Ferguson-Miller S (1999) Definition of the interaction domain for cytochrome *c* on cytochrome *c* oxidase. I. Biochemical, spectral, and kinetic characterization of surface mutants in subunit II of *Rhodobacter sphaeroides* cytochrome aa_3. J Biol Chem 274: 38032–38041

Chapter 9
Bioenergetics in a Primordial Cyanobacterium *Gloeobacter violaceus* PCC 7421

Mamoru Mimuro, Tohru Tsuchiya, Kohei Koyama and Günter A. Peschek

9.1 Introduction

Photosynthesis is a light-driven energy conversion reaction from physical energy to chemical energy through formation of chemical bonds. It sustains almost all organisms on the earth by supplying two important compounds, i.e. oxygen and carbohydrates; the former is the terminal electron acceptor for oxygenic respiration and the latter is a high energy compounds for a catabolic reaction through the TCA cycle and the respiratory electron transfer system (Blankenship 2001). In this sense, photosynthesis is called "safety- or fire-net of the earth".

Photosynthesis consists of a chain of plural reaction systems, and analysis of the overall reactions is important for general understanding of the bioenergetics. There are plenty of oxygenic photosynthetic organisms, and it is not easy to select a limited number of species for analysis. Among those, cyanobacteria are one of suitable targets for study, because several useful techniques are available on these organisms, for example, molecular biology, structure biology, ultrafast spectroscopy, physiology, and molecular evolution. Comparative analysis among several species of cyanobacteria is also useful. In addition, the whole genome information is available on many species of cyanobacteria. Even though actual experimental analysis cannot be conducted, a comparative genomic analysis will help for the general understanding by supplying information from various aspects.

We have studied the primary processes of photosynthesis on several species of cyanobacteria and understood an importance of comparative study. Recently we adopted unique species of cyanobacteria, i.e. *Gloeobacter violaceus* PCC 7421 and *Acaryochloris marina* MBIC 11017 (Mimuro et al. 2008a). The former lacks thylakoid membranes and the latter contain chlorophyll (Chl) *d* as a major pigment. The machineries for photosynthesis vary in these two organism, and those supply

M. Mimuro (✉)
Graduate School of Human and Environmental Studies, Kyoto University,
Kyoto 606-8501, Japan
e-mail: mamo-mi@mm1.mbox.media.kyoto-u.ac.jp

G. A. Peschek et al. (eds.), *Bioenergetic Processes of Cyanobacteria,*
DOI 10.1007/978-94-007-0388-9_9, © Springer Science+Business Media B.V. 2011

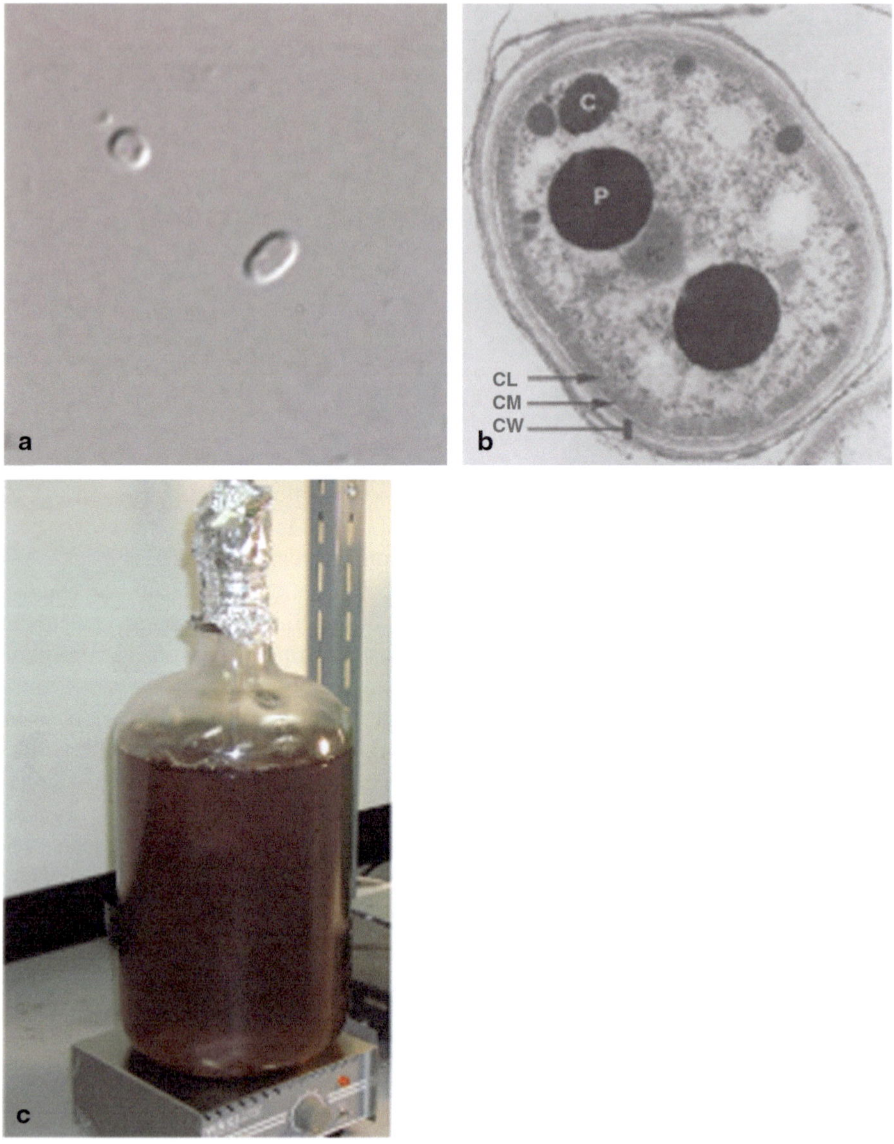

Fig. 9.1 Micrographs of intact cells of *Gloeobacter violaceus* PCC 7421. **a** Optical microscope image of cells, **b** an electron micrograph of cells (Rippka et al. 1974), and **c** an 8-liter culture of cells in our laboratory

intriguing view points for the general understanding. In this article, we mainly describe the properties of *G. violaceus*, because variations were much more evident in *G. violaceus*.

G. violaceus (Fig. 9.1a) was isolated in the early 1970s in Switzerland (Rippka et al. 1974). This species is a unique cyanobacterium that does not have thylakoid membranes (Rippka et al. 1974; Guglielmi et al. 1981) (Fig. 9.1b), the locus for photosynthesis in other cyanobacteria. Lack of thylakoid membranes in this species is the only one exception so far studied in all oxygenic photosynthetic organisms, therefore is attractive in terms of the inherent properties of cyanobacteria and evolutionary processes of oxygenic photosynthetic organisms. Usually, a new species is used for analysis to convey a new property; however *G. violaceus* has not been used frequently only because it grows very slowly. Furthermore, a mass culture is very difficult (Fig. 9.1c), and by this critical difficulty, this species is not suitable for biochemical analysis. On the contrary, in 2003 its complete genome sequence was reported (Nakamura et al. 2003), and very unique properties were suggested by its gene compositions. This can be called "a mine of gold" in terms of diversity of genome contents. Even if this species is not used for experimental analyses, its genome information is very useful for a comparative genomics and for comparative physiology. In this review, we are going to describe unique properties of *G. violaceus* for a comprehensive understanding of the bioenergetics of cyanobacteria.

9.2 Discovery of the Organism

G. violaceus PCC 7421 was discovered on a calcareous rock in Switzerland (Rippka et al. 1974). Since phycoerythrin was the major photosynthetic pigment and chlorophyll content was low (Rippka et al. 1974) (Fig. 9.2a), a color of cells looked violet. Electron microscopic observation revealed that this species is unique in lack of thylakoid membranes. According to phylogenetic analysis based on the small subunit (SSU) rDNA sequence (Fig. 9.3), it branched off from the main cyanobacterial tree in a very early stage (Nelissen et al. 1995), before branching to the species that could be an ancestor of chloroplasts. This suggests that *G. violaceus* still keeps ancestral properties of cyanobacteria in part. The complete genome sequence of this species was reported in 2003 (Nakamura et al. 2003); based on the sequence information, many unique properties of this species were shown, which leads to an evolutionary approach of photosynthetic organisms.

9.3 Genome Information and Features Revealed by Genome

The whole genome sequences of *G. violaceus* PCC 7421 (Nakamura et al. 2003) revealed unique properties of this organism, and gave important information in terms of comparative genomics, leading to a new idea from the evolutionary aspect.

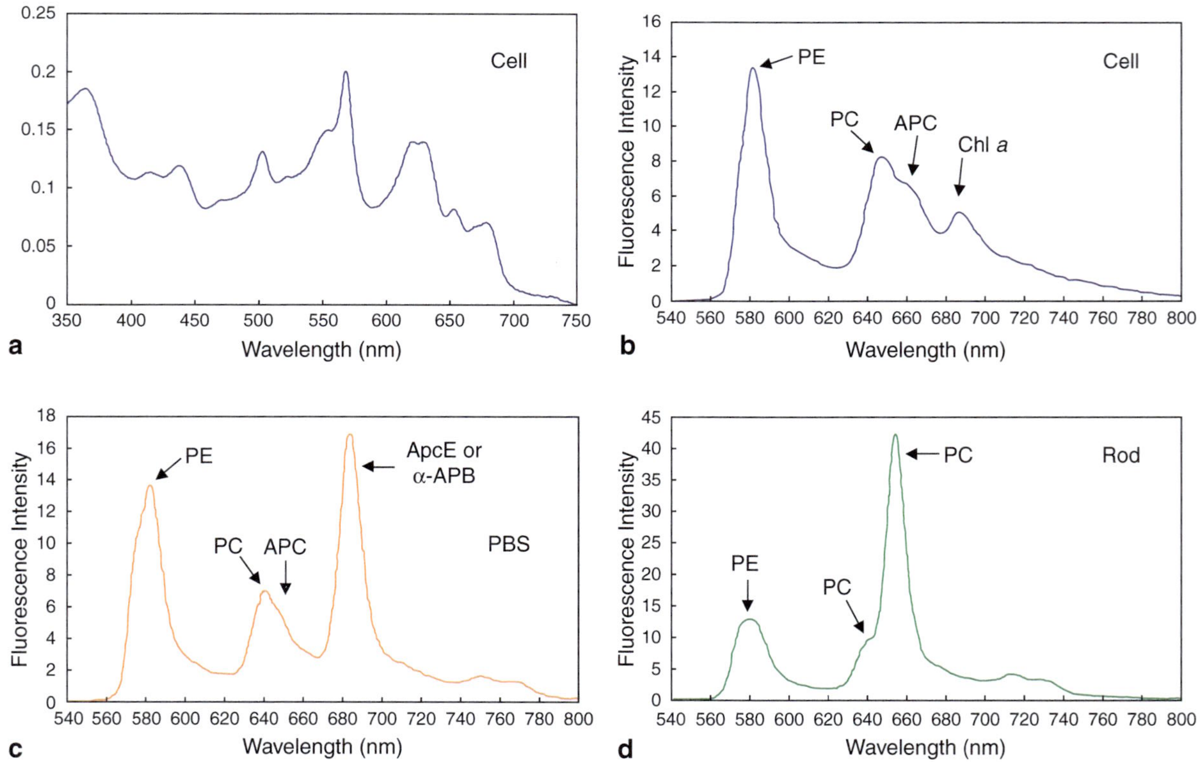

Fig. 9.2 Low temperature absorption and fluorescence spectra in *G. violaceus*. (**a**) Absorption spectrum of intact cells at 80 K, (**b**) fluorescence spectrum of intact cells at 77 K, (**c**) fluorescence spectrum of PBS at 77 K, and (**d**) fluorescence spectrum of rod at 77 K. Fluorescence spectra were measured by the excitation at 600 nm

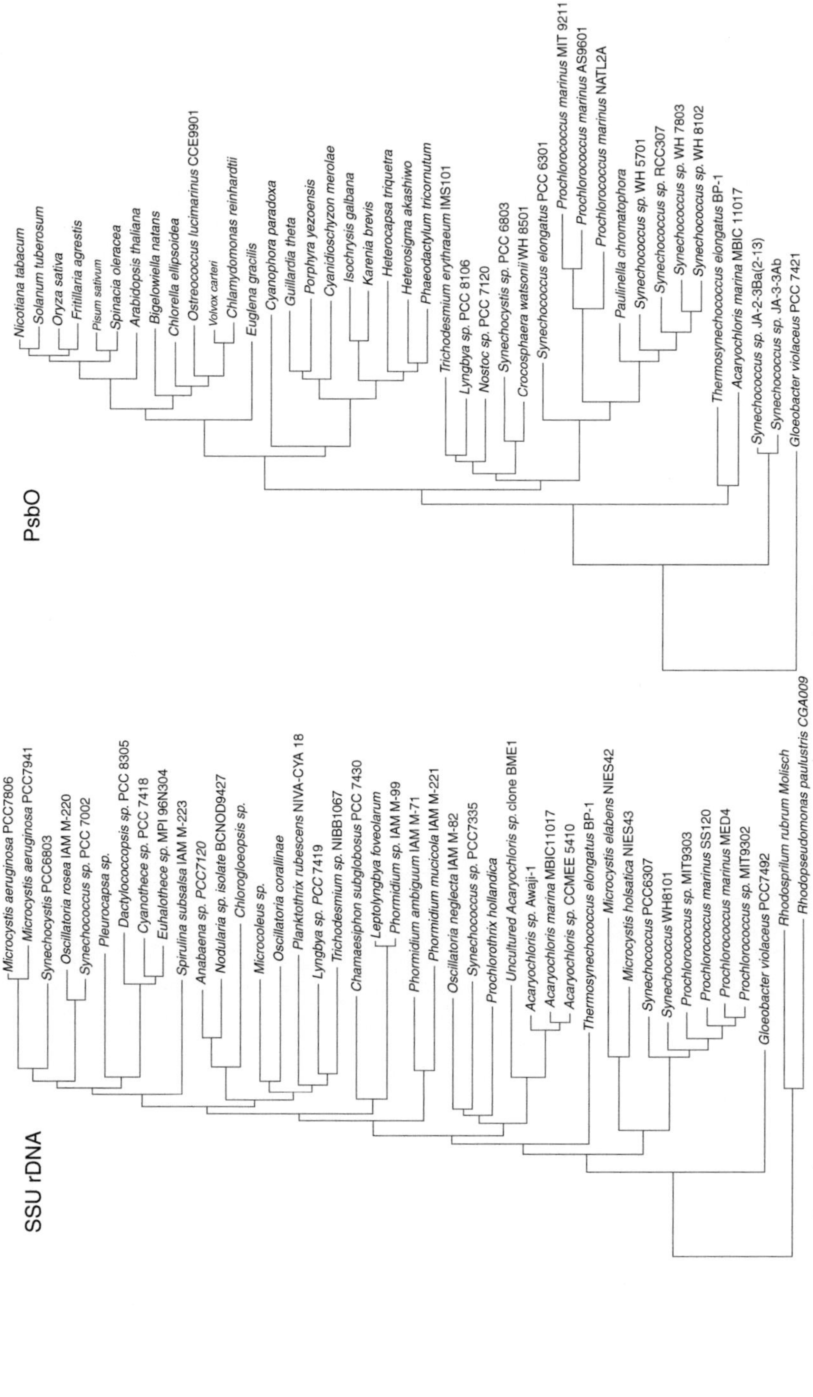

Fig. 9.3 Phylogenetic tree of *G. violaceus* drawn by the sequence similarity of SSU rDNA and a specific protein, PsbO

9.3.1 Structural Feature of the Genome

The genome of *G. violaceus* was a single circular chromosome and plasmid was not detected. A total size of genome was 4,659,019 bp with an average GC content of 62%, a very high value in cyanobacteria. The chromosome comprises 4,430 potential protein-coding regions; one additional coding region was annotated afterwards (Inoue et al. 2004). Among potential protein-coding regions, 41% (1,836) genes showed sequence similarities to genes of known function, 37% (1,635) to hypothetical genes, and remaining 22% (959) has no similarities to reported genes. A fraction of the last component was very high among cyanobacteria, indicating unique properties of this organism. As structural RNAs, one copy of rRNA gene cluster, 45 tRNA genes, and a single gene for tmRNA were assigned. Potential genes for small RNAs (B subunit of RNase P, SRP RNA, and 6Sa RNA) were also assigned.

9.3.2 Loss and/or Gain of Genes in Relation to Bioenergetics

Among 4,430 potential protein-coding regions, 158 genes were annotated to photosynthesis and respiration, and 102 genes, to energy metabolism (Tables 9.1–9.3). As for PS I, there was a set of genes for PsaA, PsaB, PsaC, PsaD, PsaE, PsaF, PsaL, and PsaM, but those for PsaI, PsaJ, PsaK, and PsaX were missing (Table 9.1). Recently, a novel subunit, PsaZ was discovered in the purified PS I complexes (Inoue et al. 2004). PsaF is a trans-membrane protein and its N-terminal region is believed to interact with soluble electron transfer catalysts such as plastocyanin (PetE) and cytochrome c_{553} or cyt c_6 (PetJ) (Table 9.1). These two electron transfer components are expected to be localized in the periplasmic space, therefore their similarities were low to other cyanobacteria. The presumptive product of *psaF* in *G. violaceus* (181 amino acid residues) was longer than those of other cyanobacteria (approximately 164 amino acid residues), which is attributed to the longer N-terminal region of the less conserved sequence.

As for PS II, five *psbA* genes, *psbB*, *psbC*, *psbD*, *psbE*, *psbF*, *psbH*, *psbI*, *psbJ*, *psbK*, *psbL*, *psbM*, *psbN*, *psbO*, *psbP*, *psbT*, *psbU*, two *psbV* genes, *psbX*, and two *psb28* genes were assigned to the genome, but genes for PsbQ, PsbY, PsbZ, and Psb27 seemed to be missing (Table 9.1). PsbO, PsbU, and PsbV, components of oxygen-evolving complex, are localized in the periplasmic space, and their sequence identities to corresponding proteins in other cyanobacteria were very low, approximately 31%. On the contrary, among *Thermosynechococcus*, *Synechocystis* sp. PCC 6803 (hereafter referred to as *Synechocystis*), and *Anabaena*, their identities were kept high (45–62%); this implied that localization of components in cells determined the sequence homology probably through the stable structures of those proteins.

Genes for the cytochrome b_6f complex, *petA*, *petB*, *petC*, *petD*, *petG*, and *petN* were assigned to the genome (Table 9.1). There was another set of *petB* (*gll1870*) and *petD* (*gll1869*) with lower sequence similarity. *petM* and *petL* could not be de-

Table 9.1 Genes for photosynthetic reaction centers and electron transfer system in *G. violaceus* PCC 7421

	Gene name	Gene no.	Function/subunit name
PS I complex	*psaA*	*glr3438*	Photosystem I P700 chlorophyll A apoprotein A1
	psaB	*glr3439*	Photosystem I P700 chlorophyll A apoprotein A2
	psaC	*gsl3287*	Photosystem I subunit VII
	psaD	*glr3701*	Photosystem I reaction center subunit II
	psaE	*gsl3408*	Photosystem I protein E
	psaF	*glr2732*	Photosystem I subunit III
	psaL	*glr2236*	Photosystem I subunit XI
	psaM	*gsl2401*	Photosystem I subunit XII
	psaZ	*gsr5001*	Novel subunit
PS II complex	*psbA*	*glr0779, glr1706, glr2322, glr2656, gll3144*	Photosystem II protein D1
	psbB	*glr2999*	Photosystem II CP47 protein
	psbC	*glr2324*	Photosystem II CP43 protein
	psbD	*glr2323*	Photosystem II protein D2
	psbE	*gsr0856*	Cytochrome b_{559} alpha-subunit
	psbF	*gsr0857*	Cytochrome b_{559} beta subunit
	psbH	*gsr3002*	Photosystem II PsbH protein
	psbI	*gsl3634*	Photosystem II PsbI protein
	psbJ	*gsr0859*	Photosystem II PsbJ protein
	psbK	*gsr2807*	photosystem II PsbK protein
	psbL	*gsr0858*	Photosystem II PsbL protein
	psbM	*gsl2997*	Photosystem II PsbM protein
	psbN	*gsl3001*	Photosystem II PsbN protein
	psbO	*glr3691*	Photosystem II manganese-stabilizing protein
	psbP	*gll1440*	Photosystem II oxygen-evolving complex 23 K protein
	psbT	*gsr3000*	Photosystem II PsbT protein
	psbU	*gll2873*	Photosystem II 12 kDa extrinsic protein
	psbV	*gll2337, gll2338*	Cytochrome c_{550}
	psbX	*gsr1874*	Photosystem II PsbX protein
	psb28	*glr1041, gsl0928*	Photosystem II protein W
Cyt b_6f complex	*petA*	*glr3039*	Apocytochrome *f*
	petB	*gll1870/gll1919*	Cytochrome b_6
	petC	*glr3038*	Plastoquinol-plastocyanin reductase (Rieske iron sulfur protein)
	petD	*gll1869/gll1918*	Cytochrome b_6f complex subunit 4
	petE	*glr2276/gll2341*	Plastocyanin
	petF	*gll3182/gsr3623*	Ferredoxin
	petG	*gsl0511*	Cytochrome b_6f complex subunit 5
	petH	*gll2295*	Ferredoxin-$NADP^+$ reductase
	petJ	*glr1906/gll1980*	Cytochrome c_{553}/cytochrome c_6
	petN	*gsl3700*	Cytochrome b_6f complex subunit 8

tected, but this may be due to a low degree of conservation for small polypeptides. A low sequence homology of PetJ and PetE has already been described above.

G. violaceus had all genes for Chl biosynthesis only except for *gun4* that stimulates Mg chelatase activity (Table 9.2, Chl specific pathway). In *Synechocystis*, *gun4* mutant exhibited the strongly decreased Chl biosynthesis and cannot grow under photoautotrophic condition (Sobotka et al. 2008). Although a gene (*glr3962*) for Gun4-like protein was found in the genome, the function of the gene is not elucidated. From the viewpoint of evolution, cyanobacteria acquired light-dependent protochlorophyllide oxidoreductase (Por) in addition to oxygen-sensitive protochlorophyllide oxidoreductase (ChlLNB). *Gloeobacter* had genes for both light-independent (*chlLNB*) and light-dependent (*por*) enzymes for protochlorophyllide reduction (Table 9.2). This observation indicates that cyanobacteria have acquired the light-dependent enzyme in the very early phase of its evolution because *G. violaceus* is believed to have diverged the earliest within the radiation of cyanobacteria.

The candidate genes for carotenoid biosynthesis were summarized in Table 9.2. Among them, four genes, *crtB* (*glr1744*), *crtI* (*glr0867*), *crtO* (*gll0394*) and *crtW* (*gll1728*), were functionally expressed in *Escherichia coli* (Steiger et al. 2005; Tsuchiya et al. 2005). On the other hand, genes corresponding to other known carotenoid biosynthetic enzymes, CrtP (phytoene desaturase), CrtQ (ζ-carotene desaturase), CrtL (lycopene cyclase), CrtR (β-carotene hydroxylase) and CrtZ

Table 9.2 Genes for synthesis of co-factors, pigments, and lipids in *G. violaceus* PCC 7421

	Gene name	Gene no.	Function/subunit name
Chl biosynthesis	*chlB*	*glr0215*	Protochlorophyllide reductase ChlB subunit
	chlD	*glr0870*	Magnesium protoporphyrin IX chelatase ChlD subunit
	chlG	*glr1809*	Chlorophyll *a* synthase
	chlH	*gll2622, glr3328*	Magnesium protoporphyrin IX chelatase ChlH subunit
	chlI	*glr1714*	Magnesium protoporphyrin IX chelatase ChlI subunit
	chlL	*gll2370*	Protochlorophyllide reductase iron-sulfur ATP-binding protein
	chlM	*glr4402*	Mg-protoporphyrin IX methyl transferase
	chlN	*gll2369*	Protochlorophillide reductase ChlN subunit
	chlP	*glr4377*	Geranylgeranyl hydrogenase
	acsF	*gll3625*	Magnesium-protoporphyrin IX monomethyl ester aerobic cyclase
	Por	*glr2486*	Protochlorophyllide oxidoreductase
		gll0878	8-Vinyl reductase

Table 9.2 (continued)

	Gene name	Gene no.	Function/subunit name
Carotenoid biosynthesis	*crtB*	*glr1744*	Phytoene synthase
	crtD	*gll2874*	3,4-Desaturase like
	crtE	*gll0416*	Geranylgeranyl pyrophosphate synthase
	crtH	*gll2133*	Carotene isomerase like
	crtI	*glr0867*	Phytoene desaturase
	crtO	*gll0394*	Beta-carotene ketolase
	crtW	*gll1728*	Beta-carotene ketolase
	cruA	*gll3598*	Lycopene cyclase
	cruP	*gll2484*	Lycopene cyclase
Phycobiliprotein biosynthesis	*cpeA*	*gll1189*	Phycoerythrin alpha subunit
	cpeB	*gll1190*	Phycoerythrin beta subunit
	cpeC	*glr1263*	Linker protein
	cpeD	*glr1264*	Linker protein
	cpeE	*glr1265*	Linker protein
	cpeG	*glr1262*	Novel linker protein
	cpeS	*glr1192*	S/U-type lyase
	cpeT	*glr1193*	T-type lyase
	cpeY	*gll1188*	Putative lyase subunit
	cpeZ	*gll1187*	Putative lyase subunit
	cpcA	*glr1185/glr3218*	Phycocyanin alpha subunit
	cpcB	*glr1184/glr3217*	Phycocyanin beta subunit
	cpcC	*glr0950/gll3219*	Phycocyanin-associated rod linker protein
	cpcD	*gsr1266/gsr1267*	Rod capping linker protein
	cpcE	*glr1268*	E/F-type lyase subunit CpcE
	cpcF	*glr1269*	E/F-type lyase subunit CpcF
	cpcJ	*glr2806*	Novel rod-core linker
	apcA	*glr1246*	Allophycocyanin alpha subunit
	apcB	*glr1247*	Allophycocyanin beta subunit
	apcC	*gsr1248*	Core linker
	apcD	*glr1181*	Allophycocyanin-B
	apcE	*glr1245*	Core-membrane linker
	apcF	*glr1930*	Core component
	pcyA	*glr2589*	Phycocyanobilin:ferredoxin oxidoreductase
	pebA	*glr1260*	Dihydrobiliverdin:ferredoxin oxidoreductase
	pebB	*glr1261*	Phycoerythrobilin:ferredoxin oxidoreductase
Menaquinone biosynthesis	*menA*	*gll1578*	1,4-Dihydroxy-2-naphtoic acid prenyltransferase
	menG	*gll0127*	2-Phytyl-1,4-benzoquinone methyltransferase

Table 9.3 Genes for respiration in *G. violaceus* PCC 7421

	Gene name	Gene no.	Function/subunit name
Complex I (NDH-1 component)	*ndhA*	*gll1584*	NADH dehydrogenase subunit 1
	ndhB	*glr3120*	NADH dehydrogenase subunit 2
	ndhC	*glr0748*	NADH dehydrogenase subunit 3
	ndhD	*glr0219/glr1004/gll2082/gll2536*	NADH dehydrogenase subunit 4
	ndhE	*gll0652*	NADH dehydrogenase subunit 4L
	ndhF	*glr0218/glr1003/gll2083/glr2599*	NADH dehydrogenase subunit 5
	ndhG	*gll0653*	NADH dehydrogenase subunit 6
	ndhH	*glr2372/gll2383*	NADH dehydrogenase subunit 7
	ndhI	*gll0654*	NADH dehydrogenase subunit I
	ndhJ	*glr0750/glr2380*	NADH dehydrogenase subunit J
	ndhK	*glr0749/glr2379*	NADH dehydrogenase subunit K
	ndhM	*glr1522*	NADH dehydrogenase subunit M
	ndhN	*gll3770*	NADH dehydrogenase subunit N
Complex II	–	*glr2944*	Succinate dehydrogenase iron sulfur protein
	–	*glr2945*	Succinate dehydrogenase chain C
	–	*glr2988*	Succinate dehydrogenase flavoprotein
bd type quinol oxidase	*cydA*	*gll1197*	Ubiquinol oxidase chain I
	cydB	*gll1196*	Ubiquinol oxidase chain I I
aa_3 type cytochrome *c* oxidase	*ctaB*	*glr2166*	Cytochrome *c* oxidase folding protein
	ctaC	*gll2164/glr0739*	Cytochrome *c* oxidase subunit I
	ctaD	*gll2163/glr0740*	Cytochrome *c* oxidase subunit II
	ctaE	*gll2162*	Cytochrome *c* oxidase subunit III
Cytochrome c_M	–	*gll2782*	Cytochrome c_M
PTOX		*gll0601*	Plasotoquinol terminal oxidase

(β-carotene hydroxylase) were not found. Usually the plant-type desaturases (CrtP and CrtQ) are used in cyanobacteria for lycopene synthesis from phytoene, however in *G. violaceus*, CrtI, the bacterial-type phytoene desaturase is functioning instead of CrtP and CrtQ (Steiger et al. 2005; Tsuchiya et al. 2005). This is the only one exception in carotenoid biosynthesis in cyanobacteria, and assigned to a primordial property of this organism. Although echinenone, one of carotenoids in this species, can be synthesized by CrtW and CrtO, the function of these two β-carotene ketolases *in vivo* is still controversial (Steiger et al. 2005; Tsuchiya et al. 2005). Thus, carotenoid biosynthesis in this species is also unique.

Biosynthesis of phycobiliproteins is carried out by many proteins. In the genome of *G. violaceus*, all known genes for apoproteins, linker proteins, biosynthetic enzymes (Frankenberg et al. 2001), and bilin lyases (Scheer and Zhao 2008) were found except for CpcG (rod-core linker) (Table 9.2). CpcG forms a multi-gene family, and disruption of all *cpcG* genes caused complete loss of highly assembled PBS in *Synechocystis* (Kondo et al. 2005). On the other hand, PBS prepared from *Gloeobacter* showed a unique morphology (Guglielmi et al. 1981). Recently, two novel linker proteins, CpeG and CpcJ, were discovered (Koyama et al. 2006; Krogmann et al. 2007). Other specific feature was found in two CpcC proteins (CpcC1 and CpcC2). CpcC proteins had N-terminal extension that was similar to both CpcD2 and the C-terminal domain of CpeC. In contrast, C-terminal domain of CpcC1 and CpcC2 was truncated by comparison with CpcC proteins in other cyanobacteria (Gutiérrez-Cirlos et al. 2006). These specific linker proteins may be involved in the formation of the unique bundle-shape structure of PBS.

Phylloquinone and menaquinone (MQ) are electron transfer components found in not only photosynthetic organisms but also many other bacteria. Especially, phylloquinone is the secondary electron acceptor of PS I in oxygenic phototroph; in the case of *G. violaceus*, phylloquinone is replaced with MQ. These naphtoquinones are synthesized from chorismate by seven enzymes (MenA to MenG). However, comparative genomic analyses revealed that only two of seven *men* genes, i.e. *menA* and *menG* that catalyze last two steps in MQ biosynthetic pathway, were found in the *G. violaceus* genome (Mimuro et al. 2005) (cf. Table 9.2). The lack of several *men* genes was also observed among several bacteria, and recently, an alternative MQ biosynthetic pathway was proved (Hiratsuka et al. 2008). Thus, we surveyed novel MQ biosynthetic genes in the *G. violaceus* genome, however no candidate genes were found. The MQ biosynthetic pathway in *Gloeobacter* is still enigmatic.

G. violaceus lacks sulfoquinovosyl diacylglycerol (SQDG) (Selstam and Campbell 1996), one of the major lipids in cyanobacteria. Whereas two genes, *sqdB* and *sqdX*, are required for SQDG biosynthesis, neither of them was found in the *G. violaceus* genome (Nakamura et al. 2003). These results strongly indicate that *Gloeobacter* does not synthesize SQDG. Although the lack of SQDG in mutant of *Synechococcus* sp. PCC 7942 did affect neither the growth nor photosynthetic activity under normal condition, reduced growth of the mutant was observed under phosphate-limiting condition (Güler et al. 1996). In the case of *Gloeobacter*, phosphatidylglycerol is the only anionic lipid like SQDG mutant. Under normal condition, this lipid composition may be enough to grow by photosynthesis for *Gloeobacter*.

9.4 Physiological Phenotype

9.4.1 Growth

G. violaceus grows very slowly under the autotrophic condition; its doubling time was 73 h at the light intensity of 50 foot candle in the BG 11 medium (Rippka et al. 1974). During culture, cells tend to stick together and form large aggregates because of the formation of multiple sheaths. It can grow under the autotrophic condition up to 30 μmol photons m^{-2} s^{-1}, however a light intensity higher than 10 μmol photons m^{-2} s^{-1} usually inhibits a growth of cells (Koyama et al. 2008). Since a growth rate is low, a high concentration of CO_2 is not necessary. We usually supply air to the medium without any addition of CO_2. Since the content of PE is high (Fig. 9.2a), a fluorescent lamp is useful for a growth. It was reported that growth inhibition by DCMU was not relieved by the addition of the common organic carbon sources including glucose (Rippka et al. 1974). Thus, the organism is regarded as obligate photoautotroph.

9.4.2 Carbon and Nitrogen Metabolism

It might be expected that this species shows unique carbon metabolic pathways however there were almost no reports on this aspect. We are able to trace the carbon metabolism based on the genome information.

This species does not grow heterotrophically, suggesting lack of sugar transporter such as *glcP* in *Synechocystis* (*sll0771*). However, artificial uptake of exogenous glucose caused metabolic disequilibrium, and introduced gene could not be maintained (Zhang et al. 1998). As for carbon metabolism, genes for pentose phosphate pathway, Calvin-Benson cycle, and carbon concentration mechanism are basically conserved (Fig. 9.4). The lack of a gene for phosphofructokinase in glycolysis was observed in the genome of *G. violaceus*, and a gene for fructose bisphosphatase that catalyzes a reverse reaction of phosphofructokinase was found. Although such gene constitution is also found not only in certain cyanobacteria but also in certain bacteria, glycolysis in *G. violaceus* is different from that in *Synechocystis* that has both genes. In general, TCA cycle in cyanobacteria was thought to be incomplete due to the absence of 2-ketoglutarate dehydrogenase, and genes for the enzyme and 2-ketoglutarate ferredoxin oxidoreductase were missing (Huynen et al. 1999). The lack of these genes was also found in the case of *G. violaceus*, and fumarate hydratase was also missing.

As for nitrogen metabolism, all genes for nitrogen uptake and enzymes for the main metabolism (nitrate reductase, nitrite reductase, glutamine synthetase and glutamate synthase) were found. The genes for nitrogenase were missing. This gene constitution is consistent with the observation that *G. violaceus* is incapable of aerobic nitrogen fixation (Rippka et al. 1974).

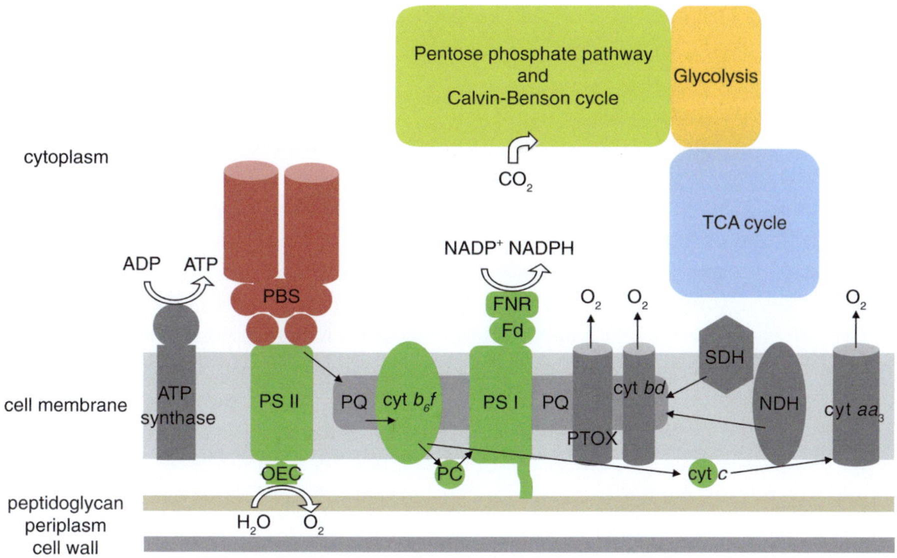

Fig. 9.4 A schematic model of the main energetic processes in relation to photosynthesis and respiration in *G. violaceus*. Major protein complexes on cell membranes and metabolic pathways in cytoplasm were schematically shown. *PTOX* plastoquinol terminal oxidase, *cyt* aa_3 terminal cytochrome *c* oxidase, *cyt bd* cytochrome *bd* type quinol oxidase, *Fd* ferredoxin, *FNR* Fd-$NADP^+$-oxidoreductase, *NDH* NAD(P)H dehydrogenase, *OEC* oxygen evolving complex, *PBS* phycobilisome, *PC* plastocyanin, *PQ* plastoquinone, *PS I* photosystem I, *PS II* photosystem II, *SDH* succinate dehydrogenase

9.5 Photosynthesis

Photosynthetic system of *G. violaceus* is essentially identical to that of other cyanobacteria on the basis of whole genome analysis (Nakamura et al. 2003). Four complexes, PS I, PS II, cyt b_6f and ATP synthase are present in cell membranes, and the electron transfer reactions proceed normally as evidence by the oxygen evolution. However, cells cannot grow under the high light condition, even though the maximum oxygen evolution activity (approximately 230 μmol O_2 (mg Chl)$^{-1}$ hr^{-1}) is comparable to other cyanobacteria. Unknown defect(s) is present in its photosynthetic system. Hereafter, characteristics of several photosynthetic components are considered.

9.5.1 PS I

PS I complexes purified from *G. violaceus* are mostly recovered as a trimer; the monomer form is present less than 10%, significantly lower than that of other cyanobacteria (Inoue et al. 2004). The complexes consist of nine subunits, PsaA, PsaB, PsaC, PsaD, PsaE, PsaF, PsaL, PsaM, and the recently discovered PsaZ (Inoue et al. 2004) (Table 9.1). PsaZ is a putative membrane-spanning small polypeptide consisting of 35 amino acids. PsaZ was found by a biochemical investigation; once the whole genome information was obtained, biochemical analysis was not necessarily performed, however this attitude was not proper to investigate the reaction system. PsaI, PsaJ, and PsaK are generally conserved in oxygenic photosynthetic organisms but are absent from this species. PsaZ may function as a substitute for one of these missing subunits. PsaB has a long C-terminal extension that has a putative peptidoglycan-binding domain in eubacteria and localizes to the periplasm (Fig. 9.4), but its actual physiological function is still unknown. In addition, the purified PS I complexes represented the novel T-shaped morphology by electron microscopy (Mangels et al. 2002), this result also supports the existence of the peptidoglycan-binding domain in the complexes. Since RC subunits possessing this functional domain have not been reported in cyanobacteria or photosynthetic bacteria, this feature is unique. These features may be related to the functional alterations and/or to the composition of polypeptides specific for *G. violaceus*.

The Chl *a*/Chl *a′* heterodimer is the primary electron donor of PS I in *G. violaceus* (Mimuro et al. 2005), although its redox potential was reported to be slightly lower (approximately 400 mV, Kato et al. 2008) than that of other cyanobacteria (415–455 mV). This difference might be due to the structure of the P700 dimer or to the amino acid environment around P700. The secondary electron acceptor of PS I (A_1) has been identified as MQ-4, rather than phylloquinone as in most oxygenic photosynthetic organisms (Mimuro et al. 2005). Since MQ-4 is frequently found in photosynthetic bacteria, leading to the idea that *G. violaceus* retains a primordial property. This is also the case for a key enzyme for carotenoid biosynthesis, phytoene desaturase, which catalyzes the conversion of phytoene into lycopene (Tsuchiya et al. 2005). *G. violaceus* uses a bacterial-type desaturase (CrtI) like many purple bacteria for lycopene biosynthesis, whereas other cyanobacteria and plants require two plant-like desaturases (CrtP and CrtQ), for which the corresponding genes are not found in the *G. violaceus* genome.

The electron donor for P700 is most probably plastocyanin (PetE). The amino acid sequence of *G. violaceus* is very different from that of other cyanobacteria. The *Synechocystis* and *G. violaceus* plastocyanins share 30–35% amino acid identity, which might reflect differences in the redox potential of P700 and its localization, i.e., in the periplasm in *G. violaceus*. The ionic environment in the periplasm is very different from that of the cytosol, which may have resulted in changes in amino acid sequences to adapt to molecular environment in the periplasm. The plastocyanin-docking protein (PsaF) is less conserved, which probably reflects its localization. An alternative electron transfer component, cyt c_6, is also found in the genome sequence, however it is not clear whether this component is expressed and functional under the normal growth condition.

9.5.2 PS II

Since almost no biochemical or biophysical analyses have been done on *G. violaceus* PS II complexes, there is very limited information on the composition and reaction processes. PS II complexes are most probably consist of polypeptides that have shown by the genome analysis (Table 9.1). Among those, five PsbA genes are present in this species; their expression levels are different depending on the growth condition (Sicora et al. 2008). On the other hand, only one PsbD gene was found, therefore D2 protein was kept the same, irrespective of composition of complexes. Another major components of PS II, i.e. PsbB (CP47) and PsbC (CP43), are present in the complexes. Three extrinsic proteins for water oxidation, i.e. PsbO, PsbU, and PsbV, are present in the periplasmic space; their sequences were less conserved among cyanobacteria (Koyama et al. 2008). When their sequences were extensively investigated, many differences were found in the regions that were important for interactions between two polypeptides in the water oxidation system. For example, there were remarkable differences between the PsbO sequences of *G. violaceus* and the other two species, including two deletions (residues 7–11 and 50–60) and a short insertion (residues 174–176). In *G. violaceus*, the N-terminus (residues 1–14) was poorly conserved, although this region is known to be functionally significant for binding and oxygen-evolving activity in spinach, and is conserved in the other cyanobacteria. Residues 56–70 form a β1–β2 loop in other cyanobacteria (De Las Rivas and Barber 2004), but residues 56–60 were absent in *G. violaceus*, probably leading to the effect of formation or structure of the β1–β2 loop. In addition, most residues in the DPKGR region are involved in binding to D1, D2 and CP47 (De Las Rivas and Barber 2004). Nevertheless, there was an insertion (residues 174–176) detected within this region in *G. violaceus*, and in addition, approximately half of the residues (13 out of 29) differed from the consensus. Since a one-residue insertion between Leu159 and Asp160 had deleterious effects on binding to PS II (Motoki et al. 2002), the short insertion (Gln174-Thr176), unique to *G. violaceus*, should lead to significant modifications of the PsbO structure, and possibly reduced binding ability to PS II. It has been reported that substitution of Lys162 with Arg reduces binding affinity and oxygen evolution (Motoki et al. 2002) and *G. violaceus* PsbO has Arg162 in fact. This residue also would affect the ability of binding to PS II. Corresponding alternation in the sequences were found in D1, D2, CP43 and CP47, leading to modification in the binding patterns (Koyama et al. 2008). Additional feature of PS II is incomplete inhibition of oxygen evolution by DCMU. A degree of insensitivity in *G. violaceus* was much enhanced, up to 40% of the total oxygen evolution activity.

9.5.3 Phycobiliproteins and Phycobilisomes

A major phycobiliprotein is phycoerythrin (PE) with three peaks, so-called "R-PE" (Fig. 9.2a). This is not common in cyanobacteria, but common in red algae. Other two molecular species of phycobiliproteins are phycocyanin (PC) and allophycocyanin (APC) (Bryant et al. 1981). A relative content of PC to PE was high, showing a clear peak in absorption spectrum; it was also the case for APC. Due to a small amount of Chl *a* relative to phycobiliproteins, a color of cells was determined by the major PE content, and looked violet in a liquid culture medium. Minor components of PBS, i.e. an anchor protein (ApcE) and APC-B (ApcD), are present in genome, and at least presence of the former (ApcE) was confirmed by fluorescence spectrum of isolated PBS (Koyama et al. 2006). The presence of the latter was suggested by the second derivative spectrum of the main fluorescence bands (see Energy transfer); the peak of the latter was somewhat shorter than the peak of ApcE (Mimuro et al. 1986).

An overall morphology of PBS in *G. violaceus* is very different from those well-known hemi-discoidal shape or hemi-spherical shape (Gantt 1981; Mimuro et al. 2008b); it is described as a bundle-like shape (Bryant et al. 1981; Koyama et al. 2006). PBS consists of two parts; one is a pillar consisting of PE and PC, and the second part is APC core, as similar to the core in other cyanobacteria containing a hemi-discoidal shape of PBS. This unique morphology might be a reflection of a relatively abundant content of phycobiliproteins and lack of thylakoid membranes in this species as a locus for the PBS attaching site. Phycobiliprotein content per unit area of membranes was very limited, thus PBS in hemi-discoidal shape might not be realized in this species. Two new linker polypeptides were identified after isolation of PBS and PMF (peptide mass fingerprinting) method. Those are assigned to the products of *glr1262* and *glr2806*; by comparing the gene sequence with partial amino acid sequences by PMF, whole amino acid sequences of the two were resolved. These two contained a linker motif three in tandem in individual proteins, and the *glr1262* product most probably functions as a rod linker (L_R) connecting PE and PC, while the *glr2806* product may function as a rod-core linker (L_{RC}). We have designated the Glr1262 as CpeG and the Glr2806 as CpcJ (Koyama et al. 2006). Locations and functions of the two new linker polypeptides were consistent with a unique morphology of PBS in this species.

9.5.4 Carotenoids

Three carotenoid species were found; β-carotene, (2S,2′S)-oscillol 2,2′-di(α-L-fucoside) and echinenone (Tsuchiya et al. 2005). Among those, a relative content of oscillol-fucoside was high; it was solubilized when cell membranes were isolated, indicating that it is loosely bound to membranes. Echinenone is a minor component among the three; its localization was not necessarily clear. By the analogy of crystal structure of thermophilic cyanobacteria (Kamiya and Shen 2003; Ferreira et al. 2004; Loll et al. 2005; Guskov et al. 2009), β-carotene, a major species, is expected to be bound to PS I and PS II core, however β-carotene is not involved in the energy transfer to PS II Chl *a*. As for PS I, β-carotene is expected to be functional in energy transfer to PS I Chl *a*, however due to lack of PS I fluorescence at 77 K, this was not yet proved.

9.5.5 Energy Transfer in the Antenna System

Energy transfer processes in this species were monitored by steady-state and time-resolved fluorescence spectrum. The most prominent feature in fluorescence spectrum of this species is lack of PS I Chl *a* band at −196°C (Koenig and Schmidt 1995); even by the time-resolved spectroscopy with a time resolution of 3 ps, this component was not detected (Mimuro et al. 2002). This property made an analysis of energy transfer difficult, because there is no index of PS I. Energy transfer from PBS to PS II Chl *a* in membranes was clearly resolved by the steady-state spectra; when PE was excited, we could detect fluorescence from PE, PC, APC and PS II Chl *a* (Fig. 9.2b), indicating a sequential energy transfer from PE to PS II Chl *a*. On the other hand, when Chl *a* was excited, an additional minor PS II Chl *a* band was observed at 695 nm, indicating that energy flow from the 685-nm component to the 695-nm component was not necessarily high.

Time-resolved fluorescence spectra and global analysis showed the energy transfer processes in details (Yokono et al. 2008). We resolved four spectral components in PE, three in PC, two in APC, and two in PS II Chl *a*. The bundle-like PBSs of *G. violaceus* showed multiple energy transfer pathways; fast (≈10 ps) and slow (≈100 and ≈500 ps) pathways were found in rods consisting of PE and PC. Energy transfer time from PE to PC was two times slower in *G. violaceus* than in *Fremyella diplosiphon* grown under green light. In *G. violaceus*, phycourobilin is present in PE and it works as an energy donor with the highest energy level; the energy transfer rate from phycourobilin to phycoerythrobilin in PE was very high compared with the rate between phycoerythrobilins. On the other hand, energy transfer from PE to PC was slow, and this was consistent with a high fluorescence intensity from PE in the steady-state spectrum. Energy transfer processes within PC and from PC to APC core were similar to those in PBSs of the hemi-discoidal shape. A unique morphology of PBS in *G. violaceus* induced an alternation in the energy transfer in PE (Fig. 9.2c, d).

9.5.6 Electron Transfer

A rate of photosynthetic oxygen evolution in this species is potentially comparable to other cyanobacteria; under the condition of 300 μmol photon m^{-2} s^{-1}, an oxygen evolution activity was approximately 230 μmol O_2 (mg Chl *a*)$^{-1}$ h^{-1} (Koyama et al. 2008). This was not low however this species cannot grow under this light condition due to unknown reason(s); there might be the other fatal reasons for a slow cell growth under the high light condition. If a partial reaction of photosynthetic electron flow is blocked, this oxygen evolution activity is not necessarily realized therefore this rate indicates the electron flow from water to $NADP^+$ is sound as that in other cyanobacteria.

A partial activity was measured only for PS I, because the complexes were isolated. Under light-saturating conditions (2,600 μmol photon m^{-2} s^{-1}), the electron transfer activities of the complexes measured by the Mehler reaction were found to be 360 ± 35 and 750 ± 70 µmol O_2 (mg Chl *a*)$^{-1}$ h^{-1} for *Gloeobacter* and *Synechocystis*, respectively (Mimuro et al. 2005). Under light-limiting conditions (130 µmol photon m^{-2} s^{-1}), the activities were 160 and 240 µmol O_2 (mg Chl *a*)$^{-1}$ h^{-1}, respectively. Even though the activity was almost half of that of *Synechocystis*, these values were enough to prove intactness in the structure and function of isolated PS I complexes.

On the other hand, there was no data on activity of the isolated PS II complexes. Isolation of the complexes has not yet been successful, therefore an activity or a property of PS II has not yet been analyzed. Instead, an oxygen-evolving activity has been measured on cells; it could be a good index for PS II activity. As mentioned above, the activity on cell basis was comparable to that in other cyanobacteria, therefore PS II activity was enough to sustain a full activity of cells.

9.6 Respiration

9.6.1 Components Deduced from Genome Information

The whole genome analysis revealed that there are at least seven components for respiration in *G. violaceus*, i.e. NAD(P)H dehydrogenase (type-1 NDH), succinate dehydrogenase (SDH), cyt $b_6 f$, plastocyanin, cyt c_6, cyt *bd* quinol oxidase, and cyt *c* oxidase (cyt aa_3) in *G. violaceus* (Fig. 9.4). Two genes for cyt $b_6 f$, *petM* and *petL*, could not be detected. Presence of an additional component, i.e. plasotoquinol terminal oxidase (PTOX), is also suggested by the genome analysis and the CN^--insensitive respiration (McDonald and Vanlerberghe 2006). Including PTOX, there might be three pathways from NADH to O_2 through PQ pool; these are NADH or SDH → PQ → cyt $b_6 f$ → plastocyanin or cyt c_6 → cyt *c* oxidase → O_2, NADH or SDH → PQ → PTOX → O_2, and NADH or SDH → PQ → cyt *bd* quinol oxidase → O_2 (Paumann et al. 2005). A contribution or electron flows of individual pathways will be surveyed more in detail near in future. Presence of the three electron transfer pathways is rare in cyanobacteria; *Nostoc* sp. PCC 7120 was reported to contain the three, however other cyanobacteria contain either of cyt *bd* quinol oxidase or alternative oxidase in addition to cyt *c* oxidase (Paumann et al. 2005).

9.6.2 Respiratory Electron Transfer Activity

A rate of respiration in *G. violaceus* is very high (50–100 µmol O_2 (mg Chl *a*)$^{-1}$ h^{-1}) (Koyama et al. 2008); this was at least ten times larger than that of other cyanobacterial species. This might be a sum of the three pathways, even though an activity of individual pathways is not known. Due to this high rate, a light compensation point raised to 35 μmol photon m^{-2} s^{-1}, an exceptionally high value in oxygenic photosynthetic organisms. This respiration rate indicates that the major energy yielding system in this species is the respiration. A unique property in respiration was the CN-insensitive respiration. Under the presence of 1 mM KCN, approximately 40% of the total respiration was not inhibited, indicating the presence of PTOX (Koyama et al. unpublished). Oxygen consumption by cyt *bd* type quinol oxidase was estimated to be far lower than that by cyt *c* oxidase in the *Synechocystis* mutant grown heterotrophically (Berry et al. 2002; Mogi and Miyoshi 2009).

9.6.3 Coupling of Respiratory Electron Transfer Chain with Photosynthetic Electron Transfer Chain on Cell Membranes

Cyanobacteria, in general, have two sites for respiration, i.e. on thylakoid membranes and cell membranes (Schmetterer 1994; Paumann et al. 2005; Peschek 2008). A respiratory activity is higher in the latter than in the former (Peschek 2008). However, since *G. violaceus* contains only cell membranes, respiratory and photosynthetic electron transfer systems are co-localized on cell membranes (Fig. 9.4). Therefore it is inferred that at least six complexes are localized on the membranes, i.e. PS I, PS II, cyt b_6f, ATP synthase, NDH or SDH, and terminal cyt *c* oxidase. In addition, there are many complexes on the membranes, which are engaged in transport or signal transduction. In this sense, cell membranes might be full of protein-complexes, which secondary affects a diffusion rate of components in membranes. If the attachment site of PBS is assumed to be the localization site of PS II, they are distributed uniformly on the membranes based on the electron micrograph observation (Guglielmi et al. 1981). Unfortunately, there is no direct information on the distribution of respiratory components in membranes. However, it is reasonable to assume that coupling of respiratory and photosynthetic electron transfer systems is a probability-based phenomenon by sharing common components.

The photosynthetic electron flow is perturbed by the respiratory electron flow when the two are mutually affected each other (Fig. 9.5). One typical example is the oxygen flash yield; it is kept constant under a low repetition rate, but is lowered under a high repetition rate, because the turnover of PQ pool will limit the electron transfer between the two PSs. In the case of *Synechocystis*, the oxygen flash yield was constant up to 10 Hz, on the other hand, in *G. violaceus*, it was lowered from 3 Hz (Koyama et al. 2008). This clearly indicated that a turnover rate of PQ pool was much slower in *G. violaceus*. A probable reason for this slow rate was decrease in active PQ molecules by an electron flow from the respiratory pathway. This was also shown by the fluorescence induction on cell level.

Coupling of photosynthetic and respiratory electron transfer chain is indirectly monitored by a fluorescence induction curve. Since the curve represents the redox state of Q_A molecule in PS II and Q_A is equilibrated with Q_B and PQ molecules in the pool, a fluorescence induction curve is affected by various factors that deter-

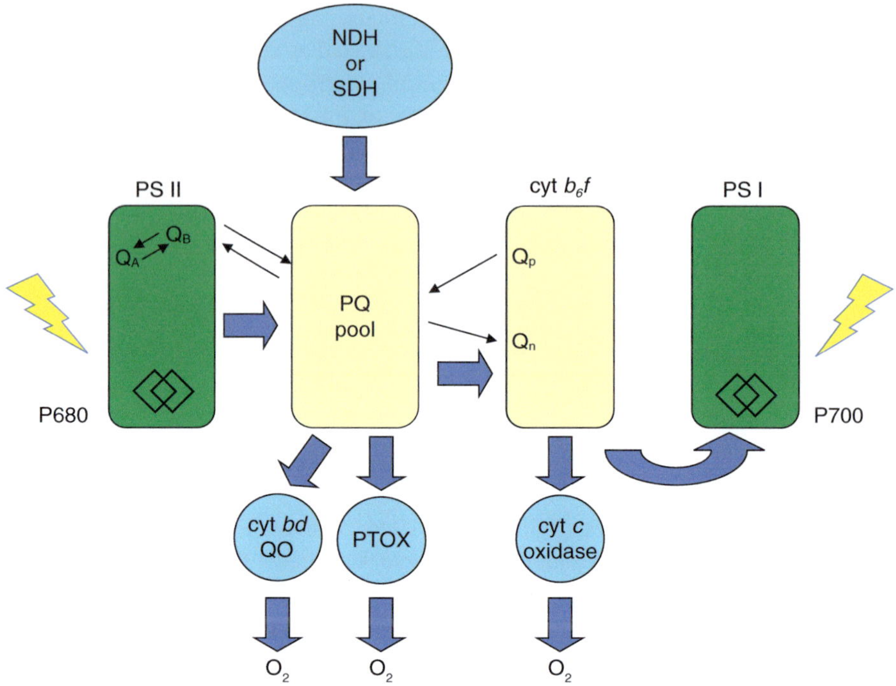

Fig. 9.5 Interconnection of photosynthetic and respiratory electron transfer systems on cell membranes in *G. violaceus*. *Thick arrows* stand for a flow of electron and *thin arrows*, a flow of quinone molecules in the two electron transfer pathways

mine the redox level of PQ pool. Respiration is one of those factors. *G. violaceus* showed an induction curve that is similar to other cyanobacteria (Fig. 9.6a). However, there was a significant difference, i.e. a low variable fluorescence, suggesting a low content of active PQ molecules in a pool and a high PE fluorescence. When far-red light is illuminated, a fraction of a variable fluorescence was increased to almost twice, whereas the increase was observed only 10% in *Synechocystis* (Koyama et al. unpublished) (Fig. 9.6). The far-red light preferentially excites PS I and accelerates an electron flow to PS I, leading to increase in an oxidized form of PQ pool. The observed low variable fluorescence is explained by a low content of an oxidized form of PQ molecules in a pool. This also proved an interconnection of the two electron transfer systems in the same membrane.

Q_A^- re-oxidation kinetics also indicates the redox state of PQ pool by monitoring a content of Q_A^- that is equilibrated with other quinone components, i.e. Q_B and PQ pool. The kinetics observed on *G. violaceus* cells were very similar to that observed on *Synechocystis* cells, but an amplitude was even low in the former. Far-red light illumination increased a magnitude of the total decay amplitude, indicating that PQ pool was significantly reduced even in the dark-adapted state. This was consistent with the observation of the fluorescence induction kinetics, and again indicates a supply of electron from the respiratory chain.

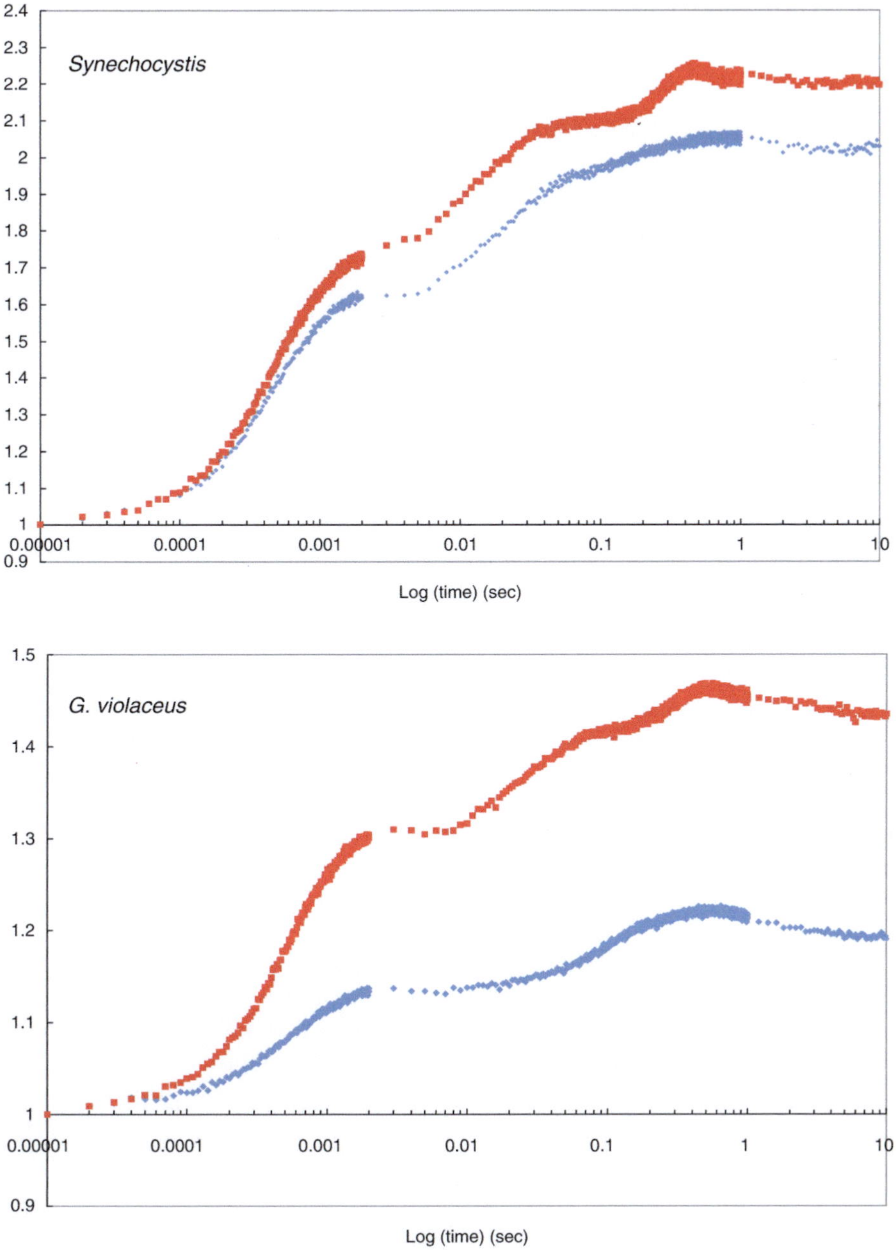

Fig. 9.6 Fluorescence induction kinetics of intact cells of *G. violaceus* at room temperature. Fluorescence induction kinetics of two species of cyanobacteria. *Blue lines* (*Diamonds*) show the kinetics of the dark-adapted cells (30-min dark incubation) and *red* (*squares*) ones, dark-adapted cells followed by far-red light illumination (>700 nm, 15 μmol photon m^{-2} s^{-1}). Note a difference in magnitudes of ordinates

9.7 Uniqueness of the Organism in the World of Cyanobacteria

9.7.1 A Unique Bioenergetic Pathway

Proteorhodopsin is a proteobacterial light-driven proton pump homologous to archaeal bacteriorhodopsin (Béjà et al. 2000) and it was found that the proteorhodopsin-like protein is widely distributed. In the genome of *G. violaceus*, a gene for proteorhodopsin-like protein (*gll0198*) was found, and a recombinant protein showed light-driven proton transport activity *in vitro* (Miranda et al. 2009). Although the expression level of this novel proton pump in the cells has not been reported, it would be worthwhile to evaluate the contribution of the proteorhodopsin-like protein to bioenergetics in *Gloeobacter*, even though absorption peak of proteorhodopsin-like protein overlaps with that of phycoerythrin.

9.7.2 Membrane Architecture

The most remarkable feature of the energy-yielding system in *G. violaceus* is a membrane system; it has only cytoplasmic membranes on which two electron transfer systems are co-localized (Fig. 9.4). In general, there has been a long discussion on the physical connection between thylakoid membranes and cytoplasmic membranes in cyanobacteria, however this was not detected by the electron microscopic observation on *G. violaceus*. Instead, invagination of cytoplasmic membranes will become important to increase the area for photosynthesis, however the electron microscopic studies did not give any evidence of the invagination of the membranes in *G. violaceus*. The cytoplasmic membranes looked smooth and are arranged along with cell walls.

A relative content of phycobiliproteins to Chl *a* is high in *G. violaceus* (Fig. 9.2a). Since this organism usually grows under a low light intensity, a large antenna size is required to absorb photons enough for photosynthesis. A bundle-like shape of PBSs (Fig. 9.7) might be an adaptive change for occupation of a relatively small area on the membranes with a large amount of chromophores. As shown by the fluorescence spectrum (Fig. 9.2b–d), PBSs are physically and energetically attached to PS II Chl *a*, indicating that the locus for PBSs is the locus for PS II RC. Based on the sequence similarities of CP47 and CP43 to other cyanobacteria, a dimer structure of PS II core complexes is expectable. Even though the protein interaction for connection between PBSs and PS II core complexes is not yet resolved, the anchor polypeptide of PBSs is associated with dimeric core complexes to realize energy transfer (Fig. 9.2c). This connection might be the same as that of other cyanobacteria through an anchor protein of PBS (ApcE).

As stated before, one of the subunits of PS I core complexes, PsaB, has a long extension in its C-terminal (Fig. 9.4); it contains a peptidoglycan-binding motif,

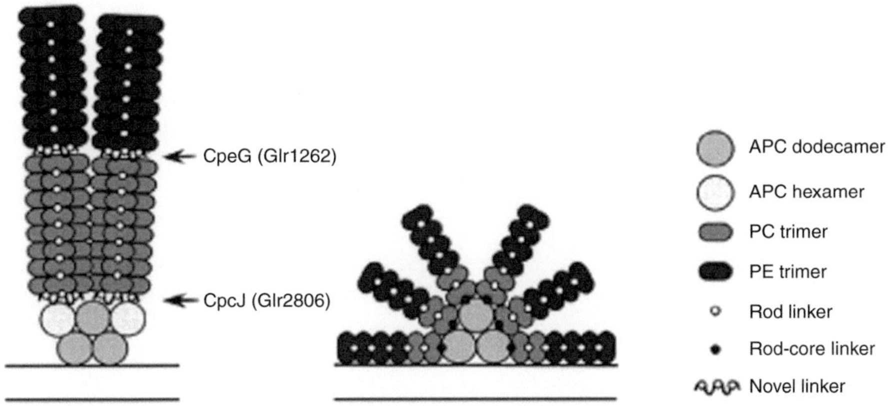

Fig. 9.7 A schematic model of phycobilisomes in *G. violaceus*. (*Left*) PBS of *G. violaceus* and (*right*) PBS of typical hemi-discoidal shape. (Koyama et al. 2006)

however a function of this extension was not yet resolved. It is interesting to note that this organism contains the other gene (*glr1828*) whose amino acid sequences were very similar to those of the C-terminal extension of PsaB; it is expected that a gene product of the *glr1828* connects the outer layer of cell wall and peptidoglycan, however its expression and function were not resolved. The ancestral form is not clear between these two genes, however these two are probably related to each other.

As for biogenesis of the membrane system, *G. violaceus* shows unique properties. *Synechocystis* have two similar genes, *pspA* for phage shock protein A and *vipp1* (Westphal et al. 2001), both of which are related to membrane biogenesis. *Vipp1* is known to be essential for formation of thylakoid membranes in *Synechocystis* and in a higher plant, *Arabidopsis thaliana* (Kroll et al. 2001). The C-terminal region of Vipp1, approximately 30 amino acid residues, is well conserved and is believed to encode a domain related to the above function, and PspA lacks this region. *glr0898* in *Gloeobacter* showed higher sequence similarity to that of *vipp1* than *pspA*, but a region corresponding to the C-terminal conserved region was missing. *glr0898* may reflect the intermediate state of a gene during the evolutional process between *vipp1* and *pspA*.

9.7.3 Evolutionary Aspect

It is suggested that based on the 16S rDNA sequence, *G. violaceus* branched off at the earliest stage of evolutionary pathway in cyanobacteria (Nelissen et al. 1995) (Fig. 9.3a). This branching pattern was also obtained when a specific protein was used as an index for comparison (Fig. 9.3b). The branching preceded the primary symbiosis, therefore features of this organism were not necessarily inherited to chloroplasts. Comparative analysis of the genome of this species revealed deletions, modifications, and sometimes redundancy of genes involved in photochemical reactions (Nakamura et al. 2003). A combination of this information with biochemical analyses will yield information on reaction processes. However due to its slow growth rate, this organism is not particularly suitable for biochemical analysis, thus discussion on an evolutionary aspect will be proceeded based on the genome information.

It is widely accepted that the major components for RCs are inherited from photosynthetic bacteria, i.e. PS I components from green-sulfur bacteria and PS II components from purple bacteria and/or green-filamentous bacteria (Blankenship 2001; Olson and Blankenship 2004). However, origins of other subunits for individual RCs are not necessarily determined. Some might be inherited from bacteria, and some might be invested or acquired in cyanobacteria. Comparative genomics may clearly state the origins or inherit pathways of those components. A comparison of the constituents of *G. violaceus* with those of other cyanobacteria suggests that ancestral cyanobacteria had PS I with a simple subunit composition and that they acquired small subunits during evolution. It was also the case for PS II. This hypothesis is consistent with the idea of the appearance of an oxygenic photosynthetic system because the two PSs were inherited from the RCs of photosynthetic bacteria with a simple composition and the number of subunits increased afterwards.

A missing link is an invention of the water oxidation system. Since ancestor of cyanobacteria, i.e. photosynthetic bacteria, do not possess the water oxidation system, it is reasonable to assume that this system was acquired after cyanobacteria were differentiated or evolved. However there was no way to trace this evolutionary process(es), because there are no intermediary organisms between photosynthetic bacteria and the first cyanobacterial species. *G. violaceus* might be the first cyanobacterial species, and its oxygen evolution system might still keep primordial properties (Fig. 9.8). However the observed oxygen-evolving properties of *G. violaceus* show that the system is properly established, almost no defect in the system.

The oxygen-evolving system consists of four Mn atoms, Ca^{2+}, Cl^{-}, and several peripheral proteins of PS II (PsbO, PsbU, and PsbV). Ligands to the Mn atoms are supplied from the two polypeptides consisting of the PS II core, i.e. D1-protein and CP43, and this active center for water oxidation is called the oxygen-evolving complex (OEC) (Kamiya and Shen 2003; Ferreira et al. 2004; Loll et al. 2005; Guskov et al. 2009). Even though an extensive study of the structure of this OEC by the crystallographic method, its actual structure is not yet established (Yano et al. 2006; Yano and Yachandra 2008). This is the main reason for a slow progress of analysis of OEC. The electrochemical potential required for water oxidation is high, i.e. higher than 1.0 V, and this potential may destroy cellular components if the reaction intermediates are not properly sealed from the other parts of the system. Components, reaction processes and reaction mechanism(s) of water oxidation are still under debate. This reaction is the largest topics to be solved in the field of photosynthesis. A step-by-step progress is the nearest way to get to the final goal, even if we need a little long time to get to that stage.

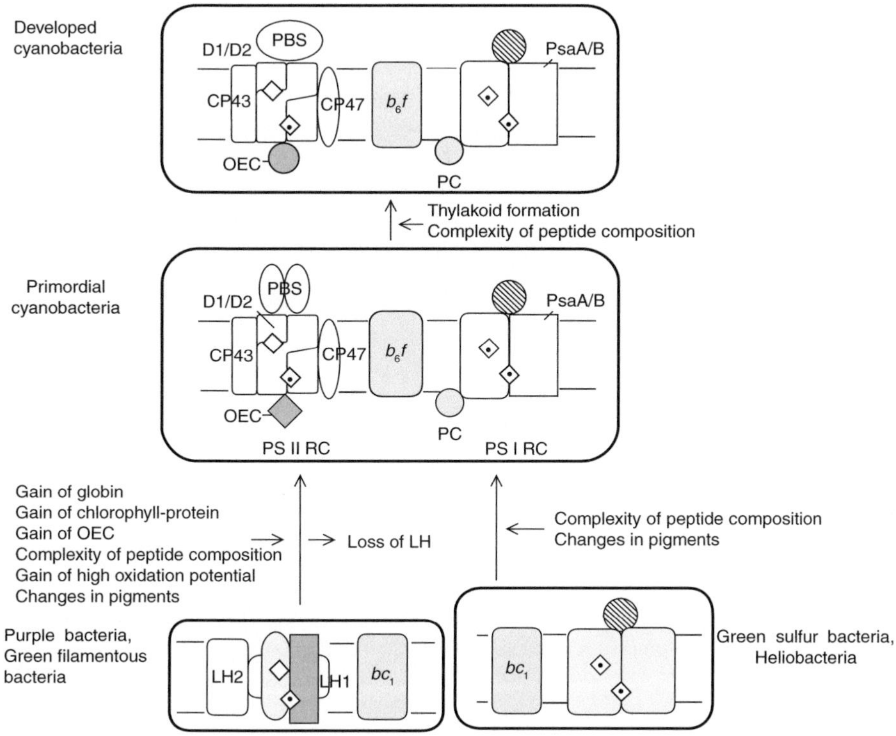

Fig. 9.8 Changes in photosynthetic components along with evolution from anoxygenic photosynthetic bacteria to oxygenic cyanobacteria. Figure was drawn on the assumption that the evolution from photosynthetic bacteria to current cyanobacteria was brought about by two steps; one from anoxygenic photosynthetic bacteria to a primordial cyanobacteria, and the second step, from a primordial cyanobacteria to current cyanobacteria. Component(s) or phenomena necessary for evolution were described. Rhombus with a dot stands for chlorophyll or bacteriochlorophyll molecules, and that without a dot, pheophytin molecules. bc_1, cytochrome bc_1; b_6f, cytochrome b_6f; CP43, 43 kDa chlorophyll-protein; CP47, 47 kDa chlorophyll-protein; D1/D2, reaction center II; LH1, light-harvesting complex 1; LH2, light-harvesting complex 2; OEC, oxygen-evolving complex; PsaA/B, reaction center I; PBS, phycobilisome; PC, plastocyanin; PS I, photosystem I; PS II, photosystem II

9.7.4 Uniqueness of G. violaceus as a Tool for Comprehensive Understanding of Bioenergetics of Cyanobacteria

G. violaceus is a unique cyanobacterium in many senses as we have described above. These properties are useful for comprehensive understanding of bioenergetics of cyanobacteria, however one important point is not yet satisfied, i.e. transformation. Since the whole genome information is known on this species, combination of genome information and transformation is the minimum requirement for driving progress of study. Introduction of plasmid DNA into *G. violaceus* was once reported

(Guo and Xu 2004), however there is no successive report. This point is a breakthrough for future progress when it is established.

Acknowledgments The authors would like to express sincere thanks to Prof. G. A. Peschek, University of Vienna, for his giving us a chance to write this article. This work was supported in part by the Grant-in-Aids for the Creative Research from the Japanese Society for Promotion of Science (JSPS) to MM (Grant No. 17GS0314). We also thank Mr. H. Inoue for his work in the early stage of our experiments.

References

Béjà O, Aravind L, Koonin EV, Suzuki MT, Hadd A, Nguyen LP, Jovanovich SB, Gates CM, Feldman RA, Spudich JL, Spudich EN and DeLong EF (2000) Bacterial rhodopsin: evidence for a new type of phototrophy in the sea. *Science* 289: 1902–1906

Berry S, Schneider D, Vermaas WFJ and Rögner M (2002) Electron transport route in whole cells of *Synechocystis* sp. strain PCC 6803: the route of the cytochrome *bd*-type oxidase. *Biochemistry* 41: 3422–3429

Blankenship RE (2001) Molecular evidences for evolution of photosynthesis. *Trends Plant Sci* 6: 4–6

Bryant DA, Cohen-Bazire G and Glazer AN (1981) Characterization of the biliproteins of *Gloeobacter violaceus*. Chromophore content of a cyanobacterial phycoerythrin carrying phycourobilin chromophore. *Arch Microbiol* 129: 190–198

De Las Rivas J and Barber J (2004) Analysis of the structure of the PsbO protein and its implications. *Photosynth Res* 81: 329–343

Ferreira K, Iverson T, Maghlaoui K, Barber J and Iwata S (2004) Architecture of the photosynthetic oxygen-evolving canter. *Science* 303: 1831–1838

Frankenberg N, Mukougawa K, Kohchi T and Lagarias JC (2001) Functional genomic analysis of the HY2 family of ferredoxin-dependent bilin reductases from oxygenic photosynthetic organisms. *Plant Cell* 13: 965–978

Gantt E (1981) Phycobilisomes. *Annu Rev Plant Physiol* 32: 327–347

Guglielmi G, Cohen-Bazire G and Bryant DA (1981) The structure of *Gloeobacter violaceus* and its phycobilisomes. *Arch Microbiol* 129: 181–189

Güler S, Seeliger A, Härtel H, Renger G and Benning C (1996) A null mutant of *Synechococcus* sp. PCC7942 deficient in the sulfolipid sulfoquinovosyl diacylglycerol. *J Biol Chem* 271: 7501–7507

Guo H and Xu X (2004) Broad host range plasmid-based gene transfer system in the cyanobacterium *Gloeobacter violaceus* which lacks thylakoids. *Prog Natl Sci* 14: 31–35

Guskov A, Kern J, Gabdulkhakov A, Broser M, Zouni M and Saenger W (2009) Cyanobacterial photosystem II at 2.9-Å resolution and the role of quinones, lipids, channels and chloride. *Nat Struct Mol Bio* 16: 334–342

Gutiérrez-Cirlos EB, Pérez-Gómez B, Krogmann DW and Gómez-Lojerob C (2006) The phycocyanin-associated rod linker proteins of the phycobilisome of *Gloeobacter violaceus* PCC 7421 contain unusually located rod-capping domains. *Biochim Biophys Acta* 1757: 130–134

Hiratsuka T, Furihata K, Ishikawa J, Yamashita H, Itoh N, Seto H and Dairi T (2008) An alternative menaquinone biosynthetic pathway operating in microorganisms. *Science* 321: 1670–1673

Huynen MA, Dandekar T and Bork P (1999) Variation and evolution of the citric-acid cycle: a genomic perspective. *Trends Microbiol* 7: 281–291

Inoue H, Tsuchiya T, Satoh S, Miyashita H, Kaneko T, Tabata S, Tanaka A and Mimuro M (2004) Unique constitution of photosystem I with a novel subunit in the cyanobacterium *Gloeobacter violaceus* PCC 7421. *FEBS Lett* 578: 275–279

Kamiya N and Shen J-R (2003) Crystal structure of oxygen-evolving photosystem II from *Thermosynechococcus vulcanus* at 3.7 Å resolution. *Proc Natl Acad Sci U S A* 100: 98–103

Kato Y, Nakamura A, Suzawa T and Watanabe T (2008) In Allen J, Gantt E, Golbeck J and Osmond B (eds) *Photosynthesis Energy from the Sun*, Springer, New York 109–112

Koenig F and Schmidt M (1995) *Gloeobacter violaceus*-investigation of an unusual photosynthetic apparatus, absence of the long wavelength emission of photosystem I in 77 K fluorescence spectra. *Physiol Plant* 94: 621–628

Kondo K, Geng XX, Katayama M and Ikeuchi M (2005) Distinct roles of CpcG1 and CpcG2 in phycobilisome assembly in the cyanobacterium *Synechocystis* sp. PCC 6803. *Photosynth Res* 84: 269–273

Koyama K, Tsuchiya T, Akimoto S, Yokono M, Miyashita H and Mimuro M (2006) New linker proteins in phycobilisomes isolated from the cyanobacterium *Gloeobacter violaceus* PCC 7421. *FEBS Lett* 580: 3457–3461

Koyama K, Suzuki H, Noguchi T, Akimoto S, Tsuchiya T and Mimuro M (2008) Oxygen evolution activities in the thylakoid-lacking cyanobacterium *Gloeobacter violaceus* PCC 7421. *Biochim Biophys Acta* 1777: 369–378

Krogmann DW, Pérez-Gómez B, Gutiérrez-Cirlos EB, Chagolla-López A, de la Vara LG and Gómez-Lojero C (2007) The presence of multidomain linkers determines the bundle-shape structure of the phycobilisome of the cyanobacterium *Gloeobacter violaceus* PCC 7421. *Photosynth Res* 93: 27–43

Kroll D, Meierhoff K, Bechtold N, Kinoshita M, Westphal S, Vothknecht UC, Soll J and Westhoff P (2001) *VIPP*1, a nuclear gene of *Arabidopsis thaliana* essential for thylakoid membrane formation. *Proc Nat Acad Sci U S A* 98: 4238–4242

Loll B, Kern J, Saenger W, Zouni A and Biesiadka J (2005) Towards complete cofactor arrangement in the 3.0 Å resolution structure of photosysterm II. *Nature* 438: 1040–1044

McDonald AE and Vanlerberghe GC (2006) Origins, evolutionary history, and taxonomic distribution of alternative oxidase and plastoquinol terminal oxidase. *Comparative Biochem Physiol*, Part D 1: 357–364

Mangels D, Kruip J, Berry S, Rögner M, Boekema EJ and Koenig F (2002) Photosystem I from the unusual cyanobacterium *Gloeobacter violaceus. Photosynth Res* 72: 307–319

Mimuro M, Lipschultz CA and Gantt E (1986) Energy flow in the phycobilisome core of *Nostoc* sp. (MAC): two independent terminal pigments. *Biochim Biophys Acta* 852: 126–132

Mimuro M, Ookubo T, Takahashi D, Sakawa T, Akimoto S, Yamazaki I and Miyashita H (2002) Unique fluorescence properties of a cyanobacterium *Gloeobacter violaceus* PCC 7421: reasons for absence of the long-wavelength PSI Chl *a* fluorescence at –196°C. *Plant Cell Physiol* 43: 587–594

Mimuro M, Tsuchiya T, Inoue H, Sakuragi Y, Itoh Y, Gotoh T, Miyashita H, Bryant DA and Kobayashi M (2005) The secondary electron acceptor of photosystem I in *Gloeobacter violaceus* PCC 7421 is menaquinone-4 that is synthesized by a unique but unknown pathway. *FEBS Lett* 579: 3493–3496

Mimuro M, Tomo T and Tsuchiya T (2008a) Two unique cyanobacteria lead to a new view on the appearance of oxygenic photosynthesis. *Photosynth Res* 97: 167–176

Mimuro M, Kobayashi M, Murakami A, Tsuchiya T and Miyashita H (2008b) Structure and function of antenna systems: oxygen evolving cyanobacteria. In Renger G (ed) *Primary Processes of Photosynthesis: Basic Principles and Apparatus*, Part 1, pp 261–299. RSC Publishing, Cambridge

Miranda MRM, Choi AR, Shi L, Bezerra Jr AG, Jung K-H and Brown LS (2009) The photocycle and proton translocation pathway in a cyanobacterial ion-pumping rhodopsin. *Biophys J* 96: 1471–1481

Mogi T and Miyoshi H (2009) Properties of cytochrome *bd* plastoquinol oxidase from the cyanobacterium *Synechocystis* sp. PCC 6803. *J Biochem* 145: 395–401

Motoki A, Usui M, Shimazu T, Hirano M and Katoh S (2002) A domain of the manganese stabilizing protein from *Synechococcus elongatus* involved in functional binding to photosystem II. *J Biol Chem* 277: 14747–14756

Nakamura Y, Kaneko T, Sato S, Mimuro M, Miyashita H, Tsuchiya T, Sasamoto S, Watanabe A, Kawashima K, Kishida Y, Kiyokawa C, Kohara M, Matsumoto M, Matsuno A, Nakazaki N, Shimpo S, Takeuchi C, Yamada M and Tabata S (2003) Complete genome structure of *Gloeobacter violaceus* PCC 7421, a cyanobacterium that lacks thylakoids. *DNA Res* 10: 137–145

Nelissen B, Van de Peer Y, Wilmotte A and De Wachter R (1995) An early origin of plastids within the cyanobacterial divergence is suggested by evolutionary trees based on complete 16S rRNA sequences. *Mol Biol Evol* 12: 1166–1173

Olson JM and Blankenship RE (2004) Thinking about the evolution of photosynthesis. *Photosynth Res* 80: 373–386

Paumann M, Regelsberger G, Obinger C and Peschek GA (2005) The bioenergetic role of dioxygen and the terminal oxidase(s) in cyanobacteria. *Biochim Biophys Acta* 1707: 231–253

Peschek GA (2008) Electron transport chains in oxygenic cyanobacteria. In Renger G (ed) *Primary Processes of Photosynthesis: Principles and Applications*, pp 383–415. RSC Publishing, Cambridge

Rippka R, Waterbury J and Cohen-Bazire G (1974) A cyanobacterium which lacks thylakoids. *Arch Microbiol* 100: 419–436

Scheer H and Zhao KH (2008) Biliprotein maturation: the chromophore attachment. *Mol Microbiol* 68: 263–276

Selstam E and Campbell D (1996) Membrane lipid composition of the unusual cyanobacterium *Gloeobacter violaceus* sp. PCC 7421, which lacks sulfoquinovosyl diacylglycerol. *Arch Microbiol* 166: 132–135

Schmetterer G (1994) Cyanobacterial respiration. *The Molecular Biology of Cyanobacteria,* pp 409–435. Kluwer Academic, Dordrecht

Sicora CI, Brown CM, Cheregi O, Vass I and Campbell DA (2008) The psbA gene family responds differentially to light and UVB stress in *Gloeobacter violaceus* PCC 7421, a deeply divergent cyanobacterium. *Biochim Biophys Acta* 1777: 130–139

Sobotka R, Dühring U, Komenda J, Peter E, Gardian Z, Tichy M, Grimm B and Wilde A (2008) Importance of the cyanobacterial Gun4 protein for chlorophyll metabolism and assembly of photosynthetic complexes. *J Biol Chem* 283: 25794–25802

Steiger S, Jackisch Y and Sandmann G (2005) Carotenoid biosynthesis in *Gloeobacter violaceus* PCC4721 involves a single crtI-type phytoene desaturase instead of typical cyanobacterial enzymes. *Arch Microbiol* 184: 207–214

Tsuchiya T, Takaichi S, Misawa N, Maoka T, Miyashita H and Mimuro M (2005) The cyanobacterium *Gloeobacter violaceus* PCC 7421 uses bacterial-type phytoene desaturase in carotenoid biosynthesis. *FEBS Lett* 579: 2125–2129

Westphal S, Heins L, Soll J and Vothknecht UC (2001) *Vipp*1 deletion mutant of *Synechocystis*: A connection between bacterial phage shock and thylakoid biogenesis? *Proc Nat Acad Sci U S A* 98: 4243–4248

Yano J, Kern J, Sauer K, Latimer MJ, Pushkar Y, Biesiadka J, Loll B, Saenger W, Messinger J, Zouni A and Yachandra VK (2006) Where water is oxidized to dioxygen: structure of the photosynthetic Mn4Ca cluster. *Science* 314: 821–825

Yano J and Yachandra VK (2008) Where water is oxidized to dioxygen: structure of the photosynthetic Mn4Ca cluster from X-ray spectroscopy. *Inorg Chem* 47: 1711–1726

Yokono M, Akimoto S, Koyama K, Tsuchiya T and Mimuro M (2008) Energy transfer processes in *Gloeobacter violaceus* PCC 7421 that possesses phycobilisomes with a unique morphology. *Biochim Biophys Acta* 1777: 55–65

Zhang CC, Jeanjean R and Joset F (1998) Obligate phototrophy in cyanobacteria: more than a lack of sugar transport. *FEMS Microbio Lett* 161: 285–292

Chapter 10
ATP Synthase: Structure, Function and Regulation of a Complex Machine

Dirk Bald

10.1 Introduction

Adenosin-triphosphate (ATP) is regarded as the currency of energy in living organisms used for endergonic cellular processes such as transport against concentration gradients, biosynthesis of metabolites, or cellular movement. ATP synthase (also referred to as F_oF_1-ATP synthase or F_oF_1-ATPase) synthesizes ATP from adenosin-diphosphate (ADP) and inorganic phosphate. It is an evolutionary conserved enzyme found in plasma membranes of bacteria, thylakoid membranes in plant or algae chloroplasts and inner membranes of mitochondria. The energy required for the ATP synthesis reaction is derived from the flow of coupling ions (mostly protons) across the bio-membrane, along the gradient of an electrochemical potential. In the case of protons as coupling ions, this potential difference is often referred to as the Proton Motive Force (PMF). The difference of electrochemical potentials on both sides of the membrane is established by the enzyme complexes of the respiratory chain in case of mitochondrial inner membranes or bacterial plasma membranes and the photosynthetic electron transport chain in chloroplast thylakoids and photosynthetic bacteria. In cyanobacteria, photosynthetic as well as respiratory electron transport occur in the same membrane system, the cytoplasmatic membrane, thus ATP synthase can utilize the PMF established by both electron transfer systems.

Synthesis of ATP is reversible, under conditions of a low PMF ATP can be hydrolyzed and protons can be transported across the membrane. Whereas some bacteria use this reversed reaction to maintain the PMF under conditions of low oxygen tensions, in ATP synthase from chloroplasts and mitochondria elaborate regulatory mechanisms suppress hydrolysis of ATP.

ATP synthase is a multi-subunit enzyme complex of roughly 15 nm length and 10 nm width with a molecular mass of ~600 kDa (Fig. 10.1), it can be reversibly

---

D. Bald (✉)
Faculty of Earth- and Life Science, Department of Molecular Cell Biology, VU University Amsterdam, De Boelelaan 1085, 1081 HV Amsterdam, The Netherlands
e-mail: dirk.bald@falw.vu.nl

G. A. Peschek et al. (eds.), *Bioenergetic Processes of Cyanobacteria*,
DOI 10.1007/978-94-007-0388-9_10,

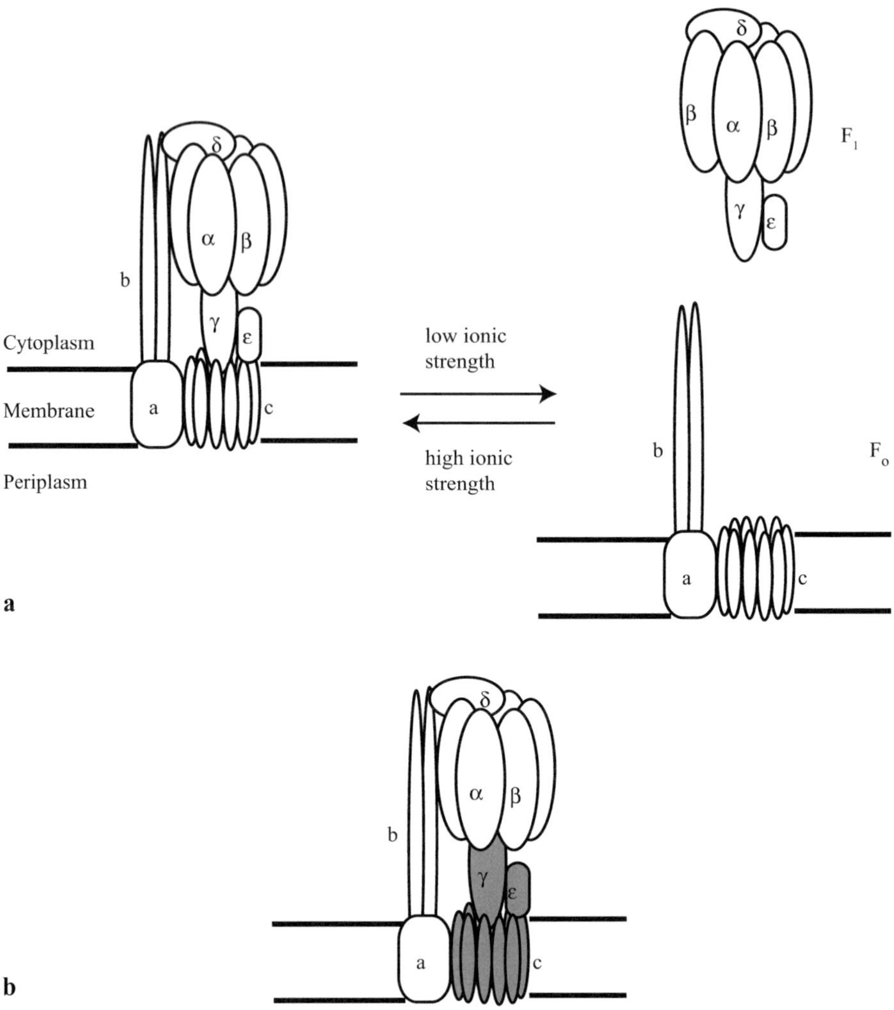

Fig. 10.1 Subunit scheme of ATP synthase. **a** Reversible separation of ATP synthase in the F_1 part and F_o part. **b** Functional classification in rotor subunits (grey) and stator subunits (white)

separated into a membrane-embedded F_o part and a peripheral F_1 part. Whereas the F_o part acts as a proton channel, the F_1 part houses the nucleotide binding sites where synthesis or hydrolysis of ATP is carried out. In most bacterial species, the F_o part consists of three different subunits: a single subunit a, two subunits b and several subunits c (10–15 mer). As an exception to this rule, in cyanobacteria along with the plant chloroplast ATP synthase a heterodimeric subunit b is found, made up by two homologous subunits called subunit b and subunit b′. The hydrophilic F_1 part consists of five different subunits with the stoichiometry $\alpha_3\beta_3\gamma\delta\varepsilon$. ATP synthase from mitochondria comprises, in addition to the eight different core subunits also found in bacteria, a number of additional peripheral subunits.

An alternative grouping of ATP synthase subunits is based on the enzyme's mechanism as a rotary molecular machine: during catalytic activity the subunits γεc ("rotor", Fig. 10.1b) rotate relative to subunits αβδab ("stator"). As such, ATP synthase is a fascinating molecular machine, which carries out two energy conversion steps at nanometer scale: in a first step energy stored in an electrochemical potential difference is converted into mechanical energy of subunit motion, which in the second step is again stored as phosphate-group transfer potential of ATP.

In the following chapters, structure, function and regulation of ATP synthase will be discussed in terms of general features of this enzyme found across domain borders as well as specific properties of cyanobacterial ATP synthase.

10.2 Core Structure and Mechanism

10.2.1 Structure of the Catalytic Core

For no ATP synthase holoenzyme any high-resolution structure is available as of today. In negative-stained electron microscopy images the enzyme appears as a slightly elongated, dumbbell-shaped object (Wilkens and Capaldi 1998; Rubinstein et al. 2003). More detailed structural information has been gained for purified subcomplexes, mostly the F_1 part, which due to its hydrophilic properties crystallizes more readily. High-resolution structures are now known for F_1 from bovine heart mitochondria (Abrahams et al. 1993; Gibbons et al. 2000), yeast mitochondria (Stock et al. 1999; Kabaleeswaran et al. 2006), rat liver mitochondria (Bianchet et al. 1998), spinach chloroplast (Groth and Pohl 2001) and the thermophilic *Bacillus* PS3 (Shirakihara et al. 1997). As may be expected in view of the high degree of sequence similarity, all structures determined so far reveal the same basic construction principle for the catalytic core of ATP synthase:

As first determined for F_1 from bovine heart mitochondria, the three subunits α and β are arranged in an alternating manner, like the segments of an orange, while subunit γ penetrates into the center of the $\alpha_3\beta_3$ ring (Fig. 10.2, Abrahams et al. 1993; Gibbons et al. 2000). Both subunits α and β consist of an N-terminal β-barrel domain, thought to be important for protein stability, a central nucleotide-binding domain and a C-terminal helical bundle, which plays a dynamic role in catalysis. The amino- and carboxy-terminal portions of subunit γ form an elongated coil-coil that for about half of its length is embedded in the central cavity of the $\alpha_3\beta_3$ ring. The other half protrudes in the direction of the F_o part, connecting to the membrane-bound subunit c (Stock et al. 1999). The central part of subunit γ displays smaller β-sheet and helical domains. There are a total of six nucleotide-binding sites located

Fig. 10.2 Structure of the catalytic core of ATP synthase. High-resolution structure of the F_1 part reveals a ring made up by subunits-α (*pink*) and subunit β (*yellow*) with subunit γ (*blue*) in the center (**a**), side-view showing subunit γ penetrating into the $\alpha_3\beta_3$ ring (**b**), *top* view indicating how the non-central position of subunit γ creates asymmetry in the three $\alpha\beta$ pairs. (PDB ID 1E79; Abrahams et al. 1993; Gibbons et al. 2000). The location of the three catalytic sites in the empty (E), ADP (DP) and AMP-PNP (TP) conformations as well as the three non-catalytic sites (NC) are indicated

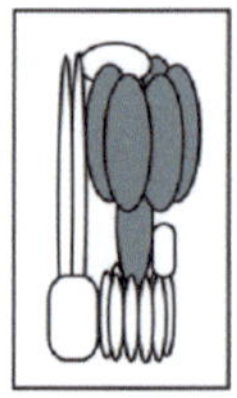

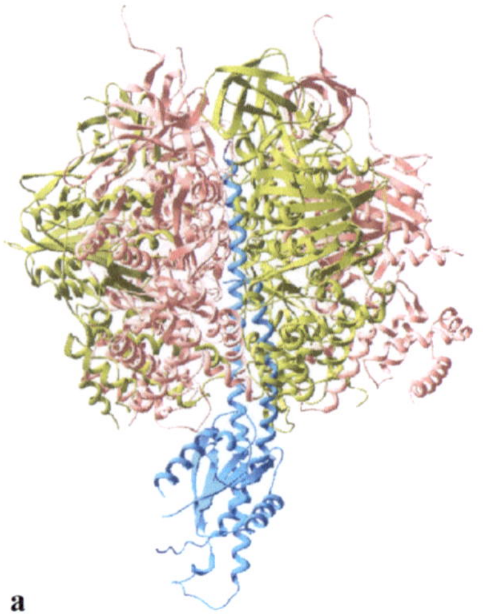

a

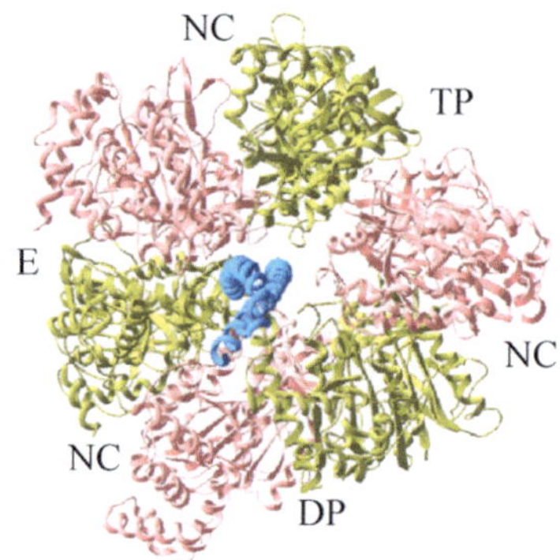

b

at the six $\alpha\beta$ interfaces, however, only three of them, separated by 120° angles, are active in catalysis ("catalytic sites"). The nucleotide binding sites at the three remaining $\alpha\beta$ interfaces sites do not catalyze ATP hydrolysis or synthesis ("non-catalytic sites"), but play a regulatory role.

In the first crystal structures reported for F_1 the three catalytic nucleotide-binding sites differ in nucleotide occupancy with one site binding ADP, a second site binding the ATP analogue AMP-PNP and the third site being empty (Fig. 10.2b, Abrahams et al. 1993). Connected with this difference in nucleotide occupancy were three distinct conformations of the three subunits β, imposed by the asymmetric environment in the presence of subunit γ. In the "empty" site the helical bundle domain was swung out, allowing the nucleotide to dissociate. The "ADP" site was closed by a hinge-like displacement of the helical bundle, whereas the AMP-PNP site was even more tightly closed. As detailed in the following chapter(s), this structural information supported the "rotary catalysis" mechanism of ATP synthase first proposed by P. Boyer.

10.2.2 ATP Synthase as Rotary Motor

A function of ATP synthase as rotary machine was first proposed by Boyer based on isotope exchange experiments (for review see Boyer 1993). These data indicated that all three subunits β behaved identical and were indistinguishable in spite of the asymmetric environment created by the smaller subunits γ, δ, and ε. Boyer postulated that rotation of subunit γ causes a change of conformation in the three subunits β, leading to nucleotide binding and release from the nucleotide binding sites. As stated above, in the first high-resolution x-ray crystal structure each subunit β displayed its own distinct conformation and the whole structure appeared like a snapshot of the enzyme active in rotary catalysis. Movement of subunit γ within the $\alpha_3\beta_3$ structure subsequently was shown for *E. coli* ATP synthase by cross-linking experiments during ATP hydrolysis (Duncan et al. 1995) and ATP synthesis (Zhou et al. 1997). Spectroscopic investigation of F_1-ATPase from spinach chloroplast immobilized on a surface with subunit γ labeled by a single dye molecule provided further evidence for the rotary mechanism (Sabbert et al. 1996). Finally, rotation was directly observed by attachment of a fluorescent-labeled actin filament to immobilized F_1-ATPase from the thermophilic *Bacillus PS3* (Noji et al. 1997). The filaments were found rotating in a unidirectional manner, interrupted by pauses of varying lengths. At low ATP concentrations, the quasi-continuous rotational movement of the filament was resolved into discrete 120° steps, as expected when subunit γ moves from one αβ pair to the next (Yasuda et al. 1998). The rotational torque of a single 120° step was calculated to about 60–80 pNm, indicating close to 100% efficiency for the transformation of the free energy of ATP hydrolysis reaction into the mechanical energy of rotary motion. Using polystyrene beads and nano-gold particles as probes combined with a high-speed CCD-camera, a single 120° step was resolved into one 80° and one 40° sub-step. These two sub-steps are thought to represent movement of subunit γ upon binding of ATP and release of inorganic phosphate, respectively (Yasuda et al. 2001; Nishizaka et al. 2004).

A close connection of the regulatory subunit ε with subunit γ was demonstrated for *E. coli* F_1-ATPase by cross-linking experiments (Aggeler et al. 1995). Rotational movement of subunit ε along with subunit γ in *Bacillus* F_1-ATPase was shown by attachment of fluorescent-labeled actin filaments (Kato-Yamada et al. 1998).

As further component of the "rotor " part of ATP synthase the oligomeric subunit c was identified by the actin filament technique (Sambongi et al. 1999; Pänke et al. 2001) as well as by cross-linking experiments fixing subunit c to the stator subunit a (Tsunoda et al. 2001b). The existence of a complex rotor extending from the nucleotide binding sites all the way down into the membrane is consistent with the "Brownian machine" mechanism for coupling of proton transport to torque generation (Vik and Antonio 1994; Junge et al. 1997). According to this mechanism, protons enter from the periplasmic space via an entry-channel in subunit a and are then transferred to an essential acidic residue in the membrane spanning part of subunit c (Asp61 in the *E. coli* enzyme). After close-to-360° rotation of the cylindrical subunit c oligomer relative to subunits a and b the protons are released on the opposite side of the membrane via a second half-channel in subunit a (Fig. 10.3).

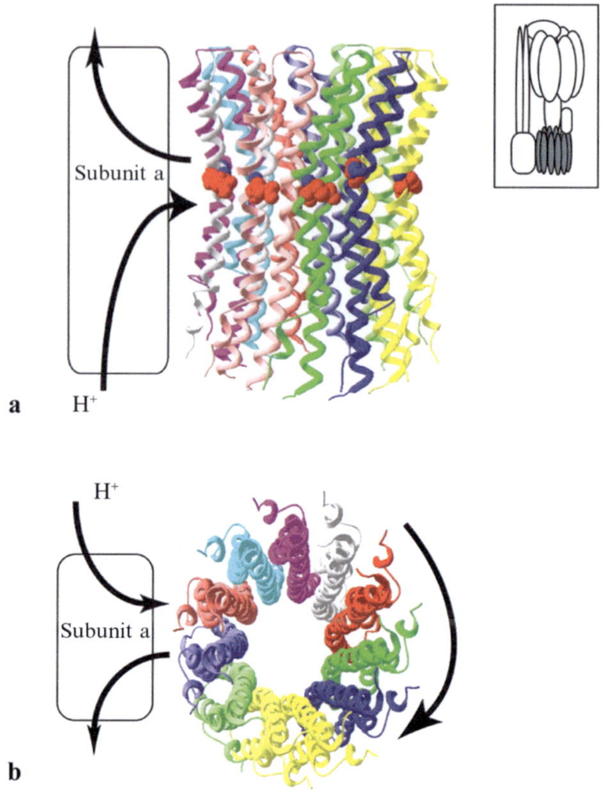

Fig. 10.3 The subunit c oligomer. Structure of the subunit c oligomer: **a** side view showing the monomers which each forms an α-helical hairpin. The route of ions crossing the membrane via subunit a and subunit c is indicated, the acidic residue essential for ion transport is displayed as space-filling model. **b** *Top* view indicating the the rotary movement of the oligomer relative to subunit a upon flow of ions (subunit c oligomer structure PDB ID: 1YCE). (Meier et al. 2005)

The movement of the ATP synthase rotor has also been demonstrated in a membrane environment by immobilizing the rotor subunit c to a solid support and observing an actin filament probe attached to the stator subunits $\alpha_3\beta_3$ (Nishio et al. 2002). Furthermore, in isolated ATP synthase from *E. coli* incorporated in liposomes Foerster Resonance Energy Transfer (FRET) was measured two organic fluorophors bound to the rotor subunit γ and the stator subunit b, respectively (Börsch et al. 2002). Three distinct levels of FRET efficiency were observed during ATP hydrolysis, consistent with the three distinct distances between the rotor and stator when subunit γ rotates from one αβ pair to the next. Similarly, periodic FRET changes were also detected by confocal microscopy for the ATP synthesis reaction using ATP synthase incorporated in lipsosomes energized by bacteriorhodopsin (Diez et al. 2004). However, the order of FRET efficiencies observed was found to be reversed compared to the ATP hydrolysis reaction, indicating that the γεc rotor actually moves in the opposite, clockwise, direction during synthesis of ATP (Diez et al. 2004).

In a landmark experiment, ATP synthesis was even achieved using the isolated F_1 part, separated from the F_0 proton channel that in vivo provides the energy for ATP synthesis. F_1 from *Bacillus* PS3 was immobilized in micro-reaction chambers on pattered surfaces and magnetic beads were bound to subunit γ. Subsequently, the beads were actively rotated in a clock-wise manner by external magnetic fields. This enforced rotary movement, imposed with magnetic tweezers, provided the energy required to synthesize ATP (Itoh et al. 2004). This experiment provides the first example of chemical synthesis driven by externally applied mechanical force. As a fascinating follow-up, a molecular "wind-up car" was constructed. After synthesis of ATP by enforced, clock-wise rotation of subunit γ, the magnetic fields were switched off, leading to counter-clockwise rotational movement due to hydrolysis of the produced ATP (Rondelez et al. 2005). Control of energy conversion at nanometer scale may prove highly useful for a potential application of nanomachines for future technological purposes, such as controlled catalysis of chemical reactions.

10.3 Connecting Ion Flow to ATP Synthesis

10.3.1 Subunit c Stochiometry

Subunit c of ATP synthase is a highly hydrophobic subunit that forms an α-helical hairpin, which is nearly completely embedded in the membrane. A conserved acidic residue of subunit c located in the central part of the membrane layer is essential for ion transport in rotational catalysis (glutamic acid 61 in the *E. coli* enzyme), and ATP synthase activity can be blocked by inhibitors binding to subunit c (von Ballmoos et al. 2004; Koul et al. 2007). In all species investigated so far, subunits c of ATP synthase form an oligomeric cylinder that constitutes a major part of the rotor (Fig. 10.3). However, reports differ considerably concerning how many subunits c comprise this oligomer: 10–15 copies of this subunit have been found for different species. X-ray analysis and genetic fusion experiments indicated 10 subunits c in ATP synthase from *Saccharomyces cerevesiae* (Stock et al. 1999), *E. coli* (Jiang et al. 2001), and *Bacillus PS3* (Mitome et al. 2004). Electron micoscopy and single-particle analysis showed 11 copies for ATP synthase from *Ilyobacter tatraricus, Propiogenium modestum* and *Clostridium paradoxum* (Stahlberg et al. 2001; Meier et al. 2003; Meier et al. 2006). 13 copies were determined in the cyanobacterium *Synochococcus elongatus* and *Bacillus* sp. strain TA2.A1 complexes by Atomic Force Microscopy and Mass Spectrometry (Meier et al. 2007; Pogoryelov et al. 2007). In chloroplasts from spinach and *Chlamodymonas rheinhardii* Atomic Force Microscopy revealed 14 subunits c in the oligomer (Seelert et al. 2000; Meier zu Tittingdorf et al. 2004). The highest number reported up to now, 15 copies of subunit c have been found in ATP synthase from the alkalophilic cyanobacterium *Arthrospira PCC9438* using Atomic Force Microscopy (Pogoryelov et al. 2005). The subunit c oligomer is part of the rotor and closely connected to subunit γ, whose rotation from one αβ pair to the next accounts for synthesis of ATP. It can thus be speculated that the number of subunits c in the oligomeric ring may be directly related to the number of protons flowing across the membrane for synthesis of one molecule of ATP. A single 120° step of subunit γ is thought to lead to synthesis of one molecule of ATP (Yasuda et al. 1998) thus a full 360° rotation produces three molecules of ATP. On the other hand, rotation of the subunit c oligomer is thought to occur in smaller steps of 360° divided by the number of subunits c per oligomer. For several species, which display a large number of subunits c in the oligomer, indeed also high H+/ATP values have been reported. In the *E. coli* and *Bacillus* PS3 enzymes, with 10 subunits c, the H+/ATP ratio is close to three, whereas for spinach chloroplast ATP synthase, with 14 subunits c, a H+/ATP ratio close to four has been found (van Walraven et al. 1996; Turina et al. 2003). If this relation between the number of subunits c and the H+/ATP ratio turns out to be generally applicable, a low number of subunits c would support ATP hydrolysis and active proton transport, whereas a larger number of subunits c would favor passive proton flow and synthesis of ATP. It seems thus logical that species/organelles which predominantly use ATP synthase to produce ATP, such as chloroplasts, display a larger number of subunits c. Furthermore, species living in an alkaline environment may depend on a large number of subunits c to survive under conditions of a low PMF. Indeed, the alkalophilics *Bacillus* sp. strain TA2.A1 and *Arthrospira* PCC9438 display oligomers of 13 and 15 subunits c, respectively (Meier et al. 2007; Pogoryelov et al. 2007). Many bacteria, on the other hand, can reverse the direction of ATP synthase depending on the PMF and cellular energy charge, so a lower number of this subunit may be of advantage. V-type ATPases, whose function purely is to hydrolyse ATP and actively transport protons, show only six subunits c in the oligomer. They are thought to have originated from a fusion of two subunit c genes followed by deletion of one of the proton binding sites in the fusion protein (for review see Cipriano et al. 2008).

However, it is not clear yet if the number of subunits c in a given species is constant or can change depending on external factors, e.g. the carbon source, as reported for the *E. coli* enzyme (Schemidt et al. 1998). In *Bacillus* PS3 the number of subunits c seems to be restricted to 10, as ATP synthase with 10 genetically fused copies of subunit c was active in ATP synthesis, but only very activity was found for the enzyme with 9, 11 or 12 fused copies (Mitome et al. 2004). Structural data indicate that in the case of *I. tartraricus*, spinach chloroplast and *Arthrospira* subunit c oligomers lacking of one or more monomers do not form a full cylinder (Müller et al. 2001). Varying the growth conditions for *Chlamydomonas reinhardtii* did not influence the oligomeric state of subunit c (Meyer zu Tittingdorf et al. 2004). Apparently, for these organisms the intrinsic structure of the subunit c monomer determines the oligomeric state with little or no flexibility to adapt to different stoichiometries. It is not known which regions of subunit c determine the size of the oligomer. A recent comparison of eight cyanobacterial species revealed that, although amino acid sequences of the investigated subunits c were highly similar with >80% sequence identity, one species displayed a subunit c 13 mer, four species 14 mers and two species 15 mers (Pogoryelov et al. 2007). Hence, cyanobacterial ATP synthase subunit c may prove a valuable model system to pinpoint factors that are critical for shaping the oligomer.

10.3.2 Stator Composition

Rotation of the $\gamma\delta c$ rotor can only lead to synthesis of ATP if the non-rotating (stator) subunits are rigidly connected, otherwise transport of protons would cause non-functional rotation of the whole ATP synthase complex. In the F_1 part, the $\alpha_3\beta_3$ ring is in contact with subunit γ (Fig. 10.2), whereas in F_o subunit a is in close vicinity of the c-ring (Fig. 10.3) (Weber 2007). The stator subunits of the F_1 and the F_o part are connected by an extended hydrophilic structure observed in electron microscopy images of ATP synthase from *E. coli* (Wilkens et al. 1998), spinach chloroplast (Böttcher et al. 1998) and bovine heart (Karrasch et al. 1999). This structure is referred to as "stator stalk" or "second stalk", as opposed to the central stalk made up by subunits γ and ε. Compared to the catalytic core of ATP synthase, the stator stalk shows considerable variety among species. In bacteria, it consists of subunit δ, a predominantly helical subunit bound to the top of the $\alpha_3\beta_3$ ring (Rodgers and Capaldi 1998; Wilkens et al. 2005) and two copies of subunit b, forming an extended dimer (Dunn 1992). In ATP synthase from cyanobacteria, other photosynthetic bacteria and chloroplasts, two different, but homologous subunits, substitute for the subunit b dimer. These two subunits are called subunit b and subunit b′ in cyanobacteria and subunit I and II in chloroplasts (Cozens and Walker 1987; Lill and Nelson 1991). In the mitochondrial enzyme, subunit δ is replaced by the homologous oligomycin-sensitivity-conferring protein (OSCP). In addition, only a single subunit b is found, accompanied by two smaller subunits called subunit d and F6 (Weber et al. 2007).

Recently, an x-ray structure of a "second-stalk-complex" from bovine heart mitochondria, comprising large portions of subunits b, d, and F6 (Dickson et al. 2006), revealed the single subunit b as an extended α-helix (Fig. 10.4a). The mitochondria-specific subunits d and F6 were wrapped around subunit b, in this way probably

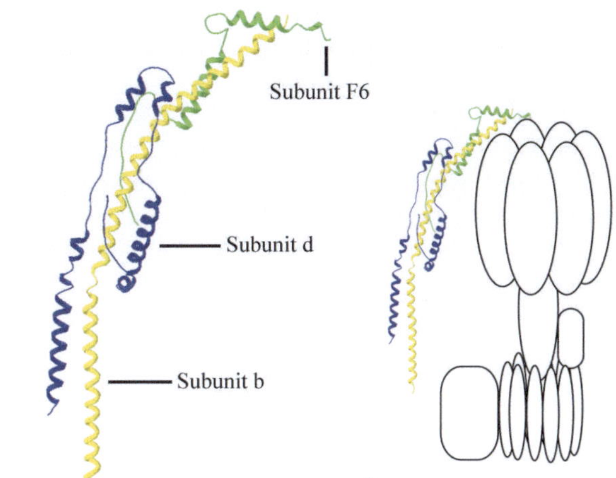

Fig. 10.4 The stator stalk. Structure of the "stator stalk" complex of ATP synthase from bovine heart mitochondria: **a** Subunit b forms an extended bent, α-helix, around which subunit d and F6 are wrapped (PDB ID: 2CLY). **b** Suggested location of the structure of the stator stalk in the ATP synthase holoenzyme: the stalk can reach from the membrane up to the top of F_1. (Model made by eye based on Dickson et al. 2006 and Weber et al. 2007)

fulfilling the same function as the second subunit b in the bacterial enzyme. When modeled into a low-resolution structure of mitochondrial ATP synthase (Rubinstein et al. 2003), the structure of this second stalk complex nicely fits into the region attributed to the stator stalk (Dickson et al. 2006). Subunit b can reach from the membrane all the way up to the top of the $\alpha_3\beta_3$ ring (Fig. 10.4b).

According to secondary structure prediction models bacterial subunit b is entirely helical and anchored in the membrane with a single N-terminal trans-membrane helix. The hydrophilic portion consist of three domains: a tether domain reaching from the membrane approximately up to the lower part of the $\alpha_3\beta_3$ ring, a central domain responsible for dimerisation of subunit b in solution, and a C-terminal part that is essential for contact with subunit δ (Rodgers and Capaldi 1998; Revington et al. 1999; McLachlin et al. 2000). In *E. coli* ATP synthase, subunit b tolerates insertions of up to 11 residues or deletions of maximal 14 residues in the tether domain without profound effect on ATP synthase activity (Sorgen et al. 1998, 1999). Also, the two subunits b may differ in length by up to 14 residues (Grabar and Cain 2003). These findings indicate that the subunit b does not function as a rigid rod, but may display some flexible, rope-like properties (Sorgen et al. 1999). A crystal structure of a monomeric peptide corresponding to the central dimerisation domain revealed a linear helical structure (DelRizzo et al. 2002). Amino acid sequence prediction and cross-linking data suggest an unusual right-handed coiled-coil structure for this domain (DelRizzo et al. 2006). A coiled-coil in the dimerisation domain was also predicted by modeling studies based on the heptad-repeat motif typical for coiled-coils, however, these theoretical studies favored the conventional left-handed coil (Wise and Vogel 2008).

For the subunit b′ found in cyanobacterial ATP synthase no high-resolution structural data are available, secondary structure prediction models and Circular Dicroism studies suggest a helical structure as in subunit b (Lill et al. 1994). Deletion of subunit b′ from *Synechocystis PCC6803* was lethal, but its hydrophilic part could be truncated down to the membrane spanning α-helical part without affecting ATP synthase assembly. This finding indicates no essential structural role of this subunit, however, truncation decreased ATP hydrolysis activity (Lill et al. 1994). Expressed and purified subunit b and subunit b′ from *Synochocystis PCC6803* formed heterodimers upon chemical crosslinking, but no homodimers, indicating a specific interaction between b and b′ in the cyanobacterial enzyme (Dunn et al. 2001). These heterodimers displayed a coiled-coil structure and an elongated shape, similar to the properties expected for subunits bb′ incorporated in ATP synthase in vivo (Dunn et al. 2001). In *E. coli* ATP synthase, nearly one third of one copy of subunit b (b_{39-86}) could be substituted by the homologous part of subunit b or subunit b′ from *Thermosynechococcus elongatus* (Claggett et al. 2007). Introduction of the *Thermosynechococcus* b′-segments in the hydrophilic region of both *E. coli* subunits b was tolerated, although the resulting chimeric enzyme was clearly less stable (Claggett et al. 2007). Recently, a bioinformatics comparison of chloroplast ATP synthase subunits I and II indicated a difference in secondary structure between these two subunits, especially in the number and lengths of helical parts (Poetsch et al. 2007).

Utilization of two different b-type subunits is not completely restricted to photosynthetic organisms, as a b′-type subunit recently was also found in the hyperthermophilic bacterium *Aquifex aeolicus* (Peng et al. 2006). It is not clear which potential advantages an asymmetric stator stalk may have, but it has been hypothesized that subunit b′ may be an adaptation to optimize binding to F_1 or F_o. The regions of subunits b and b′, which are not involved in dimer formation, may adopt different conformations and in this way improve interaction with the asymmetric subunit a and/or δ (Dunn et al. 2001). A similar reason may also account for the presence of subunits d and F6, replacing the second copy of subunit b in mitochondrial ATP synthase. The availability of purified, stable cyanobacterial ATP synthase and its subunits, provided for by new purification methods or by overexpression systems as reported for *Thermosynochococcus elongatus* (Konno et al. 2006; Suhai et al. 2008), may help to shed light on the function of the stator stalk in cyanobacteria.

10.3.3 Elastic Power Transmission?

In addition to its function as a connector of the F_1 and F_o stator subunit, the stator stalk may also serve for intermediate storage of energy during catalysis. For synthesis of ATP, the rotary stepping angles of the c ring in F_o have to align with the movement of subunit γ in F_1. Whereas subunit γ moves in steps of 120°, divided in two sub-steps of 80° and 40° (Yasuda et al. 1998, 2001), the c-ring is thought to move in steps of 360° divided by the number of subunits c in the oligomer. For all species with 10, 11, 13 or 14 copies of subunit c in the oligomer the step size ratio compared to subunit γ is a non-integer value, requiring some elastic property of the enzyme complex to effectively couple the two rotors (Cherepanov et al. 1999; Junge et al. 2001; Weber 2007). The flexible, rope-like properties proposed for the subunit b dimer in *E. coli* based on insertion and deletion mutant studies (Sorgen et al. 1998, 1999) make subunit b a promising candidate for this transient storage of elastic energy. A kink observed in the N-terminal domain of subunit b, directly adjacent to the membrane and close to the tether domain, may act as a hinge (Dmitriev et al. 1999), allowing twisting movements of the hydrophilic portion. Nucleotide-dependent shortening of the whole ATP synthase complex from *E. coli* was observed by electron microscopy of single particles, consistent with a twisting or piston-like motion of the stalk during catalysis (Capaldi et al. 2000). On the contrary, EPR spectroscopy of two spin-labels, introduced at equal height in the tether-domains of both subunits b in *E. coli*, did not show significant nucleotide-dependent changes in distance between the two labels. This result indicates no or only minor structural changes in this region during catalysis (Steigmiller et al. 2005). However, in case of a twisting or piston-like motion of the stator stalk, the most prominent changes in distance can be expected between the part of subunit b near to the membrane compared to the C-terminal part located close to the top of F_1, only to a lesser extend between to labels introduced at the same height.

It may well be possible that parts of the rotor, e.g. the extended coiled-coil found in subunit γ, are involved in energy storage as well. Coiled-coils may respond to mechanical stress by untwisting or unfolding of several turns and have been investigated as important determinants for elasticity in a number of proteins, such as myosin (Schwaiger et al. 2002), fibrin (Lim et al. 2008) and the DNA repair protein RAD50 (van Noort et al. 2003).

10.4 Regulation

10.4.1 Inhibition by Mg-ADP

Living organisms constantly have to adapt their metabolism to environmental needs and, as expected for an enzyme of such a central position in metabolism, the activity of ATP synthase is subject to a number of regulatory mechanisms. In particular, the activity of the enzyme in the presence of a low PMF needs to be tightly controlled to prevent excess hydrolysis of ATP.

An intrinsic regulatory mechanisms found in ATP synthase from bacteria, chloroplasts and mitochondria is inhibition by Mg-ADP, which occasionally becomes entrapped in a catalytic nucleotide binding site, thus blocking cooperative catalysis (Fitin et al. 1979; Drobinskaya et al. 1985; Fromme and Gräber 1990; Milgrom and Boyer 1990). Inhibition by Mg-ADP can be relieved by the PMF (Junge 1970; Strotmann et al. 1976; Fischer et al. 2000), consequently ATP synthesis is significantly less affected than ATP hydrolysis (Syroeshkin et al. 1995; Bald et al. 1998; 1999). Inhibition by Mg-ADP can arrest subunit γ during rotational catalysis at an intermediate position: as each 120° step of subunit γ consists of an 80° sub-step (ATP binding) and a subsequent 40° step (Mg-ADP release), entrapped Mg-ADP occasionally stops the rotational movement at the intermediate 80° position (Hirono-Hara et al. 2001). It has been suggested that the PMF stimulates the 40° step, releasing the bound Mg-ADP and moving subunit γ to the ATP waiting position thus allowing ATP to bind again (Feniouk et al. 2007). Consistent with this hypothesis, individual molecules of F_1-ATPase from *Bacillus* PS3 were relieved from Mg-ADP inhibition by forced rotation of about 40° with magnetic tweezers (Hirono-Hara et al. 2005).

Inhibitory Mg-ADP can also be released after binding of ATP to a regulatory non-catalytic nucleotide-binding site (Jault et al. 1995; Weber et al. 1995; Matsui et al. 1997). The non-catalytic sites lack an acidic residue essential for hydrolysis of ATP and for this reason are catalytically inactive (Abrahams et al. 1993). Their main function may be the regulation of Mg-ADP inhibition. The underlying mechanism of how binding of ATP to the non-catalytic sites can influence the affinity of Mg-ADP for the catalytic sites remains to be clarified.

10.4.2 Subunit ε

A second intrinsic mechanism of regulation found in ATP synthase from bacteria and chloroplasts, but not from mitochondria, is the inhibition by subunit ε, the smallest subunit of the hydrophilic F_1 part (reviewed in Feniouk et al. 2007).

Subunit ε is located in the central stalk of the enzyme in close association with subunit γ and the oligomeric c-ring. It plays an essential role for the holo-enzyme assembly in bacteria (Yoshida et al.1977) and for ATP synthesis activity in chloroplasts (Patrie et al. 1984). In mitochondria, a homologous subunit (unfortunately for historic reasons called subunit δ) is important for ATP synthase assembly, but apparently not for regulation of activity (Giraud et al. 1997). Subunit ε consists of two major domains, an N-terminal β-barrel and two C-terminal α-helices (Fig. 10.5) (Wilkens et al. 1995; Uhlin et al. 1997). Whereas the β-barrel is in close contact with the membrane-bound oligomeric subunit c, the α-helical domain plays a crucial role in activity regulation. The inhibitory effect is best seen in the isolated F_1 part, where subunit ε decreases activity ~5-fold in case of *E. coli* F_1 (Dunn and Tozer 1987) and ~15-fold in case of spinach chloroplast F_1 (Richter et al. 1984). In the ATP synthase holo-enzymes from *Bacillus* PS3 and spinach chloroplast, truncated subunit ε deficient of the C-terminal α-helices caused significantly enhanced

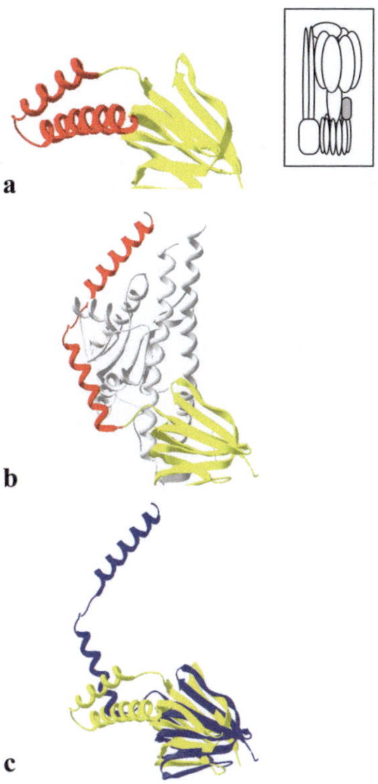

Fig. 10.5 Subunit ε **a** Structure of subunit ε of ATP synthase from *E. coli*, displaying the N-terminal β-barrel (*yellow*) and the C-terminal helices (*red*). The two helices from a contracted hairpin (PDB ID: 1AQT, Uhlin et al. 1998). **b** Structure of a γε sub-complex of *E. coli* ATP synthase with the N-terminal β-barrel of subunit ε in *yellow*, the C-terminal helices in *red* and subunit γ in *grey* (PDB ID: 1FS0, Rodgers and Wilce 2000). **c** Overlay of the two subunit ε structures from (**a**) and (**b**) with subunit ε in the contracted conformation in *yellow* and in the extended conformation in *blue*

ATP hydrolysis activity. These findings indicate that the role of this domain may be to prevent wasteful hydrolysis of ATP (Kato-Yamada et al. 1999; Nowak and McCarty 2004).

In spinach chloroplast thylakoids, conformational changes of the C-terminal domain of subunit ε were detected on illumination, indicating mobility for this domain during catalysis (Richter and McCarty 1987; Komatsu-Takaki 1989). Consistent with this result, in *E. coli* F_1, the C-terminal α-helices were crosslinked to the conserved DELSEED region of subunit β in a nucleotide dependent manner (Aggeler and Capaldi 1996). For the *Bacillus* enzyme, time-dependent activation of ATP hydrolysis activity was observed without dissociation of the inhibitory subunit ε, indicating two distinct conformations of this subunit (Kato et al. 1997). The structure determined for subunit ε in solution displayed the C-terminal α-helices as a contracted hairpin (Fig. 10.5a) (Wilkens et al. 1995) and the dimensions of the subunit were too small to reach all the way from the c-ring up to the DELSEED region of subunit β in F_1. Furthermore, x-ray structures determined for a large part of bovine heart mitochondrial F_1 and for an ATP synthase sub-complex from *Saccharomyces cerevesiae* comprising F_1 and the subunit c oligomer revealed a similar structure of the homologous mitochondrial subunit δ (Stock et al. 1999; Gibbons et al. 2000). In these structures, the C-terminal helices were located too far away from subunit β to explain the cross-linking results obtained for the bacterial enzyme. In contrast, in a subunit γε sub-complex from *E. coli* ATP synthase, the two C-terminal helices of subunit ε displayed an extended conformation, reaching up in the direction of the F_1 part (Fig. 10.5b, c) (Rodgers and Wilce 2000). This conformation can account for the observed cross-linking to subunit β as well as for the inhibitory effect. Subsequently, cross-linking experiments showed that transition between the two conformations can be induced and trapped by the presence of nucleotides and by the PMF in *E. coli* and *Bacillus* PS3 (Tsunoda et al. 2001a; Suzuki et al. 2003). Interestingly, when the helical domain of subunit ε was fixed in the contracted hairpin conformation, the enzyme was fully active in both ATP synthesis and ATP hydrolysis, however, when entrapped in the extended conformation, the ATP hydrolysis activity was strongly inhibited without significant influence on ATP synthesis.

Conformational changes of subunit ε related to activation/inactivation of ATP synthase were also observed in real-time using a FRET approach. The distances measured for two fluorophors introduced in subunit b and subunit ε, respectively, differed significantly between ATP synthase molecules actively in catalysis compared to resting enzymes (Zimmermann et al. 2005). Furthermore, two fluorophors introduced in N-terminus and C-terminus of subunit ε, respectively, approached each other in the presence of ATP, whereas the distance increased again after the ATP was hydrolysed (Iino et al. 2005).

Surprisingly, the isolated subunit ε can bind ATP although it does not contain standard ATP-binding motifs, suggesting a role for this subunit as ATP sensor (Kato et al. 2007) However, it remains an open question if binding of ATP directly to subunit ε may account for the conformation transition or if subunit ε can sense occupancy of the nucleotide binding sites in F_1. Single-molecule experiments investigating rotation of F_1 from *Thermosynechococcus elongatus* revealed that inhibition by subunit ε stops the rotary movement of subunit γ at the 80° intermediate position, corresponding to a catalytic cycle intermediate with Mg-ADP bound to the catalytic sites (Konno et al. 2006). It can be speculated that binding of the extended C-terminal helices of subunit ε to the subunit β DELSEED sequence in this way enhances the inhibitory effect of Mg-ADP (Feniouk 2007).

Subunit ε has emerged as a key regulator of ATP synthase activity. Its most significant features may be the conformational transition in response the PMF and cellular energy charge as well as the ratchet-like inhibition of ATP hydrolysis, but not ATP synthesis. Differences between species in primary structure of subunit ε may determine the extent of inhibition and allow each species to adapt to its specific environmental conditions. Subunit ε in cyanobacteria and chloroplasts, compared to its bacterial counterparts, exerts a stronger inhibitory effect and acts independent of the ATP concentration (Richter et al. 1984; Konno et al. 2006). This feature may reflect the need of photosynthetic organisms to efficiently respond to changes in light intensity. The molecular basis for this strong inhibitory effect in the cyanobacterial and chloroplast enzymes is not known, but critical residues in the C-terminal helices may be responsible (Feniouk 2007). As an example for adaptation to utilizing ATP synthase in ATP hydrolysis direction to maintain the PMF, several anaerobic bacteria such a *Chlorobium* lack the C-terminal helical domain of subunit ε (Xie et al. 1993). In mitochondrial ATP synthase the homologous subunit (δ) appears to be kept in the contracted hairpin conformation by the presence of an additional subunit (mitochondrial subunit ε), which is not found in the bacterial or chloroplast enzyme (Stock et al. 1999; Gibbons et al. 2000). Mitochondrial ATP synthase is regulated by the Inhibitor Protein (IF_1) that under conditions of low pH forms a dimer, which deeply penetrates between a catalytic αβ interface (Harris 1997; Cabezon et al. 2000, 2003), thus blocking wasteful hydrolysis of ATP.

10.4.3 Redox-Regulation

In photosynthetic organisms the PMF decreases in the dark and ATP synthase may act in reverse, hydrolysing ATP and pumping protons across the membrane. As discussed above, one mechanism to prevent this wasteful consumption of ATP may be a strongly inhibitory subunit ε, common to cyanobacteria and plant chloroplasts.

However, plants and green algae have invented an additional regulatory mechanism that allows adaptation to changes in light intensity. The activity of ATP synthase in these organisms is modulated by oxidation/reduction of two cysteine residues of subunit γ (Nalin and McCarty 1984). These cysteines are located in an insert of ~40 amino acid resides in subunit γ, which is not found in the homologous bacterial and mitochondrial subunits (Miki et al. 1988). This insert is located in the part of subunit γ close to the contact area with the c-ring. Unfortunately, for this region of the enzyme no high-resolution structural data are available, as an x-ray structure determined for F_1 from spinach chloroplast only resolved the $\alpha_3\beta_3$ hexagon (Groth and Pohl 2001). Cyanobacterial subunit γ displays a similar, although slightly shorter, insert but not the two specific cysteine residues (Lill and Nelson 1991). Consequently cyanobacterial ATP synthase does not show redox-regulation. Cyanobacteria may not need an additional regulatory mechanism, as the PMF across their plasma membranes is not only set-up by the photosynthetic, but also by the respiratory electron transport chain.

Formation of a disulfide bond upon oxidation of the cysteines suppresses ATP synthesis and ATP hydrolysis activity, cleavage of the bond catalysed by thioredoxin leads to activation (Nalin and McCarty 1984). Thioredoxin in turn is reduced by Photosystem I via Ferredoxin and the Ferredoxin-NADPH-Oxidoreductase (for review: Schürmann and Buchanan 2008). Redox-regulation has been observed for the ATP synthesis and ATP hydrolysis activity of the ATP synthase holoenzyme as well as for isolated F_1-ATPase. The mechanism of inactivation upon disulfide bridge formation is unknown, it may be speculated that oxidation leads to a structural change of the insert, which may then, in analogy to subunit ε, interact with the F_1 part, thus blocking rotation. Clearly, structural data, especially on the oxidized enzyme, are highly desired to gain deeper mechanistic insight.

Genetic introduction of the cysteine residues into ATP synthase of *Synechocystis PCC6803* yielded an enzyme susceptible to redox–regulation (Werner-Grüne et al. 1994; Krenn et al. 1997), indicating the basic capability of cyanobacteria to adopt redox-regulation for modulation of ATP synthesis activity. In another approach, the whole regulatory region of spinach chloroplast ATP synthase subunit γ was genetically transferred to *Bacillus PS3* F_1. The resulting enzyme was sensitive to redox-regulation of activity, indicating that the regulatory region can be regarded as a transferrable module, which can be inserted into homologous enzymes (Bald et al. 2000). Observation of rotation of this redox-sensitive *Bacillus* enzyme revealed long stops of activity that could be reversibly imposed upon oxidation (Bald et al. 2001). Such a modulation of motor movement by external chemical signals may provide an example of how to control a molecular machine for potential applications in an artificial environment.

Redox-regulation is closely related to the inhibitory action of subunit ε as the affinity of subunit ε for F_1 strongly decreases upon reduction of subunit γ from spinach chloroplast (Soteropoulos et al. 1992) and upon deletion of the regulatory insert from a chimeric F_1-ATPase consisting of $\alpha_3\beta_3$ from *Bacillus* PS3 and subunit γ from spinach chloroplast (Hisabori et al. 2002; Konno et al. 2004). Interestingly, deletion of the insert from *Thermosynechococcus elongatus* F_1-ATPase subunit γ also decreased inhibition by subunit ε, although this cyanobacterial insert does not confer redox-regulation (Konno et al. 2006). The insert may thus have first developed in cyanobacteria as a means to increase inhibition by subunit ε. In a second step introduction of the cysteine residues may have refined this regulation mechanism (Konno et al. 2006). For the thermophilic cyanobacterium *Spirulina platensis* two cysteine residues were found in the sequence of subunit γ, however, these two residues were not homologous to the cysteine residues responsible for redox-regulation in the plant chloroplast enzyme (Steinemann and Lill 1995). Although the role of these cysteine residues in cyanobacterial subunit γ is unclear, they may represent an evolutionary intermediate on the way to redox-regulation.

It is not fully understood yet why different classes of organisms use different strategies to regulate ATP synthase activity. It may turn out that inhibition by Mg-ADP, susceptible to changes in the PMF and the cellular energy charge, is the key intrinsic inhibitory mechanism of ATP synthase. Subunit ε in the extended conformation may act as a brake and increase inhibition by Mg-ADP by fixing subunit γ in an angular position typical for Mg-ADP binding. The insert of subunit γ found in cyanobacteria may have evolved as a loop to further strengthen action of subunit ε. Finally, in chloroplasts, redox-regulation of cysteine residues in the insert may in turn have provided a means to control the effect of the insert on subunit ε.

Further structural and mechanistic investigations will clarify if regulation by Mg-ADP, subunit ε, redox-state and IF_1 are just variations of a common theme or constitute evolutionary distinct strategies to adapt energy metabolism to changing environmental conditions.

Acknowledgement The author wishes to thank H. Lill (VU University Amsterdam) for critically reading the manuscript. Financial support from the European Commission (Marie Curie Project MRTN-CT-2005-019481) is gratefully acknowledged.

References

Abrahams JP, Leslie AGW, Lutter R and Walker JE (1993) Structure at 2.8A resolution of F_1-ATPase from bovine heart mitochondria. Nature 370: 621–8

Aggeler R and Capaldi RA (1996) Nucleotide-dependent movement of the epsilon subunit between alpha and beta subunits in the Escherichia coli F1F0-type ATPase. J Biol Chem. 271: 13888–91

Aggeler R, Weinreich F, and Capaldi RA (1995) Arrangement of the epsilon subunit in the Escherichia coli ATP synthase from the reactivity of cysteine residues introduced at different positions in this subunit. Biochim Biophys Acta 1230: 62–8

Bald D, Amano T, Muneyuki E, Pitard B, Rigaud JL, Kruip J, Hisabori T, Yoshida M and Shibata M (1998) ATP synthesis by F0F1-ATP synthase independent of noncatalytic nucleotide binding sites and insensitive to azide inhibition. J Biol Chem 273: 865–70

Bald D, Muneyuki E, Amano T, Kruip J, Hisabori T and Yoshida M (1999) The noncatalytic site-deficient alpha3beta3gamma subcomplex and FoF1-ATP synthase can continuously catalyse ATP hydrolysis when Pi is present. Eur J Biochem 262: 563–8

Bald D, Noji H, Stumpp MT, Yoshida M and Hisabori T (2000) ATPase activity of a highly stable alpha(3)beta(3)gamma subcomplex of thermophilic F(1) can be regulated by the introduced regulatory region of gamma subunit of chloroplast F(1). J Biol Chem 275: 12757–62

Bald D, Noji H, Yoshida M, Hirono-Hara Y and Hisabori T (2001) Redox regulation of the rotation of F(1)-ATP synthase. J Biol Chem 276: 39505–507

Bianchet MA, Hullihen J, Pedersen PL and Amzel LM (1998) The 2.8-A structure of rat liver F1-ATPase: configuration of a critical intermediate in ATP synthesis/hydrolysis. Proc Natl Acad Sci U S A 95: 11065–70

Börsch M, Diez M, Zimmermann B, Reuter R and Gräber P (2002) Stepwise rotation of the gamma-subunit of EF(0)F(1)-ATP synthase observed by intramolecular single-molecule fluorescence resonance energy transfer. FEBS Lett 527: 147–52

Böttcher B, Schwarz L and Gräber P (1998) Direct indication for the existence of a double stalk in CF0F1. J Mol Biol 281: 757–62

Boyer PD (1993) The binding change mechanism for ATP synthase—some probabilities and possibilities. Biochim Biophys Acta 1140: 215–50

Cabezon E, Butler PJ, Runswick MJ and Walker JE (2000) Modulation of the oligomerization state of the bovine F1-ATPase inhibitor protein, IF1, by pH. J Biol Chem 275: 25460–4

Cabezón E, Montgomery MG, Leslie AG and Walker JE (2003) The structure of bovine F1-ATPase in complex with its regulatory protein IF1. Nat Struct Biol 10: 744–5

Capaldi RA, Schulenberg B, Murray J and Aggeler R (2000) Cross-linking and electron microscopy studies of the structure and functioning of the Escherichia coli ATP synthase. J Exp Biol 203: 29–33

Cherepanov DA, Mulkidjanian AY and Junge W (1999) Transient accumulation of elastic energy in proton translocating ATP synthase. FEBS Lett 449: 1–6

Cipriano DJ, Wang Y, Bond S, Hinton A, Jefferies KC, Qi J and Forgac M (2008) Structure and regulation of the vacuolar ATPases. Biochim Biophys Acta 1777: 599–604

Claggett SB, Grabar TB, Dunn SD and Cain BD (2007) Functional incorporation of chimeric b subunits into F1Fo ATP synthase. J Bacteriol 189: 5463–71

Cozens AL and Walker JE (1987) The organization and sequence of the genes for ATP synthase subunits in the cyanobacterium Synechococcus 6301. Support for an endosymbiotic origin of chloroplasts. J Mol Biol 194: 359–83

Del Rizzo PA, Bi Y, Dunn SD and Shilton BH (2002) The "second stalk" of Escherichia coli ATP synthase: structure of the isolated dimerization domain. Biochemistry 41: 6875–84

Del Rizzo PA, Bi Y and Dunn SD (2006) ATP synthase b subunit dimerization domain: a right-handed coiled coil with offset helices. J Mol Biol 364: 735–46

Dickson VK, Silvester JA, Fearnley IM, Leslie AG and Walker JE (2006) On the structure of the stator of the mitochondrial ATP synthase. EMBO J 25: 2911–8

Diez M, Zimmermann B, Börsch M, König M, Schweinberger E, Steigmiller S, Reuter R, Felekyan S, Kudryavtsev V, Seidel CA and Gräber P (2004) Proton-powered subunit rotation in single membrane-bound F0F1-ATP synthase. Nat Struct Mol Biol 11: 135–41

Dmitriev O, Jones PC, Jiang W and Fillingame RH (1999) Structure of the membrane domain of subunit b of the Escherichia coli F0F1 ATP synthase. J Biol Chem 274: 15598–604

Drobinskaya IY, Kozlov IA, Murataliev MB and Vulfson EN (1985) Tightly bound adenosine diphosphate, which inhibits the activity of mitochondrial F1-ATPase, is located at the catalytic site of the enzyme. FEBS Lett 182: 419–24

Duncan TM, Bulygin VV, Zhou Y, Hutcheon ML and Cross RL (1995) Rotation of subunits during catalysis by Escherichia coli F1-ATPase. Proc Natl Acad Sci U S A 92: 10964–8

Dunn SD (1992) The polar domain of the b subunit of Escherichia coli F1F0-ATPase forms an elongated dimer that interacts with the F1 sector. J Biol Chem 267: 7630–6

Dunn SD and Tozer RG (1987) Activation and inhibition of the Escherichia coli F1-ATPase by monoclonal antibodies which recognize the epsilon subunit. Arch Biochem Biophys 25373–80

Dunn SD, Kellner E and Lill H (2001) Specific heterodimer formation by the cytoplasmic domains of the b and b′ subunits of cyanobacterial ATP synthase. Biochemistry 40: 187–92

Feniouk BA, Rebecchi A, Giovannini D, Anefors S, Mulkidjanian AY, Junge W, Turina P and Melandri BA (2007) Met23Lys mutation in subunit gamma of F(O)F(1)-ATP synthase from Rhodobacter capsulatus impairs the activation of ATP hydrolysis by protonmotive force. Biochim Biophys Acta 1767: 1319–30

Fischer S, Graber P, and Turina P (2000) The activity of the ATP synthase from Escherichia coli is regulated by the transmembrane proton motive force. J Biol Chem 275: 30157–62

Fitin AF, Vasilyeva EA and Vinogradov AD (1979) An inhibitory high affinity binding site for ADP in the oligomycin-sensitive ATPase of beef heart submitochondrial particles. Biochem Biophys Res Commun 86: 434–9

Fromme P and Gräber P (1990) Uni-site catalysis in thylakoids. The influence of membrane energization on ATP hydrolysis and ATP-Pi exchange. FEBS Lett 269: 247–51

Gibbons C, Montgomery MG, Leslie AG and Walker JE (2000) The structure of the central stalk in bovine F(1)-ATPase at 2.4 A resolution. Nat Struct Biol 7: 1055–61

Giraud MF and Velours J (1997) The absence of the mitochondrial ATP synthase delta subunit promotes a slow growth phenotype of rho- yeast cells by a lack of assembly of the catalytic sector F1. Eur J Biochem 245: 813–8

Grabar TB and Cain BD (2003) Integration of b subunits of unequal lengths into F1F0-ATP synthase. J Biol Chem 278: 34751–6

Groth G and Pohl E (2001) The structure of the chloroplast F1-ATPase at 3.2 A resolution. J Biol Chem 276: 1345–52

Harris DA (1997) Functional regions of the H(+)-ATPase inhibitory protein from ox heart mitochondria. Biochim Biophys Acta 1320: 8–16

Hirono-Hara Y, Noji H, Nishiura M, Muneyuki E, Hara KY, Yasuda R, Kinosita K Jr and Yoshida M (2001) Pause and rotation of F(1)-ATPase during catalysis. Proc Natl Acad Sci U S A 98: 13649–54

Hirono-Hara Y, Ishizuka K, Kinosita K Jr, Yoshida M and Noji H (2005) Activation of pausing F1 motor by external force. Proc Natl Acad Sci U S A 102: 4288–93

Hisabori T, Konno H, Ichimura H, Strotmann H and Bald D (2002) Molecular devices of chloroplast F(1)-ATP synthase for the regulation. Biochim Biophys Acta 1555: 140–6.

Iino R, Murakami T, Iizuka S, Kato-Yamada Y, Suzuki T and Yoshida M (2005) Real-time monitoring of conformational dynamics of the epsilon subunit in F1-ATPase. J Biol Chem 280: 40130–4

Itoh H, Takahashi A, Adachi K, Noji H, Yasuda R, Yoshida M and Kinosita K (2004) Mechanically driven ATP synthesis by F1-ATPase. Nature 427: 465–8

Jault JM, Matsui T, Jault FM, Kaibara C, Muneyuki E, Yoshida M, Kagawa Y and Allison WS (1995) The alpha 3 beta 3 gamma complex of the F1-ATPase from thermophilic Bacillus PS3 containing the alpha D261N substitution fails to dissociate inhibitory MgADP from a catalytic site when ATP binds to noncatalytic sites. Biochemistry 34: 16412–8

Jiang W, Hermolin J and Fillingame RH (2001) The preferred stoichiometry of c subunits in the rotary motor sector of Escherichia coli ATP synthase is 10. Proc Natl Acad Sci U S A 98: 4966–71

Junge W (1970) The critical electric potential difference for photophosphorylation. Its relation to the chemiosmotic hypothesis and to the triggering requirements of the ATPase system. Eur J Biochem 14: 582–92

Junge W, Lill H and Engelbrecht S (1997) ATP synthase: an electrochemical transducer with rotatory mechanics. Trends Biochem Sci 22: 420–3

Junge W, Pänke O, Cherepanov DA, Gumbiowski K, Müller M and Engelbrecht S (2001) Intersubunit rotation and elastic power transmission in F0F1-ATPase. FEBS Lett 504: 152–60

Kabaleeswaran V, Puri N, Walker JE, Leslie AG and Mueller DM (2006) Novel features of the rotary catalytic mechanism revealed in the structure of yeast F1 ATPase. EMBO J 25: 5433–42

Karrasch S and Walker JE (1999) Novel features in the structure of bovine ATP synthase. J Mol Biol 290: 379–84

Kato Y, Matsui T, Tanaka N, Muneyuki E, Hisabori T and Yoshida M (1997) Thermophilic F1-ATPase is activated without dissociation of an endogenous inhibitor, epsilon subunit. J Biol Chem 272: 24906–12

Kato S, Yoshida M and Kato-Yamada Y (2007) Role of the epsilon subunit of thermophilic F1-ATPase as a sensor for ATP. J Biol Chem 282: 37618–23

Kato-Yamada Y, Noji H, Yasuda R, Kinosita K Jr and Yoshida M (1998) Direct observation of the rotation of epsilon subunit in F1-ATPase. J Biol Chem 273: 19375–7

Kato-Yamada Y, Yoshida M and Hisabori T (2000) Movement of the helical domain of the epsilon subunit is required for the activation of thermophilic F1-ATPase. J Biol Chem 275: 35746–50

Komatsu-Takaki M (1989) Energy-dependent conformational changes in the epsilon subunit of the chloroplast ATP synthase (CF0CF1). J Biol Chem 264: 17750–3

Konno H, Suzuki T, Bald D, Yoshida M and Hisabori T (2004) Significance of the epsilon subunit in the thiol modulation of chloroplast ATP synthase. Biochem Biophys Res Commun 318: 17–24

Konno H, Murakami-Fuse T, Fujii F, Koyama F, Ueoka-Nakanishi H, Pack CG, Kinjo M and Hisabori T (2006) The regulator of the F1 motor: inhibition of rotation of cyanobacterial F1-ATPase by the epsilon subunit. EMBO J 25: 4596–604

Koul A, Dendouga N, Vergauwen K, Molenberghs B, Vranckx L, Willebrords R, Ristic Z, Lill H, Dorange I, Guillemont J, Bald D and Andries K (2007) Diarylquinolines target subunit c of mycobacterial ATP synthase. Nat Chem Biol 3: 323–4

Krenn BE, Strotmann H, Van Walraven HS, Scholts MJ and Kraayenhof R (1997) The ATP synthase gamma subunit provides the primary site of activation of the chloroplast enzyme: experiments with a chloroplast-like Synechocystis 6803 mutant. Biochem J 323: 841–5

Lill H and Nelson N (1991) The atp1 and atp2 operons of the cyanobacterium Synechocystis sp. PCC 6803. Plant Mol Biol 17: 641–52

Lill H, Steinemann D and Nelson N (1994) Mutagenesis of the b′-subunit of Synechocystis sp. PCC 6803 ATP-synthase. Biochim Biophys Acta 1184: 284–90

Lim BB, Lee EH, Sotomayor M and Schulten K (2008) Molecular basis of fibrin clot elasticity. Structure 16: 449–59

Matsui T, Muneyuki E, Honda M, Allison WS, Dou C and Yoshida M (1997) Catalytic activity of the alpha3beta3gamma complex of F1-ATPase without noncatalytic nucleotide binding site. J Biol Chem 272: 8215–21

McLachlin DT, Coveny AM, Clark SM and Dunn SD (2000) Site-directed cross-linking of b to the alpha, beta, and a subunits of the Escherichia coli ATP synthase. J Biol Chem 275: 17571–7

Meier T, Matthey U, von Ballmoos C, Vonck J, Krug von Nidda T, Kühlbrandt W and Dimroth P (2003) Evidence for structural integrity in the undecameric c-rings isolated from sodium ATP synthases. J Mol Biol 325: 389–97

Meier T, Polzer P, Diederichs K, Welte W and Dimroth P (2005) Structure of the rotor ring of F-Type Na+-ATPase from Ilyobacter tartaricus. Science 308: 659–62

Meier T, Ferguson SA, Cook GM, Dimroth P and Vonck J (2006) Structural investigations of the membrane-embedded rotor ring of the F-ATPase from Clostridium paradoxum. J Bacteriol 188: 7759–64

Meier T, Morgner N, Matthies D, Pogoryelov D, Keis S, Cook GM, Dimroth P and Brutschy B (2007) A tridecameric c ring of the adenosine triphosphate (ATP) synthase from the thermoalkaliphilic Bacillus sp. strain TA2.A1 facilitates ATP synthesis at low electrochemical proton potential. Mol Microbiol 65:1181–92

Meyer Zu Tittingdorf JM, Rexroth S, Schäfer E, Schlichting R, Giersch C, Dencher NA and Seelert H (2004) The stoichiometry of the chloroplast ATP synthase oligomer III in Chlamydomonas reinhardtii is not affected by the metabolic state. Biochim Biophys Acta 1659: 92–9

Milgrom YM and Boyer PD (1990) The ADP that binds tightly to nucleotide-depleted mitochondrial F1-ATPase and inhibits catalysis is bound at a catalytic site. Biochim Biophys Acta 1020: 43–8

Miki J, Maeda M, Mukohata Y and Futai M (1988) The gamma-subunit of ATP synthase from spinach chloroplasts. Primary structure deduced from the cloned cDNA sequence. FEBS Lett 232: 221–6

Mitome N, Suzuki T, Hayashi S and Yoshida M (2004) Thermophilic ATP synthase has a decamer c-ring: indication of noninteger 10:3 H+/ATP ratio and permissive elastic coupling. Proc Natl Acad Sci U S A 101: 12159–64

Müller DJ, Dencher NA, Meier T, Dimroth P, Suda K, Stahlberg H, Engel A, Seelert H and Matthey U (2001) ATP synthase: constrained stoichiometry of the transmembrane rotor. FEBS Lett 504: 219–22

Nalin CM and McCarty RE (1984) Role of a disulfide bond in the gamma subunit in activation of the ATPase of chloroplast coupling factor 1. J Biol Chem 259: 7275–80

Nishio K, Iwamoto-Kihara A, Yamamoto A, Wada Y and Futai M (2002) Subunit rotation of ATP synthase embedded in membranes: a or beta subunit rotation relative to the c subunit ring. Proc Natl Acad Sci U S A 99: 13448–52

Nishizaka T, Oiwa K, Noji H, Kimura S, Muneyuki E, Yoshida M and Kinosita K Jr (2004) Chemomechanical coupling in F1-ATPase revealed by simultaneous observation of nucleotide kinetics and rotation. Nat Struct Mol Biol 11:142–8

Noji H, Yasuda R, Yoshida M and Kinosita K Jr (1997) Direct observation of the rotation of F1-ATPase. Nature 386: 299–302

Nowak KF and McCarty RE (2004) Regulatory role of the C-terminus of the epsilon subunit from the chloroplast ATP synthase. Biochemistry 43: 3273–9

Pänke O, Gumbiowski K, Junge W and Engelbrecht S (2001) F-ATPase: specific observation of the rotating c subunit oligomer of EF(o)EF(1). FEBS Lett 472: 34–8

Patrie WJ and McCarty RE (1984) Specific binding of coupling factor 1 lacking the delta and epsilon subunits to thylakoids. J Biol Chem 259: 11121–8

Peng G, Bostina M, Radermacher M, Rais I, Karas M and Michel H (2006) Biochemical and electron microscopic characterization of the F1F0 ATP synthase from the hyperthermophilic eubacterium Aquifex aeolicus. FEBS Lett 580: 5934–40

Poetsch A, Berzborn RJ, Heberle J, Link TA, Dencher NA and Seelert H (2007) Biophysics and bioinformatics reveal structural differences of the two peripheral stalk subunits in chloroplast ATP synthase. J Biochem 141: 411–20

Pogoryelov D, Yu J, Meier T, Vonck J, Dimroth P and Muller DJ (2005) The c15 ring of the Spirulina platensis F-ATP synthase: F1/F0 symmetry mismatch is not obligatory. EMBO Rep 6: 1040–4

Pogoryelov D, Reichen C, Klyszejko AL, Brunisholz R, Muller DJ, Dimroth P and Meier T (2007) The oligomeric state of c rings from cyanobacterial F-ATP synthases varies from 13 to 15. J Bacteriol 189: 5895–902

Revington M, McLachlin DT, Shaw GS and Dunn SD (1999) The dimerization domain of the b subunit of the Escherichia coli F(1)F(0)-ATPase. J Biol Chem 274: 31094–101

Richter ML and McCarty RE (1987) Energy-dependent changes in the conformation of the epsilon subunit of the chloroplast ATP synthase. J Biol Chem 262: 15037–40

Richter ML, Patrie WJ and McCarty RE (1984) Preparation of the epsilon subunit and epsilon subunit-deficient chloroplast coupling factor 1 in reconstitutively active forms. J Biol Chem 259: 7371–3

Rodgers AJ and Capaldi RA (1998) The second stalk composed of the b- and delta-subunits connects F0 to F1 via an alpha-subunit in the Escherichia coli ATP synthase. J Biol Chem 273: 29406–10

Rodgers AJ and Wilce MC (2000) Structure of the gamma-epsilon complex of ATP synthase. Nat Struct Biol 7: 1051–4

Rondelez Y, Tresset G, Nakashima T, Kato-Yamada Y, Fujita H, Takeuchi S and Noji H (2005) Highly coupled ATP synthesis by F1-ATPase single molecules. Nature 433: 773–7

Rubinstein JL, Walker JE and Henderson R (2003) Structure of the mitochondrial ATP synthase by electron cryomicroscopy. EMBO J 22: 6182–92

Sabbert D, Engelbrecht S and Junge W (1996) Intersubunit rotation in active F-ATPase. Nature 381: 623–5

Sambongi Y, Iko Y, Tanabe M, Omote H, Iwamoto-Kihara A, Ueda I, Yanagida T, Wada Y and Futai M (1999) Mechanical rotation of the c subunit oligomer in ATP synthase (F0F1): direct observation. Science 286: 1722–4

Schemidt RA, Qu J, Williams JR and Brusilow WS (1998) Effects of carbon source on expression of F0 genes and on the stoichiometry of the c subunit in the F1F0 ATPase of Escherichia coli. J Bacteriol 180: 3205–8

Schürmann P and Buchanan BB (2008) The ferredoxin/thioredoxin system of oxygenic photosynthesis. Antioxid Redox Signal 10: 1235–74

Schwaiger I, Sattler C, Hostetter DR and Rief M (2002) The myosin coiled-coil is a truly elastic protein structure. Nat Mater 1: 232–5

Seelert H, Poetsch A, Dencher NA, Engel A, Stahlberg H and Müller DJ (2000) Structural biology. Proton-powered turbine of a plant motor. Nature 405: 418–9

Shirakihara Y, Leslie AG, Abrahams JP, Walker JE, Ueda T, Sekimoto Y, Kambara M, Saika K, Kagawa Y and Yoshida M (1997) The crystal structure of the nucleotide-free alpha 3 beta 3 subcomplex of F1-ATPase from the thermophilic Bacillus PS3 is a symmetric trimer. Structure 5: 825–36

Sorgen PL, Caviston TL, Perry RC and Cain BD (1998) Deletions in the second stalk of F1F0-ATP synthase in Escherichia coli. J Biol Chem 273: 27873–8

Sorgen PL, Bubb MR and Cain BD (1999) Lengthening the second stalk of F(1)F(0) ATP synthase in Escherichia coli. J Biol Chem 274: 36261–6
Soteropoulos P, Süss KH and McCarty RE (1992) Modifications of the gamma subunit of chloroplast coupling factor 1 alter interactions with the inhibitory epsilon subunit. J Biol Chem 267: 10348–54
Stahlberg H, Müller DJ, Suda K, Fotiadis D, Engel A, Meier T, Matthey U and Dimroth P (2001) Bacterial Na(+)-ATP synthase has an undecameric rotor. EMBO Rep 2: 229–33
Steigmiller S, Börsch M, Gräber P and Huber M (2005) Distances between the b-subunits in the tether domain of F(0)F(1)-ATP synthase from E. coli. Biochim Biophys Acta 1708: 143–53
Steinemann D and Lill H (1995) Sequence of the gamma-subunit of Spirulina platensis: a new principle of thiol modulation of F0F1 ATP synthase? Biochim Biophys Acta 1230: 86–90
Stock D, Leslie AG and Walker JE (1999) Molecular architecture of the rotary motor in ATP synthase. Science 286: 1700–5
Strotmann H, Bickel S and Huchzermeyer B (1976) Energy-dependent release of adenine nucleotides tightly bound to chloroplast coupling factor CF1. FEBS Lett 61: 194–8
Suhai T, Dencher NA, Poetsch A and Seelert H (2008) Remarkable stability of the proton translocating F1FO-ATP synthase from the thermophilic cyanobacterium Thermosynechococcus elongatus BP-1. Biochim Biophys Acta 1778: 1131–4
Suzuki T, Murakami T, Iino R, Suzuki J, Ono S, Shirakihara Y and Yoshida M (2003) F0F1-ATPase/synthase is geared to the synthesis mode by conformational rearrangement of epsilon subunit in response to proton motive force and ADP/ATP balance. J Biol Chem 278: 46840–6
Syroeshkin AV, Vasilyeva EA and Vinogradov AD (1995) ATP synthesis catalyzed by the mitochondrial F1-F0 ATP synthase is not a reversal of its ATPase activity. FEBS Lett 366: 29–32
Tsunoda SP, Rodgers AJ, Aggeler R, Wilce MC, Yoshida M and Capaldi RA (2001a) Large conformational changes of the epsilon subunit in the bacterial F1F0 ATP synthase provide a ratchet action to regulate this rotary motor enzyme. Proc Natl Acad Sci U S A 98: 6560–4
Tsunoda SP, Aggeler R, Yoshida M, Capaldi RA (2001b) Rotation of the c subunit oligomer in fully functional F1Fo ATP synthase. Proc Natl Acad Sci U S A 98: 898–902
Turina P, Samoray D and Gräber P (2003) H+/ATP ratio of proton transport-coupled ATP synthesis and hydrolysis catalysed by CF0F1-liposomes. EMBO J 22: 418–26
Uhlin U, Cox GB and Guss JM (1997) Crystal structure of the epsilon subunit of the proton-translocating ATP synthase from Escherichia coli. Structure 5: 1219–30
Uhlin U, Cox GB and Guss JM (1998) Crystal structure of the epsilon subunit of the proton-translocating ATP synthase from Escherichia coli. Structure 15: 1219–30
van Noort J, van Der Heijden T, de Jager M, Wyman C, Kanaar R and Dekker C (2003) The coiled-coil of the human Rad50 DNA repair protein contains specific segments of increased flexibility. Proc Natl Acad Sci U S A 100: 7581–6
Van Walraven HS, Strotmann H, Schwarz O and Rumberg B (1996) The H+/ATP coupling ratio of the ATP synthase from thiol-modulated chloroplasts and two cyanobacterial strains is four. FEBS Lett 379: 309–13
Vik SB and Antonio BJ (1994) A mechanism of proton translocation by F1F0 ATP synthases suggested by double mutants of the a subunit. J Biol Chem 269: 30364–9
von Ballmoos C, Brunner J and Dimroth P (2004) The ion channel of F-ATP synthase is the target of toxic organotin compounds. Proc Natl Acad Sci U S A 101: 11239–4
Weber J (2007) ATP synthase—the structure of the stator stalk. Trends Biochem Sci 32: 53–6
Weber J, Bowman C, Wilke-Mounts S and Senior AE (1995) alpha-Aspartate 261 is a key residue in noncatalytic sites of Escherichia coli F1-ATPase. J Biol Chem 270: 21045–9
Werner-Grüne S, Gunkel D, Schumann J and Strotmann H (1994) Insertion of a "chloroplast-like" regulatory segment responsible for thiol modulation into gamma-subunit of F0F1-ATPase of the cyanobacterium Synechocystis 6803 by mutagenesis of atpC. Mol Gen Genet 244: 144–50
Wilkens S and Capaldi RA (1998) ATP synthase's second stalk comes into focus. Nature 393:29
Wilkens S, Dahlquist FW, McIntosh LP, Donaldson LW and Capaldi RA (1995) Structural features of the epsilon subunit of the Escherichia coli ATP synthase determined by NMR spectroscopy. Nat Struct Biol 2: 961–7

Wilkens S, Borchardt D, Weber J and Senior AE (2005) Structural characterization of the interaction of the delta and alpha subunits of the Escherichia coli F1F0-ATP synthase by NMR spectroscopy. Biochemistry 44: 11786–94
Wise JG and Vogel PD (2008) Subunit b-dimer of the Escherichia coli ATP synthase can form left-handed coiled-coils. Biophys J 94: 5040–5
Xie DL, Lill H, Hauska G, Maeda M, Futai M and Nelson N (1993) The atp2 operon of the green bacterium Chlorobium limicola. Biochim Biophys Acta 1172: 267–73
Yasuda R, Noji H, Kinosita K Jr and Yoshida M (1998) F1-ATPase is a highly efficient molecular motor that rotates with discrete 120 degree steps. Cell 93: 1117–24
Yasuda R, Noji H, Yoshida M, Kinosita K Jr and Itoh H (2001) Resolution of distinct rotational substeps by submillisecond kinetic analysis of F1-ATPase. Nature 410: 898–904
Yoshida M, Okamoto H, Sone N, Hirata H and Kagawa Y (1977) Reconstitution of thermostable ATPase capable of energy coupling from its purified subunits. Proc Natl Acad Sci USA 74: 936–40
Zhou Y, Duncan TM and Cross RL (1997) Subunit rotation in Escherichia coli FoF1-ATP synthase during oxidative phosphorylation. Proc Natl Acad Sci U S A 94: 10583–7
Zimmermann B, Diez M, Zarrabi N, Gräber P and Börsch M (2005) Movements of the epsilon-subunit during catalysis and activation in single membrane-bound H(+)-ATP synthase. EMBO J 24: 2053–63

Part III
Cyanobacteria and Light: Oxygenic Photosynthesis (and Other Uses of Light)

Chapter 11
The Evolution of Cyanobacteria and Photosynthesis

Gerhart Drews

11.1 Introduction

Cyanobacteria are in the nineteenth century referred to as blue-green algae or phycochromaceae, because they are considered to be algae and not bacteria. Cohn (1867, 1875) concluded that the blue-green algae and bacteria were closely related and proposed that the blue-greens (Schizophyceae) and bacteria (Schizomycetes) be grouped in the one division Schizophyta. Although it will be shown that cyanobacteria are true bacteria it was proposed by a compromise proposal from August 1981: Cyanobacteria (blue-green algae) may be described following either the Botanical or the Bacteriological Code (Staley et al. 1989).

Cyanobacteria are the only members of the domain Bacteria with the ability of oxygenic photosynthesis. They are Gram-negative bacteria with peptidoglycan as a cell wall constituent. Their photosynthetic apparatus (PSA) contains two photosystems (PS I and PS II) each with a unique reaction center (RC) and chlorophyll a (Chl*a*) and phycobilisomes (consisting of phycobilins covalently bound to phycobiliproteins) as characteristic light-harvesting systems. In contrast to most cyanobacteria, prochlorophytes contain Chl*a* and Chl*b* and lack phycobilins (Partensky et al. 1999; Tomitani et al. 1999) and the cyanobacterium *Acayochloris marina* harvests far-red light with Chl*d* for photosynthesis underneath minute coral-reef invertebrates (Kühl et al. 2005). Most cyanobacteria fix molecular nitrogen (Zehr et al. 2001). Many filamentous but also some unicellular cyanobacteria move by a gliding motility directed by light, i.e. phototaxis (Drews and Nultsch 1962; Nultsch 1962; Nultsch and Häder 1974; Bhaya 2004).

During a long life history, cyanobacteria settled in nearly all habitats on Earth that are exposed to light. They thrive from the arctic tundra to the desert, in warm springs, and in all marine and freshwater habitats. All species grow under obligate or facultative photoautotrophic conditions. They are globally important primary

G. Drews (✉)
Institute of Biology 2, Microbiology, Albert-Ludwigs-Universität, Schänzlestrasse 1, 79104 Freiburg, Germany
e-mail: gerhart.drews@biologie.uni-freiburg.de

G. A. Peschek et al. (eds.), *Bioenergetic Processes of Cyanobacteria*,
DOI 10.1007/978-94-007-0388-9_11, © Springer Science+Business Media B.V. 2011

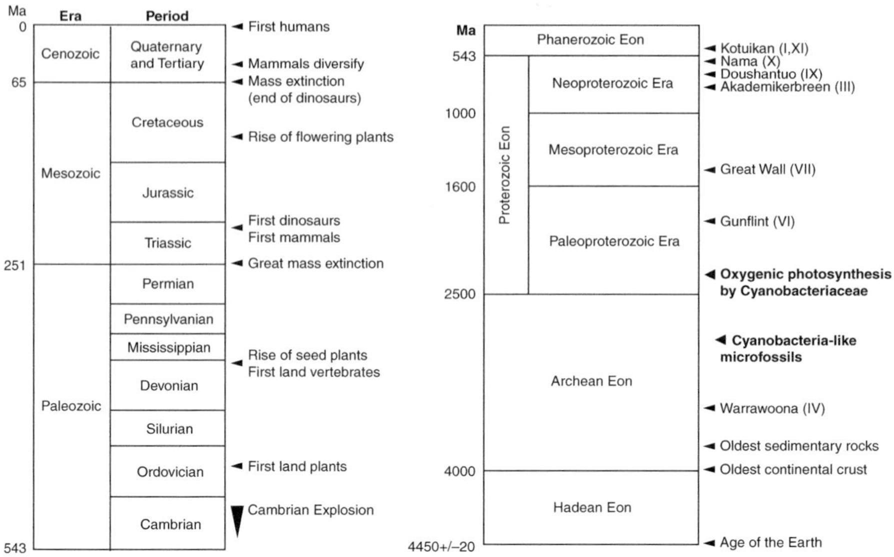

Fig. 11.1 The geologic timescale and Eras are indicated which are important for the evolution of photosynthesis and cyanobacteria. *Ma* million years before present. (From Knoll (2003))

producers today and during much of our planet's history (Capone et al. 1997). They are free living or in symbiosis with fungi (lichens and *Geosiphon pyriformis*), plants (e.g. Cycads), ferns, aquatic ferns (Azolla), mosses (*Anthoceros*), or animals (e.g. corals, zooxanthelles, ascidians (*Prochloron didemni*).

Cyanobacteria are monophyletic but morphologically and physiologically diverse. They were one of the earliest organisms on this planet (Fig. 11.1) and there is general agreement that they played a key role in the formation of atmospheric oxygen (Knoll 2003; Bekker et al. 2004). The early occurrence of cyanobacteria on Earth has been concluded from molecular-phylogenetic analyses (Gupta et al. 1999; Cavalier-Smith 2006; Tomitani et al. 1999, 2006), studies on the rise of atmospheric oxygen (Bekker et al. 2004) and micro-paleontological investigations (Schopf and Packer 1987; Schopf 1993; Knoll 2003). Recent results in these research fields indicate that oxygenic photosynthesis by cyanobacteria originated about 2,340 million years before present (Ma). Presumably, anoxygenic photosynthesis with one photosystem was a precursor of oxygenic photosynthesis with two photosystems. There is, however, no agreement on the first anoxygenic bacteria. Three major lineages of bacteria (actinobacteria, *Deinococcus*, and cyanobacteria) contributed to an early colonization of land (Battistuzzi et al. 2004).

11.2 Paleontological History of Cyanobacteria

Microfossil structures in 3.465 billion-year-old rock, the Apex cherts (microcrystalline silica) of the Warrawoona Group in Western Australia, that resemble preserved cyanobacteria and other bacteria have been considered as the oldest morphological evidence for life on Earth and the beginning of oxygen-producing (oxygenic) photosynthesis. It has been proposed that the criteria of authentic Archean microfossils are met by 11 types of filamentous prokaryotic cells (Schopf and Packer 1987; Schopf 1993, 1994, 1999). Carbon isotopes in organic matter from the chert veins in Warrawoona sediments and sulfur isotopes in sedimentary pyrite and barite likewise suggest that microbial life was present at this time. Filaments of biological origin were observed in hydrothermal mineral deposits 3.2-billion-year-old hydrothermal mineral deposits (Rasmussen 2000).

Petrographic and geochemical studies on the 3.416-billion-year-old Buck Reef Chert, South Africa indicated that most carbonaceous matter was formed by photosynthetic mats within the euphotic zone of a stratified early ocean (Tice and Lowe 2004). However, this carbonaceous matter is nearly a billion years older than putative cyanobacterial biomarkers, the rise of atmospheric oxygen and molecular-phylogenetic data (Summons et al. 1999; Catling et al. 2001; Battistuzzi et al. 2004). Serious questions about the presence of cyanobacteria in the 3.465 billion-year-old Warrawoona rocks have also been raised in recent studies. The purported microfossil-like structures have been reinterpreted as secondary artifacts formed from amorphous graphite within multiple generations of a metalliferous hydrothermal chert vein that precipitated from volcanically heated fluids that percolated through Warrawoona sediments, replacing original minerals soon after deposition (Brasier et al. 2002; Knoll 2003, pp. 59–62). The existence of life 3.5 billion years ago in the Warrawoona sediments is still questioned by many geologists (Knoll 2003, p. 64). But as Knoll (2003, p. 68) points out "it would be hazardous to interpret the absence of evidence as evidence for absence".

The interpretation of microfossil-like structures in very old rocks is always difficult because the biological and mineral materials underwent chemical and physical conversions during the long periods of Earth history. The history of rocks and their secondary modification are important in determining the significance of the observed microfossils. Moreover, because most cyanobacteria have simple shapes, the morphological similarity between ancient and modern forms should not automatically allow conclusions to be made about the physiology of the ancient forms.

Cyanobacterial microfossils have been found encased in chert of the Akademikerbreen Group, a thick stack of limestone, formed before the dawn of the Cambrian Period (about 800–600 Ma) in coastal environments at the edge of a tropical ocean. The Akademikerbreen Group is the backbone of the Spitsbergen rocks, and was produced by carbonate sedimentation along a tidal flat that bordered a subtropical to tropical climate. Northeastern Spitsbergen drifted by plate tectonic processes poleward through the Paleozoic and Mesozoic eras reaching the current latitude more than 100 Ma (Knoll 2003, p. 39). These processes exposed the rock beautifully, and the surface is little altered by weathering. The microfossils in the rocks consist of small black crusts dotting a tidal flat of lime mud laced with cyanobacterial mats. The crusts were formed in the upper part of the intertidal zone, built by small spheroidal cyanobacteria that secreted extracellular sheaths (Knoll 2003, pp. 39–47). The chert, encasing the microfossils, is an extraordinarily hard substance made up of microcrystalline silica (SiO_2). The chert and the enclosed microfossils together form black nodules. Chert in carbonate rocks contains tubes about 10 μm across which are interpreted as the extracellular sheaths of filamentous cyanobacteria.

Fortunately, the ancient process of crust formation by cyanobacteria embedded in a tidal flat of lime mud can be observed today on various tropical coasts, such as Bahama Banks and the Arabic Sea. These crusts are built by small spheroidal cyanobacteria that secreted sheaths. They bore into and lived within ooid grains. Fine-scale comparisons between mat-building cyanobacteria found today and fossils formed in comparable environments 2000 Ma show a remarkable similarity. The close similarity between the environments of ancient and living cyanobacteria suggests that the microorganisms that distributed across Spitsbergen tidal flats were very similar to modern forms in morphology, life style, and physiology. Like cyanobacteria living today in coastal habitats the chert nodules in Proterozoic carbonates center on coastal environments where silica was precipitated much like salt from evaporating seawater (Knoll 2003, pp. 42–44). The Akademikerbreen rocks contain conspicuous stromatolites, which are wavy-laminated, domed, and candelabrum-like structures. These stromatolites, formed in carbonate rocks in Precambrian oceans, and their structures can be explained by cyanobacterial movement and activities. Cyanobacteria spread across sediment surfaces and form coherent mats. They trap and bind fine particles supplied by waves and currents. Accumulation of mud and sand and growth of the population results in an upward curving. Biochemical processes deeper in the mat result in carbonate crystallization. Biofilm formation, trapping of material, and carbonate precipitation are discontinuous and result in fine layers of limestone accretion, one atop the other. In one locality of the Spitsbergen tidal flats, the cyanobacterial architects of an offshore stromatolitic reef were preserved by fine carbonate cement that encrusted individual filaments (Knoll 2003, p. 44).

Cyanobacteria, as photosynthetic organisms, leave chemical traces in the carbonate minerals and organic matter deposited beneath the sea. When photosynthetic organisms take up carbon dioxide for the synthesis of organic molecules they incorporate $^{12}CO_2$ more readily than $^{13}CO_2$. Organic carbon is depleted in ^{13}C, while inorganic carbonate is not. The geological record shows a continuous value of +27 ± 7‰ for organic carbon and a continuous value of +0.4 ± 2.5‰ for carbonate carbon. As a consequence, the ratio of ^{13}C–^{12}C, as determined by isotope fractionation, in organic matter made through photosynthesis is lower than in carbonate minerals. In the Spitsbergen rocks the ratios of carbon isotopes in carbonates and organic matter consistently differ by about 28‰. This record is consistent with biological CO_2 fixation by photosynthesis, but is also consistent with a non-photosynthetic mechanism. Presumably, photosynthesis was active in the Proterozoic sediments that formed the Spitsbergen rocks (Knoll 2003, p. 47). These Spitsbergen rocks contain no fossils of eukaryotic organisms, such as bones or shells, but do contain traces from the Neoproteozoic era that show that cyanobacteria were active about 600–800 Ma (Knoll 2003, p. 49).

The 3.8-billion-year-old rocks on Akilia island off the coast of south-western Greenland have been severely altered metamorphically. Tiny grains of mineral phosphate occur within the rocks, and these grains contain inclusions of reduced carbon, i.e. graphite. The isotopic composition of the carbon is greatly depleted in ^{13}C, consistent with the formation through biological processes. However, some researchers believe that the graphite in these grains formed through chemical reaction of hot metamorphic fluids with iron carbonate in the rock (Knoll 2003, pp. 69, 70).

11.3 The Rise of Atmospheric Oxygen

Several lines of geological and geochemical evidence indicate that the level of atmospheric oxygen was extremely low before 2450 Ma (Bekker et al. 2004). Sedimentary deposits ≥2.45 billion years old include detrital uraninite, siderite and pyrite (England et al. 2002), reduced shallow-water facies of iron formations (Beukes and Klein 1990), and highly carbonaceous shales that are not enriched in redox-sensitive elements (Holland 1994; Yang and Holland 2002). The sulfate content of the seawater was <200 μM. In contrast, sedimentary layers younger than 2,220 Ma contain red beds (Chandler 1980), oxidized iron formations (Beukes and Klein 1992), and $CaSO_4$-rich evaporated substances (El Tabakh et al. 1999).

These changes indicate an increase of the O_2 content of the atmosphere between 2,450 and 2,220 Ma from about 10^{-5} to 10^{-2}–10^{-1} of the present atmospheric level, but this may be related to different tectonic settings and post-depositional alterations (Clemmey and Badham 1982; Ohmoto 1996). The method of mass-independent fractionation (MIF) of sulfur isotopes is a new tool for tracing changes in the oxygen content of the atmosphere. The material for such studies is pyrite from highly organic-rich layers, which occurred as nodules, mineralized microbial mats, disseminated grains, and laminated seams. Pyrite has been taken from early paleoproterozoic sedimentary successions of the 2.32 billion-year-old Rooihoogte and Timeball Hill formations in South Africa. The large increase in atmospheric oxygen was determined to occur by 2,320 Ma during the great oxidation event (GOE). The level of atmospheric oxygen during the depositions of the pyrite was $>10^{-5}$ times that of the present atmospheric level. The presence of sideritic iron formation at the base of the Rooihoogte Formation and the ironstone layer consisting of haematitic pisolite and oolites in the upper Timeball Hill Formation indicate that atmospheric oxygen rose significantly (Bekker et al. 2004).

The time-series history of elemental and isotope variations was investigated in a scientific core drilled through the 2.5-billion-year-old Mount McRae organic-rich shale of northwestern Australia (Kaufman et al. 2007). Two depositional cycles were identified; each sequence starts in carbonate or siliciclastic turbidite or breccia and deepens upwards to either pelagic shale or banded iron formation. The succession has experienced only mild regional metamorphism and minimal deformation. Siderite dominates in the lower half of the formation, which is consistent with the absence of O_2 in deeper depositional environments. Calcite is a primary carbonate phase in the upper Mount McRae shale, which indicates the general absence of soluble iron in the shallow water column at this time. Carbonate and total organic ^{13}C values increase progressively up the core. The lower half of the Mount McRae shale from the Archean drilling project was accumulated in a deep, anoxic environment insofar as sediments are dominated by sideritic shale and banded iron formation. The dominance of positive $\Delta^{33}S$ through this stratigraphic interval is interpreted as indicating the preferential incorporation of atmospheric S into marine sediments facilitated by microbial S reduction. The S isotope signatures in the upper Mount McRae shale reflect the establishment of a widespread oxidative S cycle. In the late Archean ocean O_2 would accumulate in highly productive regions along continental margins and perhaps to a lesser degree in distal settings where nutrient levels were high enough to support oxygenic photosynthesis. To test whether the change in atmospheric composition is widespread, S isotope analyses of samples from the broadly equivalent Transvaal Basin in South Africa were undertaken. The time-series records of mineralogic, elemental and S isotopic change provide clues to coupled changes in the redox state of the shallow ocean, largely before the atmosphere became oxygenated, in relation to biological innovation before the Archean/Proterozoic boundary. Oxygenation of the surface ocean preceded pervasive and persistent atmospheric oxygenation by 50 million years or more (Kaufman et al. 2007). The rapid rise in the partial pressure of atmospheric O_2 during the GOE between 2,450 and 2,220 Ma has been interpreted as an immediate consequence of the evolution of oxygenic photosynthesis. Alternatively, the GOE was a consequence of an abiotic shift in the balance of oxidants and reductants at the Earth's surface. Oxidative weathering of Mo-bearing sulfide minerals in crustal rocks leads to the accumulation of Mo in the ocean, where it is the most abundant transition element with MoO_4 concentrations of about 105 nM. On an anoxic earth, Mo would be retained in unoxidized crustal sulfide minerals during weathering. Therefore, Mo concentrations in the oceans would be low. High-resolution chemostratigraphy reveals an episode of enrichment of the redox-sensitive transition metals molybdenum and rhenium in the late Archean Mount McRae shale. Correlations with organic carbon indicate that these metals were derived from contemporaneous seawater. Rhenium/osmium geochronology demonstrates that the enrichment in a primary sedimentary feature dates to 2,501 ± 8 Ma. Mo and rhenium were probably supplied to the Archean ocean by oxidative weathering of crustal sulfide minerals. It has been suggested that low amount of O_2 were present in the environment more than 50 Ma before the great increase of oxygen in the atmosphere by oxygenic photosynthesis (Anbar et al. 2007).

From recent studies on the speciation of Fe in well-preserved Neoproteozoic sedimentary rocks it was concluded that through much of the later Neoproteozoic (<742 ± 6 Ma) anoxia remained widespread beneath the mixed layer of the oceans (Canfield et al. 2008). Deeper water masses were sometimes sulfidic but mainly Fe^{2+}-enriched. In the modern world, and through much of the Phanerozoic Eon, marine anoxia produces sulfidic conditions. Neoproteozoic ferruginous deep-water chemistry may result from either limited S input to the ocean or increased input of Fe (Canfield et al. 2008).

Non-photosynthetic filamentous microfossils dated at 3,200 Ma have been found in a massive sulfide deposit from the Pilbarra Craton in Australia and show that filamentous structures should not automatically be assumed to be photosynthetic (Rasmussen 2000).

The Gunflint Formation in northwestern Ontario include shales, a little bit of carbonate, and some sandstone. It most prominently features iron formation. Volcanic material near the top of the formation contains zircons dated at 1,878 ± 2 Ma. The lowermost layer of the Gunflint Formation contains tiny fossils, iron-coated tubes 1–2 µm across, in finger-like stromatolites, which are similar to sheaths of iron bacteria. Fossilized cyanobacteria have been found in nodules of black chert in carbonate beds from the Belcher Islands near the Hudson Bay. Microbial carbon and sulfur cycles operated 2,000 Ma, and iron deposits frequently formed. But 1,800 Ma the iron formations were gone (Knoll 2003, pp. 90–93). Iron formations provide geological evidence for oxygen scarcity in the early-Archean atmosphere and oceans. Pyrite is common in sediments enriched in organic matter, formed when H_2S produced by sulfate-reducing bacteria react with iron in oxygen-depleted waters. Pyrite, siderite ($FeCO_3$) and uraninite (UO_2) have been found in deposits older than about 2.2 billion years. Exposed in rock faces, they are stripped away by weathering and erosion and destroyed by oxygen. When the oxygen-sensitive minerals disappeared from the rocks oxygen-requiring rock types increased in prominence. For example, when the groundwater that washes surface sands contain oxygen, iron oxides form. Iron oxide coats sand grains, giving them red color, as in the red sandstone beds in the Grand Canyon. Red beds are common in sedimentary successions deposited after about 2,200 Ma (Knoll 2003, pp. 95–98).

Cyanobacterial membranes contain 2-methylbacteriohopanepolyols which can be used as specific biomarkers for these organisms. In the sediment these compounds are converted to 2-methylhopanes and persist. indefinitely as molecular fingerprints of ancient cyanobacteria. 2-Methylhopanes have been found in 2.7-billion-year-old shales, which are well preserved deposits and rich in organic matter. Thus, cyanobacteria have an Archean origin.

A carbon isotope fractionation as large as 60‰ has been found in late Archean and earliest Proterozoic rocks. It is thought that between 2,800 and 2,200 Ma methanogenic archaea played a prominent role in the global carbon cycle. The oxygen content of the atmosphere was very low and CO_2 was converted to CH_4 by methanogenic archaea. The increase in the oxygen content in the atmosphere by cyanobacteria enabled methane-oxidizing bacteria to assimilate methane. The archaea diverged mostly between 4,100 and 3,100 Ma, and methanogenesis originated probably around 4,100 and 3,800 Ma, which is compatible with highly depleted carbon isotopic values found in rocks dated to be 2.6–2.8 billion years old (Battistuzzi et al. 2004).

An early origin of phototrophy is consistent with the earliest bacterial mats and structures identified as stromatolites, but a 2,600 Ma origin of cyanobacteria suggests that the Archean structures, if biologically produced, were made by anoxygenic photosynthesis (Battistuzzi et al. 2004). Sulfate reducing bacteria emerged later, after the reduced sulfur compounds were oxidized. The fractionation level of sulfur increased in the Proterozoic Eon (Knoll 2003, pp. 100–103).

The amount of oxygen generated by photosynthesis in the late Archean Eon was used up in the ocean to oxidize iron and sulfur salts before it was enriched in the atmosphere; otherwise hydrogen sulfide might have swept iron from the sea. Besides the abiotic processes that reduced the oxygen level in the atmosphere and in the ocean, the increase of organic carbon burial and a decrease in reductant input might have triggered the GOE (Goldblatt et al. 2006; Kasting 2006). The geological and geochemical data indicate that the cyanobacteria or their ancestors originated in the Archean eon, approximately 2,700 Ma. The increase of the oxygen level after the beginning of oxidative photosynthesis remained very low, about 10^{-5} of the present atmospheric level, for about 400 million years (Goldblatt et al. 2006). A significant increase in oxygen to $\geq 5 \times 10^{-3}$ of the present atmospheric level was dated to 2,320 Ma. By 1,800 Ma iron formation and oxygen-sensitive minerals disappeared from the rocks, which indicated a large increase of oxygen in the atmosphere. The deep ocean was anoxic and ferruginous until 580 Ma and became oxic afterward in the late-Neoproterozoic era, which enabled the rise of animal life (Canfield et al. 2007).

11.4 Anoxygenic and Oxygenic Photosynthesis

The oxygenic photosynthetic apparatus (PSA) of cyanobacteria and plants produces metabolic power in the form of ATP, reducing equivalents for the synthesis of sugar from CO_2, and oxygen by the water-splitting complex of photosystem (PS) II at the expence of absorbed light energy. The two types of reaction centers (RCs), of PS I (plastocyanine:ferredoxin oxidoreductase) and PS II (water:plastoquinone oxidoreductase), each have their own characteristic structure and cofactors, which have been preserved during evolution in cyanobacteria and plants, which indicates that they are homologous, i.e. have a common origin. The photosynthetic electron transport of PS II is connected to PS I. But PS I can work independently from PS II to equilibrate the formation of ATP and reducing power. PS II and PS I are integrated in the cytoplasmic and intracytoplasmic membranes. Both PSs are surrounded by light-harvesting or antenna systems that collect light energy and transfers it to an RC. Light-harvesting systems characteristic of cyanobacteria and rhodophyta are phycobilisomes. They are accessory pigment proteins attached to the PS II. The phycobiliproteins of the phycobilisomes owe their intense visible absorption properties to the presence of open-chain tetrapyrroles, the phycobilins, which are covalently bound to their apoproteins. The phycobilins are allophycocyanin, phycocyanin and phycoerythrin. The light-harvesting systems of the organisms belonging to different taxa differ considerably in their absorption spectra, structure, and organization.

Sequence analysis and biophysical measurements of the RC complexes found in all classes of organisms indicate that all RCs fall into two basic groups: those with pheophytin and a pair of quinones (type II), and those with iron sulfur (type I). All anoxygenic phototrophs contain only one RC in their PSA. The sulfur and the nonsulfur purple bacteria and the aerobic anoxygenic phototrophs, all are Proteobacteria and have an RC that is similar to that found in Chloroflexaceae (nonsulfur filamentous green bacteria). These organisms have an RC whose core is similar to the PS II RC of oxygenic organisms but lacks the oxygen-evolving complex.

These RCs contain bacteriopheophytin as an intermediate electron acceptor and two quinones as terminal electron acceptor. The spectroscopic, kinetic and magnetic resonance properties and the protein composition and structure of all these pheophytin-quinone type II RCs are very similar which suggests that they share a common ancestor. There is also evidence that a similar evolutionary relationship exists among the RCs of the Chlorobiaceae (green sulfur bacteria), Heliobacteria (Gram-positive bacteria, bacteriochlorophyll g) and the PS I of oxygenic phototrophs. These RCs contain low-potential iron-sulfur centers as early electron acceptor (Fe-S type; Blankenship 1992).

The antenna system collects light energy and transfers it to an RC, where the excitation energy is transformed into charge separation and a redox potential difference across the membrane. A cyclic electron transport via the cytochrome b/c_1 complex contributes to the generation of an electrochemical proton gradient across the membrane. This proton gradient is used to produce ATP at the ATP synthase. In the photoautotrophic green bacteria reducing equivalents are formed by a noncyclic electron transport.

PS II has a modular structure comparable to that of PS I. Apart from a similar set of constituent cofactors of the electron transfer system, the analogous features include a comparable cofactor arrangement and a corresponding secondary structure motif of the RC cores. Despite differences in distances and orientation of individual cofactors and protein composition, the great significance of analogous structures and functional similarities support the hypothesis that RC types I and II share a common evolutionary origin (Schubert et al. 1998; Heathcote et al. 2002).

11.5 Molecular Phylogenetics of the Evolution of Cyanobacteria and Photosynthesis

A remarkable variety of both eukaryotic and prokaryotic photosynthetic organisms are known to day. The photosynthetic processes in all these organisms have many similarities, and there is some comparative biochemical evidence that all (bacterio) chlorophyll-based photosynthesis is ultimately derived from a single photosynthetic ancestor. The only other group of organisms with a system completely different from the chlorophyll-based photosynthetic systems are the halobacteria, in which the retinal-containing protein bacteriorhodopsin pumps protons across a membrane upon light excitation. This system represents an independent evolutionary development (Blankenship 1992).

The photosynthetic bacteria have traditionally been divided into families and genera based on pigment content, morphology and metabolic capabilities (Staley et al. 1989). A phylogenetic classification based on sequence similarities of the 16S ribosomal RNA has been introduced by Woese (1987). The rRNA genes apparently have not undergone lateral transfers. Other widely conserved genes of the nucleic-acid-based information processing system also track with the rRNA tree.

Consequently, the rRNA-based tree represents the evolutionary flow of the genetic machinery (Pace 2006). The results of 16S rRNA analyses have led to the proposal that all organisms can be grouped into three kingdoms, the Eukarya, the Bacteria and the Archaea.

All available evidence suggests that (bacterio)chlorophyll-based photosynthesis arose in the kingdom Bacteria and was followed by subsequent endosymbiotic transfer from cyanobacteria into eukaryotic organisms (Raymond et al. 2002). On the basis of phylogenetic analyses and the complexity of the PSAs of present-day photosynthetic organisms it was concluded that mechanistically and structurally simpler anoxygenic photosynthesis preceded and was ancestral to oxygenic photosynthesis (Xiong et al. 2000; Blankenship 2002). Therefore the cyanobacteria were probably preceded by more primitive anoxygenic phototrophs. The extant anoxygenic phototrophs belong to various subgroups of the α-Proteobacteria and Gram-positive bacteria but never to the archaea. There is some evidence that core components of photosynthesis have been subject to lateral gene transfer over the course of evolution (Blankenship 1992; Raymond et al. 2002; Sato 2002; Olson and Blankenship 2004; Zhaxybayeva et al. 2004; Mix et al. 2005). Different components of the PSA might therefore have different original sources, making reliable conclusions on the revolutionary pathway problematic (Blankenship 1992).

A comparative analysis of gene maps of representatives from the five bacterial phyla cyanobacteria, proteobacteria (purple bacteria), green sulfur bacteria, green filamentous bacteria and Gram-positive heliobacteria have revealed 188 orthologs common to the genomes of *Synechocystis* sp., *Rhodobacter capsulatus*, *Chlorobium tepidum*, *Chloroflexus aurantiacus* and *Heliobacillus mobilis*. These genes include housekeeping genes involved in protein synthesis, DNA replication, transcription and components of various metabolic pathways. A comparison of photosynthetic genes supports the idea that the evolution of these genes has been disconnected from divergence and speciation in these organisms. Genetic components of the PSA have crossed species lines nonvertically, confirming the extensive role that horizontal gene transfer has played in bacterial evolution (Raymond et al. 2002). However, horizontal gene transfer does not occur with equal probability for all genes. Those genes involved in transcription, translation, and related processes thought to have more macromolecular interactions than operational genes for housekeeping are postulated to be seldom transferred (Jain et al. 1999). Identification of core genes that are resistant to horizontal gene transfer allows the separation of true phylogenetic signals from "noise". For example 359 of 682 orthologs from 13 cyanobacterial genomes (≥52%) are susceptible to horizontal gene transfer within the cyanobacterial phylum (Shi and Falkowki 2008). The remaining 323 orthologs show broad phylogenetic agreement. This core set is comprised of key photosynthetic and ribosomal proteins. Many of the key proteins are encoded by small gene clusters, which often are indicative of protein-protein, protein-prosthetic group, and protein-lipid interactions. These macromolecular interactions in complex protein structures retard the tempo of evolution of the core genes and restrict horizontal gene transfer of components of the core. Analysis of the core gene set suggests that the ancestral cyanobacterium did not fix nitrogen and probably was a thermophilic organism (Shi and Falkowki 2008). A new approach of analysis integrates clustering and sequence analysis with resolution of an integrated phylogeny of 24 cyanobacterial genomes spanning multiple taxonomic levels occupying a wide variety of environmental niches (Swingley et al. 2008). Following the initial clustering, 613 protein families fit the criterion of being absent in not more than two cyanobacteria and having not more than a total of two paralogs in all organisms. The common ancestor of all cyanobacteria is inferred to have had a conserved core of 361 protein families as these are present in the full set of 26 genomes analyzed. Other proteins are also common to all analyzed genomes but they contain paralogs and therefore were excluded. These families represent a widely conserved core of housekeeping proteins common not only across all known cyanobacterial lineages but also present to some extent in other bacterial genomes. The core genome at the base of the cyanobacterial phylum encompasses most of the major proteins of the PSA, which suggests that oxygenic photosynthesis evolved prior to or early in the cyanobacterial radiation. In contrast, the ability to fix nitrogen is found paraphyletically throughout the cyanobacterial tree. Gene loss, horizontal gene transfer or some combination of these processes may be responsible for this distribution of nitrogen fixation (Raymond et al. 2004). The NifD (nitrogen fixation catalytic subunit) tree shows evidence for several gene duplications and horizontal gene transfer, as suggested by the position of *Trichodesmium erythraeum* comprising the earliest cyanobacterial branch among NifD proteins (Swingley et al. 2008). It has been concluded that the core components of photosynthesis represent mosaic genes with very different evolutionary histories (Raymond et al. 2002; Zhaxybayeva et al. 2004).

Proteins encoded in 15 complete cyanobacterial and *Prochlorococcus* genomes have been compared to define the minimal set of genes common to all cyanobacteria and *Prochlorococcus* and to trace the conservation of these genes among other taxa (Mulkidjanian et al. 2006). It has been shown that cyanobacteria and plants share numerous photosynthesis-related genes that are lacking in genomes of other phototrophs. Of 3,188 cyanobacterial clusters of orthologous groups of proteins (CyOGs) 892 are encoded in each cyanobacterial genome and an additional 162 are encoded in 14 of 15 genomes. The combined set of 1,054 CyOGs that are lacking in no more than one cyanobacterial genome is referred to as the coreCyOGs. Certain components of PS I and PS II as well as many other polypeptides are not within the core set. Core RC 1 subunit PsaA/PsaB is found in all cyanobacteria, in plants, in *Chlorobium tepidum* and *Heliobacillus mobilis*, but not in the genomes of *Rhodopseudomonas palustris* and *Chloroflexus aurantiacus*. The enzymes for the chlorophyll biosynthesis (ChlB, ChlN, ChlD, ChlH, ChlI, ChlM, ChlG, and ChlP) are present in all of these photosynthetic organisms.

The evolution of the tetrapyrrol biosynthesis resulting in the formation of (bacterio)chlorophylls (BChl) is described by the Granick hypothesis, which proposes that biosynthetic pathways recapitulate their evolution (Granick 1957; Mauzerall 1992). It is speculated that each step fulfils a useful function. Thus, in a primitive RC, protochlorophyll *a* might have functioned at some time before chlorophyll *a* existed (Olson and Pierson 1987a, b), and protoporphyrin IX and Mg protoporphyrin IX might have served as RC pigments before protochlorophyll *a* existed (Xiong et al. 1998, 2000). The proteobacterial lineage of BChl *a* is more deeply rooted than either the heliobacterial lineage (BChl *g*) or the cyanbacterial lineage (Chl *a*) (Xiong et al. 1998, 2000). The genes encoding the protochlorophyllide and chlorine reductases originated by an ancient gene duplication (Raymond et al. 2003). The pathway for the synthesis of protoporphyrin IX probably evolved in response to evolutionary pressure for efficient electron transport cofactors such as heme in cytochrome *b* (Olson and Blankenship 2004).

The evolution of photosynthesis and RCs was proposed to begin in the prebiotic phase, using the water soluble uroporphyrin in a cyclic photoreaction and forming hydrogen from various electron donors (Mauzerall 1992; Hartman 1998). The evolution of a primitive cytochrome *b*/Fe–S complex might have occurred in a common ancestor of bacteria and archaea before the evolution of photosynthesis in bacteria (Nitschke et al. 1998). The first photochemical RC would have been integrated into an existing respiratory electron transport chain. The cytochrome *b* subunit of the cytochrome b/c_1 complex and the core polypeptides L and M of the RC 2 show sequence and structural similarity in the membrane-spanning region that contains ligands for binding of the cofactors heme and quinone (cytochrome *b*) and BChl, bacteriopheophytin and non-heme iron (RC 2), respectively (Xiong and Bauer 2002a, b).

The RC 2 subunit PsbN is found in the genomes of all plants and cyanobacteria with the exception of one *Prochlorococcus marinus*, but is lacking in all anoxygenic phototrophs. Of the 1,054 coreCyOGs genes, 936 are found in other bacteria: these genes include those encoding housekeeping proteins involved in DNA replication and repair, transcription, translation, key metabolic pathways and energy metabolism. But the vast majority of cyanobacterial photosynthetic genes have no detectable homologs in anoxygenic phototrophic bacteria (Mulkidjanian et al. 2006). As mentioned in Sect. 4 anoxygenic photosynthesis is found in the Chlorobi group (e.g. *Chlorobium tepidum*), Firmicutes (e.g. *Helicobacillus mobilis*), α-Proteobacteria (e.g. *Rhodopseudomonas palustris*), β-Proteobacteria (e.g. *Rubrivivax gelatinosum*), γ-Proteobacteria (e.g. *Chromatium vinosum*), and Chloroflexi (e.g. *Chloroflexus aurantiacus*). The Chlorobi group and Firmicutes have type 1 RCs, the Proteobacteria and Chloroflexi have type 2 RCs. It is generally accepted that the evolution of photosynthetic genes was accompanied by their dissemination by way of horizontal gene transfer between different groups of bacteria. This idea is supported by the apparent presence of non-photosynthetic organisms in all of these phyla except for cyanobacteria and the observation that photosynthetic genes are often localized on a single contiguous chromosomal region (superoperon) (Choudhary and Kaplan 2000; Xiong et al. 2000). In addition to enzymes involved in (bacterio)chlorophyll synthesis, an extremely small number of photosynthetic proteins are found in different photosynthetic bacteria. Therefore a phylogenetic tree based on the (bacterio)chlorophyll biosynthesis enzymes is not representative of the evolution of the photosynthetic machinery as a whole, although the biosynthetic pathway of (bacterio)chlorophyll seems to be monophyletic and a connecting link in the evolution of photosynthesis.

As mentioned above, all photosynthetic RCs can be placed into one of two groups. These two groups share a comparable cofactor arrangement and a corresponding secondary structure motif of their core. It seems likely that a primordial RC existed, which subsequently diverged into the ancestors of the two classes of contemporary RCs. However, there are no clear indications of any intermediate stages in the development of the oxygen-evolving system. Similarities between core antenna proteins of PS II of cyanobacteria and chloroplasts and RCs and light-harvesting polypeptides of RC 2 and RC 1 suggest that a small number of one-helix ancestral proteins may have given rise to the various RC and antenna proteins (Vermaas 1994; Mulkidjanian and Junge 1997; Schubert et al. 1998). It is speculated that the first major evolutionary accomplishment was the creation of a homodimeric RC 1 in which the two chlorophylls could join to form a special pair and four cysteines could bind a single Fe–S center. This RC 1 is assumed to be the ancestral RC for all subsequent forms (Olson and Pierson 1987a, b, Vermaas 1994; Olson and Blankenship 2004). The next major step was a gene duplication, the loss of the Fe–S center, and the splitting of one large subunit gene into two smaller genes, one coding for the core protein of PS and the other for the homodimeric RC 2. In the next steps, gene duplication and divergence resulted in light-harvesting core proteins and heterodimeric RC 1 and RC 2 and PS I and PS II, respectively. The homodimeric RC 1 of heliobacteria and green sulfur bacteria may have evolved from the ancestral RC 1. The heterodimeric RC 2 of proteobacteria and green filamentous bacteria may have evolved from the ancestral RC 2. The light-harvesting complexes of proteobacteria are not related to the core proteins of PS II but are thought to have evolved from helix D of RC 2 together with the loss of photochemical activity (Fig. 2 in Olson and Blankenship 2004). The ancestral RC 1 was adapted to photoautotrophic photosynthesis in a low redox environment functional in a phototrophic bacterium using reduced sulfur compounds as electron donor for CO_2 fixation. The evolution of RC 2 was a response to the increase in the redox level of the environment as the best electron donors for CO_2 fixation were used up (Olson and Blankenship 2004). The branching sequence for the evolution of chlorophyll synthesis has been proposed to be proteobacteria → green bacteria → heliobacteria + cyanobacteria (Xiong et al 2000). Another proposed branching sequence for photosynthetic organisms is: heliobacteria → green filamentous bacteria → cyanobacteria → green sulfur bacteria → proteobacteria (Gupta 2003). In spite of the good evidence that RC 1 and RC 2 have a common origin, several authors propose an independent evolution of RC 1 and RC 2 in different organisms and that they were brought together in the cyanobacterial line of photosynthesis with photosystems in series (Mathis 1990; Blankenship 1992; Meyer 1994; Meyer et al. 1996; Xiong et al. 1998; Xiong and Bauer 2002a, b). Thus, the cyanobacteria are the descendants of the fusion organism, and the proteobacteria are the descendants of the line containing RC 2. The green sulfur bacteria and heliobacteria are descended from the line containing RC 1. The fusion model might be rational from a genetic point of view but does not take into consideration the evolutionary pressures driving the various steps. Independent from the different hypotheses about the origin of the PSA the genetic data suggest that the evolutionary history of RC proteins differs from the history of pigment-synthesizing enzymes (Raymond et al. 2002, 2003; Mulkidjanian et al. 2006). The phycobiliproteins are pigment proteins that have evolved by multiple gene duplications and divergences from an ancestral structure that could form short rods (Grossman et al. 1995).

11.6 What was the First Photosynthetic Organism?

The process of photosynthesis is extremely complex and it is difficult to envisage how natural selection could have produced the intermediate stages in its evolution. In a model based on the premise that pre-photosynthetic life was associated with hydrothermal systems, certain early organisms developed phototaxis using near-infrared radiation, and that photosynthesis arose subsequently from bacteriochlorophyll-like molecules used for phototaxis (Nisbet et al. 1995). The results of comparative phylogenetic analysis of photosynthetic and non-photosynthetic organisms bolster the idea that the evolution of photosynthetic genes has been disconnected from divergence and speciation in these organisms, confirming the extensive role that horizontal gene flow has played in bacterial evolution.

The placement of *Chloroflexus* at the base of the bacterial radiation using 16S ribosomal RNA has been the reason for its designation as the earliest phototroph (Woese 1987; Oyaizu et al. 1987; Valadez 2000; Cavalier-Smith 2006). Considering the importance of horizontal gene transfer it is possible that *Chloroflexus* has acquired phototrophy largely through lateral gene transfer (Raymond et al. 2002).

Despite the variety of accessory pigments in photosynthetic organisms, it is generally accepted that all chloroplasts are derived from a single cyanobacterial ancestor (Wolfe et al. 1994; Bhattacharya and Medlin 1998). Prochlorophytes are bacteria that perform oxygenic photosynthesis using chlorophyll *b* and have been proposed to be the ancestors of chlorophyte chloroplasts (Lewin 1976). However, three prochlorophytes have been shown not to be the specific ancestors of chloroplasts, but only diverged members of the cyanobacteria, that contain phycobilins and lack chlorophyll *b* (Palenik and Haselkorn 1992; Urbach et al. 1992). Phylogenetic analyses show that the genes for chlorophyll *b* synthesis share a common evolutionary origin. This indicates that the progenitors of oxygenic photosynthetic bacteria, including the ancestors of chloroplasts, had both chlorophyll b and phycobilins (Tomitani et al. 1999). The members of the Heliobacteriaceae containing an Fe–S type RC 1 and represent the sole photosynthetic phylum from the Gram-positive bacteria. They are thought to be the most ancestral of the photosynthetic lineages on the basis of Hsp60, Hsp70, and other protein genes (Gupta et al. 1999; Gupta 2003; Vermaas 1994). In addition to *Heliobacillus*, *Chloroflexus* and proteobacterial lineages *Chlorobium* has been proposed to be the first photosynthetic organism (Buttner et al. 1992). Cyanobacteria are usually not explicitly considered as a lineage in which photosynthesis could have emerged because of the far greater complexity of their PSA and the higher number of photosynthetic genes. But ancestors of cyanobacteria, the procyanobacteria, should also be considered as candidates for the role of the first phototrophs. Sequence data alone do not allow the first phototroph to be determined. Geological data have shown that partially filamentous cyanobacteria were present in the Archean Eon or at least in the Paleoprotozoic Era, which, according to the carbon isotope composition, used the Calvin cycle to fix CO_2. Because the Calvin cycle is absent in *Heliobacillus* and *Chlorobium* it is unlikely that their ancestors were the phototrophic inhabitants of the Buck Reef Chert (South Africa). The absence of oxidized iron (II) and sulfur in the 3.4 billion-year-old Buck Reef Chert indicated that neither iron (II) nor sulfide but possibly atmospheric hydrogen had been used as electron donors (Tice and Lowe 2006). RC type 2, which is the only RC type in purple bacteria and Chloroflexi, would hardly be useful in a hydrogen-driven metabolism. It would be over-reduced and kinetically incompetent under these conditions. Because there is sufficient evidence that anoxygenic photosynthesis preceded oxygenic photosynthesis and was already taking place in the period between 3,500 and 2,300 Ma, it was proposed that the first phototrophs were procyanobacteria, i.e. anoxygenic ancestors of the extant cyanobacteria: these organisms would be responsible for the presence of the 2-methylhopanoid biomarkers in the 2.7-billion-year-old sediments (Summons et al. 1999; Mulkidjanian et al. 2006). These anoxygenic procyanobacteria might have relied on RC I to reduce $NADP^+$. It is speculated that in an early phase, the anoxygenic phototrophs pro-Heliobacillus and pro-Chlorobi received the photosynthetic genes of PS 1 and pigments by lateral gene transfer from the procyanobacteria and in a later stage when PS 2 was developed, the genes of PS 2 and cofactors were transferred to *Chloroflexus* and purple bacteria. In this hypothesis, modern cyanobacteria inherited their photosynthetic apparatus from ancestral phototrophs (Procyanobacteria), whereas other bacterial lineages obtained theirs by way of lateral gene transfer (Mulkidjanian et al. 2006).

Phylogenetic analyses support the hypothesis that cyanobacteria are monophyletic. They are morphologically diverse, and these morphological differences have been used to divide the group into five subsections: (I) the Chroococcales which are unicellular coccoids that divide by binary fission; (II) the Pleurocapsales which are also unicellular but undergo multiple fission to produce small, easily dispersed cells, called baeocytes; (III) the Oscillatoriales, which are filaments that have only vegetative cells; (IV) the Nostacales, which are also filaments in which vegetative cells differentiate into heterocysts, specialized in nitrogen fixation, or akinetes, resting cells that survive environmental stresses; (V) the Stigonematales, which are filaments with complicated branching patterns and in which vegetative cells differentiate into heterocysts or akinetes. Akinete fossils have been identified in 1.65–1.4-billion-year-old cherts from Siberia, China, and Australia. The earliest known akinetes are preserved in approximately 2.1 billion-year-old chert from West Africa. Putative microfossils of subsection V microfossils with complex branching filaments and possible heterocysts are preserved in the lower Devonian (approximately 400 Ma). Heterocyst fossils have not been reliable identified in Precambriam rocks. Akinetes can be traced back into the Paleoproterozoic Era. Oxygen first reached levels that would compromise nitrogen fixation and hence select for heterocyst differentiation 2,300 Ma. Cyanobacterial 16S rRNA genes, *rbcL* (gene encoding the large subunit of ribulose 1,5-biphosphate carboxylase/oxygenase) and *hetR* (gene essential for heterocyst differentiation) have been phyllogenetically analyzed. 16S rRNA gene sequence analyses of 20 cyanobacterial strains distributed among 16 genera show that cyanobacteria of groups I–III are mixed, but heterocyst- and akinete-bearing taxa (IV and V) form a monophyletic clade. Phylogenetic trees based on the *rbcL* genes support the monophyly of heterocystous cyanobacteria. The members of subsection V form a monophyletic clade in the *hetR* tree. It has been suggested that the clade of cyanobacteria marked by cell differentiation diverged between 2,450 and 2,100 Ma ago (Tomitani et al. 2006). The genomes from 24 strains of 10 genera of cyanobacteria living in freshwater, marine environments and hot springs have been analyzed by different methods to reconstruct an evolutionary history of the phylum (Swingley et al. 2008): 583 protein families of orthologs were selected and phylogenies were constructed.

11.7 Discussion

There is no doubt that the origin and subsequent evolution of the complex oxygenic photosynthetic apparatus of present-day cyanobacteria must have occurred in multiple steps under constant selective pressure. The selective pressure could come from the necessity of cells to gain energy in an environment of decreasing atmospheric hydrogen and other substrates and damaging effects of solar UV. RC 1 could have evolved by way of multiple duplication events from simpler chlorophyll-binding membrane proteins (Mulkidjanian and Junge 1997; Olson and Blankenship 2004). Upon gradual oxidation of the atmosphere, the need for further sources of redox equivalents could have driven the formation of the small, high-potential RC 2. Further depletion of electron donors upon oxidation of the available Fe(II) could have driven the evolution of RC 2 into the water-oxidizing PS II (Mulkidjanian et al. 2006). The first steps in the evolution of photosynthesis and cyanobacteria remains a topic of speculation. The formation of the pigment molecules carotenoids and (bacterio)chlorophylls, generally speaking tetrapyrrols, which are light absorbers with a very high absorption cross section owing to their network of alternating single and double bonds may have been selected as sensory molecules (Nisbet et al. 1995) or protector molecules. Model experiments with tetrapyrrols inserted in artificial membranes resulted in formation of potential differences across the membrane (Steinberg-Yfrach et al. 1997). Similar arrangements may have been formed in earlier times as a first system for producing energy. The large increase in the oxygen content of the atmosphere about 2,300 Ma was traced back to the oxygenic photosynthesis of cyanobacteria. The existence of cyanobacteria at this time is demonstrated by microfossils, biomarkers, the decreased $^{13}C/^{12}C$ isotope ratio and geochemical data. The ancestors of these oxygenic phototrophs were presumably present in the Archean Eon. Microfossils in rocks 2.7-billion years old and older were postulated to be remainders of cyanobacteria (Schopf 1999). It might be possible that these structures are the remains of the anoxygenic procyanobacteria. Heterocyst differentiation diverged once between 2,450 and 2,100 Ma, and akinete fossils have been identified in 1.65- to 1.4-billion-year-old cherts from Siberia, China, and Australia (Tomitani et al. 2006).

Acknowledgement I thank Karen Brune for editing the manuscript.

References

Anbar AD, Duan Y, Lyons TW, Arnold GL, Kendall B, Creaser RA, Kaufman AJ, Gordon GW, Scott C, Garvin J and Buick R (2007) A whiff of oxygen before the great oxidation event? Science 317:1903–1906

Battistuzzi FA, Feijao A, and Hedges SB (2004) A genomic timescale of prokaryote evolution: insights into the origin of methanogenesis, phototrophy, and the colonization of land. BMC Evol Biol 4(44):1–14

Bekker A, Holland AD, Wang PL, Rumble D, Stein HJ, Hannah JL, Coetzee IL and Beukes NJ (2004) Dating the rise of atmospheric oxygen. Nature (London) 427:117–126

Beukes NJ and Klein C (1990) Geochemistry and sedimentology of a facies transition from micro-banded to granular iron formation-in the early proterozoic Transvaal supergroup, South Africa. Precambr Res 47:99–139

Beukes KJ and Klein C (1992) In: Schopf JW and Klein C (eds) Proterozoic biosphere, pp. 147–151. Cambridge University Press, Cambridge

Bhattacharya D and Medlin L (1998) Algal phylogeny and the origin of land plants. Plant Physiol 116:9–15

Bhaya D (2004) Light matters: phototaxis and signal transduction in unicellular cyanobacteria. Mol Microbiol 53:745–754

Blankenship RE (1992) Origin and early evolution of photosynthesis. Photosynth Res 33:91–111

Blankenship RE (2002) Molecular mechanisms of photosynthesis. Blackwell Science, Oxford

Brasier MD, Green OR, Jephcoat AP, Kleppe AK, Van Kranendonk MJ, Lindsay JF, Steele A and Grassineau NV (2002) Questioning the evidence for earth's oldest fossils. Nature (London) 416:76–81

Buttner M, Xie DL, Nelson H, Pinther W, Hauska G and Nelson N (1992) Photosynthetic reaction center genes in green sulphur bacteria and in photosystem I are related. Proc Natl Acad Sci U S A 89:8135–8139

Canfield DE, Poulton SW and Narbonne GM (2007) Late-neoproterzoic deep-ocean oxygenation and the rise of animal life. Science 315:92–94

Canfield DE, Poulton SW, Knoll AH, Narbonne GM, Ross G, Goldberg T and Strauss H (2008) Ferruginous conditions dominated later neoproteozoic deep-water chemistry. Science 321: 949–952

Capone DG, Zehr JP, Paerl HW, Bergman B and Carpenter EJ (1997) *Trichodesmium*, a globally significant marine cyanobacterium. Science 276:1221–1229

Catling D, Zahnle KJ and McKay CP (2001) Biogenic methane, hydrogen escape, and the irreversible oxidation of early earth. Science 293:839–843

Cavalier-Smith T (2006) Cell evolution and earth history: stasis and revolution. Phil Trans R Soc B 361:969–1006

Chandler FW (1980) Proterozoic redbed sequences of Canada. Can Geol Surv Bull 311:1–53

Choudhary M and Kaplan S (2000) DNA sequence analysis of the photosynthetic region of *Rhodobacter sphaeroides* 2.4.1. Nucleic Acids Res 28:862–867

Clemmey H and Badham N (1982) Oxygen in the Precambrian atmosphere: an evaluation of the geological evidence. Geology 10:141–146

Cohn F (1867) Beiträge zur Physiologie der Phycochromaceen und Florideen. Arch Mikrosk Anat III:1–60
Cohn F (1875) Untersuchungen über Bakterien II. Beitr Biol Pfl 1(3):141–207
Drews G and Nultsch W (1962) Spezielle Bewegungsmechanismen von Einzellern. In: Ruhland W (ed) Handbuch der Pflanzenphysiologie, vol 17/2, pp. 887–894. Springer, Berlin
El Tabakh M, Grey K, Pirajon F and Schreiber BC (1999) Pseudomorphs after evaporitic minerals interbedded with 2.2 Ga stromatolites of the Yerriba Basin, Western Australia: origin and significance. Geology 27:871–874
England GL, Rasmussen B, Krapez B and Groves DL (2002) Paleoenvironmental significance of rounded pyrite in siliciclastic sequences of the late Archean Witwatersrand basin: oxygen-deficient atmosphere or hydrothermal alteration? Sedimentology 49:1133–1156
Goldblatt C, Lenton TM and Watson AJ (2006) Bistability of atmospheric oxygen and the great oxidation. Nature (London) 443:683–686
Granick S (1957) Speculations on the origin and evolution of photosynthesis. Ann N Y Acad Sci 69:292–308
Grossman AR, Bhaya D, Apt KE and Kehoe DM (1995) Light-harvesting complexes in oxygenic photosynthesis. Diversity, control and evolution. Ann Rev Genet 29:231–288
Gupta RS (2003) Evolutionary relationship among photosynthetic bacteria. Photosynth Res 76:173–183
Gupta RS, Mukhtar T and Singh B (1999) Evolutionary relationships among photosynthetic prokaryotes (*Heliobacterium chloroum, Chloroflexus aurantiacus*, cyanobacteria, *Chlorobium tepidum* and protobacteria): implications regarding the origin of photosynthesis. Mol Microbiol 32:893–906
Hartman H (1998) Photosynthesis and the origin of life. Origins Life Evol Biosphere 28:515–521
Heathcote P, Fyfe PK and Jones MR (2002) Reaction centres; the structure and evolution of biological solar power. TIBS 27:79–87
Holland HD (1994) In: Bengston S (ed) Early Life on Earth, pp. 237–244. Columbia University Press, New York
Jain R, Rivera MC and Lake JA (1999) Horizontal gene transfer among genomes: the complexity hypothesis. Proc Natl Acad Sci U S A 96:3801–3806
Kasting JF (2006) Ups and downs of ancient oxygen. Nature (London) 443:643–645
Kaufman AJ, Johnston DT, Farquhar J, Masterson AL, Lyons TW, Bates S, Anbar AD, Arnold GL, Garvin J and Buick R (2007) Late Archean biosphere oxygenation and atmospheric evolution. Science 317:1900–1903
Knoll AH (2003) Life on a young planet. Princeton University Press, Princeton
Kühl M, Chen M, Ralph PJ, Schreiber U and Larkum AWD (2005) A niche for cyanobacteria containing chlorophyll d. Nature (London) 433:820
Lewin RA (1976) Prochlorophyta as a proposed new division of algae. Nature 261:697–698
Mathis P (1990) Compared structure of plant and bacterial photosynthetic reaction centers. Evolutionary implications. Biochim Biophys Acta 1018:163–167
Mauzerall D (1992) Light, iron, Sam Granick and the origin of life. Photosynth Res 33:163–170
Meyer TE (1994) Evolution of photosynthetic reaction centers and light harvesting chlorophyll proteins. BioSystems 33:167–175
Meyer TE, van Beeumen JJ, Ambler RP and Cusanovich MA (1996) The evolution of electron transfer proteins in photosynthetic bacteria and denitrifying pseudomonads. In: Baltscheffsky H (ed) Origin and Evolution of Biological Energy Conversion, pp. 71–108. VCH, New York
Mix LJ, Haig D and Cavanaugh CM (2005) Phylogenetic analyses of the core antenna domain: investigating the origin of photosystem I. J Mol Evol 60:153–163
Mulkidjanian AY and Junge W (1997) On the origin of photosynthesis as inferred from sequence analysis. Photosynth Res 51:27–42
Mulkidjanian AY, Koonin EV, Makaarova KS, Mekhedov SL, Sorokin A, Wolf YI, Dufresne A, Partensky F, Burd H, Kaznadzey D and Haselkorn R (2006) The cyanobacterial genome core and the origin of photosynthesis. Proc Natl Acad Sci U S A 103:13126–13131
Nisbet EG, Cann JR and van Dover CL (1995) Origins of photosynthesis. Nature 373:479–480

Nitschke W, Mühlenhoff U and Liebl U (1998) Evolution. In: Raghavendra A (ed) Photosynthesis: A Comprehensive Treatise, pp. 285–304. Cambridge University Press, Cambridge
Nultsch W (1962) Der Einfluß des Lichtes auf die Bewegung der Cyanophyceen: Photophobotaxis bei Phormidium uncinatum. Planta 58:647–663
Nultsch W and Häder DP (1974) Über die Rolle der beiden Photosysteme in der Photophobo-taxis von *Phormidium uncinatum*. Ber Dtsch Bot Ges 87:83–97
Ohmoto H (1996) Evidence in pre-2.2 Ga paleosols for the early evolution of the atmospheric oxygen and terrestrial biota. Geology 24:1135–1138
Olson JM and Blankenship RE (2004) Thinking about the evolution of photosynthesis. Photosynth Res 80:373–386
Olson JM and Pierson BK (1987a) Evolution of reaction centers in photosynthetic prokaryotes. Ann Rev Cytol 108:209–248
Olson JM and Pierson BK (1987b) Origin and evolution of photosynthetic reaction centers. Orig Life 17:419–430
Oyaizu H, Debrunner-Vossbrinck B, Mandelco L, Studier JA and Woese CR (1987) The green non-sulfur bacteria: a deep branching in the eubacterial line of descent. Syst Appl Microbiol 9:47–53
Pace NR (2006) Time for a change. Nature (London) 441:46–51
Palenik B and Haselkorn R (1992) Multiple evolutionary origins of prochlorophytes, the chlorophyll b-containing prokaryotes. Nature (London) 355:265–267
Partensky F, Hess WR and Vanlot D (1999) *Prochlorococcus*, a marine photosynthetic prokaryote of global significance. Microbiol Mol Biol Rev 63:106–127
Rasmussen B (2000) Filamentous microfossils in a 3,235-million-years-old volcanogenic massive sulphide. Nature 405:676–679
Raymond J, Zhaxybayeva O, Gogarten JP, Gerdes SY and Blankenship RE (2002) Whole-genome analysis of photosynthetic prokaryotes. Science 298:1616–1620
Raymond J, Zhaxybayeva O, Gogarten JP and Blankenship RE (2003) Evolution of photo-synthetic prokaryotes: a maximum likelihood mapping approach. Phil Trans R Soc B 358:223–230
Raymond J, Siefert J, Staples C and Blankenship RE (2004) The natural history of nitrogen fixation. Mol Biol Evol 21:541–554
Sato N (2002) Comparative analysis of the genomes of cyanobacteria and plants. Genome Inform 13:173–182
Schopf JW (1993) Microfossils of the early Archean apex chert: new evidence of the antiquity of life. Science 260:640–646
Schopf JW (1994) In: Bengtson S (ed) Early Life on Earth, pp. 193–206. Columbia University Press, New York
Schopf JW (1999) The Cradle of Life. Princeton University Press, New York
Schopf JW and Packer BM (1987) Early Archean (3.3 billion to 3.5 billion year-old) microfossils from Warrawoona group, Australia. Science 237:70–73
Schubert WD, Klukas O, Saenger W, Witt HT, Fromme P and Krauß N (1998) A common ancestor for oxygenic and anoxygenic photosynthetic systems: a comparison based on the structural model of photosystem I. J Mol Biol 280:297–314
Shi T and Falkowski PG (2008) Genome evolution in cyanobacteria: the stable core and the variable shell. Proc Natl Acad Sci U S A 105:2510–2515
Staley JT, Bryant MP, Pfennig N and Holt JG (eds) (1989) Bergey's Manual of Systematic Bacteriology, vol. 3. Williams & Wilkins, Baltimore
Steinberg-Yfrach G, Liddell PA, Hung S-C, Moore AL, Gust D and Moore TA (1997) Conversion of light-energy to proton potential in liposomes by artificial photosynthetic reaction center. Nature (London) 385:239–241
Summons RE, Jahnke LL, Hope M and Logan GA (1999) 2-methylhopanoids as biomarker for cyanobacterial oxygenetic photosynthesis. Nature (London) 400:554–557
Swingley WD, Blankenship RE and Raymond J (2008) Integrating Markov clustering and molecular phylogenetics to reconstruct the cyanobacterial species tree from conserved protein families. Mol Biol Evol 25:643–654.

Tice MM and Lowe DR (2004) Photosynthetic microbial mats in the 3,416-Myr-old ocean. Nature (London) 431:549–551

Tice MM and Lowe DR (2006) Hydrogen based carbon fixation in the earliest known photosynthetic organism. Geology 34:37–40

Tomitani A, Okada K, Miyashita H, Matthija, HCP, Ohno T and Tanaka A (1999) Chlorophyll b and phycobilins in the common ancestor of cyanobacteria and chloroplasts. Nature (London) 400:159–162

Tomitani A, Knoll AH, Cavanaugh CM and Ohno T (2006) The evolutionary diversification of cyanobacteria: molecular-phylogenetic and paleontological perspectives. Proc Natl Acad Sci U S A 103:5442–5447

Urbach E, Robertson DL and Chisholm SW (1992) Multiple evolutionary origin of prochlorophytes within the cyanobacterial radiation. Nature 355:267–270

Valadez R (2000) In: Crockford SJ (ed) Dogs Through Time: An Archaeological Perspective, pp. 193–204. British Archaeological Reports, Oxford

Vermaas WF (1994) Evolution of heliobacteria: implications for photosynthetic reaction Center complexes. Photosynth Res 41:285–294

Woese CR (1987) Bacterial evolution. Microbiol Rev 51:221–271

Wolfe GR, Cunningham FX, Durnford D, Green BR and Gantt E (1994) Evidence for a common origin of chloroplasts with light-harvesting complexes of different pigmentation. Nature (London) 367:566–568

Xiong J and Bauer CE (2002a) A cytochrome b origin of photosynthetic reaction centers: an evolutionary link between respiration and photosynthesis. J Mol Biol 322:1025–1037

Xiong J and Bauer CE (2002b) Complex evolution of photosynthesis. Annu Rev Plant Biol 53:503–521

Xiong J, Inoue K and Bauer CE (1998) Tracking molecular evolution of photosynthesis by characterization of a major photosynthesis gene cluster from *Heliobacillus mobilis*. Proc Natl Acad Sci U S A 95:14851–14856

Xiong J, Fischer WM, Inoue K, Nakahara M and Bauer CE (2000) Molecular evidence for the early evolution of photosynthesis. Science 289:1724–1730

Yang W and Holland HD (2002) The redox-sensitive trace elements. Mo, U, and Re in Precambrian carbonaceous shales: indicators of the great oxidation event. Geol Soc Am Abstr Programs 34:381

Zehr JP, Waterbury JB, Turner PJ, Montoya JP, Omoregie E, Steward GF, Hansen A and Karl DM (2001) Unicellular cyanobacteria fix N_2 in the subtropical north pacific ocean Nature (London) 412:635–638

Zhaxybayeva O, Hamel L, Raymond J and Gogarten JP (2004) Visualization of the phylogenetic content of five genomes using dekapentagonal maps. Genome Biol 5:R20

Chapter 12
Structure of Cyanobacterial Photosystems I and II

Petra Fromme and Ingo Grotjohann

12.1 Overview of the Function of Photosystem I and II

Photosystem I and II are the key players in the energy conversion from the light of the sun into chemical energy. They have changed the atmosphere of our planet Earth from anoxygenic to oxygenic. Photosystem II (PS II) is unique and of global importance for the biosphere. All molecular dioxygen that can be presently found on Earth was produced by this protein complex with dramatic consequences for life on this planet due to formation of the oxygenic atmosphere about 2.5 billion years ago. (Lane 2003; Larkum 2008). The oxygenic respiration provides the main energy source for all higher heterotrophic organisms, including human beings. The oxygen that we breathe is exclusively produced by oxygenic photosynthesis.

Plant, algae and cyanobacteria also have a respiratory chain. The great advantage of O_2 as substrate for highly efficient Gibbs free energy extraction from food is only one side of the "coin", the other is the inherent danger for the cells due to its potential to cause oxidative degradation of biological material (see Sect. 12.2.2.1 and chapter by Renger and Ludwig).

Photosystem I and II are large multi-subunit membrane proteins located in the photosynthetic membrane (thylakoid membrane) of higher plants, green and red algae, and cyanobacteria. Ancestors of the current cyanobacteria have "invented" oxygenic photosynthesis and the cores of both Photosystem I and II are still very similar in all oxygenic photosynthetic organisms. Photosystem I (PS I) and II (PS II) catalyze the light-induced charge separation across the photosynthetic membrane. PS I and PS II capture the solar radiation by pigment protein complexes which act as antenna systems and funnel the electronic excitation energy to the reaction centers (RCs) where it drives charge separation. The anisotropic arrangement of the RCs within the thylakoid membrane gives rise to a vectorial electron transfer from

P. Fromme (✉)
Department of Chemistry and Biochemistry, Arizona State University, PO Box 871604, Tempe, AZ 85287-1604, USA
e-mail: pfromme@asu.edu

G. A. Peschek et al. (eds.), *Bioenergetic Processes of Cyanobacteria*,
DOI 10.1007/978-94-007-0388-9_12, © Springer Science+Business Media B.V. 2011

the lumenal to the cytosolic/stroma side[1]. The antenna systems consist of a peripheral/proximal antenna system, which was drastically changed, and a core antenna, which remained virtually invariant during evolution from the level of cyanobacteria up to higher plants. In cyanobacteria, large membrane associated peripheral antenna complexes called phycobilisomes serve as peripheral antenna complexes. They contain covalently bound phycobilines (open chain tetrapyrroles) as major pigments. On the contrary, green algae and higher plants lack phycobilisomes but contain membrane-integral light harvesting complexes (LHCs) (for reviews, see chapters in Green and Parson 2003). Red algae contain in addition also phycobilisomes, and this fact is one indication that they may have emerged from a second endosymbiotic event. In marine cyanobacteria (prochlorophytes), membrane intrinsic antenna complexes, the Pcb proteins (LaRoche et al. 1996), can be found. Under iron deficiency, the cyanobacterial protein IsiA surrounds PS I in the membrane and serves as its peripheral antenna (Leonhardt and Straus 1992, 1994; Boekema et al. 2001).

In PS II of all oxygen evolving organisms the electronic excitation energy captured by the peripheral/proximal antenna is transferred to the core complex, which harbors the RC with the photoactive pigment P680. The subsequent charge separation leads to formation of the stable radical ion pair $P680^{+\bullet} Q_A^{-\bullet}$ (for a recent review, see Renger and Renger 2008). Four sequential steps driven by $P680^{+\bullet}$ give rise to the extraction of four electrons from two water molecules at the water oxidizing complex (WOC)[2], resulting in the formation of molecular dioxygen and the release of four protons into the thylakoid lumen, thus contributing to the electrochemical potential difference of protons across the membrane. It should be noticed that cyanobacteria also contain a respiratory chain, located in the plasma membrane (see Chap. 20) so that they can also utilize the oxygen for respiration.

Two sequential electron transfer steps with $Q_A^{-\bullet}$ as reductant lead to formation of plastoquinol (PQH_2) at the Q_B^- site. This reaction is coupled with the uptake of two protons from the cytosolic/stroma side. The PQH_2 is replaced by a plastoquinone molecule from the PQ pool. The most recent crystal structure of PS II (Guskov et al. 2009) revealed that the Q_B binding site is connected to the pool of plastoquinones in the membrane by two channels (see Sect. 12.2.3.1).

PS I catalyzes the light driven transmembrane electron transfer from plastocyanin (PC) or cytochrome c_6 (Cyt c_6) at the lumenal side to the soluble electron carrier ferredoxin or flavodoxin at the cytosolic/stromal side of the membrane. It thereby finally provides the electrons for the reduction of protons to hydrogen in form of NADPH by the ferredoxin: $NADP^+$ reductase (FNR). PS I contains a much larger internal core antenna system than PS II. The structure of cyanobacterial PS I (Jordan et al. 2001) reveals that 96 Chlorophylls (Chls) and 22 carotenoids (Cars) capture the sunlight and transfer the electronic excitation energy to the center of the

[1] The outer aqueous phase of thylakoids is the cytosol in cyanobacteria and the stroma of chloroplasts on algae and higher plants.

[2] The water oxidizing complex is also designated oxygen evolving complex (OEC). In this book the symbol WOC will be used.

complex, where charge separation takes place at P700. After the primary charge separation, one electron is transported across the membrane by a chain of electron carriers. The organic cofactors of the ET chain (6 chlorophylls, 2 phylloquinones) are arranged in two branches, followed by three 4Fe4S clusters: F_X, F_A and F_B. From F_B the electron is transferred to ferredoxin, which docks to Photosystem I at the stromal hump that protrudes into the cytosol/stroma by approximately 40 Å. Flavodoxin replaces ferredoxin as mobile electron carrier under iron deficiency (Sandmann and Malkin 1983; Goni et al. 2009). Ferredoxin or Flavodoxin finally transfer the electron to FNR, which reduces $NADP^+$ to NADPH.

PS I and PS II are functionally coupled by the cytochrome b_6f (Cyt b_6f) complex, which releases 2 protons to the inside of the thylakoids (lumen), and subsequently reduces 2 molecules of plastocyanin or Cyt c_6 which act as electron donors to $P700^{+\bullet}$.

In addition to the 2 protons released upon PQH_2 oxidation, the Cyt b_6f complex pumps an additional proton across the membrane, which thereby further contributes to the extent of the electrochemical potential difference of protons across the photosynthetic membrane. The structure of the Cyt b_6f complex has been determined both from green algae (Stroebel et al. 2003) and cyanobacteria (Kurisu et al. 2003; Yamashita et al. 2007). For details on the Cyt b_6f complex, see chapter of Bernát and Rögner.

The photosynthetic electron transport chain, driven by the charge separation events in Photosystem I and II, establishes an electrochemical potential difference ($\Delta pH + \Delta\psi$) across the membrane (pmf) which drives synthesis of ATP by the ATP-Synthase (Mitchell 1961; Jagendorf 2002).

This chapter is focused on the structures of cyanobacterial Photosystem I and II and therefore the function will be only briefly described, in particular for PS II which is a major topic of the chapter by Renger and Ludwig.

12.2 Photosystem II

12.2.1 Overview of the Structure and Function of Photosystem II

PS II is a large multimeric membrane protein complex that catalyzes the light driven electron transfer from water to plastoquinone.

The photosynthetic process is initiated by the capture of light which leads to population of electronically excited singlet states. This excitation energy is transferred to the reaction center (RC), where the electron transfer chain (ETC) is located that performs the charge separation. Both Photosystems contain a core antenna. The pigment content of the PS II core antenna system is less than half in size compared to the antenna of PS I. The linear ETC of PS II is more complex than the electron transfer chain of PS I (see Sects. 12.2.3 and 12.3.3)

One major problem that PS II has to cope with is photodamage. Mechanistic and molecular details of photodamage in PS II are topics of current research activities (see Sect. 12.2.2.1).

In the native membrane of plants, algae, and cyanobacteria, PS II is probably a dimer (but see Takahashi et al. 2009). The monomer of PS II consists of 19–20 protein subunits in cyanobacteria.

PS I and PS II were identified in the 1950s and 1960s by their spectroscopic properties. It was concluded that there are two photosynthetically active Photosystems in the thylakoid membrane, shown to be anisotropically incorporated into the thylakoid membrane [for review, see (Witt 1971) and references therein]. It was concluded that the photoactive pigment P680 of PS II is located near to lumenal side and the acceptor Q_A on the stroma side of thylakoids in chloroplasts. Progress in the preparative methodology led to isolation of PS II core (Franzen et al. 1986; Haag et al. 1990; Tsiotis et al. 1999) and D1/D2/cytb-559 complexes (Nanba et al. 1987). These achievements were the prerequisite for advanced studies on the structure of PS II by image analyses of EM micrographs (Rögner et al. 1987; Dekker et al. 1988). As a result, the overall shape of PS II core complexes was resolved and a first crude picture obtained on the extrusion due to extrinsic PsbO protein (Haag et al. 1990). More structural details could be gathered from using the method of electron diffraction on two dimensional crystal lattices of D1/D2/CP47/cytb-559 preparation (Nakazato et al. 1996; Rhee 2001). For a detailed discussion of the history of structural analysis of Photosystems I and II, see Witt (2004) and references therein.

The first crystals of PS II that were able to split water had been grown from the thermophilic cyanobacterium *Thermosynechococcus*(*T.*) *elongatus* (Zouni et al. 2000) and led to the first X-ray structural model of the intact PS II complex at 3.8 Å resolution (Zouni et al. 2001). This model provided a deeper insight into the structure of PS II and revealed the 3 + 1 organization of the four manganese in the WOC. All further improved crystal structures and refinements were based on this type of crystals from *T. elongatus*, but one exception, i.e. the 3.7 Å crystal structure, which was determined from *T. vulcanus* (Kamiya and Shen 2003).

12.2.2 Arrangement of Protein Subunits in Cyanobacterial PS II

23 protein subunits have been assigned in structures of PS II from *T. elongatus:* PsbA to PsbF, PsbH to PsbO, PsbT to PsbZ, as well as the PsbZ like protein Psb27 and subunit ycf12 (Psb30). At least 20 proteins of these subunits have been identified in the PS II crystals by detailed mass spectroscopic analysis (Kern et al. 2005, Guskov et al. 2009). Seventeen subunits are integral membrane proteins, whereas PsbO, PsbU and PsbV do not contain transmembrane α-helices and are located at the lumenal side of the complex. In the first structural model of the intact PS II at 3.8 Å resolution (Zouni et al. 2001), 36 transmembrane helices were identified. Later structures of PS II have been published at 3.7 to 2.9 Å resolution, revealing more details. These include assignments of most of the amino acid side chains

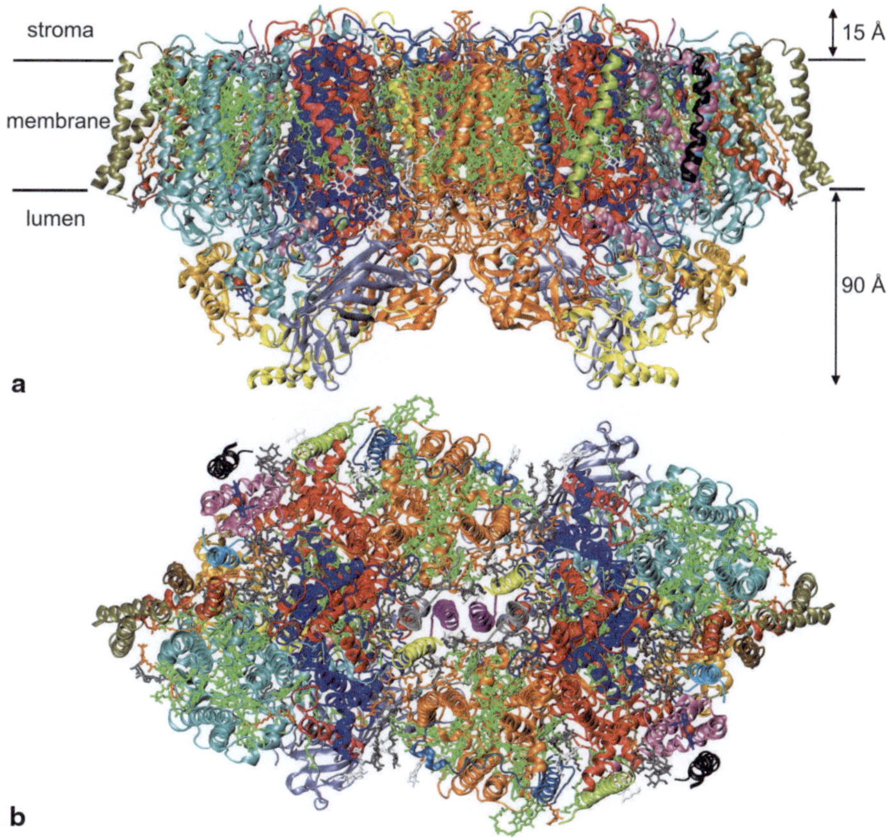

Fig. 12.1 Dimeric photosystem II from *Thermosynechococcus elongatus*. **a** View from inside the membrane plane. All extrinsic subunits of PS II can be found on the lumenal side, where the protein complex extends 90 Å out of the membrane, whereas loops reach maximally 15 Å into the cytosol. **b** View of the dimer from the cytosolic side. This face of the protein complex is mostly flat. All Photosystem II models in this chapter were made from pdb files 3bz1 and 3bz2. All images in this chapter were made with the program VMD (Humphrey et al. 1996)

and identification of the small membrane intrinsic subunits. The initial number and location of the 36 transmembrane helices in PS II were confirmed by the recent crystal structure of PS II at 3.0 Å resolution (Loll et al. 2005), whereas one helix is missing in the 3.5 Å structure of PS II (Ferreira et al. 2004). The general arrangement of subunits in PS II is shown in Figs. 12.1 and 12.2, based on the structural model of PS II at 2.9 Å structure of PS II (Guskov et al. 2009).

In analogy to the RCs of anoxygenic purple bacteria (PBRCs), the heterodimer consisting of subunits D1 (PsbA) and D2 (PsbD) is often called reaction center but this assignment should be considered with care (for a discussion, see Renger 2008). The D1/D2 heterodimer is flanked on both sides by the core antenna proteins CP47 (PsbB) and CP43 (PsbD). This core of PS II shows strong similarities to the

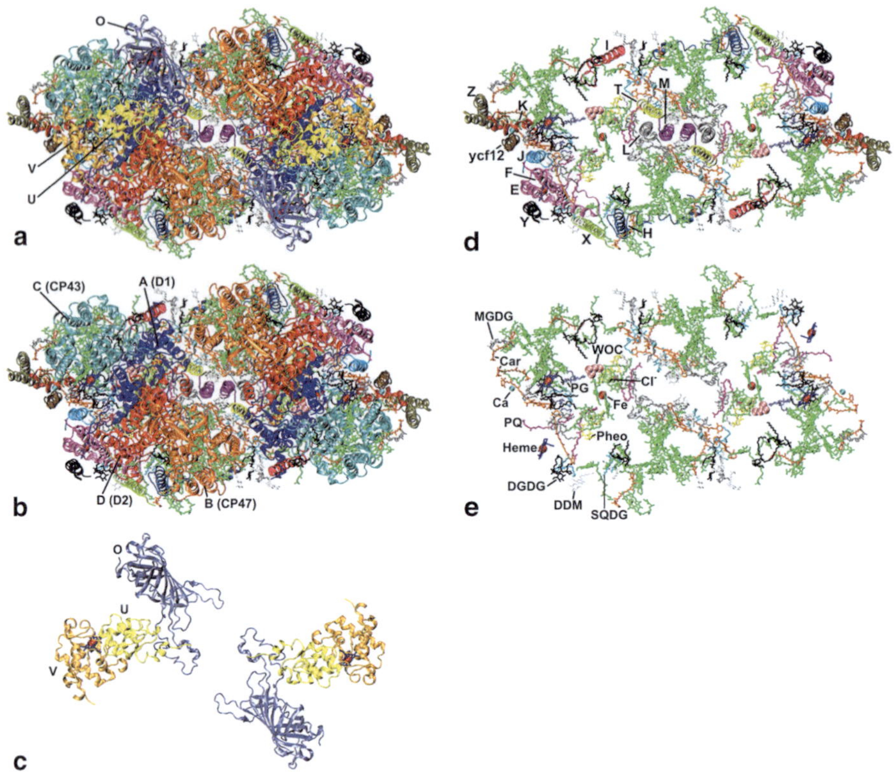

Fig. 12.2 Dimeric photosystem II from *Thermosynechococcus elongatus* as seen from the lumenal side. **a** All subunits and co-factors. The extrinsic subunits, PsbO, PsbU, and PsbV, are labeled. The prefix "Psb" has been omitted from all names for brevity. **b** Same view, with the extrinsic subunits removed. The reaction center core and core antenna subunits are labeled. The oxygen-evolving complex is now openly visible. **c** The extrinsic subunits on their own. The heme-containing PsbV is also called Cytochrome c_{550}. **d** All membrane-intrinsic small protein subunits. The core proteins (A, B, C, D) and the extrinsic subunits (O, V, U) have been omitted, but all other cofactors are still shown. **e** All non-protein cofactors. Chlorophylls are green and not labeled. *Pheo* pheophytin, *PQ* plastoquinone, *Car* carotenoid, *DGDG* digalactosyldiacylglycerole, *MGDG* monogalactosyldiacylglycerole, *SQDG* sulfoquinovosyldiacylglycerole, *PG* phosphatidylglycerole; *WOC* water-oxidizing complex, *Cl*$^-$ chloride, *Fe* iron, *Ca* calcium. Although technically no cofactors, tightly bound detergent molecules are also shown, as they might replace other tightly bound lipids: *DDM* Dodecyl-β-D-maltoside

arrangement of the transmembrane helices in the two major proteins in PS I (PsaA and PsaB), indicating that PS I and PS II have evolved from a common ancestor, as suggested by Schubert and coworkers (Schubert et al. 1998). None of the extrinsic or small integral subunits of PS I and PS II exhibit any similarities, which suggests that they became part of PS I and PS II after the evolutionary split between the two Photosystems had occurred. As evident from Figs. 12.1a and 12.3a, the most striking difference between both Photosystems is the location of the membrane extrinsic

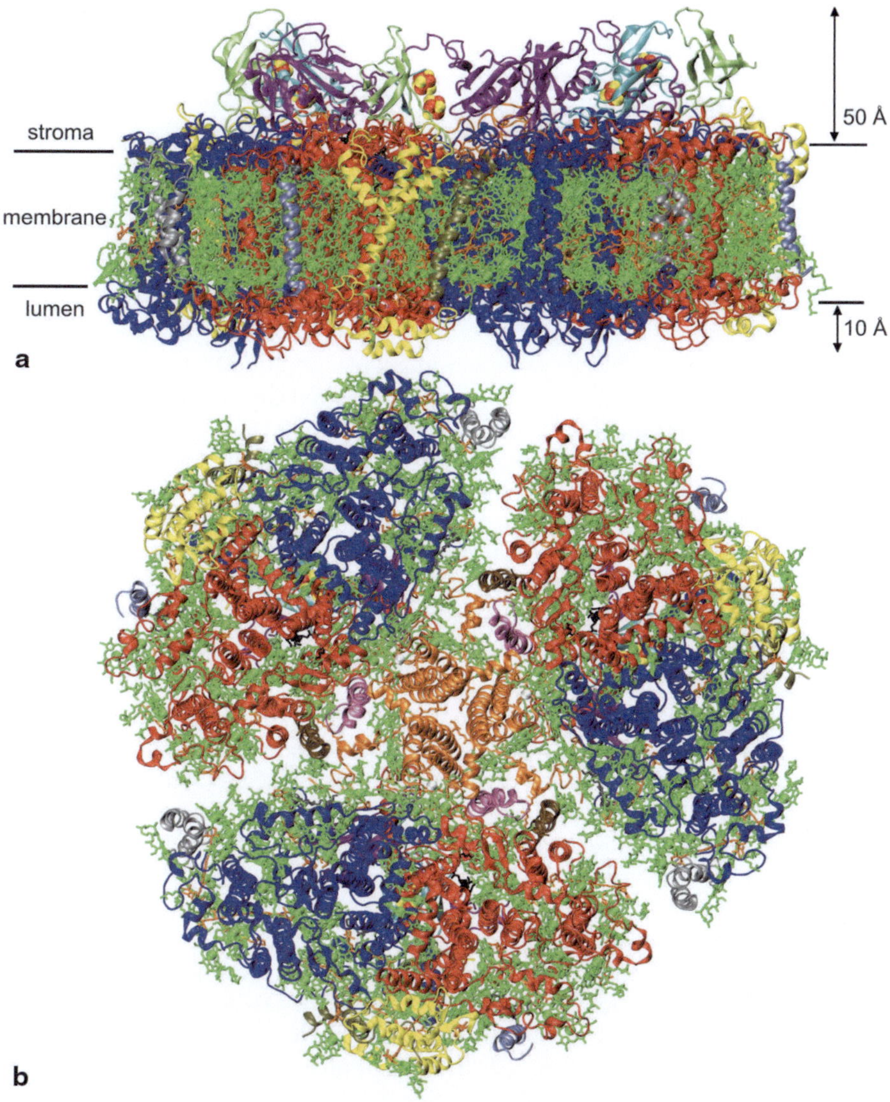

Fig. 12.3 Trimeric photosystem I from *Thermosynechococcus elongatus*. **a** View from inside the membrane plane. All extrinsic subunits of PS I can be found on the cytosolic side, where the protein complex extends 40 Å out of the membrane, whereas loops reach maximally 10 Å into the lumen. **b** View of the trimer from the lumenal side. This face of the protein complex is mostly flat. One of the central P_{700} chlorophyll dimers is marked. This is also the binding site for cytochrome c_6. All Photosystem I models in this chapter were made from pdb file 1jb0

proteins. Whereas PS I contains three extrinsic proteins at the cytrosolic/stromal side, which are involved in the docking of ferredoxin/flavodoxin, PS II does not extend more than 10 Å into the stroma. Most of the extrinsic mass of PS II is located at the lumenal site. Here, the large lumenal loops of CP43, CP47, which are unique and essential structural motifs (Glu 354 of CP 43 is a direct ligand to the Mn_4O_xCa cluster, see Sect. 12.2.3.2), and the 3 lumenal proteins PsbO, PsbU and PsbV stabilize the WOC in PS II.

12.2.2.1 Subunits D1 and D2

The subunits D1 and D2 were identified in gel electrophoresis as diffusive bands (see Baumgartner et al. 1993 and references therein) and later shown to bind Chl (Irrgang et al. 1986). D1 and D2 form a heterodimer which binds the cofactors of the ETC of the water-plastoquinone:oxido-reductase (for details of cofactor arrangement, see Fig. 12.4 of this chapter and also Fig. 12.2 in chapter of Renger and Ludwig). PS II crystal structure analyses revealed that subunits D1 (PsbA) (depicted in blue) and D2 (PsbD) (red) form a cluster of 2×5 transmembrane helices (TMHs) in an S-type arrangement (see Fig. 12.2b). This motif resembles the structure of the L and M subunits of the reaction center of purple bacteria (PBRC) (Deisenhofer et al. 1985; Deisenhofer and Michel 1991). A structural and functional comparison between PS II and PBRC suggests that the D1 protein is related to the L subunit, while the D2 protein is related to the M subunit of the PBRC. The overall structure of the D1 and D2 proteins resembles to a less pronounced extent also the C-terminal domains of the large subunits PsaA and PsaB of Photosystem I (Jordan et al. 2001). This similarity reveals that *all* existing RCs might have evolved from a common ancestor, as previously proposed (Blankenship and Kindle 1992; Schubert et al. 1998).

In contrast to PS I, where the C-terminal parts of PsaA and PsaB coordinate 28 antenna Chls, the hetrodimer of D1 and D2, coordinates only two "antenna" chlorophylls: $ChlZ_{D1}$ and $ChlZ_{D2}$ which probably play a protective role (see Lince and Vermaas 1998). Furthermore, two Cars are bound, forming hydrophobic interactions with the outermost transmembrane helix of D1 and D2 (helix a). The carotenoids (Cars) are located in close proximity to $ChlZ_{D1}$ and $ChlZ_{D2}$. A role of these Cars in the non-photochemical quenching process in PS II was proposed for the Cars associated to D2 (Telfer 2002, 2005; Telfer et al. 2003).

The D1 protein has to pay a high price for coordinating most of the ETC cofactors. It is subject to photodamage and exhibits a half-life of only 30 minutes: in plants in bright sunlight. Therefore D1 has to be constantly replaced via a repair mechanism (see Sect. 12.2.3 and Fig. 12.4).

The cyanobacterial genome contains three copies of the D1 gene, called *psbA1*, *psbA2* and *psbA3*. While expression of *psbA1* has not been shown so far, the expression patterns of psbA2 and psbA3 suggest an adaptation mechanism to different environmental conditions (Soitamo et al. 1998; Salih and Jansson 1997; Sippola and Aro 2000). The photodamage of the D1 protein leads to a complex degradation and replacement cycle. This complicated process is the subject of a

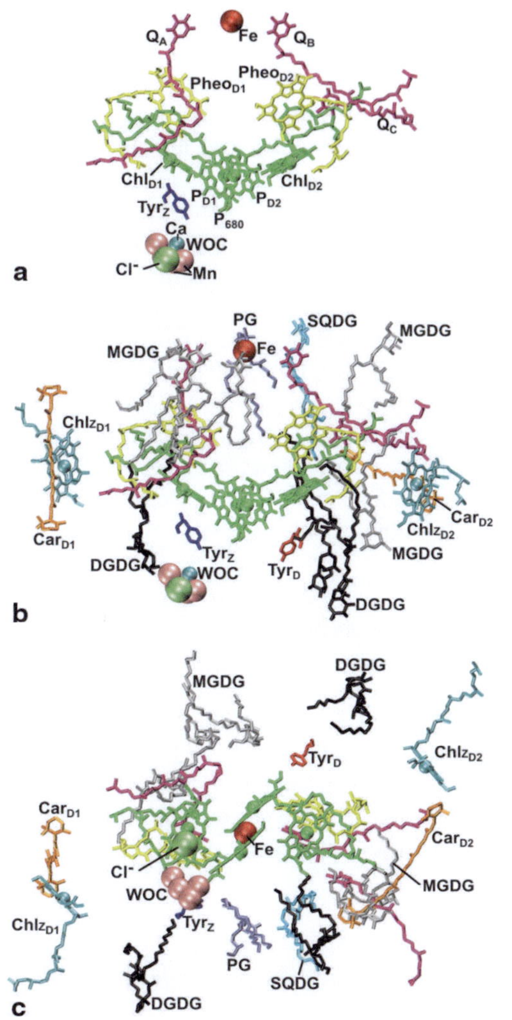

Fig. 12.4 The electron transfer chain of PS II and nearby cofactors. **a** The electron transfer chain itself. Included are four chlorophylls P_1–P_4, two pheophytins, three plastoquinones (including the newly discovered Q_C), the redox-active tyrosine TyrZ, and the water-oxidizing complex (WOC) with its four manganese and one calcium atoms, plus the associated chloride ion. The iron (Fe) is shown for orientation. **b** In addition to all components from **a**, all molecules within 15 Å from the central chlorophyll head groups are shown. This includes the inactive TyrD, the only two other chlorophylls coordinated by the subunits D1 and D2 ($Chlz_{D1}$, $Chlz_{D2}$), two carotenoids and several lipids (see Fig. 12.2 for abbreviations). In case of photodamage, the triplet state is localized on Chl P_3 (accessory Chl on the D1 side), and this chlorophyll has a distance of 21 Å to the next carotenoid, which is Car_{D1}. On the other side Chl P_4 has a much shorter distance to Car_{D2} of only 11.6 Å. **c** Same as **b**, as seen from the lumenal side

very active field of current research on PS II and has been best studied in plants so far [see Baena-Gonzalez and Aro 2002; Rokka et al. 2005; Nowaczyk et al. 2006; Mohanty et al. 2007; Vass et al. 2007; Vass and Aro 2008 and references therein]. It includes phosphorylation of the damaged D1 protein, monomerization of the PS II dimer, de-attachment of the CP43 protein and the three lumenal proteins, proteolytic degradation of D1, synthesis of D1 by membrane-bound ribosomes, cleavage of the N-terminal signal sequence of D1 in the lumen, assembly of the Mn cluster by photo-activation, and assembly of all subunits and the re-formation of PS II dimers. To illustrate the central role and location of the D1 protein this repair cycle could be compared to a heart transplantation taking place every half an hour.

12.2.2.2 CP47 and CP43

The polypeptides of CP47 and CP43 are PsbB and PsbC, respectively, which are the two largest subunits of PS II. They bind the pigments thus forming the core antenna of PS II and are in addition indispensable for stabilizing the Mn_4O_xCa cluster (see Sect. 12.2.3). These subunits are located on both sides of the central D1/D2 core of PS II, with CP47 flanking the D2 protein and CP43 being located in close proximity to D1. Each of these subunits consists of six transmembrane helices which form three pairs of dimers, consisting of helices 1/2, 3/4 and 5/6. CP47 is located close to the dimer-dimer interface, while CP43 is located at the periphery of the PS II dimer. This arrangement explains why CP43 can be more easily removed from the PS II core than CP47 (Bricker and Frankel 2002). The peripheral location of CP43 allows disassembly of this subunit from PS II in the process of D1 turnover during the photodamage repair cycle (see Sect. 12.2.3).

The arrangement of the transmembrane helices in CP47 and CP43 is very similar to the arrangement of the transmembrane helices in the N-terminal part of the PsaA/PsaB in Photosystem I (see Figs. 12.2 and 12.5) (Grotjohann et al. 2004).

CP43 and CP47 coordinate 13 and 16 chlorophylls, respectively. While the structures at 3.7 Å and 3.5 Å reported a 14th chlorophyll in CP43, the 3.0 Å and 2.9 Å structures identified a lipid (DGDG2, a digalactosyl diacylglycerol) at this site.

One of the striking differences between the N-terminal region of the PsaA/B subunits in PS I and CP47 and CP43 is the existence of extended loops at the lumenal side in case of the PS II subunits. The large loop of CP 47 interacts with loops of the D2 protein, the PsbO protein and the PsbU protein, thereby stabilizing the manganese cluster. The long loop of CP 43 interacts with loops of the D1 protein and all three extrinsic proteins, PsbO, PsbU and PsbV. Furthermore, amino acid Glu 354 provides a ligand for one manganese of the Mn_4O_xCa cluster (see Sect. 12.2.3).

12.2.2.3 Small Integral Subunits

Small integral proteins of PS II have been assigned to 14 transmembrane helices that are located peripherally to the central core. They can be subdivided into two groups, consisting of subunits located at the dimerization domain and subunits located at the periphery of the PS II dimer. Three helices are located close to the local twofold symmetry axis between the dimers, representing PsbM, PsbT and PsbL. One helix is sandwiched between D1 and CP43, which is assigned to PsbI. A field of 10 helices is located at the membrane-exposed periphery of PS II, which constitutes of PsbE, PsbF, PsbH, PsbJ, PsbK, PsbX, PsbY, PsbZ, and ycf12. The electron density is much better defined at the dimer interface than at the periphery of PS II, with the consequence that the subunit-assignment of three of the peripheral helices was ambiguous until recently (Loll et al. 2005). This question could be finally settled with the 2.9 Å structure published in 2009 (Guskov et al. 2009). One of these transmembrane helices, ycf12, had been erronously assigned

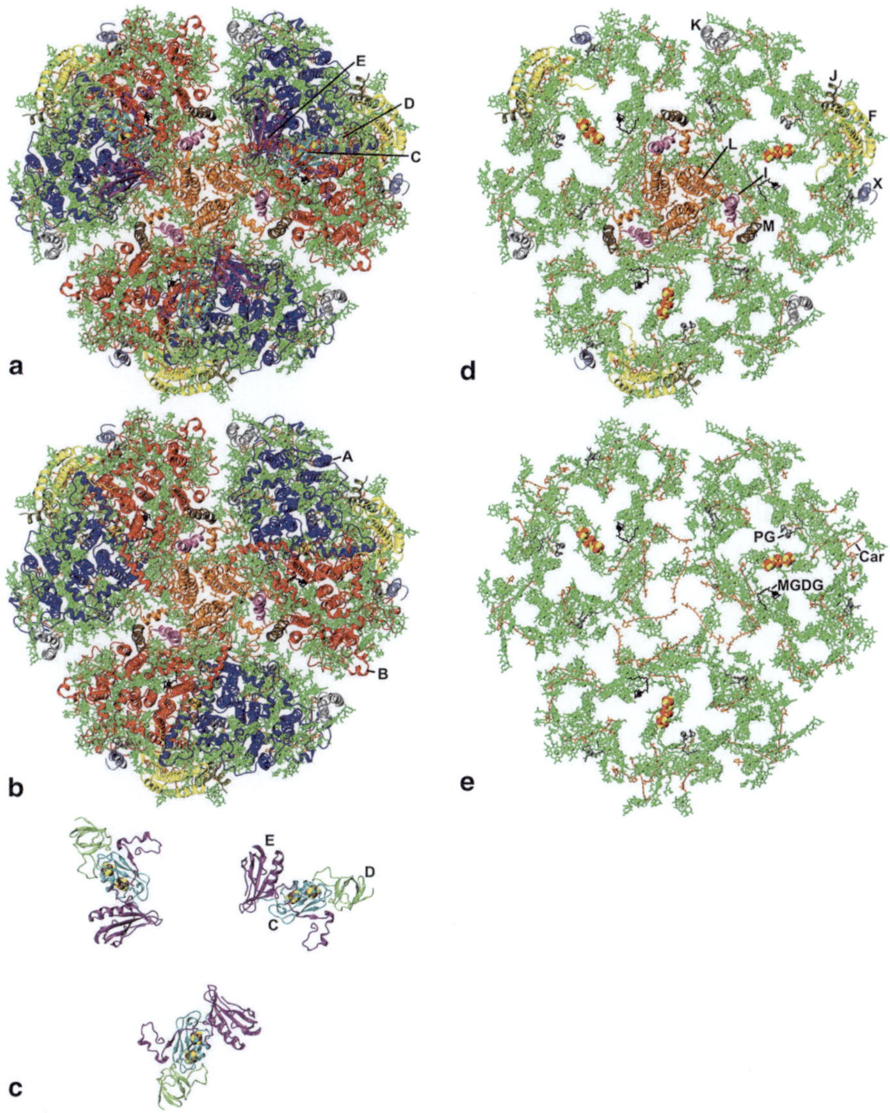

Fig. 12.5 Trimeric photosystem I from *Thermosynechococcus elongatus* as seen from the cytosolic side. **a** All subunits and co-factors. The extrinsic subunits, PsaC, PsaD, and PsaE, are labeled. The prefix "Psa" has been omitted from all names for brevity. **b** Same view, with the extrinsic subunits removed. The reaction center subunits are labeled, and some of the electron transfer chain, sandwiched between subunits PsaA and PsaB, becomes visible. **c** The extrinsic subunits on their own. Subunit PsaC contains the iron sulfur clusters F_A and F_B, which are the last molecules of the electron transfer chain. **d** All membrane-intrinsic small protein subunits. L, I, and M are found in the trimerization domain, J, F, and X are a group at the outermost edge of the protein complex, and subunit K can be found at the outer end of the monomer-monomer interface. **e** All non-protein cofactors. Chlorophylls are green and not labeled. *Car* carotenoids, *MGDG* monogalactosyldiacylglycerole, *PG* phophatidylglycerole. The carotenoids build an extensive protective network for the chlorophylls

to PsbN before (Ferreira et al. 2004), but MS data showed that this subunit is not present in PS II crystals (for a recent review on the small integral subunits, see Müh et al. 2008).

Subunits at the Periphery of PS II

Cytochrome b_{559} (PsbE and PsbF)

Cytochrome b_{559} (Cyt b_{559}) consists of two subunits: PsbE and PsbF which form a cross-shaped heterodimer where the heme group is coordinated by His 34 of PsbE and His 24 of PsbF. The cytochrome is located in close vicinity to helix 1 of D2. Cytb_{559} is an essential constituent of PS II because a deletion of either PsbE or PsbF is lethal for the assembly of a functionally competent PS II (Suorsa et al. 2004). The heme group can attain at least three different forms which are characterized by midpoint potentials of +50 to +120 mV (low potential LP form), +230 to +260 mV (intermediate IP form) and +385 to +400 mV (the unusual high potential HP form). The population probability of these forms is variable in both cyanobacteria and higher plants and also depends on the type of sample preparation (Stewart and Brudwig 1998; Kaminskaya et al. 1999, 2005; Roncel et al. 2003). The functional role of this cytochrome and its heme cofactor is still under investigation. The structure implies that the heme might be involved in the non-radiative charge recombination pathway between the singly reduced $Q_B^{-\bullet}$ and $P680^{+\bullet}$ as part of a prevention of excessive photodamage, with the help of Car_{D2} that is placed between the D2 protein and cytochrome b_{559} (Bondarava et al. 2003; Vasil'ev et al. 2003; Lakshmi et al. 2003). A redox active role of Cars has also been proposed by (Tracewell and Brudvig 2003). It is a constituent of one side of the putative quinone diffusion channel in PS II, as suggested by Loll et al. (2005).

The midpoint potential was recently shown to be affected by several compounds and a new binding site Q_c proposed for PQ/PQH_2 in the neighbourhood of Cyt b_{559} (Kaminskaya et al. 2007). This idea has been confirmed by the recent crystal structure data (Guskov et al. 2009) (see Sect. 12.2.3.1).

Subunits Close to Cytochrome b_{559}

Two further helices are located in the close vicinity of PsbE and PsbF. The helix closest to PsbF has been assigned to PsbJ, which seems to act in tandem with Cytb_{559} in forming an access gate to the putative quinone channel in PS II (Loll et al. 2005; Guskov et al. 2009). The helix close to PsaE [named X2 in (Loll et al. 2005)] was present in the 3.7 Å model (Kamiya and Shen 2003) and the 3.2 Å model (Biesiadka et al. 2004) of PS II, but is missing in the 3.5 Å model (Ferreira et al. 2004). Because of its characteristic break predicted from its secondary structure, this helix could finally be unambigously assigned to PsbY (Guskov et al. 2009). The loss of this subunit in the 3.5 Å model may have been caused by the higher detergent concentration used for the isolation of PS II in the work of Ferreira et al. (2004), and its poor electron density even in the 2.9 Å model suggests that this subunit is only loosely bound to the PS II core.

Small Subunits Close to CP 43

A group of four transmembrane helices is located at the periphery of PS II in close vicinity to CP43. The helix closest to the interface between CP43 and PsbJ has been assigned to PsbK, while two transmembrane helices that are in the vicinity of the N-terminal region of CP43 have been assigned to PsbZ. One further helix in between PsbZ and PsbK has been assigned to PsbN in the 3.5 Å structure. However, this assignment has not been confirmed at 3.0 Å resolution and PsbN could not be identified in the crystals by mass spectrometry (Kern et al. 2005). The 2.9 Å model finally allowed for assignment of the unknown helix to the protein ycf12. The subunits PsbJ and PsbK shield a field of lipids between the D1 protein and CP43 from the membrane, which in turn shield the cofactors of the D2 branch, including Q_A. The important role of PsbJ has been shown in biochemical studies: A psbJ deletion mutant is impaired in PS II electron flow to plastoquinone (Regel et al. 2001), which is in accordance with its suggested role as one of the gate posts of the putative quinone diffusion channel (Loll et al. 2005). PsbK may stablilize a cluster of chlorophylls in CP43. A tight interaction of CP43 with PsbK has also been shown by biochemical studies, where PsbK was co-purified with CP43 during ion exchange chromatography (Sugimoto and Takahashi 2003). The stabilization effect may be further enhanced by the interaction with a Car.

Small Subunits Close to CP47

Only one helix is in close contact to chlorophylls of CP 47 at the membrane-exposed periphery, and this helix has been assigned to PsbH. It forms hydrophobic contacts with several chlorophylls in CP47 and with one carotenoid, thereby stabilizing the antenna system of CP 47. PsbH seems to play a role in the repair of photodamage (Komenda et al. 2005) because it has the second highest turnover rate of all PS II subunits under illumination (Rokka et al. 2005). The exact nature of its role is unknown so far.

Small Subunits Close to $ChlZ_{D1}$ *and* $ChlZ_{D2}$

Close to $ChlZ_{D1}$ and $ChlZ_{D2}$, two symmetry-related single helices can be found. The helix close to $ChlZ_{D1}$ has been assigned to PsbI, while the helix close to $ChlZ_{D2}$ has been identified as PsbX which was previously tentatively assigned to PsbZ in the 3.5 Å structure, but this assignment was not confirmed at 2.9 Å resolution.

Deletion mutants of PsbI are strongly affected. They are still able to grow photoautotrophically in dim light, but not in high light, and the amounts of PS II complexes and the oxygen evolving activity are both reduced to 10–20% of wild-type levels (Kunstner et al. 1995). Similar to PsbH, PsbI seems to play a role in D1 turnover, and furthermore in CP43 stability (Dobakova et al. 2007). Deletion of PsbI leads to increased D1 turnover, but to decreased efficiency of the repair process, probably due to destabilization of CP43.

Subunits of the Dimerization Domain

In the dimerization domain, three helices are located, which are assigned to the subunits PsbL, PsbM and PsbT. PsbM is closest to the dimerization domain and forms interactions with the PsbM subunit of the neighboring dimer. This is the only protein to protein interaction known within homodimeric PS II. PsbT is also close to the dimer-dimer interface and to the first transmembrane helix of the D1 protein. PsbL is located between PsbM and PsbT. All three subunits form contacts with several lipids that fill the space between the dimerization domain and the opening between D1 and D2 protein. This means that all three subunits and the lipids are in close vicinity to the ETC and may even stabilize the Q_A binding site. The deletion of each of the three subunits has severe effects on the function of PS II, which are not all in agreement with the proposed location of these subunits. Deletion of PsbM in tobacco leads to high-light sensitivity of PS II, reduced phosphorylation of D1 and D2 and alteration of the Q_B site properties. The latter finding is difficult to understand with the present assignment of PsbM in *T. elongatus*. In this mutant the PS II repair may be impaired (Umate et al. 2007).

Mutants lacking PsbL are highly sensitive to photoinhibition and it was suggested that PsbL may be important to prevent reduction of PS II by back electron flow from plastoquinol (Ohad et al. 2004). A possible explanation for this discrepancy between biochemical and structural evidence might be that deletion of one of the subunits PsbM, PsbL, or PsbT could result in the loss of all three subunits, as they form a helix bundle, so the effect might be indirect or generally disturb the PS II structure.

In the case of PsbT, its location in the dimerization domain is also strongly supported by biochemical evidence, as results from a deletion mutant of PsbT suggest. PsbT seems to be involved in the stabilization and repair of the primary electron acceptor Q_A of PS II during photoinhibition. Furthermore, it may structurally stabilize the Q_A binding site, since half of the Q_A was lost during purification from the PS II core complex that lacks PsbT (Ohnishi et al. 2007).

A general problem that has to be taken into consideration is the possibility that slight differences exist in the polypeptide pattern of cyanobacteria and higher plants. Therefore a comparison of results from different species does not always permit to draw straightforward conclusions.

12.2.2.4 Lumenal Subunits PsbO, PsbV and PsbU

Cyanobacterial PS II contains three extrinsic subunits that are located at the lumenal side of the core complex: the 33 kDa protein (PsbO), the 12 kDa protein (PsbU) and the cytochrome c_{550} (PsbV) (see Fig. 12.4b).

PsbU and PsbV are unique to cyanobacteria, whereas PsbO is an essential constituent of all PS II complexes (cyanobacteria, algae, higher plants; for review, see (Bricker and Burnap 2005)). In the structure at 3.8 Å resolution, the main body of PsbO was identified as a ß-barrel structure, which was confirmed in all later structures. PsbO forms various contacts with CP43, CP47, D1 and D2 and is thereby important for the stabilization of the Mn_4O_xCa cluster, even if it does not provide a direct metal ligand. The protein subunit PsbV (Cyt c_{550}) is located at the side of the lumenal hump that faces away from the dimerization domain and is in close contact to the lumenal loops of CP43. It contains a heme with a midpoint potential of -80–90 mV (Roncel et al. 2003; Kaminskaya et al. 2005), but its function is unclear, as the reduction of the heme has not been reported. PsbV has strong structural similarity to Cyt c_6, the electron donor to PS I, and might represent the old electron donor to the non-oxygenic ancestor of PS II, which was trapped during evolution and became an extrinsic subunit of PS II that now stabilizes the oxygen evolving complex (Grotjohann et al. 2004). PsbU is a small protein that is located at the outermost lumenal tip of PS II. It may further stabilize the PS II complex.

12.2.3 Electron Transport Chain of PS II

The electron transfer chain of PS I contains the following cofactors (from the lumen to the cytosol/stroma): the 4MnCa cluster, two redox active tyrosines (Tyr_Z and Tyr_D), the primary electron donor P680, two accessory chlorophylls, 2 pheophytins, a tightly bound plastoquinone (Q_A) and a mobile plastoquinone (Q_B). PS II belongs to the larger class of type II RCs, where a mobile quinone serves as terminal electron acceptor of the ETC and the electron transfer is unidirectional in spite of a rather symmetric array of the pigments and quinones within the heterodimeric protein matrix. In PS II this heterodimer consists of the two central core proteins D1 (PsbA) and D2 (PsbD). The D1 subunit binds the Mn_4O_xCa cluster, Tyr_Z, P_{D1}, Chl_{D1}, Pheo $_{D1}$ and harbors the Q_B site of the mobile quinone Q_B. It also forms the quinone exchange pocket where Q_C is located. The protein D2 coordinates the Chls P_{D2} and Chl_{D2}, Pheo $_{D2}$ and the tightly bound quinone Q_A. Chl_{D2} and Pheo $_{D2}$ are located in the "inactive" branch and are not directly involved in the forward electron transfer.

12.2.3.1 Acceptor Side of PS II

The acceptor side of PS II is remarkably similar in all type II RCs.. The electron is transferred through the active branch from $^1P680^*$ to the tightly bound plastoquinone Q_A via pheophytin at the D1 site ($Pheo_{D1}$). From Q_A the electron is transferred to the mobile quinone Q_B. A non-heme iron is located between Q_A and Q_B, that has been shown not to be directly involved in electron transfer from $Q_A^{-\bullet}$ to $Q_B/Q_B^{-\bullet}$ but probably playing a structural function (for a discussion, see Renger and Renger

2008). Two sequential electron transfer steps with $Q_A^{-\bullet}$ as donor lead to double reduction of Q_B, followed by the binding of two protons (from the cytosolic/stroma side) to form PQH_2 which then leaves the binding pocket in exchange with a PQ molecule. The proposed existence of a third quinone binding site Q_C (Kaminskaya et al. 2007) was confirmed by the most recent X-ray structure (Guskov et al. 2009).

The head group of Q_C is located at a 17 Å distance to Q_B's head group. Eight lipids, which are forming a "mini-bilayer", are located in close vicinity to the Q_B and Q_C sites. The surrounding of these two quinones is shown in Figs. 12.4b, c. Two access portals have been proposed for exchange of the quinones with PQ pool. Based on the structure three different models have been discussed for the function of the quinone Q_C and the mechanism of the exchange of PQH_2 for a quinone from the PQ pool: the alternating, wriggling, and the single channel scenario.

In the alternating channel scenario, PQH_2 and PQ enter into the Q_B site via alternating pathways: PQH_2 leaves PS II through channel II, and the "waiting" plastoquinone in channel I takes its place in the Q_B binding site. This scenario would predict that the isoprenoid tail would still reside in channel I, while at the same time another PQ from the PQ pool moves into the channel II, waiting for the next turnovers where then PQH_2 leaves the Q_B site now via channel I. The fact that the tails of the Q_B and Q_C are well defined and Q_B is located in both monomers of PS II in channel II makes this scenario relatively unlikely, as it would require that all PS II molecules in the crystal have been trapped in the same conformation.

The second mechanism discussed is the wriggling mechanism. In this scenario, channel I is the entry channel and channel II is the exit channel. This means that plastoquinone PQ always enters through channel I and plastoquinol PQH_2 always leaves the Q_B site through channel II. This scenario would require the movement of the isoprenoid tail from channel I to channel II, and consequently the tail has to wriggle around two MGDG lipids and the phytyl tail of Chl_{D1}.

The third, "single channel" mechanism excludes the quinone in the Q_C site from PQH_2/PQ exchange. Only channel II is active and a single PQ enters the channel II after PQH_2 has left. This scenario is possible but does not provide a function for Q_C and channel I in PQH_2/PQ exchange. However it is in line with a regulatory role of Cyt b_{559} proposed by Kaminskaya et al. (2007).

At the moment one can not decide unambiguously between the 3 scenarios. Site directed mutagenesis in combination with functional studies may provide the results to unravel the mechanism of the PQH_2/PQ exchange.

12.2.3.2 Donor Site of Photosystem II

While the acceptor site of PS II is homologous in all type II RCs, the ability to use water as a donor for the electron transport chain makes the ETC of the PS II donor site unique among all enzymes on earth.

The current state of knowledge on the donor side of PS II has been described in several recent reviews (see Renger and Renger 2008 and references therein) and book chapters in Bricker and Burnap (2005) and Renger (2008). Therefore we will here only briefly summarize the general function and the structural features of the PS II donor side (for further details, see chapter of Renger and Ludwig).

P680

The cation radical of the primary donor in PS II, $P680^{+\bullet}$ is one of the most oxidizing cofactors found in biological systems (Ishikita et al. 2006) and the indispensable energetic prerequisite for the unique function of PS II: the oxidative water splitting. The midpoint potential of $P680/P680^{+\bullet}$ cannot be directly determined by redox titration. Therefore its value has been estimated by indirect methods and recently numbers of about +1.25 V were reported (Rappaport and Diner 2009). PS II has to pay a very high price for this high midpoint potential: the D1 protein in Photosystem II is easily damaged and has to be degraded and replaced in a plant in bright sunshine every half an hour. It is remarkable that Nature has not been able to solve this photodamage problem of PS II in the 2.5 billion years since PS II has been performing its function on our planet. Most of the current evidence points in the direction of the triplet state of $^3P680$ being the major contributor to the photodamage problem. Chl triplet states can be most efficiently quenched by carotenoids (Cars) (<1 ns) as nicely illustrated for the LHC II complex of plants (Schödel et al. 1998). This unique mechanism requires close proximity between P680 and Car. In the PS II RCs, however, the Cars have to be sufficiently far apart from $P680^{+\bullet}$ in order to prevent very rapid and dissipative oxidation. As a consequence quenching of $^3P680$ is significantly less efficient.

When looking at the structure of four Chls in PS II, the question arises which of the four chlorophylls represent(s) $^1P680^*$, $P680^{+\bullet}$ and $^3P680$?

Based on our current state of knowledge it is most likely that $^1P680^*$ is delocalized over the ensemble of the four Chls and two Pheo's, while $P680^{+\bullet}$ attains two different states, i.e. the inially formed $Chl_{D1}^{+\bullet}$ followed by very rapid electron transfer from P_{D1} (Barber 2002; Fromme et al. 2002; see also Di Donato et al. 2008).

We discussed above that $^3P680$ may be the primary cause of the photodamage of the D1 protein in PS II. Therefore it is interesting to know on which of the chlorophylls the triplet state of P680 is located. Spectroscopic evidence revealed that the triplet state of the primary donor $^3P680$ is located on a Chl that is tilted 30° against the membrane plane (van Mieghem et al. 1995 and references therein). This condition is only satisfied for Chl_{D1} and Chl_{D2}. As the triplet state is formed by recombination of the electron from one of the quinones to $P680^{+\bullet}$ and location of $P680^{+\bullet}$ has been shown to be on P_{D1}, the Chl_{D2} can be excluded, thus restricting the location of $^3P680$ on Chl_{D1}. It should be noted that no central ligand for this chlorophyll is visible in any of the X-ray structures, which makes it very likely that Chl_{D1} is coordinated by a water molecule.

The four chlorophylls are arranged in two symmetrically related pairs, P_{D1}/P_{D2} and Chl_{D1}/Chl_{D2}. The dimer P_{D1}/P_{D2} is oriented perpendicular to the membrane plane. A finding that confuses non-crystallographers is that the center-to-center distance between P_{D1} and P_{D2} as determined from the X-ray structures of PS II has decreased from 10 Å in the first structure of PS II at 3.8 Å resolution (Zouni et al. 2001), to 9.56 Å in the 3.7 Å structure (Kamiya and Shen 2003), to 8.6 Å in the 3.6 Å structure (Fromme et al. 2002). This value further reduced to 8.2 Å in the 3.5 Å structure (Ferreira et al. 2004) and 8.3 Å in the 3.2 Å (Biesiadka et al. 2004) structure and 8.0 Å in the structure at 3.0 Å resolution (Loll et al. 2005). The recent 2.9 Å structure of PS II showed 8.13 Å in monomer 1 (pdb code 3BZ1) and 8.11 in monomer 2 (pdb code 3BZ2) (Guskov et al. 2009). The decrease in distance with increase in resolution may be caused by a clearer assignment of the chlorin system at higher resolution, so that the center-to-center distance of 8.1 Å can now be stated with great confidence. It is very interesting to note that the chlorophylls P_{D1}/P_{D2} are more separated from each other in PS II than the Chls in P700 or the BChls in the special pair of PBRC. This larger distance implies a weaker excitonic coupling, so the Chls in P_{D1}/P_{D2} of PS II may have more "monomeric" character than the two Chls in P700 or the two BChls in P870 in the PBRC (for a review, see Renger and Renger 2008).

Water-Oxidizing Complex

Oxidative water splitting is catalyzed by the Mn_4O_xCa cluster which is bound to the protein D1. The PS II structure at 3.8 Å resolution unravelled for the first time the location of the manganese cluster and the arrangement of the four manganese in a 3+1 organization. This 3+1 arrangement of the four manganese was confirmed by all further improved crystal structures (Fromme et al. 2002; Kamiya and Shen 2003; Ferreira et al. 2004; Loll et al. 2005; Guskov et al. 2009) and is also in agreement with EPR and EXAFS studies (Kern et al. 2007; Peloquin and Britt 2001; Haddy 2007; Sauer and Yachandra 2004; Britt et al. 2004; see also Yano and Yachandra 2009 and references therein). The first evidence for the location of the Ca was provided by the 3.5 Å structure (Ferreira et al. 2004). Three manganese and the Ca^{2+} form a distorted cubane with the fourth manganese being more distal to the distorted cubane. The general structural arrangement of the Mn_4O_xCa cluster is in agreement with conclusions gathered from EPR and EXAFS data [for a more detailed discussion see (Sauer and Yachandra 2004; Yano and Yachandra 2009; Haddy 2007) and references therein]. The arrangement of the metal cluster based on the 2.9 Å structure (Guskov et al. 2009) is shown in Fig. 12.4.

Chloride has long time been suggested to be an essential part of the WOC (for a review, see van Gorkom and Yocum 2005) but is has only recently been identified in the X-ray structures of PS II (Murray et al. 2008; Kawakami et al. 2009; Guskov et al. 2009). These studies revealed that Cl– is not a direct ligand to manganese but probably part of a functionally relevant hydrgen bond network, as originally proposed by Olesen and Andréasson (2003).

Despite marked improvements in the overall resolution of the PS II structure and the protein environment, the resolution of the electron density of the Mn_4O_xCa cluster has not significantly improved since the very first crystal structure of PS II at 3.8 Å resolution in 2001. It should also be noted that an omit map with the model of the centers of the Mn_4O_xCa cluster, published in the supplement info of the 3.5 Å structure of PS II, implies an accuracy of the position of the Mn atoms which is strongly misleading. At this resolution any atom placed in the electron density (ED) map of the Mn_4O_xCa cluster at totally arbitrary position will show up as a spherical density in the omit map (Berry, personal communication). Therefore interpretations of the omit maps at this resolution should be considered with great caution and neither be published nor be used for any modeling.

The limited resolution of the crystal structure, however, is only one part of the problem of the still fragmentary knowledge on the structural details of the Mn_4O_xCa cluster. Even more important are the effects due to damage of the Mn_4O_xCa cluster by X-rays during data collection. It has been shown that the manganese in the cluster becomes rapidly damaged and reduced to Mn^{2+} during X-ray exposure (Yano et al. 2005). The distal Mn4 (often also designed Mn_A) might be the most sensitive manganese to disassemble from the cluster during X-ray damage, as it shows less than 30% occupancy in the 3.0 Å X-ray structure (Loll et al. 2005).

The determination of an undisturbed structure of the cluster by overcoming the X-ray damage problem and further improvement in resolution is absolutely essential for the determination of the structure of the Mn_4O_xCa cluster to levels that permit reliable mechanistic considerations. It must be clearly stated that none of the current structures resolves the position of individual manganese centers or provides evidence for their mode of bridging by oxygen. Likewise, the mechanistically most important ligation by water molecules is not visible in any of the X-ray structures, except of one water molecule which has so far been identified in the neighbourhood of Cl^- and Mn4 in the 2.9 Å structure of PS II (Guskov et al. 2009).

Potential ligands to the Mn_4O_xCa cluster include Asp 170, Glu 189, His 190, Asp 342, His 332, Glu 333 and the N-terminus of D1 (Ala 344). In addition Glu 354 of CP43 also provides a ligand to the cluster. Although similar amino acids are close to the Mn_4O_xCa cluster in all structures, the positioning of the amino acids differs slightly. Recently, FTIR spectroscopy studies have been performed on mutants where nearly all putative ligands to the Mn_4O_xCa cluster are replaced by other amino acid residues. Almost all mutants remained still functionally competent and showed no significant difference in the FTIR spectra (Strickler et al. 2006, 2007; Debus 2008), except the mutation CP43-Glu 354 to Gln (Strickler et al. 2008). This finding caused a lot of confusion. One reasonable explanation for this puzzle might be offered by the recent resonant inelastic X-ray scattering (RIXS) results which show that the charge is likely to be delocalized over the cluster (Yano and Yachandra 2008). Therefore, the change of one of the ligands may cause only a slight redistribution of charge and spin among the manganese centers of the cluster.

The mechanism of water splitting and the oxidation states of the Mn_4O_xCa cluster are a very hot topic of discussion and so far no model exists that can explain all experimental data. A more detailed description of current models on the mechanism is presented in chapter Renger and Ludwig. It is the hope that a higher resolution X-ray structure of the undamaged Mn_4O_xCa cluster in combination with spectroscopic results and computational modelling may be able to solve the secrets of the process of light-driven oxidative water splitting leading to formation of molecular dioxygen. This mechanism is still the "holy grail" of photosynthesis that waits to be fully discovered.

12.2.4 Antenna System of PS II

The core antenna system of the PS II monomer from *T.elongatus* contains 29 Chl *a* and 10 Car molecules in CP43/CP47. The antenna system is very well separated from the reaction center domain. There is no clear "entrance gate" of the excitation energy transfer into the PS II core. Six Chls are located in moderate distance of about 18–22 Å to the Chls of the electron transfer chain. Interestingly the two Chls coordinated by D1 and D2, $ChlZ_{D1}$ and $ChlZ_{D2,}$ are not the Chls that are located at the closest distance to the electron transfer chain. Indeed they are quite isolated and may not interact strongly with the clusters of Chls coordinated by CP43 and CP47. The clear spatial separation between the pigments of the antenna system and the RC core, which is necessary to prevent rapid oxidation by $P680^{+\bullet}$, explains why excitation energy transfer in the PS II antenna system is probably "transfer to the trap" limited. However, this question is still a matter of controversial discussion (Miloslavina et al. 2006; Broess et al. 2008; for a review, see Renger and Renger 2008).

12.2.4.1 Chlorophylls

The PS II core complex of *T.elongatus* contains 35 structurally identified Chls; 29 of these are coordinated by the proteins PsbB and PsbC which provide the axial ligands for 16 and 13 Chl *a* molecules, respectively (Loll et al. 2005; Müh et al. 2008)]. Position and location of these Chls is very well defined in the X-ray structures and all except three of them had already been identified in the first 3.8 Å resolution structure of PS II (Zouni et al. 2001).

The 16 Chls bound to CP47 and the 13 Chls bound to CP43 are arranged in two layers, close to the stromal and lumenal side of the membrane. This layer structure can also be found in the peripheral part of the PS I antenna (see Fig. 12.6). Thirteen of the 16 Chls assigned to CP47 are coordinated by His ligands, one is coordinated by the backbone carbonyl atom of a tryptophan residue, one is coordinated by the ester carbonyl group of a lipid (MGDG) via a potential water molecule, for one Chl no ligand has been assigned and it is assumed that it is coordinated by a water molecule (Müh et al. 2008). Of the 13 chlorophylls bound to CP43, 10 are coordinated by His ligands, one is coordinated by a Ser residue (possibly involving a water molecule), one is coordinated by the carbonyl side chain group of an Asn residue, and one chlorophyll is coordinated by the ester carbonyl group of a lipid (DGDG) (Müh et al. 2008). It is very astonishing that—in contrast to PS I—none of the small membrane intrinsic subunits of PS II directly coordinates any of the antenna chlorophylls.

PS I and PS II exhibit some homologies of the position and location of the Chls bound by CP47 and CP43, confirming the suggestion that the N-terminal part of PS I and CP47/CP43 may have evolved from a common ancestor (Schubert et al. 1998). The major difference of the antenna systems in PS II and PS I is the lack of the central antenna in PS II (see Fig. 12.6 for comparison of the antenna systems in

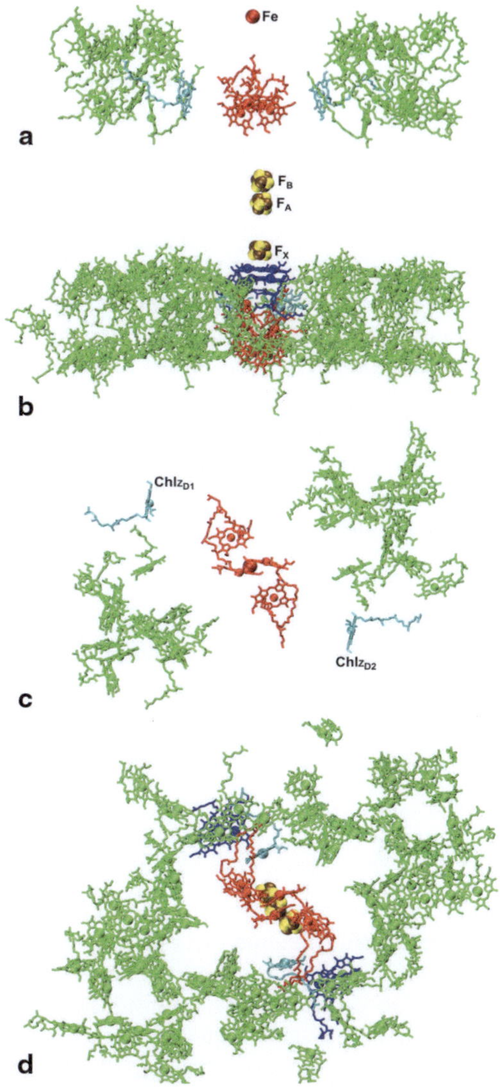

Fig. 12.6 Comparison of the antenna systems of monomeric photosystems I and II from *TS elongatus*. **a**, **c** Photosystem II; **b**, **d** Photosystem I; **a**, **b** views from within the membrane plane; **c**, **d** views from the lumenal side. Only chlorophylls are shown, with the PS II non-heme iron and the PS I iron sulfur clusters included to facilitate orientation. Chlorophylls in electron transfer chains are red, so-called "connecting chlorophylls" in PS I and $Chlz_{D1}$ and $Chlz_{D2}$ in PS II are cyan, "connecting dimers" in PS I blue. The comparison shows that PS I contains an extensive, uninterrupted chlorophyll network, whereas the electron transfer chain of PS II is isolated by a big gap, possibly to prevent the high redox potential of P_{680} from damaging other chlorophylls. In both protein complexes, the chlorophylls are clearly organized in two layers near the membrane surfaces, bridged by only a few chlorophylls and of course the ETC's

PS I and PS II). In PS I, 50 antenna Chls are located in this central domain surrounding the ET chain. In contrast, the central PS II core contains only two chlorophylls ($ChlZ_{D1}$ and $ChlZ_{D2}$), which are coordinated by the D1 and D2 proteins.

The lower efficiency of the excitation energy transfer in PS II compared to PS I is caused by the lack of the central antenna. In PS II, the quantum efficiency of excitation energy transfer is only ~90% (Hou et al. 2001; Vasil'ev et al. 2004) compared to >97% efficiency of excitation energy transfer and trapping in PS I (Gobets et al. 2001; Damjanovic et al. 2002; Sener et al. 2002). The lower efficiency of the antenna system in PS II is the reason for the observation that the chlorophyll fluorescence spectrum of photosynthetic organisms at room temperature is dominated by fluorescence from PS II (for a review, see Papageorgiou and Govindjee 2004).

Why does PS II not contain a central antenna domain? Concerning the antenna system, PS II may represent an earlier evolutionary state, where antenna and reaction center were not tightly coupled. In PS I the antenna domain and the reaction center domains are fused to form a more efficiently joint reaction center and antenna system. The reason why PS II has not developed a core antenna system analogous to PS I is very likely due to the process of photodamage and repair. PS II has to pay a high price for its ability to use water as an unlimited electron source, as it is sensitive to photodamage (see Sect. 12.2.3.2). The photodamage of the D1 protein requires the frequent replacement of the protein in a complex process that involves monomerization of the PS II dimer, phosphorylation of the D1 protein, plus disassembly of CP43, all three membrane extrinsic proteins, as well as several of the small intrinsic membrane proteins of the PS II core complex. This constant need for repair of the D1 protein may have hindered the development of a closely coupled core antenna system in PS II. At present, nothing is known about the fate of the Chls bound to D1 during the repair cycle, but if PS II would contain a central antenna domain, all central Chls bound to D1 would have also to be replaced together with the D1 protein.

The repair cycle of D1 is one, but not the only possible explanation for the lack of a tightly coupled antenna system. Another important reason for a clear separation of the antenna and reaction center domain is the high midpoint potential of $P680/P680^{+\bullet}$. The location of many antenna chlorophylls close to the highly oxidizing P_{680} could potentially caused severe damage of the antenna pigments.

12.2.4.2 Carotenoids

The 3.0 Å structure of Photosystem II (Loll et al. 2005; Müh et al. 2008) unravelled the position and location of 11 Cars, and one additional Car has been located at the D2 site of PS II in the 2.9 Å structure of PS II (Guskov et al. 2009). All Cars were modeled as β-Car, which do not mean that all carotenoids in PS II are ß-Car molecule. They were only modelled as β-car molecules due to the limitations in resolution of the X-ray structure. The same is true for PS I, where so far all Cars have been modelled as β-Car. There are no striking homologies between the arrangements of the carotenoids in PS I and PS II. In PS II, the arrangement of Cars is much more symmetrical than in PS I.

All Cars in PS II are located in the antenna domain of PS II and only two Cars are associated with the D1 and D2 core proteins. The top view of the cofactor arrangement (Fig. 12.2e) shows that none of the Cars is in close distance to any of the Chls of the ETC. This lack of Cars in the RC domain is in clear contrast to the PBRC, where a carotenoid molecule is located in close distance to the primary donor (Deisenhofer et al. 1995) so that it can efficiently quench dangerous ^3P triplet states. PS I also contains Cars in the central RC domain (Jordan et al. 2001), see also Fig. 12.5e. The lack of any Cars in the central RC domain of PS II may be the reason for the photodamage of PS II caused by ^{3}P680. This raises the question why PS II lost these essential Cars during evolution, a lack which makes it now so vulnerable to photodamage. The reason for the lack of Cars in the RC of PS II is very likely caused by the high midpoint potential of P680/P680$^{+\bullet}$ as outlined in Sect. 12.2.3.2.

Most of the Cars in the antenna system of PS II are oriented roughly perpendicular to the membrane plane, only three are oriented parallel to the membrane plane (Müh et al. 2008; Guskov et al. 2009). The Cars thereby may serve in the functional connection of the lumenal and cytosolic/stromal layers of the antenna system.

Quenching of Chl triplet states is one of the main functions of carotenoids in PS II. They can also function as an additional antenna pigments. In addition Cars may also serve as stabilizing elements in PS II together with the lipids. Carotenoids bridge CP43 with the cluster of small subunits at the periphery of PS II (PsbK and PsbZ) as well as PsbI at the inner site of the PS II dimer. Carotenoids at the monomer-monomer interface may contribute to the stabilization of the dimer by "bridging" CP 47 with PsbT and the first transmembrane helix of the D1 protein from the other monomer.

The two Cars that are associated with D1 and D2 are at symmetry-related positions, but have opposite orientations. Car_{D1} is located between helix 1 of the D1 protein and the dimer-dimer interface. It is orientated nearly perpendicular to the membrane plane. Car_{D2}, is located between helix 1 and the Cyt b_{559}, in close vicinity to $ChlZ_{D2}$, and is orientated nearly parallel to the membrane plane. These two Chls may represent the species that have been spectroscopically identified by (Kwa et al. 1992). Based on the 3.0 Å structure of PS II (Loll et al. 2005), the authors proposed that Car_{D1} may represent Car489 and Car_{D2} may correspond to Car507. It has been speculated that Car_{D1} may have the function of quenching triplet states of $ChlZ_{D1}$.

Cars cannot only quench Chl triplets but also reactive singlet dioxygen which is formed via their sensitized reaction with triplet ground state dioxygen.

Furthermore, Car_{D2} was also proposed to play a role in the photoprotective processes of charge recombination between singly reduced Q_B and P680; however this hypothesis may be questioned due to the large distance to the Chls of the RC core.

12.2.4.3 Lipids

Photosystem II has the highest lipid content from any of the membrane proteins, for which a structure has been determined so far (Guskov et al. 2009). Already the 3.0 Å structure revealed that lipids are constituents of PS II. The lipids are bound at the interface between the D1/D2 core, the antenna proteins CP43 and CP 47 and two domains of transmembrane helices from small subunits (PsbM, L and T at the dimerization domain) and (PsbF, J and K) at the membrane exposed surface of PS II. They may provide the lubricant for the smooth replacement of the D1 protein (Loll et al. 2005, 2007). The removal of damaged D1 proteins and subsequent insertion of the newly synthesized D1 proteins may be facilitated by allowing them to slide easily in and out of their place.

The extremely high lipid content of PS II is one of the most important differences between PS I and PS II. 25 lipid molecules have been identified in the 2.9 Å structure of PS II (Guskov et al. 2009), i.e. 11 molecules in addition to the 14 lipids identified in the structure of PS II at 3.0 Å resolution (Loll et al. 2005). So far only four lipids have been identified in PS I, based on the structure at 2.5 Å resolution, even if it cannot be excluded that more lipids might be visible in an X-ray structure of PS I at higher resolution.

The role of lipids as essential cofactors in PS II has been known for a long time and lipids have been assigned to play a vital role in the function and structure of PS II (Dörmann and Hölzl 2009; Wada and Mizusawa 2009). It was shown by mutagenesis studies that phosphoglycerol (PG) plays a role in electron transfer between Q_A and Q_B (Gombos et al. 2002) and digalactosyldiacylglycerol (DGDG) affects the reactions on the donor side of PS II (Steffen et al. 2005). Furthermore, lipids were also found to be absolutely essential for the dimerization of PS II (see Jones 2007; Fyfe and Jones 2005 and references therein).

The most recent structure of PS II at 2.9 Å resolution unveiled that the most prominent lipids in PS II are the glycolipids MGDG and DGDG (Guskov et al. 2009). PS II contains 11 MGDG molecules, and 7 DGDG molecules. All the head groups of the 7 DGDG molecules are located at the lumenal side of PS II, while the head groups of MGDG are found to be localized on both sides of the membrane. In addition to the glycoclipids, PS II contains 7 charged lipids, 5 of them are the sulfolipid sulfoquinovosyldiacylglycerol (SQDG) and only 2 are phospholipids. The head groups are located at the stromal side of the membrane for all seven charged lipids.

The lipid composition of PS II roughly reflects the distribution of lipids in the native membrane of the closely related thermophilic cyanobacterium *T. vulcanus*, which contain 45% MDGD, 25% DGDG, 15–25% SQDG and 1–15% PG (Sakurai et al. 2006). It is remarkable that the lipids are not interchangeable as they can be unambiguously identified at unique positions in the X-ray structure.

The lipids form 2 clusters at the monomer-monomer interface and a large cluster at the Q_B/Q_C binding pockets, which may represent the PQ/PQH_2 exchange pocket. Other lipid clusters fill a pocket at the interface between CP47 and D2/PsbH/PsbX and a pocket at the interface between CP43 and PsbJ/K. Another cluster is located at the periphery of PS II at the interface between PsbX, PsbY, and Cyt559 (PsbE/F). The most remarkable feature of all lipid clusters can be found for the cluster of 8 lipids that is located at the PQ-PQH_2 exchange pocket. This cluster forms a membrane like arrangement, which may be very important for the PQ-PQH_2 exchange in the catalytic cycle of PS II.

As the lipids of the thylakoid membrane play an important role in the repair cycle of PS II (Kanervo et al. 1995, 1997), the lipids at the interface to CP43 and at the monomer to monomer interface may have a similar function, as they may provide a “lubricant” that allows for fast disassembly/assembly of PS II during the repair cycle. This repair cycle involves monomerization and detachment of CP43 and peripheral subunits from the core of PS II to allow for degradation of the damaged D1 protein and the co-translational re-insertion of a new D1 protein into the

D2-CP47-Cyt559 core complex of PS II. The importance of lipids for the PQ/PQH_2 exchange of PS II and the repair process would also explain the much lower lipid content of PS I, as PS I does not have to undergo excessive repair in the living cell and the electron transfer is mediated by tightly bound quinones that do not exchange during the catalytic cycle.

12.3 Photosystem I

12.3.1 Overview of the Structure of Cyanobacterial Photosystem I[3]

The prevalent oligomeric form of PS I in cyanobacteria is a trimer (Ford and Holzenburg 1988; Rögner et al. 1990), but under high light intensities or stress, PS I can also monomerize in the membrane. In plants PS I is always monomeric connected with four LHC I complexes serving as the peripheral antenna (Amunts et al. 2007).

It has been argued that the trimeric PS I of cyanobacteria is the "ur" form of PS I and the trimeric structure has been disrupted later in plants to allow the attachment of the LHCs (Amunts and Nelson 2008; Nelson 2009). An alternate scenario might be more likely, where the "ur" PS I in ancient cyanobacteria was monomeric and the trimeric form is a more recent evolutionary "improvement" that allows the cyanobacterial PS I to establish a larger coupled antenna system. An indication for this idea comes from the recent findings on PS I from an ancient thermophilic and acidophilic red alga. This red alga contains a monomeric PS I that shows chimerical features of both, cyanobacterial and plant PS I (Vanselow et al. 2009).

So far, the structure of cyanobacterial PS I has only been determined from one organism, the thermophilic cyanobacterium *T. elongatus* (Jordan et al. 2001; Fromme et al. 2001). The trimeric PS I has a molecular weight of 1,080,000 Da and, thereby, is the largest and most complex membrane protein that has been crystallized so far (Fromme 1998). The structure has been improved from the first model at 6 Å resolution (Krauss et al. 1993) to 4 Å resolution (Krauss et al. 1996; Klukas et al. 1999a, b). The currently best model is based on the structure of PS I at 2.5 Å resolution (Jordan et al. 2001; Fromme et al. 2001). The structure of PS I can be described as a clover leaf with a diameter of 220 Å. This structural model in a top view from the lumenal side is shown in Fig. 12.3b and from the stromal side in Fig. 12.5a. Each monomer "leaflet" contains one electron transport chain and consists of 12 proteins, 9 of which are membrane intrinsic. Three peripheral protein subunits extend the surface of PS I by 40 Å at the stromal side and provide the docking site for ferredoxin/flavodoxin. At its lumenal side, PS I is relatively "flat" and extends the membrane by only about 15 Å. An indentation at the lumenal side in close proximity to P700

[3] For a historical review of structural investigations of Photosystem I that precede the X-ray structure, see Witt (2004).

provides the docking site for PC or Cyt c_6. A unique feature of PS I is its high cofactor content: one monomeric unit of PS I contains 127 cofactors, which thereby contribute approx 30% to the total mass of PS I. All cofactors are non-covalently bound to the protein subunits.

PS I can be regarded as a joint form of a reaction center and core antenna complex. The core of PS I formed by PsaA/B heterodimer is surrounded by 7 small membrane intrinsic proteins (PsaF, PsaI, PsaJ, PsaK, PsaL, PsaM and PsaX), which contain only 1–3 transmembrane helices each. The stromal hump is built by three subunits (PsaC, PsaD and PsaE) that provide the docking site for ferredoxin/flavodoxin.

12.3.2 Protein Subunits of PS I

12.3.2.1 Large Subunits PsaA and PsaB

PsaA and PsaB bind the organic cofactors of the ETC and 79 of the 90 antenna Chls. While all type II RCs (including PBRCs, non green sulphur bacterial RCs and PS II) are hetero-dimeric, type I RCs in non-oxygenic bacteria (heliobacteria and green sulphur bacteria) are all homodimeric (Raymond et al. 2006). This leads to the conclusion that the hetero-dimeric nature of PS I is very likely a relatively recent evolutionary event (Mix et al. 2005). PsaA and PsaB show a large sequence homology to each other, and the membrane intrinsic core of PsaA/PsaB is very symmetric, while larger differences are found in the loop regions.

PsaA and PsaB form 11 transmembrane helices (TMHs) each (see Figs. 12.3b and 12.5b). The majority of the cofactors of the electron transport chain (P700, A_0, A_1, and F_X) are coordinated by PsaA and PsaB. In addition, they provide the central ligand for 79 of the 90 antenna chlorophylls in PS I and form hydrophobic interactions with most of the Cars. PsaA and PsaB can be structurally divided into two distinct domains: The c-terminal domain of PsaA and PsaB consist of 5 transmembrane helices each and encircles the electron transport chain like a fence (see Figs. 12.3b and 12.5b). The N-terminal domains of PsaA/PsaB consist of six transmembrane helices each. This twinned pair of six helices each flanks the core of the reaction center on both sides and provide the core antenna system of PS I. The arrangement of the C-terminal domain resembles the D1/D2 core of PS II (see Sect. 12.2.2.1), while the N-terminal 6 helices of PsaA and PsaB show structural and functional homologies to the CP43 and CP47 antenna proteins of PS II (see Sect. 12.2.2.2). The similarity of the helix arrangements strongly supports the hypothesis that PS I and PS II have evolved from a common ancestor (Schubert et al. 1998).

C-terminal Domain of PsaA/PsaB

The 5 transmembrane helices of the C-terminal domain surround the cofactors of the ETC. The C-terminal domain coordinates the 6 chlorophylls of the ETC as well as the two phylloquinones and the first FeS center, F_X. The helices of the C-terminal domain of PsaA and PsaB are arranged in two semi-circles. The TMH 10 is located in the center of the complex and provides the central ligands for the primary donor P700 and the first stable electron acceptor A_0, located in the middle of the membrane. Following the semi-circle from the center to the periphery the next helix in line is TMH 11, followed by TMHs 9, 8 and 7. The phylloquinones are bound in two hydrophobic pockets formed by the loops between TMHs 10 and 11. The coordination of the first FeS cluster F_X is provided by four cysteines, two from PsaA and two from PsaB. F_X is thereby a rare example of an inter-protein FeS cluster. The loop region that coordinates F_X is located between TMHs 8 and 9 and represents the region with the highest sequence conservation between PsaA and PsaB.

Antenna Domain of PsaA/PsaB

A unique feature of PS I that also differs from PS II is the tight structural and functional coupling of the reaction center core and the integral antenna system, which indicates that PsaA/PsaB in PS I form a joint entity of RC and core antenna system. The six N-terminal TMHs of PS I coordinate the 40 Chls in the peripheral part of the core antenna domain of PS I. The six helices are arranged in form of a "trimer of dimers" where pairs of helices are formed by TMHs 1/2, TMHs 3/4, and TMHs 5/6. The helix pair 5/6 provides protein-protein interactions with the C-terminal core domain of PS I, and the helix pair 1/2 is located at the periphery of PS I. This arrangement of helices is very similar to PsbC and PsbB in PS II (see Figs. 12.1b and 12.2b). The N-terminal domain of PS I also shows interactions with Cars and four lipids. A second function of the N-terminal domain is the interaction with the small membrane intrinsic subunits. However, most of these interactions do not involve direct protein-protein contacts, but are pigment-mediated, thereby better described as protein-cofactor-protein interactions. The C-terminal domain is also involved in the antenna function of PS I. It has a dual function: in addition to the coordination of most cofactors of the ETC, it also coordinates 25 out of 90 antenna chlorophylls in PS I. A further unique feature is that also the lumenal and stromal loops of PsaA and PsaB contribute to the coordination of chlorophylls in PS I.

Loop Regions of PsaA/PsaB

The TMHs of PsaA and PsaB are highly homologous between PsaA and PsaB and show a nearly perfect two-fold symmetry. The center of this symmetry is located at an axis that crosses the first FeS cluster F_X, which indicates that F_X may also play an important role in the assembly of the core of PS I. The high symmetry of the transmembrane part of PsaA and PsaB is contrasted by large deviations from the twofold symmetry in cytosolic/stromal and lumenal loop regions of PsaA and PsaB, which differ in sequence, length and also contain varying secondary structural elements. Thereby, the loops induce elements of asymmetry into the system. These differences are especially important for the docking of the three stromal subunits, PsaC, PsaD and PsaE. Simulations of the docking of PsaC to the PS I core have indicated that they might provide the clue for the unique orientation of PsaC with regard to the core of PS I (Jolley et al. 2006; Jagannathan and Golbeck 2009a, b).

PsaA and PsaB also play a key role in the docking of the soluble electron donors to PS I. The docking site for plastocyanin and Cyt c_6 is located at the lumenal site of PS I at a shallow cleft, which allows access of the electron donor to P700. The major interaction site of PS I with PC/Cyt c_6 is provided by two helices in the loop between TMHs 9 and 10 (Grotjohann et al. 2004). The helices are the only barrier between P700 and the aqueous surface. They are mainly hydrophobic and contain two tryptophan residues that "stick out" into the aqueous phase and may directly interact with the soluble electron carriers PC/Cyt c_6 (Sommer et al. 2004). Whether they are directly involved in electron transfer is still under discussion. In cyanobacteria, only PsaA and PsaB are involved in the docking of PC/Cyt c_6, however in plants also PsaF interacts electrostatically with PC (Hippler et al. 1996, 1998; Ben-Shem et al. 2003).

12.3.2.2 Peripheral Integral Small Subunits in PS I

PS I contains seven small membrane intrinsic subunits. These protein subunits contain between one and three TMHs and surround the core of the transmembrane domain of Photosystem I.

The arrangement of the small subunits in PS I and PS II show no homology. The three subunits PsaI, PsaL and PsaM are located in vicinity to the center of the PS I trimer, thereby shielded from the membrane in the PS I trimer and form the trimerization domain at the monomer-momomer interface. These subunits are easily lost when PS I monomerizes and are present in sub-stochiometric amounts in PS I monomers (Jekow et al. 1995, 1996). Four subunits, PsaF, PsaJ, PsaK and PsaX, are located at the membrane exposed surface of the trimeric PS I.

Monomer-Monomer Interface Subunits: PsaL, PsaI and PsaM

PsaL is located at the center of the trimerization domain and forms most of the contacts between the monomers. There are some hydrophobic interactions involved, but the majority of the contact sites between the monomers in the trimerization domain are provided by hydrogen bonds and electrostatic interactions within the loop regions of the three adjunct PsaL subunits. In addition, three Ca^{2+} binding sites are located at the trimer interface. The ligands for Ca^{2+} are provided by PsaL subunits from neighbouring monomers and PsaA. The Ca^{2+} binding may further stabilize the PS I trimer. The protein interactions in the trimerization domain of PS I are completely different from protein interactions in the dimerization domain of PS II, where a leucine/isoleucine "zipper" is formed by the two PsbM subunits of adjunct monomers and provides the only protein-protein interaction in the dimerization domain of PS II (see section "Subunits of the Dimerization Domain"). The further stabilization of the trimerization domain in PS I by electrostatic interactions and H-bonds may contribute to the much higher stability of the PS I trimer compared to the PS II dimer in cyanobacteria.

PsaI is involved in stabilization of the PS I trimer (Schluchter et al. 1996). It contains only one TMH and is located between PsaL and PsaM. It has been shown that the deletion of PsaI destabilizes the trimer, but the deletion does not completely hinder the trimer formation (Schluchter et al. 1996), which is supported by the structure, which shows that it forms few contacts with the adjacent monomer. PsaI does not coordinate Chl *a*, but forms hydrophobic interactions with carotenoid molecules.

PsaM contains only one transmembrane α-helix. With a MW of 3.4 kDa, it is the smallest subunit of PS I. It is located at the monomer/monomer interface, and forms interactions with PsaI and PsaB. PsaM coordinates one Chl that belongs functionally to the next monomer and may be important for the excitation energy transfer between monomers (Sener et al. 2004). Until recently, PsaM was regarded to be a protein subunit that is unique to cyanobacteria. Some plants contain an open reading frame for PsaM, but the protein was not expressed and has not been identified in protein preparations of PS I from plants or green algae. However, PsaM was recently indentified in the thermophilic, acidophilic red alga *Galdieria sulphuraria* and is present in the PS I-LHC I complex from this organism (Thangaraj et al. 2011), which indicates that PsaM can also be found in eukaryotic PS I complexes.

It is interesting to note that PsaI and PsaL are also present in plant PS I (Amunts et al. 2007). They are located at the same position in plant and cyanobacterial PS I, despite the fact that they show only moderate sequence homology. This indicates that the arrangement of the small subunits is a motif that was conserved during evolution (Andersen and Scheller 1993; Janson et al. 1996), despite the fact that they serve different functions in plants and cyanobacteria. While PsaL forms the trimerization domain in cyanobacteria, in higher plants, PsaI and PsaL interact with PsaH and may form interactions with the Light Harvesting Complex II (LHC II) under favorable light conditions (Andersen and Scheller 1993; Janson et al. 1996; Scheller et al. 2001; Kouril et al. 2005a, c). In the red alga *Galdieria sulphuraria,* PsaL and PsaI exhibit a chimeric structure that lacks the C-terminal helix which contributes to the interactions between the PsaL subunits in the trimer, and it also lacks any conservation of the subunits that interact with PsaH in plant PS I (Vanselow et al. 2009). These findings further support the idea that the "ur" from of cyanobacterial PS I was a monomer and that PS I in the red alaga Galdieria may represent one of the most ancient forms of PS I that still exist today on Earth. In this scenario, the PS I-LHC complex found still in plants today would be more ancient than the PS I trimer that is dominant in the present cyanobacteria. PsaI, PsaL, and PsaM may have just served in the stabilization of the antenna system in the ancestor of the current cyanobacteria and have resumed their function in the trimerization domain. The trimerization domain might have evolved at a time when cyanobacteria had to fight competition by eukaryotic green and red algae and plants in the aquatic environment.

Membrane Exposed Integral Subunits: PsaF, PsaJ, PsaK and PsaX

Four small hydrophobic protein subunits (PsaF, PsaJ, PsaK and PsaX) are located at the detergent exposed surface of PS I (see Fig. 12.5a, d). These four proteins may have two functions: stabilization of the core antenna system of PS I and formation of interactions with the integral peripheral antenna system, the IsiA ring (Kouril et al. 2005a, b; Bibby et al. 2001; Boekema 2001). PsaF may also be involved in the interaction with the phycobilisomes, and deletion of PsaF may alter the redox potential and oxidation kinetics of the phylloquinone at the A-branch (Yang et al. 1998; Van der Est et al. 2004).

PsaF, PsaJ and PsaK are present in PS I from all oxygenic phototrophic organisms, but PsaX is unique to thermophilic cyanobacteria. The gene for PsaX has so far only been identified in thermophilic cyanobacteria. PsaF and PsaJ are located at the periphery of the PS I trimer at the opposite side of the trimerization domain. PsaK is only loosely bound to the PS I core and is located at the "peripheral tip" of the PS I monomer at the site of PsaA.

PsaF is the largest of the small membrane intrinsic subunits. It can be structurally divided into three domains: an N-terminal region that is membrane extrinsic and located in the lumen, the transmembrane domain, and the C-terminal region that is located in the cytosol/stroma and forms interactions with PsaA and the subunit PsaE.

The N-terminal domain undergoes processing, as the primary sequence contains the target sequence for the thylakoid lumen. This domain contains two α-helices, which are hydrophilic and are oriented parallel to the membrane plane. In plants, these helices are extended by 25 amino acids and are actively involved in docking of plastocyanin via electrostatic interactions (Fischer et al. 1998). It is very interesting to notice that the red alga *Galdieria sulphuraria* contains this plant type extra domain despite the fact that it uses Cyt c_6 as the sole electron carrier. In contrast to higher plants and green algae, where the gene for PsaF is located in the nucleus, the *psa*F gene is located in the chloroplast in *Galdieria*. This finding implies that this domain may have already been developed in cyanobacteria for the Cyt c_6 interaction and is not really a "plastocyanin binding domain", but may have originally been introduced by cyanobacteria for Cyt c_6 docking before or shortly after the second endosymbiotic event that led to the development of red algae.

However, in all currently identified cyanobacteria, PsaF plays no role in docking of cytochrome c_6 or plastocyanin, because the large distance of 15 Å to the putative docking site of cytochrome c_6 does not favour a direct role of PsaF in the docking process of the lumenal electron carriers. Site directed mutagenesis also supports this conclusion, as deletion of PsaF has no influence on the electron transfer between PC and PS I (Xu et al. 1994; Hippler et al. 1999).

The transmembrane part of PsaF has a very unusual fold, consisting of a single transmembrane helix and two short hydrophobic helices forming a V-shaped structure that penetrates the membrane by 10 Å. The two V-shaped helices are the most peripheral part of PS I; they are embedded into the outer leaflet of the membrane.

It is very remarkable that they contain 2 lysine residues which stick into the membrane. PsaF is in contact with 6 Cars and 10 Chls via hydrophobic interactions, but does not provide a central ligand for Chl. There is evidence that PsaF may also play a role in the coupling of peripheral antenna systems to PS I both in cyanobacteria and in plants. A *T. elongatus* mutant deficient in PsaF (Muhlenhoff and Chauvat 1996) was not able to grow under low light (P.Fromme unpublished). This mutant expressed large amounts of allophycocyanin under low light conditions, which gave indications for the speculation that PsaF is important for docking of phycobilisomes to PS I (Fromme et al. 2003). PsaF may also be important but is not absolutely essential for the interaction with the IsiA complex, as the PS I-IsiA supercomplex is less stable in the PsaF deletion mutants and comprises 17 instead of 18 IsiA subunits (Kouril et al. 2005a, b). The crystal structure of plant PsaF shows that it interacts with the LHC I proteins (Ben-Shem et al. 2003; Amunts et al. 2007) and it has also been shown that it can be isolated as a Chl–protein complex with LHC I proteins (Anandan et al. 1989).

PsaJ contains only one transmembrane α-helix and is located in close vicinity to PsaF at the periphery of the PS I trimer. It coordinates three Chls and is in hydrophobic contact with several Cars. PsaJ is important for the stabilization of the pigment clusters, located at the interface between PsaJ/PsaF and the PsaA/PsaB core. Furthermore, the Chls coordinated by PsaJ may directly couple with IsiA proteins under iron deficiency.

PsaK contains 2 transmembrane helices. It is located at the periphery of the PS I complex, in close vicinity to PsaA. The two transmembrane α-helices are connected in the stroma, so that both the C- and N-terminus point into the lumen. PsaK coordinates two Chls and forms various contacts with Cars. PsaK is loosely bound to the PS I core and may even shift its position to facilitate the interaction with IsiA in the IsiA-PS I complex formed under iron deficiency (Chauhan et al. 2011).

PsaX contains only one transmembrane helix. It may be unique to thermophilic cyanobacteria (Koike et al. 1989). No gene for PsaX has been identified in mesophilic cyanobacteria yet. PsaX is located in close vicinity to one lipid, which forms various hydrophobic contacts with PsaX. In addition, PsaX coordinates one chlorophyll and forms hydrophobic contacts with several Car molecules. Potential roles for PsaX may include the stabilization of thermophilic PS I and it may also play a role in the interaction of PS I with the IsiA antenna ring in thermophilic cyanobacteria.

12.3.2.3 Cytosolic/Stromal Subunits of PS I: PsaC, PsaD and PsaE

PS I contains three cytosolic/stromal subunits, PsaC, PsaD and PsaE, which form the docking site for ferredoxin and flavodoxin.

PsaC is the most important small subunit of PS I, as it coordinates the two terminal FeS clusters F_A and F_B. PsaC is the best conserved protein subunit in PS I, with high sequence homology in all oxygenic photosynthetic organisms. The structure of PsaC is identical in plants and cyanobacteria (Jordan et al. 2001; Ben-Shem et al. 2003; Jolley et al. 2005). The functionally important core of PsaC contains two 4Fe4S clusters, which are coordinated by 8 cysteine residues from two conserved sequence motives CXXCXXCXXXC. Structurally, the two FeS clusters are connected by two short α-helices. PsaC shows striking structural homologies with bacterial ferredoxins, which indicates that PsaC may have evolved from a soluble ferredoxin that already contained 2 4Fe4S clusters. EPR spectroscopy in combination with site directed mutagenesis has shown that F_A is proximal to the membrane and F_X, while F_B is the terminal FeS cluster that transfers the electron to ferredoxin (Fischer et al. 1999).

The unique parts of PsaC, which do not exhibit homologies to ferredoxins, are the C- and N-terminus of PsaC. They are providing a very important symmetry-breaking feature that might be essential for the selection of a unique conformation for the docking of PsaC to the PS I core formed by PsaA/PsaB (Antonkine et al. 2003; Jolley et al. 2006).

PsaD is essential for the proper orientation of PsaC and, thereby, important for the tuning of the electron transfer from PS I to ferredoxin (Setif 2001; Setif et al. 2002). It is located at the “inner part” of the cytosolic/stromal hump, close to the “connecting domain”.

Co-crystals between PS I and ferredoxin (Fromme et al. 2002) serve as a basis for a structure of the PS I ferredoxin complex. The cocrystal structure of PS I with ferredoxin has been determined recently and unveiled that ferredoxin docks directly to PsaD (H. Q. Yu, R. Fromme, H. Bottin, P. Setif and P. Fromme, unpublished).

In the N-terminal domain, PsaD contains a large antiparallel four stranded ß-sheet. An α-helix follows in the sequence, which forms interactions with PsaC and with PsaA. A unique feature of PsaD is a clamp that wraps around PsaC and stabilizes the electron acceptor site in PS I (Li et al. 1991; Hanley et al. 1996) by forming numerous contacts between PsaD, PsaC and PsaE.

PsaE consists of five anti-parallel stranded ß-sheets. It should be noticed that it was the first subunit of PS I for which a solution structure had been determined by NMR (Falzone et al. 1994). PsaE stabilizes the stromal hump and plays a role in cyclic electron transport (Yu and Vermaas 1993). Functional studies indicate that it may be involved in anchoring the ferredoxin (Sonoike et al. 1993; Strotmann and Weber 1993; Rousseau et al. 1993). The role of PsaE in the docking of ferredoxin and flavodoxin, (Meimberg et al. 1998; Muhlenhoff and Chauvat 1996) was questioned by the finding that PsaE deletion mutants were still able to grow photoautotrophically. van Thor et al. (1999) showed that PsaE deletion mutants increased the level of ferredoxin in the cells by orders of magnitude to compensate for deficits caused by the lack of PsaE. However the role of PsaE in ferredoxin docking is very likely indirect by stabilization of the cytosolic/stromal hump, as no direct contacts between PsaE and ferredoxin have been detected in the recent cocrystal structure.

12.3.3 The Electron Transfer Chain in Photosystem I

The ETC of PS I is located in the center of the PS I monomer. It is arranged in two branches and consists of six chlorophylls, two phylloquinones and three 4Fe4S clusters, F_X, F_A and F_B. F_X is located at the twofold symmetry axis while F_A and F_B are coordinated by PsaC and break the twofold symmetry of the ET chain. The reaction sequence of photochemical charge separation leads to oxidation of the pigment dimer P700 and reduction of F_B. It is now well established by the structure and spectroscopy that F_B serves as the terminal FeS cluster and transfers the electron to ferredoxin.

12.3.3.1 P700

P700 is a pigment pair which consists of two chemically distinct Chl molecules and is located close to the lumenal surface of PS I. A Chl *a* molecule is present at the B-branch, while the A-Branch contains the C13 epimer of Chl *a*, a Chl *a*' molecule. The existence of at least one Chl *a*' molecule was first suggested by Watanabe and co-workers (Watanabe et al. 1985) on the basis of Chl extraction experiments and it has now been well established that in PS I complexes from all oxygenic photosynthetic kingdoms, P700 is represented by a Chl *a*/Chl *a*' dimer (Watanabe et al. 1985; Yoshida et al. 2003). In addition to the chemical differences, strong differences in the hydrogen bonding network of the two Chls are also evident. While the B-branch Chl *a* does not form hydrogen bonds with the protein, three hydrogen bonds are formed between the A-branch Chl *a*' and the protein. The distance between the central Mg^{2+} ions of the two chlorophylls in P700 is 6.3 Å, which is much shorter than the distance between the two corresponding Chls of P680 in PS II (see Sect. 12.2.3.2). The two Chls in P700 are tightly coupled, as revealed by molecular orbital studies (Plato et al. 2003). Despite the tight coupling the spin density on the cation radical is unevenly distributed between the two Chl s moieties in $P700^{+\bullet}$ as unravelled by ENDOR studies showing that more than 85% of the spin density is located on the B-branch Chl *a* (Käss et al. 2001). The question of the functional and structural reason for the asymmetry in P700 is still open.

The fact that Chl *a'* is a constituent in cyanobacterial, algal and plant PS I gives an indication that Chl *a'* may play a key role for the function of PS I. One possible role might be the "gating" of the electron along the two cofactor branches. However site directed mutagenesis of the side chains that mediate the H-bonding to the Chl *a'* showed no strong influence on the PS I function, nor did these mutations lead to a replacement of the Chl *a'* by Chl *a* (Wang et al. 2003; Li et al. 2004; Petrenko et al. 2004). It is also unknown how Chl *a'* is synthesized and what guides the assembly of the Chl *a'* into the binding site at the A-branch. To answer these questions more research results are needed.

For a long time P700 has been assumed to be the primary electron donor but recent studies suggest that the charge separation may start from the electronically excited singlet state of the first monomeric Chl at the B branch rather than from $^1P700^*$ (Müller et al. 2003; Giera et al. 2010).

12.3.3.2 Primary Electron Acceptor A

It is generally assumed that the second pair of Chls of the ET chain serves as the initial electron acceptor. These two chlorophylls are not ligated by a protein side chain, but water provides the fifth ligand to the central Mg^{2+} ion. However, this pair of Chls is difficult to detect by spectroscopy, as the first steps of electron transport occur in less than 3 ps.

12.3.3.3 First Stable Electron Acceptor A_0

The first stable electron acceptor A_0 is probably one of the two Chls of the third pair of Chls. These two Chl molecules are located in the middle of the membrane and their position mirrors the position of the two pheophytins in Photosystem II (Ferreira et al. 2004) and the PbRC (Deisenhofer and Michel 1991). The third pair of Chls in PS I has very unusual ligands, as sulphur atoms of two methionine residues provide the fifth ligand to the Mg^{2+} ion. Methionine ligandation is common in heme-proteins but is unusual for Chl coordination. Met is an extremely weak ligand as only weak interactions can be established between the hard acid Mg^{2+} and methionine sulphur, which is a soft base. It has been suggested that weak interaction might be important for the tuning of the very negative redox potential of A_0. Site directed mutagenesis experiments of the methionine have shown that the replacement of the Met by strong ligands like His slow down the electron transfer chain along the corresponding branches (Fairclough et al. 2003; Santabarbara et al. 2005; Ramesh et al. 2004, 2007)

The question is still open if A_0 is really only formed by the third pair of Chls. The two Chls of the third pair of Chls may form strong interactions with the Chls of the second pair of chlorophylls (Hastings et al. 1995a, b), which leads to the suggestion that all four Chls of the second and third pair contribute to the physical-chemical properties of the electron acceptor A_0 (but see former note on the possible role of P700).

12.3.3.4 Phylloquinone A_1

Two phylloquinones QK_A and QK_B represent the electron acceptor "A_1". The phylloquinones are located in a hydrophobic pocket at the stromal side of the membrane. They are π-stacked with a tryptophan residue. Only one oxygen atom in the quinone is involved in H-bonding. The unique H-bond is formed to an NH backbone group (from a Ser) in both PsaA and PsaB. The asymmetric H-bonding leads to an asymmetry in the distribution of the unpaired electron in the radical state $A_1^{-\bullet}$ (Srinivasan et al. 2009) and may be responsible for the extremely negative reduction potential

(−770 mV) of A_1. Weakening of the H-bond by replacement of the Leu722 with a bulky tryptophan side chain leads to a weakening of the H-bond and a significant double reduction of the phylloquinone at the A-branch, thereby blocking ET beyond A_1. These results show that the H-bond is absolutely essential for the electron transfer in PS I. It prevents the protonation and lowers the probability of double reduction during periods of high light intensity (Srinivasan et al. 2009). The electron transfer from A_1 to the FeS cluster, F_X, is the rate limiting step of the electron transfer in PS I. The rate of electron transfer from Q_{KA} and Q_{KB} to F_X may depend on the organism. In the green alga *Chlamydomonas reinhardtii*, the electron transfer proceeds along both branches of the ET chain (Guergova-Kuras et al. 2001). The electron transfer is about a factor of 50 slower on the A- than on the B-branch (Guergova-Kuras et al. 2001), indicating that there might be a higher activation energy barrier on the A- compared to the B-branch. In cyanobacteria the picture is more complex, and most evidence points in the direction that most of the electrons follow the slower A-branch (Dashdorj et al. 2005).

What is the origin of this asymmetry in the rates of electron transfer from Q_{KA} and Q_{KB}? The direct protein environment of both phylloquinones is very similar, but two lipid molecules, which are located close to the phylloquinones, could contribute to the asymmetry. One hypothesis assumes that the negatively charged phospholipid, which is located on the slower A-branch, increases the reorganisation energy on this branch. At the B-branch, a neutral galactolipid is present instead of the phospholipid, which may lower the reorganization energy on the B-branch (Ishikita and Knapp 2003). The question of the involvement and the gating along the two branches is still a controversial topic and more experiments are needed to clearly demonstrate what factors determine the gating and speed of the ET along the two branches.

12.3.3.5 First FeS Cluster F_X

F_X is the first of the three 4Fe4S clusters. F_X is a rare example of an inter-protein FeS cluster; it is formed by 2 cysteines from PsaA and 2 cysteines from PsaB, which are located in the loop between the transmembrane helices 8 and 9. The F_X binding motif is highly conserved between plants, algae and cyanobacteria.

F_X is not only an essential cofactor of the electron transfer chain, but it also plays an important role in the assembly of the PS I complex. The stromal subunits PsaC, PsaD and PsaE can only be assembled on the PsaA/PsaB core of PS I after F_X has been assembled. Several proteins including rubredoxin are involved in the assembly of F_X (Shen et al. 2002a, b).

12.3.3.6 The Terminal FeS Clusters F_A and F_B

The two terminal FeS clusters, F_A and $F_{B,}$ are the only cofactors of the electron transfer chain that are not coordinated by PsaA/PsaB. F_A and F_B are embedded in the extrinsic subunit PsaC, which is located at the center of the stromal hump of PS I. The cluster in close proximity to F_X has been assigned to F_A by EPR and mutagenesis studies (Zhao et al. 1992; Mehari et al. 1995), while the distal cluster, which transfers the electrons to ferredoxin, is F_B (Kanervo et al. 1995; Yu et al. 1995a, b; Fischer et al. 1999; Golbeck 1999; Lakshmi et al. 1999).

F_A and F_B exhibit the same distance and orientation as the three 4Fe4S clusters in ferredoxin, which further supports the hypothesis that PsaC evolved from a soluble bacterial ferredoxin (Fromme et al. 2001). F_B is the terminal FeS cluster in Photosystem I and might therefore play an important role in the future for the use of Photosystem I as a building block for devices for solar energy conversion. In a recent, elegant work, Golbeck and coworkers have directly "wired" the F_B cluster to carbon-nanotubes and Platinum nano-balls (Grimme et al. 2008), using the original cysteine mutant that has been used for identification of the F_B cluster in 1995 (Yu et al. 1995a, b). The wired complexes showed very promising rates of H_2 production and might form the basis for new technologies for bioenergy production in the future.

12.3.4 Antenna System of PS I

Photosystem I contains the largest core antenna system of all photosynthetic reaction centers. One monomeric unit contains 90 antenna Chl *a* molecules and 22 Cars, i.e. more than twice as many Chls compared to the PS II core (see Sect. 12.2.4.1). This implies that the antenna system is much more tightly coupled to the ETC in PS I than in PS II. The excitation energy is transferred from the antenna pigments to P700, with a quantum yield of 99.5% at room temperature, while the quantum efficiency of the excitation energy transfer in PS II is only about 80%. The Cars serve dual functions in PS I: they act as additional light harvesting pigments and also play an important role in photo-protection, as they can quench dangerous Chl triplet states.

12.3.4.1 Antenna Chls and Mechanism of Excitation Energy Transfer in PS I

The arrangement of antenna Chls in PS I is complex (see Fig. 12.7e). The Chls are densely packed and form a clustered network to allow very fast and efficient excitation energy transfer. This arrangement shows no homologies to the ring of LH1 proteins that surround the RC core in purple bacteria (McDermott et al. 1995; Roszak et al. 2003). One analogy is the comparison of the Chl arrangement in PS I with the network of neurons in the brain where fast interconnections of the neurons in a clustered network allows very fast signal transduction. In PS I, the average center to center distance between the Chls is less than 15 Å, and each of the Chls has several neighbors that can mediate fast excitation energy transfer, which explains the high

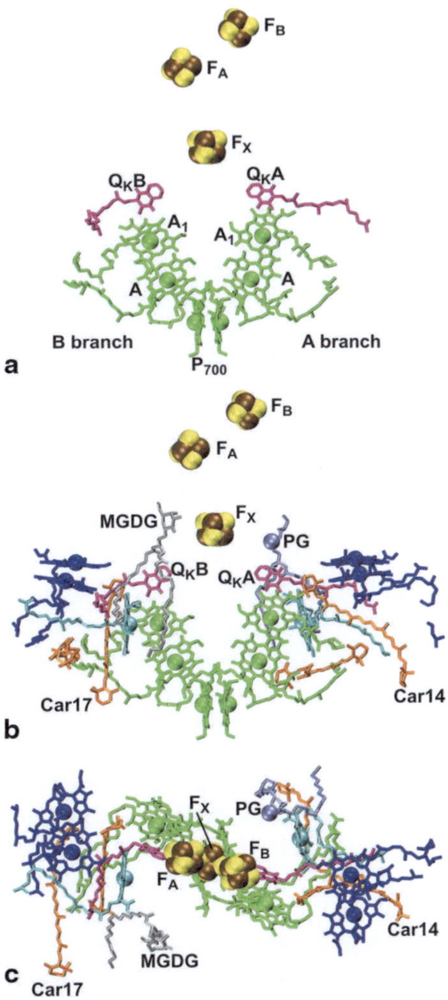

Fig. 12.7 The electron transfer chain of photosystem I from *Thermosynechococcus elongatus*. **a** The components of the electron transfer chain. This includes six chlorophylls, two phylloquinones, and 3 4Fe4S iron sulfur clusters. **b** Molecules in the vicinity of the electron transfer chain. The so-called "connecting chlorophylls" are shown in cyan, the "connecting dimers" in blue. Carotenoids Car 14 and Car 17 are near enough to play a protecting role for the chlorophylls in the electron transfer chain, a feature that is missing in PS II. Furthermore, two lipids in the vicinity of F_X and the phylloquinones are shown, with one side having a neutral lipid (MGDG), whereas the other side contains the charged lipid PG, which might have functional implications. **c** Same as **b**, but seen from the cytosolic side

efficiency and robustness of the energy transfer in the core antenna of PS I (Sener et al. 2002, 2003). This system does not require all Chls to be highly optimized concerning the orientation and transition dipole moments, but the antenna chlorophylls that come closest to the RC core must be highly optimized. The antenna system of PS I may look at the first glance arbitrary, but the comparison of the plant and cyanobacterial core antenna system of PS I shows that the position and orientation of 85 out of the 90 antenna Chls have been conserved over 1.5 Billion years of evolution (Jordan et al. 2001; Amunts et al. 2007).

The very fast excitation energy transfer in PS I cannot be described by random "hopping" from one Chl to the next and is best explained by the mechanism of "quantum beating" which has been proposed based on the experimental results from ultrafast spectroscopy of the soluble FMO protein, which contains 7 bacteriochlorophylls (Engel et al. 2007; Savikhin et al. 1998). The idea of the quantum beating is that the excitation energy moves in the antenna system in form of a wave, which is partially delocalized over several antenna pigments. The movement of the excitation energy in form of a wave allows the exploration of the energy of a large number of Chls in a very short time scale, which may allow for a more directed energy transfer from the antenna Chls to P700. It would also solve the problem of the finding that excitation energy transfer is trap-limited in PS I. It has been shown that the excitation energy reaches P700 in about 10 ps, but charge separation takes place only in about 20–30% of the events, which increases the time frame for trapping to 30–50 ps. This means that the exciton must "visit" P700 on the average 3–4 times before charge separation takes place. The quantum beating mechanisms would nicely explain how the system could reach the 99.99% quantum efficiency of excitation energy transfer and trapping despite the fact that P700 is performing charge separation only in 25% of the excitation events. Just assume we would be able to follow the way of the excitation energy after an unsuccessful "visit" to P700 in the random walk and the quantum beating mechanism. In the mechanism of random walk from Chl to Chl, the excitation energy could easily be moving back to the peripheral antenna pigments, from where they would require another long random walk to P700, which would increase the chance of the excitation being lost by fluorescence. However, this is not observed. In contrast, the quantum beating mechanism is able to explain the fast and highly efficient excitation energy transfer that is experimentally observed, as the wave is always delocalized over several chlorophylls and goes back and forth between the lowest energy Chls in a fashion that can be compared to a heart beating (therefore the name "quantum beating"). Thereby the excitation energy wave would just "wobble" back and forward between P700 and its nearest neighbour Chls until charge separation occurs.

The core antenna system of PS I can be structurally divided into two domains: the central domain surrounds the ETC, while the peripheral domains flank the core of PS I on both sides. The antenna Chls of the peripheral domains show some similarities to the core antenna system of the Chls in the antenna proteins CP47 and CP43 in PS II. Most of the Chls at the periphery are arranged in two layers. One layer is located close to the cytosolic/stromal surface of the membrane while the other layer is more closely located to the lumenal surface of the membrane. In contrast, the Chls in the central domain are distributed over the full depth of the membrane and may thereby be able to facilitate the excitation energy transfer between the two layers.

A unique feature of PS I is that it contains two Chls (named "connecting chlorophylls"), which are located in the semi-circle of 2×5 TMHs surrounding the electron transfer chain. Mutagenesis experiments on the His ligand of these two Chls showed an increase of the time for trapping of the excitation energy. The authors estimated that about 30% of the excitation energy is transferred in the WT PS I via the connecting Chls, however the mutants were still highly efficient in excitation energy transfer (Gibasiewicz et al. 2003). This shows the high robustness of the antenna system in PS I, where even the alteration of antenna Chls in closest vicinity of P700 does not greatly diminish the efficiency of PS I.

Another unique feature of PS I is the existence of "red" Chls that absorb at wavelengths $\lambda > 700$ nm in PS I. These red Chls have even led to the discovery of the existence of two light reactions [see (Rumberg et al. 1964; Allen et al. 1961; Bertsch 1962) and references therein]. The number of long wavelength Chls is not identical in PS I from all cyanobacteria. For example, PS I from *Synechocystis PCC 6804* has only a small pool of long wavelength Chls, compared to *T. elongatus.* Why does PS I, in contrast to PS II, contain long-wavelength Chls? The most compelling reason is that they are increasing the spectral width of the light absorbed by PS I. Another explanation would be that they may be involved in the funneling of the excitation energy to the center of the complex. There is so far not a complete agreement on the exact location of the red Chls, and this topic is still controversially discussed (Byrdin et al. 2002; Damjanovic et al. 2002; Vasil'ev and Bruce 2004).

12.3.4.2 Carotenoids

PS I contains 22 Cars and thereby nearly twice as many Cars as PS II, where one monomer contains only 12 Cars. The major role of the Cars is photo-protection, but they also play a structural role and may function as additional antenna pigments.

Prevention from photodamage is essential for the function of the system as over-excitation caused by excess light can lead to photoinhibition. The photoprotective role is critical to the function of PS I, and the higher Car content of PS I compared to PS II might also contribute to the much higher stability of PS I compared to PS II.

The major function of the carotenoids in PS I is to quench Chl triplet states ^{3}Chl. Chl triplet states are very dangerous and can lead to severe damage of the proteins and other biomolecules in the cell. Triplet Chls can easily interact with oxygen, which is in the ground state in the triplet state, to form singlet oxygen $^1\Delta_g O_2$ that acts as a very potent cell poison. The Cars in PS I are in close vicinity to the Chls, and they are distributed over the whole antenna system, to allow transfer of the energy from the triplet Chls to the Cars. The Car triplet state, ^{3}Car, which is formed by this process, dissipates the excess energy as heat, thereby acting essentially as a "sunscreen" for PS I and PS II. We want to mention that a common misconception can be found now even in textbooks about the chemical nature of the Cars in PS I. All Cars were modeled as β-carotene in the 2.5 Å crystal structure of PS I only because detailed features of the ring head groups were not visible at this resolution. Biochemical evidence shows that PS I (and the crystals) contains xanthophylls, but a crystal structure of PS I at higher resolution is essential to unveil the nature of the individual Cars in PS I.

12.3.4.3 Lipids

PS I has with four lipids much a smaller lipid content than PS II. The PS I structure at 2.5 Å resolution unravelled the position of 3 molecules of phosphatidylglycerol (PG) and one molecule of monogalactosyldiacylglycerol (MGDG), which confirmed the lipid content determined by previous biochemical studies (Kruse and Schmid 1995; Kruse et al. 2000). Two of the lipids (one PG and the MGDG) are located in the vicinity of the phylloquinones of the electron transfer chain, and it has been speculated that they may play an important role in the difference in the rates of electron transfer between the two different branches (Fromme et al. 2001; Ishikita and Knapp 2003).

The head groups of these two lipids are deeply embedded into the PS I complex and form specific interactions with the stromal loops of PsaA and PsaB as well as the stromal subunits PsaC, PsaD and PsaE. Their location and interactions suggest that they must be incorporated into PS I at a very early stage of the assembly process.

12.4 Concluding Remarks and Future Perspectives

The first decade of this century saw a vast increase in our understanding of both, Photosystems I and II, with the advent of crystal structures with up to 2.5 Å and 2.9 Å resolution, respectively. In both protein complexes, this allowed to see the positions of the molecules participating in the respective electron transfer chains, the Chls and Cars of the inner antennas, and of many other cofactors, like lipids or metal ions. This structural information allowed explanations of kinetic data and construction of better models for the many different physico-chemical processes that take place during excitation, charge separation, or electron transfer.

However, many open questions remain. In PS I, we still do not know with absolute certainty where the initial charge separation takes place, what exactly the role of the Chl *a'* epimer is, which role the A and B branches play, where the phycobilisomes bind, how these actually look like on a molecular level, whether there are more lipids in the molecule and what their functional role is, to name just a few points. For solving all these questions, better structural models with even higher resolution would be very helpful.

For PS II, this problem is even more severe, because the "heart" of the complex is still awfully under-determined. The exact structure of the water-oxidizing complex, the position of the manganese atoms, the mode of the oxygen bridging, and the important water ligation are at the current resolution still mostly based on speculation. In order to shine light into this central question, new methods of data acquisition might have to be developed. Whatever the solution will be, we can be sure that there will still be structural revelations to come within the coming years.

Acknowledgements The authors want to thank Prof. Gernot Renger for a thorough review of this chapter and for his valuable comments and suggestions, including a paragraph about the history of the structural investigation of Photosystem II. The authors' work on Photosystems I and II is supported by the National Science Foundation award MCB-0417142 and the National Institute of Health award 1 R01 *GM71619*-01.

References

Allen MB, Piette LH and Murchio JC (1961) Observation of two photoreactions in photosynthesis. Biochem Biophys Res Commun 4: 271–274

Amunts A and Nelson N (2008) Functional organization of a plant Photosystem I: evolution of a highly efficient photochemical machine. Plant Physiol Biochem 46: 228–237

Amunts A, Drory O and Nelson N (2007) The structure of a plant photosystem I supercomplex at 3.4 A resolution. Nature 447: 58–63

Anandan S, Vainstein A and Thornber JP (1989) Correlation of some published amino acid sequences for photosystem I polypeptides to a 17 kDa LHCI pigment-protein and to subunits III and IV of the core complex. FEBS Lett 256: 150–154

Andersen B and Scheller HV (1993) Structure, function and assembly of photosystem I. In: Sundqvist C, Rydberg M (eds) Chloroplast Pigment-Protein Complexes, Academic Press, Orlando, pp 383–418

Antonkine ML, Jordan P, Fromme P, Krauss N, Golbeck JH and Stehlik D (2003) Assembly of protein subunits within the stromal ridge of photosystem I. Structural changes between unbound and sequentially PS I-bound polypeptides and correlated changes of the magnetic properties of the terminal iron sulfur clusters. J Mol Biol 327: 671–697

Baena-Gonzalez E and Aro EM (2002) Biogenesis, assembly and turnover of photosystem II units. Philos Trans R Soc Lond B Biol Sci 357: 1451–1459

Barber J (2002) P680: what is it and where is it? Bioelectrochemistry 55: 135–138

Baumgartner BJ, Rapp JC and Mullet JE (1993) Plastid genes encoding the transcription/translation apparatus are differentially transcribed rarly in barley (Hordeum vulgare) chloroplast development (Evidence for selective stabilization of psbA mRNA). Plant Physiol 101: 781–791

Ben-Shem A, Frolow F and Nelson N (2003) Crystal structure of plant photosystem I. Nature 426: 630–635

Bertsch WF (1962) Two photoreactions in photosynthesis: evidence from the delayed light emission of chlorella. Proc Natl Acad Sci U S A 48: 2000–2004

Bibby TS, Nield J and Barber J (2001) Three-dimensional model and characterization of the iron stress-induced CP43'-photosystem I supercomplex isolated from the cyanobacterium synechocystis PCC 6803. J Biol Chem 276: 43246–43252

Biesiadka J, Loll B, Kern J, Irrgang KD and Zouni A (2004) Crystal structure of cyanobacterial photosystem II at 3.2 angstrom resolution: a closer look at the Mn-cluster. Phys Chem Chem Phys 6: 4733–4736

Blankenship JE and Kindle KL (1992) Expression of chimeric genes by the light-regulated cabII-1 promoter in Chlamydomonas reinhardtii: a cabII-1/nit1 gene functions as a dominant selectable marker in a nit1- nit2- strain. Mol Cell Biol 12: 5268–5279

Boekema EJ, Hifney A, Yakushevska AE, Piotrowski M, Keegstra W, Berry S, Michel KP, Pistorius EK and Kruip J (2001) A giant chlorophyll-protein complex induced by iron deficiency in cyanobacteria. Nature 412: 745–758

Bondarava N, De Pascalis L, Al-Babili S, Goussias C, Golecki JR, Beyer P, Bock R and Krieger-Liszkay A (2003) Evidence that cytochrome b559 mediates the oxidation of reduced plastoquinone in the dark. J Biol Chem 278: 13554–13560

Bricker TM and Burnap RL (2005) The extrinsic proteins of Photosystem II. In: Wydrzynski TJ, Satoh K (eds) Advances in photosynthesis and respiration: the light-driven water: plastoquinone oxidoreductase, Springer, Dordrecht, pp 95–120

Bricker TM and Frankel LK (2002) The structure and function of CP47 and CP43 in photosystem II. Photosynth Res 72: 131–146

Britt RD, Campbell KA, Peloquin JM, Gilchrist ML, Aznar CP, Dicus MM, Robblee J and Messinger J (2004) Recent pulsed EPR studies of the photosystem II oxygen-evolving complex: implications as to water oxidation mechanisms. Biochim Biophys Acta 1655: 158–171

Broess K, Trinkunas G, van Hoek A, Croce R and van Amerongen H (2008) Determination of the excitation migration time in photosystem II consequences for the membrane organization and charge separation parameters. Biochim Biophys Acta 1777: 404–409

Byrdin M, Jordan P, Krauss N, Fromme P, Stehlik D and Schlodder E (2002) Light harvesting in photosystem I: modeling based on the 2.5-A structure of photosystem I from Synechococcus elongatus. Biophys J 83: 433–457

Chauhan D, Folea IM, Jolley CC, Kouřil R, Lubner CE, Lin S, Kolber D, Wolfe-Simon F, Golbeck JH, Boekema EJ and Fromme P (2011) A novel photosynthetic strategy for adaptation to low-iron aquatic environments. Biochemistry 50: 686–692

Damjanovic A, Vaswani HM, Fleming GR and Fromme P (2002) Chlorophyll excitations in photosystem I as revealed by semi-empirical ZINDO/CIS calculations. Biophys J 82: 293–298

Dashdorj N, Xu W, Cohen RO, Golbeck JH and Savikhin S (2005) Asymmetric electron transfer in cyanobacterial photosystem I: charge separation and secondary electron transfer dynamics of mutations near the primary electron acceptor A0. Biophys J 88: 1238–1249

Debus RJ (2008) Protein ligation of the photosynthetic oxygen-evolving center. Coord Chem Rev 252: 244–258

Deisenhofer J and Michel H (1991) Structures of bacterial photosynthetic reaction centers. Annu Rev Cell Biol 7: 1–23

Deisenhofer J, Epp O, Miki K, Huber R and Michel H (1985) Structure of the protein subunits in the photosynthetic reaction center of rhodopseudomonas-viridis at 3a resolution. Nature 318: 618–624

Deisenhofer J, Epp O, Sinning I and Michel H (1995) Crystallographic refinement at 2.3 A resolution and refined model of the photosynthetic reaction centre from Rhodopseudomonas viridis. J Mol Biol 246: 429–457

Dekker JP, Boekema EJ, Witt HT and Rögner M (1988) Refined purification and further characterization of oxygen-evolving and tris-treated photosystem II particles from the thermophylic cyanobacterium *Synechococcus sp.* Biochim Biophys Acta 936: 307–318

Di Donato M, Cohen RO, Diner BA, Breton J, van Grondelle R and Groot ML (2008) Primary charge separation in the photosystem II core from Synechocystis: a comparison of femtosecond visible/midinfrared pump-probe spectra of wild-type and two P680 mutants. Biophys J 94: 4783–4795

Dobakova M, Komenda J and Tichy M (2007) The PsbI protein stabilizes the newly synthesized D1 protein and assists its stable incorporation into photosystem II in the cyanobacterium Synechocystis PCC. Photosynth Res 91: 209–219

Dörmann P and Hölzl G (2009) The role of glycolipids in photosynthesis. In: Wada H, Murata N (eds) Lipids in photosynthesis: essential and regulatory functions, Springer, Dordrecht, pp 265–282, CP13

Engel GS, Calhoun TR, Read EL, Ahn TK, Mancal T, Cheng YC, Blankenship RE and Fleming GR (2007) Evidence for wavelike energy transfer through quantum coherence in photosynthetic systems. Nature 446: 782–786

Fairclough WV, Forsyth A, Evans MCW, Rigby SEJ, Purton S and Heathcote P (2003) Bidirectional electron transfer in photosystem I: electron transfer on the PsaA side is not essential for phototrophic growth in Chlamydomonas. Biochim Biophys Acta 1606: 43–55

Falzone CJ, Kao YH, Zhao J, Bryant DA and Lecomte JT (1994) Three-dimensional solution structure of PsaE from the cyanobacterium Synechococcus sp. strain PCC 7002, a photosystem I protein that shows structural homology with SH3 domains. Biochemistry 33: 6052–6062

Ferreira KN, Iverson TM, Maghlaoui K, Barber J and Iwata S (2004) Architecture of the photosynthetic oxygen-evolving center. Science 303: 1831–1838

Fischer N, Hippler M, Setif P, Jacquot JP and Rochaix JD (1998) The PsaC subunit of photosystem I provides an essential lysine residue for fast electron transfer to ferredoxin. Embo J 17: 849–858

Fischer N, Setif P and Rochaix JD (1999) Site-directed mutagenesis of the PsaC subunit of photosystem I. F(b) is the cluster interacting with soluble ferredoxin. J Biol Chem 274: 23333–23340

Ford RC and Holzenburg A (1988) Investigation of the structure of trimeric and monomeric photosystem I reaction center complexes. Embo J 7: 2287–2293

Franzen LG, Styring S, Etienne AL, Hansson Ö and Vernotte C (1986) Spectroscopic and functional characterization of a highly oxygen-evolving photosystem II reaction center complex from spinach. Photobiochem Photobiophys 13: 15–28

Fromme P (1998) Crystallization of photosystem I for structural analysis. Habilitation. Technical University Berlin, Berlin

Fromme P, Jordan P and Krauss N (2001) Structure of photosystem I. Biochim Biophys Acta 1507: 5–31

Fromme P, Kern J, Loll B, Biesiadka J, Saenger W, Witt HT, Krauss N and Zouni A (2002) Functional implications on the mechanism of the function of photosystem II including water oxidation based on the structure of photosystem II. Philos Trans R Soc Lond B Biol Sci 357: 1337–1344

Fromme P, Melkozernov A, Jordan P and Krauss N (2003) Structure and function of photosystem I: interaction with its soluble electron carriers and external antenna systems. FEBS Lett 555: 40–44

Fyfe PK and Jones MR (2005) Lipids in and around photosynthetic reaction centres. Biochem Soc Trans 33: 924–930

Gibasiewicz K, Ramesh VM, Lin S, Redding K, Woodbury NW and Webber AN (2003) Excitonic interactions in wild-type and mutant PS I reaction centers. Biophys J 85: 2547–2559

Giera W, Ramesh VM, Webber AN, van Stokkum I, van Grondelle R and Gibasiewicz K (2010) Effect of the P700 pre-oxidation and point mutations near A0 on the reversibility of the primary charge separation in photosystem I from Chlamydomonas reinhardtii. Biochim Biophys Acta 1797: 106–112

Gobets B and van Grondelle R (2001) Energy transfer and trapping in photosystem I. Biochim Biophys Acta 1507: 80–99

Golbeck JH (1999) A comparative analysis of the spin state distribution of in vitro and in vivo mutants of PsaC. Photosynth Res 61: 107–144

Gombos Z, Várkonyi Z, Hagio M, Iwaki M, Kovács L, Masamoto K, Itoh S, and Wada H (2002) Phosphatidylglycerol requirement for the function of electron acceptor plastoquinone Q_B in the photosystem II reaction center. Biochemistry 41: 3796–3802

Goni G, Herguedas B, Hervas M, Peregrina JR, De la Rosa MA, Gomez-Moreno C, Navarro JA, Hermoso JA, Martinez-Julvez M and Medina M (2009) Flavodoxin: a compromise between efficiency and versatility in the electron transfer from photosystem I to Ferredoxin-NADP(+) reductase. Biochim Biophys Acta 1787: 144–154

Green B and Parson WW (2003) Light harvesting antennas. Advances in photosynthesis and respiration, Vol. 13, Springer, Dordrecht

Grimme RA, Lubner CE, Bryant DA and Golbeck JH (2008) Photosystem I/molecular wire/metal nanoparticle bioconjugates for the photocatalytic production of H2. J Am Chem Soc 130: 6308–6309

Grotjohann I, Jolley C and Fromme P (2004) Evolution of photosynthesis and oxygen evolution: implications from the structural comparison of photosystems I and II. Phys Chem Chem Phys 6: 4743–4753

Guergova-Kuras M, Boudreaux B, Joliot A, Joliot P and Redding K (2001) Evidence for two active branches for electron transfer in photosystem I. Proc Natl Acad Sci U S A 98: 4437–4442

Guskov A, Kern J, Gabdulkhakov A, Broser M, Zouni A and Saenger W (2009) Cyanobacterial photosystem II at 2.9-A resolution and the role of quinones, lipids, channels and chloride. Nat Struct Mol Biol 16: 334–342

Haag E, Irrgang KD, Boekema EJ and Renger G (1990) Functional and structural analysis of photosystem II core complexes from spinach with high oxygen evolution capacity. Eur J Biochem 189: 47–53

Haddy A (2007) EPR spectroscopy of the manganese cluster of photosystem II. Photosynth Res 92: 357–368

Hanley J, Setif P, Bottin H and Lagoutte B (1996) Mutagenesis of photosystem I in the region of the ferredoxin cross-linking site: modifications of positively charged amino acids. Biochemistry 35: 8563–8571

Hastings G, Hoshina S, Webber AN and Blankenship RE (1995a) Universality of energy and electron transfer processes in photosystem I. Biochemistry 34: 15512–15522

Hastings G, Reed LJ, Lin S and Blankenship RE (1995b) Excited stae dynamics in photosystem I: effects of detergent and excitation wavelength. Biophys J 69: 2044–2055

Hippler M, Reichert J, Sutter M, Zak E, Altschmied L, Schroer U, Herrmann RG, and Haehnel W (1996) The plastocyanin binding domain of photosystem I. EMBO J 15: 6374–6384

Hippler M, Drepper F, Haehnel W and Rochaix JD (1998) The N-terminal domain of PsaF: precise recognition site for binding and fast electron transfer from cytochrome c6 and plastocyanin to photosystem I of Chlamydomonas reinhardtii. Proc Natl Acad Sci U S A 95: 7339–7344

Hippler M, Drepper F, Rochaix JD and Mühlenhoff U (1999) Insertion of the N-terminal part of PsaF from Chlamydomonas reinhardtii into photosystem I from Synechococcus elongatus enables efficient binding of algal plastocyanin and cytochrome c_6. J Biol Chem 274: 4180–4188

Hou JM, Boichenko VA, Diner BA, and Mauzerall D (2001) Thermodynamics of electron transfer in oxygenic photosynthetic reaction centers: volume change, enthalpy, and entropy of electron-transfer reactions in manganese-depleted photosystem II core complexes. Biochemistry 40: 7117–7125

Humphrey W, Dalke A and Schulten K (1996) VMD: visual molecular dynamics. J Mol Graph 14: 33–38

Irrgang K-D, Renger G and Vater J (1986) Identification of Chl-binding proteins in a PS II preparation from spinach. FEBS Lett 204: 67–75

Ishikita H and Knapp EW (2003) Redox potential of quinones in both electron transfer branches of photosystem I. J Biol Chem 278: 52002–52011

Ishikita H, Saenger W, Biesiadka J, Loll B and Knapp EW (2006) How photosynthetic reaction centers control oxidation power in chlorophyll pairs P680, P700, and P870. Proc Natl Acad Sci U S A 103: 9855–9860

Jagannathan B and Golbeck JH (2009a) Breaking biological symmetry in membrane proteins: the asymmetrical orientation of PsaC on the pseudo-C2 symmetric photosystem I core. Cell Mol Life Sci 66: 1257–1270

Jagannathan B and Golbeck JH (2009b) Understanding of the binding interface between PsaC and the PsaA/PsaB heterodimer in photosystem I. Biochemistry 48: 5405–5416

Jagendorf AT (2002) Photophosphorylation and the chemiosmotic perspective. Photosynth Res 73: 233–241

Janson S, Andersen B and Scheller HV (1996) Nearest-neighbor analysis of higher-plant photosystem I holocomplex. Plant Physiol 112: 409–420

Jekow P, Fromme P, Witt H and Saenger W (1995) Photosystem I from *Synechococcus elongatus*: preparation and crystallization of monomers with varying subunit compositions. Biochim Biophys Acta 1229: 115–120

Jekow P, Fromme P, Witt H and Saenger W (1996) Crystallisation of intact and subunit L-deficient monomers from Synechocystis PCC 6803 photosystem I. Z Naturforsch 51c: 195–199

Jolley C, Ben-Shem A, Nelson N and Fromme P (2005) Structure of plant photosystem I revealed by theoretical modeling. J Biol Chem 280: 33627–33636

Jolley CC, Wells SA, Hespenheide BM, Thorpe MF and Fromme P (2006) Docking of photosystem I subunit C using a constrained geometric simulation. J Am Chem Soc 128: 8803–8812

Jones MR (2007) Lipids in photosynthetic reaction centres: structural roles and functional holes. Prog Lipid Res 46: 56–87

Jordan P, Fromme P, Klukas O, Witt HT, Saenger W and Krauß N (2001) Three-dimensional structure of cyanobacterial photosystem I at 2.5 Å resolution. Nature 411: 909–917

Kaminskaya O, Kurreck J, Irrgang K-D, Renger G and Shuvalov VA (1999) Redox and spectral properties of cytochrome b559 in different preparations of photosystem II. Biochemistry 38: 16223–16235

Kaminskaya O, Kern J, Shuvalov VA and Renger G (2005) Extinction coefficients of cytochromes b559 and c550 of Thermosynechococcus elongates and Cyt b559/PS II stoichiometry of higher plants. Biochim Biophys Acta 1708: 333–341

Kaminskaya O, Shuvalov VA and Renger G (2007) Evidence for a novel quinone binding site in the Photosystem II (PS II) complex which regulates the redox potential of Cyt b559. Biochemistry 46: 1091–1105

Kamiya N and Shen JR (2003) Crystal structure of oxygen-evolving photosystem II from Thermosynechococcus vulcanus at 3.7-A resolution. Proc Natl Acad Sci U S A 100: 98–103

Kanervo E, Aro EM and Murata N (1995) Low unsaturation level of thylakoid membrane lipids limits turnover of the D1 protein of photosystem II at high irradiance. FEBS Lett 364: 239–242

Kanervo E, Tasaka Y, Murata N and Aro EM (1997) Membrane lipid unsaturation modulates processing of the photosystem II reaction-center protein D1 at low temperatures. Plant Physiol 114: 841–849

Käss H, Fromme P, Witt HT and Lubitz W (2001) Orientation and electronic structure of the primary donor radical cation P-700(+center dot) in photosystem I: a single crystals EPR and ENDOR study. J Phys Chem B 105: 1225–1239

Kawakami K, Umena Y, Kamiya N and Shen JR (2009) Location of chloride and its possible functions in oxygen-evolving photosystem II revealed by X-ray crystallography. Proc Natl Acad Sci U S A 106: 8567–8572

Kern J, Loll B, Luneberg C, DiFiore D, Biesiadka J, Irrgang KD and Zouni A (2005) Purification, characterisation and crystallisation of photosystem II from Thermosynechococcus elongatus cultivated in a new type of photobioreactor. Biochim Biophys Acta 1706: 147–157

Kern J, Biesiadka J, Loll B, Saenger W and Zouni A (2007) Structure of the Mn(4)-Ca cluster as derived from X-ray diffraction. Photosynth Res 234: 356–362

Klukas O, Schubert WD, Jordan P, Krauss N, Fromme P, Witt HT and Saenger W (1999a) Localization of two phylloquinones, Q_K and Q_K', in an improved electron density map of photosystem I at 4 Å resolution. J Biol Chem 274: 7361–7367

Klukas O, Schubert WD, Jordan P, Krauss N, Fromme P, Witt HT and Saenger W (1999b) Photosystem I, an improved model of the stromal subunits PsaC, PsaD, and PsaE. J Biol Chem 274: 7351–7360

Koike K, Ikeuchi M, Hiyama T and Inoue Y (1989) Identification of photosystem I components from the cyanobacterium Synechococcus vulcanus by N-terminal sequencing. FEBS Lett 253: 257–263

Komenda J, Tichy M and Eichacker LA (2005) The PsbH protein is associated with the inner antenna CP47 and facilitates D1 processing and incorporation into PS II in the cyanobacterium Synechocystis PCC 6803. Plant Cell Physiol 46: 1477–1483

Kouril R, Arteni AA, Lax J, Yeremenko N, D'Haene S, Rogner M, Matthijs HC, Dekker JP and Boekema EJ (2005a) Structure and functional role of supercomplexes of IsiA and photosystem I in cyanobacterial photosynthesis. FEBS Lett 579: 3253–3257

Kouril R, Yeremenko N, D'Haene S, Oostergetel GT, Matthijs HC, Dekker JP and Boekema EJ (2005b) Supercomplexes of IsiA and photosystem I in a mutant lacking subunit PsaL. Biochim Biophys Acta 1706: 262–266

Kouril R, Zygadlo A, Arteni AA, de Wit CD, Dekker JP, Jensen PE, Scheller HV and Boekema EJ (2005c) Structural characterization of a complex of photosystem I and light-harvesting complex II of *Arabidopsis thaliana*. Biochemistry 44: 10935–10940

Krauss N, Hinrichs W, Witt I, Fromme P, Pritzkow W, Dauter Z, Betzel C, Wilson KS, Witt HT and Saenger W (1993) Three-dimesional structure of system I of photosynthesis at 6 Å resolution. Nature 361: 326–361

Krauss N, Schubert WD, Klukas O, Fromme P, Witt HT and Saenger W (1996) Photosystem I at 4 A resolution represents the first structural model of a joint photosynthetic reaction centre and core antenna system. Nat Struct Biol 3: 965–973

Kruse O and Schmid GH (1995) The role of phosphatidylglycerol as a functional effector and membrane anchor of the D1-core peptide from photosystem II-particles of the cyanobacterium Oscillatoria chalybea. Z Naturforsch C 50: 380–390

Kruse O, Hankamer B, Konczak C, Gerle C, Morris E, Radunz A, Schmid GH and Barber J (2000) Phosphatidylglycerol is involved in the dimerization of photosystem II. J Biol Chem 275: 6509–6514

Kunstner P, Guardiola A, Takahashi Y and Rochaix JD (1995) A mutant strain of Chlamydomonas reinhardtii lacking the chloroplast photosystem II psbI gene grows photoautotrophically. J Biol Chem 270: 9651–9654

Kurisu G, Zhang H, Smith JL and Cramer WA (2003) Structure of the cytochrome b6f complex of oxygenic photosynthesis: tuning the cavity. Science 302: 1009–1014

Kwa SLS, Newell WR, Vangrondelle R and Dekker JP (1992) The reaction center of photosystem-II studied with polarized fluorescence spectroscopy. Biochim Biophys Acta 1099: 193–202

Lakshmi KV, Eaton SS, Eaton GR and Brudvig GW (1999) Orientation of the tetranuclear manganese cluster and tyrosine Z in the O(2)-evolving complex of photosystem II: an EPR study of the S(2)Y(Z)(*) state in oriented acetate-inhibited photosystem II membranes. Biochemistry 38: 12758–12767

Lakshmi KV, Poluektov OG, Reifler MJ, Wagner AM, Thurnauer MC and Brudvig GW (2003) Pulsed high-frequency EPR study on the location of carotenoid and chlorophyll cation radicals in photosystem II. J Am Chem Soc 125: 5005–5014

Lane N (2003) Oxygen—the molecule that made the World. Oxford University Press, Oxford

Larkum AWD (2008) The evolution of photosynthesis. In: Renger G (ed) Primary processes of photosynthesis: priniciples and apparatus. Part II: Reaction centers/photosystems, electron transport chains, photophosphorylation and evolution. Royal Society Chemistry, Cambridge, pp 491–521

LaRoche J, vanderStaay GWM, Partensky F, Ducret A, Aebersold R, Li R, Golden SS, Hiller RG, Wrench PM, Larkum AWD and Green BR (1996) Independent evolution of the prochlorophyte and green plant chlorophyll a/b light-harvesting proteins. Proc Natil Acad Sci U S A 93: 15244–15248

Leonhardt K and Straus NA (1992) An iron stress operon involved in photosynthetic electron transport in the marine cyanobacterium Synechococcus sp. PCC 7002. J Gen Microbiol 138: 1613–1621

Leonhardt K and Straus NA (1994) Photosystem II genes isiA, psbDI and psbC in Anabaena sp. PCC 7120: cloning, sequencing and the transcriptional regulation in iron-stressed and iron-repleted cells. Plant Mol Biol 24: 63–73

Li N, Zhao JD, Warren PV, Warden JT, Bryant DA and Golbeck JH (1991) PsaD is required for the stable binding of PsaC to the photosystem I core protein of Synechococcus sp. PCC 6301. Biochemistry 30: 7863–7872

Li Y, Lucas MG, Konovalova T, Abbott B, MacMillan F, Petrenko A, Sivakumar V, Wang R, Hastings G, Gu F, van Tol J, Brunel LC, Timkovich R, Rappaport F and Redding K (2004) Mutation of the putative hydrogen-bond donor to P700 of photosystem I. Biochemistry 43: 12634–12647

Lince MT and Vermaas W (1998) Association of His117 in the D2 protein of photosystem II with a chlorophyll that affects excitation: energy transfer efficiency to the reaction center. Eur J Biochem 256: 595–602

Loll B, Kern J, Saenger W, Zouni A and Biesiadka J (2005) Towards complete cofactor arrangement in the 3.0 A resolution structure of photosystem II. Nature 438: 1040–1044

Loll B, Kern J, Saenger W, Zouni A and Biesiadka J (2007) Lipids in photosystem II: interactions with protein and cofactors. Biochim Biophys Acta 1767: 509–519

McDermott G, Prince A, Freeer A, Hawthornwhite-Lawless M, Papiz M, Cogdell R and Asaacs N (1995) Crystal structure of an integral membrane ligth-harvesting complex from photosynthetic bacteria. Nature 374: 517–521

Mehari T, Qiao F, Scott MP, Nellis DF, Zhao J, Bryant DA and Golbeck JH (1995) Modified ligands to F_A and F_B in photosystem I. I. Structural constraints for the formation of iron-sulfur clusters in free and rebound PsaC. J Biol Chem 270: 28108–28117

Meimberg K, Lagoutte B, Bottin H and Muhlenhoff U (1998) The PsaE subunit is required for complex formation between photosystem I and flavodoxin from the cyanobacterium Synechocystis sp. PCC 6803. Biochemistry 37: 9759–9767

Miloslavina Y, Szczepaniak M, Muller MG, Sander J, Nowaczyk M, Rogner M and Holzwarth AR (2006) Charge separation kinetics in intact photosystem II core particles is trap-limited. A picosecond fluorescence study. Biochemistry 45: 2436–2442

Mitchell P (1961) Coupling of photophosphorylation to electron and hydrogen transfer by a chemiosmotic type of mechanism. Nature 191: 144–148

Mix LJ, Haig D and Cavanaugh CM (2005) Phylogenetic analyses of the core antenna domain: investigating the origin of photosystem I. J Mol Evol 60: 153–163

Mohanty P, Allakhverdiev SI and Murata N (2007) Application of low temperatures during photoinhibition allows characterization of individual steps in photodamage and the repair of photosystem II. Photosynth Res 94: 217–224

Müh F, Renger T and Zouni A (2008) Crystal structure of cyanobacterial photosystem II at 3.0 Å resolution: a closer look at the antenna system and the small membrane-intrinsic subunits. Plant Physiol Biochem 46: 238–264

Muhlenhoff U and Chauvat F (1996) Gene transfer and manipulation in the thermophilic cyanobacterium Synechococcus elongatus. Mol Gen Genet 252: 93–100

Müller MG, Niklas J, Lubitz W and Holzwarth AR (2003) Ultrafast transient absorption studies on photosystem I reaction centers from Chlamydomonas reinhardtii. I. A new interpretation of the energy trapping and early electron transfer steps in photosystem I. Biophys J 85: 3899–3922

Murray JW, Maghlaoui K, Kargul J, Ishida N, Lai TL, Rutherford AW, Sugiura M, Boussac A and Barber J (2008) X-ray crystallography identifies two chloride binding sites in the oxygen evolving centre of photosystem II. Energy Environ Sci 1: 161–166

Nakazato K, Toyoshima C, Enami I and Inoue Y (1996) Two-dimensional crystallization and cryoelectron microscopy of photosystem II. J Mol Biol 257: 225–232

Nanba O and Satoh K (1987) Isolation of a photosystem II reaction center consisting of D-1 and D-2 polypeptides and cytochrome b-559. Proc Natl Acad Sci U S A 84: 109–112

Nelson N (2009) Plant photosystem I—the most efficient nano-photochemical machine. J Nanosci Nanotechnol 9: 1709–1713

Nowaczyk MM, Hebeler R, Schlodder E, Meyer HE, Warscheid B and Rogner M (2006) Psb27, a cyanobacterial lipoprotein, is involved in the repair cycle of photosystem II. Plant Cell 18: 3121–3131

Ohad I, Dal Bosco C, Herrmann RG and Meurer J (2004) Photosystem II proteins PsbL and PsbJ regulate electron flow to the plastoquinone pool. Biochemistry 43: 2297–2308

Ohnishi N, Kashino Y, Satoh K, Ozawa S and Takahashi Y (2007) Chloroplast-encoded polypeptide PsbT is involved in the repair of primary electron acceptor QA of photosystem II during photoinhibition in Chlamydomonas reinhardtii. J Biol Chem 282: 7107–7115

Olesen K and Andréasson LE (2003) The function of the chloride Ion in photosynthetic oxygen evolution. Biochemistry 42: 2025–2035

Papageorgiou GC (2004) In: Govindjee (ed) Chlorophyll a fluorescence: A signature of hotosynthesis. Advances in photosynthesis and respiration, Vol. 19, Springer, Dordrecht

Peloquin JM and Britt RD (2001) EPR/ENDOR characterization of the physical and electronic structure of the OEC Mn cluster. Biochim Biophys Acta 1503: 96–111

Petrenko A, Maniero AL, van Tol J, MacMillan F, Li Y, Brunel LC and Redding K (2004) A high-field EPR study of P700+ in wild-type and mutant photosystem I from Chlamydomonas reinhardtii. Biochemistry 43: 1781–1786

Plato M, Krauss N, Fromme P and Lubitz W (2003) Molecular orbital study of the primary electron donor P700 of photosystem I based on a recent X-ray single crystal structure analysis. Chem Phys 294: 483–499

Ramesh VM, Gibasiewicz K, Lin S, Bingham SE and Webber AN (2004) Bidirectional electron transfer in photosystem I: accumulation of A0- in A-side or B-side mutants of the axial ligand to chlorophyll A0. Biochemistry 43: 1369–1375

Ramesh VM, Gibasiewicz K, Lin S, Bingham SE and Webber AN (2007) Replacement of the methionine axial ligand to the primary electron acceptor A0 slows the A0- reoxidation dynamics in photosystem I. Biochim Biophys Acta 1767: 151–160

Rappaport F and Diner BA (2009) Primary photochemistry and energetics leading to the oxidation of the (Mn)4Ca cluster and to the evolution of molecular oxygen in photosystem II. Coord Chem Rev 252: 259–272

Raymond J and Blankenship RE (2006) Evolutionary relationships among Type I photosynthetic reaction centers. In: Golbeck JH (ed) Advances in photosynthesis and respiration: the light-driven plastocyanin: ferredoxin oxicoreductase. Springer, Dordrecht, pp 669–681

Regel RE, Ivleva NB, Zer H, Meurer J, Shestakov SV, Herrmann RG, Pakrasi HB and Ohad I (2001) Deregulation of electron flow within photosystem II in the absence of the PsbJ protein. J Biol Chem 276: 41473–41478

Renger G (2008) Functional pattern of photosystem II in oxygen evolving organisms. In: Renger G (ed) Primary processes of photosynthesis: priniciples and apparatus. Part II: Reaction centers/ photosystems electron transport chains, photophosphorylation and evolution. Royal Society Chemistry, Cambridge, pp 237–290

Renger G and Renger T (2008) Photosystem II: the machinery of photosynthetic water splitting. Photosynth Res 98: 53–80

Rhee KH (2001) Photosystem II: the solid structural era. Annu Rev Biophys Biomol Struct 30: 307–328

Rögner M, Dekker JP, Boekema EJ and Witt HT (1987) Size, shape and mass of the oxygen-evolving photosystem-Ii complex from the Thermophilic Cyanobacterium Synechococcus Sp. FEBS Lett 219: 207–211

Rögner M, Mühlenhoff U, Boekema EJ and Witt HT (1990) Mono-, di-, and trimeric PS I reaction center complexes isolated from the thermophilic cyano-bacterium Synechoccus sp. Size, shape and activity. Biochim Biophys Acta 1015: 415–424

Rokka A, Suorsa M, Saleem A, Battchikova N and Aro EM (2005) Synthesis and assembly of thylakoid protein complexes: multiple assembly steps of photosystem II. Biochem J 388: 159–168

Roncel M, Boussac A, Zurita JL, Bottin H, Sugiura M, Kirilovsky D and Ortega JM (2003) Redox properties of the photosystem II cytochromes b559 and c550 in the cyanobacterium Thermosynechococcus elongatus. J Biol Inorg Chem 8: 206–216

Roszak AW, Howard TD, Southall J, Gardiner AT, Law CJ, Isaacs NW and Cogdell RJ (2003) Crystal structure of the RC-LH1 core complex from Rhodopseudomonas palustris. Science 302: 1969–1972

Rousseau F, Setif P and Lagoutte B (1993) Evidence for the involvement of PS I-E subunit in the reduction of ferredoxin by photosystem I. EMBO J 12: 1755–1765

Rumberg B, Schmidt-Mende P and Witt HT (1964) Different demonstrations of the coupling of two light reactions in photosynthesis. Nature 201: 466–468

Sakurai I, Shen JR, Leng J, Ohashi S, Kobayashi M and Wada H (2006) Lipids in oxygen-evolving photosystem II complexes of cyanobacteria and higher plants. J Biochem 140: 201–209

Salih GF and Jansson C (1997) Activation of the silent psbA1 gene in the cyanobacterium Synechocystis sp strain 6803 produces a novel and functional D1 protein. Plant Cell 9: 869–878

Sandmann G and Malkin R (1983) Iron-sulfur centers and activities of the photosynthetic electron transport chain in iron-deficient cultures of the blue-green Alga Aphanocapsa. Plant Physiol 73: 724–728

Santabarbara S, Kuprov I, Fairclough WV, Purton S, Hore PJ, Heathcote P and Evans MCW (2005) Bidirectional electron transfer in photosystem I: determination of two distances between P-700(+) and A(1)(-) in spin-correlated radical pairs. Biochemistry 44: 2119–2128

Sauer K and Yachandra VK (2004) The water-oxidation complex in photosynthesis. Biochim Biophys Acta 1655: 140–148

Savikhin S, Buck DR and Struve WS (1998) Toward level-to-level energy transfers in photosynthesis: the fenna-matthews-olson protein. J Phys Chem B 102: 5556–5565

Scheller HV, Jensen PE, Haldrup A, Lunde C and Knoetzel J (2001) Role of subunits in eukaryotic photosystem I. Biochim Biophys Acta 1507: 41–60

Schluchter WM, Shen G, Zhao J and Bryant DA (1996) Characterization of psaI and psaL mutants of Synechococcus sp. strain PCC 7002: a new model for state transitions in cyanobacteria. Photochem Photobiol 64: 53–66

Schödel R, Irrgang K-D, Voigt J and Renger G (1998) Rate of carotenoid triplet formation in solubilized light-harvesting complex II (LHCII) from spinach. Biophys J 75: 3143–3153

Schubert WD, Klukas O, Saenger W, Witt HT, Fromme P and Krauss N (1998) A common ancestor for oxygenic and anoxygenic photosynthetic systems: a comparison based on the structural model of photosystem I. J Mol Biol 280: 297–314

Sener M, Park S, Lu DY, Damjanovic A, Ritz T, Fromme P and Schulten K (2003) Excitation transfer dynamics in monomeric and trimeric forms of cyanobacterial photosystem I. Biophys J 84: 274–279

Sener MK, Lu DY, Ritz T, Park S, Fromme P and Schulten K (2002) Robustness and optimality of light harvesting in cyanobacterial photosystem I. J Phys Chem B 106: 7948–7960

Sener MK, Park S, Lu D, Damjanovic A, Ritz T, Fromme P and Schulten K (2004) Excitation migration in trimeric cyanobacterial photosystem I. J Chem Phys 120: 11183–11195

Setif P (2001) Ferredoxin and flavodoxin reduction by photosystem I. Biochim Biophys Acta 1507: 161–179

Setif P, Fischer N, Lagoutte B, Bottin H and Rochaix JD (2002) The ferredoxin docking site of photosystem I. Biochim Biophys Acta 1555: 204–209

Shen G, Antonkine ML, van der Est A, Vassiliev IR, Brettel K, Bittl R, Zech SG, Zhao J, Stehlik D, Bryant DA and Golbeck JH (2002a) Assembly of photosystem I. II. Rubredoxin is required for the in vivo assembly of F(X) in Synechococcus sp. PCC 7002 as shown by optical and EPR spectroscopy. J Biol Chem 277: 20355–20366

Shen G, Zhao J, Reimer SK, Antonkine ML, Cai Q, Weiland SM, Golbeck JH and Bryant DA (2002b) Assembly of photosystem I. I. Inactivation of the rubA gene encoding a membrane-associated rubredoxin in the cyanobacterium Synechococcus sp. PCC 7002 causes a loss of photosystem I activity. J Biol Chem 277: 20343–20354

Sippola K and Aro EM (2000) Expression of psbA genes is regulated at multiple levels in the cyanobacterium Synechococcus sp. PCC 7942. Photochem Photobiol 71: 706–714

Soitamo AJ, Sippola K and Aro EM (1998) Expression of psbA genes produces prominent 5′ psbA mRNA fragments in Synechococcus sp. PCC 7942. Plant Mol Biol 37: 1023–1033

Sommer F, Drepper F, Haehnel W and Hippler M (2004) The hydrophobic recognition site formed by residues PsaA-Trp651 and PsaB-Trp627 of photosystem I in Chlamydomonas reinhardtii confers distinct selectivity for binding of plastocyanin and cytochrome c6. J Biol Chem 279: 20009–20017

Sonoike K, Hatanaka H and Katoh S (1993) Small subunits of Photosystem I reaction center complexes from Synechococcus elongatus. II. The psaE gene product has a role to promote interaction between the terminal electron acceptor and ferredoxin. Biochim Biophys Acta 1141: 52–57

Srinivasan N, Karyagina I, Bittl R, van der Est A and Golbeck JH (2009) Role of the hydrogen bond from Leu722 to the A1A phylloquinone in photosystem I. Biochemistry 48: 3315–3324

Steffen R, Kelly AA, Huyer J, Doermann P and Renger G (2005) Investigations on the reaction pattern of photosystem II in leaves from *Arabidopsis thaliana* wild type plants and mutants with genetically modified lipid content. Biochemistry 44:3134–3142

Strickler MA, Hillier W and Debus RJ (2006) No evidence from FTIR difference spectroscopy that glutamate-189 of the D1 polypeptide ligates a Mn ion that undergoes oxidation during the S0 to S1, S1 to S2, or S2 to S3 transitions in photosystem II. Biochemistry 45: 8801–8811

Strickler MA, Walker LM, Hillier W, Britt RD and Debus RJ (2007) No evidence from FTIR difference spectroscopy that aspartate-342 of the D1 polypeptide ligates a Mn ion that undergoes oxidation during the S0 to S1, S1 to S2, or S2 to S3 transitions in photosystem II. Biochemistry 46: 3151–3160

Strickler MA, Hwang HJ, Burnap RL, Yano J, Walker LM, Service RJ, Britt RD, Hillier W and Debus RJ (2008) Glutamate-354 of the CP43 polypeptide interacts with the oxygen-evolving Mn4Ca cluster of photosystem II: a preliminary characterization of the Glu 354Gln mutant. Philos Trans R Soc Lond B Biol Sci 363: 1179–1187; discussion 1187–1178

Stroebel D, Choquet Y, Popot JL and Picot D (2003) An atypical haem in the cytochrome b(6)f complex. Nature 426: 413–418

Strotmann H and Weber N (1993) On the function of PsaE in chloroplast photosystem I. Biochim Biophys Acta 1143: 204–210

Stewart DH and Brudvig GW (1998) Cytochrome b559 of photosystem II. Biochim Biophys Acta 1367: 63–87

Sugimoto I and Takahashi Y (2003) Evidence that the PsbK polypeptide is associated with the photosystem II core antenna complex CP43. J Biol Chem 278: 45004–45010

Suorsa M, Regel RE, Paakkarinen V, Battchikova N, Herrmann RG and Aro EM (2004) Protein assembly of photosystem II and accumulation of subcomplexes in the absence of low molecular mass subunits PsbL and PsbJ. Eur J Biochem 271: 96–107

Takahashi T, Inoue-Kashino N, Ozawa S, Takahashi Y, Kashino Y and Satoh K (2009) Photosystem II complex *in vivo* is a monomer. J Biol Chem 284: 15598–15606

Telfer A (2002) What is beta-carotene doing in the photosystem II reaction centre? Philos Trans R Soc Lond B Biol Sci 357: 1431–1439

Telfer A (2005) Too much light? How beta-carotene protects the photosystem II reaction centre. Photochem Photobiol Sci 4: 950–956

Telfer A, Frolov D, Barber J, Robert B and Pascal A (2003) Oxidation of the two beta-carotene molecules in the photosystem II reaction center. Biochemistry 42: 1008–1015

Thangaraj B, Jolley CC, Sarrou I, Bultema JB, Greyslak J, Whitelegge JP, Lin S, Kouřil R, Subramanyam R, Boekema EJ and Fromme P (2011) Efficient light harvesting in a dark, hot, acidic environment: the structure and function of PSI-LHCI from Galdieria sulphuraria. Biophys J 100: 135–143

Tracewell CA and Brudvig GW (2003) Two redox-active beta-carotene molecules in photosystem II. Biochemistry 42: 9127–9136

Tsiotis G, Psylinakis M, Woplensinger B, Lustig A, Engel A and Ghanotakis D (1999) Investigation of the structure of spinach photosystem II reaction center complex. Eur J Biochem 259: 320–331

Umate P, Schwenkert S, Karbat I, Dal Bosco C, Mlcochova L, Volz S, Zer H, Herrmann RG, Ohad I and Meurer J (2007) Deletion of PsbM in tobacco alters the QB site properties and the electron flow within photosystem II. J Biol Chem 282: 9758–9767

van der Est A, Valieva AI, Kandrashkin YE, Shen G, Bryant DA and Golbeck JH (2004) Removal of PsaF alters forward electron transfer in photosystem I: evidence for fast reoxidation of QK-A in subunit deletion mutants of Synechococcus sp. PCC 7002. Biochemistry 43: 1264–1275

van Gorkom HJ and Yocum CF (2005) The calcium and chloride cofactors. In: Wydrzynski T, Satoh K (eds) Photosystem II: the water/plastoquinone oxido-reductase in photosynthesis. Advances in photosynthesis and respiration, Vol 22. Springer, Dordrecht, pp 307–327

van Mieghem F, Brettel K, Hillmann B, Kamlowski A, Rutherford AW and Schlodder E (1995) Charge recombination reactions in photosystem II. I. Yields, recombination pathways, and kinetics of the primary pair. Biochemistry 34: 4798–4813

van Thor JJ, Geerlings TH, Matthijs HC and Hellingwerf KJ (1999) Kinetic evidence for the PsaE-dependent transient ternary complex photosystem I/Ferredoxin/Ferredoxin:NADP(+) reductase in a cyanobacterium. Biochemistry 38: 12735–12746

Vanselow C, Weber AP, Krause K and Fromme P (2009) Genetic analysis of the photosystem I subunits from the red alga, Galdieria sulphuraria. Biochim Biophys Acta 1787: 46–59

Vasil'ev S and Bruce D (2004) Optimization and evolution of light harvesting in photosynthesis: the role of antenna chlorophyll conserved between photosystem II and photosystem I. Plant Cell 16: 3059–3068

Vasil'ev S, Brudvig GW and Bruce D (2003) The X-ray structure of photosystem II reveals a novel electron transport pathway between P680, cytochrome b559 and the energy-quenching cation, ChlZ+. FEBS Lett 543: 159–163

Vasil'ev S, Shen JR, Kamiya N and Bruce D (2004) The orientations of core antenna chlorophylls in photosystem II are optimized to maximize the quantum yield of photosynthesis. FEBS Lett 561: 111–116

Vass I and Aro E (2008) Photoinhibition of photosynthetic electron transport. In: Renger G (ed) Primary processes of photosynthesis: priniciples and apparatus. Part I: Photophysical principles, pigments and light harvesting/adaptation/stress. Royal Society Chemistry, Cambridge, pp 393–425

Vass I, Cser K and Cheregi O (2007) Molecular mechanisms of light stress of photosynthesis. Ann N Y Acad Sci 1113: 114–122

Wada H and Mizusawa N (2009) The role of phosphatidylglycerol in photosynthesis. In: Wada H, Murata N (eds) Lipids in photosynthesis: essential and regulatory functions, Springer, Dordrecht, pp 243–263

Wang R, Sivakumar V, Li Y, Redding K and Hastings G (2003) Mutation induced modulation of hydrogen bonding to P700 studied using FTIR difference spectroscopy. Biochemistry 42: 9889–9897

Watanabe T, Kobayashi M, Hongu A, Nakazato M and Hiyama T (1985) Evidence, that a chlorophyll a′ dimer constitutes the photochemical reaction centre 1 (P700) in photosynthetic apparatus. FEBS Lett 235: 252–256

Witt HT (1971) Coupling of quanta, electrons, field ions and phosphorylation in the functional membrane of photosynthesis. Quart Res Biophys 4: 365–477

Witt HT (2004) Steps on the way to building blocks, topologies, crystals and X-ray structural analysis of Photosystems I and II of water-oxidizing photosynthesis. Photosynth Res 80: 86–107

Xu Q, Yu L, Chitnis VP and Chitnis PR (1994) Function and organization of photosystem I in a cyanobacterial mutant strain that lacks PsaF and PsaJ subunits. J Biol Chem 269: 3205–3211

Yamashita E, Zhang H and Cramer WA (2007) Structure of the cytochrome b6f complex: quinone analogue inhibitors as ligands of heme cn. J Mol Biol 370: 39–52

Yang F, Shen G, Schluchter WM, Zybailov BL, Ganago AO, Vassiliev IR, Bryant DA and Golbeck JH (1998) Deletion of the PsaF polypeptide modifies the environment of the redox active phylloquinone (A_1). Evidence for unidirectionality of electron transfer in photosystem I. J Phys Chem B 102: 8288–8299

Yano J, Kern J, Irrgang KD, Latimer MJ, Bergmann U, Glatzel P, Pushkar Y, Biesiadka J, Loll B, Sauer K, Messinger J, Zouni A and Yachandra VK (2005) X-ray damage to the Mn4Ca complex in single crystals of photosystem II: a case study for metalloprotein crystallography. Proc Natl Acad Sci U S A 102: 12047–12052

Yano J and Yachandra VK (2008) Where water is oxidized to dioxygen: structure of the photosynthetic Mn4Ca cluster from X-ray spectroscopy. Inorg Chem 47: 1711–1726

Yano J and Yachandra VK (2009) X-ray absorption spectroscopy. Photosynth Res 102: 241–254

Yoshida E, Nakamura A and Watanabe T (2003) Reversed-phase HPLC determination of chlorophyll a′ and naphthoquinones in photosystem I of red algae: existence of two menaquinone-4 molecules in photosystem I of Cyanidium caldarium. Anal Sci 19: 1001–1005

Yu J and Vermaas WF (1993) Synthesis and turnover of photosystem II reaction center polypeptides in cyanobacterial D2 mutants. J Biol Chem 268: 7407–7413

Yu L, Bryant DA and Golbeck JH (1995a) Evidence for a mixed-ligand [4Fe-4S] cluster in the C14D mutant of PsaC. Altered reduction potentials and EPR spectral properties of the F_A and F_B clusters on rebinding to the P700-F_X core. Biochemistry 34: 7861–7868

Yu L, Vassiliev IR, Jung YS, Bryant DA and Golbeck JH (1995b) Modified ligands to F_A and F_B in photosystem I. II. Characterization of a mixed ligand [4Fe-4S] cluster in the C51D mutant of PsaC upon rebinding to P700-F_x cores. J Biol Chem 270: 28118–28125

Zhao J, Li N, Warren PV, Golbeck JH and Bryant DA (1992) Site-directed conversion of a cysteine to aspartate leads to the assembly of a [3Fe-4S] cluster in PsaC of photosystem I. The photoreduction of F_A is independent of F_B. Biochemistry 31: 5093–5099

Zouni A, Jordan R, Schlodder E, Fromme P and Witt HT (2000) First photosystem II crystals capable of water oxidation. Biochim Biophys Acta 1457: 103–105

Zouni A, Witt HT, Kern J, Fromme P, Krauß N, Saenger W and Orth P (2001) Crystal structure of photosystem II from Synechococcus elongatus at 3.8 Å resolution. Nature 409: 739–743

Chapter 13
Mechanism of Photosynthetic Production and Respiratory Reduction of Molecular Dioxygen: A Biophysical and Biochemical Comparison

Gernot Renger and Bernd Ludwig

13.1 Introduction

In terms of thermodynamics all biological organisms represent open systems that are far from equilibrium by an estimated average value of 20–25 kJ/mol (Morowitz 1978). Therefore the development and sustenance of living matter requires fluxes of Gibbs free energy[1] as indispensable prerequisite (Nicolls and Prigogine 1977; Morowitz 1978). This condition is satisfied by the electro-magnetic near UV, visible and near infrared radiation from the extraterrestrial nuclear fusion reactor, the sun which acts as unique Gibbs free energy source of the biosphere. This fundamental principle can be summarized in a single sentence "life is bottled sunshine" (Wynword Read Matterdon of Man 1924), in spite of the existence of organisms which live in a special environment (e.g. in "the black smokers") and use another form(s) of Gibbs free energy as driving force. The underlying thermodynamic principles of the exploitation of solar radiation through photosynthesis were outlined in 1845 by R. J. Mayer (first law of thermodynamics) and in 1886 by L. Boltzmann (second law of thermodynamics) (see Boltzmann 1905).

Among living matter two basically different types of organisms are to be distinguished: (1) photoautotrophs which can directly use solar radiation as the driving force for their living processes and (2) photoheterotrophs which are unable to directly exploit solar energy and therefore need the uptake of energy rich substances (food) that are eventually synthesized by photoautrotrophs and serve as Gibbs free energy source through catabolic digestions.

[1] The Gibbs free energy is defined as $G=H-TS$, where H and S are the enthalpy and entropy, respectively, and T is the absolute temperature (see Atkins 2001)

G. Renger (✉)
Max-Volmer-Laboratorium für Biophysikalische Chemie,
Technische Universität Berlin, Berlin, Germany
e-mail: gernot.renger@mailbox.tu-berlin.de

B. Ludwig
Institute Biochemistry, Biocenter of Goethe University Inst,
Max-von-Laue-Str. 9, 60438 Frankfurt, Germany

G. A. Peschek et al. (eds.), *Bioenergetic Processes of Cyanobacteria*,
DOI 10.1007/978-94-007-0388-9_13,

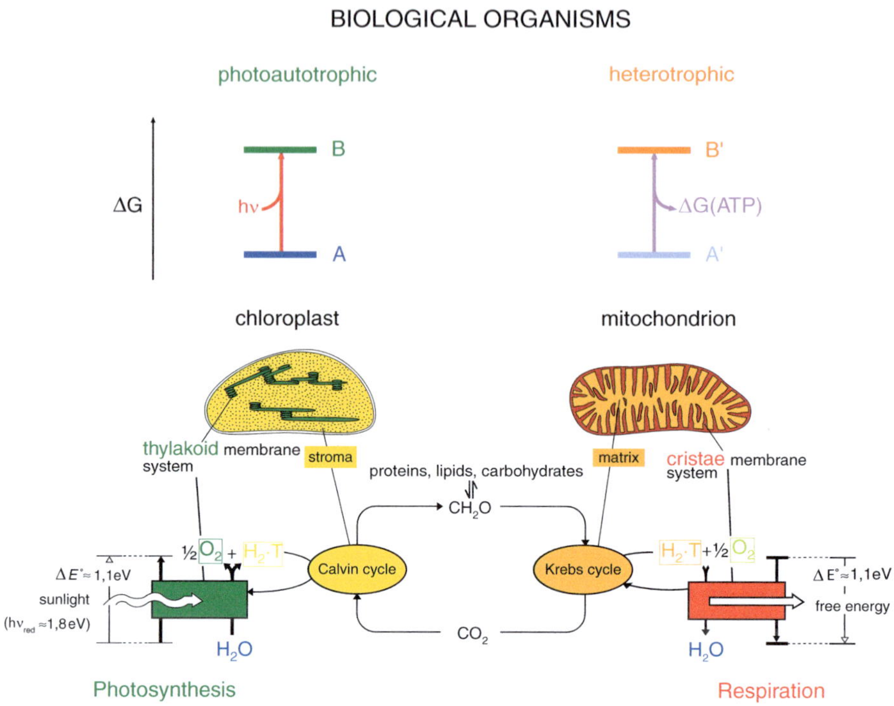

Fig. 13.1 *Top*: General scheme of bioenergetic transformation of solar radiation and food catabolism by photoautotrophic (*left side*) and heterotrophic (*right side*) organisms, respectively. *Middle*: Overall organization scheme of bioenergetics in higher eukaryotic organisms. The thermodynamically essential processes take place in chloroplasts (*left side*) and mitochondria (*right side*). *Bottom*: The reactions of photosynthetic water splitting and respiratory water formation are symbolized by *green* and *red* boxes, respectively; the complex anabolic and catabolic reaction sequences leading to CO_2 fixation and release, respectively, by colored ellipses. For the sake of simplicity the multitude of processes which are responsible for the great variety of biological material (*proteins*, *lipids*, *carbohydrates*) are not explicitly shown (for further details, see text)

The thermodynamics of these two types of direct and indirect solar energy exploitation can be simply summarized by a scheme that is shown in the top panel of Fig. 13.1. In photoautrophic organism the light flux increases the Gibbs free energy content of a suitable storage system by driving the transition from a level A to a higher level of state B. This process is referred to as photosynthesis and the specific amount of stored Gibbs free energy symbolized by $\Delta G^{\circ}_{\text{store}}(h\nu)$ depends on the properties of the components that are involved in the $A \xrightarrow{h\nu} B$ transition. In the heterotrophic organism the analogous reaction takes place in the opposite direction where food molecules with a higher Gibbs free energy content B′ are catabolized into a state A′ with a lower level and the Gibbs free energy drop is made available in form of a universal "bioenergetic currency", the ATP molecule. This process is denoted respiration (for the sake of simplicity no distinction will be made between anaerobic fermentation and aerobic respiration, see biochemistry textbooks for details).

Based on the hypothesis of van Niel (1941) the transition $A \rightarrow B$ in photosynthetic organism can be described by the overall reaction:

$$H_2X + T \xrightarrow{h\nu} X + TH_2 \tag{13.1}$$

where H_2X is a hydrogen donor and T a hydrogen acceptor substance (in most cases $NADP^+$) with the thermodynamic relation of $\Delta G°(H_2X/X) < \Delta G°(TH_2/T)$. The process (1) is coupled to the formation of ATP via a mechanism where directed proton flux through the enzyme ATP synthase provides the driving force for the endergonic reaction:

$$H^+ + ADP^{3^-} + HPO_4^{2^-} \rightarrow ATP^{4^-} + H_2O \tag{13.2}$$

(Mitchell 1961, for a review, see Junge 2008).

The products TH_2(NADPH) and ATP give rise to assimilation of the primordial elements carbon, nitrogen and sulphur. The most important among these processes is the CO_2 fixation into carbohydrates via the Calvin-Benson cycle (Calvin 1989; Benson 2002 and references therein) which is by far the quantitatively most important process on the earth surface with an estimated annual turnover rate of 200–300 billion tons of fixed CO_2 (Zamaraev and Permon 1980).

In heterotrophic organisms analogous hydrogen transfer reactions take place but in this case, in the opposite direction, i.e. from state B′ ($T'H_2$) to A′ (H_2X') with $\Delta G°(T'H_2/T') > \Delta G°(H_2X'/X')$:

$$T'H_2 + X' \rightarrow T' + H_2X' \tag{13.3}$$

and the available Gibbs free energy is transiently stored as ATP formed according to Eq. (13.2).

The efficiency of the fundamental bioenergetic processes Eqs. (13.1) and (13.3) depends on two parameters: (1) the specific storage capacity $\Delta G^{\circ}_{store}(h\nu)$ and $\Delta G°(ATP)$, respectively, and (2) the nature of the chemical compounds that enable a large scale performance of transitions $A \xrightarrow{h\nu} B$ and $B' \longrightarrow A'$ through both availability in sufficient amounts and control of the reactivity in order to avoid dissipative side reactions.

In the early stages of evolution the photoautotrophic organisms used substances like H_2, H_2S and small organic molecules (formate, acetate, methanol) that are potent electron donors However this mode of solar energy exploitation that is still existing in present day "primitive" living forms (for recent reviews, see Parson 2008 Vermeglio 2008) comprises limitations by both, substrate abundance and specific storage capacity of Gibbs free energy, i.e. comparatively small values of $\Delta G^{\circ}_{store}(h\nu)$.

From a Darwinian point of view it is not surprising that during a long lasting evolutionary process the fundamental bioenergetics of life have been optimized to the highest possible level. This development was restricted *per se* by the thermodynamic limit of water splitting into the elements H_2 and O_2 since living matter is inevitably connected with bulk water phases. Thus the maximum possible value of $\Delta G^{\circ}_{store,\ max}(h\nu)$ is 237.17 kJ/mol (Atkins 2001). In reality the hydrogen is chemically bound to molecules T like $NADP^+$ and therefore $\Delta G^{\circ}_{store}(h\nu)$ is lower by about 20 kJ/mol.

13.2 Redox Couple O_2/H_2O: The Key System of Gibbs Free Energy Transduction in the Biosphere

The bioenergetic limit of photosynthetic water splitting into molecular dioxygen and bound hydrogen (NADH) was reached about 2–3 billion years ago with the "invention" of a molecular machinery that enabled ancient procaryotic cyanobacteria to perform this process (Buick 1992; de Marais 2000; Xiong and Bauer 2002, for further reading, see Peschek 2008; Larkum 2008). This evolutionary step was the "big bang" in the biosphere which had two consequences of paramount importance: (a) the huge water pool of the earth surface became available as hydrogen source for living organisms and (b) the molecular dioxygen released as the "waste product" of photosynthetic water splitting led to the formation of an aerobic atmosphere (Kasting and Seifert 2002; Lane 2002) thus increasing by more than a factor of 10 the amount of Gibbs free energy that is extractable from the same food molecule (e.g. glucose) through aerobic respiration (for thermodynamic considerations, see Nicholls and Fergusson 1982; Renger 1983). Furthermore the oxygen in the atmosphere enabled the formation of the ozone layer in the stratosphere that established the protective "umbrella" against deleterious UV-B radiation (Worrest and Caldwell 1986).

The ability of the biosphere in using the redox couple H_2O/O_2 for solar energy storage by "loading" the system through photosynthetic water splitting and "discharging" it via aerobic respiration was the energetic prerequisite for the development and existence of all higher forms of life on our planet with the wonderful facets of flora and fauna. The middle panel of Fig. 13.1 shows the typical machineries that exist in all higher eukaryotic plants and animals, i.e. the cell organelle **chloroplasts** for photosynthesis and **mitochondria** for aerobic respiration. The overall reaction pattern of the bioenergetics of these organisms is schematically depicted in the bottom panel of Fig. 13.1.

An inspection of Fig. 13.1 (middle panel) reveals that both chloroplasts and mitochondria are characterized by the subdivision into a highly structured membranous system and a hydrophilic aqueous space of high viscosity. The key steps of Gibbs free energy transduction take place in the thylakoid membranes of chloroplasts and the cristae membrane of mitochondria whereas the great variety of metabolic and catabolic reactions are localized in the stroma and matrix "vessels" where the enzyme catalyzed Calvin-Benson and Krebs cycle, respectively, take place.

These considerations show that use and handling of the redox couple H_2O/O_2 is the "inner sanctum" of the bioenergetics of all higher forms of life on earth. Therefore the unravelling of the structural and mechanistic details of the molecular machinery of these most important processes in bioenergetics is a topic of world wide research activities.

Accordingly, the following description of this chapter is entirely focussed on the topics of water splitting in photosynthesis and "cold" combustion of metabolically bound hydrogen in the process of aerobic respiration.

Before going into details of these processes and their performance in biological systems, a few general properties of the redox chain $T(T')H_2/H_2O/O_2$ are worth considering.

The overall processes (13.1) and (13.3) with X and X′=oxygen comprise two partial reactions sequences:

$$2H_2O \rightleftarrows O_2 + 4H^+ + 4\ominus \tag{13.4}$$

and

$$2T(T') + 4H^+ + 4\ominus \rightleftarrows 2T(T')H_2 \tag{13.5}$$

Among these two "half" reactions the sequence (13.4) is the key of efficient Gibbs free energy exploitation in living matter (vide supra).

The sequence of the individual steps of the redox chemistry can be represented by Eq. (13.6).

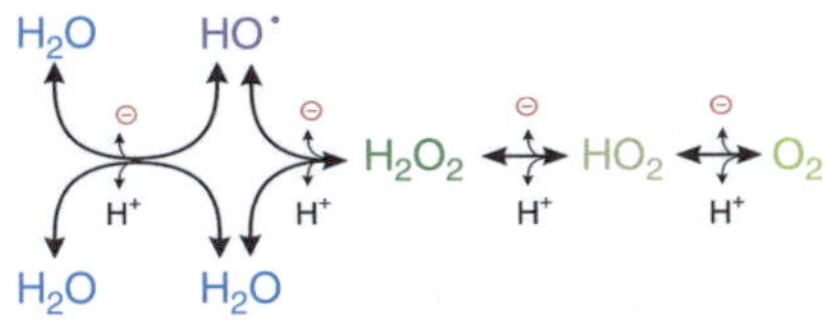

(13.6)

This reaction sequence comprises intermediates that are extremely detrimental species for biological organisms. The OH• radical is one of the most reactive species known to chemistry (it is isoelectronic with the fluorine atom) and gives rise to rapid destruction of the system. Therefore, the reaction pathways in both directions have to be properly tuned in order to avoid the formation and release of these species and to suppress undesired side reactions. This goal is achieved by "taming" the process through confinement of the chemistry of the H_2O/O_2 couple to structurally well defined catalytic sites containing redox active transition metal centers as will be outlined in Sects. 13.3.3 and 13.4.3.

The modulation of the reaction coordinate also implies a striking balancing of the energetics. A comparison of the sequences of the four 1-electron redox steps taking place in solution and in the enzyme catalyzed reactions shows the drastic differences between the individual steps in the former case which are characterized by quite different Gibbs free energy gaps (Anderson and Albu 1999) while in marked contrast those occurring in the enzymes exhibit nearly the same energetics as is illustrated for the oxidative pathway leading to dioxygen formation (see Sect. "Reaction Coordinate of Oxidative Water Cleavage").

An inspection of Eq. (13.6) reveals another property that is of high mechanistic relevance. The overall chemical reaction is coupled to the release or uptake of four protons in the forward or reverse direction, respectively. Regardless of the mechanistic details this feature shows that electron transfer (ET) and proton transfer (PT) steps are closely interrelated. The mode of coupling between ET and PT is a general problem of high relevance for many redox reactions (for reviews, see Cukier 2002; Hammes-Schiffer 2006; Huyn and Meyer 2007) and in particular for "dealing" with the redox couple H_2O/O_2 in bioenergetics.

In this respect it is important to note that the protein matrix plays a key role and that marked differences exist between the pathways in the forward and backward direction: in the highly endergonic photosynthetic water oxidation four protons are released per O_2 molecule formed (see Sect. 13.3.4) while the corresponding exergonic reduction of O_2 to H_2O is, in all canonial cases, energetically coupled to "uphill" H^+ pumping, thus giving rise to an overall ratio of 8 H^+/O_2 taken up from the bacterial cytoplasm or the mitochondrial matrix, respectively, when taking into account the "chemical" protons used for the formation of two water molecules (4 H^+/O_2) (see Sect. 13.4).

When considering the great advantages of the redox couple H_2O/O_2 for bioenergetics, the "other side of the coin" must not be ignored, i.e. the potential danger which is inhenrent in this system. It is important to note that biological organisms can only survive in an aerobic atmosphere because molecular dioxygen is characterized by a unique property: the electronic ground state of the O_2 molecule is the triplet state $^3\Sigma_g O_2$ (see Atkins 2001) whereas the vast majority of molecules that are constituents of a biological organism (proteins, lipids, carbohydrates) attain a singlet ground state. These properties imply that the chemical reaction of these molecules with O_2 comprises a transition state that is spin state "forbidden" and therefore the process is very sluggish, thus preventing rapid destruction of the organic material in an aerobic atmosphere. However, a transition into the excited singlet state $^1\Delta_g O_2$ (see Sect. 13.3.2.3) and/or formation of radicals like $O_2^{-\bullet}$ and $OH^\bullet$ eliminates this restriction and the reactive oxygen species (ROS) give rise to rapid degradation of biomolecules. As a consequence, molecular dioxygen acts in the biosphere as a "Janus-faced" species: it is indispensable for nourishing most of the life on earth through efficient Gibbs free energy extraction from food via aerobic respiration, but on the other hand it is a poison which destroys life (for further reading, see Gilbert 1981). The regulation of the reactivity of O_2 in oxygen enzymology is an important topic that will be outlined in more detail in Sect. 13.4.

13.3 Photosynthetic O_2-Production in Photosystem II (PS II)

The photosynthetic water splitting in all oxygen evolving organisms takes place in a multimeric operational unit referred to as Photosystem II (PS II) that is characterized by a striking complexity of polypeptide composition. The PS II core complex consists of at least 20 subunits (for a review, see Shi and Schröder 2004) and therefore markedly contrasts the conventional two or three subunit composition of the reaction centers (RCs) of anoxygenic photosynthetic bacteria (for a review see Parson 2008 and articles in Hunter et al. 2008). It is very intriguing that this PS II core complex evolved in a narrow geological period and that the composition of cofactors and polypeptides remained surprisingly invariant to evolutionary development (Raymond and Blankenship 2004). Interestingly, the largely invariant polypeptide patterns of the PS II core complex from ancient cyanobacteria up to higher plants also sharply differ from the marked changes of cytochrome *c* oxidase that catalyzes the reaction in opposite direction, i.e. the exergonic O_2 reduction to water. The polypeptide composition of these enzymes increased during evolution from three subunits in bacteria to thirteen in mammalians (see Sect. 13.4.2).

13.3.1 General Reaction Pattern of PS II

PS II acts as a light driven water:plastoquinone-oxido:reductase. The overall process

$$2PQ + 2H_2O \xrightarrow{h\nu} 2PQH_2 + O_2 \tag{13.7}$$

is coupled with the uptake of four protons from the cytoplasm in cyanobacteria (stroma in chloroplasts) and the release of four protons into the lumenal space of thylakoids. It comprises three different types of reaction sequences (for a review, see Renger 2008).

1. light induced formation of the radical pair $P680^{+\bullet}Q_A^{-\bullet}$

$$P680PheoQ_A \xrightarrow{\langle\varepsilon\rangle} {}^1P680^*PheoQ_A \xrightarrow{k_{PC}} P680^{+\bullet}Pheo^{-\bullet}Q_A \xrightarrow{k_{stab}} P680^{+\bullet}PheoQ_A^{-\bullet} \tag{13.8}$$

2. oxidative water splitting

$$4\oplus + 2H_2O \rightarrow O_2 + 4H^+_{lumen} \tag{13.9}$$

3. reductive plastoquinol formation

$$4H^+_{stroma} + 4\ominus + 2PQ \rightarrow PQH_2 \tag{13.10}$$

where $\langle\varepsilon\rangle$ symbolizes an electronic excitation (created by light absorption) that gives rise to excited singlet state formation of the photoactive chlorophyll *a* (Chl *a*) complex P680, Pheo is pheophytin *a* and Q_A a special noncovalently bound plastoquinone–9 molecule; $\oplus$ and $\ominus$ represent redox equivalents (including the cofactors involved, see Sects. 13.3.3 and 13.3.4) that provide the driving force for oxidative water splitting and reductive plastoquinol formation, respectively, H^+_{lumen} and H^+_{stroma} describe proton release into the lumen and proton uptake from the stroma (in cyanobacteria from cytoplasma).

13.3.2 Structure of PS II

The structure has been resolved by X-ray diffraction crystallography (XRDC) of PS II core complexes from thermophilic cyanobacteria *Thermosynechococcus* (T) *elongatus* (Zouni et al. 2001; Ferreira et al. 2004; Loll et al. 2005) and *T. vulcanus* (Kamiya and Shen 2003). At present the highest resolution available is at 2.9 Å (Guskov et al. 2009)[2]. The structure information are described in a recent book chapter (Zouni 2008) and review articles (Kern and Renger 2007; Müh and Zouni 2011). Therefore only a short summary on this topic will be presented here.

Figure 13.2 shows the structural details obtained for the dimeric PS II core complexes from *T. elongatus* (Loll et al. 2005; PDB identifier 2AXT). The top panel presents a side view of the arrangement of the transmembrane helices (TMHs) in

[2] A X-ray diffraction crystallographic (XRDC) structure model of 1.9 Å resolution was reported by J.-R. Shen, et al. (2010) at the 15th International Congress of Photosynthesis in Beijing (poster PS 6.5).

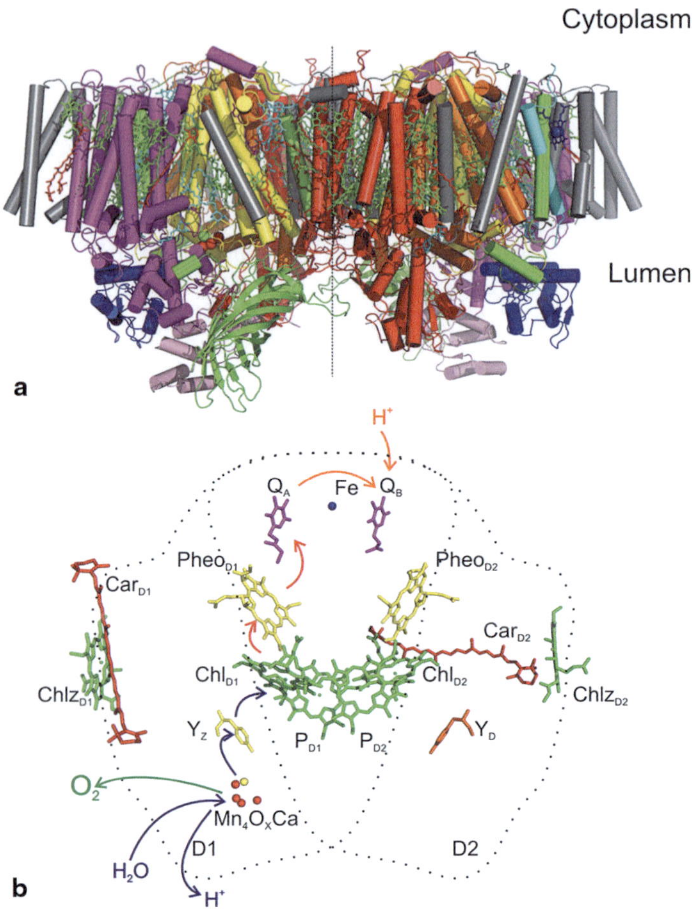

Fig. 13.2 View of dimeric PS II from *T. elongatus* and schematic representation of cofactor arrangement in the reaction center core. **a.** Side view along the membrane plane of a dimer of PS II from *T. elongatus* as derived from the 3.0 Å resolution electron density (Loll et al. 2005), Helices are depicted as cylinders, cofactors are drawn in stick models. The C2 axis, relating the two monomers in the dimer is indicated by a *black dashed line*. The large subunits *D1* (*yellow*) and *D2* (*orange*), forming the reaction center, are surrounded by CP43 (*magenta*) and CP47 (*red*) as well as several membrane intrinsic low molecular weight subunits (*grey*). The α and β subunits of the membrane intrinsic heterodimeric cyt b559 are shown in *green* and *cyan*, respectively. At the lumenal side the three extrinsic subunits PsbO (*green*), PsbV (*blue*) and PsbU (*light pink*) are interacting with large loop regions of D1/D2/CP43/CP47. Organic cofactors are coloured in *green* (*Chl*), *yellow* (*Pheo*), *magenta* (PQ9), *red* (carotenoids), *cyan* (lipids), and *blue* (heme), respectively, Ca^{2+} (*yellow*) *Fe* (*blue*) and *Mn* (*red*) are shown as spheres, figure was generated using Pymol. **b.** Cofactor arrangement in the D1/D2 heterodimer of the PS II core (view is along the membrane normal), colored in *green* (*Chl*), *yellow* (*Pheo*), *magenta* (PQ-9) and *red* (carotenoids), respectively, Ca^{2+} (*yellow*), *Fe* (*blue*) and *Mn* (*red*) are shown as spheres, the coordinating protein subunits D1 and D2 are indicated by a *dotted line*. *Red colored arrows* symbolize the direction of light induced charge separation (sequence 1), *blue* and *green arrows* the pathway of oxidative water splitting (sequence 2) and *orange arrows* the steps leading to plastoquinol formation (sequence 3). The figure was kindly provided by Jan Kern

the thylakoids membrane and the overall structures of the attached extrinsic PsbO, PsbU and PsbV subunits. Inspection of this structure reveals that the PS II core complex is rather flat on the cytosolic/stroma side whereas bulky extensions exist in the lumen, including the large loops of CP43 and CP47. These characteristics of PS II which are in marked contrast to PS I with a flat luminal side but pronounced extensions into the cytoplasm (cyanobacteria, see Fromme et al. 2008, see also chapter by Fromme and Grotjohann) or stroma (plants, see Amunts et al. 2007) are important for grana formation in chloroplasts (see Danielson et al. 2006).

The bottom panel of Fig. 13.2 presents the general arrangement of the cofactors that are involved in the three reaction sequences summarized by Eqs. (13.8)–(13.10). These are marked by arrows in red (sequence Eq. (13.8)), blue and green (sequence Eq. (13.9)) and yellow (sequence Eq. (13.10)). It shows that all cofactors participating in this reaction are embedded into a heterodimeric protein matrix consisting of polypeptides D1 and D2. Interestingly, in cytochrome *c* oxidase (COX) which catalyzes the reverse process, i.e. O_2 reduction to H_2O, all cofactors are also bound to a protein matrix consisting of only two polypeptides (see Sect. 13.4.2.2).

A highly symmetric array is seen with respect to the vertical axis crossing the nonheme iron (NHFe) center but with one marked exception, i.e. the manganese containing catalytic site of the WOC (see Sect. "Structure of the WOC") is largely displayed to the side of D1 subunit. Furthermore this Mn_4O_XCa cluster (marked as red spheres) is in addition ligated by at least one residue (Glu 354) of CP43.

13.3.3 Light Induced Charge Separation

We have recently described details on the reaction sequence (13.8) in two book chapters (Renger and Holzwarth 2005; Renger 2008) and a review article (Renger and Renger 2008) and therefore will restrict the description to the essential features of the process.

13.3.3.1 Nature of the Cofactors

The overall functional pattern and the cofactor array of reaction sequence (13.8) closely resembles that of the corresponding process in reaction centers of anoxygenic purple bacteria (PBRCs) but one drastic difference is most important: the properties of the photochemically active pigment P680 of PS II are markedly different from that of the "special pair" P_{RC} in PBRCs. The energetics of oxidative water splitting indispensably requires the generation of electron holes with sufficient oxidizing power. This goal has been achieved by incorporation of Chl *a* into a proteinaceous microenvironment with a low dielectric constant (Hasegawa and Noguchi 2005). In this way the midpoint potential of the redox couple $P680/P680^{+\bullet}$ exceeds the corresponding value of Chl *a*/Chl $a^{+\bullet}$ in solution (Kobayashi et al. 2007) by about 400 mV and reaches an estimated number of about −1.25 V (Rappaport and Diner 2008).

The nature of $P680^{+\bullet}$ is a matter of debate. In contrast to PBRCs where the photochemically active pigment is a well defined special pair P_{RC} of two bacteriochlorophyll (BChl) molecules (Parson 2008) the constitution of P680 is more complex. The cofactor array in the bottom panel exhibits a spatial array of four Chl *a* and two Pheo *a* molecules that closely resembles that of the four BChls and BPheos in PBRCs (see Lancaster 2008). However the excitonic coupling is markedly different. In PBRCs the two BChls forming the special pair P_{RC} are strongly coupled thus giving rise to a lower exciton level with a marked red shifted absorption. Accordingly P_{RC} is clearly separated spectrally from the two other BChls and BPheo's. On the other hand, the excitonic coupling between two Chl *a* molecules (P_{D1} and P_{D2}) forming a "special pair" like configuration in PS II is still slightly larger than the interaction with the other four chlorins but this is not sufficient for a significant spectral band shift. A theoretical analysis reveals that the lowest exciton state is predominantly located on Chl_{D1} rather than on the "P_{D1}–P_{D2} dimer" (Raszewski et al. 2008, see also Renger and Renger 2008). These differences have an important consequence: in PBRCs the states $^1P^*_{RC}$, $^3P_{RC}$ and $P^{+\bullet}_{RC}$ are confined to the same special pair P_{RC} (see Parson 2008 and chapters in Hunter et al. 2008) but this is definitely not the case for $^1P680^*_{RC}$, $^3P680_{RC}$ and $P680^{+\bullet}_{RC}$ and therefore the "historical" symbol P680 can be used only as operational term in order to avoid misinterpretations.

A critical survey of the literature leads to the conclusion that P680 is best described as a pigment protein complex (Chl $a)_4$ $(Pheo)_X$ with x = 0. 1 or 2 when considering the state $^1P680^*$ (Renger and Holzwarth 2005; Renger and Renger 2008). In the case of $P680^{+\bullet}$ a semantic problem emerges because two different forms exist, $P680^{+\bullet}_1$ and $P680^{+\bullet}_2$ with the electron hole located on Chl_{D1} and on P_{D1}–P_{D2}, respectively (see Sect. 13.3.3.2), whereas $^3P680$ is most likely $^3Chl_{D1}$ (Schlodder et al. 2008).

The electron acceptor of the primary charge separation is the Pheo *a* molecule bound to polypeptide D1. This $Pheo_{D1}$ forms a hydrogen bond with Gln 130 in cyanobacteria that is replaced by Glu in plants. The possible role of this hydrogen bond is not yet clarified (for discussion, see Renger 2008).

Cofactor Q_A is a tightly, but noncovalently bound PQ-9 molecule that acts as a 1-electron acceptor under physiological conditions. The binding pocket in polypeptide D2 is rather hydrophobic and formed by several amino acid residues (for structural details of this binding site, see Loll et al. 2005; Kern and Renger 2007; Zouni 2008; Guskov et al. 2009).

The functional connection of $P680^{+\bullet}$ with the WOC occurs via the redox active Tyr161 of polypeptide D1 (Y_Z) (see Sect. 13.3.4.2).

In addition to the cofactors of the reaction sequence (13.8) the D1/D2 heterodimeric protein matrix contains additional redox active groups: two carotenoids (Car_{D1} and Car_{D2}), two chlorophylls ($Chlz_{D1}$ and $Chlz_{D2}$) and the tyrosine residue Y_D (see Fig. 13.2, bottom panel). Furthermore cytochrome *b*559 as a constituent of all PS II core complexes (not shown in Fig. 13.3) participates in light induced electron transfer reactions that probably are part of a protective mechanism under stress conditions (see Kamiskaya et al. 2007 and references therein).

13.3.3.2 Energetics and Kinetics of Formation of the Radical Ion Pairs $P680^{+\bullet}$ $Pheo^{-\bullet}$ and $P680^{+\bullet}Q_A^{-\bullet}$

The energetics of the light induced charge separation are determined by the level of the lowest excited singlet state $^1P680^*$ and the midpoint potentials E_m of the redox couples $P680/P680^{+\bullet}$, $Pheo^{-\bullet}/Pheo$ and $Q_A^{-\bullet}/Q_A$. For the sake of clarity, at first a general comment is required. The *midpoint potentials* E_m determined via equilibrium redox titrations provide only limited information because these are thermodynamic parameters that reflect *per definitionem* time independent values. Equilibrium conditions, however, are rarely attained in systems at work. The redox active cofactors are embedded into a protein matrix that responds to charge transfer reactions. As a consequence, relaxation processes of the protein take place that can significantly affect the overall energetics. Likewise the dynamics of charged redox groups in the microenvironment modulate the energetics through electrostatic interactions. In order to avoid misleading discussions and incorrect conclusions a clear distinction between two parameters is indispensable: "*working potential*" E_m^{work} and "*midpoint potential*" E_m. In general E_m^{work} is a time dependent parameter that describes the energetics of a redox group under the functional conditions. The values of E_m^{work} and E_m can differ by tens to hundreds of mV. The energy gap of the pair $^1P680^*/P680$ is 1.83 eV assuming a 680 nm electronic transition. Redox titration of the couple $P680/P680^{+\bullet}$ cannot be performed because the strongly oxidizing conditions that must be applied to determine the E_m value modifies all chlorophyll and carotenoid molecules. Therefore only indirect estimations are available (for a review see Renger and Holzwarth 2005). A value of about +1.25 V is currently discussed to be a reliable number (Rappaport and Diner 2008).

Direct redox titrations can be performed for both couples $Pheo^{-\bullet}/Pheo$ and $Q_A^{-\bullet}/Q_A$. As the reduction of Pheo necessarily comprises the formation of $Q_A^{-\bullet}$ the E_m value of −610 mV determined for $Pheo^{-\bullet}/Pheo$ by redox titrations (Klimov et al. 1979) contains the contribution due to the electrostatic effect of $Q_A^{-\bullet}$. After its subtraction a value of −525 mV was obtained for E_m of $Pheo^{-\bullet}/Pheo$ in PS II core complexes where Q_A is oxidized, i.e. of a functionally competent ("open") reaction center (RC) (Gibasiewicz et al. 2001). Similar values were recently reported for PS II from cyanobacteria (Kato et al. 2009, Allakhverdiev et al. 2010) but these numbers were not corrected for the affect of $Q_A^{-\bullet}$.

For the E_m of $Q_A^{-\bullet}/Q_A$ a collection of data from numerous experiments shows a clustering of values at around −300, −100, 0 mV and +50 to +100 mV (Krieger et al. 1995). Theoretical calculations strongly depend on the underlying XRDC structure data. Values of −148 and −10 mV were reported for E_m when using the models of 3.5 and 3.0 Å resolution, respectively (Ishikita and Knapp 2005). This variance reflects intrinsic uncertainties due to limited information on details of the microenvironment, in particular of hydrogen bonding to the *p*-benzoquinone head group of Q_A. Two further complications emerge: at first studies on mutants of thermophilic cyanobacteria led to the proposal of a dynamic equilibrium between two conformational states with different $E_m(Q_A^{-\bullet}/Q_A)$ values (Rappaport et al. 2005) and secondly the possibility cannot be excluded that the midpoint potentials of $Q_A^{-\bullet}/Q_A$ are not the same in different species (mesophilic and thermophilic cyanobacteria, green algae and higher plants, see Fufezan et al. 2007). Latest spectro-electrochenical studies on PS II core complexes from *Thermosynechococcus* (*T.*) *elongatus* in the presence of redox mediators lead to E_m values of −140 mV (Shibamoto et al. 2009). This finding reveals that no drastic differences exist between *T. elongatus* and spinach. Further experiments are required to clarify this point.

The E_m was also found to exhibit a marked upshift of about 150–160 mV in PS II lacking an intact WOC in samples from both, *T. elongatus* (Shibimoto et al. 2009) and spinach (Krieger et al. 1995). A smaller shift is observed when the native PQ-9 bound to the Q_B pocket is replaced by different herbicides (Krieger-Lisztay and Rutherford 1998). Direction and magnitude of this shift depend on the nature of the herbicide molecule (Fufezan et al. 2007).

Based on these findings it is clear that straightforward and unambiguous general energetics cannot be presented for the redox couple $Q_A^{-\bullet}/Q_A$ in PS II. It seems reasonable to assume that an $E_m(Q_A^{-\bullet}/Q_A)$ value of the order of −100 mV properly reflects the reduction behaviour of $Q_A^{-\bullet}$ in "open" PS II complexes with an intact WOC and of about +50 mV in systems lacking a intact WOC (for further discussion, see Renger and Renger 2008).

As a summary of our consideration on the energetics of PS II it turns out that the radical pair $P680^{+\bullet}Q_A^{-\bullet}$ stores about 1.3 eV, i.e. 70% of the energy of a red photon of 680 nm.

Kinetics and mechanism of the reaction sequence (13.8) are not fully resolved because the individual steps leading to the radical pair $P680^{+\bullet}$ $Pheo^{-\bullet}$ cannot be directly monitored due to spectral congestion of the cofactors that are involved. Based on our current state of knowledge (for a review, see Renger and Renger 2008)[3] it appears most likely that the primary step is the electron ejection from $^1Chl_{D1}^*$ with $Pheo_{D1}$ acting as acceptor followed by very rapid electron transfer from P_{D1} to Chl_{D1} thus forming $P_{D1}^{+\bullet}Pheo_{D1}^{-\bullet}$. Accordingly the reaction sequence can be summarized by:

$$^1P680^*\,PheoQ_A \overset{k_{PC}}{\longleftrightarrow} P680_1^{+\bullet}\,Pheo_{D1}^{-\bullet}Q_A \overset{k_1}{\longleftrightarrow} P680_2^{+\bullet}\,Pheo_{D1}^{-\bullet}Q_A \overset{k_{stab}}{\longleftrightarrow} P680_2^{+\bullet}\,Pheo_{D1}Q_A^{-\bullet} \tag{13.11}$$

where $P680_1^{+\bullet} = Chl_{D1}^{+\bullet}\,P_{D1}\,P_{D2}$ and $P680_2^{+\bullet} = Chl_{D1}\,P_{D1}^{+\bullet}\,P_{D2}$ (actually the probability of the hole electron localization is not an integer and the symbols are a simplified description to illustrate the sequence of redistribution within $P680^{+\bullet}$).

Direct kinetic measurements were only performed for the last step of $P680^{+\bullet}Q_A^{-\bullet}$ formation. The population of the reduced $Q_A^{-\bullet}$ was shown to occur with a rise time of 300±50 ps (Bernarding et al. 1994; Renger et al. 1995). The other rate con-

[3] Alternative (Shelaev et al. 2008) and complementary (Romeo et al. 2010) pathways with P_{D1} as primary electron donor have also been discussed.

stants of the sequence Eq. (13.11) were gathered from model based evaluation of the results of time resolved fluorescence (Miloslavina et al. 2006) and transient absorption changes in the visible (VIS) (Holzwarth et al. 2006) and mid infrared/VIS region (Groot et al. 2005). The conclusions on the mode of rate limitation are conflicting and the values of the rate constant for the excited energy transfer (EET) from CP43 and CP43 leading to $^1P680^*$ formation (k_{EET}) and for primary charge separation (k_{PC}) are quite different. It is a controversial debate as to whether the overall process is "transfer to the trap limited" or "trapping limited". In general, the rate constant k_{PC} is much larger in the former (order of $(100–300\ fs)^{-1}$ than in the latter case (order of a $(few\ ps)^{-1}$ (for a recent review, see Renger and Renger 2008). At present a straightforward answer cannot be offered. In this respect it should be kept in mind that a photochemically successful trapping of excitation energy can be only achieved if the radical pair $P680^{+\bullet}Pheo_{D1}^{-\bullet}$ is stabilized by electron transfer to Q_A. Taking this as the key criterion, then the overall transformation process of light into storable Gibbs free energy in PS II is certainly "trapping limited" with a rate constant of 300 ± 100 ps.

13.3.3.3 Reactions of the Radical Pair $P680^{+\bullet}Q_A^{-\bullet}$

In intact PS II complexes the radical ion pair $P680^{+\bullet}Q_A^{-\bullet}$ provides the driving forces for the reaction sequences of oxidative water splitting (see Sect. 13.3.4) and plastoquinol formation (see Sect. 13.3.5), respectively. However, in systems lacking an intact WOC or a functionally competent Q_B-site dissipative recombination reactions take place. Among these the recombination of the radical pair in the triplet configuration $^3[P680^{+\bullet}Pheo^{-\bullet}]$ is most critical because it leads to population of $^3P680$ which acts as sensitizer for singlet oxygen formation:

$$^3P680 + {}^3\Sigma_g O_2 \rightarrow P680 + {}^1\Delta_g O_2 \tag{13.12}$$

The singlet oxygen $^1\Delta_g O_2$ is rather reactive and gives rise to oxidative degradation reactions. The implication of $^3P680$ formation for photodamage is beyond the scope of this chapter (for reviews, see Pogson et al. 2005; Chow and Aro 2005; Vass and Aro 2008).

13.3.4 Oxidative Water Splitting

13.3.4.1 Overall Reaction Pattern: Kok Cycle

The oxidative splitting of two water molecules into molecular dioxygen and four protons requires the cooperation of four oxidizing redox equivalents (see Eq. (13.6)) that are generated as a result of the light-induced charge separation (see Eq. (13.11)). The mode of this coupling has been unravelled by the pioneering work of Pierre Joliot, Bessel Kok and their coworkers [see Renger (pp. 351–370) and

Joliot (pp. 371–378) in Govindjee et al. (2005)]. Joliot et al. (1969) found that dark adapted samples (green algae, isolated chloroplasts) illuminated by a train of single turnover flashes exhibit a characteristic period four oscillation of the oxygen yield per flash with maxima after the 3rd, 7th, 11th etc flash. This pattern is virtually independent of the relative content of functionally fully competent PS II complexes (Kok et al. 1970). These findings unambiguously show that the indispensable cooperation is achieved via a sequence of 1-electron transfer steps which take place in a functional unit referred to as water oxidizing complex (WOC) where the oxidizing equivalents are transiently stored as intermediates until after the accumulation of four, dioxygen is formed and released. The experimental result are consistently described by the scheme proposed by Kok et al. (1970) referred to as the Kok cycle.

Figure 13.3 shows a typical trace of oxygen yield measurements (top panel) and a modified version of the Kok cycle (bottom panel) comprising two essential extensions of the original scheme: (a) the stepwise oxidation of the WOC by $P680^{+\bullet}$ is mediated by the redox active tyrosine Y_Z (see Fig. 13.2b) and (b) "super reduced" redox states S_{-i} exist below the level of S_0 that are probably intermediates in the photoassembly of the WOC (see Renger 2001 and references therein). The storage states of the WOC are symbolized by S_i where i is the number of oxidizing redox equivalents above the lowest level S_0 that is populated under steady state turnover conditions.

Although PS II contains two redox active tyrosines (Y_Z in polypeptide D1 and Y_D in polypeptide D2) that are symmetrically arranged around P680, only Y_Z is sufficiently close to the Mn_4O_XCa cluster (see bottom panel of Fig. 13.2b) to mediate a rapid turnover of the WOC. In addition, the midpoint potentials of Y_Z and Y_D are quite different at physiological pH-values so that only Y_Z^{OX} provides the driving force for the Kok cycle while Y_D^{OX} can solely oxidize S_0 to S_1 via a rather slow reaction and Y_D actually acts as a reductant for the decay of S_2 and S_3 (see Fig. 13.3, for further details, see Renger 2007 and references therein). Therefore, in the following only the turnover of Y_Z will be described.

13.3.4.2 Y_Z Oxidation by $P680^{+\bullet}$

At first glance the cofactor arrangement might suggest that Y_Z oxidation by $P680^{+\bullet}$ is a simple first order reaction. However, the kinetics of this process in PS II complexes with an intact WOC are surprisingly complex and depend on the redox state S_i. The time course of each individual oxidation step of Y_Z in the Kok cycle can be approximately described by triphasic kinetics with half lifetimes of 25–70 ns ("fast" ns kinetics), 300–600 ns ("slow" ns kinetics) and 15–50 μs (slow μs kinetics). This pattern of multiphasic kinetics, which is basically the same in all sample types from ancient cyanobacteria up to higher plants (see Renger 2007 and references therein), is explained by relaxation processes within the protein environment (Kühn et al. 2004). For energetic reasons the oxidation of Y_Z by $P680^{+\bullet}$ is coupled to proton transfer (PT). Kinetic analyses (activation energies, kinetic

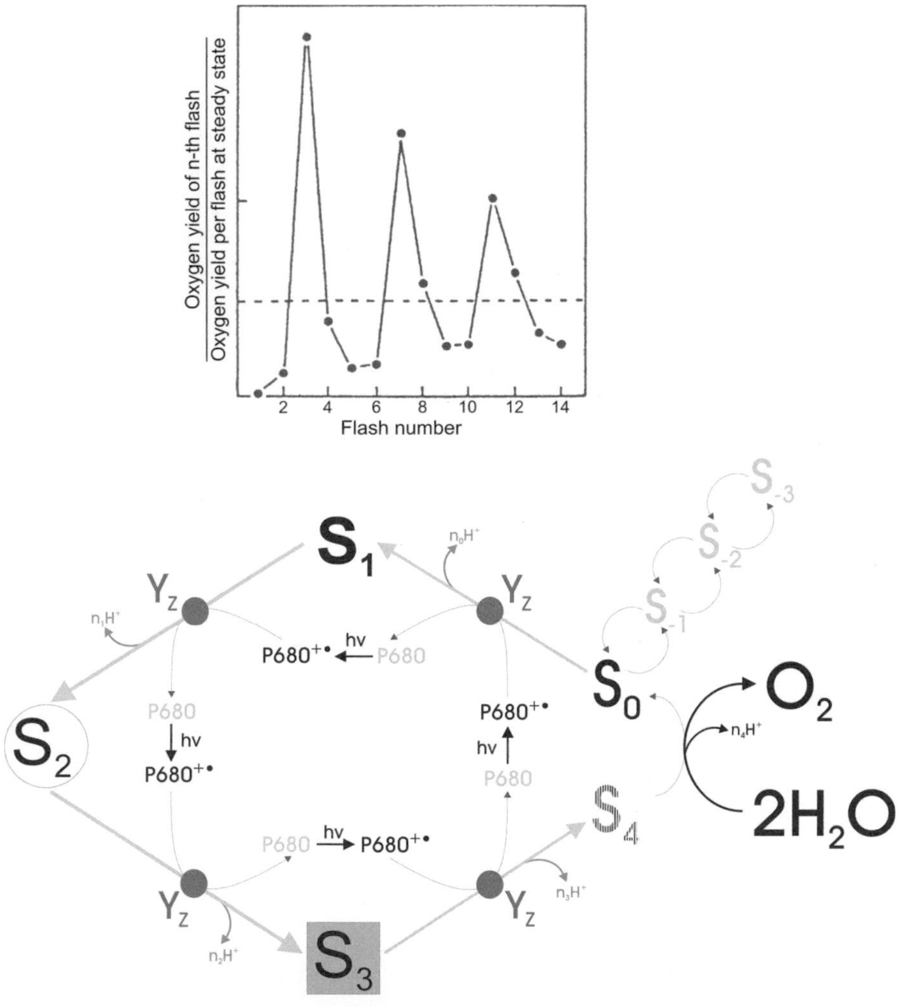

Fig. 13.3 Typical period four oscillation pattern of *oxygen yield* induced by a train of single turnover flashes in dark-adapted spinach thylakoids (*top panel*, data redrawn from Messinger et al. 1991) and extended Kok-cycle of oxidative water splitting (*bottom panel*). The *dashed line* in the *top panel* represents the normalized steady state oxygen yield per flash. *Bottom panel*: extended Kok-cycle of oxidative water cleavage. The photo oxidation of $P680^{+\bullet}$ is marked by *arrows* (*hμ bold italics*), the intermediary redox component is symbolized by a *dark grey* dot, the S_i states are symbolized in the following way: the *dark* stable redox state S_1 by a *bold capital*, the metastable redox state S_2 and S_3 by capitals encircled and *dark grey* background, respectively, and the transient "elusive" state S_4 by a *dashed symbol*, super reduced sates are marked in *grey*. For the sake of simplicity the reactions with the redox couple Y_D^{OX}/Y_D and the slow dark relaxation reactions of S_2 and S_3 are omitted [for a review, see Renger 1999; Messinger and Renger 2008]

H/D isotope exchange effects, pH dependence) revealed that the "fast" ns step requires a structurally well defined hydrogen bond between the OH group of Y_Z and His 190 (see Fig. 13.4 upper panel) and that different orbitals participate in the coupled ET/PT reactions (see Renger and Kühn 2007 and references therein). This type of multiple site electron and proton transfer (MS-EPT) is of general relevance for many biological redox processes (see Meyer et al. 2007). The normalized amplitude of the "fast" ns reaction is determined by the equilibrium constant of $P680^{+\bullet}Y_Z \rightleftharpoons P680Y_Z^{OX}$ (Kühn et al. 2004) and its kinetics are limited by the rate constant of the nonadiabatic electron transfer step (Marcus and Sutin 1985) which is characterized by a reorganization energy of about 0.5 eV (Renger et al. 1989). The slower kinetics leading to a higher extent of Y_Z oxidation ("slow" ns and μs kinetics) are ascribed to thermodynamic shifts of the equilib-

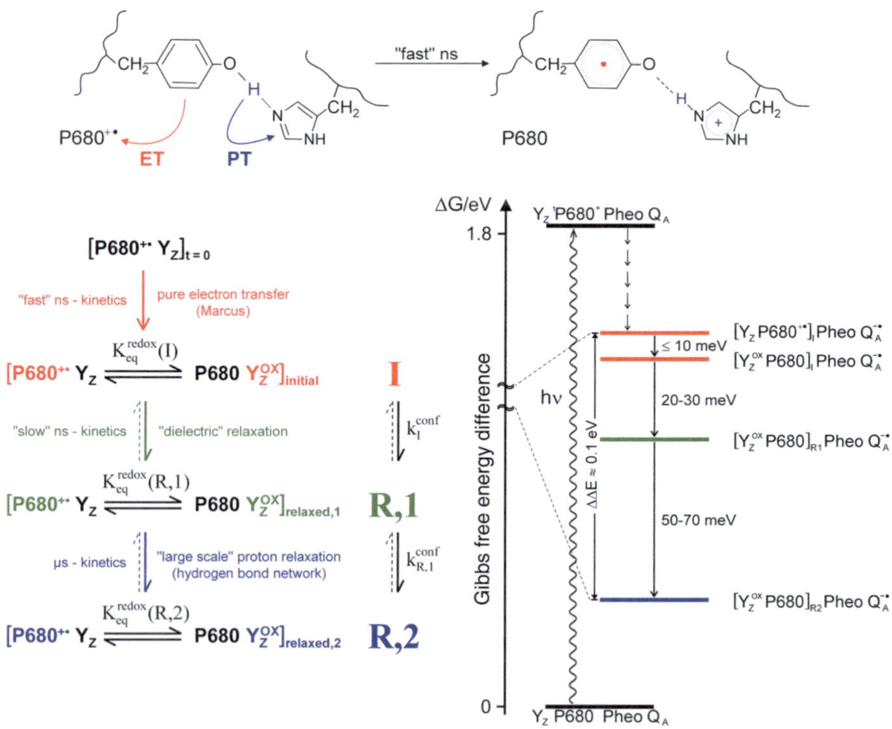

Fig. 13.4 Scheme of proton coupled electron transfer of P 680$^+$ reduction by Y_Z with His as hydrogen bonding partner (*top panel*) and reaction scheme (*bottom panel, left side*) and energetics (*bottom panel, right side*) of $P680^{+\bullet}$ reduction by Y_Z in PS II complexes with an intact water-oxidizing complex in redox state S_1. The initial state I and the two relaxed states R, 1 and R, 2 are marked in *red* (I), *green* (R, 1) and *blue* (R, 2). For the sake of simplicity the panel on the right side presents only ΔΔE values for the different relaxation states because the absolute energy gap between $[Y_Z{}^1P680^*PheoQ_A]$ and $[Y_Z^{OX}P\ 680PheoQ_A^{-\bullet}]$ is not exactly known. Likewise, energetic relaxations around the $Q_A^{-\bullet}$ site and the energy loss due to partial reoxidation of $Q_A^{-\bullet}$ by $Q_B(Q_B^{-\bullet})$ in the μs time domain are omitted

rium $P680^{+\bullet}Y_Z \rightleftharpoons P680Y_Z^{OX}$ towards the right side due to relaxations of the protein environment (local response in the "slow" ns time domain and subsequent proton rearrangement within hydrogen bond network(s) in the µs-time domain) as schematically illustrated in the bottom panel of Fig. 13.4 (left side). The relaxation processes significantly contribute to the energetics of the overall reaction (Fig. 13.4 bottom panel, right side) (for detailed discussion, see Kühn et al. 2004).

Interestingly, the extent of the relaxation in PS II core complexes from *T. elongatus* increases upon replacement of Ca^{2+} by Sr^{2+} but remains virtually unaffected by Cl^-/Br^- exchange (Ishida et al. 2008). The effect of Ca^{2+} on the $P680^{+\bullet}$ reduction kinetics has been shown in earlier studies on mildly trypsinized PS II membrane fragments from spinach (Renger et al. 1989) and seems to be a general phenomenon of PS II in cyanobacteria and higher plants (see also Miqyass et al. 2008).

In samples deprived of the WOC the oxidation of Y_Z is coupled to a stoichiometric proton release into the lumen (Renger and Völker 1982) and the properties of the reaction markedly change as reflected by higher activation energy and a significantly larger H/D exchange effect (for further details, see Renger and Kühn 2007). Interestingly, the reorganization energy increases by a factor of about 3 in these samples. The value of 1.6 eV (Renger et al. 1998; Tommos and Babcock 2000) closely resembles that of 1.4 eV found for model systems in solution (Sjödin et al. 2002). This striking similarity is in line with the idea that the microenvironment of Y_D is rather hydrophilic and contains several water molecules when the WOC is destroyed but on the contrary it is devoid of "bulk" water in the presence of an intact WOC (Renger et al. 1998). This conclusion is supported by recent EPR studies (Zhang 2007).

13.3.4.3 Stepwise Oxidation of the Water Oxidizing Complex (WOC) by Y_Z^{OX}: Cofactors, Assembly and Structure of the WOC, Reaction Coordinate

The Kok cycle *per se* provides a description of the overall four step reaction sequence leading to oxidative water splitting in the WOC but a deeper understanding of the process requires information on: (1) structure of the catalytic site including the water molecules, (2) electronic configuration and nuclear geometry of the catalytic site in each redox state S_i, (3) reaction coordinates of the individual redox step, and (4) pathways for substrate (H_2O) uptake and product (O_2, H^+) release.

In the following our current knowledge on these points will be briefly summarized and finally an attempt presented to cast this information into a proposed model.

Structure of the WOC

Several biochemical and spectroscopic studies revealed that the catalytic site of the WOC is a Mn_4O_XCa cluster, where x denotes the number of µ-oxo-bridges (for reviews, see Yachandra 2005; Messinger and Renger 2008). Accordingly, the structure of the WOC is defined by (a) the spatial arrangement of the four manganese centers, the bridging oxygens and the Ca^{2+} ion, (b) the coordination sphere of the Mn_4O_XCa cluster and (c) the surrounding protein matrix.

Information on the structure of the Mn_4O_XCa cluster and the nature and geometry of its coordination sphere have been gathered from a combination of results obtained by applying different techniques including XRDC, extended X-ray absorption fine structure (EXAFS), Fourier transform infrared (FTIR) spectroscopy and site directed mutagenesis (for various biophysical techniques in photosynthesis, see Aartsma and Matysik 2008). The current state of knowledge and the limitations of structure determination (in particular radiation damage in XRDC structure analysis) are described in a review of Kern and Renger (2007). This structure information is summarized in Fig. 13.5 (see also chapter of Fromme and Grotjohann).

The ligand sphere of the metal centers is formed by (1) the oxo-bridges, (2) amino acid residues and (3) water including the substrate molecules.

The oxo-bridges were assigned by EXAFS (see Yachandra 2005). Several potential amino acid ligands have been identified by site directed mutagenesis, FTIR difference spectroscopy and XRDC (see references in Kern and Renger 2007; Müh and Zouni 2011), including Asp 170, Glu 189, His 332, Glu 333, Asp 342 and the C-terminus Ala 344 of polypeptide D1 and Glu 354 of CP43. Consensus is lacking so far for His 337. One interesting conclusion emerging from the XRDC data is the idea that the Mn_4O_XCa cluster is probably stabilized by metal ligation via bridging carboxylate groups and that this mode of coordination could provide the scaffold for fixing and for fine tuning of functionally important distances between metal centers and ligands (including the substrate) during the catalytic cycle.

At present there is no evidence for a direct ligation by inorganic cofactors: bicarbonate can be excluded (Ulas et al. 2008; Sheleva et al. 2008) and chloride seems to be part of a hydrogen bond network close to the Mn_4O_XCa cluster as originally proposed by Olesen and Andreasson (2003) (for further discussion, see Renger and Renger 2008). The binding of chloride via hydrogen bonding is highly supported by new XRDC data at 2.9 A resolution (Guskov et al. 2009).

Fig. 13.5 Structure of the donor side of PS II and EXAFS derived model for arrangement of manganese in the Mn_4O_XCa cluster. **Panel a:** Structural model where the Chl P_{D1} (*green*), the redox active Tyr_Z and the metal ions of the Mn_4O_XCa cluster are shown as derived from the 3.0 Å structure (Loll et al. 2005), the view is approximately along the membrane plane with lumenal side on *bottom*. Mn (*red*) and Ca (*green*) are shown as spheres, amino acids and structural elements from D1 (*yellow*) and CP43 (*magenta*) in the vicinity of the Mn_4O_XCa cluster are labelled, figure was generated using Pymol. **Panel b:** Schematic representation of the ligation of the Mn_4O_XCa cluster derived from the X-ray structure. *Grey*, *blue* and *green* lines indicate 2.7 Å, 3.3 Å and 3.4 Å distances, respectively, *red dotted lines* symbolize possible ligand-metal interactions with distances given in angstrom units (adapted from Loll et al. 2005). **Panel c:** The most likely arrangements of the four manganese in the Mn_4O_XCa cluster of PS II as derived from polarized EXAFS measurements on PS II single crystals. Manganese (*red*) and calcium (*green*) are shown as spheres, the distances Mn_A-Mn_B, Mn_B-Mn_C, and Mn_C-Mn_D are 2.7 Å, the distance Mn_B-Mn_D is 3.2 Å. For details, see (Pushkar et al. 2008).The characterization of manganese by numbers 1,2,3 and 4 in panel b corresponds with those by indices C, D, B and A in panel c: The data of panel c were kindly provided by Yulia Pushkar and Vittal Yachandra

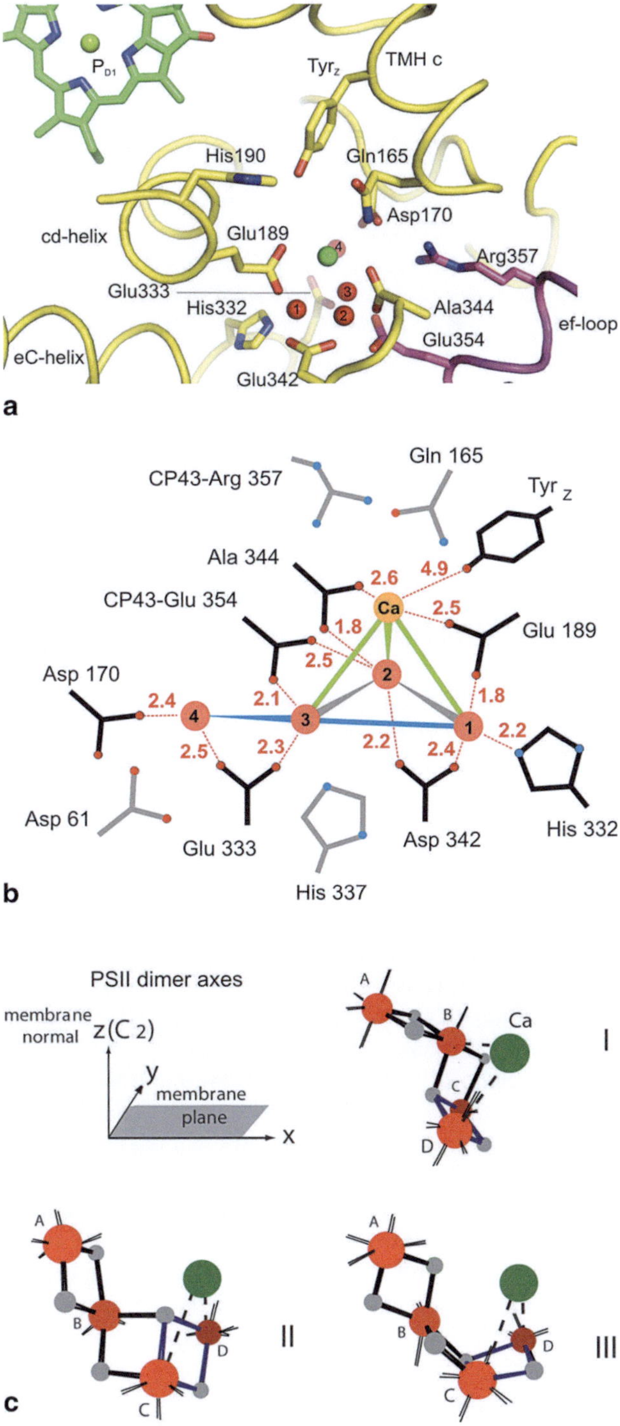

P$_{D1}$
Tyr$_Z$
TMH c
His190
Gln165
Asp170
cd-helix
Glu189
Arg357
Glu333
His332
Ala344
ef-loop
Glu354
eC-helix
Glu342
a
CP43-Arg 357
Gln 165
Tyr Z
Ala 344
CP43-Glu 354
Ca
Glu 189
Asp 170
Asp 61
Glu 333
His 337
Asp 342
His 332
b
PSII dimer axes
membrane normal
z(C 2)
y
membrane plane
x
A
B
C
D
Ca
I
II
III
c

The functionally most relevant ligands are the substrate molecules, their protonation in the different S_i-states, the mode of hydrogen bonding with the protein environment and their dynamics coupled with the S_i state transitions, in particular during oxidation of redox states S_2 and S_3.

At present direct structural information is lacking and not even the number of water molecules in the microenvironment of the Mn_4O_XCa cluster is known. Indirect evidence for the existence of a cluster of 6–12 water molecules was gathered from EPR and mass spectrometry data (Hansson et al. 1986; Bader et al. 1993; Burda et al. 2001). Measurements of $H_2{}^{18}O/H_2{}^{16}O$ exchange kinetics in the different S_i-states revealed that two substrate water molecules are bound in a different manner (Hillier and Wydrzynski 2008). It must be emphasized that in addition to the two molecules which provide the oxygen atoms that are eventually linked to form the essential O–O bond, further water molecules probably play an essential role in mediating proton transfer and affecting reaction pathways. FTIR studies of Noguchi's group revealed that changes of hydrogen bonding of H_2O take place during the S_i state transitions and that most likely at least four water molecules are involved in the overall process (see also Noguchi 2008). The key functional role of water molecules is currently unresolved and remains an exciting challenge of future research.

Electronic Configuration and Nuclear Geometry of S_i States

The S_i states of the Kok cycle (see Fig. 13.3, bottom panel) characterize the formal redox states of the WOC without discrimination between (1) the individual metal centers within the Mn_4O_XCa cluster and (2) manganese and ligands (including the substrate molecules).

The electronic configuration of the Mn_4O_XCa cluster can be best probed by X-ray absorption near edge (XANES), EPR and ^{55}Mn electron-nuclear double resonance (ENDOR) spectroscopy (for recent reviews, see Messinger and Renger 2008; Sauer et al. 2008). Interpretation of ^{55}Mn ENDOR data within the framework of a structural model of the Mn_4O_XCa cluster gathered from EXAFS studies on single crystals of *T. elongatus* (Yano et al. 2006) leads to the following conclusions on the electronic structure of the redox states (Kulik et al. 2007): $S_0=Mn_A(III)Mn_B(III)Mn_C(IV)Mn_D(III)$, $S_1=Mn_A(III)Mn_B(IV)Mn_C(IV)Mn_D(III)$ and $S_2=Mn_A(III)Mn_B(IV)Mn_C(IV)Mn_D(IV)$[4] where indices A, B, C and D correspond with those shown in Fig. 13.5, panel c. This assignment is based on experimental data obtained at 10 K on samples containing 3% methanol and tacitly implies that the manganese centers can be distinguished with respect to their valence states. However, the latter assumption is not in line with conclusions gathered from X-ray resonant Raman scattering (RIXS) experiments indicating that the electron removed from the manganese during the $S_1 \rightarrow S_2$ transition probably originates from delocalized orbital (Glatzel et al. 2004). At present, extent and mode of delocalization and its possible variation with temperature are not known.

[4] A theoretical analysis of EPR, ENDOR and ESEEM data led to the alternative conclusion that Mn_C rather than Mn_A attains the state Mn(III) in S_2 (Schinzel et al. 2010).

An even more complex picture emerges for S_3. It is not yet even clear whether or not the $S_2 \rightarrow S_3$ transition comprises a manganese centered redox step (for a review, see Messinger and Renger 2008). Furthermore, the possibility is discussed that S_3 could be a multistate redox level of the WOC (see Sect. 13.3.4.4).

The Mn_4O_XCa cluster and its protein environment undergo structural changes during the S_i state transitions. Only minor changes occur due to the $S_1 \rightarrow S_2$ transition while the $S_0 \rightarrow S_1$ transition is accompanied by a lengthening of one of the short Mn-Mn-distances from about 2.7 to 2.85 Å (Robblee et al. 2002). The most striking rearrangement of Mn_4O_XCa accompanies the oxidation of the S_2-state to S_3: the short (~2.7 Å) Mn-Mn vectors are elongated to ~2.82 Å and ~2.95 Å (Liang et al. 2000). This finding supports earlier proposals on a significant structural change during the $S_2 \rightarrow S_3$ transition that were gathered from independent lines of evidence (for a review, see Messinger and Renger 2008). Likewise, new data obtained on a *T. elongatus* mutant suggest that His 332 hardly affects the S_2 state while significant changes are observed for S_3 (Sugiura et al. 2009). This result likely reflects a marked structural rearrangement within the WOC coupled with the oxidation of S_2 to S_3. Recent information gathered from EXAFS measurements on Sr^{2+} containing PS II core complexes from *T. elongatus* (Pushkar et al. 2008) revealed that the $S_2 \rightarrow S_3$ transition is not only accompanied by a Mn-Mn distance increase but also comprises a concomitant decrease of the Mn-Ca distance. Based on this finding the Ca^{2+} is inferred to be actively involved in the process of oxidative water splitting (Pushkar et al. 2008).

Reaction Coordinate of Oxidative Water Cleavage

The reaction coordinate of oxidative water splitting in the WOC is determined by the energetics of the individual S_i-states and the activation barriers for the stepwise forward transitions between these states of the Kok-cycle.

Absolute values are not known for the Gibbs free energy levels $G^{0'}(S_i)$ of the individual S_i states and even the $\Delta G^{o'}$ values for the redox transitions $S_i \rightarrow S_{i+1}$ can be only indirectly estimated from experimental data (for a detailed description, see Messinger and Renger 2008). On the basis of currently available information the following values reflect the energetics of the WOC within limited ranges:

$$\Delta G^{0'}(S_{i+1}/S_i)\text{: } 0.70 - 0.85 \text{ eV } (i = 0), 1.0 - 1.10 \text{ eV } (i = 1), \\ 1.0 - 1.15 \text{ eV } (i = 2) \text{ and about } 1.0 \text{ eV } (i = 3)$$

Regardless of the uncertainties of the exact numbers, this pattern shows that the energetics of oxidative water cleavage in the WOC entirely differs from those of the analogous process in solution via the intermediates hydroxyl radical, hydrogen peroxide and superoxide radical (see Eq. (13.6)).

The time course of the individual redox steps of the WOC can be described by virtually exponential kinetics, except of the reduction of Y_Z^{OX} by S_3 which exhibits a lag phase of variable extent (for a discussion, see Renger 2011). This feature of the

Table 13.1 Kinetics of redox transitions (half life times, activation energies and kinetic H/D isotope exchange effects) of the reaction $Y_Z^{OX}S_i \rightarrow Y_Z S_{i+1-4\delta_{i3}} + \delta_{i3}O_2 + n_i H^+$ where $\delta_{i3}=1$ for $i=3$ and zero for $i \neq 3$[a]

i	$t_{\frac{1}{2}}(\mu s)$[b]	E_A(kJ/mol)[c]	k_H/k_D[b]
0	(30)[d]	5	n.d.
1	85	12.0	1.3
2	240	36.0	1.3
3	1300	20.0 (46.0)	1.4

[a] for the sake of simplicity protolytic reactions are symbolized by n_iH^+ without distinction between extrinsic and intrinsic proton release (Renger 1978)
[b] data from Karge et al. (1997)
[c] data from Renger and Hanssum (1992)
[d] the value taken from van Leeuwen et al. (1993) differs by almost one order of magnitude from the value (250 μs) reported by Rappaport et al. (1994) which appears to be more realistic (Renger 2011)

S_i state transitions markedly differs from that of the multi-phasic kinetics of Y_Z^{OX} formation (see Sect. 13.3.4.2). Experimental data on half life times, activation energies and kinetic H/D isotope exchange effects of the S_i state transitions are compiled in Table 13.1.

Based on the findings on energetics and kinetics the generalized reaction coordinate can be constructed for the oxidative water splitting of oxygenic photosynthesis. The result is shown in Fig. 13.6. In this scheme the S_3 state is assumed to comprise

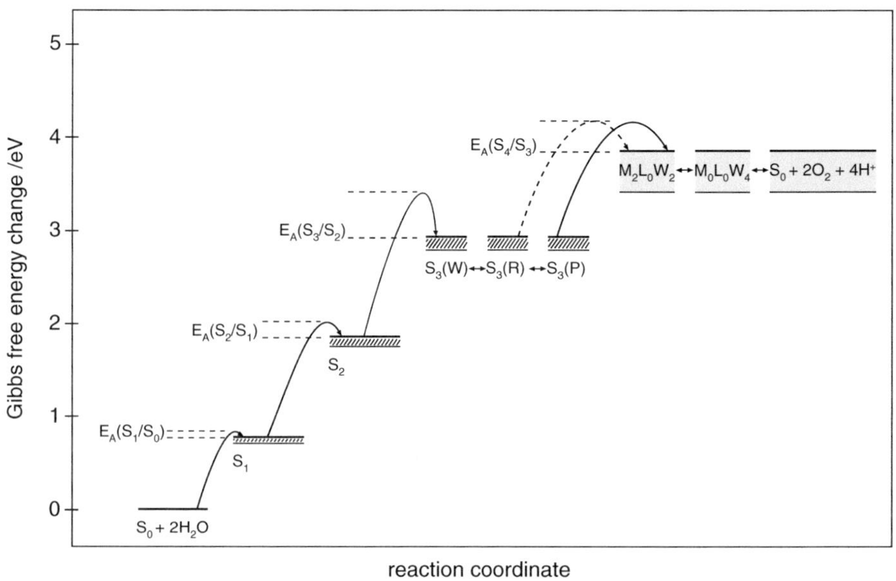

Fig. 13.6 Generalized reaction coordinates of the four step oxidative water cleavage in the photosynthetic apparatus (for details, see text)

three distinguishable electronic configurations (redox isomerism) with different nuclear geometry (including proton tautomerism) as will be outlined in the next Sect. 13.3.4.4.

Two important conclusions emerge from these studies: (1) the characteristic kinetic features—in particular the values for the rate constants and activation energies—appear to be unique for the WOC and to persist without significant change during the evolution from cyanobacteria up to the level of higher plants because the absolute values were found to be virtually the same in PS II preparations from *Thermosynechococcus vulcanus* Copeland (Koike et al. 1987) and from spinach (Karge et al. 1997), and (2) all redox steps in the WOC are triggered reactions, probably by proton and/or conformational gating because an application of the empirical rate constant distance relationship for NET (Page et al. 1999) leads to results that are in marked conflict with the well known structure data for the distance between Y_Z and the Mn_4O_XCa cluster (for further details, see Renger and Renger 2008; Renger 2011).

Substrate/Product Pathways

The Mn_4O_XCa is shielded from bulk water in the lumen by both the large loop of CP43 protein and extrinsic proteins, in particular PsbO, (see Fig. 13.2a) so that a sufficiently fast substrate transport is required in order to sustain high rates of oxygen evolution under saturating light intensities. The turnover numbers exceed the slow phase of $H_2{}^{18}O/H_2{}^{16}O$ exchange kinetics at the WOC (Hillier and Messinger 2005) by at least one order of magnitude. Therefore the latter kinetics neither reflect the substrate transport nor the rapid replacement of newly formed O_2 by two water molecules. Based on measurement of the interaction of small hydrophilic surrogate electron donors (NH_2NH_2 or NH_2OH) the substrate water transport to the catalytic site was estimated to be of the order of a few milliseconds (Messinger et al. 1991).

Stimulated by the postulation of a putative pathway for oxygen substrate transport in cytochrome *c* oxidase (see Sect. 13.4) by Riistama et al. (1996) a complementary pathway was speculated to exist for the release of the product dioxygen from the catalytic site of the WOC (Renger 1999; Anderson 2001). Progress in PS II structure resolution in combination with theoretical calculations led to proposals for pathways of both, substrate water and product dioxygen (Murray and Barber 2007; Ho and Styring 2008). A modelling of new structure data (Guskov et al. (2009) revealed nine potential channels for H^+ and O_2 release from the Mn_4O_XCa cluster and substrate entry (Gabdulkhakov et al. (2009). The most important conclusion from this study is the proposal of a strict spatial separation of pathways for substrate and products. Furthermore it seems likely that the dioxygen is transported through the PS II complex to the cytosolic side (Guskov et al. 2009) as intuitively assumed in earlier studies (Renger 2001).

The "chemical" protons released due to oxidative water splitting are probably funnelled via Asp 61 (McEvoy and Brudvig 2006; Meyer et al. 2007) and Glu 65

(Guskov et al. 2009) of polypeptide D1 into a pathway of the extrinsic PsbO protein that is a constituent of PS II in all oxygen evolving organisms. Based on the unusual titration behaviour of soluble PsbO, Shutova et al. (1997) concluded that Glu and Asp residues in this protein provide an exit pathway of H^+ into the lumen; a more refined analysis supported this idea (Shutova et al. 2007). Likewise theoretical calculations favor the proposal of H^+ transport pathways in the PsbO protein (Ishikita et al. 2006).

13.3.4.4 Mechanism of Oxidative Water Splitting

Any mechanistic considerations on oxidative water splitting in the WOC must take into account the following characteristic properties: (1) water is the direct substrate for dioxygen formation (Clausen et al. 2004, 2005; Hillier et al. 2006) without any catalytic function of bicarbonate (for a proposal, see Castelfranco et al. 2007) that is not a constituent of the WOC (see Sect. “Structure of the WOC”), (2) the S_i-state transitions up to the redox state S_3 are virtually first order reactions with comparatively small S_i state dependent activation energies (see Sect. “Reaction Coordinate of Oxidative Water Cleavage”) and are coupled with proton release in a MS–EPT manner (for a discussion, see Meyer et al. 2007), and (3) the oxidation steps of the WOC by Y_Z^{OX} are triggered by proton transfer (PT) and/or conformational changes (see Sect. “Reaction Coordinate of Oxidative Water Cleavage”) and become blocked below threshold temperatures (for compilation of data, see Renger 2001, 2004) and hydration levels (Noguchi and Sugiura 2002) that both depend on the redox state S_i.

The mechanism of oxidative water splitting has been discussed in several recent review articles (Evoy and Brudvig 2006; Renger 2007; Meyer 2007; Renger and Renger 2008) and book chapters (Hillier and Messinger 2005; Messinger and Renger 2008). Therefore only a short summary and some new data will be presented.

The essential steps of the process are summarized in Fig. 13.7. The S_i-States are characterized by the triple symbol $M_jL_kW_l$ (i=j+k+l) in order to distinguish between formal redox states of manganese (M_j), ligand (L_k) and substrate (W_l); in some cases a distinction between L_k and W_l is not possible, i.e. $L_kW_l=(LW)_{k+l}$ (for a review on this nomenclature, see Messimger and Renger 2008).

There is consensus that the transitions $S_0 \rightarrow S_1$ and $S_1 \rightarrow S_2$ are metal centered redox steps (see Sect. “Electronic Configuration and Nuclear Geometry of S_i States” and Renger 2011). However the nature of the following steps of S_2 and S_3 oxidation are a matter of controversy. Essentially three types of models are currently under debate (see Renger and Renger 2008; Renger 2011 and references therein):

(a) the “Mn only” model where both steps are metal centered (see McEvoy and Brudvig 2006),
(b) the “oxo-radical” model where an oxo-radical is formed in S_3 (see Pushkar et al. 2008)
(c) the “multiple S_3 state” model (see Renger 2004) where S_3 comprises three states with different electronic configuration and nuclear geometry.

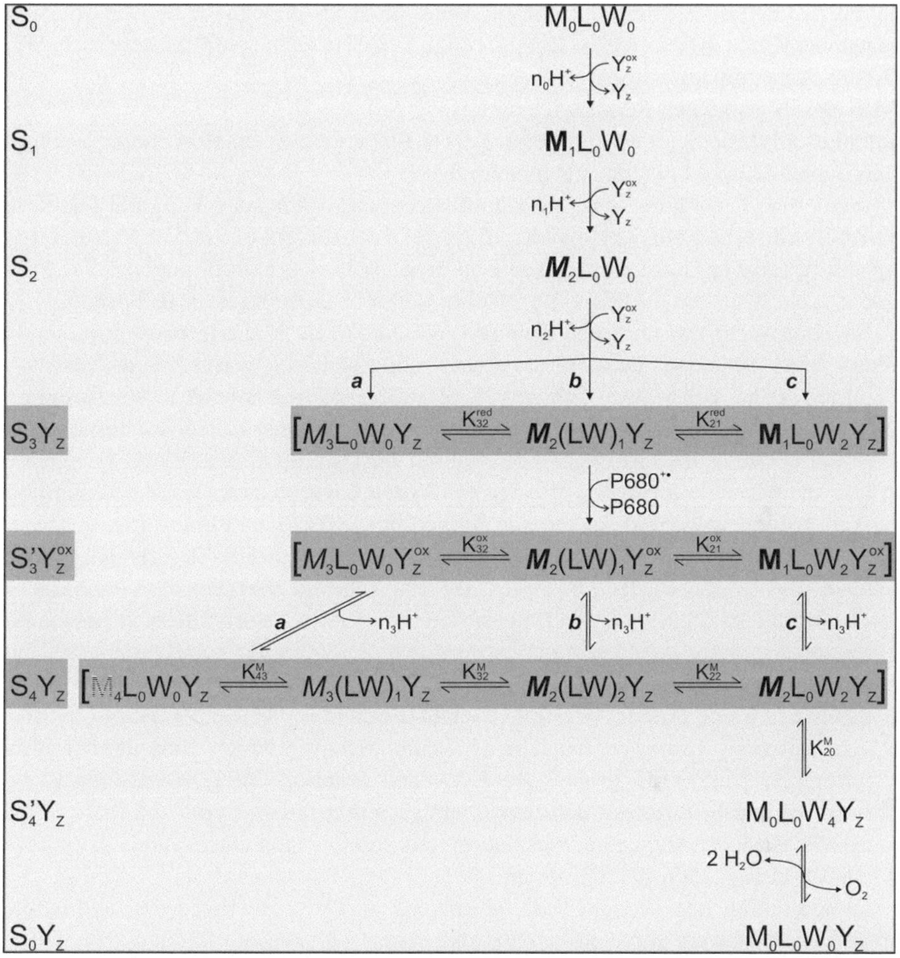

Fig. 13.7 Simplified mechanism of oxidative water splitting. The redox states S_i of the WOC are symbolized by $M_jL_kW_l/M_j(LW)_m$. Different redox states of the manganese are distinguished by using different types of capital letters M and index j=0,…4. Likewise, L_k and W_l describe the redox states of ligand and substrate (water), respectively. In the case of $(LW)_m$ no distinction is made between the redox levels of L and W (for details, see text); ***a***, ***b*** and ***c*** represent different possible pathways for oxidation of S_2 by Y_Z^{OX}; K_{32}^{red}, K_{21}^{red} and K_{32}^{OX}, K_{21}^{OX} are the overall equilibrium constants between the three different states $M_3L_0W_0$, $M_2(LW)_1$ and $M_1L_0W_2$ of S_3 and of $Y_Z^{OX}S_3$, respectively, including redox isomerism and proton tautomerism, K_{43}^M, K_{32}^M and $K_{22}^M$are the equilibrium constants for the possible states of S_4 and K_{20}^M describes the equilibration between S_4 configuration $M_2L_0W_2$ (complexed peroxide) and S_4' (complexed dioxygen), respectively. Oxidation of Y_Z by $P680^{+\bullet}$ is explicitly shown only for the transition $S_3Y_ZP680^{+\bullet} \rightarrow S_3Y_Z^{OX}P680$, otherwise only oxidation of the WOC by $Y_Z^{OX}$is presented

In Fig. 13.7 the "Mn-only" model is represented by pathway ***a*** and equilibrium constants $K_{32}^{red} = K_{32}^{OX} = K_{21}^{red} = K_{21}^{OX} = 0$, i.e. S_3 exists exclusively as state $M_3L_0W_0$; in the "oxo-radical" model $K_{32}^{red} = K_{32}^{OX} \rightarrow \infty$, $K_{31}^{red} = K_{21}^{OX} = 0$, i.e. only state $M_2(LW)_1$ is populated through formation via pathway ***b***; the "multistate S_3" model involves all three S_3 states and pathways (in this case a distinction between metal- and ligand-centered reactions is meaningless).

State $Y_Z^{OX}S_3'$ contains four redox equivalents above the state Y_ZS_0 and therefore formally correspond to S_4. However, Y_Z is not a constituent of the first coordination sphere of the Mn_4O_XCa cluster (see Loll et al. 2005; Kern and Renger 2007; Müh and Zouni 2011) and therefore $Y_Z^{OX}S_3'$ should not be considered as an S_4 state.

In contrast to the unresolved nature of S_3 as to be a single or multiple state redox level consensus seem to exist that redox states S_4 comprises at least two different states, where one of these is a peroxidic intermediate $M_2L_0W_2$. Based on effects of oxygen pressure on UV absorption changes that reflect the turnover of $Y_Z^{OX}S_3'$ (Velthuys 1981; Renger and Weiss 1982), an equilibrium was postulated where the population of $M_2L_0W_2$ due to oxygen back pressure leads to inhibition of Y_Z^{OX} reduction by S_3 (Clausen and Junge 2004). This idea implies that the reaction $[S_2(H_XO_2)] \rightleftharpoons [Y_ZS_0(O_2)] \rightleftharpoons Y_ZS_0 + O_2 + n_3H^+$ is only slightly exergonic. The O_2 back pressure effect, however, has been recently questioned on the basis of flash induced fluorescence yield measurements (Kolling et al. 2009). A new mass spectrometry study unambiguously shows that O_2 evolution is not suppressed by high O_2 pressure (Sheleva et al. 2011). This finding supports our earlier proposal of a highly exergonic product/substrate exchange reaction (Renger 1978).

The formerly proposed existence of state $M_4L_0W_0$ which contains a highly electrophilic Mn(V)=O group (see Evoy and Brudvig 2006) is unlikely to be formed according to recent theoretical analyses that rather favour the oxo-radical state Mn(VI)=O$^{\bullet}$ (Siegbahn 2008; Sproviero 2008) which corresponds to symbol $M_3(LW)_1$ in Fig. 13.7, i.e. $K_{43}^{M} \rightarrow \infty$.

When taking into account that the kinetics of Y_Z^{OX} reduction by S_3 and of the dioxygen release are very similar (Strzalka et al. 1990; Razeghifard and Pace 1999; Clausen and Junge 2004) it turns out that the transition $S_3'Y_Z^{OX} \rightarrow S_4Y_Z$ must be highly endergonic in formation of either $M_4L_0W_0$ or $M_3(LW)_1$ (pathway (***a***) or $M_2(LW)_2$ (pathway (***b***) so that the transient populations of these states are rather low (Siegbahn 2009). On the other hand no such a restriction is required for pathway ***c*** where $M_1L_0W_2$ is assumed to be a complexed peroxide which acts as the entatic state for O_2 formation (Renger 2001 and references therein). The population probability of state $M_1L_0W_2$ was postulated to be dependent on the redox state of Y_Z, i.e. $K_{32}^{red}\,K_{21}^{red} < K_{32}^{OX}\,K_{21}^{OX}$, and estimated to attain values not higher than 10–20% (see Renger and Renger 2008).

The scheme of Fig. 13.7 does not explicitly distinguish between the experimentally detected "extrinsic" proton release pattern and the "chemical" deprotonation steps at the Mn_3O_XCa cluster with its associated substrate molecules (for a discussion, see Renger 1978, 1983) because the molecular details (in particular the structure of hydrogen bond networks) are not known.

Likewise we refrain from any detailed mechanistic proposals because the experimental data are not sufficient to provide a sound basis for theoretical calculations. The currently discussed models gathered from DFT calculations (Sproviero et al. 2008; Siegbahn 2008, 2009; Zein et al. 2008) offer interesting ideas but are suffering from strong limitations due to insufficient structure information (in particular, at present only limited information is available about the water molecules, see Sect. "Structure of the WOC"). In this respect it is important to note that DFT calculations performed on a much simpler and well characterized binuclear ruthenium catalyst for water oxidation to molecular dioxygen led to a mechanism that turned out to be incorrect when isotope labelling experiment with $H_2{}^{18}O$ were performed (Romain et al. 2009). This example nicely illustrates that utmost caution is required when using DFT calculations for mechanistic considerations on the WOC (see also Orio et al. 2009).

Of key relevance for O–O bond formation is the local proton activity and its change due to the redox transitions of the WOC. Information on this essential parameter are lacking and therefore at present only a formal description can be presented by using a time dependent local proton gradient ∇ $(\vec{r}, t, S_i)$ at the catalytic site (for further discussion, see Renger and Renger 2008).

The mechanistic considerations show that the protein matrix is an essential active part of the system and therefore the classical cofactor-approtein concept must be set aside, i.e. PS II with its WOC should be rather considered as a tailored molecular machine of nanoscale dimensions.

13.3.5 Formation of Metabolically Bound Hydrogen in Form of PQH_2

In marked contrast to the unique process of oxidative water splitting with no analogy in biology, the basic principles of PQH_2 formation are more widely used and closely resemble those of the corresponding reactions of ubiquinone (UQ) reduction in anoxygenic purple bacteria as outlined in recent book chapters (see Petrouleas and Crofts 2005; Parson 2008; Renger 2008) and therefore will be only briefly summarized in this chapter.

13.3.5.1 Cofactor(s) and Binding Site(s)

The two step reduction of PQ via the semiquinone state as intermediate takes place in a special pocket referred to as Q_B-site with $Q_A^{-\bullet}$ acting as the direct reductant without involvement of additional cofactor(s) (see next Sect. 13.3.5.2 and Fig. 13.8). The Q_B-site is exclusively formed by polypeptide D1 and comprises three structural elements: the C-terminal part of transmembrane helix (TMH) d, the two cytosolic helixes and the N-terminal part of TMH e (for a review, see Kern and Renger 2007). The headgroup of the substrate molecule PQ-9 is connected via hydrogen bonds to amino acid residues Ser 264 and His 215 and the backbone amide of Phe 265. Structure data at 2.9 Å resolution unraveled the orientation of the isoprenoid tail of PQ-9 which is accommodated in the newly discovered PQ/PQH_2 exchange channel (Guskov et al. 2009).

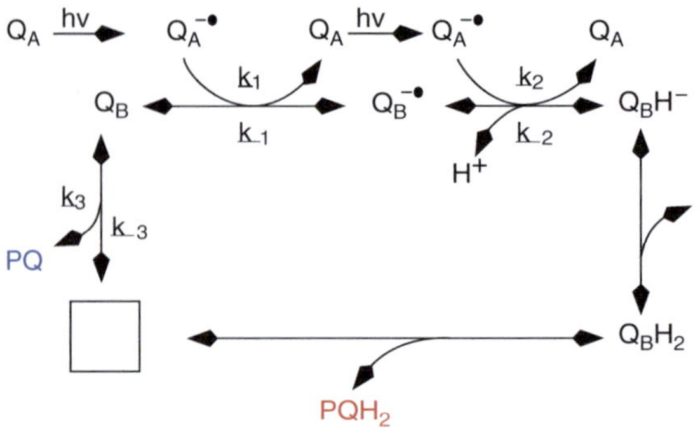

Fig. 13.8 Simplified reaction scheme of PQ reduction to PQH_2 at the Q_B site by the 1-electron reductant $Q_A^{-\bullet}$ that is formed as the result of light induced charge separation. The open square symbolizes an empty Q_B site. For the sake of clarity, protolytic steps are explicitly shown only for the PQ-9 molecule in its different redox states, i.e. protonation/deprotonation reactions of amino acid residues are omitted. (For further details, see text)

In addition to the binding site of Q_A in polypeptide D2 and the Q_B-site in polypeptides D1 a third binding site of PQ was predicted to exist in the neighborhood of cytochrome *b559* and symbolized by Q_C (Kaminskaya et al. 2007). The existence of the Q_C-site is now confirmed by the XRDC data. Guskov et al. (2009) propose that this site is involved in the PQ/PQH_2 exchange mechanism.

13.3.5.2 Reaction Pattern and Mechanism

PQH_2 is formed via a reaction sequence of two 1-electron redox steps that is schematically summarized in Fig. 13.8. For the sake of simplicity, the reactions of light induced charge separation (see Eq. (13.8) and Fig. 13.2) and of the donor side (see Eqn. (13.9)) are omitted and the formation of $Q_A^{-\bullet}$ is symbolized by an arrow labeled with $h\nu$ to indicate the driving force for this reaction, Q_B, $Q_B^{-\bullet}$, Q_BH^- and Q_BH_2 represent the PQ-9 molecule in different redox and protonation states bound to the Q_B site.

Inspection of Fig. 13.8 shows that the energetics of the reaction sequence depend on the midpoint potentials of the redox couples $Q_A^{-\bullet}/Q_A$, $Q_B^{-\bullet}/Q_B$ and $Q_BH^-/Q_B^{-\bullet}$ and the protonation constants of Q_BH^- and $Q_B^{-\bullet}$. The pK_a values of Q^{2-} and Q_BH^- are expected to be significantly larger than physiological pH values so that the population of these forms is negligibly small under *in vivo* conditions. In comparison to the oxidized (Q_B) and fully reduced (Q_BH_2) forms of PQ-9 the semiquinone form $Q_B^{-\bullet}$ is highly stabilized and firmly bound (for details on thermodynamics and kinetics, see Petrouleas and Crofts 2005; Renger and Renger 2008).

The time course of the reduction of PQ bound to the Q_B site is characterized by rate constants of 200–400 μs for Q_B reduction by $Q_A^{-\bullet}$ and somewhat slower kinetics (500–800 μs) in the second redox step (see Wijn and van Gorkum 2001 and references therein).

Eq. (13.10) shows that the overall reaction is coupled with the uptake of two protons. It is known that the second reduction step requires the protonation of $Q_B^{-\bullet}$ but the exact pathway of proton transfer is not yet clarified (Suzuki et al. 2005).

A striking feature of $Q_A^{-\bullet}$ reoxidation by Q_B is the pronounced dependence of the reaction on temperature and hydration level of the sample (for a review, see Renger and Renger 2008). Neutron scattering data revealed that these phenomena correlate with protein flexibility (Pieper and Renger 2009). This finding clearly illustrates the functional importance of protein dynamics.

13.4 Respiratory O_2 Reduction

Cytochrome *c* oxidase (COX) is the terminal member of the respiratory electron transport chain of mitochondria and many bacteria. Reduced cofactors such as NADH or $FADH_2$ are oxidized by complex-I (NADH:ubiquinone oxidoreductase) or complex-II (succinate:ubiquinone oxidoreductase), the reduced ubiqinol cofactor then transfers its electrons via complex-III (ubiquinol:cytochrome *c* oxidoreductase) to cytochrome *c*, the ultimate substrate for COX (also termed complex-IV). A unique situation is encountered in the intracytoplasmic membranes of cyanobacteria where photosynthetic and respiratory electron transport share a series of common components. This membrane contains both, the oxygen evolving WOC of PS II and the oxygen consuming COX (for reviews, see Paumann et al. 2005; Peschek 2008, and chapters in this book).

COX catalyzes the highly exergonic reduction of molecular oxygen to water via a concerted four-electron transfer step, and transforms the Gibbs free energy available from this reaction (see Sect. 13.1) into an electrochemical proton gradient across the membrane, which then provides the proton motive force for driving ATP synthesis (Mitchell 1961, for a review, see Junge 2008). As in the case of WOC (see Sect. 13.3.3.4), COX may be considered a perfect molecular machine specifically tailored to couple electron transfer to "uphill" transmembrane proton translocation. General agreement has been reached that the stoichiometry of this pump activity is unity (i.e. 1 proton translocated per electron). Thus, during a complete O_2 reduction cycle, out of 8 protons taken up from the inside (mitochondrial matrix or bacterial cytoplasm), four are translocated across the bilayer (vectorial or "pumped" protons), while another four, designated as scalar or "chemical" protons, are directed to the active site for water formation:

$$4\text{cyt}c^{2+} + 8\text{H}^+{}_{\text{in}} + \text{O}_2 \rightarrow 2\text{H}_2\text{O} + 4\text{cyt}c^{3+} + 4\text{H}^+{}_{\text{out}}$$

Despite early 3-D structural information, the molecular mechanism of the actual coupling process in the integral membrane protein is still not fully understood, and a matter of intense research activities (see Sect. 13.4.3).

In recent years, it became evident that apart from the endosymbiotic aspect, nature has evolved a large number of divergent (bacterial) oxidase specificities, varying in substrate, heme cofactor, and physiological niche conditions, to qualify their grouping into a still growing super-family of heme/copper terminal oxidases.

13.4.1 General Reaction Pattern of Cytochrome c Oxidase (COX)

Electrons received from its single-electron donor cytochrome *c* are transferred to the primary acceptor of COX, the Cu_A center in subunit II (SU II), a homo-binuclear copper center of mixed valence. From there, electrons reach redox centers in SU I: first a heme *a*, and then the binculear center (BNC) made up of heme a_3 and a copper ion termed Cu_B, both closely spaced and electronically coupled. This latter arrangement represents the active site of the enzyme, where not only oxygen reduction occurs but also most likely the essential steps of coupling to proton translocation take place. Importantly, this BNC is located by about one-third into the depth of the hydrophobic core of the membrane, with its shielded location making any charge movements energetically highly unfavourable without invoking a compensating countercharge, thus following the electroneutrality principle.

The mitochondrial enzyme suffers a number of drawbacks when studying mechanistic details: it consists of a large number of subunits, up to 13 in higher organisms such as mammalia, and it is the product of two genetic systems: the three largest subunits are encoded by the mitochondrial genome, while all the other structural genes appear randomly scattered across the nuclear genome, and their gene products, often even present as tissue-specific isoforms, require organelle import, undergo complex assembly steps, and appear to be key targets for cytoplasma/organelle regulation. Therefore, the approach to address the much simpler bacterial counterparts has been followed, and largely proven successful to yield both structural and functional results applicable to the understanding of general oxidase mechanisms.

Based on the first 3-D structure of a terminal oxidase (Iwata et al. 1995), it became obvious that any core cytochrome *c* oxidase complex requires the presence of at least those two integral membrane protein subunits which carry the relevant redox centers. This very framework has been confirmed in any of the subsequent structures solved. Basic principles of electron transfer and energy transduction can thus be addressed and analyzed with much more experimental ease in any of the simpler prokaryotic systems available. With the exact location of all amino acid side chains at hand, extensive site-directed mutagenesis approaches became feasible on a much more rational basis to study cytochrome *c* interaction sites on either partner protein, tentative internal electron transfer paths, and the molecular events at the active site during a full dioxygen reduction cycle.

An even more challenging point of interest in this respiratory complex is the mechanism of coupling electron transfer to "uphill" proton pumping. Without substantial redox-related conformational changes between redox states, structure alone does not immediately provide any direct functional clues. Current and future activities aim at clarifying details of the electron transfer reaction cycle, the precise timing of each of the energy transduction steps, and its molecular mechanism(s) of coupling both these processes, making the enzyme such an efficient molecular machine. Moreover, there is increasing evidence that oxidase, along with other complexes of the respiratory chain in bacteria, and even more so in mitochondria, (1) is organized in a higher-hierarchy arrangement of supercomplexes for more efficient electron transfer interactions, (2) represents a target for regulatory cross-talk to modulate cellular energy levels and demands, and (3) provides an experimental challenge in studying and understanding its biosynthesis, for its ever rising number of essential chaperone functions identified over the past years, with defects eventually also eliciting diseases in humans. This complexity applies not only to subunit translocation and membrane insertion/protein assembly steps, but also to the concerted transfer of cofactors, heme *a* as well as copper ions, by co- or post-translational mechanisms.

This review does not attempt to extensively cover the oxidase field, but rather focusses on some of the above cited highlights, with a particular preponderance on bacterial systems and their advantages; for excellent recent in-depth reviews on specialized topics over the past few years, see Wikström (2004); Hosler et al. (2006); Olsson et al. (2007); Shimokata et al. (2007); Wikström and Verkhovsky (2007); Belevich and Verkhovsky (2008); Brzezinski and Gennis (2008); Fadda et al. (2008); Fee et al. (2008); Pereira et al. (2008); Pisliakov et al. (2008); Sharpe and Ferguson-Miller (2008); Siegbahn and Blomberg (2008); Kaila et al. (2010).

13.4.2 Structure of COX

13.4.2.1 Overall Structure in Bacteria and Mammalia

COX is an integral membrane protein located in the mitochondrial inner (cristae) membrane (see Fig. 13.1), or the bacterial cytoplasmic membrane resp. Any COX complex described so far shares a common denominator consisting of SU I and II that provides the ligands to all redox cofactors (for comparison, see PS II cofactor arrangement in a protein matrix consisting of two subunits described in 13.3.2). Many preparations carry an additional subunit III, all three of which in eucaryotes are encoded on the mitochondrial genome. Mammalian oxidase is characterized by the presence of up to 10 additional subunits: their genes are spread across the nuclear genome, expressed in the cytoplasm and the (precursor) proteins subsequently imported and assembled with the mitochondrial protein subunits into a functional complex in the organelle (see e.g. Wickner and Schekman 2005; Neupert and Herrmann 2007).

13.4.2.2 Core Structure and Cofactor Arrangement

Subunit I, a protein of 62.4 kDa in the soil bacterium *Paracoccus* (*P.*) *denitrificans* is characterized by a highly hydrophobic amino acid composition, and like SU I

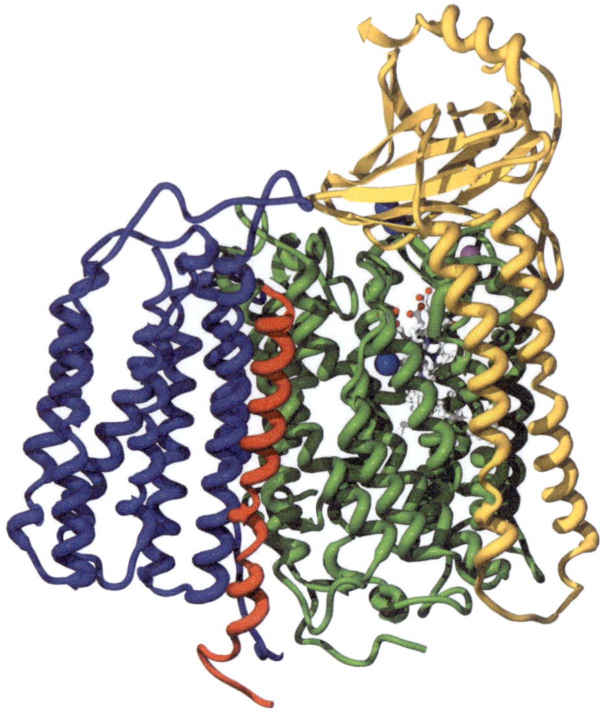

Fig. 13.9 3-D structure of the four subunit COX isolated from *P. denitrificans*. Based on the PDB structure (pdb1QLE) the four subunits are presented as α-traces: green, SU I with its 12 transmembrane helices and its redox centers, hemes *a* as well as a_3 and Cu_B (see also Fig. 13.10 for a more detailled view, taken from a similar perspective); SU II in *yellow*, with its two transmembrane helices and the hydrophilic domain housing the Cu_A center (*copper ions in blue*); in *magenta*, the Mg/Mn site at the interface of SU I/II; SU III and IV in *dark blue/red*

from most other sources, traverses the membrane 12 times, providing a rather rigid framework to harbour all but one of the redox-active metal centers. The transmembrane helices (TMHs) in top view (Iwata et al. 1995), form a three-winged propeller structure enclosing three apparent pores which, however, are blocked by the cofactor moieties, and by hydrophobic side chains.

Both hemes (*a* and a_3) are buried equally deep down into the membrane interior, to roughly one third of the membrane depth down from the P-side (Figs. 13.9 and 13.10); while being aligned perpendicular to the membrane, their interplanar angle is 108°. The low-spin iron of heme *a* is liganded by two His residues, while the high-spin iron of heme a_3 has only one His as the proximal ligand, leaving its sixth position unoccupied. This iron ion is within close distance of approx. 5 Å to a copper ion (termed Cu_B) giving rise to electronic coupling between both metals. Cu_B itself is liganded by three His residues, one of which (H276 in *P. denitrificans*) shows the remarkable feature of being covalently crosslinked to Tyr280 (Buse et al. 1990), thus forming a highly rigid side chain ring structure (Pinakoulaki et al. 2002). This copper ion and the heme a_3 moiety, together designated the binuclear center (BNC) of SU I, establish the site of oxygen binding and reduction, and moreover represent the most likely site of coupling oxygen chemistry to energy transduction, see below. Thus, electrons reach SU I to first reduce heme *a*, then heme a_3 at a metal:metal distance of 13.2 Å, and finally equilibrate with Cu_B within the BNC. It should be noted at this point already that nature has evolved a large number of different terminal

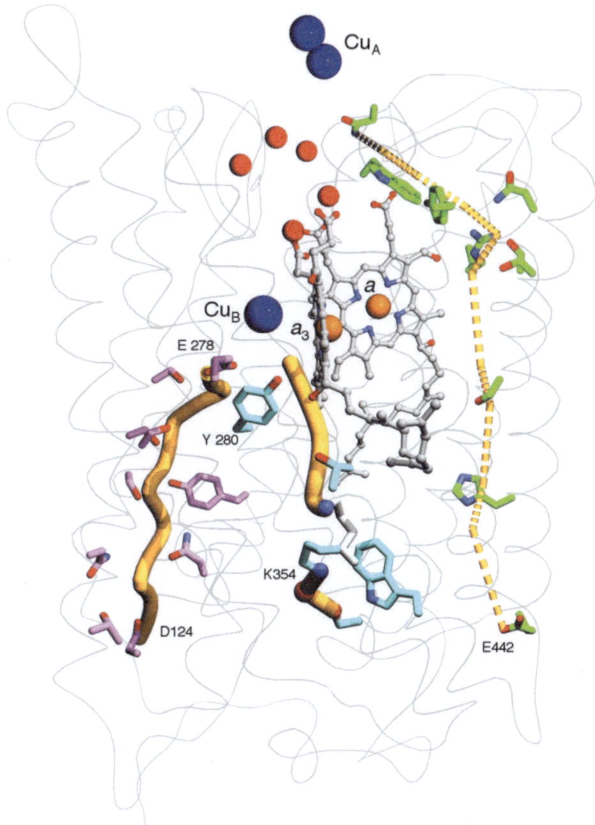

Fig. 13.10 Redox centers, proton pathways, and crucial residues in subunits I and II of bacterial COX. With secondary structure elements of SU I depicted in *grey* in the background for general orientation, redox cofactors of oxidase are indicated (Cu_A located in SU II, and heme *a*, heme $a_3 \cdot Cu_B$ in SU I); *yellow* pathway elements indicate presumed protonic conduits: *left*, the D pathway originating at Asp124 (D124) extending to Glu278 (E278); *center*: the K pathway with Lys 354 (K354); *right*: a potential path ("H") postulated for the mammalian enzyme, but not confirmed for bacterial oxidases; *red* spheres, defined water molecules observed above the hemes

oxidases that accomodate other types of heme species both in their low- and in their high-spin sites, see also below and chapter of Teixiera et al. in this book.

Subunit II accomodates the first redox center for ET to the oxidase complex, Cu_A. Being processed both on its N- as well on its C-terminus (Steinrücke et al. 1987), the mature 28 kDa *P. denitrificans* protein spans the membrane twice, exposing the extended hydrophilic Cu_A domain of its C-terminal half right above the membrane surface on the periplasmic side of the membrane, see Fig. 13.9. Two copper ions, 2.6 Å apart and thus close enough to allow a mixed-valence electronic state, are liganded by two Cys (with both Cu and the two S atoms in plane), two His, one Met, and the carbonyl oxygen of a Glu residue (being a ligand to a Mn ion, see below). Electrons donated by cytochrome *c* (see below) reach the Cu_A center located around 5 Å below the surface, and from there are transmitted to heme *a* in SU I, spanning a center to center distance of 19.5 Å.

Subunit III represents the most hydrophobic SU of the oxidase complex, spanning the membrane seven times (Fig. 13.9); the TMHs are organized in two bundles in a V-shaped arrangement in this 30.6 kDa protein in *P. denitrificans*. While lacking any redox centers, this SU has been assigned a stabilizing function in some bacterial oxidase complexes under specific conditions (Haltia et al. 1991; Bratton et al. 1999), but the originally isolated *P. denitrificans* complex which consists of only two subunits and lacks this SU III (Ludwig and Schatz 1980) has been characterized as fully competent both in its structural features (Ostermeier et al. 1997) as well as in terms of energy transduction (Hendler et al. 1991).

Subunit IV in the *P. denitrificans* complex, easily overlooked in gels due to its small size of only 49 amino acids, shows a single TMH topology. Its role in the complex is still unknown, and a specific deletion of this small peptide SU did not exhibit a particular oxidase phenotype (Witt and Ludwig 1997).

Other ion binding sites identified in (bacterial) oxidases: (1) a Mn^{2+}/Mg^{2+} site is located between SU I and II and liganded by residues of both subunits at the periplasmic side of the oxidase complex (Witt et al. 1997). While its occupancy can be shifted by the type of divalent metal prevailing in the growth medium, the Mn^{2+} ion is not redox-active, yet communicates the redox state of the Cu_A site (Seelig et al. 1981), due to its proximity to the copper center. This observation makes the Mn^{2+} an ideal EPR-sensitive redox reporter for distance measurements (Käß et al. 2000) and for studying potential protonic and water exit pathways (Florens et al. 2001; Sharpe et al. 2009 and see below). A structural role for this site has also been considered in keeping the surfaces of SU I and SU II in tight contact for inter-subunit ET (Witt et al. 1997).

Furthermore, a Ca^{2+}-binding site has been identified both in the *P. denitrificans* and *Rhodobacter* (*Rb.*) *sphaeroides* enzymes as well as in the mammalian oxidase subunit I, close to the P-side of the membrane (Pfitzner et al. 1999; Lee et al. 2002; Kirichenko et al. 2005), but a functional role for this site has not yet been established with certainty.

13.4.3 Molecular Mechanism of COX

13.4.3.1 Electron Transfer: Pathways from Cytochrome *c* to Oxygen

Cytochrome *c* as the Electron Donor

Early studies with the mitochondria system had made clear that the interaction of COX with its substrate cytochrome *c* is guided by ionic interactions (Margoliash and Bosshard 1983), where the same highly positive surface charge density on the mitochondrial cytochrome *c* is rationalized to allow for interactions both with complex-III and with oxidase. Later, cross-linking data identified SU II as the main docking site on COX (e.g. Bisson et al. 1982), pointing at surface-exposed acidic residues in an area that was later characterized as the hydrophilic Cu_A domain. However, on looking at other terminal oxidases, it turned out that not all cytochrome *c* interactions follow this same principle, but a different scenario is encountered for thermophilic ET reactions, see below. At present, little is known on ET reactions and domain mobility in cases where the electron donating *c*-type cytochrome domain is fused to SU II, like in the heme caa_3 cytochrome oxidases (e.g. Mather et al. 1993).

A comprehensive survey of interaction sites on either ET reaction partner has been performed on the *P. denitrificans* system, where a membrane-bound cytochrome c_{552} is the main ET donor to the aa_3 COX (Turba et al. 1995; Baker et al. 1998; Otten et al. 2001). Both interaction partners, mostly expressed and studied as soluble domains (Reincke et al. 1999, 2001) to specifically address the actual redox reaction without interference by subsequent steps, were extensively modified by site-directed mutagenesis and analyzed for their interaction properties and redox kinetics (Witt et al. 1995 1998a, b; Drosou et al. 2002a, b): A specific cluster of surface-exposed acidic residues on SU II was identified, along with a set of Lys residues positioned around the heme cleft in cytochrome c_{552} (Harrenga et al. 2000). Yet, no particular side chain mutation on either partner protein stood out in its catalytic parameter phenotype, in line with the previous assumption of pseudospecificity (e.g. Tiede et al. 1993; Williams et al. 2005) that the overall surface potential, and not any individual charge pair(s), would drive the long-range interaction/preorientation. At the same time this idea also explains the fact that many of the bacterial oxidases show excellent ET activity with, e.g., the mitochondrial cytochrome *c*, being almost identical in its surface potential properties to the bacterial donor (Harrenga et al. 2000; Reincke et al. 2001).

Following this first orientation step through space, several experimental indications suggested a subsequent fine-tuning between both proteins to reach the final, but short-lived conformation for efficient ET. A set of mutants in hydrophobic residues exposed around the interaction site on SU II were hardly affected in their steady-state K_M parameters, but severely decreased for their k_{cat} (Drosou et al. 2002b), while NMR chemical shift mapping and EPR relaxation studies confirmed the merely transient nature of such an encounter (Wienk et al. 2003; Lyubenova et al. 2007). Dynamic ensembles have been observed for the *Rb. sphaeroides* aa_3 oxidase interaction as well (Roberts and Pique 1999), and are also predicted by docking simulations (Flöck and Helms 2002; Bertini et al. 2005). One particular surface-exposed amino acid, Trp121, positioned approx. 5 Å above the solvent-shielded Cu_A center, showed an unparalleled phenotype upon mutation: any replacement introduced for this amino acid resulted in an almost complete loss of ET activity (Witt et al. 1998b), assigning this position a crucial role in ET between the *c*-type cytochrome donor and the Cu_A center of SU II.

A different scenario for the interaction of a *c*-cytochrome with oxidase is encountered for *Thermus* (*T.*) *thermophilus*. As was noted early on, canonical charged residues are missing on the hypothetical docking surfaces of both the ET substrate, the soluble cytochrome c_{552}, and on the hydrophilic domain of SU II in the heme ba_3-type oxidase (Than et al. 1997; Soulimane et al. 2000). Moreover, kinetic data (Giuffré et al. 1999) showed that ET activity remains high even at very low ionic strength conditions. A systematic Brønsted analysis revealed that the interaction between the Cu_A binding domain is largely governed by hydrophobic forces, favoured over electrostatic interactions at such elevated growth temperatures; unlike for the corresponding *P. denitrificans* ET couple where between two and three effective charges on either interacting surface appear to be responsible. In this thermophilic organism, however, less than one nominal charge on either molecule contributes to complex formation (Maneg et al. 2003, 2004). Based on a combined NMR and computational approach, a model for the transient docking complex and a suggestion for an ET pathway between the two redox centers has been put forward recently for *T. thermophilus* (Muresanu et al. 2006).

A rather complex situation may arise whenever the cytochrome *c* substrate domain is covalently fused to SU II of COX, such as in the *caa*$_3$ terminal oxidase of *T. thermophilus* (Mather et al. 1993). Only little is known as to whether such a domain still requires sufficient flexibility in receiving the electron and then donating it to the Cu_A site, assuming a single docking interface on the *c*-domain for both interactions. Moreover, a direct encounter of the electron donating cytochrome *bc* complex (complex III) with the *caa*$_3$ oxidase has been suggested for the *T. thermophilus* ET chain, based on fast ET reactivities between corresponding domains from either complex (Janzon et al. 2007).

Internal ET Pathways

Rates of internal ET from Cu_A via heme *a* to the BNC (see Fig. 13.10) have been determined in great detail for the mitochondrial enzyme (see Hill 1994; Van Eps et al. 2000), even with approaches to differentiate between various kinetic forms (e.g. Brand et al. 2007). Corresponding studies on bacterial oxidases fully confirmed the sequence of ET steps and, in most cases, even their rates are virtually the same within narrow numerical margins (Ludwig and Gibson 1981; Farver et al. 2006; Belevich et al. 2007, Belevich and Verkhovsky 2008). Based on the main distance criterion for NET efficiency (Page et al. 1999; Moser et al. 2006, 2008), the slightly shorter edge to edge distance between Cu_A and heme *a* (of around 16 Å) favours this redox pathway kinetically compared to that from Cu_A to heme a_3 because the distance between these cofactors is longer by ≈2.5 Å (Iwata et al. 1995). Both heme moieties, positioned at the same depth within the membrane (Fig. 13.10), are only 4.7 Å apart, thus invoking a fast ET. The expected rate of electron equilibration in the ns range has only recently been demonstrated in a reverse ET approach (Pilet et al. 2004) and the much slower rates initially measured for the oxidized COX in the forwad reaction is explained by a slow shift of the heme redox potential due to a coupled proton uptake (Belevich et al. 2007).

The Oxygen Reaction Cycle

For reduction of dioxygen to water, a total of four electrons and four "chemical" protons (see Eq. (13.4) need to reach the BNC in a stepwise manner (see Eq. (13.6)). Fig. 13.11 depicts the major catalytic intermediates identified mostly on a spectroscopic/kinetic basis over the last 50 years (for relevant references, see reviews cited in the last paragraph of Sect. 13.4.1). Starting with the oxidized form of COX

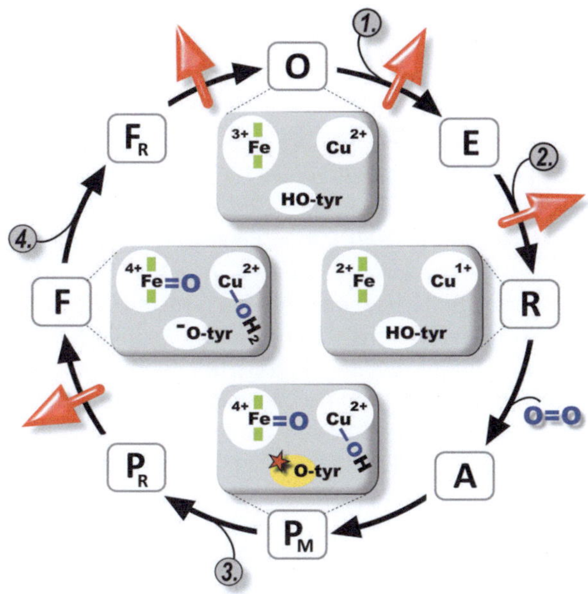

Fig. 13.11 Simplified scheme of a full dioxygen reduction cycle catalyzed by COX. *Outer circle:* spectroscopically discernible reaction intermediates (*boxed letters*) on reducing the fully oxidized state ("O") by stepwise addition of electrons (*grey circles, numbered*) originating from cytochrome *c*. Once two electrons are present at the binuclear center ("R"), dioxygen binds, and the oxidative half-cycle eventually completes the cycle, regenerating the "O" state with the production of two water molecules; *red arrows*: proton pumping steps and their presumed timing relative to the electron entry steps. *Circle inside:* four representative states (O, R, P_M, and F) are highlighted as *grey* boxes, depicting redox-reactive sites at the binuclear center consisting of heme a_3, Cu_B, and the cross-linked tyrosine residue which may initially contribute the fourth electron required for fast O=O bond splitting, thus transiently forming a radical (*). The fate of the two oxygen atoms derived from O_2 is indicated by *blue* symbols, while no reference is given to other oxygenous ligands, e.g. hydroxyl, water, presumably present at this site during turnover; likewise, scalar protons entering *via* the K- and D-pathways are omitted for clarity as well; for further details, see text

("O"), both redox-active metal cofactors in the BNC (symbolized within the grey rectangle(s)) are present in their oxidized forms, $Fe_{heme\ a3}^{3+} \bullet Cu^{2+}$. With the first two electrons entering one by one from cytochrome *c* and transferred via Cu_A and heme *a*, states "E" and "R" are reached. Only at this stage with both heme a_3 and Cu_B reduced, an oxygen molecule is able to bind the binuclear center to get trapped between both metal ions as an adduct ("A"). In a unique and very fast reaction sequence, four electrons are transferred to the dioxygen molecule to split the O–O bond and leave both oxygen atoms at a formal oxidation state of water. This "P_M" mixed valence state, historically thought to house a peroxi form of oxygen, develops within less than 100 μs (Proshlyakov et al. 1998) and is characterized by its spectroscopic signature as an oxo-ferryl (Fe(IV)=O) state of the heme. A striking analogy was noted with respect to the reverse reaction during the water-splitting reaction catalyzed by photosystem II (Tommos and Babcock 2000). The fast kinetics of this concerted step during oxidase turnover has also been taken as the major safeguard why COX is not a source of reactive oxygen species (ROS), leaving no potentially damaging oxygen intermediates free to diffuse to the outside.

What are the sources of all four electrons required for this almost spontaneous O_2 reduction? Two electrons are available from the Fe(II) to Fe(IV) transition of heme a_3, another one is donated due to Cu_B oxidation (see Fig. 13.11). As suspected earlier (Proshlyakov et al. 1998), the side chain of a close-by aromatic residue contributes the fourth electron, transforming itself into a (neutral) radical and concomitantly provides also a proton to the chemistry at the BNC (see Fig. 13.11). The most likely residue for this role is thought to be the prominent tyrosine (Y280 in *P. denitrificans)* which is crosslinked in its side chain to one of the histidines which are ligands to the Cu_B (see above): its unique chemical properties do influence both the proton and the electron affinity in the BNC, and have been discussed to be properly positioned to donate both an electron and a proton for oxygen bond splitting (Gorbikova et al. 2008a, b; Kaila et al. 2009). Direct evidence for this particular Tyr forming a radical that is kinetically competent to participate in the O_2 cycle is, however, still lacking. Evidence for both a Tyr radical (e.g. Fabian and Palmer 1995; Macmillan et al. 1999) or Tyr167 in *P. denitrificans* (Budiman et al. 2004)) and for a Trp radical (W272 in *P. denitrificans*) has been presented (Wiertz et al. 2004; MacMillan et al. 2006). This latter position has been deduced from microsecond freeze-hyperquenching experiments under single turnover conditions by analyzing the formation of this radical signal in a catalytically relevant, sub-millisecond time window (Wiertz et al. 2007). Taken together, these data support the view that a radical of state P_M may well be initially localized at Tyr280, but may easily equilibrate with suitable residues in the vicinity such as Trp272, or even migrate to more distant positions such as Y167 which may represent a thermodynamic sink under artificial experimental conditions lacking further electron input.

Concluding the oxygen cycle (Fig. 13.11), the third and the fourth electron input steps first reduce the radical to form the P_R/F state, and then the oxoferryl state to recover the O state. It should be noted at this point that recent data suggest that the tyrosinate present in F will become protonated only during re-entry into the reductive phase of the cycle (Gorbikova et al. 2008b), and that the occupation of the BNC may be more complex than depicted in Fig. 13.11, because recent crystal structure densities provide evidence for the existence of a peroxide ligand already present in the O state (Aoyama et al. 2009; Koepke et al. 2009).

13.4.3.2 Proton Pumping Mechanisms

While general agreement has been reached on the overall stoichiometry of one proton translocated per electron, the precise timing of each pump step during a full reduction cycle, the molecular mechanism of the coupling process, or even the proton pathways and their utilization still pose challenging problems for current COX research.

Figure 13.11 summarizes our present consensus view on proton translocation steps and their timing, gathered from a large number of experimental studies on both the mitochondrial oxidase and various bacterial aa_3-type oxidases, combined with insights from extensive theoretical calculations and modelling (e.g. Michel 1999; Cukier 2004; Wikström and Verkhovsky 2007; Belevich and Verkhovsky 2008; Fadda et al. 2008; Fee et al. 2008; Pisliakov et al. 2008; Sharpe and Ferguson-Miller 2008; Siegbahn and Blomberg 2008; Chakrabarty et al. 2011).

Many different hypothetical schemes have been put forward in the past (and later dismissed) to describe the coupling behaviour of COX. Three unifying features emerge from the more recent findings: (1) at least during continuous oxidase turnover (e.g. Belevich et al. 2007), proton pumping steps seem to occur both during the reductive and the oxidative phase of the oxygen cycle, as indicated by bold red arrows in Fig. 13.11 (see e.g. Michel 1999; Ruitenberg et al. 2000; Wikström and Verkhovsky 2007); (2) every pump step appears to be immediately preceded by an ET step (e.g. Faxen et al. 2005; Belevich et al. 2006) and as a consequence, (3) each individual step may be considered largely identical at least in its basic setup (see e.g. Belevich et al. 2006; Fadda et al. 2008).

Coupling as such is best understood by the fact that any charge, be it an electron or a proton, introduced into a confined hydrophobic environment such as the heme sites (and the BNC itself), is energetically unfavourable, and will immediately invoke counter-ion movements, or expel an ion of the same charge. This electroneutrality principle (Rich 1995) has guided the design of more elaborate schemes for the sequence of events leading to "uphill" proton pumping (see e.g. Belevich et al. 2006, 2007; Belevich and Verkhovsky, 2008; Fadda et al. 2008):

(a) An electron arriving at heme *a* increases the pK value of a "pump site" residue acting as proton loading site (PLS) located above both hemes and most likely being the D-propionate of heme a_3 (Kaila et al. 2008a; Belevich et al. 2010), This pK shift givies rise to the uptake of a proton supplied via Glu278 at the end of the D pathway (this key Glu residues is conseverved in most oxidases and corresponds to Glu 286 in *Rb. sphaeroides* and Glu 242 in mitochondrial COX);
(b) this first step paves the way for fast electron equilibration between heme *a* and the binuclear center;
(c) in a subsequent charge compensatory step, a "chemical" proton is taken up and directed to the BNC where it is used in oxygen chemistry reactions;
(d) the resulting charge repulsion exerted on the preloaded proton still residing at the pump site causes its extrusion to the P side of the membrane, thus completing a formal "one proton per electron" pump cycle.

With four such unit steps required for a full dioxygen reduction, it is tempting to assume that basically the same mechanism is responsible for the pumping of every proton at each separate electron input event. While this may hold true for the immediate ET steps between heme *a* and the binuclear center along with their respective charge-compensatory steps a-d as outlined above, neither the oxygen chemistry at the BNC (see Fig. 13.11, and above) nor the channel usage at each step are identical (see also Sect. 13.4.3).

13.4.3.3 Channel Usage and Gating Concepts

In a first appraisal, structure-derived information for potential proton pathways was assessed, and merged with early mutagenesis data on conspicuous hydrophilic and even charged amino acid residues found along the hydrophobic transmembrane helices of SU I (e.g. Hosler et al. 1993; Thomas et al. 1993; Pfitzner et al. 1998, 2000; Iwata et al. 1995; Tsukihara et al. 1996). The finding of two separate protonic conduits was initially interpreted as one of them serving the four protons for water formation at the binuclear center, the K pathway (for the crucial residue Lys354), and the D pathway (Asp124 residue) being responsible for pumped protons (see Fig. 13.10). Time-resolved electrometric analyses on specific channel mutants, however, established that the first two protons reach the BNC through the K pathway, while a hybrid pathway usage in the D channel for the remaining six hydrogen ions taken up from the inside during a complete oxygen cycle (see Konstantinov et al. 1997; Ruitenberg et al. 2000, 2002; Tuukkanen et al. 2006; and see above). As an immediate corollary of this finding, a distinguishing device in the D pathway for directing every single proton to its relevant target site, either the BNC or the pump (exit) site, had to be invoked to avoid energetic short-cuts that would give rise to uncoupling of the oxidase. All present evidence suggests that this gating device is, directly or indirectly, associated with a glutamate side chain at the end of the well-established part of the D-pathway (Glu278, see Fig. 13.10). The motional flexibility of this residue had attracted attention early-on, using infrared-spectroscopic and other approaches (Hellwig et al. 1998; Karpefors et al. 2000; Kaila et al. 2008a, b). Replacements in this position rendered mutant oxidases catalytically inactive. This mobile side chain has been observed in crystal structures only in a position facing "down" the D pathway (see e.g. Iwata et al. 1995 for the *P. denitrificans* structure), while an "up" position is functionally mandatory for donating protons either to the binuclear center, or to the PLS (see Sect. 13.4.3.2).

This key role of Glu278, in particular in its anionic form after proton delivery, has recently been acknowledged by molecular dynamics simulations (Tuukkanen et al. 2007; Kaila et al. 2008a, b) as a kinetically gated valve that prevents backflow of protons from the P side of the membrane, due to its discrete equilibrium distribution.

Interestingly, a recent *P. denitrificans* oxidase mutation in a distant asparagine, N131D close to the lower entrance of the D pathway in SU I, has shed some light on this presumed gating mechanism: its phenotype is that of an ideally uncoupled COX with full ET capacity, but lacking any proton pump activity (Pfitzner et al. 2000; Dürr et al. 2008); similar results have been obtained also for a related COX (Vakkasoglu et al. 2006; Bränden et al. 2006; Lepp et al. 2008a). The 3-D structure of COX from the *P. denitrificans* mutant is virtually identical to that of the wild type, but reveals both the "up" and the "down" densities for the Glu278 side chain and a modified water chain (Dürr et al. 2008). This finding is indicative of a restricted mobility of its carboxylate position, obviously locked in a conformational equilibrium exclusively serving the binuclear center supply route, but unable to deliver protons to the pump site. Various studies (e.g. Wikström et al. 2003; Belevich et al. 2007) suggest that this latter decision in gating, still poorly understood, may be exerted by specifically positioned water molecules within hydrophobic cavities at or around the BNC, sensing and relaying the electronic state of the metal centers.

Proton movements through both the K and the D pathways are thought to be mediated mainly by a Grotthuss type of mechanism (de Grotthuss 2006) along extensive chains of localized water molecules, either experimentally observed in the recent crystallographic structures or inferred from modelling approaches (Wikström et al. 2003; Olsson et al. 2007; Shimokata et al. 2007; Dürr et al. 2008; Koepke et al. 2009). A certain degree of "plasticity" in coordinating this water chain (Namslauer et al. 2007) would explain the general observation that many residues potentially lining these pathways may be exchanged in the enzyme without functional loss (e.g. Pfitzner et al. 1998, 2000).

The further path for pumped protons, once they reached the PLS above the hemes (see Sec. 13.4.3.2), is not clear at present, nor is the fate of the reaction product water known after leaving the BNC. However, a number of indications (Florens et al. 2001; Sharpe et al. 2009) point at an exit channel at or near the Mg/Mn binding site at the interface between SU I and II (see Fig. 13.9, and above).

The intricate interplay of proton pathway usage in the reductive/oxidative phase of the catalytic cycle, as outlined above, has been explained recently by the distinct feature of the (charged) Lys354 residue in the K pathway (Lepp et al. 2008b; Kaila et al. 2009), providing dielectric buffering in the oxidative part, and net proton movement during the reductive phase. It is interesting, though, that some of the more distant members within the family of heme-copper oxidases, notably the *T. thermophilus* ba_3 oxidase, are devoid of this distinct pathway architecture, as judged both from the lack of key residues as well as their 3-D structure (Soulimane et al. 2000; Fee et al. 2008; Smirnova et al. 2008), yet appear functionally competent (though not fully; see also Kannt et al. 1998) in their energy transduction properties.

For the mitochondrial oxidase, a third protonic conduit has been suggested as the main pathway for pumped protons, invoking also a different mechanism of energy coupling for the beef heart enzyme (Yoshikawa et al. 1998; Muramoto et al. 2007; Shimokata et al. 2007). However, the presence of such a pathway (see Fig. 13.10, "H" pathway, spotted trace) operating in bacteria could not be experimentally confirmed by extensive mutagenesis studies (Pfitzner et al. 2000; Lee et al. 2000; Salje et al. 2005).

Apart from proton pathways into and across the enzyme, both the substrate dioxygen and the product water need to gain access/exit to/from the BNC. An oxygen channel has been detected early-on in the *P. denitrificans* COX structure, and was later characterized by site-directed mutagenesis (Riistama et al. 1996, 2000), Differences in the extent of this oxygen cavity between different terminal oxidases has recently been visualized by Xe binding to protein crystals (Luna et al. 2008). Both experimental data (Florens et al. 2001; Sharpe et al. 2009) and MD simulations provided evidence for water exit pathways to the periplasmic side, passing by the Mg/Mn site close to Cu_A.

13.4.4 Related Topics

With the extensive search for bacterial counterparts of the mitochondrial COX started some three decades ago (Ludwig and Schatz 1980), it became obvious that today's versions of oxidases not only evolved from Nature's "toolbox" of structural domains originally developed for denitrification enzymes such as N_2O and NO reductase (Saraste and Castresana 1994). The variability of bacterial pendants extends to different electron input substrates (either *c*-type cytochromes or quinols), different internal routes of ET within the complex such as in the cbb_3 oxidases lacking the Cu_A site, and the bewildering variations of different heme moieties found both in the low- and in the high-spin site. The common denominator, the presence of a BNC composed of a heme group and the Cu_B site, discriminates this large group of oxidase enzymes from the only distantly (but evolutionarily) related NO reductases, and assumes, in a unifying concept, similar strategies to handle and make best use of the terminal substrate dioxygen in energy transduction (Pereira et al. 2008; see contribution by M. Teixera in this book).

13.4.4.1 COX as a Constituent in Supercomplex Assemblies

Previous attempts carefully focussed on isolating and characterizing enzyme units, such as the mitochondrial 13-SU COX, that were defined by a minimum structural complexity to sustain all the (presumed) catalytic properties; on an operational basis this often meant to rely on an efficient detergent for solubilization, potentially splitting membrane protein associations into small(er) units. However, contradicting the long-held random collision model, supramolecular assemblies were observed upon gentle detergent treatment followed by appropriate separation of these high molecular weight complexes (see Schägger and Pfeiffer 2001; Krause et al. 2004; Stroh et al. 2004; Vonck and Schäfer 2009; Genova et al. 2008). These supercomplexes did not only show up when the highly specialized (and therefore specifically enriched) inner mitochondrial membrane was addressed, but were also isolated from a bacterial cytoplasmic membrane in distinct stoichiometries for complexes I, III, and IV (e.g. Stroh et al. 2004); such respirasomes may be optimally poised for substrate channelling (for a recent discussion, see Genova et al. 2008).

13.4.4.2 COX Biogenesis Mechanisms

This topic receives increasing attention for the fact that the mitochondrial enzyme poses a considerable challenge due to its complexity in its number of protein subunits, its cofactor insertion mechanisms, and the regulation of nuclear-cytoplasmic communication required for gene expression (see e.g. Hamza and Gitlin 2002; Wickner and Schekman 2005; Fontanesi et al. 2006; Neupert and Herrmann 2007; Khalimonchuk and Winge 2008). Genetic defects originating from either of the two genetic systems involved in COX assembly are increasingly recognized as direct cause of diseases (e.g. Sacconi et al. 2003)). With a number of advantages outlined above, bacteria often offer a suitable model system to address the most elementary steps of biogenesis common to both the bacterial as well as the eukaryotic/organellar world (Hiser et al. 2000; Smith et al. 2005; Bundschuh et al. 2008; Greiner et al. 2008).

13.4.4.3 Regulation of Cellular Energy Demands

The mitochondrial oxidative phosphorylation system has previously been viewed largely as that of an self-employed organelle. Rising evidence suggests that organelles are indeed linked and embedded either way into the cell's general regulatory networks, making individual redox complexes targets for post-translational modifications and cellular signalling cascades (e.g. Fukuda et al. 2007; Vogt et al. 2007; Wagner et al. 2008).

13.5 Concluding Remarks

This chapter is an attempt to summarize our current state of knowledge on the use of the redox system H_2O/O_2 as the core of the bioenergetics found in all higher forms of life.

Water splitting into dioxygen and metabolically bound hydrogen in photosynthesis and the reversal of this process in respiration, are mediated by multimeric, integral membrane protein complexes. Elementary steps are the light-driven oxidative water splitting into O_2 and four protons catalysed by PS II, and the reduction of O_2 to water mediated by COX.

Both reactions comprise a sequence of one-electron redox steps at catalytic centers containing transition metals as reactive sites. It must be emphasized that in both cases the protein matrix is of paramount functional relevance, i.e. WOC and COX have to be considered as molecular machines that are especially tailored for optimal process performance either in the forward or the backward direction. Specially tuned reaction steps avoid the generation of harmful ROS species.

Despite considerable progress in unravelling the structural and functional organization, important questions on the mechanism of key reaction steps still remain to be answered: in particular the pathway of O–O bond formation in the WOC and molecular details of gating in proton pumping of COX are challenging topics of current research activities. In this respect the unravelling of (1) the mode of coupling between proton and electron transfer steps, (2) of the key functional role of single water molecules and (3) the tuning of the reaction coordinates by protein dynamics are of utmost importance.

The tremendous achievements by using genetic engineering and significant progress in monitoring protein dynamics in combination with highly advanced theoretical approaches in the field of quantum chemistry and molecular mechanics will pave the way to a higher level of understanding of biocatalysts in general and the WOC and COX in particular.

Acknowledgements G.R. would like to thank Susanne Renger for preparing the electronic version of Fig. 13.1, J. Kern for Figs. 13.2 and 13.4 a, b and P. Kühn for Fig. 13.8, Fig. 13.4c was kindly provided by V. Yachandra. The financial support by Deutsche Forschungsgemeinschaft (SFB 429 TPA1) is gratefully acknowledged. B.L. thanks O. Richter for providing Fig. 13.10; work from his lab was supported by long-term funding by Deutsche Forschungsgemeinschaft (SFB 472 P8, and Cluster of Excellence "Macromolecular Complexes", Project EXC 115).

References

Aartsma TJ, Matysik J (eds) (2008) Biophysical techniques in photosynthesis, Volume II. Advances in photosynthesis and respiration, Vol. 26, Springer, Dordrecht

Allakhverdiev SI, Tomo T, Shimada Y, Kindo H, Nagao R, Klimov VV and Mimuro M (2010) Redox potential of pheophytin a in photosystem II of two cyanobacteria having the different special pair chlorophylls. Proc Natl Acad Sci 107: 3924–3929

Amunts A, Drory O and Nelson N (2007) The structure of a plant Photosystem I supercomplex at 3.4 Å resolution. Nature 447: 58–63

Anderson JM (2001) Does functional Photosystem II complex have an oxygen channel? FEBS Lett 488: 1–4

Anderson AB and Albu TV (1999) Ab initio determination of reversible potentials and activation energies for outer-sphere oxygen reduction to water and the reverse oxidation reaction. J Am Chem Soc 121: 11855–11863

Aoyama H, Muramoto K, Shinzawa-Itoh K, Hirata K, Yamashita E, Tsukihara T, Ogura T and Yoshikawa S (2009) A peroxide bridge between Fe and Cu ions in the O_2 reduction site of fully oxidized cytochrome *c* oxidase could suppress the proton pump. Proc Natl Acad Sci U S A 106: 2165–2169

Atkins PW (2001) Physical Chemistry, Oxford University Press, Oxford

Bader KP, Renger G and Schmidt GH (1993) A mass spectroscopic analysis of the water-splitting reaction. Photosynth Res 38: 355–361

Baker SC, Ferguson SJ, Ludwig B, Page MD, Richter O-MH and van Spanning RJM (1998) Molecular genetics of the genus *Paracoccus*—metabolically versatile bacteria with bioenergetic flexibility. Microbiol Mol Biol Rev 62: 1046–1078

Belevich I and Verkhovsky MI (2008) Molecular mechanism of proton translocation by cytochrome *c* oxidase. Antioxid Redox Sign 10: 1–29

Belevich I, Verkhovsky MI and Wikström M (2006) Proton-coupled electron transfer drives the proton pump of cytochrome *c* oxidase. Nature 440: 829–832

Belevich I, Bloch DA, Belevich N, Wikström M and Verkhovsky MI (2007) Exploring the proton pump mechanism of cytochrome *c* oxidase in real time. Proc Natl Acad Sci U S A 104: 2685–2690

Belevich I, Gorbikova E, Belevich NP, Rauhamäki V, Wikström M and Verkhovsky MI (2010) Initiation of the proton pump of cytochrome c oxidase. Proc Natl Acad Sci U S A 107: 18469–18474

Benson AA (2002) Following the path of carbon in photosynthesis: A personal story. Photosynth Res 73: 29–49

Bernarding J, Eckert H-J, Eichler H-J, Napiwotzki A and Renger G (1994) Kinetic studies on the stabilization of the primary radical pair $P680^{+}Pheo^{-}$ in different Photosystem II preparations from higher plants. Photochm Photobiol 59: 566–573

Bertini I, Cavallaro G and Rosato A (2005) A structural model for the adduct between cytochrome *c* and cytochrome *c* oxidase. J Biol Inorg Chem 10: 613–624

Bisson R, Steffens GCM, Capaldi RA and Buse G (1982) Mapping of the cytochrome *c* binding site on cytochrome *c* oxidase. FEBS Lett 144: 359–363

Boltzmann L (1905) Der zweite Hauptsatz der mechanischen Wärmetheorie. In: Populäre Schriften, J. A. Barth, Leipzig (in German)

Brand SE, Rajagukguk S, Ganesan K, Geren L, Fabian M, Han D, Gennis RB, Durham B and Millett F (2007) A new ruthenium complex to study single-electron reduction of the pulsed O(H) state of detergent-solubilized cytochrome oxidase. Biochemistry 46: 14610–14618

Bränden G, Pawate AS, Gennis RB and Brzezinski P (2006) Controlled uncoupling and recoupling of proton pumping in cytochrome *c* oxidase. Proc Natl Acad Sci U S A 103: 317–322

Bratton MR, Pressler MA and Hosler JP (1999) Suicide inactivation of cytochrome *c* oxidase: Catalytic turnover in the absence of subunit III alters the active site. Biochemistry 38: 16236–16245

Brzezinski P and Gennis RB (2008) Cytochrome *c* oxidase: Exciting progress and remaining mysteries. J Bioenerg Biomembr 40: 521–531

Budiman K, Kannt A, Lyubenova S, Richter O-MH, Ludwig B, Michel H and MacMillan F (2004) Tyrosine-167: The origin of the radical species observed in the reaction of cytochrome *c* oxidase with hydrogen peroxide in *Paracoccus denitrificans*. Biochemistry 43: 11709–11716

Buick R (1992) The antiquity of oxygenic photosynthesis; evidence from stromatolites in sulphate-deficient Archaen lakes. Science 255: 74–77

Bundschuh FA, Hoffmeier K and Ludwig B (2008) Two variants of the assembly factor Surf1 target specific terminal oxidases in *Paracoccus denitrificans*. Biochim Biophys Acta 1777: 1336–1343

Burda K, Bader KP and Schmid GH (2001) An estimation of the size of the water cluster present at the cleavage site of the water splitting enzyme. FEBS Lett 491: 81–84

Buse G, Soulimane T, Dewor M, Meyer HE and Blüggel M (1990) Evidence for a copper-coordinated histidine-tyrosine cross-link in the active site of cytochrome oxidase. Protein Sci 8: 985–990

Calvin M (1989) Forty years of photosynthesis and related activities. Photosynth Res 21: 3–16

Castelfranco PA, Lu Y-K and Stemler AJ (2007) Hypothesis: The peroxydicarbonic acid cycle in photosynthetic oxygen evolution. Photosynth Res 94: 235–246

Chakrabarty S, Namslauer I, Brzezinski P and Warshel A (2011) Exploration of the cytochrome *c* oxidase pathway puzzle and examination of the origin of elusive mutational effects. Biochim Biophys Acta 1807: 413–426

Chow WS and Aro EM (2005) Photoinactivation and mechanism of recovery In: Wydrzynski T and Satoh K (eds) Photosystem II: The water/plastoquinone oxido-reductase in photosynthesis, advances in photosynthesis and respiration, Vol. 22, Springer, Dordrecht, pp 627–648

Clausen J and Junge W (2004) Detection of an intermediate of photosynthetic water oxidation. Nature 430: 480–483

Clausen J, Debus RJ and Junge W (2004) Time-resolved oxygen production by PS II: Chasing chemical intermediates. Biochim Biophys Acta 1655: 184–194

Clausen J, Beckmann K, Junge W and Messinger J (2005) Evidence that bicarbonate is not the substrate in photosynthetic oxygen evolution. Plant Physiol 139: 1444–1450

Cukier RI (2002) A theory that connects proton-coupled electron-transfer and hydrogen-atom transfer reactions. J Phys Chem B 106: 1746–1757

Cukier RI (2004) Quantum molecular dynamics simulation of proton transfer in cytochrome *c* oxidase. Biochim Biophys Acta 1656: 189–202

Danielsson R, Suorsa M, Paakkarinen V, Albertsson P-Å, Styring S, Aro E-M, and Mamedov F (2006) Dimeric and monomeric organization of Photosystem II. Distribution of five distinct complexes in the different domains of the thylakoids membrane. J Biol Chem 281: 14241–14249

de Grotthuss CJT (2006) Memoir on the decomposition of water and of the bodies that it holds in solution by means of galvanic electricity. Biochim Biophys Acta 1757: 871–875

de Marais DJ (2000) Evolution. When did photosynthesis emerge on Earth? Science 289: 1703–1705
de Wijn R and van Gorkom HJ (2001) Kinetics of electron transfer from Q_A to Q_B in Photosystem II. Biochemistry 40: 11912–11922
Drosou V, Malatesta F and Ludwig B (2002a) Mutations in the docking site for cytochrome *c* on the *Paracoccus* heme aa_3 oxidase: Electron entry and kinetic phases of the reaction. Eur J Biochem 269: 2980–2988
Drosou V, Reincke B, Schneider M and Ludwig B (2002b) Specificity of interaction between the *Paracoccus denitrificans* oxidase and its substrate cytochrome *c*: Comparing the mitochondrial to the homologous bacterial cytochrome c_{552}, and its truncated and site-directed mutants. Biochemistry 41: 10629–10634
Dürr K, Koepke J, Hellwig P, Müller H, Angerer H, Peng G, Olkova E, Richter O-MH, Ludwig B and Michel H (2008) A D-pathway mutation decouples the *Paracoccus denitrificans* cytochrome *c* oxidase by altering the side chain orientation of a distant, conserved glutamate. J Mol Biol 384: 865–877
Fabian M and Palmer G (1995) The interaction of cytochrome oxidase with hydrogen peroxide: The relationship of compounds P and F. Biochemistry 34: 13802–13810
Fadda E, Yu CH and Pomès R (2008) Electrostatic control of proton pumping in cytochrome *c* oxidase. Biochim Biophys Acta 1777: 277–284
Farver O, Grell E, Ludwig B, Michel H and Pecht I (2006) Rates and equilibrium of Cu_A to heme *a* electron transfer in *Paracoccus denitrificans* cytochrome *c* oxidase. Biophys J 90: 2131–2137
Faxen K, Gilderson G, Ädelroth P and Brzezinski P (2005) A mechanistic principle for proton pumping by cytochrome *c* oxidase. Nature 437: 286–289
Fee JA, Case DA and Noodleman L (2008) Toward a chemical mechanism of proton pumping by the B-Type cytochrome *c* oxidases: Application of density functional theory to cytochrome ba_3 of *Thermus thermophilus*. J Am Chem Soc 130: 15002–15021
Ferreira K, Iverson TM, Maghlouni K, Barber J and Iwata S (2004) Architecture of the photosynthetic oxygen-evolving center. Science 303: 1831–1838
Flöck D and Helms V (2002) Protein–protein docking of electron transfer complexes: Cytochrome *c* oxidase and cytochrome *c*. Proteins 47: 75–85
Florens L, Schmidt B, McCracken J and Ferguson-Miller S (2001) Fast deuterium access to the buried magnesium/manganese site in cytochrome *c* oxidase. Biochemistry 40: 7491–7497
Fontanesi F, Soto IC, Horn D and Barrientos A (2006) Assembly of mitochondrial cytochrome *c*-oxidase, a complicated and highly regulated cellular process. Am J Physiol Cell Physiol 291: 1129–1147
Fromme R, Grotjohann I and Fromme P (2008) Structure and function of photosystem I. In: Renger G (ed) Primary processes of photosynthesis: Principles and apparatus, Part II reaction centers/photosystems, electron transport chains, photophosphorylation and evolution, Royal Society Chemistry, Cambridge, pp 111–146
Fufezan C, Gross CM, Sjödin M, Rutherford AW, Krieger-Liszkay A and Kirilovsky D (2007) Influence of the redox potential of the primary quinone electron acceptor on photoinhibition in Photosystem II. J Biol Chem 282: 12492–12502
Fukuda R, Zhang H, Kim, J, Shimoda L, Dang CV and Semenza GL (2007) HIF-1 Regulates cytochrome oxidase subunits to optimize efficiency of respiration in hypoxic cells. Cell 129: 111–122
Gabdulkhakov A, Guskov A, Broser M, Kern J, Müh F, Saenger W and Zouni A (2009) Probing the accessibility of the Mn_4Ca cluster in Photosystem II: Channels calculation, noble gas derivatization, and cocrystallization with DMSO. Structure 17: 1223–1234
Genova ML, Baracca A, Biondi A, Casalena G, Faccioli M, Falasca AI, Formiggini G, Sgarbi G, Solaini G and Lenaz G (2008) Is supercomplex organization of the respiratory chain required for optimal electron transfer activity? Biochim Biophys Acta 1777: 740–774
Gibasiewicz K, Dobek A, Breton J and Leibl W (2001) Modulation of primary radical pair kinetics and energetics in Photosystem II by the redox state of the quinone electron acceptor QA. Biophys J 80: 1617–1630

Gilbert DL (ed) (1981) Oxygen and living processes: An interdisciplinary approach, Springer, New York

Giuffrè A, Forte E, Antonini A, D'Itri E, Brunori M, Soulimane T and Buse G (1999) Kinetic properties of *ba*$_3$ oxidase from thermus thermophilus: Effect of temperature. Biochemistry 38: 1057–1065

Glatzel P, Bergmann U, Yano J, Visser H, Robblee JH, Gu WW, de Groot FMF, Christou G, Pecoraro VL, Cramer SP and Yachandra VK (2004) The electronic structure of Mn in oxides, coordination complexes, and the oxygen-evolving complex of Photosystem II studied by resonant inelastic X-ray scattering. J Am Chem Soc 126: 9946–9959

Gorbikova EA, Belevich I, Wikström M and Verkhovsky MI (2008a) The proton donor for O–O bond scission by cytochrome *c* oxidase. Proc Natl Acad Sci U S A 105: 10733–10737

Gorbikova EA, Wikström M and Verkhovsky MI (2008b) The protonation state of the cross-linked tyrosine during the catalytic cycle of cytochrome *c* oxidase. J Biol Chem 283: 34907–34912

Govindjee, Beatty JT, Gest H and Allen JF (eds) (2005) Discoveries in photosynthesis. Advances in photosynthesis and respiration, Vol. 20, Springer, Dordrecht

Greiner P, Hannappel A, Werner C and Ludwig B (2008) Biogenesis of cytochrome *c* oxidase—bacterial approaches to study cofactor insertion into subunit I. Biochim Biophys Acta 1777: 904–911

Groot ML, Pawlowicz NP, van Wilderen LJGW, Breton J, van Stokkum IHM and van Grondelle R (2005) Initial electron donor and acceptor in isolated photosystem II reaction centers identified with femtosecond mid-IR spectroscopy. Proc Natl Acad Sci U S A 102: 13087–13092

Guskov A, Kern J, Gabdulkhakov A, Broser M, Zouni A and Saenger W (2009) Cyanobacterial Photosystem II at 2.9 Å resolution and the role of quinones, lipids, channels and chloride. Nature Struct Mol Biol 16: 334–342

Haltia T, Saraste M and Wikström M (1991) Subunit III of cytochrome *c* oxidase is not involved in proton translocation: A site-directed mutagenesis study. EMBO J 10: 2015–2021

Hamza I and Gitlin JD (2002) Copper chaperones for cytochrome *c* oxidase and human disease. J Bioenerg Biomembr 34: 381–388

Hammes-Schiffer S (2006) Hydrogen tunneling and protein motion in enzyme reactions. Acc Chem Res 39: 93–100

Hansson Ö, Andreasson LE and Vänngard T (1986) Oxygen from water is coordinated to manganese in the S_2 state of Photosystem II. FEBS Lett 195: 151–154

Harrenga A, Reincke B, Rüterjans H, Ludwig B and Michel H (2000) Structure of the soluble domain of cytochrome c_{552} from *Paracoccus denitrificans* in the oxidized and reduced states. J Mol Biol 295: 667–678

Hasegawa K and Noguchi T (2005) Density functional theory calculations on the dielectric-constant dependence of the oxidation potential of chlorophyll: Implication for the high potential of P680 in Photosystem II. Biochemistry 44: 8865–8872

Hellwig P, Behr J, Ostermeier C, Richter O-MH, Pfitzner U, Odenwald A, Ludwig B, Michel H and Mäntele W (1998) Involvement of Glutamic Acid 278 in the Redox Reaction of the Cytochrome *c* Oxidase from *Paracoccus denitrificans* investigated by FTIR Spectroscopy. Biochemistry 37: 7390–7399

Hendler RW, Pardhasaradhi K, Reynafarje B and Ludwig B (1991) Comparison of energy-transducing capabilities of the two- and three-subunit cytochromes *aa*$_3$ from *Paracoccus denitrificans* and the 13-subunit beef heart enzyme. Biophys J 60: 415–423

Hill BC (1994) Modeling the sequence of electron transfer reactions in the single turnover of reduced, mammalian cytochrome *c* oxidase with oxygen. J Biol Chem 269: 2419–2425

Hillier W and Messinger J (2005) Mechanism of photosynthetic oxygen production In: Wydrzynski T and Satoh K (eds) Photosystem II. The light-driven water: Plastoquinone oxidoredutase, Advances in photosynthesis and respiration, Vol. 22, Springer, Dordrecht, pp 567–608

Hillier W, McConnell I, Badger MR, Boussac A, Klimov VV, Dismukes GC and Wydrzynski T (2006) Quantitative assessment of intrinsic carbonic anhydrase activity and the capacity for bicarbonate oxidation in Photosystem II. Biochemistry 45: 2094–2102

Hillier W and Wydrzynski T (2008) ^{18}O-water exchange in Photosystem II: Substrate binding and intermediates of the water splitting cycle. Coord Chem Rev 252: 306–317

Hiser L, Di Valentin M, Hamer AG and Hosler JP (2000) Cox11p is required for stable formation of the Cu_B and magnesium centers of cytochrome *c* oxidase. J Biol Chem 275: 619–623

Ho FM and Styring S (2008) Access channels and methanol binding site to the $CaMn_4$ cluster in Photosystem II based on solvent accessibility simulations, with implications for substrate water access. Biochim Biophys Acta 1777: 140–153

Holzwarth AR, Müller MG, Reus M, Nowaczyk M, Sander J and Rögner M (2006) Kinetics and mechanism of electron transfer in intact photosystem II and in the isolated reaction center: Pheophytin is the primary electron donor. Proc Natl Acad Sci U S A 103: 6895–6900

Hosler JP, Ferguson-Miller S, Calhoun MW, Thomas JW, Hill J, Lemieux L, Ma J, Georgiou C, Fetter J, Shapleigh J, Tecklenburg MMJ, Babcock GT and Gennis RB (1993) Insight into the active site structure and function of cytochrome oxidase by analysis of site-directed mutants of bacterial cytochrome aa_3 and cytochrome *bo*. J Bioenerg Biomembr 25: 121–136

Hosler JP, Ferguson-Miller S and Mills DA (2006) Energy Transduction: Proton transfer through the respiratory complexes. Ann Rev Biochem 75: 165–187

Hunter CN, Daldal F, Thurnauer MC and Beatty JT (eds) (2008) The purple phototrophic bacteria. Advances in photosynthesis and respiration, Vol. 28, Springer, Dordrecht

Huynh MHV and Meyer TJ (2007) Proton-coupled electron transfer. Chem Rev 107: 5004–5064

Ishida N, Sugiura M, Rappaport F, Lai T-L, Rutherford AW and Boussac A (2008) Biosynthetic exchange of bromide for chloride and strontium for calcium in the Photosystem II oxygen-evolving enzymes. J Biol Chem 283: 13330—13340

Ishikita H, Knapp E-W (2005) Control of quinone redox potentials in Photosystem II: Electron transfer and photoprotection. J Am Chem Soc 127: 14714–14720

Ishikita H, Saenger W, Loll B, Biesiadka J and Knapp EW (2006) Energetics of a possible proton exit pathway for water oxidation in Photosystem II. Biochemistry 45: 2063–2071

Iwata S, Ostermeier C, Ludwig B and Michel H (1995) Structure at 2.8 Å resolution of cytochrome *c* oxidase from *Paracoccus denitrificans*. Nature 376: 660–669

Janzon J, Ludwig B and Malatesta F (2007) Electron transfer kinetics of soluble fragments indicate a direct interaction between complex III and the caa_3 oxidase in *Thermus thermophilus.* IUBMB Life 59: 563–569

Joliot P, Barbieri G and Chabaud R (1969) Un nouveau modele des centres photochimiques du système II. Photochem Photobiol 10: 309–329

Junge W (2008) Evolution of photosynthesis. In: Renger G (ed) Primary processes of photosynthesis: Priniciples and apparatus, Part II reaction centers/photosystems, electron transport chains, photophosphorylation and evolution, Royal Society Chemistry, Cambridge, pp 447–487

Kaila VRI, Verkhovsky M, Hummer G and Wikström M (2008a) Prevention of leak in the proton pump of cytochrome *c* oxidase. Biochim Biophys Acta 1777: 890–892

Kaila VRI, Verkhovsky M, Hummer G and Wikström M (2008b) Glutamic acid 242 is a valve in the proton pump of cytochrome *c* oxidase. Proc Natl Acad Sci U S A 105: 6255–6259

Kaila VRI, Johansson MP, Sundholm D, Laakkonen L and Wikström M (2009) The chemistry of the Cu_B site in cytochrome *c* oxidase and the importance of its unique His–Tyr bond. Biochim Biophys Acta 1787: 223–233

Kaila VR, Verkhovsky MI and Wikström M (2010) Proton-coupled electron transfer in cytochrome oxidase. Chem Rev 110: 7062–7081

Kaminskaya O, Shuvalov VA and Renger G (2007) Evidence for a novel quinone binding site in the PS II (PS II) complex which regulates the redox potential of Cyt b559. Biochemistry 46: 1091–1105

Kamiya N and Shen J-R (2003) Crystal structure of oxygen-evolving Photosystem II from *Thermosynechococcus vulcanus* at 3.7 Å resolution. Proc Natl Acad Sci U S A 100: 98–103

Kannt A, Soulimane T, Buse G, Becker A, Bamberg E and Michel H (1998) Electrical current generation and proton pumping catalyzed by the ba_3-type cytochrome *c* oxidase from *Thermus thermophilus*. FEBS Lett 434: 17–22

Karge M, Irrgang K-D and Renger G (1997) Analysis of the reaction coordinate of photosynthetic water oxidation by kinetic measurements of 355 nm absorption changes at different temperatures in Photosystem II preparations suspended in either H_2O or D_2O. Biochemistry 36: 8904–8913

Karpefors M, Ädelroth P and Brzezinski P (2000) Localized control of proton transfer through the D-pathway in cytochrome *c* oxidase: application of the proton-inventory technique. Biochemistry 39: 6850–6856

Kasting JF and Seifert JF (2002) Life and the evolution of earth's atmosphere. Science 296: 1066–1067

Käß H, MacMillan F, Ludwig B and Prisner TF (2000) Investigation of the Mn binding site in cytochrome *c* oxidase from *Paracoccus denitrificans* by high-frequency EPR. J Phys Chem 104: 5362–5371

Kato Y, Sugiura M, Oda A and Watanabe T (2009) Spectroelectrochemical determination of the redox potential of pheophytin *a*, the primary electron acceptor in Photosystem II. Proc Nat Acad Sci U S A 106: 17365–17370

Kern J and Renger G (2007) Photosystem II: structure and mechanism of the water:plastoquinone oxidoreductase. Photosynth Res 94: 183–202

Khalimonchuk O and Winge DR (2008) Function and redox state of mitochondrial localized cysteine-rich proteins important in the assembly of cytochrome *c* oxidase. Biochim Biophys Acta 1783: 618–628

Kirichenko AV, Pfitzner U, Ludwig B, Soares CM, Vygodina TV and Konstantinov AA (2005) Cytochrome *c* oxidase as a calcium binding protein. Studies on the role of a conserved aspartate in helices XI–XII cytoplasmic loop in cation binding. Biochemistry 44: 12391–12401

Klimov VV, Allakhverdiev SI, Demeter S and Krasnovsky AA (1979) Photoreduction of pheophytin in Photosystem II of chloroplasts as a function of redox potential of the medium. Dokl Adad Nauk SSSR 249: 227–230

Kobayashi M, Ohashi S, Iwamoto K, Shiraiwa Y, Kato Y and Watanabe T (2007) Redox potential of chlorophyll *d in vitro*. Biochim Biophys Acta 1767: 596–602

Koepke J, Olkhova E, Angerer H, Müller H, Peng G and Michel H (2009) High resolution crystal structure of *Paracoccus dentrificans* cytochrome c oxidase: New insights into the active site and the proton transfer pathways. Biochim Biophys Acta 1787(6): 635–645

Koike H, Hanssum B, Inoue Y and Renger G (1987) Temperature dependence of the S-state transitions in a thermophilic cyanobacterium, *Synechococcus vulcanus* Copeland measured by absorption changes in the ultraviolet region. Biochim Biophys Acta 893: 524–533

Kok B, Forbush B and McGloin M (1970) Cooperation of charges in photosynthetic O_2 evolution. Photochem Photobiol 11: 457–476

Kolling DRJ, Brown TS, Ananyev G and Dismukes GC (2009) Photosynthetic oxygen evolution is not reversed at high oxygen pressures: Mechanistic consequences for the water-oxidizing complex. Biochemistry 48: 1381–1389

Konstantinov AA, Siletsky S, Mitchell D, Kaulen A and Gennis RB (1997) The roles of the two proton input channels in cytochrome *c* oxidase from *Rhodobacter sphaeroides* probed by the effects of site-directed mutations on time-resolved electrogenic intraprotein proton transfer. Proc Natl Acad Sci U S A 94: 9085–9090

Krause F, Reifschneider NH, Vocke D, Seelert H, Rexroth S and Dencher NA (2004) "Respirasome"-like supercomplexes in green leaf mitochondria of spinach. J Biol Chem 279: 48369–48375

Krieger-Liszkay A and Rutherford AW (1998) Influence of herbicide binding on the redox potential of teh quinone acceptor in photosystem II: Relevance to photodamage and phototoxicity. Biochemistry 37: 17339–17344

Krieger A, Rutherford AW and Johnson GN (1995) On the determination of redox midpoint potential of the primary quinone electron-acceptor, Q(a), in Photosystem II. Biochim Biophys Acta 1229: 193–201

Kühn P, Eckert H-J, Eichler H-J and Renger G (2004) Analysis of the $P680^{+\bullet}$ reduction pattern and its temperature dependence in oxygen-evolving PS II core complexes from thermophilic cyanobacteria and higher plants. Phys Chem Chem Phys 6: 4838–4843

Kulik LV, Epel B, Lubitz W and Messinger J (2007) Electronic structure of the Mn_4O_XCa cluster in the S_0 and S_2 states of the oxygen-evolving complex of Photosystem II based on pulse ^{55}Mn-ENDOR and EPR spectroscopy. J Am Chem Soc 129: 13421–13435

Lancaster R (2008) Structures of reaction centers in anoxygenic bacteria. In: Renger G (ed) Primary processes of photosynthesis: Priniciples and apparatus, Part II reaction centers/photosystems, electron transport chains, photophosphorylation and evolution, Royal Society Chemistry, Cambridge, pp 5–56

Lane N (2002) Oxygen—the molecule that made the World. Oxford University Press, Oxford

Larkum AWD (2008) The evolution of photosynthesis. In: Renger G (ed) Primary processes of photosynthesis: Priniciples and apparatus, Part II reaction centers/photosystems, electron transport chains, photophosphorylation and evolution, Royal Society Chemistry, Cambridge, pp 491–521

Lee A, Kirichenko L, Vygodina T, Siletsky SA, Das TK, Rousseau DL, Gennis R and Konstantinov AA (2002) Ca(2+)-binding site in *Rhodobacter sphaeroides* cytochrome C oxidase. Biochemistry 41: 8886–8898

Lee H, Das TK, Rousseau DL, Mills D, Ferguson-Miller S and Gennis RB (2000) Mutations in the putative H-channel in the cytochrome *c* oxidase from *Rhodobacter sphaeroides* show that this channel is not important for proton conduction but reveal modulation of the properties of heme *a*. Biochemistry 39: 2989–2996

Lepp H, Salomonsson L, Zhu JP, Gennis RB and Brzezinski P (2008a) Impaired proton pumping in cytochrome *c* oxidase upon structural alteration of the D pathway. Biochim Biophys Acta 1777: 897–903

Lepp H, Svahn E, Faxén K and Brzezinski P (2008b) Charge transfer in the K proton pathway linked to electron transfer to the catalytic site in cytochrome *c* oxidase. Biochemistry 47: 4929–4935

Liang W, Roelofs TA, Cinco RM, Rompel A, Latimer MJ, Yu WO, Sauer K, Klein MP and Yachandra VK (2000) Structural change of the Mn cluster during the S_2 to S_3 state transition of the oxygen evolving complex of Photosystem II. Does it reflect the onset of water/substrate oxidation? Determination by Mn x-ray absorption spectroscopy. J Am Chem Soc 122: 3399–3412

Loll B, Kern J, Saenger W, Zouni A and Biesiadka J (2005) Towards complete cofactor arrangement in the 3.0 Å resolution structure of Photosystem II. Nature 438: 1040–1044

Ludwig B and Gibson QH (1981) Reacton of oxygen with cytochrome *c* oxidase from *Paracoccus denitrificans*. J Biol Chem 256: 10092–10098

Ludwig B and Schatz G (1980) A Two-Subunit Cytochrome *c* Oxidase (Cytochrome aa_3) from *Paracoccus denitrificans.* Proc Natl Acad Sci USA 77: 196–200

Luna VM, Chen Y, Fee JA and Stout CD (2008) Crystallographic studies of Xe and Kr binding within the large internal cavity of cytochrome ba_3 from *Thermus thermophilus*: structural analysis and role of oxygen transport channels in the heme-Cu oxidases. Biochemistry 47: 4657–4665

Lyubenova S, Siddiqui K, Penning de Vries M, Ludwig B and Prisner TF (2007) Protein-protein interactions studied by EPR relaxation measurements: cytochrome *c* and cytochrome *c* oxidase. J Phys Chem B 111: 3839–3846

MacMillan F, Kannt A, Behr J, Prisner T and Michel H (1999) Direct evidence for a tyrosine radical in the reaction of cytochrome *c* oxidase with hydrogen peroxide. Biochemistry 38: 9179–9184

MacMillan F, Budiman K, Angerer H and Michel H (2006) The role of tryptophan 272 in the *Paracoccus denitrificans* cytochrome *c* oxidase. FEBS Lett 580: 1345–1349

Maneg O, Ludwig B and Malatesta F (2003) Different interaction modes of two cytochrome *c* oxidase soluble Cu_A fragments with their substrates. J Biol Chem 278: 46734–46740

Maneg O, Malatesta F, Ludwig B and Drosou V (2004) Interaction of cytochrome *c* with cytochrome oxidase: Two different docking scenarios. Biochim Biophys Acta 1655: 274–281

Marcus RA and Sutin N (1985) Electron transport in chemistry and biology. Biochim Biophys Acta 811: 265–322

Margoliash E and Bosshard HR (1983) Guided by electrostatics, a textbook protein comes of age. Trends Biochem Sci 8: 316–320

Mather MW, Springer P, Hensel S, Buse G, and Fee JA (1993) Cytochrome oxidase genes from *Thermus thermophilus*. Nucleotide sequence of the fused gene and analysis of the deduced primary structures for subunits I and III of cytochrome caa_3. J Biol Chem 268: 5395–5408

McEvoy JP and Brudvig GW (2006) Water-splitting chemistry of Photosystem II. Chem Rev 106: 4455–4483

Messinger J and Renger G (2008) Photosynthetic water splitting. In: Renger G (ed) Primary processes of photosynthesis: Basic priniciples and apparatus, Volume II: Reaction centers/photosystems, electron transport chains, photophosphorylation and evolution, Royal Society Chemistry, Cambridge, pp 291–349

Messinger J, Wacker U and Renger G (1991) Unusual low reactivity of the water oxidase in redox state S_3 toward exogenous reductants. Analysis of the NH_2OH and NH_2NH_2-induced modifications of flash-induced oxygen evolution in isolated spinach thylakoids. Biochemistry 30: 7852–7862

Meyer TJ, Hang M, Huynh V and Thorp HH (2007) The role of proton coupled electron transfer (PCET) in water oxidation by photosystem II. Wiring for protons. Angew Chem Int Ed 46: 5284–5304

Michel H (1999) Cytochrome *c* oxidase: Catalytic cycle and mechanisms of proton pumping—A discussion. Biochemistry 38: 15129–15140

Miloslavina Y, Szczepaniak M, Müller MG, Sander J, Nowaczyk M, Rögner M and Holzwarth AR (2006) Charge separation kinetics in intact photosystem II core particles is trap-limited. A picosecond fluorescence study. Biochemistry 45: 2436–2442

Miqyass M, Marosvölgyi MA, Nagel Z, Yocum CF and van Gorkom HJ (2008) S-State dependence of the calcium requirement and binding characteristics in the oxygen-evolving complex of photosystem II. Biochemistry 47: 7915–7924

Mitchell P (1961) Coupling of photophosphorylation to electron and hydrogen transfer by a chemiosmotic type of mechanism. Nature 191: 144–148

Morowitz HJ (1978) Foundations of bioenergetics, Academic Press, New York, p 344

Moser CC, Farid TA, Chobot SE and Dutton PL (2006) Electron tunneling chains of mitochondria. Biochim Biophys Acta 1757: 1096–1109

Moser CC, Chobot SE, Page CC and Dutton PL (2008) Distance metrics for heme protein electron tunneling. Biochim Biophys Acta 1777: 1032–1037

Muramoto K, Hirata K, Shinzawa-Itoh K, Yoko-o S, Yamashita E, Aoyama H, Tsukihara T and Yoshikawa S (2007) A histidine residue acting as a controlling site for dioxygen reduction and proton pumping by cytochrome *c* oxidase. Proc Natl Acad Sci U S A 104: 7881–7886

Muresanu L, Pristovšek P, Löhr F, Maneg O, Mukrasch MD, Rüterjans H, Ludwig B and Lücke C (2006) The electron transfer complex between cytochrome c_{552} and the Cu_A domain of the *Thermus thermophilus* ba_3 oxidase—a combined NMR and computational approach. J Biol Chem 281: 14503–14513

Müh F and Zouni A (2011) Light-induced water oxidation in Photosystem II. Front Biosci (in press)

Murray JW and Barber J (2007) Structural characteristics of channels and pathways in photosystem II including the identification of an oxygen channel. J Struct Biol 159: 228–237

Namslauer A, Lepp H, Brändén M, Jasaitis A, Verkhovsky MI and Brzezinski P (2007) Plasticity of proton pathway structure and water coordination in cytochrome *c* oxidase. J Biol Chem 282: 15148–15158

Neupert W and Herrmann JM (2007) Translocation of proteins into mitochondria. Annu Rev Biochem 76: 723–749

Nicholls DG and Fergusson SJ (1982) Bioenergetics, vol. 2, Academic Press, London, pp 215–218

Nicolls G, Prigogine I (1977) Self organization in nonequilibrium systems, Wiley, New York

Noguchi T (2008) FTIR detection of water reactions in the oxygen-evolving centre of photosystem II. Phil Trans R Soc B 363: 1189–1195

Noguchi T and Sugiura M (2002) Flash-induced FTIR difference spectra of the water oxidizing complex in moderately hydrated Photosystem II core films: Effect of hydration extent on S-state transitions. Biochemistry 41: 2322–2330

Olesen K and Andréasson L-E (2003) The function of the chloride ion in photosynthetic oxygen evolution. Biochemistry 42: 2025–2035

Olsson MHM, Siegbahn PEM, Blomberg MRA and Warshel A (2007) Exploring pathways and barriers for coupled ET/PT in cytochrome *c* oxidase: A general framework for examining energetics and mechanistic alternatives. Biochim Biophys Acta 1767: 244–260

Orio M, Pantazis DA and Neese F (2009) Density functional theory. Photosynth Res 102(2–3): 443–453

Ostermeier C, Harrenga A, Ermler U and Michel H (1997) Structure at 2.7 Å resolution of the *Paracoccus denitrificans* two-subunit cytochrome *c* oxidase complexed with an antibody Fv fragment. Proc Natl Acad Sci U S A 94: 10547–10553

Otten MF, van der Oost J, Reijnders WNM, Westerhoff HV, Ludwig B and van Spanning RJM (2001) Cytochromes c_{550}, c_{552} and c_1 in the electron transport network of *Paracoccus denitrificans*: Redundant or subtly different in function? J Bacteriol 183: 7017–7026

Page CC, Moser CC, Chen X and Dutton PL (1999) Natural engineering principles of electron tunneling in biological oxidation-reduction. Nature 402: 47–52

Parson WW (2008) Functional patterns of reaction centers in anoxygenic photosynthetic bacteria. In: Renger G (ed) Primary processes of photosynthesis: Priniciples and apparatus, Part II reaction centers/photosystems, electron transport chains, photophosphorylation and evolution, Royal Society Chemistry, Cambridge, pp 57–109

Paumann M, Regelsberger G, Obinger C and Peschek GA (2005) The bioenergetic role of dioxygen and the terminal oxidase(s) in cyanobacteria. Biochim Biophys Acta 1777: 231–253

Pereira MM, Sousa FL, Verissimo AF and Teixiera M (2008) Looking for the minimum common denominator in haem–copper oxygen reductases: Towards a unified catalytic mechanism. Biochim Biophys Acta 1777: 929–934

Peschek G (2008) Electron transport chains in oxygenic cyanobacteria. In: Renger G (ed) Primary processes of photosynthesis: Priniciples and apparatus, Part II reaction centers/photosystems, electron transport chains, photophosphorylation and evolution, Royal Society Chemistry, Cambridge, pp 383–415

Petrouleas V and Crofts AR (2005) The iron-quinone acceptor complex. In: Wydrzynski T and Satoh K (eds) Photosystem II: The water/plastoquinone oxido-reductase in photosynthesis, Springer, Dordrecht, pp 177–206

Pfitzner U, Odenwald A, Ostermann T, Weingard L, Ludwig B and Richter O-MH (1998) Cytochrome *c* oxidase (heme aa_3) from *Paracoccus denitrificans*: analysis of mutations in putative proton channels of subunit I. J Bioenerg Biomembr 30: 89–97

Pfitzner U, Kirichenko AV, Konstantinov AA, Mertens M, Wittershagen A, Kolbesen BO, Steffens GCM, Harrenga A, Michel H and Ludwig B (1999) Mutations in the Ca^{2+} binding site of the *Paracoccus denitrificans* cytochrome *c* oxidase. FEBS Lett 456: 365–369

Pfitzner U, Hoffmeier K, Harrenga A, Kannt A, Michel H, Bamberg E, Richter O-MH and Ludwig B (2000) Tracing the D-pathway in reconstituted site-directed mutants of cytochrome *c* oxidase from *Paracoccus denitrificans*. Biochemistry 39: 6756–6762

Pieper J and Renger G (2009) Protein dynamics investigated by neutron scattering. Photosynth Res 102: 281–293

Pilet E, Jasaitis A, Liebl U and Vos MH (2004) Electron transfer between hemes in mammalian cytochrome *c* oxidase. Proc Natl Acad Sci U S A 101: 16198–16203

Pinakoulaki E, Pfitzner U, Ludwig B and Varotsis C (2002) The role of the cross-link his-tyr in the functional properties of the binuclear center in cytochrome *c* oxidase. J Biol Chem 277: 13563–13568

Pisliakov AV et al (2008) Electrostatic basis for the unidirectionality of the primary proton transfer in cytochrome *c* oxidase. Proc Natl Acad Sci U S A 105: 7726–7731

Pogson JB, Rissler HM and Frank HA (2005) The role of carotenoids in energy quenching. In: Wydrzynski T and Satoh K (eds) Photosystem II: The water/plastoquinone oxido-reductase in photosynthesis, Springer, Dordrecht, pp 515–537

Proshlyakov DE, Pressler MA and Babcock GT (1998) Dioxygen activation and bond cleavage by mixed-valence cytochrome *c* oxidase. Proc Natl Acad Sci U S A 95: 8020–8025

Pushkar Y, Yano J, Sauer K, Boussac A and Yachandra VK (2008) Structural changes in the Mn_4Ca cluster and the mechanism of photosynthetic water splitting. Proc Nat Acad Sci U S A 105: 1879–1884

Rappaport F and Diner BA (2008) Primary photochemistry and energetics leading to the oxidation of the $(Mn)_4Ca$ cluster and to the evolution of molecular oxygen in Photosystem II. Coord Chem Rev 252: 259–272

Rappaport F, Blanchard-Desce M and Lavergne J (1994) Kinetics of the electron transfer and electrochromic change during the redox transitions of the photosynthetic oxygen-evolving complex. Biochim Biophys Acta 1184: 178–192

Rappaport F, Cuni A, Xiong L, Sayre R and Lavergne J (2005) Charge recombination and thermoluminescence in photosystem II. Biophys J 88: 1948–1958

Raszewski G, Diner BA, Schlodder E and Renger T (2008) Spectroscopic properties of reaction center pigments in photosystem II core complexes: Revision of the multimer model. Biophys J 95(1): 105–119, doi: 10.1529/biophysj.107.123935

Raymond J and Blankenship RE (2004) The evolutionary development of the protein complement of Photosystem II. Biochim Biophys Acta 1655: 133–139

Razeghifard MR and Pace RJ (1999) EPR kinetic studies of oxygen release in thylakoids in PS II membranes: A kinetic intermediate in the S_3 to S_0 transition. Biochemistry 38: 1252–1257

Reincke B, Thöny-Meyer L, Dannehl C, Odenwald A, Aidim M, Witt H, Rüterjans H and Ludwig B (1999) Heterologous expression of soluble fragments of cytochrome c_{552} acting as electron donor to the *Paracoccus denitrificans* cytochrome *c* oxidase. Biochim Biophys Acta 1411: 114–120

Reincke B, Pérez C, Pristovsek P, Lücke C, Ludwig C, Löhr F, Rogov VV, Ludwig B and Rüterjans H (2001) Solution structure and dynamics of the functional domain of *Paracoccus denitrificans* c_{552} in both redox states. Biochemistry 40: 12312–12320

Renger G (1978) Theoretical studies about the functional and structural organization of the photosynthetic oxygen evolution. In: Metzner H (ed) Photosynthetic oxygen evolution, Academic Press, London, pp 229–248

Renger G (1983) Biological energy conservation. In: Hoppe W, Lohmann W, Markl H and Ziegler H (eds), Biophysics, Springer, Berlin, pp 347–371

Renger G (1999) Molecular mechanism of water oxidation. In: Singhal GS, Renger G, Govindjee, Irrgang KD and Sopory SK (eds) Concepts in photobiology: Photosynthesis and photomorphogenesis, Kluwer Academic Publishers (now Springer), Dordrecht and Narosa Publishing Co., Delhi, pp 292–329

Renger G (2001) Photosynthetic water oxidation to molecular oxygen: Apparatus and mechanism. Biochim Biophys Acta 1503: 210–228

Renger G (2004) Coupling of electron and proton transfer in oxidative water cleavage in photosynthesis. Biochim Biophys Acta 1655: 195–204

Renger G (2007) Oxidative photosynthetic water splitting: Energetics, kinetics and mechanism. Photosynth Res 92: 407–425

Renger G (2008) Functional pattern of Photosystem II in Oxygen Evolving Organisms. In: Renger G (ed) Primary processes of photosynthesis: principles and apparatus, Part II reaction centers/photosystems, electron transport chains, photophosphorylation and evolution. Royal Society Chemistry, Cambridge, pp 237–290

Renger G (2011) Light induced oxidative water splitting in photosynthesis: energetics, kinetics and mechanism. Photobiochem Photobiophys B: Biol, doi: 10.1016/j.jphotobiol.2011.01.023

Renger G and Hanssum B (1992) Studies on the reaction coordinates of the water oxidase in PS II membrane fragments from spinach. FEBS Lett 299: 28–32

Renger G and Holzwarth AR (2005) Primary electron transfer In: Wydrzynski T and Satoh K (eds) Photosystem II: The water: Plastoquinone oxido-reductase in photosynthesis, Springer, Dordrecht, pp 139–175
Renger G and Kühn P (2007) Reaction pattern and mechanism of light induced oxidative water splitting in photosynthesis. Biochim Biophys Acta 1767: 458–471
Renger G and Renger T (2008) Photosystem II: The machinery of photosynthetic water splitting. Photosynth Res 98: 53–80
Renger G and Völker M (1982) Studies on the proton release of the donor side of system II. Correlation between oxidation and deprotonization of donor D_1 in Tris-washed inside-out thylakoids. FEBS Lett 149: 203–207
Renger G and Weiss W (1982) The detection of intrinsic 320 nm absorption changes reflecting the turnover of the water splitting enzyme system Y which leads to oxygen formation in trypsinized chloroplats. FEBS Lett 137: 217–221
Renger G, Eckert HJ and Völker M (1989) Studies on the electron transfer from Tyr-161 of polypeptide D-1 to $P680^+$ in PS II membrane fragments from spinach. Photosynth Res 22: 247–256
Renger G, Eckert H-J, Bergmann A, Bernarding J, Liu B, Napiwotzki A, Reifarth F, and Eichler JH (1995) Fluorescence and spectroscopic studies on exciton trapping and electron transfer in Photosystem II of higher plants. Austr J Plant Physiol 22: 167–181
Renger G, Christen G, Karge M, Eckert H-J and Irrgang K-D (1998) Application of the Marcus theory for analysis of the temperature dependence of the reactions leading to photosynthetic water oxidation—results and implications. J Bioinorg Chem 3: 360–366
Rich PR (1995) Towards an understanding of the chemistry of oxygen reduction and proton translocation in the iron-copper respiratory oxidases. Aust J Plant Physiol 22: 479–486
Riistama S, Puustinen A, García-Horsman A, Iwata S, Michel H and Wikström M (1996) Channelling of dioxygen into the respiratory enzyme. Biochim Biophys Acta 1275: 1–4
Riistama S, Puustinen A, Verkhovsky MI, Morgan JE and Wikström M (2000) Binding of O_2 and its reduction are both retarded by replacement of valine 279 by isoleucine in cytochrome *c* oxidase from *Paracoccus denitrificans*. Biochemistry 39: 6365–6372
Robblee JH, Messinger J, Cinco RM, McFarlane KL, Fernandez C, Pizarro SA, Sauer K and Yachandra VK (2002) The Mn cluster in the S_0 state of the oxygen evolving complex of Photosystem II studied by EXAFS spectroscopy: are there three di-μ-oxo-bridged Mn_2 moieties in the tetranuclear Mn complex? J Am Chem Soc 124: 7459–7471
Roberts VA and Pique ME (1999) Definition of the interaction domain for cytochrome *c* on cytochrome *c* oxidase—III. Prediction of the docked complex by a complete, systematic search. J Biol Chem 274: 38051–38060
Romain S, Bozoglian F, Sala X and Llobet A (2009) Oxygen−oxygen bond formation by the Ru-Hbpp water oxidation catalyst occurs solely via an intramolecular reaction pathway. J Am Chem Soc 131: 2768–2769
Romero E, van Stokkum IHM, Novoderezhkin VI, Dekker JP and van Grondelle R (2010) Two different charge separation pathways in Photosystem II. Biochemistry 49: 4300–4307
Ruitenberg M, Kannt A, Bamberg E, Michel H, Ludwig B and Fendler K (2000) Single-electron reduction of the oxidized state is coupled to proton uptake via the K-pathway in *Paracoccus denitrificans* cytochrome *c* oxidase. Proc Natl Acad Sci U S A 97: 4632–4636
Ruitenberg M, Kannt A, Bamberg E, Fendler K and Michel H (2002) Reduction of cytochrome *c* oxidase by a second electron leads to proton translocation. Nature 417: 99–102
Sacconi S, Salviati L, Sue CM, Shanske S, Davidson MM, Bonilla E, Naini AB, De Vivo DC and DiMauro S (2003) Mutation screening in patients with isolated cytochrome *c* oxidase deficiency. Pediatr Res 53: 224–230
Salje J, Ludwig B and Richter O-MH (2005) Is a third proton-conducting pathway operative in bacterial cytochrome *c* oxidases? Biochem Soc Trans 33: 829–831
Saraste M and Castresana J (1994) Cytochrome oxidase evolved by tinkering with denitrification enzymes. FEBS Lett 341: 1–4
Sauer K, Yano J and Yachandra VK (2008) X-ray spectroscopy of the photosynthetic oxygen-evolving complex. Coord Chem Rev 252: 318–335

Schägger H and Pfeiffer K (2001) The ratio of oxidative phosphorylation complexes I–V in bovine heart mitochondria and the composition of respiratory chain supercomplexes. J Biol Chem 276: 37861–37867

Schinzel S, Schraut J, Arbuznikov AV, Siegbahn PEM and Kaipp M (2010) Density functional calculations of ^{55}Mn, ^{14}N and ^{13}C electron paramagnetic resonance parameters support an energetically feasible model system for the S_2 state of the oxygen-evolving complex of Photosystem II. Chem Eur J 16: 10424–10438

Schlodder E, Renger T, Raszewski G, Coleman WJ, Nixon PJ, Cohen RO and Diner B (2008) Site-directed mutations at D1-Thr179 of Photosystem II in *Synechocystis* sp. PCC 6803 modify the spectroscopic properties of the accessory chlorophyll in the D1-branch of the reaction center. Biochemistry 47: 3143–3154

Seelig A, Ludwig B, Seelig J and Schatz G (1981) Copper and Manganese Electron Spin Resonance Studies of Cytochrome *c* Oxidase from *Paracoccus denitrificans*. Biochim Biophys Acta 636: 162–167

Sharpe MA and Ferguson-Miller S (2008) A chemically explicit model for the mechanism of proton pumping in heme–copper oxidases. J Bioenerg Biomembr 40: 541–549

Sharpe MA, Krzyaniak MD, Xu S, McCracken J and Ferguson-Miller S (2009) EPR Evidence of Cyanide Binding to the Mn(Mg) Center of Cytochrome *c* Oxidase: Support for Cu_A -Mg Involvement in Proton Pumping. Biochemistry 48: 328–335

Shelaev IV, Gostev FE, Nadtochenko VA, Shkuropatov AY, Zabelin AA, Mamedov MD, Semenov AY, Sarkisov OM and Shuvalov VA (2008) Primary light-energy conversion in tetrameric chlorophyll structure of Photosystem II and bacterial reaction centers: II. Femto- and picosecond charge separation in PS II D1/D2/Cyt b559 complex. Photosynth Res 98: 95–103

Shevela D, Su J-H, Klimov V and Messinger J (2008) Hydrogencarbonate is not a tightly bound constituent of the water oxidizing complex in photosystem II. Biochim Biophys Acta 1777: 532–539

Sheleva D, Beckmann K, Clausen J, Junge W and Messinger J (2011) Photosynthetic oxygen-evolution at elevated oxygen pressure: direct detection by membrane inlet mass spectrometry. Proc Natl Acad Sci 108: 3602–3607

Shi L-X and Schröder WP (2004) The low molecular mass subunits of the photosynthetic supracomplex, Photosystem II. Biochim Biophys Acta 1608: 75–96

Shibamoto T, Kato Y, Sugiura M and Watanabe T (2009) Redox potential of the primary plastoquinone electron acceptor QA in Photosystem II from *Thermosynechococcus elongatus* determined by spectroelectrochemistry. Biochemistry 48: 10682–10684

Shimokata K, Katayama Y, Murayama H, Suematsu M, Tsukihara T, Muramoto K, Aoyama H, Yoshikawa S and Shimada H (2007) The proton pumping pathway of bovine heart cytochrome *c* oxidase. Proc Natl Acad Sci U S A 104: 4200–4205

Shutova T, Irrgang K-D, Shubin V, Klimov VV and Renger G (1997) Analysis of pH-induced structural changes of the isolated extrinsic 33 kDa protein of Photosystem II. Biochemistry 36: 6350–6358

Shutova T, Klimov VV, Andersson B and Samuelsson G (2007) A cluster of carboxylic groups in PsbO protein is involved in proton transfer from the water oxidizing complex of Photosystem II. Biochim Biohys Acta 1767: 434–440

Siegbahn PEM (2008) Theoretical studies of O–O bond formation in Photosystem II. Inorg Chem 47: 1779–1786

Siegbahn PEM (2009) Structures and energetics for O_2 formation in Photosystem II. Acc Chem Res 42: 1871–1880

Siegbahn PEM and Blomberg MRA (2008) Proton pumping mechanism in cytochrome *c* oxidase. J Phys Chem A 112: 12772–12780

Sjödin M, Styring S, Åkermark B, Sun L and Hammarström L (2002) The mechanism for proton coupled electron transfer from tyrosine in a model complex and comparison with tyrosine Z oxidation in Photosystem II. Phil Trans B 357: 1471–1478

Smirnova IA, Zaslavsky D, Fee JA, Gennis RB and Brzezinski (2008) Electron and proton transfer in the ba_3 oxidase from *Thermus thermophilus*. J Bioenerg Biomembr 40: 281–287

Smith D, Gray J, Mitchell L, Antholine WE and Hosler JP (2005) Assembly of cytochrome-*c* oxidase in the absence of assembly protein Surf1p leads to loss of the active site heme, J Biol Chem 280: 17652–17656

Soulimane T, Buse G, Bourenkov GP, Bartunik HD, Huber R, and Than ME (2000) Structure and mechanism of the aberrant ba(3)-cytochrome *c* oxidase from *Thermus thermophilus*. EMBO J 19: 1766–1776

Sproviero EM, Gascon JA, McEvoy JP, Brudvig GW and Batista VS (2008) Quantum mechanics/molecular mechanics study of the catalytic cycle of water splitting in Photosystem II. J Am Chem Soc 130: 3428–3442

Steinrücke P, Steffens GCM, Panskus G, Buse G and Ludwig B (1987) Subunit II of cytochrome *c* oxidase from *Paracoccus denitrificans*: DNA sequence, gene expression, and the protein. Eur J Biochem 167: 431–439

Stroh A, Anderka O, Pfeiffer K, Yagi T, Finel M, Ludwig B and Schägger H (2004) Assembly of respiratory complexes I, III, and IV into NADH oxidase supercomplex stabilizes complex I in *Paracoccus denitrificans*. J Biol Chem 279: 5000–5007

Strzalka K, Walczak T, Sarna T and Swartz HM (1990) Measurement of time-resolved oxygen concentration changes in photosynthetic systems by nitroxide-based EPR oximetry. Arch Biochem Biophys 281: 312–318

Sugiura M, Rappaport F, Hillier W, Dorlet P, Ohno Y, Hayashi H and Boussac A (2009) Evidence that D1-His332 in Photosystem II from *Thermosynechococcus elongatus* interacts with the S_3-State and not with the S_2-State. Biochemistry 48: 7856–7866

Suzuki H, Nagasaka M, Sugiura M and Noguchi T (2005) Fourier transform infrared spectrum of the secondary quinone electron Acceptor Q_B in Photosystem II. Biochemistry 44: 11323–11328

Than ME, Hof P, Huber R, Bourenkov GP, Bartunik HD, Buse G and Soulimane T (1997) Thermus thermophilus Cytochrome-c552: A new highly thermostable Cytochrome-c structure obtained by MAD Phasing. J Mol Biol 271: 629–644

Thomas JW, Lemieux LJ, Alben JO and Gennis RB (1993) Site-directed mutagenesis of highly conserved residues in helix VIII of subunit I of the cytochrome *bo* ubiquinol oxidase from *Escherichia coli*: an amphipathic transmembrane helix that may be important in conveying protons to the binuclear center. Biochemistry 32: 11173–11180

Tiede DM, Vashishta AC and Gunner MR (1993) Electron-transfer kinetics and electrostatic properties of the *Rhodobacter sphaeroides* reaction center and soluble *c*-cytochromes. Biochemistry 32: 4515–4531

Tommos C and Babcock GT (2000) Proton and hydrogen currents in photosynthetic water oxidation. Biochim Biophys Acta 1458: 199–299

Tsukihara T, Aoyama H, Yamashita E, Tomizaki T, Yamaguchi H, Shinzawa-Itoh K, Nakashima R, Yaono R and Yoshikawa S (1996) The whole structure of the 13-subunit oxidized cytochrome *c* oxidase at 2.8 Å. Science 272: 1136–1144

Turba A, Jetzek M and Ludwig B (1995) Purification of the cytochrome c_{552} of *Paracoccus denitrificans* and sequence analysis of the gene. Eur J Biochem 231: 259–265

Tuukkanen A, Kaila VRI, Laakkonen L, Hummer G and Wikström M (2007) Dynamics of the glutamic acid 242 side chain in cytochrome *c* oxidase. Biochim Biophys Acta 1767: 1102–1110

Tuukkanen A, Verkhovsky MI, Laakkonen L and Wikström M (2006) The K-pathway revisited: A computational study on cytochrome *c* oxidase. Biochim Biophys Acta 1757: 1117–1121

Ulas G, Olack G, Brudvig GW (2008) Evidence against bicarbonate bound in the O_2-evolving complex of Photosystem II. Biochemistry 47: 3073–3075

Vakkasoglu AS, Morgan JE, Han D, Pawatea AS and Gennis RB (2006) Mutations which decouple the proton pump of the cytochrome *c* oxidase from *Rhodobacter sphaeroides* perturb the environment of glutamate 286. FEBS Lett 580: 4613–4617

Van Eps N, Szundi I and Einarsdottir Ó (2000) A new approach for studying fast biological reactions involving dioxygen: The reaction of fully reduced cytochrome *c* oxidase with O_2. Biochemistry 39: 14576–14582

van Leeuwen PJ, Heimann C, Gast P, Dekker JP and Gorkom HJ (1993) Flash-induced redox changes in oxygen-evolving spinach Photosystem II core particles. Photosynth Res 38: 169–176 van Niel CB (1941) The bacterial photosynthesis and their importance of the general problem of photosynthesis. Act Enzymol 1: 263–328

Vass I and Aro EM (2008) Photoinhibition of photosynthetic electron transport. In: Renger G (ed) Primary processes of photosynthesis: Basic priniciples and apparatus, Part I photophysical principles, pigments and light harvesting/adaptation/stress, Royal Society Chemistry, Cambridge, pp 393–425

Velthuys B (1981) Spectroscopis studies of the S-state transitions of Photosystem II and of the interacxtion of its charged donorchains with lipid-soluble anions. In: Akoyunoglou G (ed) Photosynthesis II. Electron transport and photophosphorylation. Balaban, Philadelphia, pp 75–85

Vermeglio A (2008) Electron transport chain and phosphorylation. In: Renger G (ed) Primary processes of photosynthesis: Priniciples and apparatus, Part II reaction centers/photosystems, electron transport chains, photophosphorylation and evolution, Royal Society Chemistry, Cambridge, pp 353–382

Vogt S, Riehl A, Koch V and Kadenbach B (2007) Regulation of oxidative phosphorylation by inhibition of its enzyme complexes via reversible phosphorylation. Curr Enzyme Inhib 3: 189–206

Vonck J and Schäfer E (2009) Supramolecular organization of protein complexes in the mitochondrial innermembrane. Biochim Biophys Acta 1793: 117–124

Wagner BK, Kitami T, Gilbert TJ, Peck D, Ramanathan A, Schreiber SL, Golub TR and Mootha VK (2008) Large-scale chemical dissection of mitochondrial function. Nat Biotech 26: 343–351

Wickner W and Schekman R (2005) Protein translocation across biological membranes. Science 310: 1452–1456

Wienk H, Maneg O, Lücke C, Pristovsek P, Löhr F, Ludwig B and Rüterjans H (2003) Interaction of cytochrome *c* with cytochrome *c* oxidase: An NMR study on two soluble fragments derived from *Paracoccus denitrificans*. Biochemistry 42: 6005–6012

Wiertz FGM, Richter O-MH, Cherepanov AV, MacMillan F, Ludwig B and de Vries S (2004) An oxo-ferryl tryptophan radical catalytic intermediate in cytochrome *c* and quinol oxidases trapped by microsecond freeze-hyperquenching (MHQ). FEBS Lett 575: 127–130

Wiertz FGM, Richter O-MH, Ludwig B and de Vries S (2007) Kinetic resolution of a tryptophan-radical intermediate in the reaction of *Paracoccus denitrificans* cytochrome *c* oxidase. J Biol Chem 282: 31580–31591

Wikström M (2004) Cytochrome *c* oxidase: 25 years of the elusive proton pump. Biochim Biophys Acta 1655: 241–247

Wikström M and Verkhovsky MI (2007) Mechanism and energetics of proton translocation by the respiratory heme-copper oxidases. Biochim Biophys Acta 1767: 1200–1214

Wikström M, Verkhovsky MI and Hummer G (2003) Water-gated mechanism of proton translocation by cytochrome *c* oxidase. Biochim Biophys Acta 1604: 61–65

Williams PA, Fülöp V, Leung YC, Chan C, Moir JW, Howlett G, Ferguson SJ, Radford SE and Hajdu J (2005) Pseudospecific docking surfaces on electron transfer proteins as illustrated by pseudoazurin, cytochrome c_{550} and cytochrome cd_1 nitrite reductase. Nat Struct Biol 2: 975–982

Witt H, Zickermann V and Ludwig B (1995) Site-directed mutagenesis of cytochrome *c* oxidase reveals two acidic residues involved in the binding of cytochrome *c*. Biochim Biophys Acta 1230: 74–76

Witt H, Wittershagen A, Bill A, Kolbesen BA and Ludwig B (1997) Asp-193 and Glu-218 of subunit II are involved in the Mn^{2+}-binding of *Paracoccus denitrificans* cytochrome *c* oxidase. FEBS Lett 409: 128–130

Witt H and Ludwig B (1997) Isolation, analysis, and deletion of the gene coding for subunit IV of cytochrome *c* oxidase in *Paracoccus denitrificans*. J Biol Chem 272: 5514–5517

Witt H, Malatesta F, Nicoletti F, Brunori M and Ludwig B (1998a) Cytochrome *c* binding site on cytochrome oxidase in *Paracoccus denitrificans*. Eur J Biochem 251: 367–373

Witt H, Malatesta F, Nicoletti F, Brunori M and Ludwig B (1998b) Tryptophan 121 of subunit II is the electron entry site to cytochrome *c* oxidase in *Paracoccus denitrificans*—involvement of a hydrophobic patch in the docking reaction. J Biol Chem 273: 5132–5136

Worrest RC and Caldwell MM (eds) (1986) Stratospheric ozone reduction, solar ultraviolet radiation and plant life, Springer, Berlin, pp 171–184

Xiong J and Bauer CE (2002) Complex evolution of photosynthesis. Annu Rev Plant Biol 53: 503–521

Yachandra VK (2005) The catalytic manganese cluster: organisation of the metal ions. In: Wydrzynski T and Satoh K (eds) Photosystem II. The Light-Driven Water:Plastoquinone Oxidoreductase, Springer, Dordrecht, pp 235–260

Yano J, Kern J, Sauer K, Latimer JM, Pushkar Y, Biesiadka J, Loll B, Saenger W, Messinger J, Zouni A and Yachandra VK (2006) Where water is oxidized to dioxygen: Structure of the photosynthetic Mn_4Ca Cluster. Science 314: 821–825

Yoshikawa S and Shinzawa-Itoh K (1998) Redox-coupled crystal structural changes in bovine heart cytochrome *c* oxidase. Science 280: 1723–1729

Zamaraev KI and Parmon VN (1980) Potential methods and perspectives of solar energy conversion via photocatalytic processes. Rev Sci Eng 22: 261–324

Zein S, Kulik LV, Yano J, Kern J, Pushkar Y, Zouni A, Yachandra VK, Lubitz W, Neese F and Messinger J (2008) Focusing the view on nature's water-splitting catalyst. Phil Trans R Soc B 363: 1167–1177

Zhang C (2007) Low-barrier hydrogen bond plays key role in active Photosystem II — A new model for photosynthetic water oxidation. Biochim Biophys Acta 1767: 493–499

Zouni A (2008) From cell growth to the 3.0 Å resolution crystal structure of cyanobacterial Photosystem II. In: Renger G (ed) Primary processes of photosynthesis: Priniciples and apparatus, Part II reaction centers/photosystems, electron transport chains, photophosphorylation and evolution, Royal Society Chemistry, Cambridge, pp 193–236

Zouni A, Witt HT, Kern J, Fromme P, Krauß N, Saenger W and Orth P (2001) Crystal structure of Photosystem II from *Synechococcus elongatus* at 3.8 Å resolution. Nature 409: 739–743

Chapter 14
Photoprotection in Cyanobacteria: The Orange Carotenoid Protein and Energy Dissipation

Cheryl A. Kerfeld and Diana Kirilovsky

14.1 Introduction

Photosynthetic organisms have developed physiological mechanisms allowing acclimation and survival in a wide range of environmental conditions. By harvesting solar energy and converting it into chemical energy, plants, algae and cyanobacteria provide organic carbon molecules and oxygen that are essential for life on earth. However, changes in light quality and, especially, light intensity can be harmful to photosynthetic organisms. Changes in light quality could cause an imbalance of energy absorption by the photosystems resulting in a decrease in photosynthetic efficiency. In addition, excess light can be lethal because harmful reactive oxygen species are generated in the photochemical reaction centers. Most of the mechanisms induced in response to changes in light quality and light intensity are controlled by photosynthesis-mediated changes in cellular redox potential and lumenal pH. For example, in order to counterbalance the changes in light quality, the redox state of the plastoquinone pool (and/or of the cytochrome b_6f complex) regulates the energy re-distribution between the two photosystems by changing the relationship between the antennae and the reaction centers (state transitions; van Thor et al. 1998; Wollman 2001) and by regulating the transcription of photosynthetic genes (Allen 1993; Allen et al. 1995; Foyer and Allen 2003). In plants, the low lumenal pH, generated during photosynthesis under high irradiance, induces a photoprotective mechanism involving energy dissipation as heat in the antenna (resulting in a detectable quenching of the chlorophyll fluorescence). This mechanism decreases the energy arriving at the PSII reaction center, thus reducing the generation of oxygen reactive species and the probability of PSII damage (Horton et al. 1996; Niyogi 1999).

C. A. Kerfeld (✉)
United States Department of Energy, Joint Genome Institute, 2800 Mitchell Drive, Walnut Creek, CA 94598, USA
e-mail: ckerfeld@lbl.gov

Department of Plant and Microbial Biology, University of California, Berkeley, CA 94720, USA

G. A. Peschek et al. (eds.), *Bioenergetic Processes of Cyanobacteria*,
DOI 10.1007/978-94-007-0388-9_14, © Springer Science+Business Media B.V. 2011

Cyanobacteria, like plants carry out oxygenic photosynthesis; due to their numeric abundance, they play a key role in global carbon cycling. For cyanobacteria, only recently was a light-induced process for decreasing the energy transfer between the antenna and the reaction centers described (El Bissati et al. 2000; Rakhimberdieva et al. 2004; Scott et al. 2006; Wilson et al. 2006). In contrast to its counterpart in plants, the cyanobacterial process is not triggered by a lowering of the lumenal pH or a change in the redox state of the plastoquinone pool (Rakhimberdieva et al. 2004; Scott et al. 2006; Wilson et al. 2006). Instead, the decrease of energy reaching the reaction centers and the quenching of the antenna fluorescence is induced by light activation of the Orange-Carotenoid-Protein (OCP) (Wilson et al. 2006), a soluble 35kD protein containing a single-non covalently bound carotenoid (Holt and Krogmann 1981; Wu and Krogmann 1997; Kerfeld 2004).

14.2 Protective Antenna-Related Energy Dissipation Mechanisms

The photosynthetic apparatus of the organisms that carry out oxygenic photosynthesis is composed of two photosystems, Photosystem I (PSI) and Photosystem II (PSII). In each photosystem, the light harvesting antennae collect and concentrate light energy for the photochemistry in the reaction centers and for conversion light energy to chemical energy by photosynthesis. While the structure of the reaction center complexes appears to be conserved in all the oxygenic photosynthetic organisms, there is a wide variety of light harvesting complexes of distinctive pigment and protein composition. In plants and green algae, light is principally harvested by membrane embedded complexes non-covalently binding chlorophyll and carotenoid molecules, known as light-harvesting-complexes (LHCs). These complexes are structurally and functionally flexibile, allowing plants and algae to adapt to changes in environmental conditions. The LHCs, which are very efficient in the collection of light energy under low light conditions, can reversibly switch to a photoprotected state in which the excess energy is converted into heat under high light conditions (Horton et al. 1996; Niyogi 1999; Pascal et al. 2005; Ruban et al. 2007). The switch is induced by the ΔpH that builds up across the thylakoid membrane during electron transport. The drop of the lumen pH induces the interconversion of LHC carotenoids by the xanthophyll cycle (violaxanthin to zeaxanthin) and the protonation of a PSII subunit (PsbS), a member of the LHC superfamily. Conformational changes in the LHCII, which are also induced by the acidification of the lumen, are crucial for this mechanism. These changes modify the interaction between chlorophylls and carotenoids to allow thermal dissipation (Horton et al. 1996; Niyogi 1999; Pascal et al. 2005; Ruban et al. 2007). Energy dissipation is accompanied by a diminution of PSII-related fluorescence emission, known as non-photochemical-quenching fast (NPQ_f) or qE. This NPQ process is rapidly reversible (within seconds) in the dark, in the absence of protein synthesis. Decrease of fluorescence emission usually serves as a measure of the dissipation process.

To harvest light, cyanobacteria (with the exception of the Prochlorophytes) use a distinctive type of photosynthetic antenna, a large membrane extrinsic complex called the phycobilisome (PB), composed of several types of chromophorylated phycobiliproteins and of linker peptides needed for the structural organization and functioning of the PBs (for reviews, see Glazer 1984; Grossman et al. 1993; MacColl 1998; Tandeau de Marsac 2003; Adir 2005) Phycobilisomes are composed of a core from which rods radiate. The major core protein is allophycocyanin (APC), while in most of the freshwater cyanobacteria the rods contain only phycocyanin (PC) and, in many marine cyanobacteria, phycoerythrin (PE) or phycoerythrocyanin (PEC) are found in the distal end of the rods. These complexes are attached to the outer surface of the thylakoid membranes (Gantt and Conti 1966a, b) via the large, chromophorylated, core membrane linker protein, L_{CM} (Redlinger and Gantt 1982). This protein and APC^{B} serve as PB terminal energy acceptors. Harvested light energy is transferred from them to the chlorophylls (chls) of PS II and PSI (Mullineaux 1992; Rakhimberdieva et al. 2001).

Since key elements involved in the qE mechanism of plants such as the LHC, the enzymes of the xanthophyll cycle and the PsbS protein are absent in cyanobacteria it has been long assumed that they do not have an antenna-related NPQ photoprotective mechanism. Only relatively recently, in 2000, appeared the first report establishing that in cyanobacteria there exists a photoprotective NPQ-antenna-related process (El Bissati et al. 2000). Exposure of *Synechocystis* PCC 6803 (hereafter referred to as *Synechocystis*) cells to strong intensities of blue light induced PSII fluorescence quenching under conditions in which the plastoquinone (PQ) pool was largely oxidized and the oxygen evolving activity was not saturated. Moreover, the recovery of the fluorescence occurred in the presence of inhibitors of translation (El Bissati et al. 2000). These results demonstrated that the blue-green light induced quenching was not related to photoinhibition or state transitions. Under high intensities of white light, the decrease of fluorescence is associated with the inactivation of PSII and the damage of the D1 protein (photoinhibition) (for reviews see Prasil et al. 1992; Aro et al. 1993; Vass and Aro 2008; Tyystjärvi 2008). Recovery of fluorescence requires the replacement of the damaged D1 protein and it cannot occur in the presence of inhibitors of protein synthesis. In state transitions, the decrease of fluorescence is induced by the reduction of the PQ pool caused by a preferential illumination of PSII by orange light (or green light when PE is present) principally absorbed by the phycobilisomes (for review Williams and Allen 1987; van Thor et al. 1998; Wollman 2001). In contrast, blue light, principally absorbed by PSI, induces the oxidation of the PQ pool and increases the fluorescence associated with the state 1 transition. Moreover, while state transitions are largely inhibited by lowering the temperature (increasing the rigidity of the thylakoids), the fluorescence quenching induced by strong blue light is not affected by the rigidity of the membrane and has a lower temperature dependence (El Bissati et al. 2000). The investigators proposed that, as in plants, in cyanobacteria high intensities of white or blue-green light induce energy dissipation at the level of the PSII antenna: high light intensities induce phycobilisome fluorescence quenching accompanied by a decrease of the energy transfer from the phycobilisome to the photosystems. This hypothesis was soon substantiated. Rakhimberdieva et al. (2004) first showed that in a *Synechocystis* mutant lacking reaction center II and the chl antennae CP43 and CP47, blue-light induced a reversible quenching of the phycobilisome emission. These results were confirmed by two other laboratories (Scott et al. 2006; Wilson et al. 2006). In these mutants, a specific decrease of the phycobilisome-related-fluorescence was observed and there was no concurrent change in the chlorophyll-related-fluorescence emission (Scott et al. 2006; Wilson et al. 2006). In contrast, blue-light was unable to induce fluorescence quenching in a phycobilisome-deficient mutant or in a mutant containing only the phycocyanin rods of the phycobilisome (Wilson et al. 2006). Phycobilisomes are thus the essential components of the blue-light-induced NPQ mechanism. The presence of only the core of the phycobilisome is sufficient to induce the quenching (Wilson et al. 2006). Picosecond time-resolved fluorescence decay data were consistent with a mechanism of blue-light-induced quenching localized in the phycobilisome core that decreased the excited-state population (Scott et al. 2006). Recently, it was also proposed that in conditions in which all the PSII centers are closed, as in the presence of DCMU, chl excitation could be quenched via the phycobilisome quenching mechanism. In these conditions, excited chl molecules could be sufficiently long-lived to allow the exciton be transferred from PSII chl to the APC in the core of the phycobilisome where it is quenched (Rakhimberdieva et al. 2007b)

In cyanobacteria the fluorescence changes associated with energy dissipation in the antenna are no faster than those associated with state transitions; both are on the time-scale of minutes. Thus, the name NPQ_{fast} is not appropriate for the cyanobacterial mechanism. qE is more appropriate because the fluorescence quenching is also associated with energy dissipation in cyanobacteria. However, the cyanobacteria and the plant qE are completely different mechanisms. To differentiate both mechanisms we call the cyanobacterial mechanism qE_{cya}. Additional details about this mechanism are described in (Karapetyan 2007; Kerfeld 2009; Kirilovsky 2007, 2010; Bailey and Grossman 2008).

14.3 The Inducer of the qE_{cya} Mechanism: The Orange Carotenoid Protein

In higher plants, the antenna-associated qE or NPQ_f is induced by a low lumenal pH (Horton et al. 1996; Niyogi 1999). In contrast, the phycobilisome-associated qE_{cya} is not dependent on the presence of a transthylakoid ΔpH (Wilson et al. 2006). The induction of the quenching in cyanobacteria appears to be independent of excitation pressure on PS II or of changes in the redox state of the plastoquinone pool (El Bissati et al. 2000; Wilson et al. 2006). A specific characteristic of the qE_{cya} mechanism is that its induction depends on the quality of light; only blue-green light triggers it. The action spectrum for the phycobilisome emission quenching in PSII-deficient mutants of *Synechocystis* PCC 6803, has peaks at 496, 470 and 430 nm; this suggested that a carotenoid molecule could be involved (Rakhimberdieva et al. 2004). Wilson et al. (2006) then demonstrated that indeed a carotenoid

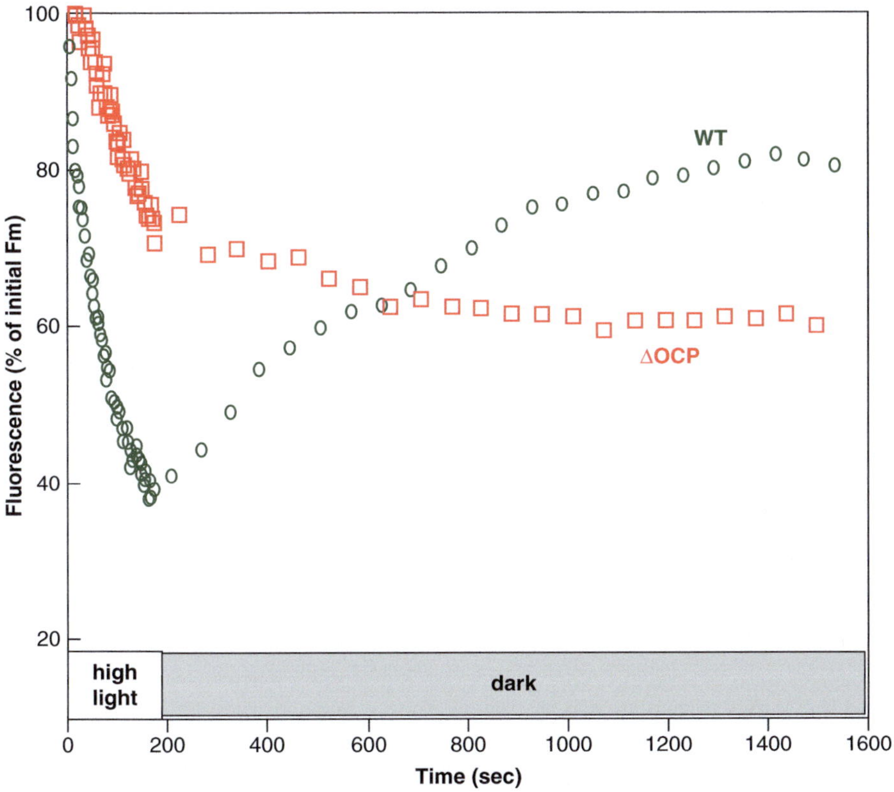

Fig. 14.1 The OCP is essential for the reversible fluorescence quenching induced by high light intensities. Changes in maximal fluorescence in dark-adapted WT (*green circles*) and ΔOCP (*red squares*) cells (in the presence of chloramphenicol) illuminated by high intensities of white light (3 min 1500 μmol photons. m^{-2} s^{-1}) and then incubated in darkness. High irradiance induced a larger fluorescence quenching in WT cells than in ΔOCP cells. However, while the fluorescence quenching observed in WT cells was largely reversible, the quenching induced in ΔOCP cells was irreversible. Thus, in WT cells fluorescence quenching is associated to the photoprotective mechanism and in ΔOCP to photoinhibition. More details in (Wilson et al. 2006)

protein, the Orange Carotenoid Protein (OCP), is specifically involved in the qE_{cya} mechanism (Fig. 14.1). In the absence of the OCP, the fluorescence quenching induced by strong blue-green light in *Synechocystis* cells is completely inhibited and the cells are more sensitive to high light intensities. This is manifested as a faster decrease in PSII activity in a mutant lacking the OCP (Wilson et al. 2006). Studies by immuno-gold labeling and electron microscopy showed that most of the OCP is present in the inter-thylakoid cytoplasmic region, on the phycobilisome side of the membrane (Wilson et al. 2006). The existence of an interaction between the OCP and the phycobilisomes and thylakoids was supported by the co-isolation of the OCP with the phycobilisome-associated membrane fraction (Wilson et al. 2006, 2007; Boulay et al. 2008b). A possible direct interaction between the OCP and the

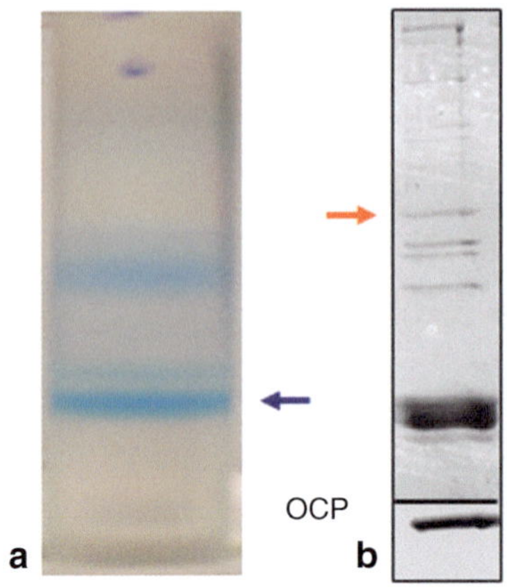

Fig. 14.2 Relationship between the OCP and the phycobilisome. **a** Reconstituted phycobilisomes in the presence of OCP in a non-linear sucrose gradient (0.25 M, 0.5 M, 0.75 M and 1.5 M sucrose from up to bottom). The *blue arrow* shows the band containing the reconstituted phycobilisomes. **b** Coomassie-stained gel electrophoresis of the phycobilisome (*upper panel*) and OCP immuno-blot detection (*lower panel*)**.** The *orange arrow* shows the 35 kDa OCP band. The OCP co-migrated with the reconstituted phycobilisomes

phycobilisomes was suggested by phycobilisome reconstitution experiments in the presence of the OCP, followed by co-migration in a sucrose gradient, separation and isolation of a phycobilisome-OCP complex (Fig. 14.2; Boulay et al. 2008b). Figure 14.3 summarizes the working model of the OCP-phycobilisome-related photoprotective mechanism.

In *Synechocystis*, the strain used for almost all of the studies on the OCP-related-photoprotective mechanism, the OCP is encoded by the *slr1963* gene (Wu and Krogmann 1997) and it is constitutively expressed. It is present even in *Synechocystis* mutants lacking phycobilisomes (Wilson et al. 2007). High light increases levels of OCP transcripts in *Synechocystis* (Hihara et al. 2001): recent data suggests this is transient, increasing after 15 min of exposure to high light, but then returning to baseline levels after the elapse of one hour (Singh et al. 2008). Other stress conditions (salt stress, iron starvation) increase the levels of OCP transcripts and proteins (Hihara et al. 2001; Fulda et al. 2006; Wilson et al. 2007). Under iron starvation (Cadoret et al. 2004; Joshua et al. 2005; Rakhimberdieva et al. 2007b; Wilson et al. 2007), and in the absence of PSI (Kirilovsky, unpublished), blue-green light induces a large fluorescence quenching. It was proposed that the chlorophyll-binding protein IsiA, a stress-induced protein of the LHC family, was the responsible for this quenching (Cadoret et al. 2004; Joshua et al. 2005). However, more recently it was clearly demonstrated that the iron-depletion induced qE_{cya} was dependent on the OCP, not IsiA (Rakhimberdieva et al. 2007b; Wilson et al. 2007). Moreover, the higher OCP concentration (relative to chl and phycobiliproteins), the larger fluorescence quenching (Wilson et al. 2007) indicating that the OCP plays an essential

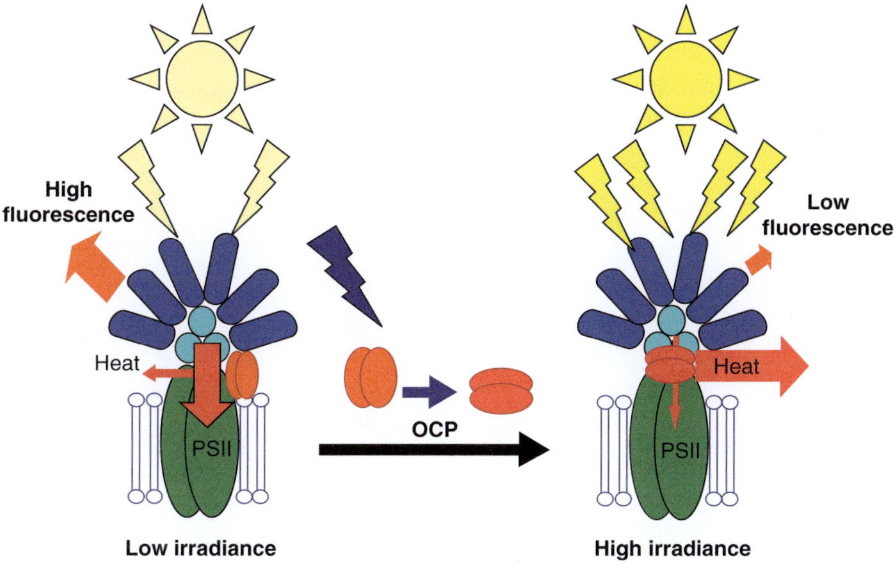

Fig. 14.3 The photoprotective mechanism. The ubiquitous OCP, which has strong interactions with the phycobilisomes and the thylakoids, is activated by absorption of high intensities of *blue-green* light inducing conformational changes of the carotenoid and the protein. The activated protein induces quenching of phycobilisome emission and decrease of energy transfer from the phycobilisomes to the reaction centers. We hypothesize that the activated protein is able to absorb the energy absorbed by the phycobilisomes via interaction with the phycobilisome core and convert this energy into heat

role in the acclimation of *Synechocystis* to environmental changes. The relationship between the concentration of the OCP and energy dissipation was later confirmed using a *Synechocystis* mutant overexpressing OCP, in which large quantities of the OCP were present and a very large fluorescence quenching was observed (Wilson et al. 2008).

14.4 Distribution of the OCP

A survey of the currently available cyanobacterial genome sequences shows that genes homologous to *slr1963* in *Synechocystis* are present in most phycobilisome-containing cyanobacteria (Table 14.1). The primary structure of the OCP is highly conserved (Fig. 14.4); 129 of 318 residues in the primary structure of the OCP are absolutely conserved among the orthologs. A phylogenetic tree (Fig. 14.5) of full length OCP sequences and its N-terminal fragments (discussed below) based on pairwise alignments with *slr1963*, shows the most distant relatives to *slr1963* to be those from *Gloeobacter violaceus* (50%) and *Nodularia spumigena* (56% identical). The marine *Synechococcus* OCP sequences cluster tighly and are ~65% identi-

Table 14.1 Occurrence of the ocp, its paralogs and fragments (*Nostoc punctiforme* ORFs in Bold are known to be expressed based on a proteomic analysis (Anderson et al. 2006))

Organism	OCP*	OCP-Nter*	OCP C-ter*
Acaroychloris marina	AMI_5842		
Anabaena varaibilis ATCC29413	Ava_3843	Ava_2052 Ava_2230 Ava_4694	Ava_2231
Arthrospira maxima CS-328	ZP_03274607		
Crocosphaera watsonii WH 8501		CWATdraft_0985	CWATdraft_5349
Cyanobium PCC7001			JCVI_CPCC7001_1352 (86)
Cyanothece ATCC51142	cce_1649	cce_1537 cce_1991	cce_0742
Cyanothece CCY0110	CY0110_09677	CY0110_08696	CY0110_8806
Cyanothece PCC7424	ZP_02975753	ZP_02976887 ZP_02974930 ZP_02975513	ZP_02976888 ZP_02974026
Cyanothece PCC7425		ZP_03136620 ZP_03137335 ZP_03140186	ZP_03140187
Cyanothece PCC7822		ZP_03155220 ZP_03155118 ZP_03157201	ZP_03157200
Cyanothece PCC8801(2)		ZP_02940387	ZP_02940386
Gloeobacter violaceus PCC7421	glr0050 glr3935 (274)	gll0259 gll0260	gll2503
Lyngbya sp PCC8106	L8106_29210	L8106_04666 L8106_29395	L8106_29390
Microcoleus chthonoplastes PCC7420	MC7420_3617 (282)	MC7420_905 *P30 and 33*	
Microcystis aeruginosa NIES 843	MAE_18910		
Microcystis aeruginosa PCC7806	IPF_5686		
Nodularia spumigena CCY9414	N9414_13085	N9414_12098 N9414_22258	N9414_22253
Nostoc punctiforme PCC73102	NpR5144 NpR0404	**NpF6242** NpF6243 **NpF5913** NpR5130	NpF5133
Nostoc sp PCC7120	all3149	all1123 alr4783 all4941 all3221	all4940
Synechococcus CC9311	Sync_1803		
Synechococcus CC9902	syncc9902_0973		

Table 14.1 (continued)

Organism	OCP*	OCP-Nter*	OCP C-ter*
Synechococcus PCC7002	SYNPCC7002_ A2810		
Synechococcus PCC7335	EDX85092	EDX87112	EDX87269
	EDX83307	EDX84507	
	EDX83475		
Synechococcus BL107	BL107_14105		
Synechococcus RCC307	RCC307_1992		
Synechococcus RS9917	RS9917_00692		
Synechococcus WH5701	WH5701_04010		
	WH5701_00210 (219)		
Synechococcus WH7803	SynWH7803_0929		
Synechococcus WH7805	WH7805_01202		
Synechococcus WH8102	SYNW1367		
Thermosynechococcus elongates BP-1		tll1269	tll1268

*Genes/ORFs are listed by locus tag (if given in the Genbank record or in Integrated Microbial Genomes http://img.jgi.doe.gov/) or Genbank gi/protein record Identifier.
For truncated forms of the full-length OCP, the number of amino acids is given in parenthesis

cal to the OCP of *Synechocystis*. Seven of these strains have been examined for the expression of the full-length OCP; it is constitutively present in each of them (Boulay et al. 2008a). These seven strains are able to perform blue-green light induced fluorescence quenching, substantiating the suggestion that the OCP-related photoprotective mechanism is widespread in cyanobacteria and plays an important role in the physiology of these organisms (Boulay et al. 2008a). Furthermore *Arthrospira maxima* was examined for its OCP response to iron starvation: as in *Synechocystis*, iron starvation induces an increase in OCP concentration and a large fluorescence quenching (Boulay et al. 2008a), consistent with the general observation from microarray analyses (Hihara et al. 2001; Fulda et al. 2006; Wilson et al. 2007) that the OCP functions in response to stress. The presence of a larger qE_{cya} protects the cells from the presence of "dangerous" functionally disconnected phycobilisomes under iron starvation conditions. The transfer of excess absorbed energy from the phycobilisomes to the thylakoids could induce oxidative damage (e.g. lipid peroxidation) if it is not thermally dissipated at the level of the phycobilisomes.

Most of the cyanobacterial genomes sequenced to-date contain only a single copy of the OCP gene. Exceptions include *Gloeobacter violaceus* PCC 7421 and *Synechococcus* WH5701 which contain additional truncated paralogs (274 and 219 amino acids, respectively). *Synechococcus* PCC7335 is unusual in that it contains 3 full length OCP genes. A subset of the genomes also contain, in addition to full length OCP gene, one or more open reading frames (ORFs) encoding homologs to the N-terminal domain (Fig. 14.5) while containing a single copy of the C-terminal domains (with the exception of *Microcoleus chthonplastes* which does not contain a C-terminal domain; Table 14.1). The primary structures of these fragments

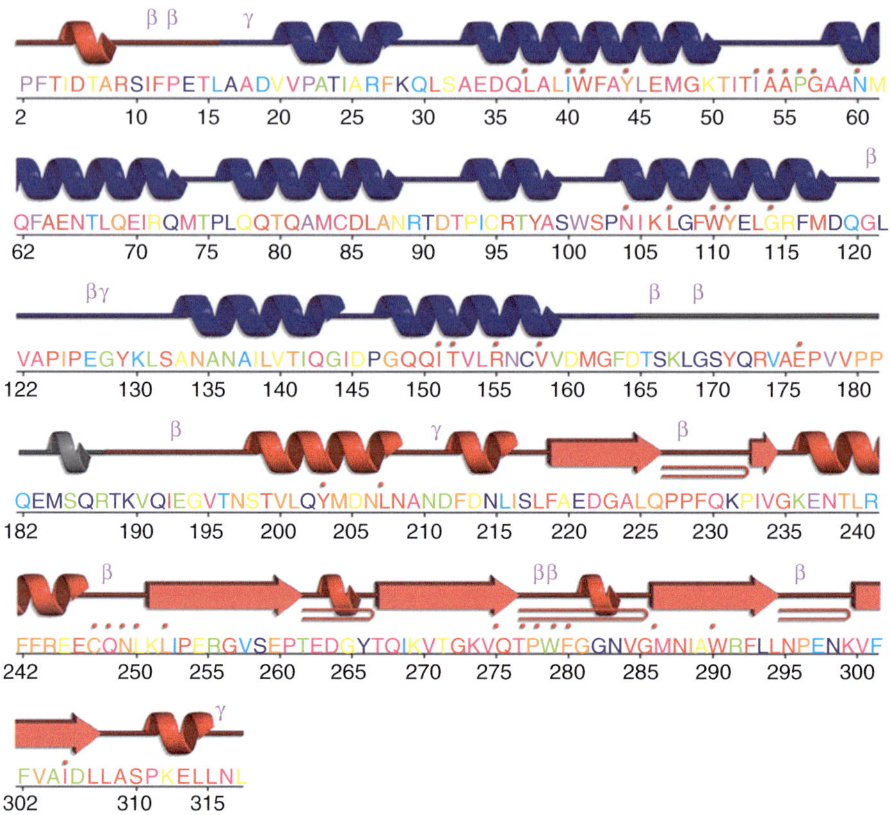

Fig. 14.4 Primary and secondary structure of the OCP. The primary structure of *A. maxima* OCP is shown with secondary structure elements (*arrows* for beta strands, *coils* for helices; N-and C-terminal domain secondary structure is shown in *blue* and *red*, respectively). The loop joining the two domains is *grey*. The types of turns present are given in Greek symbols: β for beta turn, γ for gamma turn. Beta hairpins are shown as the sideways letter u. Residues are colored according to conservation among OCP orthologs using a gradient from *red* (high) to *blue* (low). Residues marked with a *red dot* are within van der waals contact of the carotenoid. (Figure adapted from PDBsum Laskowsi 2001)

show a significant amount of conservation (Fig. 14.6), including residues that are in the immediate environment of the carotenoid in the OCP structure (Fig. 14.7). The N-terminal homologs are cluster in groups across species, rather than collectively by organism (Fig. 14.5), suggesting their origin is relatively ancient gene duplication. The function of these fragments is unknown. When they were first noted (Kerfeld et al. 2003), it was suggested that they could combine in modular fashion to produce OCP variants for specific environmental conditions; this remains to be tested in those organisms (many of which occupy environments with large fluctuations in conditions) which the fragments are redundant to the full-length OCP. More

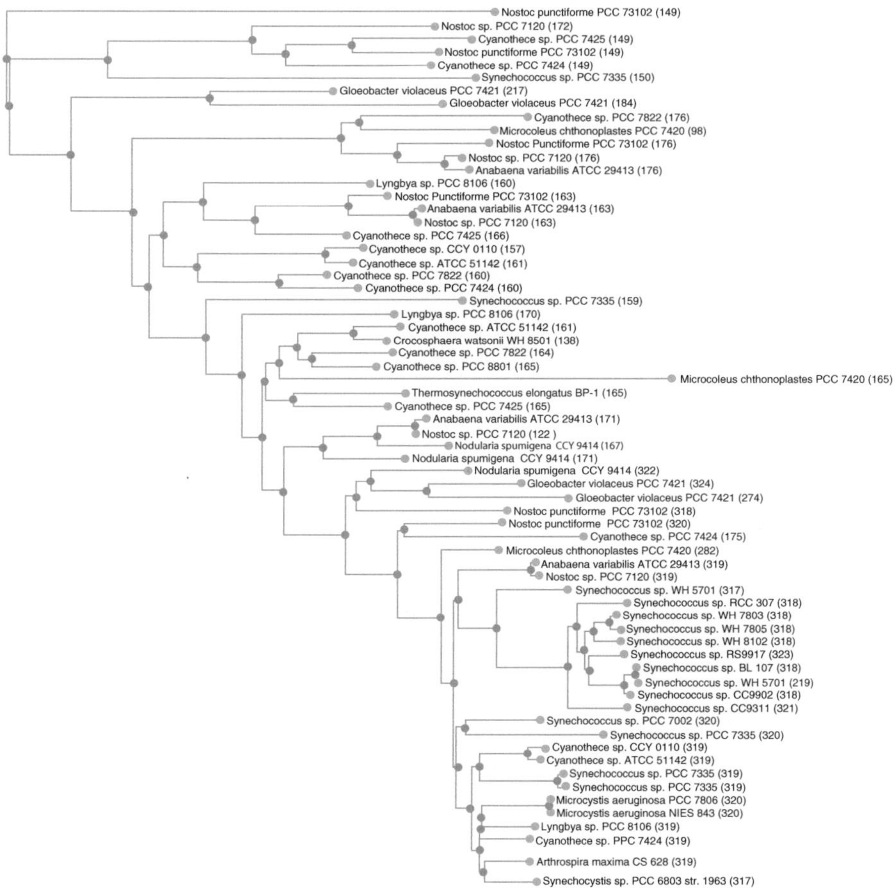

Fig. 14.5 Phylogenetic tree of the full length OCP and N-terminal OCP fragments. Tree based on pairwise alignment the protein sequences retrieved via a BLAST search using the primary structure of *Synechocystis* OCP using Fast Minimum evolution (Desper and Gascuel 2004) as implemented as a Tree View in NCBI BLAST. ORFs encoding the N-terminal domain of the OCP are also shown. The length of the (putative) protein sequence is given in parentheses

recently evidence for their expression has been reported. In *Cyanothece* 51142 a unicellular diazathroph that temporally separates photosynthesis and nitrogen fixation, a whole-genome DNA microarray was used to detect oscillating light-dark transcripts; one of the N-terminal (cce_1537) and the C-terminal (cce_0742) OCP fragment ORFs cycle similarly to the full length OCP, peaking in the light phase (Stockel et al. 2008). Cluster analysis grouped the OCP gene in a subnetwork with photosynthesis, cell division, glycogen biosynthesis and carbon fixation genes. In *Nostoc punctiforme*, proteomic analysis provides additional evidence for a func-

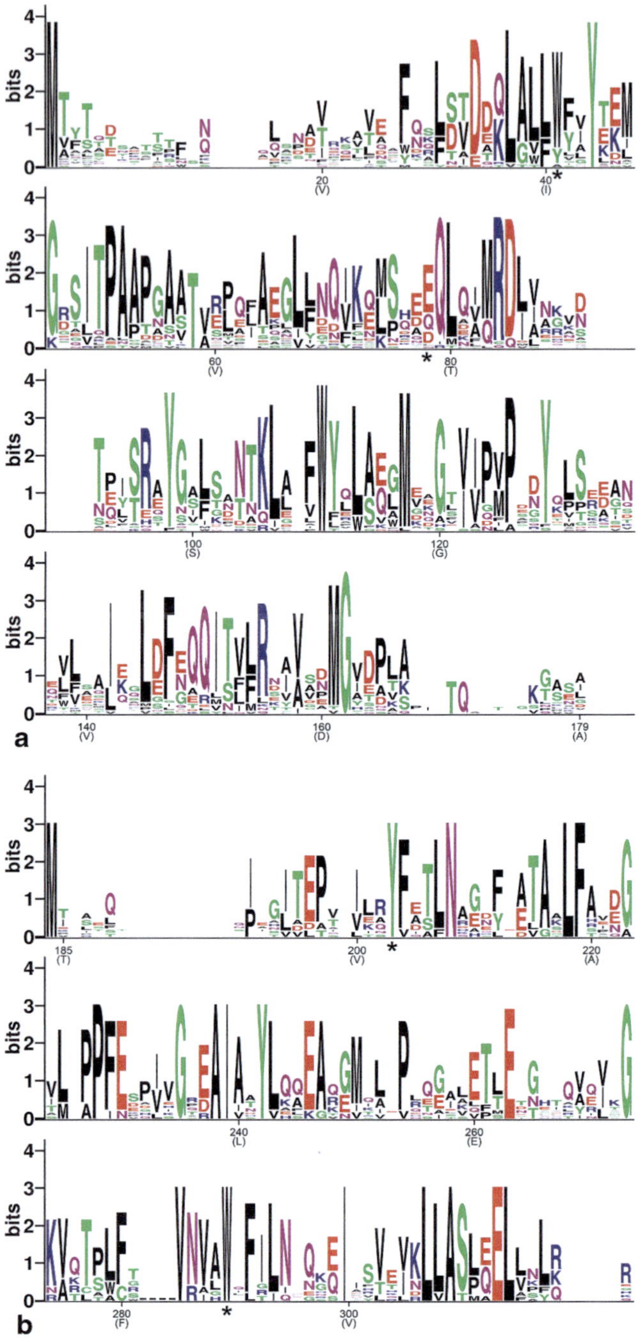

Fig. 14.6 Sequence logos for the **a** N- and **b** C-terminal fragments of the OCP identified from genomic sequences. Residues are numbered according to alignment with the *A. maxima* OCP. The height of each letter is proportional to the observed frequency of the amino acid in that position. The overall height of each stack of letters is proportional to sequence conservation as measured in bits (Schneider and Stephens 1990). A sample correction for underestimation of entropy was included in the generation of the Logo. The amino acids are colored according to their chemical properties: *polar, green*; *basic, blue*; *acidic, red*; *hydrophobic, black*. In the C-terminal fragment logo, the positions of the Tyr and Trp residues that hydrogen bond to the carotenoid in the full-length OCP are marked with an *. (Figure prepared with WEBLOGO Crooks et al. 2004)

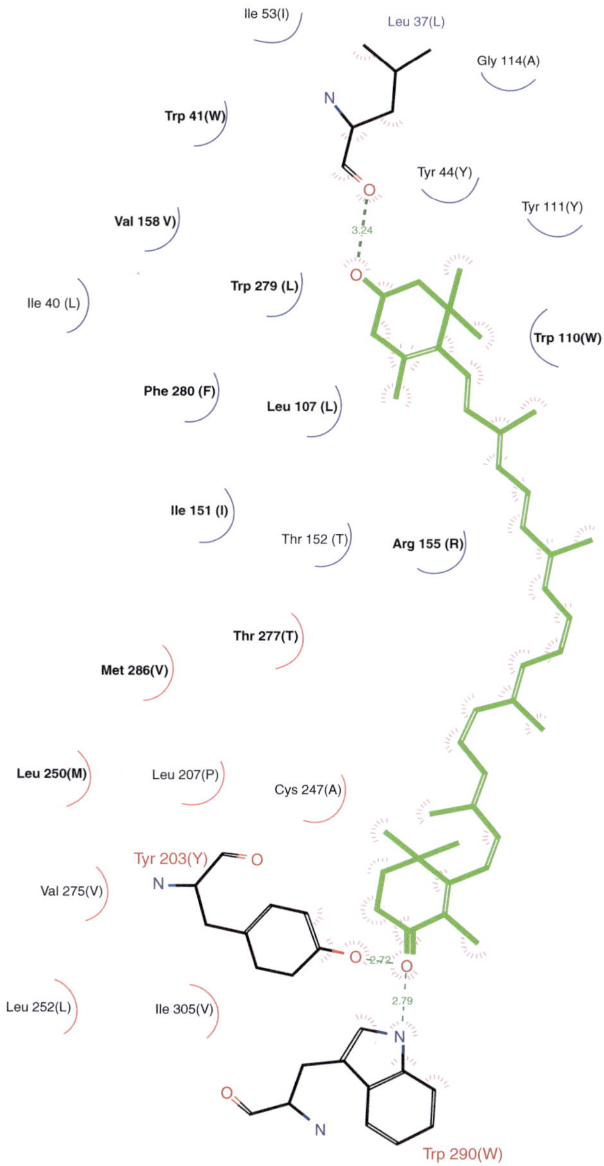

Fig. 14.7 Two-dimensional view of protein-pigment interactions (hydrogen bonding and hydrophobic contacts) in the *A. maxima* OCP structure. Residues in bold are absolutely conserved among OCP orthologs. The residue most commonly found in each position in the isolated N- and C-terminal fragments (see Fig. 14.6) of the OCP is shown in parentheses. (Figure prepared with LIGPLOT, Wallace et al. 1995)

tional role for the OCP fragments; several of the N-terminal paralogs are known to be expressed (Anderson et al. 2006).

Among phycobilisome-containing cyanobacteria, only a few strains lack a full length OCP gene homolog, including the freshwater *Synechococcus elongatus* PCC 7942 and PCC 6301, the thermophile *Thermosynechococcus elongatus*, *Synechococcus* sp. A, *Synechococcus* sp. B', the marine *Synechococcus* sp. CC9605 and *Crocosphaera watsonii* WH8501. *T. elongatus*, *C. watsonii* WH8501, *Cyanothece*

PCC7425, *Cyanothece* PCC7822 and *Cyanothece* PCC8801, however, contain two ORFs one for each of the N- and the C-terminal domain of the OCP (Table 14.1). In *T. elongatus, Cyanothece* PCC7425, *Cyanothece* PCC7822 and *Cyanothece* PCC8801, they occur as separate but adjacent genes (and *Cyanothece* PCC7425 and *Cyanothece* PCC7822 also contain isolated N-terminal domains ORFs) whereas in *C. watsonii*, the ORFs for the N- and C-terminal domains are in different parts of the genome. In *S. elongatus* PCC 7942 and *T. elongatus*, the qE_{cya} mechanism appears to be absent confirming the correlation between the blue-green light induced qE_{cya} and the presence of the full length OCP, (Boulay et al. 2008a), and that the ORFs coding for the N- and C-terminal of the OCP do not function to induce qE_{cya} in *T elongatus*. The strains lacking the OCP are more sensitive to episodes of high light irradiance, similar to the phenotype of the ΔOCP *Synechocystis* mutant (Boulay et al. 2008a) Interestingly, *S elongatus* PCC 7942 and *T elongatus*, which lack the OCP-related energy dissipation mechanism, use a different strategy to cope with excess light energy. Under iron starvation conditions, these organisms protect themselves by quickly decreasing the phycobiliprotein cell content to avoid the accumulation of the "dangerous" functionally disconnected phycobilisomes (Boulay et al. 2008a).

14.5 The Structure and Photoactivity of the OCP

14.5.1 The Three-dimensional Structure of the OCP

The structure of the OCP isolated from *Arthrospira maxima* (Kerfeld et al. 2003) is known (Fig. 14.8); given the sequence homology between the *Synechocystis* and *A. maxima* OCP (82% identical), the structure provides a useful model for interpreting data from functional studies.

The OCP consists of two domains: an all α-helical N-terminal domain and a mixed α/β C-terminal domain that resembles the fold of nuclear transport factor 2 (NTF2; Pfam 02136). The embedded ketocarotenoid, 3'-hydroxyechinenone (hECN), is almost completely buried within the protein with its keto terminus deep in the hydrophobic pocket of the NTF2 domain (Figs. 14.7 and 14.8); only 3.4% of the pigment surface is accessible to solvent. The N-terminal domain of the OCP is an orthogonal alpha helical bundle, subdivided into two four-helix bundles (Figs. 14.4 and 14.8). To-date, the OCP N-terminal domain is the only known protein with this particular fold (Pfam 09150). The hydroxyl terminus of the hECN is nestled between the two bundles. The absolutely conserved Tyr 44 and Trp110 sidechains make hydrophobic contacts with the hydroxyl terminal end of the carotenoid, whereas two other absolutely conserved residues, Tyr203 and Trp290, form hydrogen bonds to the carbonyl moiety at the keto terminus of the hECN (Kerfeld et al. 2003 and Fig. 14.7). Interestingly, most of these residues are conserved among the ORFs that potentially encode isolated N- and C-terminal fragments of OCP

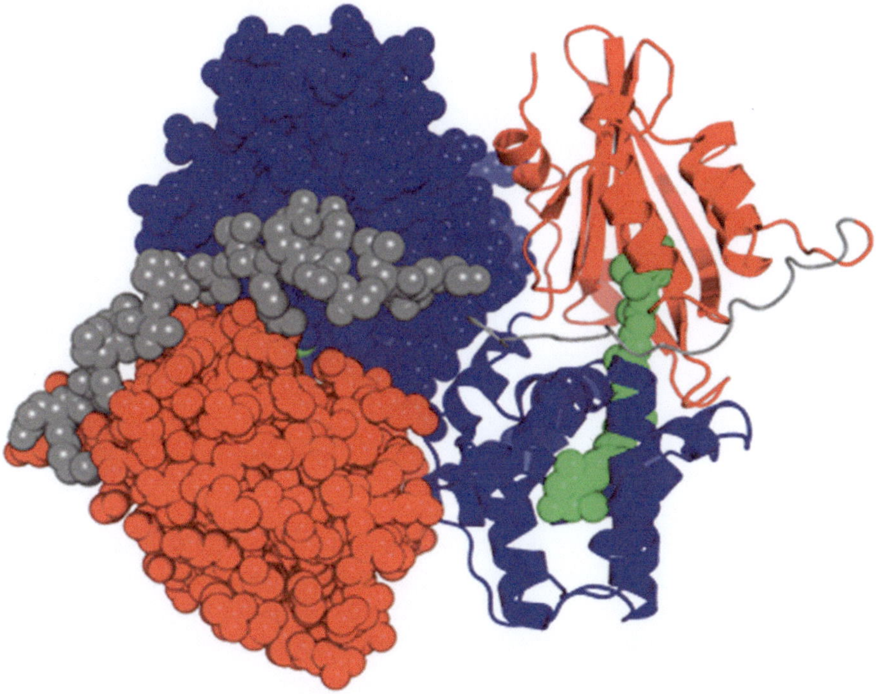

Fig. 14.8 The structure of the OCP. The dimer of the OCP is shown, the left molecules is shown in space-filling representation, the right molecules as a ribbon diagram. The N-terminal domain is *colored blue*, the C-terminal domain is *red* and the linker joining the domains is *colored gray* (as in Fig. 14.4). The 3′ hydroxyechinenone is shown in green space-filling representation

that are found in some cyanobacterial genomes; including absolute conservation of W290 and Y203 (Fig. 14.6).

14.5.2 The Photoactivity of the OCP in the Context of its Structure

The protein has a large effect on the spectroscopic characteristics of the carotenoid: In organic solvents, hECN is yellow (λmax=450 nm). In the OCP, in darkness or dim light, it appears orange and its absorption spectrum presents a typical carotenoid shape, reflecting the strongly allowed S_0–S_2 transition with three distinct vibrational bands; the 0–0 vibrational peak located at 496 nm (Kerfeld et al. 2003; Polivka et al. 2005; Wilson et al. 2008 and Fig. 14.9). The position and shape of this transition corresponds to that of hECN locked into an all-*trans* conformation by the surrounding protein (Polivka et al. 2005), in agreement with the crystallographic

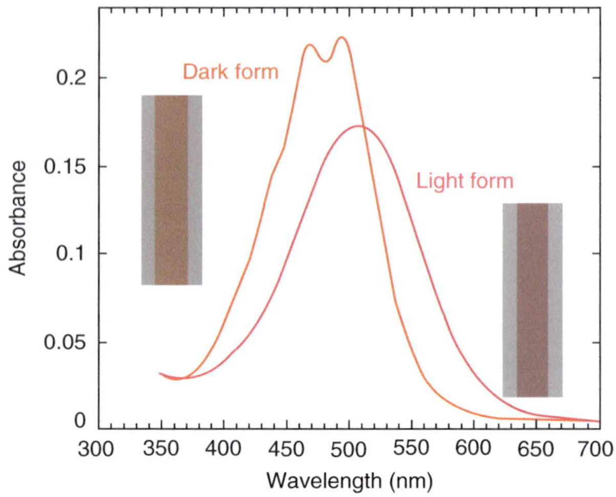

Fig. 14.9 The OCP is a photoactive protein. To obtain the *red form* the isolated OCP was illuminated with *blue-green* light at 12°C for 2 min. Absorbance spectra (and photograph) of the *dark orange* form (*orange*) and the *light red form* (*red*)

structure (Kerfeld et al. 2003): the protein forces the carotenoid to have the C = C bond at the terminal ring containing the conjugated carbonyl group in *trans* position with respect to the conjugated backbone (Kerfeld et al. 2003; Polivka et al. 2005). The Raman spectra of the OCP from *A. maxima* and *Synechocystis*, which also indicate that the hECN is in an all-*trans* conformation, are nearly identical, implying that the carotenoid-peptide interactions in these two proteins are remarkably similar (Wilson et al. 2008).

Upon illumination with blue-green light (at 10°C), the OCP (which appears orange) is completely photoconverted to a red form (Wilson et al. 2008 and Fig. 14.9). The red-shifted spectrum of the light-activated form, with a maximum at 510 nm, loses the resolution of the vibrational bands (Fig. 14.9). The initial rate of the photoconversion strongly depended on light intensity but it was temperature-independent. The conversion occurs with a very low quantum yield (about 0.03); this can be explained by the fact that the OCP is involved in a photoprotective mechanism that must be induced only under high irradiance. In darkness, the red form of the OCP spontaneously reverts to the orange form (Wilson et al. 2008). This step, which is not accelerated by illumination, shows large temperature dependence.

Resonance Raman spectroscopy and light-induced Fourier transform infrared (FTIR) difference spectra demonstrated that light absorbance by the OCP induces structural changes not only in the carotenoid but also in the apo-protein (Wilson et al. 2008). Upon illumination of the *A. maxima* and *Synechocystis* OCP, the apparent conjugation length of hECN increased by about one conjugated bond, and hECN reaches a less distorted, more planar structure. These changes are in agreement with the observed red-shift of the absorbance spectrum. The results strongly suggested that a *trans-cis* isomerisation of the hECN did not occur upon illumination. It was concluded that that hECN is still all-*trans* in red OCP form. Although the specific changes in the carotenoid of the OCP remain to be elucidated, they are clearly different from those occurring in retinal, a carotenoid derivative, which is the chromophore of rhodopsins (reviewed in Spudich et al. 2000). In rhodopsins, the retinal is covalently bound to the protein via a protonated Schiff-base linkage to a Lys residue; light induces an isomerization of the retinal, initiating the photocycle. The photocycle involves proton transfer between the Schiff-base and a carboxylate residue. In the OCP not only there is no isomerization detected but also the hECN is not covalently bound to the protein.

Large changes were also observed in the FTIR spectra upon OCP illumination, suggesting a major rearrangement of the protein backbone. The changes observed corresponded to a less rigid helical structure and a compaction of the β-sheet domain upon conversion to the red form. Further studies will be necessary to identify the specific nature of the changes in the carotenoid and the protein.

14.5.3 The Red Form of the OCP is the Active Form

After it was demonstrated that the OCP is a photoactive protein, the link between the light-induced red form and the qE_{cya} mechanism required elucidation. The first results suggesting the relationship between the induction of fluorescence quenching and the formation of the red form of the OCP came from absorbance spectra of a *Synechocystis* strain overexpressing OCP, measured before and after illumination with strong blue light. The calculated light-minus-dark difference spectra strongly suggest that the red form of the OCP accumulates *in vivo* under conditions inducing energy dissipation and fluorescence quenching (strong blue-green light) (Wilson et al. 2008).

The observed strong dependency on light intensity of the blue-light induced qE_{cya} in whole cells (Wilson et al. 2006) correlates with that of the kinetics of the formation of the red OCP form *in vitro* (Wilson et al. 2008). The back–reaction kinetics (red to orange) show a large temperature dependence, the half-time of recovery varying from 30 s at 32°C to about 45 min at 11°C (Wilson et al. 2008). *In vivo,* the recovery qE_{cya} kinetics also show a large temperature dependence, however the kinetics are slower (El Bissati et al. 2000; Wilson et al. 2006; Rakhimberdieva et al. 2007a) than the red-to-orange dark conversion, suggesting that the red OCP form is more stable *in vivo* than *in vitro* or that the fluorescence quenching remains longer than the red form. In contrast, the initial rate of the light photoconversion was temperature-independent. Thus, at a fixed light intensity, the steady-state concentration of the red form depended on the temperature. This is reflected *in vivo*, where a larger fluorescence quenching was observed at 7°C than at 33°C (Wilson et al. 2008). Thus, the ratio of the two forms of the OCP under illumination depends on light intensity and temperature: more red form at high irradiance and low temperatures, conditions in which photoprotection is needed. The OCP is a sensor of photoinhibitory conditions inducing a greater fluorescence quenching (photoprotection) under increasing irradiance and lowering temperature.

Collectively these results already strongly suggested that the red form of the OCP is the active form. This was confirmed by the study of qE_{cya} in three *Synechocystis* mutants: an overexpressing C-terminal His-tagged OCP mutant, an overexpressing C-terminal His-tagged W110S OCP mutant and an overexpressing OCP ΔcrtO double mutant (Fig. 14.10). In the latter mutant, echinenone and

Fig. 14.10 The red form is the active OCP form. This activated form is essential for the photoprotective mechanism. Absorbance spectra of the *dark* (*black*) and the *light* (*red*) form of the OCP isolated from an overexpressing W110S OCP mutant **a** and an overexpressing WT-OCP ΔCrtO mutant **b** This mutant lacks hECN and echinenone and the OCP binds mostly zeaxanthin. The isolated proteins were illuminated 15 min at 12°C. **c** Decrease of maximal fluorescence (Fm') during exposure of overexpressing OCP (*green circles*), WT (*open green circles*), overexpressing W110S OCP (*red triangles*) and overexpressing WT-OCP ΔCrtO mutant (*blue squares*) cells

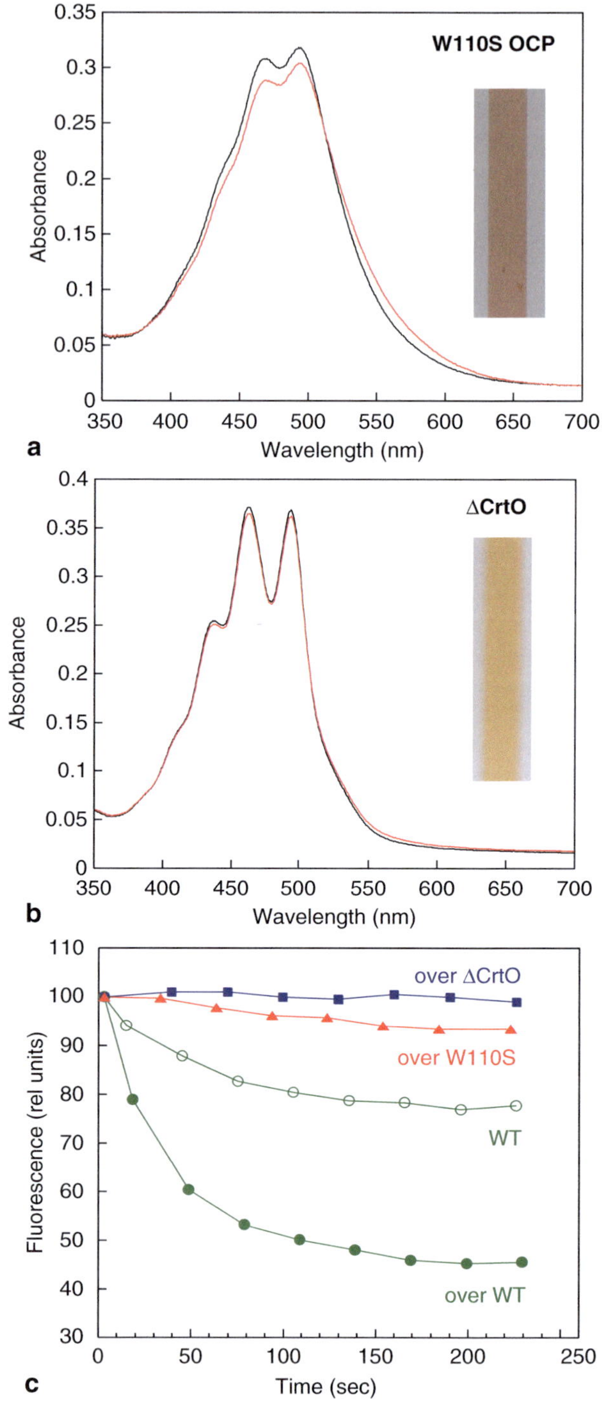

hECN are lacking due to the deletion of the *crtO* gene coding for the β-carotene monoketolase which catalyzes the formation of echinenone and hECN. In this mutant the OCP binds mostly zeaxanthin and it is yellow (Punginelli et al. 2009). In contrast, the W110S OCP point mutant is orange and contains mostly hECN. The WT OCP isolated from the overexpressing His-tagged OCP mostly contains echinenone (50–85%) and hECN (35–15%) in different proportions depending on the preparation. In all three mutants, large quantities of the OCP were present. However, while high intensities of blue-green light induced a large fluorescence quenching in the overexpressing His-tagged OCP strain, no (or almost no) fluorescence quenching was induced in the other two mutants. Only a very small portion of the W110S mutant OCP and none of the zeaxanthin-OCP were converted to the red form, even after prolonged illumination with strong blue-green light at 10°C (Wilson et al. 2008; Punginelli et al. 2009) (Fig. 14.10). The red shift of the WT-hECN spectrum is caused by the increase of the apparent conjugation length of the hECN by about one conjugated bond, due to a more planar conformation of the carotenoid in the light form (Wilson et al. 2008) and probably a change in the position of the carbonyl group. The absence of the carbonyl group in the zeaxanthin prevents the increase of the apparent conjugation length and/or the stabilization of a more planar structure of the zeaxanthin with a red-shifted spectrum. The inactive W110S mutant OCP binds hECN, indicating that the presence of the carbonyl is not sufficient for the photoactivity and for the stabilization of the red form. The hydrophobic interaction between the hydroxyl β-ring and the tryptophan sidechain seems to be also important to form and/or stabilize the hECN in the red form. Interestingly, in contrast to the W110S mutant, a W110F mutant OCP is able to stabilize the red form and to induce the photoprotective mechanism (Wilson et al. 2010). On the other hand, the absence of the hydroxyl group of the hECN appears not to be essential because the echinenone-OCP was completely converted to the red form (Punginelli et al. 2009). In conclusion, the stabilization of the red form is essential for the induction of the fluorescence quenching.

14.6 Hypotheses and Perspectives

By comparing the OCP with other blue-light responsive photoactive proteins in the context of the phenotype of various OCP mutants, it is possible to speculate about the role of the OCP in the qE_{cya} mechanism.

Blue-green light is an important biological signal for many organisms; blue-light receptors are found in both prokaryotes and eukaryotes. For example, BLUF domains (Blue light using FAD; Pfam 04940) are widespread in bacteria in which they function to relay light signals to diverse outputs. Many other blue-light responsive proteins found in algae and plants as well as bacteria are members of the PAS-fold superfamily (Pfam clan CL0183). These include the flavin-binding phototropins

(Crosson and Moffat 2001; Briggs 2006; Takahashi et al. 2007). The phototropins contain the well-characterized light-oxygen-voltage (LOV) domains. In addition, photoactive yellow protein (PYP) which is involved in chemotaxis is also a member of the PAS superfamily. Structural and spectroscopic data are available for several of these proteins in both the resting and the photoactive state. In addition to responsiveness to blue light, they share some structural similarity with the OCP; these blue-light responsive proteins contain a core consisting of a 4–5 stranded beta sheet similar to the C-terminal domain of the OCP.

The lack of the activity of the zeaxanthin-OCP and of the hECN-W110S OCP, both of which are unable to stabilize the red OCP form, could have several underlying causes: (1) these proteins cannot accept the energy arriving from the phycobilisome to dissipate it; (2) the interaction with the phycobilisome is disrupted; (3) they are unable to propagate the signal.

We outline four hypotheses about the qEcya mechanism that require testing in the coming years in order to elucidate the OCP's photoprotective mechanism.

14.6.1 The Red Form is the Active Form Because it is Able to Absorb the Energy Coming from the Phycobilisomes and Dissipate It

Carotenoids are able to quench excitation energy from tetrapyrrole molecules through energy transfer to their optically forbidden S1 state, coupled to an intramolecular charge-transfer state (Berera et al. 2006). This seems to be the mechanism at the origin of qE in the plants (Ruban et al. 2007). Hydroxyechinenone and echinenone, which contain a conjugated carbonyl group, clearly have such an intramolecular charge transfer state. Indeed, it was shown that excited state of hECN in the OCP has strong charge-transfer character and, consequently, its lifetime is about half that in solution (Polivka et al. 2005). In the OCP the excited-state-absorption band (ESA) of the carotenoid is extremely red-shifted, having its maximum at 650 nm. This band seems to be due to a stabilization of an intramolecular charge transfer (and/or a breaking of hydrogen bonds upon excitation) (Polivka et al. 2005). Therefore, we propose that the OCP is not only the sensor of light but also the quencher of the excitation energy. In principle, the orange-hECN and zeaxanthin are able to quench excitation energy like the red-hECN; however only the red-hECN is able to induce the quenching of phycobilisome emission. We hypothesize that the red shift of the hECN spectrum is necessary to tune the optically forbidden S1 state of hECN to a position allowing the energy transfer from the phycobilisome to the OCP where excitation energy can then be quenched. The zeaxanthin-OCP and the W110S mutant OCP, due to the lack of a metastable "red" form are unable to absorb the energy coming from the phycobilisome and quench this energy.

14.6.2 The OCP Interacts with the APC Core of the Phycobilisome. Light-Induced Changes in the C-terminal Domain are Essential for Energy and Fluorescence Quenching

Absolutely conserved Tyr203 and Trp290 residues form hydrogen bonds with the carbonyl moiety at the keto terminus of the hECN (Kerfeld et al. 2003). These residues belong to the C-terminal domain of the OCP, which has some structural similarity to the phycobilisome 7.8 kD core linker protein (Wilson et al. 2008). We had hypothesized that the C-terminal OCP domain, by interacting with the center of an APC trimer in the phycobilisome core, may bring the carotenoid into proximity of the APC chromophores (Wilson et al. 2008). Perturbations in the secondary structure of the protein, especially in the C-terminal domain, could be induced by the binding of zeaxanthin, instead of hECN, due to the lack of the carbonyl group. These changes could prevent optimal contact between the APC trimers and the OCP. In addition light induced changes in the carbonyl-protein interactions could possibly modify the interaction between the carotenoid and the APC chromophore favoring energy transfer under high light conditions. In the zeaxanthin-OCP, in which there are no hydrogen bonds between the carbonyl and the protein, the modifications required to induce energy transfer and dissipation are probably not induced.

The crystal structure of the *A. maxima* OCP is assumed to represent the dark-state adapted form of the protein. Given the almost complete concealment of the carotenoid by the protein, it is difficult to envision how the carotenoid could mediate quenching of energy absorbed by the phycobilisome. However at the time of the structure elucidation the number of cavities and protrusions on the surface of the protein was noted to be exceptional (Kerfeld et al. 2003). These could play a role in interaction with other proteins. In addition, the two domains of the structure are joined by a long flexible linker (Fig. 14.4 and Fig. 14.8); conformational changes in the protein that alter the disposition of the two domains with respect to each other could unmask sites for protein-protein interactions and allow increased accessibility to the carotenoid. A precedent for this is found in PYP; in activating the PYP alterations in hydrogen bonding result in a displacement of a conserved Arg residue. This movement increases solvent accessibility to the chromophore. In the OCP an invariant Arg residue, Arg155, is positioned at the interface between the N- and C-terminal domains (Fig. 14.7). A change in the conformation of the sidechain of this residue could increase the solvent accessibility of the carotenoid.

14.6.3 The Signal is Propagated Through Protein Conformational Changes

A role for conformational changes in transducing a signal through the OCP and/or in altering inter-protein or intra-protein (domain-domain) interactions is supported by the observation that glutaraldehyde and high concentrations of sucrose and glycerol abolish NPQ formation in *Synechocystis* (Scott et al. 2006; Rakhimberdieva et al. 2007a). Functionally important conformational changes in plant LOV domains have been shown to be substantially reduced by similar treatments (Iwata et al. 2007). These experiments also revealed the important role of structural water molecules; with additional structures of the OCP we will be able to begin to determine if there are similarly important structurally conserved water molecules that may be important in the photoresponse of the OCP.

In PYP and LOV domains, blue light causes changes in the chromophore that induce protein conformational changes that are required for the light-responsive function (Rubinstenn et al. 1998; Harper et al. 2003). These conformational changes have their basis in the dynamics of the hydrogen bonding networks (as in BLUF domains) within the LOV domain and PYP (Anderson et al. 2004; Kort et al. 2004; Jung et al. 2006). The OCP contains a substantial number of hydrogen bonds; several could serve as relays between the carotenoid and the protein surface via interaction with Tyr203 and Trp290. For example, Tyr203 is hydrogen-bonded to the main chain atoms of Thr 199 and Leu 207; Thr 199 is surface exposed. Likewise Trp 290 is hydrogen bonded to the absolutely conserved residues Val271 and Phe292; these residues are part of the beta sheet and are surface exposed. Light triggered conformational changes of hECN could alter the strength of the hydrogen bond between the carbonyl and Tyr203 and Trp290. This, as in LOV domains (Nozaki et al. 2003; Harper et al. 2003; Halavaty and Moffat 2007), could alter the conformation of the beta sheet C-terminal domain and its interaction with a short alpha-helix from the N-terminus, affording signal propagation pathway from the carotenoid to the surface of the OCP. At the opposite end of the carotenoid molecule, the hydroxyl group has the potential to form an hydrogen bond to the mainchain of Leu 37 which is hydrogen bonded to the mainchain of Ala 33 and Trp 41; these residues are also surface exposed.

When the structure of the OCP was first described, the large number of Met residues within 6.5 Å of the carotenoid was noted. These may be important in protection (via oxidation to methioinine sulfoxide and methionine sulfone) of the carotenoid from harmful reactive oxygen species. The Met residues could also be involved in signal propagation. BLUF domains provide an example for a role of Met residues in signal propagation via sidechain conformational changes (Jung et al. 2006). The sulphur atom in Met sidechains is polarizeable and Met sidechains adopt distinctive conformations with electrophiles and with nucleophiles which may also facilitate signalling. All of the conserved Met residues in the OCP make multiple hydrogen bonds to residues that are surface-exposed.

14.6.4 Is the Oligomeric State of the OCP is Functionally Important?

In the Phototropin 1 LOV domain of *Arabidopsis*, alterations in hydrogen bonding patterns induced by light, also changes the monomer-dimer equilibrium (Nakasone et al. 2006). This process is temperature dependent, underscoring the importance with the conformational changes that are known to be important for biological activity (Nakasone et al. 2008). The OCP is a dimer in the crystal (Kerfeld et al. 2003; Wilson et al. 2010) (Fig. 14.8). The intermolecular interactions are largely mediated by hydrogen bonding among the N-terminal 30 amino acids. There is no indication suggesting that the orange to red light-induced conversion involves a dissociation of the OCP into monomers. On the other hand, it is easier to envision an efficient energy transfer from the chromophores of the APC to the carotenoid if the OCP monomer, rather than the dimer, interacts with the PBS. For the moment, there is no data indicating if OCP is a monomer or a dimer *in vivo* and if both the orange and the red form interact with the phycobilisome or only the red form. However the OCP is a monomer in solution. If *in vivo* the orange OCP is a dimer having low affinity for the phycobilisome, light-induced protein conformational changes and hydrogen bonding patterns could affect the inter-protein interactions in the dimer, possibly leading to its dissociation and facilitating the binding of the OCP monomer to the PBS core. In the zeaxanthin-OCP and the W110S mutant OCP, the changes in the hydrogen bonding patterns is not induced; perhaps an associated failure to dissociate into monomers could underlie the lack of activity of these proteins.

14.7 Conclusion

The recent discovery of an energy dissipating photoprotective mechanism in cyanobacteria is very exciting because this mechanism is completely distinct from that known in plants. It involves a new photoactive protein, the OCP, which contains a carotenoid as the active chromophore and is essential for this mechanism. The function of the OCP as the light sensor that mediates energy dissipation and fluorescence quenching is unequivocal. We propose that, in addition, the OCP may be the energy dissipator and the quencher of fluorescence but this hypothesis remains to be verified.

The OCP also constitutes an ideal model system *in vitro* for studies of the electronic and photochemical properties of protein-bound carotenoids. These properties are still poorly understood due to the fact that carotenoids are most often embedded in integral membrane proteins where they are surrounded by other cofactors that complicate spectroscopic studies and mutagenesis experiments. The OCP is a water soluble protein that binds only a single carotenoid per polypeptide chain.

The construction of OCP and phycobilisome mutants and the associated characterization of their phenotypes by spectroscopic and structural studies will be essential to elucidate which components of the PBS core are involved in the interaction with the OCP and in energy dissipation. Likewise this will inform more reductionistic studies that dissect the role of specific amino acids that are essential for photoactivity and for energy dissipation, and to understand which sidechains influence the properties of the carotenoid and the affinity of the carotenoid for the protein.

The study of the OCP provides not only fundamental knowledge about the influence of proteins on carotenoid properties and about a new photoprotective mechanism involving a new type of photoactive protein but the OCP also has applications in biomass production using cyanobacteria. The importance of stress protection of cyanobacteria that are used for biofuel, food, or chemical production is becoming apparent to the biotechnology industry. Light is known to be the principal limiting factor of photobioreactor efficiency; it is too low or quasi-null in the depth of culture, and too high in the vicinity of the light source. Cultivating strains with modifications in the antenna and in the antenna-associated photoprotective mechanism could potentially lead to an increase of the efficiency of photobioreactor-based biotechnology The confluence of structural, spectroscopic and functional studies in cyanobacterial photoprotection has been extraordinarily fruitful in providing insight into a novel mechanism of photoprotection that may have important implications for bioengineering.

Acknowledgements We thank David Krogmann for helpful discussions and Edwin Kim for assistance with the figures. DK thanks her students Adjélé Wilson, Clémence Boulay and Claire Punginelli who largely contributed to the results described in this work. The research of CAK is performed under the auspices of the US Department of Energy's Office of Science, Biological and Environmental Research Program, and by the University of California, Lawrence Berkeley National Laboratory under contract number DE-AC02-05CH11231, Lawrence Livermore National Laboratory under contract number DE-AC52-07NA27344. The research of DK and her group is supported by the "Commissariat à l'Energie Atomique", the "Center National de la Recherche Scientifique" and the "Agence Nationale de la Recherche" (projet CAROPROTECT).

References

Adir N (2005) Elucidation of the molecular structures of components of the phycobilisome: reconstructing a giant. Photosynth Res 85: 15–32

Allen JF (1993) Control of gene expression by redox potential and the requirement for chloroplast and mitochondrial genomes. J Theor Biol 165: 609–631

Allen JF, Alexciev K, Hakansson G (1995) Photosynthesis. Regulation by redox signalling. Curr Biol 5: 869–872

Anderson DC, Campbell EL, Meeks JC (2006) A soluble 3D LC/MS/MS proteome of the filamentous cyanobacterium *Nostoc punctiforme*. J Proteome Res 5: 3096–3104

Anderson S, Crosson S, Moffat K (2004) Short hydrogen bonds in photoactive yellow protein. Acta Crystallogr D60: 1008–1016

Aro EM, Virgin I, Andersson B (1993) Photoinhibition of photosystem II, inactivation, protein damage and turnover. Biochim Biophys Acta 1143: 113–134

Bailey S, Grossman A (2008) Photoprotection in cyanobacteria: regulation of light harvesting. Photochem Photobiol 84: 1410–1420

Berera R, Herrero C, van Stokkum IH, Vengris M, Kodis G, Palacios RE, van Amerongen H, van Grondelle R, Gust D, Moore TA, Moore AL, Kennis JT (2006) A simple artificial light-harvesting dyad as a model for excess energy dissipation in oxygenic photosynthesis. Proc Natl Acad Sci U S A 103: 5343–5348

Boulay C, Abasova L, Six C, Vass I, Kirilovsky D (2008a) Occurrence and function of the orange carotenoid protein in photoprotective mechanisms in various cyanobacteria. Biochim Biophys Acta 1777: 1344–1354

Boulay C, Wilson A, Kirilovsky D (2008b) Orange carotenoid protein (OCP) related NPQ in *Synechocystis* PCC 6803 OCP: phycobilisomes interactions. In: Allen JF, Gantt E, Golbeck JH, Osmond B (eds) Photosynthesis 2007. Energy from the sun. Proceedings of the 14th international congress on photosynthesis. Springer, Heidelberg, pp 997–1001

Briggs WR (2006) Photomorphogenesis in plants and bacteria. Springer, Dordrecht

Cadoret JC, Demouliere R, Lavaud J, van Gorkom HJ, Houmard J, Etienne AL (2004) Dissipation of excess energy triggered by blue light in cyanobacteria with CP43′ (IsiA). Biochim Biophys Acta 1659: 100–104

Crooks GE, Hon G, Chandonia J-M, Brenner SE (2004) WEBLOG: a sequnce logo generator. Genome Res 14: 1188–1190

Crosson S, Moffat K (2001) Structure of a flavin-binding plant photoreceptor domain: Insights into light-mediated signal transduction. Proc Natl Acad Sci U S A 98: 2995–3000

Desper R, Gascuel O (2004) Fast minimum evolution. Mol Biol Evol 21: 587–598

El Bissati K, Delphin E, Murata N, Etienne A, Kirilovsky D (2000) Photosystem II fluorescence quenching in the cyanobacterium *Synechocystis* PCC 6803: involvement of two different mechanisms. Biochim Biophys Acta 1457: 229–242

Foyer CH, Allen JF (2003) Lessons from redox signaling in plants. Antioxid Redox Signal 5: 3–5

Fulda S, Mikkat S, Huang F, Huckauf J, Marin K, Norling B, Hagemann M (2006) Proteome analysis of salt stress response in the cyanobacterium Synechocystis sp. strain PCC 6803. Proteomics 6: 2733–2745

Gantt E, Conti SF (1966a) Granules associated with chloroplast lamellae of *Porphyridium cruentum*. J Cell Biol 29: 423

Gantt E, Conti SF (1966b) Phycobiliprotein localization in algae. Brookhaven Symp Biol 19: 393–405

Glazer AN (1984) Phycobilisome—a macromolecular complex optimized for light energy-transfer. Biochim Biophys Acta 768: 29–51

Grossman AR, Schaefer MR, Chiang GG, Collier JL (1993) The phycobilisome, a light-harvesting complex responsive to environmental-conditions. Microbiol Mol Biol Rev 57: 725–749

Halavaty AS, Moffat K (2007) N- and C-terminal flanking regions modulate light-induced signal transduction in the LOV2 domain of the blue light sensor phototropin 1 from Avena sativa. Biochemistry 46(49):14001–14009

Harper SM, Neil LC, Gardner KH (2003) Structural basis of a phototropin light switch. Science 301(5639): 1541–1544

Hihara Y, Kamei A, Kanehisa M, Kaplan A, Ikeuchi M (2001) DNA microarray analysis of cyanobacterial gene expression during acclimation to high light. Plant Cell 13: 793–806

Holt TK, Krogmann DW (1981) A carotenoid-protein from cyanobacteria. Biochim Biophys Acta 637: 408–414

Horton P, Ruban AV, Walters RG (1996) Regulation of light harvesting in green plants. Annu Rev Plant Physiol Plant Mol Biol 47: 655–684

Iwata T, Yamamoto A, Tokutomi S, Kandori H (2007) Hydration and temperature similarly affect light-induced protein structural changes in the chromophoric domain of phototropin. Biochemistry 46: 7016–7021

Joshua S, Bailey S, Mann NH, Mullineaux CW (2005) Involvement of phycobilisome diffusion in energy quenching in cyanobacteria. Plant Physiol 138: 1577–1585

Jung A, Reinstein J, Domratcheva T, Shoeman RL, Schlichting I (2006) Crystal structures of the AppA BLUF domain photoreceptor provide insights into blue light-mediated signal transduction. J Mol Biol 362: 717–732

Karapetyan NV (2007) Non-photochemical quenching of fluorescence in cyanobacteria. Biochemistry (Mosc) 72: 1127–1135

Kerfeld CA (2004) Structure and function of the water-soluble carotenoid-binding proteins of cyanobacteria. Photosynth Res 81: 215–225

Kerfeld CA, Sawaya MR, Brahmandam V, Cascio D, Ho KK, Trevithick-Sutton CC, Krogmann DW, Yeates TO (2003) The crystal structure of a cyanobacterial water-soluble carotenoid binding protein. Structure 11: 55–65

Kerfeld CA, Alexandre M, Kirilovsky D (2009) The orange carotenoid protein in cyanobacteria. In: J. Landrum (ed) Carotenoids: physical, chemical and biological functions and properties. Taylor and Francis group, pp 3–19

Kirilovsky D (2007) Photoprotection in cyanobacteria: the orange carotenoid protein (OCP)-related non-photochemical-quenching mechanism. Photosynth Res 93: 7–16

Kirilovsky D (2010) The photoactive orange carotenoid protein and photoprotection in cyanobacteria. Adv Exp Med Biol 675: 139–159

Kort R, Hellingwerf KJ, Ravelli RBG (2004) Initial events in the photocycle of photoactive yellow protein. J Biol Chem 279: 26417–26424

Laskowsi RA (2001) PDBsum: summaries and analyses of PDB structures. Nucleic Acids Res 29: 221–222

MacColl R (1998) Cyanobacterial phycobilisomes. J Struct Biol 124: 311–334

Mullineaux CW (1992) Excitation energy transfer from phycobilisomes to photosystem-I in a cyanobacterium. Biochim Biophys Acta 1100: 285–292

Nakasone Y, Eitoku T, Matsuoka D, Tokutomi S, Terazima M (2006) Kinetic measurement of transient dimerization and dissociation reactions of *Arabidopsis* phototropoin 1 LOV2 domain. Biophys J 91: 645–653

Nakasone Y, Eitoku T, Zikihara K, Matsuoka D, Tokutomi S, Terazima M (2008) Stability of dimer and domain-domain interaction of arabidopsis phototropin 1 LOV2. J Mol Biol 383: 904–913

Niyogi KK (1999) Photoprotection revisited: genetic and molecular approaches. Annu Rev Plant Physiol Plant Mol Biol 50: 333–359

Nozaki IT, Tokutomi D, Kagawa S, Wada M, Kandori H. (2003) Light-induced structural changes in the LOV2 domain of Adiantum phytochrome 3 studied by low-temperature FTIR and UV-visible spectroscopy. Biochemistry 42(27): 8183–8191

Pascal AA, Liu ZF, Broess K, van Oort B, van Amerongen H, Wang C, Horton P, Robert B, Chang WR, Ruban A (2005) Molecular basis of photoprotection and control of photosynthetic light-harvesting. Nature 436: 134–137

Polivka T, Kerfeld CA, Pascher T, Sundström V (2005) Spectroscopic properties of the carotenoid 3′-hydroxyechinenone in the orange carotenoid protein from the cyanobacterium *Arthrospira maxima*. Biochemistry 44: 3994–4003

Prasil O, Adir N, Ohad I (1992) Dynamics of photosystem II: mechanism of photoinhibition and recovery processes. In: Barber J (ed) The photosystems: structure, function and molecular biology. Elsevier Science, Amsterdam, pp 295–348

Punginelli C, Wilson A, Routaboul J-M, Kirilovsky D (2009) Influence of zeaxanthin and echinenone binding on the activity of the orange carotenoid protein. Biochim Biophys Acta 1787(4): 280–288

Rakhimberdieva MG, Boichenko VA, Karapetyan NV, Stadnichuk IN (2001) Interaction of phycobilisomes with photosystem II dimers and photosystem I monomers and trimers in the cyanobacterium *Spirulina platensis*. Biochemistry 40: 15780–15788

Rakhimberdieva MG, Stadnichuk IN, Elanskaya IV, Karapetyan NV (2004) Carotenoid-induced quenching of the phycobilisome fluorescence in photosystem II-deficient mutant of *Synechocystis* sp. FEBS Lett 574: 85–88

Rakhimberdieva MG, Bolychevtseva YV, Elanskaya IV, Karapetyan NV (2007a) Protein-protein interactions in carotenoid triggered quenching of phycobilisome fluorescence in *Synechocystis* sp. PCC 6803. FEBS Lett 581: 2429–2433

Rakhimberdieva MG, Vavilin DV, Vermaas WF, Elanskaya IV, Karapetyan NV (2007b) Phycobilin/chlorophyll excitation equilibration upon carotenoid-induced non-photochemical fluorescence quenching in phycobilisomes of the cyanobacterium *Synechocystis* sp. PCC 6803. Biochim Biophys Acta 1767: 757–765

Redlinger T, Gantt E (1982) A M(r) 95,000 polypeptide in *Porphyridium cruentum* phycobilisomes and thylakoids: possible function in linkage of phycobilisomes to thylakoids and in energy transfer. Proc Natl Acad Sci U S A 79: 5542–5546

Ruban AV, Berera R, Ilioaia C, van Stokkum IHM, Kennis JTM, Pascal AA, van Amerongen H, Robert B, Horton P, van Grondelle R (2007) Identification of a mechanism of photoprotective energy dissipation in higher plants. Nature 450: U575–U522

Rubinstenn G, Vuister GW, Mulder FA, Dux PE, Boelens R, Hellingwerf KJ, Kaptein R (1998) Structural and dynamic changes of photoactive yellow protein during its photocycle in solution. Nat Struct Biol 5: 568–570

Schneider TD, Stephens RM (1990) Sequence logos: a new way to display consensus sequences. Nucleic Acids Res 18: 6097–6100

Scott M, McCollum C, Vasil'ev S, Crozier C, Espie GS, Krol M, Huner NP, Bruce D (2006) Mechanism of the down regulation of photosynthesis by blue light in the Cyanobacterium *Synechocystis* sp. PCC 6803. Biochemistry 45: 8952–8958

Singh AK, Thanura E, Bhattacharyya-Pakrasi M, Aurora R, Ghosh B, Pakrasi HB (2008) Intergration of carbon and nitrogen metabolism with energy production is crucial to light acclimation in the cyanobacterium Synechocystis. Plant Physiol 148: 467–478

Spudich JL, Yang CS, Jung KH, Spudich EN (2000) Retinylidene proteins: structures and functions from archaea to humans. Annu Rev Cell Dev Biol 16: 365–392

Stockel J, Welsh EA, Liberton M, Kunnvakkam R, Aurora R, Pakrasi HB (2008) Global transcriptomic analysis of Cyanothece 51142 reveals robust diurnal oscillation of central metabolic processes. Proc Natl Acad Sci U S A 105: 6156–6161

Takahashi F, Yamagata D, Ishikawa M, Fukamatsu Y, Ogura Y, Kasahara M, Kiyosue T, Kikuyama M, Wada M, Kataoka H (2007) AUREOCHROME, a photoreceptor required for photomorphogenesis in stramenopiles. Proc Natl Acad Sci U S A 104: 19625–19630

Tandeau de Marsac N (2003) Phycobiliproteins and phycobilisomes: the early observations. Photosynth Res 76: 197–205

Tyystjärvi E (2008) Photoinhibition of photosystem II and photodamage of the oxygen evolving manganese cluster. Coord Chem Rev 252: 361–376

van Thor JJ, Mullineaux CW, Matthijs HCP, Hellingwert KJ (1998) Light harvesting and state transitions in cyanobacteria. Bot Acta 111: 430–443

Vass I, Aro EM (2008) Photoinhibition of Photosystem II electron transport. In: Renger G (ed) Primary processes of photosynthesis: basic principles and apparatus. Royal Society of Chemistry, Cambridge, pp 393–411

Wallace AC, Laskowski RA, Thornton JM (1995) LIGPLOT: a program to generate schematic diagrams of protein-ligand interactions. Protein Eng 8: 127–134

Williams WP, Allen JF (1987) State-1/State-2 changes in higher-plants and algae. Photosynth Res 13: 19–45

Wilson A, Ajlani G, Verbavatz JM, Vass I, Kerfeld CA, Kirilovsky D (2006) A soluble carotenoid protein involved in phycobilisome-related energy dissipation in cyanobacteria. Plant Cell 18: 992–1007

Wilson A, Boulay C, Wilde A, Kerfeld CA, Kirilovsky D (2007) Light-induced energy dissipation in iron-starved cyanobacteria: roles of OCP and IsiA proteins. Plant Cell 19: 656–672

Wilson A, Kinney JN, Zwart PH, Punginelli C, D'Haene S, Perreau F, Klein MG, Kirilovsky D, Kerfeld CA (2010) Structural determinants underlying photoprotection in the photoactive orange carotenoid protein of cyanobacteria. J Biol Chem 285:18364–18375

Wilson A, Punginelli C, Gall A, Bonetti C, Alexandre M, Routaboul JM, Kerfeld CA, van Grondelle R, Robert B, Kennis JT, Kirilovsky D (2008) A photoactive carotenoid protein acting as light intensity sensor. Proc Natl Acad Sci U S A 105: 12075–12080

Wollman FA (2001) State transitions reveal the dynamics and flexibility of the photosynthetic apparatus. EMBO J 20: 3623–3630

Wu YP, Krogmann DW (1997) The orange carotenoid protein of *Synechocystis* PCC 6803. Biochim Biophys Acta 1322: 1–7

Chapter 15
Mn Transport and the Assembly of Photosystem II

Eitan Salomon, Gernot Renger and Nir Keren

15.1 Transition Metals in Photosynthetic Cells

Transition metals play an important role in many biological processes (for reviews, see Kaim and Schwederski 1994). In the photosynthetic apparatus, transition metals are required as cofactors in electron transport processes (Merchant and Sawaya 2005). Among these are Fe containing cofactors such as iron-sulfor clusters, cytochromes and non-heme iron (see Chaps. 12 and 13), Cu in plastocyanin (see Chap. 21) and Mn in Photosystem II (see Chaps. 12 and 13). As a result, the demand for some of these metals in cyanobacteria and plants far exceeds that of other organisms. A most illustrative example is the evolutionary transition from non-oxygenic to oxygenic photosynthesis which resulted in a ~100 times higher internal Mn quota in the oxygen evolving cyanobacterium *Synechocystis sp*. PCC 6803 compared to that of the photosynthetic bacterium *Rhodobacter capsulatus* which is unable to evolve oxygen (Keren et al. 2002).

The "invention" of oxygenic photosynthesis was the key step for the evolution of the whole biosphere because it provided the indispensable prerequisites for the development and sustenance of all higher forms of life on earth (Des Marais 2000). Nevertheless, oxygenic photosynthesis generates severe risks for living matter due to the production of molecular dioxygen with its peculiar properties (for further reading, see articles in Gilbert 1981). Unique problems emerge from the necessity to transport, accumulate and assemble transition metal ions into functional cofactors because these cations catalyze the formation of the extremely reactive hydroxyl radicals ($OH^{\bullet}$) from superoxide and/or hydrogen peroxide via the Haber-Weiss reaction (Haber and Weiss 1934). Reactive oxygen species (ROS) can be formed at a number of junctions along the course of the photosynthetic electron transport chain.

N. Keren (✉)
Department of Plant and Environmental Sciences, The Silberman Institute of Life Sciences, The Hebrew University of Jerusalem, Edmond Safra Campus, Givat-Ram, Israel
e-mail: nirkeren@vms.huji.ac.il

G. A. Peschek et al. (eds.), *Bioenergetic Processes of Cyanobacteria,*

DOI 10.1007/978-94-007-0388-9_15,

Potential candidates for ROS generation are (1) the sensitized reaction of chlorophyll triplets with molecular dioxygen in its triplet ground state ($^3\sum_g O_2$) leading to its reactive singlet state ($^1\Delta_g O_2$), and (2) redox reactions on the donor (and acceptor) side of PS II that give rise to formation of superoxide and/or hydrogen peroxide (reviewed in Shcolnick and Keren 2006). In order to overcome inherent dangers from radical chain reactions between transition metal cations and intermediates in photosynthetic catalysis, the handling of these metals is tightly regulated (Pittman 2005; Shcolnick and Keren 2006).

15.2 Mn Availability in the Environment

Manganese is a common element on earth's surface (Kaim and Schwederski 1994). In the photosynthetic apparatus Mn plays a unique role because it is the essential constituent of the catalytic Mn_4O_xCa cluster in the water oxidizing complex (WOC). This element cannot be replaced by any other transient metal, in marked contrast to the redox inert Ca^{2+} which can be substituted by Sr^{2+} (for a review, see Van Gorkom and Yocum 2005). The evolutionary pathway leading to the "construction" of the WOC is not yet resolved. Current hypotheses assume that suitable bicarbonate-manganese adducts were used as templates for the evolution of the catalytic site of the WOC (Dismukes et al. 2001). Details on the assembly of the unique Mn_4O_xCa cluster will be discussed in Sect. 15.3.

15.2.1 Mn Transport in Cyanobacteria

While Mn is not considered as a limiting factor for primary productivity like Fe, its concentration is nonetheless in the low nM range in open oceans (Morel 2008) and in fresh water bodies (Sterner et al. 2004). For comparison, the Mn concentration in standard growth media like BG11 (Allen 1968) or A+ (Stevens and Porter 1980) is in the micromolar range. In the analysis of Mn transport and homeostasis in cyanobacteria we should consider two conditions; the Mn deficient condition which prevails in open water bodies and the Mn sufficient condition existing for brief periods following wind-driven dust depositions (Guieu et al. 1994; Baker et al. 2006) and in culture media.

15.2.1.1 Mn Transport Under Deficient Conditions: The MntABC Transporter

Metal ions are charged species that cannot freely cross hydrophobic membrane cores for energetic reasons. Therefore, the transport of transition metals through biological membranes requires active transport. In cyanobacteria, only one Mn transporter is known, MntABC (Bartsevich and Pakrasi 1995). Interestingly enough, this transporter was identified by a screen for photosynthetic mutants that took advantage of the glucose sensitive strain of *Synechocystis sp.* PCC 6803 (Williams 1988). In this strain, only mutants with impaired photosynthetic abilities can survive on plates containing glucose in the light. This method has yielded a number of interesting mutants, including one that was mapped to the *mntCAB* operon encoding for the 3 subunits of the MntABC transporter (Bartsevich and Pakrasi 1995). The mutant exhibited low oxygen evolution rates and could be rescued by the addition of excess Mn. MntABC was shown to be a member of the ATP requiring ABC transporter family (Linton 2007). Bacterial ABC type transporters contain three protein components: a solute (ion) binding protein in the periplasm (or linked to the inner plasma membrane), a transmembrane permease and an ATP binding protein (Linton 2007). The function of MntABC mutants was further studied by ^{54}Mn transport assays (Bartsevich and Pakrasi 1996). A difference in the transport rate between wild type and mutant cultures was observed only under Mn deficient conditions, leading to the suggestion that MntABC is a high affinity transporter which is expressed under conditions of limited Mn supply.

The structure of the periplasmic ion binding protein, denoted MntC, of the Mn transporter MntABC, was resolved by X-ray crystallography, which discovered a surprising feature (Rukhman et al. 2005). The unit cell of the crystal contained two forms of the protein; one with a disulfide bond and one in which these cysteines are reduced (Fig. 15.1). The existence of a redox active cysteine pair in a periplasmic solute binding protein is unexpected. Rukhman et al. (2005) suggested a regulatory role for this feature, changing the protein from an active into an inactive state depending on the redox poise of the periplasm. Support for the existence of such a

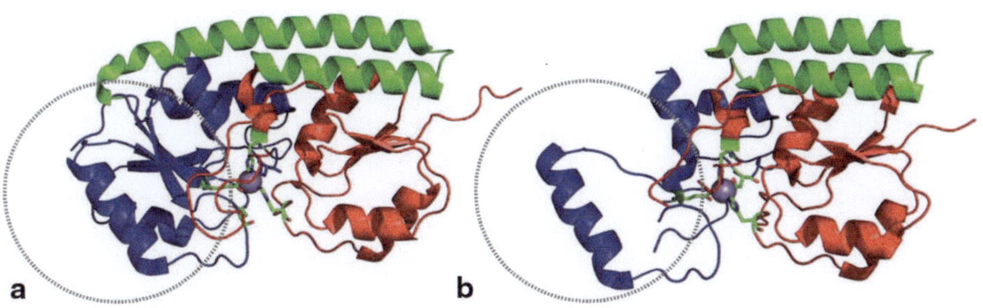

Fig. 15.1 The two redox dependant configuration of the MntC protein. **a** MntC monomer A, the oxidized disulfide bond conformation. A disulfide bond is found between Cys219 and Cys268. This enables the correct folding pattern of the metal binding site resulting in a high Mn^{2+} affinity. **b** MntC monomer C, the reduced disulfide bond conformation. The disulfide bond is broken, which lead to conformational changes in the metal binding site and results in a lower Mn binding capacity. The variable region is marked by a dashed circle. (Courtesy of Prof. Noam Adir, data redrawn from Rhukhman et al. 2005)

mechanism comes from a recent work by Singh and coworkers demonstrating that a plasma membrane protein is involved in disulfide bond formation in the extracytoplasmic space of cyanobacteria (Singh et al. 2008).

15.2.1.2 Mn Transport Under Sufficient Conditions: Mn Accumulation in an Outer Membrane Bound Pool

Measurements of ^{54}Mn transport rates clearly demonstrated that, in addition to MntABC, other transport systems function in *Synechocystis sp.* PCC6803 (Bartsevich and Pakrasi 1996). While Mn transport was severally inhibited in $\Delta mntC$ mutants under Mn deficient conditions, no inhibition could be measured under Mn sufficient conditions. Furthermore, the rate of Mn uptake exhibited a bi-phasic dependence on Mn concentration, with increasing transport rates in the millimolar range, both in wild type and in $\Delta mntC$ mutant cells. V_{max} values could not be reached even in the presence of 2 mM Mn (Bartsevich and Pakrasi 1996).

Further analysis of the uptake under Mn sufficient conditions showed that *Synechocystis sp.* PCC 6803 cells, in the early log phase of growth, have the ability to bind large amounts of Mn from fresh BG11 media (Keren et al. 2002). Fractionation of these cells indicated that most of the bound Mn is associated with the outer membrane (Keren et al. 2002). The face of the membrane to which Mn is bound remains to be determined. X-ray absorption spectroscopy studies confirmed that this pool contains Mn^{2+} and extended X-ray absorption fine structure (EXAFS) studies revealed that Mn interacts with the membrane. However, the nature of the binding sites could not be resolved by using this technique (Keren et al. 2002). Recent work from the Robinson group suggested that a protein from the cupin family, MncA, is the major Mn binding protein in the periplasm of *Synechocystis sp.* PCC 6803 cells (Tottey et al. 2008). The relation of this protein to Mn storage has not been determined.

Release of Mn from the outer membrane pool could be achieved by incubation with EDTA at concentrations higher than 2 mM. However, high EDTA concentrations can poke holes in the outer membrane of Gram-negative bacteria in addition to chelating released transition metal ions (Vaara 1992). Using this method it was possible to establish that an excess of $\sim 1 \times 10^7$ atoms/cell can be stored in the pool bound to the outer membrane (OM) of *Synechoccystis sp.* PCC 6803 cells in the early log phase of growth, i.e. an amount of approximately 10 times higher than the Mn concentration inside the plasma membrane (PM). Along the progression of growth in the culture, Mn from the outer membrane pool is gradually utilized, keeping the internal Mn concentration constant (Keren et al. 2002). This mode of transport is not affected by inactivation of MntABC transporter (Keren et al. 2002). The maintenance of a Mn concentration difference of such a large extent requires the supply of Gibbs free energy. Indeed, accumulation of Mn in this outer membrane pool was found to be coupled to photosynthetic activity (Keren et al. 2002). The mechanism by which photosynthesis in the thylakoid membrane supports Mn accumulation on the outer membrane is still unknown.

15.2.1.3 Mn Sensing and Regulation of Transport

In order to regulate different Mn transport pathways, cyanobacteria require a Mn sensor. Ogawa and coworkers and Yamaguchi and coworkers have independently identified the components of such a system. A mutant of the ManS sensory kinase was identified in a screen that utilized the *mntCAB* promoter linked to a reporter gene (Ogawa et al. 2002). In addition to ManS the response regulator ManR has been discovered by Microarray analysis of mutants in genes coding for two-component regulators (Yamaguchi et al. 2002). The ManS/R system represses the expression of the *mntCAB* operon and its inactivation results in constitutive transcription of its mRNA. Interestingly, inactivation of either *manS* or *manR* did not significantly affect the transcription of any other genes apart from those of the *mntABC* operon (Yamaguchi et al. 2002).

The ManS protein is embedded into the plasma membrane, like many other two-component sensors. The sensor domain of the protein is exposed to the periplasm and the kinase domain to the cytoplasm (Ogawa et al. 2002; Yamaguchi et al. 2002). The topology of the protein should have an effect on its mode of sensing. Lowering the external Mn concentration will result in the depletion of the outer membrane bound pool, which will activate the ManS/R system in advance of an internal Mn deficiency.

15.3 Mn and PSII Function

PSII is one of only a few enzymes that utilize Mn in cyanobacterial cells. It was estimated that 10–50% of the cellular Mn quota is bound to PSII complexes (Keren et al. 2002). The link between Mn availability and PSII function in cyanobacteria was originally observed by Cheniae and Martin (Cheniae and Martin 1967). Removal of Mn from the medium of *Anacystis nidulans* (now designated *Synechococcus elongatus* PCC 7942) results in inactivation of oxygen evolution, which can be reversed by addition of Mn. The close correlation between Mn availability and PSII function is clear but the pathway of Mn cluster assembly is still a matter of debate. Three major questions regarding Mn cluster assembly will be discussed in this review: the order of events in the assembly of the PSII protein complex, the photoactivation sequence and the topology of the assembly process.

15.3.1 Assembly of the Mn Cluster: Structural and Functional Requirements

The donor side of PSII is localized near the lumenal side of the thylakoid membrane. The catalytic component of this site is the Mn_4O_xCa- cluster (Yachandra

2005; Barber 2008; Messinger and Renger 2008). Structural studies indicate that this cluster is mainly coordinated by residues of the D1 protein and by one residue (Arg 357) of the CP43 protein (Kamiya and Shen 2003; Ferreira et al. 2004; Loll et al. 2005), which are integral subunits of the PSII complex. The Mn_4O_xCa cluster is secluded from the aqueous environment of the thylakoid lumen by several extrinsic proteins. In cyanobacteria these proteins were found to be PsbO, PsbU, PsbV (cytochrome c_{550}), PsbQ and PsbP (Thornton et al. 2004). Only PsbO/U/V were resolved in the X-ray crystallographic structure of PS II core complexes from thermophilic cyanobacteria (Kamiya and Shen 2003; Ferreira et al. 2004; Loll et al. 2005). It is important to note that the *psbO*, *psbV* and *psbU* genes are found in all fully sequenced cyanobacterial genomes, while *psbP* and *psbQ* are found only in some (Thornton et al. 2004). The distribution and evolution of the extrinsic proteins among cyanobacterial species is described in a recent review (Roose et al. 2007).

A more complex picture arises from functional studies of deletion mutants in genes coding for PSII subunits. Deletions of D1, D2, cytochrome b_{559}, CP43 or CP47 results in complete loss of PSII water splitting activity. In the case of cytochrome b_{559} (Pakrasi et al. 1989; Shukla et al. 1992), D1 (Jansson et al. 1987) and D2 (Yu and Vermaas 1990) deletions, assembled PSII complexes could not be detected. In the case of CP43 and CP47 deletion mutants a core complex of PSII did accumulate but was unable to oxidize water and grow photoautotrophically (Vermaas et al. 1986; Rogner et al. 1991). Deletion of the extrinsic subunits (PsbO/U/V/P/Q) affects the function of PSII complexes, however these mutants were all capable of a limited extent of oxygen evolution and photoautotrophic growth (Vermaas et al. 1986; Burnap and Sherman 1991; Rogner et al. 1991; Burnap et al. 1992; Shen et al. 1995a, b, 1997; Thornton et al. 2004; Inoue-Kashino et al. 2005; Summerfield et al. 2005, 2007; Balint et al. 2006; Kashino et al. 2006). It is interesting to mention that deletion of *psbO* leads to more severe effects in eukaryotic plants (Mayfield et al. 1987).

15.3.2 Assembly of the Mn Cluster: Donor Side Protein Association

The assembly of the Mn_4O_xCa cluster occurs for the first time during *de novo* synthesis of PSII complexes and then repeatedly throughout the photodamage/repair cycle (reviewed in Adir et al. 2003; Aro et al. 2005). Considering the function and the unique properties of this cluster, it is obvious to expect that the assembly process will be tightly regulated and that stop-gap measures will be put into place to avoid its untimely activation. So far two control mechanisms have been identified.

The D1 subunit of PSII protein is translated as a precursor containing a C-terminal extension. Cleavage of the C' terminus, a process that is catalyzed by the CtpA protease, trims the D1 to the correct size and position for formation and ligation of the Mn_4O_xCa cluster (Diner et al. 1988; Anbudurai et al. 1994; Roose and Pakrasi 2004). While the C- terminal extension can be found in the vast majority of photosynthetic organisms, replacement of pD1 by a mature D1 does not result in any observable problems in the assembly process (Ivleva et al. 2000). Nevertheless, in mixed culture competition experiments between *Synechocystis sp.* PCC6803 strains containing either pD1 or mature D1, cells containing pD1 prevailed (Ivleva et al. 2000).

An additional protein, Psb27, was found to be attached to PSII centers lacking Mn and extrinsic proteins (Nowaczyk et al. 2006; Roose and Pakrasi 2008). These complexes were found to lack oxygen evolution capacity and to exhibit impaired forward electron transfer from $Q_A^{-\bullet}$ to Q_B. (Mamedov et al. 2007). Based on these studies it was suggested that transient formation of Psb27-PSII assemblies play a role in the repair cycle of PSII (Nowaczyk et al. 2006; Roose and Pakrasi 2008).

The studies performed so far suggest that the order of events for the Mn cluster assembly is as follows: the pD1 protein is inserted into a partially assembled PSII core complex. After cleavage of its C′ terminus by CtpA, the mature D1 protein can attach the Mn cluster. It is not clear whether the cluster is assembled sequentially from the ligation of individual Mn atoms, or whether a part or the entire complex is assembled and then transferred to PSII by a Mn chaperon. Interestingly, *in vitro* experiments revealed that synthetic binuclear manganese complexes were more efficient in photo-activation of the oxygen evolution activity compared to Mn^{2+} ions (Allakhverdiev et al. 1994). Psb27 prevents Mn attachment by associating with the lumenal side of PSII, blocking the docking sites of the extrinsic proteins (Nowaczyk et al. 2006; Roose and Pakrasi 2008). This prevents immature binding and interference with the sequence of the photosystem assembly process. Following the completion of the D1 processing step, Psb27 is detached (Roose et al. 2007; Roose and Pakrasi 2008) and the extrinsic regulatory proteins of the WOC can bind on the luminal side of the PS II complex.

A prediction for the order of binding of the extrinsic proteins can be derived from biochemical and structural studies. The extrinsic protein, PsbO, is attached to the lumenal side of CP47 (Kamiya and Shen 2003; Ferreira et al. 2004; Loll et al. 2005), followed by PsbV, which binds to CP43 (Shen and Inoue 1993; Kamiya and Shen 2003; Ferreira et al. 2004; Loll et al. 2005) and PsbU, which interacts with CP43, CP47, PsbO and PsbV (Eaton-Rye 2005). In *Synechcocystis sp.*PCC 6803 the assembly process of extrinsic proteins culminates with the binding of PsbQ. PSII complexes containing PsbQ were more active and stable as compared with PSII units lacking this protein. (Kashino et al. 2006; Roose et al. 2007). PsbP was found to be associated only with a fraction of PSII complexes and is not assumed to be a functional component of the extrinsic proteins, but rather to have a regulatory role (Thornton et al. 2004).

The study of the order of events leading to the assembly of a functional Mn complex is hampered by the lack of kinetic data. Mutational studies, commonly used for establishing the function of unknown proteins, provide only static results.

Establishing the sequence of events requires the ability to track the individual steps. Recent advances in quantitative peptide mass spectrometry are expected to solve this problem. Using these techniques, Nowachick and coworkers (Nowaczyk et al. 2006) were able to track incorporation of ^{15}N tagged D1 into PSII complexes at different maturation states, ranging from photosynthetically inactive monomers to fully operational dimers.

15.3.3 Assembly of the Mn Cluster: Photoactivation

Cheniae and co-workers showed that the oxygen evolution capacity of Mn depleted *Synechococcus elongatus sp.* PCC 7942 cells can be restored by addition of $MnCl_2$ to the suspending medium when the cells are illuminated (Cheniae and Martin 1967). The overall process comprises the uptake of Mn in the dark (this process is stimulated by illumination) and a subsequent reaction which is strictly dependent on light. Therefore this process is designated **photoactivation**. Numbers of about 6 Mn on the oxidizing side of each PS II were gathered from the data on photoactivation (Cheniae and Martin 1969). Later studies revealed that the center of the WOC is a Mn_4O_xCa cluster (for reviews, see Yachandra 2005; Messinger and Renger 2008 and references therein).

In enzymology, numerous examples are known for metalloenzymes in which the catalytic metal center can be reversibly removed from the apoprotein and reinserted or even replaced by another metal (see textbooks on metalloenzymes, e.g. Kaim and Schwederski 1994). In marked contrast to these systems, however, the formation of the Mn-containing functionally-competent WOC through ligation of manganese (and Ca^{2+}, *vide infra*) is an endergonic process that can be only achieved via the light-driven reaction sequence of photoactivation.

This phenomenon of photoactivation is not restricted to cyanobacteria but is a general intrinsic feature of all oxygen evolving organisms as was shown for algae and higher plants. The WOC formation in the green alga *Chlorella pyrenoidosa* grown in the dark which is not able to evolve oxygen must be preceded by the formation of PS II complexes that are capable of stable light induced charge separation. The assembly of the WOC *per se* is not coupled with a *de novo* protein synthesis (Cheniae and Martin 1973). Likewise, bean leaves grown under special conditions of illumination with short light flashes separated by long dark times were shown to assemble a PS II complex which is able to perform light-induced charge separation, but lacking the capability of oxygen evolution. A subsequent illumination with continuous light was required to establish a functionally competent WOC (Strasser and Sironval 1972).

More detailed information on the mechanism of the assembly of the Mn_4O_xCa cluster was obtained by studies on manganese-deprived samples. Several treatments (incubation in suspensions of high pH or with solutions containing compounds like Tris- or hydroxylamine at sufficiently high concentrations) lead to a release of manganese from the sample without severe destruction of the apoprotein matrix (Cheniae and Martin 1973; for review see Debus 1992). The overall process of photoactivation in hydroxylamine treated PS II membrane fragments from spinach was shown to comprise two light reactions separated by a stabilization step taking place in the dark within the order of 100 ms (see Tamura and Cheniae 1987 and references therein). It was found that the reconstitution of a functionally competent WOC is absolutely specific for Mn, which cannot be replaced by any other transition metal ion. On the other hand, the redox inert Ca^{2+} can be replaced by Sr^{2+} as a surrogate (Ba^{2+} and Mg^{2+} are much less efficient, see Tamura et al. 1989). Under standard conditions, the O_2-evolving activity of WOCs containing a Mn_4O_xSr cluster is about 50% of that with a Mn_4O_xCa cluster (Ghanotakis et al. 1984; Westphal et al. 2000; Boussac et al. 2004; Cinco et al. 2004; Strickler et al. 2005).

Figure 15.2 summarizes our current knowledge on the reaction pattern of the photoassembly of the Mn_4O_xCa cluster in manganese deprived samples. After occupation of the high affinity binding site in the apo-WOC by a Mn^{2+} (reaction state A) this cation becomes oxidized to Mn^{3+} in the first light induced step (reaction A $\rightarrow$ B) where $P680^{+\bullet}$ acts as oxidant and Y_Z as intermediary redox carrier (Tamura and Cheniae 1986; Hoganson et al. 1989). The kinetics of this reaction is accelerated in the presence of bicarbonate (Baranov et al. 2004). It was postulated that a ternary complex is formed between Mn^{2+}, (bi)carbonate and the apo-WOC which is thermodynamically easier to oxidize than without the (bi)carbonate (Kozlov et al. 2004) which was shown to coordinate with the tightly bound Mn^{3+} (Dasgupta et al. 2007). In the absence of (bi)carbonate a water molecule is assumed to coordinate this site in the proposed ternary complex. Regardless of the presence of (bi)carbonate the formation of state C requires the indispensable redox inert cation Ca^{2+} (or Sr^{2+}). The dark reaction B $\rightarrow$ C comprises a conformational

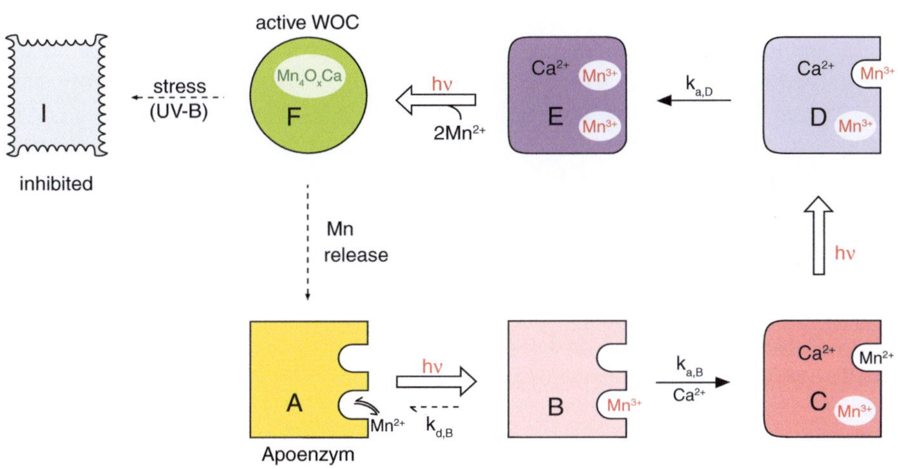

Fig. 15.2 Proposed model for photoactivation

change and gives rise to an unstable intermediate which completely disappears within the time domain of a few seconds in the dark when the second light induced reaction into state D is missing. (Tamura and Cheniae 1987). State C is probably a μ-oxo bridged complex $[Mn^{3+}(O^{2-})Ca^{2+}]$ (Dasgupta et al. 2010). The subsequent intermediates (D and E) of Mn incorporation and the coupled oxidation steps are not yet fully characterized. It must be emphasized that (bi)carbonate is not a constituent of the first coordination sphere of the mature Mn_4O_xCa cluster (Shevela et al. 2008; Ulas et al. 2008). Recent structure analyses are in line with this finding (Loll et al. 2005).

The different redox states of the Mn_4O_xCa cluster[1] in the WOC can be specifically manipulated by several exogenous compounds. Among these agents, hydrophilic exogenous reductants give rise to the population of "reduced" S_i states (S_{-i} states). So far, a stable population could be achieved for states S_{-1}, S_{-2} and S_{-3} (see Messinger et al. 1997). Indirect evidence for a possible existence of redox states S_{-4} and S_{-5} has also been presented (Messinger et al. 2001). It was speculated that the reduced states are transients in the photoactivation process (Renger 1997; Messinger et al. 2001; Ono 2001; Hillier and Messinger 2005). This idea is supported by the finding that the states $S_{-3} - S_{-1}$ are populated *in vivo* (Higuchi et al. 2003; Quigg et al. 2003). As an extension of this postulate it appears attractive to assign the reduced states S_{-5} and S_{-4} to the intermediates A (one Mn^{2+} bound at the high affinity site) and B/C (one Mn^{3+} bound at the high affinity site) of the scheme in Fig. 15.2 because the stable redox state of the WOC in the dark is S_1 which was shown to attain the electronic configuration $Mn_2(III)Mn_2(IV)$ (for reviews, see Yachandra 2005; Messinger and Renger 2008 and references therein). Accordingly, S_{-5} corresponds with a cluster where all four Mn are reduced to Mn(II) and in redox state S_{-4} one Mn is oxidized to Mn(III). The low quantum yield of the $A \rightarrow B$ and $C \rightarrow D$ transitions would explain why the apparent population of S_{-5} and S_{-4} is rather small and hardly resolvable in the patterns of flash-induced oxygen yield detected in NH_2OH treated thylakoids (Messinger et al. 2001).

In conclusion, the photoassembly of the WOC implies that the Mn_4O_xCa cluster is a non-equilibrium state, which needs to be stabilized by the protein environment. This notion is supported by the finding that the WOC is the primary target of UV-B light- (Renger et al. 1989) and temperature- (Thompson et al. 1989; Enami et al. 1994) induced destruction of PS II. It has been suggested that the WOC is also prone to destruction by visible light thus leading to photoinhibition (Ohnishi et al. 2005; Tyystjarvi 2008) but this idea is a matter of discussion (see Vass and Czer 2009 and references therein).

[1] It has to be kept in mind that the actual charge of the manganese centers in the redox states Mn(II), Mn(III) and Mn(IV) are not intgers (+2, +3 and +4. respectively). In fact density functional theory (DFT) calculations on synthetic mixed-valent oxomanganese complexes (Sproviero et al. 2006a) and a charge population analysis reveals that actually Ca^{2+} carries the highest positive charge within the Mn_4O_XCa cluster although manganese attains the redox states III and IV (Sproviero et al. 2006b).

15.3.4 Assembly of the Mn Cluster: Topological Constraints

The structure of the cyanobacterial cell is more complex than that of other prokaryotic cells. On top of a periplasmatic and intracellular space, defined by the outer (OM) and plasma membrane (PM), which are common to all Gram negative bacteria, cyanobacteria contain a third lumenal space enclosed by the thylakoid membrane (TM). The thylakoid membranes take up a large fraction of the intracellular space, creating a maze that hampers the transport processes inside the cell (Liberton et al. 2006; Nevo et al. 2007). From an evolutionary standpoint, thylakoid membranes are considered to be the descendents of the photosynthetic membranes of purple bacteria (Cogdell et al. 2006). However, while in purple bacteria the photosynthetic membranes are connected to the plasma membrane (Oelze and Drews 1972), in most cyanobacterial species there is no clear evidence for such a connection (Liberton et al. 2006; Nevo et al. 2007). Nevertheless, in *Gloeobacter violaceus* PCC 7421, which is considered as a member of an early branching cyanobacterial linage, TM is a direct invagination from the PM (Rippka et al. 1974).

While it is clear that PSII functions in the thylakoid membranes, there are conflicting reports on the site of its biogenesis. Smith and Howe (1993) were able to detect chlorophyll and D1 proteins in PM membranes of *Synechococcus sp.* PCC 7942. In this study, membranes were isolated by sucrose density gradient centrifugation. Using a two-phase separation technique (Albertsson 1970), Zak and coworkers were able to detect D1, Cytochrome b_{559} and CtpA, but not CP43, CP47 or PsbO in PM of *Synechocystis sp.* PCC 6803 (Zak et al. 2001). The PSII proteins present in the PM preparation formed a complex that migrated as one band in native gels. Based on these data Zak et al. suggested that PSII biogenesis occurs in the PM rather than in TM. This hypothesis represents a departure from the accepted view of the location of PSII assembly in the chloroplast thylakoids of eukaryotic organisms (Aro et al. 2005). Keren et al. (2005) demonstrated, by using the same preparation technique, that a limited extent of charge separation can be measured in partially assembled PSII complexes localized in PM. On a chlorophyll basis the concentration of Mn attached to PM was found to be ~20% of that in TM (Keren et al. 2005). The lack of stable charge separation and the lower Mn content in PM is not surprising considering the role of the CP47 and CP43 extra-membranal loops in stabilizing the oxygen evolving complex (Gleiter et al. 1995). Jansen et al. (2002) combined the two techniques, performing sucrose gradient density centrifugation followed by two-phase separation and were still able to detect D1, D2 and cytochrome b_{559} in PM preparations of *Synechocystis* sp. PCC 6803. In a recent study from the same group (Srivastava et al. 2006), right-side out and inside out vesicles of PM were isolated from *Synechocystis sp.* PCC 6803. The additional fractionation step revealed heterogeneity with respect to protein distribution within the PM: D1 and pD1 were enriched in inside out as compared to right-side out vesicles on a total protein basis.

While the results of the studies mentioned above suggest that the PSII assembly pathway is in the PM, there are a number of studies that did not detect chlorophyll in PM. These reports include work on membranes isolated by sucrose gradient centrifugation from *Synechococcus elongatus* 7942 (Murata et al. 1981), *Synechocystis* 6714 (Omata and Murata 1984; Jurgens and Weckesser 1985), or by hyperspectral confocal fluorescence imaging of *Synechocystis* 6803 cells (Vermaas et al. 2008). Peschek et al. (1989), analyzing membrane fractions isolated from *Synechococcus sp.* PCC7942 cells, could detect precursors of the chlorophyll biosynthesis pathway but very little or no chlorophyll in PM.

15.4 The Relations Between Mn Transport and PSII Biogenesis

Regardless of the mechanistic details, the first step of Mn uptake by cyanobacteria is the transport through the outer membrane (OM). The details of this transport step are still unknown. Beyond that point, when taking into consideration the different routes suggested for Mn transport under deficient and sufficient conditions, and the conflicting views on the location of PSII assembly events, four different possible scenarios can be discussed for the assembly of the Mn cluster as is schematically illustrated in Fig. 15.3.

1. **Mn sufficient conditions, PSII assembly in the Plasma membrane**
   Under sufficient conditions there will be plentiful Mn in the periplasmic space. After assembly of the Mn complex, PSII needs to be transferred to the thylakoid membrane where the process will be completed by the addition of CP43 and CP47. Currently, there is no clear evidence either for a direct contact or for vesicle transport between the two membrane systems.
2. **Mn sufficient conditions, PSII assembled in the Thylakoid membrane**
   According to this scenario Mn is transported through the OM, PM and finally through the TM to reach the site of assembly. The identity of the participant(s) in all three transport steps is still unknown.
3. **Mn deficient conditions, PSII assembly in the plasma membrane**
   Under Mn deficient conditions the MntABC transporter is expressed. If the TM is a topological extension of PM (as in the case of *Gloeobacter violaceus PCC 7421*) the MntABC will transport Mn away from the site of precursor complex or Mn cluster assembly in the PM. Under these conditions the redox active cysteine pair of the MntC subunit (Fig. 15.1) can have a role in controlling the allocation of the limited Mn for PSII assembly in the periplasm or for the assembly of enzymes such as Mn-SOD in the cytoplasmatic space.
4. **Mn deficient conditions, PSII assembly in the thylakoid membrane**
   According to this scenario, MntABC function supports the assembly of Mn enzymes in the cytoplasm directly and, following transport through the TM, the assembly of the Mn_4O_xCa cluster.

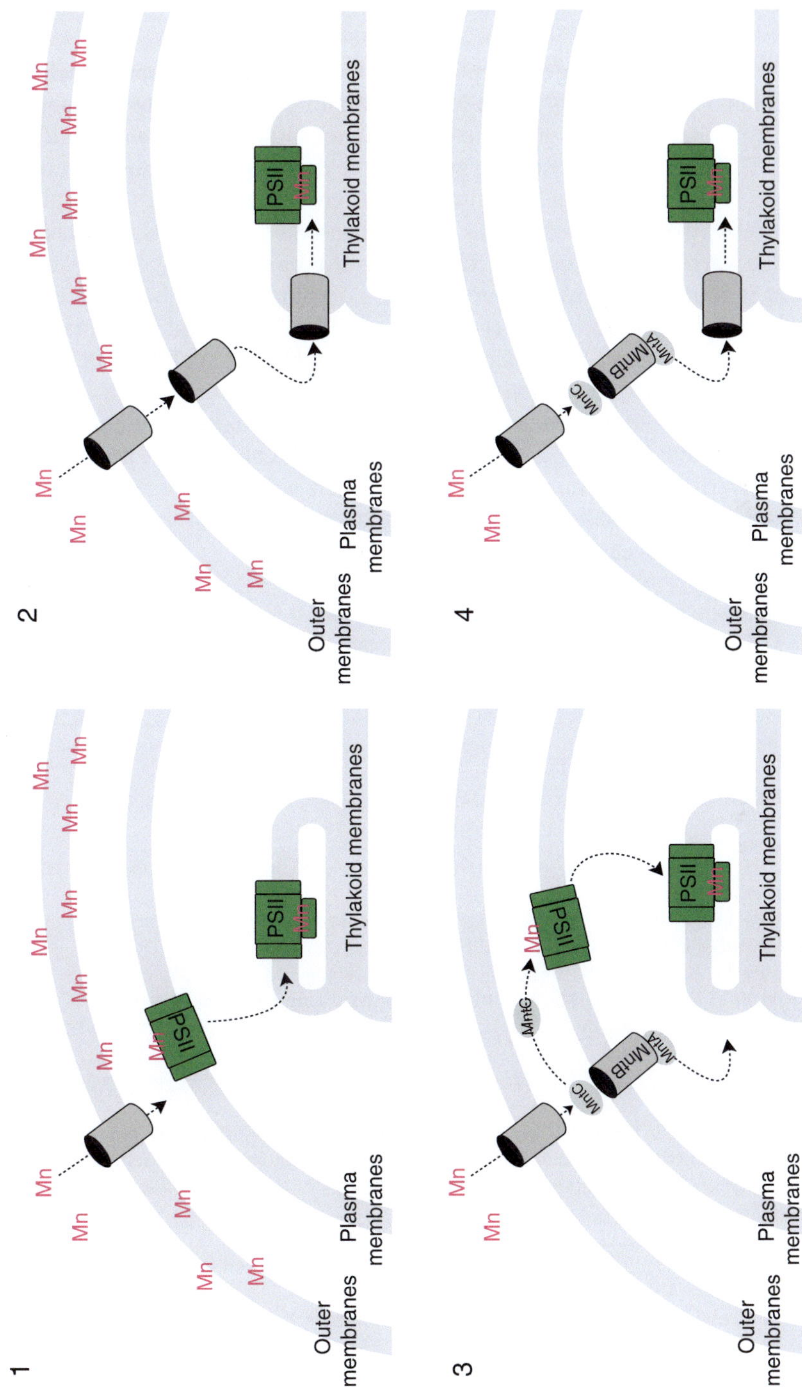

Fig. 15.3 Possible scenarios for Mn transport and the Mn cluster assembly process

15.5 Concluding Remarks and Future Perspectives

The current state of knowledge does not permit to draw straightforward conclusions on the details of the pathways leading to the assembly of a functionally competent WOC in cyanobacteria. It is possible that PSII assembly takes place in different topological locations during *de novo* synthesis or during repair, when Mn concentrations are high or low or in response to other environmental cues. Any attempt to further resolve Mn transport routes and PSII assembly pathways must take into account the variability imposed by changes in the bioavailability of Mn. In this respect, it is important to remember that Mn concentrations in oceans and lakes are in the nanomolar range (Chester and Stoner 1974; Sunda and Huntsman 1988; Sterner et al. 2004; Morel 2008). However, in different recipes of the standard cyanobacterial growth media, Mn concentrations are in the micromolar range, approximately three orders of magnitude higher. While it is hard to control transition metal bioavailability, and even harder to precisely control the physiological status of the cells, it is nevertheless essential that these parameters will be measured and reported so that experimental results from different growth conditions and from different cyanobacterial strains can be compared.

Acknowledgments This work was supported by the Israeli Science Foundation (grant No. 1168/07), G. R. Gratefully acknowledges financial support by Deutsche Forsacknschungsgemeinschaft (Sfb 429).

References

Adir N, Zer H, Shochat S, Ohad I (2003) Photoinhibition—A historical perspective. Photosynthesis Research 76: 343–370

Albertsson PA (1970) Partition of cell particles and macromolecules in polymer two-phase systems. Advances in Protein Chemistry 24: 309–341

Allakhverdiev SI, Karacan MS, Somer G, Karacan N, Khan EM, Rane SY, Padhye S, Klimov VV, Renger G (1994) Reconstitution of the water-oxidizing complex in manganese depleted potosystem II complexes by using synthetic binuclear manganese complexes. Biochemistry 33: 12210–12214

Allen MM (1968) Simple conditions for growth of unicellular blue-green algae on plates. Journal of Phycology 4: 1–4

Anbudurai PR, Mor TS, Ohad I, Shestakov SV, Pakrasi HB (1994) The Ctpa gene encodes the C-terminal processing protease for the D1 protein of the photosystem II reaction center complex. Proceedings of the National Academy of Sciences of the United States of America 91: 8082–8086

Aro EM, Suorsa M, Rokka A, Allahverdiyeva Y, Paakkarinen V, Saleem A, Battchikova N, Rintamaki E (2005) Dynamics of photosystem II: A proteomic approach to thylakoid protein complexes. Journal of Experimental Botany 56: 347–356

Baker AR, Jickells TD, Witt M, Linge KL (2006) Trends in the solubility of iron, aluminium, manganese and phosphorus in aerosol collected over the Atlantic Ocean. Marine Chemistry 98: 43–58

Balint I, Bhattacharya J, Perelman A, Schatz D, Moskovitz Y, Keren N, Schwarz R (2006) Inactivation of the extrinsic subunit of photosystem II, PsbU, in *Synechococcus* PCC 7942 results in elevated resistance to oxidative stress. FEBS Letters 580: 2117–2122

Baranov SV, Tyryshkin AM, Katz D, Dismukes GC, Ananyev GM, Klimov VV (2004) Bicarbonate is a native cofactor for assembly of the manganese cluster of the photosynthetic water oxidizing complex. Kinetics of reconstitution of O_2 evolution by photoactivation. Biochemistry 43: 2070–2079

Barber J (2008) Photosynthetic generation of oxygen. Philosophical Transactions of the Royal Society B—Biological Sciences 363: 2665–2674

Bartsevich VV, Pakrasi HB (1995) Molecular identification of an ABC transporter complex for manganese—Analysis of a cyanobacterial mutant strain impaired in the photosynthetic oxygen evolution process. EMBO Journal 14: 1845–1853

Bartsevich VV, Pakrasi HB (1996) Manganese transport in the cyanobacterium *Synechocystis* sp. PCC 6803. Journal of Biological Chemistry 271: 26057–26061

Boussac A, Rappaport F, Carrier P, Verbavatz JM, Gobin R, Kirilovsky D, Rutherford AW, Sugiura M (2004) Biosynthetic Ca2+/Sr2 +exchange in the photosystem II oxygen evolving enzyme of *Thermosynechococcus elongatus*. Journal of Biological Chemistry 279: 22809–22819

Burnap RL, Sherman LA (1991) Deletion mutagenesis in *Synechocystis* sp. PCC6803 indicates that the Mn stabilizing protein of photosystem II is not essential for O_2 evolution. Biochemistry 30: 440–446

Burnap RL, Shen JR, Jursinic PA, Inoue Y, Sherman LA (1992) Oxygen yield and thermoluminescence characteristics of a cyanobacterium lacking the manganese stabilizing protein of photosystem II. Biochemistry 31: 7404–7410

Cheniae GM, Martin IF (1967) Photoreactivation of manganese catalyst in photosynthetic oxygen evolution. Biochemical and Biophysical Research Communications 28: 89–95

Cheniae GM, Martin IF (1969) Photoreactivation of manganese catalyst in photosynthetic oxygen evolution. Plant Physiology 44: 351–360

Cheniae GM, Martin IF (1973) Absence of oxygen evolving capacity in dark grown *Chlorella*—Photoactivation of oxygen evolving centers. Photochemistry and Photobiology 17: 441–459

Chester R, Stoner JH (1974) The distribution of zink, nickel, manganese, cadmium, copper and iron in some surface waters from the world ocean. Marine Chemistry 2: 17–32

Cinco RM, Robblee JH, Messinger J, Fernandez C, Holman KLM, Sauer K, Yachandra VK (2004) Orientation of calcium in the Mn_4Ca cluster of the oxygen evolving complex determined using polarized strontium EXAFS of photosystem II membranes. Biochemistry 43: 13271–13282

Cogdell RJ, Gall A, Kohler J (2006) The architecture and function of the light harvesting apparatus of purple bacteria: from single molecules to *in vivo* membranes. Quarterly Reviews of Biophysics 39: 227–324

Dasgupta J, Tyryshkin AM, Dismukes GC (2007) ESEEM spectroscopy reveals carbonate and an N-donor protein ligand binding to Mn^{2+} in the photoassembly reaction of the Mn_4Ca cluster in photosystem II. Angewandte Chemie-International Edition 46: 8028–8031

Dasgupta J, Tyryshkin AM, Baranov SV, Dismukes GC (2010) Bicarbonate coordinates to Mn^{3+} during photo assembly of the catalytic Mn_4Ca core of photosynthetic water oxidation: EPR characterization. Applied Magnetic Resonance 37: 137–150

Debus RJ (1992) The manganese and calcium ions of photosynthetic oxygen evolution. Biochimica Et Biophysica Acta 1102: 269–352

Des Marais DJ (2000) Evolution—When did photosynthesis emerge on earth? Science 289: 1703–1705

Diner BA, Ries DF, Cohen BN, Metz JG (1988) COOH -terminal Processing of polypeptide D1 of the photosystem II reaction center of *Scenedesmus obliquus* is necessary for the assembly of the oxygen evolving complex. Journal of Biological Chemistry 263: 8972–8980

Dismukes GC, Klimov VV, Baranov SV, Kozlov YN, Dasgupta J, Tyryshkin A (2001) The origin of atmospheric oxygen on Earth: The innovation of oxygenic photosynthesis. Proceedings of the National Academy of Sciences of the United States of America 98: 2170–2175

Eaton-Rye JJ (2005) Requirements for different combinations of the extrinsic proteins in specific cyanobacterial photosystem II mutants. Photosynthesis Research 84: 275–281
Enami I, Kitamura M, Tomo T, Isokawa Y, Ohta H, Katoh S (1994) Is the primary cause of thermal inactivation of oxygen evolution in spinach PSII membranes release of the extrinsic 33 Kda protein or of Mn. Biochimica Et Biophysica Acta-Bioenergetics 1186: 52–58
Ferreira KN, Iverson TM, Maghlaoui K, Barber J, Iwata S (2004) Architecture of the photosynthetic oxygen evolving center. Science 303: 1831–1838
Ghanotakis DF, Babcock GT, Yocum CF (1984) Calcium reconstitutes high rates of oxygen evolution in polypeptide depleted photosystem II preparations. FEBS Letters 167: 127–130
Gilbert DL (ed) (1981) Oxygen and living processes: an interdisciplinary approach, Springer, New York
Gleiter HM, Haag E, Shen JR, Eaton-Rye JJ, Seeliger AG, Inoue Y, Vermaas WF and Renger G (1995) Involvement of the CP47 protein in stabilization and photoactivation of a functional wateroxidizing complex in the cyanobacterium Synechocystis sp. PCC 6803. Biochemistry 34: 6847–6856
Guieu C, Duce R, Arimoto R (1994) Dissolved input of manganese to the ocean aerosol source. Journal of Geophysical Research-Atmospheres 99: 18789–18800
Haber F, Weiss JJ (1934) The catalytic decomposition of H_2O_2 by iron salts. Proceedings of the Royal Society of London for Biology A 147: 332–351
Higuchi M, Noguchi T, Sonoike K (2003) Over reduced states of the Mn cluster in cucumber leaves induced by dark-chilling treatment. Biochimica Et Biophysica Acta-Bioenergetics 1604: 151–158
Hillier W, Messinger J (2005) Mechanism of photosynthetic oxygen production. In T Wydrzynski, K Satoh, eds, Photosystem II. The Light-Driven Water: Plastoquinone Oxidoredutase. Springer, Dordrecht, pp 567–608
Hoganson CW, Ghanotakis DF, Babcock GT, Yocum CF (1989) Mn^{2+} reduces Y_z^+ in manganese depleted photosystem II preparations. Photosynthesis Research 22: 285–293
Inoue-Kashino N, Kashino Y, Satoh K, Terashima I, Pakrasi HB (2005) PsbU provides a stable architecture for the oxygen evolving system in cyanobacterial photosystem II. Biochemistry 44: 12214–12228
Ivleva NB, Shestakov SV, Pakrasi HB (2000) The carboxyl-terminal extension of the precursor D1 protein of photosystem II is required for optimal photosynthetic performance of the cyanobacterium *Synechocystis* sp. PCC 6803. Plant Physiology 124: 1403–1411
Jansen T, Kanervo E, Aro EM, Maenpaa P (2002) Localisation and processing of the precursor form of photosystem II protein D1 in *Synechocystis* 6803. Journal of Plant Physiology 159: 1205–1211
Jansson C, Debus RJ, Osiewacz HD, Gurevitz M, Mcintosh L (1987) Construction of an obligate photoheterotrophic mutant of the cyanobacterium *Synechocystis* 6803—Inactivation of the *psbA* gene family. Plant Physiology 85: 1021–1025
Jurgens UJ, Weckesser J (1985) Carotenoid containing outer membrane of *Synechocystis* sp. strain PCC 6714. Journal of Bacteriology 164: 384–389
Kaim W, Schwederski B (1994) Bioinorganic chemistry: Inorganic elements in the chemistry of life, an introduction and guide. Inorganic Chemistry: A Textbook Series. John Wiley & Sons, Chichester
Kamiya N, Shen JR (2003) Crystal structure of oxygen evolving photosystem II from *Thermosynechococcus vulcanus* at 3.7-angstrom resolution. Proceedings of the National Academy of Sciences of the United States of America 100: 98–103
Kashino Y, Inoue-Kashino N, Roose JL, Pakrasi HB (2006) Absence of the PsbQ protein results in destabilization of the PsbV protein and decreased oxygen evolution activity in cyanobacterial photosystem II. Journal of Biological Chemistry 281: 20834–20841
Keren N, Kidd MJ, Penner-Hahn JE, Pakrasi HB (2002) A light dependent mechanism for massive accumulation of manganese in the photosynthetic bacterium *Synechocystis* sp. PCC 6803. Biochemistry 41: 15085–15092

Keren N, Liberton M, Pakrasi HB (2005) Photochemical competence of assembled photosystem II core complex in cyanobacterial plasma membrane. Journal of Biological Chemistry 280: 6548–6553

Kozlov YN, Zharmukhamedov SK, Tikhonov KG, Dasgupta J, Kazakova AA, Dismukes GC, Klimov VV (2004) Oxidation potentials and electron donation to photosystem II of manganese complexes containing bicarbonate and carboxylate ligands. Physical Chemistry Chemical Physics 6: 4905–4911

Liberton M, Berg RH, Heuser J, Roth R, Pakrasi HB (2006) Ultrastructure of the membrane systems in the unicellular cyanobacterium *Synechocystis* sp. strain PCC 6803. Protoplasma 227: 129–138

Linton KJ (2007) Structure and function of ABC transporters. Physiology 22: 122–130

Loll B, Kern J, Saenger W, Zouni A, Biesiadka J (2005) Towards complete cofactor arrangement in the 3.0 angstrom resolution structure of photosystem II. Nature 438: 1040–1044

Mamedov F, Nowaczyk MM, Thapper A, Rogner M, Styring S (2007) Functional characterization of monomeric photosystem II core preparations from *Thermosynechococcus elongatus* with or without the Psb27 protein. Biochemistry 46: 5542–5551

Mayfield SP, Bennoun P, Rochaix JD (1987) Expression of the nuclear encoded Oee1 protein is required for oxygen evolution and stability of photosystem II particles in *Chlamydomonas reinhardtii*. EMBO Journal 6: 313–318

Merchant S, Sawaya MR (2005) The light reactions: A guide to recent acquisitions for the picture gallery. Plant Cell 17: 648–663

Messinger J, Renger G (2008) Photosynthetic water splitting. In G Renger, ed, Primary Processes of Photosynthesis: Basic Priniciples and Apparatus, Vol. II: Reaction Centers/Photosystems, Electron Transport Chains, Photophosphorylation and Evolution. Royal Society Chemistry, Cambridge, pp 291–349

Messinger J, Seaton G, Wydrzynski T, Wacker U, Renger G (1997) S_3 state of the water oxidase in photosystem II. Biochemistry 36: 6862–6873

Messinger J, Robblee JH, Bergmann U, Fernandez C, Glatzel P, S. I, Hanssum B, Renger G, Cramer SP, Sauer K, Yachandra VKMosipI (2001) Manganese oxidation states in photosystem II. In Proceedings of 12th International Congress on Photosynthesis, Brisbane www.publish.csiro.au/ps2001, S10-019

Morel FMM (2008) The co-evolution of phytoplankton and trace element cycles in the oceans. Geobiology 6: 318–324

Murata N, Sato N, Omata T, Kuwabara T (1981) Separation and characterization of thylakoid and cell envelope of the blue green alga (cyanobacterium) *Anacystis nidulans*. Plant and Cell Physiology 22: 855–866

Nevo R, Charuvi D, Shimoni E, Schwarz R, Kaplan A, Ohad I, Reich Z (2007) Thylakoid membrane perforations and connectivity enable intracellular traffic in cyanobacteria. EMBO Journal 26: 1467–1473

Nowaczyk MM, Hebeler R, Schlodder E, Meyer HE, Warscheid B, Rogner M (2006) Psb27, a cyanobacterial lipoprotein, is involved in the repair cycle of photosystem II. Plant Cell 18: 3121–3131

Oelze J, Drews G (1972) Membranes of photosynthetic bacteria. Biochimica Et Biophysica Acta 265: 209–239

Ogawa T, Bao DH, Katoh H, Shibata M, Pakrasi HB, Bhattacharyya-Pakrasi M (2002) A two-component signal transduction pathway regulates manganese homeostasis in *Synechocystis* 6803, a photosynthetic organism. Journal of Biological Chemistry 277: 28981–28986

Ohnishi N, Allakhverdiev SI, Takahashi S, Higashi S, Watanabe M, Nishiyama Y, Murata N (2005) Two step mechanism of photodamage to photosystem II: Step 1 occurs at the oxygen evolving complex and step 2 occurs at the photochemical reaction center. Biochemistry 44: 8494–8499

Omata T, Murata N (1984) Isolation and characterization of 3 types of membranes from the cyanobacterium (blue green alga) *Synechocystis* PCC 6714. Archives of Microbiology 139: 113–116

Ono T (2001) Metallo radical hypothesis for photoassembly of Mn_4 cluster of photosynthetic oxygen evolving complex. Biochimica Et Biophysica Acta-Bioenergetics 1503: 40–51

Pakrasi HB, Diner BA, Williams JGK, Arntzen CJ (1989) Deletion mutagenesis of the cytochrome b_{559} protein inactivates the reaction center of photosystem II. Plant Cell 1: 591–597

Peschek GA, Hinterstoisser B, Wastyn M, Kuntner O, Pineau B, Missbichler A, Lang J (1989) Chlorophyll precursors in the plasma membrane of a cyanobacterium, *Anacystis nidulans*—Characterization of protochlorophyllide and chlorophyllide by spectrophotometry, spectrofluorimetry, solvent partition, and high-performance liquid chromatography. Journal of Biological Chemistry 264: 11827–11832

Pittman JK (2005) Managing the manganese: molecular mechanisms of manganese transport and homeostasis. New Phytologist 167: 733–742

Quigg A, Beardall J, Wydrzynski T (2003) Photoacclimation involves modulation of the photosynthetic oxygen evolving reactions in *Dunaliella tertiolecta* and *Phaeodactylum tricornutum*. Functional Plant Biology 30: 301–308

Renger G (1997) Mechanistic and structural aspects of photosynthetic water oxidation. Physiologia Plantarum 100: 828–841

Renger G, Volker M, Eckert HJ, Fromme R, Hohmveit S, Graber P (1989) On the mechanism of photosystem II deterioration by UV-B irradiation. Photochemistry and Photobiology 49: 97–105

Rippka R, Waterbury J, Cohenbazire G (1974) Cyanobacterium which lacks thylakoids. Archives of Microbiology 100: 419–436

Rogner M, Chisholm DA, Diner BA (1991) Site directed mutagenesis of the *psbC* gene of photosystem II—Isolation and functional characterization of CP43 less photosystem II core complexes. Biochemistry 30: 5387–5395

Roose JL, Pakrasi HB (2004) Evidence that D1 processing is required for manganese binding and extrinsic protein assembly into photosystem II. Journal of Biological Chemistry 279: 45417–45422

Roose JL, Pakrasi HB (2008) The Psb27 protein facilitates manganese cluster assembly in photosystem II. Journal of Biological Chemistry 283: 4044–4050

Roose JL, Wegener KM, Pakrasi HB (2007) The extrinsic proteins of photosystem II. Photosynthesis Research 92: 369–387

Rukhman V, Anati R, Melamed-Frank M, Adir N (2005) The MntC crystal structure suggests that import of Mn^{2+} in cyanobacteria is redox controlled. Journal of Molecular Biology 348: 961–969

Shcolnick S, Keren N (2006) Metal homeostasis in cyanobacteria and chloroplasts. Balancing benefits and risks to the photosynthetic apparatus. Plant Physiology 141: 805–810

Shen JR, Inoue Y (1993) Cellular localization of cytochrome c_{550}—Its specific association with cyanobacterial photosystem II. Journal of Biological Chemistry 268: 20408–20413

Shen JR, Burnap RL, Inoue Y (1995a) An independent role of cytochrome c_{550} in cyanobacterial photosystem II as revealed by double deletion mutagenesis of the *psbO* and *psbV* genes in *Synechocystis* sp. PCC 6803. Biochemistry 34: 12661–12668

Shen JR, Vermaas W, Inoue Y (1995b) The role of cytochrome c_{550} as studied through reverse genetics and mutant characterization in *Synechocystis* sp. PCC 6803. Journal of Biological Chemistry 270: 6901–6907

Shen JR, Ikeuchi M, Inoue Y (1997) Analysis of the *psbU* gene encoding the 12 kDa extrinsic protein of photosystem II and studies on its role by deletion mutagenesis in *Synechocystis* sp. PCC 6803. Journal of Biological Chemistry 272: 17821–17826

Shevela D, Su JH, Klimov V, Messinger J (2008) Hydrogencarbonate is not a structural part of the Mn_4O_xCa cluster in photosystem II. Biochimica Et Biophysica Acta 1777: 532–539

Shukla VK, Stanbekova GE, Shestakov SV, Pakrasi HB (1992) The D1 protein of the photosystem II reaction center complex accumulates in the absence of D2—Analysis of a mutant of the cyanobacterium *Synechocystis* sp. PCC 6803 lacking cytochrome b_{559}. Molecular Microbiology 6: 947–956

Singh AK, Bhattacharyya-Pakrasi M, Pakrasi HB (2008) Identification of an atypical membrane protein involved in the formation of protein disulfide bonds in oxygenic photosynthetic organisms. Journal of Biological Chemistry 283: 15762–15770

Smith D, Howe CJ (1993) The distribution of photosystem I and photosystem II polypeptides between the cytoplasmic and thylakoid membranes of cyanobacteria. FEMS Microbiology Letters 110: 341–347

Sproviero EM, Gascon JA, McEvoy JP, Brudvig GW, Batista VS (2006a) Characterization of synthetic oxomanganese complexes and the inorganic core of the O_2 evolving complex in photosystem II: Evaluation of the DFT/B3LYP level of theory. Journal of Inorganic Biochemistry 100: 786–800

Sproviero EM, Gascon JA, McEvoy JP, Brudvig GW, Batista VS (2006b) QM/MM models of the O2 evolving complex of photosystem II. Journal of Chemical Theory and Computation 2: 1119–1134

Srivastava R, Battchikova N, Norling B, Aro EM (2006) Plasma membrane of *Synechocystis* PCC 6803: A heterogeneous distribution of membrane proteins. Archives of Microbiology 185: 238–243

Sterner RW, Smutka TM, McKay RML, Qin XM, Brown ET, Sherrell RM (2004) Phosphorus and trace metal limitation of algae and bacteria in Lake Superior. Limnology and Oceanography 49: 495–507

Stevens SE, Porter RD (1980) Transformation in *Agmenellum quadruplicatum*. Proceedings of the National Academy of Sciences of the United States of America-Biological Sciences 77: 6052–6056

Strasser RJ, Sironval C (1972) Induction of photosystem II activity in flashed leaves. FEBS Letters 28: 56–60

Strickler MA, Walker LM, Hillier W, Debus RJ (2005) Evidence from biosynthetically incorporated strontium and FTIR difference spectroscopy that the C-terminus of the D1 polypeptide of photosystem II does not ligate calcium. Biochemistry 44: 8571–8577

Summerfield TC, Shand JA, Bentley FK, Eaton-Rye JJ (2005) PsbQ (Sll1638) in *Synechocystis* sp. PCC 6803 is required for photosystem II activity in specific mutants and in nutrient limiting conditions. Biochemistry 44: 805–815

Summerfield TC, Eaton-Rye JJ, Sherman LA (2007) Global gene expression of a delta *psbO*:Delta *psbU* mutant and a spontaneous revertant in the cyanobacterium *Synechocystis* sp. strain PCC 6803. Photosynthesis Research 94: 265–274

Sunda WG, Huntsman SA (1988) Effect of sunlight on redox cycles of manganese in the southwestern Sargasso sea. Deep Sea Research Part A -Oceanographic Research Papers 35: 1297–1317

Tamura N, Cheniae GM (1986) Requirements for the photoligation of Mn^{2+} in PSII membranes and the expression of water oxidizing activity of the polynuclear Mn catalyst. FEBS Letters 200: 231–236

Tamura N, Cheniae G (1987) Photoactivation of the water oxidizing complex in photosystem II membranes depleted of Mn and extrinsic proteins I. Biochemical and kinetic characterization. Biochimica Et Biophysica Acta 890: 179–194

Tamura N, Inoue Y, Cheniae GM (1989) Photoactivation of the water oxidizing complex in photosystem II membranes depleted of Mn, Ca and extrinsic proteins. II. Studies on the functions of Ca^{2+}. Biochimica Et Biophysica Acta 976: 173–181

Thompson LK, Blaylock R, Sturtevant JM, Brudvig GW (1989) Molecular basis of the heat denaturation of photosystem II. Biochemistry 28: 6686–6695

Thornton LE, Ohkawa H, Roose JL, Kashino Y, Keren N, Pakrasi HB (2004) Homologs of plant PsbP and PsbQ proteins are necessary for regulation of photosystem II activity in the cyanobacterium *Synechopystis* 6803. Plant Cell 16: 2164–2175

Tottey S, Waldron KJ, Firbank SJ, Reale B, Bessant C, Sato K, Cheek TR, Gray J, Banfield MJ, Dennison C, Robinson NJ (2008) Protein folding location can regulate manganese binding versus copper or zinc binding. Nature 455: 1138–1142

Tyystjarvi E (2008) Photoinhibition of photosystem II and photodamage of the oxygen evolving manganese cluster. Coordination Chemistry Reviews 252: 361–376

Ulas G, Olack G, Brudvig GW (2008) Evidence against bicarbonate bound in the O_2 evolving complex of photosystem II. Biochemistry 47: 3073–3075

Vaara M (1992) Agents that increase the permeability of the outer membrane. Microbiological Reviews 56: 395–411

Van Gorkom HJ and Yocum CF (2005) The calcium and chloride cofactors. In: Wydrzynski T and Satoh K (eds) Photosystem II. The Light-Driven Water: Plastoquinone Oxidoredutase. Springer, Dordrecht, pp 307–327

Vass I, Czer K (2009) Janus-faced charge recombinations in photosystem II photoinhibition. Trends in Plant Sciences 14: 200–205

Vermaas WFJ, Williams JGK, Rutherford AW, Mathis P, Arntzen CJ (1986) Genetically engineered mutant of the cyanobacterium *Synechocystis* 6803 lacks the photosystem II chlorophyll binding protein CP47. Proceedings of the National Academy of Sciences of the United States of America 83: 9474–9477

Vermaas WFJ, Timlin JA, Jones HDT, Sinclair MB, Nieman LT, Hamad SW, Melgaard DK, Haaland DM (2008) In vivo hyperspectral confocal fluorescence imaging to determine pigment localization and distribution in cyanobacterial cells. Proceedings of the National Academy of Sciences of the United States of America 105: 4050–4055

Westphal KL, Lydakis-Simantiris N, Cukier RI, Babcock GT (2000) Effects of Sr^{2+} substitution on the reduction rates of Y_z in PSII membranes—Evidence for concerted hydrogen atom transfer in oxygen evolution. Biochemistry 39: 16220–16229

Williams JGK (1988) Construction of specific mutations in photosystem II photosynthetic reaction center by genetic engineering methods in *Synechocystis* 6803. Methods in Enzymology 167: 766–778

Yachandra VK (2005) The catalytic manganese cluster: Organisation of the metal ions. In T Wydrzynski, K Satoh, eds, Photosystem II. The Light Driven Water: Plastoquinone Oxidoreductase. Springer, Dordrecht pp 235–260

Yamaguchi K, Suzuki L, Yamamoto H, Lyukevich A, Bodrova I, Los DA, Piven I, Zinchenko V, Kanehisa M, Murata N (2002) A two-component Mn^{2+} sensing system negatively regulates expression of the mntCAB operon in *synechocystis*. Plant Cell 14: 2901–2913

Yu JJ, Vermaas WFJ (1990) Transcript levels and synthesis of photosystem II components in cyanobacterial mutants with inactivated photosystem II genes. Plant Cell 2: 315–322

Zak E, Norling B, Maitra R, Huang F, Andersson B, Pakrasi HB (2001) The initial steps of biogenesis of cyanobacterial photosystems occur in plasma membranes. Proceedings of the National Academy of Sciences of the United States of America 98: 13443–13448

Part IV
Electron Entry (Dehydrogenation)

Chapter 16
Structure and Physiological Function of NDH-1 Complexes in Cyanobacteria

Natalia Battchikova, Eva-Mari Aro and Peter J. Nixon

16.1 Introduction

Complex I (NADH:ubiquinone oxidoreductase, EC 1.6.5.3) is the first of the three energy-transducing enzyme complexes of the mitochondrial respiratory chain and its role is to transfer electrons from the electron donor NADH to a ubiquinone molecule, with concomitant generation of a proton motive force used for ATP synthesis. Related complexes are found throughout nature, but the type of electron donor and electron acceptor each uses can differ (Yagi et al. 1998; Bäumer et al. 2000; Weerakoon and Olsen 2008). Complex I in eubacteria is considered to be composed of three structural modules or sub-complexes: an electron-input device or activity module involved in NADH-binding and oxidation, an interconnecting amphipathic fragment containing a number of Fe-S centers and a hydrophobic membrane fragment involved in proton translocation and possibly quinone binding (Friedrich et al. 1995; Friedrich and Scheide 2000; Sazanov 2007). The size of complex I varies considerably: the 'minimal' complex I found in bacteria (designated NDH-1) is approximately 550 kDa and consists of 13–15 subunits (reviewed by Sazanov 2007) whereas bovine complex I contains 45 subunits and is close to 1 MDa (Carroll et al. 2006). The crystal structure of the intact holenzyme from *Thermus thermophilus* was recently solved at 4.5 Å resolution (Efremov et al. 2010) and the soluble hydrophilic domain, which encompasses the NADH-oxidizing module and interconnecting fragment, and contains 8 subunits, 1 FMN and 9 Fe-S centers has been determined at 3.3 Å resolution (Sazanov and Hinchcliffe 2006). How proton pumping is coupled to electron transfer is currently unknown with both redox-linked and conformationally linked mechanisms proposed (Sazanov 2007).

Early DNA sequencing experiments, reinforced by analysis of cyanobacterial genome sequences, lent support to previous conclusions that cyanobacteria contained a respiratory complex I involved in NADH or perhaps NADPH oxidation (reviewed by Schmetterer 1994). Cyanobacteria, as well as chloroplasts, were found to encode

P. J. Nixon (✉)
Department of Life Sciences, Biochemistry Building, Imperial College London,
S. Kensington Campus, London, SW7 2AZ, UK
e-mail: p.nixon@imperial.ac.uk

G. A. Peschek et al. (eds.), *Bioenergetic Processes of Cyanobacteria,*
DOI 10.1007/978-94-007-0388-9_16,

Table 16.1 Ndh subunits and NDH-1 complexes in *Synechocystis* sp. PCC 6803

Ndh subunits (CyanoBase designation)[a]	Homologous *E. coli* subunits[b]	Location of Ndh subunits in NDH-1[d]	NDH-1 (sub) complexes[e]	Redox centers and cofactor binding sites[f]
–	NuoE	–		Fe-S: N1a
–	NuoF	–		NADH-binding site, FMN, Fe-S: N3
–	NuoG	–		Fe-S: N1b, N4, N5, N7
NdhA (Sll0519)	NuoH	Membrane (core)	L, M, MS	Q binding?
NdhB (Sll0223)	NuoN	Membrane (core)	L, M, MS	
NdhC (Slr1279)	NuoA	Membrane (core)	L, M, MS	
NdhD1 (Slr0331)	NuoM	Membrane (D1F1)	L	Q binding?
NdhD2 (Slr1291)		Membrane (D2F1)	L'	
NdhD3 (Sll1733)		Membrane (CUP-A)	S, MS	
NdhD4 (Sll0027)		Membrane (CUP-B)	MS'	
NdhD5 (Slr2007)		Not detected	–	
NdhD6 (Slr2009)		Not detected	–	
NdhE (Sll0522)	NuoK	Membrane (core)	L, M	
NdhF1 (Slr0844)	NuoL	Membrane (D1F1)	L	Q binding?
NdhF3 (Sll1732)		Membrane (CUP-A)	S, MS	
NdhF4 (Sll0026)		Membrane (CUP-B)	MS'	
NdhG (Sll0521)	NuoJ	Membrane (core)	L, M, MS	
NdhH (Slr0261)	NuoD[c]	Interconnecting	L, M, MS	Q binding?
NdhI (Sll0520)	NuoI	Interconnecting	L, M, MS	Fe-S: N6a, N6b
NdhJ (Slr1281)	NuoC[c]	Interconnecting	L, M, MS	
NdhK (Slr1280)	NuoB	Interconnecting	L, M, MS	Fe-S: N2 Q binding?
NdhL (Ssr1386)	–	Membrane (core)	L, M, MS	
NdhM (Slr1623)	–	Interconnecting	L, M, MS	
NdhN (Sll1262)	–	Interconnecting	L, M, MS	
NdhO (Ssl1690)	–	Interconnecting	L, M, MS	
CupA (Sll1734)	–	Membrane (CUP-A)	S, MS	
CupB (Slr1302)	–	Membrane (CUP-B)	MS'	
CupS (Sll1735)	–	Membrane (CUP-A)	S, MS	

[a] Cyanobase is at http://genome.kazusa.or.jp/cyanobase/

[b] The subunits of NDH-1 from *E. coli* are included for comparison

[c] NuoCD are fused in *E. coli*

[d] The subunits are assigned to the interconnecting and membrane fragments of NDH-1 as described in Fig. 16.2. The membrane fragment is further subdivided into a core domain, the NdhD1NdhF1 (D1F1) module, the putative NdhD2NdhF1 module (D2F1), the CUP-A module and the CUP-B module

[e] Confirmed and predicted NDH-1 complexes that contain the Ndh subunit. See abbreviations and text for details

[f] Predicted location of redox centers and cofactor-binding sites in *E. coli* NDH-1 based on Sazanov and Hinchliffe (2006)

homologues (designated NdhA-G) of the 6 *E. coli* subunits found in the hydrophobic membrane fragment and homologues (NdhH-K) of the 4 subunits found in the interconnecting fragment (Table 16.1). Notably absent from the cyanobacterial genomes was a conserved set of genes encoding homologues of the NADH-oxidizing electron-input module, consisting of NuoE, F and G in *E. coli.* Hence, although the cyanobacterial complex is annotated as the NDH-1 complex, or type I NAD(P)H dehydrogenase, there is still considerable uncertainty as to the nature of the electron donor. Indeed a role for NDH-1 as a ferredoxin:plastoquinone oxidoreductase cannot be excluded. Despite this gap in our knowledge, the NDH-1 complexes found in cyanobacteria and chloroplasts appear to possess certain conserved features which allow them to be categorized as a separate subclass of the family of complex I enzymes, specific for organisms capable of oxygenic photosynthesis. In this chapter we will present a critical overview of the structure and function of cyanobacterial NDH-1 complexes.

16.2 Physiological Studies on *ndh* Mutants

16.2.1 The ndh Genes

Analysis of the first cyanobacterial genome sequence, that of *Synechocystis* sp. PCC 6803, led to the identification of 11 potential Ndh subunits (NdhA-K) based on sequence comparisons with characterized complex I subunits in other systems (Table 16.1) (Kaneko et al. 1996). Unlike *E. coli,* where all the *nuo* genes, encoding the NDH-1 subunits, are located in one cluster (Weidner et al. 1993), the annotated *ndh* genes of *Synechocystis* 6803 are dispersed throughout the genome usually as single copies, sometimes in operons (e.g., *ndhAIGE, ndhCJK*). Unusually, however, *ndhD* and *ndhF* were found to be present as small gene families, six in the case of *ndhD* (*ndhD1-6*) and three in the case of *ndhF* (*ndhF1, 3, 4*). NdhD and NdhF (as well as NdhB) are homologous to monovalent cation/proton antiporters (Mathiesen and Hägerhäll 2003); consequently a role for members of the *ndhD* and *ndhF* gene families outside NDH-1 function could not be discounted. The *ndhD1* and *ndhF1* genes are found in all of the cyanobacteria so far sequenced whereas the presence of the remaining *ndhD* and *ndhF* genes is more variable (Ogawa and Mi 2007). To help resolve the physiological importance of the *ndh* genes, substantial effort has been directed towards the generation and characterization of defined mutants, created mainly in *Synechocystis* sp. PCC 6803. Except for *ndhH,* which has been reported as essential (Pieulle et al. 2000), *ndh* null mutants can be isolated, usually with a high-CO_2-requiring phenotype (discussed in Battchikova and Aro 2007).

16.2.2 Role in Respiration

Cyanobacteria have the coding potential to synthesise a number of different plastoquinone reductases and plastoquinol oxidases, which could participate in a proton-pumping respiratory chain (Peschek 1980; Wastyn et al. 1988; Howitt and Vermaas 1998; Kuntz 2004; McDonald and Vanlerberghe 2006). By extrapolation from other systems it was hypothesized that one physiological role for the NDH-1 complex would be to contribute to respiratory electron flow. Indeed early studies showed that the *ndhB* insertion mutant, M55, of *Synechocystis* 6803 was impaired in dark respiration (Mi et al. 1992a). However, it has been argued that the impaired rates of respiration seen in this and other *ndh* mutants (Ohkawa et al. 2000a) might be indirect and actually due to low levels of succinate in these strains, which would limit the respiratory activity of the succinate:plastoquinone oxidoreductase which has been proposed to be the main route for dark reduction of plastoquinone (Cooley and Vermaas 2001).

16.2.3 Role in Cyclic Electron Flow

Given that plastoquinone is a common electron carrier in both respiratory and photosynthetic electron flow (Scherer et al. 1988), a role for NDH-1 in cyclic electron flow around Photosystem I has long been suspected (Mi et al. 1992b). Indeed, the M55 mutant lacking NdhB was found to display characteristics expected of impaired cyclic electron flow around PSI. These included enhanced accumulation of $P700^+$ in the light, a reduced rate of reduction of $P700^+$ in the dark following illumination (Mi et al. 1992a) and reduced energy storage upon far-red illumination as assessed by photoacoustic spectroscopy (Yeremenko et al. 2005). Chloroplast mutants lacking the chloroplast NDH-1 (or Ndh/NDH) complex also show perturbed cyclic electron flow (Burrows et al. 1998; Shikanai and Endo 2000; Joët et al. 2002; Peltier and Cournac 2002; Munekage et al. 2004; Rumeau et al. 2007). Consequently NDH-1 appears to have a conserved role in cyclic electron flow in both chloroplasts and cyanobacteria. Whether this role is direct or indirect, such as by controlling the redox poise of the plastoquinone pool, is currently not clear. A second parallel pathway of cyclic electron flow involves PGR5 in chloroplasts (Munekage et al. 2002, 2004) and its homologue, Ssr2016, in *Synechocystis* 6803 (Yeremenko et al. 2005). Cyclic electron flow serves to increases the ratio of ATP/NADPH produced in photosynthetic electron flow and plays an important physiological role when cyanobacteria are exposed to environmental stress and require additional ATP to be synthesised (reviewed by Nixon and Mullineaux 2001). This might explain why a *Synechocystis* mutant lacking NDH-1 is sensitive to salt shock (Tanaka et al. 1997).

16.2.4 Role in Uptake of Inorganic Carbon

A distinguishing feature of many, but not all, cyanobacterial *ndh* mutants is the impaired uptake of inorganic carbon, specifically CO_2 (Ogawa 1991a, b). Most of the organic carbon in autotrophically grown cells is obtained through CO_2 fixation

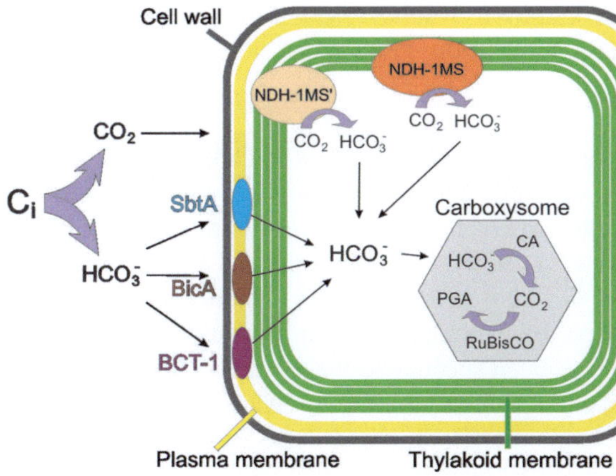

Fig. 16.1 The types of carbon-concentrating mechanism (CCM) found in a cyanobacterial cell. Two systems specialized in CO_2 uptake are the constitutively expressed NDH-1MS' complex (low-affinity CO_2 uptake) and the low CO_2-inducible NDH-1MS complex (high-affinity CO_2 uptake). Three components of HCO_3^- uptake include the Na^+-dependent bicarbonate transporter SbtA, the ABC-type transporter BCT-1 encoded by the *cmpABCD* operon, and the Na^+-dependent BicA transporter. For details, see Badger et al. (2002, 2006) and Kaplan et al. (2007). (The figure is re-drawn from Kaplan et al. 2007)

by Rubisco. Aquatic photosynthetic organisms face severe problems in carbon acquisition since the affinity of their Rubisco enzymes for CO_2 is low. To compensate, cyanobacteria have developed multiple types of carbon concentrating mechanism (CCM, Fig. 16.1) to build up an internal inorganic carbon (C_i) concentration much higher than the external one (reviewed by Kaplan and Reinhold 1999). As a source of carbon, both CO_2 and HCO_3^- can be utilized by cyanobacteria (Volokita et al. 1984). Bicarbonate uptake, which requires assistance of special transporters to transfer a HCO_3^- molecule across the membrane, is beyond the scope of this chapter and is described elsewhere (for example, Price et al. 2008). Briefly, three types of bicarbonate transporters shown in Fig. 16.1 have been discovered in the plasma membrane of cyanobacteria. They include the ABC-type transporter BCT-1 encoded by the *cmp* operon (Omata and Ogawa 1985; Omata et al. 1999), the Na^+-dependent bicarbonate transporter SbtA found in *Synechocystis* (Shibata et al. 2002a), and the Na^+-dependent BicA transporter of the sulfate transporters/permeases (SulP) family which was identified in marine cyanobacteria (Price et al. 2004).

In contrast to HCO_3^- ions, CO_2 enters the cells by diffusion via water-channels and the uptake of CO_2 is strongly inhibited by a water-channel blocker (Tchernov et al. 2001). Inside cells the inorganic carbon is accumulated in the form of bicarbonate, so CO_2 uptake in cyanobacteria involves a step of intracellular conversion of CO_2 to HCO_3^- which should be mediated by a carbonic anhydrase (CA)-like enzyme (Volokita et al. 1984; Kaplan and Reinhold 1999). Finally, the bicarbonate accumulated in the cytosol of the cells is transported into the carboxysome where

CA converts it to CO_2 in close proximity to Rubisco (Volokita et al. 1984; Kaplan and Reinhold 1999; Badger and Spalding 2000; Price et al. 2002).

There are two systems specialized in CO_2 uptake (Fig. 16.1), a constitutively expressed low-affinity CO_2 uptake system, and a high-affinity CO_2 uptake system induced at limiting CO_2 conditions (Klughammer et al. 1999; Ohkawa et al. 2000a; Shibata et al. 2001). Inactivation of NdhB completely abolishes the acquisition of C_i in the form of CO_2 indicating the involvement of NDH-1 in both systems (Ogawa 1991a, b; Marco et al. 1993). How NDH-1 might facilitate CO_2 uptake is still unclear but based on analysis of different *ndh* mutants it does not seem to be an indirect effect stemming from impaired cyclic photophosphorylation or respiration (Ohkawa et al. 2000a). Instead it has been speculated that after CO_2 enters the cells by diffusion, NDH-1 complexes catalyse its hydration to HCO_3^- at a site containing a reactive metal-bound hydroxide ion (Badger and Price 2003) or in a special alkaline pocket (Kaplan and Reinhold 1999; Tchernov et al. 2001).

Extensive reverse genetic studies performed in several cyanobacterial species (*Synechocystis* 6803, *Synechococcus* 7942 and *Synechococcus* 7002) have implicated specific members of the *ndhD* and *ndhF* gene families in CO_2 uptake (Schluchter et al. 1993; Klughammer et al. 1999; Ohkawa et al. 2000a; Shibata et al. 2001; Maeda et al. 2002). For instance, a Δ*ndhD3*/Δ*ndhD4* double mutant grows well under photoheterotrophic conditions, shows normal rates of respiration and cyclic electron flow but is impaired in CO_2 uptake (Ohkawa et al. 2000a; Shibata et al. 2001), whereas the single Δ*ndhD3* and Δ*ndhD4* mutants are still able to grow at air levels of CO_2 indicating effective CO_2 uptake (Ohkawa et al. 2000a). In contrast, a Δ*ndhD1*/Δ*ndhD2* double mutant of *Synechocystis* 6803, like the M55 mutant lacking NdhB, grows normally at air levels of CO_2, but is unable to grow under photoheterotrophic conditions and shows a low rate of respiration (Ohkawa et al. 2000a).

Interpretation of the phenotype of *ndhD3* and *ndhD4* insertion mutants is complicated by the fact they are actually part of two operons in *Synechocystis* 6803: the *ndhF3-ndhD3-cupA* operon which is induced under low CO_2, and the *ndhF4-ndhD4* operon (in many strains *ndhF4, ndhD4 and cupB* are found in one operon) which is expressed constitutively. Nevertheless, analysis of various mutants has led to the proposal that these two operons define two distinct CO_2-uptake systems: NdhD3/NdhF3/CupA are responsible for low CO_2-inducible, high-affinity CO_2 acquisition, and NdhD4/NdhF4/CupB are involved in a constitutive, low-affinity CO_2 uptake (Ohkawa et al. 2000a, b; Maeda et al. 2002; Shibata et al. 2002b). The physiological roles of NdhD5 and NdhD6 currently remain unclear.

16.3 Biochemical Studies

16.3.1 Isolation and Activity of NDH-1 Complexes

Traditional methods to study complex enzymes include their purification from cellular extracts to homogeneity by various chromatographic approaches with concomitant monitoring of an increase in the specific activity. This approach has proved to have limited success in the case of cyanobacterial NDH-1 complexes. Similar to complex I from other bacteria and mitochondria, NDH-1 of cyanobacteria is a membrane-located complex containing hydrophilic subunits together with highly hydrophobic ones (in *Synechocystis* 6803, all NdhF and NdhD proteins as well as NdhB contain 12–17 transmembrane helixes as predicted by PRED-TMR2, Pasquier and Hamodrakas 1999). Thus one potential problem is that membrane solubilization using detergent might cause extensive fragmentation of cyanobacterial NDH-1 complexes resulting in a loss of activity and/or hydrophobic subunits. Another problem concerns the presence of several types of NADH dehydrogenase activity in cyanobacteria (Matsuo et al. 1998a; Yagi et al. 1998; Howitt et al. 1999) and several NADPH-dependent enzymes including FNR and DrgA (Matsuo et al. 1998b), which may mask or contaminate the activity of the NDH-1 complex. Since mitochondrial and bacterial complex I is an NADH-dependent oxidoreductase (Friedrich et al. 1995; Yagi et al. 1998; Brandt 2006), in earlier publications the isolation and characterization of cyanobacterial NDH-1 was based on using NADH as a substrate (Alpes et al. 1989; Berger et al. 1993). At present current evidence suggests that the cyanobacterial NDH-1 complex might be an NADPH-dependent, rather than an NADH-dependent, enzyme (Matsuo et al. 1998a; Deng et al. 2003a, b; Ma et al. 2006). However, without knowledge of the active subunits this assignment remains open to debate. Instead other methods (such as detection of subunits with antibodies or by mass spectrometry) have been used to monitor the purification of the enzyme (Berger et al. 1993; Matsuo et al. 1998b; Deng et al. 2003a; Ma et al. 2006).

Berger et al. (1993) first described the isolation of an NDH-1 subcomplex, from Triton X-100 treated *Synechocystis* 6803 cell extracts, by immunoaffinity chromatography using an NdhK-specific antibody coupled to the Protein A sepharose. The isolated subcomplex contained several hydrophilic Ndh subunits (I, J, K, H, as well as later discovered M) and a few unknown polypeptides which appeared to be contaminants. The authors monitored the purification with NdhK- and NdhJ-specific antibodies and measurement of NADH-oxidizing activity and concluded that the subcomplex was functionally inactive. No hydrophobic subunits were detected. Considering the tendency of complex I from mitochondria and *E. coli* to dissociate into three subcomplexes (activity domain, interconnecting fragment and the membrane fragment) in the presence of chaotropic anions, it can be concluded that the authors had isolated part or all of the interconnecting fragment of cyanobacterial NDH-1.

Later Matsuo et al. (1998b) described an active hydrophilic NDH-1 subcomplex of 380 kDa from a CHAPS-treated *Synechocystis* 6803 cell extract. The subcomplex reacted with a polyclonal antibody against NdhH but not against NdhA or NdhB. The kinetic properties were determined for the subcomplex after purification by anion exchange and gel filtration chromatography. The enzyme appeared to be specific for NADPH and had high affinities for the artificial electron acceptors ferricyanide and dichloroindophenol. K_m for NADPH and V_{max} were determined as 5.1 μM and 5.4 units (mg protein)$^{-1}$, respectively, in a NADPH-$K_3Fe(CN)_6$ oxidoreductase reaction. The absence of hydrophobic subunits indicated that the membrane domain of the NDH-1 complex had dissociated during the isolation procedure, which would help explain why the analyzed subcomplex had a low affinity for plastoquinone and was not inhibited by rotenone, a classic inhibitor of Complex I in mitochondria.

Another NADPH-active NDH-1 subcomplex, of about 230–250 kDa, was purified by conventional chromatographic methods by Deng et al. (2003a) from *Synechocystis* 6803 cell extracts treated with n-dodecyl β-D-maltoside (DM). Using antibodies, the authors demonstrated the presence of the hydrophobic NdhA subunit in this subcomplex. The NDH-1 subcomplexes isolated by Matsuo et al. (1998b) and Deng et al. (2003a) were possible to separate by native PAGE and to stain "in-gel" for NADPH-nitroblue tetrazolium oxidoreductase activity. Further Deng et al. (2003b) showed that the activity of the band in the native gel, proposed to correspond to the NADPH-active NDH-1 subcomplex, increased if cells were grown in low CO_2 conditions. Unfortunately, the proteins contributing to the activity were not identified and remain unknown at present.

A more successful approach for the isolation of NDH-1 complexes has come from the exploitation of His-tagging technology in both *Synechocystis* 6803 (Prommeenate et al. 2004) and *Thermosynechococcus elongatus* (Zhang et al. 2005). In the case of *Synechocystis* 6803, His-tagged NdhJ was used to isolate a major complex of approximate size 460 kDa containing NdhA, B, C, D1, F, G, H, I, J, K, a large C-terminal breakdown fragment of NdhF1 and two new Ndh subunits assigned as NdhM and NdhN. One or more protein components found within the low molecular mass region were unable to be identified. The complex did not display any NAD(P)H:FeCN oxidoreductase activity.

For the thermophilic cyanobacterium, *T. elongatus,* a His-tagged derivative of NdhL, which had earlier been shown to be a component of the *Synechocystis* 6803 NDH-1 complex (Battchikova et al. 2005), was used (Zhang et al. 2005). The thermophilic nature of *T. elongatus* makes it an excellent organism to isolate complexes that are often more unstable in mesophilic cyanobacteria. Again the isolated NDH-1 preparation displayed no detectable NAD(P)H:FeCN oxidoreductase activity (Zhang et al. 2005). A number of distinct complexes were characterized following size selection by BN/SDS-PAGE. A 450-kDa complex, designated NDH-1L on the basis of earlier proteomics experiments (described below), contained 15 subunits (Table 16.1) and is likely to be equivalent to the largest complex (complex A) isolated by Prommeenate et al. (2004). Importantly, a second complex, of 490 kDa, was also isolated in which NdhD1 and NdhF1 had been replaced by NdhD3, NdhF3, CupA and Tll0220 (CupS homologue). This isolated complex, designated NDH-1MS (described in detail below), provided the crucial experimental evidence to show that NdhD3/NdhF3/CupA were actually components of a larger NDH-1 complex rather than an independent complex that was evolutionarily related. Interestingly, NDH-1L could also be isolated from wild type *T. elongatus* in the absence of a His-tag (Zhang et al. 2005), possibly because a short histidine-rich region in the *T. elongatus* NdhF1 protein served as a "natural His-tag" strong enough to bind the NDH-1L complex to Ni^{2+} resin.

16.3.2 Identification of New Ndh Subunits

Analysis of His-tagged complexes together with proteomic studies has so far identified 4 new cyanobacterial NDH-1 subunits (NdhM, NdhN, NdhO and NdhL) (Prommeenate et al. 2004; Battchikova et al. 2005) with nucleus-encoded homologues in the chloroplast NDH-1 complex (Rumeau et al. 2005; Shimizu et al. 2008). NdhL had earlier been identified as being important for the uptake of CO_2 in *Synechocystis* 6803 and originally designated *ictA* (Ogawa 1991b, 1992). NdhM and NdhN knock-out mutants also require high levels of CO_2 for growth whereas NdhO does not seem to be critical for NDH-1 function (R. deVries and P. J. Nixon, unpublished). In contrast all 4 genes are needed for accumulation of the chloroplast NDH-1 complex (Rumeau et al. 2005; Shimizu et al. 2008). None of the new subunits possesses obvious sequence motifs that give clues to the function of the subunit. Apart from NdhM, which shows some sequence similarity to the B13 subunit of bovine complex I (Berger et al. 1993; Prommeenate et al. 2004), the newly discovered NDH-1 subunits appear to be specific to the NDH-1 complex of cyanobacteria and chloroplasts. Very recently, two additional small subunits (designated NdhP and NdhQ) were detected in the NDH-1 complex isolated from *T. elongatus* (Nowaczyk et al. 2011). NdhP shows some sequence similarity to the chloroplast NDF6 subunit but as yet the importance of NdhP and NdhQ for NDH-1 function is unclear.

16.3.3 Proteomic Studies

The presence of several distinct NDH-1 complexes with different sets of hydrophobic subunits was first revealed by proteomic studies of the membrane protein complexes from *Synechocystis* 6803 (Herranen et al. 2004). The proteomic approach included separation of the DM-solubilized membrane compartment by 2-D BN/SDS-PAGE which allowed separation of high molecular mass protein complexes in the 1st dimension followed by dissociation of complexes into subunits in the 2nd dimension. Identities of subunits were clarified by mass spectrometry supplemented with detection using specific antibodies. Three complexes containing Ndh subunits were detected and named NDH-1L (*L*arge), NDH-1M (*M*edium) and NDH-1S (*S*mall) according to their sizes of 490, 350 and 200 kDa, respectively (Herranen et al. 2004). The identity and functions of these three complexes were further clarified by detailed analysis of their subunit composition combined with proteomic studies of several specific *ndh* gene knockout mutants grown under different CO_2 and pH conditions (Zhang et al. 2004; Battchikova et al. 2005). The scheme of cyanobacterial NDH-1 complexes is shown in Fig. 16.2.

Expression of the NDH-1L complex appeared relatively unaffected by different growth conditions including high and low CO_2, iron deficiency, salt stress, photoauto-, mixo-, or photoheterotrophy (Herranen et al. 2004). In contrast, the NDH-1M and NDH-1S complexes were strongly expressed in cells cultured photoautotrophi-

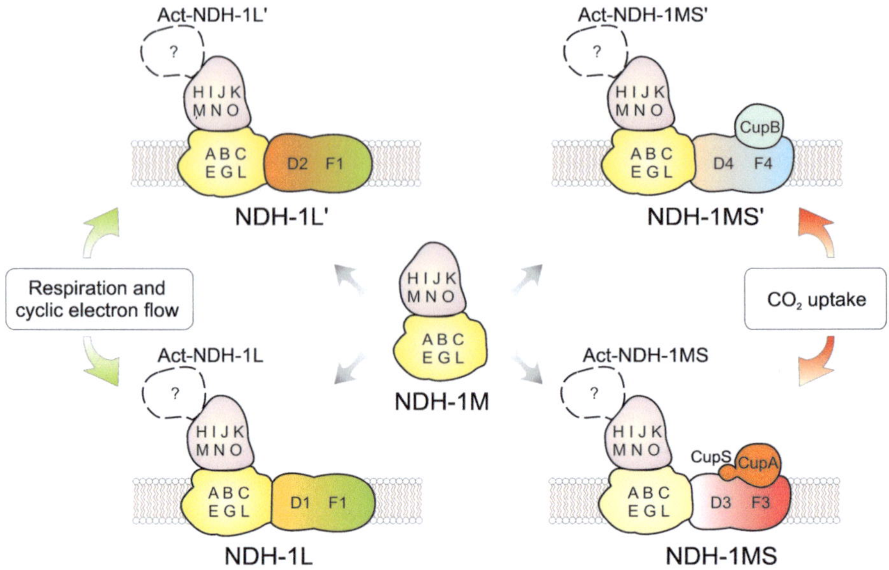

Fig. 16.2 Cyanobacterial NDH-1 complexes. Four NDH-1 complexes are predicted by reverse genetics. *NDH-1L* and *NDH-1L'* participate in respiration and cyclic electron flow around PSI while *NDH-1MS* and *NDH-1MS*' are involved in CO_2 uptake. The *NDH-1M* complex in the center of the picture is a basic subcomplex present in *NDH-1L* and *NDH-1MS* and probably in other two complexes as well. Membrane modules containing NdhD and NdhF subunits differ among the four NDH-1 complexes. Moreover, NDH-1 complexes, functional in CO_2 uptake, contain specific subunits associated with the corresponding membrane NdhD/F module (CupA and CupS in *NDH-1MS* and CupB in *NDH-1MS'*). The as-yet unidentified electron input device or activity module associated with NDH-1 is indicated by the question mark. (Modified from Battchikova and Aro 2007)

cally under air levels of CO_2. An increase in CO_2 concentration drastically reduced the expression of both the NDH-1M and NDH-1S complexes (Herranen et al. 2004; Battchikova et al. 2005; Zhang et al. 2005).

The NDH-1M complex is composed of the same 13 subunits present in NDH-1L except that NdhF1 and NdhD1 are absent (Battchikova et al. 2005). No other proteins of the NdhD and NdhF families were found in the NDH-1M complex.

The NDH-1S complex comprises NdhD3, NdhF3, CupA and Sll1734 (Herranen et al. 2004), the latter protein later designated CupS by Ogawa and Mi (2007). These proteomic results provided important evidence that the CupA protein was associated with NdhD3 and NdhD4, in agreement with conclusions made from reverse genetic studies (Shibata et al. 2001; Maeda et al. 2002). Furthermore, the CupS protein appeared to be a novel subunit of the same complex, despite the fact that no changes in the phenotype were found in a *cupS* deletion mutant of *Synechococcus* 7002 (Klughammer et al. 1999). The physiological function of CupS remains unknown.

Proteomic studies of NDH-1 complexes from the thermophilic cyanobacterium *T. elongatus* showed further that NDH-1M and NDH-1S formed a distinct complex, designated NDH-1MS, of 490 kDa (Zhang et al. 2005). NDH1-MS was observed only when low concentrations of detergent were used to solubilize the membranes and dissociated into NDH-1M and NDH-1S upon an increase in DM concentration (Zhang et al. 2005). The DM concentration of 1–1.5% (w/v), which is necessary to solubilize the NDH-1 complex from the *Synechocystis* 6803 membranes, nearly completely dissociated the NDH-1MS complex from *T. elongatus*. It is thus likely that the NDH-1MS complex is also the functional form of the NDH-1S complex in *Synechocystis* 6803. Additionally, the relatively high level of DM required for solubilization of *Synechocystis* membranes probably led to fragmentation of the NDH-1S complex into a smaller NDH-1S subcomplex (140 kDa, designated as NDH-1S_2) composed of only the NdhD3 and NdhF3 subunits (Herranen et al. 2004).

Functional proteomic analysis of various cyanobacterial mutants containing disruptions of *ndh* genes verified at the protein level the conclusions made from reverse genetic studies concerning the structural and functional multiplicity of NDH-1 complexes in cyanobacteria. The knock-out of *ndhB* resulted in complete disappearance of NDH-1L and NDH-1M complexes indicating that the hydrophobic NdhB subunit is vital for their assembly (Zhang et al. 2004; Battchikova et al. 2005). The observed inability of the *ndhB* mutant to form the NDH-1L and NDH-1M complexes explained the significance of the NdhB subunit for all NDH-1 functions: cyclic electron flow, respiration and CO_2 uptake (Ogawa 1991a; Mi et al. 1992a; Ohkawa et al. 2000b, 2002). In contrast, the assembly of NDH-1L and NDH-1M was not compromised in the *ndhL* mutant (Battchikova et al. 2005). Thus the importance of NdhL for the CO_2 uptake, which was explicitly demonstrated by Ogawa (1992), might be due to a more subtle functional influence of this small hydrophobic subunit.

In the *ndhD1* and *ndhD1/D2 Synechocystis* null mutants the NDH-1L complex is missing while NDH-1M and NDH-1S, which probably comprise the fragile NDH-1MS complex, are abundant in low CO_2 conditions (Zhang et al. 2004). Taking into account the functional studies performed with these mutants, it is possible that NDH-1L participates in cyclic electron flow and respiration and is important for heterotrophic growth. Conversely, the *ndhD3* and *ndhD3/D4* insertion mutants exhibit wild-type levels of NDH-1L. In both latter mutants the NDH-1S complex is absent at low CO_2 conditions (Zhang et al. 2004), and the function of CO_2 uptake is impaired. Thus, the putative NDH-1MS complex participates in low-CO_2 uptake. In the *ndhD3/D4* double mutant CO_2 acquisition is completely abolished in low CO_2 conditions at close to neutral pH (<7.5). However, the ability of the double mutant to grow at low CO_2 can be restored at elevated pH (>8.3). Proteomic analysis revealed strong induction of the SbtA bicarbonate transporter under this latter growth condition (Herranen et al. 2004; Zhang et al. 2004) indicating that the impairment in CO_2 uptake was compensated for by activation of mechanisms involved in bicarbonate acquisition, which agrees well with reverse genetic studies (Shibata et al. 2002b; Badger et al. 2006).

Recently Xu et al. (2008) introduced His_6 and cMyc tags into the C-terminus of *Synechocystis* CupB and demonstrated the presence of the tagged CupB protein in a membrane-bound protein complex of 440 kDa. The size of this complex is similar to that of the NDH-1MS complex containing the CUP-A module. The association of CupB with NdhD4 and NdhF4 was supported by the fact that the CupB protein was not found integrated into a membrane protein complex in *ndhD4* and *ndhF4* disruption mutants. The complex, designated NDH-1MS', appeared to be highly unstable during the purification procedure which prevented detailed characterization of the subunit composition. However, it is logical to predict that the discovered NDH-1MS' complex would be similar to NDH-1MS but with the CUP-B complex (NdhD4, NdhF4 and the CupB protein), instead of the CUP-A module, in the distal part of the membrane arm. The NDH-1MS' complex had not been detected previously in WT and mutant thylakoid proteomes possibly because of its low abundance, its instability, and/or difficulty in solubilization (in *Synechocystis,* the NdhF4 subunit is predicted to contain 17 transmembrane segments, the highest value for all proteins in this organism) (Zhang et al. 2004). These reasons might also explain why the NdhD2 subunit has not yet been detected in proteomic experiments. Based on reverse genetics experiments, it is predicted that this subunit is part of an analogous complex to NDH-1L, termed the NDH-1L' complex, except that it contains NdhD2 rather than NdhD1 (Fig. 16.2).

It is important to re-emphasize the fact that no proteins homologous to the activity subunits of the *E. coli* complex I (NuoE, F or G) have been found in NDH-1L and NDH-1MS (or NDH-1M and NDH-1S) complexes in proteomics experiments, and neither has NADPH nor NADH dehydrogenase activity been detected in isolated complexes. It is possible that both NDH-1L and NDH-1MS represent the connecting and membrane domains of NDH-1 left after dissociation of the components responsible for the enzymatic activity. However the current data cannot yet rule out the absence of a dehydrogenase complex and the use of some other electron input device or activity module.

Given that NDH-1 complexes appear to fragment during detergent solubilization, it also remains possible that NDH-1 complexes might exist as even larger complexes *in vivo.* Ma et al. (2006) observed that a minor amount of NDH-1 from high CO_2 grown *Synechocystis* cells migrates in native PAGE as a supercomplex of about 1000 kDa, in addition to complexes of smaller sizes. The corresponding band showed NADPH oxidation activity with nitroblue tetrazolium. NdhH, NdhI, NdhK as well as hydrophobic NdhA and NdhB subunits were detected with specific antibodies in this region of the gel while possible association with FNR was ruled out. The corresponding supercomplex was absent in protein extracts from *ndhD1/D2* and M55 *Synechocystis* mutants indicating that it is an NDH-1L-like complex. However, evidence showing the direct association of NADPH dehydrogenase activity with the NDH-1 complex, through affinity purification, for example, is lacking. Since the molecular mass of the complex was more than twice that of NDH-1L the authors suggested that the observed supercomplex (designated as Act-NDH-1Sup, *Act*ive *Sup*er-complex) could be a dimer of NDH-1L with additional unknown subunits. Further Ma and Mi (2008) reported that the activity of Act-NDH-1Sup (as determined by the intensity of in-gel staining) is proportional to the rate of cyclic PSI electron flow reflected by the initial rate of $P700^+$ reduction in the dark and is inhibited by addition of exogenous glucose. It is, however, intriguing that the active supercomplex almost completely disappeared when *Synechocystis* cells were grown at low CO_2. Instead, another NADPH-active band was observed corresponding to a NDH-1 complex of smaller molecular mass (~380 kDa) designated as Act-NDH-1M (*Act*ive *M*edium complex). The appearance of the smaller complex might be explained by the fact that the low CO_2 inducible NDH-1MS complex of *Synechocystis* was more susceptible to degradation than NDH-1L (Herranen et al. 2004, Zhang et al. 2004). Act-NDH-1M appears to be similar to the active NDH-1 complex identified by Matsuo et al. (1998a) and Deng et al. (2003a). Nevertheless, a strong decrease in the expression of NDH-1L in low CO_2 grown cells compared to high CO_2 grown ones has not been reported before. Besides, the *ndhD1* and *ndhF1* genes are considered to be constitutively expressed (Wang et al. 2004; Eisenhut et al. 2007), therefore disappearance of Act-NDH-1Sup at low CO_2 needs further investigation.

16.3.4 Cellular Location of NDH-1 Complexes

All variants of cyanobacterial NDH-1, similar to complex I of any origin, are membrane-embedded complexes. In non-photosynthetic bacteria, for example *E. coli,* the complex is found in the plasma membrane. Cyanobacterial cells contain in addition to the plasma membrane an extensive internal thylakoid membrane system as an alternative site for NDH-1. The exact location of cyanobacterial NDH-1 complexes remains at present a subject of debate and a dual location of some NDH-1 complexes in both the thylakoid membrane and the plasma membrane remains a possibility.

Data in the literature are controversial. Berger et al. (1991) and Pieulle et al. (2000) found NdhJ and NdhK subunits of *Synechocystis* 6803 complex in both the plasma membrane and the thylakoid membrane. The same result was described by Dworsky et al. (1995) for *Anacystis nidulans* NDH-1. Howitt et al. (1993) localized the NDH-1 complex in *Anabaena* sp. PCC 7120 explicitly to the plasma membrane. In contrast, Ogawa (1992) found the NdhL subunit only in the thylakoid membrane of *Synechocystis* 6803. In agreement with the latter observation, Zhang et al. (2004) detected NdhJ, NdhK, NdhD3 and NdhF3 subunits in the thylakoid membrane of *Synechocystis* and showed nearly complete absence of these proteins in the plasma membrane. Xu et al. (2008) obtained analogous results for the CupB subunit of *Synechocystis.* The thylakoid localization of NDH-1 was also supported by Mi et al. (1995) who demonstrated that the donation of electron from NADPH to plastoquinone occurs in the thylakoid membrane of wild type *Synechocystis* 6803.

It has been suggested that various NDH-1 complexes having different functions might reside in different membranes (Ohkawa et al. 1998; Price et al. 1998). However, the results of Ogawa (1992), Zhang et al. (2004) and Xu et al. (2008) imply that NDH-1L, NDH-1MS and NDH-1MS' localize largely, if not exclusively, to the thylakoid membrane of *Synechocystis* 6803. Moreover, since NdhJ and NdhK should belong (although it is not yet proven) to all four predicted variants of *Synechocystis* NDH-1, including the still evasive NDH-1L', this conclusion should hold true for all four NDH-1 complexes. The inconsistency in conclusions might arise from variations in the purity of the membrane preparations obtained with different isolation methods. The more recent experiments showing that in *Synechocystis* 6803 the NDH-1 complexes are confined in the thylakoid membrane (Ohkawa et al. 2001, 2002; Zhang et al. 2004, Xu et al. 2008) were performed with thylakoid and plasma membranes prepared by the two-phase partitioning method (Norling et al. 1998), whose purity has been validated using antibodies specific to widely accepted protein markers of the thylakoid (CP43 of Photosystem II) and plasma membranes (NrtA/SbtA), respectively. However, it still remains unclear whether the two membrane fractions are contaminated by specific membrane domains from the other membrane type.

In the case of studies on *Anabaena* sp. PCC 7120 (Howitt et al. 1993), antibodies raised against complex I from red beet mitochondria detected the cyanobacterial NDH-1 complex in the cytoplasmic membrane and not the thylakoid membrane. However, given the considerable differences between the cyanobacterial NDH-1 complexes and plant mitochondrial complex I, both in subunit complexity and primary structure (Heazlewood et al. 2003), it is important to confirm that the cross-reactions obtained in cyanobacteria were indeed due to *bona fide* cyanobacterial NDH-1 subunits.

16.3.5 Structural Studies

Both the mitochondrial and bacterial NDH-1 complexes have been shown by electron microscopy to have a characteristic L-shaped structure in which the hydrophobic module is embedded in the membrane and the hydrophilic peripheral arm of the complex protrudes into the aqueous phase (Sazanov 2007). Similar 'L-shaped' structures have also been detected in His-tagged NDH-1 complexes isolated from *T. elongatus* (Arteni et al. 2006) although there are some interesting differences.

The subunit interactions have been examined in His-tagged NDH-1 complexes of *Synechocystis* 6803 following exposure to the detergent Triton X-100 to fragment the complex into smaller sub-complexes (Prommeenate et al. 2004). It was found that the hydrophobic NdhD1 and NdhF1 subunits could be easily detached as expected for a position at the exposed distal end of the complex. In line with current models of complex I, sub-complexes composed of hydrophilic (NdhH, I, J, K and M) and hydrophobic (NdhA, B, C, E, G) subunits could be generated. More unexpectedly was the finding that NdhI and NdhA could be selectively removed from these sub-complexes indicating a peripheral location. Of the other cyanobacterial-specific Ndh subunits, the hydrophilic NdhN and NdhO subunit are likely to be part of the interconnecting fragment, although evidence is currently lacking, and the hydrophobic NdhL subunit is most probably part of the membrane domain (Battchikova et al. 2005).

16.4 Comparison with the Chloroplast NDH Complex

The closely related plastid NDH-1 complex (or NDH/Ndh complex) is also thought to play a role in the reduction of plastoquinone, either through cyclic electron flow around PSI or through a chlororespiratory activity involving oxidation of a stromal reductant (reviewed recently by Nixon and Rich 2006; Endo et al. 2008). Like the cyanobacterial complex, attempts to isolate the complex have been frustrated by its low abundance (approximately 1% of the levels of PSII in chloroplasts) (Burrows et al. 1998) and its lability (Sazanov et al. 1998; Rumeau et al. 2005). A His-tagging approach has been used to isolate a NDH-1 subcomplex from tobacco (Rumeau et al. 2005) so it is likely that the use of a suitable affinity tag and appropriate solubilisation conditions might allow the isolation of the intact complex. Activity assays using isolated NDH-1 preparations suggest that the chloroplast NDH-1 complex acts as an NADH:plastoquinone oxidoreductase (Sazanov et al. 1998; Rumeau et al. 2005) but, like cyanobacterial NDH-1, this assignment must remain tentative until the nature of the electron input module has been characterized. Careful experiments using BN-PAGE have recently revealed that the NDH-1 complex found in etioplasts exists as a 550-kDa monomeric complex whereas in chloroplasts it forms a PSI/NDH-1 supercomplex of mass greater than 1000 kDa, possibly to facilitate cyclic electron flow around PSI (Peng et al. 2008). Whether a similar PSI/NDH-1 supercomplex also occurs in cyanobacteria is currently unknown. Proteomic studies on bundle sheath chloroplasts in maize have also raised the interesting possibility that chloroplasts, like cyanobacteria, might contain different types of NDH-1 complex, some of which might be involved in CO_2 uptake (Majeran et al. 2008). However, much more work is required to substantiate this interesting possibility.

The presence of a functional chloroplast NDH complex is associated with a post-illumination increase in the yield of chlorophyll fluorescence from leaves following a period of illumination (Burrows et al. 1998; Shikanai et al. 1998). A number of *Arabidopsis* mutants have now been isolated that lack this fluorescence signature, some of which show reduced amounts of the chloroplast NDH complex when assessed by immunoblotting (Table 16.2). Some of the mutants are affected in mRNA processing in the chloroplast (e.g., Hashimoto et al. 2003) but others appear to be either assembly factors or structural components of the NDH-1 complex. Interestingly several subunits do not appear to have cyanobacterial homologues (Table 16.2) and so would appear to be specific features of the chloroplast NDH-1 complex, whereas several have potential cyanobacterial homologues which might play some role in NDH-1 assembly and function in cyanobacteria. Of greatest interest are subunits that might have a redox role such as NDF4, which is known to contain a Fe-S

Table 16.2 Candidate Ndh subunits identified in *A. thaliana* and their *Synechocystis* 6803 homologues

Arabidopsis protein	Possible function	Closest homologue in *Synechocystis* 6803 (% sequence identity)
CRR1 (At5g52100) (Shimizu and Shikanai 2007)	NAD(P)H-binding?	Sll1058/*dapB* (43%)
CRR3 (At2g01590) (Muraoka et al. 2006)		
CRR6 (At2g47910) (Munshi et al. 2006)		Slr1097 (46%)
CRR7 (At5g39210) (Munshi et al. 2005)		Ssl3829 (33%)
NDH45/NDF2 (At1g64770) (Sirpiö et al. 2009; Takabayashi et al. 2009)		
NDH48/NDF1 (At1g15980) (Sirpiö et al. 2009; Takabayashi et al. 2009)	Glycosyltransferase?	Slr0606 (22%)
NDF4 (At3g16250) (Takabayashi et al. 2009)	Binds 4Fe-4S cluster	Ssl3044 (47%)
NDF5 (At1g55370) (Ishida et al. 2009)		
NDF6 (At1g18730) (Ishikawa et al. 2008)		
PPL2 (At2g39470) (Ishihara et al. 2007)		Sll1418/PsbP2 (27%)

center and which has a homologue in *Synechocystis* 6803 (Ssl3044) and in other cyanobacteria (Takabayashi et al. 2009). In addition CRR1 is potentially an NADH-binding protein that might be part of an electron-input module in the chloroplast NDH complex (Shimizu and Shikanai 2007). There is a homologue in *Synechocystis* 6803 but it has been argued that it is unlikely to play a role in NDH-1 function (Shimizu and Shikanai 2007).

16.5 Conclusions and Future Directions

Overall the current data suggest that NDH-1 complexes in cyanobacteria and chloroplasts are an important sub-class of the complex I superfamily with specialized functions important for oxygenic photosynthesis. Much progress has been made in characterizing their subunit composition in *Synechocystis* 6803 and *T. elongatus*. The common structural element of NDH-1 complexes from these cyanobacteria is the NDH-1M complex composed of a hydrophilic connecting domain (containing the NdhH-K and NdhM-O subunits) and a hydrophobic domain consisting of the NdhA-C, NdhE, NdhG and NdhL subunits (Fig. 16.2). The diversity seen in the cyanobacterial NDH-1 complexes seems to be generated by combining the NDH-1M core complex with specific NdhD/NdhF modules: for example NdhD1/NdhF1 in NDH-1L, NdhD3/NdhF3/CupA/CupS (CUP-A module) in NDH-1MS, and NdhD4/NdhF4/CupB (CUP-B module) in NDH-1MS' (Table 16.1). Based on reverse genetics experiments, an additional NDH-1L complex (termed NDH-1L'), containing NdhD2 instead of NdhD1, is predicted, but its existence continues to remain elusive. Given their physiological importance, a key area of future research will be to understand how the NDH-1MS and NDH-1MS' complexes facilitate CO_2 uptake.

A longstanding question that still remains to be answered is the identity of the electron donors and activity modules used by the NDH-1 complexes of cyanobacteria and chloroplasts. The NdhI and NdhK proteins are homologous to NuoI and NuoB of *E. coli* complex I, respectively, which bind the three Fe-S clusters (N6a, N6b and N2) (Table 16.1) that form the terminal part of the electron transfer chain from NADH to ubiquinone (Sazanov and Hinchliffe 2006). It remains unclear how electrons are delivered to the proposed N6a cluster in the NdhI subunit. Over the years a number of candidates have been suggested to act as the electron-input module, including the bi-directional hydrogenase (Appel and Schulz 1996) and a homologue of the coenzyme $F_{420}H_2$ dehydrogenase (Prommeenate et al. 2004). However there is little evidence to support either of these suggestions. The recent detection of large NDH-1 supercomplexes, apparently retaining NADPH dehydrogenase activity, is a promising development that might lead to clarification of the unknown sub-complex (Ma et al. 2006). However, given the modular nature of complex I (Friedrich et al. 1995; Friedrich and Scheide 2000), it is quite feasible that the chloroplast and cyanobacterial NDH-1 complexes have evolved to use a different electron donor to NAD(P)H. Indeed the longer the dehydrogenase module remains unidentified the greater the suspicion that the chloroplast and cyanobacterial NDH-1 complexes actually function as ferredoxin:plastoquinone oxidoreductases as proposed by Friedrich et al. (1995) and for which there is some experimental evidence (Mi et al. 1995). It should also be noted that not all members of the complex I family of respiratory enzymes oxidize NADH. For instance, complex I in both *Helicobacter pylori* and *Campylobacter jejuni* lack homologues of NuoE and NuoF and instead contain two subunits, not conserved in other complex I enzymes, that might provide a docking site for an alternative electron donor to the complex (Finel 1998). In the case of *C. jejuni* the donor has recently been identified as flavodoxin rather than ferredoxin or NAD(P)H (Weerakoon and Olson 2008). It is therefore tempting to speculate that NdhM, NdhN and NdhO might provide a docking site(s) for the, as yet, unidentified electron input module(s) of the cyanobacterial and chloroplast NDH-1 complex.

If one wants to understand the enzymology of the NDH-1 complexes at a molecular level, it is important to obtain high resolution structures of the various complexes. The use of thermophilic cyanobacteria such as *T. elongatus* offers a promising route for the isolation of stable NDH-1 complexes, although low yield and sample heterogeneity might well provide formidable obstacles. However, these might be overcome with the use of multiple affinity tags and optimization of solubilization conditions.

Acknowledgements PJN is grateful to the Biotechnology and Biological Sciences Research Council for financial support. EMA acknowledges the Academy of Finland (Project No. 118637), the EU project Solar-H2 (FP7 contract no 212508), Nordic Energy and the Maj and Tor Nessling Foundation for support of research.

References

Alpes I, Scherer S and Böger P (1989) The respiratory NADH dehydrogenase of the cyanobacterium *Anabaena variabilis:* purification and characterization. Biochim Biophys Acta 973: 41–46

Appel J and Schulz R (1996) Sequence analysis of an operon of a NAD(P)-reducing nickel hydrogenase from the cyanobacterium Synechocystis sp. PCC 6803 gives additional evidence for direct coupling of the enzyme to NAD(P)H-dehydrogenase (complex I). Biochim Biophys Acta 1298: 141–147

Arteni AA, Zhang P, Battchikova N, Ogawa T, Aro E-M, Boekema EJ (2006) Structural characterization of the NDH-1 complexes of *Thermosynechococcus elongatus* by single particle electron microscopy. Biochim Biophys Acta 1757: 1469–1475

Badger MR and Price GD (2003) CO_2 concentrating mechanisms in cyanobacteria: molecular components, their diversity and evolution. J Exp Bot 54: 609–622

Badger MR and Spalding MH (2000) CO_2 acquisition, concentration and fixation in cyanobacteria and algae. In: Leegood RC, Sharkey TD and von Caemmerer S (eds) Photosynthesis: Physiology and Metabolism, pp 369–397. Kluwer Academy Publishers, Dordrecht, The Netherlands

Badger MR, Hanson D and Price GD (2002) Evolution and diversity of CO_2 concentrating mechanisms in cyanobacteria. Funct Plant Biol 29: 161–173

Badger MR, Price GD, Long BM and Woodger FJ (2006) The environmental plasticity and ecological genomics of the cyanobacterial CO_2 concentrating mechanism. J Exp Bot 57: 249–265

Battchikova N and Aro E-M (2007) Cyanobacterial NDH-1 complexes: multiplicity in function and subunit composition. Physiol Plant 131: 22–32

Battchikova N, Zhang P, Rudd S, Ogawa T and Aro E-M (2005) Identification of NdhL and Ssl1690 (NdhO) in NDH-1L and NDH-1 M complexes of *Synechocystis* sp. PCC 6803. J Biol Chem 280: 2587–2595

Bäumer S, Ide T, Jacobi C, Johann A, Gottschalk G and Deppenmeier U (2000) The $F_{420}H_2$ dehydrogenase from *Methanosarcina mazei* is a redox-driven proton pump closely related to NADH dehydrogenases. J Biol Chem 275: 17968–17973

Berger S, Ellersiek U and Steinmüller K (1991) Cyanobacteria contain a mitochondrial complex I-homologous NADH-dehydrogenase. FEBS Lett 286: 129–132

Berger S, Ellersiek U, Kinzelt D and Steinmüller K (1993) Immunopurification of a subcomplex of the NAD(P)H-plastoquinone-oxidoreductase from the cyanobacterium *Synechocystis* sp. PCC6803. FEBS Lett 326: 246–250

Brandt U (2006) Energy converting NADH:quinone oxidoreductase (Complex I). Annu Rev Biochem 75: 69–92

Burrows PA, Sazanov LA, Svab Z, Maliga P and Nixon PJ (1998) Identification of a functional respiratory complex in chloroplasts through analysis of tobacco mutants containing disrupted plastid *ndh* genes. EMBO J 17: 868–876

Carroll J, Fearnley IM, Skehel JM, Shannon RJ, Hirst J and Walker JE (2006) Bovine complex I is a complex of 45 different subunits. J Biol Chem 281: 32724–32727

Cooley JW and Vermaas WFJ (2001) Succinate dehydrogenase and other respiratory pathways in thylakoid membranes of *Synechocystis* sp. strain PCC 6803: capacity comparisons and physiological function. J Bacteriol 183: 4251–4258

Deng Y, Ye J, Mi H and Shen Y (2003a) Separation of hydrophobic NAD(P)H dehydrogenase subcomplexes from cyanobacterium *Synechocystis* PCC6803. Acta Biochim Biophys Sinica 35: 723–727.

Deng Y, Ye J and Mi H (2003b) Effects of low CO_2 on NAD(P)H dehydrogenase, a mediator of cyclic electron transport around photosystem I in the cyanobacterium *Synechocystis* PCC 6803. Plant Cell Physiol 44: 534–540

Dworsky A, Mayer B, Regelsberger G, Fromwald S and Peschek GA (1995) Functional and immunological characterization of both "mitochondria-like" and "chloroplast-like" electron/proton transport proteins in isolated and purified cyanobacterial membranes. Bioelectrochem Bioenerg 38: 35–43

Eisenhut M, von Wobeser EA, Jonas L, Schubert H, Ibelings BW, Bauwe H, Matthijs HCP and Hagemann M (2007) Long-term response toward inorganic carbon limitation in wild type and glycolate turnover mutants of the cyanobacterium *Synechocystis* sp. Strain PCC 6803. Plant Physiol 144: 1946–1959

Efremov RG, Baradaran R and Sazanov LA (2010) The architecture of complex I. Nature 465: 441–447

Endo T, Ishida S, Ishikawa N and Sato F (2008) Chloroplastic NAD(P)H dehydrogenase complex and cyclic electron transport around Photosystem I. Mol Cells 25: 158–162

Finel M (1998) Does NADH play a central role in energy metabolism in *Helicobacter pylori?* Trends Biochem Sci 23: 412–414

Friedrich T and Scheide D (2000) The respiratory complex I of bacteria, archaea and eukarya and its module common with membrane-bound multisubunit hydrogenases. FEBS Lett 479: 1–5

Friedrich T, Steinmüller K and Weiss H (1995) The proton-pumping respiratory complex I of bacteria and mitochondria and its homologue in chloroplasts. FEBS Lett 367: 107–111

Hashimoto M, Endo T, Peltier G, Tasaka M and Shikanai T (2003) A nucleus-encoded factor, CRR2, is essential for the expression of chloroplast *ndhB* in *Arabidopsis.* Plant J 36: 541–549

Heazlewood JL, Howell KA and Millar AH (2003) Mitochondrial complex I from Arabidopsis and rice: orthologs of mammalian and fungal components coupled with plant-specific subunits. Biochim Biophys Acta 1604: 159–169

Herranen M, Battchikova N, Zhang P, Graf A, Sirpiö S, Paakkarinen V and Aro E-M (2004) Towards functional proteomics of membrane protein complexes in *Synechocystis* sp. PCC 6803. Plant Physiol 134: 470–481

Howitt CA and Vermaas WFJ (1998) Quinol and cytochrome oxidases in the cyanobacterium *Synechocystis* sp. PCC 6803. Biochemistry 37: 17944–17951

Howitt CA, Smith GD and Day DA (1993) Cyanide-insensitive oxygen uptake and pyridine nucleotide dehydrogenases in the cyanobacterium *Anabaena* PCC 7120. Biochim Biophys Acta 1141: 313–320

Howitt C, Udall PK and Vermaas W (1999) Type 2 NADH dehydrogenases in the cyanobacterium Synechocystis sp. strain PCC 6803 are involved in regulation rather than respiration. J Bacteriol 181: 3994–4003

Ishida S, Takabayashi A, Ishikawa N, Hano Y, Endo T and Sato F (2009) A novel nuclear-encoded protein, NDH-dependent cyclic electron flow 5, is essential for the accumulation of chloroplast NAD(P)H dehydrogenase complexes. Plant Cell Physiol 50: 383–393

Ishihara S, Takabayashi A, Ido K, Endo T, Ifuku K and Sato F (2007) Distinct functions for the two PsbP-like proteins PPL1 and PPL2 in the chloroplast thylakoid lumen of *Arabidopsis.* Plant Physiol 145: 668–679

Ishikawa N, Takabayashi A, Ishida S, Endo T and Sato F (2008) NDF6: a thylakoid protein specific to terrestrial plants is essential for activity of chloroplastic NAD(P)H dehydrogenase in *Arabidopsis*. Plant Cell Physiol 49: 1066–1073

Joët T, Cournac L, Peltier G and Havaux M (2002) Cyclic electron flow around photosystem I in C3 plants. In vivo control by the redox state of chloroplasts and involvement of the NADH-dehydrogenase complex. Plant Physiol 128: 760–769

Kaneko T, Sato S, Kotani H, Tanaka A, Asamizu E, Nakamura Y, Miyajima N, Hirosawa M, Sugiura M, Sasamoto S, Kimura T, Hosouchi T, Matsuno A, Muraki A, Nakazaki N, Naruo K, Okumura S, Shimpo S, Takeuchi C, Wada T, Watanabe A, Yamada M, Yasuda M and Tabata S (1996) Sequence analysis of the genome of the unicellular cyanobacterium Synechocystis sp. strain PCC6803. II. Sequence determination of the entire genome and assignment of potential protein-coding regions. DNA Res 3: 109–136

Kaplan A and Reinhold L (1999) The CO_2 concentrating mechanisms in photosynthetic microorganisms. Annu Rev Plant Physiol Plant Mol Biol 50: 539–570

Kaplan A, Hagemann M, Bauwe H, Kahlon S and Ogawa T (2007) Carbon acquisition by cyanobacteria: mechanisms, comparative genomics, and evolution. In: Herrero A and Flores E (eds) The Cyanobacteria: Molecular Biology, Genomics and Evolution, pp 305–334. Horizon Scientific Press, Norwich, UK

Klughammer B, Sültemeyer D, Badger MR and Price GD (1999) The involvement of NAD(P)H dehydrogenase subunits, NdhD3 and NdhF3, in high-affinity CO_2 uptake in *Synechococcus* sp. PCC7002 gives evidence for multiple NDH-1 complexes with specific roles in cyanobacteria. Mol Microbiol 32: 1305–1315

Kuntz M (2004) Plastid terminal oxidase and its biological significance. Planta 218: 896–899

Ma W and Mi H (2008) Effect of exogenous glucose on the expression and activity of NADPH dehydrogenase complexes in the cyanobacterium *Synechocystis* sp. strain PCC 6803. Plant Physiol Biochem 46: 775–779

Ma W, Deng Y, Ogawa T and Mi H (2006) Active NDH-1 complexes from the cyanobacterium *Synechocystis* sp. strain PCC 6803. Plant Cell Physiol 47: 1432–1436

Maeda S, Badger MR and Price GD (2002) Novel gene products associated with NdhD3/D4-containing NDH-1 complexes are involved in photosynthetic CO_2 hydration in the cyanobacterium, *Synechococcus* sp. PCC7942. Mol Microbiol 43: 425–435

Majeran W, Zybailov B, Ytterberg AJ, Dunsmore J, Sun Q and van Wijk KJ (2008) Consequences of C4 differentiation for chloroplast membrane proteomes in maize mesophyll and bundle sheath cells. Mol Cell Proteomics 7: 1609–1638

Marco E, Ohad N, Schwarz R, Lieman-Hurwitz J, Gabay C and Kaplan A (1993) High CO_2 concentration alleviates the block in photosynthetic electron transport in an *ndhB*-inactivated mutant of *Synechococcus* sp. PCC 7942. Plant Physiol 101: 1047–1053

Mathiesen C and Hägerhäll C (2003) The 'antiporter module' of respiratory chain complex I includes the MrpC/NuoK subunit—a revision of the modular evolution scheme. FEBS Lett 549: 7–13

Matsuo M, Endo T and Asada K (1998a) Properties of the respiratory NAD(P)H dehydrogenase isolated from the cyanobacterium *Synechocystis* PCC6803. Plant Cell Physiol 39: 263–267

Matsuo M, Endo T and Asada K (1998b) Isolation of a novel NAD(P)H-quinone oxidoreductase from the cyanobacterium synechocystis PCC6803. Plant Cell Physiol 39: 751–755.

McDonald AE and Vanlerberghe GC (2006) Origins, evolutionary history, and taxonomic distribution of alternative oxidase and plastoquinol terminal oxidase. Comp Biochem Biophys Part D1: 357–364

Mi H, Endo T, Schreiber U, Ogawa T and Asada K (1992a) Electron donation from cyclic and respiratory flows to the photosynthetic intersystem chain is mediated by pyridine nucleotide dehydrogenase in the cyanobacterium *Synechocystis* PCC 6803. Plant Cell Physiol 33: 1233–1237

Mi H, Endo T, Schreiber U and Asada K (1992b) Donation of electrons from cytosolic components to the intersystem chain in the cyanobacterium *Synechococcus* sp. PCC 7002 as determined by reduction of P700+. Plant Cell Physiol 33: 1099–1105

Mi H, Endo T, Ogawa T and Asada K (1995) Thylakoid membrane-bound, NADPH-specific pyridine nucleotide dehydrogenase complex mediates cyclic electron transport in the cyanobacterium *Synechocystis* sp. PCC 6803. Plant Cell Physiol 36: 661–668

Munekage Y, Hojo M, Meurer J, Endo T, Tasaka M and Shikanai T (2002) PGR5 is involved in cyclic electron flow around photosystem I and is essential for photoprotection in *Arabidopsis*. Cell 110: 361–371

Munekage Y, Hashimoto M, Miyake C, Tomizawa K, Endo T, Tasaka M and Shikanai T (2004) Cyclic electron flow around photosystem I is essential for photosynthesis. Nature 429: 579–582

Munshi MK, Kobayashi Y and Shikanai T (2005) Identification of a novel protein, CRR7, required for the stabilization of the chloroplast NAD(P)H dehydrogenase complex in *Arabidopsis*. Plant J 44: 1036–1044

Munshi MK, Kobayashi Y and Shikanai T (2006) CHLORORESPIRATORY REDUCTION 6 is a novel factor required for accumulation of the chloroplast NAD(P)H dehydrogenase complex in *Arabidopsis*. Plant Physiol 141: 737–744

Muraoka R, Okuda K, Kobayashi Y and Shikanai T (2006) A eukaryotic factor required for accumulation of the chloroplast NAD(P)H dehydrogenase complex in *Arabidopsis*. Plant Physiol 142: 1683–1689

Nixon PJ and Mullineaux CW (2001) Regulation of photosynthetic electron transport. In: Aro E-M and Andersson B (eds) Regulation of Photosynthesis. Advances in Photosynthesis and Respiration, vol 11, pp 533–555. Kluwer Academic Publishers, Dordrecht, The Netherlands

Nixon PJ and Rich PR (2006) Chlororespiratory pathways and their physiological significance. In: Wise RR and Hoober JK (eds) The Structure and Function of Plastids. Advances in Photosynthesis and Respiration, vol 23, pp 237–251. Kluwer Academic Publishers, Dordrecht, The Netherlands

Norling B, Zak E, Andersson B and Pakrasi H (1998) 2D-isolation of pure plasma and thylakoid membranes from the cyanobacterium *Synechocystis* sp. PCC 6803. FEBS Lett 436: 189–192

Nowaczyk MM, Wulfhorst H, Ryan CM, Souda P, Zhang H, Cramer WA and Whitelegge JP (2011) NdhP and NdhQ: Two novel small subunits of the cyanobacterial NDH-1 complex. Biochemistry (in press)

Ogawa T (1991a) A gene homologous to the subunit-2 gene of NADH dehydrogenase is essential to inorganic carbon transport of *Synechocystis* PCC6803. Proc Natl Acad Sci USA 88: 4275–4279

Ogawa T (1991b) Cloning and inactivation of a gene essential to inorganic carbon transport of *Synechocystis* PCC6803. Plant Physiol 96: 280–284

Ogawa T (1992) Identification and characterization of the *ictA/ndhL* gene product essential to inorganic carbon transport of *Synechocystis* sp. strain PCC6803. Plant Physiol 99: 1604–1608

Ogawa T and Mi H (2007) Cyanobacterial NADPH dehydrogenase complexes. Photosynth Res 93: 69–77

Ohkawa H, Sonoda M, Katoh H and Ogawa T (1998) The use of mutants in the analysis of the CO_2-concentrating mechanism in cyanobacteria. Can J Bot 76: 1035–1042

Ohkawa H, Pakrasi HB and Ogawa T (2000a) Two types of functionally distinct NAD(P)H dehydrogenases in *Synechocystis* sp. strain PCC 6803. J Biol Chem 275: 31630–31634

Ohkawa H, Price GD, Badger MR and Ogawa T (2000b) Mutation of *ndh* genes leads to inhibition of CO_2 uptake rather than HCO_3^- uptake in *Synechocystis* sp. strain PCC 6803. J Bacteriol 182: 2591–2596

Ohkawa H, Sonoda M, Shibata M and Ogawa T (2001) Localization of NAD(P)H dehydrogenase in the cyanobacterium *Synechocystis* sp. strain PCC 6803. J Bacteriol 183: 4938–4939

Ohkawa H, Sonoda M, Hagino N, Shibata M, Pakrasi HB and Ogawa T (2002) Functionally distinct NAD(P)H dehydrogenases and their membrane localization in *Synechocystis* sp. PCC6803. Funct Plant Biol 29: 195–200

Omata T and Ogawa T (1985) Changes in the polypeptide composition of the cytoplasmic membrane in the cyanobacterium *Anacystis nidulans* during adaptation to low CO_2 conditions. Plant Cell Physiol 26: 1075–1081

Omata T, Price GD, Badger MR, Okamura M, Gohta S and Ogawa T (1999) Identification of an ATP-binding cassette transporter involved in bicarbonate uptake in the cyanobacterium *Synechococcus* sp. strain PCC 7942. Proc Natl Acad Sci USA 96: 13571–13576

Pasquier C and Hamodrakas SJ (1999) An hierarchical artificial neural network system for the classification of transmembrane proteins. Protein Eng 12: 631–634

Peltier G and Cournac L (2002) Chlororespiration. Annu Rev Plant Biol 53: 523–550

Peng L, Shimizu H and Shikanai T (2008) The chloroplast NAD(P)H dehydrogenase complex interacts with Photosystem I in *Arabidopsis*. J Biol Chem 283: 34873–34879

Peschek GA (1980) Electron transport reactions in respiratory particles of hydrogenase-induced *Anacystis nidulans*. Arch Microbiol 125: 123–131

Pieulle L, Guedeney G, Cassier-Chauvat C, Jeanjean R, Chauvat F and Peltier G (2000) The gene encoding the NdhH subunit of type 1 NAD(P)H dehydrogenase is essential to survival of *Synechocystis* PCC6803. FEBS Lett 487: 272–276

Price GD, Sültemeyer D, Klughammer B, Ludwig M and Badger MR (1998) The functioning of the CO_2 concentrating mechanism in several cyanobacterial strains: a review of general physiological characteristics, genes, proteins, and recent advances. Can J Bot 76: 973–1002

Price GD, Maeda S, Omata T and Badger MR (2002) Modes of active inorganic carbon uptake in the cyanobacterium, *Synechococcus* sp. PCC7942. Funct Plant Biol 29: 131–149

Price GD, Woodger FJ, Badger MR, Howitt SM and Tucker L. (2004) Identification of a SulP-type bicarbonate transporter in marine cyanobacteria. Proc Natl Acad Sci USA 101: 18228–18233

Price GD, Badger MR, Woodger FJ and Long BM (2008) Advances in understanding the cyanobacterial CO_2-concentrating-mechanism (CCM): functional components, Ci transporters, diversity, genetic regulation and prospects for engineering into plants. J Exp Bot 59: 1441–1461

Prommeenate P, Lennon AM, Markert C, Hippler M and Nixon PJ (2004) Subunit composition of NDH-1 complexes of *Synechocystis* sp. PCC 6803: identification of two new *ndh* gene products with nuclear-encoded homologues in the chloroplast Ndh complex. J Biol Chem 279: 28165–28173

Rumeau D, Becuwe-Linka N, Beyly A, Louwagie M, Garin J and Peltier G (2005) New subunits NDH-M, -N, and -O, encoded by nuclear genes, are essential for plastid Ndh complex functioning in higher plants. Plant Cell 17: 219–232

Rumeau D, Peltier G and Cournac L (2007) Chlororespiration and cyclic electron flow around PSI during photosynthesis and plant stress response. Plant Cell Environment 30: 1041–1051

Sazanov LA (2007) Respiratory complex I: mechanistic and structural insights provided by the crystal structure of the hydrophilic domain. Biochemistry 46: 2275–2288

Sazanov LA and Hinchliffe P (2006) Structure of the hydrophilic domain of respiratory complex I from Thermus thermophilus. Science 311: 1430–1436

Sazanov LA, Burrows PA and Nixon PJ (1998) The plastid *ndh* genes code for an NADH-specific dehydrogenase: isolation of a complex I analogue of from pea thylakoid membranes. Proc Natl Acad Sci USA 95: 1319–1324

Scherer S, Almon H and Böger P (1988) Interaction of photosynthesis, respiration and nitrogen fixation in cyanobacteria. Photosynth Res 15: 95–114

Schluchter WM, Zhao J and Bryant DA (1993) Isolation and characterization of the *ndhF* gene of *Synechococcus* sp. strain PCC 7002 and initial characterization of an interposon mutant. J Bacteriol 175: 3343–3352

Schmetterer G (1994) Cyanobacterial respiration. In: Bryant DA (ed) The Molecular Biology of Cyanobacteria. Advances in Photosynthesis and Respiration, vol 1, pp 409–435. Springer, Dordrecht, The Netherlands

Shibata M, Ohkawa H, Kaneko T, Fukuzawa H, Tabata S, Kaplan A and Ogawa T (2001) Distinct constitutive and low-CO_2-induced CO_2 uptake systems in cyanobacteria: genes involved and their phylogenetic relationship with homologous genes in other organisms. Proc Natl Acad Sci USA 98: 11789–11794

Shibata M, Katoh H, Sonoda M, Ohkawa H, Shimoyama M, Fukuzawa H, Kaplan A and Ogawa T (2002a) Genes essential to sodium-dependent bicarbonate transport in cyanobacteria: function and phylogenetic analysis. J Biol Chem 277: 18658–18664

Shibata M, Ohkawa H, Katoh H, Shimoyama M and Ogawa T (2002b) Two CO_2 uptake systems in cyanobacteria: four systems for inorganic carbon acquisition in *Synechocystis* sp. strain PCC 6803. Funct Plant Biol 29: 123–129

Shikanai T and Endo T (2000) Physiological function of a respiratory complex, NAD(P)H dehydrogenase in chloroplasts: dissection by chloroplast reverse genetics. Plant Biotech 17: 79–86

Shikanai T, Endo T, Hashimoto T, Yamada Y, Asada K and Yokota A (1998) Directed disruption of the tobacco *ndhB* gene impairs cyclic electron flow around photosystem I. Proc Natl Acad Sci USA 95: 9705–9709

Shimizu H and Shikanai T (2007) Dihydrodipicolinate reductase-like protein, CRR1, is essential for chloroplast NAD(P)H dehydrogenase in *Arabidopsis*. Plant J 52: 539–547

Shimizu H, Peng L, Myouga F, Motohashi R, Shinozaki K and Shikanai T (2008) CRR23/NdhL is a Subunit of the Chloroplast NAD(P)H Dehydrogenase Complex in Arabidopsis. Plant Cell Physiol 49: 835–842

Sirpiö S, Allahverdiyeva Y, Holmstrom M, Khrouchtchova A, Haldrup A, Battchikova N and Aro E-M (2009) Novel nuclear-encoded subunits of the chloroplast NAD(P)H dehydrogenase complex. J Biol Chem 284: 905–912

Takabayashi A, Ishikawa N, Obayashi T, Ishida S, Obokata J, Endo T and Sato F (2009) Three novel subunits of *Arabidopsis* chloroplastic NAD(P)H dehydrogenase identified by bioinformatic and reverse genetic approaches. Plant J 57: 207–219

Tanaka Y, Katada S, Ishikawa H, Ogawa T and Takabe T (1997) Electron flow from NAD(P)H dehydrogenase to photosystem I is required for adaptation to salt shock in the cyanobacterium *Synechocystis* sp. PCC 6803. Plant Cell Physiol 38: 1311–1318

Tchernov D, Helman Y, Keren N, Luz B, Ohad I, Reinhold L, Ogawa T and Kaplan A (2001) Passive entry of CO_2 and its energy-dependent intracellular conversion to HCO_3^- in cyanobacteria are driven by a photosystem I-generated $\Delta\mu H^+$. J Biol Chem 276: 23450–23455

Volokita M, Zenvirth D, Kaplan A and Reinhold L (1984) Nature of the inorganic carbon species actively taken up by the cyanobacterium *Anabaena variabilis*. Plant Physiol 76: 599–602

Wang H-L, Postier BL and Burnap RL (2004) Alterations in global patterns of gene expression in *Synechocystis* sp. PCC 6803 in response to inorganic carbon limitation and the inactivation of ndhR, a LysR family regulator. J Biol Chem 279: 5739–5751

Wastyn M, Achatz A, Molitor V and Peschek GA (1988) Respiratory activities and aa_3-type cytochrome oxidase in plasma and thylakoid membranes from vegetative cells and heterocysts of the cyanobacterium *Anabaena* ATCC 29413. Biochim Biophys Acta 935: 217–224.

Weerakoon DR and Olson JW (2008) The *Campylobacter jejuni* NADH:ubiquinone oxidoreductase (Complex I) utilizes flavodoxin rather than NADH. J Bacteriol 190: 915–925

Weidner U, Geier S, Ptock A, Friedrich T, Leif H and Weiss H (1993) The gene locus of the proton-translocating NADH:ubiquinone oxidoreductase in *Escherichia coli*. J Mol Biol 233: 109–122

Xu M, Ogawa T, Pakrasi HB and Mi H (2008) Identification and Localization of the CupB Protein Involved in Constitutive CO_2 Uptake in the Cyanobacterium, *Synechocystis* sp. Strain PCC 6803. Plant Cell Physiol 49: 994–997

Yagi T, Yano T, Di Bernardo S and Matsuno-Yagi A (1998) Procaryotic complex I (NDH-1), an overview. Biochim Biophys Acta 1364: 125–133

Yeremenko N, Jeanjean R, Prommeenate P, Krasikov V, Nixon PJ, Vermaas WFJ, Havaux M and Matthijs HCP (2005) Open reading frame ssr2016 is required for antimycin A-sensitive Photosystem I-driven cyclic electron flow in the cyanobacterium *Synechocystis* sp. PCC 6803. Plant Cell Physiol 46: 1433–1436

Zhang P, Battchikova N, Jansen T, Appel J, Ogawa T and Aro E-M (2004) Expression and functional roles of the two distinct NDH-1 complexes and the carbon acquisition complex NdhD3/NdhF3/CupA/Sll1735 in *Synechocystis* sp. PCC 6803. Plant Cell 16: 3326–3340

Zhang P, Battchikova N, Paakkarinen V, Katoh H, Iwai M, Ikeuchi M, Pakrasi HB, Ogawa T and Aro E-M (2005) Isolation, subunit composition and interaction of the NDH-1 complexes from *Thermosynechococcus elongatus* BP-1. Biochem J 390: 513–520

Chapter 17
The Superfamily of Succinate:Quinone Oxidoreductases and its Implications for the Cyanobacterial Enzymes

C. Roy D. Lancaster

17.1 Introduction

Cyanobacteria, also known as blue-green algae, are the paradigmatic organisms of oxygenic photosynthesis and aerobic respiration in that they uniquely accommodate both a photosynthetic electron transport chain and a respiratory electron transfer chain within a single prokaryotic cell (Jones and Myers 1963). Between 20 and 30% of the 10^{11} t of carbon in CO_2 estimated to be converted per year into biomass by plant-type photosynthesis are assigned to cyanobacteria, in particular to small unicellular marine *Synechococcus* and *Prochlorococcus* species (Waterbury et al. 1979; Chisholm et al. 1988; Partensky et al. 1999; Scanlan et al. 2009). Cyanobacterial respiration is much less well-understood than photosynthesis, and, in spite of the availability of gene-derived amino acid sequence information (cf. Table 17.1 below) and (comparatively few) publications reporting the analysis of cell extracts (e.g. Leach and Carr 1970) or the comparative analysis of deletion mutants (Cooley et al. 2000, 2001), this applies in particular to cyanobacterial succinate:quinone oxidoreductases.

Succinate:quinone oxidoreductases (SQORs; EC 1.3.5.1; Hägerhäll 1997; Lancaster 2002a, b) are enzymes that couple the two-electron oxidation of succinate to fumarate (reaction 17.1) to the two-electron reduction of quinone to quinol (reaction 17.2).

$$\text{succinate} \rightleftarrows \text{fumarate} + 2\text{H}^+ + 2\text{e}^- \qquad (17.1)$$

$$\text{quinone} + 2\text{H}^+ + 2\text{e}^- \rightleftarrows \text{quinol} \qquad (17.2)$$

They can also catalyse the opposite reaction, the coupling of quinol oxidation to quinone to the reduction of fumarate to succinate (Lemma et al. 1991). The *cis*-configuration isomer of fumarate, maleinate, is neither produced in the oxidation reaction, nor is it consumed as a substrate in the reduction reaction, i.e. the reaction is stereospecific

C. R. D. Lancaster (✉)
Faculty of Medicine, Department of Structural Biology, Institute of Biophysics, Saarland University, Building 60, 66421 Homburg, Germany
e-mail: roy.lancaster@structural-biology.eu

G. A. Peschek et al. (eds.), *Bioenergetic Processes of Cyanobacteria,*
DOI 10.1007/978-94-007-0388-9_17, © Springer Science+Business Media B.V. 2011

Table 17.1 List of abbreviations for the respective cyanobacterial strains, their derived succinate dehydrogenase subunit compositions, and access codes of all genes within the scope of this review

Order/Genus	Strain	Abbreviation	Accession numbers for gene encoding succinate dehydrogenase…				
			Subunit A	Subunit B	Subunit C (Type B Membrane anchor)	Subunit E (Type E Membrane anchor)	Reference
Chroococcales							
Crocosphaera							
Crocosphaera watsonii	WH 8501	Cro.wat	ZP_00515966	ZP_00518809		ZP_00518019[a]	Copeland et al. unpublished
Cyanothece							
Cyanothece sp.	CCY0110	Cya0110	ZP_01728536	ZP_01731813		ZP_01731991[b]	Stal et al. unpublished
Cyanothece sp.	ATCC 51142	Cya51142	YP_001802080	YP_001804658		YP_001804642[c]	Welsh et al. (2008)
Cyanothece sp.	PCC 7424	Cya7424	YP_002376991	YP_002380545		YP_002379309[c]	Lucas et al. unpublished
Cyanothece sp.	PCC 7425	Cya7425	YP_002482300	YP_002482022		YP_002483244[c]	Lucas et al. unpublished
Cyanothece sp.	PCC 7822	Cya7822	ZP_03157073	ZP_03153901		ZP_03157932[c]	
Cyanothece sp.	PCC 8801	Cya8801	YP_002372055	YP_002370649		YP_002373097[c]	Lucas et al. unpublished
Cyanothece sp.	PCC 8802	Cya8802	YP_003137614	YP_003136199		YP_003138825[c]	Lucas et al. unpublished
Microcystis							
Microcystis aeruginosa	NIES-843	Mic.aer843	YP_001660639	YP_001656381		YP_001659575[c]	Kaneko et al. (2007)
Synechococcus							
Synechococcus elongatus	PCC 6301	Syn6301	YP_171594	YP_173229	YP_171909		Sugita et al. (2007)

Table 17.1 (continued)

Order/Genus	Strain	Abbreviation	Accession numbers for gene encoding succinate dehydrogenase…				
			Subunit A	Subunit B	Subunit C (Type B Membrane anchor)	Subunit E (Type E Membrane anchor)	Reference
Synechococcus elongatus	PCC 7942	Syn7942	YP_399660	YP_400550	YP_399333		Copeland et al. unpublished
Synechococcus sp.	PCC 7002	Syn7002	YP_001735801	YP_001734352		YP_001734095[c]	Li et al. unpublished
Synechococcus sp.	BL107	Syn107	ZP_01468469	ZP_01468470	ZP_01468468		
Synechococcus sp.	WH 5701	Syn5701	ZP_01085403	ZP_01085404	ZP_01085402		
Synechococcus sp.	WH8102	Syn8102	NP_896684	NP_896683	NP_896685		Palenik et al. (2003)
Synechococcus sp.	JA-2-3B'a(2-13)	SynJA23	YP_478336	YP_478702	YP_376598	YP_477584	Bhaya et al. (2007)
Synechocystis							Kaneko et al. (1995)
Synechocystis sp.	PCC 6803	Syc6803	NP_440839	NP_441091		NP_442350[c]	Kaneko et al. (1996)
Thermosynechococcus							
Thermosynechococcus elongatus	BP-1 The.elo		NP_682167	NP_682544		NP_681769	Nakamura et al. (2002)
Gloeobacteria							
Gloeobacter							Nakamura et al. (2003a)
Gloeobacter violaceus	PCC 7421	Glo.vio	NP_925934	NP_925890		NP_925891	Nakamura et al. (2003b)

Table 17.1 (continued)

Order/Genus	Strain	Abbreviation	Accession numbers for gene encoding succinate dehydrogenase…				
			Subunit A	Subunit B	Subunit C (Type B Membrane anchor)	Subunit E (Type E Membrane anchor)	Reference
Nostocales							
Anabaena							
Anabaena variabilis	ATCC 29413	Ana.var	YP_321455	YP_321042		YP_324149	Copeland et al. unpublished
Nostoc							
Nostoc punctiforme	ATCC 29133	Nos.pun	YP_001869117	YP_001869179		YP_001867864[b]	Copeland et al. unpublished
Oscillatoriales							
Arthrospira maxima							
Arthrospira maxima	CS-328	Art.max	ZP_03276500	ZP_03276258		ZP_03272563[c]	Lucas et al. unpublished
Lyngbya							
Lyngbya sp.	PCC 8106	Lyn8106	ZP_01623494	ZP_01622068		ZP_01619713[b]	Stal et al. unpublished
Microcoleus							
Microcoleus chthonoplastes	PCC 7420	Mio.cht	ZP_05029125	ZP_05026392		ZP_05029470[d]	Tandeau de Marsac et al. unpublished
Trichodesmium							
Trichodesmium erythraeum	IMS101	Tri.ery	YP_722156	YP_723255		YP_720385	Copeland et al. unpublished

Table 17.1 (continued)

Order/Genus	Strain	Abbreviation	Accession numbers for gene encoding succinate dehydrogenase…				
			Subunit A	Subunit B	Subunit C (Type B Membrane anchor)	Subunit E (Type E Membrane anchor)	Reference
Prochlorales							
Prochlorococcus							
Prochlorococcus marinus subsp. marinus str.	CCMP1375	Pro1375	NP_875091	NP_875092	NP_875090		Dufresne et al. (2003)
Prochlorococcus marinus str.	MIT 9303	Pro9303	YP_001016189	YP_001016190	YP_001016188		Kettler et al. (2007)
Prochlorococcus marinus str.	MIT 9313	Pro9313	NP_893965	NP_893966	NP_893964		Rocap et al. (2003)
unclassified Cyanobacteria							
Acaryochloris							
Acaryochloris marina	MBIC11017	Aca.mar	YP_001517100	YP_001517973		YP_001519799[c]	Swingley et al. (2008)

[a] annotated as "protein of unknown function"
[b] annotated as "hypothetical protein"
[c] annotated as "heterodisulfide reductase subunit B"
[d] annotated as "cysteine-rich domain protein"

in both directions. Depending on the direction of the reaction catalysed *in vivo*, the members of the superfamily of succinate:quinone oxidoreductases can be classified as either succinate:quinone reductases (SQR) or quinol:fumarate reductases (QFR). SQR and QFR can be degraded to form succinate dehydrogenase and fumarate reductase (both EC 1.3.99.1), which no longer react with quinone and quinol, respectively.

SQR and QFR complexes are anchored in the cytoplasmic membranes of archaebacteria, eubacteria and in the inner mitochondrial membrane of eukaryotes with the hydrophilic domain canonically extending into the cytoplasm and the mitochondrial matrix, respectively. SQR (respiratory complex II) is involved in aerobic metabolism (Fig. 17.1a) as part of the citric acid cycle and of the aerobic respiratory chain (Saraste 1999). QFR participates in anaerobic respiration with fumarate as the terminal electron acceptor (Fig. 17.1b, Kröger 1978; Kröger et al. 2002; Lancaster 2004), and is part of the electron transport chain catalysing the oxidation of various donor substrates (e.g. NADH, H_2, or formate) by fumarate. These reactions are coupled via an electrochemical proton potential (Δp) to ADP phosphorylation with inorganic phosphate by ATP synthase (Mitchell 1979).

Succinate:quinone oxidoreductases generally contain four protein subunits, referred to as A, B, C, and D. Subunits A (64–79 kDa) and B (27–31 kDa) are hydrophilic, whereas the canonical subunits C (13–18 kDa) and D (11–16 kDa) are integral membrane proteins. Among species, subunits A and B have high sequence homology, while that for the hydrophobic subunits is much lower. Most of the SQR enzymes of Gram-positive bacteria, cyanobacterial SQR from *Synechococcus* and *Prochlorococcus* species (cf. Table 17.1), and the QFR enzymes from ε-proteobacteria contain only one larger hydrophobic polypeptide (C, 23–30 kDa), which, based on comparisons of sequences (Hägerhäll and Hederstedt 1996; Hederstedt 1999) and structures (Lancaster et al. 1999; Lancaster 2003d) is thought to have evolved from a fusion of the genes for the two smaller subunits C and D. While subunit A harbors the site of fumarate reduction and succinate oxidation, the hydophobic subunit(s) contain the site of quinol oxidation and quinone reduction.

Based on their hydrophobic domain and heme content (Hägerhäll and Hederstedt 1996; Hederstedt 1999), succinate:quinone oxidoreductases can be classified in five types (cf. Fig. 17.2; Lancaster 2001a, 2002b). Type A enzymes (Schäfer et al. 2002) contain two hydrophobic subunits and two heme groups, e.g. SQR from the archaea *Archaeoglobus fulgidus, Natronomonas pharaonis*, and *Thermoplasma acidophilum*. Type B enzymes contain one hydrophobic subunit and two heme groups, as is the case for SQR from the Gram-positive bacteria *Bacillus subtilis* (Hederstedt 2002), *Paenibacillus macerans*, for QFR from the ε-proteobacteria *Campylobacter jejuni*, *Helicobacter pylori*, and *Wolinella succinogenes* (Lancaster and Simon 2002), and, as will be described below, for cyanobacterial SQR from *Synechococcus* and *Prochlorococcus* species (cf. Table 17.1). Examples for type C enzymes, which possess two hydrophobic subunits and one heme group, are SQR from mammalian mitochondria and from the proteobacteria *Paracoccus denitrificans* (Hederstedt 2002) and *Escherichia coli* and QFR from the nematode *Ascaris suum* (Kita et al. 2002). The QFR of *E. coli* is an example for a type D enzyme, which contains two hydrophobic subunits and no heme group (Cecchini et al. 2002). Finally, type E enzymes (Schäfer et al. 2002; Lemos et al. 2002), such as SQRs from the archaea

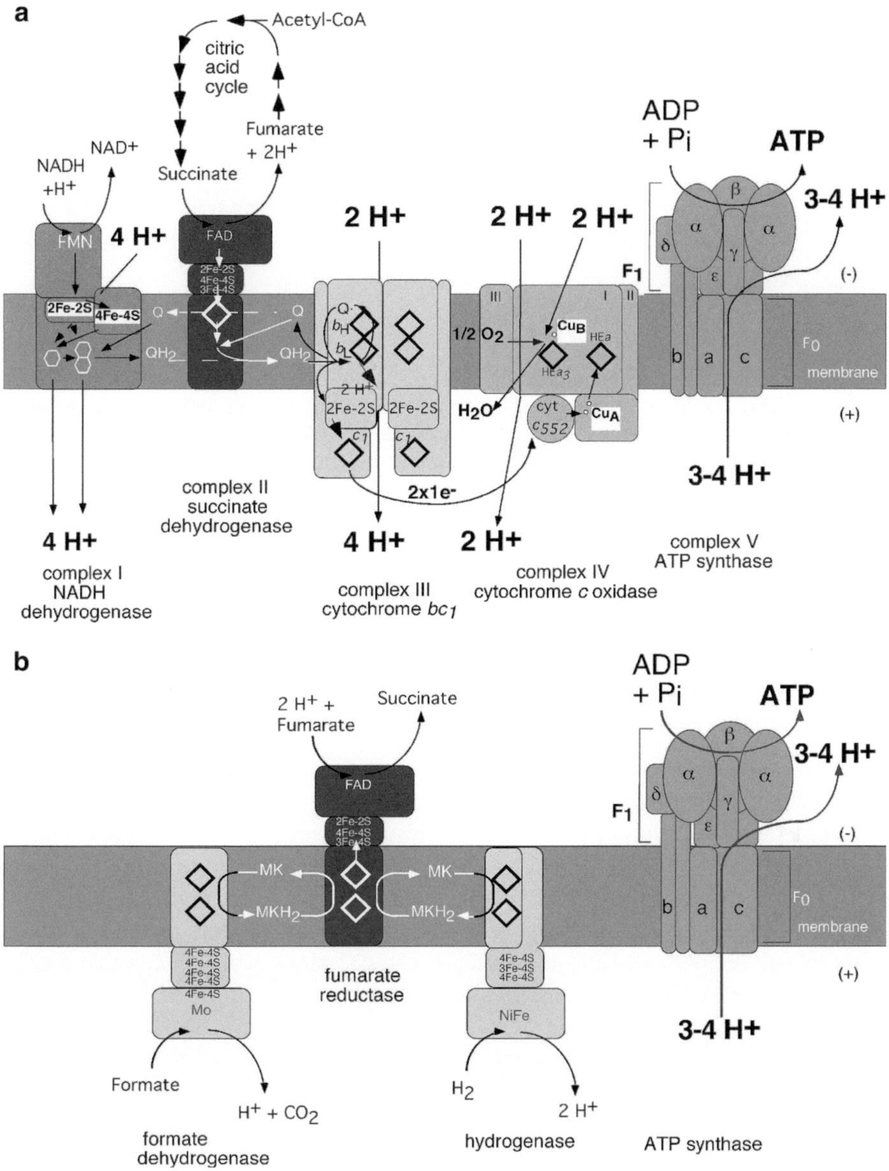

Fig. 17.1 Electron flow and the generation and utilisation of a transmembrane electrochemical potential in aerobic respiration (**a**) and anaerobic respiration (**b**). Ubiquinone is abbreviated as Q, menaquinone as MK. The positive (+) and negative (−) sides of the membrane are indicated. In bacteria, the negative side is the cytoplasm ("inside"), the positive side the periplasm ("outside"). For mitochondrial systems, these are the mitochondrial matrix and the intermembrane space, respectively. (This figure was modified from Lancaster 2001b, 2002b)

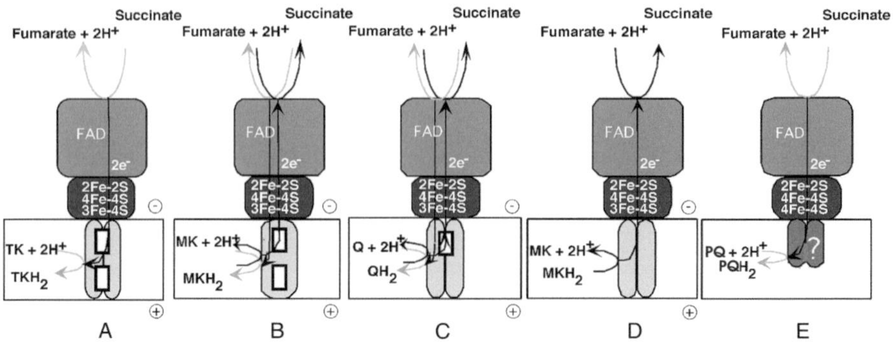

Fig. 17.2 Classification (*A* to *E*) of succinate:quinone oxidoreductases (Lancaster 2001a, b) based on their hydrophobic domain and heme content (Hägerhäll and Hederstedt 1996; Hederstedt 2003). The hydrophilic subunits *A* and *B* are drawn schematically in *grey* and *dark grey*, respectively, the hydrophobic subunits *C* and *D* in *light grey* or *grey*. Heme groups are symbolized by small rectangles. The directions of the reactions catalysed by SQR and QFR are indicated by *grey* and *black arrows*, respectively. White rectangles symbolize the respective cytoplasmic or inner mitochondrial membrane bilayer. The positive (+) and negative (−) sides of the membrane are indicated as in Fig. 17.1. The type of quinone transformed in vivo is not necessarily unique for each type of enzyme. The examples given are thermoplasma-quinone (TK), menaquinone (MK), ubiquinone (Q), and plastoquinone (PQ; Collins and Jones 1981; Hägerhäll 1997; Lancaster 2001b). Whereas most E-type enzymes are predicted to have their hydrophilic subunits canonically oriented towards the cytoplasm, those of *W. succinogenes* MFR have been shown to be oriented towards the periplasm (Juhnke et al. 2009)

Acidianus ambivalens and *Sulfolobus acidocaldarius*, but also from *Synechocystis* and most other cyanobacteria (cf. Table 17.1) and the 8-methylmenaquinol:fumarate reductase (MFR) enzymes from the epsilon proteobacteria *W. succinogenes* (Juhnke et al. 2009) and *C. jejuni* (Guccione et al. 2010) also contain no heme, but have one or two hydrophobic subunits very different from the other four types and more similar to those of heterodisulfide reductase from methanogenic archaea (Schäfer et al. 1999). The phylogenetic analyses presented (Lemos et al. 2002; Schäfer et al. 2002) corroborate the above classification scheme.

Generally, succinate:quinone oxidoreductases contain three iron-sulfur clusters, which are exclusively bound by the B subunit. Enzyme types A–D contain one $[2Fe\text{-}2S]^{2+,1+}$, one $[4Fe\text{-}4S]^{2+,1+}$, and one $[3Fe\text{-}4S]^{1+,0}$ cluster, whereas an additional [4Fe-4S] cluster apparently replaces the [3Fe-4S] in the type E enzyme (Gomes et al. 1999). With very few exceptions (e.g. Juhnke et al. 2009), the A subunit of all described membrane-bound succinate:quinone oxidoreductase complexes contains a covalently bound FAD prosthetic group (Singer and McIntire 1984). The chemical structure of the linkage as 8α-[Nε-histidyl]-FAD was first established for mammalian SQR (Walker and Singer 1970) and subsequently for the QFR enzymes of *W. succinogenes* (Kenny and Kröger 1977) and *E. coli* (Weiner and Dickie 1979).

With some interesting exceptions (Lemma et al. 1990, 1991; Madej et al. 2006b), SQR enzymes tend to catalyse the reduction of a high-potential quinone, such as ubiquinone, whereas QFR enzymes tend to oxidize a low-potential quinol such as

Table 17.2 Scores for respective sequence alignments of *Synechococcus elongatus* flavoprotein and iron-sulfur subunits to those of known three-dimensional structure

Species/strain	Subunit	ID[a]	Score[b]	E -value[b]	Sequence							
					Overlap	Identical	Percentage	Positive	Percentage	Gaps	Percentage	Resolution
Synechococcus elongatus PCC 6301	SdhA	YP_171594										
Wolinella succinogenes DSM 1740	QfrA	2BS2_A	190	1,00E-48	644	177	27%	287	45%	85	13%	1.78
Escherichia coli	QfrA	1KF6_A	189	2,00E-48	581	167	29%	260	45%	54	9%	2.7
Gallus gallus (chicken)	SdhA	2H88_A	158	7,00E-39	608	165	27%	261	43%	68	11%	1.8
Sus scrofa (pig)	SdhA	1ZOY_A	152	3,00E-37	626	170	27%	265	42%	82	13%	2.4
Escherichia coli	SdhA	1NEK_A	120	1,00E-27	602	154	26%	256	43%	89	15%	2.6
Synechococcus elongatus PCC 6301	SdhB	YP_171594										
Wolinella succinogenes DSM 1740	QfrB	2BS2_B	73.2	8,00E-14	239	66	28%	112	47%	30	13%	1.78
Escherichia coli	SdhB	1NEK_B	65.5	2,00E-11	237	64	27%	101	43%	24	10%	2.6
Sus scrofa (pig)	SdhB	1ZOY_B	42.4	1,00E-04	234	49	21%	84	36%	24	10%	2.4
Gallus gallus (chicken)	SdhB	2H88_B	40	8,00E-04	233	48	21%	82	35%	22	9%	1.8

[a] *ID* respective NCBI or PDB accession code

[b] According to Karlin and Altschul (1990, 1993), and Brenner et al. (1998)

menaquinol, thus making, in the respective physiological direction, the overall reaction favorable under standard conditions. Cyanobacteria are unusual in that they possess neither ubiquinones nor menaquinones (Collins and Jones 1981). They do contain the 1,4-naphthoquinone derivative phylloquinone (vitamin K_1) and the 1,4-benzoquinone derivative plastoquinone-9, which are indigenous to the plant kingdom, but are not normally found in bacteria.

So far, the three-dimensional structure of a cyanobacterial SQOR has not been determined, nor has that of any other representative of the type E enzymes. The first available crystal structures of succinate:quinone oxidoreductases were those of two prokaryotic quinol:fumarate reductases (cf. Table 17.2). The *E. coli* QFR, determined at 3.3 Å resolution (Iverson et al. 1999; Iverson et al. 2002), belongs to the type D enzymes, and the QFR of *W. succinogenes*, refined initially at 2.2 Å (Lancaster et al. 1999), later at 1.78 Å resolution (Madej et al. 2006a), is of type B. Subsequently, structures of type C enzymes have become available. The structure of the SQR from *E. coli* was reported at 2.6 Å resolution (Yankovskaya et al. 2003), those of the mitochondrial porcine and avian enzymes at resolutions of 2.4 Å (Sun et al. 2005) and 1.8 Å (Huang et al. 2006), respectively.

In this contribution, evidence from the SQOR crystal structure determinations, primarily that of the B-type QFR from *W. succinogenes*, in combination with multiple sequence alignments, will serve as a basis for the discussion of the cyanobacterial succinate:quinone oxidoreductases.

17.2 Subunit A, the Flavoprotein

The overall result of the phylogenetic analysis of subunit A (Fig. 17.3a) indicates two groups of cyanobacterial SQR flavoproteins flanking the SQOR subunits A of known three-dimensional structure. One group ("group I") consists of the SQR

Fig. 17.3 The flavoprotein subunit SdhA. **a** Reconstructed phylogenetic tree of 35 succinate:quinone oxidoreductase flavoprotein subunits. The tree was obtained with the MEGA4 package (Tamura et al. 2007) using the neighbor-joining method with 1000 bootstrap replications. The abbreviations for the 29 cyanobacterial species are defined in Table 17.1, the remaining six sequences (see also Table 17.2) are the A subunits of *E. coli* succinate dehydrogenase (E.coliSdhA), porcine succinate dehydrogenase (Sus.scSdhA), chicken succinate dehydrogenase (GallusSdhA), *E. coli* quinol:fumarate reductase (E.coliQfrA), *W. succinogenes* 8-methylmenaquinol:fumarate reductase (W.succMfrA), and *W. succinogenes* quinol:fumarate reductase (W.succQfrA). Group I and group II are indicated as discussed in the text. **b** Multiple sequence alignment of the first three domains of 35 succinate:quinone oxidoreductase flavoprotein subunits. The alignment was obtained with the T-coffee method (Notredame et al. 2000). The abbreviations for the 29 cyanobacterial species are defined in Table 17.1, the remaining six sequences (see also Table 17.2) are the A subunits of *E. coli* succinate dehydrogenase (E_coliSdhA), porcine succinate dehydrogenase (Sus_scSdhA), chicken succinate dehydrogenase (GallusSdhA), *E. coli* quinol:fumarate reductase (E_coliQfrA), *W. succinogenes* 8-methylmenaquinol:fumarate reductase (W_succMfrA), and *W. succinogenes* quinol:fumarate reductase (W_succQfrA). The capping domain and the helical domain are indicated by boces. The remaining regions correspond to the (bipartite) FAD-binding domain. The highlighted residues are numbered according to their positions in *W. succinogenes* QFR

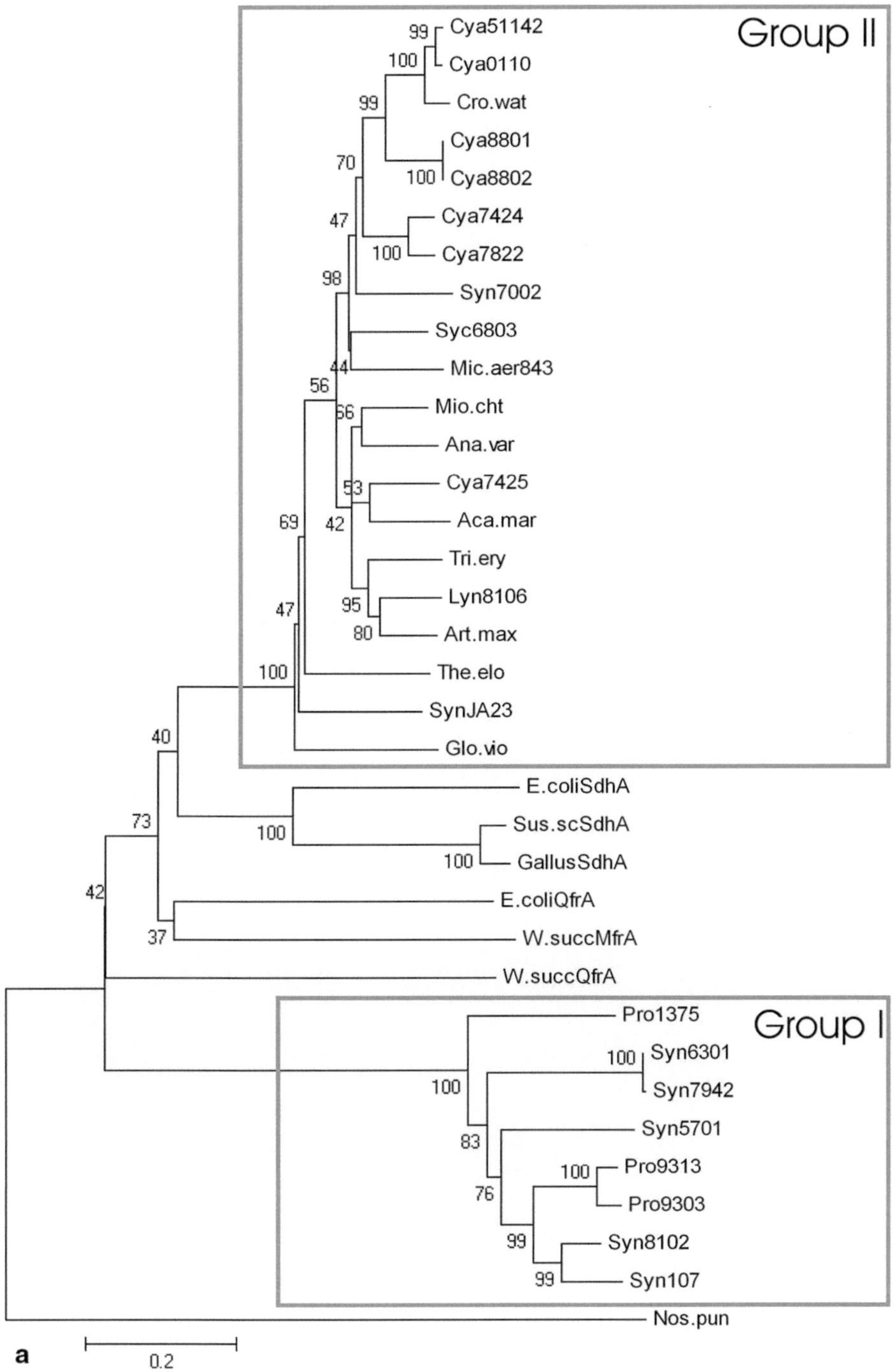
Group II
Cya51142
Cya0110
Cro.wat
Cya8801
Cya8802
Cya7424
Cya7822
Syn7002
Syc6803
Mic.aer843
Mio.cht
Ana.var
Cya7425
Aca.mar
Tri.ery
Lyn8106
Art.max
The.elo
SynJA23
Glo.vio
E.coliSdhA
Sus.scSdhA
GallusSdhA
E.coliQfrA
W.succMfrA
W.succQfrA
Group I
Pro1375
Syn6301
Syn7942
Syn5701
Pro9313
Pro9303
Syn8102
Syn107
Nos.pun
a
0.2

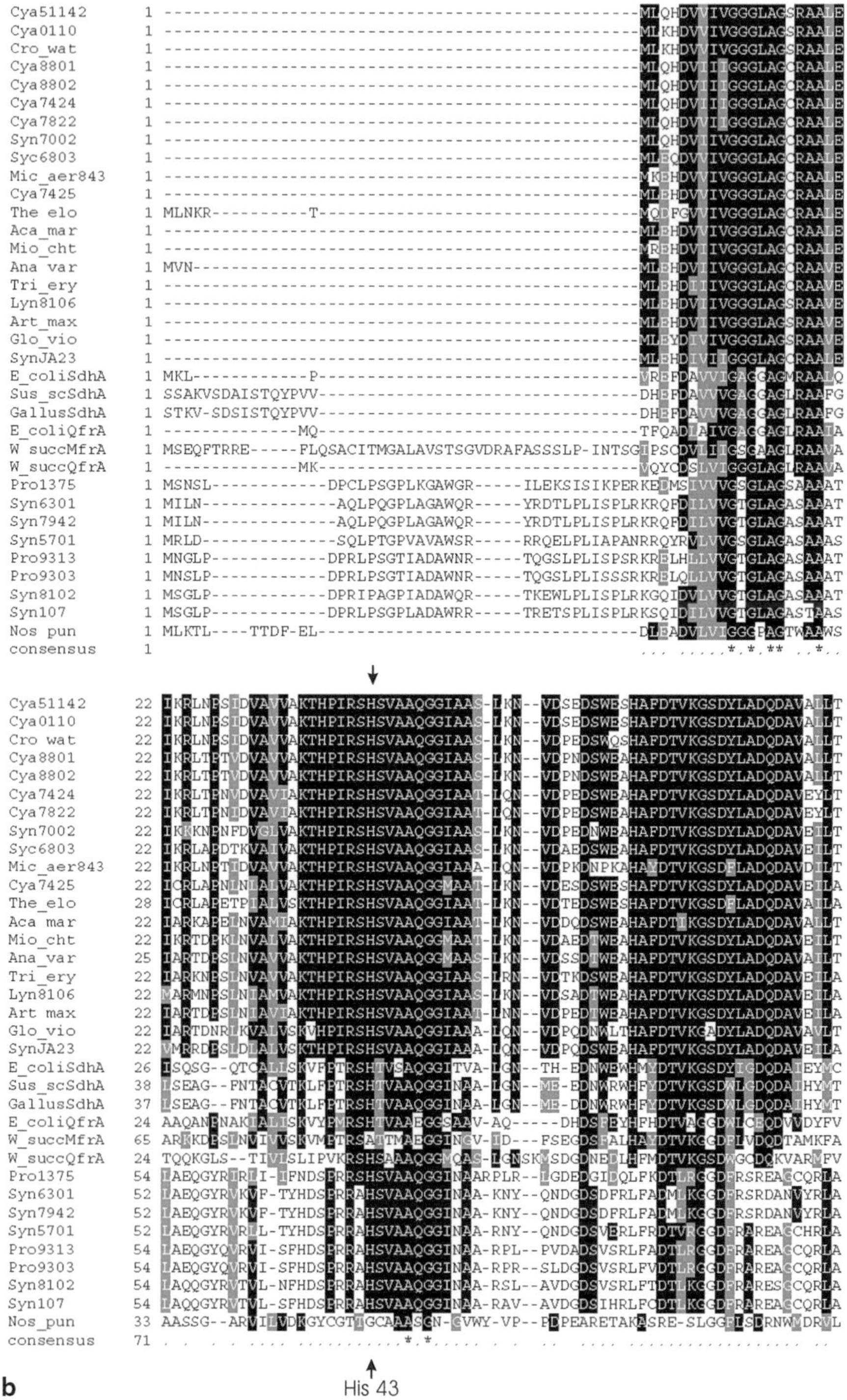

b

Fig. 17.3 (continued)

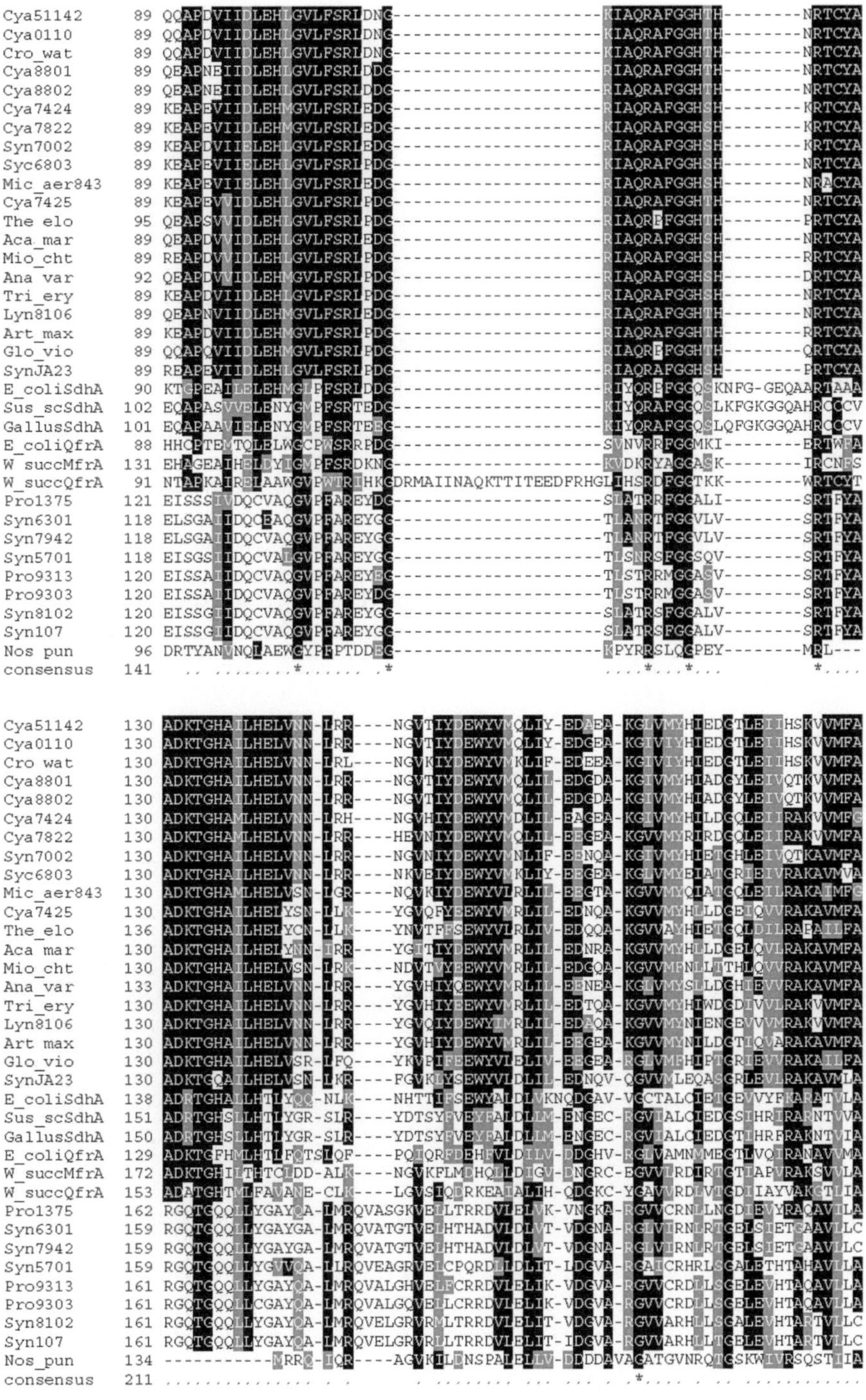

Fig. 17.3 (continued)

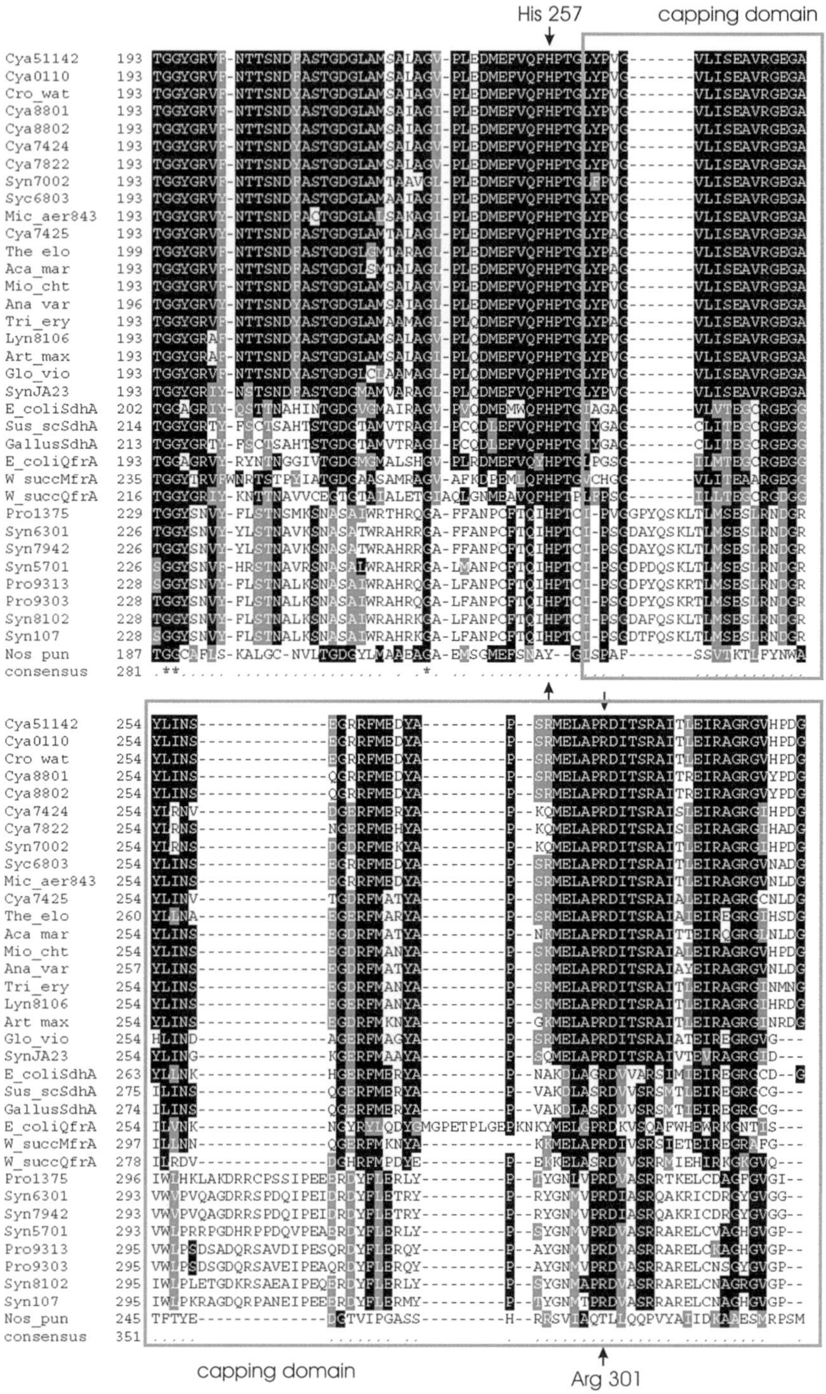

Fig. 17.3 (continued)

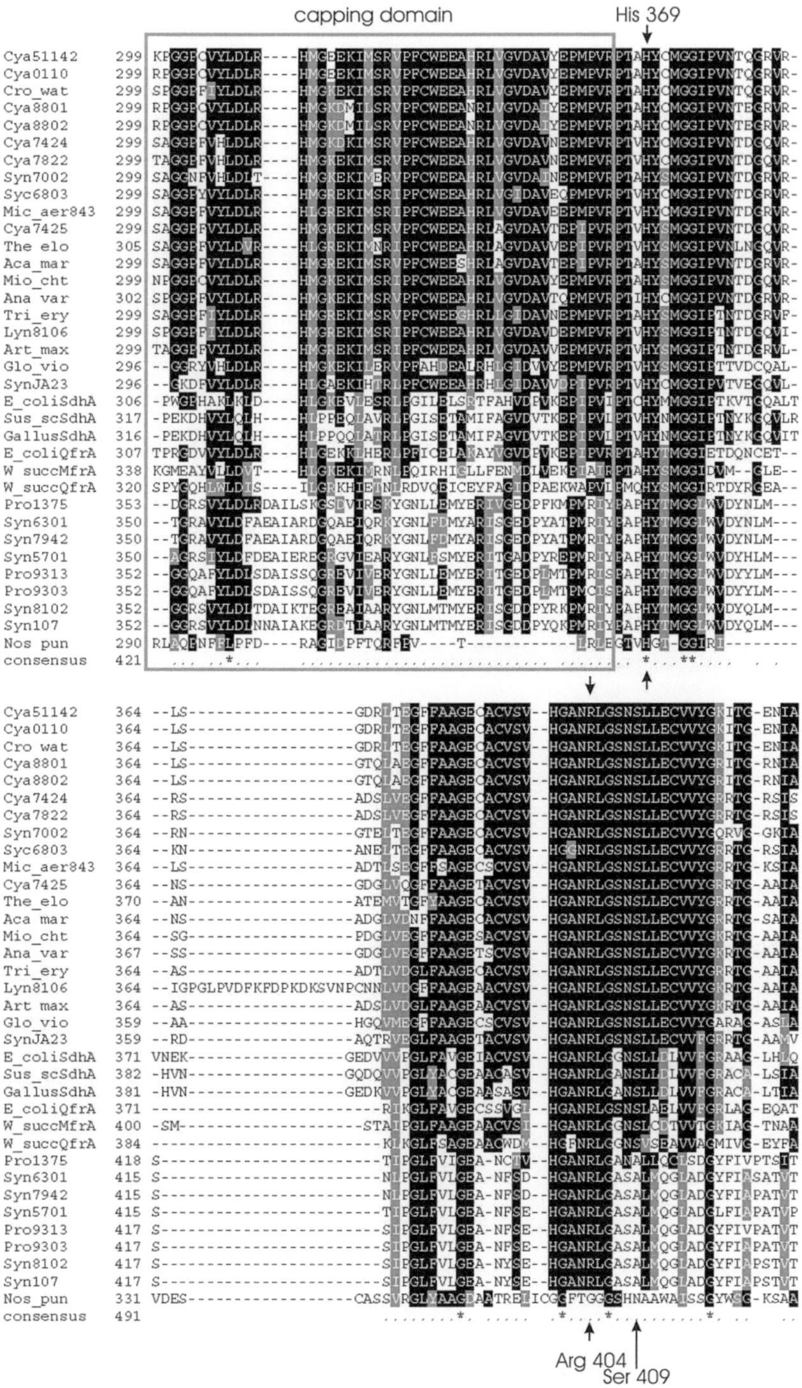

Fig. 17.3 (continued)

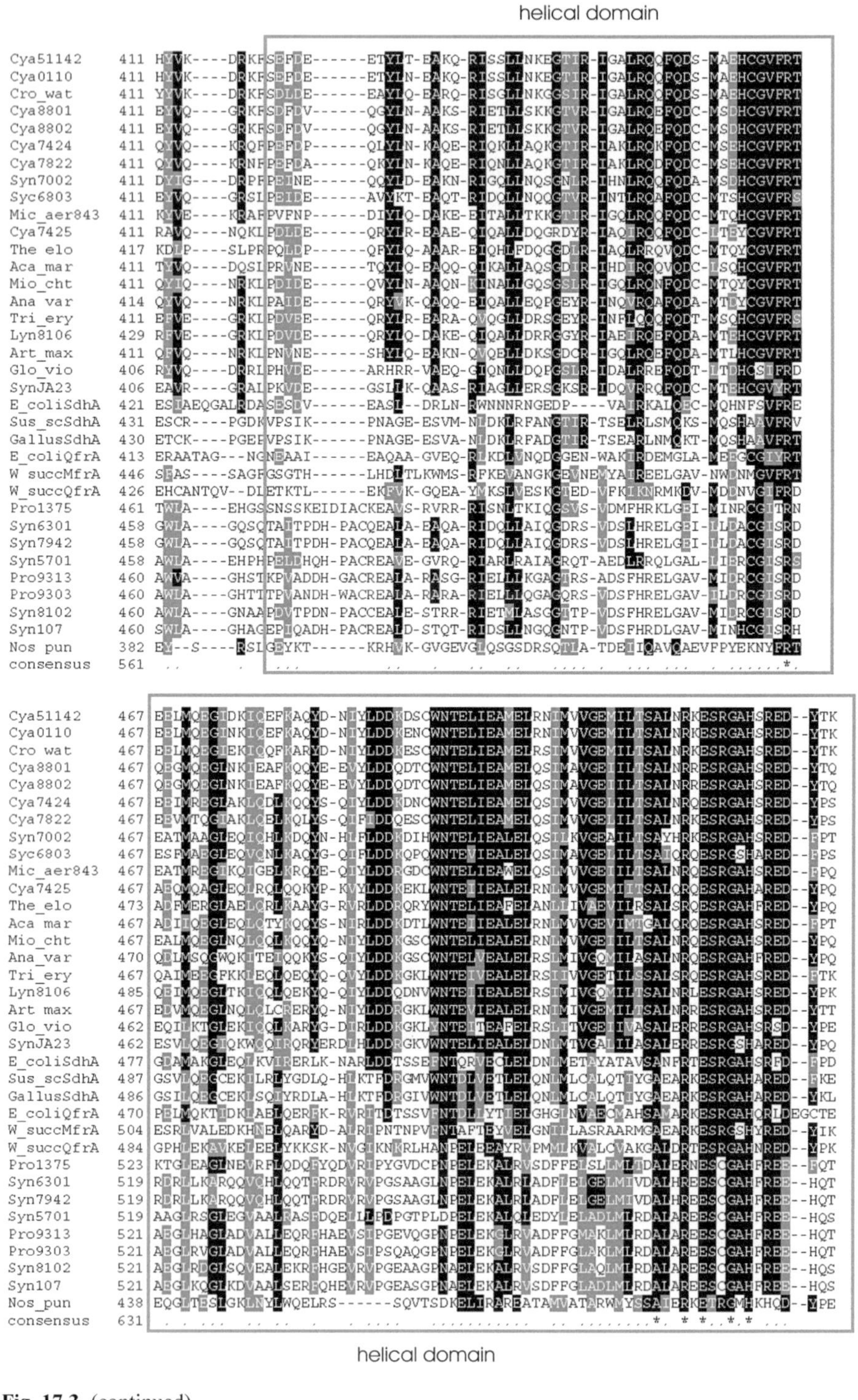

Fig. 17.3 (continued)

flavoproteins from all *Prochlorococcus* and all *Synechococcus* species considered except for *Synechococcus elongatus* PCC7002 and *Synechococcus sp.* JA-2-3B'a(2-13), the second group ("group II") includes the flavoproteins from all other species considered with the notable exception of *Nostoc punctiformae* ATCC 29133, which appears isolated. The alignments of these sequences are displayed in Fig. 17.3b and will be discussed with reference to the known structure of *W. succinogenes* QFR subunit A, a polypeptide of 73 kDa (Lauterbach et al. 1990), which is composed of four domains (Fig. 17.4a), the bipartite FAD binding domain (residues A1-260 and A366-436, with "A" indicating the A subunit), into which the capping domain (indicated by the solid-lined ellipse, A260-366) is inserted, the helical domain (indicated by dashed lines, A436-554), and the C-terminal domain (A554-656, not shown in Fig. 17.4a). The fourth domain is not conserved within the superfamily and therefore not included in the alignments in Fig. 17.3b. The FAD is covalently bound as 8α-[Nε-histidyl]-FAD (Kenny and Kröger 1977) to the *W. succinogenes* QFR residue His A43. As is apparent from Fig. 17.3b, this residue is conserved in all cyanobacterial SQR flavoproteins except for the (rather questionable sequence of the) A subunit from *Nostoc punctiformae* ATCC 29133. His A43 is also not conserved in *W. succinogenes* MFR, which has been demonstrated to lead to non-covalent attachment of the FAD (Juhnke et al. 2009). The capping domain contributes to burying the otherwise solvent-exposed FAD isoalloxazine ring from the protein surface.

A *W. succinogenes* QFR crystal grown in the presence of fumarate was found to be of crystal form 'B'. The structure was refined at 2.33 Å resolution (PDB entry 1QLB; Lancaster et al. 1999). This allowed the localization of the fumarate binding site between the FAD binding domain and the capping domain next to the plane of the FAD isoalloxazine ring (Fig. 17.4b). The structure of the enzyme in the third crystal form, 'C' (Lancaster et al. 2000) was refined at 3.1 Å resolution (PDB entry 1E7P; Lancaster et al. 2001). Compared with the previous crystal forms, the altered crystal packing (Lancaster 2003a, b) results in the capping domain being in a different arrangement relative to the FAD-binding domain (Fig. 17.4a). This leads to interdomain closure at the fumarate reducing site, suggesting that the structure encountered in this crystal form represents a closer approximation to the catalytically competent state of the enzyme (Fig. 17.4b). The *trans* hydrogenation of fumarate to succinate could occur by the combination of the transfer of a hydride ion and of a proton from opposite sides of the fumarate molecule. One of the fumarate methenyl carbon atoms could be reduced by direct hydride transfer from the N5 position of the reduced $FADH^-$, while the other fumarate methenyl carbon is protonated by the sidechain of Arg A301 (Fig. 17.4c, d). The assignment as to which of the fumarate methenyl carbon atoms accepts the hydride and which the proton is currently ambiguous (Fig. 17.4c *versus* 17.4d), because data of sufficient completeness and quality for this crystal form have so far only been obtained for the complex with malonate and not yet in the presence of fumarate. Release of the product could be facilitated by movement of the capping domain away from the dicarboxylate site (Lancaster et al. 1999; Lancaster et al. 2001). All residues implicated in substrate binding and catalysis are conserved throughout the superfamily of succinate:quinone oxidore-

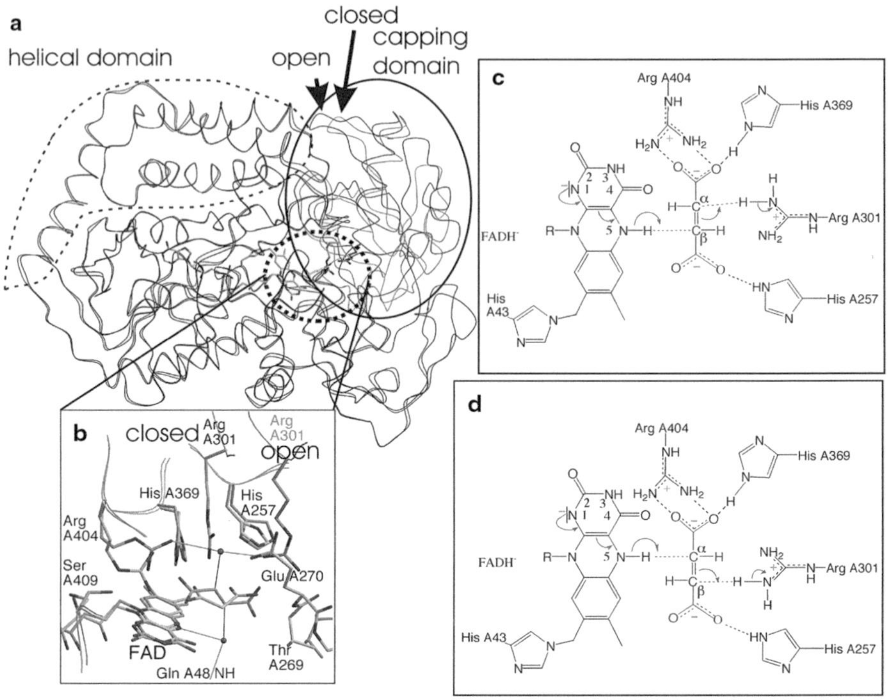

Fig. 17.4 *W. succinogenes* QFR subunit A and the possible mechanism of fumarate reduction. Figures modified from Lancaster et al. (2001) and Lancaster (2001a). **a** The Cα traces of the first three domains of subunit A and the different relative orientations of the capping domain in crystal forms B (PDB entry 1QLB (Lancaster et al. 1999), "open") and C (PDB entry 1E7P (Lancaster et al. 2001), "closed"). **b** Comparison of crystal forms 'C' (PDB entry 1E7P, "closed", complex with malonate) and 'B' (PDB entry 1QLB, "open", complex with fumarate) at the site of fumarate reduction in subunit A. The isolated spheres correspond to the oxygen atoms of two water molecules in PDB entry 1QLB. The dicarboxylate binding site in the form C crystal for which a diffraction data set could be obtained contained the smaller competitive inhibitor malonate rather than fumarate, but this structural difference is negligible compared to the large structural other differences shown here. **c** and **d** Alternate possible mechanisms of fumarate reduction in *W. succinogenes* QFR involving the residues shown in panel b for the crystal form C. Since the precise location of the bound fumarate molecule in this crystal form is not yet known, it could either be the β-methenyl group **c** or the α-methenyl group **d** which which is reduced by hydride transfer from the N5 position of $FADH^-$. This is coupled to proton transfer to the respective other methenyl group from the side chain of Arg A301

ductases (with the exception of the questionable *Nostoc punctiformae* ATCC 29133 sequence), as illustrated in Fig. 17.3b, so that this reversible mechanism is considered generally relevant for all succinate:quinone oxidoreductases.

This mechanistic interpretation of the structure is supported by the results from site-directed mutagenesis, where *W. succinogenes* QFR residue Arg A301 (conserved in all sequences except for *Nostoc punctiformae* ATCC 29133) was replaced relatively conservatively by a Lys (Lancaster et al. 2001). Strain FrdA-R301 K contained a variant enzyme, very similar to the wild type enzyme in terms of cofactor and subunit composition, in particular a fluorescence typical for FAD covalently bound to the A subunit, but which lacked succinate dehydrogenase and fumarate reductase activity (Lancaster et al. 2001). The loss in enzymatic activity is tentatively attributed to the fact that Lys ($pK_{sol}=10.8$) cannot substitute for Arg ($pK_{sol}=12.5$) in protonating the fumarate methenyl carbon, possibly because the protonating group is no longer close enough to protonate the fumarate methenyl group.

17.3 Subunit B, the Iron-Sulfur Protein

The overall result of the phylogenetic analysis of subunit B (Fig. 17.5a) again indicates two groups of cyanobacterial SQR iron-sulfur proteins flanking the SQOR subunits B of known three-dimensional structure. One group ("group I") consists of the SQR iron-sulfur proteins from all *Prochlorococcus* and all *Synechococcus* species considered except for *Synechococcus elongatus* PCC7002 and *Synechococcus sp.* JA-2-3B'a(2-13), the second group ("group II") includes the iron-sulfur proteins from all other species considered with the notable inclusion of *Nostoc punctiformae* ATCC 29133, which appears closely related to the other *Nostocales* sequence considered, that from *Anabena*. The alignments of these sequences are displayed in Fig. 17.5b and will be discussed with reference to the known structure of *W. succinogenes* QFR subunit B, the Cα trace of which is shown in Fig. 17.6. This subunit of 27 kDa (Lauterbach et al. 1990) consists of two domains, an N-terminal "plant ferredoxin" domain (B1-106), binding the [2Fe-2S] iron-sulfur cluster and a C-terminal "bacterial ferredoxin" domain (B106-239) binding the [4Fe-4S] and the [3Fe-4S] iron-sulfur clusters. The [2Fe-2S] iron-sulfur cluster is coordinated by the Cys residues B57, B62, B65, and B77 as proposed on the basis of sequence alignments (Lauterbach et al. 1990). All four Cys residues are within segments that are in contact with the A subunit. The [4Fe-4S] iron-sulfur cluster is ligated to the protein through Cys residues B151, B154, B157, and B218, and the [3Fe-4S] cluster is coordinated by Cys residues B161, B208, and B214. The latter three residues are within segments that are in contact with the C subunit. At the position corresponding to the fourth Cys of the [4Fe-4S] cluster, *W. succinogenes* QFR [3Fe-4S] cluster contains a Leu. Whereas the introduction of a Cys into *E. coli* QFR (Manodori et al. 1992) could replace the native [3Fe-4S] by a [4Fe-4S] cluster, this was not the case for *B. subtilis* SQR (Hägerhäll et al. 1995). With the exception of Leu B211, all Cys ligands discussed above are conserved thoroughout the cyanobacterial sequences considered. The residue corresponding to Leu B211 is invariably a Cys in all sequences of E-type enzymes considered, whereas it is a Leu, Ile, Gln, or Asn in the B-type and C-type enzymes (Fig 17.5b). Significantly, the B subunits of the E-type enzymes contain an approximately 80

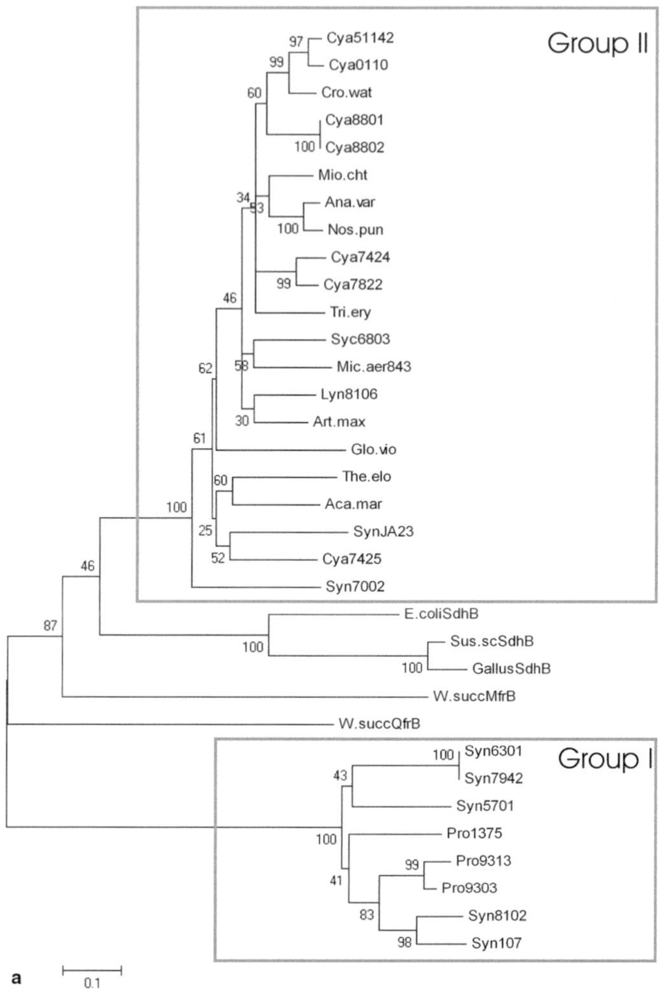

Fig. 17.5 The iron-sulfur-protein subunit SdhB. **a** Reconstructed phylogenetic tree of 34 succinate:quinone oxidoreductase iron-sulfur-protein subunits. The tree was obtained with the MEGA4 package (Tamura et al. 2007) using the neighbor-joining method with 1000 bootstrap replications. The abbreviations for the 29 cyanobacterial species are defined in Table 17.1, the remaining five sequences (see also Table 17.2) are the B subunits of *E. coli* succinate dehydrogenase (E.coliSdhB), porcine succinate dehydrogenase (Sus.scSdhB), chicken succinate dehydrogenase (GallusSdhB), *W. succinogenes* 8-methylmenaquinol:fumarate reductase (W.succMfrB), and *W. succinogenes* quinol:fumarate reductase (W.succQfrB). Group I and group II are indicated as discussed in the text. **b** Multiple sequence alignment of 34 succinate:quinone oxidoreductase iron-sulfur-protein subunits. The alignment was obtained with the T-coffee method (Notredame et al. 2000). The abbreviations for the 29 cyanobacterial species are defined in Table 17.1, the remaining six sequences (see also Table 17.2) are the B subunits of *E. coli* succinate dehydrogenase (E_coliSdhB), porcine succinate dehydrogenase (Sus_scSdhB), chicken succinate dehydrogenase (GallusSdhB), *W. succinogenes* 8-methylmenaquinol:fumarate reductase (W_succMfrB), and *W. succinogenes* quinol:fumarate reductase (W_succQfrB). The highlighted residues are numbered according to their positions in *W. succinogenes* QFR

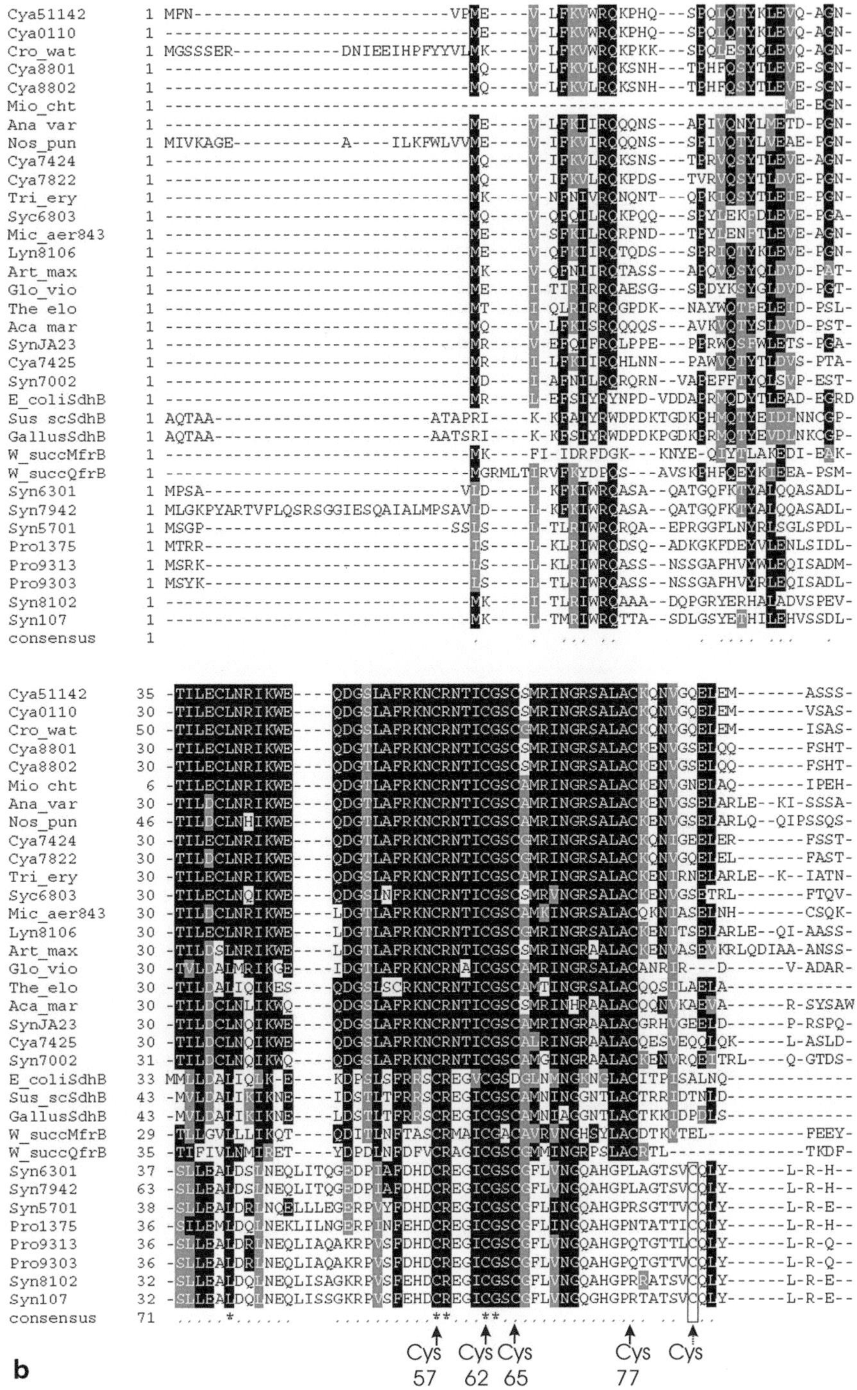

Fig. 17.5 (continued)

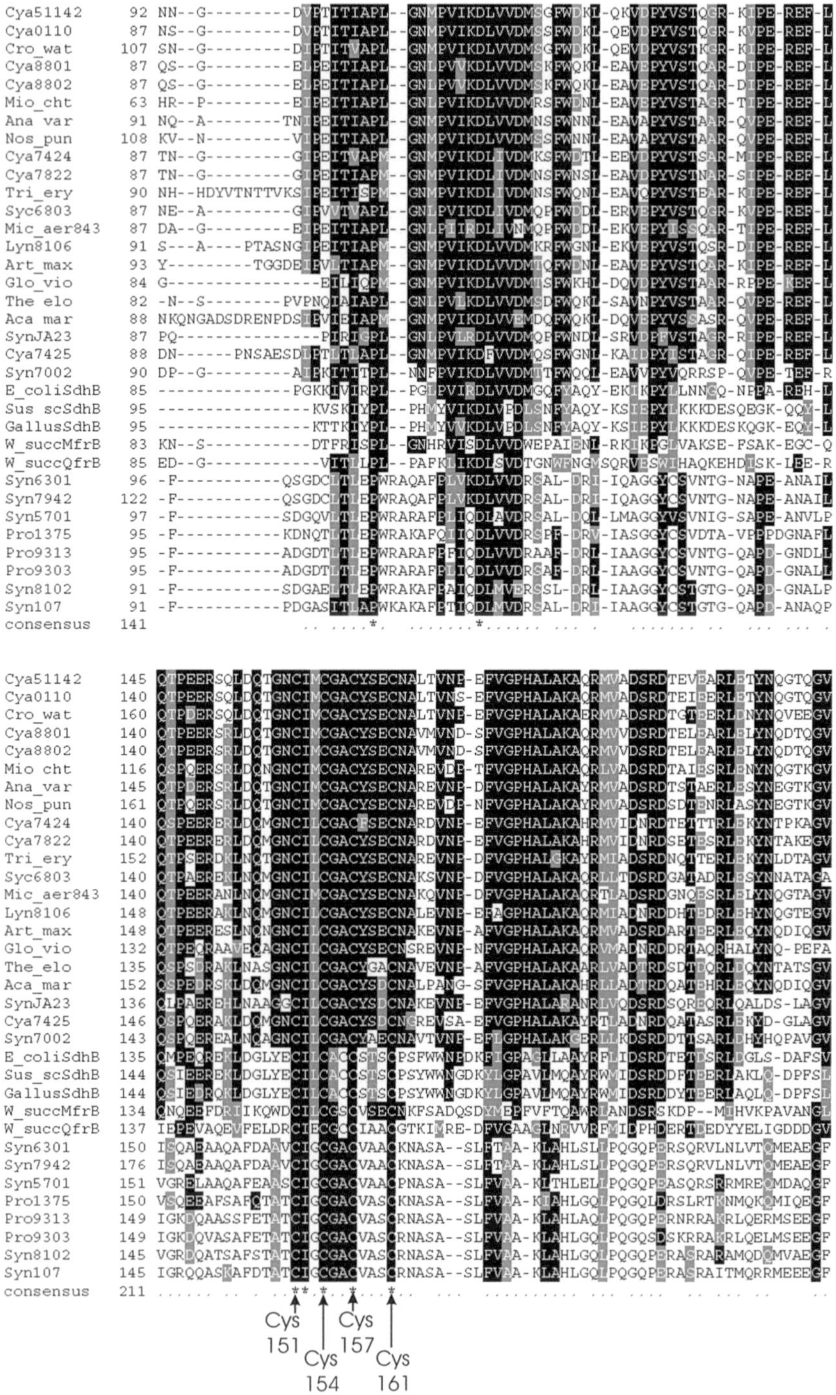

Fig. 17.5 (continued)

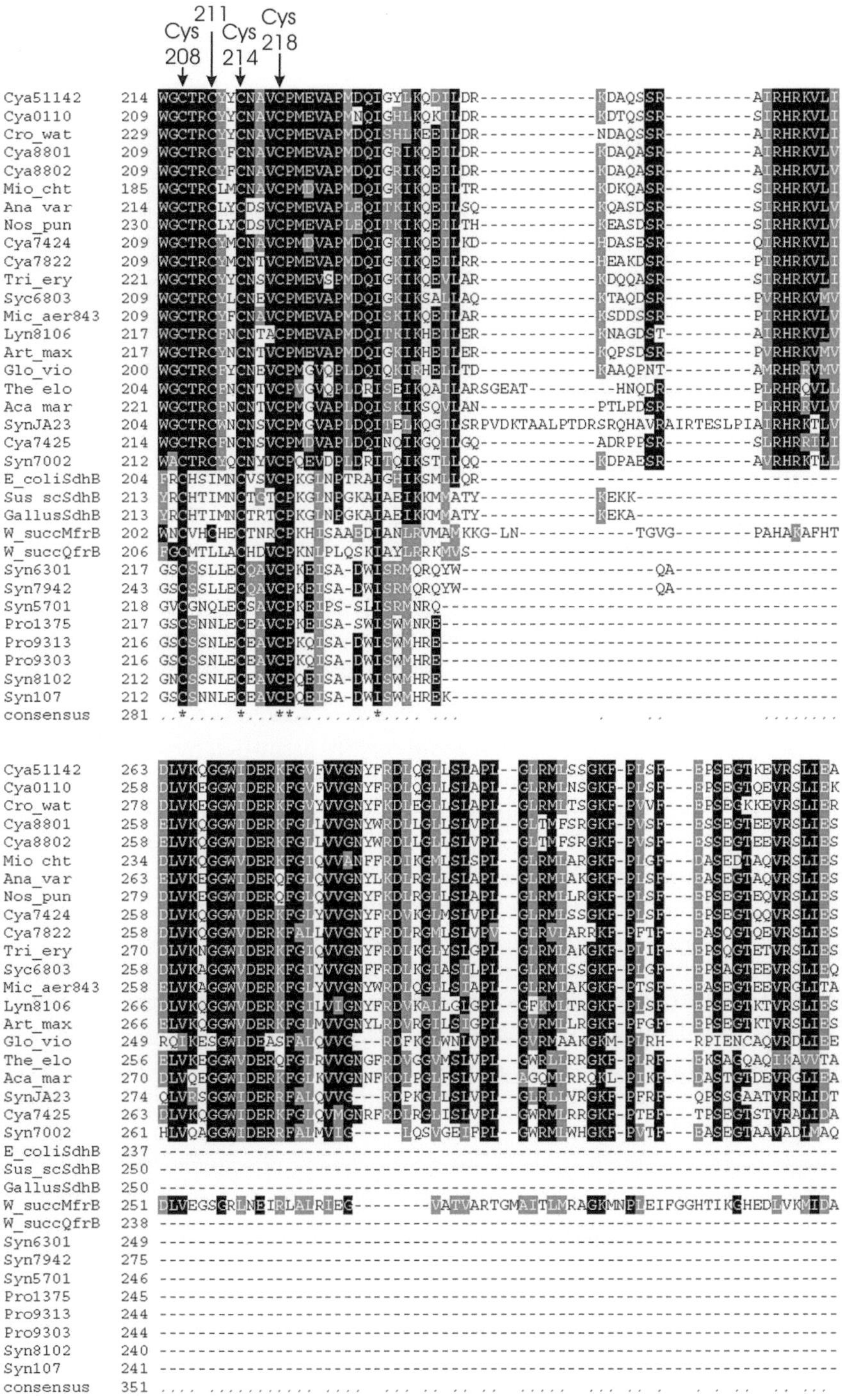

Fig. 17.5 (continued)

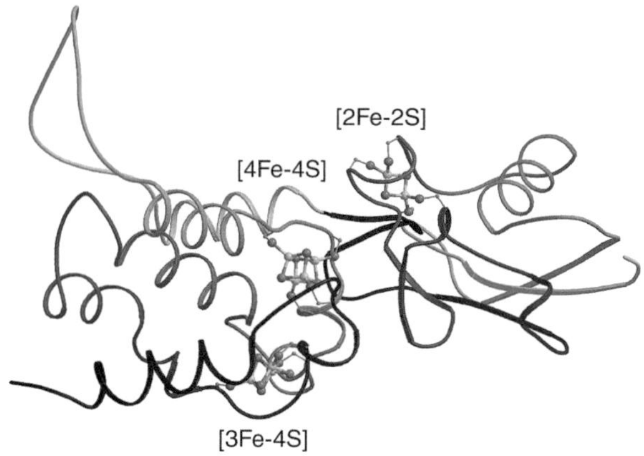

Fig. 17.6 Subunit B, the iron-sulfur protein of *W. succinogenes* QFR. The Cα trace is color-coded from *light grey* to *dark grey* for the amino-terminal [2Fe-2S] domain on the right (residues B1-106) and again from *light grey* to *dark grey* for the carboxy-terminal [7Fe-8S] domain on the left (B106–B239). (Figure modified from Lancaster et al. 1999)

amino-acid residue C-terminal extension, which is not present in those of the other types (Fig. 17.5b).

17.4 The Membrane-Associated Subunit E

While the hydrophilic subunits A and B are readily identified by sequence alignment, the assignment of the membrane-exposed subunit(s) tends to be more ambiguous. A number of possible subunit E sequences from cyanobacteria considered here have been annotated as either "protein of unknown function", "hypothetical protein", "heterodisulfide reductase subunit B" (frequently in the absence of an assignment for "heterodisulfide reductase subunit A") or as a "cysteine-rich domain protein", as detailed in Table 17.1. The overall result of the phylogenetic analysis of subunit E (Fig. 17.7a) includes only the second group encountered for the analysis of the hydrophilic subunits, i.e. all non-*Prochlorococcus* and all non-*Synechococcus* species considered plus *Synechococcus elongatus* PCC7002 and *Synechococcus sp.* JA-2-3B'a(2-13) The alignments of these cyanobacterial sequences are displayed in Fig. 17.7b and indicate sequence identities of between 61 and 92% for between 297 and 301 residues. They are compared to *W. succinogenes* MFR subunit E, revealing a high degree of sequence identity (37% over 284 residues for *Cyanothece sp.* ATCC 51142 subunit E). These subunits E have two striking characteristics (Schäfer et al. 2002; Lemos et al. 2002). First, they contain a duplicated motif rich in cysteines, $CX_{31\text{-}35}CCGX_{33\text{-}39}CX_2C$, which is thought to be essential for the interaction with quinone. Second, several of the putative helices have a predicted amphipathic nature (Fig 17.7b), which is thought to be relevant for the membrane attachment of these monotopic membrane proteins.

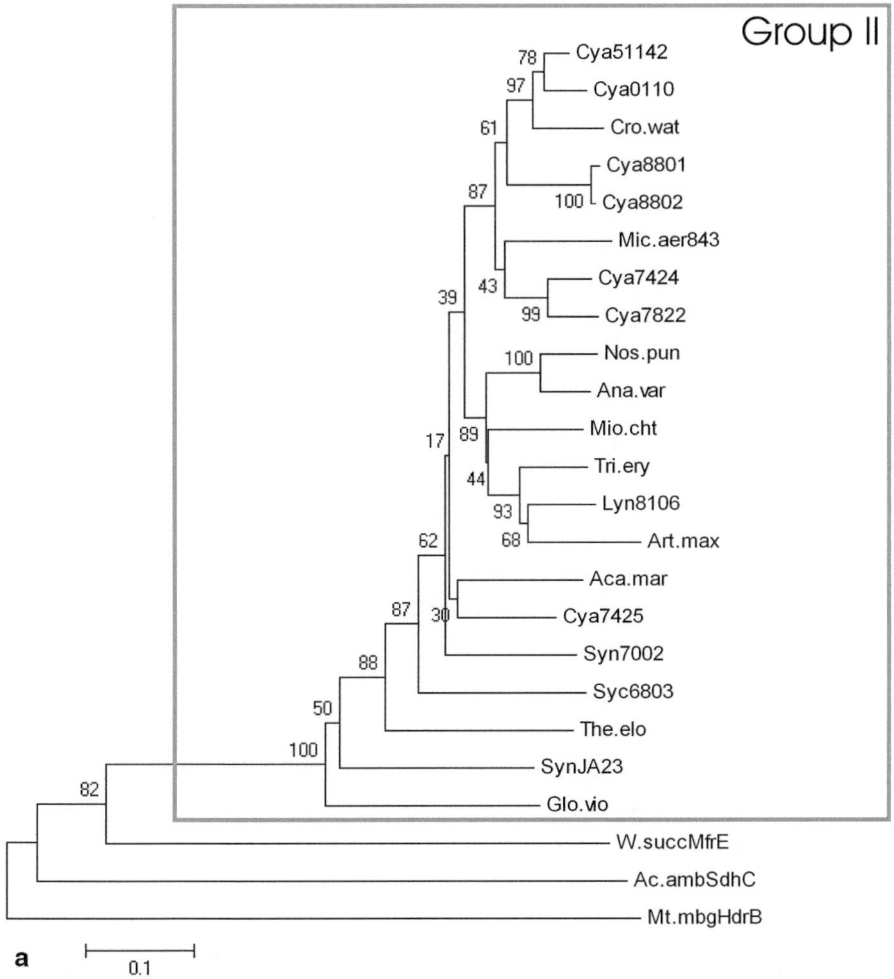

Fig. 17.7 The membrane-associated subunit SdhE. **a** Reconstructed phylogenetic tree of 22 succinate:quinone oxidoreductase membrane-associated subunits E. The tree was obtained with the MEGA4 package (Tamura et al. 2007) using the neighbor-joining method with 1000 bootstrap replications. The abbreviations for the 21 cyanobacterial species are defined in Table 17.1, the remaining sequence is *W. succinogenes* 8-methylmenaquinol:fumarate reductase subunit E (W.succMfrE). **b** Multiple sequence alignment of 22 succinate:quinone oxidoreductase membrane-associated subunits E. The alignment was obtained with the T-coffee method (Notredame et al. 2000). The abbreviations for the 21 cyanobacterial species are defined in Table 17.1, the remaining sequence is *W. succinogenes* 8-methylmenaquinol:fumarate reductase subunit E (W_succMfrE). The positions of the short helices classified as amphipathic by Lemos et al. (2002) are indicated. The highlighted residues are numbered according to their positions in *W. succinogenes* MFR

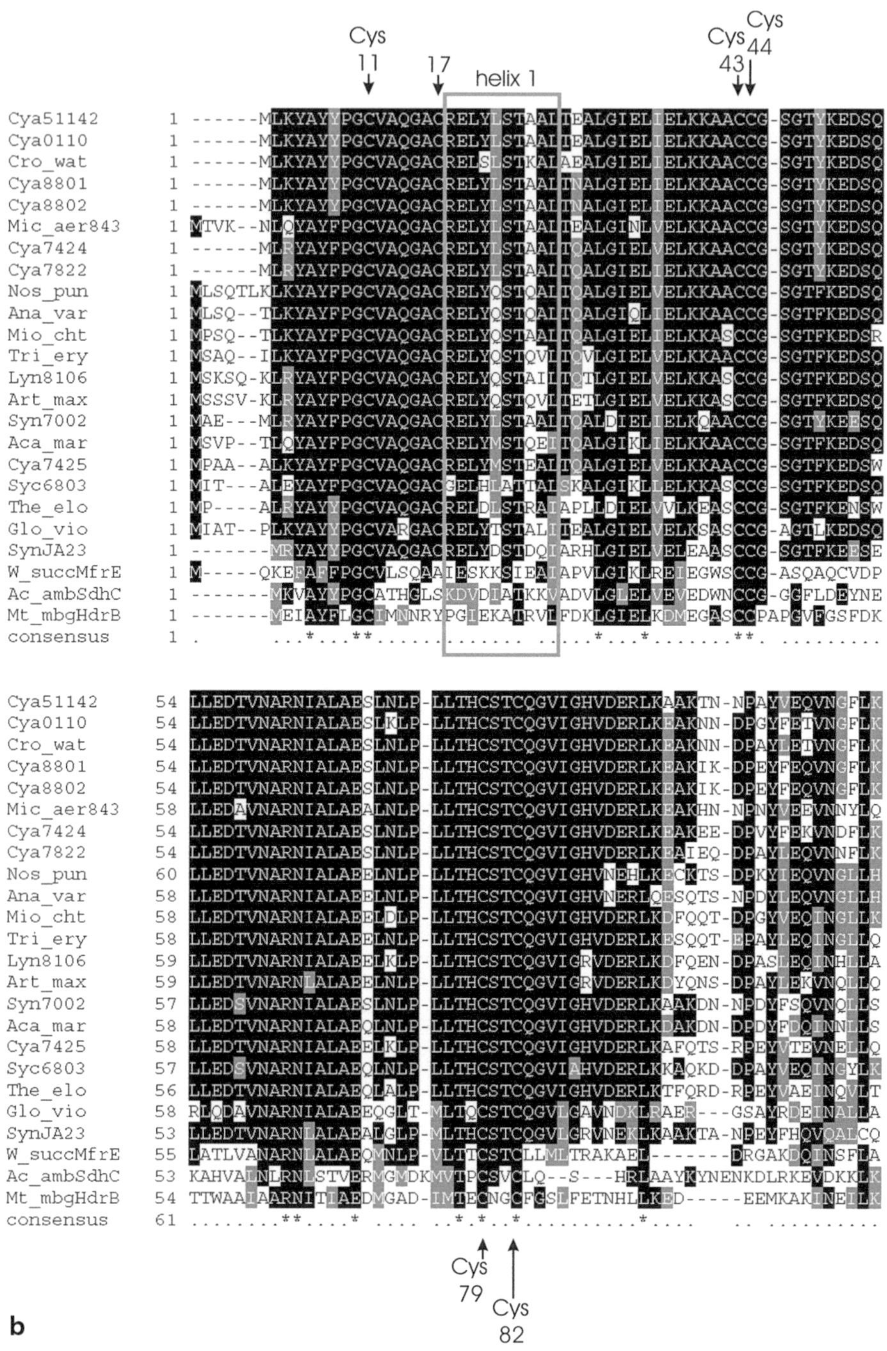

Fig. 17.7 (continued)

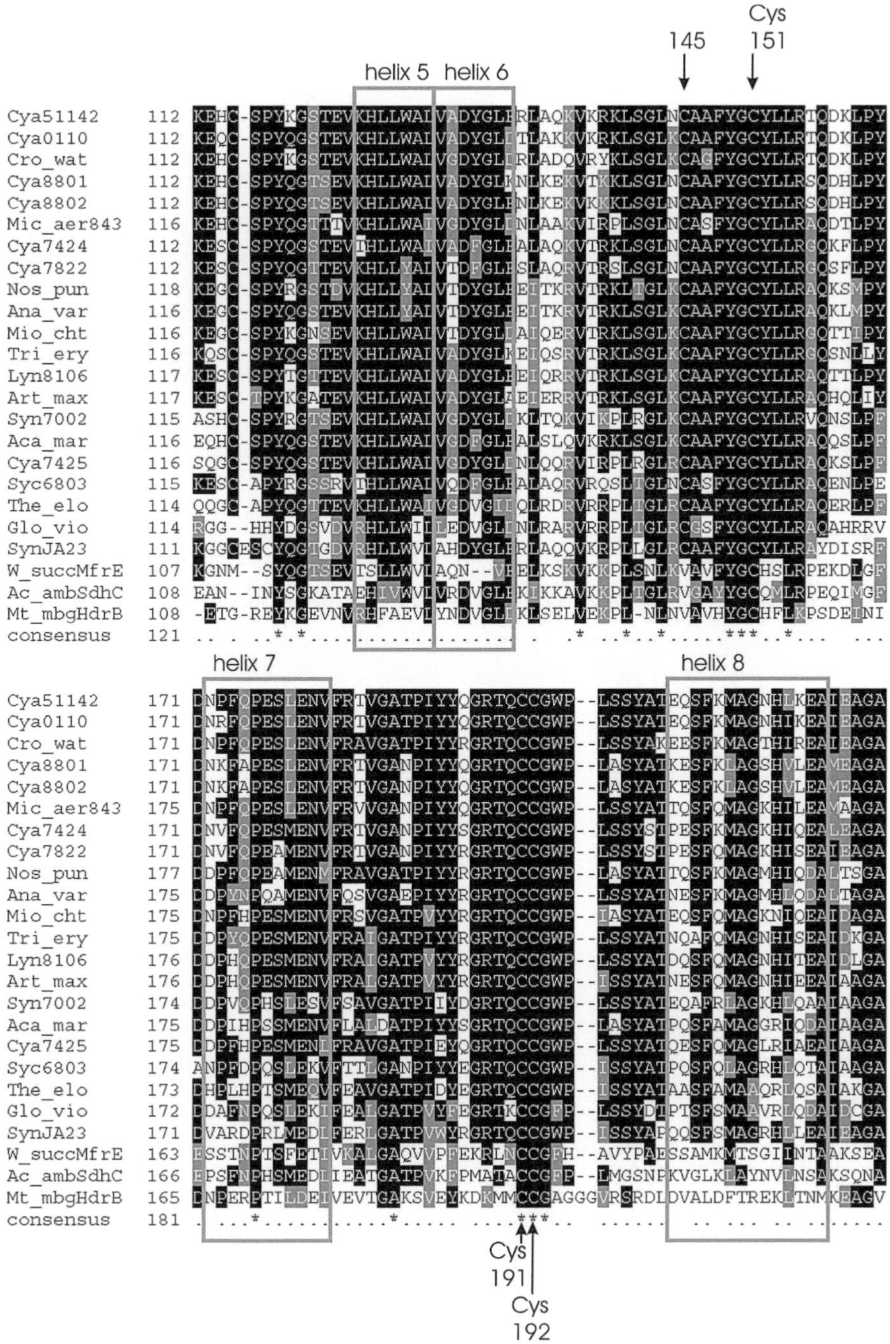

Fig. 17.7 (continued)

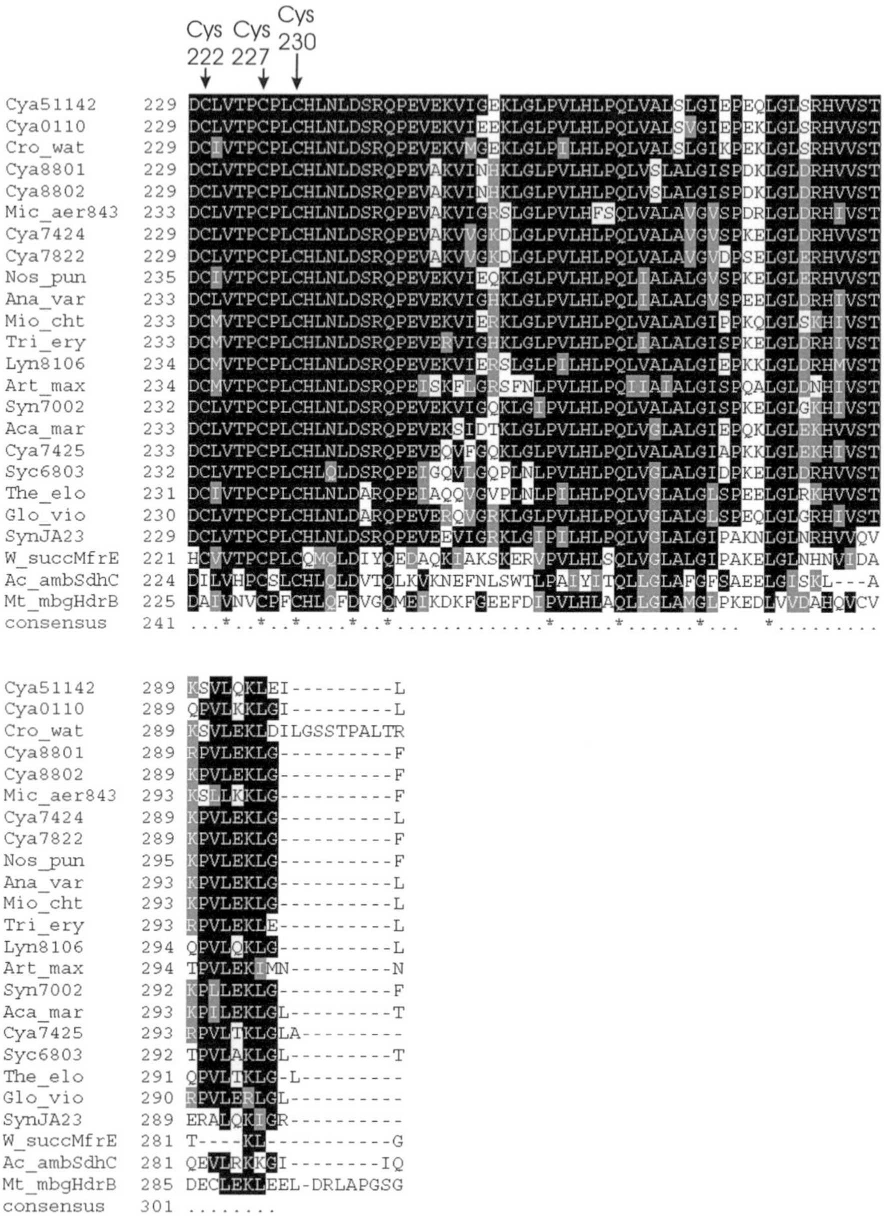

Fig. 17.7 (continued)

17.5 Subunit C, the Membrane-Integral Diheme-Cytochrome *b*

The overall result of the phylogenetic analysis of subunit C (Fig. 17.8a) includes only the first group encountered for the analysis of the hydrophilic subunits, i.e. all *Prochlorococcus* and all *Synechococcus* species except for *Synechococcus elongatus* PCC7002. An exception is *Synechococcus sp.* JA-2-3B'a(2-13), for which both a subunit E and a subunit C can be identified. The alignments of these cyanobacterial sequences are displayed in Fig. 17.8b and indicate sequence identities of between 32 and 90% for between 193 and 230 residues. They are compared to *W. succinogenes* QFR subunit C, revealing a low degree of overall sequence identity (17% over 256 residues for *Synechococcus sp. BL107* subunit C). Strikingly, however, a number of crucial residues are conserved for all sequences aligned. These will be discussed below after introduction of the structure of *W. succinogenes* subunit C.

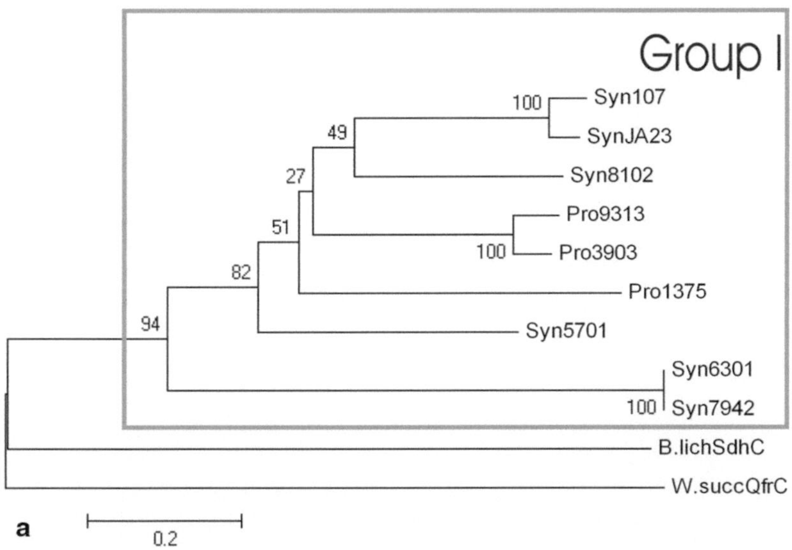

Fig. 17.8 The membrane-integral diheme-cytochrome *b* subunit SdhC. **a** Reconstructed phylogenetic tree of 11 succinate:quinone oxidoreductase diheme-cytochrome *b* subunits. The tree was obtained with the MEGA4 package (Tamura et al. 2007) using the neighbor-joining method with 1000 bootstrap replications. The abbreviations for the 9 cyanobacterial species are defined in Table 17.1, the remaining sequences (see also Table 17.2) are the C subunits of *B. licheniformis* SQR (B.lichSdhC, Veith et al. 2004) and of *W. succinogenes* QFR (W.succQfrC, Baar et al. 2003). **b** Multiple sequence alignment of 11 succinate:quinone oxidoreductase diheme-cytochrome *b* subunits. The alignment was obtained with the T-coffee method (Notredame et al. 2000). The abbreviations for the 9 cyanobacterial species are defined in Table 17.1, the remaining sequences (see also Table 17.2) are the C subunits of *B. licheniformis* SQR (B_lichSdhC) and of *W. succinogenes* QFR (W_succQfrC). The highlighted residues are numbered according to their positions in *W. succinogenes* QFR

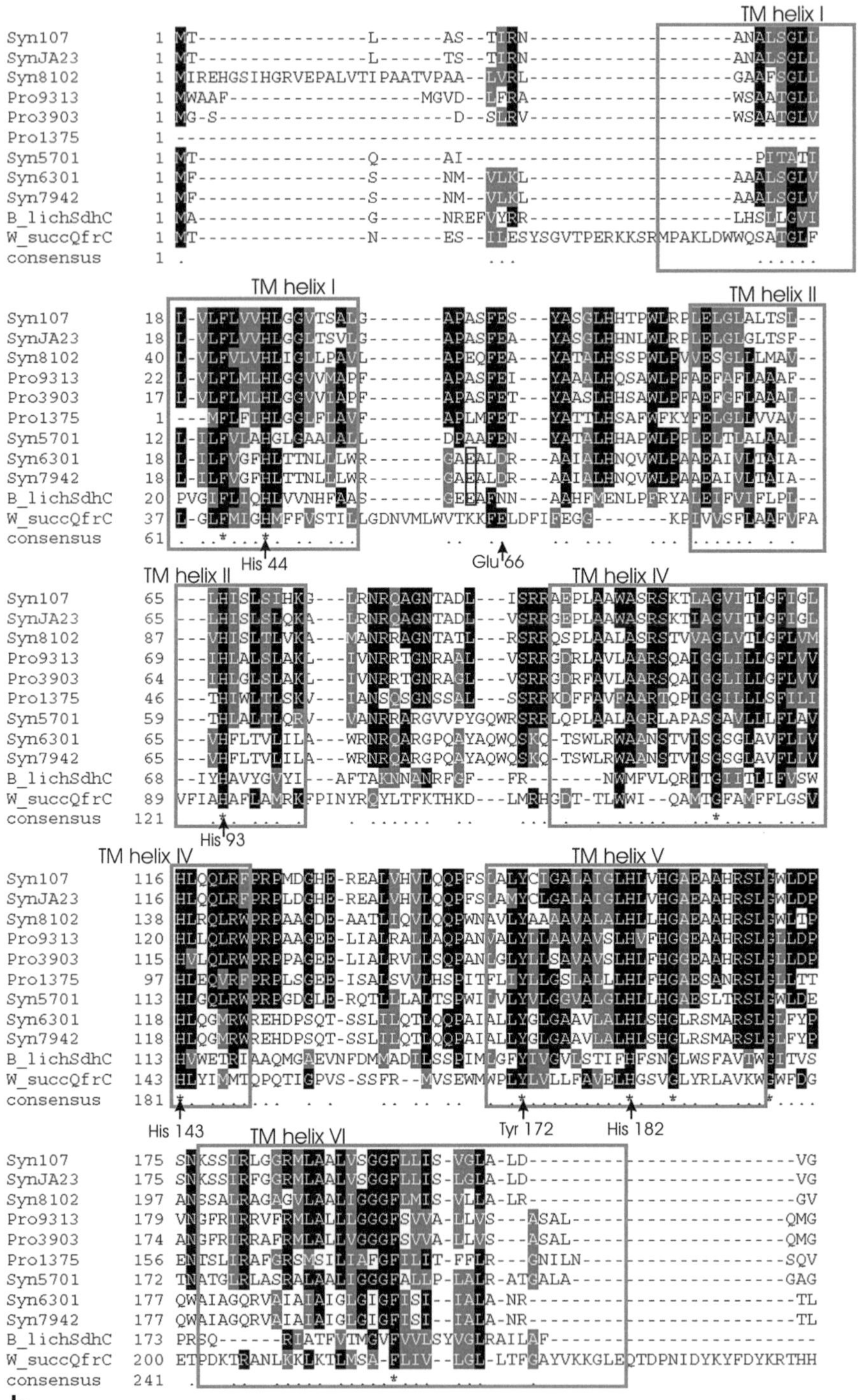

Fig. 17.8 (continued)

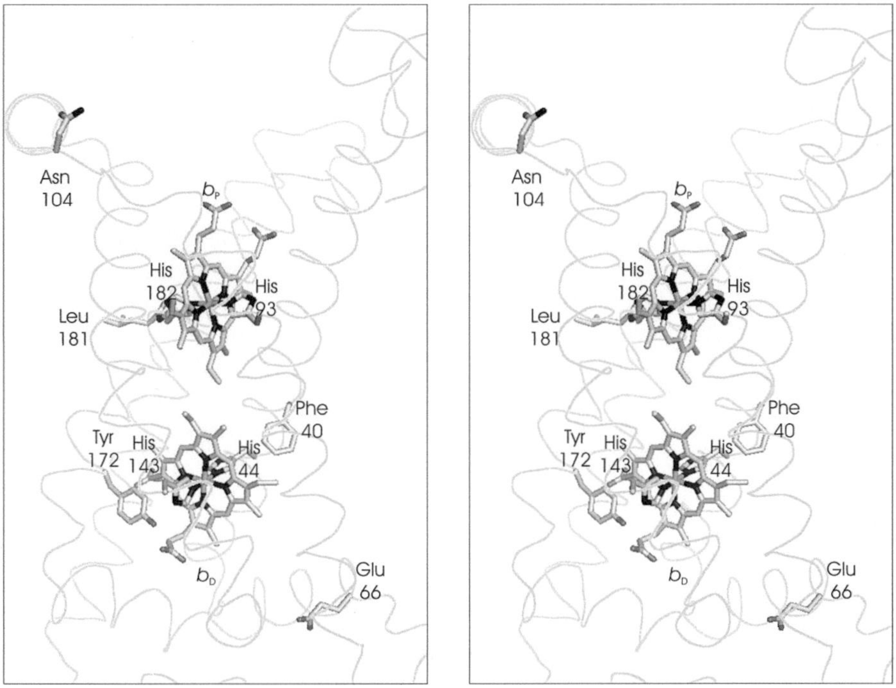

Fig. 17.9 *W. succinogenes* QFR subunit C (stereo view). Selected conserved C subunit residues are labelled

The Cα trace of *W. succinogenes* subunit C is shown in Fig. 17.9. This subunit of 30 kDa (Körtner et al. 1990) contains five membrane-spanning segments with preferentially helical secondary structure. For systematic reasons within the superfamily of succinate:quinone oxidoreductases, these segments are labelled (according to (Hägerhäll and Hederstedt 1996)) I (C22-52), II (C77-100), IV (C121-149), V (C169-194), and VI (C202-237). To a varying degree, all five transmembrane segments are tilted with respect to the membrane normal, and helix IV is strongly kinked at position C137 (Lancaster et al. 1999). This kink is stabilised by the side chain γ-hydroxyl of Ser C141, which, instead of its backbone NH, donates a hydrogen bond to the carbonyl oxygen of Phe C137. As pointed out earlier (Lancaster and Kröger 2000), this feature is very similar to that found for helix F of bacteriorhodopsin (bR, PDB entry 1C3W (Luecke et al. 1999)). However, this Ser residue is not conserved in the cyanobacterial subunits C (Fig. 17.8b).

The planes of both heme molecules bound by the *W. succinogenes* enzyme are approximately perpendicular to the membrane surface and their interplanar angle is 95° (Lancaster et al. 1999). The axial ligands to the "proximal" heme b_P are His C93 of transmembrane segment II and His C182 of transmembrane segment V (Fig. 17.7a). This causes heme b_P to be located towards the cytoplasmic surface of the membrane, and thus towards the [3Fe-4S] iron-sulfur cluster. Hydrogen bonds and salt bridges with the propionate groups of heme b_P are formed with the side chains of residues Gln C30, Ser C31, Trp C126 and Lys C193. Thus, side chains from the residues of the first four transmembrane segments are involved in the binding of heme b_P (Lancaster et al. 1999), which underscores the structural importance of the bound heme (Simon et al. 1998). Of these residues interacting with the b_P propionates, however, only Trp C126 is (partially) conserved in the cyanobacterial sequences (Fig. 17.8b). The axial ligands to the "distal" heme b_D are His C44 of transmembrane segment I and His C143 of transmembrane segment IV (Fig. 17.9), demonstrating that all four heme axial ligands had been correctly predicted by sequence alignment (Körtner et al. 1990) and site-directed mutagenesis (Simon et al. 1998). Residues of *W. succinogenes* QFR subunit C conserved among the succinate:quinone oxidoreductases from cyanobacteria (Fig. 17.8b) include all four His ligands of the heme groups, a number of Gly residues (133, 186, 195) within or at the end of transmembrane helices, and Tyr C172, which interacts with the ring D propionate of heme $b_{D.}$

The two heme groups have different oxidation-reduction potentials (Kröger and Innerhofer 1976), one is the "high-potential" heme b_H, the other the low-potential "low-potential" heme b_L. For the membrane-bound enzyme, these potentials are E_M=−20 mV and E_M=−200 mV, respectively (Kröger and Innerhofer 1976). For the detergent-solubilized QFR enzyme, the respective values are −15 and −150 mV (Lancaster et al. 2000). It has been established that the "high-potential" heme corresponds to b_P and the "low-potential" heme to b_D (Haas and Lancaster 2004).

17.5.1 General Comparison of Membrane-Integral Diheme Cytochrome b Proteins

As noted earlier (Lancaster et al. 1999), the binding of the two heme *b* molecules by an integral membrane protein four-helix bundle described here is very different from that described for the four-helix bundle of the cytochrome bc_1 complex (Xia et al. 1997). In the latter complex, only two transmembrane segments provide two axial heme *b* ligands each. This we refer to as a "two-helix motif" (Lancaster 2002c, 2003c; Fig. 17.10a). Examples for a "three-helix motif", where one transmembrane helix provides two heme *b* ligands, and two others provide one heme *b* ligand each (Fig. 17.10b), may be found in the cases of membrane-bound hydrogenases (Berks et al. 1995; Groß et al. 1998), and formate dehydrogenases (Berks et al. 1995; Jormakka et al. 2002). As described above, the axial ligands for heme binding in *W. succinogenes* QFR, are located on four different transmembrane segments ("four-helix motif"; Fig. 17.10c). One consequence of this difference is that the distance between the two heme iron centers is distinctly shorter in *W. succinogenes* QFR (15.6 Å) than it is in the mitochondrial cytochrome bc_1 complex (21 Å; Xia et al. 1997) and in *E. coli* formate dehydrogenase-N (20.5 Å; Jormakka et al. 2002).

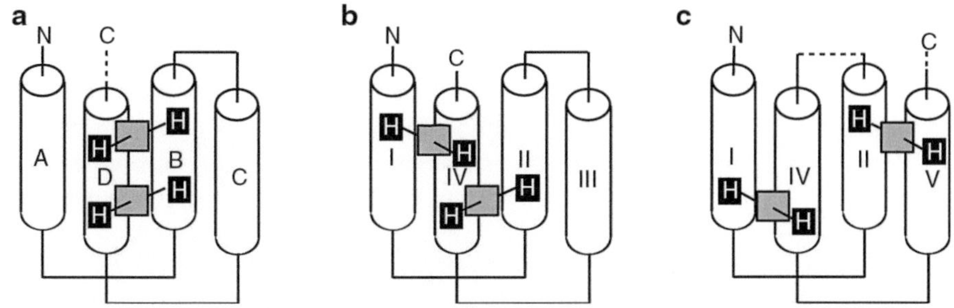

Fig. 17.10 Diheme binding by integral membrane protein four helix bundles (Lancaster 2002c). **a** "Two-helix motif": His ligands from two transmembrane helices (mitochondrial cytochrome bc_1 complex). **b** "Three-helix motif": His ligands from three transmembrane helices (e.g. hydrogenase, formate dehydrogenase). **c** "Four-helix motif": His ligands from four transmembrane helices (dihemeic succinate:quinone oxidoreductases)

17.5.2 The Site(s) of Menaquinol Oxidation and Quinone Reduction

In the *W. succingenes* QFR crystal structure, a cavity which extends from the hydrophobic phase of the membrane, close to the distal heme b_D, to the periplasmic aqueous phase could accommodate a menaquinol molecule, after minor structural alterations (Lancaster et al. 2000), which are consistent with experimentally observed structural differences for the presence and absence of a quinone substrate (Lancaster and Michel 1997). A glutamate residue (Glu C66) lines the cavity and could be involved in the acceptance of the protons liberated upon oxidation of the menaquinol (Fig. 17.9). Replacement of Glu C66 by a glutamine residue resulted in a mutant which did not catalyze quinol oxidation by fumarate, whereas the activity of fumarate reduction was not affected by the mutation (Lancaster et al. 2000). X-ray crystal structure analysis of the Glu C66 → Gln variant enzyme ruled out significant structural alterations. The midpoint potentials of the two heme groups of subunit C were not significantly affected. These results indicate that the inhibition of quinol oxidation activity in the mutant enzyme is due to absence of the carboxyl group of Glu C66. Thus it was concluded that Glu C66, which is conserved in the QFR enzymes from the ε-proteobacteria *C. jejuni* and *H. pylori*, is an essential constituent of the menaquinol oxidation site (Lancaster et al. 2000) close to heme b_D (Fig. 17.9). Glu C66 is conserved in most of the "Group I" cyanobacterial sequences (Fig. 17.8b). The remaining sequences display a conserved Glu three residues earlier (boxed in Fig. 17.8b), which emphasizes the diffulty of obtaining reliable sequence alignments with such low overall sequence identity.

17.6 Overall Description of Diheme-Containing SQOR (Type B)

In all three *W. succinogenes* QFR crystal forms, two heterotrimeric complexes of A, B, and C subunits are associated in an identical fashion, thus forming a dimer (Fig. 17.11). *W. succinogenes* QFR has an overall length of 120 Å in the direction perpendicular to the membrane. Parallel to the membrane, the maximum width is 130 Å for the dimer, and 70 Å for the monomer. Approximately 3665 $Å^2$ (8%) of the *W. succinogenes* QFR monomer surface is buried upon dimer formation. For the B-type enzymes from *W. succinogenes*, *C. jejuni*, and *H. pylori*, it has been derived from analytical gel filtration and analytical ultracentrifugation experiments, that the homodimer is apparently also present in the detergent-solubilized state of the enzyme (Mileni et al. 2006). The characterisation of crystals of *E. coli* SQR at 4.0 Å resolution (Törnroth et al. 2002) indicated that three heterotetrameric enzyme complexes form a crystallographic trimer, and that the formation of homo-oligomers is not uncommon for this type of membrane protein complexes.

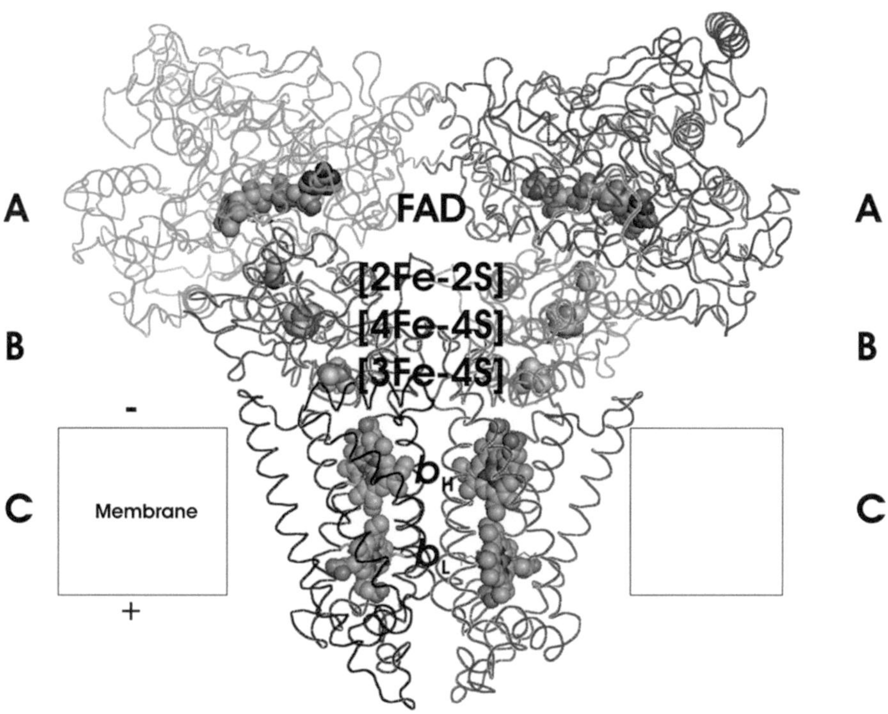

Fig. 17.11 Three-dimensional structure of the *W. succinogenes* QFR dimer of heterotrimeric complexes of *A, B,* and *C* subunits. The Cα traces and the atomic structures of the six prosthetic groups per heterotrimer are shown

17.6.1 Relative Orientation of Soluble and Membrane-Embedded QFR Subunits

The structure of D-type *E. coli* QFR can be superimposed on the structure of B-type *W. succinogenes* QFR based on the hydrophilic subunits A and B (Lancaster et al. 1999). This similarity in structure was expected based on sequence comparisons. However, in this superimposition, the membrane-embedded subunits cannot be aligned. In an alternate superimposition, the transmembrane subunits C and D of the *E. coli* enzyme can be overlayed on to the *W. succinogenes* C subunit (Lancaster et al. 1999). Compared to the former superimposition, the latter involves a rotation around the membrane normal of approximately 180° and an orthogonal 25° rotation. This immediately leads to two important conclusions (Lancaster et al. 1999). First, the structures of the transmembrane subunits carrying no hemes and two hemes, respectively, can be aligned to a significant degree, although only eleven of the aligned residues are identical. Second, the relative orientation of the soluble subunits and the transmembrane subunits is different in the QFR complexes from the two species. Interestingly, in spite of the differences in polypeptide composition and heme content, the C-type SQR from *E. coli* exhibits the same relative orientation of hydrophilic and hydrophobic subunits as the B-type QFR from *W. succinogenes* (Lancaster 2003d).

17.6.2 Electron Transfer

For the function of QFR, electrons have to be transferred from the quinol-oxidising site in the membrane to the fumarate-reducing site, protruding into the cytoplasm. The arrangement of the prosthetic groups in the QFR dimer is displayed in Fig. 17.11. It has been shown for other electron transfer proteins that physiological electron transfer occurs if edge-to-edge distances beteen redox centres are shorter than 14 Å, but not if they are longer than 14 Å (Page et al. 1999). In the case of *W. succinogenes* QFR, this indicates that physiological electron transfer can occur between the six prosthetic groups of one QFR heterotrimeric complex, but not between the two QFR complexes in the dimer (Lancaster et al. 2000).

The fumarate molecule is in van der Waals contact with the isoalloxazine ring of FAD. The linear arrangement of the prosthetic groups in one QFR complex therefore provides one straightforward pathway by which electrons could be transferred efficiently from the menaquinol oxidising site via the two heme groups, the three iron-sulfur clusters and the FAD to the site of fumarate reduction.

Because of ist very low midpoint potential ($E_m < -250$ mV; Unden et al. 1984), the [4Fe-4S] iron-sulfur cluster has been suggested not to participate in electron transfer (see Ackrell et al. (1992) and Hägerhäll (1997) for discussions). However, the determined low potential may be an artefact due to anti-co-operative electrostatic interactions between the redox centers (Salerno 1991). The position of the [4Fe-4S] cluster as revealed in the available QFR and SQR structures is highly suggestive of its direct role in electron transfer from the [3Fe-4S] cluster to the [2Fe-2S] cluster. Despite this major thermodynamically unfavorable step, the calculated rate of electron transfer is on a microsecond scale, demonstrating that this barrier can easily be overcome by thermal activation as long as the electron transfer chain components are sufficiently close to promote intrinsically rapid electron tunneling (Dutton et al. 1998).

17.6.2.1 The *W. succinogenes* Paradox and the Mechanism of Catalysis in B-type SQR

An overview of the current status of discussion of electron and proton transfer in succinate:quinone oxidoreductases is shown in Fig. 17.12a–d. In E-type cyanobacterial SQR (Fig. 17.12a), mitochondrial complex II and other C-type enzymes, such as SQR from *P. denitrificans* and *E. coli* (Fig. 17.12b), electron transfer from succinate to quinone does not lead to the generation of a transmembrane electrochemical potential Δp, since the protons released by succinate oxidation are on the same

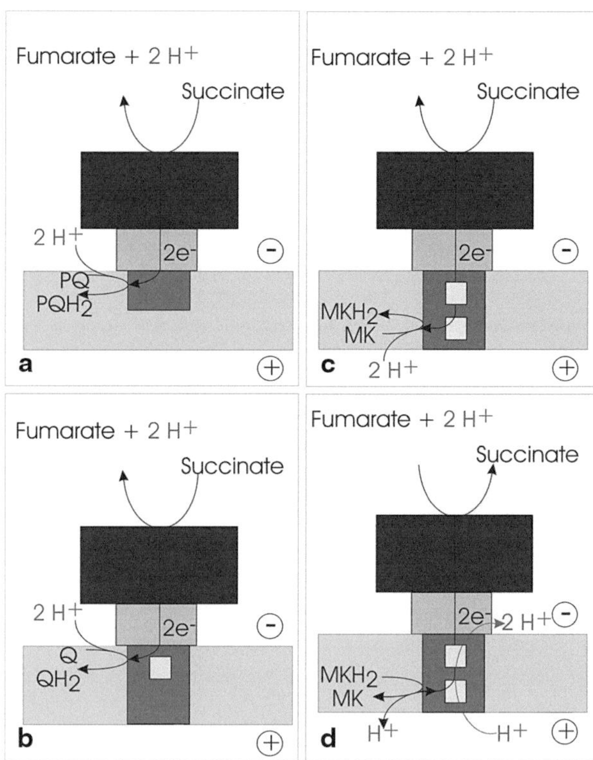

Fig. 17.12 The coupling of electron and proton flow in succinate: quinone oxidoreductases in aerobic (a, b, c) and anaerobic respiration (d), respectively. Positive and negative sides of the membrane are described for Fig. 17.1. (**a, b**) Electroneutral reactions as catalyzed by E-type cyanobacterial SQR from group II (a) and C-type SQR enzymes (b). (**c**) Utilization of a transmembrane electrochemical potential Δp as catalyzed by B-type SQR enzymes from Gram-positive bacteria (Madej et al. 2006b) and possibly also from cyanobacteria from group I. (**d**) Electroneutral fumarate reduction by B-type QFR enzymes with a proposed compensatory "E-pathway" (Lancaster 2002c; Madej et al. 2006a, 2009)

side of the membrane as those consumed by quinone reduction (see Ohnishi et al. (2000) for a review).

The orientation of the catalytic sites of fumarate reduction (Lancaster et al. 2001), associated with proton binding, and menaquinol oxidation (Lancaster et al. 2000), associated with proton release, towards opposite sides of the membrane indicated that quinol oxidation by fumarate should be an electrogenic process in *W. succinogenes*, i.e. associated directly with the establishment of an electrochemical proton potential across the membrane.

This electrogenic catalysis indeed appeared to be the case for some diheme-containing representatives of the superfamily of QFRs and succinate:menaquinone reductases (SQRs). Succinate oxidation by menaquinone, an endergonic reaction under standard conditions, had been proposed to be driven by the electrochemical proton potential in the Gram-positive bacterium *B. subtilis* and in other prokaryotes containing succinate:menaquinone reductases (Lemma et al. 1990; Schirawski and Unden 1998; Schnorpfeil et al. 2001; Zaunmuller et al. 2006). This is the analogous reaction to that suggested by the arrangement of the sites of catalysis, but in the opposite direction (Fig 17.12c). However, the experiments on the SQR had previously been performed only with whole cells and isolated membranes and it has been questioned (Azarkina and Konstantinov 2002) whether the observed effects are associated specifically with the SQR. More recently, it has been shown that the dihaem-containing succinate:menaquinone reductase, isolated from the Gram-positive bacterium *Bacillus licheniformis* and reconstituted into proteoliposomes, is unable to support the reduction of a soluble menaquinone analog unless the protonophore carbonyl cyanide *m*-chloro-phenylhydrazone (CCCP) is added (Madej et al. 2006b). Apparently the driving force of the reaction, even with a large excess of starting material over product, was insufficient to support the establishment of a Δp across the proteoliposomal membrane. In order to increase this driving force, the quinone substrate was modified to increase its oxidation/reduction midpoint potential. For an increased driving force of the reverse reaction, a lower redox midpoint potential was achieved by designing a suitable substrate. Both in the case of quinol oxidation, as well as in the case of quinone reduction, a membrane potential could be measured, thus unequivocally demonstrating electrogenic catalysis in the case of this enzyme (Madej et al. 2006b).

However, as discussed earlier (Lancaster et al. 2000), analogous experiments for isolated *W. succinogenes* QFR reconstituted into liposomes had shown that the oxidation of quinol by fumarate as catalysed by this enzyme is electroneutral (Madej et al. 2006a; Geisler et al. 1994; Kröger et al. 2002; Biel et al. 2002).

To reconcile these apparently conflicting experimental observations, the so-called "E-pathway hypothesis" was proposed (Lancaster 2002c; Fig. 17.12d). According to this working hypothesis, the transmembrane transfer of two electrons in *W. succinogenes* QFR is coupled to the compensatory, parallel translocation of one proton per electron from the periplasm to the cytoplasm. The proton transfer pathway used is transiently established during reduction of the haem groups and is closed in the oxidised enzyme. The two most prominent constituents of the proposed novel pathway were suggested to be the ring C propionate of the distal heme b_D and, in particular, the amino acid residue Glu C180, after which the "E-pathway" was named. Since the first proposal of this hypothesis, a number of theoretical (Haas and Lancaster 2004) and experimental results (Haas et al. 2005; Lancaster et al. 2005; Mileni et al. 2005) were obtained that supported it (reviewed by Lancaster et al. 2006, 2008) and it was finally proven (Madej et al. 2006a, 2009). However, Glu C180 is not conserved in the cyanobacterial B-type enzymes (Fig. 17.8b), indicating that these probably operate in analogy to the SQR enzymes of Gram-positive bacteria (Fig 17.12c).

References

Ackrell BAC, Johnson MK, Gunsalus RP and Cecchini G (1992) Structure and function of succinate dehydrogenase and fumarate reductase. In: Muller F (ed) Chemistry and biochemistry of flavoenzymes, vol 3, pp 229–297. CRC Press, Boca Raton FL

Azarkina N and Konstantinov AA (2002) Stimulation of menaquinone-dependent electron transfer in the respiratory chain of Bacillus subtilis by membrane energization. J Bacteriol 184, 5339–5347

Baar C, Eppinger M, Raddatz G, Simon Jr, Lanz C, Klimmek O, Nandakumar R, Gross R, Rosinus A, Keller H, Jagtap P, Linke B, Meyer F, Lederer H and Schuster SC (2003) Complete genome sequence and analysis of *Wolinella succinogenes*. Proc Natl Acad Sci U S A 100, 11690–11695

Berks BC, Page MD, Richardson DJ, Reilly A, Cavill A, Outen F and Ferguson SJ (1995) Sequence-analysis of subunits of the membrane-bound nitrate reductase from a denitrifying bacterium—The integral membrane subunit provides a prototype for the diheme electron carrying arm of a redox loop. Mol Microbiol 15, 319–331

Bhaya D, Grossman AR, Steunou A.-S, Khuri N, Cohan FM, Hamamura N, Melendrez MC, Bateson MM, Ward DM and Heidelberg JF (2007) Population level functional diversity in a microbial community revealed by comparative genomic and metagenomic analyses. ISME J 1, 703–713

Biel S, Simon J, Groß R, Ruiz T, Ruitenberg M and Kröger A (2002) Reconstitution of coupled fumarate respiration in liposomes by incorporating the electron transport enzymes isolated from *Wolinella succinogenes*. Eur J Biochem 269, 1974–1983

Brenner SE, Chothia C and Hubbard TJP (1998) Assessing sequence comparison methods with reliable structurally identified distant evolutionary relationships. Proc Natl Acad Sci U S A 95, 6073–6078

Cecchini G, Schröder I, Gunsalus RP and Maklashina E (2002) Succinate dehydrogenase and fumarate reductase from *Escherichia coli*. Biochim Biophys Acta 1553, 140–157

Chisholm SW, Olson RJ, Zettler ER, Goericke R, Waterbury JB and Welschmeyer NA (1988) A novel free-living prochlorophyte abundant in the oceanic euphotic zone. Nature 334, 340–343

Collins MD and Jones D (1981) Distribution of isoprenoid quinone structural types in bacteria and their taxonomic implications. Microbiol Rev 45, 316–354

Cooley JW and Vermaas WJF (2001) Succinate dehydrogenase and other respiratory pathways in thylakoid membranes of *Synechocystis sp.* strain PCC 6803: capacity comparisons and physiological function. J Bacteriol 183, 4251–4258

Cooley JW, Howitt CA and Vermaas WJF (2000) Succinate:quinol oxidoreductases in the cyanobacterium *Synechocystis sp.* strain PCC 6803: presence and function in metabolism and electron transport. J Bacteriol 182, 714–722

Dufresne A, Salanoubat M, Partensky Fdr., Artiguenave Fo., Axmann IM, Barbe Vr, Duprat S, Galperin MY, Koonin EV, Le Gall F, Makarova KS, Ostrowski M, Oztas S, Robert C, Rogozin IB, Scanlan DJ, de Marsac NT, Weissenbach J, Wincker P, Wolf YI and Hess WR (2003) Genome sequence of the cyanobacterium *Prochlorococcus marinus* SS120, a nearly minimal oxyphototrophic genome. Proc Natl Acad Sci U S A 100, 10020–10025

Dutton PL, Chen X, Page CC, Huang S, Ohnishi T and Moser CC (1998) Respiratory electron transfer proteins. In: Canters GW, Vijgenboom E (eds) Biological electron transfer Chains: genetics, composition and mode of operation, pp 3–8. Kluwer Academic Publishers, Dordrecht

Geisler V, Ullmann R and Kröger A (1994) The direction of proton-exchange associated with the redox reactions of menaquinone during electron transport in *Wolinella succinogenes*. Biochim Biophys Acta 1184, 219–226

Gomes CM, Lemos RS, Teixeira M, Kletzin A, Huber H, Stetter KO, Schäfer G and Anemüller S (1999) The unusual iron sulfur composition of the Acidianus ambivalens succinate dehydrogenase complex. Biochim Biophys Acta 1411, 134–141

Groß R, Simon J, Lancaster CRD and Kröger A (1998) Identification of histidine residues in *Wolinella succinogenes* hydrogenase that are essential for menaquinone reduction by H_2. Mol Microbiol 30, 639–646

Guccione E, Hitchcock A, Hall SJ, Mulholland F, Shearer N, Vliet AHM v and Kelly DJ (2010) Reduction of fumarate, mesaconate and crotonate by Mfr, a novel oxygen-regulated periplasmic reductase in *Campylobacter jejuni*. Environ Microbiol 12, 576–591

Haas AH and Lancaster CRD (2004) Calculated coupling of transmembrane electron and proton transfer in dihemic quinol:fumarate reductase. Biophys J 87, 4298–4315

Haas AH, Sauer US, Gross R, Simon J, Mantele W and Lancaster CRD (2005) FTIR difference spectra of Wolinella succinogenes quinol:fumarate reductase support a key role of Glu C180 within the "E-pathway hypothesis" of coupled transmembrane electron and proton transfer. Biochemistry 44, 13949–13961

Hägerhäll C (1997) Succinate:quinone oxidoreductases. Variations on a conserved theme. Biochim Biophys Acta 1320, 107–141

Hägerhäll C and Hederstedt L (1996) A structural model for the membrane-integral domain of succinate:quinone oxidoreductases. FEBS Lett 389, 25–31

Hägerhäll C, Sled V, Hederstedt L and Ohnishi T (1995) The trinuclear iron-sulfur cluster S3 in *Bacillus subtilis* succinate:menaquinone reductase—Effects of a mutation in the putative cluster ligation motif on enzyme activity and EPR properties. Biochim. Biophys Acta 1229, 356–362

Hederstedt L (1999) Respiration without O_2. Science 284, 1941–1942

Hederstedt L (2002) Succinate: quinone oxidoreductase in the bacteria *Paracoccus denitrificans* and *Bacillus subtilis*. Biochim Biophys Acta 1553, 74–83

Hederstedt L (2003) Complex II is complex too. Science 299, 671–672

Huang L-S, Shen JT, Wang AC and Berry EA (2006) Crystallographic studies of the binding of ligands to the dicarboxylate site of Complex II, and the identity of the ligand in the "oxaloacetate-inhibited" state. Biochim Biophys Acta 1757, 1073–1083

Iverson TM, Luna-Chavez C, Cecchini G and Rees DC (1999) Structure of the *Escherichia coli* fumarate reductase respiratory complex. Science 284, 1961–1966

Iverson TM, Luna-Chavez C, Croal LR, Cecchini G and Rees DC (2002) Crystallographic studies of the *Escherichia coli* quinol:fumarate reductase with inhibitors bound to the quinol-binding site. J Biol Chem 277, 16124–16130

Jones LW and Myers J (1963) A common link between photosynthesis and respiration in a blue-green alga. Nature 199, 670–672

Jormakka M, Törnroth S, Byrne B and Iwata S (2002) Molecular basis of proton motive force generation: structure of formate dehydrogenase-N. Science 295, 1863–1868

Juhnke HD, Hiltscher H, Nasiri HR, Schwalbe H and Lancaster CRD (2009) Production, characterization and determination of the real catalytic properties of the putative 'succinate dehydrogenase' from *Wolinella succinogenes*. Mol Microbiol 71, 1088–1101

Kaneko T, Tanaka A, Sato S, Kotani H, Sazuka T, Miyajima N, Sugiura M and Tabata, S (1995) Sequence analysis of the genome of the unicellular cyanobacterium *Synechocystis sp.* strain PCC6803. I. Sequence features in the 1 Mb region from map positions 64% to 92% of the genome. DNA Res 2, 153–166

Kaneko T, Sato S, Kotani H, Tanaka A, Asamizu E, Nakamura Y, Miyajima N, Hirosawa M, Sugiura M, Sasamoto S, Kimura T, Hosouchi T, Matsuno A, Muraki A, Nakazaki N, Naruo K, Okumura S, Shimpo S, Takeuchi C, Wada T, Watanabe A, Yamada M, Yasuda M and Tabata, S (1996) Sequence analysis of the genome of the unicellular cyanobacterium *Synechocystis sp.* Strain PCC6803. II. Sequence determination of the entire genome and assignment of potential protein-coding regions. DNA Res 3, 109–136

Kaneko T, Nakajima N, Okamoto S, Suzuki I, Tanabe Y, Tamaoki M, Nakamura Y, Kasai F, Watanabe A, Kawashima K, Kishida Y, Ono A, Shimizu Y, Takahashi C, Minami C, Fujishiro T, Kohara M, Katoh M, Nakazaki N, Nakayama S, Yamada M, Tabata S and Watanabe MM (2007) Complete genomic structure of the bloom-forming toxic cyanobacterium *Microcystis aeruginosa* NIES-843. DNA Res 14, 247–256

Karlin S and Altschul SF (1990) Methods for assessing the statistical significance of molecular sequence features by using general scoring schemes. Proc Natl Acad Sci U S A 87, 2264–2268

Karlin S and Altschul SF (1993) Applications and statistics for multiple high-scoring segments in molecular sequences. Proc Natl Acad Sci U S A 90, 5873–5877

Kenny WC and Kröger A (1977) The covalently bound flavin of *Vibrio succinogenes* succinate dehydrogenase. FEBS Lett 73, 239–243

Kettler GC, Martiny AC, Huang K, Zucker J, Coleman ML, Rodrigue S, Chen F, Lapidus A, Ferriera S, Johnson J, Steglich C, Church GM, Richardson P and Chisholm SW (2007) Patterns and Implications of gene gain and loss in the evolution of *Prochlorococcus*. PLoS Genet 3, e231

Kita K, Hirawake H, Miyadera H, Amino H and Takeo S (2002) Role of complex II in anaerobic respiration of the parasite mitochondria from *Ascaris suum* and *Plasmodium falciparum*. Biochim Biophys Acta 1553, 123–139

Körtner C, Lauterbach F, Tripier D, Unden G and Kröger A (1990) *Wolinella succinogenes* fumarate reductase contains a diheme cytochrome *b*. Mol Microbiol 4, 855–860

Kröger A (1978) Fumarate as terminal electron acceptor of phosphorylative electron transport. Biochim Biophys Acta 505, 129–145

Kröger A and Innerhofer A (1976) The function of the *b* cytochromes in the electron transport from formate to fumarate of *Vibrio succinogenes*. Eur J Biochem 69, 497–506

Kröger A, Biel S, Simon J, Gross R, Unden G and Lancaster CRD (2002) Fumarate respiration of *Wolinella succinogenes*: enzymology, energetics and coupling mechanism Biochim Biophys Acta 1553, 23–38

Lancaster CRD (2001a) Succinate:quinone oxidoreductases—What can we learn from *Wolinella succinogenes* quinol:fumarate reductase? FEBS Lett 504, 133–141

Lancaster CRD (2001b) Succinate:quinone oxidoreductases. In: Messerschmidt A, Huber R, Poulos T, Wieghardt K (eds) Handbook of metalloproteins, pp 379–401. Wiley, Chichester

Lancaster CRD (ed) (2002a) Special issue on fumarate reductases and succinate dehydrogenases. Biochim Biophys Acta 1553, 1–176

Lancaster CRD (2002b) Succinate:quinone oxidoreductases: an overview. Biochim Biophys Acta 1553, 1–6

Lancaster CRD (2002c) *Wolinella succinogenes* quinol:fumarate reductase—2.2 Å resolution crystal structure and the E-pathway hypothesis of coupled transmembrane proton and electron transfer. Biochim Biophys Acta 1565, 215–231

Lancaster CRD (2003a) Crystallization of *Wolinella succinogenes* quinol:fumarate reductase. In: Hunte C, Schägger H and von Jagow G (eds) Membrane protein purification and crystallization: a practical guide, 2nd edn, pp 219–228. Academic, San Diego

Lancaster CRD (2003b) Crystallization of *Wolinella succinogenes* quinol:fumarate reductase in three crystal forms. In: Iwata S (ed) Methods and results in membrane protein crystallization, pp 177–192. University Line, La Jolla

Lancaster CRD (2003c) The structure of wolinella succinogenes quinol:fumarate reductase and its relevance to the superfamily of succinate:quinone oxidoreductases. Adv Protein Chem 63, 131–149

Lancaster CRD (2003d) Wolinella succinogenes quinol:fumarate reductase and its comparison to Escherichia coli succinate:quinone reductase. FEBS Lett 555, 21–28

Lancaster CRD (2004) Structure and function of succinate:quinone oxidoreductases and the role of quinol:fumarate reductases in fumarate respiration. In: Zannoni D (ed) Respiration in archaea and bacteria volume 1: diversity of prokaryotic electron transport carriers, pp 57–85. Kluwer Scientific, Dordrecht,

Lancaster CRD and Kröger A (2000) Succinate:quinone oxidoreductases: new insights from X-ray crystal structures. Biochim Biophys Acta 1459, 422–431

Lancaster CRD and Michel H (1997) The coupling of light-induced electron transfer and proton uptake as derived from crystal structures of reaction centres from *Rhodopseudomonas viridis* modified at the binding site of the secondary quinone, QB. Structure 5, 1339–1359

Lancaster CRD and Simon J (2002) Succinate:quinone oxidoreductases from ε-proteobacteria. Biochim Biophys Acta 1553, 84–101

Lancaster CRD, Kröger A, Auer M and Michel H (1999) Structure of fumarate reductase from *Wolinella succinogenes* at 2.2 Å resolution. Nature 402, 377–385

Lancaster CRD, Groß R, Haas A, Ritter M, Mäntele W, Simon J and Kröger A (2000) Essential role of Glu-C66 for menaquinol oxidation indicates transmembrane electrochemical potential generation by *Wolinella succinogenes*. Proc Natl Acad Sci U S A 97, 13051–13056

Lancaster CRD, Groß R and Simon J (2001) A third crystal form of *Wolinella succinogenes* quinol:fumarate reductase reveals domain closure at the site of fumarate reduction. Eur J Biochem 268, 1820–1827

Lancaster CRD, Sauer US, Gross R, Haas AH, Graf J, Schwalbe H, Mantele W, Simon J and Madej MG (2005) Experimental support for the "E pathway hypothesis" of coupled transmembrane e(-) and H+ transfer in dihemic quinol:fumarate reductase. Proc Natl Acad Sci U S A 102, 18860–18865

Lancaster CRD, Haas AH, Madej MG and Mileni M (2006) Recent progress on obtaining theoretical and experimental support for the "E-pathway hypothesis" of coupled transmembrane electron and proton transfer in dihaem-containing quinol:fumarate reductase. Biochimi Biophys Acta 1757, 988–995

Lancaster CRD, Herzog E, Juhnke HD, Madej MG, Müller FG, Paul R and Schleidt PG (2008) Electroneutral and electrogenic catalysis by dihaem-containing succinate:quinone oxidoreductases. Biochem Soc Trans 36, 996–1000

Lauterbach F, Körtner C, Albracht SP, Unden G and Kröger A (1990) The fumarate reductase operon of *Wolinella succinogenes*. Sequence and expression of the *frdA* and *frdB* genes. Arch Microbiol 154, 386–393

Leach CK and Carr NG (1970) Electron transport and oxidative phosphorylation in the blue-green alga *Anabaena variabilis*. J Gen Microbiol 64, 55–70

Lemma E, Unden G and Kröger A (1990) Menaquinone is an obligatory component of the chain catalyzing succinate respiration in *Bacillus subtilis*. Arch Microbiol 155, 62–67

Lemma E, Hägerhäll C, Geisler V, Brandt U, von Jagow G and Kröger A (1991) Reactivity of the *Bacillus subtilis* succinate dehydrogenase complex with quinones. Biochim Biophys Acta 1059, 281–285

Lemos RS, Fernandes AS, Pereira MM, Gomes CM and Teixeira M (2002) Quinol:fumarate oxidoreductases and succinate:quinone oxidoreductases: phylogenetic relationships, metal centers and membrane attachment. Biochim Biophys Acta 1553, 158–170

Luecke H, Schobert B, Richter HT, Cartailler JP and Lanyi JK (1999) Structure of bacteriorhodopsin at 1.55 angström resolution. J Mol Biol 291, 899–911

Madej MG, Nasiri HR, Hilgendorff NS, Schwalbe H and Lancaster CRD (2006a) Evidence for transmembrane proton transfer in a dihaem-containing membrane protein complex. EMBO J 25, 4963–4970

Madej MG, Nasiri HR, Hilgendorff NS, Schwalbe H, Unden G and Lancaster CRD (2006b) Experimental evidence for proton motive force-dependent catalysis by the diheme-containing succinate: menaquinone oxidoreductase from the gram-positive bacterium *Bacillus licheniformis*. Biochemistry 45, 15049–15055

Madej MG, Muller FG, Ploch J and Lancaster CRD (2009) Limited reversibility of transmembrane proton transfer assisting transmembrane electron transfer in a dihaem-containing succinate:quinone oxidoreductase. Biochim Biophys Acta 1787, 593–600

Manodori A, Cecchini G, Schröder I, Gunsalus RP, Werth MT and Johnson MK (1992) [3Fe-4S] to [4Fe-4S] cluster conversion in *Escherichia coli* fumarate reductase by site-directed mutagenesis. Biochemistry 31, 2703–2712

Mileni M, Haas AH, Mantele W, Simon J and Lancaster CRD (2005) Probing heme propionate involvement in transmembrane proton transfer coupled to electron transfer in dihemic quinol:fumarate reductase by C-13-labeling and FTIR difference spectroscopy. Biochemistry 44, 16718–16728

Mileni M, MacMillan F, Tziatzios C, Zwicker K, Haas AH, Mantele W, Simon J and Lancaster CRD (2006) Heterologous production in Wolinella succinogenes and characterization of the quinol:fumarate reductase enzymes from helicobacter pylori and campylobacter jejuni. Biochem J 395, 191–201

Mitchell P (1979) Keilin's respiratory chain concept and its chemiosmotic consequences. Science 206, 1148–1159

Nakamura Y, Kaneko T, Sato S, Ikeuchi M, Katoh H, Sasamoto S, Watanabe A, Iriguchi M, Kawashima K, Kimura T, Kishida Y, Kiyokawa C, Kohara M, Matsumoto M, Matsuno A, Nakazaki N, Shimpo S, Sugimoto M, Takeuchi C, Yamada M and Tabata S (2002) Complete genome structure of the thermophilic cyanobacterium *Thermosynechococcus elongatus* BP-1. DNA Res 9, 123–130

Nakamura Y, Kaneko T, Sato S, Mimuro M, Miyashita H, Tsuchiya T, Sasamoto S, Watanabe A, Kawashima K, Kishida Y, Kiyokawa C, Kohara M, Matsumoto M, Matsuno A, Nakazaki N, Shimpo S, Takeuchi C, Yamada M and Tabata, S (2003a) Complete genome structure of *Gloeobacter violaceus* PCC 7421, a Cyanobacterium that lacks thylakoids. DNA Res 10, 137–145

Nakamura Y, Kaneko T, Sato S, Mimuro M, Miyashita H, Tsuchiya T, Sasamoto S, Watanabe A, Kawashima K, Kishida Y, Kiyokawa C, Kohara M, Matsumoto M, Matsuno A, Nakazaki N, Shimpo S, Takeuchi C, Yamada M and Tabata S (2003b) Complete genome structure of gloeobacter violaceus PCC 7421, a cyanobacterium that lacks thylakoids (Supplement). DNA Res 10, 181–201

Notredame C, Higgins DG and Heringa J (2000) T-coffee: a novel method for fast and accurate multiple sequence alignment. J Mol Biol 302, 205–217

Ohnishi T, Moser CC, Page CC, Dutton PL and Yano T (2000) Simple redox-linked proton-transfer design: new insights from structures of quinol-fumarate reductase. Structure 8, R23–R32

Page CC, Moser CC, Chen X and Dutton PL (1999) Natural engineering principles of electron tunnelling in biological oxidation-reduction. Nature 402, 47–52

Palenik B, Brahamsha B, Larimer FW, Land M, Hauser L, Chain P, Lamerdin J, Regala W, Allen EE, McCarren J, Paulsen I, Dufresne A, Partensky F, Webb EA and Waterbury J (2003) The genome of a motile marine *Synechococcus*. Nature 424, 1037–1042

Partensky F, Hess WR and Vaulot D (1999) Prochlorococcus, a marine photosynthetic prokaryote of global significance. Microbiol Mol Biol Rev 63, 106–127

Rocap G, Larimer FW, Lamerdin J, Malfatti S, Chain P, Ahlgren NA, Arellano A, Coleman M, Hauser L, Hess WR, Johnson ZI, Land M, Lindell D, Post AF, Regala W, Shah M, Shaw SL, Steglich C, Sullivan MB, Ting CS, Tolonen A, Webb EA, Zinser ER and Chisholm SW (2003) Genome divergence in two Prochlorococcus ecotypes reflects oceanic niche differentiation. Nature 424, 1042–1047

Salerno JC (1991) Electron transfer in succinate:ubiquinone reductase and quinol:fumarate reductase. Biochem Soc Trans 19, 599–605

Saraste M (1999) Oxidative phosphorylation at the fin de siècle. Science 283, 1488–1493

Scanlan DJ, Ostrowski M, Mazard S, Dufresne A, Garczarek L, Hess WR, Post AF, Hagemann M, Paulsen I and Partensky F (2009) Ecological genomics of marine picocyanobacteria. Microbiol Mol Biol Rev 73, 249–299

Schäfer G, Engelhard M, Müller V (1999) Bioenergetics of the archaea. Microbiol Mol Biol Rev 63:570–620

Schäfer G, Anemüller S and Moll R (2002) Archaeal complex II: 'classical' and 'non-classical' succinate:quinone reductases with unusual features. Biochim Biophys Acta 1553, 57–73

Schirawski J and Unden G (1998) Menaquinone-dependent succinate dehydrogenase of bacteria catalyzes reversed electron transport driven by the proton potential. Eur J Biochem 257, 210–215

Schnorpfeil M, Janausch IG, Biel S, Kröger A and Unden G (2001) Generation of a proton potential by succinate dehydrogenase of Bacillus subtilis functioning as a fumarate reductase. Eur J Biochem 268, 3069–3074

Simon J, Groß R, Ringel M, Schmidt E and Kröger A (1998) Deletion and site-directed mutagenesis of the *Wolinella succinogenes* fumarate reductase operon. Eur J Biochem 251, 418–426

Singer TP and McIntire WS (1984) Covalent attachment of flavin to flavoproteins: occurrence, assay, and synthesis. Methods Enzymol 106, 369–378

Sugita C, Ogata K, Shikata M, Jikuya H, Takano J, Furumichi M, Kanehisa M, Omata T, Sugiura M and Sugita M (2007) Complete nucleotide sequence of the freshwater unicellular cyanobacterium Synechococcus elongatus PCC 6301 chromosome: gene content and organization. Photosynth Res 93, 55–67

Sun F, Huo X, Zhai YJ, Wang AJ, Xu JX, Su D, Bartlam M and Rao ZH (2005) Crystal structure of mitochondrial respiratory membrane protein complex II. Cell 121, 1043–1057

Swingley WD, Chen M, Cheung PC, Conrad AL, Dejesa LC, Hao J, Honchak BM, Karbach LE, Kurdoglu A, Lahiri S, Mastrian SD, Miyashita H, Page L, Ramakrishna P, Satoh S, Sattley WM, Shimada Y, Taylor HL, Tomo T, Tsuchiya T, Wang ZT, Raymond J, Mimuro M, Blankenship RE and Touchman JW (2008) Niche adaptation and genome expansion in the chlorophyll d-producing cyanobacterium *Acaryochloris marina*. Proc Natl Acad Sci U S A 105, 2005–2010

Tamura K, Dudley J, Nei M and Kumar S (2007) MEGA4: molecular evolutionary genetics analysis (MEGA) software version 4.0. Mol Biol Evol 24, 1596–1599

Törnroth S, Yankovskaya V, Cecchini G and Iwata S (2002) Purification, crystallisation and preliminary crystallographic studies of succinate:ubiquinone oxidoreductase from *Escherichia coli*. Biochim Biophys Acta 1553, 171–176

Unden G, Albracht SPJ and Kröger A (1984) Redox potentials and kinetic properties of fumarate reductase complex from *Vibrio succinogenes*. Biochim Biophys Acta 767, 460–469

Veith B, Herzberg Ch, Steckel S, Feesche J, Maurer KH, Ehrenreich P, Bäumler S, Henne A, Liesegang H, Merkl R, Ehrenreich A and Gottschalk G (2004) The complete genome sequence of *Bacillus licheniformis* DSM13, an organism with great industrial potential. J Mol Biotechnol 7, 204–211

Walker WH and Singer TP (1970) Identification of covalently bound flavin of succinate dehydrogenase as 8α-histidyl flavin adenine dinucleotide. J Biol Chem 245, 4224–4225

Waterbury JB, Watson SW, Guillard RRL and Brand LE (1979) Widespread occurrence of a unicellular, marine, planktonic, cyanobacterium. Nature 277, 293–294

Weiner JH and Dickie P (1979) Fumarate reductase of *Escherichia coli*—Elucidation of the covalent flavin component. J Biol Chem 254, 8590–8593

Welsh EA, Liberton M, Stöckel J, Loh T, Elvitigala T, Wang C, Wollam A, Fulton RS, Clifton SW, Jacobs JM, Aurora R, Ghosh BK, Sherman LA, Smith RD, Wilson RK and Pakrasi HB (2008) The genome of *Cyanothece* 51142, a unicellular diazotrophic cyanobacterium important in the marine nitrogen cycle. Proc Natl Acad Sci U S A 105, 15094–15099

Xia D, Yu CA, Kim H, Xian JZ, Kachurin AM, Zhang L, Yu L and Deisenhofer J (1997) Crystal structure of the cytochrome bc1 complex from bovine heart mitochondria. Science 277, 60–66

Yankovskaya V, Horsefield R, Tornroth S, Luna-Chavez C, Miyoshi H, Leger C, Byrne B, Cecchini G and Iwata S (2003) Architecture of succinate dehydrogenase and reactive oxygen species generation. Science 299, 700–704

Zaunmuller T, Kelly DJ, Glockner FO and Unden G (2006) Succinate dehydrogenase functioning by a reverse redox loop mechanism and fumarate reductase in sulphate-reducing bacteria. Microbiology 152, 2443–2453

Part V

Connecting (−) and (+): Electron and Protein Circuits Between Dehydrogenases and Oxygen Reductases in Cyanobacteria

Chapter 18
The Water-Soluble Cytochromes of Cyanobacteria

Kwok Ki Ho, Cheryl A. Kerfeld and David W. Krogmann

18.1 Introduction

The presence of water-soluble cytochrome (cyt) in cyanobacteria was first examined and described in detail by Holton and Myers (1963) in a short paper in Science. There had been a few sightings of cyt spectral peaks in cell extracts but little pursuit of chemical details. Jack Myers, who had worked on biophysical aspects of photosynthesis, had turned to the physiology of blue- green algae (soon to be called cyanobacteria). The blue and red phycobiliproteins and the small spherical cells of *Anacystis nidulans* (now called *Synechocystis* sp. PCC 6803) held promise for biophysical experiments. R. W. Holton joined Jack and brought skills in protein purification and characterization. Holton and Myers (1963) outlined the isolation and distinctive character of three cyts in their short Science paper. In 1967, they published two substantial papers. One described in further detail on the extraction and spectral properties of three c-type cyts (Holton and Myers 1967a) and the other told of their physicochemical properties (Holton and Myers 1967b). An important distinction came from the apparent abundance of each cyt. Cyt c554 was a thousand-fold and cyt c550 a hundred-fold more abundant than cyt c552. Cyt c554 had a redox potential of +0.35 V, and cyt c550 of −0.355 V. Cyt c552 is auto-oxidizable. Its redox potential is 0.151 V. Cyts c554 and c550 showed acidic character in ion-exchange chromatography while cyt c552 was weakly basic. In contemporary study of cyanobacteria, cyts c552 and c554 are frequently referred to as cyts cM and c_6, respectively. These current names will be used hereinafter.

Cyt c_6 is functionally equivalent to a copper protein, plastocyanin. Either protein can act as an electron carrier between the cytochrome b_6f complex and Photosystem I (PSI) in some cyanobacteria and green algae (Wood 1978; Sandmann et al. 1983). This cyt serves well. There is often a greater availability of iron and a scarcity of copper in natural waters. In response to copper deficiency, many species of

D. W. Krogmann (✉)
Department of Biochemistry, Purdue University, 175 S. University Street,
West Lafayette, IN 47907-2063, USA
e-mail: krogmann@purdue.edu

G. A. Peschek et al. (eds.), *Bioenergetic Processes of Cyanobacteria*,
DOI 10.1007/978-94-007-0388-9_18, © Springer Science+Business Media B.V. 2011

cyanobacteria were found to replace plastocyanin by increasing the content of cyt c_6 (Sandmann 1986). The gene encoding cyt c_6 or plastocyanin has been inactivated by genetic manipulations in *Synechocystis* sp. PCC 6803 (Duran et al. 2004) and *Synechococcus* sp. PCC7942 (Clarke and Campbell 1996). In *Synechocystis* sp. PCC 6803, the normal functioning of photosynthesis and respiration was found to be firmly dependent on the availability of either cyt c_6 or plastocyanin based on the results obtained with the deletion mutants. In *Synechococcus* sp. PCC7942, inactivation of the PetE gene encoding plastocyanin resulted in a large increase of messenger RNA for cyt c_6. However, it led to a loss in both photosynthetic capacity and tolerance to cold-induced photoinhibition. Attempts to inactivate the PetJ gene encoding cyt c_6 in *Synechococcus* sp. 7002 (Nomura and Bryant 1998) were not successful. This gene could not be replaced by the one encoding cyt c_6, plastocyanin or cyt cM from *Synechocystis* sp. PCC 6803. The observation that different cyanobacteria responded differently to inactivation of the gene encoding plastocyanin or cyt c_6 suggests that the extent of functional equivalence between the two electron components varies among cyanobacteria.

Cyt c_6 can act as an electron carrier between cytochrome b_6f complex and cyt oxidase during cyanobacterial respiration (Pils and Schmetterer 2001). Figure 18.1a–e provides striking evidence of structural similarities between catalysts of photosynthesis and respiration. For instance, both structures of cyt c_6 from a cyanobacterium *Arthrospira maxima* and horse-heart mitochondrial cyt c are characterized by the presence of four helices and a number of loops such that three of the helices cluster around one edge of the heme. The spatial arrangements of these three helices are similar for both structures when superimposed. Figure 18.1e shows that the N-terminal and C-terminal helices of *A. maxima* cyt c_6 closely superimpose on those of horse-heart cyt c and that the sequences of the two helices are similar for both cyts. Horse-heart cyt c has four tyrosines, of which Tyr 97 and Tyr 74 are thought to be target sites for phosphorylation and nitration, respectively. When either site was modified, there was a marked effect on either protein structure or electron-transfer kinetics with cyt oxidase (Lee et al. 2006; Abriata et al. 2009). *A. maxima* cyt c_6 has three tyrosines, of which Tyr 80 occupies the same spatial position as Tyr 97 of horse-heart cyt c and Tyr 53 is close in space to Tyr 74 of horse-heart cyt c. Since cyt c_6 is thought to interact with cyt oxidase, it will be of interest to see if any of the *A. maxima* tyrosines can be modified by phosphorylation or nitration.

Cyt cM appears to be the same protein as cyt c552. The name was introduced by Malakhov et al. (1994) to identify a small water-soluble c-type cyt in *Synechocytsis* PCC 6803. It is relatively short, about 76 amino acids when compared to cyt c_6 and cyt c550 which have about 90 and 130–140 amino acids, respectively. Cyt cM has yet to be crystallized. The name cyt cM might be used until more is learned about the function and structure of this protein. The gene for cyt cM has been identified in many cyanobacterial genomes. It is not clear if any of these genomes have additional homologs to the cyt cM gene. Of the 36 cyanobacterial genomes fully available from National Center for Biotechnology Information (NCBI) database, the *Prochlorococcus marinus* subsp. Pastoris str. CCMP1986 genome appears to be the only one that is without a gene for either cyt c_6 or c550. Since this organism has

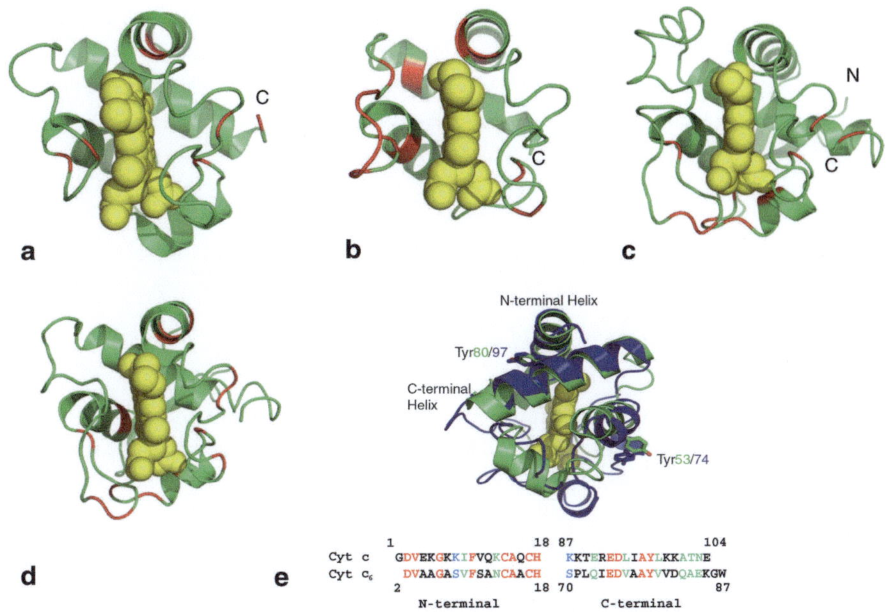

Fig. 18.1 Ribbon diagrams of **a** Cyt c_6, **b** CytM, **c** Cyt c550, **d** PsbV2, and **e** a superposition of *A. maxima* cytochrome c_6 (PDB ID 1f1f; Sawaya et al. 2001) and horse-heart cytochrome c (PDB ID 1hrc; Bushnell et al. 1990). The view in **e** is rotated ~180° about an axis parallel to the page relative to that shown in **a**–**d**. In **a**–**d**, the positions of absolutely conserved amino acids (with the exception of the heme ligands) are colored in red. The heme is shown in space-filling and oriented with the propionate oxygens pointing downwards in the cyts shown in **a**–**d**. In **e** horse-heart cyt c is shown in blue. Tyr 53 and Tyr 80 of cyt c_6 and Tyr 74 and 97 of horse-heart cyt c are shown in sticks. The sequence alignment was conducted with LALIGN (Gasteiger et al. 2003) at ExPASy server (http://www.expasy.ch). This figure and figure 18.2 were prepared with Pymol (DeLano 2002)

both cyt cM and plastocyanin genes, it will be interesting to inactivate the plastocyanin gene and study its effect as there are suggestions that cyt cM can substitute for plastocyanin or cyt c_6 during photosynthesis (Malakhov et al. 1999). The inability to construct double mutants for cyt c_6 and plastocyanin in any conditions has so far made the assessment of cyt cM's role in photosynthesis difficult.

Cyt c550 is sometimes referred to as Low Potential (L.P.) cyt c550. This water-soluble protein contains a single heme with bihistidine coordination. It is one of the extrinsic Photosystem II (PS II) subunits involved in oxygen evolution. While deletion of the gene for cyt c550, *psbV*, leads to impaired growth and diminished oxygen evolution, the role of the heme remains unclear. The presence of homologs to *psbV* in the same genome can complicate the study of cyt c550. An additional homolog to *psbV*, *psbV2*, has been cloned from *Thermosynechococcus elongatus* and *T. vulcans*. It is not clear how widespread is *psbV2* among cyanobacterial genomes sequenced so far.

In previous surveys of Ho (2005), Bialek et al. (2008) and Bernroitner et al. (2008), multiple but not identical genes encoding cyt c_6s were found in some cyanobacterial genomes. Here we have extended the study of cyanobacterial genomes to include all three water-soluble cyts. Our study includes an identification of the cyts' coding sequences in more genomes from the NCBI database and an analysis of their deduced primary structures, aiming towards a better understanding of their roles in cyanobacteria.

18.2 Gene Content and Arrangements

18.2.1 Cytochromes c_6, cM and c550

Most of the cyanobacteria studied here have multiple but not identical gene sequences for cyt c_6. The ones that don't provide a gene sequence for cyt c_6 include 4 strains of *Prochlorococcus marinus*, CCMP1986, AS9601, MIT9215 and MIT9515. These unicellular cyanobacteria have a gene for plastocyanin which is a functional equivalent of cyt c_6. With a size of about 1.7 Mbp, their genomes are the smallest found among the cyanobacteria. Within 45 genomes, we found 49 sequences that are annotated as cyt c_6 and 44 other sequences listed as cyt cs. The latter group of cyts was accepted here as cyt c_6, based on their close similarity in sequence to the annotated ones. Table 18.1 summarizes the distribution of 93 cyt c_6 genes among 36 unicellular and 9 filamentous cyanobacterial genomes. In general, the number of cyt c_6 genes for individual genomes is small, varying from 1 to 4. Despite a larger genome size, the highest number among the filamentous genomes is 3. A study of how cyt c_6 genes distribute within individual genomes reveals some interesting patterns that may be related to gene function and evolution. First, with one exception, the multiple cyt c_6 genes found within the same genome are not in close proximity to each other. The only exception is in *Nodularia spumigena CCY9414* genome, where 2 cyt c_6 genes are about 400 bp apart. Second, 16 out of 36 unicellular genomes have a cyt c_6 gene located next to a plastocyanin gene. This pattern of arrangement of the cyt c_6 and plastocyanin genes suggests that in these organisms the synthesis of cyt c_6 is regulated in coordination with the synthesis of plastocyanin. This suggestion is consistent with earlier studies that in some species of cyanobacteria, syntheses of these two proteins were somehow coordinate regulated by the presence of copper ions (Sandmann and Boger 1980; Bovy et al. 1992; Ghassemian et al. 1994; Zhang et al. 1992). Of the 36 unicellular genomes, 17 come from different *Synechococcus* strains, which include 12 marine, 4 freshwater and 1 intertidal ones. The above gene arrangement appears common among the marine strains of *Synechococcus*, occurring in 10 of the 12 strains. The remaining 2 marine strains, PCC7002 and RCC307, don't seem to have a gene for plastocyanin. As for the freshwater strains, JA-2–3B'a(2–13), JA-3–3Ab, PCC6301 and PCC7942, and the intertidal one, PCC7335, a cyt c_6 sequence is quite far away from a plastocyanin sequence. An understanding of how the cyt c_6 and plastocyanin genes are regulated is very much needed and the marine strains of *Synechococcus* may be useful for

Table 18.1 Distribution of cytochrome c_6, cM, c550, and c550-like genes among different cyanobacterial genomes

Cyanobacteria	Gene			
	cyt c_6	cyt cM	cyt c550	cyt c550-like
Synechococcus				
BL107	2	1	1	
CC9311	4	1	1	
CC9605	2	1	1	
CC9902	2	1	1	
JA-2-3B'a(2-13)	1	1	1	1
JA-3-3Ab	1	1	1	1
PCC6301	3	1	1	
PCC7002	2	1	1	
PCC7335	1		2	
PCC7942	3	1	1	
RCC307	2	1	1	
RS9916	4	1	1	
RS9917	4	1	1	
WH5701	2	1	1	
WH7803	3	1	1	
WH7805	4	1	1	
WH8102	2	1	1	
non-*Synechococcus*				
Acaryochloris marina MBIC 11017	2		1	1
Crocosphaera watsonii 8501	3	1	1	
Cyanobium sp. PCC 7001	2	1	1	
Cyanothece sp. ATCC 51142	2	1	1	
Cyanothece sp. CCY 0110	2	1	1	
Cyanothece sp. PCC 7424	2	1	1	
Cyanothece sp. PCC 7425	2	1	1	1
Cyanothece sp. PCC 7822	1	1	1	
Cyanothece sp. PCC 8802	2	1	1	
Gloeobacter violaceus PCC7421	2		1	1
Microcystis aeruginosa NIES-843	1	1	1	
Prochlorococcus marinus MIT 9211	1	1		
Prochlorococcus marinus MIT 9301	1	1		
Prochlorococcus marinus MIT 9303	2	1	1	
Prochlorococcus marinus MIT 9312	1	1		
Prochlorococcus marinus MIT 9313	2	1	1	
Prochlorococcus marinus NATL1A	1	1		
Prochlorococcus marinus NATL2A	1	1		
Synechocystis sp. PCC 6803	1	1	1	
Filamentous				
Anabaena sp. PCC7120	3	1	1	
Anabaena sp. PCC7119	2			
Anabaena variabilis ATCC29413/PCC7937	3	1	1	
Nostoc punctiforme ATCC29133/PCC73102	3	1	1	
Nodularia spumigena CCY9414	2	1	1	
Arthrospira maxima CS-328	2	1	1	1
Trichodesmium erythraeum IMS101	2	1	1	1
Lyngbya sp. PCC8106	2	1	1	1
Microcoleus chthonoplastes PCC7420	1	1	1	1

such study. Lastly, of the 9 filamentous cyanobacteria, 5 form heterocysts and 4 don't. The 5 that form heterocysts are *Anabaena* sp. 7120, *Anabaena variabilis* 29413/7937, *Nostoc punctiforme* 29133/73102, *Anabaena* sp. *7119* and *N. spumigema* CCY9414. The first three organisms have a plastocyanin gene located between a cyt c_6 gene and a c550 gene. *N. spumigema* CCY9414 has a plastocyanin gene located next to a cyt c_6, and its genome like the one from *Anabaena* sp. 7119 is not complete. The cytc_6/plastocyanin/cyt c550 gene arrangement may have some importance to the filamentous cyanobacteria that form heterocysts. Incidentally, it is not found in other filamentous cyanobacteria that don't form heterocysts. These include *Microcoleus chthonoplates* 7420, *A. maxima* CS-328, *Lyngbya* sp. 8106 and *Trichodesmium erythraeum* IMS101.

Unlike cyt c_6, there is no additional homolog to cyt cM in the same genome. We obtained one cyt cM gene sequence each from 42 out of the 45 cyanobacterial genomes (Table 18.1). The complete genomes from *Synechococcus* sp. PCC7335 and *Gloeobacter violaceus* PCC7421, and the incomplete one from *Anabaena* sp. PCC7119 provide no cM gene sequences. Of the 42 cyt cM coding sequences, 21 are found to locate next to a gene for the cyt b_6f complex subunit petG. This pattern of gene arrangement spreads across both unicellular and filamentous cyanobacterial genomes.

We obtained 40 c550 gene (*psbV*) sequences from 39 cyanobacterial genomes, of which only one, *Synechococcus* sp. PCC7335 has 2 sequences (Table 18.1). In this organism, the 2 cyt c550 gene sequences are in close proximity. *M. chthonoplates* 7420, *A. maxima* CS-328, *Lyngbya* sp. 8106 and *T. erythraeum* IMS101 are filamentous cyanobacteria that lack heterocysts. These 4 together with 5 other unicellular cyanobacteria have an additional homolog sequence to cyt c550, *psbV2*. In these organisms, the psbV2 gene is located immediately downstream of the *psbV*.

18.3 Structure and Properties

The deduced amino acid sequences of cyts c_6, cM and c550 were all retrieved from the Protein Knowledgebase server (http://www.uniprot.org). The sequences were divided into 3 sets based on their origins (*Synechococcus*, non-*Synechococcus* and filamentous). Non-*Synechococcus* refers to a group of unicellular cyanobacteria that are not *Synechococcus*. Alignments of cyts c_6 and c550 sequences were performed on the mature portions of the proteins. With cyt cM, the full-length of the sequence was used. All alignments were conducted using ClustalW (Thompson et al. 1994) at Network Protein Sequence Analysis server (http://npsa-pbil.ibcp.fr/). Identical, strongly similar and weakly similar residues were indicated by (*), (:) and (.) respectively. Within the alignment tables, individual sequences were numbered and identified by an accession code. Next to the C-terminal end of the alignment are columns for the isoelectric point [pI] of the mature sequence, letter identifying the homologs from the same organism [L] and name of the organism. The pI values for cyt c550 from different cyanobacteria were all in the acidic range and so were not shown in Tables 18.2, 18.3 and 18.4. The leader sequence and pI value of individual sequences were determined using SignalP and Compute pI/Mw (Gasteiger et al. 2003) respectively at the ExPASy server (http://www.expasy.ch).

Table 18.2 An alignment of cytochrome c550 from different *Synechococcus* strains

```
                    10        20        30        40        50        60        70
                     |         |         |         |         |         |         |
 [1]Q31LM9   AELTAETRTVKLNPQGDNVTLSLKQVAEGKQLFAYACGQCHVGGITKTDPNVGLDPEALALATPPRDSVE
 [2]Q5N095   AELTAETRTVKLNPQGDNVTLSLKQVAEGKQLFAYACGQCHVGGITKTDPNVGLDPEALALATPPRDSVE
 [3]B4WI32   AELDANLRTIPLNEAGDTVTLSLEEFTRGQQKFNNACAICHLGGITKTNPNVGLDTESLAGAFPDRNNLE
 [4]P0A386   AELTPEVLTVPLNSEGKTITLTEKQYLEGKRLFQYACASCHVGGITKTNPSLDLRTETLALATPPRDNIE
 [5]P0A387   AELTPEVLTVPLNSEGKTITLTEKQYLEGKRLFQYACASCHVGGITKTNPSLDLRTETLALATPPRDNIE
 [6]Q3AZH8   VQWDAETLTVPVNPDGAEVTFTDREINAGRKVFNTSCGTCHAGGITKTNQNVGLDPETLALATPARDNVD
 [7]Q064V6   VQWDAETLTVPVNPDGAEVTFTDREINAGRKVFNTSCGTCHAGGITKTNQNVGLDPETLALATPARDNVE
 [8]Q3AHP7   AQWDAETLTVPADPSGTEVTFSDREIDSGRKVFNTSCGTCHAGGITKTNHNVGLDPETLALATPARDNVE
 [9]Q7U8S5   VRWDAETLTVPGNPEGTQVTFSEQEINTGRKVFNTSCGTCHAGGITKTNQNVGLDPETLALATPARDNVD
[10]A4CRA5   ASWDAETLTVPADSEGSTVTFTEQQVKAGRKVFNTSCGTCHAGGITKTNHNVGLDPETLALATPARDNVA
[11]A5GN88   ASWDAETLTVPADAEGSPVTFSEQQVKAGRKVFNTSCGTCHAGGITKTNHNVGLDPETLALATPARDNVA
[12]Q0I7X4   AQWDAETLTVPADGDGALVTFSEQEIKTGRKVFNVSCGTCHAGGITKTNQNVGLDTETLALATPARDNVA
[13]Q05U25   AQWDADTLTVNADASGTEVTFSEQEIKTGRKVFNTSCGTCHAGGITKTNHNVGLDPETLALATPARNSVE
[14]A3YVH4   AQWDTATLTVPGDPDGNLVTFSEAEIKGGRKLFNDSCGTCHAGGVTKTNYNVGLDPETLSLATPSRDSVE
[15]A3Z698   AQWDAATLTVPADASGDQVTFSEQEIKTGRKIFNTSCGTCHAGGITKTNHNVGLDPETLALATPARDNVD
[16]A5GV34   VQWDAETLTVPADASGSTVTFTESEIKAGRKVFNNSCGECHAGGITKTNQNVGLDPETLALATPPRDSVD
[17]B4WKJ4   SEIDPASRTVSLNETGEQVTFSTRQLMQGQRKFNGSCAQCHIDGASKTNPDVDLGLETLANATPARDSVE
[18]Q55210   TALREVDRTVNLNET-ETVVLSDQQVAKGERIFINTCSTCHNSGRTKSNPNVTLSLVDLEGAEPRRDNIL
[19]Q2JTY6   ---ARQIPPVALSPT-ESITFTEAQLARGKQLFNRACAQCHVGGQTYPNPDVTLKLSDLEGATPPRDNVL
[20]Q2JIS8   ---ARDIPAVALSPT-ESITFTEAQLAKGKKLFNTACAQCHVGGQTYPNPDVSLKLSDLEGATPPRDNVL
                     .:  .       .:  :   *.: *   *  **  * : .: .: *    *  * * *:.:
  Consensus  --------TV--N-------FS--Q---GRK-F---C--CH--G-T-TN-NV-L----L--A-P-RDNV-

                    80        90       100       110       120       130       140
                     |         |         |         |         |         |         |
 [1]Q31LM9   SLVDYLHNPTTYDGEREISELHPSTKSTDIFPKMRNLTEDDLVAISGHILLQPKIVGTKWGGGKIYY--------
 [2]Q5N095   SLVDYLHNPTTYDGEREISELHPSTKSTDIFPKMRNLTEDDLVAISGHILLQPKIVGTKWGGGKIYY--------
 [3]B4WI32   GLVDYLHNPTTYDGLTEISELHPSTKSSDIYPKMRNLTEDDLRAISGHILLMPKIRGDQWGAGKTKSL-------
 [4]P0A386   GLVDYMKNPTTYDGEQEIAEVHPSLRSADIFPKMRNLTEKDLVAIAGHILVEPKILGDKWGGGKVYY--------
 [5]P0A387   GLVDYMKNPTTYDGEQEIAEVHPSLRSADIFPKMRNLTEKDLVAIAGHILVEPKILGDKWGGGKVYY--------
 [6]Q3AZH8   ALVDYMKDPTSYDGEYSISDVHPSMRSAELYPAMRDLDDEDLRLMAGYILVSPKVQGSSWGGGKIYF--------
 [7]Q064V6   ALVDYMKDPTSYDGEYSISDVHPSMRSAELYPAMRDLDDEDLRLMAGYILVSPKVQGSSWGGGKIYF--------
 [8]Q3AHP7   ALVDYMKDPTSYDGEYSIADLHPSMRDAELYPAMRDLDDEDLRLMAGYILVSPKVQGSAWGGGKIYF--------
 [9]Q7U8S5   ALVDYMKDPTSYDGEYSIADLHPSMRDAELYPAMRDLTDEELRLMAGYILVAPKVQGTAWGGGKIYF--------
[10]A4CRA5   ALVDYLQDPTSYDGEYSIADLHPSMRSRDLYPAMRDLTDEDLRLMSAYILVAPKVLGAEWGGGKIYF--------
[11]A5GN88   ALVDYLQDPTSYDGEYSIADLHPSMRSRDLYPAMRDLTDEDLRLMSAYILVAPKVLGQEWGGGKIYF--------
[12]Q0I7X4   SLVDYLQDPTSYDGEYSIADLHPSMRSRDLYPAMRDLTDEDLRLMSGYILVAPKVLGVDWGGGKIYF--------
[13]Q05U25   GLVDYLNDPTSYDGEYSIADLHPSIRSRDLYPAMRDLTDEDLRLMAGYILVAPKVLGEEWGGGKIYF--------
[14]A3YVH4   ALVDFIKDPTSYDGEYSIADVHPSMRSSDVFVKMRDLDDDDLRLIAGYILVAPKVSGTQWGGGKVYF--------
[15]A3Z698   SLVDYMKDPTSYDGEYSIADVHPSMRSSDLFVQMRDLTDDDLRLIAGYILVAPKVQGNQWGGGKIYF--------
[16]A5GV34   ALVDYMKDPTSYDGEISIAEFHPSMRSADVFVKMRDLNDEDLRRMAGYILTSPKAQGTQWGGGKIYF--------
[17]B4WKJ4   AIVDYLHDPTTYDGLRSLAELHPSTIRADLFPRMRDLGDDDLSAIAGYILAQPNIIGDQWAGGKPKR--------
[18]Q55210   AMVDYLKNPTSYDGELDLSQLHPNTVRADIWSSMRNLNEEDLQNVSGYVLVQAQVRGVAWGGGKTVN--------
[19]Q2JTY6   ALVDYIKNPVTYDGVESLVEYHPNTQLLSEYPRLRNLTDEDLKLIAGYILVQAKTV-PGWGGTKSESHSDLSAYL
[20]Q2JIS8   ALVDYIKNPVTYDGVESLLEYHPNTQLTSEYPRLRNLTDEDLKLIAGYILVQAKTV-PGWGGTKNESHSDLSAYL
             .::*:::.*.:***  .  : * .    . :  :*:: :.:*  ::.::*  .:     *..
  Consensus  ALVDYLKDPTSYDG--S--E-H-S----D-Y--MRDL-DEDL--IAGYIL--PK-----WGG-------------
```

[1]Q31LM9, *Synechococcus* sp. PCC 7942; [2]Q5N095, *Synechococcus* sp. PCC 6301; [3]B4WI32, *Synechococcus* sp. PCC 7335; [4]P0A386, *Thermosynechococcus elongatus*; [5]P0A387, *Thermosynechococcus vulcanus*; [6]Q3AZH8, *Synechococcus* sp. CC9902; [7]Q064V6, *Synechococcus* sp. BL107 [8]Q3AHP7, *Synechococcus* sp. CC9605; [9]Q7U8S5, *Synechococcus* sp. WH8102; [10]A4CRA5, *Synechococcus* sp. WH7805; [11]A5GN88, *Synechococcus* sp. WH7803; [12]Q0I7X4, *Synechococcus* sp. CC9311; [13]Q05U25, *Synechococcus* sp. RS9916; [14]A3YVH4, *Synechococcus* sp. WH 5701; [15]A3Z698, *Synechococcus* sp. RS9917; [16]A5GV34, *Synechococcus* sp. RCC307; [17]B4WKJ4, *Synechococcus* sp. PCC 7335; [18]Q55210, *Synechococcus* sp. PCC 7002; [19] Q2JTY6; *Synechococcus* sp. JA-3-3Ab; [20]Q2JIS8, *Synechococcus* sp. JA-2-3B'a(2-13.

18.3.1 Cytochrome c_6

18.3.1.1 Amino Acid Sequences

In general, the alignments corresponding to different sets of cyanobacteria (Tables 18.5, 18.6 and 18.7) show that the cyt c_6 sequences from the same organism are less similar to each other than they are to the sequences from other organisms and that the sequences appear to divide into groups such that each sequence in the same group comes from a different organism. Also, the heme attachment site of Cys-xx-

Table 18.3 An alignment of cytochrome c550 from non-*Synechococcus*, unicellular cyanobacteria

```
                    10        20        30        40        50        60        70
                     |         |         |         |         |         |         |
[1]B2F867    LELDKDTLTIPLNEGGDTTTLTSKQATNGQRLFNLECTQCHLQGKTKTNNNVSLGLEDLSKADPRRDNVL
[2]B4AYY6    LELDKDTLTIPLNEQGETTTLTTKQYANGQKLFNLECTQCHLQGKTKTNNNVSLGLSDLSKAEPPRDNLL
[3]B0JSW0    LELNEKTLTITLNDAGESVTLTSEQATEGQKLFVANCTKCHLQGKTKTNNNVSLGLGDLAKAEPPRDNLL
[4]P19129    LELDEKTLTITLNDAGESVTLTSEQATEGQKLFVANCTKCHLQGKTKTNNNVSLGLGDLAKAEPPRDNLL
[5]Q55013    VELTESTRTIPLDEAGGTTTLTARQFTNGQKIFVDTCTQCHLQGKTKTNNNVSLGLADLAGAEPRRDNVL
[6]A3INJ0    LELTEETRTVPLTEEGETTTLSSEQITNGQRLFIRECTQCHLQGKTKTNNNVSLGLEDLSLAEPPRDNVL
[7]B1WVX0    LELTEETRTVPLNEEGETITLTSEQITNGQRLFIRECTQCHLQGKTKTNNNVSLGLEDLSLAEPPRDNVL
[8]Q4C8X2    LELNEKTRTIPLAEDGQETTLSDQQLANGNRLFIRECTQCHLQGKTKTNNNVSLGLEDLAGAEPRRDNLL
[9]B8HUH9    LELDEATRTVALDETGKTTIVTSKQITNGQRLFVQECTQCHLQGKTKTNNNVSLGLDDLAGAEPPRNNVL
[10]B0C731   AELDDATRTVALNE-GSTVTLSTQQAKEGQRLFNFACANCHIGGDTKTNPSINLSSASLAGANPRRDNVE
[11]B4CB52   AQLDATTRTVALNDQGKNITLSRRQVEDGQRLFNEACASCHAGGITKTNPSLDLSPETLALATPRRDSVA
[12]A2C723   AQWDAETLTVPAGSGGQQVTFSESEIKSASKLFKSNCATCHNQGVTKTNQNVGLDLEALSLASPARDNVD
[13]Q7V5W0   AQWDAETLTVPAGSGGQQVTFTESEIKSASKLFKSNCATCHNQGVTKTNQNVGLDLEALSLASPARDNVD
[14]B5IMK6   AQWTAEQLTVPASPDGATVTFTETEIKAGRKLFNSSCGECHAGGITKTNQNVGLDPETLALATPARDNVE
              :      *:.    *    .:  :   . ::*   *  **  * **** .:.*.   *: * * *:.:
Consensus    -E------TVP----G----LT--Q---G-KLF---C--CH--G-TKTN-NVSLG---LA-A-P-RDNV-

                    80        90       100       110       120       130
                     |         |         |         |         |         |
[1]B2F867    ALVDYLKNPTSYDGEDDYTELHPNVSRPDLFPELKNFTEDDVFDVAGYMLIAPKLD-TRWGG-TIYF
[2]B4AYY6    ALVDYLKNPTSYDGEDDYSELHPNVSRPDLYPELKNFTEDDVFDVSAYVLIAPKLD-TRWGG-TIYF
[3]B0JSW0    ALIDYLEHPTSYDGEDDLSELHPNVSRPDIFPELRNLTEDDVYNVAAYMLVAPRLD-ERWGG-TIYF
[4]P19129    ALIDYLEHPTSYDGEDDLSELHPNVSRPDIYPELRNLTEDDVYNVAAYMLVAPRLD-ERWGG-TIYF
[5]Q55013    ALVEFLKNPKSYDGEDDYSELHPNISRPDIYPEMRNYTEDDIFDVAGYTLIAPKLD-ERWGG-TIYF
[6]A3INJ0    GLVDYLKYPTSYDGEDDYTELHVNVSRPDIYPELRDFTEDDLYDVSGYILVAPKLD-SYWGG-SIYF
[7]B1WVX0    GLVDYLKYPTSYDGEDDYTELHVNVSRPDIYPELRDFTEDDLYDVSGYILVAPKLD-SYWGG-SIYF
[8]Q4C8X2    AIVDYLKQPTSYDGEDDYTEFHVNVNRPDIYPELRDFTEEDLYDVSGYVLVAPKLD SYWGG-SIYF
[9]B8HUH9    ALVDYLKHPTSYDGEDNYEELHVNVTRPDLFPELRNFTEDDLYDVAGYVLVAPKLD-AYWGG-SIYF
[10]B0C731   GLVDYMNNPTTYDGFDTISEVHPSTQSTDVFPLMRNLSDEDLFDIAGHILIQPSVIGDQWGGGKANR
[11]B4CB52   GLVDYMKDPTTYDGEDSIAEIHPSIKSADIYPKMRNLSEQDLTAIAGYILVQPKVQGEKWGGGKIYF
[12]A2C723   GLVEFLKNPMSYDGEYSIADTHPGISSSDVYVQMRTLNDEDLRLIAGYILTAEKVQGDQWGGGKIYF
[13]Q7V5W0   GLVEFLKNPMSYDGEYSIADTHPGISSSDVYVQMRTLNDEDLRLIAGYILTAEKVQGDQWGGGKIYF
[14]B5IMK6   ALVDYMKDPTSYDGEYSIADIHPSMRSSDIFVKMRDLDDDDLRLMAGYILVAPKAQGVQWGGGKIYF
             .*::::: * :***     : * .   .*::  ::   ::*:  ::.: *         *** .
Consensus    ALVDYLK-P-SYDG-----E-H-N---PDIY--LR---EDDL--VAGY-L---------WGG-T---
```

[1]B2F867, *Cyanothece* sp. PCC 7424;[2]B4AYY6, *Cyanothece* sp. PCC 7822;[3]B0JSW0, *Microcystis aeruginosa* NIES-843;[4]P19129, *Microcystis aeruginosa*;[5]Q55013, *Synechocystis* sp. PCC 6803;[6]A3INJ0, *Cyanothece* sp. CCY 0110; [7]B1WVX0,*Cyanothece* sp. ATCC 51142;[8]Q4C8X2, *Crocosphaera Watsonii* 8501; [9] B8HUH9, *Cyanothece* sp. PCC 8802; [10]B0C731, *Acaryochloris marina* MBIC 11017;[11]B4CB52, *Cyanothece* sp. PCC 7425;[12]A2C723, *Prochlorococcus marinus* MIT 9303;[13]Q7V5W0, *Prochlorococcus marinus* MIT 9313;[14]B5IMK6, *Cyanobium* sp. PCC 7001.

Cys His and the sixth heme ligand, Met, appear to be conserved in all sequences. Table 18.5 has 43 cyt c_6 sequences from different strains of *Synechococcus*, of which 42 are derived from gene sequences and 1 from strain PCC6312 (#27) is determined by protein sequencing (Aitken 1979). There are 17 strains of *Synechococcus* corresponding to the 42 sequences. This group includes 4 sequences each for 4 strains (e, f, g, h), 3 sequences each for 3 strains (a, b, i), 2 sequences each for 7 strains (c, d, j, k, l, m, n) and 1 sequence each for 3 strains. There are 13 identical residues conserved in all sequences. The pI column shows that only 8 sequences have pIs above 8. Note that 11 sequences have pIs from 4.5 to 9.2 (Group I), 4 sequences have pIs from 4.3 to 4.9 (Group II), 14 sequences from 4.4 to 5.2 (Group III) and 14 sequences from 5 to 8.7 (Group IV). In Group I and Group III, there are more acidic than basic sequences. Nearly all the sequences in which each is derived from a gene sequence next to a plastocyanin gene sequence cluster together in Group III. These include sequences number 16–24. The only one exception is sequence number 5 of *Synechococcus* PCC5701, which appears in Group I.

Table 18.4 An alignment of cytochrome c550 from different filamentous cyanobacteria

```
                        10        20        30        40        50        60        70
                         |         |         |         |         |         |         |
[1]Q8Z044       LELDEATRTVPLNAQGDTVTLSLKQVKEGKRLFQYACAQCHVGGVTKTNQNVGLEPEALALATPNRNNIE
[2]Q3M9H7       LELDEATRTVPLNAQGDTVTLSLKQVKEGKRLFQYACAQCHVGGVTKTNQNVGLEPEALALATPNRNNIE
[3]B2IVD8       VELDKATRTVPLNAEGETTVLSLKQVKEGKRLFNYACAQCHAGGVTKTNQNVGLTPEDLALATPNRNNIE
[4]A0ZF54       LELDEAIRTVPLNDQGDTVVLSLKQVKEGKRLFQFACAQCHAGGVTKTNQNVGLEPETLALASPNRNNIE
[5]B5W3Z1       LELTEELRTLPINAQGDTAVLSLKEIKKGQQVFNAACAQCHALGVTKTNPDVNLSPEALALATPPRDNIA
[6]A0YIA2       IELTEDLRTFPLNEQGEPTVLSLKEIKQGKQLFNYACAQCHALGVTKTNPDVNLSPQSLEGAVPSRNNIG
[7]Q111E9       AKLDDNVRTLPLNEDKE-VVLTIKEYTQGKREFTNVCSQCHVGGITKTNPDVSLDPETLALAYPARDNIE
[8]B4VPE5       LELSEDIRTVPVNEQGNTTVLSLEEVAKGKRLFNDSCSQCHIGGRTKTNPNVTLRLEDLKLAFPPRDNIA
                  * .  **.    : : ..:: ::  .*:: *   *:***  * *:** :* *  : *  * * *:**
 Consensus      --L-E--RTV----Q-D-VVLS-KE--EGKR-F---CAQCH--G-TKTN-NV-L--E-L--A-P-RDNI-

                        80        90       100       110       120       130
                         |         |         |         |         |         |
[1]Q8Z044       GLVDYMKNPTTYDGVEEISEIHPSLKSADIFTAMRNLTDKDLESIAGHILLQPKILGDKWGGGKIYY
[2]Q3M9H7       GLVDYMKNPTTYDGVEEISEIHPSIKSADIFTAMRNLTDKDLESIAGHILLQPKILGDKWGGGKIYY
[3]B2IVD8       GLVDYLKNPTTYDGEEEISEIHPSIKSADIFTAMRNLTDEDLEAIAGHILIQPKIVGTKWGGGKIYY
[4]A0ZF54       GLVDYMKNPTTYDGEMEISEIHPSLKSADIFTEMRIFTEDDLVATAGHILLQPKIVGDKWGGGKIYY
[5]B5W3Z1       ALVDYVKNPTTYDGFIEIYELHPSLKSADIFPKMRNLSEDDLYNVAAYILLQPKVRGEQWGGGKYLR
[6]A0YIA2       ALIDYMQNPTTYDGFYEISELHPSTKSSDLYPKMRNLTEDDLYSIAGYILLEPKIRGEQWGGGKYLR
[7]Q111E9       GLIDYMQNPTTYDGFIEISEFHPSIKSADIYPEMRNLTEDDLYAIAGYILVQPKVLGKQWGGGKIFR
[8]B4VPE5       ALVDYMKNPTTYDGFTEIYEFHPSTRSADIFPEMRNLTEDDLKAIAGYILVQPKVQGIMWGGGKVYN
                .::**:::**:***  *  *:* .   .*: .  * :::.**   :.::*  **     **
  Consensus     GLVDYMKNPTTYDG--E--EIH-S---ADI-P--R-LTEDDL---AGYIL--PK-----WG------
```

[1]Q8Z044, *Anabaena* sp. PCC 7120 [2]Q3M9H7, *Anabaena variabilis* ATCC 29413/PCC 7937; [3]B2IVD8, *Nostoc punctiforme* ATCC 29133/PCC 73102; [4]A0ZF54, *Nodularia spumigena* CCY 9414; [5]B5W3Z1, *Arthrospira maxima* CS-328; [6]A0YIA2, *Lyngbya* sp. PCC 8106; [7]Q111E9, *Trichodesmium erythraeum* IMS101; [8]B4VPE5, *Microcoleus chthonoplastes* PCC 7420.

Table 18.6 is a miscellaneous collection of non-*Synechococcus* cyanobacterial sequences. It has 31 deduced cyt c_6 sequences from 19 strains of 8 different genera. This collection includes 3 sequences for 1 strain (h), 2 sequences each for 10 strains (a–g, i–k) and 1 sequence each for 8 strains. The pI column shows that 14 sequences have values ranging from 4.2 to 9.5 (Group I) and 17 sequences from 5.6 to 9.6 (Group II). Group I is comprised of mostly acidic sequences (9 acidic, 2 neutral and 3 basic) whereas Group II has mostly basic sequences (12 basic, 3 neutral and 2 acidic). All the sequences (#3, 5 and 9–12), in which each is derived from a cyt c_6 gene next to a gene for plastocyanin, are clustered together in group I.

In Table 18.7, there are 23 cyt c_6 sequences from different species of filamentous cyanobacteria, of which 20 are deduced from gene sequences and 3 from amino acid sequencing. The 3 protein sequences number 9, 12 and 16 belong to *Plectonema boryanum* (Aitken 1977), *Aphanizomenon flos-aquae* (Ulrich et al. 1982) and *Anabaena variabilis* (Aitken 1976), respectively. There are 9 different species providing the 20 deduced sequences. These groups include 3 sequences each for 3 species (a, c, d), 2 sequences each for 5 species (b, e–h) and 1 sequence for 1 species. The pI column shows that 8 sequences in Group I have mainly basic values ranging from 8.0 to 9.5. In Group II, all the sequences fall within a narrow range of pI values from 9.0 to 9.4 except one at pH 6.8. Group III has 6 sequences, of which 1 has a basic pI value of 9.0, 2 have the same value of 6.8 close to neutral and 3 have acidic values between 5.3 and 5.6. The 3 cyt c_6 sequences (# 19–21) in which each is located next to a plastocyanin and a cyt c550 gene appears in Group III and all are acidic.

Table 18.5 An alignment of cytochrome c_6 from different *Synechococcus* strains

	10 20 30 40 50 60 70 80
[1]Q2JLC6	----------PDLALGAKVFQAKCVGCHLNGRNTLVAAKNLSLAALHEYHVDT---P----ELIQAQVRNGKGAMPAFAK
[2]Q2JSC5	----------PDLALGAKVFQAKCVGCHLNGRNTLVAAKNLSLAALHEYHVDT---P----ELIQAQVRNGKGAMPAFGK
[3]Q31RP8	----------ADLQVGARLFQSNCTTCHLNGGNVINGQKTLRQEALRRYGMDS-------VAAIQKQVTYGKNAMPAFGQ
[4]Q5N2K6	----------ADLQVGARLFQSNCTTCHLNGGNVINGQKTLRQEALRRYGMDS-------VAAIQKQVTYGKNAMPAFGQ
[5]A3Z188	----------ADAAHGGQIFSANCAACHMGGGNVVNAERTLKADALTAYLANYSSDH---EAAIAAQVTVGKNAMPAFLG
[6]A5GSG1	----------ADVAHGGQIFSANCVACHMGGGNVVNGERTLKAEALDAYLANYGDGH---ESAIAYQVTNGKNAMPAFGG
[7]A3Z4W1	----------ADAAHGAQIFSANCAACHIGGGNVVNAERTLKQADLDAYLSGYGEGH---EAAIAAQVTKGKNAMPSFLG
[8]A4CU10	----------ADSAHGEQIFSANCAACHMGGGNVVNGQRTLQQDDLKAYLANYNEGH---EEAIAYQVTNGKNGMPAFGG
[9]Q05TV5	----------ADAAHGAQVFSANCAACHIGGGNVVNGQRTLQQDDLKAYLANYNDGH---EEAIAYQVTNGKNGMPAFGG
[10]Q0I9C6	----------ADVAHGEQVFSANCAACHMGGGNVVNGQRTLKQDDLKAYLSDYNDGH---ESAIAHQVTNGKNGMPAFGS
[11]A5GKE5	----------TDTVRGGQIFTTNCAACHAGGGNIIKAERTLQEADLKAHLPNYLGAH---ESAIVAQVTYGRNAMPAFVD
[12]A4CTU5	----------TDAVRGGQIFTTNCAACHAGGGNIIKAERTLREADLKAHLPNYLGAH---ESAIVAQVTYGRNAMPAFVD
[13]Q0I9B5	----------SDTVRGGQIFNTNCAACHAGGGNIVKSERTLRQDDLEAFLPNYLTGH---ETGIVAQVTYGRNAMPAFLD
[14]A3Z4X0	----------ADAALGAKVFSSNCAACHAGGGNIVNGERTLRQADLQDYLAQYSNGH---ESAIVAQVTYGRNAMPAFID
[15]Q05R02	----------GDAVNGAKVFSANCAACHAGGGNVVNGEHTLQQDDLKAYLAQYNNGH---ESAIVAQVTYGLNEMPAFIG
[16]A5GJR3	--------P--DPEHGAQLFSANCAACHMGGGNVISASRTLSQTDLQAHLESYGMDP---LEAIEHQIENGKNAMPAYEG
[17]A4CXE7	--------P--DPEHGAQLFSANCAACHMGGGNVISASRTLSQTDLQAHLESYGMDP---LEAIEHQIENGKNAMPAYEG
[18]Q05RA2	-------NP--DSGNGSQIFSTNCAACHMGGGNVIRASRTLSEADLQAHLDSYSQDH---LEAIEHQIEAGKNAMPPYQG
[19]Q0I8X8	-----ASSPSGDLDHGGQLFSANCAACHMGGGNVISASRTLSQSDLQAHLNEYGDDH---LEAIEHQIEKGKNAMPSFVG
[20]Q3AYE6	-----FDSP---LEQGEQIFSQNCAACHMGGGNVIRASRTLNIRDLNAHLTEYPQDP---LEAIEHQIEDGKNAMPSYAG
[21]Q065X1	-----FDSA---VEQGEQIFSQNCAACHMGGGNVIRASRTLNIRDLNAHLEEYPQDP---LEAIEHQIEDGKNAMPSYAG
[22]Q7U642	-----LESAL--IEQGEQIFSSNCAACHMGGGNVIRANRSLKIRDLNAHLEEYQQDP---LEAIEHQIEAGKNAMPSYEG
[23]Q3AKW0	-----LKSSA--LERGEQIFNSNCAACHMGGGNVIRANRTLKISDLNAHVEAYSSSP---LEALEHEIEDGLNAMPSYAD
[24]A3Z568	-----TDPVPANLENGSQVFSAQCAACHMGGGNVIRASRTLCQSDLQAHLAAYRSDH---LEAIEDQVEHGKNAMPAFAS
[25]O30881	----------ADAAAGAQVFAANCAACHAGGNNAVMPTKTLKADALKTYLAGYKDGSKSLEEAVAYQVTNGQGAMPAFGG
[26]B4WN82	----------GDIAQGKQVFATNCVACHAGGRNVVQADKTLKQDALESYLENY--GAEHNISAIVYQVTNGKNAMPAFSG
[27]P00115	----------ADIADGAKVFSANCAACHMGGGNVVMANKTLKKEALEQFGMNS-------ADAIMYQVQNGKNAMPAFGG
[28]P07497	----------ADLAHGGQVFSANCAACHLGGRNVVNPAKTLQKADLDQYGMAS-------IEAITTQVTNGKGAMPAFGS
[29]P25935	----------ADLAHGGQVFSANCAACHLGGRNVVNPAKTLQKADLDQYGMAS-------IEAITTQVTNGKGAMPAFGS
[30]A5GUE2	-----------AAADGAALFEAHCVGCHLNGGNIIRRGKTLKLKALERNGIEG---P----LEIAMIATEGRGQMSAYGE
[31]A3YZ99	---------ESPAASGESLFANHCAGCHINGGNILRRGKTLKLKALERNGITG---P----GAIAAIASAGLGQMSGYGE
[32]A5GJP6	-----------LSSDGAQLFDLHCAGCHPNGGNIIRRGRSLKLKDLSKRGLDN---P----EAIAAIAKEGIGQMSGYGD
[33]A4CXC9	-----------LAPDGEQLFDLHCAGCHPNGGNIIRRGRSLKLKDLSKRGLDN---P----EAIAAIAREGIGQMSGYGH
[34]Q05RC9	TLAIFPGVSPVLATSGQELFSNHCAGCHVNGGNIIRRGRTLKLAALKKNGLDD---P----DAIARVAREGIGQMSGYGD
[35]A3Z583	-----------SAADGEQLFNNHCAGCHPHGGNIIRRGRTLKLRALEQRQLNN---P----DAIARIAREGIGQMGGYAE
[36]Q3AYG2	------------SSEGAALFQQHCAACHINGGNIIRRGKTLKLKALERQGIAS---M----DAIAQIAREGVGQMSGYGD
[37]Q065V4	------------SNEGAALFQQHCAACHINGGNIIRRGKTLKLKALERQGIAS---V----DAIAQIAREGVGQMSGYGD
[38]Q3AKY1	-----------DASTGALLFEQHCAGCHINGGNIIRRGKNLKLKTLEREQIAT---V----DAIAAIAREGRGQMSGYAD
[39]Q7U624	-----------TSGEGAVLFGQHCAGCHVNGGNIIRRGKNLKLATLKRQGLDS---T----EAIASIARKGIGQMSGYGD
[40]Q0I8W0	------------VETGASLFQNNCASCHPNGENIIRRGRTLKMKALTKRGLDS---S----EAIAQVAREGIGQMSGYAD
[41]Q31K47	----------AAPATGAALFEAQCVGCHIGGGNIIRRGKTLQLKALKKNGYDT---A----EAIATIVSQGKNNMSAYSD
[42]Q5N1R2	----------AAPATGAALFEAQCVGCHIGGGNIIRRGKTLQLKALKKNGYDT---A----EAIATIVSQGKNNMSAYSD
[43]Q8KX15	----------ADLDQGAQIFEAHCAGCHLNGGNIVRRGKNLKKRAMAKNGYTS---V----EAIANLVTQGKGNMSAYGD
	* :* :*. ** * * : :.* : : * * :
Consensus	---------------G--LF--NCA-CH--G-N-I---RTL----L-----------------I------G---M--F--

	90 100 110	pI	L	Strain	
[1]Q2JLC6	ILKPEEIEAVAAYVLDRAEHN-WSRG----	7.2		JA-2-3B'a(2-13)	
[2]Q2JSC5	LLKPEEIEAVAAYVLDRAEHN-WSKG----	7.2		JA-3-3Ab	
[3]Q31RP8	RLSPEQIEAVATYVFDRAERG-WTAL----	9.2	a	PCC7942	
[4]Q5N2K6	RLSPEQIEAVATYVFDRAERG-WTAL----	9.2	b	PCC6301	
[5]A3Z188	KLTETDIADVSAYVEEMAAKG-WA------	4.8	c	WH5701	
[6]A5GSG1	KLSDGDIADVAAYVEDMASKG-WA------	4.6	d	RCC307	Group I
[7]A3Z4W1	KLSDTDIADVAAYVEQQAGNG-WT------	4.8	e	RS9917	
[8]A4CU10	KLGPDDIADVAAYVESQSING-WA------	4.5	f	WH7805	
[9]Q05TV5	KLSSDDIADVAAYVESQSVNG-WA------	4.5	g	RS9916	
10]Q0I9C6	KLGSDDISDVAAYVESQSVKG-WA------	5.0	h	CC9311	
11]A5GKE5	VLSESEIADVAAYVEEQSSQG-WS------	4.7	i	WH7803	
12]A4CTU5	VLSESEIADVAAYVEEQSSQG-WS------	4.9	f	WH7805	
13]Q0I9B5	VLSENEIADVAAYVEDQASHG-WS------	4.6	h	CC9311	Group II
14]A3Z4X0	TLTETQISDVAAYVEEQASKG-WS------	4.6	e	RS9917	
15]Q05R02	TLTENQISDVAAYVEEQASNG-WG------	4.3	g	RS9916	
16]A5GJR3	KLSDQDIADVAAYVESQAEQG-WSR-----	4.4	i	WH7803	
17]A4CXE7	KLSDQDIADVAAYVERQAEHG-WSR-----	4.7	f	WH7805	
18]Q05RA2	KLSDDDIADVAAYVEEQAERG-WQR-----	4.5	g	RS9916	
19]Q0I8X8	KLSEQDILDVAAYVELKAEKG-WQR-----	5.0	h	CC9311	
20]Q3AYE6	KLSESEIVAVATYVEQQAEMG-W-------	4.8	j	CC9902	
21]Q065X1	KLSESEIIAVATYVEQQAEMG-W-------	4.4	k	BL107	Group III
22]Q7U642	KLTEAEIIAVATYVEQQAELG-W-------	4.6	l	WH8102	
23]Q3AKW0	TLSDEEIMAVATYVEQRAELG-WSRR----	4.7	m	CC9605	
24]A3Z568	KLSERDIADVAAFVEEQAERG-WGR-----	5.2	e	RS9917	
25]O30881	RLSDADIANVAAYIADQAENNKW-------	4.7	n	PCC7002	
26]B4WN82	RLTADQIEDVAAYVNDQAESG-WTS-----	5.1		PCC7335	
27]P00115	RLSEAQIENVAAYVLDQSSNK-WAG-----	5.7		PCC6312	
28]P07497	KLSADDIADVASYVLDQSEKG-WQG-----	5.0	b	PCC6301	
29]P25935	KLSADDIADVASYVLDQSEKG-WQG-----	5.0	a	PCC7942	
30]A5GUE2	VLGEEGADAVAEFIWQQAQEG-WS------	4.9	d	RCC307	
31]A3YZ99	VLGPGGPEAVAGWVWQQALGN-WSAEAKPV	8.1	c	WH 5701	
32]A5GJP6	ALGEGNEQVVGDWVWLQAQNA-WIQG----	5.5	i	WH7803	
33]A4CXC9	VLGEGNDNVVGDWVWLQAQNA-WIQE----	5.5	f	WH7805	
34]Q05RC9	VLGVGGDELVAAWIWEQAQNA-WIQG----	5.5	g	RS9916	
35]A3Z583	VLGPNGDQRVAEWIWDQAQKA-WVQG----	8.1	e	RS9917	
36]Q3AYG2	VLGPGGDQLVAEWIWEQAQNA-WIQG----	6.1	j	CC9902	Group IV
37]Q065V4	VLGPGGDQLVAEWIWDQAQNA-WIQG----	6.0	k	BL107	
38]Q3AKY1	VLGSEGDQLVAEWVLMQAQNA-WTQG----	6.1	m	CC9605	
39]Q7U624	KLGEGGDQLVAGWILEQAQNA-WTQG----	8.0	l	WH8102	
40]Q0I8W0	ALGEDGDVFVAEWVWQQAQKA-WTQG----	5.7	h	CC9311	
41]Q31K47	RLSPEEIAIVSAYVLEQANAG-WPRS----	8.7	a	PCC7942	
42]Q5N1R2	RLSPEEIAIVSAYVLEQANAG-WPRS----	8.7	b	PCC6301	
43]Q8KX15	KLSSEEIQAVSQYVLQQSQTD-WKS-----	8.7	n	PCC 7002	
	* *. :: : *				
Consensus	-L-------VA-YV---A--- W-------				

Table 18.6 An alignment of cytochrome c_6 from non-*Synechococcus*, unicellular cyanobacteria

	10 20 30 40 50 60 70 80
[1]A2CAJ0	----ADSAHGGQVFSSTCAACHAGGGNIVD-PAKTLQKAALEATLSNYGSGHEEAIVAQVTNGKGGMPSFA--DVLSAAD
[2]Q7TV20	----ADSAHGGQVFSSTCAACHAGGGNIVD-PAKTLQKAALEATLSNYGSGHEEAIVAQVTNGKGGMPSFA--DVLSAAD
[3]B5IL30	----ADVAHGGQLFSANCAACHMGGGNVVN-AERTLKQDALEAYLANYSSDHEAAIAYQVTNGKNAMPAFG--GKLSEGD
[4]P46445	----ADLAHGKAIFAGNCAACHNGGLNAIN-PSKTLKMADLEAN----GKNSVAAIVAQITNGNGAMPGFK--GRISDSD
[5]Q7NJ52	---AADLAQGEKVFKANCAMCHAGGRNTVN-PAKTLKIEDLKKY----KMDTAAAISAQLYNGKGAMPAFGKNGKLKQDQ
[6]B4CB54	-----DAAAGAKVFSANCAACHMGGNNVIM-ANKTLKKEALEQF----GMNSADAIIYQVQHGKNAMPAFG--GRLSDEQ
[7]Q7NJC6	---QPDLAAGEKIFKANCAACHAGGNNIVE-PEKTLKKEALAHF----GMGSPAAIIQQVTGGKNAMPAFG--GELSTEE
[8]B0C733	----ADAGAGAGVFNANCAACHAGGNNVVQ-ADKTLKADALSAN----GMDSADAIINQVTNGKGGMPAFG--ASLSPAD
[9]B0JHN8	--------DGASIFSANCASCHMGGKNVVN-AAKTLKKEDLVKY----GKDSVEAIVTQVTKGMGAMPAFG--GRLSAED
[10]A3IMH4	----GDASAGKGIFTANCASCHMGGGNVVAGASKGLAKDALEKN----GVDTLEKIVYQVTNGKNAMPAFQ--GRLNAQQ
[11]Q4C0X0	----GDAGAGKGIFTANCNACHMGGGNVVAGPSKGLAKNYLEKN----GVDTLEKIVYQVTNGKNAMPAFG--GRLDAQK
[12]A1KYG5	----GDAAAGKGIFTANCASCHMGGGNVVAGASKGLTKDALENN----KMLSEEAIIAQVTKGKAAMPAFL--GRLTDTQ
[13]B4C4K8	----GDAAAGKTVFTAKCAQCHLGGKNLVN-PAKTLSKADLEAN----GMASLDAIITQVTNGKAAMPAFG--KLLTAEQ
[14]B2F961	----GDAGNGSKVFSANCNACHLGGKNVVN-AAKTLNKSDLEKY----AMLDLEAIKTQVTNGKGAMPAFG--KRLTPDQ
[15]A3PCC0	----SDIRDGETIFRNVCAGCHVRGGSVVLKGSKSLKLSDLEKR----GIADVNSITIIANEGIGFMKGYK--NKLNDGE
[16]Q31BG9	----SDIRNGETIFRNVCAGCHVRGGSVVLKGSKSLKLSDLEKR----GIADEISIAKIANEGIGYMKGYK--KKLKDGE
[17]Q7TV27	----LDTDAGGSLFKQHCSGCHVNGGNIIR-RNKTLRLKALERN----GLDNPQAIARVAREGIGQMSGYE--DVLGDSG
[18]A2CAQ1	----LDIDSGGSLFKQHCSGCHVNGGNIIR-RNKTLRLKALERN----GLDNPQAIARVAREGIGQMSGYE--DVLGDSG
[19]A9BEJ9	----LSEMDGENLFIKHCSGCHINGGNIIR-RNKTLKLKDLQRN----GIDTPEAISKIAREGIGIMSGYE--EVLGEGR
[20]Q46LX6	----IEADSGRTLFNHNCAGCHINGGNIIR-RSKNLKISSLKRN----GIDNPEAIAKIARQGVGIMSGYE--DELGDNG
[21]A2C131	----IEADSGKNLFNHNCAGCHINGGNIIR-RSKNLKISSLKRN----GIDNPEAIANIARQGVGIMSGYE--DELGDNG
[22]B5IHT4	ASAALPAADGARLFENHCAGCHLNGGNIIR-RRKTLRLAALERQ----GVASEQAIAAIAAAGVGQMGGYG--DVLGDDG
[23]B0C975	----ASPGDGAVLFETYCAGCHPKGGNIIR-RRKTLKLKSLQRD----GYDTLAPVINLITQGQNNMPGFA--DQLNDTQ
[24]B4CIR5	----ATGGEGAQLFDLHCAGCHLNGGNIVR-RGKTLKLPALQRN----GYDTVEAITAIVSNGKGNMSAYK--ERLTPDQ
[25]A3IW71	----EEVTDGSKIFTVHCAGCHPNGNNIIR-RGKNLKMRALKRN----KVDTLDSIIHLVTYGKNNMSAYE--EKLTQNE
[26]B1WTZ1	----EEATDGAKIFTVHCAGCHPNGNNIIR-RGKNLKMKALKRN----KVDTLDAVINIVTYGKNNMSAYE--ERLTQDE
[27]Q4C6T3	----QDISDGSKIFTIHCAGCHPNGKNIIR-RGKNLKLKALKKN----KVDTLDSIMNLVSYGKNNMSAYE--ERLTQNE
[28]Q4BXE0	----EDINDGAKIFSIHCVGCHPQGKNIIR-RAKNLKLRALKRN----KVDSLDAIINLVTYGKNNMSAYE--DKLTKEQ
[29]B4BTF3	----VDLNHGEEIFSIHCAGCHPNGSNIIR-RGKNLKLKALQKN----KVDSLEAIAALVTNGKNNMSAYS--DRLTPEE
[30]B2F001	--LTEEIGQGEQLFQTYCVGCHPNGNNIIR-RGKTLKSQALKRY----KMDSLEAITNLVTYGKNNMSAFQ--DRLSQPE
[31]B4AYG9	----SEIGEGAKIFNTHCAGCHPNGNNIIR-RGKTLQARALKKY----KMDSLEAISNLVANGKNNMSAFK--ERLSEQE
	* :* * ** * . : * * * : * * .: :
Consensus	---------G--IF---C--CH--G-N-2----K-L----L-------------I------G---M-AF-----L----

	90	pI	L	Cyanobacteria	
[1]A2CAJ0	IADVAAYVEAQASSGW---	4.5	a	*Prochlorococcus marinus* MIT 9303	
[2]Q7TV20	IADVAAYVEAQASSGW---	4.5	b	*Prochlorococcus marinus* MIT 9313	
[3]B5IL30	IADVAAYVEDMASKGWA--	5.1	c	*Cyanobium* sp. PCC 7001	
[4]P46445	MEDVAAYVLDQAEKGW---	5.5		*Synechocystis* sp. PCC 6803	
[5]Q7NJ52	IDSVTAYVLDQANKGWKK-	9.5	d	*Gloeobacter violaceus*	Group I
[6]B4CB54	IQDVAAYVLEQSEKGWKG-	6.1	e	*Cyanothece* sp. PCC 7425	
[7]Q7NJC6	IRQVASYVLEMADKDWQK-	5.5	d	*Gloeobacter violaceus*	
[8]B0C733	IENVAAYVLDQADK-W---	4.2	f	*Acaryochloris marina* MBIC 11017	
[9]B0JHN8	IEAVANYVLAQAEKGW---	7.7		*Microcystis aeruginosa* NIES-843	
[10]A3IMH4	IEDVATYVLSQAETGW---	5.7	g	*Cyanothece* sp. CCY 0110	
[11]Q4C0X0	IEDVATYVLSQAEAGW---	6.8	h	*Crocosphaera watsonii* 8501	
[12]A1KYG5	IEDVAAYVLSQAEAGW---	5.2	i	*Cyanothece* sp. ATCC 51142	
[13]B4C4K8	IENVATYVLAQAEADWK--	6.8	j	*Cyanothece* sp. PCC 8802	
[14]B2F961	IEDVATYVLEKAEKGW---	8.0	k	*Cyanothece* sp. PCC 7424	
[15]A3PCC0	DKVLAQWIIQNAEKGWK--	9.0		*Prochlorococcus marinus* MIT 9301	
[16]Q31BG9	DKVLAQWIIQNAENGWE--	8.9		*Prochlorococcus marinus* MIT 9312	
[17]Q7TV27	DQLVAAWIWAQAQNAWTQG	6.9	b	*Prochlorococcus marinus* MIT 9313	
[18]A2CAQ1	DQLVAAWIWAQAQNAWTQG	6.9	a	*Prochlorococcus marinus* MIT 9303	
[19]A9BEJ9	DEVLAHWIWKQSQKAWVQG	7.0		*Prochlorococcus marinus* MIT 9211	
[20]Q46LX6	DQIVANWVWEQAQKAWVQE	6.1		*Prochlorococcus marinus* NATL2A	
[21]A2C131	DQIVANWVWEQAQKAWVQE	5.6		*Prochlorococcus marinus* NATL1A	
[22]B5IHT4	VVAVARWVWEQAQAGWPRS	8.1	c	*Cyanobium* sp. PCC 7001	
[23]B0C975	IETVAQFVLEQASNNWK--	8.0	f	*Acaryochloris marina* MBIC 11017	Group II
[24]B4CIR5	IAIVADYVLQQATQNWH--	8.0	e	*Cyanothece* sp. PCC 7425	
[25]A3IW71	IKVVSQYVLEQAKNNWHNP	9.2	g	*Cyanothece* sp. CCY 0110	
[26]B1WTZ1	IKMVSQYVLEQANNNWRNP	9.0	i	*Cyanothece* sp. ATCC 51142	
[27]Q4C6T3	IKAVSQYILDKAQNNWRN-	9.6	h	*Crocosphaera watsonii* 8501	
[28]Q4BXE0	IESVSKYVLQQAQNNWHT-	9.4	h	*Crocosphaera watsonii* 8501	
[29]B4BTF3	IQAVSVYVLEQAQNNWTK-	8.0	j	*Cyanothece* sp. PCC 8802	
[30]B2F001	IESVARYVLEQAEKDWR--	8.0	k	*Cyanothece* sp. PCC 7424	
[31]B4AYG9	IKTVAQYVLEQAPKNWR--	9.6		*Cyanothece* sp. PCC 7822	
	:: :: : *				
Consensus	---VA-YV---A---W---				

Table 18.7 An alignment of cytochrome c_6 from filamentous cyanobacteria

	10 20 30 40 50 60 70	80 90	pI	L	Cyanobacterial Species	
[1]Q8Z0D7	--ENTINGEQIFSVHCAGCHINGSNIIR-RGKNLQKKTLKKYGM----DSLEAIEAIVTNGKNNMSAYKD	RLSEQEIQDVAAYVLEQAEKGWR------	8.0	a	*Anabaena* sp. PCC7120	
[2]B0BF00	--ENTINGEQIFSVHCAGCHINGSNIIR-RGKNLQKKTLKKYGM----DSLEAIEAIVTNGKNNMSAYKD	RLSEQEIQDVAAYVLEQAEKGWR------	8.0	b	*Anabaena* sp. PCC7119	
[3]Q3MCY4	--VDTANGEQIFSVHCAGCHINGSNIIR-RGKNLQKKALKKYGM----DSLEAVAAIVTNGKNNMSAYKD	RLSEQEIQNVAAYVLEQAEKGWR------	9.0	c	*Anabaena variabilis* ATCC29413/PCC7937	
[4]B2J3S0	--ADIVNGEQIFSLHCAGCHINGSNIVR-RGKNLKKQALKKYGM----DSIEAVTSIVTNGKNNMSAYKD	RLTEQQITDVAAYVLEQAEKDWR------	8.6	d	*Nostoc punctiforme* ATCC29133/PCC73102	
[5]A0ZCV9	--AETNHGAEVFSVHCAGCHINGGNIIR-RGKNLKKPALKRYGM----DTIEAVTSIITNGKNNMSAYQD	RLTPQEIQEVATYVLEQAQTGWR------	8.7	e	*Nodularia spumigena* CCY9414	I
[6]B5W5T9	DNLNIEKGGEIFSVHCVGCHAGGGNIIR-RGKNLKQRALQRNGY----DSIEAIAEIVANGKSNMSAYSD	RLSVSEIQQVAAYVLDQAEKNWR------	8	f	*Arthrospira maxima* CS-328	
[7]Q10VY3	--TEITQGAEVFQIHCAGCHAKGGNIVK-WWKNLKIRTLKRNKL----DSVEAIAYLVKNGKNNMSAYKD	RLTEIEIQTVSAYVLKQAENRWN------	9.5	g	*Trichodesmium erythraeum* IMS101	
[8]A0YJE3	--ANSQNGAKVFNQQCAGCHAGGGNIVR-WGKTLKKNALNRNGY----DTIEAISYLVTQGKGNMPAYKE	RLTPKEIQDVSEYVMKQAEKNWKS-----	9.5	h	*Lyngbya* sp. PCC8106	
[9]P00117	--ADAAAGGKVFNANCAACHASGGGQIN-GAKTLKKNALTANGK----DTVEAIVAQVTNGKGAMPAFKG	RLSDDQIQSVALYVLDKAEKGW-------	9.0		*Plectonema boryanum*	
[10]Q111R3	--ADIASGKGVFQGNCAACHIGGKNNIN-PAKTLQKSDLEKYGM----FAAEKIIYQVTNGKNAMPAFGR	RLKPQQIENVAAYVMAQAEGGWK------	9.4	g	*Trichodesmium erythraeum* IMS101	
[11]B2IWE9	--GDAVSGAKVFSANCASCHAGGKNLVQ-AAKNLKKEALEKYGL----YSAEAIIAQVTNGKNAMPAFGK	RLKADQIENVAAYVLSQADKGWK------	9.4	d	*Nostoc punctiforme* ATCC29133/PCC73102	
[12]P00116	--ADTVSGAALFKANCAQCHVGGGNLVN-RAKTLKKEALEKYNM----YSAKAIIAQVTHGKGAMPAFGI	RLKAEQIENVAAYVLEQADNGWKK-----	9.4		*Aphanizomenon flos-aquae*	
[13]A0ZCV7	--ADTVNGAKIFGANCAACHAGGRNLVQ-AQKTLKKDALEKYGL----YSAEAIISQVTKGKNAMPAFKG	RLKSEQIENVAAYVLEQADKGW-------	9.2	e	Nodularia spumigena CCY9414	II
[14]P0A3X8	--ADSVNGAKIFSANCASCHAGGKNLVQ-AQKTLKKADLEKYGM----YSAEAIIAQVTNGKNAMPAFKG	RLKPEQIEDVAAYVLGKADADWK------	8.9	b	*Anabaena* sp. PCC7119	
[15]P0A3X7	--ADSVNGAKIFSANCASCHAGGKNLVQ-AQKTLKKADLEKYGM----YSAEAIIAQVTNGKNAMPAFKG	RLKPEQIEDVAAYVLGKADADWK------	8.9	a	*Anabaena* sp. PCC7120	
[16]P0C180	--ADSVNGAKIFSANCASCHAGGKNLGV-AQKTLKKADLEKYGA----YSAMAIGAQVTNGKNAMPAFKG	RLKPEEIXXVAAYVLGKAEAEWK------	9.4		*Anabaena variabilis*	
[17]Q3MDW2	--ADVANGAKIFSANCASCHAGGKNLVQ-AQKTLKKEDLEKFGM----YSAEAIIAQVTNGKNAMPAFKG	RLKPDQIEDVAAYVLGQADKSWK------	9.0	c	*Anabaena variabilis* ATCC 29413/PCC7937	
[18]B4VU02	--ADTAKGAKIFSANCAACHIGGNNIVM-AQKTLKKDALEKYGM----DSIEKIVYQAKNGKGAMPAFIG	RLSDSDIEDVAAYVIEQAEKGW-------	6.8		*Microcoleus chthonoplastes* PCC7420	
[19]Q3M9H9	--AELPTGAKIFNNNCASCHIGGGNILI-SEKTLKKEALLKYLEDYETNSIQAIIHQVQYGKNAMPAFKD	KLSTEEILEVAAYIFQKAEKDWSNLEKEG	5.3	c	Anabaena variabilis ATCC29413/PCC7937	
[20]Q8Z046	--AELPTGAKIFNNNCASCHIGGGNILI-SEKTLKKEALLKYLEDYETNSIQAIIHQIQYGKNAMPAFKD	KLSTEEILEVAAYIFQKAEKDWSNLEKEG	5.3	a	Anabaena sp. PCC7120	III
[21]B2IVD6	--AETSNGAKIFEANCASCHIGGGNILI-SQKTLKKEALSKYLENYNSDSIEAIIHQVQNGKNAMPAFKG	KLSAEEILDVAAYVFQNAEQGW-------	5.6	d	*Nostoc punctiforme* ATCC29133/PCC73102	
[22]B5VVZ5	--ADAAAGASVFSANCAACHMGGRNVIV-ANKTLSKSDLAKYLKGFDEDAVASVAYQVTNGKNAMPAFNG	RLSPKQIEDVAAYVVDQAEKGW-------	6.8	f	*Arthrospira maxima* CS-328	
[23]A0YKW1	-DGDPATGSQVFAANCNACHMGGKNVIM-SNKTLSKADLAKYLKGFNDDPQAAIAYQITKGKNAMPAFKG	RLSPQQIEDVSAYVFSKADKGW-------	9.0	h	*Lyngbya sp.* PCC8106.	
	: * :* :* ** * . * * * : : **. *.. :	:* .* *: : . :* *				
Consensus	---D---G--IF--NC--CH--G-N------K-L----L----------S---I------GKN-MPAF--	RL---QI--VA-YVL-QAE--W				

In an earlier paper, Ho and Krogmann (1984) commented on the abundance of high pI values in the cyt c_6s from filamentous species and the abundance of the low pIs in the cyt c_6s of unicellular species. However, the data sets presented above show that a wider spread of pI values seemed common among both unicellular and filamentous cyanobacteria. We have proposed that the different cyts within the same organism might be identified with a Roman number, for example I–IV, as in case of *Synechococcus* sp. RS9917 which has 4 cyt c_6 homologs. The Roman number I refers to the cyt that is found in cells grown under optimal conditions for photosynthesis. The meanings of the other numbers will emerge when more is known about the circumstances that make the other homologs appear, as discussed later in the concluding section.

18.3.1.2 Structural Properties

Beyond the amino acid sequence of cyt c_6, there are many interesting properties of this molecule. One is its reversible denaturation. Krogmann found that impure cyt c_6 could be mixed with trifluoroacetic acid to 5% and centrifuged to remove the denatured proteins. The yellow supernatant was loaded on a Sephacryl S 200 column and eluted with 50 mM Tris buffer. In a few hours the trifluoroacetic acid had evaporated and the cyt containing tubes had the normal pink color. The spectrum was that of pure native cyt c_6. Ho (2005) had published a similar purification. This is a simple method to isolate highly purified fractions of cyt c_6 from cell samples and it makes use of mass spectrometry for protein identification. In Ho's procedure a sample of cells with approximately 5 µg of chlorophyll per ml is used. The cells are broken by 2 cycles of freezing and thawing. The particles are removed by centrifugation of the extract. The supernatant is brought to 40% ammonium sulfate saturation and centrifuged to remove the phycobiliproteins. The ammonium sulfate concentration is raised to 80% saturation and this is centrifuged to concentrate the cyt c_6 and some other proteins. The salt is removed from the dissolved precipitate and the sample is mixed with 2 parts of chloroform and 3 parts of ethanol at −21°C. This is centrifuged and the supernatant can be put in the mass spectrometer. There is one sharp peak with the mass of cyt c_6. The procedure is simple and indicates the strong tendency of this polypeptide to fold into the native 3-D form.

A very interesting behavior of cyt c_6 was found by Kerfeld et al. (2002). She found that cyt c_6I from *A. maxima* is present in isoforms that can be separated by size exclusion chromatography. One of these isoforms, lacking any post-translational modifications, crystallized in a rare cubic space group. Elucidation of the crystal structure revealed that the proteins assemble themselves into a 24 molecule cluster (Fig. 18.2). The cluster is hollow and has a diameter at various points of 75–120 Å. The edge to edge distance of hemes is 17.4 Å which is close enough for electron transfer. There are six pores of about 13 Å diameter. If there is a function of this assembly, it is yet to be known.

The 24 cyt c cluster turned up in another line of experiments. A very pure sample of *A. maxim* cyt c_6 was used to examine the small bands of cyt that appear during elution of a DEAE cellulose column (Kerfeld et al. 1998). When a very shallow gradient of NaCl was used, 4 pink bands of cyt preceded the main band and another 4 pink bands followed it. When samples of each band were subjected to time-of-flight mass spectroscopy, 5–8 mass peaks of from 9,850 to 10,003 Da appeared. The increments between peaks varied from 15 to 21 Da and most frequently the increment was 16. The calculated mass of cyt c_6 is 9,855 Da. The protein is negatively charged and it appears that there are some molecules that are more attracted to the ion-exchange cellulose. *A. maxima* cyt c_6 has two methionine residues. One

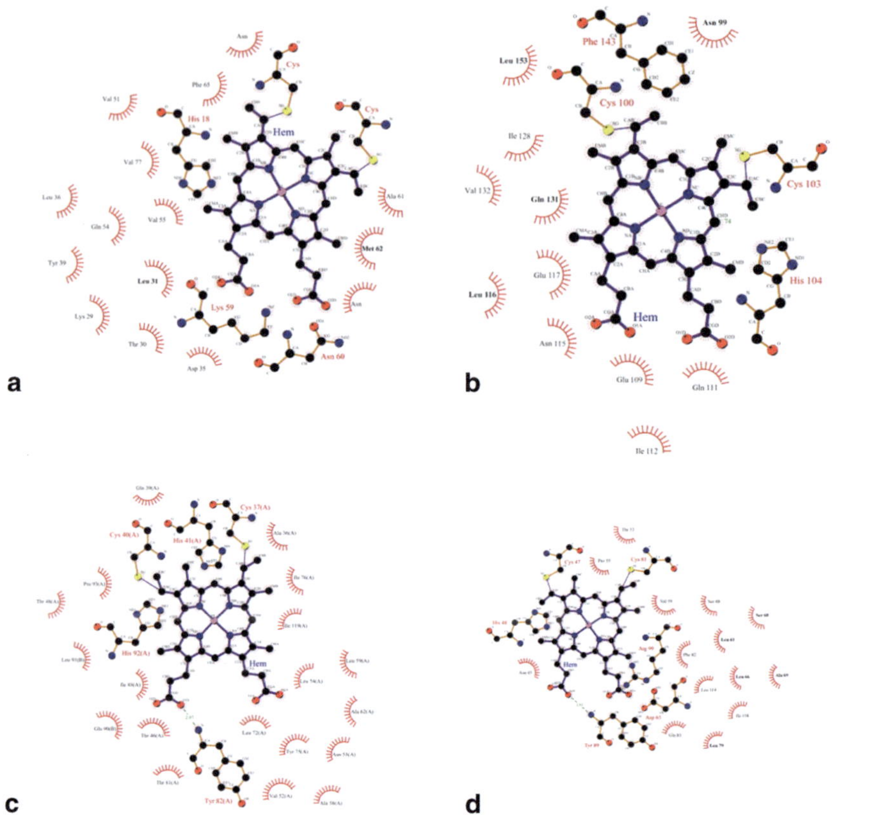

Fig. 18.2 Two-dimensional ligand-binding plots showing hydrogen-bonding interactions and distances, and hydrophobic contacts for the heme in **a** Cyt c_6, **b** Cyt cM, **c** Cyt c550 and **d** PsbV2. (Putative) hydrogen bonds are shown in dashed lines with distances indicated. Residues labed in bold are absolutely conserved in the primary structure of the proteins. In c the two protein chains of the dimer are distinguished by **a** and **b**. (Figures were prepared with LIGPLOT, Wallace 1995)

is residue 44 and the other is residue 65. The sulfur in the latter residue serves as a ligand to the iron in heme. If the methionines were both oxidized to sulfoxide, this would increase the total mass by 32 Da and if both were further oxidized to sulfones the increase would be 64 Da. One might guess that there would be some disturbance to the heme. When Jung et al. (1995) used some of this batch of cyt c_6 in their development of a high affinity assay for cyanobacterial PSI reduction, they found that the specific activity of the cyt was doubled by pretreating the cyt with β-mercaptoethanol which would surely reduce the sulfone and sulfoxide on the methionines (Jung, personal communication). The varying clumps of mass fragments in each eluted band could be the result of the cyt c_6 forming the 24 molecule clusters. Ho et al. (in preparation) have found a yellow form of cyt c_6 in extracts of an exotic, filamentous cyanobacterium, *Lyngbya wollei*, at a stage of growth where

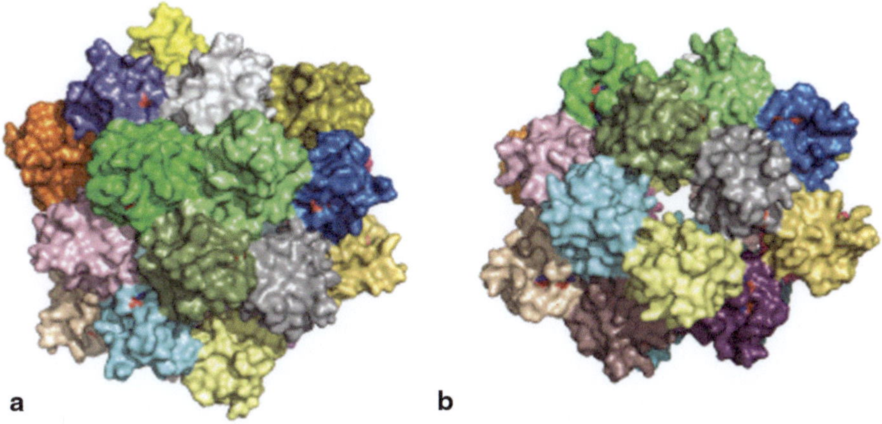

Fig. 18.3 Surface representation of the *A. maxima* cyt c_6 cluster. **a** View down the three-fold and **b** pseudo four-fold symmetry axis

sulfide is driving anoxyegenic photosynthesis. The normal reduced cyt c_6 has peaks at 279,520 and 552 nm. The yellow cyt c_6 which is reduced and cannot be oxidized has peaks at 433, 533 and 579 nm. This might be an accidental chemical product since there are sulfides, thiols, sulfones, sulfoxones, photons and active oxygen available.

While cyt c_6 is a cyanobacterial-specific protein, recently structure elucidation of a putative outer membrane c-type cyt (OmcF) from the dissimilatory metal-reducing bacterium *Geobacter sulfurreducens* revealed that it was most structurally similar to the cyanobacterial cyt c_6 (Lukat et al. 2008). Cyt c_6 and OmcF have similar redox spectra. However, the E_m of OmcF (+121 mV) is remarkably lower than that of c_6. The observation of a Tyr residue, the structural counterpart of Gln 54 (Fig. 18.3a) in cyt c_6, is suggested to be an important factor in reducing the E_m of OmcF relative to that of cyt c_6. The importance of this position in tuning redox potential was first noted in the structure of *Arabidopsis thaliana* cyt c6A (where it is a small hydrophobic residue; Worrall et al. 2007). Among the cyt c_6 paralogs, OmcF is most similar to petJ2 from *Synechococcus* PCC7002. Proposed roles for OmcF include redox sensing or involvement in iron reduction (Weigel et al. 2003).

18.3.2 Cytochrome cM

18.3.2.1 Amino Acid Sequences

Table 18.8 shows the full-length sequences of cyt cM from16 different strains of *Synechococcus*. All have an Asp-Pro-Tyr trio (residues 69–71) prior to a predicted cleavage site (↓) of the leader sequence. This trio of amino acids was con-

Table 18.8 An alignment of cytochrome cM from different *Synechococcus* strains

```
                       10        20        30        40        50        60        70
                        |         |         |         |         |         |         |
[1]Q2JNK7     ----------------MQRFFT-------------SVWAQKTRLGLALLGLGLAILTWAWLSAEARLDPY
[2]Q2JRK7     ---------------------------------------------MAVVGL--AILTWIWLAAGAGLDPY
[3]Q3AYZ2     ----------------MIEPSSTAAE--------QPERGRGLITALVVLAAAACVVLVLWVLGNARQDPY
[4]Q067U7     ----------------MIEPSSTAAE--------QPEQGRGLITALVVLAAAACVVLVLWVWGNARQDPY
[5]Q7U8A2     ----------------MIEPTSTATE--------QQESGRGLITALVVLAASACAVLLIWVINGAQQDPY
[6]Q3AI93     ----------------MVQPSRIAAESASADANDSSDRGRGLIAALVVSAATACVVLVLWVLGSAQQDPY
[7]A3Z7S0     -----------------MAPSSTAAAP--------AGRHRGLITGLVILAAVACIVLVVWVMGNASRDPY
[8]Q05XQ9     --------MFHGLDAIVTSPSSTAAAP--------SPKRRGLIAGLVTVAAITCIVLVAWVMTSTNRDPY
[9]A4CVQ9     --------MFHATLASVTAPSS-TAAS--------PDRKRGLVSALIAVAGATALVMLIWLFGVNRLDPY
[10]A5GM57    -------------MASVTVPSS-TAAP--------ADRKRGLVSALVAVAGVTALVMLVWLFGVNRLDPY
[11]Q0IBJ3    ----------------MTAPSSNAAAI--------PERSRGLIAALTVLAAMACIVLLVWMLGNTRQDPY
[12]A3YY40    ---------------MTGDPSTILEATP------APVQRQGLITALVLLAAAGCILLLLLVLPAARSDPY
[13]A5GRX7    MSRNLRGDTPMLGEVWPRRHPLTGSELLD------EEPRKRGLITALVVLAVVALVAMSALVASANSRDPY
[14]Q5MZ66    MILLRHVLQTTDSLSPAIASAVEQSLSKS--------IESRAAQTLGWLLATAVMVAIGLVVTLIRPADPY
[15]Q9Z3G2    MIILRHVLQTTDSLSPAIASAVEQSLSKS--------IESRAAQTLGWLLATAVMVAIGLVVTLIRPADPY
[16]B1XNK7    ----------------MANSSELSVPN----------FSKLLFVIILVLAIAALGIFGVYST--QASDPY
                                                                .                 ***
 Consensus    -------------------------------------------------A-----------------DPY

                 ↓81        91       101       111       121       131       141
                   |         |         |         |         |         |         |
[1]Q2JNK7     TEAVLSLQGQVSHGASLFALNCAACHGEEADGRVGPSLRGVSSRRSDRFIIQQVTGGKTPPMPQFQPDPQE
[2]Q2JRK7     TKAVLSLEGQPSHGASLFALNCAACHGEEADGRVGPSLRGVSNRRSDRFIIHQVTSGKTPPMPQFQPDPQE
[3]Q3AYZ2     VRASLDLPGSADHGGQLFRINCAGCHGLAGQGLVGPKLVGISEQRNDPVLVHQIISGETPPMPSFQMEPQS
[4]Q067U7     VRASLDLQGSTEHGGQLFRINCAGCHGLAGQGLVGPKLVGITEQRNDPVLVHQIISGETPPMPSFQMEPQS
[5]Q7U8A2     VRASLELQGDPDHGGQLFRINCAGCHGLAGQGLLAPKLAGISERMRDPALIHQIVSGDTPPMPSFQMEPAS
[6]Q3AI93     IKASLELQGAVDHGGQLFRINCAGCHGLAGQGLVGPQLQGVSNQMKDPALVHQIISGETPPMPSFEMEPQS
[7]A3Z7S0     VQATRMLNGDADHGGQVFRINCAGCHGIAAQGLVGPNLHGVADRRSDPSLIHQVVSGATPPMPRFEVEPQT
[8]Q05XQ9     VVATRSLDGDAQHGGLLFRINCAGCHGIAGQGLVGPSLQGVSTRLSDPQIIQQVVSGQTPPMPRFEIEPQG
[9]A4CVQ9     SRATLSLSGDVQHGGQLFRINCAGCHGIAGQGLVGPSLQGVAGKRSDRSIIHQIVSGETPPMPRFEIEPQG
[10]A5GM57    SKATLSLSGEVSHGGQLFRINCAGCHGIAGQGLVGPNLQGVSAKRSDRSIIHQIVSGETPPMPRFEIEPQG
[11]Q0IBJ3    TKATLALEGSEQHGGQMFRINCAGCHGIAGQGLVGPSLKGVSDRRKDMKIIHQVVSGDTPPMPRFEIEPQN
[12]A3YY40    TRRTLELSGSLETGGRLFRMNCAGCHGIAAQGLVGPSLHGVSQRRSDRQLIRQVVSGRTPPMPRFQPEPQA
[13]A5GRX7    VTSSLEKVGDQEHGSKLFLINCAGCHGIAAQGVVGPNLHGVSQRKNDSALIQQVISGKTPPMPSFQPDPQS
[14]Q5MZ66    VSTVLNLPGNAERGQAIFQINCAGCHGPEGRGLVGPDLANVSNRKSRKDLIRQVTTGETPPMPKFQPSPET
[15]Q9Z3G2    VSTVLNLPGNAERGQAIFQINCAGCHGPEGRGLVGPDLANVSNRKSRKDLIRQVTTGETPPMPKFQPSPET
[16]B1XNK7    IQQVLALQGDELRGNAIFQINCAGCHGPQADGNVGPSLRAVAQRKSDVRLIQQVISGKTPPMPKFQPAPQE
                      *    *  :* :***.***  . * :.*.*  :: :     :::*:  * ***** *:  *
 Consensus    --------G----G--LF-INCAGCHG--G-G-VGPSL--VS-R-----LIHQV--G-TPPMP-FQ--P--

                   151          pI        Strain
                    |
 [1]Q2JNK7     MADLLSYLKTL----     7.0     Synechococcus sp. JA-2-3B'a(2-13)
 [2]Q2JRK7     MADLLSYLKTL----     6.3     Synechococcus sp. JA-3-3Ab
 [3]Q3AYZ2     MADLLAYLHDLS---     5.3     Synechococcus sp. CC9902
 [4]Q067U7     MADLLAYLHDLS---     5.0     Synechococcus sp. BL107
 [5]Q7U8A2     MADLLSHLHKLS---     6.2     Synechococcus sp. WH8102
 [6]Q3AI93     MADLLAYLHTLS---     5.2     Synechococcus sp. CC9605
 [7]A3Z7S0     MADLLAYLKSLE---     5.7     Synechococcus sp. RS9917
 [8]Q05XQ9     MADLLAHLHSFSDAE     5.0     Synechococcus sp. RS9916
 [9]A4CVQ9     MADLLSYLKTVT---     7.0     Synechococcus sp. WH7805
[10]A5GM57     MADLLSYLKTVT---     7.0     Synechococcus sp. WH7803
[11]Q0IBJ3     MADLLAYLKTLS---     7.0     Synechococcus sp. CC9311
[12]A3YY40     MADLLAHLHSLE----   10.2     Synechococcus sp. WH 5701
[13]A5GRX7     MADLLAYLHELQ---     6.4     Synechococcus sp. RCC307
[14]Q5MZ66     MADLLRYLETL----     8.0     Synechococcus sp PCC6301
[15]Q9Z3G2     MADLLRYLETL----     8.0     Synechococcus sp PCC7942
[16]B1XNK7     MADLLSYLRTL----     8.8     Synechococcus sp. PCC7002
               ***** :*. .
 Consensus     MADLL-YLH-L----
```

served in all the 8 cyt cM sequences from filamentous cyanobacteria (residues 76–78, Table 18.9) but not in all the 18 sequences from non-*Synechococcus* ones (residues 68–70, Table 18.10). The mature portion of all the cyt cM sequences is short, averaging 70–80 amino acids. Apart from sharing a highly conserved heme binding site, all the sequences are characterized by the presence of a conserved

Table 18.9 An alignment of cytochrome cM from filamentous cyanobacteria

```
                      10        20        30        40        50        60        70
                       |         |         |         |         |         |         |
[1]A0YWH9     ----------------------------------------------------------MLLILLWVFGLN
[2]B5W431     MPSTQESIKLYSEQVKFFVDSCIRVALDN--------HIANTHITEIIQRVALVALGLMVVILLWVWGVS
[3]Q3M5W1     --------------------------MDN----------QITKPEILIQRIALVALVILLAIPLGFFGVQ
[4]Q9ZB68     --------------------------MDN----------QITKPEILIQRIALVALVILLAIPLGFFGVQ
[5]B2J1A5     --------------------------MDN----------QITKPEILIQRIVLLTLAILLSVPLGFFGIQ
[6]A0ZBG1     --------------------------MDN----------QITKPETRIQWITLLALVLLLAAPLGIFGVQ
[7]B4W441     ---------------------MAGDPLEN----------QLDQPQVLVRRIIMVVLALLFVIGIGIFGFQ
[8]Q113F0     ----------------------------------------------------------------------

 Consensus    ----------------------------------------------------------------------

                      80  ↓     90       100       110       120       130       140
                       |         |         |         |         |         |         |
[1]A0YWH9     LLQVSDPYIEQVLSLDGNPVKGHAIFEMNCAGCHFQKASDQVGPSLLDVSQRKSPVGIIHQVTSGKTPPM
[2]B5W431     SVKISDPYIQKVLSLTGDPTRGHAIFEMNCAGCHVSQVEHQIGPNLEDISQRKSQISIIQQVVSGKTPPM
[3]Q3M5W1     LVKASDPYVKSVLAMKGDPIQGHAIFQINCAGCHGLEADGRVGPSLQAVSKRKSKYGLIHQVISGDTPPM
[4]Q9ZB68     LVKASDPYVKSVLAMKGDPIQGHAIFQINCAGCHGLEADGRVGPSLQAVSKRKSKYGLIHQVISGDTPPM
[5]B2J1A5     LVQASDPYVKSVLSLTGNPVQGHAIFQINCAGCHGLEADGRVGPSLQAVSKHKSQYGLIHQVISGETPPM
[6]A0ZBG1     MVRSADPYAKNVLSLKGDQVQGHAIFQINCAGCHGLEADGLVGPSLQDVSKRKSRYKLIHQVISGETPPM
[7]B4W441     VLQVSDPYIQKVLSLNGDPVRGQAIFEINCADCHGMHADGKVGPSLQHISKHRSPSGLIEQVISGQTPPM
[8]Q113F0     ---MSDPYSKDVLSLEADPIKGEDIFQINCAGCHKSSMDSQVGPRLEGVSQHKSDLAIIKQVVSGQTPPM
                   ***  .*::: .:  :*. :*  ***.**    .  :** *  :.:::*   :*.** ** ****
 Consensus    -----DPY--SVLSL-GD--QGH-IF--NCAGCH----D--VGP-L--VSKRKS---LIHQV-SG-TPPM

                     150       160   pI     Cyanobacteria
                       |         |
[1]A0YWH9     PQFQPTPQEMADLLSYLQKL   7.0    Lyngbya sp. PCC 8106
[2]B5W431     PQFQPNPQEMADLLNYLEQI   5.0    Arthrospira maxima CS-328
[3]Q3M5W1     PKFQPNTQEMADLLSFLETL   6.8    Anabaena variabilis ATCC 29413/PCC 7937
[4]Q9ZB68     PKFQPNTQEMADLLSFLETL   7.0    Anabaena sp. PCC 7120
[5]B2J1A5     PKFQPSTQEMADLLSYLESL   6.0    Nostoc punctiforme ATCC 29133/PCC 73102
[6]A0ZBG1     PKFQPSVQEMSDLLSYLESL   6.3    Nodularia spumigena CCY9414
[7]B4W441     PKFQPSPQEMADLLTYLEKL   6.0    Microcoleus chthonoplastes PCC 7420
[8]Q113F0     PQFQPSEQEMADLLSYLKKI   5.1    Trichodesmium erythraeum IMS101
              *:***. ***:*** :*: :
 Consensus    PKFQPS-QEMADLL-YLE-L
```

region, Thr-Pro-Pro-Met-Pro, which is about 20 amino acids away from the N-terminus. Tables 18.8 and 18.10 each shows that the mature portions of the cyt cM sequences from unicellular cyanobacteria have pI values ranging from 5 to 10. As for the filamentous cyanobacteria, the pI values have a narrower range from 5 to 7 (Table 18.9).

18.3.2.2 Structural Properties

Cyt cM gene of *Synechocystis* PCC6803 has been cloned and expressed in bacteria (Malakhov et al. 1994). The gene product shows striking differences from both cyt c_6 and cyt c550. The coding sequence from *Synechocystis* PCC 6803 corresponds to 107 amino acids including a Cys-xx-Cys His heme binding site at residue 49–53 out from the N terminus. There are 2 methionines at residues 82 and 91, one of which can be a sixth ligand to the heme. A hydrophobic patch is centered at residue 30. Malakhov et al. (1994) commented that the basic cyt c552 of *Anacystis nidulans* with a molecular mass of 9.0 kDa and a pI value of 7.0 was most similar to the cyt cM gene product which had an 8.3 kDa mass and a pI value of 7.3. After a long pause, cyt 552 became cyt cM and was known in much better detail. Six year later,

Table 18.10 An alignment of cytochrome cM from non-Synechococcus, unicellular cyanobacteria

	10 20 30 40 50 60 70
[1]A2C850	--------------------------MTGPSSN--NQSKNGLIGALVVLVAAVCIAILLWVLGHSQRDPY
[2]Q7TUV4	------------------MFHAPELTVTGPSSN--NQRKNGFIGPLVVLAAAVCIAILLWVLGHSQRDPY
[3]B5INH8	--------------------------MNSQIVQ--PAMPRGLVAALVLAAATACVVLALLVLPAARTDPY
[4]B7KHU8	--------------------------MDSQLSQ--PKALASRLMIAFVAIALIMAIIILGIYISRVSDPY
[5]B4B4R2	--------------------------MDSQLTQ--PKVLVSRLIIAVIAIGAIVAVIFLGLYISRVYDPY
[6]B0JQE2	--------------------------MNSQLFK--PKFGLYRFLMVIIVVLFLISTYLITFHLAKMSDPY
[7]B1WY28	------------------------MTVNEELIQ--PKDAIQRLILIILGMIFLAIVVIVGWYLHQISDPY
[8]A3IXY6	--------------------------MTEQLIQ--PKDAIQRIILIILGMIFLAIIIVVGWYLHEISDPY
[9]Q4C7D0	MIGLTFYSQKIAKCKEKDNTIPPEIIVTEQLIQ--PKDAIQRVILSLLGIIFLAIVVVVGWYFHQISDPY
[10]B4C4X3	--------------------------MTDQFAQ--RKDGIGRFALTLLGIIIIALVLVMGWYLHQLSDPY
[11]B8HUQ6	---MLIALAIGLVTGLTLLGIHQVQMSDPY
[12]B0CC81	--------------------MEDPALKSDQTLN--PGLSIQKALLVVLGWLLVIAVIIGVVYQARLSDPY
[13]P42351	--MAGIVSLVILAVALFSFMN---FDPY
[14]A2C3M6	-----------------------MNIPSSKALIK-ENQEISSWRIFLLTSVTLILLILFWRMGDLKQDPF
[15]Q46JZ2	-----------------------MNLPSSKALIK-ENQEISSWRIFLLTSATLILLIFFWRMGDFKQDPF
[16]Q31AG8	-----------------------MSTSSSSAAE--RDIKREFFKKFFIVFIVSLLCFSIFFLNNHENNKY
[17]A3PDG1	---MICFSIIFVNHHDNNKY
[18]A9BB52	-----------------------MKRTSPIGIEDKRNPTKGLLALIIISATSCLIILSWLFIYLEKGSPY
	. :
Consensus	---D-Y

	↓ 80 90 100 110 120 130 140
[1]A2C850	INATLDLEGSLEQGGRLFRINCAGCHGITAQGNLGPNLLDVSERRNDAQLIRQVVSGNTPPMPRFQLEPQ
[2]Q7TUV4	VNATLDLEGSLEQGGRLFRINCAGCHGITAQGNLGPNLLDVSERRNDAQLIRQVVSGNTPPMPRFQLEPQ
[3]B5INH8	TRQTLELRGSVEHGARLFRLNCAGCHGIAAQGLVGPDLHGVSQRKSQRQLIQQVVSGRTPPMPRFQPEPQ
[4]B7KHU8	IQEVLTLEGDLARGYEIFQLNCAGCHGQQADGSVGPSLHHVPKRKSKVRLIEQVISGKTPPMPKFQPSAQ
[5]B4B4R2	VQEVLSLKGDLARGYEIFQINCAGCHGQQADGSVGPSLHHVPKRKSKVRLIEQVISGDTPPMPKFQPSPQ
[6]B0JQE2	IKEVLSLQGNVSRGYEIFQINCAACHGQFADGIVGPSLEDVSHRKSRISLIEQVISGKTPPMPKFQPDTQ
[7]B1WY28	IHEVLSRQGDVSRGQAIFEINCAGCHGLEADGIVGPSLHHVQKHKSKISLIQQVISGNTPPMPKFQPSPQ
[8]A3IXY6	INEVLSRQGDVSRGQAIFEINCAGCHGIEADGIVGPSLHRVQKHKSKVSLIQQVISGKTPPMPKFQPSSQ
[9]Q4C7D0	MGEVLSLQGDVTRGQAMFEINCAGCHGLNADGIVGPSLHHVQQHKSKISLIKQVTSGKTPPMPKFQPNPQ
[10]B4C4X3	VKDVLSLKGNITQGEAIFQINCAGCHGLKADGSVGPSLHHVHQHKSKISLINQVISGKTPPMPKFQPTAQ
[11]B8HUQ6	VRQVMELVGDPEQGKAIFVMNCSGCHGLDGAGEVGPSLWRVSDRRSQISLIHQITSGRTPPMPKFQASPA
[12]B0CC81	VQDVLHHEGTVAQGNAIFQLNCSGCHGTSGDGKVGPSLRRVTTHRSQVGLIYQITSGKTPPMPKFQANPQ
[13]P42351	VSQVLALKGDADRGRAIFQANCAVCHGIQADGYIGPSLWGVSQRRSQSHIIHQVVSGQTPPMPQFEPNPQ
[14]A2C3M6	ITETLSLQGESLSGSKLFKINCVGCHGISAQGFVGPDLHEATQEMSDKKIINQVIRGLTPPMPSFEIEPQ
[15]Q46JZ2	ITETLSIQGDSLSGSKLFKINCVGCHGISAQGFVGPDLHEATQEMSDKKIINQVIRGLTPPMPSFEIEPQ
[16]Q31AG8	IIDTLELNGSAKEGDALFKINCVGCHGITARGLVGPDLHSITQRLNDKEIIKQVTGGLTPPMPSFEIDPV
[17]A3PDG1	IIETLELNGSVEEGDALFKINCVGCHGITARGLVGPDLHSITQRLNDKEIIKQVTGGLTPPMPSFEIDPV
[18]A9BB52	IQNSLELNGSIEKGGNLFRMNCVGCHGISAQGLVGPDLNNVTNQLSDERIINQVIKGLTPPMPSFEMDPQ
	: * * :* ** *** . * :**.* . . :* *: * ***** *: .
Consensus	----L---G----G--LF--NC--CHG--A-G-VGPDL-----R-S---II-QV--G-TPPMP-FE--P-

	150	pI	Cyanobacteria
[1]A2C850	EMADLLAYLNSLK--	5.7	*Prochlorococcus marinus* MIT 9303
[2]Q7TUV4	EMADLLAYLNSLN--	5.1	*Prochlorococcus marinus* MIT 9313
[3]B5INH8	AMADLLAYLNGLV--	10.0	*Cyanobium* sp. PCC 7001
[4]B2F479	EMADLLTYLEGL---	8.0	*Cyanothece* sp. PCC 7424
[5]B4B4R2	EMADLLNYLERL---	8.0	*Cyanothece* sp. PCC 7822
[6]B0JQE2	AMADLLVYLENISKN	6.1	*Microcystis aeruginosa* NIES-843
[7]B1WY28	DMADLLSYLEKL---	7.1	*Cyanothece* sp. ATCC 51142
[8]A3IXY6	EMADLLSYLEKL---	8.7	*Cyanothece* sp. CCY 0110
[9]Q4C7D0	EMADLLIYLEGL---	7.1	Crocosphaera Watsonii 8501
[10]B4C4X3	EMADLLSYLESL---	8.0	*Cyanothece* sp. PCC 8802
[11]B4CAX3	EMADLLSFLNTL---	6.0	*Cyanothece* sp. PCC 7425
[12]B0CC81	EMADLLEYLKTL---	8.8	Acaryochloris marina MBIC 11017
[13]P42351	EMADLLNYLKTL---	7.0	*Synechocystis* sp. PCC 6803
[14]A2C3M6	SMADLLEYMHSLN--	5.1	*Prochlorococcus marinus* NATL1A
[15]Q46JZ2	SMADLLEYMHSLN--	5.1	*Prochlorococcus marinus* NATL2A
[16]Q31AG8	NMSNLLKYLHSLE--	6.3	*Prochlorococcus marinus* MIT 9312
[17]A3PDG1	NMSNLLKYLHSLE--	5.5	*Prochlorococcus marinus* MIT 9301
[18]A9BB52	SMADLLSYLHSLN--	5.0	*Prochlorococcus marinus* MIT 9211
	*::** ::. :		
Consensus	-MADLL-YLH-L---		

Cho et al. (2000) prepared and expressed the cyt cM of *Synechocystis* PCC6803 in *Escherichia coli*. They reported that cyt cM had an alpha peak at 550 nm and a measured pI value of 5.6 which was lower than the calculated one of 7.2. In 2002, Dr. Whitmarsh gave the cyt cM gene to Dr. Krogmann who hoped to provide purified cyt cM to Dr. Kerfeld for crystallography. For cyt cM purification, an extract of pink cells was loaded on a DEAE column and eluted with a gradient of increasing NaCl concentration. There appeared two distinct bands, one the oxidized and the other the reduced cyt cM. This cyt is autoxidizable and affords a special opportunity in purification. The extract can be oxidized by potassium ferricyanide and run through the column. The eluted cyt can be reduced by dithionate and run through the column a second time. The result is a pure cyt with an ultraviolet peak that is lower than the reduced alpha peak. Cyts c552 and c554 crystallize at this level of purity. However, the recovery of cyt cM was poor. Each trip through the column gave a 50% loss of the cyt. The few milligrams of cyt cM left would be enough. Unfortunately the cyt's color disappears in a vast array of solvents and even in crystallization attempts done under argon. It is only stable while frozen. One fears another long pause in cyt cM research. Dr. Krogmann found a band of cyt with a cyt cM spectrum in the extract of *L. wollei*. It was low in concentration and was weakly held on DEAE indicating a nearly neutral pI. The cyt faded away as it was eluted from the column. This behavior is a unique identifier of cyt cM.

Because cyt cM has been refractory to crystallization due to the instability of the protein, we constructed a structural model of the *A. maxima* cyt cM using automated threading (Schwede et al. 2003); the algorithms choose horse-heart cyt c as the best template, despite only 21.4% sequence identity between the two proteins. The N-terminal residue of the model (Fig. 18.1b) is residue 86 in Table 18.9. Because of the lack of predicted secondary structure in the region of residues near the heme, it is not possible to ascertain which amino acid is the sixth axial ligand to the heme (there are two conserved Met residues in this region, Table 18.9) but neither is predicted to be close to the heme in the structure (Fig. 18.3b). Earlier, Shuvalov et al. (2001) published spectroscopic studies on *Synechocystis* PCC 6803 cyt cM. They found, in whole cell studies, that cyt cM was oxidized by light absorption of PSI in cells acclimated to low temperature. This suggests that cyt cM might be donating electrons to photo-oxidized P700. The cyt cM is synthesized under conditions of low temperature stress. Under sulfur deprivation, in contrast to other electron carriers, the expression of cyt cM is observed to increase (Zhang et al. 2008). Interestingly, most of the other genes that are expressed at a late stage in sulfur deprivation conditions are hypothetical proteins of unknown function.

There is one more cyt that is vaguely related by name to cyt cM. Fier et al. (1999) reported a soluble cyt 552 and its participation in sulfur metabolism of *Oscillatoria* strain Bo 32. The cells were grown in artificial sea water and could be grown under conditions of anaerobic hydrogen sulfide in a gas tight flask in the culture medium at 1 mM hydrogen sulfide. The purification and characterization is well done. The unique property is a molecular weight of 23,500 Da as determined by SDS gel electrophoresis. At this time, there is genome information for at least two members of the *Oscillatoriales*, *T. erythraeum* IMS101 and *A. maxima* CS-328. Finally there is a well done paper by Molina-Heredia et al. (2002) describing a flash induced kinetic analysis of PSI reduction by plastocyanin, cyt c_6 and cyt cM. The bimolecular rate constants are one hundred times lower with cyt cM and the redox and surface electrostatic potential are distant from those of cyt cM, cyt c_6 and plastocyanin. Cyt cM is not a likely participant in PSI reduction.

In a recent paper, Bernroitner et al. (2009) reported that cyt cM and cyt c_6 were more efficient electron carriers to cyt oxidase when compared with plastocyanin.

18.3.3 Cytochrome c550

18.3.3.1 Amino Acid Sequences

Table 18.2 has 20 cyt c550 sequences, 18 of which are deduced from *Synechococcus* gene sequences. The remaining 2, number 4 and 5, are deduced from gene sequences of *T. elongatus* and *T. vulcans*, respectively (Katoh et al. 2001). We included these two thermophiles which are closely related to the *Synechococcus*. It is interesting to note that only one strain, PCC7335, has 2 cyt c550 sequences which have about 54% identity. These two sequences are more similar to the other cyanobacterial cyt c550 sequences than they are to cyt c550-like (PsbV2) protein sequences that will be discussed later, suggesting that these two c550 paralogs reflect a relatively recent gene duplication. The length of all the mature sequences is about 137 amino acids, except for the ones from 2 strains, JA-3–3Ab and JA-2–3B'a(2–13), which are about 143. Table 18.3 consists of 14 sequences from a collection of non-*Synechococcus* cyanobacteria and each has a similar length of 137–138 amino acids. The one from *Microcystis aeruginosa* (#4) was obtained by protein sequencing (Cohn et al. 1989). There are 2 cyt c550 gene sequences from *G. violaceus* PCC7421 that have very low sequence identity when compared with the other cyt c550 sequences. Table 18.4 has 8 sequences from filamentous cyanobacteria and all have about 137 amino acids. Interestingly, all the cyt c550 sequences were predicted to have an acidic pI value. From protein work, cyt c550 is an acidic protein which is slightly larger than cyt c_6 (100 amino acids) but is likely diverged from the same gene.

18.3.3.2 Structural Properties

Despite only 32% sequence identity, cyts c_6 and c550 of *A. maxima* have remarkable structural similarity (Fig. 18.1a, c); the two proteins superimpose over 263 backbone atoms with an root-mean-square deviation of only 0.7 Å (Sawaya et al. 2001). This is especially interesting in the context of their midpoint potentials which differ by almost 600 mV. Through a structural comparison of cyts c_6 and c550, several differences can be identified that may be responsible for tuning E_m in these proteins. In addition to a difference in the type of amino acid providing the sixth axial ligand to the heme (His in cyt c550, Met in cyt c_6), there are striking differences in the hydrogen partners to the fifth axial ligand (Sawaya et al. 2001) and the environment of the propionate oxygen atoms (Fig. 18.3a, c). In addition, the heme in cyt c550 was significantly more solvent exposed in the structure, which is likewise expected to favor a lower E_m. Also, the *A. maxima* cyt c550 formed a dimer in the crystal that is based on structural arguments as well as a calculation of the binding energy of dimerization (Kerfeld, unpublished), would be expected to be stable in solution. The dimerization reduces the solvent exposure of the heme, but introduces a Glu side-chain (Glu 90 in Fig. 18.3c) into the heme environment; both of these alterations would be expected to affect the midpoint potential. The co-existence of monomer and dimeric forms of cyt c550 could explain why two redox potentials have been measured for it (Hogansen et al. 1990). Although the association of monomeric cytochrome c550 as a component of PSII is well known, several functions for soluble (dimeric?) cytochrome c550 in cyanobacteria have been proposed including participating in cyclic photo-phosphorylation (Kienzel and Peschek 1983), oxidation of NADPH (Pulich 1977) or reduction of hydrogenase (Krogmann 1991; Morand et al. 1994). Furthermore, the abundance of cyt c550 appears to vary with growth conditions, being especially abundant in natural blooms. The distribution of cyt c550 in cyanobacterial genomes also adds to the enigma of its function; although this highly conserved protein is found in most cyanobacterial genomes (Tables 18.2, 18.3 and 18.4) the psbV gene which encodes this cyt is not found in any of the *Prochlorococcus* species with the exception of two low-light adapted strains, MIT9303 and MIT9313.

18.4 Cyt c550-like Protein

In a subset of the cyanobacterial genomes sequenced to date, there is an additional homolog sequence to cyt c550, *psbV2*. This homolog was found in both unicellular and filamentous cyanobacteria (Table 18.11). In these organisms the PsbV2 gene is located immediately downstream of the gene for cyt c550. The amino acid sequence of psbV2 is most closely related to cyt c550 (psbV2 and cyt c550 of *T. elongatus* are 43.9% identical). Small amounts of this protein can be purified from *T. elongatus* (A. Boussac, personal communication) and the midpoint is apparently extremely low (Kerfeld et al. 2003). A structural model for residues 11–129 of psbV2, based on the structure of *T. elongatus* c550, is shown in Fig. 18.1d. Based on EPR data and the structural model, the presumed sixth axial ligand to the heme was proposed to be Tyr 96 but this remains speculative. The heme in the psbV2 model appears to be more solvent accessible than the heme in cyt c550, consistent with its lower midpoint potential (Fig. 18.3c, d).

Table 18.11 An alignment of cytochrome c550-like protein from different cyanobacteria

```
                     10        20        30        40        50        60        70
                      |         |         |         |         |         |         |
[1]Q8DJE2    AAGVDNYVIQYLKVTDTVELPVNDRGETKTFTAVDLTRGKRLFEENCKNCHVGGSTLPNPLVSLSLKDLKGATPP
[2]Q9ET95    AAGVDNYVIQYLKVTDTVELPVNDRGETKTFTAVDLTRGKRLFEENCKNCHVGGSTLPNPLVSLSLKDLKGATPP
[3]B8HUH8    -ARVDEYVYRFL-AKEPVTIPINSAGETQLFSPEELTLGKQLFDQNCKNCHVGGATLPNPAESLALADLRAANPP
[4]B5W3Z0    AQNVDPYVLRYL-TDQPVEIQSDRQGGTQTFSVEDLSAGKVLFTDSCLNCHAGGVTIPYPTVSLSLEDLKGATPP
[5]A0YIA1    ADGVDPYVRRYL-ASQPVEIPADNQGHTQQFLPEDINTGKMLFEASCLNCHAGGVNIPYPTVSLSLEDLKAATPA
[6]Q111E8    -AAVDSYVKRYLDAEIPVGIKLNKQGELKNFSAEDLSEGKQTFAKNCLNCHVGGANLVNPSVSLSLEKLKGATPP
[7]B4VPE3    AYAIDDYVKRFLKVTEPIALDLDGQGQTKAFTPGDLAEGKELFKTHCINCHVGGTTLQDPRVSLGLETLAGATPP
[8]B0C732    AGDIEDYIVRYLKVTEPVPIVMDASGETQLFSPEELSDGKRLFDDNCKNCHVGGATLPNPVQSLSMKNLQGATPS
[9]Q2JTY7    AGRIDPYVSHYLRTTEPIELPWDAEGHMLTFTPEQLTDGKNRFQSACLNCHVGGSTLPAPNISLSLKDLRGATPP
[10]Q2JIS7   AARIDPYVNQYLRVSGPVELPLDAAGHTLTFTPEQLTDGKNRFQSACLNCHVGGATLPAPNISLSLKDLRGATPP
                :: *: ::* .  .: :  :  *    *   ::  **  *   * ***.** .:
Consensus:   ---VD-YV-RYL-V--PV-L--D--G----F---DL--GK--F---C-NCHVGG-TL--P--SLSL--L-GATPP

                       85        95       105       115       125       135       145
                        |         |         |         |         |         |         |
[1]Q8DJE2     RDTIASLVAFQRSPKSYDGSEESYSCRRVSEDWLTTEQLETLAAFILRAAAVAPGWGVESFPDSAP-----
[2]Q9ET95     RDTIASLVAFQRSPKSYDGSEESYSCRRVSEDWLTTEQLETLAAFILRAAAVAPGWGVESFPDSAP-----
[3]B8HUH8     RDNVAALVAYQRSPMLQDGSELSYLCRRVSEDWLTNDQLQDLAAFLLRAAEKSPGWGTDALELSF------
[4]B5W3Z0     RDTIKDLVAYMRHPVSYDGSDTNYWCREIPENWLDDAEIDNLAAFILQSAIKAPGWAKK------------
[5]A0YIA1     RDNIKNLVSYFRYPLSFDGSDTNYWCRQISESWMSQVQAEKLAAYIIRSAQAAPGWGKTQEANELPPEFRM
[6]Q111E8     RDDLNNLVAFLRDPMIYDGSSYTLFCRQITENWMSQQEVENIAAFILRAAQKAPYWGVENVR---------
[7]B4VPE3     RDNVDGIISFLREPMTYDGSAPSFWCRQVPDTWMPQEQIEKLAAFVLRAAQKAPGWGGKASS---------
[8]B0C732     RDNVANLIEFQREPTIYDGSDVSFWCRQVPDTWLPDKELANLEAFILRAAEVARGWGTEDVADGIDFT---
[9]Q2JTY7     RDTIQALVEYQRDPRSYDGTEVSYGCRPVPPSWMDDEALRNLAAFILRAAQVAPGWGSNAIGGG-------
[10]Q2JIS7    RDTLQALMEYQRDPRSYDGSDSSYGCRPVPPSWMDDEAVN-------YPAPTAKRYGAGFQ----------
              ** :  :: : * *   **:  .  ** :.  *:              .*  :  :.
Consensus:    RD-I--LV-2-R-P---DGS--S--CR-VP--WM--------------AA--A--WG--------------
```

[1]Q8DJE2, *Synechococcus elongates*;[2]Q9ET95, *Thermosynechococcus vulcanus*; [3]B8HUH8, *Cyanothece sp. PCC 7425*; [4]B5W3Z0, *Arthrospira maxima* CS-328; [5] A0YIA1, *Lyngbya sp.* PCC 8106,[6]Q111E8, *Trichodesmium erythraeum* IMS101;[7]B4VPE3, *Microcoleus chthonoplastes* PCC 7420; [8]B0C732, *Acaryochloris marina* MBIC 11017;[9]Q2JTY7, *Synechococcus sp.* JA-3-3Ab; [10]Q2JIS7, *Synechococcus sp.* JA-2-3B'a(2-13). 2 =Y/F.

18.5 A Possible Expression of Silent Genes

Some years ago, Krogmann had obtained a few small samples—5 to 10 gm wet weight of the *L. wollei*, that were handfuls taken from fresh water lakes. The filaments are weakly attached to the lake bottom and tend to intertwine to form bundles. The cells are four times larger than most bacteria and are disc shaped. They are sheathed in a polysaccharide layer and they contain gas vesicles whose buoyancy raises the filaments in vertical columns up to 5 ft lengths. The cells in the small samples were very hard to break but the small amounts of extract obtained were put through a fractionation ending in column chromatography and there were too many bands of cyt c_6. One can easily harvest large amounts of *L. wollei*, wash it with a garden hose and dry it in the sun. Caution is needed since some strains have toxins that can be brushed off to skin and cause painful inflammation (Carmichael et al. 1997). Our collections at two sites proved harmless. Dr. Jerry Brand found a lake near Marble Falls, Texas that had an abundance of *L. wollei* and large collections were made. In the year of bloom appearance, the bundles reached lengths of 5 ft and were green. Microscopic examination of the washed filaments revealed no other bacteria. In the second year, the bundles were blue-green with some areas of easily washed off black and a few black mats were on the bottom. There was a strong smell of hydrogen sulfide which is often found near *L. wollei* due to the presence of *Desulfovibrio* bacteria. When washed, these filaments showed no sign of other bacteria, no odor of hydrogen sulfide and no sign of *Desulfovibrio* cyts which have spectra that are distinctive from cyanobacterial cyts. When the water cools in the fall, the bundles form thick mats that are completely blackened with sulfides that don't wash off. The mats coil into cylinders and float to the surface. In the next year the bloom is more abundant, even after ice layers cover the surface in northern locations. The green cells show only cyt c_6I after extraction and chromatography as do the thoroughly blackened cells in the late second year cylinders. The collection made between these two stages showed three bands of cyt c_6. Cyt c_6I was the most abundant and the other two were present at about 5% of the concentration of c_6I. In the early green cells, photosynthesis is driving the growth and c_6I might well be reducing PSI. In the second phase, phycocyanin has increased and the hydrogen sulfide could be used for anoxygenic photosynthesis in which hydrogen sulfide supplies electrons to drive into PSI as described by Cohen et al. (1975). Cyts c_6II and c_6III appear and might contribute to the anoxygenic process. In the third phase, when there is no light reaching the cells, energy could be supplied by cyt c_6I through respiration. Obinger et al. (1990) demonstrated that cyt c_6I is the most efficient electron carrier to the plasma membrane cyt c oxidase.

Acknowledgements The work of CAK is performed under the auspices of the US Department of Energy's Office of Science, Biological and Environmental Research Program, and by the University of California, Lawrence Berkeley National Laboratory under contract number DE-AC02–05CH11231, and Lawrence Livermore National Laboratory under contract number DE-AC52–07NA27344.

References

Abriata LA, Cassina A, Tórtora V, Marín M, Souza JM, Castro L, Vila AJ and Radi R (2009) Nitration of solvent-exposed tyrosine 74 on cytochrome c triggers heme iron-methionine 80 bond disruption: nuclear magnetic resonance and optical spectroscopy studies. J Biol Chem 284: 17–26

Aitken A (1976) Protein evolution in cyanobacteria. Nature 263: 793–796

Aitken A (1977) Purification and primary structure of cytochrome f from the cyanobacterium, *Plectonema boryanum*. Eur J Biochem 78: 273–279

Aitken A (1979) Purification and primary structure of cytochrome c-552 from the cyanobacterium, *Synechococcus* PCC 6312. Eur J Biochem 101: 297–308

Bernroitner M, Zamocky M, Pairer M, Furtmüller PG, Peschek GA and Obinger C (2008) Heme-copper oxidases and their electron donors in cyanobacterial respiratory transport. Chem Biodivers 5: 1927–1961

Bernroitner M, Tangl D, Lucini C, Furtmüller PG, Peschek GA and Obinger C (2009) Cytochrome cM: probing its role as electron donor for CuA of cyt c oxidase. Biochim Biophys Acta 1787: 135–143

Bialek W, Nelson M, Tamiola K, Kallas T and Szczepaniak A (2008) Deeply branching c6-like cytochromes of cyanobacteria. Biochemistry 47: 5515–5522

Bovy A, De Vrieze G, Borrias M and Weisbeek P (1992) Transcriptional regulation of the plastocyanin and cytochrome c-553 genes from the cyanobacterium *Anabaena* species PCC 7937. Mol Microbiol 6: 1507–1513

Bushnell GW, Louie GV and Brayer GD (1990) High-resolution three-dimensional structure of horse heart cytochrome c. J Mol Biol 214: 585–595

Carmichael WW, Evans WR, Yin QQ, Bell P and Moczydlowski E (1997) Evidence for paralytic shellfish poisons in the freshwater cyanobacterium *Lyngbya wollei* (Farlow ex Gomont) comb. nov. Appl Environ Microbiol 63: 3104–3110

Cho YS, Pakrasi HB and Whitmarsh J (2000) Cytochrome cM from *Synechocystis* 6803. Detection in cells, expression in *Escherichia coli*, purification and physical characterization. Eur J Biochem 267: 1068–1074

Clark AK and Campbell D (1996) Inactivation of the petE gene for plastocyanin lowers photosynthetic capacity and exacerbates chilling-induced photoinhibition in the cyanobacterium *Synechococcus*. Plant Physiol 112: 1551–1561

Cohen Y, Padan E and Shilo M (1975) Facultative anoxygenic photosynthesis in the cyanobacterium *Oscillatoria liminetica*. J Bact 123: 855–861

Cohn CL, Sprinkle JR, Alam J, Hermodson M, Meyer T and Krogmann DW (1989) The amino acid sequence of low-potential cytochrome c550 from the cyanobacterium *Microcystis aeruginosa*. Arch Biochem Biophys 270: 227–235

DeLano WL (2002) The PyMOL Molecular Graphics System. California, USA: DeLano Scientific, San Carlos

Duran RV, Hervas M, De la Rosa M and Navarro JA (2004) The efficient functioning of photosynthesis and respiration in *Synechocystis* sp. PCC6803 strictly requires the presence of either cytochrome c6 or plastocyanin. J Biochem 279: 7229–7333

Fier I, Rethmeier J and Fischer U (1999) Molecular properties of soluble cytochrome c-552 and its participation in sulfur metabolism of *Oscillatoria* strain Bo32. In: Peschek GA, Loeffelhardt W and Schmetterer G (eds) The Phototrophic Prokaryotes, pp 275–280, Kluwer Academic, New York

Gasteiger E, Gattiker A, Hoogland C, Ivanyi I, Appel RD and Bairoch A (2003) ExPASy: the proteomics server for in-depth protein knowledge and analysis. Nucleic Acids Res 31: 3784–3788

Ghassemian M, Wong B, Ferreira F, Markley JL and Straus NA (1994) Cloning, sequencing and transcriptional studies of the genes for cytochrome c-553 and plastocyanin from *Anabaena* sp. PCC 7120. Microbiology 140: 1151–1159

Hogansen CW, Lagenfelt G, Andréasson L-E and Vanngard T (1990) EPR spectra of cytochrome c549 of *Anacystis nidulans*. In: Baltscheffsky H (ed) Current Research in Photosynthesis, Vol 3, pp 319–322, Kluwer Academic, New York

Ho KK (2005) Cytochrome c_6 genes in cyanobacteria and higher plants. In: Pessarakli M (ed) Handbook of Photosynthesis, 2nd ed, pp 273–284, Marcel Dekker, New York

Ho KK and Krogmann DW (1984) Electron donors to P700 in cyanobacteria and algae. An instance of unusual genetic variability. Biochim Biophys Acta 766: 310–316

Holton RW and Myers J (1963) Cytochromes of a blue-green alga: extraction of a c-type cytochrome with a strongly negative redox potential. Science 142: 234–235

Holton RW and Myers J (1967a) Water-soluble cytochromes from a blue-green alga. I. Extraction, purification and spectral properties of cytochromes C (549, 552, and 554, *Anacystis nidulans*). Biochim Biophys Acta 131: 362–374

Holton RW and Myers J (1967b) Water-soluble cytochromes from a blue-gree alga. II. Physicochemical properties and quantitative relationships of cytochromes C (549, 552, and 554 *Anacystis nidulans*). Biochim Biophys Acta 131: 375–384

Jung YS, Yu L and Golbeck JH (1995) Reconstitution of iron-sulfur center FB results in complete restoration of $NADP^+$ photoreduction in Hg-treated photosystem I complexes from *Synechococcus* sp. PCC 6301. Photosynth Res 46: 249–255

Katoh H, Itoh S, Shen JR and Ikeuchi M (2001) Functional analysis of psbV and a novel c-type cytochrome gene psbV2 of the thermophilic cyanobacterium *Thermosynechococcus elongatus* strain BP-1. Plant Cell Physiol 42: 599–607

Kerfeld CA, Ho KK and Krogmann DW (1998) The cytochromes c of cyanobacteria. In: Peschek GA, Loeffelhardt W and Schmetterer G (eds) The Phototrophic Prokaryotes, pp 259–268, Plenum Press, New York

Kerfeld CA, Sawaya MR, Krogmann DW and Yeates TO (2002) Structure of cytochrome c_6 from *Arthrospira maxima*: an assembly of 24 subunits in a nearly symmetric shell. Acta Crystallogr D58: 1104–1110

Kerfeld CA, Sawaya MR, Bottin H, Tran KT, Suqiura M, Cascio D, Desbois A, Yeates TO Kirilovsky D and Boussac A (2003) Structural and EPR characterization of the soluble form of cytochrome c-550 and of the psbV2 gene product from the cyanobacterium *Thermosynechococcus elongatus*. Plant Cell Physiol 44: b697–b706

Kienzel F and Peschek GA (1983) Cytochrome c-549: an endogenous cofactor of cyclic photophosphorylation in the cyanobacterium *Anacystis nidulans*. FEBS Lett 162: b76–b80

Krogmann DW (1991) The low-potential cytochrome c of cyanobacteria and algae. Biochim Biophys Acta 1058: 35–37

Lee I, Salomon AR, Yu KB, Doan JW, Grossman LI and Hutterman M (2006) New prospects for an old enzyme: mammalian cytochrome c is tyrosine-phosphorylated in vivo. Biochemistry 45: 9121–9128

Lukat P, Hoffmann M and Einsle O (2008) Crystal packing of the c_6-type cytochrome OmcF from Geobacter sulfurreducens is mediated by an N-terminal strep-tag II. Acta Crystallogr D 64: 919–926

Malakhov MP, Wada H, Los DA, Sakamoto VE and Murata N (1994) A new type of cytochrome c from *Synechocystis* PCC 6803. J Plant Physiol 144: 259–264

Malakhov MP, Malakhova OA and Murata N (1999) Balanced regulation of expression of the gene for cytochrome cM and that of genes for plastocyanin and cytochrome c6 in *Synechocystis*. FEBS Lett 448: 281–284

Molina-Heredia FP, Balme A, Hervás M, Navarro JA and De la Rosa MA (2002) A comparative structural and functional analysis of cytochrome cM, cytochrome c6 and plastocyanin from the cyanobacterium *Synechocystis* sp. PCC 6803. FEBS Lett 517: 50–54

Morand LZ, Cheng RH and Krogmann DW (1994) Soluble electron transfer catalysts of cyanobacteria. In: Bryant DA (ed) The Molecular Biology of Cyanobacteria, Vol 1, pp 243–249, Kluwer Academic, Dordrecht

Nomura C and Bryant DA (1998) Cytochrome c6 from Synechococcus PCC 7002. In: Peschek GA, Loeffelhardt W and Schmetterer G (eds) The Photorophic Prokaryotes, pp 269–273, Kluwer Academic, New York

Obinger C, Knepper JC, Zimmermann U and Peschek GA (1990) Identification of a periplasmic c-type cytochrome as electron donor to the plasma membrane-bound cytochrome oxidase of the cyanobacterium *Nostoc* Mac. Biochem Biophys Res Commun 169: 492–501

Pils D and Schmetterer G (2001) Characterization of three bioenergetically active respiratory terminal oxidases in the cyanobacterium Synechocystis sp. strain PCC 6803. FEMS Microbiol Lett 203: 217–222

Pulich W Jr (1977) Cytochrome c548 in Nostoc sp. (Cyanophyceae): an electron acceptor from reduced NADP in the dark. J Phycol 13: 40–45

Sandmann G (1986) Formation of plastocyanin and cytochrome c-553 in different species of blue-green algae. Arch Microbiol 145: 76–79

Sandmann G and Boger P (1980) Copper induced exchange of plastocyanin and cytochrome c553 in cultures of *Anabaena variabilis*. Plant Sci Lett 17: 417–424

Sandmann GH, Kessler RE and Boger P (1983) Distribution of plastocyanin and soluble plastidic cytochrome c in various classes of algae. Arch Microbiol 134: 23–27

Sawaya MR, Krogmann DW, Serag A, Ho KK, Yeates TO and Kerfeld CA (2001) Structures of cytochrome c-549 and cytochrome c6 from the cyanobacterium *Arthrospira maxima*. Biochemistry 40: 9215–9225

Schwede T, Kopp J, Guex N and Peitsch MC (2003) SWISS-MODEL: an automated protein homology-modeling server. Nucleic Acids Res 31: 3381–3385

Shuvalov VA, Allakhverdiev SI, Sakamoto A, Malakov M and Murata N (2001) Optical study of cytochrome cM formation in *Synechocystis*. IUBMB Life 51: 93–97

Thompson JD, Higgins DG and Gibson TJ (1994) CLUSTAL W: improving the sensitivity of progressive multiple sequence alignment through sequence weighting, position-specific gap penalties and weight matrix choice. Nucleic Acids Res 22: 4673–4680

Ulrich EL, Krogmann DW and Markley JL (1982) Structure and heme environment of ferrocytochrome c553 from 1H NMR studies. J Biol Chem 257: 9356–9364

Wallace AC, Laskowski RA and Thornton JM (1995) LIGPLOT: a program to generate schematic diagrams of protein-ligand interactions. Protein Eng 8: 127–134

Weigel M, Varotto C, Pesaresi P, Finazzi G, Rappaport F, Salamini F and Leister D (2003) Plastocyanin is indispensible for photosynthic electron flow in *Arabidopsis thaliana*. J Biol Chem 278: 31286–31289

Wood PM (1978) Interchangeable copper and iron proteins in algal photosynthesis. Studies on plastocyanin and cytochrome c-552 in *Chlamydomonas*. Eur J Biochem 87: 9–19

Worrall JA, Schlarb-Ridley BG, Reda T, Marcaida MJ, Moorlen RJ, Wastl J, Hirst J, Bendall DS, Luisi BF and Howe CJ (2007) Modulation of heme redox potential in the cytochrome c6 family. J Am Chem Soc 129: 9468–9475

Zhang L, McSpadden B, Pakrasi HB and Whitmarsh J (1992) Copper-mediated regulation of cytochrome c553 and plastocyanin in the cyanobacterium *Synechocystis* 6803. J Biol Chem 267: 19054–19059

Zhang Z, Pendse ND, Phillips KN, Cotner JB and Khodursky A (2008) Gene expression patterns of sulfur starvation in *Synechocystis* PCC 6803. BMC Genomics 9: 344

Chapter 19
Transient Interactions Between Soluble Electron Transfer Proteins. The Case of Plastocyanin and Cytochrome *f*

Derek S. Bendall, Beatrix G. Schlarb-Ridley and Christopher J. Howe

19.1 Introduction

Electron transfer is the central process of biological energy metabolism, in which energy is provided by exploitation of the energy potential of non-equilibrium situations in the environment. By far the most important processes by which organisms do this are respiration and photosynthesis, both of which depend on a series of electron transfers to make energy available to the cell in a chemically useful form. There are two outstanding features of these reactions: they must be fast, and they depend upon a chain of reactions so that each component, frequently but not invariably a protein, has to react with at least two other components. The biological response to the latter requirement depends on either a fixed complex of proteins with their embedded redox centers, or the collisional interaction between proteins by diffusion. In this chapter we will first discuss the physical properties of such reactions, which impose a framework within which the biological processes must operate. We will then take as an example the reaction between cytochrome *f* and plastocyanin, which is central to photosynthetic electron transfer in oxygenic organisms.

19.2 The Physical Framework for Electron Transfer

We will concentrate here on reactions in which at least one of the partners is a soluble, diffusible protein. This imposes restraints additional to those occurring with the fixed scaffold of a protein complex such as a photosynthetic reaction center. In a transient complex brought together by diffusion the two redox centers must be sufficiently close for rapid electron transfer, but there is a conflicting requirement for the interaction not to be so strong or specific as to inhibit the reaction of the donor with

D. S. Bendall (✉)
Department of Biochemistry, University of Cambridge, Downing Site,
Cambridge CB2 1QW, UK
e-mail: dsb4@mole.bio.cam.ac.uk

G. A. Peschek et al. (eds.), *Bioenergetic Processes of Cyanobacteria,*
DOI 10.1007/978-94-007-0388-9_19, © Springer Science+Business Media B.V. 2011

another donor, and the acceptor with another acceptor. In other words, the primary requirement is for rapid turnover of the whole chain. The experimental investigation of this physical framework has mostly involved studying the interactions between purified components in solution, but the question of how this framework relates to the physiological system *in vivo* must not be neglected.

19.2.1 Transient Interactions and Rapid Turnover

Electron transfers are the simplest of chemical reactions, which are capable of achieving very high rates. For biological processes to take advantage of this fact the lifetime of a reaction complex formed by diffusion must be very short, no more than about 1 ms. This is only achievable because of the inherently rapid nature of electron transfer. The situation implies that tight, specific complexes are not formed, so that binding constants usually lie within the range 10^2–10^6 M^{-1}. The problem, therefore, is how a weak, transient interaction can bring the redox centers (metal or organic cofactors) into a relative configuration that satisfies the physical requirements for rapid electron transfer. These requirements are what we must consider next.

19.2.2 Physical Requirements for Electron Transfer

Electron transfer is best regarded as a quantum mechanical process. It occurs because there is overlap of the orbitals (wave functions) of the electron when it is associated with either the donor or the acceptor molecule. In biological systems the redox centers are usually separated by quite large distances (say 10–20 Å), so the amount of orbital overlap is very small. By contrast, in 'ordinary' chemical reactions, where bonds are broken and made, atoms approach closely to each other and there is much stronger interaction and distortion of orbitals.

The basic equation to describe the rate of electron transfer is simple enough. The rate of transition between two quantum mechanical states arises out of time-dependent perturbation theory and is often known as Fermi's Golden Rule:

$$k_{et} = \frac{2\pi}{\hbar} H_{\mathrm{AB}}^2 (FC)$$

where $\hbar = h/2\pi$, h being Planck's constant (DeVault 1980, 1984; Marcus and Sutin 1985). This equation can be thought of as made up of two parts, that is

$$k_{et} = (\text{electronic factor}).(\text{nuclear factor})$$

The corresponding classical equation derived from transition state theory is

$$k_{et} = \kappa B \exp(-\Delta G^{\ddagger}/k_{\mathrm{B}}T)$$

where B is a frequency factor which depends on whether the reaction is monomolecular or bimolecular (DeVault 1980, 1984); the transmission coefficient, κ, is essentially an electronic factor and the exponential term containing the activation free energy, $\Delta G^{\ddagger}$, is the nuclear factor.

The electronic factor in the Golden Rule equation (the H_{AB}^2 term) represents the electronic coupling between reactant and product wavefunctions, and can only be understood in quantum mechanical terms. On the other hand the nuclear, or Franck-Condon factor (*FC*) can often be treated classically provided the frequency of the nuclear motions connected with the electron transfer (which governs the spacing of vibrational energy levels) is not too high in relation to the temperature. In practice this means that for the type of reaction considered here, occurring around room temperature or above, a classical description is adequate.

19.2.2.1 Electronic Coupling

The electronic factor is most simply described by Gamow's tunnelling equation according to which the electronic coupling, H_{AB}^2, decays exponentially with distance. If we write H_r for a tunnelling distance of r and H_0 for a distance of zero, then

$$H_{\mathrm{r}}^2 = H_0^2 \mathrm{e}^{-\beta \mathrm{r}}$$

where $\beta = (2\pi/\hbar)(2mV)^{1/2}$, m being the mass of the electron and V the height of the potential energy barrier through which the electron has to tunnel. In this way of looking at the problem the tunnelling barrier between the two redox centers, that is the chemical structure between donor and acceptor, is treated as a continuum with uniform properties. This is manifestly not the case, but has turned out to be a remarkably useful approximation.

The value of β is often taken to be 1.4 Å^{-1} for protein, based upon the measured values in model systems (Moser et al. 1995). This leads to the important conclusion that k_{et} will decline exponentially by a factor of 10 for every increase in distance between redox centers of about 1.6 Å. As a result, rapid rates (say $\geq 10^3$ s^{-1}) require the distance to be not more than about 14 Å. In fact, this is found to be the upper limit of separation for most biological electron transfer systems (Page et al. 1999). Longer pathways involve intermediate redox centers. The distance limitation is enough to explain, for example, the lack of an unwanted cycling between plastocyanin and P680 within the thylakoid lumen. On the other hand, when considering diffusional interaction between a pair of redox proteins, it is evident that a distance between redox centers of 14 Å will remove some of the need for a precise fit between the two protein surfaces, and there may be redundancy in the relative configurations that allow effective electron transfer.

Another useful way of treating the problem of electronic coupling is to assume that the electron follows discrete pathways through the covalent structure of the protein (Onuchic et al. 1992). The coupling decays by a constant factor for each covalent bond traversed. If the pathway necessitates a jump between atoms that are not covalently linked the coupling is weaker and decays exponentially with the distance 'through-space' between the two atoms. Recognition of the more rapid decay 'through-space' than 'through-bond' has led to a modification of the simple distance/continuum model by including the packing density of the material between donor and acceptor groups, that is by introducing a different value of β for through-atom ($\beta = 0.9$ Å^{-1}) and through-space ($\beta = 2.8$ Å^{-1}) transfer (Page et al. 1999). Dutton and colleagues have argued that the validity of a simple distance model implies that evolutionary pressure has not been exerted on the detailed structure of the protein bridge between redox centers (Page et al. 2003).

In the case of electron transfer by diffusional interaction between two proteins, at least one through space jump is almost inevitable. An important recent development, however, is the finding in several systems that bound water molecules can provide significant electronic coupling when they form a hydrogen bond network between the proteins (Lin et al. 2005; Miyashita et al. 2005). Even so, high rates are likely to involve expulsion of water from the interface and hydrophobic contact between the two surfaces.

19.2.2.2 The 'Energy Gap' Law

The nuclear or Franck-Condon factor in the Golden Rule equation is an exponential term containing the activation free energy, and is related to the nuclear rearrangement that occurs in response to movement of the charge of the electron. The classical treatment first worked out by Marcus shows that $\Delta G^{\ddagger}$ depends on two formal energy terms, ΔG°, the driving force or standard free energy change for the reaction, and λ, the reorganization energy defined as the energy required to change the atomic configuration from that in the reactants to that in the products (without actually moving the electron). The 'Energy Gap' law deduced by Marcus (Marcus and Sutin 1985) is

$$\Delta G^{\ddagger} = \frac{(\lambda + \Delta G^{\circ})^2}{4\lambda}$$

which is incorporated into the full semi-classical equation (the term $(2\pi/\hbar)H_{AB}^2$ is still required) to give

$$k_{et} = \frac{2\pi}{\hbar} H_{AB}^2 (4\pi\lambda k_B T)^{-1/2} \exp\left[-(\lambda + \Delta G^{o})^2/4\lambda k_B T\right]$$

The important point to note here is that a small value for λ favours fast electron transfer. A rigid coordination sphere for the metal ion, as provided by the porphyrin in cytochromes, contributes to this, but λ also depends on the rearrangement of any solvent molecules close enough to be influenced by the charge transfer. Thus fast rates are likely to depend on a hydrophobic environment for the major electron pathway at the contact point between the two proteins.

19.2.3 The Intracellular Environment

Plastocyanin is a soluble protein which occurs in the thylakoid lumen. It is presumably able to diffuse, even if it spends much of its time bound to protein complexes embedded in the membrane. Cytochrome *f* is a component of the cytochrome *bf* complex with its functional group containing the haem on the luminal side of the membrane. The interaction between cytochrome *f* and plastocyanin has been extensively studied in solution with proteins purified from plants, algae and cyanobacteria, or from heterologous gene expression in *E. coli*. This approach allows the application of powerful tools to define the relevant physical and chemical properties of the system, but the relevance of the conclusions to the situation *in vivo* must be established. For example, studies *in vitro* have shown that electron transfer between cytochrome *f* and plastocyanin of the green alga *Chlamydomonas reinhardtii* is strongly influenced by an electrostatic attraction between the acidic patches on plastocyanin and a group of basic residues on cytochrome *f*, as is the case with higher plant proteins. However, when a mutant strain of *C. reinhardtii* was prepared in which the relevant basic residues of cytochrome *f* had been neutralized, there was only a small effect on the growth rate and rate of oxygen evolution, much smaller than was expected from the behaviour *in vitro* (Soriano et al. 1996; Soriano et al. 1998). This result is most readily explained by assuming that the rate limiting step *in vivo* is diffusion of plastocyanin from its oxidation site on Photosystem I to cytochrome *f* bound to the cytochrome *bf* complex, or its release from Photosystem I (Finazzi et al. 2005), but this leaves open the question of why the basic patch on cytochrome *f* should have been conserved throughout evolution of green algae and higher plants.

The intracellular environment contains high concentrations of protein, which can have marked effects on protein-protein interactions compared with the behaviour of the same proteins in simple buffer solutions (Ellis 2001a, b; Minton 2001).The effects are not readily predictable, however, as different effects can oppose each other. Attempts have been made to mimic the effect of a high protein concentration by the use of artificial polymers such as Dextran 70 or Ficoll 70. Osmotic pressure measurements have shown that in a solution of 7% (w/v) Ficoll the Ficoll molecules occupy about 60% of the total volume (Wenner and Bloomfield 1999), so one would expect this excluded volume effect to raise the apparent binding constant of a pair of proteins by a factor of $1/0.4 = 2.5$. A significant increase, although not as large as this prediction, has been observed in the case of the binding of superoxide dismutase to xanthine oxidase (Zhou et al. 2006). On the other hand, high concentrations of macromolecules increase the viscosity of a solution, which may have an adverse effect on the kinetics of electron transfer between two proteins.

The reaction between cytochrome *f* and plastocyanin occurs in the thylakoid lumen. Current estimates suggest that the lumen contains a concentration of soluble protein of about 20 mg mL^{-1}, whereas higher concentrations (50–400 mg mL^{-1}) prevail in cell cytoplasm (Minton 2001). The thylakoid lumen is a tightly confined space and it may be that effects of confinement are more important than macromolecular crowding, but they are less amenable to experimental investigation. In fact, concentrations of Ficoll 70 up to 7% or Dextran 70 up to 5% have little or no effect on the rate of reaction between the proteins from the cyanobacterium *Phormidium laminosum*, even though low molecular weight viscogens such as glycerol cause a marked inhibition over a similar range of macroviscosity (Schlarb-Ridley et al. 2005). There is the further possibility that in the narrow, confined space of the lumen two-dimensional diffusion over the membrane surface is more important than three-dimensional liquid phase diffusion, so providing a substantial rate enhancement.

19.3 The Reaction Between Cytochrome *f* and Plastocyanin

19.3.1 Structures of the Individual Proteins

This is the starting point for an understanding of the reaction in molecular terms. How does the appropriate site for electron transfer on one protein find the complementary sites on its various reaction partners? How does it do this without getting stuck with only one partner? To answer these questions we need first to know the chemical characteristics of each protein surface and then the structure of the complex formed between each pair of reactants. Furthermore, there is the possibility that formation of the reaction complex involves conformational change, and also, in the extreme case, that the complex is no single structure but rather a dynamic state in which multiple configurations are suitable for electron transfer. If a single structure does exist we can compare it with those of the individual components, but here we come up against the problem that complexes of electron transfer proteins are not readily crystallized because of the low binding constants and only a few such complexes have been successfully crystallized. NMR has given invaluable information about the interfaces between reactants, but interpretations of NMR data usually depend heavily on X-ray crystal structures of individual components.

The first redox complex to have its structure determined by X-ray crystallography was that between cytochrome *c* and cytochrome *c* peroxidase from yeast (Pelletier and Kraut 1992). The backbone structures of the individual proteins can be superimposed on the complex with root mean square deviations of 0.57 Å and 0.39 Å for cytochrome *c* and cytochrome *c* peroxidase respectively. Thus there is no conformational change in the backbones as the free proteins combine to form the complex. Rapid electron transfer has been shown to occur in the crystals, so the crystal structure defines the active complex, although weaker binding sites are also important.

Publication of the first X-ray crystal structure of plastocyanin in 1978 was a major event for understanding of blue copper proteins (Colman et al. 1978). The first crystals were obtained with protein from the leaves of black poplar (*Populus nigra* var. *italica*), and subsequently crystal structures have been reported from plastocyanin from spinach (Xue et al. 1998; Jansson et al. 2003), white campion (*Saline pratensis*) (Fraústa da Silva and Williams 1991), a fern (*Dryopteris crassirhizoma* (Inoue et al. 1999b; Kohzuma et al. 1999), green algae (*Ulva pertusa* (Shibata et al. 1999), *Chlamydomonas reinhardtii* (Redinbo et al. 1993) and *Enteromorpha prolifera* (Collyer et al. 1990)) and several cyanobacteria (*Phormidium laminosum* (Bond et al. 1999), *Anabaena* sp. (Schmidt et al. 2006), *Synechococcus* sp. PCC 7942 (Inoue et al. 1999a) and *Synechocystis* sp. PCC 6803 (Romero et al. 1998)). NMR structures have also been obtained for parsley (*Petroselinum crispum*) (Bagby et al. 1994) and French bean (*Phaseolus vulgaris*) (Moore et al. 1991) among higher plants and *Synechocystis* sp. PCC 6803 (Bertini et al. 2001), *Anabaena variabilis* (Badsberg et al. 1996) and *Prochlorothrix hollandica* (Babu et al. 1999) among cyanobacteria. The tertiary structure is a highly conserved Greek key β-sandwich with a fold which is related to many other classes of protein. Conservation is much lower when it comes to the surface properties of the molecule. The plant proteins are consistently acidic; they have a well-conserved electrostatic surface showing a characteristeric patch of acidic residues on one face. The cyanobacterial proteins, on the other hand, are much more variable. The acidic patch is either weakened or may even be basic. Thus, whereas the pI for spinach plastocyanin is 3.0, that for *Anabaena variabilis* is 8.4. We will see that such variations have important consequences for the way plastocyanin interacts with cytochrome *f*.

In the case of cytochrome *f* only three high resolution X-ray structures are available, but being from a higher plant (turnip, *Brassica rapa*) (Martinez et al. 1994, 1996), a green alga (*Chlamydomonas reinhardtii*) (Chi et al. 2000) and a cyanobacterium (*Phormidium laminosum*) (Carrell et al. 1999) they effectively span the range of oxygenic organisms. These structures (which refer to the whole molecule except for the transmembrane helix and a short stromal sequence) are remarkably highly conserved. The tertiary structures are almost identical, and the luminal domain is acidic in all cases. The one major difference is that the plant protein possesses a cluster of basic residues which complements the acidic patch of plastocyanin. These residues are acidic in the cyanobacterial case. Cytochrome *f* is unique amongst *c*-type cytochromes in being largely β-sheet, but its fold is related to some other β-sandwich Greek key proteins which include some transcription factors and diphtheria toxin.

19.3.2 Structure of the Complex

The complex between cytochrome *f* and plastocyanin has not been crystallized, but the interaction between the two proteins has been extensively studied using NMR techniques (Ubbink et al. 1998; Ejdebäck et al. 2000; Bergkvist et al. 2001; Crowley et al. 2001, 2002, 2004; Díaz-Moreno et al. 2005b, c; Lange et al. 2005; Hulsker et al. 2008). The interaction has also been simulated by Brownian dynamics (Pearson and Gross 1998; De Rienzo et al. 2001; Gross and Pearson 2003; Gross 2004; Haddadian and Gross 2006). The advantage of NMR is that the proteins are studied in solution and so it is likely to give results closer to the situation *in vivo* compared with X-ray crystallography. The two methods are complementary, however; whereas crystallography can give precise atomic coordinates within a complex, NMR can give information on the dynamics of a complex. In the well-studied case of yeast cytochrome *c* and cytochrome *c* peroxidase NMR has shown that the dominant form of the complex in solution is very similar to that observed in crystallography (Worrall et al. 2001; Volkov et al. 2006). In addition, however, paramagnetic relaxation enhancement experiments have revealed the existence of alternative configurations within the complex, although >70% of the lifetime of the complex is spent in the one, dominant form.

The interaction between plastocyanin and cytochrome *f* has been studied by NMR with proteins from several different organisms, ranging from plants to cyanobacteria, and some proteins with crucial residues modified by mutagenesis. A remarkable variability in the nature of the complex has emerged from these studies. The results are important for understanding inter-protein electron transfer, although their full significance is not yet clear. The classic example is that of the higher plant complex with cytochrome *f* from turnip (*Brassica rapa*) and plastocyanin from spinach (Ubbink et al. 1998), for which a structure was obtained by restrained rigid body molecular dynamics using restraints from diamagnetic chemical shift changes and intermolecular paramagnetic pseudocontact shifts. Although a homologous system from higher plants has not yet been studied experimentally, a model of the homologous spinach complex has been simulated by computational docking and molecular dynamics (Musiani et al. 2005). In most cases a single orientation complex could be defined in which the hydrophobic patch of plastocyanin containing the copper ligand His87 (higher plant numbering) makes contact with a region of cytochrome *f* containing the haem ligand Tyr1 (Prudêncio and Ubbink 2004; Díaz-Moreno et al. 2005b, c). An important feature of the higher plant complexes, which are all very similar although not identical, is that an electrostatic attraction between the acidic patch of plastocyanin and a group of basic residues in cytochrome *f* pulls plastocyanin into a sideways-on approach to cytochrome *f*. This allows a good electron-transfer pathway between the haem and the copper through the haem ligand Tyr1 and the copper ligand His87 to give an Fe–Cu distance of about 11 Å with turnip cytochrome *f* and spinach plastocyanin.

Electrostatics thus play a consistent role in the interaction *in vitro* between cytochrome *f* and plastocyanin throughout the plants and algae. Sequence comparisons confirm that the conclusions from NMR studies on a small number of species is likely to be general. This is no longer the case when considering extant cyanobacteria. The complex from *Phormidium laminosum* perhaps represents an extreme case (Crowley et al. 2001). Although the hydrophobic face of plastocyanin presents itself to cytochrome *f* it does so in a head-on fashion, that is to say electrostatics have very little influence, as shown by the insensitivity to the ionic strength of the medium. The *Nostoc* complex forms an interesting intermediate case. An electrostatic attraction pulls plastocyanin round to a position between those adopted by plant and *Phormidium* proteins, but the charges are reversed (Díaz-Moreno et al. 2005c). Cytochrome *f* presents a consistently negatively charged surface to a positively charged patch on plastocyanin. The lack of any positively charged patch on cyanobacterial cytochrome *f* seems to be a general phenomenon, but the charge pattern of plastocyanin is more varied, although there is no known case of a strong acidic patch such as occurs in higher plants. The apparently neutral character of *Phormidium* plastocyanin is the result of charge balance between a pair of acidic residues (D44 and D45) and three basic residues (predominantly R93 with weaker contributions from K46 and K53) (Schlarb-Ridley et al. 2002; Hart et al. 2003).

NMR can provide a rich source of information regarding the dynamics of protein-protein interactions. A general line-broadening effect, due to an increase in rotational correlation time, is evidence for complex formation. Most electron transfer proteins show chemical shift changes for residues in the interface of a complex, which implies that the new chemical shift is a weighted mean between those for the free and the bound states, even though the free proteins may still be present. This situation is indicative of fast exchange between the two states, consistent with formation of a transient complex. In cases where a dominant single orientation of the complex can be defined, its formation is often aided by a specific electrostatic attraction, as in the higher plant reaction between cytochrome *f* and plastocyanin, although the same reaction with the *Phormidium* proteins shows that hydrophobic effects, which must involve solvent exclusion, can be equally effective. On the other hand a more diffuse electrostatic attraction can encourage the non-specific encounter complex, even when no single orientation predominates. The latter situation (which is associated with very small chemical shift changes) has been found with both non-physiological and physiological complexes, and often the electron transfer rates are surprisingly fast, despite the lack of a specific complex (Ubbink and Bendall 1997; Worrall et al. 2002, 2003; Hulsker et al. 2008). In some cases this might be explained by good electronic coupling over a relatively wide region of the interface, for example when haem groups are involved as in the reaction between cytochrome b_5 and myoglobin, but an easy explanation is not always forthcoming.

The complex of cytochrome *f* and plastocyanin from the cyanobacterium *Prochlorothrix hollandica* provides a remarkable example of a shift between formation of a single orientation complex and a purely dynamic one. The wild type proteins show a weak electrostatic attraction, and a single orientation of the complex can be defined although, as is also true for *Phormidium laminosum*, it is a relatively dynamic one compared to *Nostoc* and higher plants, as judged by the smaller chemical shift changes and a less tightly confined ensemble of possible structures. The hydrophobic patch of *Prochlorothrix* plastocyanin, which forms part of the interface with cytochrome *f*, contains two unusual residues, Y12 and P14. If these are mutated to the consensus residues, G12 and L14, the complex with cytochrome *f* becomes much more dynamic and no single orientation predominates; despite this the overall electron transfer rate changes surprisingly little (Hulsker et al. 2008). The physical basis for this effect is unclear, and it contrasts with the behaviour of the complex formed between myoglobin and cytochrome b_5. The wild type complex is highly dynamic and electron transfer occurs at only a modest rate, as one might intuitively expect. If the negative charges on the haem propionates of myoglobin are neutralized electron transfer increases by two orders of magnitude, even though the binding constant is little affected (Wheeler et al.

2007). This can be explained by a much greater concentration of contacts in the region of the exposed haem edge, although the total number of contacts remains much the same. Observations of this kind emphasize the flexibility of electron transfer reactions and the lack of highly specific structural requirements.

19.3.3 Kinetics of Interaction

If we measure electron transfer between two soluble proteins what is it we are measuring? The dynamics of the overall process suggests that several steps are involved, any one of which may be the rate-limiting one for the second-order rate constant, k_2. A widely accepted model for the formation of a specific protein-protein complex is represented in Fig. 19.1. This model can be taken to apply to redox proteins when a single orientation complex is formed. The free proteins first interact by diffusion to form a relatively loose, dynamic encounter complex, often under the influence of an electrostatic attraction between complementary charges in the region of the reaction site of each protein. At this stage the complex can form several different configurations of roughly equivalent energy so that rapid interchange between them occurs. At least one layer of water molecules tends to remain between the two protein surfaces; any direct contact is momentary and insufficient to direct the proteins to their final configuration in the complex. Eventually an appropriate configuration is found and rearrangement takes the complex through a transition state to its relatively static single orientation. Short range forces now dominate, especially the entropic benefit of water exclusion and van der Waals attraction (the hydrophobic effect). This scheme can be written as a series of chemical reactions, for plastocyanin being reduced by cytochrome *f*, as follows:

$$\mathrm{Pc_{ox}} + \mathrm{Cyt}f_{\mathrm{red}} \underset{k_{-a}}{\overset{k_a}{\rightleftarrows}} [\mathrm{Pc_{ox}Cyt}f_{\mathrm{red}}]_1 \underset{k_{-r}}{\overset{k_r}{\rightleftarrows}} [\mathrm{Pc_{ox}Cyt}\,f_{\mathrm{red}}]_2$$

$$\overset{k_{\mathrm{et}}}{\longrightarrow} [\mathrm{Pc_{red}Cyt}\,f_{\mathrm{ox}}] \longrightarrow \mathrm{Pc_{red}} + \mathrm{Cyt}\,f_{\mathrm{ox}}$$

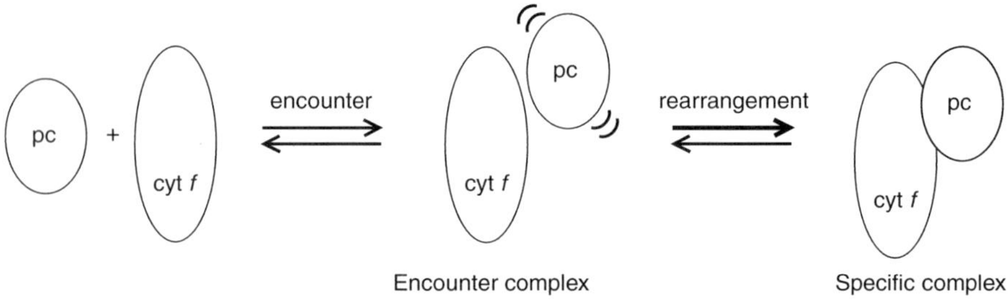

Fig. 19.1 The two-step model for protein-protein binding. In the encounter complex plastocyanin is shown as capable of both rotational and translational motion relative to the surface of cytochrome *f*

The general equation for this scheme is

$$k_2 = \frac{k_a k_r}{k_{-a}(k_{et} + k_{-r}) + k_r k_{et}} k_{et}$$

which in its reciprocal form can be rearranged to show how the overall reaction time can be represented as the sum of the transit times of the three individual steps

$$\frac{1}{k_2} = \frac{1}{k_a} + \frac{1}{K_A k_r} + \frac{1}{K_A K_R k_{et}}$$

K_A and K_R being the equilibrium constants for the first two steps, association (k_a/k_{-a}) and rearrangement (k_r/k_{-r}) (Schlarb-Ridley et al. 2005). The second order rate constant, k_2, is normally measured by stopped flow spectrophotometry as the proportionality constant between k_1, the observed first order rate constant and the concentration of the reagent being used in excess, usually plastocyanin in this case (Bendall 1996). Several different simplifying assumptions are possible. If both rearrangement and the intrinsic rate of electron transfer, k_{et}, are very fast then the reaction becomes truly diffusion limited and $k_2 = k_a$. If rearrangement is rate limiting, sometimes referred to as a gated reaction, $k_2 = K_A k_r$. A third possibility, the activation-controlled regime, occurs when the electron transfer step is limiting, and then $k_2 = K_A K_R k_{et}$.

The above equation shows that k_{et}, the intrinsic rate of electron transfer, is not measured directly. In some systems, although not cytochrome *f*/plastocyanin, rate saturation can be observed. The maximum value of k_1 then gives k_{et} directly only when k_r is very fast, so that rearrangement cannot be observed. Thus in general further kinetic information is required to understand the physical factors contributing to k_2. Measurement of the binding constant, $K_A K_R$, is helpful, but this is often surprisingly difficult with highly transient interactions. The most successful method for binding constants in the range 10^3–10^6 M^{-1} seems to be the titration of chemical shift changes by NMR. A minimum value for k_{et} is given by $k_2/K_A K_R$. Flash photolysis provides a means of measuring k_{et} directly for electron transfer from cytochrome *f* to plastocyanin in the preformed complex (Meyer et al. 1993; Qin and Kostic 1993; Navarro et al. 2004; Albarrán et al. 2005). The flash induces photoreduction of 5-deazariboflavin to the semiquinone which rapidly and selectively reduces cytochrome *f*. A significant proportion of the proteins are present as complex only at very low ionic strength and measurements under these conditions have given estimates of $k_{et} = 2.8 \times 10^3$ s^{-1} at $I = 4$ or 5 mM (Meyer et al. 1993) for the higher-plant proteins. This should be compared with an estimated minimum value of $k_{et} \geq 2.6 \times 10^4$ s^{-1} at $I = 100$ mM from the second-order rate constant measured by stopped-flow spectrophotometry and the binding constant measured by NMR (Kannt et al. 1996). In the case of the homologous proteins from *Nostoc* sp. (a cyanobacterium) a minimum value $k_{et} \geq 1.3 \times 10^4$ s^{-1} at $I \approx 30$ mM has been measured by laser flash absorption spectroscopy (Albarrán et al. 2005).

Further useful information regarding the interpretation of k_2 can be obtained by measuring the effect of temperature, so that the activation parameters, $\Delta G^\ddagger$, $\Delta H^\ddagger$, $\Delta S^\ddagger$, can be calculated, and also the redox potential to give the driving force, ΔG°, especially in conjunction with mutagenesis of specific residues. The activation parameters are composite ones for the overall reaction, but ΔG° will affect k_{et} alone.

So what can we say about the interaction between cytochrome *f* and plastocyanin? For the proteins from higher plants and green algae reacting in solution a distinct pattern is emerging. The diffusion equation gives a rate constant for collision between two small, uncharged proteins in the range 10^9–10^{10} $M^{-1}\ s^{-1}$. The expected rate constant for reaction, however, is orders of magnitude smaller than this because reaction only occurs when relatively small areas on the surface of each protein come into contact. The solvent cage effect prolongs the lifetime of the diffusional complex and so increases the probability that the initial collison will give rise to the final specific structure. Brownian dynamics simulations have shown that the expected second order rate constant for formation of a specific complex is 10^5–10^6 $M^{-1}\ s^{-1}$ (Northrup and Erickson 1992). In fact $k_2 = 2 \times 10^6\ M^{-1}\ s^{-1}$ has been observed for turnip cytochrome *f* reacting with pea plastocyanin at very high ionic strength (3 M) (Schlarb-Ridley et al. 2003). At the more physiological value of $I = 100$ mM, however, the reaction is 100 times faster ($k_2 = 1.5 \times 10^8\ M^{-1}\ s^{-1}$). The rate enhancement is brought about by a strong electrostatic attraction, which is screened out at high ionic strength, between a basic patch of residues on cytochrome *f* close to the haem and an acidic patch on plastocyanin adjacent to the hydrophobic patch containing the histidine copper ligand. Judging by the high degree of conservation of the relevant residues of cytochrome *f* and plastocyanin, this pattern is likely to be a general feature of the interaction for most, perhaps all higher plants. It also applies to *Chlamydomonas reinhardtii* and at least some other green algae.

Electrostatic attraction has two effects. In the first place it enhances the rate of diffusional interaction as predicted by Debye-Hückel theory and so gives larger values of k_a in the kinetic scheme above, and at the same time it would probably diminish the rate of separation, k_{-a}. Secondly, complementary charges exert an electrostatic steering effect which helps to define an encounter complex favourable for formation of the final single orientation complex (Selzer and Schreiber 2001). The overall effect is to increase the binding constant and the occupancy of the encounter complex, which in turn increase the rate of complex formation. The electrostatic attraction can be modulated by adjusting the salt concentration. Thus, increasing the ionic strength decreases k_a and K_a and hence k_2. The effect on K_r, however, can be the opposite of that on K_a because electrostatic attraction stabilises the encounter complex and decreases k_r (Lange et al. 2005). As a result the overall rate of reaction, k_2, can show an optimum as a function of ionic strength and such an effect has been observed with the reaction between cytochrome *f* and plastocyanin and with several other electron transfer reactions that are sensitive to ionic strength (Meyer et al. 1993). Once a specific complex has been formed an effect of ionic strength on the intrinsic rate of electron transfer, k_{et}, would not be expected. However, if the observed value of k_{et} is an ensemble average (Lin and Beratan 2005) electrostatics may distort the distribution within that ensemble. We noted above that k_{et} decreases by at least a factor of ten between $I=100$ mM and $I=4$-5 mM. At $I=100$ mM the reaction is very fast and we have previously suggested that it is close to being diffusion limited (Gong et al. 2000); in response to very low ionic strength there may be a change in kinetic regime towards activation control.

The above pattern of behaviour of the reaction between cytochrome *f* and plastocyanin seems reasonably consistent for all green plants. This is certainly no longer true for cyanobacteria in which there is considerable variability and no known case corresponds to the green plant model. As with the plant proteins, the electron transfer rate between those from *Nostoc* shows a strong dependence on ionic strength, indicative of electrostatic attraction. In this case, however, mutational analysis (Albarrán et al. 2005; Albarrán et al. 2007) and determination of the structure of the complex by paramagnetic NMR (Díaz-Moreno et al. 2005) show that the polarity of the interface is reversed, with K11, K35, R93, K62 and K51 of plastocyanin interacting with D100, E108, E165, E189, D64, and D190 of cytochrome *f*. In contrast with *Nostoc*, electron transfer with the proteins from *Phormidium laminosum* shows only weak electrostatic attraction, and the structure of the complex and binding constant studied by NMR show no ionic strength dependence. As one might expect, at 27°C the higher plant proteins give a higher value for k_2 (1.5×10^8 M^{-1} s^{-1}) than those from *P. laminosum* (4.7×10^7 M^{-1} s^{-1}) (Schlarb-Ridley et al. 2002). *P. laminosum*, however, is a moderate thermophile and from the measured effect of temperature on k_2 (Schlarb-Ridley et al. 2005) it can be predicted that at the organism's normal growth temperature of 45°C k_2 would be 1.4×10^8 M^{-1} s^{-1}, comparable with the value for higher plants at 27°C. It should be noted that neither *P. laminosum* nor *Nostoc* are electrostatically optimized for fast electron transfer. *Phormidium* plastocyanin contains two acidic residues, D44 and D45, in the region that corresponds to the acidic patch of plant plastocyanin. These exert an electrostatic repulsion from the negatively charged surface of cytochrome *f*, and if changed to neutral residues allow a much stronger attraction exerted by the basic residues of plastocyanin (predominantly R93) and an increased value of k_2 (Schlarb-Ridley et al. 2002). In the case of *Nostoc* k_2 can be increased by the mutation D54K in plastocyanin (Albarrán et al. 2005).

A common view of rapid and specific protein-protein interactions, such as are observed in electron transfer reactions, is that long-range electrostatic attraction accelerates formation of the encounter complex in which both proteins remain hydrated. The proteins are then brought into direct contact in a specific configuration by short-range forces, which are dominated by the hydrophobic effect involving desolvation of the interface. Although this model is appropriate for the cytochrome *f*/plastocyanin reaction of plants, green algae and *Nostoc*, the *Phormidium* proteins illustrate the fact that hydrophobic effects alone can give rise to rapid rates. At high ionic strength, which removes the residual electrostatic attraction, a rate

$k_2 = 3.2 \times 10^7$ M^{-1} s^{-1} is still observed, which is 10 times faster than with higher plant proteins (Schlarb-Ridley et al. 2002). Fast hydrophobic protein-protein binding has also been observed in other systems, for example the binding of the inhibitor ovomucoid third domain to α-chymotrypsin (Camacho et al. 1999, 2000). Camacho et al. have concluded that partial desolvation provides a slowly varying attractive force over a region of the protein surface which leads to a low free energy attractor defining the encounter complex (precursor state in their terminology). How far the hydrophobic attraction extends from the protein surface as a result of desolvation is uncertain, but it seems to be of the order of two water molecules. A weak attraction between hydrophobic surfaces in water has been detected out to approximately 25 Å by atomic force microscopy (Tsao et al. 1993) where an entropic effect of desolvation has not yet come into play. The origin of this attraction is thought to be electrostatic, resulting from the polarization field of water molecules oriented at the hydrophobic surface (Despa and Berry 2007).

We have previously suggested that the reaction between cytochrome *f* and plastocyanin with plant proteins is close to being diffusion controlled (Gong et al. 2000), that is to say that $k_{et} > k_{-r}$ and $k_r > k_{-a}$ and the initial binding is rate limiting. This is plausible in view of the high value of k_2 and its enhancement by a strong electrostatic attraction (Kannt et al. 1996), but has not been critically tested. Characteristic features of diffusion controlled reactions are a low activation enthalpy and a sensitivity to solvent viscosity. The measured activation enthalpy for the *Phormidium* reaction is 45.3 kJ mol^{-1}, considerably higher than would be expected for a simple diffusion controlled reaction. Although k_2 is sensitive to viscosity, the dependence is unusual. According to Kramers' theory one would expect an inverse dependence on viscosity (Kramers 1940) whereas the non-linear dependence on reciprocal viscosity suggested a power law dependence according to the equation:

$$k_2^0/k_2 = \mathrm{A}(\eta/\eta^0)^x$$

with the exponent $x = 1.8$ (Schlarb-Ridley et al. 2005). This is incompatible with simple diffusion control but could be explained if k_2 depends on two viscosity-sensitive reactions in series (k_a and k_r), with the reverse reactions (k_{-a} and k_{-r}) little affected. Thus the rate limiting step could be either rearrangement (a gated reaction) for which $k_2 = K_A k_r$, or k_{et} (activation control) with $k_2 = K_A K_R k_{et}$. In the *Phormidium* complex His92, the copper ligand of plastocyanin, makes van der Waals contact with Phe3 of cytochrome *f* and one might expect the best coupling pathway to be less effective than one through the copper ligand Tyr1, as in the plant case, because of the need for two through-space jumps rather than one. The latter possibility would suggest rearrangement of the stable complex observed by NMR to an unstable configuration in which His92 contacts Tyr1. However, the electronic coupling calculated for the Phe3 pathway in *Phormidium* using the Pathways program is almost as high as for the Tyr1 pathway in plants (Díaz-Moreno et al. 2005). Nevertheless, some contribution to k_2 from k_{et} cannot be ruled out so that at this stage we should conclude that both rearrangement and electron transfer have some controlling influence over k_2.

19.4 Structure and Variability of the Interface

The interfaces of four protein-protein complexes are illustrated in Figs. 19.2 and 19.3. Figure 19.2 compares that of the electron-transfer complex between yeast *iso*-1-cytochrome *c* with cytochrome *c* peroxidase (ccpc, lower panel) and that of

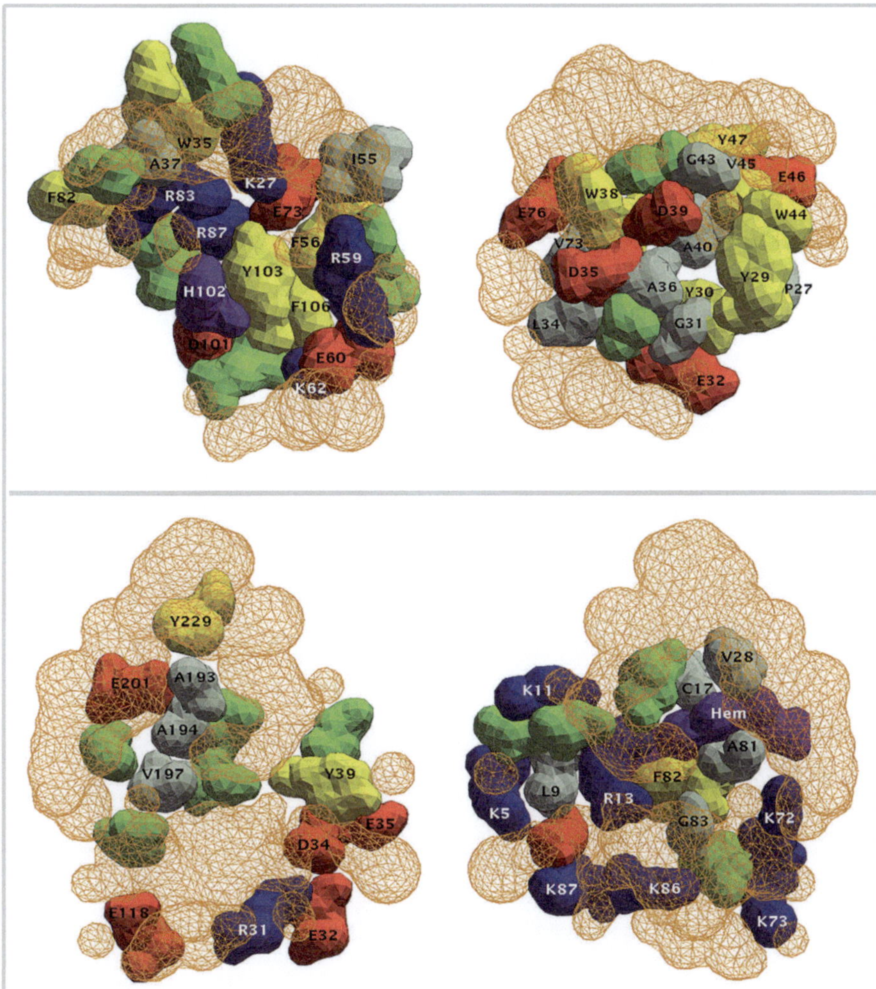

Fig. 19.2 Interface of complexes displayed with Surfnet and ImageMagick. Upper panel, complex of barnase and barstar (brs), barnase left, barstar right; lower panel, complex of yeast cytochrome *c* peroxidase and *iso*-1-cytochrome *c* (ccpc), cytochrome *c* peroxidase left, cytochrome *c* right. In each complex the lefthand protein has been rotated −90° and the righthand protein +90°. Red, acidic residues; blue basic residues; cyan hydrophobic residues; green polar residues; yellow, Phe, Tyr or Trp; purple, His; magenta, haem; orange mesh, volume penetrated by a test sphere minimum radius 1.4 Å, maximum radius 4.0 Å

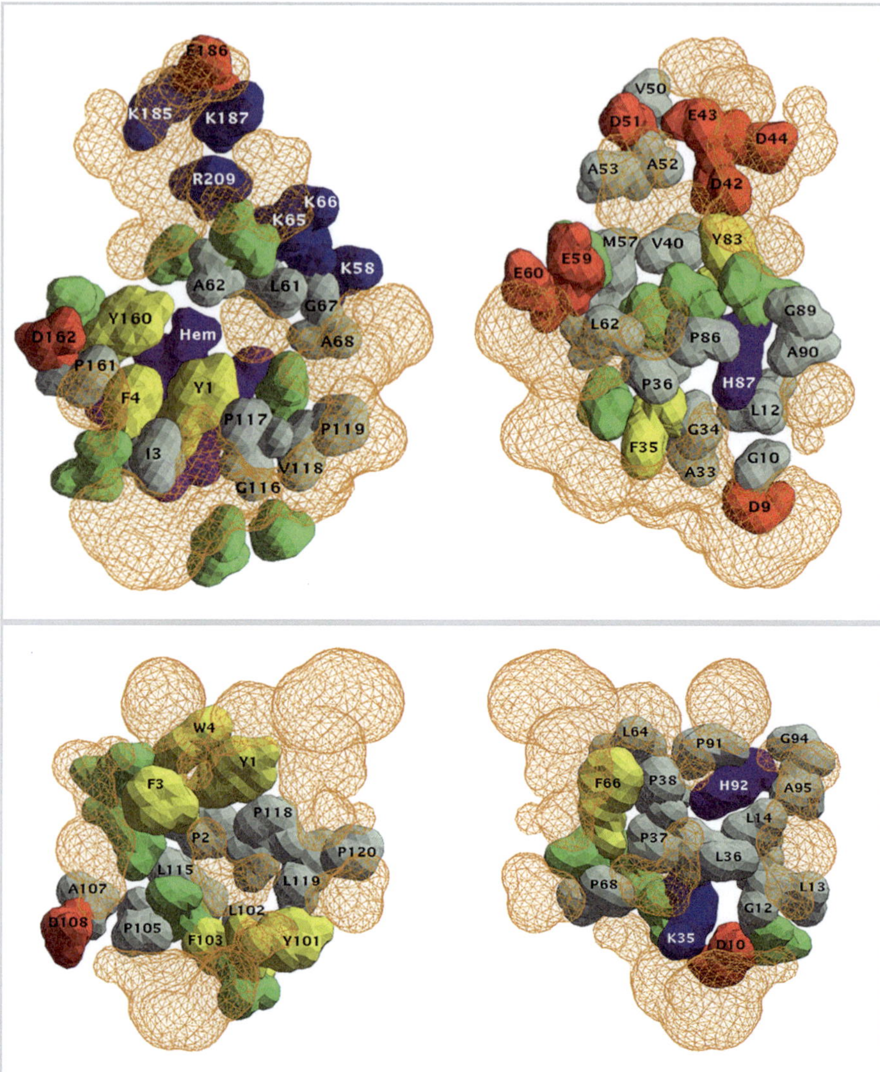

Fig. 19.3 Interfaces of complexes. Upper panel, complex of cytochrome *f* and plastocyanin from higher plants (pcf); lower panel, complex of cytochrome *f* and plastocyanin from the cyanobacterium *Phormidium laminosum* (plpcf); cytochrome *f* left, plastocyanin, right. Other details as in Fig. 19.2

barnase, the bacterial RNase, with its natural inhibitor protein, barstar (brs, upper panel). Both are predominantly electrostatic in nature and have been extensively studied on the basis of their X-ray crystal structures, and with the aim of understanding the kinetics and energetics of binding. The validity of the crystal structure of the cytochrome *c*/cytochrome *c* peroxidase complex in solution has been confirmed

Table 19.1 Association constants and interface geometry of four protein-protein complexes

Complex	Association constant, K_A $(M^{-1})^a$	ΔG_A (kcal mol^{-1})	Interface area $(Å^2)^b$	Gap Index[c]	Inner Gap Index[d]	Shape correlation, S_c[e]
brs	10^{14}	−19.1	1590	1.67	0.143	0.721
ccpc	5.6×10^{5}	−7.8	1262	4.34	0.147	0.574
pcf	10^{4}	−5.5	1852	2.57	0.157	0.584
plpcf	10^{3}	−4.1	1219	3.45	0.176	0.522

[a] Approximate values

[b] Sum of the two interface surfaces calculated with Surfcv (Sridharan et al. 1992)

[c] Measured with Surfnet (Jones and Thornton 1996), calculated as interface volume divided by interface solvent-accessible area. Interface volume is defined by a test sphere with minimum and maximum radii of 1.0 Å and 5.0 Å respectively

[d] Calculated as interface volume divided by interface area when the interface volume is defined by the difference in volume occupied by test spheres with minimum and maximum radii of 1.0/3.0 and 1.4/3.0 Å respectively

[e] Calculated with Sc in the CCP4 package (Lawrence and Colman 1993)

by NMR (Pelletier and Kraut 1992; Volkov et al. 2006). Figure 19.3 compares the two extreme types of complex shown between cytochrome *f* and plastocyanin, the higher plant complex in the upper panel (pcf) and the cyanobacterial complex from *Phormidium laminosum* in the lower panel (plpcf). The plant complex is highly electrostatic, but electrostatics plays only a small part in the *Phormidium* complex.

At first sight the two complexes shown in Fig. 19.2 are not obviously very different, and yet the association constants, K_A, differ by many orders of magnitude (Table 19.1). This reflects the physiological need for very rapid and tight binding between barnase and barstar, and for rapid but highly unstable binding between the peroxidase and cytochrome *c*. Closer inspection of the figure, however, reveals an important difference in the role of water in the interface. This is shown by the orange mesh which represents the ability of a test probe of radius 1.4 Å (a water molecule) to penetrate the interface. The interface of brs is tight and compact, with a well-defined area from which water is excluded. By contrast, the interface of ccpc is looser, with water able to penetrate deeply. The two electron transfer complexes shown in Fig. 19.3 are also relatively loose. The interface of plpcf is the smallest of the four and that of pcf the largest. The difference is due to the presence in pcf of a group of charged residues which remain partially solvated.

Some properties of the four interfaces are shown in Table 19.1. The challenge is to explain the difference in binding constant between the three electron-transfer complexes ($K_A = 10^3 - 10^6\ M^{-1}$) and brs ($K_A = 10^{14}\ M^{-1}$). The area of the interface is potentially important because it provides the first crude measure of the magnitude of the hydrophobic effect (mainly entropic at room temperature) which arises from the exclusion of water from the hydrophobic surfaces of the interface. Table 19.1 shows that area is clearly not a dominant factor in explaining the difference because that of brs lies in the middle of the range defined by the four complexes. The gap index, as defined by Jones and Thornton (1996) is the ratio of the gap volume to the interface area and provides a quantitative measure of the looseness of the interface. The gap volume is the volume traced out by a test sphere of adjustable radius, minimum 1.0 Å, maximum 5.0 Å (Laskowski 1995). The gap index of brs is substantially lower than those for the three electron transfer complexes, consistent with the qualitative comments about the interface in the previous paragraph. The gap index is dominated by the looseness of the edges of the interface. In order to get information about the less exposed parts of the interface we have calculated an inner gap index which is defined by a probe sphere of maximum radius 1.4 Å (see Table 19.1 for details). The differences in the gap index largely disappear when the inner gap index is calculated, showing that there is close contact between the two proteins in the inner regions in all four cases. Nevertheless, the shape complementarity statistic, S_c, of Lawrence and Colman (1993) shows that the fit of the electron transfer complexes is significantly less good than for brs, which is typical of protein-inhibitor complexes. S_c takes account not only of the distance between the two surfaces, point by point, but also of their relative orientation and is only calculated after removal of the peripheral part of the interface. A perfect fit would give $S_c = 1$. This difference is important because it probably means that the van der Waals contribution to binding energy is relatively low for the electron transfer complexes.

19.4.1 Energetics of Binding

To form a well-defined complex in solution two proteins have to overcome an entropic free energy barrier of about 15 kcal mol^{-1} (Finkelstein and Janin 1989) because of the loss of translational and rotational freedom of motion. For the association constant to exceed this by every factor of 10 requires an additional $\Delta G_A = -1.4$ kcal mol^{-1}. While this degree of accuracy has occasionally been achieved with some stronger protein-protein complexes, it has only been done by suitable choice of several adjustable parameters. For the highly transient, weak binding, complexes between electron transfer proteins the central problem is how to achieve rapid electron transfer in a complex between promiscuous proteins that has a very short lifetime (rapid k_{off}). The discussion in previous sections shows that a variety of solutions have been adopted, taking advantage of the variety of physical contributions to ΔG_A.

Electrostatics and the hydrophobic effect are major contributors to binding energy, but there may also be uncompensated changes in van der Waals energy when protein-water contacts are replaced by protein-protein contacts at the interface. A complete inventory should also include additional entropic factors, particularly the loss of sidechain mobility at the interface. The hydrophobic effect is the result of desolvation of hydrophobic surfaces on complex formation and is largely entropic at room temperature. It always makes a stabilizing contribution to binding energy (negative ΔG_A) and approximate calculations can be made based on the assumption that the free energy contribution is proportional to the area of the interface. Honig and colleagues have used a value of 58.2 cal Å^{-2} (Froloff et al. 1997; Polticelli et al. 1999) which, together with the difference in interfacial area of 328 Å^2, is enough to explain the marked difference in binding energy between brs and ccpc (Table 19.1).

On the other hand it clearly cannot explain the difference between brs and pcf because the latter has the larger interfacial area. Although both interfaces are highly electrostatic (as is that of ccpc) it is to the details of the electrostatic interactions that we must look for the explanation.

19.4.2 Electrostatics of Binding

The principle of electrostatic interactions is simple, that is the application of Coulomb's law to the interactions between every charged atom in the system, but detailed calculations are huge and complicated for proteins in solution because of the very large numbers involved and the difficulty of knowing the precise spatial location of each atom at any instant in time. A common method of keeping the computation within bounds is to treat the polar solvent (water) and protein interior as separate continua with uniform dielectric constants. Electrostatic potentials of the polar atoms of the protein can then be calculated by the finite difference Poisson-Boltzmann method (Warwicker and Watson 1982; Sharp and Honig 1990). Although the calculation itself can be made reasonably accurate, the absolute values obtained for binding energies depend on several assumptions. Firstly, the atomic structure of the complex must be known, and then it is often assumed that the proteins can be treated as rigid bodies which do not change conformation on complex formation. This is usually a fair approximation for small, redox proteins. Secondly, it is necessary to decide how to construct the boundary between the high dielectric solvent and the low dielectric protein interior, and the value to assign to the dielectric constant of the latter. Ideally a value of 1 should be used, but 2 or 4 are common because they allow for atomic polarizabilites and 4 accounts for the dielectric response to conformational changes that are not treated explicitly (Froloff et al. 1997). The interface between these two dielectrics is defined by the surface traced out by a probe sphere of radius 1.4 Å (to imitate a water molecule) either as the solvent accessible surface (given by the center of the sphere) or the molecular surface (given by the contact point between probe and protein (Vijayakumar and Zhou 2001; Dong and Zhou 2002, 2006). All these factors can have a significant effect on the final result, but comparative calculations under constant conditions should be less affected.

The results of comparative calculations with the program Delphi for the four different complexes discussed above (Table 19.1) are shown in Table 19.2. The electrostatic contribution to binding energy, ΔG_{es}, is obtained as the difference between the electrostatic energy of the complex and the sum of those of the individual proteins. A thermodynamic cycle gives ΔG_{es} as the sum of the coulombic interaction energy, ΔG_{c}, and the reaction field or solvation energy, ΔG_{rf} (Gilson and Honig 1988; Froloff et al. 1997). The Coulombic energy is calculated with Coulomb's law when the whole system is assigned a single dielectric constant, usually that of the protein interior, and the solvation energy is that required for changing the dielectric constant of the external medium for that of the solvent (water). From Table 19.2

Table 19.2 Electrostatic binding energies of four protein-protein complexes

Complex	Electrostatic binding energy (kcal mol^{-1})[a]			Electrostatic interaction energies (kcal mol^{-1})[b]			
	ΔG_c	ΔG_{rf}	ΔG_{es}	if/bb	if/other	bb/other	overall
brs	−158.5	165.6	7.1	−28.1	−3.3	−11.5	−43.2
ccpc	−498.8	484.7	−14.1	−9.3	−13.2	−1.2	−22.7
pcf	−124.5	162.4	37.9	2.8	−3.6	−8.0	−6.6
plpcf	87.1	−70.9	16.2	−7.7	−2.4	−3.2	−13.3

[a] Calculated with Delphi (Gilson et al. 1987; Gilson and Honig 1988; Sharp and Honig 1990). The Parse parameters (Sitkoff et al. 1994) were used for atomic charges and radii, the ionic strength was 0.1 M, and inner and outer dielectric constants were 4.0 and 78.5 respectively

[b] Calculated with thermodynamic cycles (Serrano et al. 1990; Di Cera 1998). Interface charges (if) included the formal, but not partial, charges of interface residues; backbones charges (bb) were the partial charges of the peptide backbone of the whole protein, and charges at the N and C termini; other charges (other) included all other charges of the protein

it is immediately apparent that ΔG_{es} is the sum of two large numbers, one positive and one negative. The balance between the two depends much upon the magnitude of ΔG_{rf}, which in turn depends on the details of the charge distribution within the complex and the individual proteins. The net result is that even with complementary charges on the reaction partners, electrostatics can have a destabilizing effect on the complex (Elcock et al. 1999; Sheinerman et al. 2000; Miyashita et al. 2003; Brock et al. 2007). The major advantage of such a charge pattern is then the long-distance attraction between the two proteins (before desolvation has taken place), which accelerates the formation of the complex (increases k_{on}) and the possibility of electron transfer (Selzer et al. 2000).

The difference between coulombic and desolvation energies is influenced by a number of factors. The distribution of interface sidechain charges is important: partial burial in the monomer, or partial exposure to solvent in the complex help to minimize the desolvation penalty, but interprotein hydrogen bonds and salt bridges also have a stabilizing tendency and networks of charge-charge interactions are especially favourable (Albeck et al. 2000; Lebbink et al. 2002; Sheinerman and Honig 2002). The results in Table 19.2 show that cooperative interactions between different groups of charges, including those not in the interface, may make significant contributions to the overall electrostatic component of binding energy. Thus the interaction between formal interface charges and the partial charges of the backbone make an important contribution to stabilizing the complex in brs and to a lesser extent in ccpc. One factor contributing to this interaction in brs is the α–helix of barstar which is a prominent feature of the interface. The most important stabilizing influence for ccpc is the cooperativity between the interface charges and all other sidechain charges, including partial charges of all sidechain polar residues and formal sidechain charges from outside the interface. Cooperative interactions between the backbone and other charges are also significant for brs, and to a lesser extent for pcf. The variability of these different interactions demonstrates that the arrangement of the backbone and other charges needs to be taken into account in the assessment of the structural origins of the electrostatic binding energy. A major contributor to the large difference in binding energy between brs and the electron-transfer complexes, especially pcf and plpcf, is the cooperativity between the interface charges and backbone and other charges. This is despite the fact that the coulombic attraction between cytochrome *f* and plastocyanin in pcf (but not plpcf) is much larger than in brs.

19.5 Promiscuity of Electron Transfer

The commonest electron carrier in the thylakoid lumen of present day oxygenic organisms is plastocyanin; but there is an alternative, now known as cytochrome c_6. Cytochrome c_6 is prevalent in red algae and possibly in those algae containing chlorophyll *c* (brown algae), whereas many green algae and cyanobacteria are capable of synthesizing both proteins. Normally a choice is made between them, depending on the availability of copper and iron in the growth medium, plastocyanin being preferred unless the supply of copper is limited. Conditions on the early earth were very different. Before the development of oxidizing conditions following the advent of oxygenic photosynthesis in cyanobacteria, the oceans were anaerobic and rich in sulphides. Copper sulphides are extremely insoluble in water, whereas ferrous sulphide is sufficiently soluble to allow organisms to develop the use of both iron-sulphur complexes and haem in electron-transfer proteins. Evolution of cyanobacteria led first to oxidation of this huge pool of sulphide, and later to free dioxygen in the atmosphere. Oxidizing conditions and the disappearance of free sulphide caused the availabilities of copper and iron to be reversed, with ferric hydroxide being extremely insoluble and cupric hydroxide showing a moderate solubility (Fraústa da Silva and Williams 1991). The potential presence of both plastocyanin and cytochrome c_6 implies that not only does plastocyanin have to be able to react with both cytochrome *f* and photosystem I, but also the latter two complexes have to be capable of reacting with both plastocyanin and cytochrome c_6.

Plastocyanin and cytochrome c_6 are structurally very different proteins with different evolutionary origins. Plastocyanin is a typical cupredoxin with a Greek key arrangement of β-strands as a β-sandwich. Cytochrome c_6 is a typical class I *c*-type cytochrome, largely α-helix in structure. Nevertheless, from any one organism they tend to have a similar surface charge pattern, similar redox potentials and they perform the same function within the thylakoids. The kinetics of reaction between cytochromes *f* and c_6 have not been studied because of the supposed difficulty of distinguishing the two spectroscopically, although in fact they give a substantial difference spectrum in the γ-band region. On the other hand, NMR studies have shown a similar mode of binding between cytochrome c_6 or plastocyanin with cytochrome *f* from *Nostoc* sp. PCC 7119 (Díaz-Moreno et al. 2005a, c). As described above, plastocyanin binds in such a way that the copper-binding H92 and its surrounding hydrophobic patch bind close to the haem region of cytochrome *f*, and an adjacent basic patch binds close to some acidic groups on cytochrome *f*. Cytochrome c_6 binds in an analogous fashion, with the exposed haem edge taking the part of H92 and an adjacent group of basic residues showing electrostatic interaction. An interesting feature is that involved in these interactions is an arginine residue adjacent to the haem edge in cytochrome c_6, or H92 in plastocyanin, which appears to be highly conserved in both proteins throughout all cyanobacteria (Kerfeld et al. 1995; Molina-Heredia et al. 2001). An arginine occurs in this position in some, but not all, algae; the equivalent position in higher-plant plastocyanin is occupied by a glutamine. With proteins from *Phormidium laminosum* R93 has been shown to be important in the reaction with cytochrome *f*, interacting specifically with D63 (Schlarb-Ridley et al. 2002; Hart et al. 2003), even though the interaction of the proteins is predominantly hydrophobic.

By contrast with the reaction with cytochrome *f*, the kinetics of electron donation by both plastocyanin and cytochrome c_6 to photosystem I have been extensively studied by flash absorption spectroscopy. In preparations from plants and green algae the photooxidation of plastocyanin or cytochrome c_6 shows kinetic behaviour which is called type III. This can be described by the following scheme

$$\mathrm{PSI_{ox}} + \mathrm{Pc_{red}} \underset{k_{-a}}{\overset{k_a}{\rightleftarrows}} [\mathrm{PSI_{ox}Pc_{red}}]_1 \underset{k_{-r}}{\overset{k_r}{\rightleftarrows}} [\mathrm{PSI_{ox}Pc_{red}}]_2 \xrightarrow{k_{et}} [\mathrm{PSI_{red}Pc_{ox}}]$$
$$\rightarrow \mathrm{PSI_{red}} + \mathrm{Pc_{ox}}$$

which is similar to that shown above in section IIIC for reduction by cytochrome *f*. In cyanobacteria types I and II predominate. These refer to situations in which one of the three steps in the reaction is rate limiting. Thus type II occurs when the electron transfer step is rate limiting, but rearrangement is fast (referred to as 'activation controlled' above), and type I when the slowest step is the initial binding (diffusion controlled). The type I case gives a single exponential, the rate constant for which increases with reduced plastocyanin concentration until saturation is reached, when k_{obs} corresponds to k_{et}.

The promiscuity of plastocyanin and cytochrome c_6 in cyanobacteria extends further, because of the presence of both photosynthetic and respiratory chains in the thylakoid membranes. These two systems are not kept separate, as the cytochrome b_6f complex serves them both. The major oxidase is a cytochrome *c* oxidase which is remarkably similar to its eukaryotic counterpart localized in mitochondria (Hart et al. 2005; Paumann et al. 2005). These are the classical cytochrome aa_3 type of oxidase with a haem-copper binuclear center located in subunit I. The surprising feature of the cyanobacterial enzyme is that mammalian cytochrome *c*, which has a ring of basic residues surrounding the exposed haem edge, is a good substrate. This is a reflection of the fact that electron donation to cytochrome oxidase is through the Cu_A site on subunit II and that the binding site is essentially acidic in almost all known cytochrome oxidases (Zhen et al. 1999). An interesting exception to this is the cytochrome ba_3 oxidase of the bacterium *Thermus thermophilus*, which is an extreme thermophile (Maneg et al. 2004). This may be a consequence, as mentioned above when discussing *Phormidium laminosum*, of the fact that electrostatic interactions are weakened as the temperature increases. It seems likely that the native donors in cyanobacteria are plastocyanin and cytochrome c_6, despite the varying charge properties of these proteins in different species, but a thorough study has been made only in the case of *Nostoc* sp. PCC 7119 and *Synechocystis* sp. PCC 6803 (Durán et al. 2004; Navarro et al. 2005). The highly conserved arginine residue on the binding surface may be significant. Why the negative charges on the surface of subunit II are so widely conserved is unclear, but a consequence may be that when cytochrome oxidase was lost from the thylakoid membranes with the formation of eukaryotic oxyphototrophs, plastocyanin and cytochrome c_6 became free to evolve into the highly acidic proteins we now see in plants and green algae.

19.6 Conclusions

There are three aspects to the study of the role of plastocyanin and cytochrome c_6 in electron transfer. First and foremost is the need to understand the physical and structural properties of the proteins concerned that allow rapid electron transfer and turnover of the system as a whole, that is to say the ability to react equally well with more than one partner. This aspect is the most highly developed. It is complicated by the need to take into account the atomic level structure of each of the proteins, but we have tried to show above that underlying this complexity are some relatively simple principles.

When the physical basis of the functional requirements are understood it becomes possible to consider the second aspect, which is the evolutionary pressures to which the system has been exposed and the evolutionary basis of what we actually observe in the modern world. The most striking feature of our present understanding is that the need to balance the rate of electron transfer with the specifity of protein-protein binding allows considerable flexibility of interface design. The variability of surface properties of plastocyanin and cytochrome c_6 in cyanobacteria provides an excellent system with which to study the underlying requirements. As a result certain broad evolutionary conclusions seem reasonably well established, in particular how the availability of copper and iron determines the choice between plastocyanin and cytochrome c_6, and the fact that the three-dimensional structure of the protein is irrelevant provided it is small and has suitable surface properties – in other words we have an example of evolutionary *bricolage* in the expression of Jacob (Jacob 1977; Howe et al. 2006). However, the system also provides a good opportunity, which has not been adequately exploited, to investigate coevolution and conservation of interface residues in weak protein-protein interactions of limited specifity (Teichmann 2002; Lee et al. 2007; Parson 2007). An important distinguishing feature of cyanobacterial thylakoid membranes is the presence of cytochrome oxidase. The endosymbiotic incorporation of a cyanobacterial cell into a eukaryotic cell that already possessed a cytochrome oxidase in mitochondria removed a significant evolutionary constraint, the need for plastocyanin and cytochrome c_6 to be capable of reacting with both photosystem I and cytochrome oxidase.

The least developed aspect of the system is the relation between the physical properties of the proteins concerned and the behaviour *in vivo*, partly because of the experimental difficulties. The concentration of free protein in the thylakoid lumen, at least in chloroplasts (Kieselbach et al. 1998), suggests that macromolecular crowding such as is expected to occur in cell cytoplasm (Ellis 2001b; Minton 2001) is not likely to modify seriously the behaviour of plastocyanin in the lumen (Schlarb-Ridley et al. 2005). However, the thylakoid lumen is a very narrow, flattened sac, with a bumpy internal surface. If plastocyanin needs to diffuse from the cytochrome b_6f complex to photosystem I it could be that confinement is more important than macromolecular crowding, as normally understood. There is also some evidence that at functional pHs, that is pH < 7 for the lumen, plastocyanin is not free in solution (Bendall and Wood 1978; Lockau 1979), so that the diffusion is two-dimensional on the membrane surface rather than in bulk solution. The nature of the diffusion path and the location of the cytochrome b_6f complex in the cyanobacterial membranes have not been established. In *Synechocystis* sp. PCC 6803 photosystem I complexes are located between rows of photosystem II complex (Mullineaux 1999), so the distance between cytochrome *f* and the plastocyanin binding site on photosystem I is likely to be small. For the purple bacterium *Rhodobacter sphaeroides* a supercomplex involving the reaction center, the cytochrome bc_1 complex and cytochrome c_2 has been described (Joliot et al. 2005), but there is no evidence for a corresponding supercomplex in any oxygenic organism.

The study of electron transfer reactions involving soluble proteins has proved particularly valuable for developing our current understanding of the more general problem of the structural requirements for weak, transient interactions between proteins, which is relevant to many signalling systems as well as to electron transfer. These processes can be characterized as requiring a high value for k_{off}. For electron transfer reactions the functional need for an appropriate balance of specificity against rapid turnover and limited promiscuity can be understood in physical terms as the establishment of an appropriate distance for good electronic coupling (roughly 10–14 Å), and a small reorganization energy, by a balance of attractive and repulsive forces. These forces acting at the interface include electrostatics, hydrophobic interactions and desolvation effects. Desolvation is crucial for determining the net binding energy because it can make stabilizing and destabilizing contributions which are both very large. Long distance electrostatic attraction between complementary charges is important for enhancement of k_{on} but at short range attraction is much weakened, or even reversed, by a heavy desolvation penalty. On the other hand direct contact between hydrophobic surfaces makes a large, stabilizing contribution to binding energy because of the larger entropy of bulk water compared with that of water molecules bound to a hydrophobic surface. Even so, the lack of good surface complementarity in the hydrophobic contact areas of at least some electron transfer complexes indicates that there can be a destabilizing contribution from incomplete compensation for the loss of the van der Waals energy of solvent–solvent interactions.

The above considerations lead to a conclusion about the overall structure of the interface that is required for electron transfer systems. In the first place the interface must be small to medium in size because a large interface would be likely to provide too large a hydrophobic stabilization. A survey of 11 crystal structures of electron transfer complexes in the Protein Databank (Crowley and Carrondo 2004) confirms this conclusion, and agrees with the NMR structures for plastocyanin/cytochrome *f* complexes described above. Water is inimical to tight binding, because the hydrophobic effect depends on solvent exclusion and electrostatic contributions from hydrogen bonding and salt bridges depend on close contact. The interface of redox complexes is thus relatively permeable to water, but nevertheless contains a close-contact hydrophobic core which includes the dominant pathway for electron transfer in order to minimize reorganization energy and distance between redox centers.

Cyanobacteria have been important in reaching these conclusions, especially in the detailed studies of the interaction between cytochrome *f* and plastocyanin, because they have shown most clearly that the physiological requirements have led to no single solution in structural terms. Electrostatic attraction is frequently an important factor in enhancing the rate of association, but we have seen that there is no need for a strong charge or surface complementarity. This helps to keep binding weak by minimizing the contribution of inter-protein hydrogen bonds and salt bridges. The importance of hydrophobic interactions has become clearer, not only because they provide the conditions necessary for fast electron transfer, and also a minimum of binding energy but, surprisingly perhaps, they can also enhance k_{on} almost to the exclusion of any need for long-distance electrostatic attraction.

Acknowledgements Work from our own laboratory has been supported by the Biotechnology and Biological Sciences Research Council, the Oppenheimer Fund, University of Cambridge, and Corpus Christi College, Cambridge. We are grateful to Barry Honig for making available the program Delphi, to Edward Solomon for providing atomic partial charges for the copper center of plastocyanin and to Peter Crowley for providing the coordinates of the complex between cytochrome *f* and plastocyanin from *Phormidium laminosum*.

References

Albarrán C, Navarro JA, Molina-Heredia FP, Murdoch PS, De la Rosa MA, Hervás M (2005) Laser flash-induced kinetic analysis of cytochrome *f* oxidation by wild-type and mutant plastocyanin from the cyanobacterium *Nostoc* sp. PCC 7119. Biochemistry 44:11601–11607

Albarrán C, Navarro JA, De la Rosa MA, Hervás M (2007) The specificity in the interaction between cytochrome *f* and plastocyanin from the cyanobacterium *Nostoc* sp. PCC 7119 is mainly determined by the copper protein. Biochemisry 46:997–1003

Albeck S, Unger R, Schreiber G (2000) Evaluation of direct and cooperative contributions towards the strength of buried hydrogen bonds and salt bridges. J Mol Biol 298:503–520

Babu CR, Volkman BF, Bullerjahn GS (1999) NMR solution structure of plastocyanin from the photosynthetic prokaryote, *Prochlorothrix hollandica*. Biochemistry 38:4988–4995

Badsberg U, Jorgensen AMM, Gesmar H, Led JJ, Hammerstad JM, Jespersen L-L, Ulstrup J (1996) Solution structure of reduced plastocyanin from the blue-green alga *Anabaena variabilis*. Biochemistry 35:7021–7031

Bagby S, Driscoll PC, Harvey TS, Hill HAO (1994) High-resolution solution structure of reduced parsley plastocyanin. Biochemistry 33:6611–6622

Bendall DS (1996) Interprotein electron transfer. In: Bendall DS (ed) Protein electron transfer. Bios Scientific Publishers, Oxford, pp 43–68

Bendall DS, Wood PM (1978) Kinetics of electron transfer through higher-plant plastocyanin. In: Hall DO et al (eds) Proceedings of the fourth international congress on photosynthesis 1977. The Biochemical Society, London, pp 771–775

Bergkvist A, Ejdebäck M, Ubbink M, Karlsson BG (2001) Surface interactions in the complex between cytochrome *f* and the E43/D44N and E59K/E60Q plastocyanin double mutants as determined by ^{1}H-NMR chemical shift analysis. Protein Sci 10:2623–2626

Bertini I, Ciurli S, Dikly A, Fernández CO, Luchinat C, Safarov N, Shumilin S, Vila AJ (2001) The first solution structure of a paramagnetic copper(II) protein: the case of oxidized plastocyanin from the cyanobacterium *Synechocystis* PCC6803. J Am Chem Soc 123:2413

Bond CS, Bendall DS, Freeman HC, Guss JM, Howe CJ, Wagner MJ, Wilce MCJ (1999) The structure of plastocyanin form the cyanobacterium *Phormidium laminosum*. Acta Cryst D55:414–421

Brock K, Talley K, Coley K, Kundrotas P, Alexov E (2007) Optimization of electrostatic interactions in protein-protein complexes. Biophys J 93:3340–3352

Camacho CJ, Weng ZP, Vajda S, Delisi C (1999) Free energy landscapes of encounter complexes in protein-protein association. Biophys J 76:1166–1178

Camacho CJ, Kimura SR, Delisi C, Vajda S (2000) Kinetics of desolvation-mediated protein-protein binding. Biophys J 78:1094–1105

Carrell CJ, Schlarb BG, Bendall DS, Howe CJ, Cramer WA, Smith JL (1999) Structure of the soluble domain of cytochrome *f* from the cyanobacterium, *Phormidium laminosum*: convergent evolution of the membrane-bound *c*-type cytochromes from photosynthetic and respiratory cytochrome *bc* complexes. Biochemistry 38:9590–9599

Chi Y-I, Huang L-S, Zhang Z, Fernández-Velasco JG, Berry EA (2000) X-ray structure of a truncated form of cytochrome *f* from *Chlamydomonas reinhardtii*. Biochemistry 39:7689–7701

Collyer CA, Guss JM, Sugimura Y, Yoshizaki F, Freeman HC (1990) Crystal structure of plastocyanin from a green alga, *Enteromorpha prolifera*. J Mol Biol 211:617–632

Colman PM, Freeman HC, Guss JM, Murata M, Norris VA, Ramshaw JAM, Venkatappa MP (1978) X-ray crystal structure analysis of plastocyanin at 2.7 A resolution. Nature 272:319–324

Crowley PB, Carrondo MA (2004) The architecture of the binding site in redox protein complexes: implications for fast dissociation. Proteins 55:603–612

Crowley PB, Otting G, Schlarb-Ridley BG, Canters GW, Ubbink M (2001) Hydrophobic interactions in a cyanobacterial plastocyanin-cytochrome *f* complex. J Am Chem Soc 123: 10444–10453

Crowley PB, Vintonenko N, Bullerjahn GS, Ubbink M (2002) Plastocyanin-cytochrome *f* interactions: the influence of hydrophobic patch mutations studied by NMR spectroscopy. Biochemistry 41:15698–15705

Crowley PB, Hunter DM, Sato K, McFarlane W, Dennison C (2004) The parsley plastocyanin-turnip cytochrome *f* complex: a structurally distorted but kinetically functional acidic patch. Biochem J 378:45–51

De Rienzo F, Gabdoulline RR, Menziani MC, De Benedetti PG, Wade RC (2001) Electrostatic analysis and brownian dynamics simulation of the association of plastocyanin and cytochrome *f*. Biophys J 81:3090–3104

Despa F, Berry RS (2007) The origin of long-range attraction between hydrophobes in water. Biophys J 92:373–378

DeVault D (1980) Quantum mechanical tunnelling in biological systems. Q Rev Biophys 13: 387–564

DeVault D (1984) Quantum-mechanical tunnelling in biological systems. Cambridge University Press, Cambridge

Di Cera E (1998) Site-specific analysis of mutational effects in proteins. Adv Protein Chem 51: 59–119

Díaz-Moreno I, Díaz-Quintana A, Ubbink M, De la Rosa MA (2005a) An NMR-based docking model for the physiological transient complex between cytochrome *f* and cytochrome c_6. FEBS Lett 579:2891–2896

Díaz-Moreno I, Díaz-Quintana A, De la Rosa MA, Crowley PB, Ubbink M (2005b) Different modes of interaction in cyanobacterial complexes of plastocyanin and cytochrome *f*. Biochemistry 44:3176–3183

Díaz-Moreno I, Díaz-Quintana A, De la Rosa MA, Ubbink M (2005c) Structure of the complex between plastocyanin and cytochrome *f* from the cyanobacterium *Nostoc* sp. PCC 7119 as determined by paramagnetic NMR. J Biol Chem 280:18908–18915

Dong F, Zhou H-X (2002) Electrostatic contributions to T4 lysozyme stability: solvent-exposed charges versus semi-buried salt bridges. Biophys J 83:1341–1347

Dong F, Zhou H-X (2006) Electrostatic contribution to the binding stability of protein-protein complexes. Proteins 65:87–102

Durán RV, Hervás M, De la Rosa MA, Navarro JA (2004) The efficient functioning of photosynthesis and respiration in *Synechocystis* sp. PCC 6803 strictly requires the presence of either cytochrome c_6 or plastocyanin. J Biol Chem 279:7229–7233

Ejdebäck M, Bergkvist A, Karlsson BG, Ubbink M (2000) Side-chain interactions in the plastocyanin-cytochrome *f* complex. Biochemistry 39:5022–5027

Elcock AH, Gabdoulline RR, Wade RC, McCammon JA (1999) Computer simulation of protein-protein association kinetics: acetylcholinesterase-fasciculin. J Mol Biol 291:149–162

Ellis RJ (2001a) Macromolecular crowding: an important but neglected aspect of the intracellular environment. Curr Opin Struct Biol 11:114–119

Ellis RJ (2001b) Macromolecular crowding: obvious but underappreciated. Trends Biochem Sci 26:597–604

Finazzi G, Sommer F, Hippler M (2005) Release of oxidized plastocyanin from photosystem I limits electron transfer between photosystem I and cytochrome b_6f complex *in vivo*. Proc Natl Acad Sci U S A 102:7031–7036

Finkelstein AV, Janin J (1989) The price of lost freedom: entropy of bimolecular complex formation. Protein Eng 3:1–3

Fraústa da Silva JJR, Williams RJP (1991) The biological chemistry of the elements: the inorganic chemistry of life. Clarendon Press, Oxford

Froloff N, Windemuth A, Honig B (1997) On the calculation of binding free energies using continuum methods: application to MHC class I protein-peptide interactions. Protein Sci 6: 1293–1301

Gilson MK, Sharp KA, Honig B (1987) Calculating the electrostatic potential of molecules in solution: method and error assessment. J Comput Chem 9:327–335

Gilson MK, Honig B (1988) Calculation of the total electrostatic energy of a macromolecular system: solvation energies, binding energies, and conformational analysis. Proteins 4:7–18

Gong X-S, Wen JQ, Fisher NE, Young S, Howe CJ, Bendall DS, Gray JC (2000) The role of individual lysine residues in the basic patch on turnip cytochrome *f* for electrostatic interactions with plastocyanin *in vitro*. Eur J Biochem 267:3461–3468

Gross EL (2004) A brownian dynamics study of the interaction of *Phormidium laminosum* plastocyanin with *Phormidium laminosum* cytochrome *f*. Biophys J 87:2043–2059

Gross EL, Pearson DC (2003) Brownian dynamics simulations of the interaction of *Chlamydomonas* cytochrome *f* with plastocyanin and cytochrome c_6. Biophys J 85:2055–2068

Haddadian EJ, Gross EL (2006) A brownian dynamics study of the effects of cytochrome *f* structure and deletion of its small domain in interactions with cytochrome c_6 and plastocyanin in *Chlamydomonas reinhardtii*. Biophys J 90:566–577

Hart SE, Schlarb-Ridley BG, Delon C, Bendall DS, Howe CJ (2003) Role of charges on cytochrome *f* from the cyanobacterium *Phormidium laminosum* in its interaction with plastocyanin. Biochemistry 42:4829–4836

Hart SE, Schlarb-Ridley BG, Bendall DS, Howe CJ (2005) Terminal oxidases of cyanobacteria. Biochem Soc Trans 33:832–835

Howe CJ, Schlarb-Ridley BG, Wastl J, Purton S, Bendall DS (2006) The novel cytochrome c_6 of chloroplasts: a case of evolutionary *bricolage*? J Exp Bot 57:13–22

Hulsker R, Baranova MV, Bullerjahn GS, Ubbink M (2008) Dynamics in the transient complex of plastocyanin-cytochrome *f* from *Prochlorothrix hollandica*. J Am Chem Soc 130:1985–1991

Inoue T, Sugawara H, Hamanaka S, Tsukui H, Suzuki E, Kohzuma T, Kai Y (1999a) Crystal structure determinations of oxidized and reduced plastocyanin from the cyanobacterium *Synechococcus* sp PCC 7942. Biochemistry 38:6063–6069

Inoue T, Gotowda M, Sugawara H, Kohzuma T, Yoshizaki F, Sugimura Y, Kai Y (1999b) Structure comparison between oxidized and reduced plastocyanin from a fern, *Dryopteris crassirhizoma*. Biochemistry 38:13853–13861

Jacob F (1977) Molecular tinkering in evolution. In: Bendall DS (ed) Evolution from Molecules to Men. Cambridge University Press, Cambridge, pp 131–144

Jansson H, Ökvist M, Jacobson F, Ejdebäck M, Hansson Ö, Sjölin L (2003) The crystal structure of the spinach plastocyanin double mutant G8D/L12E gives insight into its low reactivity towards photosystem I and cytochrome *f*. Biochim Biophys Acta 1607:203–210

Joliot P, Joliot A, Verméglio A (2005) Fast oxidation of the primary electron acceptor under anaerobic conditions requires the organization of the photosynthetic chain of *Rhodobacter sphaeroides* in supercomplexes. Biochim Biophys Acta 1706:204–214

Jones S, Thornton JM (1996) Principles of protein-protein interactions. Proc Natl Acad Sci U S A 93:13–20

Kannt A, Young S, Bendall DS (1996) The role of acidic residues of plastocyanin in its interaction with cytochrome *f*. Biochim Biophys Acta 1277:115–126

Kerfeld CA, Anwar HP, Interrante R, Merchant S, Yeates TO (1995) The structure of chloroplast cytochrome c_6 at 1.9 Å resolution: evidence for functional oligomerization. J Mol Biol 250:627–647

Kieselbach T, Hagman Å, Andersson B, Schröder WP (1998) The thylakoid lumen of chloroplasts—Isolation and characterization. J Biol Chem 273:6710–6716

Kohzuma T, Inoue T, Yoshizaki F, Sasakawa Y, Onodera K, Nagatomo S, Kitagawa T, Uzawa S, Isobe Y, Sugimura Y, Gotowda M, Kai Y (1999) The structure and unusual pH dependence of plastocyanin from the fern *Dryopteris crassirhizoma*. J Biol Chem 274:11817–11823

Kramers HA (1940) Brownian motion in a field of force and the diffusion model of chemical reactions. Physica 7:284–304

Lange C, Cornvik T, Díaz-Moreno I, Ubbink M (2005) The transient complex of poplar plastocyanin with cytochrome *f*: effects of ionic strength and pH. Biochim Biophys Acta 1707:179–188

Laskowski RA (1995) SURFNET: a program for visualizing molecular surfaces, cavities and intermolecular interactions. J Mol Graph 13:323–330

Lawrence MC, Colman PM (1993) Shape complementarity at protein/protein interfaces. J Mol Biol 234:946–950

Lebbink JHG, Consalvi V, Chiaraluce R, Berndt KD, Ladenstein R (2002) Structural and thermodynamic studies on a salt-bridge triad in the NADP-binding domain of glutamate dehydrogenase from *Thermotoga maritima*: cooperativity and electrostatic contribution to stability. Biochemistry 41:15524–15535

Lee H, Cheng Y-C, Fleming GR (2007) Coherence dynamics in photosynthesis: protein protection of excitonic coherence. Science 316:1462–1465

Lin J, Beratan DN (2005) Simulation of electron transfer between cytochrome c_2 and the bacterial photosynthetic reaction center: brownian dynamics analysis of the native proteins and double mutants. J Phys Chem B 109:7529–7534

Lin J, Balabin IA, Beratan DN (2005) The nature of aqueous tunneling pathways between electron-transfer proteins. Science 310:1311–1313

Lockau W (1979) The inhibition of photosynthetic electron transport in spinach chloroplasts by low osmolarity. Eur J Biochem 94:365–373

Maneg O, Malatesta F, Ludwig B, Drosou V (2004) Interaction of cytochrome *c* with cytochrome oxidase: two different docking scenarios. Biochim Biophys Acta 1655:274–281

Marcus RA, Sutin N (1985) Electron transfers in chemistry and biology. Biochim Biophys Acta 811:265–322

Martinez SE, Huang D, Szczepaniak A, Cramer WA and Smith JL (1994) Crystal structure of chloroplast cytochrome *f* reveals a novel cytochrome fold and unexpected heme ligation. Structure 2:95–105

Martinez SE, Huang D, Ponomarev M, Cramer WA, Smith JL (1996) The heme redox center of chloroplast cytochrome *f* is linked to a buried five-water chain. Protein Sci 5:1081–1092

Meyer TE, Zhao ZG, Cusanovich MA, Tollin G (1993) Transient kinetics of electron transfer from a variety of *c*-type cytochromes to plastocyanin. Biochemistry 32:4552–4559

Minton AP (2001) The influence of macromolecular crowding and macromolecular confinement on biochemical reactions in physiological media. J Biol Chem 276:10577–10580

Miyashita O, Onuchic JN, Okamura MY (2003) Continuum electrostatic model for the binding of cytochrome c_2 to the photosynthetic reaction center from *Rhodobacter sphaeroides*. Biochemistry 42:11651–11660

Miyashita O, Okamura MY, Onuchic JN (2005) Interprotein electron transfer from cytochrome c_2 to photosynthetic reaction center: tunneling across an aqueous interface. Proc Natl Acad Sci U S A 102:3558–3563

Molina-Heredia FP, Hervás M, Navarro JA, De la Rosa MA (2001) A single arginyl residue in plastocyanin and cytochrome c_6 from the cyanobacterium *Anabaena* sp. PCC 7119 is required for efficient reduction of photosystem I. J Biol Chem 276:601–605

Moore JM, Lepre CA, Gippert GP, Chazin WJ, Case DA, Wright PE (1991) High-resolution solution structure of reduced French bean plastocyanin and comparison with the crystal structure of poplar plastocyanin. J Mol Biol 221:533–555

Moser CC, Page CC, Farid RS, Dutton PL (1995) Biological electron transfer. J Bioenerg Biomembr 27:263–274

Mullineaux CW (1999) The thylakoid membranes of cyanobacteria: structure, dynamics and function. Aus J Plant Physiol 26:671–677

Musiani F, Dikiy A, Semenov AY, Ciurli S (2005) Structure of the intermolecular complex between plastocyanin and cytochrome *f* from spinach. J Biol Chem 280:18833–18841

Navarro JA, Lowe CE, Amons R, Kohzuma T, Canters GW, De la Rosa MA, Ubbink M, Hervás M (2004) Functional characterization of the evolutionarily divergent fern plastocyanin. Eur J Biochem 271:3449–3456

Navarro JA, Durán RV, De la Rosa MA, Hervás M (2005) Respiratory cytochrome *c* oxidase can be efficiently reduced by the photosynthetic redox proteins cytochrome c_6 and plastocyanin in cyanobacteria. FEBS Lett 579:3565–3568

Northrup SH, Erickson HP (1992) Kinetics of protein-protein association explained by Brownian dynamics computer simulation. Proc Natl Acad Sci U S A 89:3338–3342

Onuchic JN, Beratan DN, Winkler JR, Gray HB (1992) Pathway analysis of protein electron-transfer reactions. Annu Rev Biophys Biomol Struct 21:349–377

Page CC, Moser CC, Chen XX, Dutton PL (1999) Natural engineering principles of electron tunnelling in biological oxidation-reduction. Nature 402:47–52

Page CC, Moser CC, Dutton PL (2003) Mechanism for electron transfer within and between proteins. Curr Opin Chem Biol 7:551–556

Parson WW (2007) Long live electronic coherence! Science 316:1438–1439

Paumann M, Regelsberger G, Obinger C and Peschek GA (2005) The bioenergetic role of dioxygen and the terminal oxidase(s) in cyanobacteria. Biochim Biophys Acta 1707:231–253

Pearson DC Jr., Gross EL (1998) Brownian dynamics study of the interaction between plastocyanin and cytochrome *f*. Biophys J 75:2698–2711

Pelletier H, Kraut J (1992) Crystal structure of a complex between electron transfer partners, cytochrome c peroxidase and cytochrome c. Science 258:1748–1755

Polticelli F, Ascenzi P, Bolognesi M, Honig B (1999) Structural determinants of trypsin affinity and specificity for cationic inhibitors. Protein Sci 8:2621–2629

Prudêncio M, Ubbink M (2004) Transient complexes of redox proteins: structure and dynamic details from NMR studies. J Mol Recognit 17:524–539

Qin L, Kostic NM (1993) Importance of protein rearrangement in the electron-transfer reaction between the physiological partners cytochrome *f* and plastocyanin. Biochemistry 32:6073–6080

Redinbo MR, Cascio D, Choukair MK, Rice D, Merchant S, Yeates TO (1993) The 1.5-Å crystal structure of plastocyanin from the green alga *Chlamydomonas reinhardtii*. Biochemistry 32:10560–10567

Romero A, De la Cerda B, Varela PF, Navarro JA, Hervás M, De la Rosa MA (1998) The 2.15 Å crystal structure of a triple mutant plastocyanin from the cyanobacterium *Synechocystis* sp. PCC 6803. J Mol Biol 275:327–336

Schlarb-Ridley BG, Bendall DS, Howe CJ (2002) Role of electrostatics in the interaction between cytochrome *f* and plastocyanin of the cyanobacterium *Phormidium laminosum*. Biochemistry 41:3279–3285

Schlarb-Ridley BG, Bendall DS, Howe CJ (2003) Relation between interface properties and kinetics of electron transfer in the interaction of cytochrome *f* and plastocyanin from plants and the cyanobacterium *Phormidium laminosum*. Biochemistry 42:4057–4063

Schlarb-Ridley BG, Mi H, Teale WD, Meyer VS, Howe CJ, Bendall DS (2005) Implications of the effects of viscosity, macromolecular crowding, and temperature for the transient interaction between cytochrome *f* and plastocyanin from the cyanobacterium *Phormidium laminosum*. Biochemistry 44:6232–6238

Schmidt L, Christensen HEM, Harris P (2006) Structure of plastocyanin from the cyanobacterium *Anabaena variabilis*. Acta Cryst D Biol Crystallogr 62:1022–1029

Selzer T, Albeck S, Schreiber G (2000) Rational design of faster associating and tighter binding protein complexes. Nature Struct Biol 7:537–541

Selzer T, Schreiber G (2001) New insights into the mechanism of protein-protein association. Proteins 45:190–198

Serrano L, Horovitz A, Avron B, Bycroft M, Fersht AR (1990) Estimating the contribution of engineered surface electrostatic interactions to protein stability by using double-mutant cycles. Biochemistry 29:9343–9352

Sharp KA, Honig B (1990) Electrostatic interactions in macromolecules: theory and applications. Annu Rev Biophys Biophys Chem 19:301–332

Sheinerman FB, Honig B (2002) On the role of electrostatic interactions in the design of protein-protein interfaces. J Mol Biol 318:161–177

Sheinerman FB, Norel R, Honig B (2000) Electrostatic aspects of protein-protein interactions. Curr Opin Struct Biol 10:153–159

Shibata N, Inoue T, Nagano C, Nishio N, Kohzuma T, Onodera K, Yoshizaki F, Sugimura Y, Kai Y (1999) The crystal structure of *Ulva pertusa* plastocyanin at 1.6 Å resolution, structural basis for regulation of the copper site by residue 88. J Biol Chem 274:4225

Sitkoff D, Sharp KA, Honig B (1994) Accurate calculation of hydration free energies using macroscopic solvent models. J Phys Chem 98:1978–1988

Soriano GM, Ponamarev MV, Tae GS, Cramer WA (1996) Effect of the interdomain basic region of cytochrome *f* on its redox reactions *in vivo*. Biochemistry 35:14590–14598

Soriano GM, Ponamarev MV, Piskorowski RA, Cramer WA (1998) Identification of the basic residues of cytochrome *f* responsible for electrostatic docking interactions with plastocyanin in vitro: relevance to the electron transfer reaction in vivo. Biochemistry 37:15120–15128

Sridharan S, Nicholls A, Honig B (1992) A new vertex algorithm to calculate solvent accessible surface areas. FASEB J 6:A174

Teichmann SA (2002) The constraints protein-protein interactions place on sequence divergence. J Mol Biol 324:399–407

Tsao Y-H, Evans DF, Wennerström H (1993) Long-range attractive force between hydrophobic surfaces observed by atomic force microscope. Science 262:547–550

Ubbink M, Bendall DS (1997) Complex of plastocyanin and cytochrome *c* characterized by NMR chemical shift analysis. Biochemistry 36:6326–6335

Ubbink M, Ejdebäck M, Karlsson BG, Bendall DS (1998) The structure of the complex of plastocyanin and cytochrome *f*, determined by paramagnetic NMR and restrained rigid-body molecular dynamics. Structure 6:323–335

Vijayakumar M, Zhou H-X (2001) Salt bridges stabilize the folded structure of barnase. J Phys Chem B 105:7334–7340

Volkov AN, Worrall JAR, Holtzmann E, Ubbink M (2006) Solution structure and dynamics of the complex between cytochrome *c* and cytochrome *c* peroxidase determined by paramagnetic NMR. Proc Nat Acad Sci U S A 103:18945–18950

Warwicker J, Watson HC (1982) Calculation of the electric potential in the active site cleft due to α-helix dipoles. J Mol Biol 157:671–679

Wenner JR, Bloomfield VA (1999) Crowding effects on *Eco*RV kinetics and binding. Biophys J 77:3234–3241

Wheeler KE, Nocek JM, Cull DA, Yatsunyk LA, Rosenzweig AC, Hoffman BM (2007) Dynamic docking of cytochrome b_5 with myoglobin and α-hemoglobin: heme-neutralization "squares" and the binding of electron-transfer-reactive configuration. J Am Chem Soc 129:3906–3917

Worrall JAR, Kolczak U, Canters GW, Ubbink M (2001) Interaction of yeast iso-1-cytochrome *c* with cytochrome *c* peroxidase investigated by [^{15}N,^{1}H] heteronuclear NMR spectroscopy. Biochemistry 40:7069–7076

Worrall JAR, Liu Y, Crowley PB, Nocek JM, Hoffman BM, Ubbink M (2002) Myoglobin and cytochrome b_5: a nuclear magnetic resonance study of a highly dynamic protein complex. Biochemistry 41:11721–11730

Worrall JAR, Reinle W, Bernhardt R, Ubbink M (2003) Transient protein interactions studied by NMR spectroscopy: the case of cytochrome *c* and adrenodoxin. Biochemistry 42:7068–7076

Xue YF, Ökvist M, Hansson Ö, Young S (1998) Crystal structure of spinach plastocyanin at 1.7 Å resolution. Protein Sci 7:2099–2105

Zhen Y, Hoganson CW, Babcock GT, Ferguson-Miller S (1999) Definition of the interaction domain for cytochrome *c* on cytochrome *c* oxidase. I. Biochemical, spectral, and kinetic characterization of surface mutants in subunit II of *Rhodobacter sphaeroides* cytochrome aa_3. J Biol Chem 274:38032–38041

Zhou Y-L, Liao J-M, Chen J, Liang Y (2006) Macromolecular crowding enhances the binding of superoxide dismutase to xanthine oxidase: implications for protein-protein interactions in intracellular environments. Int J Biochem Cell Biol 38:1986–1994

Chapter 20
Center of the Cyanobacterial Electron Transport Network: The Cytochrome b_6f Complex

Gábor Bernát and Matthias Rögner

20.1 Introduction

Major aim of this chapter is to provide an overview on electron and proton transport processes in Cytochrome (Cyt) b_6f complex including related regulatory aspects. While functional and regulatory aspects of the Cyt b_6f structure are thoroughly discussed in several current reviews (Allen 2004; Cape et al. 2006; Cramer and Zhang 2006; Cramer et al. 2006, 2008a, b; Baniulis et al. 2008), we focus on the following topics:

Section 20.2 focuses on the role of non-essential subunits (i.e., PetL, PetM, PetP) as well as on the multiplicity of the Rieske 2Fe–2S protein with data essentially obtained from studies on cyanobacteria. Section 20.3 discusses the role of certain amino acid residues or small domains in the vicinity of catalytic sites and/or prosthetic groups. Here, due to the limited amount of data on cyanobacteria, we also include data from the green alga *Chlamydomonas reinhardtii* or—in some cases—from higher plant mutants. Especially the high structural and functional similarity of cyanobacterial and plant Cyt b_6f complex as revealed by X-ray structure analysis allows this extension. (For simplicity, *Mastigocladus laminosus* numbering is used.) As complementary component of the plastoquinol (PQH_2) oxidation which is highly dependent on the Cyt b_6f activity, the potential role of Cyt *bd* oxidase in the electron transport processes is also shortly reviewed.

G. Bernát (✉)
Plant Biochemistry, Ruhr-University Bochum, 44780 Bochum, Germany
e-mail: gabor.bernat@rub.de

G. A. Peschek et al. (eds.), *Bioenergetic Processes of Cyanobacteria,* 573
DOI 10.1007/978-94-007-0388-9_20, © Springer Science+Business Media B.V. 2011

20.2 Overall Structure and Function of the Cyanobacterial Cytochrome b_6f Complex

20.2.1 Role of Cytochrome b_6f in the Cyanobacterial Electron Transport Network Including Cytochrome bd Oxidase

Central features of oxygenic photosynthesis are the sequentially coupled two photosytsems, photosystem II (PS II, see Chapter Renger and Ludwig) and photosystem I (PS I, see Chapter Fromme and Grotjohann) located in the thylakoid membrane (TM) and connected by the intersystem electron transport chain (ETC) (Fig. 20.1). This downhill ETC chain consists of two mobile electron carriers, plastoquinone (PQ) and plastocyanin (PC)—of which the latter can be substituted by cytochrome c_6 (Cyt c_6) in cyanobacteria and algae—and the integral membrane cytochrome b_6f (Cyt b_6f) complex, the plastoquinone:plastocyanin oxidoreductase (see Fig. 20.2)

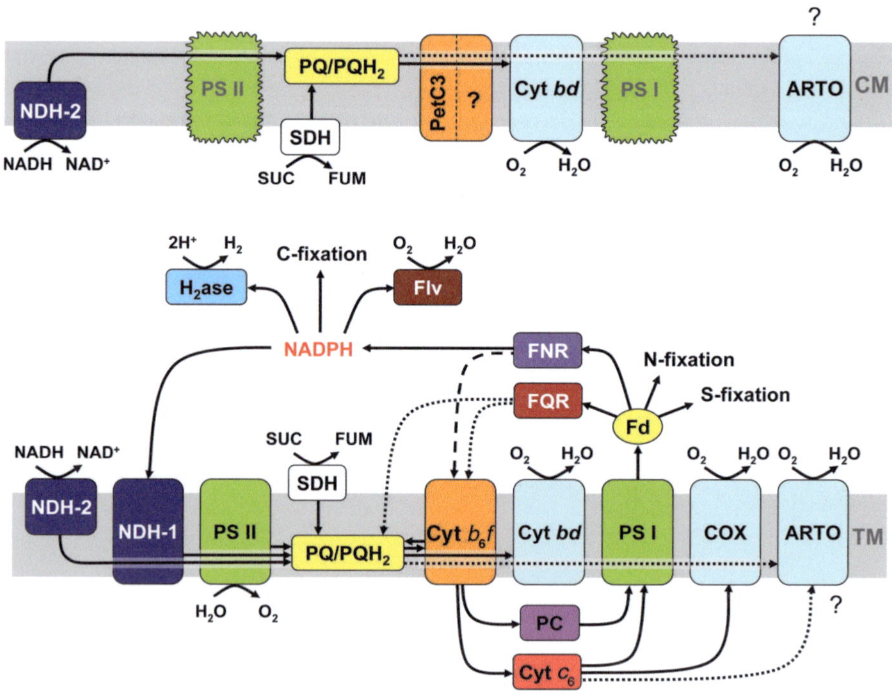

Fig. 20.1 Major-components and -routes of the cynobacterial ET network in the CM and TM of the cyanobacterium *Synechocystis* sp. PCC 6803. ET processes in the CM (*upper part*) must be limited due to incomplete or missing components. Question marks and dotted lines indicate that localization and substrate of ARTO as well as the ET route of PQH_2 reduction *via* FQR are still to be identified. The *dashed line* shows the proposed direct Cyt b_6f-reduction *via* Fd/FNR. *H_2-ase* hydrogenase, *Flv* Flavoprotein-1 and -3 (Helman et al. 2003). For more abbreviations, further details and references, see text

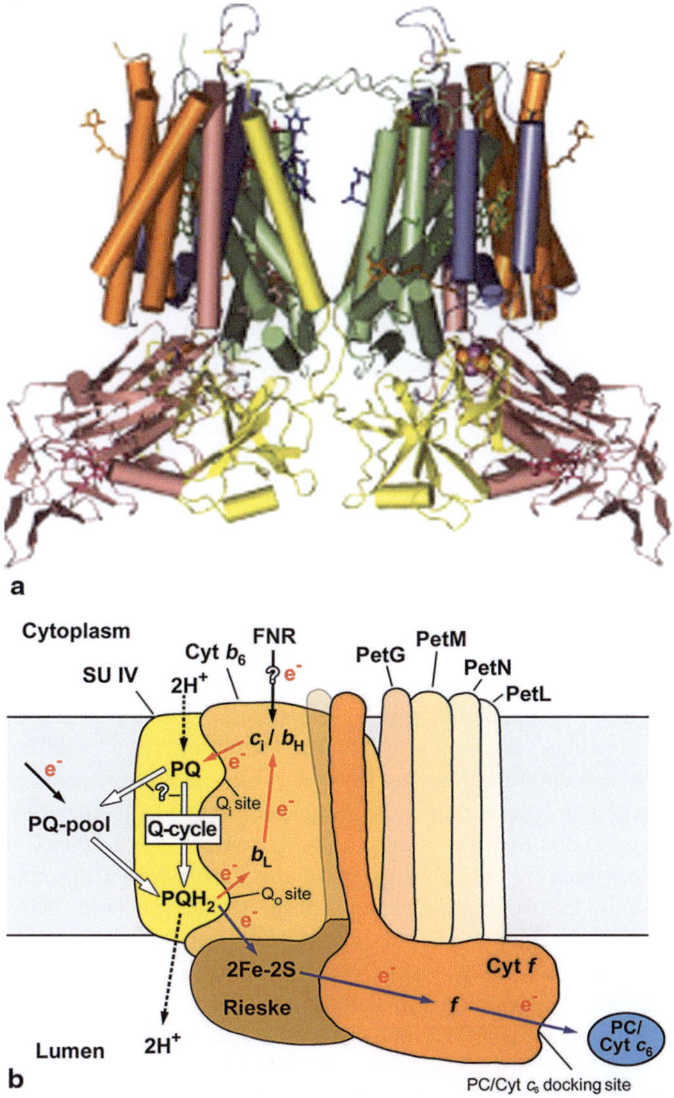

Fig. 20.2 **a** Structure of the dimeric Cyt b_6f complex (Reproduced from Cramer and Zhang 2006; copyright ©: Elsevier B.V.) with cytoplasmic side up and lumenal side down. Color codes of subunits: Cyt *f*, *pink*; Cyt b_6, *green*; Rieske 2Fe–2S subunit, *yellow*; SU IV, *blue*; small subunits, *green*. Non-protein components are shown as spheres (2Fe–2S center) or sticks (others). Color codes of these components: heme *f*, *pink*; hemes b_L and b_H, *purple*; heme c_i, *pink*; Chl, *green*; Car, *orange*; PQ in the Q_i quinone binding site, *blue*; artificial quinone analogue TDS in the Q_o quinone binding site, *orange*. **b** Schematic view of the electron transport routes and components of the Cyt b_6f complex. For clarity only the model of one monomer (instead of the functional dimer) is shown. *Solid*, *dashed* and *open arrows* represent electron and proton transfer and plastoquinol diffusion, respectively. ET in low- and high-potential chain is marked by *red* and *blue arrows*, respectively. Domain swapping of the Rieske 2Fe–2S subunit is indicated by the shadow of its membrane-spanning alpha-helical anchor. For details see text

[for recent reviews see (Allen 2004; Cramer and Zhang 2006; Cramer et al. 2006, 2008a, b; Cape et al. 2006; Baniulis et al. 2008), for book chapters see also (Kallas 1994; Hauska and Büttner 1996; Wollman 1998; Ke 2001). Beyond being an oxido-reductase, the Cyt b_6f complex (or the homologous Cyt bc_1 and Cyt bc in mitochondria and bacteria, respectively) operates as a proton pump and couples the downhill, vectorial electron transport with an effective proton translocation (Fig. 20.2b).

The Cyt b_6f complex is structurally well conserved as revealed by crystallographic structures at 3.1 Å and 3.0 Å resolution of isolated complexes from the unicellular green alga *Chlamydomonas reinhardtii* and two cyanobacteria, *Mastigocladus laminosus* and *Nostoc* sp. PCC 7120, respectively (Kurisu et al. 2003; Stroebel et al. 2003; Yamashita et al. 2007; Baniulis et al. 2009). It contains two quinone/quinol binding sites, Q_o (=Q_p) and Q_i (=Q_n), located close to the luminal and stromal/cytoplasmic surface of the membrane, respectively, and a PC/Cyt c_6 binding/docking site on the luminal surface of the complex (Fig. 20.2b). Besides the transiently bound quinone/quinol molecules it contains also seven non-protein components {four hemes, one chlorophyll (Chl) *a*, one 2Fe–2S center, and one carotenoid [Car, 9-*cis* β-carotene or echinenone, see (Boronowsky et al. 2001; Wenk et al. 2005)]} and lipid molecules. The Cyt b_6f complex consists of four large (17.5–32 kDa) subunits [Cyt *f* (encoded by the *petA* gene), Cyt b_6 (*petB*), the Rieske 2Fe–2S protein (*petC*) and subunit IV (*petD*)] and four small subunits (3.3–4.1 kDa; PetG, -L, -M and -N) at unit stoichiometry. With the exception of Car (being inserted between the PetL and PetM helices) all components are attached to the large subunits. All but two subunits, Cyt b_6 and subunit (SU) IV with four and three transmembrane helices, respectively, span the membrane once, resulting in a complex with 13 transmembrane helices. Besides their membrane spanning parts, Cyt *f* and Rieske 2Fe–2S proteins possess large hydrophilic cofactor binding domains which are extended into the lumen. Also, in contrast to all other Cyt b_6f subunits which are encoded by only a single gene, the Rieske 2Fe–2S protein (PetC) is encoded by a gene family in most cyanobacteria [(Schneider et al. 2002, 2004a, b); for reviews see (Schneider and Schmidt 2005) and Sect. 20.2.2.1]. While all three available X-ray structures show eight Cyt b_6f subunits, there is evidence for other protein components interacting transiently with the Cyt b_6f complex: ferredoxin:NADP$^+$ oxidoreductase (FNR), PetO, and PetP. Cramer et al. co-isolated Cyt b_6f and FNR from spinach chloroplast (Zhang et al. 2001), and the functional coupling of a small phosphoprotein, PetO (Hamel et al. 2000) to Cyt b_6f has also been reported for higher plants. This polypeptide might be analogous to the PetP, which has been proposed as a new cyanobacterial Cyt b_6f subunit [(Volkmer et al. 2007; Gendrullis et al. 2008); for details and model see Sect. 20.2.2.3)].

The photosynthetic protein complexes often form oligomers: while PS II and ATP-synthase can form dimers (Barber 2003; Rexroth et al. 2004; Nowaczyk et al. 2006), PS I forms trimers in cyanobacterial membranes (Jordan et al. 2001) and the Cyt b_6f complex is dimeric both in cyanobacteria and higher plants. However, in contrast to oligomerized photosynthetic supercomplexes and ATP-synthase, monomeric Cyt b_6f complex (Rögner et al. 1990) is inactive (Huang et al. 1994; Breyton et al. 1997) due to the so-called "domain swapping": X-ray crystallog-

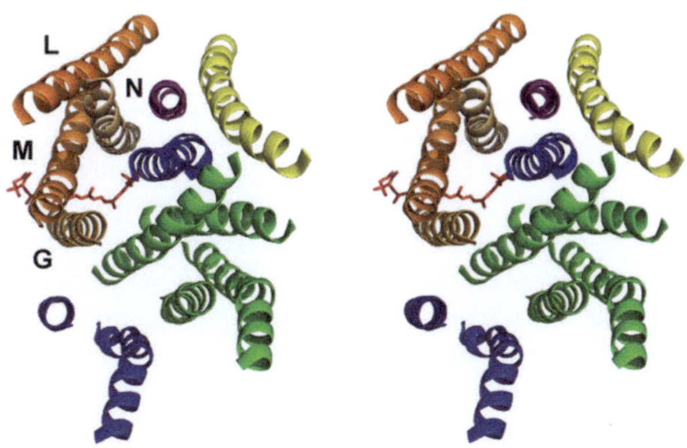

Fig. 20.3 Stereographic view of the membrane-spanning part of the monomeric Cyt b_6f complex as viewed from the lumenal side. Color coding of the subunits is as in Fig. 20.2a. Localization of the Car molecule (*red* sticks) is also shown

raphy revealed, that the membrane-spanning alpha-helical anchor and the cofactor binding extrinsic domain of the 2Fe–2S Rieske protein are parts of opposing monomers within the homodimer, respectively. For this reason, monomeric Cyt b_6f complex shows no fast electron transport between PQH_2 and PC/Cyt c_6 (see below and Fig. 20.2b.). The two Cyt b_6f monomers form two central, interconnected inner cavities which harbor the redox reactions of the (plasto)quinone species [Fig. 20.3a, (Kurisu et al. 2003; Cramer et al. 2006)]. The wall of each cavity is formed by six transmembrane helices (three from each monomer). The inner cavities may provide a higher local quinone/quinol concentration and a conduit for a transmembrane PQ/PQH_2 transport between the two quinone sites (Cramer et al. 2006). In combination with the proposed inter-monomer electron transfer (cross-talk) *via* the quinone/quinol pool inside the cavity, this may result in a more efficient ET (see the Q-cycle below) as "side effect" of dimerization. Also, the local quinone concentration may have implications in the quinone chemistry (see below).

Photosynthetic electron transport comprises a series of one-, two-, and four electron gated processes. While water-splitting at the manganese cluster in PS II is a four-electron gated process (see Chapter Renger and Ludwig), the mobile electron carriers PQ and FNR transport two electrons. In general, transporting electrons from a four- or two-electron carrier to a one-electron carrier is chemically difficult and needs complex machineries. The Cyt b_6f complex carries electrons from the two-electron carrier PQH_2 to the one-electron carrier PC/Cyt c_6 and solves this problem by transferring only one of the electrons directly to PC/Cyt c_6 *via* the Cyt b_6f complex while the other one is recycled in a cyclic process (the so-called Q-cycle; see Fig. 20.2b and below). "Recycling" the second electron of PQH_2 in the Q-cycle is part of the proton pumping mechanism of Cyt b_6f. Details of this complex scheme of electron and proton transfer reactions in Cyt b_6f are still not completely understood (see below).

Another central feature of the Cyt b_6f complex are the two branches of electron transport bifurcated at the PQH_2 binding site Q_o, formed by SU IV and Cyt b_6 residues and the Rieske 2Fe–2S center (Figs. 20.2b and 20.3) (Stroebel et al. 2003; Cramer et al. 2006; Yamashita et al. 2007). One of the two available electrons is transferred to the energetically favored, so-called high-potential chain (Fig. 20.2b, blue arrows) which gives rise to a a downhill electron transport towards PC/Cyt c_6 and PS I *via* the Rieske 2Fe–2S cluster and Cyt *f*. These electron transport components are reduced in darkness and become oxidized by $P700^{+\bullet}$ after charge separation in PS I. Due to their midpoint redox potentials, the Rieske 2Fe–2S protein (300–320 mV) and Cyt *f* (300–370 mV) are able to oxidize PQH_2 to $PQ^{\bullet-}$. {Noteworthy, the midpoint redox potential of Cyt *f* in cyanobacteria is more negative than in higher plants [300–330 mV *vs*. 355–370 mV (Ponamarev et al. 2000; Albarrán et al. 2005; Alric et al. 2005)]; however, due to an accompanying shift in the P700/$P700^{+\bullet}$ redox potential (Hamacher et al. 1996; Nakamura et al. 2005) the redox gap between Cyt *f*/Cyt f^{+} and P700/$P700^{+\bullet}$ is almost identical resulting the same driving force for the electron transfer.} Also, the ET from Rieske cluster to Cyt *f* is facilitated by a movement ("tethered diffusion") of the Rieske 2Fe–2S protein towards Cyt *f*.

The midpoint redox potential of the PQH_2/PQ couple (100 mV) and the Rieske 2Fe–2S center (300–320 mV) allows a 400–440 mV (two times the difference between the midpoint redox potentials given in parentheses) split between the PQH_2/$PQ^{\bullet-}$ and the $PQ^{\bullet-}$/PQ couple [see e.g. (Chobot et al. 2008)]. This in turn enables the ("oxidant induced") reduction of the other, so-called low potential chain which consists of two *b*-type hemes [Cyt b_L ($E_m \approx -50$–-150 mV) and Cyt b_H (0–-50 mV)] in a vectorial, membrane-spanning arrangement (Fig. 20.2b, red arrows) (Widger et al. 1984; Cramer et al. 1985). Surprisingly, X-ray structure determination of the Cyt b_6f complex (Kurisu et al. 2003; Stroebel et al. 2003) revealed a new redox-active component, a *c*-type heme [heme c_i; $E_m \approx 100$ mV, (Alric et al. 2005)] located in between the heme b_L and the second quinone-binding niche, Q_i. The Q_i pocket, located close to the stromal/cytoplasmic surface of the membrane is formed, similarly to the Q_o pocket, by SU IV and Cyt b_6 residues as well as by the heme c_i, which apparently coordinates the quinone directly (Kurisu et al. 2003; Stroebel et al. 2003; Yamashita et al. 2007; Cramer et al. 2008a). The Q_i site operates as a two-electron gate. Full reduction of the bound PQ molecule (to PQH_2) requires two electrons and the b_H/c_i couple may act as a two-electron donor in this reaction (Cramer et al. 2004; Cramer and Zhang 2006; Cramer et al. 2006, 2008a; Baniulis et al. 2008). As the low-potential chain receives only one electron from the quinol of the Q_o site, two subsequent quinol oxidations at the Q_o site are required for the PQ reduction at the Q_i site—at least according to the "classical view" when this is the only way to reduce the components of the low-potential chain (see below). PQH_2 at the Q_i pocket can either enter the PQ/PQH_2-pool or diffuse directly to the Q_o site (Cramer et al. 2006; Baniulis et al. 2008) for oxidation which closes the cycle (Fig. 20.2b, open arrows). Thus, in summary, this "Q-cycle" (Mitchell 1975) recycles halves of the electrons transported to the Q_o site. Apparently, this recycling has two major functions:

First, it is essential for converting the two-electron gated PQH_2 oxidation into two sequential reductions of the mobile one-electron carriers (PC or Cyt c_6). Second, it is crucial for an effective proton translocation resulting in an electrochemical proton gradient (proton motive force) across the membranes which in turn is the driving force of ATP-synthesis (Mitchell 1961, for a review, see Junge 2008). PQH_2 oxidation and PQ reduction at the Q_o and Q_i site, respectively, are accompanied with the translocation of two protons from the cytoplasm/stroma to the lumen, i.e. in a continuously running Q-cycle (Sacksteder et al. 2000) two protons per electron are transported. In combination with the water-oxidation at PS II (1 H^+/e^-) this yields an overall H^+/e^--stoichiometry of 3:1. The H^+/ATP ratio is determined by the oligomeric ratio of the c subunit and α/β dimers of the ATP synthase (Junge 2008). The c ring stoichiometry in plant and cyanobacterial ATP synthase is varied from c_{13} to c_{15} (Seelert et al. 2000, Pogoryelov et al. 2007; Junge 2008) which results in H^+/ATP ratios of 13/3, 14/3 or 15/3, respectively. These ratios is equivalent with the ATP/NAD(P)H ratios of 18:13, 9:7 and 6:5 (as a theoretical minimum), respectively, which are close to the ratio (3:2) required for CO_2 fixation, but not yet sufficient. The remaining ATP demand (e.g. 3/2–9/7 = 3/14 ATP/NAD(P)H for CO_2 fixation at c_{14} stoichiometry) is provided by a cyclic electron transport around PS I (Bendall and Manasse 1995; Allen 2003a; Joliot et al. 2006; Shikanai 2007).

This electron flow around PS I [which, actually, was discovered before the linear ETC (Arnon et al. 1954)] transfers electrons from reduced ferredoxin (Fd) back to PQ, yielding ATP (cyclic photophosphorylation), but not NAD(P)H. By varying the ratio of cyclic and linear electron flow, the ATP/NAD(P)H ratio can be adjusted according to the actual physiological conditions within a wide range, i.e. from 18:13/9:7/6:5 (only linear ET) to 1:0 (only cyclic ET). Without such a control mechanism an imbalance in the ATP/NAD(P)H ratio would lead to a feedback inhibition of the ET processes [see (Avenson et al. 2005)]. In some cases, cyclic electron transport provides a better utilization of the absorbed light, for instance in cyanobacteria which often show high PS I/PS II ratios (Fujita et al. 1994) or in shade plants/leaves which are mainly exposed to PS I light. Cyclic flow enables to maintain ET (and ATP synthesis), when PS II is damaged or down-regulated. Redox poise of the PQ-pool *via* cyclic/linear ET ratio adjustment is essential not only for maximal light energy utilization but also for the protection against photodamage. [For PS I/PS II ratio and redox-control see (Allen 2003b; Pfannschmidt 2003); for cyclic *vs.* linear ET and state transitions, see (Wollman 2001; Finazzi et al. 2002)].

While in higher plants two major cyclic electron transport routes are considered—the FNR- and the FQR- (ferredoxin-plastoquinone oxidoreductase) mediated route [(Bendall and Manasse 1995; Bukhov and Carpentier 2004; Munekage et al. 2004; Joliot et al. 2006; Rumeau et al. 2007), see also (Shikanai 2007; DalCorso et al. 2008; Endo et al. 2008)]—the situation in cyanobacteria is more complicated. In higher plants, photophosphorylation and oxidative phosphorylation are located in two separate organelles, the chloroplasts and mitochondria, respectively (see Chapter Renger and Ludwig). Cyanobacteria, the phylogenetical ancestors of plants' chloroplasts, perform both processes in parallel in the same compartment with PQ-pool and Cyt b_6f complex being shared components of both photosynthetic and respiratory ET chains. [This is partly also true for the Cyt b_6f complexes in chloroplasts, which retained some respiration, the so-called chlororespiration (Rumeau et al. 2007)]. Thus, the cyanobacterial Cyt b_6f-complex—in combination with the PQ-pool—plays a central role in the complex cyanobacterial ET network. With very few exceptions like the unique *Gloeobacter violaceus* (Rippka et al. 1974), cyanobacteria contain two distinct membrane systems: the cytoplasmic (CM) and the thylakoid membranes (TM). Most recent data indicate that they are non-connected, separate entities, which possibly communicate by vesicle formation and transport (Liberton et al. 2006). While both photosynthetic and respiratory ET occurs in the TM's, ET processes in the CM's must be limited due to incomplete or missing components. Importantly, the lack of all major Cyt b_6f subunits in CM—with the exception of the PetC3 Rieske isoform (for details see Sect. 20.2.2.1)—has been reported (Aldridge et al. 2008; Schultze et al. 2009), which strongly suggests the absence of a functional Cyt b_6f complex in the CM. This lack excludes a linear or cyclic ET in the CM and allows only a truncated ET chain *via* the PQ-pool [Fig. 20.1, (Schultze et al. 2009)], which is much more restricted than previously assumed (Pils and Schmetterer 2001; Teuber et al. 2001; Peschek et al. 2004; Paumann et al. 2005). In line with this, some major ET components such as type I NADPH dehydrogenase (NDH-1) and the aa_3-type Cyt *c* oxidase (COX) seem to be completely absent (Pils and Schmetterer 2001; Ohkawa et al. 2002; Zhang et al. 2004b) in CM. [(For the role of CM in PS II and PS I assembly and turnover see (Zak et al. 2001; Nowaczyk et al. 2006)].

Cyanobacterial cyclic electron transport routes can be classified into direct- and indirect cyclic pathways (Yeremenko et al. 2005): While in the direct routes FQR and FNR guide ("photosynthetic") electrons from PS I back to the PQ-pool *via* redox intermediates, the indirect routes guide ("respiratory") electrons from metabolism *via* NDH-1 or succinate dehydrogenase (SDH) to the PQ-pool (Fig. 20.1). However, reality is in between these arbitrary categories: While FNR is certainly a member of both (direct- and indirect) cyclic and linear ET, NDH-1 shows a large heterogeneity in subunit composition, function and involvement in ET processes for reviews see (Battchikova and Aro 2007; Ogawa and Mi 2007). This complex network of cyclic ET routes (together with pseudocyclic routes, see below) enables the rapid adaptation of cyanobacteria to changing environmental factors (light, temperature, nutrient availability, etc.), although quantitative experimental data are mainly lacking. For instance, it is still an open question whether NDH-1 or SDH is the main electron donor under certain environmental and metabolic conditions (Cooley and Vermaas 2001).

Although the direct involvement of Cyt b_6f in the cyclic ET from PS I back to PQ has been proposed for a long time, only recent structural data imply the possibility of a direct redox "wire" between Fd and the heme b_H/c_i couple: The narrow shielding of the b_H/c_i couple from the cytoplasm/stroma and the report of a purified FNR-Cyt b_6f complex—although up to now not from cyanobacteria (Zhang et al. 2001)—support this view (Cramer et al. 2004; Cramer and Zhang 2006; Cramer et al. 2006, 2008a; Baniulis et al. 2008).

Pseudocyclic routes connect water oxidation at PS II and concomitant reduction of oxygen to water. This latter typically occurs at the PS I acceptor side *via* the so-called Mehler reaction (Mehler 1951; Asada 1999) or at terminal oxidases—in cyanobacteria usually cytochrome *c* oxidase (COX), the alternative terminal oxidase (ARTO) and the cytochrome *bd* oxidase (Cyt *bd*) (Howitt and Vermaas 1998;

Pils and Schmetterer 2001; Berry et al. 2002; Hart et al. 2005; Paumann et al. 2005, see Chapter Bernroitner et al.). Although quinol oxidase activity of Cyt *bd* in cyanobacteria is still under debate (Peschek et al. 2004; Paumann et al. 2005), recent data with cyanobacterial Cyt *bd* overexpressed in *Escherichia coli* clearly show that this enzyme, indeed, functions as plastoquinol oxidase (Mogi and Miyoshi 2009). Moreover, chlorophyll fluorescence data clearly show an increased Cyt *bd* activity (a) in cyanobacterial wild type (WT) strains exposed to conditions which result in a highly reduced PQ pool (Berry et al. 2002) and (b) in various mutants with impaired ET through Cyt b_6f complex [(Schneider et al. 2001, 2004a), see also Sects. 20.2.2, and 20.3.1]. In accordance with this, an enhanced Cyt *bd* protein level was recently detected in the *Synechocystis* sp. PCC 6803 *petC1* strain with a rather low Cyt b_6f population size (Tsunoyama et al. 2009). These reports clearly show that Cyt *bd* has an important (protective) role in the PQ redox poise by taking over (partly) the role of the Cyt b_6f in PQH_2 re-oxidation. Some cyanobacteria—such as *G. violaceus* and *Anabaena variabilis*—use the plastid terminal oxidase (PTOX) as an alternative oxidase (McDonald et al. 2003). In contrast to Cyt *bd* and PTOX, which accept electrons from PQH_2 (i.e. "before" the Cyt b_6f), COX is reduced by Cyt c_6 or PC (i.e."after" Cyt b_6f), while the substrate of ARTO is not yet identified. Terminal oxidases generate a membrane potential by ET-coupled proton translocation. In photoautotrophs they primarily seem to function as security valves to prevent over-reduction of the linear electron transport chain (e.g. under limited CO_2 fixation or other conditions described above). However, similarly to the cyclic routes (see above), activation/regulation and usage of these processes is still largely unknown.

Beyond its very central role in the coupled proton/electron transport processes, Cyt b_6f also plays an important role in the redox regulation, notably as key component of signal cascades. Especially for state transitions, the short-term regulation of light energy distribution between the two photosystems, a kinase is phosphorylated at the Cyt b_6f complex after its conformational change, induced by the occupation of the Q_o site (Vener et al. 1995, 1997, 1998; Zito et al. 1999; Finazzi et al. 2001; Mao et al. 2002). For reviews see (Wollman 2001; Finazzi and Forti 2004); for the putative role of Rieske-heterogeneity and PetP subunits in regulatory processes, see Sects. 20.2.2.1 and 20.2.2.3, respectively.

20.2.2 Role of Cytochrome b_6f Subunits for Assembly and Function

The minimal functional Cyt b_6f should consist of six essential subunits: Cyt *f*, Cyt b_6, the Rieske 2Fe–2S protein, subunit IV, PetG, and PetN. Deletion of the other two small subunits, PetL and PetM,—although they have an impact on the ET properties—keep the Cyt b_6f complex functional [(Schneider et al. 2001, 2007); see Sect. 20.2.2.2]. Also, deletion of some (but not all) Rieske-encoding *petC* isoforms is possible—without impairing the function of Cyt b_6f complexes completely—(at least) in *Synechocystis* PCC 6803 (see Sect. 20.2.2.1).

The central role of this complex in the electron transport network (see Sect. 20.2.1) is the major reason why the essential cytochrome b_6f subunits cannot be deleted. The green alga *Chlamydomonas reinhardtii* is the only known example for viable mutants without functional Cyt b_6f complexes: These strains are still able to grow heterotrophically and to generate ATP and NAD(P)H by oxidative phosphorylation in mitochondria. Rieske- (de Vitry et al. 1999), Cyt b_6- (Kuras and Wollman 1994; Kuras et al. 1997), Cyt *f* (Kuras and Wollman 1994), and SU IV (Kuras and Wollman 1994; Zito et al. 1999) knock-out mutants have already been created and characterized. These mutants provided important data on the biogenesis of the Cyt b_6f complex and revealed that its accumulation in the membrane is a concerted process in which the stability or synthesis of certain subunits highly depends on the presence of other Cyt b_6f subunits (Kuras and Wollman 1994).

20.2.2.1 Multiple Rieske-Proteins and Cytochrome b_6f Heterogeneity

The availability of multiple sets of gene isoforms is a strategy of evolution for acclimation to changing environmental conditions. These isoforms encode two or more functionally distinct proteins from the same protein families. Under normal conditions usually only one form is active while the silent gene(s) are expressed under certain physiological/stress conditions. Among the eight *bona fide* Cyt b_6f proteins the Rieske protein is the only one which is encoded by multiple genes (i.e., a *petC* family) in some cyanobacterial genomes. While in the genome of *T. elongatus* and *G. violaceus* (like in all eukaryotes) the Rieske protein is encoded by a single *petC* gene, in *Nostoc punctiforme*, *Synechococcus* PCC 7002 and *Synechocystis* PCC 6803, three Rieske isoforms exist [for a review see (Schneider and Schmidt 2005)]. Some other cyanobacteria like *Nostoc* PCC 7120 even contain an additional fourth Rieske protein (Schneider and Schmidt 2005).

If an important protein is encoded by three genes, and if the gene products can partially substitute each other in a vital phenotype, six different mutants should be created: three single-, and three double mutants. In case there are combinations where substitution is not possible, the number of vital mutants will be lower. The cyanobacterium *Synechococcus* PCC 7002 (containing 3 Rieske genes) was unable to survive a *petC1* deletion (Yan and Cramer 2003), suggesting that only *petC1* encodes a functional Rieske protein and that *petC2* and *petC3* are silent in function (although Δ*petC2*, Δ*petC3* and Δ*petC2/3* strains were not constructed). In contrast, in *Synechocystis* PCC 6803 *petC1* alone can be deleted: In this organism five Δ*petC* strains have been generated of which only the double deletion mutant Δ*petC1*/Δ*petC2* was lethal (Schneider et al. 2004a). This finding reveals that the presence of either PetC1 or PetC2 is essential and they can partly replace each other, which is also in good correlation with their high sequence identity.

An in depth *in vivo* analysis of the various *petC* gene deficient mutant strains indicated that PetC1, which is phylogenetically close to the group of chloroplast

Rieske proteins (Schneider et al. 2002), is the major Rieske iron-sulfur protein in *Synechocystis* PCC 6803. Its deletion results in a remarkably low electron transport rate through the Cyt b_6f complex—in spite of redox and functional similarities between proteins PetC1 and PetC2. This may be due to a lower expression of the Cyt *f* encoding *petA* gene which is organized with *petC1* in the *petCA* operon (Kallas et al. 1988) as is apparent from the low amount of functional Cyt b_6f complexes (PetC2-Cyt b_6f) in this strain (Schultze et al. 2009; Tsunoyama et al. 2009). In consequence, sufficient PQH_2-oxidation is prevented, which results in the inability to perform state transitions, high light stress sensitivity, a decreased PS II/PS I ratio, and enhanced Cyd *bd* activity (Schneider et al. 2004a; Tsunoyama et al. 2009).

Although PetC2 exhibits low abundance in WT *Synechocystis* sp. PCC 6803 under all examined conditions, significantly increased transcript accumulation was observed under high light (HL) conditions or in absence of PetC1 (see above). This phenomenon, in combination with the unusually high growth rate under HL conditions of the Δ*petC2* strains, suggests a role of PetC2 in long-term light adaptation (Tsunoyama et al. 2009).

PetC3 can neither substitute PetC1 nor PetC2 due to its unique features. At first: (a) its N-terminal end lacks more than ten residues in comparison with PetC1/PetC2, (b) the sequence of its putative transmembrane helix is poorly conserved and (c) the connected hinge region is absent (Schneider et al. 2002). Second, its midpoint redox potential of 135 mV (*vs.* 300–320 mV for PetC1/PetC2) excludes oxidation of PQH_2, i.e. a role in the linear electron transport chain (Schneider et al. 2002, 2004b). Third, in contrast to PetC1 and PetC2 with exclusive location in the thylakoid membrane, PetC3 is exclusively located in the cytoplasmic membrane with all other major Cyt b_6f subunits missing (Aldridge et al. 2008; Schultze et al. 2009). In summary, these properties strongly indicate that PetC3 is not a component of the "classical" Cyt b_6f complex which apparently is missing completely in the cytoplasmic membrane. Nevertheless, current data show a regulatory impact of PetC3 on the photosynthetic electron transport and long-term light adaptation, for instance by activation of the cyclic electron transport around PS I under high light conditions (Tsunoyama et al. 2009). Details of this regulation still have to be elucidated.

In *Nostoc* PCC 7120, the PetC4 protein, which has a high sequence similarity to PetC1 and PetC2, apparently plays a role in nitrogen fixation and/or heterocyst formation and is also proposed to be a redox sensor (Schneider and Schmidt 2005).

20.2.2.2 Small Cytochrome b_6f Subunits

The four low molecular weight Cyt b_6f subunits PetG, -L, -M and -N are located at the periphery of the dimeric complex and form a so-called "picket fence" structure. According to the current structure determinations (Yamashita et al. 2007; Baniulis et al. 2009), PetM and PetL are positioned (very) peripherally and PetG and PetN deeper inside the complex (Fig. 20.3) as was found also in the *C. reinhardtii* structure (Stroebel et al. 2003). Noteworthy, homologous small subunits are absent in the corresponding Cyt bc_1 structures [(Xia et al. 1997; Iwata et al. 1998; Zhang et al. 1998), see also (Berry et al. 2004)].

In order to clarify the role of these small subunits, knock-out lines have been generated in *Synechocystis* [all four knock-outs, (Schneider et al. 2001, 2007)], in *C. reinhardtii* [Δ*petG*, (Berthold et al. 1995); Δ*petL*, (Takahashi et al. 1996; Zito et al. 2002)] and in tobacco [Δ*petG*, (Schwenkert et al. 2007), Δ*petL*, (Schöttler et al. 2007; Schwenkert et al. 2007), Δ*petN*, (Hager et al. 1999; Schwenkert et al. 2007). Also, a chimeric SU IV-PetL fusion protein was designed in *C. reinhardtii* (Zito et al. 2002). In summary, in agreement with the structural data, these mutants showed that the deletion of *petG*- or *petN* genes causes dramatic effects, while deletion of *petM* or *petL* gene results only in moderate phenotypes. However, disruption of *petL* results in impaired growths in *C. reinhardtii* which is probably due to the relatively high divergence of PetL (Schwenkert et al. 2007). Also, similar to PetG and PetN, the presence of PetL seems to be important for Cyt b_6f stability (Schöttler et al. 2007; Schwenkert et al. 2007), while its role as a sensor is still controversial [see (Zito et al. 2002; Schwenkert et al. 2007) *vs.* (Schöttler et al. 2007)]. In *Synechocystis* sp. PCC 6803, a regulatory role of PetM has been proposed (Schneider et al. 2001).

20.2.2.3 The Newly Discovered PetP Subunit

Highly purified Cyt b_6f complexes from *Synechocystis* sp. PCC 6803 and *T. elongatus* contained a potential new Cyt b_6f subunit with a molecular mass of 7.2 kDa which was coined PetP, i.e. the 9th Cyt b_6f subunit (Volkmer et al. 2007; Gendrullis et al. 2008). This polypeptide is highly conserved in cyanobacteria and encoded by the open reading frames (ORFs) *ssr2998* and *tsr0524* in *Synechocystis* sp. PCC 6803 and *T. elongatus*, respectively (Volkmer et al. 2007). Sequence analysis predicted PetP as a soluble protein. The absence of this subunit in all up to now published cyanobacterial structures suggests, that PetP is loosely/temporarily associated with the complex, similar to PetO of the plastid Cyt b_6f complex (Hamel et al. 2000). However, in contrast to PetO, PetP is apparently not a phosphoprotein and not (directly) involved in the regulation of state transitions. The fact, that the PetP-deletion mutant of *Synechocystis* sp. PCC 6803 shows a slower ET kinetics through the Cyt b_6f complex (Volkmer et al. 2007) indicates a regulatory role of PetP in ET processes of cyanobacteria.

20.3 Key Functional Sites of the Cytochrome b_6f Complex

During the last 4–5 decades several *C. reinhardtii* mutants with partially active Cyt b_6f complex have been generated by *function-directed* random mutagenesis which provided valuable information on the organization, assembly, structure-function relationships and regulatory roles of the Cyt b_6f complex [see e.g. (Finazzi et al. 1997;

de Vitry et al. 1999; Shao et al. 2006)]. However, the lack of direct structural information strongly limited the conclusions drawn from these experiments. The currently available 3D structures based on X-ray crystallography (Kurisu et al. 2003; Stroebel et al. 2003; Yamashita et al. 2007; Baniulis et al. 2009) combined with the availability of (directed) mutants from *Synechococcus* PCC 7002, *Synechocystis* sp. PCC 6803 and *C. reinhardtii* (de Vitry et al. 1999; de Lavalette et al. 2008a) allows in depth examinations of structure-function relationships on the level of target residues, i.e., a new dimension of gaining knowledge. In future, these techniques should be combined with focus on one organism, from which both 3D-structure and function of WT and directed mutants will be available. Up to now such an approach was hampered by the fact that directed mutants of *C. reinhardtii* are difficult to obtain and that the cyanobacterium *M. laminosus* is not transformable (see also Concluding remarks and future perspectives).

One drawback of site-directed mutagenesis is that only non-lethal mutations can be investigated. Site-directed mutagenesis of essential amino acids impairs the Cyt b_6f complex [see e.g. (Kuras et al. 1997; Lee et al. 2001)] and results in merodiploid cells (in cyanobacteria) instead of viable segregants. On the other hand, many cyanobacterial mutants with altered—but still functional—electron transport across Cyt b_6f are still vital. For this reason, site-directed mutagenesis is an efficient and appropriate tool for the investigation of the structure, function and regulation of this complex as will be outlined—also for *C. reinhardtii*—in the next sections.

20.3.1 The Q_o Quinone Binding Site

Electrons are transported to/from the Cyt b_6f complex *via* mobile carriers, PQH_2, PC (and possibly FNR) which are transiently attached to the complex at their docking/binding sites. The Cyt b_6f complex contains two quinone/quinol binding sites, Q_o (=Q_p) and Q_i (=Q_n), located close to the luminal and stromal/cytoplasmic surface of the membrane, respectively. The PQH_2 binding site Q_o, the "Q_o pocket", where oxidation and deprotonation of PQH_2 occurs (see Sect. 20.2.1), is an antechamber which is connected with the quinone exchange cavity *via* a small portal and which is formed under participation of SU IV, Cyt b_6 and the Rieske subunit/center. The structure of this chamber is shown in Fig. 20.4a, b where binding of PQ(H_2) is visualized by the quinone-analog tridecylstigmatellin (TDS, orange).

The first site-directed cyanobacterial Cyt b_6f mutants with modified Q_o quinone binding niche were reported by Kallas et al. (Lee et al. 2001) for *Synechococcus* PCC 7002. These mutations (Asp141Gly, Ala147Gly, Ser152Ala; *Synechococcus* numbering: Asp148, Ala154, Ser159) effected the lumen-extended *cd1*-helix of Cyt b_6 (Fig. 20.4), which, according to the crystal structure, forms the *p*-side "floor" of the quinone exchange cavity together with the *cd2* helix in the *cd* loop (Cramer et al. 2006). These mutants showed no remarkable phenotype unless specific quinone-analogs such as 2,5-dibromo-3-methyl-6-isopropyl-*p*-benzoquinone (DBMIB), stigmatellin and myxothiazol were applied in low concentrations (Lee et al.

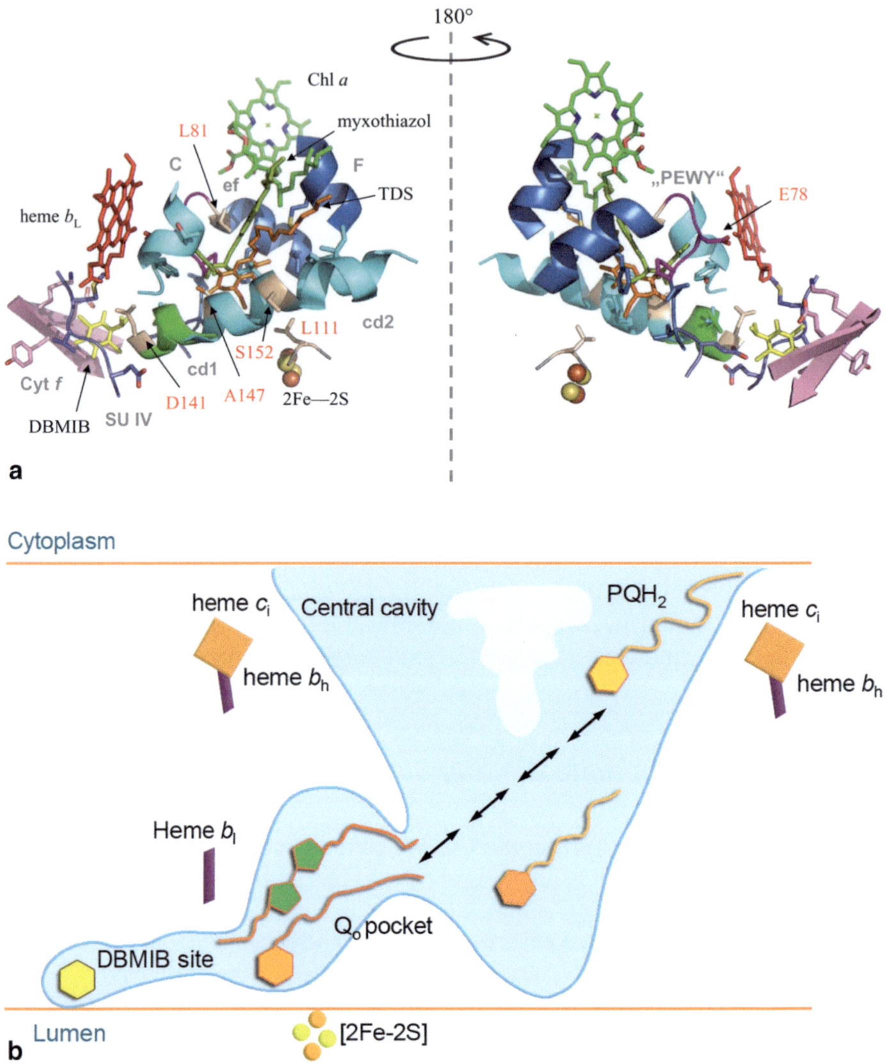

Fig. 20.4 **a** Structure of the cyanobacterial Q_o quinone binding pocket in presence of TDS (PDB, 2e76; Yamashita et al. 2007) or DBMIB (PDB, 2d2c; Yan et al. 2006). The quinone binding pocket, with bound quinone analogs TDS (*orange*) and myxothiazol [(*olive*—taken from the bc_1 structure (PDB, 1sqp; Esser et al. 2004)], is surrounded by residues from Cyt b_6 (*cyan*) and SU IV (*sky blue*/*magenta*), the Rieske 2Fe–2S center (*yellow*/*orange spheres*), and heme b_L (*red*). The phytyl tail of Chl *a* (*green*) extends into the chamber. Entry of the Q_o pocket is close to the “tail” parts of TDS and myxothiazol (*upper right* in the left scheme). The high affinity DBMIB (*yellow*) binding site, connected to the Q_o site, is surrounded by Cyt *f* (*pink*), SU IV (*stale*), and Cyt b_6 (*green*) residues. Amino acid residues located at ≤4 Å from the inhibitors are shown as sticks. The conserved “PEWY” sequence on the ef loop of SU IV is shown in *magenta*, while the substituted amino acids (*wheat*) in the vicinity of the Q_o pocket are shown with their identifier (*red*). **b** Cartoon of the quinone-exchange cavity and the Q_o quinone-binding pocket (Reproduced from Cramer et al. 2006 with modifications; copyright ©: Annual Reviews) with bound quinone/quinone analogs. For details see text

2001). The observed binding affinities of these agents to the mutant complexes indicated a different binding mode (location, interacting partners) of different agents in the same binding niche, which was later supported by the structural data in presence of bound quinone analogs [Fig. 20.4; (Stroebel et al. 2003; Cramer et al. 2006; Yamashita et al. 2007; Yan et al. 2006)]. Although native quinones have not been detected in the Q_o pocket (neither in b_6f nor in bc_1), these inhibitors helped to define their coordinates. In this context, the surprisingly "deep" location of DBMIB in the chamber (Yan et al. 2006) may be explained by a movement from the Q_o (inhibitory) site to its high-affinity site (shown in Fig. 20.4) *via* an internal passage after its reduction (Roberts et al. 2004; Cramer et al. 2006; Yan et al. 2006).

Figure 20.4a shows TDS in the native "ring-in" conformation as found both in *C. reinhardtii* and the recent *M. laminosus* structure (Stroebel et al. 2003; Yamashita et al. 2007). Previously, to explain some alterations between the structures [(Kurisu et al. 2003) *vs.* (Stroebel et al. 2003)] a bimodal binding of the quinone, i.e. in "ring-in" and "ring-out" conformation was proposed (Yan and Cramer 2004; Cramer et al. 2006). This (later neglected) hypothesis inspired the generation of a set of *Synechococcus* PCC 7002 mutants. They revealed that replacement of

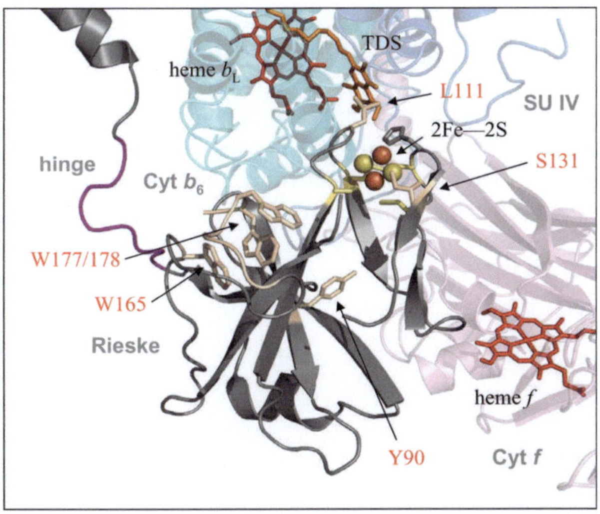

Fig. 20.5 The 2Fe–Fe cluster and its molecular environment (PDB, 1vf5 and 2e76; Kurisu et al. 2003; Yamashita et al. 2007, respectively). The quinone binding pocket is shown with the bound quinone analog TDS (*orange*, see also Fig. 20.4a). While subunits Cyt b_6, SU IV, and Cyt f are marked by *transparent cyan*, *sky blue*, and *pink*, respectively, the Rieske 2Fe–2S is colored in *grey*. For clarity, residues 138–146 in the Rieske protein are omitted. The following color codes have been used for the cofactors and residues: Rieske 2Fe–2S *center* (*yellow/orange spheres*), hemes b_L and f (*red* sticks), four cystein residues around the 2Fe–2S cluster (*yellow* sticks), His129 (*grey* sticks), sites of mutagenesis in the water soluble Rieske domain (*wheat* sticks/cartoon, *red* identifier), hinge region between water soluble domain and transmembrane helix of the Rieske 2Fe–2S protein (*magenta*). Transmembrane helix (*upper left*) and water soluble part of the Rieske protein (*center*) belong to opposing monomers. For details see text

Rieske Leu111 (Figs. 20.4a and 20.5) by alanine or tyrosine, results in complete loss or more than tenfold decrease in sensitivity against TDS and stigmatellin, respectively. This indicates the critical role of Leu111 in inhibitor binding and/or proper docking of the Rieske water soluble domain to Cyt b_6 and SU IV (Yan and Cramer 2004); see also the next paragraph and Sect. 20.3.2.1]. In contrast, substitution of SU IV-Leu81 on the small *ef* helix at the immediate entrance of the Q_o site (Fig. 20.4) by the bulkier phenylalanine highly increased the sensitivity, while substitution by the less bulkier alanine showed no strong effect (Yan and Cramer 2004). Possibly, the smaller size of the quinone entrance portal in the Leu81Phe mutant stabilizes TDS at the inhibitory site in proper orientation and conformation (Yamashita et al. 2007).

These data clearly indicate that inhibitor binding properties of the Q_o site can be modified without compromising catalytic activity. Model calculations reveal how changes in the PQH_2 binding affinity affect the ET rate through the Cyt b_6f complex. When assuming $K_M \approx 2.0 \times 10^4$ M^{-1} (Kramer et al. 1994), 1 mM PQ/PQH_2 in the thylakoid membranes (Rich 1982) and a tenfold excess of PQ/PQH_2 compared to Cyt b_6f (Finazzi et al. 1997), a fivefold decrease in the binding affinity would result only in a 20% decrease of the electron transport rate [for a calculation, see (Finazzi et al. 1997)], which is still within range of the experimental error gained with *cd1* helix mutants (Lee et al. 2001). For the Rieske-Leu111 mutants (Yan and Cramer 2004), the same calculation resulted in a 10- to 15-fold decrease in the PQH_2-binding affinity. (The accelerated electron transport kinetics in the SU IV-Leu81Phe mutant cannot be modeled solely by changes in the binding affinity.)

In this context, the FUD2-mutant of *C. reinhardtii* with a duplication of 12 amino acids in the *cd2* helix domain of the Cyt b_6 subunit should be mentioned (Bennoun et al. 1978; Finazzi et al. 1997): This mutant exhibited an 8-times slower ET through the Cyt b_6 than WT, due to a 100-times slower binding affinity of the Q_o site for plastoquinols (Finazzi et al. 1997).

[For Rieske hinge mutants effecting sensitivity towards Q_o site inhibitors see (Yan and Cramer 2003) and Sect. 20.3.2.1].

20.3.2 *The High Potential Chain*

20.3.2.1 The Rieske Protein and 2Fe–2S Center

The Rieske 2Fe–2S protein consists of an N-terminal transmembrane helix and a large (140 residue) mobile soluble domain which extends into the lumen and is connected by a flexible hinge region (Fig. 20.5). This soluble extrinsic domain is subdivided into a small and a large subdomain [Fig. 20.5, Carrell et al. 1997]. PQH_2 is oxidized to semiquinone by the Rieske 2Fe–2S center which is localized in the small subdomain, 8–10 Å (edge-to-edge) away from the bound quinone with Rieske-Leu111 and His129 in between (see above, Figs. 20.4a and 20.5).

While site-directed hinge mutants have been generated from both *Chlamydomonas* and cyanobacteria (de Vitry et al. 1999; Yan and Cramer 2003; de Vitry et al. 2004), mutations effecting the environment of the 2Fe–2S center were mostly carried out with *Chlamydomonas,* which tolerates—in contrats to cyanobacteria—deletion of essential amino acids (de Vitry et al. 1999). Some lethal substitutions of key residues such as Cys113 and Cys128 (*C. reinhardtii* numbering: Cys110, -125) of the Rieske disulfide bridge were investigated by overexpressing a corresponding fusion protein in *E. coli* (Ouyang et al. 2004). Generally, the larger the distance from the 2Fe–2S cluster in the available mutants [Ser131-Tyr90-Trp177/178-Trp165 (*C. reinhardtii*: Ser128, Tyr87, Trp176, -177, -163)] the smaller the effect of mutations on structure and catalytic activity of the cluster [Fig. 20.5, (de Vitry et al. 1999)]. For instance, Ser131, which is in direct contact with the 2Fe–2S cluster (Fig. 20.5), tolerates only the conservative substitution with threonine under phototrophic growth conditions, while mutations Ser131Ala and Ser131Cys are lethal. Replacement of the polar serine residue with such nonpolar residues unable to form H-bond may impair the catalytic site and/or shift the redox potential of the 2Fe–2S cluster which, in turn, prevents an effective ET as found in the analogous Rieske mutants of the mitochondrial bc_1 complex (Denke et al. 1998). At larger distance (~10 Å) from the 2Fe–2S cluster, mutant Tyr90Asp (=mutant *794*, created by Sabeeha Merchant) shows an approximately 20-fold slower ET activity and is unable to grow phototrophically [Fig. 20.5, de Vitry et al. 1999]. Interestingly, this mutant contains only 1% Rieske-subunit of WT due to its lower stability (de Vitry et al. 1999). In summary, these observations indicate that Tyr90 is essential for the ternary structure (and catalytic activity) of the Rieske protein. Another aromatic residue, Trp165 on the large subdomain (Fig. 20.5), also seems to be important for maintaining the protein structure, while deletion of the C-terminal tail region (with Trp177 and 178, see Fig. 20.5) has no impact on the electron transport (de Vitry et al. 1999).

Oxidation of the 2Fe–2S center by Cyt *f* needs a large motion ("tethered" diffusion, see Sect. 20.2.1) of the Rieske soluble domain toward Cyt *f*. This should be highly influenced by the flexibility of the hinge region linking the α-helical anchor and the water-soluble domain (Fig. 20.5). While substitution with glycine residues and/or insertion of one or more residues increases, substitution with proline residues or shortening its length decreases its flexibility. Creation of such mutants showed, that, in contrast to the bc_1 complex, the Cyt b_6f complex is relatively insensitive to the modified length or flexibility of the hinge region (Yan and Cramer 2003; de Vitry et al. 2004). Interestingly, decreased and increased hinge flexibilities were accompanied by a higher resistance or sensitivity, respectively, against Q_o-site specific quinol analogs. While the former (truncated hinge) may alter the Q_o-site, resulting in insufficient contact between the 2Fe–2S cluster and the quinol-binding niche, the latter (elongated hinge) allows a different fit of the 2Fe–2S center to the Q_o site which, in turn, results in an increased affinity for site-specific inhibitors (Yan and Cramer 2003).

20.3.2.2 Cytochrome *f* and Mobile Carrier Binding

Cyt *f* is the largest Cyt b_6f subunit (MW≅32 kDa). It is anchored in the membrane by only one transmembrane helix, while its redox-active heme is located in the large

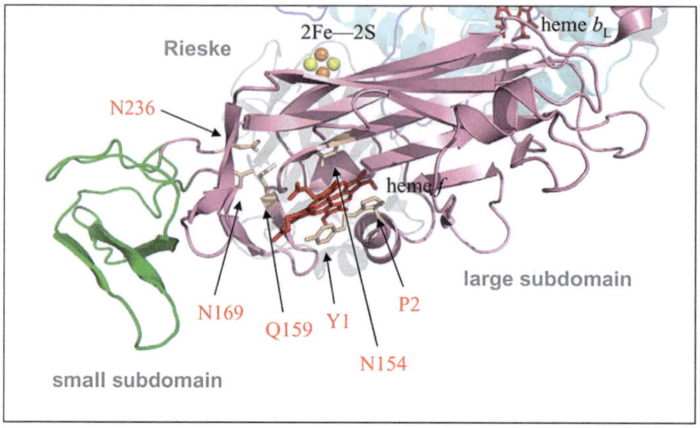

Fig. 20.6 The Cyt *f* water-soluble domain (PDB, 1vf5 and 2e76; Kurisu et al. 2003; Yamashita et al. 2007, respectively) with its large (*pink*) and small subdomain (*green*). The color coding of the Cyt b_6f subunits and co-factors is as in Fig. 20.5. For clarity, residues 69–72 and 117–120 from the large subdomain are omitted. Some sites of mutagenesis are indicated (*wheat* sticks, *red* identifier). These are Tyr1 and Pro2 at heme *f* and Asn154, Gln159, Asn169 and Asn 236 which binds the H_2O molecules of the proton channel *via* H-bonds. The PC/Cyt c_6 docking site is down-left, possibly with the participation of both small and large subdomain. For details see text

extrinsic domain with approximately 250 amino acids. Its main function is mediating electrons from the reduced Rieske 2Fe–2S center [for details, see Sects. 20.2.1 and 20.3.2.1] to the mobile water soluble carriers PC and Cyt c_6. [For a proposed 2Fe–2S—PC ET bypassing Cyt *f*, see (Fernández-Velasco et al. 2001)]. The soluble extrinsic domain of Cyt *f* has a predominant β-strand structure and can be subdivided (similar to the Rieske 2Fe–2S soluble domain) into a small and a large subdomain (Fig. 20.6.). Based on the high (1.96 Å) resolution 3D structure of the extrinsic domain (Martinez et al. 1996) many mutants have been designed, notably effecting (1) the heme *f* environment on the large subdomain (N-terminal mutants) (Zhou et al. 1996; Baymann et al. 1999; Gong et al. 2000b), (2) the whole small subdomain (Gong et al. 2001), (3) a water channel which was revealed by crystallographic analysis (Martinez et al. 1994; Ponamarev and Cramer 1998; Chi et al. 2000; Sainz et al. 2000), (4) five positively charged lysines which are responsible for the mobile carrier docking in *C. reinhardtii* (Soriano et al. 1996; Fernández-Velasco et al. 1997; Soriano et al. 1998; Gong et al. 2000a).

The high-potential *c*-type (*f*-type) heme is covalently bound *via* thioether bonds and has a histidine (His26) as fifth ligand. The sixth ligand is the α-amino group of (the N-terminal) Tyr1 (Fig. 20.6) which can be substituted without problem (Baymann et al. 1999) while substitution of the neighbouring Pro2 has more serious consequences, probably by disrupting the (helical) structure of the N-terminus which might effect the 2Fe–2S to Cyt *f* ET (Zhou et al. 1996). In agreement with this, even subtle changes in the hydrophobic patch around heme *f* alter its midpoint redox potential and the binding constant of PC which results in decreased ET rates (Gong et al. 2000b). Remarkably, the whole small subdomain can be deleted without affecting the ET to PC (Chi et al. 2000; Gong et al. 2001). According to Brownian dynamics simulations (Haddadian and Gross 2006), one function of the small domain may be to guide PC or Cyt c_6 to a uniform dock with Cyt *f*. Furthermore, intactness of the buried five-water channel ("proton wire") is required for a high rate of Cyt *f* reduction (Ponamarev and Cramer 1998). According to the structural data, a H^+-transfer from the 2Fe–2S cluster can only occur if Cyt *f* is in the proximal position (Cramer et al. 2006). Hence, a cascade of deprotonation-protonation steps in the water channel should precede the quinol-deprotonation at the Q_o site (notably, Cyt *f* and the Rieske 2Fe–2S should be oxidized first by PS I before accepting electrons from PQH_2).

20.3.3 The Low Potential Chain

20.3.3.1 The PEWY Sequence and Heme b_L

Blocking of the channel(s) for deprotonation of PQH_2 (or protonation of PQ at Q_i, see Sect. 20.3.3.2) should impair the photosynthetic electron transport, too (see Sect. 20.3.2.2). The conservative Pro-Glu-Trp-Tyr (PEWY) sequence in the *ef* loop of SU IV (Fig. 20.4a)—located in the vicinity of both the Q_o quinone binding site and heme b_L—may also be involved in the deprotonation of PQH_2. Non-conservative substitutions of Glu78 within this sequence in *C. reinhardtii* resulted in a 2- to 5-fold decrease in the Cyt *f* reduction kinetics, but neither prevented Cyt b_6f assembly nor phototrophy (Zito et al. 1998; Finazzi 2002). Glu78 was proposed to be involved in the transport of the second proton of PQH_2 (Cramer et al. 2006); however, as indicated by the slower ET, PQH_2 may also be deprotonated by another base, i.e. water or amino acid side chains. (The huge size of the quinone-binding cavity and the unknown quinone location in it keeps several possibilities open.) As proposed by Crofts (Crofts et al. 2000; Crofts 2004), the (first) electron and proton is transferred in a concerted action to the 2Fe–2S center, protonating one histidine ligand of the center (His129, Fig. 20.5). This proton can then be released to the lumen *via* an internal water bridge in Cyt *f* [(Martinez et al. 1996; Ponamarev and Cramer 1998), see Sect. 20.3.2.2]. Thus, possibly only the release-route of the second proton should be still identified. Interestingly, preventing PQH_2 deprotonation seems to impair ET less severely than protonation of PQ at the Q_i site (see next section).

Based on the Cyt b_6 primary structure and structural predictions two Gln47Ala mutants (in *Synechococcus* PCC 7002 and *Synechocystis* PCC 6803; Gln54Ala) have also been designed (Tietjens, Ambill, Schneider, Berry, Rögner, unpublished). Cyt b_6-Gln47 (in helix A) is close to and faces heme b_L (not shown). Fluorescence induction measurements indicated impaired electron transport through the Cyt b_6f and an increased Cyt *bd* oxidase activity in the partially segregated b_6-Gln47Ala mutant [(Ambill 2004); for the role of Cyt *bd* oxidase in cyanobacteria see Sect. 20.2.1]. This effect may be due to a lower Cyt b_6f population size or an impaired/blocked ET.

20.3.3.2 The Heme b_H /c_i Couple and the Q_i Quinone Binding Site

Substitution of residues with direct interaction in heme b_H and the Q_i quinone-binding site (Cyt b_6-Gly37Val, Gly37Asn and Cys35Val in helix A) in *C. reinhardtii* result phenotypes which are unable to grow phototrophically (de Lavalette et al. 2008a). Substitution of Gly37, which is only 3.7 Å away from the heme b_H porphyrin ring, with bulkier and/or more polar residues may shift the Cyt b_H midpoint potential, disrupt the protein structure and inhibit heme binding or protein folding (or a combination of them). Recently, it has been shown that binding of both hemes is crucial for the formation of the holo-cytochrome b_6 (Dreher et al. 2008). This may—together with the assumption of an improperly bound or lacking heme b_H—explain the low accumulation of the Cyt $b_6 f$ complex in these strains [see (Kuras et al. 1997; de Lavalette et al. 2008a; Saint-Marcoux et al. 2009)]. Noteworthy, the conserved Gly37 was also required in artificial proteins mimicking Cyt *b* (Rau and Haehnel 1998; Ghirlanda et al. 2004) indicating the space requirement for proper packing. Likewise, results on mutants with subsitution of the conserved Leu204 (close to the heme-binding His202 in helix D; Fig. 20.7) which impairs photosynthesis (Zito et al. 1997), underlines the importance of a well-defined protein structure for heme-binding.

Heme c_i is a unique cofactor of Cyt $b_6 f$ complexes in oxygenic photoautotrophs (Kurisu et al. 2003; Stroebel et al. 2003; Baniulis et al. 2009) and in *bc* complexes of several phyla of gram positive bacteria (Ducluzeau et al. 2008), but it does not exist in the mitochondrial bc_1 complex. Its still enigmatic role in the Cyt $b_6 f$ complex is a big challenge for future research. Heme c_i is covalently bound (*via* a thioether linkage) to Cys35 (see above and Fig. 20.7) and coupled electronically to the perpendicularly oriented Cyt b_H (edge to edge distance ≈3.5 Å; iron-iron distance ≈9.8 Å) via a water (or hydroxide) bridge (Kurisu et al. 2003; Stroebel et al. 2003; Zhang et al. 2004a; Alric et al. 2005; Cramer et al. 2006; Zatsman et al. 2006; Baymann et al.

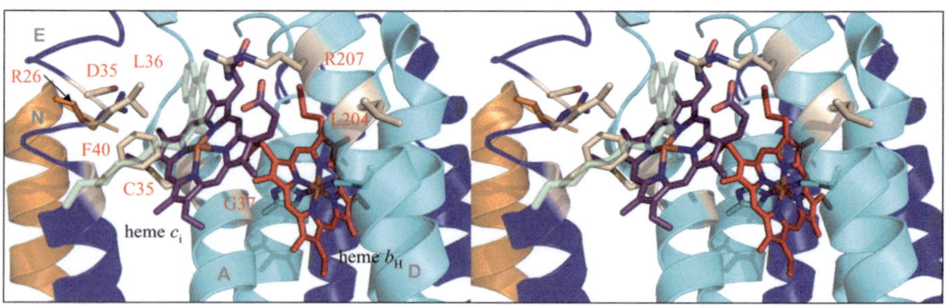

Fig. 20.7 Stereographic view of the cyanobacterial Q_i quinone binding pocket and its molecular environment in presence of NQNO (semi-apparent *pale green*) (PDB, 2e75; Yamashita et al. 2007). The quinone binding pocket is surrounded by residues from Cyt b_6 (*cyan*) and SU IV (*sky blue*), and PetN (*orange*). For clarity, subunits PetG, -L, and -M are omitted. The following color codes are used: heme b_H (*red*), heme c_i (*violet*), His100 and His202, axial ligands of heme b_H (*cyan* sticks), sites of mutagenesis (*wheat* sticks/cartoon, *red* identifier), PetN-Arg26 (*orange* sticks, *red* identifier). For details see text

2007; Baniulis et al. 2008). However, as revealed by directed mutagenesis, covalent binding of heme c_i is not a prerequisite for assembly of functional Cyt b_6f complex but contributes to the folding of the complex in a protease-resistant form (Saint-Marcoux et al. 2009). Heme c_i also interacts with Arg207 on helix D of Cyt b_6 and with Phe40 on helix E of SU IV (see below and Fig. 20.7). Its unique arrangement and the fact that it binds PQ directly without any axial ligand from polypeptides led to speculations on its role (Zhang et al. 2004a; Cramer and Zhang 2006; Cramer et al. 2006, 2008a; Baniulis et al. 2008; de Lavalette et al. 2009). Heme c_i has been proposed to prevent the formation of reactive oxygen species (ROS) at the Q_i site (Allen 2004; Cape et al. 2006; Zatsman et al. 2006; Baymann et al. 2007; Baniulis et al. 2008; Cramer et al. 2008a; de Lavalette et al. 2009)—a protection which is not required in mitochondria and anoxygenic bacteria. In the proposed model the heme b_H/c_i couple functions as a two-electron gate in which heme c_i remains reduced until heme b_H is reduced—either by heme b_L (after a second turnover) or by ferredoxin or another mobile carrier of the PS I-mediated cyclic electron transport. The latter route is suggested by the narrow shielding of c_i from the cytoplasmic space and a co-isolated cytochrome b_6f/FNR-complex (Zhang et al. 2001). Thereafter PQ is reduced in a concerted 2-electron transfer (Cramer et al. 2004, 2006, 2008a; Cramer and Zhang 2006; Baniulis et al. 2008). Noteworthy, despite huge efforts and excellent instrumentation semiquinone was never observed in the Q_i pocket of the complex (Christof Klughammer and Ron Pace, personal communication). Nevertheless, phylogenetical analysis shows that the appearance of heme c_i coincides with the split of the Cyt b into Cyt b_6 and SU IV. Evolutionary, this appeared much earlier than oxygenic photosynthesis and maybe correlated with a role of c_i in stabilizing the protein structure (i.e. connecting Cyt b_6 and SU IV) (Saint-Marcoux et al. 2009). Since early attempts to engineer a Cyt b_6–SU IV fusion protein failed (Deborah Berthold, personal communication), this problem still remains a challenging task of future research. The emergence of heme c_i in anoxygenic organisms suggests that either heme c_i does not play a role in protection against ROS formation or the prevention of ROS formation by heme c_i is only an indirect consequence of the energetically favored two-electron process at the Q_i site without formation of a stable semiquinone radical [see (Cape et al. 2006)]. However, very recently, c_i was also proposed as a putative terminal oxidase [(Twigg et al. 2009), see also below].

The involvement of Cyt b_6f and especially heme c_i in the cyclic ET processes around PS I is reinforced by heliobacteria which rely only on cyclic ET around a type I reaction center (Ducluzeau et al. 2008). In this organism, heme c_i has a protonatable axial ligand, probably Glu8 of SU IV. In Cyt b_6f complexes heme c_i does not have any axial sixth ligand (besides the quinone) and the protonatable glutamate residue (Glu8) is replaced by a phenylalanine (Phe40) which protrudes above the heme plane and is in steric clash with the bound quinone (Fig. 20.7; de Lavalette et al. 2009). Glu8 in heliobacteria can be replaced by the quinone analog 2-*n*-nonyl-4-hydroxyquinoline-*N*-oxide (NQNO) below pH 7.0 indicating some pH-dependent (self-) regulation of the ET in heliobacteria (Ducluzeau et al. 2008). Periodic changes in the redox state of the b_H/c_i couple during the Q-cycle should involve changes in the local protein structure, and control the interaction between

PQ and heme c_i (de Lavalette et al. 2009). Hence, transient orientation changes of Phe40 (as of Glu8 in heliobacteria) may play a role in the concerted 2-electron reduction of PQ. Mutagenesis studies have shown that the b_H/c_i couple may act not only an electron donor but also as a redox trigger in this process (de Lavalette et al. 2009). Also, the importance of the changing accessibility of heme c_i during the Q-cycle is indicated by light-sensitivity of the SU IV-Leu36Ala *C. reinhardtii* mutant with enlarged entrance of the Q_i site (de Lavalette et al. 2008a).

A mutation Asp35Ala, one residue away from Leu 36 (see Fig. 20.7), was detrimental for photosynthesis and accumulation of Cyt b_6f. It was proposed that this amino acid may participate in the protonation of the quinone (Ducluzeau et al. 2008). [For a thorough review on the potential *n*-side proton transport routes, see (Cramer et al. 2008a); for *p*-side proton transport see Sects. 20.3.2.2 and 20.3.3.1] Blocking of the protonation pathway would result in inhibition of the ET which, in turn, prevents phototrophic growth. As Asp35 of SU IV is salt bridged to Arg26 on the small subunit PetN (Fig. 20.7; see Sect. 20.2.2.2) site-directed mutagenesis of Arg26 may provide more information on the role of SU IV-Asp35.

Another mutant in the heme c_i/b_H moiety, b_6-Arg207His (Fig. 20.7; in *Synechococcus* PCC 7002, b_6-Arg214His), showed a blocked Cyt b_6 re-oxidation kinetics, while both b_6 reduction kinetics and midpoint redox potential remained unchanged (Nelson et al. 2005). An enhanced superoxide production was also reported (Ouyang et al. 2004) which could probably be attributed to the semiquinone formation at the Q_o site. However, this mutation removes the positively charged arginine from the vicinity of the c_i heme which should lower its midpoint potential, resulting in the possibility of oxygen reduction by heme c_i (Toivo Kallas, personal communication). Also, the obtained data indicate that Arg207 is important in modulating substrate (PQ) binding affinity and/or protonation of PQ. Arg207 is salt-bridged to the heme c_i propionate group which may provide the second proton to PQ protonation; a similar propionate-mediated protonation at heme a_3 was found in cytochrome *c* oxidase (Popovic and Stuchebrukhov 2004; Brändén et al. 2005; Hellwig et al. 2008, see also Chapter Renger and Ludwig). The presence of Arg207 should facilitate this type of proton transfer. Finally, the propionate-arginine couple with heme c_i may also form an oxygen/redox sensor as reported for the FixLH oxygen sensor from *Bradyrhizobium japonicum* (Balland et al. 2006). Current EPR detection of the O_2 surrogate NO bound to heme c_i further strengthen this hypothesis (Twigg et al. 2009). A role of Cyt b_6f in the redox signal pathway which finally leads to state transitions, has already been outlined (Vener et al. 1995); however, in this process the Q_o site rather than the Q_i site seems to play the decisive role (Vener et al. 1997; Zito et al. 1999; Finazzi et al. 2001; Wollman 2001; Mao et al. 2002).

20.3.4 Pigments Molecules and Sulfolipid in the Cytochrome b_6f Complex

As first shown by spectroscopy and later confirmed by the 3D crystal structures, the Cyt b_6f complex contains two pigment molecules, i.e. one chlorophyll *a* and one

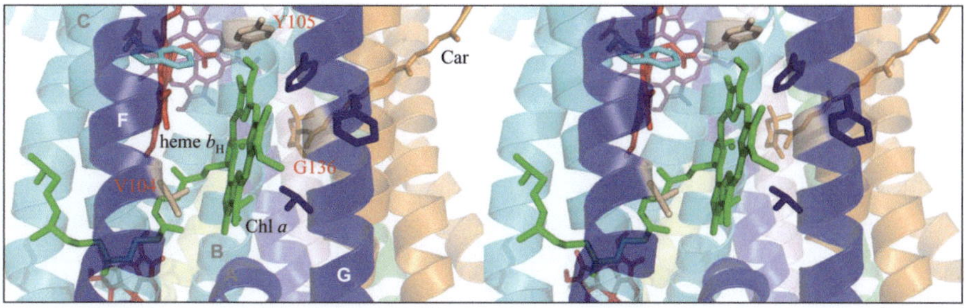

Fig. 20.8 Stereographic view of the Chl *a* molecule (*green*) and its molecular environment (PDB, 2e74; Yamashita et al. 2007). The color coding of the Cyt b_6f subunits and co-factors is as before. The chorine ring of Chl *a* is inserted between helices F and G of SU IV (*sky blue*), and its phytyl side-chain protrudes into the entrance of the Q_o site (see also Fig. 20.4a). The four aromatic resudies around the chlorine ring and SU IV-Thr140 are shown as sticks. Some sites of mutagenesis is also shown (*wheat* sticks/cartoon, *red* identifier). For details see text

carotenoid (Bald et al. 1992; Pierre et al. 1997; Peterman et al. 1998; Zhang et al. 1999; Boronowsky et al. 2001; Kurisu et al. 2003; Stroebel et al. 2003; Wenk et al. 2005). While the chlorine ring of Chl *a* is sandwiched between the F and G transmembrane helices of SU IV and probable also H-bonded to residues in helices B and C of Cyt b_6 (Fig. 20.8), its phytyl side-chain protrudes into the entrance of the Q_o site (Fig. 20.4a; Stroebel et al. 2003; de Lavalette et al. 2008b; Yan et al. 2008). The porphyrin ring of Chl *a* faces two small, non-polar residues (Val104 and Gly136 in helix F and G, respectively) and is coordinated by two water molecules, H-bonded to Phe135 and Thr140 in helix G, as a fifth (axial) ligand (of the Mg^{2+} ion). It is also surrounded by four aromatic residues [Tyr105 and Trp118 (Tyr112 and Trp125 in *Synechococus* sp. PCC 7002) on the B and C helices of Cyt b_6 and Phe133 and Phe135 from SU IV; see Fig. 20.8] which possibly stabilize its configuration *via* π-π electronic interactions (Yan et al. 2008). The Car molecule is inserted between the PetG and PetM helices (Fig. 20.3; see Sect. 20.2.2.2). This carotenoid is 9-*cis* β-carotene in *C. reinhardtii* and higher plants and echinenone in cyanobacteria, respectively (Wenk et al. 2005). Surprisingly, deletion of the *crtO* gene resulted in a phenotype with normal growth and Cyt b_6f properties in *Synechocystis* (Wenk et al. 2005). Both Chl *a* and Car is exclusively found in Cyt b_6f complexes but not in other, including heliobacterial, *bc*-type complexes (for the role of the other unique Cyt b_6f cofactor, heme c_i, in heliobacteria, see Sect. 20.3.3.2). Despite several indications, the function of these prosthetic groups is still enigmatic. It has been proposed that Chl *a* may facilitate the electron transport, regulate the Cyt b_6f assembly and/or stabilize the structure of the complex (Pierre et al. 1997; Wenk et al. 2005). Its short (~200–250 ps) singlet excited-state lifetime (Peterman et al. 1998; Dashdorj et al. 2005; Yan et al. 2008) as well as the detected long-range Chl-Car triplet energy transfer (Kim et al. 2005) may indicate their participation in protection against singlet oxygen formation (Dashdorj et al. 2007; Ma et al. 2009). Among several *Synechococcus* PCC 7002 and *C. reinhardtii* mutants in the vicinity of the Chl *a* cofactor (de Lavalette et al. 2008b; Yan et al. 2008) only the b_6-Tyr105Phe mutation (Fig. 20.8) seems to be lethal (Yan et al. 2008). Both *in vitro* and *in vivo* characterization of these mutants suggest that Chl *a* stabilizes the structure of the Cyt b_6f complex and affects the electron transport kinetics. Purified Cyt b_6f complexes from these strains showed a lowered Chl *a* content, indicating a weakened Chl *a* binding. In combination with the observed monomerization of purified mutant complexes from *C. reinhardtii* (Gly136Phe and Val104Phe/Gly136Phe) these data strongly indicate a structural role of Chl *a* (de Lavalette et al. 2008b; Yan et al. 2008). Structurally, Chl *a* was suggested to substitute the extra transmembrane helix in the bacterial *bc* complex as a stability element (de Lavalette et al. 2008b). The recent finding of protochlorophyll *a,* a chlorophyll *a* precursor, in dimeric Cyt b_6f complexes of barley etioplast, strongly suggests that phytylated tetrapyrrol is essential for the assembly of the complex (Reisinger et al. 2008). Functionally, *in vivo* characterization of SU IV mutants in *C. reinhardtii* indicates a potential role of Chl *a* in triggering state transitions, i.e. the mechanism balancing the supply of excitons between the two photosystems—possibly by the induction of a conformational change *via* Chl *a* (de Lavalette et al. 2008b). Also, for Car a structural role is proposed, which may be replaced by other Car's in case blocked synthesis pathways (Wenk et al. 2005).

The *Chlamydomonas* Cyt b_6f structure showed an endogenous sulfolipid, sulfoquinovosyl diacylglycerol (SQDG) (Stroebel et al. 2003) which is also visible in the recent *M. laminosus* and *Nostoc* structures (Yamashita et al. 2007; Baniulis et al. 2009). [For the role of SQDG in photosynthesis see also (Sato et al. 1995; Sato et al. 2003; Aoki et al. 2004; Riekhof et al. 2003)]. This SQDG molecule is inserted between Rieske- and Cyt *f*-transmembrane helices close to the stromal/cytoplasmic surface of the membrane. It is hydrogen-bonded by Rieske Arg16 and Asn20 and Cyt *f*-Lys275 residues (Arg13, Asn17 and Lys272 in *Chlamydomonas*) and mutagenesis studies suggest a participation in the autoregulation of Cyt *f* synthesis (Choquet et al. 2003; de Vitry et al. 2004).

20.4 Concluding Remarks and Future Perspectives

The X-ray structure determination of Cyt b_6f in 2003 was a milestone in understanding structure-function relations of this complex. This achievement gave rise to a very important boost to studies especially on the Q_i site of Cyt b_6f with the previously hidden heme c_i which is absent in the homologous Cyt bc_1 complex. However, many basic questions on Cyt b_6f still remain to be answered. For instance:

- Is ET in the high- and low-potential chain always concerted?
- What is the role of heme c_i in the low-potential ETC? Is it just a "wire" between heme b_H and the PQ acceptor or a decisive component for the "buffering" of electrons?
- Can heme c_i (or heme b_H) be reduced directly by FNR or other components of the PS I mediated cyclic ET?
- What is the molecular mechanism of proton and electron transfer at the Q_o and Q_i sites?

- What is the role of Cyt b_6f in signal transduction(s) including the regulation of state transitions? In which extent are these processes identical/different in cyanobacteria and higher plants?
- What is the role of PetC heterogeneity in cyanobacteria, especially PetC2, -C3 and -C4?
- What is the role of transiently or weakly bound subunits, PetO and PetP? What is the specified role of the four small *bona fide* subunits?
- What is the role of the pigment molecules in the Cyt b_6f complex? Why is echinenone found in some Cyt b_6f complexes and β-carotene in others? Is there a light-triggered effect of Chl *a* and why was its potential structural role not replaced by other components during evolution?

Most of these questions are best addressed by the combination of structural and functional analysis of WT in comparison with directed mutants. X-ray data provide in-depth information on possible structure-function relationships of the complex which can be tested by the creation and functional analysis of mutants. However, while structural analysis was mainly performed on the non-transformable thermophilic cynobacterium *Mastigocladus laminosus*, mutant analysis had to be performed using the transformable mesophilic cyanobacteria *Synechococcus* sp. PCC 7002 and *Synechocystis* sp. PCC 6803, which yield no stable Cyt b_6f crystals. The recent 3D structure at (3.0 Å resolution) of the transformable cyanobacterium *Nostoc* sp. PCC 7120 (Baniulis et al. 2009) paves the way for a straightforward combination of structural and mutagenesis investigations by using the same organism which should be an important step forward to solve the still open questions outlined above.

Acknowledgements The authors are especially grateful to Achim Trebst for many stimulating discussions. Also, the kind help by many valuable hints of Frauke Baymann, Deborah Berthold, William A. Cramer, Giovanni Finazzi, Toivo Kallas, Helmut Kirchhoff, Christof Klughammer, Ron Pace and Dirk Schneider is gratefully acknowledged. Special thanks to Dieter Wunsch for drawing Figs. 20.2 and 20.4, to Till Rudack for the first step using PyMol (GB), and to Dorothea Rexroth, Corinna Lüer and Gernot Renger for critical reading of the manuscript. Financial support by SFB 480 (project C1) was the basis for all investigations on Cyt b_6f of the authors.

References

Albarrán C, Navarro JA, Molina-Heredia FP, Murdoch P del S, De la Rosa MA and Hervás M (2005) Laser flash-induced kinetic analysis of cytochrome *f* oxidation by wild-type and mutant plastocyanin from the cyanobacterium *Nostoc* sp. PCC 7119. Biochemistry 44: 11601–11607

Aldridge A, Spence E, Kirkilionis MA, Frigerio L and Robinson C (2008) Tat-dependent targeting of Rieske iron-sulfur proteins to both the plasma and thylakoid membranes in the cyanobacterium *Synechocystis* PCC6803. Mol Microbiol 70: 140–150

Allen JF (2003a) Cyclic, pseudocyclic and noncyclic photophosphorylation: New links in the chain. Trends Plant Sci 8: 15–19

Allen JF (2003b) State transitions—A question of balance. Science 299: 1530–1532

Allen JF (2004) Cytochrome b_6f: Structure for signalling and vectorial metabolism. Trends Plant Sci 9: 130–137

Alric J, Pierre Y, Picot D, Lavergne J and Rappaport F (2005) Spectral and redox characterization of the heme c_i of the cytochrome b_6f complex. Proc Natl Acad Sci U S A 102: 15860–15865

Ambill M (2004) Structure—Funktionsbeziehungen am cyanobakteriellen Cytochrom b_6f-Komplex unter Verwendung neuer Reinungsstrategien. Ph D Thesis, Ruhr-University Bochum, Germany

Aoki M, Sato N, Meguro A and Tsuzuki M (2004) Differing involvement of sulfoquinovosyl diacylglycerol in photosystem II in two species of unicellular cyanobacteria. Eur J Biochem 271: 685–693

Arnon DI, Allen MB and Whatley FR (1954) Photosynthesis by isolated chloroplasts. Nature 174: 394–396

Asada K (1999) The water-water cycle in chloroplasts: Scavenging of active oxygens and dissipation of excess photons. Annu Rev Plant Physiol Plant Mol Biol 50: 601–639

Avenson TJ, Cruz JA, Kanazawa A and Kramer DM (2005) Regulating the proton budget of higher plant photosynthesis. Proc Natl Acad Sci U S A 102: 9709–9713

Bald D, Kruip J, Boekema EJ and Rögner M (1992) Structural investigations on cyt. b_6f-complex and PSI-complex from the cyanobacterium Synechocystis PCC6803. In: Murata N (ed) Research in Photosynthesis, Vol. I, pp 629–632. Kluwer Academic, The Netherlands

Balland V, Bouzhir-Sima L, Anxolabéhère-Mallart E, Boussac A, Vos MH, Liebl U and Mattioli TA (2006) Functional implications of the propionate 7-arginine 220 interaction in the FixLH oxygen sensor from *Bradyrhizobium japonicum*. Biochemistry 45: 2072–2084

Baniulis D, Yamashita E, Zhang H, Hasan SS and Cramer WA (2008) Structure-function of the cytochrome b_6f complex. Photochem Photobiol 84: 1349–1358

Baniulis D, Yamashita E, Whitelegge JP, Zatsman AI, Hendrich MP, Hasan SS, Ryan CM and Cramer WA (2009) Structure-function, stability, and chemical modification of the cyanobacterial cytochrome b_6f complex from *Nostoc* sp PCC 7120. J Biol Chem 284: 9861–9869

Barber J (2003) Photosystem II: The engine of life. Quart Rev Biophys 36: 71–89

Battchikova N and Aro EM (2007) Cyanobacterial NDH-1 complexes: Multiplicity in function and subunit composition. Physiol Plant 131: 22–32

Baymann F, Zito F, Kuras R, Minai L, Nitschke W and Wollman FA (1999) Functional characterization of *Chlamydomonas* mutants defective in cytochrome *f* maturation. J Biol Chem 274: 22957–22967

Baymann F, Giusti F, Picot D and Nitschke W (2007) The c_i/b_H moiety in the b_6f complex studied by EPR: A pair of strongly interacting hemes. Proc Natl Acad Sci U S A 104: 519–524

Bendall DS and Manasse RS (1995) Cyclic photophosphorylation and electron-transport. Biochim Biophys Acta 1229: 23–38

Bennoun P, Masson A, Piccioni R and Chua NH (1978) Uniparental mutants of *Chlamydomonas reinhardtii* defective in photosynthesis. In Akoyunoglou G (ed) Photosynthesis, pp 721–726. Elsevier-North/Holland, Amsterdam

Berry S, Schneider D, Vermaas WFJ and Rögner M (2002) Electron transport routes in whole cells of *Synechocystis* sp strain PCC 6803: The role of the cytochrome *bd*-type oxidase. Biochemistry 41: 3422–3429

Berry EA, Huang LS, Saechao LK, Pon NG, Valkova-Valchanova M and Daldal F (2004) X-ray structure of *Rhodobacter capsulatus* cytochrome bc_1: Comparison with its mitochondrial and chloroplast counterparts. Photosynth Res 81: 251–275

Berthold DA, Schmidt CL and Malkin R (1995) The deletion of *petG* in *Chlamydomonas-Reinhardtii* disrupts the cytochrome *bf* complex. J Biol Chem 270: 29293–29298

Boronowsky U, Wenk SO, Schneider D, Jäger C and Rögner M (2001) Isolation of membrane protein subunits in their native state: Evidence for selective binding of chlorophyll and carotenoid to the b_6 subunit of the cytochrome b_6f complex. Biochim Biophys Acta 1506: 55–66

Brändén G, Brändén M, Schmidt B, Mills DA, Ferguson-Miller S and Brzezinski P (2005) The protonation state of a heme propionate controls electron transfer in cytochrome *c* oxidase. Biochemistry 44: 10466–10474

Breyton C, Tribet C, Olive J, Dubacq JP and Popot JL (1997) Dimer to monomer conversion of the cytochrome b_6f complex—Causes and consequences. J Biol Chem 272: 21892–21900

Bukhov N and Carpentier R (2004) Alternative photosystem I-driven electron transport routes: Mechanisms and functions. Photosynth Res 82: 17–33

Cape JL, Bowman MK and Kramer DM (2006) Understanding the cytochrome *bc* complexes by what they don't do. The Q-cycle at 30. Trends Plant Sci 11: 46–55

Carrell CJ, Zhang HM, Cramer WA and Smith JL (1997) Biological identity and diversity in photosynthesis and respiration: Structure of the lumen-side domain of the chloroplast Rieske protein. Structure 5: 1613–1625

Chi YI, Huang LS, Zhang ZL, Fernández-Velasco JG and Berry EA (2000) X-ray structure of a truncated form of cytochrome *f* from *Chlamydomonas reinhardtii*. Biochemistry 39: 7689–7701

Chobot SE, Zhang HB, Moser CC and Dutton PL (2008) Breaking the Q-cycle: Finding new ways to study Q_o through thermodynamic manipulations. J Bioenerg Biomembr 40: 501–507

Choquet Y, Zito F, Wostrikoff K and Wollman FA (2003) Cytochrome *f* translation in *Chlamydomonas* chloroplast is autoregulated by its carboxyl-terminal domain. Plant Cell 15: 1443–1454

Cooley JW and Vermaas WFJ (2001) Succinate dehydrogenase and other respiratory pathways in thylakoid membranes of *Synechocystis* sp strain PCC 6803: Capacity comparisons and physiological function. J Bacteriol 183: 4251–4258

Cramer WA and Zhang HM (2006) Consequences of the structure of the cytochrome b_6f complex for its charge transfer pathways. Biochim Biophys Acta 1757: 339–345

Cramer WA, Widger WR, Herrmann RG and Trebst A (1985) Topography and function of thylakoid membrane-proteins. Trends Biochem Sci 10: 125–129

Cramer WA, Zhang HM, Yan JS, Kurisu G and Smith JL (2004) Evolution of photosynthesis: Time-independent structure of the cytochrome b_6f complex. Biochemistry 43: 5921–5929

Cramer WA, Zhang HM, Yan JS, Kurisu G and Smith JL (2006) Transmembrane traffic in the cytochrome b_6f complex. Annu Rev Biochem 75: 769–790

Cramer WA, Baniulis D, Yamashita E, Zhang H, Zatsman AI and Hendrich MP (2008a) Structure, spectroscopy, and function of the cytochrome b_6f complex: heme c_n and *n*-side electron and proton transfer reactions. In: Fromme P (ed) Photosynthetic Protein Complexes: A Structural Approach, pp 155–179. Wiley-VCH, Weinheim

Cramer WA, Zhang H, Yan J., Kurisu G, Yamashita E, Dashdorj N, Kim H and Savikhin S (2008b) Structure-function of the cytochrome b_6f complex: A design that has worked for three billion years. In: Renger G (ed) Primary Processes of Photosynthesis: Priniciples and Apparatus, Part II: Reaction Centers/Photosystems, Electron Transport Chains, Photophosphorylation and Evolution, pp 417–446. Royal Society Chemistry, Cambridge

Crofts AR (2004) Proton-coupled electron transfer at the Q_o-site of the bc_1 complex controls the rate of ubihydroquinone oxidation. Biochim Biophys Acta 1655: 77–92

Crofts AR, Guergova-Kuras M, Kuras R, Ugulava N, Li JY and Hong SJ (2000) Proton-coupled electron transfer at the Q_o site: What type of mechanism can account for the high activation barrier? Biochim Biophys Acta 1459: 456–466

DalCorso G, Pesaresi P, Masiero S, Aseeva E, Nemann DS, Finazzi G, Joliot P, Barbato R and Leister D (2008) A complex containing PGRL1 and PGR5 is involved in the switch between linear and cyclic electron flow in *Arabidopsis*. Cell 132: 273–285

Dashdorj N, Zhang HM, Kim HY, Yan JS, Cramer WA and Savikhin S (2005) The single chlorophyll *a* molecule in the cytochrome b_6f complex: Unusual optical properties protect the complex against singlet oxygen. Biophys J 88: 4178–4187

Dashdorj N, Yamashita E, Schaibley J, Cramer WA and Savikhin S (2007) Ultrafast optical pump-probe studies of the cytochrome b_6f complex in solution and crystalline states. J Phys Chem B 111: 14405–14410

de Lavalette AD, Barbagallo RP and Zito F (2008a) Why is it so difficult to construct Q_i site mutants in *Chlamydomonas reinhardtii*? Compt Rend Biol 331: 510–517

de Lavalette AD, Finazzi G and Zito F (2008b) b_6f-associated chlorophyll: Structural and dynamic contribution to the different cytochrome functions. Biochemistry 47: 5259–5265 de Lavalette AD, Barucq L, Alric J, Rappaport F and Zito F (2009) Is the redox state of the c_i heme of the cytochrome b_6f complex dependent on the occupation and structure of the Q_i site and vice versa? J Biol Chem 284: 20822–20829

de Vitry C, Finazzi G, Baymann F and Kallas T (1999) Analysis of the nucleus-encoded and chloroplast-targeted rieske protein by classic and site-directed mutagenesis of *Chlamydomonas*. Plant Cell 11: 2031–2044

de Vitry C, Ouyang Y, Finazzi G, Wollman FA and Kallas T (2004) The chloroplast Rieske iron-sulfur protein. At the crossroad of electron transport and signal transduction. J Biol Chem 279: 44621–44627

Denke E, Merbitz-Zahradnik T, Hatzfeld OM, Snyder CH, Link TA and Trumpower BL (1998) Alteration of the midpoint potential and catalytic activity of the Rieske iron-sulfur protein by changes of amino acids forming hydrogen bonds to the iron-sulfur cluster. J Biol Chem 273: 9085–9093

Dreher C, Prodöhl A, Hielscher R, Hellwig P and Schneider D (2008) Multiple step assembly of the transmembrane cytochrome b_6. J Mol Biol 382: 1057–1065

Ducluzeau AL, Chenu E, Capowiez L and Baymann F (2008) The Rieske/cytochrome *b* complex of Heliobacteria. Biochim Biophys Acta 1777: 1140–1146

Endo T, Ishida S, Ishikawa N and Sato F (2008) Chloroplastic NAD(P)H dehydrogenase complex and cyclic electron transport around photosystem I. Mol Cells 25: 158–162

Esser L, Quinn B, Li YF, Zhang MQ, Elberry M, Yu L, Yu CA and Xia D (2004) Crystallographic studies of quinol oxidation site inhibitors: A modified classification of inhibitors for the cytochrome bc_1 complex. J Mol Biol 341: 281–302

Fernández-Velasco JG, Zhou J and Malkin R (1997) The putative plastocyanin binding site in *Chlamydomonas reinhardtii* cytochrome *f*. Biophys J 72: A126

Fernández-Velasco JG, Jamshidi A, Gong XS, Zhou J and Ueng RY (2001) Photosynthetic electron transfer through the cytochrome b_6f complex can bypass cytochrome *f*. J Biol Chem 276: 30598–30607

Finazzi G (2002) Redox-coupled proton pumping activity in cytochrome b_6f, as evidenced by the pH dependence of electron transfer in whole cells of *Chlamydomonas reinhardtii*. Biochemistry 41: 7475–7482

Finazzi G and Forti G (2004) Metabolic flexibility of the green alga *Chlamydomonas reinhardtii* as revealed by the link between state transitions and cyclic electron flow. Photosynth Res 82: 327–338

Finazzi G, Buschlen S, de Vitry C, Rappaport F, Joliot P and Wollman FA (1997) Function-directed mutagenesis of the cytochrome b_6f complex in *Chlamydomonas reinhardtii*: involvement of the cd loop of cytochrome b_6 in quinol binding to the Q_o site. Biochemistry 36: 2867–2874

Finazzi G, Zito F, Barbagallo RP and Wollman FA (2001) Contrasted effects of inhibitors of cytochrome b_6f complex on state transitions in *Chlamydomonas reinhardtii*—The role of Q_o site occupancy in LHCII kinase activation. J Biol Chem 276: 9770–9774

Finazzi G, Rappaport F, Furia A, Fleischmann M, Rochaix JD, Zito F and Forti G (2002) Involvement of state transitions in the switch between linear and cyclic electron flow in *Chlamydomonas reinhardtii*. EMBO Rep 3: 280–285

Fujita Y, Murakami A, Aizawa K and Ohki K (1994) Short-term and long-term adaptation of the photosynthetic apparatus: Homeostatic properties of thylakoids. In: Bryant DA (ed) The Molecular Biology of Cyanobacteria, pp 677–692. Kluwer Academic, Dordrecht

Gendrullis M, Dyczmons N, Gomolla D, Gathmann S, Bernát G, Schneider D and Rögner M (2008) PetP, a new cytochrome b_6f subunit, and cytochrome *bd* oxidase—two potential regulatory players of cyanobacterial electron transport? In: Allen JF, Gantt JH, Golbeck JH and Osmond B (eds) Photosynthesis Energy from the Sun, pp 585–589. Springer, Dordrecht

Ghirlanda G, Osyczka A, Liu WX, Antolovich M, Smith KM, Dutton PL, Wand AJ and DeGrado WF (2004) De novo design of a D-2-symmetrical protein that reproduces the diheme four-helix bundle in cytochrome bc_1. J Am Chem Soc 126: 8141–8147

Gong XS, Jiang QW, Fisher NE, Young S, Howe CJ, Bendall DS and Gray JC (2000a) The role of individual lysine residues in the basic patch on turnip cytochrome *f* for electrostatic interactions with plastocyanin in vitro. Eur J Biochem 267: 3461–3468

Gong XS, Wen JQ and Gray JC (2000b) The role of amino-acid residues in the hydrophobic patch surrounding the haem group of cytochrome *f* in the interaction with plastocyanin. Eur J Biochem 267: 1732–1742

Gong XS, Chung S and Fernández-Velasco JG (2001) Electron transfer and stability of the cytochrome b_6f complex in a small domain deletion mutant of cytochrome *f*. J Biol Chem 276: 24365–24371

Haddadian EJ and Gross EL (2006) A Brownian dynamics study of the effects of cytochrome *f* structure and deletion of its small domain in interactions with cytochrome c_6 and plastocyanin in *Chlamydomonas reinhardtii*. Biophys J 90: 566–577

Hager M, Biehler K, Illerhaus J, Ruf S and Bock R (1999) Targeted inactivation of the smallest plastid genome-encoded open reading frame reveals a novel and essential subunit of the cytochrome b_6f complex. EMBO J 18: 5834–5842

Hamacher E, Kruip J, Rögner M and Mäntele W (1996) Characterization of the primary electron donor of photosystem I, P700, by electrochemistry and fourier transform infrared (FTIR) difference spectroscopy. Spectrochim Acta A 52: 107–121

Hamel P, Olive J, Pierre Y, Wollman FA and de Vitry C (2000) A new subunit of cytochrome b_6f complex undergoes reversible phosphorylation upon state transition. J Biol Chem 275: 17072–17079

Hart SE, Schlarb-Ridley BG, Bendall DS and Howe CJ (2005) Terminal oxidases of cyanobacteria. Biochem Soc Trans 33: 832–835

Hauska GS and Büttner M (1996) The cytochrome b_6f complex—Composition, structure and function. In: Ort DR and Yocum CF (eds) Oxygenic Photosynthesis: The light reactions, pp 377–398. Kluwer Academic, Dordrecht

Hellwig P, Böhm A, Pfitzner U, Mäntele W and Ludwig B (2008) Spectroscopic study on the communication between a heme a_3 propionate, Asp399 and the binuclear center of cytochrome *c* oxidase from *Paracoccus denitrificans*. Biochim Biophys Acta 1777: 220–226

Helman Y, Tchernov D, Reinhold L, Shibata M, Ogawa T, Schwarz R, Ohad I and Kaplan A (2003) Genes encoding A-type flavoproteins are essential for photoreduction of O_2 in cyanobacteria. Curr Biol 13: 230–235

Howitt CA and Vermaas WFJ (1998) Quinol and cytochrome oxidases in the cyanobacterium *Synechocystis* sp. PCC 6803. Biochemistry 37: 17944–17951

Huang D, Everly RM, Cheng RH, Heymann JB, Schagger H, Sled V, Ohnishi T, Baker TS and Cramer WA (1994) Characterization of the chloroplast cytochrome-b_6f complex as a structural and functional dimer. Biochemistry 33: 4401–4409

Iwata S, Lee JW, Okada K, Lee JK, Iwata M, Rasmussen B, Link TA, Ramaswamy S and Jap BK (1998) Complete structure of the 11-subunit bovine mitochondrial cytochrome bc_1 complex. Science 281: 64–71

Joliot P, Joliot A and Johnson G (2006) Cyclic electron transfer around photosystem I. In: Golbeck JH (ed) Photosystem I: The Light-driven Plastocyanin:Ferredoxin Oxidoreductase, pp 639–656. Springer, Dordrecht

Jordan P, Fromme P, Witt HT, Klukas O, Saenger W and Krauss N (2001) Three-dimensional structure of cyanobacterial photosystem I at 2.5 angström resolution. Nature 411: 909–917

Junge W (2008) Evolution of photosynthesis. In: Renger G (ed) Primary Processes of Photosynthesis: Priniciples and Apparatus, Part II: Reaction Centers/Photosystems, Electron Transport Chains, Photophosphorylation and Evolution, pp 447–487. Royal Society Chemistry, Cambridge

Kallas T (1994) The cytochrome b_6f complex. In: Bryant DA (ed) The molecular biology of cyanobacteria, pp 259–317. Kluwer Academic, Dordrecht

Kallas T, Spiller S and Malkin R (1988) Characterization of 2 operons encoding the cytochrome-b_6f complex of the cyanobacterium *Nostoc* Pcc-7906—Highly conserved sequences but different gene organization than in chloroplasts. J Biol Chem 263: 14334–14342

Ke B (2001) The interphotosystem cytochrome-b_6f complex and the related cytochrome-bc_1 complex. In: Photosynthesis Photobiochemistry and Photobiophysics, pp 635–664. Kluwer Academic, Dordrecht

Kim H, Dashdorj N, Zhang HM, Yan JS, Cramer WA and Savikhin S (2005) An anomalous distance dependence of intraprotein chlorophyll-carotenoid triplet energy transfer. Biophys J 89: L28–L30

Kramer DM, Joliot A, Joliot P and Crofts AR (1994) Competition among plastoquinol and artificial quinone/quinol couples at the quinol oxidizing site of the cytochrome *bf* complex. Biochim Biophys Acta 1184: 251–262

Kuras R and Wollman FA (1994) The assembly of cytochrome b_6/f complexes—An approach using genetic-transformation of the green-alga *Chlamydomonas-reinhardtii*. EMBO J 13: 1019–1027

Kuras R, de Vitry C, Choquet Y, Girard-Bascou J, Culler D, Büschlen S, Merchant S and Wollman FA (1997) Molecular genetic identification of a pathway for heme binding to cytochrome b_6. J Biol Chem 272: 32427–32435

Kurisu G, Zhang HM, Smith JL and Cramer WA (2003) Structure of the cytochrome b_6f complex of oxygenic photosynthesis: Tuning the cavity. Science 302: 1009–1014

Lee TX, Metzger SU, Cho YS, Whitmarsh J and Kallas T (2001) Modification of inhibitor binding sites in the cytochrome *bf* complex by directed mutagenesis of cytochrome b_6 in *Synechococcus* sp PCC 7002. Biochim Biophys Acta 1504: 235–247

Liberton M, Berg RH, Heuser J, Roth R and Pakrasi HB (2006) Ultrastructure of the membrane systems in the unicellular cyanobacterium *Synechocystis* sp strain PCC 6803. Protoplasma 227: 129–138

Ma F, Chen XB, Sang M, Wang P, Zhang JP, Li LB and Kuang TY (2009) Singlet oxygen formation and chlorophyll a triplet excited state deactivation in the cytochrome b_6f complex from *Bryopsis corticulans*. Photosynth Res 100: 19–28

Mao HB, Li GF, Ruan X, Wu QY, Gong YD, Zhang XF and Zhao NM (2002) The redox state of plastoquinone pool regulates state transitions via cytochrome b_6f complex in *Synechocystis* sp PCC 6803. FEBS Lett 519: 82–86

Martinez SE, Huang D, Szczepaniak A, Cramer WA and Smith JL (1994) Crystal-structure of chloroplast cytochrome-*f* reveals a novel cytochrome fold and unexpected heme ligation. Structure 2: 95–105

Martinez SE, Huang D, Ponomarev M, Cramer WA and Smith JL (1996) The heme redox center of chloroplast cytochrome *f* is linked to a buried five-water chain. Prot Sci 5: 1081–1092

McDonald AE, Amirsadeghi S and Vanlerberghe GC (2003) Prokaryotic orthologues of mitochondrial alternative oxidase and plastid terminal oxidase. Plant Mol Biol 53: 865–876

Mehler AH (1951) Studies on reactions of illuminated chloroplasts. 1. Mechanism of the reduction of oxygen and other hill reagents. Arch Biochem Biophys 33: 65–77

Mitchell P (1961) Coupling of phosphorylation to electron and hydrogen transfer by a chemiosmotic type of mechanism. Nature 191: 144–148

Mitchell P (1975) Protonmotive redox mechanism of cytochrome-bc_1 complex in respiratory-chain—Protonmotive ubiquinone cycle. FEBS Lett 56: 1–6

Mogi T and Miyoshi H (2009) Properties of cytochrome *bd* plastoquinol oxidase from the cyanobacterium *Synechocystis* sp PCC 6803. J Biochem 145: 395–401

Munekage Y, Hashimoto M, Miyaka C, Tomizawa KI, Endo T, Tasaka M and Shikanai T (2004) Cyclic electron flow around photosystem I is essential for photosynthesis. Nature 429: 579–582

Nakamura A, Suzawa T, Kato Y and Watanabe T (2005) Significant species-dependence of P700 redox potential as verified by spectroelectrochemistry: Comparison of spinach and *Thermosynechococcus elongatus*. FEBS Lett 579: 2273–2276

Nelson ME, Finazzi G, Wang QJ, Middleton-Zarka KA, Whitmarsh J and Kallas T (2005) Cytochrome b_6 arginine 214 of *Synechococcus* sp PCC 7002, a key residue for quinone-reductase site function and turnover of the cytochrome *bf* complex. J Biol Chem 280: 10395–10402

Nowaczyk MM, Hebeler R, Schlodder E, Meyer HE, Warscheid B and Rögner M (2006) Psb27, a cyanobacterial lipoprotein, is involved in the repair cycle of photosystem II. Plant Cell 18: 3121–3131

Ogawa T and Mi H (2007) Cyanobacterial NADPH dehydrogenase complexes. Photosynth Res 93: 69–77

Ohkawa H, Sonoda M, Hagino N, Shibata M, Pakrasi HB and Ogawa T (2002) Functionally distinct NAD(P)H dehydrogenases and their membrane localization in *Synechocystis* sp PCC6803. Funct Plant Biol 29: 195–200

Ouyang Y, Grebe R, Guo LW, Horn D, Nelson M, Wang QJ, Withmarsh J, Finazzi G, Facciotti C, Wollman FA, de Vitry C and Kallas T (2004) Mutational analysis of the Rieske iron-sulfur protein and the cytochrome *bf* complex quinone-reductase site. Cell Mol Biol Lett 9: S38–S42

Paumann M, Regelsberger G, Obinger C and Peschek GA (2005) The bioenergetic role of dioxygen and the terminal oxidase(s) in cyanobacteria. Biochim Biophys Acta 1707: 231–253

Peschek GA, Obinger C and Paumann M (2004) The respiratory chain of blue-green algae (cyanobacteria). Physiol Plant 120: 358–369

Peterman EJG, Wenk SO, Pullerits T, Palsson LO, van Grondelle R, Dekker JP, Rögner M and van Amerongen H (1998) Fluorescence and absorption spectroscopy of the weakly fluorescent chlorophyll *a* in cytochrome b_6f of *Synechocystis* PCC6803. Biophys J 75: 389–398

Pfannschmidt T (2003) Chloroplast redox signals: How photosynthesis controls its own genes. Trends Plant Sci 8: 33–41

Pierre Y, Breyton C, Lemoine Y, Robert B, Vernotte C and Popot JL (1997) On the presence and role of a molecule of chlorophyll *a* in the cytochrome b_6f complex. J Biol Chem 272: 21901–21908

Pils D and Schmetterer G (2001) Characterization of three bioenergetically active respiratory terminal oxidases in the cyanobacterium *Synechocystis* sp strain PCC 6803. FEMS Microbiol Lett 203: 217–222

Pogoryelov D, Reichen C, Klyszejko AL, BrunisholZ R, Muller DJ, Dimroth P and Meier T (2007) The oligomeric state of c rings from cyanobacterial F-ATP synthases varies from 13 to 15. J Bacteriol 189: 5895–5902

Ponamarev MV and Cramer WA (1998) Perturbation of the internal water chain in cytochrome *f* of oxygenic photosynthesis: Loss of the concerted reduction of cytochromes *f* and b_6. Biochemistry 37: 17199–17208

Ponamarev MV, Schlarb BG, Howe CJ, Carrell CJ, Smith JL, Bendall DS and Cramer WA (2000) Tryptophan-heme π-electrostatic interactions in cytochrome *f* of oxygenic photosynthesis. Biochemistry 39: 5971–5976

Popovic DM and Stuchebrukhov AA (2004) Electrostatic study of the proton pumping mechanism in bovine heart cytochrome *c* oxidase. J Am Chem Soc 126: 1858–1871

Rau HK and Haehnel W (1998) Design, synthesis, and properties of a novel cytochrome *b* model. J Am Chem Soc 120: 468–476

Reisinger V, Hertle AP, Plöscher M and Eichacker LA (2008) Cytochrome b_6f is a dimeric protochlorophyll *a* binding complex in etioplasts. FEBS J 275: 1018–1024

Rexroth S, Meyer zu Tittingdorf JMW, Schwaßmann HJ, Krause F, Seelert H and Dencher NA (2004) Dimeric H^+-ATP synthase in the chloroplast of *Chlamydomonas reinhardtii*. Biochim Biophys Acta 1658: 202–211

Rich P (1982) A physicochemical model of quinone-cytochrome *b-c* complex electron transfers. In: Trumpower BL (ed) Functions of Quinones in Energy Conserving Systems, pp 73–83. Academic, New York

Riekhof WR, Ruckle ME, Lydic TA, Sears BB and Benning C (2003) The sulfolipids 2′-O-acyl-sulfoquinovosyldiacylglycerol and sulfoquinovosyldiacylglycerol are absent from a *Chlamydomonas reinhardtii* mutant deleted in SQD1. Plant Physiol 133: 864–874

Rippka R, Waterbury J and Cohen-Bazire G (1974) A cyanobacterium which lacks thylakoids. Arch Microbiol 100: 419–436

Roberts AG, Bowman MK and Kramer DM (2004) The inhibitor DBMIB provides insight into the functional architecture of the Q_o site in the cytochrome b_6f complex. Biochemistry 43: 7707–7716

Rögner M, Nixon P and Diner B (1990) Purification and characterization of photosystem I and photosystem II core complexes from wild-type and phycocyanin-deficient strains of the cyanobacterium *Synechocystis* PCC 6803. J Biol Chem 265: 6189–6196

Rumeau D, Peltier G and Cournac L (2007) Chlororespiration and cyclic electron flow around PSI during photosynthesis and plant stress response. Plant Cell Environ 30: 1041–1051

Sacksteder CA, Kanazawa A, Jacoby ME and Kramer DM (2000) The proton to electron stoichiometry of steady-state photosynthesis in living plants: A proton-pumping Q cycle is continuously engaged. Proc Natl Acad Sci U S A 97: 14283–14288

Saint-Marcoux D, Wollman FA and de Vitry C (2009) Biogenesis of cytochrome b_6 in photosynthetic membranes. J Cell Biol 185: 1195–1207

Sainz G, Carrell CJ, Ponamarev MV, Soriano GM, Cramer WA and Smith JL (2000) Interruption of the internal water chain of cytochrome *f* impairs photosynthetic function. Biochemistry 39: 9164–9173

Sato N, Sonoike K, Tsuzuki M and Kawaguchi A (1995) Impaired photosystem-II in a mutant of *Chlamydomonas-reinhardtii* defective in sulfoquinovosyl diacylglycerol. Eur J Biochem 234: 16–23

Sato N, Aoki M, Maru Y, Sonoike K, Minoda A and Tsuzuki M (2003) Involvement of sulfoquinovosyl diacylglycerol in the structural integrity and heat-tolerance of photosystem II. Planta 217: 245–251

Schneider D and Schmidt CL (2005) Multiple Rieske proteins in prokaryotes: Where and why? Biochim Biophys Acta 1710: 1–12

Schneider D, Berry S, Rich P, Seidler A and Rögner M (2001) A regulatory role of the PetM subunit in a cyanobacterial cytochrome b_6f complex. J Biol Chem 276: 16780–16785

Schneider D, Skrzypczak S, Anemüller S, Schmidt CL, Seidler A and Rögner M (2002) Heterogeneous Rieske proteins in the cytochrome b_6f complex of *Synechocystis* PCC6803? J Biol Chem 277: 10949–10954

Schneider D, Berry S, Volkmer T, Seidler A and Rögner M (2004a) PetC1 is the major Rieske iron-sulfur protein in the cytochrome b_6f complex of *Synechocystis* sp PCC 6803. J Biol Chem 279: 39383–39388

Schneider D, Volkmer T, Berry S, Seidler A and Rögner M (2004b) Charakterisation of the PETC gene family in the cyanobacterium *Synecoicystis* PCC 6803. Cell Mol Biol Lett 9: 51–55

Schneider D, Volkmer T and Rögner M (2007) PetG and PetN, but not PetL, are essential subunits of the cytochrome b_6f complex from *Synechocystis* PCC 6803. Res Microbiol 158: 45–50

Schöttler MA, Flügel C, Thiele W and Bock R (2007) Knock-out of the plastid-encoded PetL subunit results in reduced stability and accelerated leaf age-dependent loss of the cytochrome b_6f complex. J Biol Chem 282: 976–985

Schultze M, Forberich B, Rexroth S, Dyczmons NG, Roegner M and Appel J (2009) Localization of cytochrome b_6f complexes implies an incomplete respiratory chain in cytoplasmic membranes of the cyanobacterium *Synechocystis* sp PCC 6803. Biochim Biophys Acta 1787: 1479–1485

Schwenkert S, Legen J, Takami T, Shikanai T, Herrmann RG and Meurer J (2007) Role of the low-molecular-weight subunits PetL, PetG, and PetN in assembly, stability, and dimerization of the cytochrome b_6f complex in tobacco. Plant Physiol 144: 1924–1935

Seelert H, Poetsch A, Dencher NA, Engel A, Stahlberg H and Müller DJ (2000) Structural biology—Proton-powered turbine of a plant motor. Nature 405: 418–419

Shao N, Vallon O, Dent R, Niyogi KK and Beck CF (2006) Defects in the cytochrome b_6f complex prevent light-induced expression of nuclear genes involved in chlorophyll biosynthesis. Plant Physiol 141: 1128–1137

Shikanai T (2007) Cyclic electron transport around photosystem I: Genetic approaches. Annu Rev Plant Biol 58: 199–217

Soriano GM, Ponamarev MV, Tae GS and Cramer WA (1996) Effect of the interdomain basic region of cytochrome *f* on its redox reactions in vivo. Biochemistry 35: 14590–14598

Soriano GM, Ponamarev MV, Piskorowski RA and Cramer WA (1998) Identification of the basic residues of cytochrome *f* responsible for electrostatic docking interactions with plastocyanin in vitro: Relevance to the electron transfer reaction in vivo. Biochemistry 37: 15120–15128

Stroebel D, Choquet Y, Popot JL and Picot D (2003) An atypical haem in the cytochrome b_6f complex. Nature 426: 413–418

Takahashi Y, Rahire M, Breyton C, Popot JL, Joliot P and Rochaix JD (1996) The chloroplast ycf7 (petL) open reading frame of *Chlamydomonas reinhardtii* encodes a small functionally important subunit of the cytochrome b_6f complex. EMBO J 15: 3498–3506

Teuber M, Rögner M and Berry S (2001) Fluorescent probes for non-invasive bioenergetic studies of whole cyanobacterial cells. Biochim Biophys Acta 1506: 31–46

Tsunoyama Y, Bernát G, Dyczmons NG, Schneider D and Rögner M (2009) Multiple Rieske proteins enable short- and long-term light adaptation of *Synechocystis* sp PCC 6803. J Biol Chem 284: 27875–27883

Twigg AI, Baniulis D, Cramer WA and Hendrich MP (2009) EPR detection of an O_2 surrogate bound to heme c_n of the cytochrome b_6f complex. J Am Chem Soc 131: 12536–12537

Vener AV, Vankan PJM, Gal A, Andersson B and Ohad I (1995) Activation deactivation cycle of redox-controlled thylakoid protein-phosphorylation—Role of plastoquinol bound to the reduced cytochrome *bf* complex. J Biol Chem 270: 25225–25232

Vener AV, van Kan PJM, Rich PR, Ohad I and Andersson B (1997) Plastoquinol at the quinol oxidation site of reduced cytochrome *bf* mediates signal transduction between light and protein phosphorylation: Thylakoid protein kinase deactivation by a single-turnover flash. Proc Natl Acad Sci U S A 94: 1585–1590

Vener AV, Ohad I and Andersson B (1998) Protein phosphorylation and redox sensing in chloroplast thylakoids. Curr Opin Plant Biol 1: 217–223

Volkmer T, Schneider D, Bernát G, Kirchhoff H, Wenk SO and Rögner M (2007) Ssr2998 of *Synechocystis* sp PCC 6803 is involved in regulation of cyanobacterial electron transport and associated with the cytochrome b_6f complex. J Biol Chem 282: 3730–3737

Wenk SO, Schneider D, Boronowsky U, Jäger C, Klughammer C, de Weerd FL, van Roon H, Vermaas WFJ, Dekker JP and Rögner M (2005) Functional implications of pigments bound to a cyanobacterial cytochrome b_6f complex. FEBS J 272: 582–592

Widger WR, Cramer WA, Herrmann RG and Trebst A (1984) Sequence homology and structural similarity between cytochrome-*b* of mitochondrial complex-III and the chloroplast-b_6-f complex—position of the cytochrome-*b* hemes in the membrane. Proc Natl Acad Sci U S A 81: 674–678

Wollman FA (1998) The structure, function and biogenesis of cytochrome b_6f complexes. In: Rochaix JD, Goldschmidt-Clermont M and Merchant S (eds) The Molecular Biology of Chloroplasts and Mitochondria in *Chlamydomonas*, pp 459–476. Kluwer Academic, Dordrecht

Wollman FA (2001) State transitions reveal the dynamics and flexibility of the photosynthetic apparatus. EMBO J 20: 3623–3630

Xia D, Yu CA, Kim H, Xian JZ, Kachurin AM, Zhang L, Yu L and Deisenhofer J (1997) Crystal structure of the cytochrome bc_1 complex from bovine heart mitochondria. Science 277: 60–66

Yamashita E, Zhang H and Cramer WA (2007) Structure of the cytochrome b_6f complex: Quinone analogue inhibitors as ligands of heme c_n. J Mol Biol 370: 39–52

Yan JS and Cramer WA (2003) Functional insensitivity of the cytochrome b_6f complex to structure changes in the hinge region of the Rieske iron-sulfur protein. J Biol Chem 278: 20925–20933

Yan JS and Cramer WA (2004) Molecular control of a bimodal distribution of quinone-analogue inhibitor binding sites in the cytochrome b_6f complex. J Mol Biol 344: 481–493

Yan JS, Kurisu G and Cramer WA (2006) Intraprotein transfer of the quinone analogue inhibitor 2,5-dibromo-3-methyl-6-isopropyl-p-benzoquinone in the cytochrome b_6 f complex. Proc Natl Acad Sci U S A 103: 69–74

Yan JS, Dashdorj N, Baniulis D, Yamashita E, Savikhin S and Cramer WA (2008) On the structural role of the aromatic residue environment of the chlorophyll *a* in the cytochrome b_6f complex. Biochemistry 47: 3654–3661

Yeremenko N, Jeanjean R, Prommeenate P, Krasikov V, Nixon PJ, Vermaas WFJ, Havaux M and Matthijs HCP (2005) Open reading frame ssr2016 is required for antimycin A-sensitive photosystem I-driven cyclic electron flow in the cyanobacterium *Synechocystis* sp PCC 6803. Plant Cell Physiol 46: 1433–1436

Zak E, Norling B, Maitra R, Huang F, Andersson B and Pakrasi HB (2001) The initial steps of biogenesis of cyanobacterial photosystems occur in plasma membranes. Proc Natl Acad Sci U S A 98: 13443–13448

Zatsman AI, Zhang HM, Gunderson WA, Cramer WA and Hendrich MP (2006) Heme-heme interactions in the cytochrome b_6f complex: EPR spectroscopy and correlation with structure. J Am Chem Soc 128: 14246–14247

Zhang ZL, Huang LS, Shulmeister VM, Chi YI, Kim KK, Hung LW, Crofts AR, Berry EA and Kim SH (1998) Electron transfer by domain movement in stockbroker bc_1. Nature 392: 677–684

Zhang HM, Huang DR and Cramer WA (1999) Stoichiometrically bound β-carotene in the cytochrome b_6f complex of oxygenic photosynthesis protects against oxygen damage. J Biol Chem 274: 1581–1587

Zhang HM, Whitelegge JP and Cramer WA (2001) Ferredoxin: $NADP^+$ oxidoreductase is a subunit of the chloroplast cytochrome b_6f complex. J Biol Chem 276: 38159–38165

Zhang HM, Primak A, Cape J, Bowman MK, Kramer DM and Cramer WA (2004a) Characterization of the high-spin heme x in the cytochrome b_6f complex of oxygenic photosynthesis. Biochemistry 43: 16329–16336

Zhang PP, Battchikova N, Jansen T, Appel J, Ogawa T and Aro EM (2004b) Expression and functional roles of the two distinct NDH-1 complexes and the carbon acquisition complex NdhD3/NdhF3/CupA/Sll1735 in *Synechocystis* sp PCC 6803. Plant Cell 16: 3326–3340

Zhou J, Fernández-Velasco JG and Malkin R (1996) N-terminal mutants of chloroplast cytochrome *f*. Effect on redox reactions and growth in *Chlamydomonas reinhardtii*. J Biol Chem 271: 6225–6232

Zito F, Kuras R, Choquet Y, Kossel H and Wollman FA (1997) Mutations of cytochrome b_6f in *Chlamydomonas reinhardtii* disclose the functional significance for a proline to leucine conversion by petB editing in maize and tobacco. Plant Mol Biol 33: 79–86

Zito F, Finazzi G, Joliot P and Wollman FA (1998) Glu78, from the conserved PEWY sequence of subunit IV, has a key function in cytochrome b_6f turnover. Biochemistry 37: 10395–10403

Zito F, Finazzi G, Delosme R, Nitschke W, Picot D and Wollman FA (1999) The Q_o site of cytochrome b_6f complexes controls the activation of the LHCII kinase. EMBO J 18: 2961–2969

Zito F, Vinh J, Popot JL and Finazzi G (2002) Chimeric fusions of subunit IV and PetL in the b_6f complex of *Chlamydomonas reinhardtii*—Structural implications and consequences on state transitions. J Biol Chem 277: 12446–12455

Chapter 21
The Convergent Evolution of Cytochrome c_6 and Plastocyanin Has Been Driven by Geochemical Changes

Miguel A. De la Rosa, José A. Navarro and Manuel Hervás

21.1 Introduction

The soluble metalloproteins cytochrome c_6 (Cc_6) and plastocyanin (Pc), which are both located in the thylakoidal lumen of most oxygen-evolving photosynthetic organisms, play the same physiological role, that is the transfer of electrons between the photosynthetic membrane-embedded complexes cytochrome b_6-*f* and photosystem I (PSI). In cyanobacteria, the two proteins play an extra respiratory role as they can serve as donor of electrons to cytochrome *c* oxidase (C*c*O) (Peschek 1999).

In higher plant chloroplasts, the only carrier is the copper protein Pc (Molina-Heredia et al. 2003). In many, but not all, eukaryotic algae and cyanobacteria, the carrier is also Pc, but their metabolism is so versatile as to allow them to adapt to changing environments: in copper-sufficient cells, Pc is used, but under copper deficiency this is replaced by Cc_6 (Wood 1978; Ho and Krogmann 1984). In each organism, the synthesis of either Cc_6 or Pc is controlled by copper availability, and so electrons can flow through the heme protein rather than the copper protein when the latter metal is in short supply. Finally, there are some other cyanobacteria and algae which can only synthesize Cc_6.

Some years ago, we proposed as a plausible hypothesis that there has been a gradual evolution from Cc_6 to Pc, with the first being replaced by the latter at the end of the evolutionary pathway (De la Rosa et al. 2002; Hervás et al. 2003). In this chapter, such a hypothesis will be analyzed from a geochemical point of view, so suggesting that the changing availability of the chemical elements iron and copper with time as the atmospheric dioxygen content increased along the Earth's life has been the force driving the transition from Cc_6 to Pc (cf. De la Rosa et al. 2006; Díaz-Quintana et al. 2008, for recent reviews). We will also review and summarize, from an evolutionary point of view, the most recent experimental data in the literature on the structural and functional features of the two metalloproteins.

M. A. De la Rosa (✉)
Instituto de Bioquímica Vegetal y Fotosíntesis, Universidad de Sevilla & CSIC,
Américo Vespucio 49, 41092 Sevilla, Spain
e-mail: marosa@us.es

G. A. Peschek et al. (eds.), *Bioenergetic Processes of Cyanobacteria*,
DOI 10.1007/978-94-007-0388-9_21,

21.2 Iron and Copper

Iron and copper are two of the most important elements used by proteins to play their biological function. Actually, the two metals are essential for all living cells and organisms, ranging from bacteria to plants and animals. They are even classical elements in Mankind's history as they were widely used by the human beings from the very early stages of their existence on the Earth, namely the "Bronze Age", in which the most advanced metalworking included techniques for smelting copper and tin, and the "Iron Age", in which iron was the main component of tools and weapons.

21.2.1 Two "Classical" Elements

In the Middle Ages, the alchemists linked metals to planets, so Mars (the Roman god of war) was linked to iron, and Venus (the Roman goddess of love and beauty) was linked to copper (Crichton and Pierre 2001). Mars is usually characterized as a youthful and aggressive warrior, whose attributes are usually a spear, a sword and sometimes a halberd (iron), whereas the attribute of Venus is usually a mirror (copper) (Fig. 21.1).

The two planets were also linked to male and female, so Men are from Mars and Women are from Venus. It then also links all three—namely, Iron/Mars/Men and Copper/Venus/Women—by pointing out that men have a much higher concentration of iron in their blood stream than women, and women have a much higher concentration of copper in their blood than men. Actually, the symbols ♂ and ♀, first used by the alchemists to design iron and copper, were after widely used in modern biology to distinguish male and female organs and individuals. They illustrate how an apparently simple question in science—namely, the way in which the conventional signs for iron/male and copper/female, known as the Shields of Mars and the Mirror of Venus, first came to be used—also has a long complexity history which touches upon mythology, astrology, alchemy, pharmacy, chemistry, etc. (Stearn 1962).

21.2.2 Chemistry and Biochemistry

Iron is the fourth most abundant element of the Earth's crust and the second most abundant metal. In the middle of the first transition series of the Periodic Table, iron possesses 26 electrons arranged in the following configuration: [Ar] $3d^6 4s^2$. Copper is also one of the most abundant metals in the planet; its atomic number is 29 and its electronic configuration is [Ar] $3d^{10} 4s^1$ (Fig. 21.1). Iron can exist in various oxidation states, the principal being 2+ ($3d^6 4s^0$) and 3+ ($3d^5 4s^0$). Copper can also exist in various oxidation states, the principal being 1+ ($3d^{10} 4s^0$) and 2+ ($3d^9 4s^0$).

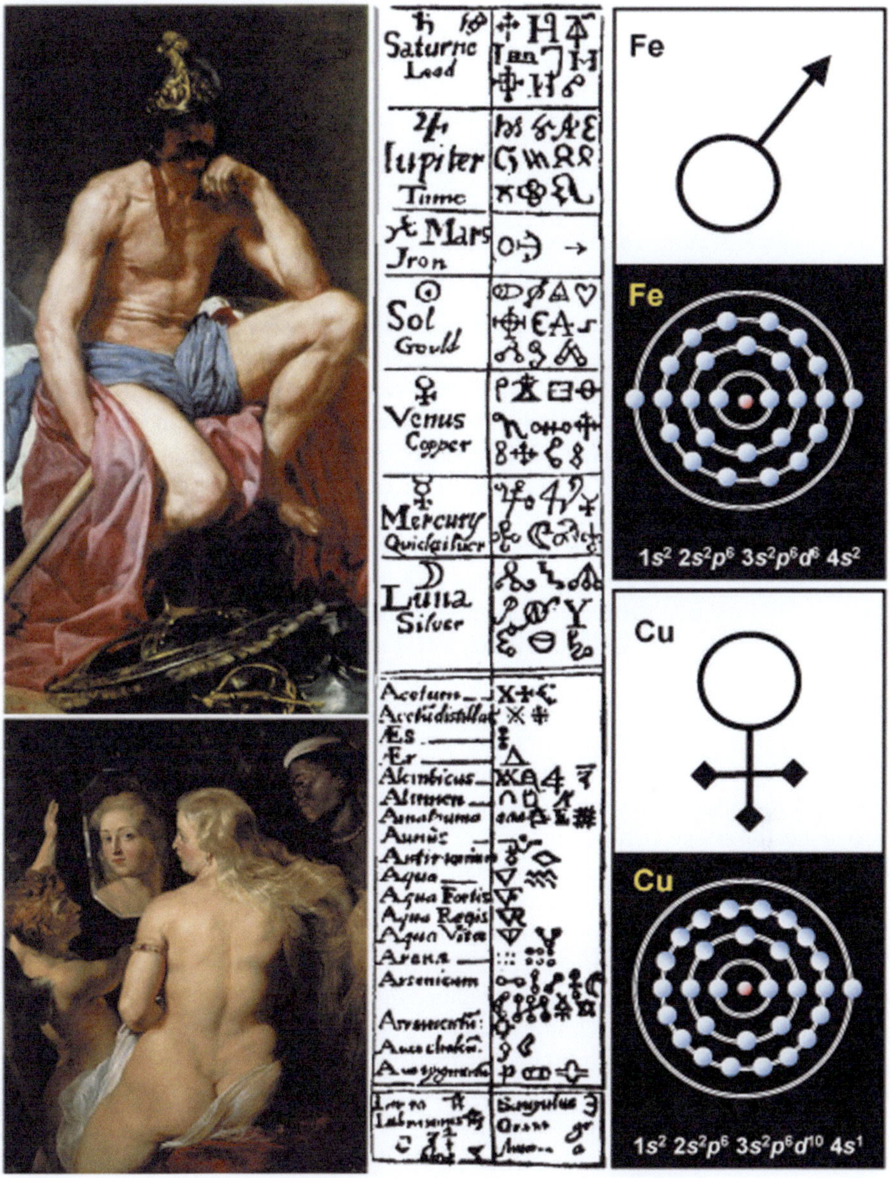

Fig. 21.1 From Mars/Iron/Men to Venus/Copper/Women. In alchemy, metals were linked to planets, so Mars was linked to iron and Venus was linked to copper. These two planets were also linked to male and female, so Men are from Mars and Women are from Venus. Therefore, the conventional signs for iron and copper, known as the Shields of Mars and the Mirror of Venus, came to be used for male and female, respectively. *Left,* Mars (*upper*) and Venus (*lower*) by Velázquez and Rubens, respectively. *Middle,* old alchemist manuscript from the 1600s. *Right,* alchemist symbols and electronic structure for iron and copper

Table 21.1 Solubility-product constant (K_{sp}) of iron and copper compounds. (Data from Williams and Fraústo da Silva 1996)

Substance	Formula	K_{sp} [a]
Copper(II)sulfide	CuS	6×10^{-36}
Iron(II)sulfide	FeS	6×10^{-16}
Copper(II)hydroxide	$Cu(OH)_2$	1.6×10^{-19}
Iron(III)hydroxide	$Fe(OH)_3$	4×10^{-38}

[a] The solubilities of the ions are in units of molarity (mol l^{-1})

The solubility of iron and copper in water strongly depends on their redox state, although many other factors—namely, formation of precipitates and formation of complexes, including acid-base equilibria—can also affect the solubility of substances and, hence, govern the main types of processes that occur in aqueous solutions. At equilibrium, the quantitative connection between the contents of a solution and the solids with which it is in contact—for example, the sea and the minerals on the sea bed—is determined by the solubility product constant (K_{sp}), which stands for the product of the concentrations of the ions in solutions. Table 21.1 shows how precipitation of hydroxides and sulfides of both iron and copper varies. Actually, the K_{sp} value for iron sulfide is much higher than that for iron hydroxide, whereas the K_{sp} value for copper sulfide is much lower than that for copper hydroxide. This point has been of extreme importance for the evolution of living organisms (see below).

Although at low concentrations, iron and copper are crucial elements for living cells, which possess a lot of proteins containing one or even more metal atoms—the so-called metalloproteins—to play their physiological role. Discussing this topic is beyond the scope of this review, but it is worth mentioning that a number of biological questions are currently being addressed, within the framework of the "omics" revolution, through the construction of genome-scale data sets for all the metal species present within a cell (metallomics) and, in particular, for metalloproteins (metalloproteomics) (Bertini and Cavallaro 2008). Among them, the iron- and copper-containing proteins are mostly abundant.

The classification of metalloproteins based on the coordination chemistry of the metal allows us to highlight the diversity of biochemical functions that metals can play viewed through the ligands, which bind it to the protein. Both Fe(II) and Fe(III) are most frequently coordinated by six ligands in octahedral geometry, but Cu(I) can form complexes with coordination numbers 2, 3 or 4, and Cu(II) prefers coordination numbers 4, 5 or 6. Whereas four-coordinate complexes of Cu(II) are square-planar, the corresponding Cu(I) complexes are tetrahedral (Crichton 2008).

In the particular case of iron, we may consider the following classification: (1) hemoproteins, in which an iron porphyrin is incorporated into different apo-proteins to give O_2 carriers, O_2 activators and electron transfer proteins; (2) iron-sulfur proteins, most of which are involved in electron transfer; and (3) non-heme, non-

Table 21.2 Iron and copper centers in metalloproteins

	Examples	Function
Iron proteins		
Heme proteins	Myoglobin, hemoglobin	O_2 storage and transport
	Cytochromes	Electron transfer
	Catalases, peroxidases, terminal oxidases	O_2 activation
	Sulfite and nitrite reductases	S and N metabolism
Fe-S proteins	Ferredoxin, rubredoxin	Electron transfer
	Oxygenases, hydrogenase, nitrogenase	Redox reaction
	Hybrid cluster proteins (HCP)	Unknown
Others	Ferritin, transferrin	Fe storage and transport
Copper proteins		
Type 1 (tetrahedral)	Blue copper proteins (plastocyanin, amycianin, azurin, etc.)	Electron transfer
Type 2 (square-planar)	Nitrite reductase, superoxide dismutase, amino and galactose oxidases	Redox reaction
Type 3 (dinuclear)	Hemocyanin	O_2 transport
	Tyrosinase	Hydroxylation
Type 4 (trinuclear)	Ceruloplasmin	Fe metabolism
	Ascorbate oxidase, laccase	Redox reaction
Type 5. Cu_A (dinuclear) and Cu_B (mononuclear)	Cytochrome *c* oxidase	O_2 reduction

iron sulfur, iron-containing proteins, which include proteins of iron storage and transport. In copper-containing proteins, three major types of metal centers can be found. Types 1 and 2 have one copper atom—which has an intense blue colour in the former and is almost colourless in the latter—whereas Type 3 has two copper atoms. Also the multicopper centers can be classified in two types, namely Type 4, which has three metal atoms, and Type 5, which consists of a dinuclear (Cu_A) and mononuclear (Cu_B) center (Table 21.2).

21.3 Geochemical Changes

During the whole Earth's existence, there have been almost continuous chemical changes of the environment mostly due to life's activities and these have had a huge effect on speciation (Williams 2007). The strong interplay of geochemical change and biological evolution is now a well accepted fact, with the evolution of biological molecules (mainly, nucleic acids and proteins) being driven by the geochemical environment and the appearance of microorganisms determining the chemical composition of the planet surface (Kasting and Siefert 2003; Macalady and Banfield 2003). Microbes affected the chemistry and distribution of many elements in the Periodic Table at the beginning of the Earth's life and, thus, directly influenced their bioavailability (Newman and Banfield 2002).

21.3.1 Increase in Atmospheric Dioxygen Content

The evolution of Earth's atmosphere is linked tightly to the evolution of its biota. The planet is ca. 4.6 billion years old, and life (biosphere) appeared ca. 3.8 billion years ago (BYA) (Fig. 21.2). At the beginning, the biosphere was exclusively unicellular and was dominated by hyperthermophiles that utilized chemical sources of energy and employed a range of metabolic pathways for CO_2 assimilation (Des Marais 1998). When the first living organism, unicellular bacteria, appeared on the primitive Earth, molecular oxygen did not exist in the atmosphere. Non-oxygenic photosynthesis arose very early, and the oxygen-evolving process arose later but still prior to 2.7 BYA. It appears that oxygenic photosynthesis—an extremely complex biochemical process—was "invented" only once, and a primitive cyanobacterium was the organism responsible. Cyanobacteria elaborated an efficient light-driven machinery and gradually developed into unicellular and multicellular photosynthetic organisms, which produced more and more molecular oxygen. The transition toward the modern global environment was paced by a decline in volcanic and hydrothermal activity (Des Marais 1998).

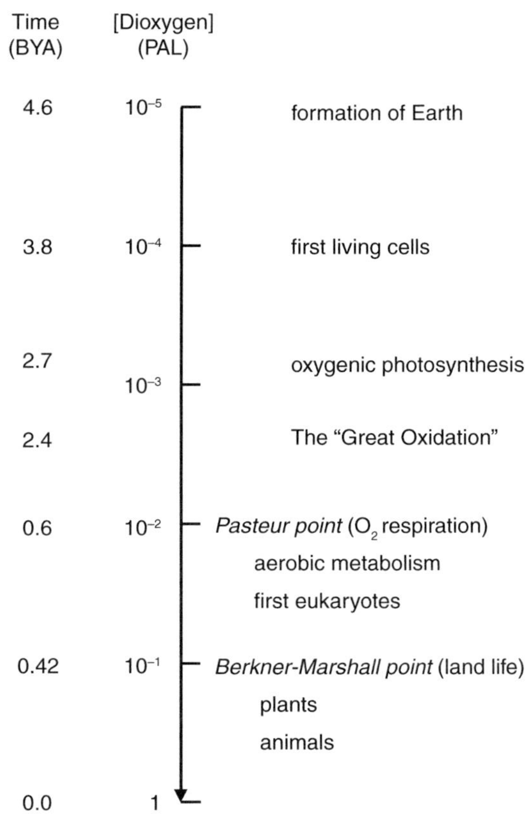

Fig. 21.2 Change in atmospheric dioxygen content and major evolutionary events along the Earth's life. Oxygen concentration is expressed as fractions of the Present Atmospheric Level (PAL). Years before present are expressed as Billions Years Ago (BYA)

So the largest chemical transition in the history of the Earth—which has been characterized by a series of major transitions separated by long periods of relative stability—was the 'Great Oxidation'. It took place ca. 2.4 BYA, when atmospheric oxygen concentrations rose from less than 10^{-5} of the present atmospheric level (PAL) to more than 0.01 PAL, and possibly to more than 0.1 PAL. This transition happened long after oxygenic photosynthesis is thought to have evolved, but the causes of this delay and of the Great Oxidation itself remain uncertain (Goldblatt et al. 2006).

It has been proposed that molecular oxygen did not enter the atmosphere during at least two billion years of photosynthetic activity, and so dioxygen concentration remained stable at a low level because it was being used for chemical oxidation of huge amounts of reduced compounds (namely sulfide and ferrous iron) as long as cyanobacteria were producing it. With the appearance of eukaryotic organisms, oxygen concentration began to increase rapidly and the limit marked by the Pasteur point (0.01 PAL) was reached ca. 0.6 BYA. The Pasteur point—which means that respiration consuming molecular oxygen can produce energy much more efficiently than anaerobic fermentation—does represent the minimum to sustain the aerobic life (Fig. 21.2).

Accompanying the buildup of O_2 was a new atmospheric development that also had an important effect on the biosphere, namely, the emergence of an ozone layer. The presence of atmospheric ozone is essential to the existence of most land life since ozone is the only important absorber of solar near-ultraviolet radiation between 200 and 300 nm. This dependence led Berkner and Marshall (1965) to link the spread of life onto land with the development of the ozone layer ca. 0.42 BYA. The so-called Berkner-Marshall point (0.1 PAL) marks the formation of a substantial ozone layer in the upper atmosphere along with the land colonization by plants and animals (Fig. 21.2) (Fischer 1965; Peschek 1999). However, the uncertainty inherent in calculations on the establishment of the ozone screen and the invasion of land leaves open the possibility that a causal relationship between the evolution of atmospheric ozone and the appearance of land life did indeed exist (Kasting and Donahue 1981).

21.3.2 Changing Availability of Iron and Copper with Rising Dioxygen

The environment played a critical role as the driving force behind the evolution of internal cellular mechanisms. The oxygen released by cyanobacteria changed the world from anaerobic to aerobic and made possible the last great advance in energy-yielding metabolism, aerobic respiration (Williams and Fraústo da Silva 1996).

The slow development of a dioxygen atmosphere had a gradual effect on many elements on the surface of the Earth and in the sea. There was a gradual rise in redox potential of the surface of the sea as the partial pressure of O_2 rose. The initial redox potential in the reducing atmosphere some 4.5 BYA could have been well below

0.0 V. If we take this value as a starting point and then introduce dioxygen up to present-day levels then the potential rose slowly and today, near the sea's surface, must be around that of the water/dioxygen couple of 0.82 V at pH 7. The change was slow and remains far from complete since the reducing components of the early surface and sea formed a large buffer, especially through iron sulfides. Dioxygen itself could only appear in quantity after most of the surface elements had been oxidized (Williams and Fraústo da Silva 1996).

Amongst the metals iron and copper, the change in availability due to oxygen advent is largely due to the difference in solubility of sulfides relative to hydroxides (see above, and Table 21.1). So copper became more available, and at the same time iron became of greatly reduced availability. The corresponding changes in availability of the two metals with the gradually increasing redox potential of surface water are shown in Fig. 21.3. In stages with the rise of redox potential, when most of the H_2S had been oxidized, there must have been first the simultaneous loss of Fe/S complexes and precipitation of $Fe(OH)_3$ (Williams and Fraústo da Silva 1996).

The redox potential value for the Fe^{2+}/Fe^{3+} couple is extremely variable but can be tuned by the ligands provided by the protein matrix, so making iron most suitable for catalysis in living organisms. Actually, the redox potential value of iron-containing proteins ranges from ca. −0.5 to ca. +0.6 V so as to encompass the whole spectrum of biologically relevant redox functions. Thus the first enzymes involved in anaerobic metabolism contained iron and were designed to act in the lower portion of the range of redox potentials. The presence of dioxygen created the need for new redox systems with redox potentials in the range from 0.0 to 0.8 V. And copper proved eminently suitable for this role. For aerobic metabolism, enzymes and other proteins with higher redox potentials came to be utilized to take advantage of the oxidizing power of dioxygen (Crichton 2008).

Iron and copper are actually metals which play an important role in the living world. From a brief consideration of their chemistry and biochemistry (see above) it can be concluded that the early chemistry of life used water soluble ferrous iron

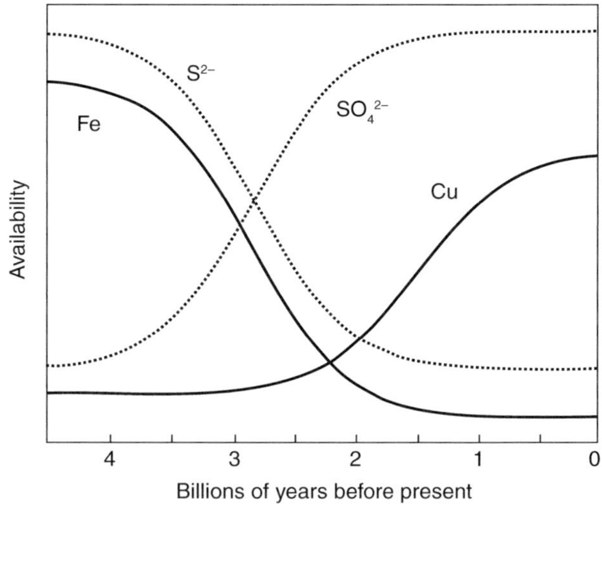

Fig. 21.3 The changing bioavailability of some chemical elements with time as atmospheric dioxygen content increases

while copper was in the water-insoluble Cu(I) state as highly insoluble sulfides. In contrast to the oxidation of iron and its loss of bioavailability as insoluble Fe(III), the oxidation of insoluble Cu(I) led to soluble Cu(II) (Crichton and Pierre 2001). In summary, life's 'addiction' to iron is thought to reflect this early evolution in an iron-rich reducing environment. Large-scale seeding experiments confirm that today iron availability limits growth of oceanic cyanobacteria possibly due to its requirement in photosynthesis and nitrogen fixation (Martin et al. 1994; Cavet et al. 2003). On the other hand, copper is contained in a number of enzymes and proteins, but most of them are found only in eukaryotes (Ochiai 1983). Thus the early evolution of life was the "iron age", but the subsequent was the "copper age", or rather the "iron-copper age", where both metals are involved together (Crichton 2008).

On the basis of the changing bioavailabilities of iron and copper throughout the Earth's life, the transition form Cc_6 to Pc can be interpreted as a consequence of microbial adaptation to environmental conditions. This means that the first O_2-generating photosynthetic organisms produced the heme protein, as iron was much more abundant than copper, to transfer electrons from cytochrome b_6-f to PSI and CcO. When copper became available as the atmosphere turn from reducing to oxidizing, the photosynthetic organisms learnt to use such a "novel" metal and incorporated it into proteins to play the same role as some of the "old" iron proteins (De la Rosa et al. 2006).

21.4 Iron-Copper Homeostasis in Cyanobacteria

Like other organisms, cyanobacteria require metal cofactors (mainly, iron and copper) for the functioning of an ample variety of proteins (Messerschmidt et al. 2001). Thus cyanobacteria have developed elaborated metal homeostasis mechanisms and regulation processes not only to ensure an adequate metal supply in a changing environment but also to avoid harmful reactions coming from free metal ions (Bryant 1994; Shcolnick and Keren 2006). Whereas iron and copper participate in essential metabolic pathways, namely photosynthesis and respiration, both free metals also promote at high concentration the generation of reactive oxygen species (ROS) causing oxidative damage, cell malfunction and even death (Halliwell and Gutteridge 1984; Crichton and Ward 2006). This is of particular relevance in the case of photosynthetic organisms, in which photosystems' redox chemistry provides several active sites where ROS can be generated (Shcolnick and Keren 2006).

Cyanobacterial photosynthetic and respiratory proteins require iron (including iron-sulfur clusters and heme groups) and/or copper (either as mono o multi-nuclear centers) cofactors (Hervás et al. 2003). As a result, the demand for both metals of photosynthetic organisms surpasses that of non-photosynthetic ones, resulting in an increased sensitivity to metal limitation and in the developing of mechanisms for metal uptake and collection (Shcolnick and Keren 2006). Thus, the occurrence of Fe(III) chelators, such as the ubiquitous ferritin iron-storage protein family, contributes to increase the iron bioavailability and to decrease the deleterious effects of metal free ions (Keren et al. 2004).

Table 21.3 Examples of iron and copper proteins that play equivalent roles. (Adapted from Merchant et al. 2006)

Function	Iron protein	Copper protein
Electron transfer	Cytochrome c_6	Plastocyanin
Oxygen transport	Hemoglobin Hemerythrin	Hemocyanin
Hydroxylation	Di-iron MMO Cytochrome P450	Particulate MMO Mononuclear tyrosinase
Oxidation	Catechol dioxygenase	Dinuclear catechol oxidase
Terminal oxidase	Di-iron alternative oxidase	Cu_ACu_B cytochrome *c* oxidase Cu_ACu_Z N_2O reductase
Anti-oxidant	Fe-SOD	Cu/Zn-SOD
Nitrite reduction	Heme nitrite reductase	Cu nitrite reductase

SOD superoxide dismutase, *MMO* methane mono-oxygenase

Since the insoluble ferric species became the prevalent form of iron for living organisms, a supplementary case of adaptation arose by developing new analogue enzymes possessing other metals and, in particular, copper (Table 21.3). In some cases, the organisms were even able to use one protein over the other depending on metal availability, thus adding an extra capability for metabolic adaptation to environmental changes (Merchant et al. 2006). In particular, as many algae and cyanobacteria can synthesize Pc or Cc_6 depending on copper availability (Sandmann et al. 1983; Ho and Krogmann 1984), iron limitation also induces the replacement of the iron-containing protein ferredoxin by the flavin-containing protein flavodoxin to accept electrons from PSI and to function in other redox pathways (Rogers 1987). In both cases, protein alternation results in maintaining the metal uses in other essential (and irreplaceable) redox proteins under metal limitation, thus suggesting that cyanobacteria can experience iron and copper-deficiency in nature (Martin et al. 1994; Morel and Price 2003).

In the particular case of Pc and Cc_6, the occurrence of a cytochrome as electron donor to the chlorophyllic system of the primitive anoxygenic photosynthetic bacteria is an extra evidence in favour of our evolutionary hypothesis that the heme protein represents the ancestral photosynthetic redox carrier (Blankenship 1992; De la Rosa et al. 2006). Pc, in its turn, is evolutionarily related to other bacterial blue-copper proteins with redox functions yet not involved in photosynthetic electron transfer, namely azurins and amicyanins (Messerschmidt et al. 2001). Thus, the replacement of Cc_6 by Pc implies the recruitment of a copper-protein as a response to iron limitations. The synthesis of either Pc or Cc_6 seems to be regulated by copper at the level of the initiation of transcription in *Synechocystis* and other cyanobacteria, thus acting on the amount of mRNA for the Pc-encoding *petE* and Cc_6-encoding *petJ* genes (Briggs et al. 1990; Zhang et al. 1992). In consequence, Pc-mRNA is detected only in the presence of copper and Cc_6-mRNA is detected only in the absence of this metal (Zhang et al. 1992).

The cyanobacterial thylakoid is the site for both photosynthetic and respiratory electron transports, with the participation of an ample variety of iron proteins, the copper protein Pc and cytochrome *c* oxidase (C*c*O)—the latter containing both iron and copper (Vermaas 2001). Pc and Cc_6 are imported into the thylakoidal lumen as unfolded protein precursors (Van der Plas et al. 1989; Bovy et al. 1992). Cyanobacterial thylakoids are currently the only location for copper proteins known to require this metal entering into bacterial cytoplasms, and Pc in particular seems to be responsible for most of copper requirements as Pc concentration in the lumen is estimated to be in the range of sub-millimolar (Finazzi et al. 2005). Copper is transferred across the plasma and thylakoidal membranes by the consecutive action of two different copper-transporting P-type ATPases, the CtaA and PacS complexes (Fig. 21.4) (Shcolnick and Keren 2006). Inside the cell, copper is associated to Atx1, a small soluble copper metallochaperone (Banci et al. 2006; Shcolnick and

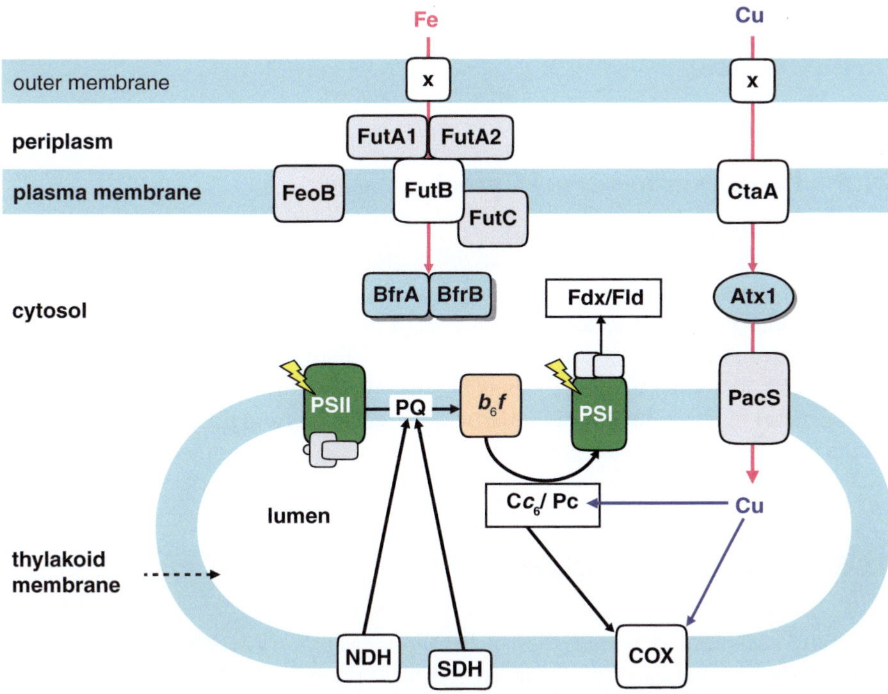

Fig. 21.4 Pathways for copper and iron transport into *Synechocystis* cells and thylakoidal proteins requiring them in photosynthesis and respiration. FutA1, FutA2, FutB and FutC are the components of the FutABC ferric iron transporter; FeoB is responsible for ferrous iron uptake; CtaA and PacS are two P-type ATPases for copper transport across the plasma and thylakoidal membranes, respectively. Iron is stored in Bacterioferritin A and B (BfrA, BfrB). Copper is associated to Atx1, a metallochaperone that delivers it to PacS. 'X' denotes transporters that are still unknown. The abbreviations for the photosynthetic and respiratory components are: PSI and PSII, photosystem I and II; PQ, plastoquinone; b_6f, cytochrome b_6-f complex; Pc, plastocyanin; Cc_6, cytochrome c_6; Fdx, ferredoxin; Fld, flavodoxin; NDH, NADH dehydrogenase; SDH, succinate dehydrogenase; C*c*O, cytochrome *c* oxidase. *Red, blue* and *black arrows* stand for metal transport, copper incorporation to redox proteins and electron transfer reactions, respectively. (The scheme has been adapted from Shcolnick and Keren (2006) and De la Cerda et al. (2008))

Keren 2006). Atx1 is supposed to obtain copper from CtaA and release it to PacS (Fig. 21.4) (Tottey et al. 2002, 2005; Banci et al. 2006), although it may also acquire copper from another unknown importer (Tottey et al. 2005).

In *Synechocystis,* the ferric iron ATP-binding cassette transporter FutABC plays a major role in iron transport across the plasma membrane (Fig. 21.4) (Katoh et al. 2001). This transporter appears to be composed of four polypeptide subunits: FutA1 and FutA2 as periplasmic proteins, and FutB and FutC, probably forming an inner-membrane bound unit (Katoh et al. 2001). Alternatively, ferrous iron transport depends on the FeoB transporter, although it is not essential for iron acquisition unless the ferric pathway is disabled (Katoh et al. 2001). Inside the cyanobacterial cell, up to 50% of the stored Fe(II) is associated to bacterioferritins, where it is oxidized to the insoluble Fe(III) form before being transferred to the different protein substrates (Keren et al. 2004; Shcolnick and Keren 2006). Cyanobacteria possess two bacterioferritins (BfrA and BfrB) that would be able to form a heterodimeric structure (Keren et al. 2004).

21.5 Plastocyanin versus Cytochrome c_6

Pc and Cc_6 are by now well-characterized proteins. A great number of crystal and solution structures are already available for the two molecules from both prokaryotes and eukaryotes, and functional analyses of their roles in differently evolved organisms have also been carried out. On the basis of such a structural and functional background, a comparative analysis is herein presented to highlight not only their main general features but also their particular similarities and differences depending on the organism from which both proteins come from.

21.5.1 Structural Features

The main structural characteristics of Pc and Cc_6 are very well conserved among photosynthetic organisms (cf. De la Rosa et al. 2006 and Díaz-Quintana et al. 2008, for recent reviews). Whereas Cc_6 is a typical Class I *c*-type cytochrome, with a heme group as redox center and a α-helix-based tertiary structure, Pc is a typical Type-1 blue copper protein, which consists of several β-strands and a short α-helix (Fig. 21.5) (De la Rosa et al. 2006; Díaz-Quintana et al. 2008).

The iron atom in the heme group of Cc_6 presents a typical histidine-methionine axial coordination (Frazão et al. 1995; Sawaya et al. 2001), with part of the heme structure (pyrrole ring C and the ring D propionic group) being accessible to the solvent as to establish the most probable electron transfer port of the protein (Frazão et al. 1995; Ubbink et al. 1998; Crowley et al. 2002; Díaz-Moreno et al. 2005a, b). In Pc, the copper atom is coordinated by two histidines, one cysteine and one methionine in a distorted tetrahedral geometry (Guss and Freeman 1983; Guss et al.

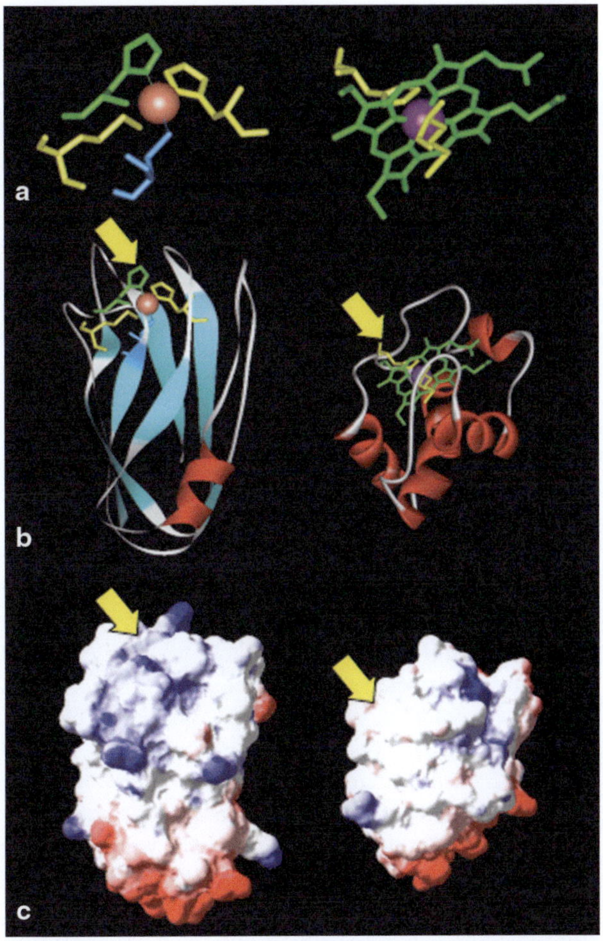

Fig. 21.5 Structural comparison between redox centers (**a**), overall foldings (**b**) and surface electrostatic potentials (**c**) of *Synechocystis* plastocyanin (PDB code, 1PCS) (*left*) and *Synechococcus* cytochrome c_6 (PDB code, 1C6S) (*right*). The two molecules are oriented with their respective site 1 at the top and site 2 in front. Calculations for the surface electrostatic potential were made at 40 mM ionic strength and pH 7, with positively and negatively charged areas being depicted in *blue* and *red,* respectively. The *arrows* point to the electron transfer port

1986; Inoue et al. 1999). As the copper atom is not accessible to the solvent, His87 (according to plant residue numbering) is the only solvent-exposed Cu-ligand, thus making this group the most probable electron transfer pathway to and from the redox cofactor (Fig. 21.5).

In summary, the main structural features—including the redox cofactor, the amino acid sequence and 3D structure—are dramatically different in Pc and Cc_6. So the following question arises: How can two proteins so different play the same physiological role and interact with three different partners, namely the membrane protein complexes cytochrome b_6-f, PSI and CcO? (see below).

Although Pc and Cc_6 are two structurally unrelated proteins, they do however share a number of critical parameters related to their functional equivalence, namely the midpoint redox potential (ca. +350 mV, at pH 7), molecular mass (ca. 9–10 kDa), and isoelectric point (pI). In particular, the pI value varies from one organism to another but is practically the same for the two proteins isolated from the same source (De la Rosa et al. 2002). As the main structural framework is strongly conserved in Pc and Cc_6 from differently evolved organisms, these variations in pI reveal changes in the acidic and basic residues at the protein surface. Actually, the pI values for cyanobacterial Pc and Cc_6 can be slightly acidic in *Synechocystis* (ca. 5.5), neutral in *Arthrospira* (ca. 7), or even strongly basic in *Nostoc* (ca. 9) (De la Rosa et al. 2006; Díaz-Quintana et al. 2008). This finding reveals a unique and intriguing feature in the evolution of these two alternative proteins. Noteworthy is that flavodoxin and ferredoxin exhibit in all cyanobacteria a strongly conserved acidic protein surface, with no significant differences from one organism to another (Rogers 1987).

The comparative structural and functional analysis of the protein surface areas of Pc and Cc_6 has made possible to identify well-defined regions that could play similar functions (De la Rosa et al. 2002, 2006). Thus, Pc contains a hydrophobic patch at the protein surface around the His87 residue coordinating the copper atom (the so-called north pole, or site 1), whereas Cc_6 exhibits a series of hydrophobic residues surrounding the solvent accessible part of the heme cofactor (Fig. 21.5) (Guss and Freeman 1983; Frazão et al. 1995; Sigfridsson 1998). In addition, Pc and Cc_6 possess an electrostatically charged region, similarly located at the surface of both proteins related to their respective hydrophobic regions and redox centers (the so-called east face, or site 2) (Fig. 21.5) (Guss and Freeman 1983; Frazão et al. 1995; Sigfridsson 1998; Díaz-Quintana et al. 2003). The electrostatic patch is negatively-charged in the eukaryotic proteins, but is formed by acid or basic residues in cyanobacterial Pc and Cc_6 so as to vary in parallel and in consonance with the pI of their respective partner (Fig. 21.5) (De la Rosa et al. 2002, 2006). From both kinetic analyses of site-directed mutants of Pc and Cc_6 and NMR data, it is now clear that site 2 is responsible to electrostatically drive the formation of the transient complex with PSI and *Cf*, whereas the hydrophobic residues at site 1 are responsible to maintain the surface complementarity with PSI and *Cf* and to establish the electron transfer pathway (Díaz-Quintana et al. 2008). However, it still remains to be established if the same areas are involved in the interaction with *Cc*O, the respiratory partner.

In agreement with the close similarity of their functional areas, identical residues can be identified in Pc and Cc_6 that are similarly located in relation to sites 1 and 2 (De la Cerda et al. 1999; Molina-Heredia et al. 2001; Albarrán et al. 2005, 2007). Actually, cyanobacterial Pc and Cc_6 possess just one arginine—the only arginyl residue in the whole amino acid sequence—between their hydrophobic and electrostatic areas. In *Nostoc,* this residue (Arg88 in Pc, Arg64 in Cc_6) has been shown to be critical for the efficient oxidation of *Cf* (Albarrán et al. 2005, 2007) and reduction of PSI (Molina-Heredia et al. 2001). In addition, the replacement of an aspartate residue (Asp49 in Pc, Asp72 in Cc_6) that is similarly located at the edge of site 2 in Pc and Cc_6 by a positively-charged amino acid accelerates PSI reduction by the two proteins (Molina-Heredia et al. 2001).

Thus Pc and Cc_6 represent a fascinating case of biological evolution at the molecular level that is both convergent and parallel. Actually, they are two genetically unrelated proteins that have evolved to acquire a number of similar structural features so as to play the same physiological role; however, such structural features can be different from one organism to another but similar within the same organism.

21.5.2 Reaction Mechanisms

The double photosynthetic and respiratory role of cyanobacterial Pc and Cc_6 as redox carriers has been unequivocally demonstrated. However, their functioning has been more in depth analyzed in photosynthesis than in respiration (Paumann et al. 2005; Díaz-Quintana et al. 2008). Actually, the functional interaction of Pc and Cc_6 with PSI has been exhaustively studied not only in cyanobacteria but also in an ample variety of eukaryotic organisms, ranging from green algae to plants (Hope 2000; De la Rosa et al. 2002; Díaz-Quintana et al. 2003; Hervás et al. 2003). As regards the interaction with C*f*, the kinetic analyses have been performed with Pc from a limited number of organisms (Kannt et al. 1996; Hope 2000; Schlarb-Ridley et al. 2002; Albarrán et al. 2005, 2007), whereas no information with Cc_6 has been attained because of the similar spectral properties of Cc_6 and C*f* and overlap of absorbance bands.

The experimental data demonstrating the involvement of the two cyanobacterial metalloproteins in respiration is comparatively scarce. A kinetic analysis using the recombinant, soluble Cu_A domain of the C*c*O subunit II from *Synechocystis* shows that it is reduced by Cc_6 and Pc with bimolecular rate constants of ca. 5×10^5 and 5×10^4 M^{-1} s^{-1}, respectively (Paumann et al. 2004a, b, c). Another time-resolved analysis using the whole enzyme reveals that both Cc_6 and Pc can be efficiently oxidized by the aa_3-type cytochrome *c* oxidase in *Nostoc* sp. PCC 7119 (Navarro et al. 2005).

To simplify the analysis of the experimental data, we have proposed (Hervás et al. 1995; De la Rosa et al. 2002; Díaz-Quintana et al. 2008) three different kinetic models: Type I, when the acceptor and donor proteins interact with each other by following a collisional reaction mechanism, with no formation of any kinetically detectable transient complex; Type II, when an intermediate transient complex between the acceptor and donor proteins can be detected; and Type III, when an additional step resulting from the structural rearrangement of the two redox partners within the complex, prior to electron transfer, can be observed. The three kinetic models are represented in the following scheme:

$$\text{I)}\quad \mathrm{A_{ox}} + \mathrm{D_{red}} \stackrel{k_2}{\rightarrow} \mathrm{A_{red}} + \mathrm{D_{ox}}$$

$$\text{II)}\quad \mathrm{A_{ox}} + \mathrm{D_{red}} \stackrel{K_\mathrm{A}}{\longleftrightarrow} [\mathrm{A_{ox}}\ldots D_\mathrm{red}] \stackrel{k_\mathrm{et}}{\rightarrow} \mathrm{A_{red}} + \mathrm{D_{ox}}$$

$$\text{III)}\quad \mathrm{A_{ox}} + \mathrm{D_{red}} \stackrel{K_\mathrm{A}}{\longleftrightarrow} [\mathrm{A_{ox}}\ldots D_\mathrm{red}] \stackrel{K_\mathrm{R}}{\longleftrightarrow} [\mathrm{A_{ox}}\ldots D_\mathrm{red}]^* \stackrel{k_\mathrm{et}}{\rightarrow} \mathrm{A_{red}} + \mathrm{D_{ox}}$$

where A and D stand for the acceptor and donor proteins, respectively; the brackets indicate formation of the transient complex; and the asterisk denotes the properly rearranged complex. Whichever the kinetic model is applied in every case will depend on the relative values of the equilibrium constants for complex association (K_A) and rearrangement (K_R), as well as of the rate constant for electron transfer itself (k_{et}). As noted by Hope (2000), the Type III model does correspond to the full description of the whole reaction in all systems, although kinetically they will appear as one of the above three models depending on the relative rate-limiting steps. This classification of models for the reaction mechanisms applies to the interaction of Pc and Cc_6 with all their known redox partners, namely PSI, *Cf* and *Cc*O.

In most Pc/PSI and Cc_6/PSI systems from different organisms analyzed so far, the kinetics of PSI reduction can be well fitted to one of such kinetic models (Hervás et al. 1995; Sigfridsson et al. 1995; Hippler et al. 1997). Actually, the two proteins react with PSI following the same kinetic model and exhibit similar rate constants when they are both isolated from the same organism but vary from one organism to another. As to the interaction with *Cc*O, both Pc and Cc_6 follow a Type II model in *Nostoc* sp. PCC 7119 (Navarro et al. 2005). And finally, Pc seems to react with *Cf* by following a Type II model in *Nostoc* and plants (Hope 2000; Albarrán et al. 2005, 2007), but the experimental data could not be assigned to any kinetic model in other organisms, such as the cyanobacterium *Phormidium laminosum* (Kannt et al. 1996; Schlarb-Ridley et al. 2002).

The comparative analysis of PSI reduction by its two donor proteins in a wide variety of organisms has shed new light on the evolution of the electron transfer mechanism, whereas the lack of data on the interaction of Pc and Cc_6 with their other redox partners prevents such evolutionary analysis. PSI reduction by Pc and Cc_6 has experienced a long evolutionary process, from the Type I to Type II and III models, in order not only to reach the maximum kinetic efficiency but also to promote the transition from the iron- (Cc_6) to the copper-protein (Pc). Thus PSI was first being adapted to react with Cc_6 so as to evolve from Type I to Type III, but the further appearance of Pc required a second adaptation of PSI to its new electron donor protein (De la Rosa et al. 2002; Hervás et al. 2003; Díaz-Quintana et al. 2008). In fact, in most cyanobacteria the reaction mechanism of Cc_6 with PSI ranges from Type I to Type II and even Type III. Pc, in turn, is always less efficient than Cc_6 and only reacts with PSI according to Type I or Type II models (Díaz-Quintana et al. 2003, 2008). In eukaryotic algae, however, the experimental data is just the opposite, with both Pc and Cc_6 following the Type III kinetic model but with Pc being even more reactive than Cc_6, so suggesting that Pc has got time enough to evolve and learn to interact with PSI by following the most efficient way. Finally, in plants, Cc_6 has lost its original function so as to be definitively replaced by the copper protein as donor to PSI (De la Rosa et al. 2006; Díaz-Quintana et al. 2008).

In summary, our proposal is that molecular recognition between partners and complex formation were not well established at the beginning of the evolutionary process, when the interaction forces were mainly hydrophobic. Later on, the subsequent evolution made the attractive electrostatic interactions the factors governing the redox reaction; the electron transfer was thus optimized within a rather stable transient complex, wherein the partners are oriented to allow the establishment of the hydrophobic interactions needed for efficient electron transfer upon complex reorganization. Finally, local structural changes in the interaction area between the two partners were designed to reach the last stage—that is, complex dissociation after electron transfer—and start a new redox cycle (De la Rosa et al. 2002; Hervás et al. 2003; Díaz-Quintana et al. 2008).

21.5.3 Knockout Mutants for Plastocyanin and Cytochrome c_6 Genes

The relevance of copper homeostasis in cyanobacteria—not only as such but also as a model system for higher plants and even for humans—has made our group to analyze the effect of copper deprivation in *Synechocystis* cells, with particular attention on the differential copper-regulated expression of Pc- and Cc_6-coding genes (Malakhov et al. 1999; Hervás et al. 2003; De la Cerda et al. 2008). First, we constructed two deletion mutants—each lacking either the *petE* or *petJ* gene, which respectively codes for Pc or Cc_6—in order to analyze their behavior, as compared with WT, when growing in media with or without copper. As expected, the WT cells do similarly grow under photoautotrophic conditions either in the absence or presence of added copper, so confirming that *Synechocystis* possesses a metabolism so versatile as to synthesize either Pc or Cc_6 as a function of copper bioavailability. The two deletion mutants grow at rates equivalent to that of WT in media that allow them to express one of the two electron donor proteins. In fact, Δ*petE* is only able to grow at a standard rate in the absence of copper, when Cc_6 is being produced, whereas Δ*petJ* grows normally when Pc is synthesized because of copper induction. When the Δ*petE* and Δ*petJ* strains are under conditions at which the expression levels of both Pc and Cc_6 are not detectable, their respective growth rate is much lower than that of WT cells (Durán et al. 2004).

Taking into account that the cyanobacterial respiratory and photosynthetic electron transfer chains share a number of redox components, such as Cc_6 and Pc (Schmetterer 1994; Pils et al. 1997; Paumann et al. 2005), the ability of the WT, Δ*petE* and Δ*petJ* strains to grow in glucose-supplemented culture media was also investigated. Under such heterotrophic conditions, glucose is used as an organic carbon source (Anderson and McIntosh 1991) and the electrons are transported from the sugar molecule to dioxygen throughout the respiratory pathway. None of the cell strains (including the WT) can grow in Cu-free medium. This is as expected because of the specific requirement for copper of C*c*O to be correctly assembled and function (Pils et al. 1997). In Cu-containing media, however, the WT and Δ*petJ* strains synthesize Pc and grow at a normal rate, but the Δ*petE* strain produces neither Cc_6 nor Pc and is thus unable to grow because of the absence of a donor of electrons to C*c*O (Durán et al. 2004).

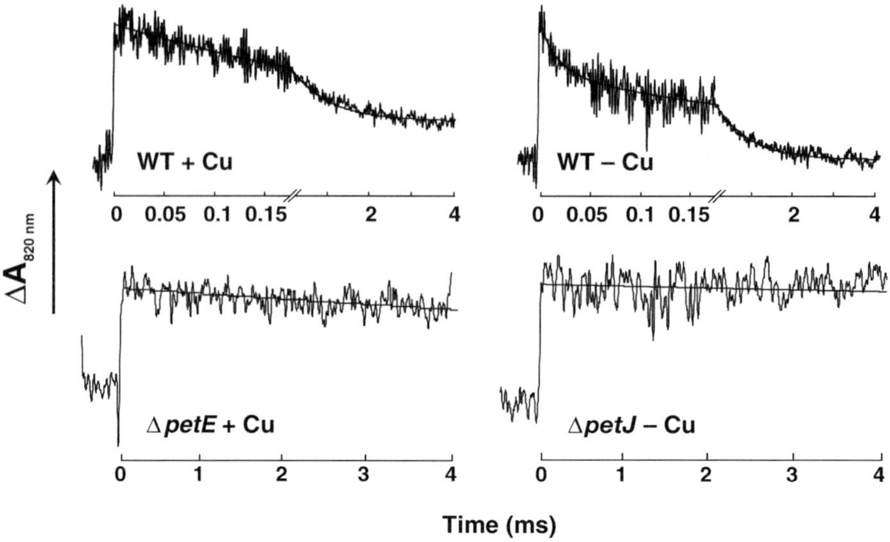

Fig. 21.6 Kinetic traces showing *in vivo* PSI reduction in WT, ΔpetE and ΔpetJ *Synechocystis* cells. The cells were previously cultured in the presence or absence of copper, as indicated. The reaction cell contained an amount of *Synechocystis* cells equivalent to a total chlorophyll content of 150–300 μg ml^{-1} in 20 mM Tricine-KOH buffer, pH 7.0, supplemented with 10% (w/v) Ficoll, to avoid cell aggregation, and 1 mM sodium ascorbate. Absorbance changes were recorded at 820 nm. The continuous lines correspond to theoretical fittings to mono or biexponential kinetics. Interferences arising from photosystem II were avoided by preillumination of the samples with white light in the presence of 10 μM 3-(3′,4′-dichlorophenyl)-1,1-dimethylurea (DCMU) and 10 mM hydroxylamine. (Adapted from Durán et al. (2004))

The *in vivo* PSI reduction kinetics of the mutants were also compared with those of the WT cells. In WT *Synechocystis* cells growing in culture media supplemented with copper, PSI reduction is only ascribed to Pc—as Cc_6 synthesis is repressed—and the kinetics are well fitted to monoexponential curves (Fig. 21.6). The value for the observed rate constant of such a monophasic kinetic is ca. 2,700 s^{-1}. In WT *Synechocystis* cells growing in Cu-depleted cultures, in which Pc synthesis is repressed, Cc_6-dependent PSI reduction follows a biphasic kinetic, with a fast phase (rate constant of ca. 45,000 s^{-1}) and a second, slower phase (rate constant of ca. 2,000 s^{-1}) (Fig. 21.6). The fast and slow phases account for 37 and 63%, respectively, of the total signal amplitude (Durán et al. 2004, 2006). The fast phase has been typically assigned to the formation of a donor-PSI complex prior to laser excitation. According to the so-called Type II or III kinetic models (see above), the transient complex is formed before electron transfer. Thus, it seems that Cc_6 interacts with PSI *in vivo* by following a more complex and efficient mechanism—involving transient complex formation—than Pc.

The results reported for the *in vivo* reduction of PSI contrast with the kinetic behaviour previously described for the isolated proteins, where monophasic kinetics corresponding to a single collisional mechanism (Type I) was reported for the interaction of both Pc and Cc_6 with PSI (Díaz-Quintana et al. 2003). The disagreement between the *in vitro* and *in vivo* data can be explained by assuming that the donor-PSI complex formation is a very precise process, which can be altered by subtle changes coming from the different donor-PSI environment existing both *in vitro* and *in vivo*. First, the ionic strength inside the thylakoid is 0.2–0.3 M; second, the presence of the thylakoidal membrane imposes additional electrostatic and hydrophobic interactions, along with a diffusion constraining factor; and third, the molecular crowding inside the cell can also impose severe limitations to the diffusion and encounter of protein partners (Durán et al. 2006).

In vivo PSI reduction in Δ*petE* and Δ*petJ* cells growing under conditions that allow the expression of one of the two electron donor proteins follows kinetics similar to those observed in the WT strain. In fact, the kinetics with the Δ*petE* mutant in the absence of Cu and with the Δ*petJ* mutant in the presence of Cu are respectively biphasic and monophasic, and their observed rate constants are equivalent to those attained with the WT cells, but no PSI reduction is detected in mutant cells in which neither Pc nor Cc_6 is expressed (Fig. 21.6).

Taken together, all this data also indicate that, contrary to previous proposals (Zhang et al. 1994; Metzger et al. 1995), there is no any other alternative redox mediator as efficient as Cc_6 and Pc in the photosynthetic and respiratory electron transport chains, and therefore *Synechocystis* cells strictly depend on the synthesis of either Cc_6 or Pc to grow not only photoautotrophically but also heterotrophically.

To further characterize the WT and mutant strains, a fractionation of cells grown either in the presence or absence of copper was performed, with the thylakoidal and plasma membranes being clearly solved on sucrose gradient. The same gradient pattern was attained with WT cells independently of copper availability, thus confirming that copper is not strictly required by *Synechocystis* cells to grow photoautotrophically. As to the mutants, both Δ*petE* in the absence of Cu and Δ*petJ* in the presence of Cu showed a pattern very similar to WT. However, in mutant cells in which neither Pc nor Cc_6 is expressed, the lack of the thylakoidal membrane band as well as a more intense blue colour band of soluble phycobiliproteins were observed, indicating that the membrane organization and pigment synthesis are significantly altered in the mutants with no electron donors to PSI and the terminal oxidase.

A preliminary proteomic study of WT and mutant *Synechocystis* cells has also been carried out. The soluble fractions of the three variants photoautotrophically grown in the presence or absence of copper were subjected to 2D electrophoresis, and some of the differentially expressed proteins were selected and analyzed by MALDI-TOF. The mutants grown under restrictive conditions showed a pattern of over-expressed proteins (tiorredoxins, superoxide dismutase, etc.) similar to that obtained with oxidatively stressed cells. This is as expected, because the absence of Pc and Cc_6 blocks the photosynthetic chain, so inducing the accumulation of oxidized PSI, the reduction of quinones and the oxidation of final acceptors. The response of mutants to oxidative stress is the over-expression of peroxiredoxin (Kobayashi et al. 2004) and DnaK (Deuerling et al. 1999). Moreover, over-expression of enzymes involved in the main metabolic pathways has also been observed, as expected from the metabolic alteration induced by oxidative stress.

It is worth to note that some of the mutants are able to revert spontaneously, at a high frequency, under restrictive conditions. Research is now in progress in our lab with such revertants—which are probably affected not only in the regulation by copper of several metabolic pathways but also in copper intake itself—as they can be a powerful model system to get a deeper insight into the complex process of copper homeostasis in cyanobacteria.

Acknowledgements Research work at the authors' laboratory has been funded by the Spanish Ministry of Science and Innovation, the Andalusian Government and the European Commission. The authors thank their co-workers and collaborators, most of them listed in the references herein cited, for their contributions to this research.

References

Albarrán C, Navarro JA, Molina-Heredia FP, Murdoch PS, De la Rosa MA and Hervás M (2005) Laser flash-induced kinetic analysis of cytochrome *f* oxidation by wild-type and mutant plastocyanin from the cyanobacterium *Nostoc* sp. PCC 7119. Biochemistry 44: 11601–11607

Albarrán C, Navarro JA, De la Rosa MA and Hervás M (2007) The specificity in the interaction between cytochrome *f* and plastocyanin from the cyanobacterium *Nostoc* sp. PCC 7119 is mainly determined by the copper protein. Biochemistry 46: 997–1003

Anderson SL and McIntosh L (1991) Light-activated heterotrophic growth of the cyanobacterium *Synechocystis* sp. PCC 6803: A blue-light-requiring process. J Bacteriol 173: 2761–2767

Banci L, Bertini I, Ciofi-Baffoni S, Kandias N, Robinson NJ, Spryoulias G, Su X-G, Tottey S and Vanarotti M (2006) The delivery of copper for thylakoid import observed by NMR. Proc Natl Acad Sci USA 103: 8320–8325

Berkner LV and Marshall LC (1965) History of major atmospheric components. Proc Natl Acad Sci USA 53: 1215–1225

Bertini I and Cavallaro G (2008) Metals in the "omics" world: Copper homeostasis and cytochrome *c* oxidase assembly in a new light. J Biol Inorg Chem 13: 3–14

Blankenship RE (1992) Origin and early evolution of photosynthesis. Photosynth Res 33: 91–111

Bovy A, de Vrieze G, Borrias M and Weisbeek P (1992) Transcriptional regulation of the plastocyanin and cytochrome c_{553} genes from the cyanobacterium *Anabaena* species PCC 7937. Mol Microl 6: 1507–1513

Briggs LM, Pecoraro VL and McIntosh L (1990) Copper-induced expression, cloning, and regulatory studies of the plastocyanin gene from the cyanobacterium *Synechocystis* sp. PCC 6803. Plant Mol Biol 15: 633–642

Bryant DA (ed) (1994) The Molecular Biology of Cyanobacteria. Advances in Photosynthesis and Respiration Series, Vol 1. Kluwer Academic Publishing, Dordrecht

Cavet JS, Borrelly GPM and Robinson NJ (2003) Zn, Cu and Co in cyanobacteria: Selective control of metal availability. FEMS Microbiol Rev 27: 165–181

Crichton RR (2008) Biological inorganic chemistry: An introduction. Elsevier, Amsterdam

Crichton RR and Pierre JL (2001) Old iron, young copper: From Mars to Venus. Biometals 14: 99–112

Crichton RR and Ward RJ (eds) (2006) Metal-based neurodegeneration: From molecular mechanisms to therapeutic strategies. John Wiley & Sons, Chichester

Crowley B, Díaz-Quintana A, Molina-Heredia FP, Nieto P, Sutter M, Haehnel W, De la Rosa MA and Ubbink M (2002) The interactions of cyanobacterial cytochrome c_6 and cytochrome *f* characterized by NMR. J Biol Chem 277: 48685–48689

De la Cerda B, Díaz-Quintana A, Navarro JA, Hervás M and De la Rosa MA (1999) Site-directed mutagenesis of cytochrome c_6 from *Synechocystis* sp. PCC 6803. The heme-protein possesses a negatively charged area that may be isofunctional with the acidic patch of plastocyanin. J Biol Chem 274: 13292–13297

De la Rosa MA, Navarro JA, Díaz-Quintana A, De la Cerda B, Molina-Heredia FP, Balme A, Murdoch PS, Díaz-Moreno I, Durán RV and Hervás M (2002) An evolutionary analysis of the reaction mechanisms of photosystem I reduction by cytochrome c_6 and plastocyanin. Bioelectrochemistry 55: 41–45

De la Rosa MA, Molina-Heredia FP, Hervás M and Navarro JA (2006) Convergent evolution of cytochrome c_6 and plastocyanin. In: Golbeck J (ed) Photosystem I. The Light-Driven Plastocyanin/Cytochrome c_6:Ferredoxin/Flavodoxin Reductase, pp 683–696. Springer, Dordrecht

De la Cerda B, Castielli O, Durán RV, Navarro JA, Hervás M and De la Rosa MA (2008) A proteomic approach to iron and copper homeostasis in cyanobacteria. Brief Funct Genomic Proteomic 6: 322–329

Des Marais DJ (1998) Earth's early biosphere. Gravit Space Biol Bull 11: 23–30

Deuerling E, Schulze-Specking A, Tomoyasu T, Mogk A and Bukau B (1999) Trigger factor and DnaK cooperate in folding of newly synthesized proteins. Nature 400: 693–696

Díaz-Quintana A, Navarro, JA, Hervás M, Molina-Heredia FP, De la Cerda B and De la Rosa MA (2003) A comparative structural and functional analysis of cyanobacterial plastocyanin and cytochrome c_6 as alternative electron donors to photosystem I. Photosynth Res 75: 97–110

Díaz-Moreno I, Díaz-Quintana A, De la Rosa MA and Ubbink M (2005a) Structure of the complex between plastocyanin and cytochrome *f* from the cyanobacterium *Nostoc* sp. PCC 7119 as determined by paramagnetic NMR. The balance between electrostatic and hydrophobic interactions within the transient complex determines the relative orientation of the two proteins. J Biol Chem 280: 18908–18915

Díaz-Moreno I, Díaz-Quintana A, Molina-Heredia FP, Nieto PM, Hansson Ö, De la Rosa MA and Karlsson BG (2005b) NMR analysis of the transient complex between membrane photosystem I and soluble cytochrome c_6. J Biol Chem 280: 7925–7931

Díaz-Quintana A, Hervás M, Navarro JA and De la Rosa MA (2008) Plastocyanin and cytochrome c_6: The soluble electron carriers between the cytochrome b_6f complex and photosystem I. In: Fromme P (ed) Structure of Photosynthetic Proteins, pp 181–200. Wiley-VCH, Weinheim

Durán RV, Hervás M, De la Rosa MA and Navarro JA (2004) The efficient functioning of photosynthesis and respiration in *Synechocystis* sp. PCC 6803 strictly requires the presence of either cytochrome c_6 or plastocyanin. J Biol Chem 279: 7229–7233

Durán RV, Hervás M, De la Rosa MA and Navarro JA (2006) A laser flash-induced kinetic analysis of *in vivo* photosystem I reduction by site-directed mutants of plastocyanin and cytochrome c_6 in *Synechocystis* sp. PCC 6803. Biochemistry 45: 1054–1060

Finazzi G, Sommer F and Hippler M (2005) Release of oxidized plastocyanin from photosystem I limits electron transfer between photosystem I and cytochrome b_6f complex *in vivo.* Proc Natl Acad Sci USA 102: 7031–7036

Fischer AG (1965) Fossils, early life, and atmospheric history. Proc Natl Acad Sci USA 53: 1205–1213

Frazão C, Soares CM, Carrondo MA, Pohl E, Dauter Z, Wilson KS, Hervás M, Navarro JA, De la Rosa MA and Sheldrick GM (1995) *Ab initio* determination of the crystal structure of cytochrome c_6 and comparison with plastocyanin. Structure 3: 1159–1169

Goldblatt C, Lenton TM and Watson AJ (2006) Bistability of atmospheric oxygen and the great oxidation. Nature 443: 683–686

Guss JM and Freeman HC (1983) Structure of oxidized poplar plastocyanin at 1.6 Å resolution. J Mol Biol 169: 521–563

Guss JM, Harrowell PR, Murata M, Norris VA and Freeman HC (1986) Crystal structure analyses of reduced (CuI) poplar plastocyanin at six pH values. J Mol Biol 192: 361–387

Halliwell B and Gutteridge JM (1984) Oxygen toxicity, oxygen radicals, transition metals and disease. Biochem J 219: 1–14

Hervás M, Navarro JA, Díaz A, Bottin H and De la Rosa MA (1995) Laser-flash kinetic analysis of the fast electron transfer from plastocyanin and cytochrome c_6 to photosystem I. Experimental evidence on the evolution of the reaction mechanism. Biochemistry 34: 11321–11326

Hervás M, Navarro JA and De la Rosa MA (2003) Electron transfer between soluble proteins and membrane complexes in photosynthesis. Accounts Chem Res 36: 798–805

Hippler M, Drepper F, Farah J and Rochaix JD (1997) Fast electron transfer from cytochrome c_6 and plastocyanin to photosystem I of *Chlamydomonas reinhardtii* requires PsaF. Biochemistry 36: 6343–6349

Ho KK and Krogmann DW (1984) Electron donors to P700 in cyanobacteria and algae. An instance of unusual genetic variability. Biochim Biophys Acta 766: 310–316

Hope AB (2000) Electron transfer amongst cytochrome *f*, plastocyanin and photosystem I: Kinetics and mechanisms. Biochim Biophys Acta 1456: 5–26

Inoue T, Sugawara H, Hamanaka S, Tsukui H, Suzuki E, Kohzuma T and Kai Y (1999) Crystal structure determinations of oxidized and reduced plastocyanin from the cyanobacterium *Synechococcus* sp. PCC 7942. Biochemistry 38: 6063–6069

Kannt A, Young S and Bendall DS (1996) The role of acidic residues of plastocyanin in its interaction with cytochrome *f*. Biochim Biophys Acta 1277: 115–126

Kasting JF and Donahue TM (1981) Evolution of oxygen and ozone in Earth's atmosphere. In: Proceedings of the Conference Life in the Universe, Moffett Field, CA, 1979, pp 149–162. MIT Press, Cambridge, MA

Kasting JF and Siefert JL (2003) Life and the evolution of Earth's atmosphere. Science 296: 1066–1068

Katoh H, Hagino N, Grossman AR and Ogawa T (2001) Genes essential to iron transport in the cyanobacterium *Synechocystis* sp. strain PCC 6803. J Bacteriol 183: 2779–2784

Keren N, Aurora R and Pakrasi HB (2004) Critical roles of bacterioferritins in iron storage and proliferation of cyanobacteria. Plant Physiol 135: 1–8

Kobayashi M, Ishizuka T, Katayama M, Kanehisa M, Bhattacharyya-Pakrasi M, Pakrasi HB and Ikeuchi M (2004) Response to oxidative stress involves a novel peroxiredoxin gene in the unicellular cyanobacterium *Synechocystis* sp. PCC 6803. Plant Cell Physiol 45: 290–299

Macalady J and Banfield JF (2003) Molecular geomicrobiology: Genes and geochemical cycling. Earth Planet Sci Lett 209: 1–17

Malakhov MP, Malakhova OA and Murata N (1999) Balanced regulation of expression of the gen for cytochrome c_M and that of genes for plastocyanin and cytochrome c_6 in *Synechocystis.* FEBS Lett 444: 281–284

Martin JH et al. (1994) Testing the iron hypothesis in ecosystems of the equatorial Pacific Ocean. Nature 371: 123–129

Merchant S, Allen MD, Kropat J, Moseley JL, Long JC, Tottey S and Terauchi AM (2006) Between a rock and a hard place: Trace element nutrition in *Chlamydomonas.* Biochim Biophys Acta 1763: 578–594

Messerschmidt A, Huber R, Poulos T and Wieghardt K (eds) (2001) Handbook of Metalloproteins. John Wiley & Sons, Chichester

Metzger SU, Pakrasi HB and Whitmarsh J (1995) Characterization of a double deletion mutant that lacks cytochrome c_6 and cytochrome c_M in *Synechocystis* 6803. In: Mathis P (ed) Photosynthesis: From Light to Biosphere, pp 823–826. Kluwer Academic Publishing, Dordrecht

Molina-Heredia FP, Hervás M, Navarro JA and De la Rosa MA (2001) A single arginyl residue in plastocyanin and cytochrome c_6 from the cyanobacterium *Anabaena* sp. PCC 7119 is required for efficient reduction of photosystem I. J Biol Chem 276: 601–605

Molina-Heredia FP, Wastl J, Navarro JA, Bendall DS, Hervás M, Howe C and De la Rosa MA (2003) Photosynthesis: A new function for an old cytochrome? Nature 424, 33–34.

Morel FMM and Price NM (2003) The biogeochemical cycles of trace metals in the oceans. Science 300: 944–947

Navarro JA, Durán RV, De la Rosa MA and Hervás M (2005) Respiratory cytochrome *c* oxidase can be efficiently reduced by the photosynthetic redox proteins cytochrome c_6 and plastocyanin in cyanobacteria. FEBS Lett 579: 3565–3568

Newman DK and Banfield JF (2002) Geomicrobiology: How molecular-scale interactions underpin biogeochemical systems. Science 296: 1071–1077

Ochiai EI (1983) Copper and the biological evolution. Biosystems 16: 81–86

Paumann M, Bernroitner M, Lubura B, Peer M, Jakopitsch C, Furtmüller PG, Peschek GA and Obinger C (2004a) Kinetics of electron transfer between plastocyanin and the soluble CuA domain of cyanobacterial cytochrome *c* oxidase. FEMS Microbiol Lett 239: 301–307

Paumann M, Feichtinger M, Bernroitner M, Goldfuhs J, Jakopitsch C, Furtmüller PG, Regelsberger G, Peschek GA and Obinger C (2004b) Kinetics of interprotein electron transfer between cytochrome c_6 and the soluble CuA domain of cyanobacterial cytochrome c oxidase. FEBS Lett 576: 101–106

Paumann M, Lubura B, Regelsberger G, Feichtinger M, Köllensberger G, Jakopitsch C, Furtmüller PG, Peschek GA and Obinger C (2004c) Soluble CuA domain of cyanobacterial cytochrome c oxidase. J Biol Chem 279: 10293–10303

Paumann M, Regelsberger G, Obinger C and Peschek GA (2005) The bioenergetic role of dioxygen and the terminal oxidase(s) in cyanobacteria. Biochim Biophys Acta 1707: 231–253

Peschek G (1999) Photosynthesis and respiration of cyanobacteria. In: Peschek G, Loffelhardt W and Schmetterer G (eds) The Phototrophic Prokaryotes, pp 201–209. Kluwer Academic Publishing, New York

Pils D, Gregor W and Schmetterer G (1997) Evidence for *in vivo* activity of three distinct respiratory terminal oxidases in the cyanobacterium *Synechocystis* sp. PCC 6803. FEMS Microbiol Lett 152: 83–88

Rogers LJ (1987) Ferredoxins, flavodoxins and related proteins: Structure, function and evolution. In: Fay P and Van Baalen C (eds) The Cyanobacteria, pp 35–67. Elsevier, Amsterdam

Sandmann G, Reck H, Kessler E and Böger P (1983) Distribution of plastocyanin and soluble plastidic cytochrome *c* in various classes of algae. Arch Microbiol 134: 23–37

Sawaya MR, Krogmann DW, Serag A, Ho KK, Yeates TO and Kerfeld CA (2001) Structures of cytochrome *c*-549 and cytochrome c_6 from the cyanobacterium *Arthrospira maxima.* Biochemistry 40: 9215–9225

Schlarb-Ridley BG, Bendall DS and Howe CJ (2002) Role of electrostatics in the interaction between cytochrome *f* and plastocyanin of the cyanobacterium *Phormidium laminosum.* Biochemistry 41: 3279–3285

Schmetterer G (1994) Cyanobacterial respiration. In: Bryant DG (ed) The Molecular Biology of Cyanobacteria, pp 409–435. Kluwer Academic Publishing, Dordrecht

Shcolnick S and Keren N (2006) Metal homeostasis in cyanobacteria and chloroplasts. Balancing benefits and risks to the photosynthetic apparatus. Plant Physiol 141: 805–810

Sigfridsson K (1998) Plastocyanin, an electron-transfer protein. Photosynth Res 57: 1–28

Sigfridsson K, Hansson Ö, Karlsson BG, Baltzer L, Nordling M and Lundberg LG (1995) Spectroscopic and kinetic characterization of the spinach plastocyanin mutant Tyr83-His: A histidine residue with a high pKa value. Biochim Biophys Acta 1228: 28–36

Stearn WT (1962) The origin of the male and female symbols of biology. Taxon 11: 109–113

Tottey S, Rondet SAM, Borrelly GPM, Robinson PJ, Rich PR and Robinson NJ (2002) A copper metallochaperone for photosynthesis and respiration reveals metal-specific targets, interaction with an importer, and alternative sites for copper acquisition. J Biol Chem 277: 5490–5497

Tottey S, Harvie DR and Robinson NJ (2005) Understanding how cells allocate metals using metal-sensors and metallochaperones. Accounts Chem Res 38: 775–783

Ubbink M, Ejdebäck M, Karlsson BG and Bendall DS (1998) The structure of the complex of plastocyanin and cytochrome *f* determined by paramagnetic NMR and restrained rigid-body molecular dynamics. Structure 6: 323–335

Van der Plas J, Bovy A, Kruyt F, de Vrieze G, Dassen E, Klein B and Weisbeek P (1989) The gene for the precursor for plastocyanin from the cyanobacterium *Anabaena* sp. PCC 7937: Isolation, sequence and regulation. Mol Microbiol 3: 275–284

Vermaas WFJ (2001) Photosynthesis and respiration in cyanobacteria. In: Encyclopedia of Life Sciences, pp 245–251. http://www.els.net, Nature Publishing Group, London

Williams RJP (2007) Life, the environment and our ecosystem. J Inorg Biochem 101: 1550–1561

Williams RJP and Fraústo da Silva JJR (1996) The Natural Selection of the Chemical Elements. Oxford University Press, Oxford

Wood PM (1978) Interchangeable copper and iron proteins in algal photosynthesis. Studies on plastocyanin and cytochrome *c*-552 in *Chlamydomonas.* Eur J Biochem 87: 9–19

Zhang L, McSpadden B, Pakrasi HB and Whitmarsh J (1992) Copper-mediated regulation of cytochrome *c*553 and plastocyanin in the cyanobacterium *Synechocystis* 6803. J Biol Chem 267: 19054–19059

Zhang L, Pakrasi HB and Whitmarsh J (1994) Photoautotrophic growth of the cyanobacterium *Synechocystis* sp. PCC 6803 in the absence of cytochrome *c*553 and plastocyanin. J Biol Chem 269: 5036–5042

Chapter 22
Flavodiiron Proteins and Their Role in Cyanobacteria

Vera L. Gonçalves, João B. Vicente, Lígia M. Saraiva and Miguel Teixeira

22.1 Introduction

The flavodiiron proteins (FDP) constitute a large family of enzymes that reduce nitric oxide (nitrogen monoxide) to nitrous oxide (N_2O) or oxygen to water (for recent reviews see Saraiva et al. (2004); Vicente et al. (2008a–d)). These enzymes were given this name because they contain a flavin mononucleotide (FMN) and a diiron centre as the common prosthetic groups. The first reported example was *r*ubredoxin:*o*xygen *o*xidoreductase, ROO, from the sulphate reducing bacterium *Desulfovibrio* (*D.*) *gigas.* This enzyme was shown to be the terminal oxidase of a three-component electron transfer chain coupling NADH oxidation (by an NADH:rubredoxin oxidoreductase, NRO) to oxygen reduction (Chen et al. 1993a, b). Electron transfer from NRO to ROO is mediated by a small iron-protein, rubredoxin (Rd), which contains a FeCys4 centre (Chen et al. 1993b; Gomes et al. 1997). This process, which occurs in the cytoplasm, was proposed to confer oxygen tolerance to this anaerobic bacterium, simultaneously enabling NAD^+ regeneration (Fareleira et al. 2003). The first structure of an enzyme of the flavodiiron family was also that obtained for the *D. gigas* ROO, that allowed to clearly identify the catalytic centre, built by two iron ions ligated by the side chains of histidines and aspartates/glutamates (H^{79}-X-E^{81}-X-D^{83}-X_{62}-H^{146}-X_{18}-D^{165}-X_{60}-H^{226}, numbering of *D. gigas* ROO), and to envisage the intramolecular electron transfer pathways. Whereas the initial studies suggested a general role in oxygen stress alleviation in anaerobes, evidence for a role in nitric oxide detoxification broadened the interest on this protein family. In fact, nitric oxide is a well recognized toxic molecule, in concentrations above the nanomolar range, being either a weapon of the immune system or an intermediate of microbial denitrification; thus, organisms need to have enzymes able to deal with NO or its derived and toxic species. Until the discovery of this function of the FDPs, NO detoxification was thought to be mainly restricted

M. Teixeira (✉)
Instituto de Tecnologia Química e Biológica, Universidade Nova de Lisboa, Av. da República, 2780-157 Oeiras, Portugal
e-mail: miguel@itqb.unl.pt

G. A. Peschek et al. (eds.), *Bioenergetic Processes of Cyanobacteria,*
DOI 10.1007/978-94-007-0388-9_22,

to the flavohemoglobin family (for a recent review see Lewis et al. (2008)), while the heme-iron membrane-bound NO reductases are mainly considered as enzymes of the denitrification pathway.

Since those early findings, a wealth of data has been accumulated in this field for different organisms. The main features of this challenging enzyme family will be discussed, with a particular focus on what is so far known on flavodiiron enzymes from cyanobacteria.

22.2 FDPs: A Family of Modular Enzymes

The analysis of the primary structure of the large number of FDPs encoded in microbial and eukaryotic genomes revealed a common flavodiiron core with ca 400 amino acids, composed by a ~250-amino acids metallo-β-lactamase domain and a ~150-amino acids flavodoxin domain, and the presence of extra structural domains fused at the C-terminus of the flavodiiron core. This latter observation led us to establish four subfamilies (Classes A–D), according to the type of structural domains present in each protein (Fig. 22.1a–d); the domain arrangement is intimately associated with the type of electron transfer chains involved in each case (e.g., Saraiva et al. 2004; Vicente et al. 2008b).

The prototype enzymes, which constitute the bulk of identified FDPs, are those that have only the core domains, i.e., the β-lactamase and the flavodoxin-like domains (Class A, Fig. 22.1a). In several organisms these enzymes receive electrons from rubredoxins, acting as the terminal oxidase of a three components chain that involves also an NADH:rubredoxin oxidoreductase. However, in most organisms the immediate electron donor is not known and, as far as it can be predicted from genomic data, it is not a rubredoxin, since genes coding for this type of proteins are absent. Distinct types of electron donors have been reported, namely the $F_{420}H_2$ cofactor in methanogenic Archaea (Seedorf et al. 2007), and the pyruvate:ferredoxin oxidoreductase system for the *Trichomonas vaginalis* hydrogenosomal FDP (Smutna et al. 2009).

The second type of enzyme to be described was the so-called flavorubredoxin, FlRd, since it contains at the C-terminus a third structural domain, similar to rubredoxins. The ca 50 amino acids rubredoxin domain is fused to the flavodiiron core by an apparently unstructured linker of about 20 amino acids. So far, this type of enzymes (Class B, Fig. 22.1b) is restricted to enterobacteria, such as *Escherichia* and *Salmonella* species. FlRd has been shown to accept electrons directly from an NADH:FlRd oxidoreductase (from the family of NADH:rubredoxin oxidoreductases), establishing a two component electron transfer chain (Gomes et al. 2000, 2002; Vicente et al. 2007). The crystallographic structure of these enzymes remains to be determined, but structural studies by small angle X-ray scattering proposed that the linker between the Rd domain and the flavodiiron core is highly flexible, and thus the Rd domain protrudes into the solvent extending from the tetrameric quaternary structure, resembling an independently-behaving "quasi" free rubredoxin (Petoukhov et al. 2008).

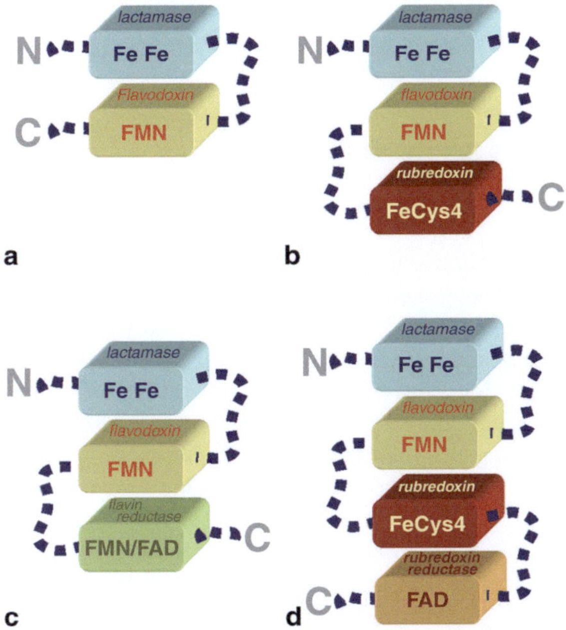

Fig. 22.1 Flavodiiron proteins are modular enzymes. Scheme depicting the modular nature of flavodiiron proteins, which in several organisms display extra C-terminal structural domains and were classified accordingly. ***a***—Class A FDPs are the structural prototype of this protein family, being composed of an N-terminal β lactamase-like domain (*light blue box*) and a C-terminal flavodoxin-like domain (*light yellow box*). ***b***—Class B FDPs, found thus far only in enterobacteria, bear an extra C-terminal rubredoxin domain (*dark red box*), harbouring a FeCys4 center. ***c***—Class C FDPs, from cyanobacteria and some eukaryotic oxygenic phototrophs, having a NAD(P)H:flavin oxidoreductase domain (*light green box*) fused at the C-terminus of the flavodiiron core, that may bind one FMN or one FAD moiety. ***d***—Class D FDPs are, so far, only found in the genomes of *Trichomonas vaginalis* and in a few *Clostridia* species. These FDPs appear to result from a fusion between a Class B FDP and its reductase partner, of the rubredoxin reductase family (*light orange box*)

Class C FDPs (Fig. 22.1c) were firstly identified in cyanobacteria (further discussed below)—these enzymes also bear a third domain at the C-terminus, homologous to NAD(P)H flavin reductases (Wasserfallen et al. 1998; Saraiva et al. 2004; Vicente et al. 2002, 2008a, b). Therefore, these FDPs have the interesting feature of condensing in a single polypeptide chain the whole electron transfer chain that couples NAD(P)H oxidation to oxygen and/or NO reduction.

The fourth class of FDPs (Class D, Fig. 22.1d) is composed by enzymes that appear to result from the fusion of a Class B FDP with the respective NADH:FlRd oxidoreductase, yielding a condensed four-domain polypeptide that is likely to directly accomplish O_2 or NO reduction at the expense of NAD(P)H oxidation. So far, the only examples were found in the genomes of the anaerobic protozoan *Trichomonas vaginalis* (Carlton et al. 2007) and of the anaerobic bacterial pathogen *Clostridium perfringens* (Shimizu et al. 2002).

22.3 Amino Acid Sequences Analyses and FDPs Distribution

A BLAST search on the sequence databases using several queries (such as the sequences of the enzymes from cyanobacteria and from anaerobic bacteria) retrieved close to 500 sequences of flavodiiron proteins, which were aligned using ClustalX (Larkin et al. 2007). A dendrogram was constructed, based only on the common FDP core (discarding the extra C-terminal domains and several types of signal peptides at the N-terminus), using the neighbour joining method implemented in ClustalX (Fig. 22.2). FDPs, which were originally thought to be present only in prokaryotes, are also found in uni- and multicellular Eukarya, including oxygenic photosynthetic organisms (Table 22.1) and anaerobic protozoa (*Trichomonas, Giardia* and *Entamoeba* species). As previously proposed, there are evidences for multiple gene transfer events, but it is also possible to distinguish groups corresponding to particular phylogenies (Andersson et al. 2003, 2006; Vicente et al. 2008b). A striking example is the observation that sequences retrieved for the cyanobacteria and for all the other oxygenic photosynthetic organisms so far known to contain genes encoding FDPs (algae, mosses, lycophytes or even the higher plant *Picea sitchensis*, see Table 22.1) form a distinct clade from the other classes of FDPs. The sequences of 82 putative FDPs of cyanobacteria were aligned separately, together with those from eukaryotic oxygenic phototrophs (Table 22.1) and with a few other prototypic FDPs, which served as a reference; a subset of this alignment is presented in Fig. 22.3, and the resulting dendrogram in Fig. 22.2b. Genes coding for FDPs are present in all complete genomes of cyanobacteria, and each cyanobacterium sequenced has always at least two FDP encoding genes; cyanobacteria are also among the organisms that have a higher number of homologues in a single genome (up to six homologues, as in *Anabaena* sp. PCC7120 and *Anabaena variabilis* ATCC 29413, see Table 22.1). As already mentioned, all cyanobacterial FDPs and those from oxygenic phototrophs are of the C-type, i.e., have the extra C-terminal flavin-reductase domain; the only exception is the FDP from *Picea sitchensis*, which, in spite of having high amino acid identity/similarity with the homologues of the remaining photosynthetic organisms, lacks the flavin reductase domain. The fact that the FDP sequences of the eukaryotic oxygenic photosynthetic organisms are closer to the cyanobacterial ones than to those from other eukaryotes (see Fig. 22.2a, b), indicates that eukaryotes acquired these genes more than once, i.e., their origin is not monophyletic. It had already been proposed that the anaerobic protozoa had acquired FDPs from multiple lateral gene transfers from anaerobic prokaryotes (Andersson et al. 2003, 2006). Although a similar in-depth analysis is out of the context of this review, at least two arguments suggest a different scenario for the origin of the FDPs in oxygenic eukaryotes: (1) a close inspection of the sub-tree obtained

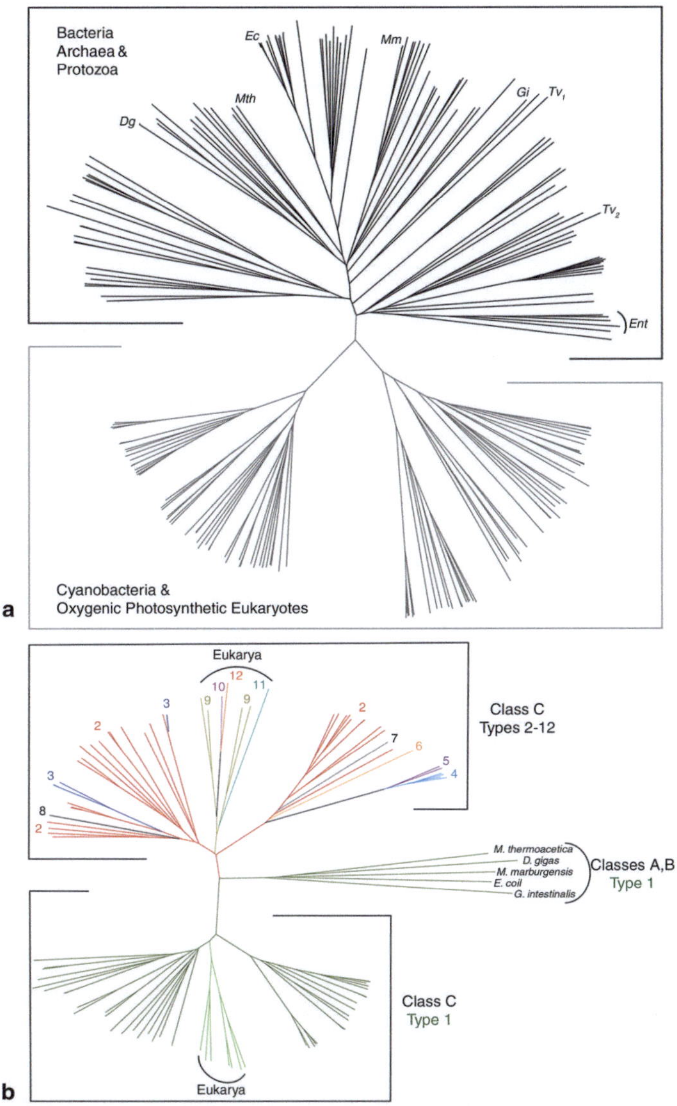

Fig. 22.2 **a** Dendrogram of selected flavodiiron proteins. A few sequences are highlighted: *D. gigas* (Dg), *Moorella thermoacetica* (Mth), *E. coli* (Ec), *Methanothermobacter marburgensis* (Mm), *Giardia intestinalis* (Gi), *Trichomonas vaginalis* (Tv_1, Tv_2), and several *Entamoeba* species (Ent). **b** Dendrogram of the flavodiiron proteins from cyanobacteria and oxygenic photosynthetic eukaryotes. The numbers correspond to the several Types of FDPs. Colors according to each type: *green*—1; *red*—2; *dark blue*—3; *light blue*—4; *magenta*—5; *light orange*—6; *gray*—7; *black*—8; *light green*—9; *pink*—10; *ocean blue*—11; *dark orange*—12. The dendrogram was obtained using ClustalX (Larkin et al. 2007), and Dendroscope, version 2.4 (Huson et al. 2007)

Table 22.1 Oxygenic photosynthetic organisms with genes coding for flavodiiron proteins

Organism	Order	No. of FDP genes
Cyanobacteria		
Acaryochloris marina (MBIC11017)	Acaryochloris	2
Anabaena (PCC7120)	Nostocales	6
Anabaena variabilis (ATCC 29413)	Nostocales	6
Cyanothece (ATCC 51142)	Chroococcales	4
Gloeobacter violaceus (PCC7421)	Gloeobacteria	2
Microcystis aeruginosa (NIES-843)	Chroococcales	4
Nostoc punctiforme (ATCC 29133)	Nostocales	5
Prochlorococcus marinus (MIT 9211, 9215, 9301, 9303, 9312, 9313, 9515; AS9601; NATL1A, 2A; MED4; SS120)	Prochlorales	2
Synechococcus (JA-2-3B'a(2-13); JA-3-3Ab; WH 7803, 8102; RCC307; CC9311, 9605, 9902, 9902; PCC7002)	Chroococcales	2
Synechococcus elongatus (PCC6301, 7942)	Chroococcales	2
Synechocystis (PCC6803)	Chroococcales	4
Thermosynechococcus elongatus (BP-1)	Chroococcales	2
Trichodesmium erythraeum (IMS101)	Oscillatoriales	2
Eukarya		
Algae		
Chlamydomonas reinhardtii	Chlamydomonadales	1
Micromonas pusilla (CCMP1545, RCC299)	Mamiellales	2
Ostreococcus lucimarinus (CCE9901)	Mamiellales	2
Ostreococcus tauri	Mamiellales	2
Paulinella chromatophora	Euglyphida	2
Lycophyte		
Selaginella moellendorffii	Selaginellale	2
Tree		
Picea sitchensis	Coniferales	1
Moss		
Physcomitrella patens (*subsp. Patens*)	Funariales	2

only for the oxygenic photosynthetic organisms (Fig. 22.2b) strongly suggests that the higher organisms obtained FDPs from an ancient, possibly non extant, cyanobacterium; (2) several putative FDPs of algae, mosses, lycophytes and of the tree *Picea sitchensis*, are nuclear encoded, and the deduced amino acid sequences reveal clear signal peptides at the N-terminus (peptides rich in aliphatic residues, as well as in basic residues, namely arginines (Gould et al. 2008)) which suggest that the FDPs will be present in cellular organelles. Quite interestingly, the FDPs of *Paulinella chromatophora* (an organism considered to have acquired its plastid from a recent endosymbiotic event that may have occurred only a few millions years ago (Bodyl et al. 2007; Yoon et al. 2009)), lack those signal peptides and, accordingly, are encoded in the plastidic genome. In a speculative way, it appears that migration of the FDP genes from the plastidic chromosome to the nucleus did not yet occur in *Paulinella chromatophora*, contrary to what already happened quite longer ago in the other photosynthetic organisms. On this basis, as well as taking into account the proposed function for FDPs in cyanobacteria (see Sect. 22.7), it is tempting to suggest localization in the chloroplasts for the FDPs of the other eukaryotic phototrophs, but experimental work will be needed to prove this hypothesis. This situation is reminiscent of that for at least one of the FDPs of the protozoan *Trichomonas vaginalis*, which possesses a signal peptide targeting the enzyme to the hydrogenosome (Smutna et al. 2009).

As recently noted by E.M. Aro and co-workers (Zhang et al. 2009), the FDP sequences of the cyanobacteria form two distinct clusters (Fig. 22.2b), which those authors named A and B. However, due to the above mentioned classification of the FDPs according to the respective modular arrangement, we prefer a different nomenclature for this division. Indeed, a detailed analysis of the cyanobacterial sequences (c.f. Fig. 22.3) reveals another remarkable characteristic, as compared to the remaining FDP-encoding organisms: while about half of the sequences have conserved residues matching the "canonical" ones known to be involved in iron coordination (H^{81}-X-E^{83}-X-$D^{85}H^{86}$-X_{62}-H^{148}-X_{18}-D^{167}-X_{60}-H^{228}, *Morella thermoacetica* FDP numbering), which correspond to what we now propose to designate as Class C, Type 1 FDPs (Cluster B in Fig. 8 of Zhang et al. (2009)), a significant variation in these residues is present in the sequences forming the other cluster (Cluster A in Fig. 8 of that reference): up to eleven possible different ligand substitution combinations could be detected (Types 1–12, see Table 22.2, Figs. 22.2b and 22.3), with the corresponding sequences scattered along that second cluster, in between the second major group, the Type 2 enzymes. It is striking that the amino acid changes involve, in general, substitutions by neutral or positively charged residues (arginines, lysines and asparagines), or even aliphatic residues, such as isoleucines or alanines. It should also be noted that almost all cyanobacterial and photosynthetic eukaryotic organisms contain at least two FDP genes, one coding for a "canonical", Type 1 enzyme, and a second, encoding one of the multiple types identified (Tables 22.1, 22.2 and Fig. 22.3). Some sub-types appear to be more common than others (c.f. Table 22.2), and Types 9–12 are, so far, restricted to eukaryotes.

These multiple putative ligand substitutions will be further discussed in the next section, regarding the possible involvement of these amino acids as ligands for the iron ions. At present, no rationale can be proposed for this diversity and its apparent occurrence only in oxygenic phototrophs, either from the Bacteria or from the Eukarya domains.

The genomic organization of the genes encoding flavodiiron proteins is, in general, very diverse. In enterobacteria, the gene coding for the FDP forms a dicistronic unit with that coding its NADH oxidoreductase; in *D. gigas, roo* forms also a dicistronic unit with the gene encoding rubredoxin, the immediate electron donor to ROO. However, in the majority of the cases, the FDP-encoding genes have in their vicinity genes coding for proteins whose functions are unrelated to oxidative or nitrosative stress responses. We analysed the genome regions surrounding the 82

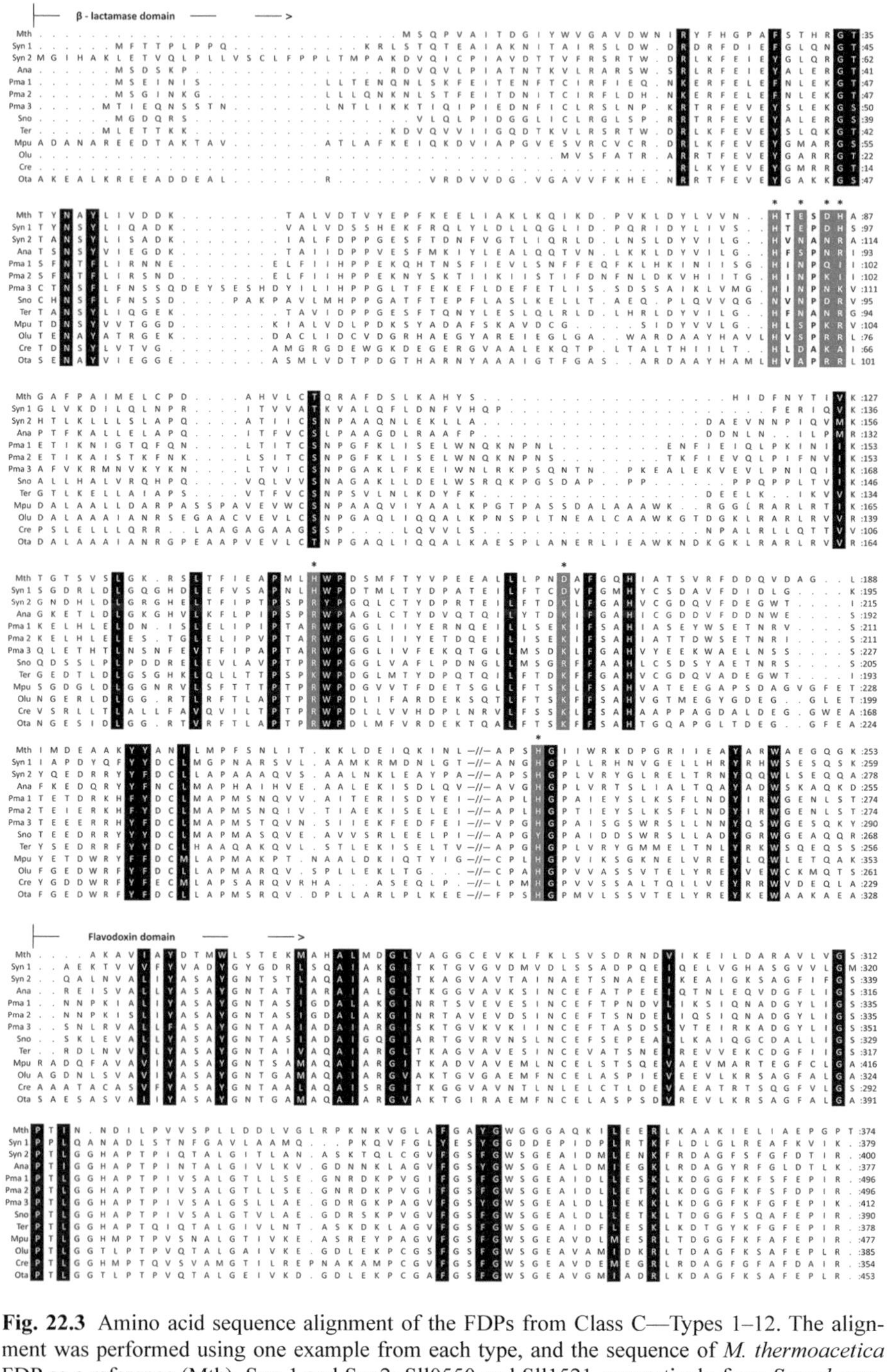

Fig. 22.3 Amino acid sequence alignment of the FDPs from Class C—Types 1–12. The alignment was performed using one example from each type, and the sequence of *M. thermoacetica* FDP as a reference (Mth). Syn 1 and Syn2: Sll0550 and Sll1521, respectively from *Synechocystis sp.* PCC6803; Ana: All0177 from *Anabaena sp.* PCC7120; Pma1: P9215_00501 from *Prochlorococcus (P.) marinus* MIT 9215; Pma2: P9515_00491 from *P. marinus* MIT 9515; Pma3: PMN2A_1375 from *P. marinus* NATL2A; Sno: SynRCC307_2387 from *Synechococcus* sp:

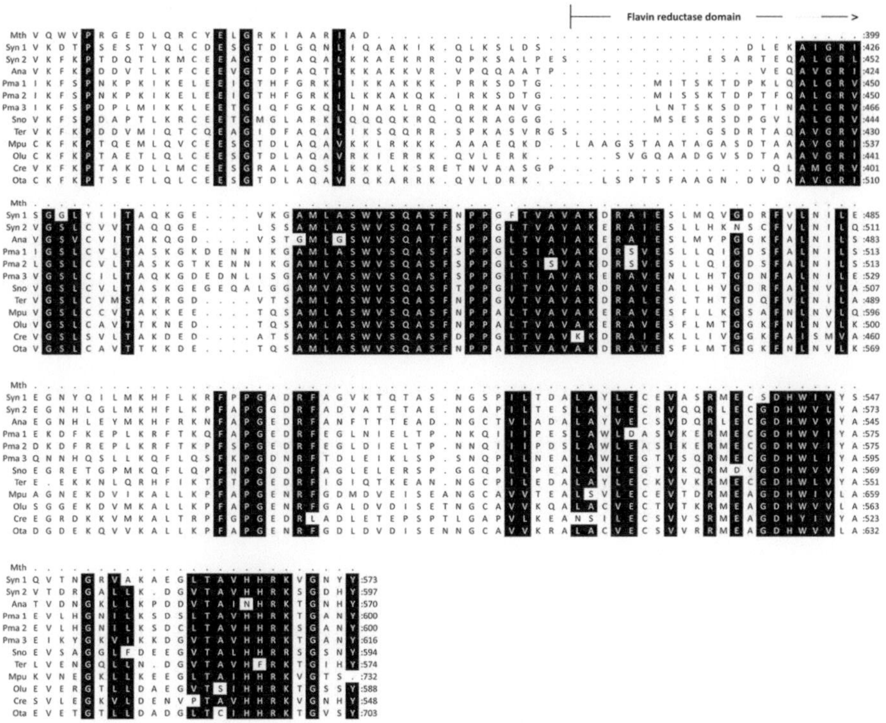

Fig. 22.3 (continued) RCC307; Ter: Tery_0770 from *Trichodesmium* erythraeum IMS101; Mpu: EEH_58658 from *Micromonas pusilla* CCMP1545; Olu: XP_001416100 from *Ostreococcus lucimarinus* CCE9901; Cre: XP_001692916 from *Chlamydomonas reinhardtii*; Ota: CAL52487 from *Ostreococcus tauri.* Several N-terminal extensions were deleted. The residues implicated in the binding of the diiron site in *Moorella thermoacetica* FDP are marked with an *asterisk* (*). *Black* shadows correspond to strictly conserved residues; *Gray* shadows correspond to the putative diiron site ligands. Too large non-conserved segments from the sequences designated by Mpu and Ota were removed for clarity and are represented by -//-

FDP-encoding genes retrieved from the Cyanobase website (http://genome.kazusa.or.jp/cyanobase/). From the 35 genomes scrutinized, 20 have genes coding for FDPs adjacently transcribed (Fig. 22.4a), where one of the genes corresponds always to a type 1 FDP, and the other to types 2–7. The flanking regions surrounding these contiguous FDP-encoding genes are variable. However, we found some examples of a common organization, as depicted in Fig. 22.4a. A particularly interesting case is found in a few *Prochlorococcus marinus* strains, where a rubrerythrin-encoding gene (a protein involved in oxidative stress response, as an H_2O_2 reductase (Kurtz 2006)) is adjacently upstream to one of the FDP coding genes and transcribed in the same direction.

In four genomes (including that of *Synechocystis* sp. PCC6803), two of the genes coding for FDPs are almost contiguous (always encoding a type 1 and a type 2 FDP enzymes), displaying only one gene (encoding a hypothetical protein) in between them, and the three genes are transcribed in the same direction (*sll0217-sll0219* in *Synechocystis* sp. PCC6803, Fig. 22.4b). That same gene is conserved among the

Table 22.2 Ligands of the diiron site in "canonical" FDPs, and corresponding amino acids substitutions in the enzymes from cyanobacteria and eukaryotic oxygenic phototrophs

Type (No. of examples)	Fe 1	Fe 2
1 (49)	H^{91}-X-E^{93}-X-D^{95}-H^{96}	H^{158}-X_{18}-D^{177}-X_{56}-H^{233}
2 (29)	H^{108}-X-N^{110}-X-N^{112}-R^{113}	R^{178}-X_{18}-K^{197}-X_{56}-H^{254}
3 (4)	H^{87}-X-S^{89}-X-N^{91}-R^{92}	R^{155}-X_{18}-K^{174}-X_{56}-H^{231}
4 (4)	H^{96}-X-N^{98}-X-Q^{100}-I^{101}	R^{174}-X_{18}-K^{193}-X_{56}-H^{250}
5 (2)	H^{96}-X-N^{98}-X-K^{100}-I^{101}	R^{174}-X_{18}-K^{193}-X_{56}-H^{250}
6 (1)	H^{105}-X-N^{107}-X-N^{109}-K^{110}	R^{190}-X_{18}-K^{209}-X_{56}-H^{226}
7 (1)	N^{89}-X-N^{91}-X-D^{93}-R^{94}	R^{168}-X_{18}-R^{187}-X_{56}-Y^{244}
8 (1)	H^{88}-X-N^{90}-X-N^{92}-R^{93}	K^{156}-X_{18}-K^{175}-X_{56}-H^{232}
9 (4)	H^{147}-X-S^{149}-X-K^{151}-R^{152}	R^{236}-X_{18}-K^{255}-X_{122}-H^{378}
10 (1)	H^{70}-X-S^{72}-X-R^{74}-R^{75}	R^{160}-X_{18}-K^{179}-X_{57}-H^{237}
11 (1)	H^{60}-X-D^{62}-X-K^{64}-A^{65}	R^{128}-X_{18}-K^{147}-X_{57}-H^{205}
12 (1)	R^{143}-X-A^{145}-X-R^{147}-R^{148}	R^{233}-X_{18}-K^{252}-X_{99}-H^{352}

Aminoacids numbering according to the sequences presented in Fig. 22.3, excluding *Morella thermoacetica* FDP. 1—*Synechocystis sp.* PCC6803 sll0550; 2—*Synechocystis sp.* PCC6803 sll1521; 3—*Anabaena sp.* PCC7120 all0177; 4—*Prochlorococcus* (*P.*) *marinus* MIT 9215 P9215_00501; 5—*P. marinus* MIT 9515 P9515_00491; 6—*P. Marinus* NATL2A PMN2A_1375; 7—*Synechococcus* sp RCC307 SynRCC307_2387; 8—*Trichodesmium* erythraeum IMS101 Tery_0770; 9—*Micromonas pusilla* CCMP1545 EEH_58658; 10—*Ostreococcus lucimarinus* CCE9901 XP_001416100; 11—*Chlamydomonas reinhardtii* XP_001692916; 12—*Ostreococcus tauri* CAL52487

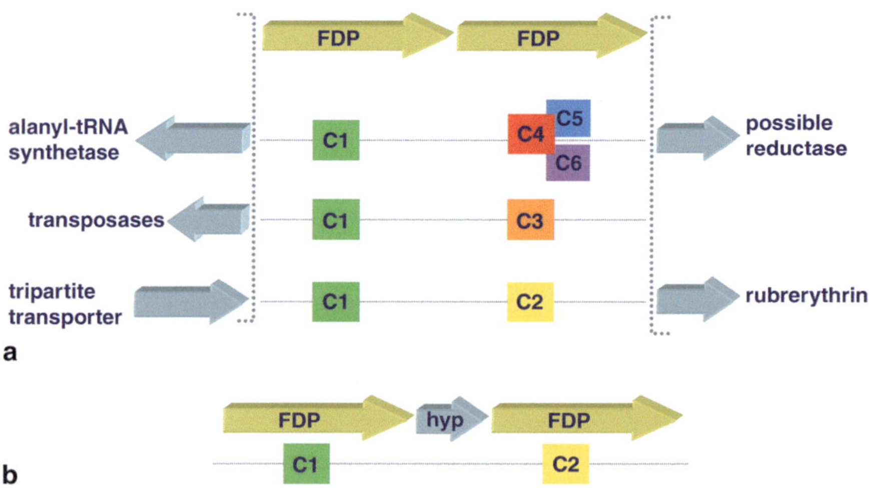

Fig. 22.4 Genomic regions surrounding cyanobacterial FDP encoding genes. The genome regions around each cyanobacterial FDP encoding gene were analyzed in the Cyanobase website (http://genome.kazusa.or.jp/cyanobase). **a** Types of genome loci that were found in at least two genomes, and the corresponding Type of Class C FDPs encoded. **b** In four genomes (including that of *Synechocystis* sp. PCC6803), genes encoding FDPs were found to be interspaced by a conserved gene encoding a hypothetical protein containing a transmembrane domain; these encoded FDPs are in every case a Type 1 and a Type 2

genomes displaying this organization, which suggests that transfer events may have included the set of FDP genes interspaced by this gene. Another interesting observation is that a secondary structure prediction (using the PSI-PRED server (Jones 1999, p. 839)) suggests the presence of transmembrane helices in that putative gene product. It should be recalled that the *Synechocystis* sp. PCC6803 *sll0217-sll0219* gene products (the type 1 and 2 FDPs in this kind of genomic organization) are found, at least partially, in the bacterial membranes, despite lacking any predictable transmembrane helices. Whether that hypothetical transmembranar protein contributes to this localization is an interesting question that requires experimental data to be answered.

22.4 Three Dimensional Structure of Flavodiiron Enzymes

The first step in the elucidation of the structural features of FDPs consisted in the resolution of the crystallographic structure of *D. gigas* FDP, rubredoxin:oxygen oxidoreductase (Frazão et al. 2000). This three dimensional structure confirmed the homodimeric quaternary arrangement determined in solution, with each ~43 kDa monomer being built by two structural domains: a diiron-containing domain, structurally analogous to zinc β-lactamases (despite the poor amino acids sequence similarity), followed by a short-chain flavodoxin-like domain, harbouring an FMN. In each monomer, the redox active cofactors are placed ca. 30 Å apart, precluding biologically active electron transfer between each other. However, the quaternary structure is such that, due to the "head-to-tail" arrangement of each monomer within the dimer, the diiron site of one monomer is almost in Van der Waals contact with the FMN from the other monomer, ensuring a fast electron transfer between the two redox centres (c.f. Fig. 22.5). The iron ions from the diiron centre are bound by histidines and glutamates/aspartates, and are bridged by a μ-hidroxo(oxo) species and one aspartate residue (Fig. 22.5c). Several FDP′s structures are now available (Silaghi-Dumitrescu et al. 2005; Seedorf et al. 2007; Di Matteo et al. 2008), revealing a conservation of the dimeric "head-to-tail" arrangement (even when the quaternary structure is a tetramer), as well as of the diiron ligand sphere.

As this revision was being written, the structure of a truncated form of *Anabaena* sp. PCC7120 FDP encoded by *all0177* (a Type 3 FDP), consisting solely of its lactamase-like domain was deposited in the Protein Data Bank (code 3HNN). The amino acids that act as iron ligands in "canonical", Type 1, FDPs, are herein replaced in several positions (Table 22.2); furthermore, the structure model does not contain any metal ion. Whereas the overall fold of this truncated domain resembles the correspondent ones in the other FDPs (Fig. 22.5a), the structure of the diiron site pocket is markedly different (Fig. 22.5d). In this structure, the side chain of a lysine (K174, all0177 numbering, equivalent to K197 in Fig. 22.5d—see sequence alignment in Fig. 22.3), which replaces the bridging aspartate in Type 1 FDPs, protrudes into the space occupied by the iron ions in the other FDPs. Moreover, the number of basic

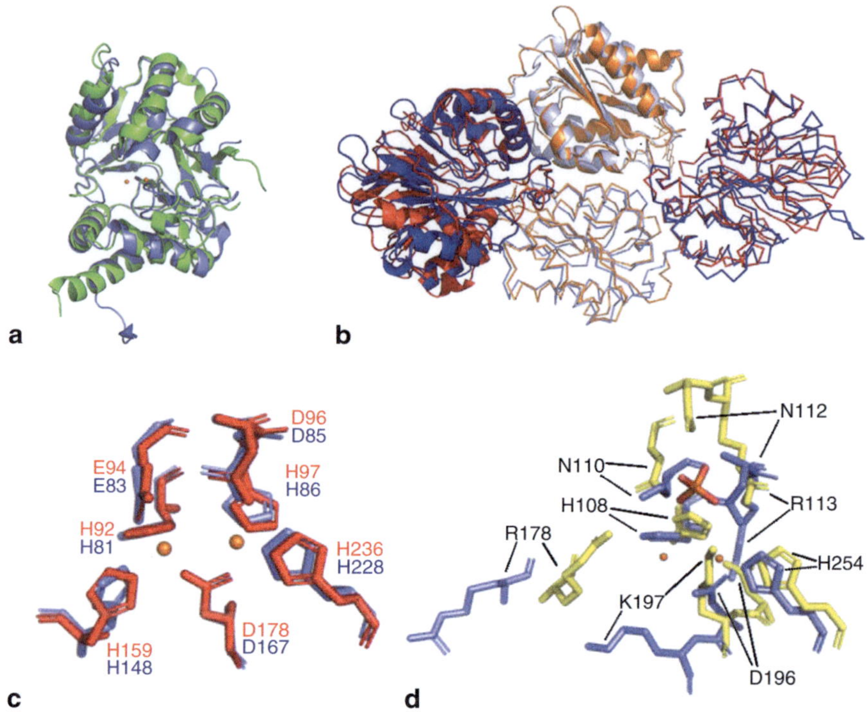

Fig. 22.5 Structural models of cyanobacterial flavodiiron proteins. **a** Structures of β-lactamase domains of a Class A FDP (*Moorella thermoacetica* FDP, PDB code 1YCG, in *blue*), and of a Class C Type 3 FDP (*Anabaena* sp. PCC7120 all0177, PDB code 3HNN, in *green*). The structures were aligned in Pymol (DeLano 2002) with an RMS of 1.364 Å for 185 aligned residues. **b** Models of the flavodiiron core of cyanobacterial FDPs (the C-terminal flavin reductase domains were removed from the modelled sequences). Two types of *Synechocystis* sp. PCC6803 FDPs were modelled using deposited FDPs structures as templates. sll550 (Type 1 FDP) was modelled using 1E5D, 1VME and 2OHH as templates (model in *blue*); sll1521 (Type 2 FDP) represented in *red* and *orange*, modelled using 2OHH as template. Each monomer is displayed either in cartoon or ribbon representation. **c** Superimposed modelled diiron site of sll550 (in *blue*) and of *Moorella thermoacetica* FDP (in *red*); iron ions as *orange* spheres. **d** Superimposition of the modelled diiron sites of sll1521 (Type 2), using different modelling strategies: *blue,* ligand substitutions modelled with structures of "canonical" Class A FDPs structures; *yellow,* the same "substituting" residues modelled with the structure of *Anabaena* sp. PCC7120 Type 3 FDP lactamase domain (PDB code 3HNN)

residues, such as asparagine and arginine (that replace the "canonical" carboxylate ligands) crowding the pocket may be the reason why a phosphate ion is stabilized inside the cavity. Since this structure was obtained for a truncated domain, one has to take into account a possible destabilization of the overall structure, in spite of the apparently retention of the lactamase fold (as described above and shown in Fig. 22.5a). This raises the question whether iron is missing because of the nature of the "alternative ligands", or due to any technical issue related with the protein production, purification or crystallization processes, which may trigger a rearrangement of the binding pocket. It should be noted that in several of the other types of enzymes (Types 2–8) the residue that replaces the bridging aspartate is preceded by another aspartate or by a glutamate, which may fulfil the same role, as will be discussed below. However, from the amino acid sequences comparison (see data on Table 22.2 and Fig. 22.3), it is clear that while some amino acid substitutions are conservative, in the sense that the amino acid substitute may bind to a metal ion through its side chain (even if representing less frequent modes of biological metal coordination), other amino acids may bind only through the backbone carbonyls, as the respective side chains are aliphatic (for example, the cases of isoleucine and alanine). Also, in many cases it is not possible to find in the sequences possible sensible amino acid substitutes, being the bridging aspartate a possible exception to this observation. These multiple amino acid substitutions raise the question whether some of those "unusual" FDPs are indeed metalloenzymes, which may be clarified only by appropriate experimental data. Of course, the possible absence of the iron ions will have unpredictable influences on the protein function.

Physiological studies on cyanobacterial FDPs have thus far focused on *Synechocystis* sp. PCC6803 enzymes (Vicente et al. 2002; Helman et al. 2003; Hackenberg et al. 2009; Zhang et al. 2009). This organism's genome codes for four FDP homologues, two of which fall into Type 1 and the other two into Type 2 subgroups. Since there appears to be no difference in terms of physiological roles between the two types of FDPs (see Sect. 22.7), we attempted to predict the structures of one of each type using deposited PDB files as templates. Since Class C FDPs have the extra C-terminal flavin reductase domain, the sequences used for modelling the structure were previously truncated, in order to model only the flavodiiron core. The protein coded by gene *sll0550* (a Type 1 enzyme) was modelled by combining the generated optimal structural alignments between sequence segments and several FDP structures (*D. gigas* ROO (1E5D), *Thermotoga maritima* FDP (1VME) and *Methanothermobacter marburgensis* $F_{420}H_2$ oxidase (2OHH)). The resulting model is thus an assembly of the best-fit models for different parts of the sll0550 sequence. The predicted structure (Fig. 22.5b) is quite similar to those of the "canonical" FDPs. Looking in detail at the binuclear site (Fig. 22.5c, red), it is observed that the conserved diiron ligands are modelled in the same positions and geometries as those of *Moorella thermoacetica* FDP (Fig. 22.5c, blue).

The Type 2 FDP, encoded by *sll1521*, was modelled by two different strategies. First, it was modelled using as template the structure of *Methanothermobacter marburgensis* $F_{420}H_2$ oxidase (2OHH). By this approach, it was possible to obtain a model where the residues substituting the "canonical" ligands display a geometry compatible with an iron coordination equivalent to Type 1 FDPs (Fig. 22.5d). Moreover, this modelling strategy resulted for sll1521 in a structure where the iron ions are present and the aspartate preceding the substituting lysine is placed in the same three dimensional position as the "canonical" aspartate. This model for sll1521 is thus compatible with a protein where the alternative amino acids indeed bind the iron ions. However, when we modelled sll1521 using the structure of the lactamase-like domain of *Anabaena* sp. PCC7120 Type 3 FDP (PDB code 3HNN) as a template, the structure of the diiron binding pocket changed dramatically, as expected

(Fig. 22.5d). As already noted for the *Anabaena* truncated enzyme, the side chain of the lysine that substitutes in the sequence of all177 the bridging aspartate protrudes into the space occupied by iron in Type 1 FDPs. Moreover, the equivalent basic residues that substitute the Type 1 ligands are placed in the same position as those in the al177 structure. Therefore, the fact that both models are plausible leaves an open question, whether the metal ions that constitute the diiron site are present or not in the native sll1521 enzyme, and which would be the effect of their presence or absence on the protein's function. As above mentioned, this question may be answered only with further structural and functional studies of the different types of FDPs.

22.5 Biochemical Properties

The FDPs thus far characterized are homodimers or homotetramers (a dimer of "head-to-tail" homodimers) in solution. Each monomer has ~45 kDa (Class A), ~54 kDa (Class B), and ~63 kDa (Class C) (no Class D FDP has been characterized), and contains two iron ions and one FMN per flavodiiron core, plus one iron per rubredoxin domain in Class B FDPs, and one flavin per flavin reductase domain in Class C FDPs. The latter extra domain is promiscuous in terms of flavin, being capable of harbouring FMN or FAD (Vicente et al. 2002). Due to the very low molar absortivity of the diiron centres, the electronic spectra of these enzymes are dominated by the flavin absorption, with maxima at ~460 and ~380 nm. In Class B enzymes, due to the presence of the FeCys4 centre the band at ~460 nm shifts to ~470 nm and an extra band appears at ~570 nm. Most Class A flavodiiron proteins thus far characterized displayed visible spectra which were broad and smooth in the band centred at ~460 nm. However, the spectra of a cyanobacterial FDP (Synechocystis sp. PCC6803 sll0550 (Vicente et al. 2002)) and of a methanogenic FDP (Wasserfallen et al. 1995) display this band with two shoulders. It has been noted that the FDPs with this kind of spectrum, lack a conserved tryptophan residue in the flavodoxin-like domain (Trp347 in *D. gigas* ROO numbering) which is coplanar with the FMN isoalloxazine ring. It was thus proposed that the presence/absence of this Trp residue was responsible for the heterogeneity in the spectral shape of the flavin moiety (Saraiva et al. 2004).

The diiron centre was first assessed by the structure of the *D. gigas* enzyme (Frazão et al. 2000). Later, it was studied in detail by EPR spectroscopy for the *E. coli* enzyme (Vicente and Teixeira 2005), and more recently for the *Giardia* and *Thrichomonas* enzymes (Vicente et al. 2009; Smutna et al. 2009). As characteristic of this type of iron centre, it can exist in three different oxidation states: diferric (Fe(III) Fe(III)), mixed valence (Fe(III) Fe(II)) and diferrous (Fe(II) Fe(II)). The diferric state is EPR silent, since the two iron ions (high-spin, $S=5/2$) are antiferromagnetically coupled, yielding a total spin $S=0$; for the *E. coli* enzyme a resonance at $g\sim 11$ for the diferrous state was detected, suggesting a $S=4$ spin ground state. The mixed valence state has a total spin $S=1/2$ and is EPR active, yielding a typical set of resonances below $g=2.0$. The diiron centre of *Moorella thermoacetica* FDP was characterized by Mössbauer spectroscopy, confirming that the iron ions in the binuclear site are magnetically coupled (Silaghi-Dumitrescu et al. 2003).

The spectroscopic signatures from the several redox cofactors of FDPs were used to study their redox properties, combining Visible and EPR spectroscopies with potentiometric titrations. The flavodoxin domain-bound FMN cofactor of several FDPs displayed reduction potentials in the −224 to +25 mV range for the $FMN_{ox} \rightarrow FMN_{sq}$ transition and −117 to +25 mV range for the $FMN_{sq} \rightarrow FMN_{red}$ transition (Gomes et al. 1997; Silaghi-Dumitrescu et al. 2003; Vicente and Teixeira 2005; Vicente et al. 2009; Smutna et al. 2009). The majority of the FDPs studied have reduction potentials that are enough separated to allow transient stabilization of the one electron-reduced semiquinone (Sq) state, which so far is always of the red, anionic type. However, recent studies on protozoan FDPs reported very close reduction potentials for the two redox transitions of the flavin, suggesting that the semiquinone state is not stabilized in these enzymes. Notably, those enzymes are only able to efficiently reduce oxygen, whereas the other FDPs, which stabilize the semiquinone flavin state, are either only NO (Vicente and Teixeira 2005) or ambivalent NO/O_2 reductases (Silaghi-Dumitrescu et al. 2003; Rodrigues et al. 2006). Until now, it has not been possible to propose a rationale for this difference in redox behaviour between O_2-reducing and NO/O_2-reducing FDPs, mainly because, at the available structural resolutions, the flavin-binding pockets are essentially identical among the several FDPs. Concerning the binuclear iron active site, the reduction potentials have been determined only for the *E. coli* flavorubredoxin (Vicente and Teixeira 2005) and for the protozoan FDPs (of *Trichomonas vaginalis* and *Giardia intestinalis*) (Di Matteo et al. 2008; Smutna et al. 2009), coupling redox potentiometry to EPR spectroscopy. The diiron sites of the protozoan enzymes have much higher reduction potentials than those of the *E. coli* FDP. However, whereas the reduction potentials in all three cases are compatible with oxygen or nitric oxide reduction, it is noteworthy that the two protozoan enzymes, that display only O_2 reductase activity, have reduction potentials markedly different from those of the NO-reducing *E. coli* enzyme. As mentioned for the flavin cofactor, the available structures do not allow yet understanding the differences in redox properties of the diiron centers.

One interesting aspect of FDPs redox properties is the observation, for the *E. coli* FlRd, that the reduction potentials of the iron center are altered in the presence of the reductase partner, which strongly suggests the formation of an electron transfer complex between the two enzymes (Vicente and Teixeira 2005).

22.6 Enzymatic Studies

Enzymatic studies to assess the putative role of FDPs as oxygen or nitric oxide reductases have been performed in Clark-type electrodes specific for each molecule. In typical experiments, *in vitro* electron transfer chains were assembled to mediate, under non rate-limiting conditions, electron delivery from NAD(P)H to FDP and determine its reductase activity towards each substrate. Fast kinetics methods, namely stopped-flow coupled to Visible spectroscopy was also used to further confirm the steady-state activity of the *Giardia* enzyme (Di Matteo et al. 2008).

The first enzymatic studies on a member of this protein family, performed for *D. gigas* ROO, led to the proposal that FDPs would be involved in the detoxification of dioxygen, which is both an inhibitor of many enzymes from anaerobes and a source of deleterious reactive oxygen species. Following the initial characterization of the *E. coli* enzyme that was shown to be capable of binding nitric oxide (Gomes et al. 2000), Gardner and co-workers proposed that *E. coli* FlRd was involved in NO detoxification under anaerobic conditions (Gardner et al. 2002). This proposal led to the *in vitro* demonstration of nitric oxide reduction by FlRd, coupled to NADH oxidation by its physiological partner, the NADH:FlRd oxidoreductase, with a turnover of 15–20 s^{-1} (Gomes et al. 2002). Since then, nitric oxide reductase activities of the same order of magnitude have been determined for several other FDPs (Gomes et al. 2002; Silaghi-Dumitrescu et al. 2003, 2005; Rodrigues et al. 2006). However, the data so far available indicate that the FDPs may have an ambivalent activity, i.e., some are able to reduce mainly NO, others mainly, or even exclusively, O_2 (Di Matteo et al. 2008; Smutna et al. 2009), and there are few cases for which comparable activities with both substrates were measured (Silaghi-Dumitrescu et al. 2003, 2005; Rodrigues et al. 2006; Hillmann et al. 2009). Concerning *Synechocystis* sp. PCC6803, preliminary data obtained for Sll0550 revealed that the enzyme has a relatively low oxygen reductase activity, which could have been hampered by the sub-stoichiometric flavin load (Vicente et al. 2002). It is important to emphasize that the data available also suggest a sluggish reaction of the reduced enzymes with hydrogen peroxide, H_2O_2, and with several reactive nitrogen species (Di Matteo et al. 2008; Smutna et al. 2009 and our own unpublished data), i.e., the enzymes appear to favour NO and O_2 as substrates. In this respect, it should be mentioned that the substrate binding cavity present in zinc β-lactamases is occluded in FDPS (e.g., Frazão et al. 2000). The dual oxygen/nitric oxide reducing activity raises the question whether both activities are physiologically relevant and what dictates the preference for each substrate. The evidences for the physiological roles of FDPs will be further discussed in the next section.

22.7 Physiological Roles of FDPs: FDPS as O_2 and/or NO Reductases

The work by Gardner and co-workers first showed that FlRd protects *E. coli* against nitric oxide; deletion of the *flrd* gene highly compromised the growth viability of *E. coli* upon exposure to NO, under anaerobic conditions (Gardner et al. 2002). Furthermore, it was shown that *flrd* transcription is regulated by NO via NorR (Gardner et al. 2003), a transcriptional regulator that in enterobacteria is located immediately upstream, and divergently transcribed of the dicistronic unit that encodes both FlRd and its NADH oxidoreductase (da Costa et al. 2003). Saraiva and co-workers (Justino et al. 2005a) further showed that the binding of the NorR trimer to three sites of the flavorubredoxin gene promoter is required for nitric oxide-dependent induction of *flrd.* Several transcriptomic studies confirmed that *flrd* transcription is indeed up-regulated by NO (Mukhopadhyay et al. 2004; Flatley et al. 2005; Justino et al. 2005b; Pullan et al. 2006), including in *E. coli* cells grown under anaerobic conditions (Justino et al. 2005b). On the contrary, transcriptional studies performed for *E. coli* under several oxidative stress conditions did not reveal a significant alteration on the expression of the FlRd-encoding gene. These *in vitro* studies established a role for flavorubredoxin (a Class B FDP) in *E. coli,* as an enzyme responsive to the deleterious action of nitric oxide, namely under anaerobic conditions. Similar evidences were later on obtained for a few other bacteria, such as *D. gigas* (Rodrigues et al. 2006). But, in apparent agreement with the fact that the *D. gigas* enzyme has comparable NO and O_2 reducing activities, ROO seems to be important also under oxidative stress conditions (Rodrigues et al. 2006). A similar situation was recently reported for *Clostridium acetobutylicum,* in which the flavodiiron proteins were shown to be up-regulated upon exposure of *C. acetobutylicum* to O_2 and important to protect this anaerobic bacterium against oxygen (Hillmann et al. 2009).

For cyanobacteria, there is no data regarding the role of FDPs in response to nitrosative stress conditions; however, as will be described below, several experimental data clearly suggest a function for their FDPs in oxidative stress protection.

A survey of DNA microarray data so far reported for *Synechocystis* sp. showed that the genes encoding FDPs have their expression modified under certain conditions. The expression of the already mentioned gene cluster *sll0217-0219* was found to be induced during *Synechocystis* acclimation from low- to high-light intensity (Hihara et al. 2001). The *sll0217* gene is also among the 36 genes encoding potential FMN-containing proteins that were induced by UV-B light (Huang et al. 2002). The induction of the FDP encoding genes observed under high energy supply is possibly related to the higher generation of reactive oxygen species under these conditions; in fact, other genes encoding ROS scavenging enzymes, such as superoxide dismutase and glutathione peroxidase, were also up-regulated under those conditions (Huang et al. 2002). Chauvat and co-workers (Houot et al. 2007) reported that the transcription of the *sll1521, sll0217* and *sll0219* genes, encoding FDPs, increases in *Synechocystis* cells exposed to 3 mM hydrogen peroxide, or grown in the presence of excess of metals such as cadmium, iron and zinc. Nevertheless, the observed variations were low (2-4-fold) and contradictory data has been reported (Hihara et al. 2003; Kobayashi et al. 2004; Li et al. 2004). In summary, the DNA array data is still too scarce and somehow ambiguous. More specific and clarifying studies were performed also for *Synechocystis* sp. PCC6803, mainly based on the analysis of the behaviour of several single and multiple deletion mutants of its FDP genes under different growth conditions, namely illumination and carbon dioxide fluxes, and combinations of both these factors (Helman et al. 2003; Hackenberg et al. 2009; Zhang et al. 2009). It was concluded that sll1521 and sll0550 (enzymes of Types 2 and 1, respectively) were essential for the Mehler reaction, i.e., for electron transport from photosystem I to oxygen, without formation of reactive oxygen species. In the same report it was proposed that the two enzymes could assemble as a heterodimer, and that sll0550 could be sufficient to sustain photoreduction of dioxygen. More recently was studied in further detail the possible role of the other two FDPs, sll0217 and sll0219 (Zhang et al. 2009). The main conclusion was that both enzymes were important to protect photosystem II against oxidative stress, caused by conditions of high illumination intensity and/or low carbon dioxide fluxes. Interestingly, it was further shown that those FDPs are membrane associated, and that the expression of *sll0217* was enhanced in the single deletion mutants of *sll1521* or *sll0550*, i.e., it appears that sll0217 may substitute also for either of those two enzymes.

These proposals are in line with an oxygen reductase activity of the cyanobacterial enzymes, as first shown for the gene product of *sll0550* (Vicente et al. 2002): the depletion of dioxygen by a NAD(P)H:oxygen oxidoreductase activity of the FDPs is fully consistent with their proposed roles in acting against oxidative stressing conditions, avoiding the formation of reactive oxygen species through the direct reduction of oxygen to water and enabling the elimination of excess reducing equivalents in the photosynthetic electron transfer chain. Nevertheless, these activities remain to be directly proven *in vitro* for the other three *Synechocystis* FDPs. This is particularly relevant, due to the fact that there is no apparent correlation between the proposed functions for the *Synechocystis* enzymes and their types, i.e., the presence or absence of the "canonical" ligands: sll1521 and sll0219 are enzymes of Type 2, lacking some of those ligands, while sll0550 and sll0217 are enzymes of the "canonical" Type 1.

22.8 Concluding Remarks

In summary, the flavodiiron proteins are a family of modular enzymes having nitric oxide and/or oxygen reductase activities. Both activities are supported either by *in vitro* experiments or by *in vivo* approaches, namely through transcriptional and phenotypic analysis of wild type and deletion strains. The abundance of homologues in cyanobacteria is particularly striking and may be related to the fact that oxygenic phototrophs are particularly prone to oxidative stress, due in special to the endogenous production of dioxygen and to the delicate balance between photonic energy supply and carbon dioxide availability. As flavodiiron enzymes appear to have evolved quite early, eventually before the split between Bacteria and Archaea, they may have been already present when oxygenic photosynthesis started, ca 3–3.5 billion years ago, as one of the first mechanisms to respond to the "oxygen paradox", i.e., to avoid its intracellular toxicity. A similar scenario has been proposed for another type of diiron proteins, the rubrerythrins (Gomes et al. 2001), which more recently were proposed to act as hydrogen peroxide reductases, forming water directly through reduction of H_2O_2 (e.g., Kurtz 2006) and which are also quite spread among cyanobacteria (our unpublished observation). A schematic representation of

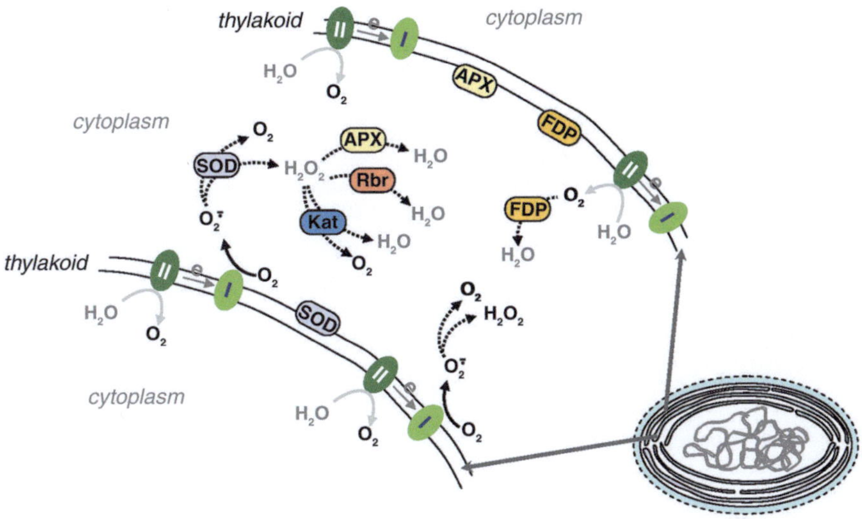

Fig. 22.6 Scheme illustrating oxidative stress and response mechanisms in cyanobacteria. Zoom-in into the thylakoid membranes in cyanobacteria and the cytoplasmatic space in between, where formation of reactive oxygen species and their breakdown by specialized enzymes are likely to take place. Oxygen is produced by water oxidation at the oxygen-evolving complex in Photosystem II (PS II, in *dark green*). Excess electrons in Photosystem I (PS I, in *light green*) generate superoxide anion from reaction with molecular oxygen; superoxide can self-dismutate to yield oxygen and hydrogen peroxide. Excess oxygen can be fully reduced to water by flavodiiron proteins (FDP, in *orange*), soluble and possibly also membrane-associated. Besides its self-dismutation, superoxide anion can be efficiently scavenged by superoxide dismutase (SOD, in *light blue*), both soluble and membrane-bound, whose products are O_2 and H_2O_2. Hydrogen peroxide can be removed by at least three different enzymatic systems detected in the genomes of cyanobacteria: both ascorbate peroxidase (APX, in *yellow*, soluble and membrane-bound forms) and rubrerythrin (Rbr, in *red*) can reduce hydrogen peroxide to water; catalase (Kat, in *darker blue*) generates oxygen and water

the diverse mechanisms involved in oxidative stress protection in cyanobacteria is presented in Fig. 22.6, where it becomes clear the importance of having enzymes, such as FDPs, that avoid the production of reactive oxygen species, like the superoxide anion (detoxified by superoxide dismutases), or hydrogen peroxide (scavenged by catalases, peroxidases, rubrerythrins or ascorbate peroxidases). It has still to be shown if in cyanobacteria FDPs play also a role in nitrosative stress defence. But whatever the actual function of the several homologues of FDPs will be, their multiplicity suggests a crucial role in cyanobacteria.

Several challenging questions remain to be answered: one relates to the evolution of FDPs and their function in higher eukaryotes; the second, the unusual diversity of putative "ligands" to the diiron site—do all types of FDPs here described contain indeed a metal site? If yes, they will have quite unprecedented ligand combinations; if no, which will then be the function of the non-metal containing enzymes? Answers to these questions will bring new insights into the structure and function of diiron proteins and into the field of oxidative stress, and will further enlarge the knowledge on the physiology of oxygenic phototrophs.

Acknowledgments Work by the authors discussed in this review was supported by projects from the Fundação para a Ciência e Tecnologia (to MT and LST). VLG and JVB are recipients of grants SFRH/BD/29428/2006 and SFRH/BPD/26895/2006. We would like to thank our collaborators in this field whose names appear in the appropriate references, and Dr. Bruno Victor (ITQB) for performing the homology model of the *Synechocystis* FDP.

References

Andersson JO, Sjogren AM, Davis LA, Embley TM and Roger AJ (2003) Phylogenetic analyses of diplomonad genes reveal frequent lateral gene transfers affecting eukaryotes. Curr Biol 13 (2): 94–104

Andersson JO, Hirt RP, Foster PG and Roger AJ (2006) Evolution of four gene families with patchy phylogenetic distributions: influx of genes into protist genomes. BMC Evol Biol 6: 27

Bodyl A, Mackiewicz P and Stiller JW (2007) The intracellular cyanobacteria of *Paulinella chromatophora:* endosymbionts or organelles? Trends Microbiol 15 (7): 295–296

Carlton JM, Hirt RP, Silva JC, Delcher AL, Schatz M, Zhao Q, Wortman JR, Bidwell SL, Alsmark UC, Besteiro S, Sicheritz-Ponten T, Noel CJ, Dacks JB, Foster PG, Simillion C, Van de Peer Y, Miranda-Saavedra D, Barton GJ, Westrop GD, Muller S, Dessi D, Fiori PL, Ren Q, Paulsen I, Zhang H, Bastida-Corcuera FD, Simoes-Barbosa A, Brown MT, Hayes RD, Mukherjee M, Okumura CY, Schneider R, Smith AJ, Vanacova S, Villalvazo M, Haas BJ, Pertea M, Feldblyum TV, Utterback TR, Shu CL, Osoegawa K, de Jong PJ, Hrdy I, Horvathova L, Zubacova Z, Dolezal P, Malik SB, Logsdon JM, Jr., Henze K, Gupta A, Wang CC, Dunne RL, Upcroft JA, Upcroft P, White O, Salzberg SL, Tang P, Chiu CH, Lee YS, Embley TM, Coombs GH, Mottram JC, Tachezy J, Fraser-Liggett CM and Johnson PJ (2007) Draft genome sequence of the sexually transmitted pathogen *Trichomonas vaginalis.* Science 315 (5809): 207–212

Chen L, Liu MY, Legall J, Fareleira P, Santos H and Xavier AV (1993a) Purification and characterization of an NADH-rubredoxin oxidoreductase involved in the utilization of oxygen by *Desulfovibrio gigas.* Eur J Biochem 216 (2): 443–448

Chen L, Liu MY, LeGall J, Fareleira P, Santos H and Xavier AV (1993b) Rubredoxin oxidase, a new flavo-hemo-protein, is the site of oxygen reduction to water by the “strict anaerobe” *Desulfovibrio gigas.* Biochem Biophys Res Commun 193 (1): 100–105

da Costa PN, Teixeira M and Saraiva LM (2003) Regulation of the flavorubredoxin nitric oxide reductase gene in *Escherichia coli:* nitrate repression, nitrite induction, and possible post-transcription control. FEMS Microbiol Lett 218 (2): 385–393

DeLano WL (2002) The PyMOL User’s Manual DeLano Scientific. Palo Alto, CA, USA

Di Matteo A, Scandurra FM, Testa F, Forte E, Sarti P, Brunori M and Giuffre A (2008) The O2-scavenging flavodiiron protein in the human parasite *Giardia intestinalis.* J Biol Chem 283 (7): 4061–4068

Fareleira P, Santos BS, Antonio C, Moradas-Ferreira P, LeGall J, Xavier AV and Santos H (2003) Response of a strict anaerobe to oxygen: survival strategies in *Desulfovibrio gigas.* Microbiology 149 (Pt 6): 1513–1522

Flatley J, Barrett J, Pullan ST, Hughes MN, Green J and Poole RK (2005) Transcriptional responses of *Escherichia coli* to S-nitrosoglutathione under defined chemostat conditions reveal major changes in methionine biosynthesis. J Biol Chem 280 (11): 10065–10072

Frazão C, Silva G, Gomes CM, Matias P, Coelho R, Sieker L, Macedo S, Liu MY, Oliveira S, Teixeira M, Xavier AV, Rodrigues-Pousada C, Carrondo MA and Le Gall J (2000) Structure of a dioxygen reduction enzyme from *Desulfovibrio gigas.* Nat Struct Biol 7 (11): 1041–1045

Gardner AM, Helmick RA and Gardner PR (2002) Flavorubredoxin, an inducible catalyst for nitric oxide reduction and detoxification in *Escherichia coli.* J Biol Chem 277 (10): 8172–8177

Gardner AM, Gessner CR and Gardner PR (2003) Regulation of the nitric oxide reduction operon (norRVW) in *Escherichia coli.* Role of NorR and sigma54 in the nitric oxide stress response. J Biol Chem 278 (12): 10081–10086

Gomes CM, Silva G, Oliveira S, LeGall J, Liu MY, Xavier AV, Rodrigues-Pousada C and Teixeira M (1997) Studies on the redox centers of the terminal oxidase from *Desulfovibrio gigas* and evidence for its interaction with rubredoxin. J Biol Chem 272 (36): 22502–22508

Gomes CM, Vicente JB, Wasserfallen A and Teixeira M (2000) Spectroscopic studies and characterization of a novel electron-transfer chain from *Escherichia coli* involving a flavorubredoxin and its flavoprotein reductase partner. Biochemistry 39 (51): 16230–16237

Gomes CM, Le Gall J, Xavier AV and Teixeira M (2001) Could a diiron-containing four-helix-bundle protein have been a primitive oxygen reductase? Chembiochem 2 (7–8): 583–587

Gomes CM, Giuffre A, Forte E, Vicente JB, Saraiva LM, Brunori M and Teixeira M (2002) A novel type of nitric-oxide reductase. *Escherichia coli* flavorubredoxin. J Biol Chem 277 (28): 25273–25276

Gould SB, Waller RF and McFadden GI (2008) Plastid evolution. Annu Rev Plant Biol 59 491–517

Hackenberg C, Engelhardt A, Matthijs HC, Wittink F, Bauwe H, Kaplan A and Hagemann M (2009) Photorespiratory 2-phosphoglycolate metabolism and photoreduction of O2 cooperate in high-light acclimation of *Synechocystis* sp. strain PCC6803. Planta 230 (4): 625–637

Helman Y, Tchernov D, Reinhold L, Shibata M, Ogawa T, Schwarz R, Ohad I and Kaplan A (2003) Genes encoding a-type flavoproteins are essential for photoreduction of O(2) in cyanobacteria. Curr Biol 13 (3): 230–235

Hihara Y, Kamei A, Kanehisa M, Kaplan A and Ikeuchi M (2001) DNA microarray analysis of cyanobacterial gene expression during acclimation to high light. Plant Cell 13 (4): 793–806

Hihara Y, Sonoike K, Kanehisa M and Ikeuchi M (2003) DNA microarray analysis of redox-responsive genes in the genome of the cyanobacterium *Synechocystis* sp. strain PCC6803. J Bacteriol 185 (5): 1719–1725

Hillmann F, Riebe O, Fischer RJ, Mot A, Caranto JD, Kurtz DM, Jr. and Bahl H (2009) Reductive dioxygen scavenging by flavo-diiron proteins of *Clostridium acetobutylicum.* FEBS Lett 583 (1): 241–245

Houot L, Floutier M, Marteyn B, Michaut M, Picciocchi A, Legrain P, Aude JC, Cassier-Chauvat C and Chauvat F (2007) Cadmium triggers an integrated reprogramming of the metabolism of *Synechocystis* PCC6803, under the control of the Slr1738 regulator. BMC Genomics 8: 350

Huang L, McCluskey MP, Ni H and LaRossa RA (2002) Global gene expression profiles of the cyanobacterium *Synechocystis* sp. strain PCC6803 in response to irradiation with UV-B and white light. J Bacteriol 184 (24): 6845–6858

Huson DH, Richter CD, Rausch C, Dezulian T, Franz M and Rupp R (2007) Dendroscope: an interactive viewer for large phylogenetic trees. BMC Bioinformatics 8: 460

Jones DT (1999) Protein secondary structure prediction based on position-specific scoring matrices. J Mol Biol 292 (2): 195–202

Justino MC, Goncalves VM and Saraiva LM (2005a) Binding of NorR to three DNA sites is essential for promoter activation of the flavorubredoxin gene, the nitric oxide reductase of *Escherichia coli.* Biochem Biophys Res Commun 328 (2): 540–544

Justino MC, Vicente JB, Teixeira M and Saraiva LM (2005b) New genes implicated in the protection of anaerobically grown *Escherichia coli* against nitric oxide. J Biol Chem 280 (4): 2636–2643

Kobayashi M, Ishizuka T, Katayama M, Kanehisa M, Bhattacharyya-Pakrasi M, Pakrasi HB and Ikeuchi M (2004) Response to oxidative stress involves a novel peroxiredoxin gene in the unicellular cyanobacterium *Synechocystis* sp. PCC6803. Plant Cell Physiol 45 (3): 290–299

Kurtz DM Jr. (2006) Avoiding high-valent iron intermediates: superoxide reductase and rubrerythrin. J Inorg Biochem 100 (4): 679–693

Larkin MA, Blackshields G, Brown NP, Chenna R, McGettigan PA, McWilliam H, Valentin F, Wallace IM, Wilm A, Lopez R, Thompson JD, Gibson TJ and Higgins DG (2007) Clustal W and Clustal X version 2.0. Bioinformatics 23 (21): 2947–2948

Lewis ME, Corker HA, Gollan B and Poole RK (2008) A survey of methods for the purification of microbial flavohemoglobins. Methods Enzymol 436: 169–186

Li H, Singh AK, McIntyre LM and Sherman LA (2004) Differential gene expression in response to hydrogen peroxide and the putative PerR regulon of *Synechocystis* sp. strain PCC6803. J Bacteriol 186 (11): 3331–3345

Mukhopadhyay P, Zheng M, Bedzyk LA, LaRossa RA and Storz G (2004) Prominent roles of the NorR and Fur regulators in the *Escherichia coli* transcriptional response to reactive nitrogen species. Proc Natl Acad Sci U S A 101 (3): 745–750

Petoukhov MV, Vicente JB, Crowley PB, Carrondo MA, Teixeira M and Svergun DI (2008) Quaternary structure of flavorubredoxin as revealed by synchrotron radiation small-angle X-ray scattering. Structure 16 (9): 1428–1436

Pullan ST, Gidley MA, Jones RA, Barrett J, Stevanin TM, Read RC, Green J and Poole RK (2006) Nitric Oxide in Chemostat-Cultured *Escherichia coli* is Sensed by Fnr and Other Global Regulators; Unaltered Methionine Biosynthesis Indicates Lack of S-Nitrosation. J Bacteriol 189 (5): 1845–55

Rodrigues R, Vicente JB, Felix R, Oliveira S, Teixeira M and Rodrigues-Pousada C (2006) *Desulfovibrio gigas* flavodiiron protein affords protection against nitrosative stress in vivo. J Bacteriol 188 (8): 2745–2751

Saraiva LM, Vicente JB and Teixeira M (2004) The role of the flavodiiron proteins in microbial nitric oxide detoxification. Adv Microb Physiol 49: 77–129

Seedorf H, Hagemeier CH, Shima S, Thauer RK, Warkentin E and Ermler U (2007) Structure of coenzyme F420H2 oxidase (FprA), a di-iron flavoprotein from methanogenic Archaea catalyzing the reduction of O2 to H2O. FEBS J 274 (6): 1588–1599

Shimizu T, Ohtani K, Hirakawa H, Ohshima K, Yamashita A, Shiba T, Ogasawara N, Hattori M, Kuhara S and Hayashi H (2002) Complete genome sequence of *Clostridium perfringens*, an anaerobic flesh-eater. Proc Natl Acad Sci U S A 99 (2): 996–1001

Silaghi-Dumitrescu R, Coulter ED, Das A, Ljungdahl LG, Jameson GN, Huynh BH and Kurtz DM Jr. (2003) A flavodiiron protein and high molecular weight rubredoxin from *Moorella thermoacetica* with nitric oxide reductase activity. Biochemistry 42 (10): 2806–2815

Silaghi-Dumitrescu R, Kurtz DM Jr., Ljungdahl LG and Lanzilotta WN (2005) X-ray crystal structures of *Moorella thermoacetica* FprA. Novel diiron site structure and mechanistic insights into a scavenging nitric oxide reductase. Biochemistry 44 (17): 6492–6501

Smutna T, Goncalves VL, Saraiva LM, Tachezy J, Teixeira M and Hrdy I (2009) Flavodiiron protein from *Trichomonas vaginalis* hydrogenosomes: the terminal oxygen reductase. Eukaryot Cell 8 (1): 47–55

Vicente JB and Teixeira M (2005) Redox and spectroscopic properties of the *Escherichia coli* nitric oxide-detoxifying system involving flavorubredoxin and its NADH-oxidizing redox partner. J Biol Chem 280 (41): 34599–34608

Vicente JB, Gomes CM, Wasserfallen A and Teixeira M (2002) Module fusion in an A-type flavoprotein from the cyanobacterium *Synechocystis* condenses a multiple-component pathway in a single polypeptide chain. Biochem Biophys Res Commun 294 (1): 82–87

Vicente JB, Scandurra FM, Rodrigues JV, Brunori M, Sarti P, Teixeira M and Giuffre A (2007) Kinetics of electron transfer from NADH to the *Escherichia coli* nitric oxide reductase flavorubredoxin. FEBS J 274 (3): 677–686

Vicente JB, Carrondo MA, Teixeira M and Frazao C (2008a) Structural studies on flavodiiron proteins. Methods Enzymol 437B: 3–19

Vicente JB, Carrondo MA, Teixeira M and Frazão C (2008b) Flavodiiron proteins: nitric oxide and/or oxygen reductases. Handbook of Metalloproteins, Ed. Albrecht Messerschmidt, John Wiley & Sons, Vol. 4: 1–19

Vicente JB, Justino MC, Goncalves VL, Saraiva LM and Teixeira M (2008c) Biochemical, spectroscopic, and thermodynamic properties of flavodiiron proteins. Methods Enzymol 437B: 21–45

Vicente JB, Scandurra FM, Forte E, Brunori M, Sarti P, Teixeira M and Giuffre A (2008d) Kinetic characterization of the *Escherichia coli* nitric oxide reductase flavorubredoxin. Methods Enzymol 437B: 47–62

Vicente JB, Testa F, Mastronicola D, Forte E, Sarti P, Teixeira M and Giuffre A (2009) Redox properties of the oxygen-detoxifying flavodiiron protein from the human parasite *Giardia intestinalis.* Arch Biochem Biophys 488 (1): 9–13

Wasserfallen A, Huber K and Leisinger T (1995) Purification and structural characterization of a flavoprotein induced by iron limitation in *Methanobacterium thermoautotrophicum* Marburg. J Bacteriol 177 (9): 2436–2441

Wasserfallen A, Ragettli S, Jouanneau Y and Leisinger T (1998) A family of flavoproteins in the domains Archaea and Bacteria. Eur J Biochem 254 (2): 325–332

Yoon HS, Nakayama T, Reyes-Prieto A, Andersen RA, Boo SM, Ishida K and Bhattacharya D (2009) A single origin of the photosynthetic organelle in different *Paulinella* lineages. BMC Evol Biol 9: 98

Zhang P, Allahverdiyeva Y, Eisenhut M and Aro EM (2009) Flavodiiron proteins in oxygenic photosynthetic organisms: photoprotection of photosystem II by Flv2 and Flv4 in *Synechocystis* sp. PCC6803. PLoS One 4 (4): e5331

Part VI
Electron Exit (Terminal Oxidation)

Chapter 23
Cyanobacterial Respiratory Electron Transport: Heme-Copper Oxidases and Their Electron Donors

Margit Bernroitner, Marcel Zamocky, Martin Pairer, Günter A. Peschek and Christian Obinger

23.1 Introduction

Cyanobacteria (blue-green algae) are the paradigmatic organisms of oxygenic (plant-type) photosynthesis and aerobic (mitochondrial) respiration in bioenergetic, evolutionary and ecological respects. They have uniquely accommodated both a photosynthetic electron transport chain (PET) and a respiratory electron transport chain (RET) within a single prokaryotic cell (Jones and Myers 1963). Since many cyanobacteria in addition are also capable of N_2-fixation (Gallon 1992; Berman-Frank et al. 2003; Zhang et al. 2006) these oxygenic phototrophic organisms are often called the bioenergetic "nonplus-ultra" among living beings. At the expense of sunlight, water, atmospheric air and a few ubiquitous minerals they cover all their needs for growth and proliferation. From the estimated 10^{11} t of carbon (in the form of CO_2) per year that are converted into biomass by plant-type photosynthesis between 20% and 30% are assigned to cyanobacteria, in particular to small unicellular marine *Synechococcus* species (Waterbury et al. 1979) and to likewise unicellular planktonic *Prochlorophytes* that are widespread in all oceans (Chisholm et al. 1988).

Cyanobacterial respiration is much less well understood than photosynthesis. Most cyanobacteria are obligate photoautotrophs which, in the dark, depend completely on respiration to maintain energy levels. In almost all cyanobacteria PET is localized in intracytoplasmatic membranes (ICM) or thylakoids, which are also the site of RET. The two processes share several components such as the cytochrome b_6f complex and the redox carriers plastoquinol (PQ), cytochrome c_6 (CYTc_6, previously known as cytochrome c_{553}) and plastocyanin (PC) (Lockau 1981; Scherer et al. 1988; Scherer 1990; Peschek et al. 2004). The cytoplasmic membrane (CM) does not contain a functional PET, but contains a second respiratory chain. The presence of RET in CM is essential to provide energy for various transport pro-

C. Obinger (✉)
Department of Chemistry, Division of Biochemistry, BOKU—University of Natural Resources and Life Sciences, Muthgasse 18, A-1190 Vienna, Austria
e-mail: christian.obinger@boku.ac.at

G. A. Peschek et al. (eds.), *Bioenergetic Processes of Cyanobacteria*,
DOI 10.1007/978-94-007-0388-9_23, © Springer Science+Business Media B.V. 2011

cesses and its importance increases under stress conditions such as high salt ("salt-respiration") and nitrogen fixation (Wolk et al. 1994; Fry et al. 1986).

Understanding of RET in cyanobacteria is hampered by the high complexity of its components and their interactions. Common to all other prokaryotes, electron-transfer chains are branched (Poole and Cook 2000) and recent experimental findings and genome analysis (Bernroitner et al. 2008) underline the presence of multiple terminal oxidases and (putative) electron donors downstream of plastoquinol. This review summarizes the present knowledge how cyanobacterial branched RET chains could help to meet the energetic and (micro)environmental demands under a variety of environmental conditions and differentiation states (e.g. heterocyst formation in some N_2-fixing cyanobacteria). It is focused on terminal heme-copper oxidases and the one-electron donors cytochrome c_6, plastocyanin and cytochrome c_M.

23.2 Terminal Heme-Copper Oxidases in Cyanobacteria

Heme-copper oxidases (or oxygen reductases) are redox driven proton pumps that couple the four-electron reduction of molecular oxygen to water to the vectorial translocation of protons across the membrane. This transmembrane proton and voltage gradient generated by heme-copper oxidases and the other complexes of RET is directly converted to more useful energy forms via energy conserving systems such as ATP synthase. Heme-copper oxidases can use cytochromes (Lockau 1981; Michel et al. 1998), type-1 (or blue-) copper proteins (Lockau 1981; Lubbern et al. 1994; Paumann et al. 2004a) or quinols (Abramson et al. 2000) as electron donors and are characterized for having a heme-copper binuclear reaction center comprising a Cu_B electronically coupled with a high-spin heme. Additionally, all heme-copper oxidases contain a low-spin heme in subunit I (Pereira and Teixeira 2004). Depending on the nature of electron donors, the superfamily of (cyanide sensitive) heme-copper oxidases is divided into two branches, namely cytochrome *c* oxidases (COX, aa_3-type cytochrome oxidase) or quinol oxidases (QOX, *bo*-type quinol oxidase) (Musser et al. 1993; Howitt and Vermaas 1998; Hart et al. 2005). In contrast, the cytochrome *bd*-quinol oxidases do not belong to this protein superfamily, since they lack the binuclear reaction center and instead contain two heme groups. Additionally, they do not pump protons. This review focuses on heme-copper oxidases, i.e. COX and QOX.

Cytochrome *c* oxidase (COX) has been found in all cyanobacteria by spectroscopic (Peschek et al. 1982), inhibitor (Peschek et al. 1982), EPR studies (Peschek et al. 1988) and immunological studies (Fry et al. 1989). In 1993 a gene cluster (*coxBAC* or *ctaCDE*) encoding subunits II, I and III of COX from *Synechocystis* PCC6803 has been cloned for the first time (Alge and Peschek 1993; Schmetterer et al. 1994). Further operons encoding subunits I-III in the same order have also been cloned from *Synechococcus vulcanus* (Sone et al. 1993), *Anabaena variabilis* (Schmetterer et al. 2001) and *Nostoc* PCC7120 (Jones and Haselkorn 2002). It has been demonstrated by analysis of completely or partially sequenced genomes (Bernroitner et al. 2008) that all cyanobacteria contain at least one complete *coxBAC* operon. *Nostocales* (heterocyst forming N_2-fixing cyanobacteria, see below) have two *coxBAC* operons (designated *coxBAC1* and *coxBAC2*) (Table 23.1). In *Gloeobacter violaceus* also two clusters are found but one lacks subunit III.

The 3D structure of a cyanobacterial COX is still unknown. Figure 23.1 presents the two-subunit structure of *P. denitrificans* COX (Iwata et al. 1995). *Paracoccus denitrificans* COX is a four-subunit enzyme, but only subunits I and II contain all redox cofactors and residues involved in O_2 reduction and proton pumping (Michel et al. 1998). Subunit III possesses seven transmembrane helices and subunit IV only one. Subunits III & IV are most likely not involved in catalysis (Michel et al. 1998). Usually, subunit I in heme-copper oxidases is strikingly-well conserved. Several high-resolution 3D structures of heme-copper oxidases are known including COX from *P. denitrificans* (Iwata et al. 1995) and *Rhodobacter sphaeroides* (Svensson-Ek et al. 2002). Similar to *Paracoccus* COX, all cyanobacterial COX subunits I have 12 (predicted) transmembrane helices with a short loop between helices I and II, similar to many other SU-I but much shorter than that of *P. denitrificans* or *R. sphaeroides* (Pereira et al. 2001; Bernroitner et al. 2008). There are no extended loops between the transmembrane helices comparable to other bacterial COXs and the C-terminal extensions of some cyanobacterial SU-I are 10–40 amino acid residues longer than that of *P. denitrificans* or *R. sphaeroides* (Bernroitner et al. 2008). The core of 12 transmembrane helices are known to form three arcs within the membrane with each arc being shaped by four transmembrane helices and together with the last segment of the previous arc pore-like arrangements are formed (Iwata et al. 1995; Svensson-Ek et al. 2002). The first pore is filled with aromatic residues, the second holds the binuclear reaction center, and the third one the low-spin heme. Additionally, the first and second pores form two uptake pathways for protons from the *N*-side, viz. the so-called D- and K-channels (see below).

Amino acid residues coordinating the three metal centers are strictly conserved in all cyanobacterial COX subunits I: His94, His276, His325, His326, His411 and His413 (*Paracoccus* numbering) (Fig. 23.1 & Bernroitner et al. 2008). Also in cyanobacterial COX SU-I His276 could be covalently linked with a conserved Tyr280, which has been reported to form a transient tyrosyl radical during catalysis (Soulimane et al. 2000). In addition to the six inner-sphere histidine ligands of metal centers (Fig. 23.1), several other amino acids are strongly conserved in cyanobacterial COX SU-I. Additional residues interacting with heme *a* are Arg54 on helix I and Trp87 (Tyr in all cyanobacterial COXs) on helix II (Iwata et al. 1995). Almost all residues in helix VI are conserved including Trp272, which interacts with Cu_B-coordinating His276 and Val279 (Bernroitner et al. 2008). The latter has been described as part of the oxygen diffusion channel (Riistama et al. 1996). In helix VIII Thr344 (Ser in *Gloeobacter*) is close to another Cu_B coordinating His. Between helices IX and X, all cyanobacteria possess Asp399, which interacts with heme a_3. Phe412 may be involved in electron transfer between the two hemes, yet it is replaced in all *Nostocales* COX1 by Leu as well as in *Gloeobacter* and *Acaryochloris marina* by Met (Bernroitner et al. 2008).

Table 23.1 Cyanobacterial heme copper oxidases and electron donors. ORFs and genes of cyanobacterial heme-copper oxidases and their (putative) electron carriers in 39 completely or partially (*) sequenced strains (including their genome size). Nitrogen-fixing cyanobacteria are highlighted in *grey*, heterocyst-forming species are highlighted in *dark-grey*

COX	QOX	Cyanobacteria	Classification	Genome size	N_2-fixation	Heterocysts	PC	CYTc_6	CYTc_M
1	0	***Prochlorococcus marinus*** **str. MIT9301**	***Pleurocapsales***	**1.6 Mb**	**no**	**no**	**1**	**1**	**1**
1	1	***Prochlorococcus marinus*** **str. MIT9515**	***Pleurocapsales***	**1.7 Mb**	**no**	**no**	**1**	**0**	**0**
1 1 1	0 90% 1 10%	**10 x** ***Prochlorococcus marinus*** [a]	***Pleurocapsales***	**1.7 - 2.7 Mb**	**no**	**no**	**1 1 1**	**0 40% 1 40% 2 20%**	**1 1 1**
1[#]	0	***Gloeobacter violaceus*** **PCC7421**	***Chroococcales***	**4.6 Mb**	**no**	**no**	**2**	**2**	**1**
1	0	***Synechococcus elongatus*** **PCC6301**	***Chroococcales***	**2.7 Mb**	**no**	**no**	**1**	**3**	**1**
1	0	***Synechococcus elongatus*** **PCC7942**	***Chroococcales***	**2.7 Mb**	**no**	**no**	**1**	**3**	**1**
1 1 1 1	0 30% 1 70%	**13 x** ***Synechococcus sp.*** [b]	***Chroococcales***	**2.2 - 3.0 Mb**	**no**	**no**	**1 0 1 1**	**1 15% 2 45% 3 10% 4 30%**	**1 1 1 1**
1	1	***Synechocystis sp.*** **PCC6803**	***Chroococcales***	**3.6 Mb**	**no**	**no**	**1**	**1**	**1**
1	0	***Thermosynechococcus elongatus*** **BP-1**	***Chroococcales***	**2.6 Mb**	**no**	**no**	**0**	**1**	**1**
1	0	***Acaryochloris marina*** **MBIC11017**	**unclassified**	**6.5 Mb**	**no**	**no**	**1**	**2**	**1**
1	1	***Crocosphaera watsonii*** **WH8501***	***Chroococcales***	**6.2 Mb**	**yes**	**no**	**1**	**3**	**1**
1	1	***Cyanothece sp.*** **CCY0110***	***Chroococcales***	**5.9 Mb**	**yes**	**no**	**1**	**2**	**1**
1	1	***Lyngbya sp.*** **PCC8106***	***Oscillatoriales***	**7.0 Mb**	**yes**	**no**	**1**	**2**	**1**
1	1	***Trichodesmium erythraeum*** **IMS101**	***Oscillatoriales***	**7.7 Mb**	**yes**	**no**	**1**	**2**	**1**
2	2[#]	***Anabaena variabilis*** **ATCC29413**	***Nostocales***	**6.3 Mb**	**yes**	**yes**	**1**	**3**	**1**
2	1	***Nodularia spumigena*** **CCY9414***	***Nostocales***	**5.3 Mb**	**yes**	**yes**	**1**	**4**	**1**
2	1[#]	***Nostoc punctiforme*** **PCC73102***	***Nostocales***	**9.0 Mb**	**yes**	**yes**	**1**	**3**	**1**
2	1	***Nostoc (Anabaena) sp.*** **PCC7120**	***Nostocales***	**6.4 Mb**	**yes**	**yes**	**1**	**2**	**1**

a) The *Prochlorococcus marinus* genus includes following strains: Pro9601, Pro9211, Pro9215, Pro9303, Pro9312, Pro9313, ProNATL1A, ProNATL2A, Pro1375, Pro1986.

b) Following strains are included in the *Synechococcus* group: Syn107*, Syn9311, Syn9605, Syn9902, SynJA23, SynJA33, Syn307, Syn9916*, Syn9917*, Syn5701*, Syn7803, Syn7805*, Syn8102.

\# Genome analysis shows the presence of additional but incomplete operons for COX or QOX (absence of subunit III)

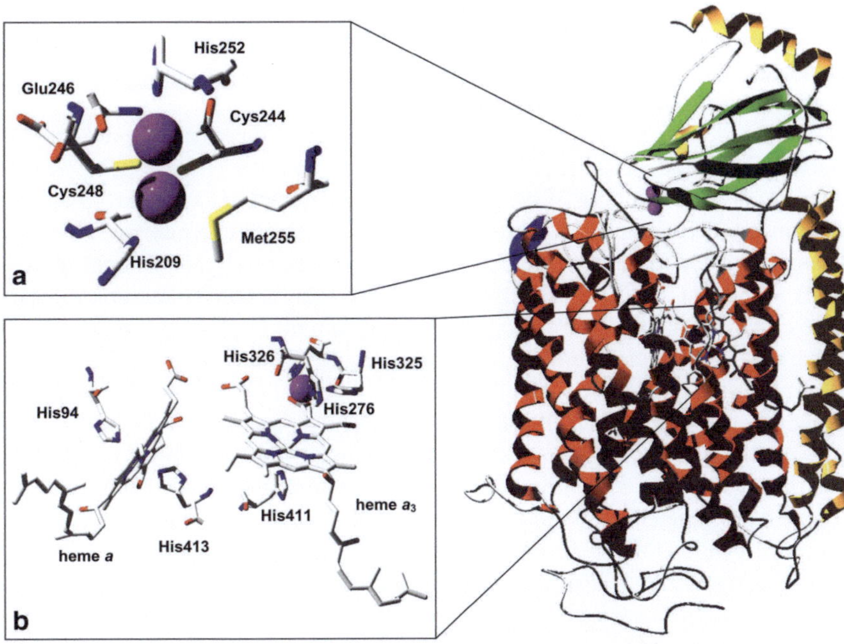

Fig. 23.1 Structure of cytochrome *c* oxidase subunit I (**a**) and subunit II (**b**) from *Paracoccus denitrificans* (PDB code: 1QLE) (Iwata et al. 1995). Figure was built by using SWISS-PDB Viewer (www.expasy.ch/spdbv). In addition the conserved complexation pattern of the metal centers is shown

According to the classification of heme-copper oxidases suggested by Pereira et al. (2001) cyanobacteria are type A_2 (YS-) enzymes having D- and K-channels for proton pumping including the highly conserved helix VI sequence -YQHX_2W-F<u>YS</u>HPAVX- (see *Paracoccus* Phe274, Ser275) (Bernroitner et al. 2008). Other residues involved in proton translocation are for the D-channel Tyr35, Asn113, Asp124, Asn131, Ser134 (Ala in most cyanobacteria), Ser193 and Asn199. The K-channel comprises Tyr280, Ser291, Thr351 and Lys354 at the second pore, which leads directly to the binuclear reaction center.

Cytochrome *c* oxidase subunit II (Fig. 23.1b) includes a typical hairpin-like structure comprising two transmembrane helices and a peripheral domain on the *P*-side of the membrane, which latter can be either CM or ICM. In case of cyanobacterial COXs, this extrinsic domain has been shown to be uniquely located in two entirely distinct environments, viz. the periplasmic space and the thylakoid lumen (Pereira et al. 2001; Peschek 1996). Besides some deviations in sequence at the N- and C-termini of cyanobacterial COXs as compared to the *P. denitrificans* or *R. sphaeroides* enzymes (Bernroitner et al. 2008), the core of the extrinsic domain shows similar structural elements, viz. a β-barrel formed by 10 β-sheets. Cyanobacterial COX subunit II sequences display an insertion located between the second transmembrane helix and the cupredoxin-like domain. This characteric insertion seems to be another typical feature of heme-copper oxidases of the A_2 subfamily (Pereira and Teixeira 2004). Furthermore, most of cyanobacterial COX subunits II have an extended C-terminus (Bernroitner et al. 2008).

Subunit II of COX participates in both electron transfer and proton pumping. All cyanobacterial SU-II have in common a glutamate residue (Glu106, *Paracoccus* numbering) at the beginning of the second transmembrane helix (Bernroitner et al. 2008). Since Glu106 is close to the conserved Lys in the K-channel (see above), it has been suggested to regulate proton entry or transit through this channel (Pereira and Teixeira 2004). In COX structures published so far, donated electrons enter the oxidase complex via a conserved tryptophan on subunit II (Trp149), which, when mutated, abolishes electron transfer (Witt et al. 1998; Zhen et al. 1999). Interestingly, the amino acid pattern in this region in cyanobacterial COX SU-II is completely different (Bernroitner et al. 2008). Finally, electrons are transferred to the Cu_A center, which is composed of two electronically coupled, mixed-valence copper ions (Michel et al. 1998). In all COXs with known structure, this center is binuclear and the ligands are located near the C-terminus. The strictly conserved residues also found in all cyanobacterial COXs are His209, Cys244, Glu246, Cys248, His252 and Met255, respectively (Fig. 23.1, Bernroitner et al. 2008). The two cysteine residues bridge the two copper atoms.

Phylogenetic analysis of COX subunits I and II revealed that cyanobacterial genes diverged early in the evolutionary history from other bacterial *cox*-genes (Bernroitner et al. 2008). Cyanobacterial COXs are rather distantly related to COX counterparts with known 3D structure (e.g. *P. denifricans* or *R. sphaeroides*) thus representing an own subfamily with above-average diversity of the *cox*-genes (in comparison to e.g. proteobacterial sequences). Gene duplication occurred frequently within this subfamily as obvious from the presence of closely related genes in branches of related species. In any case, COX subunits I expressed in heterocyst forming N_2-fixing cyanobacteria are closer related to diazotrophic *Trichodesmium* COX, whereas COXs expressed in vegetative cells are related to those of diazotrophic *Chroococcales* and *Oscillatoriales* (Bernroitner et al. 2008). It is important to note that subunits I of both COX1 and COX2 of *Nostocales* reveal a high sequence homology, whereas subunits II show some significant differences. In detail, cytochrome *c* oxidase expressed in vegetative cells (COX1, see below) have a characteristic insertion in SU-II between the second transmembrane helix and the cupredoxin domain. The C-terminal end of this insertion comprises a conserved sequence with charged residues (-NLNQEK-) neither found in other cyanobacterial COX nor in the enzymes from *P. denitrificans* or *R. sphaeroides*. This could indicate different binding patterns and/or electron donors for COX expressed in vegetative cells or N_2-fixing heterocysts (see Table 23.1).

Deletion of *coxBAC* in *Synechocystis* PCC6803 resulted in a mutant strain that was still viable under both photoautotrophic and photomixotrophic conditions. The strain respired at near wild-type rates, and this respiration was cyanide-sensitive (Schmetterer et al. 1994). Isolated membranes from the mutant were unable to oxidize reduced horse heart cytochrome *c*. This was the first evidence for a second cyanide-sensitive terminal respiratory oxidase in *Synechocystis*. Although COX appears to be the main terminal oxidase, as is reflected by its occurrence in all cyanobacteria (Table 23.1) there is also evidence of the presence and function of the other branch of the heme-copper oxidase superfamily [i.e. the quinol oxidase (QOX) or *bo*-type quinol oxidase] in many cyanobacteria. QOX is encoded by a set of genes (*qoxBAC*) very similar in sequence to the *coxBAC* genes. Other similarities between the two branches of heme-copper oxidases include the presence of one low-spin (six-coordinated heme *b* in QOX) and one high-spin (five-coordinated heme o_3 in QOX) heme, exchange coupling between high-spin heme and Cu_B in the binuclear reaction site in subunit I, heme-heme interaction as well as alignment of these hemes with respect to the membrane bilayer (Abramson et al. 2000; Musser et al. 1993). Figure 23.2 shows the two-subunit structure of bo_3 ubiquinol oxidase from *E. coli* (since no 3D-structure of a cyanobacterial QOX is available). QOX from *E. coli* is a four-subunit heme-copper oxidase that catalyzes the four-electron reduction of O_2 to water and also functions as a proton pump (Puustinen et al. 1991). QOX subunit I contains all redox cofactors and is homologous to the corresponding subunit in COX. The ligands of the two heme groups and of Cu_B have been identified as invariant His residues (Abramson et al. 2000): His106, His284, His333, His334, His419 and His421, respectively (*E. coli* numbering). These histidines are also strictly conserved in all cyanobacterial QOXs (Bernroitner et al. 2008). In contrast to cytochrome *c* oxidases, QOX subunit II has neither a binuclear Cu_A center

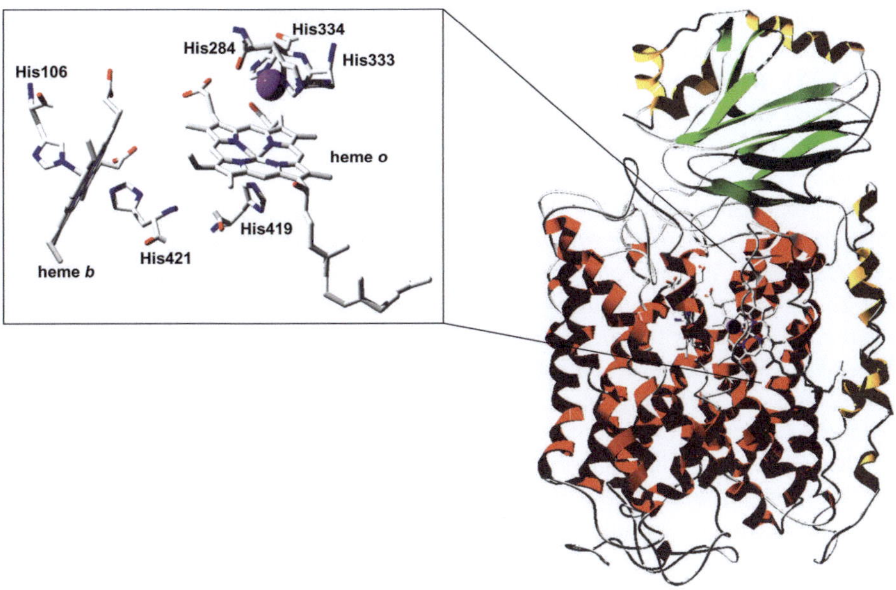

Fig. 23.2 Structure of quinol oxidase subunit I and subunit II from *E. coli* (PDB code 1FFT) (Abramson et al. 2000). Figure was built by using the SWISS-PDB Viewer (www.expasy.ch/spdbv). In addition the conserved complexation pattern of the metal centers is shown

nor a cytochrome *c* binding site (Abramson et al. 2000). In QOX the Cu_A center found in COX is completely blocked by hydrophobic residues which prevent access from the *P*-side (Abramson et al. 2000; Paumann et al. 2005). Instead, heme *b* receives electrons directly from a membrane solubilized quinol molecule (ubiquinol in case of *E. coli* QOX). The protons produced upon quinol oxidation are released on *P*-side of the membrane. In cyanobacteria plastoquinol (PQ-9) is used in both PET and RET and is functionally equivalent to ubiquinol. Cyanobacteria are devoid of any other major quinol species (Peschek 1980). It has been proposed that ubiquinol in QOX from *E. coli* binds to the membrane domain of subunit I (Abramson et al. 2000) which is defined by mostly hydrophobic α-helices (Fig. 23.2). The ubiquinol binding site lies within the lipid bilayer on the surface of helices I and II and contains a striking patch of conserved polar and charged residues. These include Arg71, Asp75, Met78, Met79, His98 and Gln101 (*E. coli* QOX numbering). These residues are highly conserved in QOXs (Abramson et al. 2000) but, interestingly, not in cyanobacterial enzymes (Bernroitner et al. 2008). In cyanobacterial QOXs only Met78 and Met79 are found. Nevertheless, also in cyanobacterial QOX the conserved hydrophobic residues Met79 and Phe103 are situated between the proposed quinol binding site and heme *b*.

The two proton transfer pathways of COX appear to be present also in subunit I of ubiquinol oxidase with a series of polar residues that are also found in the cyanobacterial enzymes. The D-channel in *E. coli* QOX begins with Asn124, Asp135, Asn142, Ser145 (Ala in most cyanobacteria), Thr201 (Ser in cyanobacteria), Asn207 and Glu286 (Ala in cyanobacteria) (Abramson et al. 2000; Bernroitner et al. 2008). Likewise, the K-channel of *E. coli* QOX contains many conserved residues like Ser299, Thr359, Lys362 and Tyr288 which are also present in cyanobacterial QOXs. In subunit II of cyanobacterial QOXs (but not in the *E. coli* enzyme) a conserved Glu (Ala90 in *E. coli* QOX) homologous to Glu106 in *P. denitrificans* SU-II of COX, which is close to the conserved Lys362 (SU-I) in the K-channel, is found.

No *qox* operon is found in the genomes of many unicellular cyanobacteria, i.e. it is absent in most of the *Prochlorococcus* species, in *Gloeobacter*, *Synechococcus elongatus*, *Thermosynechococcus elongatus* and *Acaryochloris marina* (Table 23.1). By contrast, all nitrogen fixing species have usually one *qox* operon. In *Anabaena variabilis* three operons are found but one lacks subunit III. Transcription of *qoxBAC* has been detected in *Nostoc* PCC7120 (Valladares et al. 2003), *Anabaena variabilis* (Pils et al. 2004) and *Synechocystis* PCC6803 (Howitt and Vermaas 1998; Hart et al. 2005). Two studies using deletion mutants of *Synechocystis* PCC6803 and inhibitors proposed that QOX is active and contributes to energy metabolism (Pils et al. 1997, 2001). On the other hand, in *Synechocystis* PCC6803, *qoxC* transcripts could only be detected in cells grown under low light in strains with deletions in both *coxBAC* and in those genes that encode the *bd*-quinol oxidase (Howitt and Vermaas 1998). In strains lacking PSI, deletion of *cox* but not of *qox* genes had an impact on the fluorescence decay kinetics after illumination (Berry et al. 2002). Thus, it has been suggested that COX is the major oxidase at least in

ICM (Howitt et al. 1998). The presence of a putative FNR binding site (a transcriptional regulator of the switch from aerobic to anaerobic growth in *E. coli*) upstream of the *qoxBAC* operon may indicate that QOX functions under anaerobic or microaerobic conditions. As will be discussed below, in *Nostocales*, QOX transcripts were exclusively detected in the heterocysts (Valladares et al. 2003). One major obstacle in understanding the role of QOX in cyanobacterial RET is the absence of its characterization at protein level. The operon is transcribed but the low level of mRNA (Howitt and Vermaas 1998; Pils et al. 2004) suggests that characterization at the protein level may remain problematic.

23.3 Role of Cytochrome c_6 and Plastocyanin in Cyanobacterial Respiration

It is well known that in oxygenic photosynthesis the transport of electrons from the cytochrome b_6f complex to PSI is performed by either plastocyanin (PC) or cytochrome c_6 (CYTc_6), which are both small soluble metalloproteins (Fig. 23.3) located on the *P*-side of membranes containing PET. Whereas PC is the only carrier in plants, both PC and CYTc_6 are synthesized by most cyanobacteria and green algae (Hervas et al. 2003; De la Rosa et al. 2006). Actually, despite having different structures, the two proteins can replace each other and play the same physiological role in cyanobacterial PET, but the synthesis of either one is controlled by copper availability within each organism (Wood 1978). Plastocyanin (encoded by the *petE* gene) consists of a single polypeptide chain forming a β-barrel with eight β-strands and a small α-helix, along with a type-1 blue copper center (Fig. 23.3). The metal is coordinated by two histidines (His39 and His86, *Synechocystis* numbering), one methionine (Met91) and one cysteine (Cys83) (Bertini et al. 2001; Guss and Freeman 1983; Guss et al. 1986; Bernroitner et al. 2008). His86 is the only solvent-exposed Cu-ligand, thus making this residue the most probable electron entry site of PC. On the other hand, CYTc_6 (encoded by the *petJ* gene) is a typical class I *c*-type cytochrome containing four α-helices and a covalently-linked heme group (Cys14, Cys17, *Arthrospira maxima* numbering) in which the iron atom is axially coordinated by a histidine (His18) and a methionine (Met62) (Fig. 23.3, Bernroitner et al. 2008) (Sawaya et al. 2001; Frazao et al. 1995; Kerfeld et al. 1995). The edge of pyrrole ring C and the ring D propionic groups are solvent accessible, thus establishing a tentative electron transfer pathway to and from the heme iron atom (Ubbink et al. 1998; Diaz-Moreno et al. 2005).

Despite their very different structures, both proteins share a number of physicochemical properties: their molecular masses are about 10 kDa and their midpoint redox potential values are around 350 mV at pH 7, which is in accordance with their role as electron shuttle between cytochrome b_6f and PSI (Molina-Heredia et al. 2002). Moreover, their isoelectric points are often similar within the same organism but can vary in parallel from one organism to another. Beside *Synechococcus sp*

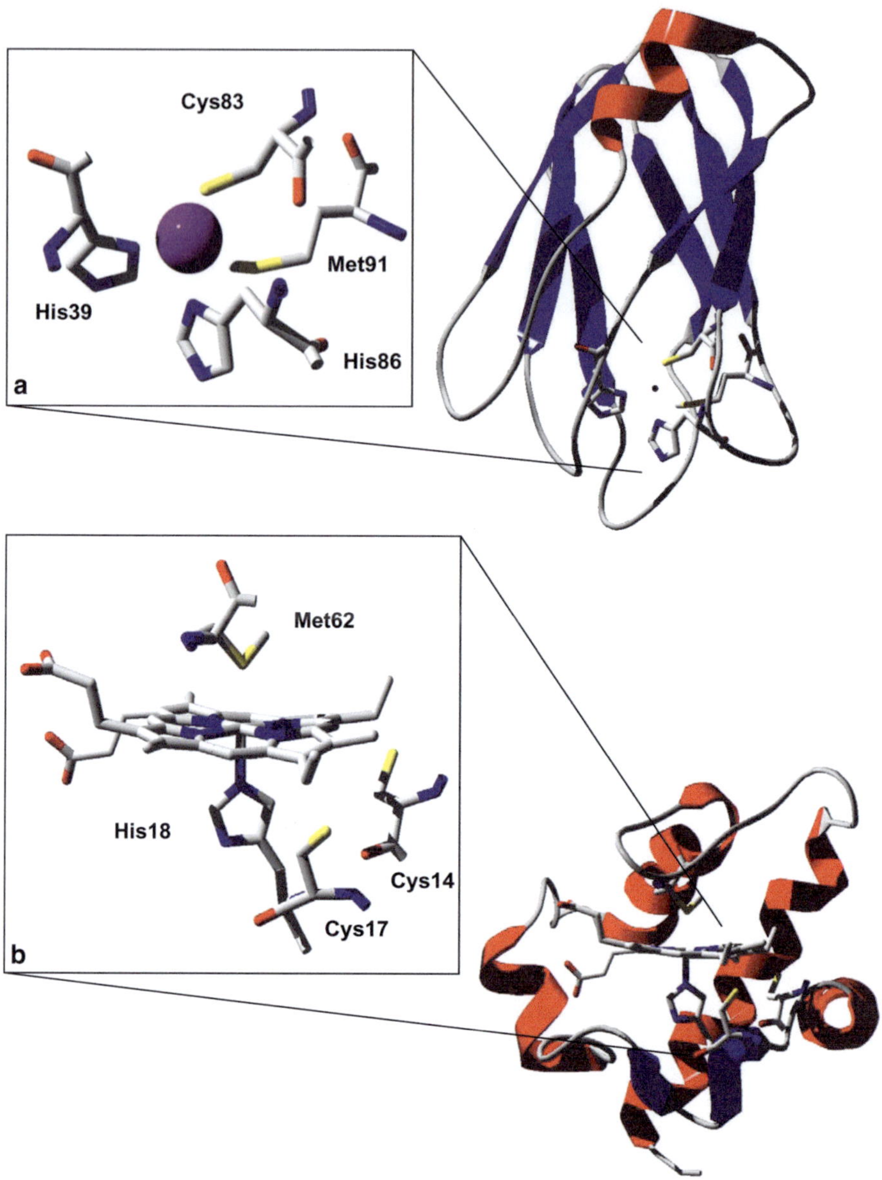

Fig. 23.3 Structure of plastocyanin from *Synechocystis* PCC6803 (PDB code 1J5D) (Bertini et al. 2001) and structure of cytochrome c_6 from *Arthrospira maxima* (PDB code 1F1F) (Sawaya et al. 2001). Figure was built by using the SWISS-PDB viewer (www.expasy.ch/spdbv). In addition the conserved complexation pattern of the metal centers and the cysteines, which participate in the covalent heme to protein linkage, are depicted

RCC307 and *Thermosynechococcus el.* BP-1, all cyanobacteria contain one *petE* gene (exception: *Gloeobacter violaceus* with two paralogs) (Table 23.1). Roughly, in filamentous cyanobacteria PC is usually positively charged at pH 7, whereas PC from unicellular species is mostly acidic (exception: *Gloeobacter*) (Table 23.2). Especially in cyanobacteria that have several (2–4) *petJ* genes, the isoelectric points of PC and CYTc_6 can also be dissimilar (compare Tables 23.1 and 23.2). Nevertheless, regarding PET the kinetic information, along with structural data, indicate that PC and CYTc_6 possess similar surface regions to interact with their redox partners in PET, but also that they are responsible for determining the reaction mechanism and the spatial conformation of the transient reaction complex (for a review of the role of PC and CYTc_6 in PET see (De la Rosa et al. 2006 and this book).

The role of both mobile electron carriers in RET is still under discussion. A recent investigation of two deletion mutants of *Synechocystis* PCC6803 (each mutant lacking either *petE* or *petJ*) and their photoautotrophic and heterotrophic growth rate in copper-free and copper-supplemented medium clearly demonstrated that efficient function of respiration requires the presence of either CYTc_6 or PC (Duran et al. 2004). These data confirm an early suggestion of Lockau (1981) that cyanobacterial RET can use both PC and Cytc_6 as electron donor. This is also underlined by investigations of the electron transfer kinetics between PC or Cytc_6 and the recombinant Cu_A domain of SU-II of COX from *Synechocystis* PCC6803 (Paumann et al. 2004a, b). Recently, it could be demonstrated that both mobile electron carriers can efficiently be oxidized by COX in *Nostoc* PCC7119 (Navarro et al. 2005).

Both CYTc_6 and PC are located inside the thylakoid lumen, and also a periplasmic CYTc_6 was identified as an electron donor to CM-bound COX (Obinger et al. 1990; Serrano et al. 1990). The occurrence of PC in the periplasmic space has not yet been proven. Regarding interchangeable CYTc_6 and PC in cyanobacterial RET, different redox carriers with acidic IPs were investigated in their interaction with respiring membranes from unicellular species showing that reaction at low ionic strength was slow but increased at higher ionic strength (Kienzl and Peschek 1982; Moser et al. 1991). By contrast, in the same organisms PC reacted effectively with PSI at low ionic strength (De la Rosa et al. 2006). It is known that in the thylakoid lumen ionic strength may rise as ions such as H^+, Mg^{2+} and Cl^- accumulate during PET activity. And it is also known that COX activity is induced in CM by growth under high ionic strength (see below). So it was hypothesized that the ionic strength can act as a "switch" for the electron transfer pathways in both CM and ICM (Nicholls et al. 1992). However, these data are contrasted with the oxidation kinetics of CYTc_6 and PC by *Nostoc* PCC7119 COX (Navarro et al. 2005). In this filamentous species the oxidation was significantly impaired by salt, a finding that reveals the existence of strong electrostatic attractive interactions between the reactants, as was reported for other COX/cytochrome *c* systems (Lappalainen et al. 1993; Wang et al. 1999; Maneg et al. 2004). These discrepancies could reflect mechanistic differences between unicellular and filamentous cyanobacteria. It is interesting to note that strong ionic effects have been also reported for the PC and CYTc_6 interaction with PSI in *Nostoc* as compared with *Synechocystis* (Diaz-Quintana et al. 2003).

Table 23.2 Isoelectric points of electron donors. Isoelectric points of plastocyanin, cytochrome c_6 and cytochrome c_M calculated from proteins without signal peptides [ProtParam (http://www.expasy.org/tools/protparam.html)]. SignalP (http://www.cbs.dtu.dk/services/SignalP/) was used to predict the signal peptide cleavage sites

Cyanobacteria	Calculated isoelectric points		
	PC	CYTc_6	CYTc_M
Acaryochloris marina MBIC11017	4.3	4.2 / 8.0	8.8
Gloeobacter violaceus PCC7421	9.1 / 9.5	5.6 / 9.6	5.8
Prochlorococcus marinus str. AS9601	4.7	—	5.2
Prochlorococcus marinus str. MIT9211	4.8	—	4.7
Prochlorococcus marinus str. MIT9215	4.7	—	5.5
Prochlorococcus marinus str. MIT9301	4.7	9.0	5.2
Prochlorococcus marinus str. MIT9303	4.7	4.5 / 6.9	5.2
Prochlorococcus marinus str. MIT9312	4.8	9.2	5.9
Prochlorococcus marinus str. MIT9313	4.5	4.5 / 6.9	4.8
Prochlorococcus marinus str. MIT9515	4.5	—	—
Prochlorococcus marinus str. NATL1A	4.7	5.6	5.1
Prochlorococcus marinus str. NATL2A	4.7	6.1	5.1
Prochlorococcus marinus subsp. marinus str. CCMP1375	4.8	8.0	4.4
Prochlorococcus marinus subsp. marinus str. CCMP1986	4.7	—	4.8
Synechococcus elongatus PCC6301	4.8	5.0 / 8.7 / 9.2	8.0
Synechococcus elongatus PCC7942	4.8	5.0 / 8.7 / 9.2	8.0
Synechococcus sp. BL107	4.5	4.4 / 6.0	5.0
Synechococcus sp. CC9311	4.7	5.0 / 4.6 / 5.7 / 5.0	7.1
Synechococcus sp. CC9605	4.8	6.0 / 4.7	5.0
Synechococcus sp. CC9902	4.6	6.1 / 4.5	4.9
Synechococcus sp. JA-2-3B'a(2-13)	7.1	7.2	6.7
Synechococcus sp. JA-3-3Ab	8.1	7.2	6.2
Synechococcus sp. RCC307	—	4.6 / 4.9	6.0
Synechococcus sp. RS9916	4.7	4.6 / 6.0 / 4.6 / 4.3	5.0
Synechococcus sp. RS9917	4.7	8.1 / 5.2 / 4.6 / 4.8	5.7
Synechococcus sp. WH5701	4.8	8.1 / 4.8	9.0
Synechococcus sp. WH7803	4.7	4.4 / 5.5 / 4.7	5.2
Synechococcus sp. WH7805	4.7	4.9 / 4.5 / 4.7 / 5.5	6.9
Synechococcus sp WH8102	4.7	4.6 / 8.0	5.9
Synechocystis sp. PCC6803	5.5	5.5	7.0
Thermosynechococcus elongatus BP-1	—	5.5	5.4
Crocosphaera watsonii WH8501	5.5	6.8 / 9.4 / 9.7	6.9
Cyanothece sp. CCY0110	5.5	5.7 / 9.2	8.7
Lyngbya sp. PCC8106	5.4	9.5 / 9.0	6.8
Trichodesmium erythraeum IMS101	8.8	9.6 / 9.4	5.1
Anabaena variabilis ATCC29413	7.1	9.0 / 9.0 / 5.4	7.0
Nodularia spumigena CCY9414	8.7	8.7 / 9.2	6.2
Nostoc punctiforme PCC73102	8.7	5.6 / 9.4 / 8.7	6.0
Nostoc (Anabaena) sp. PCC7120	7.1	8.1 / 8.9	7.0

23.4 Respiration and Stress

With respect to stress and adaptation, it is the cytoplasmatic membrane (CM) and not the intracellular thylakoid membrane (ICM) that is closest to the environment and, thus to practically any environmental stress. It will therefore also be primarily affected. Respiratory membrane energization and ATP supply are crucial to stressed cyanobacteria, which usually shut down PET first (Peschek 2001). It has been demonstrated that ATPases are located in CM (Peschek et al. 1994) and that energization of CM is a KCN-sensitive process (Pils and Schmetterer 2001).

In *A. nidulans* (*Synechococcus* PCC6301) exposed to a variety of seemingly unrelated stress conditions (e.g. high-salt or light limitation) RET in CM was shown to be significantly enhanced while ICM-bound respiratory chain was much less affected ('salt respiration') (Peschek 2001). Additionally, the amount of COX in CM (but not in ICM) was increased (Fry et al. 1986). The increased respiration was proposed to generate a proton gradient required to drive H^+/Na^+ exchange. H^+/Na^+ antiporters have been identified in *Synechocystis* PCC6803 (Inaba et al. 2001). COX was shown to be necessary and sufficient for 'salt respiration' (Pils and Schmetterer 2001).

Recently, a role of both COX and QOX in photoinhibition and high light tolerance in *Synechococcus* PCC7002 was demonstrated (Nomura et al. 2006). Both *coxA* and *qoxA* genes were reported to be dispensable for growth under normal to moderately high and stressful light conditions. In *Synechococcus* PCC7002, COX seems to be the primary oxidase under stress conditions, while QOX contributes very little to RET (Nomura et al. 2006). A *coxA* mutant in *Synechococcus* PCC7002 exhibited drastically reduced rates of oxygen consumption under stress. Usually, photoinhibition occurs under conditions of excessive light intensities or electron flow limitations due to environmental stresses. In *Synechococcus* PCC7002 the absence of *coxA* caused changes in the electron flow around the photosystems that were similar to those observed in cells grown under photoinhibitory conditions (Nomura et al. 2006). Additionally, there was also a decrease in the total amount of chlorophyll in the mutants underlining that the lack of a functioning COX causes photoinhibition. The inability of the cells to deal with an increased amount of reductive stress becomes more evident when the cells were exposed to increasing light conditions. As the light intensity increased, the cells lowered the number of active PSII complexes (Nomura et al. 2006). These findings suggest a role of COX in ICM to account for possible excessive reductive stress since both COX and photosystems are placed within the same membrane.

There have been suggestions that a small *c*-type cytochrome, CYTc_M (cytochrome c_M; Fig. 23.4) is capable of interacting with COX and/or PSI as an alternative to PC and CYTc_6 under stress conditions such as high light and low temperatures (Zhang et al. 1994; Malakhov et al. 1994; Manna and Vermaas 1997; Malakhov et al. 1999). Its association with CM and ICM has been demonstrated recently (Bernroitner et al. 2009). Table 23.1 shows that (with the exception of one *Prochlorococcus* strain) all cyanobacteria contain a gene encoding cytochrome c_M with IPs

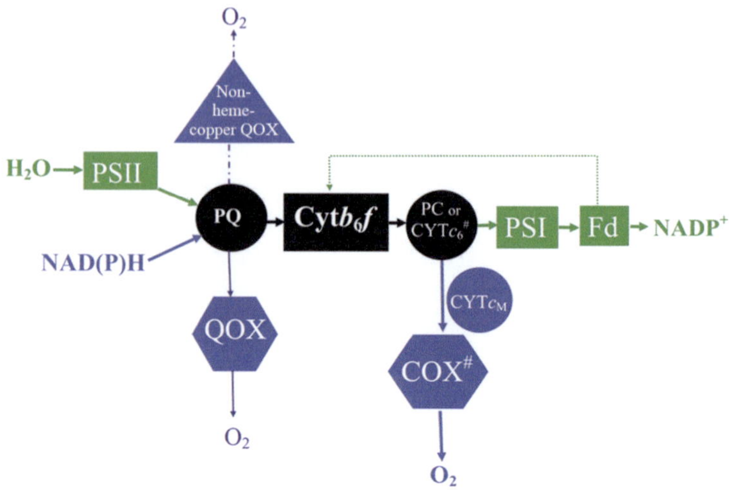

Fig. 23.4 Scheme of the branched dual-function photosynthetic (PET) -respiratory (RET) electron transport chain(s) in cyanobacteria (Bernroitner et al. 2008). Note that PET is only localized in intracytoplasmic membranes or thylakoids (ICM), whereas RET is active in both ICM and cytoplasmic membrane (CM). Electron donor and acceptor as well as protein complexes involved exclusively in PET are depicted in green, those found only in RET in blue color. Electron components used in both PET and RET are shown in black. In contrast to cytochrome *c* (aa_3-type) oxidase (COX, hexagon), which has been detected on a protein level in both CM and ICM (Wolk et al. 1994), there is no unequivocal information about the localization of heme-copper (QOX, hexagon) and non-heme-copper quinol oxidases (triangle). Depending on copper-availability plastocyanin (PC) and/or cytochrome c_6 (CYTc_6) functions as electron donor for both, photosystem I (PSI) or COX. Another mobile electron carrier exclusively found in cyanobacteria (cytochrome c_M, CYTc_M) most probably is involved in electron shuttling between cytochrome b_6f (Cytb_6f) and COX during environmental stress. COX is the main terminal oxidase in RET, but—depending on growth conditions and nitrogen supply—can partially be substituted by QOX or non-heme-copper oxidases that both use plastiquinol-9 (PQ) as electron donor. In contrast to PC and CYTc_M, up to four paralogs for CYTc_6 are found in cyanobacterial genomes (#). In *Nostocales* two paralogs for COX are found (#). Cyclic electron flow around PSI is shown in green dashed line

varying from 4.4–9.0 (Table 23.2). In contrast to CYTc_6 the IPs of cyanobacterial CYTc_M seem to be randomised between unicellular and filamentous species. *Synechocystis* cytochrome c_M has a molar mass of about 8 kDa and a midpoint redox potential value of 150 mV, which is significantly lower than that reported for PC and CYTc_6 (Molina-Heredia et al. 2002). Similar to CYTc_6 the heme is covalently linked and the heme axial ligands are histidine and methionine, but most likely due to the specific environment of the latter (-TPPMP-, see Bernroitner et al. 2008, 2009) the actual heme architecture and thus the electron density at the heme iron is different. So far no 3D-structure of a CYTc_M is available. From a thermodynamic point of view, it is difficult to see how CYTc_M with a redox potential of 150 mV could be reduced by cytochrome *f* in PET that has a redox potential of +320 mV (Metzger et al. 1995). Indeed, in *Synechocystis* PCC6803 CYTc_M reacts slowly with PSI compared to PC and CYTc_6 as was determined by laser-induced kinetic analysis of PSI reduction (Molina-Heredia et al. 2002). In an *in vitro* system its reactivity with the soluble Cu_A domain of COX from *Synechocystis* has been demonstrated to be between CYTc_6 and PC (Bernroitner et al. 2009).

A role of CYTc_M in RET is under discussion. A deletion mutant of *Synechocystis* lacking the *petJ* gene was able to grow at near wild-type rates in copper-free media, that is, when the synthesis of PC is repressed (Zhang et al. 1994). This is not amazing since, as has been outlined above, deletion of *coxBAC* in *Synechocystis* had also no effect on growth and RET rates (Howitt and Vermaas 1998), because alternative (quinol-dependent) oxidases are present. To identify the electron carriers in a PSI-less *Synechocystis* PCC6803 mutant from PSII to COX, genes encoding PC, CYTc_6 and CYTc_M were deleted and the electron flow out of PSII was monitored through chlorophyll fluorescence yields (Manna and Vermaas 1997). Loss of CYTc_6 or PC, but not of CYTc_M, decreased the rate of electron flow out of PSII. Surprisingly, CYTc_M could not be deleted in a PSI-less background strain, and also a double mutant lacking both PC and CYTc_6 could not be obtained. Due to some homology of CYTc_M with cytochrome *c*-binding regions of the *caa*$_3$-type COX from *Bacillus spp.* and *Thermus thermophilus*, it has been suggested that CYTc_M is a component of COX in cyanobacteria that serves as a redox intermediate between the soluble electron carriers and COX, and that either PC or CYTc_6 can shuttle electrons from the b_6f complex to cytochrome c_M (Manna and Vermaas 1997).

On the other hand, Northern blotting analysis revealed that CYTc_M is scarcely expressed under normal conditions in *Synechocystis* PCC6803 but its expression is enhanced when cells were exposed to low temperature or high-intensity light. By contrast, the expression of *petJ* and *petE* genes was suppressed under these stress conditions (Malakhov et al. 1999). These data suggest that PC and/or CYTc_6, which are the dominant electron donors for COX under non-stressed conditions, could be replaced by CYTc_M under stress conditions.

23.5 Respiration in Nitrogen Fixing Cyanobacteria

Cyanobacteria are the only diazotrophs (nitrogen-fixing organisms) that produce oxygen as a by-product of PET, and which must negotiate the inevitable presence of molecular oxygen with an essentially anaerobic enzyme. The biological reduction of N_2 is absolutely restricted to prokaryotes and is catalyzed by a multimeric metalloenzyme (nitrogenase), which is irreversibly inhibited by molecular oxygen and reactive oxygen species (Fay 1992; Postgate 1998). Moreover, O_2 not only affects the protein structure but also inhibits the synthesis of nitrogenase in many diazotrophs (Gallon 1992). Phylogenetic analyses of the *nifDK* (encoding for α- and β-subunits of nitrogenase) and the *nifEN* genes (encoding for the two subunits of the NifEN protein complex involved in biosynthesis of the Fe-Mo cofactor) suggest that all extant nitrogenases sequenced so far are derived from a single ancestor and that the catalytic subunits of the enzyme complex existed prior to oxygenation of Earth's atmosphere (Broda and Peschek 1983; Burgess and Lowe 1996). The change from an anoxic to an oxygenic atmosphere resulting from cyanobacterial photosynthesis, posed major challenges to diazotrophic organisms (including cyanobacteria themselves). During the course of planetary evolution, cyanobacteria have co-evolved with the changing oxidation state of the ocean and atmosphere to accommodate the machinery of PET and RET and oxygen-sensitive N_2-fixation within the same cell and/or colony of cells. These strategies have been generally described as temporal and/or spatial separation of photosynthesis and N_2-fixation.

The severe energy (ATP) requirement of biological N_2-fixation can be regarded also as a stress situation for diazotrophic organisms. Correspondingly, it was found that generation times of photoautotrophically growing *Anabaena variabilis* cells were 3.8 h on NH_4Cl, 4.9 h on $NaNO_3$ but 5.2 h on N_2 (Van Baalen 1987). Quite in accordance with what has been found about the role of respiration as a bioenergetic stress defense mechanism in cyanobacteria (see above and Winkenbach and Wolk 1972) rates of RET go steeply up when cyanobacteria grew on atmospheric N_2 as the sole source (Wastyn et al. 1988; Peschek et al. 1991; Bergman et al. 1993). Moreover, RET efficiently lowers the concentration of dissolved O_2 and may help to maintain nitrogenase in a virtually oxygen-free cellular environment (respiratory protection) (Peschek et al. 1991).

The simplest adaptation is seen for the genus *Trichodesmium* (order *Oscillatoriales)*, the major bloom-forming diazotrophic organism in modern oceans. *Trichodesmium* is a non-heterocystous filamentous cyanobacterium that fixes nitrogen during the day (Capone et al. 1997). Phylogenetic trees of diazotrophic cyanobacteria, based on *nifH* gene sequences, suggest that *Trichodesmium* branched out very early in evolution (Zehr et al. 1998). Protection against oxygen in *Trichodesmium* was found to be a complex interaction between spatial and temporal segregation of PET, RET and N_2-fixation within the photoperiod. In *Trichodesmium*, nitrogenase is compartmentalized in a fraction of the cells (10–20%) that are often arranged consecutively along the trichome. However, active photosynthetic components (PSI, PSII, Rubisco, carboxysomes) are found in all cells, even those harboring nitrogenase (Berman-Frank et al. 2001). Light initiates photosynthesis, providing energy and reductants for carbohydrate synthesis and storage, stimulating cyclic and pseudocyclic (Mehler) electron flow through PSI, and poising the PQ pool at reduced levels. Interestingly, respiration has been shown to be active also early in the photoperiod at high rates (Carpenter and Roenneberg 1995) supplying carbon skeletons for amino acid synthesis (the primary sink for fixed nitrogen). RET in *Trichodesmium erythraeum* could use two heme-copper oxidases. The present genome analysis suggests the presence of two ORFs, one encoding a cytochrome *c* oxidase (COX) and one a quinol oxidase (QOX). Putative electron donors for cytochrome *c* oxidase are PC (1 ORF; cIP=8.8) and CYTc_M (1 ORF; cIP=5.1) and CYTc_6 (2 ORFs; cIPs=9.4 and 9.6) (Tables 23.1 and 23.2). In any case, RET reduces the PQ pool further, sending negative feedback to linear PET. As a consequence, PSII becomes down-regulated and oxygen consumption exceeds oxygen production. This opens a window of opportunity for N_2-fixation during the photoperiod. As the carbohydrate pool is consumed, PET through the PQ pool diminishes. The PQ pool becomes increasingly oxidized, net O_2 production exceeds consumption and nitrogen activity is lost during the following day (Berman-Frank et al. 2001).

The combination of spatial and temporal segregation of N_2-fixation and oxygenic photosynthesis during the photoperiod represents the simplest oxygen-protective mechanism in diazotrophic cyanobacteria and could reflect their evolutionary history of N_2-fixation. Filamentous *Plectonema boryanum* (order *Oscillatoriales*) is an interesting organism, since it still remained specialized to fix N_2 only in low oxygen environments. *Plectonema* contains nitrogenase in all cells in equal amounts (Rai et al. 1992) and it will only fix nitrogen under microaerobic conditions (under 1.5% oxygen in darkness and 0.5% oxygen in light) (Tuli et al. 1996). *Plectonema* is unable to fix nitrogen when grown anaerobically, reflecting the need of oxygen for RET which is an important source of ATP and reductant. Under microaerobic environment, *Plectonema* temporally separates PET and N_2-fixation. Other *Oscillatoriales* that fix nitrogen only under microaerobic conditions are *Phormidium* and *Pseudoanabaena* (*Limnothrix*). Under contemporary oxygen levels, all these organisms are relegated to narrow environmental niches.

With the evolution of cyanobacteria and the subsequent generation of molecular oxygen, more sophisticated oxygen protective mechanisms became essential. A full temporal separation, in which nitrogen is fixed only at night, then developed in other filamentous genera of the order *Oscillatoriales* (e.g. *Lyngbya* or *Symploca*) as well as in unicellular members of the order of *Chroococcales* (e.g. *Gloeothece*, *Cyanothece* and *Crocosphaera*). Finally, in yet other filamentous organisms (order *Nostocales*), complete segregation of N_2-fixation and photosynthesis was achieved with the cellular differentiation and evolution of heterocystous cyanobacteria. Concomitantly with these developments, the number of genes encoding heme-copper oxidases and electron donors increased most probably by gene duplication (Table 23.1).

A temporal separation without the need for a microaerobic environment is observed in the filamentous cyanobacteria *Symploca* (Kumazawa et al. 2001), *Lyngbya majuscula* (Omoregie et al. 2004) and the unicellular *Gloeothece*, *Cyanothece* and *Crocosphaera* (Colon-Lopez et al. 1997; Gallon 2001). These organisms fix nitrogen at night and nitrogenase is typically found in all cells. *Cyanothece* shows extensive metabolic periodicities of photosynthesis, respiration and nitrogen fixation when grown under N_2-fixing conditions. Nitrogen fixation and respiration peaked at 24 h intervals early in the dark or subjective-dark period, whereas photosynthesis was approximately 12 h out of phase and peaked toward the end of the light or subjective-light phase. Gene regulation studies demonstrated that nitrogenase is carefully controlled at the transcriptional and posttranslational levels. Indeed, *Cyanothece* has developed an expensive mode of regulation, such that nitrogenase was synthesized and degraded each day (Colon-Lopez et al. 1997). These patterns were seen when cells were grown under either light-dark or continuous-light conditions. Nitrogenase mRNA was synthesized from the *nifHDK* operon during the first 4 h of the dark period under light-dark conditions or during the first 6 h of the subjective-dark period when grown in continuous light. It has been hypothesized that cellular O_2 levels were kept low by decreasing photosynthesis and by increasing respiration in the early dark or subjective-dark periods to permit nitrogenase activity. The subsequent increase in O_2 levels resulted in nitrogenase damage and eventual degradation (Colon-Lopez et al. 1997). The genomes of diazotrophic cyanobacteria that fix nitrogen at night contain one ORF for COX and one for QOX (Table 23.1). All so far sequenced cyanobacteria that fix nitrogen at night contain one ORF for PC (cIPs=5.4–5.5) and one for CYTc_M (IPs=6.8–8.7), respectively, and at least two ORFs for CYTc_6. In *Crocosphaera* three ORFs for CYTc_6 are found (Tables 23.1 and 23.2). At the moment it is completely unclear why these organisms have more than one CYTc_6 and how this could be related to temporal segregation of nitrogen fixation. It was interesting to see that the calculated isolelectric points of these proteins are different within the same organism: *Crocosphaera watsonii* (IPs: 6.8, 9.4, 9.7), *Cyanothece sp.* CCY0110 (IPs: 5.7, 9.2) and *Lyngbya sp.* PCC8106 (IPs: 9.5, 9.0), respectively.

A highly refined specialization is found in heterocystous cyanobacteria and this is also reflected by the increased number of genes encoding terminal heme-copper oxidases and their putative electron donors. In these organisms, nitrogenase is confined to a micro-anaerobic cell, the heterocysts, which differentiates completely and irreversibly 12–20 h after combined nitrogen sources are removed from the medium. Heterocysts are enveloped by two superimposed layers, one consisting of polysaccharides and the other of glycolipids, the latter slowing the diffusion of O_2. Heterocysts have high PSI activity but lack PSII, increased RET, and have lost the capability of cell division. Heterocysts rely on vegetative cells as source of carbon and reductant and, in return, they supply the surrounding vegetative cells with fixed nitrogen. Cyclic electron flow around PSI and RET supplies ATP. Heterocyst formation represents a striking example of prokaryotic cellular differentiation and the regulatory network has been described recently (Zhang et al. 2006; Golden and Yoon 2003). Promotors appear to be confined either to vegetative cells or to heterocysts, with genes active in both cell types possessing multiple promotors (Ramasubramanian et al. 1996). Respiration involving COX has been identified in heterocysts where it seems to provide ATP for nitrogen fixation and creates a microoxic environment (Häfele et al. 1988). Interestingly, heterocysts prefer ICM as the site of RET, while unicellular N_2-fixing species prefer CM (Wastyn et al. 1988; Peschek et al. 1991).

Nostocales possess 2 ORFs for COX (Table 23.1). It has been demonstrated that in both *Anabaena variabilis* (ATCC29413) and *Nostoc* PCC7120 expression of COX1 (*coxBAC1*) occurs in vegetative cells, whereas expression of COX2 (*coxBAC2*) is dependent on the transcription factors NtcA and HetR and limited to heterocysts (Jones and Haselkorn 2002; Valladares et al. 2003; Lopez-Gomollon et al. 2007; Jones et al. 2003). NtcA is required for the full activation of genes that are involved in nitrogen metabolism and for the early stages of heterocyst development, whereas HetR is one of the key regulators of heterocyst development. Moreover, the availability of iron has been shown to play an important role in nitrogen signaling (Lopez-Gomollon et al. 2007). In *Nostoc* PCC7120 iron uptake is controlled by FurA and, interestingly, expression of FurA is modulated by NtcA suggesting a tight link between the reserves of iron and nitrogen metabolism. Upon using a bioinformatic approach, Lopez-Gomollon and colleagues found a significant number of NtcA-regulated genes exhibiting iron boxes in their upstream regions encoding for proteins involved in photosynthesis and respiration (e.g. *coxB2*), heterocyst differentiation, transcriptional regulation or redox balance.

Interestingly, in *Nostoc* PCC7120 a *coxB2* disruption mutant was still able to grow diazotrophically. *Nostoc* PCC7120 has been shown to express *qoxBAC* (encoding a QOX) in dependence on *ntcA* and *hetR* (Valladares et al. 2003). When genes of both *coxBAC2* and *qoxBAC* operons were inactivated, nitrogenase activity was significantly reduced and cultures failed to grow diazotrophically under aerobic conditions. This suggests that both COX and QOX contribute to respiration in the heterocysts, thus generating an environment adequate for *nif* gene expression and/or nitrogenase operation. The observation that also under light conditions one of either COX2 or QOX is required to attain high nitrogenase activities (Valladares et al. 2003). indicates a role of both types of terminal heme-copper oxidases in O_2 protection of nitrogenase.

Nostoc PCC7120 is an obligatory photoautotrophic cyanobacterium, whereas *Anabaena variabilis* is able to grow also chemoheterotrophically in darkness, with combined nitrogen or with dinitrogen as nitrogen source. It has been demonstrated that COX1 is absolutely essential for fructose-dependent growth in the dark (Schmetterer et al. 2001) very similar to *Synechocystis* PCC6803, where removal of *coxBAC* also led to loss of its chemoheterotrophic growth (Pils et al. 1997). Transcription of *coxBAC1* in *Anabaena variabilis* has two starting points, one was constitutive and one was induced by fructose (Schmetterer et al. 2001). Removal of *coxBAC1* did not lead to loss of RET activity in cells grown photoautotrophically in nitrate suggesting the expression of other terminal oxidase(s). Diazotrophic growth under autotrophic conditions was also unaffected by deletion of *coxBAC1* (Schmetterer et al. 2001) since in heterocysts of *Anabaena variabilis* COX2 and at least one QOX is expressed similar to *Nostoc* PCC7120 (Pils et al. 2004). Moreover, deletion of *coxBAC2* as well as double mutants lacking both *coxBAC1* and *coxBAC2* could grow diazotrophically, suggesting a role of QOX in RET and oxygen protection in heterocysts. Nevertheless, COX2 activity was dominating in heterocysts and it was the only heme-copper oxidase that was significantly upregulated under nitrogen-fixing conditions (Pils et al. 2004). A recent work has confirmed that deletion of *qoxBAC* impaired nitrogenase activity in the heterocysts of *Nostoc* PCC7120 whereas in a double mutant (deletion of both *coxBAC2* and *qox BAC*) nitrogenase activity was undetectable (Valladares et al. 2007). Moreover, it has been proposed that both COX2 and QOX, besides protecting the N_2 fixation machinery against oxygen, are required for the normal development and performance of heterocysts (Valladares et al. 2007).

It has to be mentioned that under anaerobic conditions some heterocystous cyanobacteria, such as *Anabaena variabilis*, can synthesize alternative nitrogenase(s) (e.g. Nif2) in vegetative cells (Thiel and Pratte 2001). Nif2 is expressed shortly after nitrogen depletion but prior to heterocyst formation, and can support the fixed nitrogen needs of the filaments independently of Nif1.

As Table 23.1 shows in *Nostocales* the number of ORFs encoding QOXs varies from 1 to 3. As an example, *Anabaena variabilis* has three operons that encode putative heme-copper quinol oxidases (Table 23.1), but one operon is incomplete encoding only SU-I and -II. Further investigations have to be performed in order to establish the role of each of these oxidases in RET and especially in heterocyst metabolism. In any case, COXs expressed in heterocysts are closer related to *Trichodesmium* COX, whereas COXs expressed in vegetative cells are related to those of diazotrophic *Chroococcales* and *Oscillatoriales* (Bernroitner et al. 2008). All *Nostocales* possess one ORF for PC (cIPs 7.1–8.7) and one for CYTc_M (cIPs 6.0–7.0). Most interestingly all heterocyst forming cyanobacteria can have 2–4 ORFs encoding CYTc_6 and these cytochromes c_6 can differ in cIPs: *Anabaena variabilis*, cIPs = 5.4, 9.0, 9.0; *Nostoc* PCC7120, cIPs = 8.1, 8.9; *Nostoc punctiforme*, cIPs = 5.6, 8.7, 9.4; *Nodularia spumigena*, cIPs = 8.7 (2×), 9.2 (2×). The recently discovered cross-talk between iron and nitrogen regulatory networks in *Nostoc* PCC7120 and that expression of COX2 is dual regulated by NtcA and FurA (Lopez-Gomollon et al. 2007) might reflect also the need for cytochrome(s) c_6 as electron donor(s) for COX2. The cellular localization of at least one CYTc_6 in the heterocysts of *Anabaena variabilis* has been demonstrated (Serrano et al. 1990). Clearly more data are needed about the regulation of expression of cytochrome(s) c_6 (and plastocyanin) in vegetative cells and heterocysts and their interaction with COX1 and COX2.

23.6 Conclusion

The scheme in Fig. 23.4 presents a generalized summary about the branched dual-function PET-RET pathways downstream of plastquinol (PQ), with PET being localized exclusively on ICM. Membrane protein complexes and mobile electron carriers involved in both pathways are PQ, cytochrome b_6f, plastocyanin and cytochrome c_6. All cyanobacteria possess a *coxBAC* operon encoding subunits I-III of cytochrome *c* (aa_3-type) oxidase that catalyzes the four-electron reduction of O_2 to water and acts as a proton pump. The occurrence of COX in both CM and ICM has been demonstrated. In diazotrophic *Nostocales* two *coxBAC* paralogs are found with COX2 being expressed only in heterocysts. Depending on copper availability electron donors for COXs have been shown to be PC and/or CYTc_6. For the latter several paralogs (1–4) can be found. However, besides eventually having different activities towards COX1 and COX2, the physiological role of several CYTc_6 isoforms within one organism (often of different isoelectric points) remains to be investigated. In any case, the interplay between iron metabolism in nitrogen signaling and the fact that all diazotrophic cyanobacteria have at least two paralogs for CYTc_6 suggests that this mobile heme protein and not plastocyanin could act as electron donor for COX during nitrogen fixation. Another electron carrier exclusively found in all cyanobacteria is cytochrome c_M. Since it is a bad electron donor to PSI and upregulated under environmental stress (where PC and CYTc_6 are down-regulated) it has been proposed to act as alternative electron donor for COX. This fits with the observation that under stress conditions PET is down- and RET is upregulated. Genomes of all diazotrophic and many other (unicellular) cyanobacteria contain the *qoxBAC* operon that encodes subunits I-III of a heme-copper quinol oxidase (QOX). The present sequence analysis suggests that QOX, which uses plastoquinol-9 as electron donor, can also function as both four-electron reductant of O_2 and proton pump. Due to the absence of its characterization at protein level, its localization (CM and/or ICM) and physicochemical properties (e.g. K_M for O_2 compared to COX) are unknown. Nevertheless, transcription of QOX and its involvement in cyanobacterial metabolism, e.g. in oxygen protection in heterocysts and during reductive stress, have been demonstrated. Additionally, Fig. 23.4 reflects the fact that many cyanobacterial genomes have also the genes or ORFs for another terminal oxidase, namely the *bd*-quinol oxidase, which is not a member of the heme-copper protein superfamily since it does not contain a binuclear reaction site and is incapable of proton pumping. The role of this enzyme in cyanobacterial electron transport chains is even more speculative.

Acknowledgements This work was supported by the Austrian Science Foundation FWF (project number P17928).

References

Abramson J, Riistama S, Larsson G, Laakkonen L, Puustinen A, Iwata S and Wikström M (2000) The structure of the ubiquinol oxidase and its ubiquinone binding site. Nature 7: 910–915

Alge D and Peschek GA (1993) Identification and characterization of the ctaC (coxB) gene as part of an operon encoding subunits I, II, and III of the cytochrome *c* oxidase (cytochrome aa_3) in the cyanobacterium *Synechocystis* PCC 6803. Biochem Biophys Res Commun 191: 9–17

Bergman B, Siddiqui PJA, Carpenter EJ and Peschek GA (1993) Cytochrome oxidase: Subcellular distribution and relationship to nitrogenase expression in the nonheterocystous marine cyanobacterium *Trichodesmium thiebautii*. Appl Environ Microbiol 59: 3239–3244

Berman-Frank I, Lundgren P, Chen YB, Küpper H, Kolber Z, Bergman B and Falkowski P (2001) Segregation of nitrogen fixation and oxygenic photosynthesis in the marine cyanobacterium *Trichodesmium*. Science 294: 1534–1537

Berman-Frank I, Lundgren P and Falkowski P (2003) Nitrogen fixation and photosynthetic oxygen evolution in cyanobacteria. Res Microbiol 154:157–164

Bernroitner M, Zamocky M, Pairer M, Furtmüller PG, Peschek GA, Obinger C (2008) Heme-copper oxidases and electron donors in cyanobacterial respiratory electron transport. Chem Biodivers 5: 1927–1961

Bernroitner M, Tangl D, Lucini C, Furtmüller PG, Peschek GA, Obinger C (2009) Cyanobacterial cytochrome c_M: Probing its role as electron donor for Cu_A of cytochrome c oxidase. Biochim Biophys Acta—Bioenergetics 1787: 135–143

Berry S, Schneider D, Vermaas WF and Rogner M (2002) Electron transport routes in whole cells of *Synechocystis* sp. strain PCC 6803: The role of the cytochrome *bd*-type oxidase. Biochemistry 41: 3422–3429

Bertini I, Ciurli S, Dikiy A, Fernandez CO, Luchinat C, Safarov N, Shumilin S and Vila AJ (2001) The first solution structure of a paramagnetic copper (II) protein: The case of oxidized plastocyanin from the cyanobacterium *Synechocystis PCC6803*. J Am Chem Soc 123: 2405–2413

Broda E and Peschek GA (1983) Nitrogen fixation as evidence for the reducing nature of the early biosphere. Biosystems 16: 1–8

Burgess BK and Lowe DJ (1996) Mechanism of Molybdenum Nitrogenase. Chem Rev 96: 2983–3012

Capone DC, Zehr JP, Paerl HW, Bergman B and Carpenter EJ (1997) *Trichodesmium*, a globally significant marine cyanobacterium. Science 276 : 1221–1229

Carpenter EJ and Roenneberg T (1995) The marine planktonic cyanobacteria *Trichodesmium* spp.: Photosynthetic rate measurements in the SW Atlantic Ocean. Mar Ecol Prog Ser 118: 267–273

Chisholm SW, Olson RJ, Zettler ER, Goericke JB, Waterbury JB and Welschmeyer NA (1988) A novel free living prochlorophyte abundant in the oceanic euphotic zone. Nature 334: 340–343

Colon-Lopez M, Sherman DM and Sherman LA (1997) Transcriptional and translational regulation of nitrogenase in light-dark- and continuous-light-grown cultures of the unicellular cyanobacterium *Cyanothece* sp. strain ATCC 51142. J Bacteriol 179: 4317–4327

De la Rosa MA, Molina-Heredia FP, Hervas M and Navarro JA (2006) Convergent evolution of cytochrome c_6 and plastocyanin. In: Photosystem I: The light-driven plastocyanin: Ferredoxin oxidoreductase (Golbeck JH, ed.) Advances in photosynthesis and respiration, Vol 24. Springer, Dordrecht, pp 683–696

Diaz-Quintana A, Navarro JA, Hervas M, Molina-Heredia FP, De la Cerda B and De la Rosa MA (2003) A comparative structural and functional analysis of cyanobacterial plastocyanin and cytochrome c_6 as alternative electron donors to photosystem I. Photosynth Res 75: 97–110

Diaz-Moreno I, Diaz-Quintana A, Molina-Heredia FP, Nieto PM, Hansson Ö, De la Rosa MA and Karlsoon BG (2005) NMR Analysis of the Transient Complex between Membrane Photosystem I and Soluble Cytochrome c_6. J Biol Chem 280: 7925–7931

Duran RV, Hervas M, De la Rosa MA, Navarro JA (2004) The efficient functioning of photosynthesis and respiration in *Synechocystis* sp. PCC 6803 strictly requires the presence of either cytochrome c_6 or plastocyanin. J Biol Chem 279: 7229–7233

Fay P (1992) Oxygen relations of nitrogen fixation in cyanobacteria. Microbiol Rev 56: 340–373

Frazao C, Soares CM, Carrondo MA, Pohl E, Dauter Z, Wilson KS, Hervas M, Navarro JA, De la Rosa MA and Sheldrick GM (1995) Ab initio determination of the crystal structure of cytochrome c_6 and comparison with plastocyanin. Structure 3: 1159–1169

Fry IV, Huflejt M, Erber WW, Peschek GA and Packer L (1986) Role of respiration during adaptation of the freshwater cyanobacterium *Synechococcus* 6311 to salinityArch Biochem Biophys 244: 686–691

Fry IV, Peschek GA, Huflejt M and Packer L (1989) Biochem. EPR signals of redox active copper in EDTA washed membranes of the cyanobacterium *Synechococcus* 6311. Biochem Biophys Res Commun 129: 109–116

Gallon JR (1992) Reconciling the incompatible: N_2 fixation and O_2. New Phytologist 122: 571–609

Gallon JR (2001) N_2 fixation in phototrophs: Adaptation to a specialized way of life. Plant Soil 230: 39–48

Golden JW and Yoon H-S (2003) Heterocyst development in *Anabaena*. Curr Opin Microbiol 6: 557–563

Guss JM and Freeman HC (1983) Structure of oxidized poplar plastocyanin at 1.6 Å resolution. J Mol Biol 169: 521–563

Guss JM, Harrowell PR, Murata M, Norris VA and Freeman HC (1986) Crystal structure analyses of reduced (Cu^I) poplar plastocyanin at six pH values. J Mol Biol 192: 361–387

Häfele U, Scherer S and Böger P (1988) Cytochrome aa_3 from heterocysts of the cyanobacterium *Anabaena vanabilis*: isolation and spectral characterization. Biochim Biophys Acta 934: 186–190

Hart SE, Schlarb-Ridley BG, Bendall DS and Howe CJ (2005) Terminal oxidases of cyanobacteria. Biochem Soc Trans 33: 832–835

Hervas M, Navarro JA and De la Rosa MA (2003) Electron transfer between membrane complexes and soluble proteins in photosynthesis. Acc Chem Res 36: 798–805

Howitt CA and Vermaas WFJ (1998) Quinol and cytochrome oxidases in the cyanobacterium *Synechocystis* sp. PCC 6803. Biochemistry 37: 17 944–17 951

Inaba M, Sakamoto A and Murata N (2001) Functional expression in *Escherichia coli* of low-affinity and high-affinity $Na^+(Li^+)/H^+$ antiporters of *Synechocystis*. J Bacteriol 183: 1376–1384

Iwata S, Ostermeier C, Ludwig B and Michel H (1995) Structure at 2.8 Å resolution of cytochrome *c* oxidase from *Paracoccus denitrificans*. Nature 376: 660–669

Jones W and Myers J (1963) A common link between photosynthesis and respiration in a blue-green alga. Nature 199: 670–672

Jones KM, Haselkorn R (2002) Newly identified cytochrome *c* oxidase operon in the nitrogen-fixing cyanobacterium *Anabaena* sp. strain PCC 7120 specifically induced in heterocysts. J Bacteriol 184: 2491–2499

Jones KM, Buikema WJ and Haselkorn R (2003) Heterocyst-specific expression of *patB*, a gene required for nitrogen fixation in *Anabaena* sp. strain PCC 7120. J Bacteriol 185: 2306–2314

Kerfeld CA, Anwar HP, Interrante R, Merchant S and Yeates TO (1995) The Structure of chloroplast cytochrome c_6 at 1.9 Å resolution: Evidence for functional oligomerization. J Mol Biol 250: 627–647

Kienzl PF and Peschek GA (1982) Oxidation of *c*-Type cytochromes by the membrane-bound cytochrome oxidase (cytochrome aa_3) of blue-green algae. Plant Physiol 69: 580–584

Kumazawa S, Yumura S and Yoshisuji H (2001) Photoautotrophic growth of a recently isolated N_2-fixing marine non-heterocystous filamentous cyanobacterium, *Symploca* sp. (cyanobacteria). J Phycol 37: 482–487

Lappalainen P, Aasa R, Malmström BG and Saraste M (1993) Soluble CuA-binding domain from the *Paracoccus* cytochrome *c* oxidase. J Biol Chem 268: 26416–26421

Lockau, W (1981) Evidence for a dual role ofcytochrome c-553 and plastocyanin. Arch Microbiol 128: 336–349

Lopez-Gomollon S, Hernandez JA, Pellicer S, Espinosa A, Peleato ML and Fillat MF (2007) Cross-talk between iron and nitrogen regulatory networks in *Anabaena*(*Nostoc*) sp. PCC 7120: Identification of overlapping genes from/in FurA and NtcA regulons. J Mol Biol 37: 267–281

Lubbern M, Arnaud S, Castresana J, Warne A, Albracht SP and Saraste M (1994) A second terminal oxidase in *Sulfolobus acidocaldarius*. Eur J Biochem 224: 151–159

Malakhov MP, Wada H, Los DA, Semenenko VE and Murata N (1994) A new type of cytochrome *c* from *Synechocystis* PCC6803. J Plant Physiol 144: 259–264

Malakhov MP, Malakhova OA and Murata N (1999) Balanced regulation of expression of the gene for cytochrome c_M and that of genes for plastocyanin and cytochrome c_6 in *Synechocystis*. FEBS Lett 444: 281–284

Maneg O, Malatesta F, Ludwig B and Drosou V (2004) Interaction of cytochrome *c* with cytochrome oxidase: Two diffferent docking scenarios. Biochim Biophys Acta 1655: 274–281

Manna P and Vermaas W (1997) Lumenal proteins involved in respiratory electron transport in the cyanobacterium *Synechocystis* sp. PCC 6803. Plant Mol Biol 35: 407–416

Metzger SU, Pakrasi HB and Whitmarsh J (1995) Characterization of a double deletion mutant that lacks cytochrome c_6 and cytochrome c_M in *Synechocystis* 6803. In: Photosynthesis: From light to biosphere (Mathis P, ed.) Kluwer Academic Press, Dordrecht, pp 823–826

Michel H, Behr J, Harrenga A and Kannt A (1998) Cytochrome *c* oxidase: Structure and spectroscopy. Annu Rev Biophys 27: 329–356

Molina-Heredia FP, Balme A, Hervas A, Hervas M, Navarro JA and De la Rosa MA (2002) A comparative structural and functional analysis of cytochrome c_M, cytochrome c_6 and plastocyanin from the cyanobacterium *Synechocystis* sp. PCC 6803. FEBS Lett 517: 50–54

Moser D, Nicholls P, Wastyn M and Peschek GA (1991) Acidic cytochrome c_6 of unicellular cyanobacteria is an indispensable and kinetically competent electron donor to cytochrome oxidase in plasma and thylakoid membranes. Biochem Intern 24: 757–768

Musser SM, Stowell MHB and Chan SI (1993) Comparison of ubiquinol and cytochrome *c* terminal oxidases: An alternative view. FEBS Lett 327: 131–136

Navarro JA, Duran RV, De la Rosa MA and Hervas M (2005) Respiratory cytochrome *c* oxidase can be efficiently reduced by the photosynthetic redox proteins cytochrome c_6 and plastocyanin in cyanobacteria. FEBS Lett 579: 3565–3568

Nicholls P, Obinger C, Niederhauser H and Peschek GA (1992) Cytochrome oxidase in *Anacystis nidulans*: stoichiometries and possible functions in the cytoplasmic and thylakoid membranes. Biochim Biophys Acta 1098: 184–190

Nomura CT, Persson S, Shen G, Inoue-Sakamoto K and Bryant DA (2006) Characterization of two cytochrome oxidase operons in the marine cyanobacterium *Synechococcus* sp. PCC 7002: Inactivation of ctaDI affects the PS I:PS II ratio. Photosynth Res 87: 215–228

Obinger C, Knepper J-C, Zimmermann U and Peschek GA (1990) Identification of a periplasmic *c*-type cytochrome as electron donor to the plasma membrane-bound cytochrome oxidase of the cyanobacterium *Nostoc* Mac. Biochem Biophys Res Commun 169: 492–501

Omoregie EO, Crumbliss LL, Bebout BM and Zehr JP (2004) Determination of nitrogen-fixing phylotypes in *Lyngbya* sp. and *Microcoleus chthonoplastes* cyanobacterial mats from Guerrero Negro, Baja California, Mexico. Appl Environ Microbiol 70: 2119–2128

Paumann M, Bernroitner M, Lubura B, Peer M, Jakopitsch C, Furtmüller PG, Peschek GA and Obinger C (2004a) Kinetics of electron transfer between plastocyanin and the soluble CuA domain of cyanobacterial cytochrome *c* oxidase. FEMS Microbiol Lett 239: 301–307

Paumann M, Feichtinger M, Bernroitner M, Goldfuhs J, Jakopitsch C, Furtmüller PG, Regelsberger G, Peschek GA and Obinger C (2004b) Kinetics of interprotein electron transfer between cytochrome c_6 and the soluble Cu_A domain of cyanobacterial cytochrome *c* oxidase. FEBS Lett 576: 101–106

Paumann M, Regelsberger G, Obinger C and Peschek GA (2005) The bioenergetic role of dioxygen and the terminal oxidase(s) in cyanobacteria. Biochim Biophys Acta 1707: 231–253

Pereira MM, Teixeira M (2004) Proton pathways, ligand binding and dynamics of the catalytic site in haem-copper oxygen reductases: A comparison between the three families. Biochim Biophys Acta 1655: 340–346

Pereira MM, Santana M and Teixeira M (2001) A novel scenario for the evolution of haem-copper oxygen reductases. Biochim Biophys Acta 1505: 185–208

Peschek GA (1980) Restoration of respiratory electron-transport reactions in quinone-depleted particle preparations from *Anacystis nidulans*. Biochem J 186: 515–523

Peschek GA (1996) Structure-function relationships in the dual-function photosynthetic-respiratory electron-transport assembly of cyanobacteria (blue-green algae). Biochem Soc Trans 24: 729–733

Peschek GA (2001) Temperature stress and basic bioenergetic strategies for stress defense. In: Algal adaptation to environmental stress (Rai LC, Gaur JP, eds.) Springer-Verlag, Berlin, Heidelberg, New York, pp 203–258.

Peschek GA, Schmetterer G, Lauritsch G, Nitschmann WA, Kienzl PF and Muchl R (1982) Do cyanobacteria contain "mammalian-type" cytochrome oxidase? Arch Microbiol 131: 261–265

Peschek GA, Molitor V, Trnka M, Wastyn M and Erber W (1988) Characterization of cytochrome-*c* oxidase in isolated and purified plasma and thylakoid membranes from cyanobacteria. Methods Enzymol 167: 437–449

Peschek GA, Villgrater K and Wastyn M (1991) Respiratory protection of the nitrogenase in dinitrogen-fixing cyanobacteria. Plant Soil 137: 17–24

Peschek GA, Obinger C, Fromwald S and Bergman B (1994) Correlation between immuno-gold labels and activities of the cytochrome-*c* oxidase (aa_3-type) in membranes of salt stressed cyanobacteria. FEMS Microbiol Lett 124: 431–437

Peschek GA, Obinger C and Paumann M (2004) The respiratory chain of blue-green algae (cyanobacteria). Physiol Plant 120 358–369

Pils D and Schmetterer G (2001) Characterization of three bioenergetically active respiratory terminal oxidases in the cyanobacterium *Synechocystis* sp. strain PCC 6803. FEMS Microbiol Lett 203: 217–222

Pils D, Gregor W and Schmetterer G (1997) Evidence for activity of three distinct respiratory terminal oxidases in the cyanobacterium *Synechocystis* strain sp. PCC 6803. FEMS Microbiol Lett 152: 83–88

Pils D, Wilken C, Valladares A, Flores E, and Schmetterer G (2004) Respiratory terminal oxidases in the facultative chemoheterotrophic and dinitrogen fixing cyanobacterium *Anabaena variabilis* strain ATCC 29413: Characterization of the *cox2* locus. Biochim Biophys Acta 1659: 32–45

Poole RK and Cook GM (2000) Redundancy of aerobic respiratory chains in bacteria? Adv Microb Physiol 43: 165–224

Postgate JR (1998) Nitrogen Fixation, Cambridge University Press, Cambridge

Puustinen A, Finel M, Haltia T, Gennis RB and Wikström M (1991) Properties of the two terminal oxidases of *Escherichia coli*. Biochemistry 30: 3936–3942

Rai AN, Borthakur M and Bergman B (1992) Nitrogenase derepresssion, its regulation and metabolic changes associated with diazotrophy in the non-heterocystous cyanobacterium *Plectonema boryanum* PCC 73110. J Gen Microbiol 138: 481–491

Ramasubramanian TS, Wei T-F, Oldham AK and Golden JW (1996) Transcription of the *Anabaena* sp. strain PCC 7120 *ntcA* gene: Multiple transcripts and NtcA binding. J Bacteriol 178: 922–926

Riistama S, Puustinen A, Garcia-Horsman A, Iwara S, Michel H and Wikström M (1996) Channeling of dioxygen into the respiratory enzyme. Biochim Biophys Acta 1275: 1–4

Sawaya MR, Krogmann DW, Serag A, Ho KK, Yeates TO and Kerfeld CA (2001) Structures of cytochrome *c*-549 and cytochrome c_6 from the Cyanobacterium *Arthrospira maxima*. Biochemistry 40: 9215–9225

Scherer S (1990) Do photosynthetic and respiratory electron transport chains share redox proteins? Trends Biochem Sci 15: 458–462

Scherer S, Stürzl E and Böger P (1988) Interaction of respiratory and photosynthetic electron transport in *Anabaena variabilis* Kütz. Arch Microbiol 132: 333–337

Schmetterer G, Alge D and Gregor W (1994) Deletion of cytochrome c oxidase genes from the cyanobacterium *Synechocystis* sp. PCC6803: Evidence for alternative respiratory pathways. Photosynth Res 42: 43–50

Schmetterer G, Valladares A, Pils D, Steinbach S, Pacher M, Muro-Pastor AM, Flores E and Herrero A (2001) The *coxBAC* operon encodes a cytochrome *c* oxidase required for heterotrophic growth in the cyanobacterium *Anabaena variabilis* strain ATCC 29413. J Bacteriol 183: 6429–6434

Serrano A, Gimenez P, Scherer S and Böger P (1990) Cellular localization of cytochrome c_{553} in the N_2-fixing cyanobacterium *Anabaena variabilis*. Arch Microbiol 154: 614–618

Sone N, Tano H and Ishizuka M (1993) The genes in the thermophilic cyanobacterium *Synechococcus vulcanus* encoding cytochrome-*c* oxidase. Biochim Biophys Acta 1183: 130–138

Soulimane T, Buse G, Burenkov GP, Bartunik HT, Huber R and Than ME (2000) Structure and mechanism of the aberrant ba(3)-cytochrome c oxidase from *Thermus thermophilus*. EMBO J 19: 1766–1776

Svensson-Ek M, Abramson J, Larrson G, Tornroth S, Brezesinski P and Iwata S (2002) The X-ray crystal structures of wild-type and EQ(I-286) mutant cytochrome *c* oxidases from *Rhodobacter sphaeroides*. J Mol Biol 321: 329–339

Thiel T and Pratte B (2001) Effect on heterocyst differentiation of nitrogen fixation in *Anabaena variabilis* ATCC 29413. J Bacteriol 183: 280–286

Tuli R, Naithani S and Misra HS (1996) Cyanobacterial photosynthesis and the problem of oxygen in nitrogen-fixation: A molecular genetic view. J Sci Indust Res 55: 638–657

Ubbink M, Ejdebäck M, Karlsson BG and Bendall DS (1998) Structure of the complex of plastocyanin and cytochrome *f*, determined with paramagnetic NMR and restrained rigid-body molecular dynamics. Structure 6: 323–335

Valladares A, Herrero A, Pils D, Schmetterer G and Flores E (2003) Cytochrome *c* oxidase genes required for nitrogenase activity and diazotrophic growth in *Anabaena* sp. PCC 7120. Mol Microbiol 47: 1239–1249

Valladares A, Maldener I, Muro-Pastor AM, Flores E and Herrero AJ (2007) Heterocyst development and diazotrophic metabolism in terminal respiratory oxidase mutants of the cyanobacterium *Anabaena* sp strain PCC 7120. Bacteriol 189: 4425–4430

Van Baalen C (1987) Nitrogen fixation. In: The cyanobacteria (Fay P, Van Baalen C, eds.) Elsevier, Amsterdam, pp 187–198

Wang K, Zheng Y, Sadoski R, Grinell S, Geren L, Ferguson-Miller S, Durham W and Millett F (1999) Definition of the interaction domain for cytochrome *c* on cytochrome *c* oxidase. II.

Rapid kinetic analysis of electron transfer from cytochrome *c* to *Rhodobacter sphaeroides* cytochrome oxidase surface mutants. J Biol Chem 274: 38042–38050

Wastyn M, Achatz A, Molitor V and Peschek GA (1988) Respiratory activities and aa_3-type cytochrome oxidase in plasma and thylakoid membranes from vegetative cells and heterocysts of the cyanobacterium *Anabaena* ATCC 29413. Biochim Biophys Acta 935: 217–224

Waterbury JB, Watson SW, Guillard RRL and Brand LE (1979) Widespread occurrence of a unicellular, marine planktonic, cyanobacterium. Nature 277: 293–294

Winkenbach F and Wolk CP (1972) Activities of enzymes of the oxidative and reductive pentose phosphate pathways in heterocysts of a blue-green alga. Plant Physiol 52: 480–483

Witt H, Malatesta F, Nicoletti F, Brunori M and Ludwig B (1998) Tryptophan 121 of subunit II is the electron entry site to cytochrome *c* oxidase in *Paracoccus denitrificans.* Involvement of a hydrophobic patch in the docking reaction. J Biol Chem 273: 5132–5136

Wolk CP, Ernst A and Elhai J (1994) Heterocyst metabolism and development. In: The molecular biology of cyanobacteria (Bryant DA, ed.) Kluwer Academic Publishers, Dordrecht, The Netherlands, pp 769–823

Wood PM (1978) Interchangeable copper and iron proteins in algal photosynthesis. Eur J Biochem 87: 9–19

Zehr JP, Mellon MT and Zani S (1998) New nitrogen-fixing microorganisms detected in oligotrophic oceans by amplification of nitrogenase (*nifH*) genes. Appl Environ Microbiol 64: 3444–3450

Zhang L, Pakrasi HB and Whitmarsh J (1994) Photoautotrophic growth of the cyanobacterium *Synechocystis* sp. PCC 6803 in the absence of cytochrome c_{553} and plastocyanin. J Biol Chem 269: 5036–5042

Zhang C-C, Laurent S, Sakr S, Peng L, Bedu S (2006) Heterocyst differentiation and pattern formation in cyanobacteria: A chorus of signals. Mol Microbiol 59: 367–375

Zhen Y, Hoganson CW, Babcock GT and Ferguson-Miller S (1999) Definition of the interaction domain for cytochrome *c* on cytochrome *c* Oxidase. I. Biochemical, spectral, and kinetic characterization of surface mutants in subunit II of *Rhodobacter sphaeroides* cytochrome aa_3. J Biol Chem 274: 38032–38041

Part VII
Progress in the Genetic Manipulation of Cyanobacteria

Chapter 24
Tools for Genetic Manipulation of Cyanobacteria

Annegret Wilde and Dennis Dienst

24.1 Introduction

The stable accommodation of a cyanobacterium into a non-photosynthetic eukaryote was a milestone in the evolution of all other oxygenic photosynthetic organisms. This process, called endosymbiosis appeared only once during evolution and constitutes the origin of chloroplasts (Douglas 1998; Rodriguez-Ezpeleta et al. 2005). It is therefore not surprising that photosynthetic processes in higher plants are very similar to those in cyanobacteria and that their photosynthetic complexes have most of the subunits in common.

This is the principal reason for the extensive use of cyanobacteria to study fundamental processes in photosynthesis and the structure of photosynthetic complexes (Chitnis and Chitnis 1993; Shen and Vermaas 1994; Jordan et al. 2001; Ferreira et al. 2004; Loll et al. 2005). Moreover, cyanobacteria are fast growing prokaryotic organisms, some of which can easily be genetically modified. The sequencing of the genome of the naturally transformable cyanobacterium *Synechocystis* sp. PCC 6803 (Kaneko et al. 1996) was pioneering work in the development of this bacterium into a model organism for the research into photosynthesis and related processes. This would not have been possible without the entry of techniques and genetic tools to aid mutational experiments.

In addition, several laboratories are searching for biotechnological applications of cyanobacteria in terms of hydrogen or biofuel production (Tamagnini et al. 2007; Dismukes et al. 2008; Gressel 2008), expressing foreign proteins (Lluisma et al. 2001) or exploitation of interesting pigments or bioactive compounds from cyanobacteria (Singh et al. 2005; Eriksen 2008).

In this chapter, we will cover known methods for the genetic manipulation of cyanobacteria with a particular focus on *Synechocystis* sp. PCC 6803, depicting representative applications of these methods to study gene function. Where ap-

A. Wilde (✉)
Institute of Microbiology and Molecular Biology, Justus-Liebig-University Giessen, Heinrich-Buff-Ring 26-32, 35392 Giessen, Germany
e-mail: annegret.wilde@mikro.bio.uni-giessen.de

G. A. Peschek et al. (eds.), *Bioenergetic Processes of Cyanobacteria,*
DOI 10.1007/978-94-007-0388-9_24,

propriate, advances made in genetic engineering of other cyanobacterial strains will be considered in this review. However, since the available data concerning genetic manipulation systems for many other cyanobacteria—unicellular as well as filamentous strains—is huge, this review does not attempt to review the history of that field, but rather concentrates on the current state of the art and on potential applications. Known processes of DNA transfer in the laboratory and nature as well as genetic tools for cyanobacteria have been described in detail in recent reviews (Koksharova and Wolk 2002a; Nakasugi and Neilan 2006). Different approaches to mutagenize the chromosome of *Synechococcus elongatus* together with detailed protocols have been described recently (Clerico et al. 2007; Mackey et al. 2007).

24.2 DNA Transfer in Cyanobacteria

There have been two major processes of DNA transfer into cyanobacterial cells intensively used so far in the laboratory: transformation and conjugation. Transduction, which is the third known mechanism mediating DNA transfer, has never been demonstrated in cyanobacteria; although there is increasing evidence that a number of phages has the potential to transfer DNA via infection (Khudiakov and Gromov 1973; Waterbury and Valois 1993; McDaniel et al. 2002; Yoshida et al. 2008).

Shestakov and Khyen (1970) showed for the first time that *Synechococcus elongatus* sp. PCC 7942 (former name: *Anacystis nidulans* R2) could be transformed by exogenous DNA. This cyanobacterial strain is naturally transformable, a property that was also shown for other unicellular cyanobacteria such as *Synechococcus* sp. PCC 7002 (former name: *Agmenellum quadruplicatum* PR-6), the thermophilic *Thermosynechococcus elongatus* BP-1 and the mesophilic freshwater bacterium *Synechocystis* sp. PCC 6803. The latter one has become a very attractive model organism since the publication of its genome sequence by Kaneko et al. (1996). This was the first complete genome sequence of a photosynthetic organism thus raising not only cyanobacterial genetics and physiology to a higher level but also photosynthesis research. As *Synechocystis* sp. PCC 6803 is able to grow mixotrophically on glucose with impaired photosystems I and II, several laboratories started to use this organism for the study of photosynthesis in the eighties of the last century (Carpenter and Vermaas 1989; Smart et al. 1991; Shen and Vermaas 1994; Smart et al. 1994). In addition, electroporation has also been used to transform several cyanobacteria, e.g. *Microcystis aeruginosa*, *Thermosynechococcus elongatus* BP-1 or filamentous cyanobacteria (Thiel and Poo 1989; Chiang et al. 1992). Although transformation was also utilized for manipulation of filamentous strains, nowadays conjugation is the prevailing method to manipulate these cyanobacteria genetically (Wolk et al. 1984; Elhai and Wolk 1988; Zhu et al. 2001). In addition, conjugation is commonly used for genetic manipulation of unicellular cyanobacteria (Tsinoremas et al. 1994; Dienst et al. 2008).

24.2.1 Natural Competence and Type IV Pili (Fimbriae)

Type IV pili (Tfp) are multifunctional hair-like filaments extending from the outer membrane of diverse genera of both gram-negative and gram-positive bacteria (Mattick 2002; Nudleman and Kaiser 2004). These dynamic structures predominantly consist of a highly conserved structural protein encoded by the *pilA* gene and are known to be involved in bacterial cell adhesion. Tfp are most extensively studied in *Neisseria gonorrhoeae* and *Pseudomonas aeruginosa* and functionally range from bacteriophage adsorption (Bradley 1974) to biofilm maturation (Chiang and Burrows 2003; Jurcisek et al. 2007; Barken et al. 2008). For several pathogens, Tfp act as central host colonization factors by mediating cell adhesion and translocation across epithelial tissue (Mattick et al. 1996). In addition, they use a mechanism termed "twitching motility". This type of locomotion can also be observed in cyanobacteria as a prerequisite for phototactic movement (reviewed in Bhaya 2004; Yoshihara and Ikeuchi 2004) on solid surfaces, and has been attributed to an ATP dependent "extension and retraction mechanism" of the pilus through assembly and disassembly (reviewed recently by Burrows 2005).

Clear indications for a functional link between Tfp and natural competence—this term refers to the ability of bacteria to take up extracellular DNA—were first provided by Sparling (1966) and later by Biswas et al. (1977). Many bacteria exhibiting Tfp indeed are naturally transformable (Lorenz and Wackernagel 1994); and this competence is often dependent on intact piliation, whereupon in some cases, Tfp assembly factors rather than the pili structures itself are required for transformation (Averhoff 2004). For naturally transformable bacteria that exhibit Tfp on their surfaces, the presence of these appendages appears to be connected with competence (Sparling 1966; Stone and Kwaik 1999; Kennan et al. 2001; Yoshihara et al. 2001).

Yoshihara and co-workers (Yoshihara et al. 2001) suggested a fundamental role of type IV pili in the natural competence of unicellular cyanobacteria. The corresponding studies demonstrated that inactivation of homologs of the *pil* genes *pilA1, pilQ, pilM, pilN, pilO* and *pilB1*in *Synechocystis* sp. PCC 6803 resulted in the loss of Tfp structures that was accompanied by an impairment of motility and transformation efficiency. In fact, *pilA1* encodes the homolog of the major pilin, accounting for the main structural component of the apparatus, whereas *pilQ* encodes the homolog of the outer membrane secretin, that is likewise one of those components, that are shared by Tfp and Ψ-pili. *pilM-O* represent a cluster of homologs of the corresponding *Pseudomonas aeruginosa* genes, encoding constituents of pilus assembly with unknown function (Nudleman and Kaiser 2004). The gene product of *pilB1* exhibits NTPase activity and is regarded as a pilus extension motor that is indispensible for pilus assembly (Nudleman and Kaiser 2004; Jakovljevic et al. 2008). Thereupon, also a set of homologs of regulatory *pil* genes was shown to be involved in natural competence of *Synechocystis* sp. PCC 6803 (Yoshihara et al. 2002). Okamoto and Ohmori (2002) in turn provided genetic evidence that a homolog of *pilT1* was essential for natural competence in *Synechocystis* sp. PCC 6803. Homologs of this gene encode an AAA^+ type ATPase driving the pilus apparatus as retraction motor (Brossay et al. 1994; Merz et al. 2000; Nakasugi et al. 2007), which is indispensable for natural competence and motility (Wolfgang et al. 1998).

In addition to *pil* genes, the gene *slr0197*, encoding a putative DNA binding protein (Yura et al. 1999), appeared to be essential for natural competence but not for motility (Yoshihara et al. 2001). It was designated *comA*, since its gene product shows similarity to the DNA receptor ComEA, which is required for natural transformation of *Bacillus subtilis* (Provvedi and Dubnau 1999). Nakasugi et al. (2006) recently proposed the locus *slr0388* as an additional competence factor, encoding a protein with reasonable sequence identity to ComF proteins, which are required for natural transformation of heterotrophic bacteria. In fact in *Synechocystis* sp. PCC 6803 this designated *comF* gene emerged to be essential for competence of *Synechocystis* sp. PCC 6803 as well.

24.2.2 Maintenance of Foreign DNA

Among others, *Synechocystis* sp. PCC 6803 is known to harbor a highly efficient homologous recombination system, which facilitates the targeted inactivation of genes by genomic integration of foreign DNA via double-crossover. Note that the efficiency of such events strongly depends on the size of the homologous stretches on the DNA molecules (Williams 1988; Chauvat et al. 1989; Labarre et al. 1989).

Transformation is a widely employed method to inactivate genes by integration of genetic material into the chromosome of cyanobacteria. For this type of genetic manipulation, researchers frequently use so called suicide vectors, plasmids that do not replicate autonomously in the recipient cell. However, since uptake of extracellular (vector-) DNA into *Synechocystis* sp. PCC 6803 instantly leads to linearization and conversion of a circular plasmid into single-stranded DNA fragments (Barten and Lill 1995), natural transformation cannot be used to transfer and maintain self-replicating plasmids for expression studies or mutant complementations. In such cases, conjugation provides an alternative way to transfer recombinant DNA into cyanobacterial cells. These purposes require autonomously replicating plasmids that can be mobilized for conjugational transfer.

The ability of autonomous replication in cyanobacterial cells has been attributed to two groups of cloning vectors, which can be mobilized: the first group includes shuttle vectors that are hybrids between a native cyanobacterial plasmid and the *E. coli* plasmid pBR322 or its closely related derivatives allocating the *oriT* region. For conjugational transfer, this group depends on the conjugative plasmid RP-4 and further *trans*-acting mobilization genes, which are encoded on the plasmid ColK and its relatives (Wolk et al. 1984). Vectors of the second group harbor a replicon from plasmid RSF1010 (Marraccini et al. 1993), which belongs to the IncQ group of plasmids. These vectors are able to be mobilized by several conjugative plasmids and replicate in a very broad range of bacterial hosts including cyanobacteria (reviewed in Mermet-Bouvier et al. 1993; Rawlings and Tietze 2001). They harbor the *cis*-acting "origin of transfer site" *oriT* as well as three *trans*-acting *mob* genes encoding proteins that interact with *oriT* (Derbyshire and Willetts 1987).

The IncP (P-incompatibility) group plasmid RP-4 and its relatives are able to transfer themselves into a wide range of gram-negative bacteria, but self-replication of RP-4 within cyanobacteria has not been described yet. Certainly, these plasmids are also able to mobilize non-conjugative vectors for conjugation and in their corresponding review, Koksharova and Wolk (2002a) provide a number of examples for (RP-4 mediated) mobilization of plasmid DNA from *E. coli* to cyanobacteria. These mobilizing plasmids (e.g. RP4) in turn provide the *tra* genes that encode pili and a membrane pore, the Mfp (mating pair formation) complex as well as further components specifically required fur conjugational transfer (Lanka and Wilkins 1995).

A set of IncQ vectors that are particularly suitable for gene cloning in unicellular cyanobacteria was constructed by Zinchenko et al. (1999). These pVZ vectors, deriving from RSF1010, were equipped each with two antibiotic resistance genes, which contain and are flanked by unique restriction recognition sites facilitating efficient cloning and subsequent selection of transconjugants. In practice, conjugational gene transfer is performed with two strains of *E. coli*, harboring the

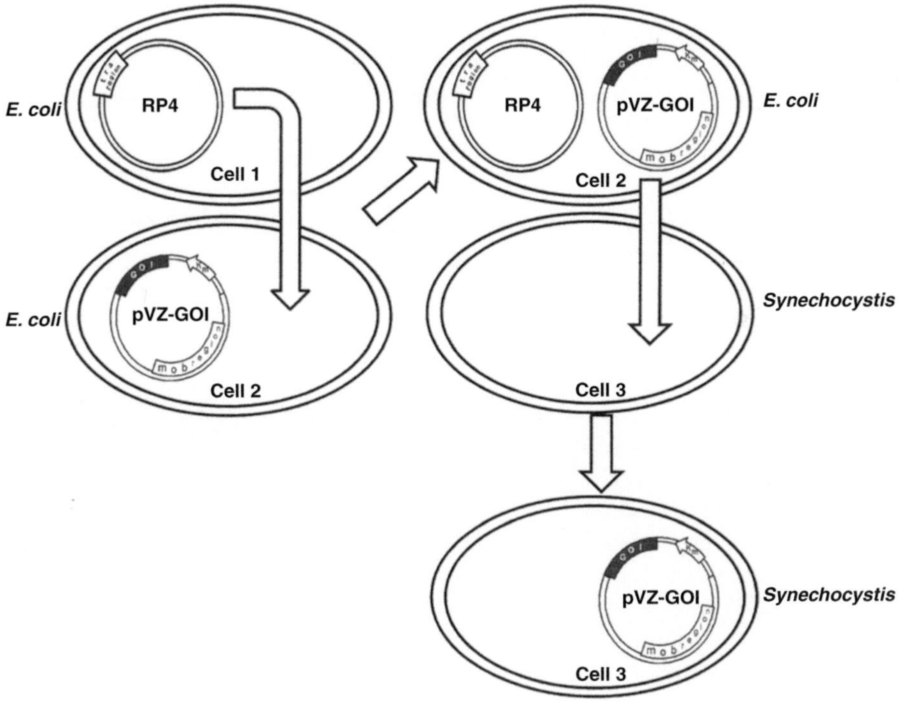

Fig. 24.1 Schematic mechanism of triparental mating for conjugational transfer of pVZ321 based plasmids from *E. coli* to *Synechocystis* sp. PCC 6803. The depicted example shows the hypothetical plasmid pVZ-GOI harboring the gene of interest, which is inserted into pVZ321 substitutional for the chloramphenicol resistance cassette. RP4 acts as the mobilizing plasmid providing the *tra* gene products (as described in the text). The following symbols are used: Km, kanamycin resistance; mob region, region harboring the mobilization genes *mobA, mob and mobC*; GOI, gene of interest

mobilizing helper plasmid (e.g. RP4) and the cargo plasmid (e.g. pVZ321), respectively and a cyanobacterial recipient strain. In this process termed "tri-parental mating", the cargo plasmid is transferred to the cyanobacterial cells, as described in Fig. 24.1.

Within the course of reverse genetics research, the pVZ321 vector has served as a reliable tool in our lab to emphasize phenotypical effects of knock-out mutations in *Synechocystis* sp. PCC 6803 by reintroduction of the corresponding gene on this self-replicating plasmid (Baier et al. 2001; Dühring et al. 2006a, b; Dienst et al. 2008).

24.3 Examples of Genetic Approaches to Study Cyanobacterial Physiology and Gene Regulation

24.3.1 Inactivation of Specific Genes—Different Strategies

The soaring amount of sequenced genomes of diverse organisms bears a multitude of putative open reading frames (ORF) with completely unknown or computational predicted functions. Since such predictions might be misleading without experimental evidence, reverse genetics and/or biochemical approaches are necessary to shed light on the functions of these genes.

Cyanobacterial genetic studies initially were based upon the isolation of spontaneous mutants on UV light as well as chemical mutagenesis. Cells were selected for specific (e.g. nutritional) requirements or loss of specific physiological characteristics like photoautotrophic growth or heterocyst formation (in diazotrophic filaments) under nitrogen limitation. Searching for the site of a mutation usually was a time consuming process, since e.g. libraries of chromosomal fragments had to be constructed for complementation of the mutation, that way facilitating recovery of the wild-type phenotype (Wolk et al. 1988; Buikema and Haselkorn 1991; Shestakov et al. 1994).

Utilization of the powerful homologous recombination system of many cyanobacterial strains has therefore developed into a major tool for directed mutagenesis of genes. Williams and Szalay (1983) were the first to publish the technique of stable integration of foreign DNA into the chromosome of *Anacystis nidulans* R2. In this process, a piece of chromosomal DNA containing the gene of interest (GOI) is nowadays amplified by PCR and ligated into a cloning vector that replicates in *E. coli*. Thereafter, part of the GOI should be replaced by an antibiotic resistance cassette, which can alternatively be ligated just into a single site, thus inactivating the respective gene. Instead, gene disruption can be achieved without cloning procedures, just by integrating the resistance marker gene into the GOI by a two-step PCR approach (Taroncher-Oldenburg and Stephanopoulos 2000; Lee et al. 2004). The amplified DNA fragment is then transferred into the wild-type strain, thereupon replacing the original gene by the mutated gene copy by homologous recombination.

Whereas in *Synechocystis* sp. PCC 6803 integration of homologous DNA usually happens by double reciprocal recombination, in many other cyanobacteria like *Nostoc* sp. PCC 7120 (former name *Anabaena* sp. PCC 7120) the second recombination event has to be selected by using the *sacB* gene (Cai and Wolk 1990). *SacB* encoding levan sucrase is a conditionally lethal gene that should be included within the suicide plasmid bearing the mutated version of the GOI. Expression of *sacB* is induced by sucrose and the gene product is lethal for many gram-negative bacteria. Because single-crossover events lead to integration of the whole plasmid, these recombinant cells harbor the *sacB* gene and thus are sucrose-sensitive. Double-crossover leads to loss of the *sacB* gene resulting in sucrose-resistant clones. Double recombinants are selected on medium containing 5% sucrose.

However, as many cyanobacteria often harbor multiple copies of their chromosome per cell (Herdman et al. 1979), the experimenter has to verify complete genomic segregation by the use of PCR and/or Southern Blot analysis. That phenomenon might on the other hand be beneficial to investigation of essential genes, whose disruption would be lethal to the cells. In such cases, a "partial knock-out" of the GOI could permit survival of the mutant cells, still providing an insight into the function of the gene.

A different approach to alter gene expression in cyanobacteria has been proposed by Collier and Grossman (1994), who demonstrated that induced overexpression of an RNA molecule with a sequence antisense to the mRNA encoding the phycobilisome degradation protein NblA actually prevented these structures from being degraded. This method is based on the ability of an antisense RNA to bind to its sense RNA counterpart (typically an mRNA), thereby disturbing translation or facilitating degradation of RNA duplexes by RNAses.

Over the past years, there has been collected increasing evidence for the existence of a significant number of regulatory non-coding RNAs (ncRNAs) that fulfill central roles in post-transcriptional regulation of gene expression in bacteria as well as in eukaryotes. These ncRNAs are a heterogeneous group of RNA molecules normally (but not always) lacking a protein-coding function. They are usually smaller than 200 nt in size and act to regulate mRNA translation/decay but can also bind to proteins and thereby modify protein function (for recent reviews see Gottesman 2004, 2005). The method described by Collier and Grossman (1994) is particularly useful when working with such RNAs. Since ncRNA genes can be found in antisense orientation within an open reading frame (Dühring et al. 2006a) or maybe within the 5′- or 3′-untranslated regions of genes (Kawano et al. 2005), it is not always reasonable to introduce a resistance cassette into the respective DNA region in order to knock out a ncRNA. Introduction of a construct into the cell that harbours a sequence antisense to the ncRNA (excluding terminator regions) together with a promoter that facilitates expression of the antisense RNA can lead to decreasing amounts of the ncRNA. For most purposes the addition of a Rho-independent terminator sequence (e.g. oop terminator from phage lambda) should be considered as described in Dühring et al. 2006a.

24.3.2 Rps12-Mediated Gene Replacement

Multiple targeted gene deletions or interruptions are often restricted by the number of available drug resistance markers. Moreover, use of cyanobacterial strains in biotechnology has a demand of mutants that are free from drug resistance markers. Matsuoka et al. (2001) have reported on a method termed *rps12*-mediated gene replacement for construction of such mutants in *Synechococcus elongatus* PCC 7942 (Fig. 24.2).

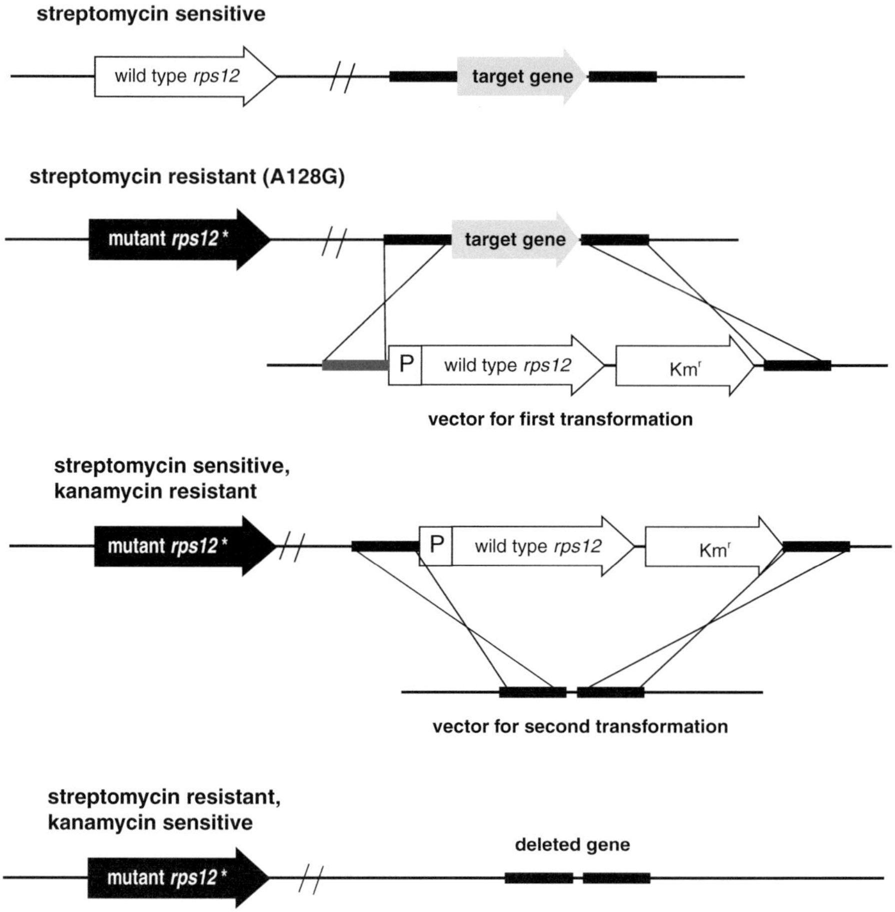

Fig. 24.2 Schematic illustration of *rps12*-mediated gene replacement in cyanobacteria. First, the wild-type *rps12* gene is a replaced by a streptomycin resistant variant using homologous recombination. Second, the gene of interest is replaced by the wild-type *rps12* gene in combination with a kanamycin resistance cassette (Km^r). Flanking regions that are use for homologous recombination are shown as wavy lines. Using a construct bearing only the flanking regions of the deleted gene and selecting on streptomycin resistance the *rps12*-Km^r cassette will be removed from the genome

There are several reports on point mutations in the S12 ribosomal protein subunit that confer resistance to streptomycin (Funatsu and Wittmann 1972; Timms et al. 1992). These changes in the amino acid sequences of mutated subunits are thought to destabilize the streptomycin-ribosome complex, resulting in resistance to this antibiotic. The system described by Matsuoka et al. (2001) is based on a streptomycin resistant *rps12-R43* allele with a lysine to arginine substitution. First, the dominant wild-type *rps12* allele is inserted into an ectopic site of a streptomycin resistant *rps12- R43* mutant using an associated antibiotic resistance cassette for selection of transformants.

The selected merodiploid clones (concerning *rps12*) are sensitive to streptomycin but resistant to the drug used for selection. Second, using the wild-type *rps12* locus any GOI can be inserted into the chromosome by homologous recombination. Streptomycin resistant clones can be selected that contain the desired gene in combination with elimination of the wild-type *rps12*/drug resistance cassette. Takahama et al. (2004) have improved the method by employing a heterologous *rps12* gene from *Synechocystis* sp. PCC 6803, thus, inhibiting gene conversion between the homologous *rps12* alleles occurring together in mutants after first transformation. Alternatively, if inactivation of a specific gene was the aim of the study, the wild-type *rps12* gene associated with an antibiotic resistance cassette should replace the GOI. Here, flanking regions of the GOI can be used for homologous recombination. After transformation with a second construct bearing these flanking regions but missing the GOI, streptomycin resistant gene knockout clones can be selected (Fig. 24.2).

24.3.3 Complementation Studies

In order to determine whether the phenotypical outcomes of a gene mutation are really due to the gene inactivation itself or rather to polar effects (e.g. by interruption of the regulatory integrity of an operon) implementation of complementation studies should be taken into consideration. The wild-type gene will therefore be re-introduced into the mutant cell either by expressing it on a self-replicating plasmid like the pVZ vectors or by inserting it into a "neutral site" on the chromosome. "Neutral sites" are thought to be sequences on the cyanobacterial chromosome where ectopic sequences can be inserted without generating a phenotype. Such a sequence might be a silent or redundant gene, whose inactivation has been proved to be without any phenotypical effects under the relevant growth conditions.

However, so-called intergenic regions, referring to chromosomal segments in between of protein encoding genes, should not be taken into consideration as "neutral sites", since there is increasing evidence for an enormous coding potential with regard to ncRNAs within these regions, also in cyanobacteria (Steglich et al. 2008; Georg et al. 2009; Mitschke et al. 2011). In addition, inserting foreign DNA into 3′-untranslated reagions can lead to destabilization of the respective transcript (Meyer et al. 1996), since the 3′-ends of bacterial mRNAs are protected against exonucleolytic digestion by secondary terminator structures (reviewed in Rauhut and Klug 1999). In either case, one should run a control approach, additionally transforming the mutant cells with the complementation vector devoid of the GOI, thereby excluding phenotypical restoration by secondary effects.

Vectors for ectopic expression of genes by means of introduction of DNA into the chromosome should contain an antibiotic-resistance marker, a controllable or constitutive promoter (unless the own promoter of the GOI is used) and a site where the GOI can be inserted. These (clustered) features are flanked by two DNA sequences that are similar to the chromosomal site of recombination. If gene expression should be controlled by an inducible promoter, one should be aware of physiological outcomes, which may accompany gene expression. P_{petJ}, P_{petE}, P_{nirA} or a constitutive promoter (e.g. P_{rnpB}) can be used (see Sect. 24.4). The definition of a site, where DNA sequences could be inserted without a phenotypic effect can be difficult.

Generally, it seems more reasonable to use self replicating plasmids for this kind of studies. But note that in this case problems might occur with respect to the copy number of the respective plasmid. Using a plasmid with a copy number higher than the amount of chromosomal copies can lead to an overexpression of the GOI with new effects on the mutant phenotype. The same holds true, when using different promoters for expression. In order to avoid such problems, expression studies have to be made to exclude additional phenotypic alterations due to non adequate expression of the respective gene.

24.3.4 Overexpression and Isolation of Proteins

The most common strategy for study of gene function is classical inactivation strategy of a gene followed by phenotypical analysis. However, this approach may not be suitable if the phenotype is complex and the contribution of the gene product should be studied in time and space. The above described methods on complementation analysis can also be used for overexpression studies in order to demonstrate the effects of abundance of protein or ncRNA on cell metabolism or to isolate native proteins from cells. Therefore, a self-replicating plasmid with an appropriate copy number is suggested in combination with a strong but inducible promoter, as overexpression of certain genes might be deleterious to the cells. Moreover, controlled induction of expression allows the evaluation of immediate cellular responses to overexpression of an ectopic gene rather than steady-state conditions.

In *Synechococcus elongatus* PCC 7942 a *lacI-lacO* repression system similar to the *E. coli* system is effective (Clerico et al. 2007). Our laboratory and others (Tous et al. 2001) frequently use the *petJ* promoter for gene expression. The *petJ* gene encodes the electron carrier protein cytochrome c_{553}, which is normally expressed under copper (Cu^{2+}) limitation, thereby replacing plastocyanin. Cu^{2+} regulates the activity of this promoter proportionally to its concentration in the medium. In standard BG-11 medium—which contains about 0.3 μM of Cu^{2+}—this promoter facilitates a basal expression level. Increasing the Cu^{2+} concentration up to 5 μM during growth leads to complete repression, whereas P*petJ* dependent expression is induced in Cu^{2+}-free medium. Basal expression (in standard BG-11 medium) has been experienced to be typically sufficient for complementation of null mutants (Peter et al. 2009). Tous et al. (2001) have used this system to create a mutant of the gene encoding the RNA subunit of ribonuclease P (RNP), which is essentially involved in tRNA processing by cleaving at the 5′ ends of tRNA precursors. The RNA subunit (*rnpB*) as well as the protein component (*rnpA*), are essential *in vivo*. In order to analyze the *in vivo* processing pathway of tRNA precursors, the authors constructed a strain, which conditionally expresses *rnpB*. The *rnpB* gene was therefore inserted downstream of the *petJ* promoter within an appropriate vector, followed by its integration into a neutral genomic locus. The wild-type *rnpB* allele was hereupon deleted from its original locus. Addition of repressing amounts of Cu^{2+} eventually enabled the detection of defects in tRNA processing.

The Cu^{2+}-inducible promoter of *petE* (encoding the copper-binding electron carrier plastocyanin) (Briggs et al. 1990; Zhang et al. 1992), on the other hand, is an established tool for controlled overexpression of heterocyst differentiation factors in *Nostoc* sp. PCC 7120 (Buikema and Haselkorn 2001; Callahan and Buikema 2001; Wu et al. 2004), but has recently been proved successful for genetic studies in *Synechocystis* sp. PCC 6803 as well (Yang et al. 2008).

Another inducible expression system was established by Qi et al. (2005), who used the nitrate-inducible promoter of *nirA* from *Synechococcus* sp. strain PCC 7942 for the expression of transgenes in *Synechocystis* sp. PCC 6803. Since this promoter is repressed by ammonium-containing medium, the authors point at its suitability especially for the engineered production of potentially toxic peptides or intermediates in *Synechocystis* sp. PCC 6803.

Expression of gene fusion products has developed into a critical need for co-purification approaches to identify components of protein complexes. Therefore, several affinity-tags turned out to be useful in cyanobacteria. The histidine tagging technique is a common approach to facilitate purification of specific proteins or protein complexes. Bricker et al (1998) have constructed a *Synechocystis* sp. PCC 6803 strain fusing an extension of six histidines to the carboxyl terminus of the CP47 protein allowing rapid purification of highly active photosystem II complexes. In addition, a couple of proteins associated with photosystem II were highly enriched in such preparations. Further work of different groups revealed the existence of several photosystem II associated proteins, which were identified after purification of His-tagged photosystem II from *Thermosynechococcus elongatus* (Nowaczyk et al. 2006) or *Synechocystis* sp. PCC 6803 (Inoue-Kashino et al. 2008).

Our group routinely uses the 3xFLAG tag (Sigma-Aldrich) as a very sensitive epitope tag for isolation of less abundant proteins. Using this tag it was shown that the putative assembly factor Ycf37 binds to photosystem I (Dühring et al. 2006b). Furthermore, the H subunit of the magnesium chelatase was co-immunopurified with the 3xFLAG tagged regulator Gun4 (Sobotka et al. 2008). Karradt et al. (2008) have used GST-tagged NblA protein expressed in *Nostoc* sp. PCC 7120 to co-precipitate ClpC a component of the Clp protease.

24.3.5 Promoter Studies

In order to improve studies on gene expression in cyanobacteria, the promoterless *luxAB* genes, which encode for the structural proteins of luciferase from *Vibrio harveyi*, were introduced as reporter system several years ago. Likewise the reporter gene *gfp*, which encodes for the light-emitting green fluorescent protein (Gfp) of the jellyfish *Aequorea victoria*, has been widely used to monitor gene expression in cyanobacteria (Yoon and Golden 1998).

The *luxAB* reporter system, which was initially used to analyze the response to nitrogen depriviation in *Anabaena* strains (Wolk et al. 1991) has been widely applied to continuative studies on nitrogen fixation in filamentous strains (e.g. Leganes et al. 1994; Cai and Wolk 1997) and on environmental- and circadian-regulated gene expression in unicellular strains (recently reviewed by Mackey et al. 2007). The groups of S. Golden and T. Kondo used the *luxAB* system to identify individual colonies exhibiting mutations in genes involved in the circadian clock system of *Synechococcus elongatus* PCC7942 (Kondo et al. 1994). In principle any promoter can be cloned upstream of a reporter gene. In *Synechococcus elongatus* bacterial luciferase (*luxAB*) has been widely used for measuring circadian activity driven by different promoters permitting assays within particular colonies. Liu et al. (1995) inserted bacterial luciferase genes randomly into the *Synechococcus* genome and screened transformed clones for bioluminescence. Using this method the authors showed that the control of gene expression by circadian clocks is very widespread and targets the majority of genes.

Other reporters like GFP were especially helpful to demonstrate transcription within particular types of cells in a filament (e.g. Yoon and Golden 1998) or to show subcellular localization of FtsZ (Sakr et al. 2006). Kunert et al. (2000) constructed promoter probe vectors for *Synechocystis* sp. PCC 6803 using *gfp* and *luxAB* genes for studies of the promoter of the *isiAB* operon.

24.3.6 Transposon Mutagenesis

Transposon mutagenesis is a powerful screening tool to identify structural and regulatory components that are important for specific cellular functions. Random mutagenesis in *Nostoc* sp. PCC 7120 using the Tn5 transposon was first described by Borthakur and Haselkorn (1989). Tn5 is a DNA-element, which inserts into any DNA-sequence and thereby causes gene inactivation. The advantage of Tn5-mutagenesis is the possibility of positive selection using transposon mediated antibiotic resistance, while the site of integration can be idientified by various molecular techniques. These techniques include the recovery of transposon containing plasmids (Katayama et al. 1999; Koksharova and Wolk 2002b), direct genomic sequencing (Bhaya et al. 2001) or inverse PCR (Huang et al. 2000; Fujimori et al. 2005). Using this method, transposon mutants with a wide variety of phenotypes were obtained in *Nostoc* sp. strain ATCC 29133 (Cohen et al. 1994) and in *Nostoc* sp. PCC 7120 (Ernst et al. 1992). Recently established strategies involve commercialized transposon-based *in vitro* systems to insert transposons randomly into a genomic library (e.g. Bhaya et al. 2001; Shibata et al. 2001; Ozaki et al. 2007). Many more studies, which are not listed here, were conducted to identify genes that are involved in nitrogen or carbon fixation, in photosynthesis, motility, differentiation, metabolism, cell division and other processes.

In addition, reporter genes such as luciferase can be inserted into genomes using transposon mutagenesis in order to identify regulatory components (Wolk et al. 1991; Cai and Wolk 1997; Schwartz et al. 1998).

24.4 Conclusion

The publication of many cyanobacterial sequences in the last 10 years has generated interest in developing tools for different organisms. Although genetic tools are available for several freshwater cyanobacteria, only very few techniques have been published for marine cyanobacteria. These have to be developed in future. Some genetic tools used with freshwater strains may also be applicable for marine cyanobacteria. The next phase of genome analysis is the elucidation of the function of open reading frames with unknown function and regulatory RNA genes that will be identified in genome sequences. The combination of genome sequencing, identification of genes and genetic tools provides the opportunity to understand how cyanobacteria survive, how they adapt to their environment and to elucidate their physiological capacities for biotechnological applications.

Acknowledgments The work of A. W. and D. D. was supported by funds from the Priority Program SPP1258 of the Deutsche Forschungsgemeinschaft and FORSYS-Partner Program of BMBF.

References

Averhoff B (2004) DNA transport and natural transformation in mesophilic and thermophilic bacteria. J Bioenerg Biomembr 36: 25–33

Baier K, Nicklisch S, Grundner C, Reinecke J and Lockau W (2001) Expression of two *nblA*-homologous genes is required for phycobilisome degradation in nitrogen-starved *Synechocystis* sp. PCC6803. FEMS Microbiol Lett 195: 35–39

Barken KB, Pamp SJ, Yang L, Gjermansen M, Bertrand JJ, Klausen M, Givskov M, Whitchurch CB, Engel JN and Tolker-Nielsen T (2008) Roles of type IV pili, flagellum-mediated motility and extracellular DNA in the formation of mature multicellular structures in *Pseudomonas aeruginosa* biofilms. Environ Microbiol 10: 2331–2343

Barten R and Lill H (1995) DNA-uptake in the naturally competent cyanobacterium, *Synechocystis* sp. PCC 6803. FEMS Microbiol Lett 129: 83–88

Bhaya D (2004) Light matters: phototaxis and signal transduction in unicellular cyanobacteria. Mol Microbiol 53: 745–754

Bhaya D, Takahashi A and Grossman AR (2001) Light regulation of type IV pilus-dependent motility by chemosensor-like elements in *Synechocystis* PCC6803. Proc Natl Acad Sci U S A 98: 7540–7545

Biswas GD, Sox T, Blackman E and Sparling PF (1977) Factors affecting genetic transformation of *Neisseria gonorrhoeae*. J Bacteriol 129: 983–992

Borthakur D and Haselkorn R (1989) Tn5 mutagenesis of *Anabaena* sp. strain PCC 7120: isolation of a new mutant unable to grow without combined nitrogen. J Bacteriol 171: 5759–5761

Bradley DE (1974) The adsorption of *Pseudomonas aeruginosa* pilus-dependent bacteriophages to a host mutant with nonretractile pili. Virology 58: 149–163

Bricker TM, Morvant J, Masri N, Sutton HM and Frankel LK (1998) Isolation of a highly active photosystem II preparation from *Synechocystis* 6803 using a histidine-tagged mutant of CP 47. Biochim Biophys Acta 1409: 50–57

Briggs LM, Pecoraro VL and McIntosh L (1990) Copper-induced expression, cloning, and regulatory studies of the plastocyanin gene from the cyanobacterium *Synechocystis* sp. PCC 6803. Plant Mol Biol 15: 633–642

Brossay L, Paradis G, Fox R, Koomey M and Hebert J (1994) Identification, localization, and distribution of the PilT protein in *Neisseria gonorrhoeae*. Infect Immun 62: 2302–2308

Buikema WJ and Haselkorn R (1991) Isolation and complementation of nitrogen fixation mutants of the cyanobacterium *Anabaena* sp. strain PCC 7120. J Bacteriol 173: 1879–1885

Buikema WJ and Haselkorn R (2001) Expression of the *Anabaena hetR* gene from a copper-regulated promoter leads to heterocyst differentiation under repressing conditions. Proc Natl Acad Sci U S A 98: 2729–2734

Burrows LL (2005) Weapons of mass retraction. Mol Microbiol 57: 878–888

Cai Y and Wolk CP (1990) Use of a conditionally lethal gene in *Anabaena* sp. strain PCC 7120 to select for double recombinants and to entrap insertion sequences. J Bacteriol 172: 3138–3145

Cai Y and Wolk CP (1997) Nitrogen deprivation of *Anabaena* sp. strain PCC 7120 elicits rapid activation of a gene cluster that is essential for uptake and utilization of nitrate. J Bacteriol 179: 258–266

Callahan SM and Buikema WJ (2001) The role of HetN in maintenance of the heterocyst pattern in *Anabaena* sp. PCC 7120. Mol Microbiol 40: 941–950

Carpenter SD and Vermaas WFJ (1989) Directed mutagenesis to probe the structure and function of photosystem II. Physiol Plantarum 77: 436–443

Chauvat F, Rouet P, Bottin H and Boussac A (1989) Mutagenesis by random cloning of an *Escherichia coli* kanamycin resistance gene into the genome of the cyanobacterium *Synechocystis* PCC 6803: selection of mutants defective in photosynthesis. Mol Gen Genet 216: 51–59

Chiang GG, Schaefer MR and Grossman AR (1992) Transformation of the filamentous cyanobacterium *Fremyela diplosiphon* by conjugation or electroporation. Plant Physiol Biochem 30: 315–325

Chiang P and Burrows LL (2003) Biofilm formation by hyperpiliated mutants of *Pseudomonas aeruginosa*. J Bacteriol 185: 2374–2378

Chitnis VP and Chitnis PR (1993) PsaL subunit is required for the formation of photosystem I trimers in the cyanobacterium *Synechocystis* sp. PCC 6803. FEBS Lett 336: 330–334

Clerico EM, Ditty JL and Golden SS (2007) Specialized techniques for site-directed mutagenesis in cyanobacteria. Methods Mol Biol 362: 155–171

Cohen MF, Wallis JG, Campbell EL and Meeks JC (1994) Transposon mutagenesis of *Nostoc* sp. strain ATCC 29133, a filamentous cyanobacterium with multiple cellular differentiation alternatives. Microbiology 140 (Pt 12): 3233–3240

Collier JL and Grossman AR (1994) A small polypeptide triggers complete degradation of light-harvesting phycobiliproteins in nutrient-deprived cyanobacteria. EMBO J 13: 1039–1047

Derbyshire KM and Willetts NS (1987) Mobilization of the non-conjugative plasmid RSF1010: a genetic analysis of its origin of transfer. Mol Gen Genet 206: 154–160

Dienst D, Dühring U, Mollenkopf HJ, Vogel J, Golecki J, Hess WR and Wilde A (2008) The cyanobacterial homologue of the RNA chaperone Hfq is essential for motility of *Synechocystis* sp. PCC 6803. Microbiology 154: 3134–3143

Dismukes GC, Carrieri D, Bennette N, Ananyev GM and Posewitz MC (2008) Aquatic phototrophs: efficient alternatives to land-based crops for biofuels. Curr Opin Biotechnol 19: 235–240

Douglas SE (1998) Plastid evolution: origins, diversity, trends. Curr Opin Genet Dev 8: 655–661

Dühring U, Axmann IM, Hess WR and Wilde A (2006a) An internal antisense RNA regulates expression of the photosynthesis gene *isiA*. Proc Natl Acad Sci U S A 103: 7054–7058

Dühring U, Irrgang KD, Lunser K, Kehr J and Wilde A (2006b) Analysis of photosynthetic complexes from a cyanobacterial *ycf37* mutant. Biochim Biophys Acta 1757: 3–11

Elhai J and Wolk CP (1988) Conjugal transfer of DNA to cyanobacteria. Methods Enzymol 167: 747–754

Eriksen NT (2008) Production of phycocyanin—a pigment with applications in biology, biotechnology, foods and medicine. Appl Microbiol Biotechnol 80: 1–14

Ernst A, Black T, Cai Y, Panoff JM, Tiwari DN and Wolk CP (1992) Synthesis of nitrogenase in mutants of the cyanobacterium *Anabaena* sp. strain PCC 7120 affected in heterocyst development or metabolism. J Bacteriol 174: 6025–6032

Ferreira KN, Iverson TM, Maghlaoui K, Barber J and Iwata S (2004) Architecture of the photosynthetic oxygen-evolving center. Science 303: 1831–1838

Fujimori T, Higuchi M, Sato H, Aiba H, Muramatsu M, Hihara Y and Sonoike K (2005) The mutant of *sll1961*, which encodes a putative transcriptional regulator, has a defect in regulation of photosystem stoichiometry in the cyanobacterium *Synechocystis* sp. PCC 6803. Plant Physiol 139: 408–416

Funatsu G and Wittmann HG (1972) Ribosomal proteins. 33. Location of amino-acid replacements in protein S12 isolated from *Escherichia coli* mutants resistant to streptomycin. J Mol Biol 68: 547–550

Georg J, Voss B, Scholz I, Mitschke J, Wilde A, Hess WR (2009) Evidence for a major role of antisense RNAs in cyanobacterial gene regulation. Mol Syst Biol 5: 305

Gressel J (2008) Transgenics are imperative for biofuel crops. Plant Sci 174: 246–263

Gottesman S (2004) The small RNA regulators of *Escherichia coli*: roles and mechanisms. Annu Rev Microbiol 58: 303–328

Gottesman S (2005) Micros for microbes: non-coding regulatory RNAs in bacteria. Trends Genet 21: 399–404

Herdman M, Janvier M, Rippka R and Stanier R (1979) Genome size of cyanobacteria. J Gen Microbiol 111: 73–85

Huang G, Zhang L and Birch RG (2000) Rapid amplification and cloning of Tn5 flanking fragments by inverse PCR. Lett Appl Microbiol 31: 149–153

Inoue-Kashino N, Takahashi T, Ban A, Sugiura M, Takahashi Y, Satoh K and Kashino Y (2008) Evidence for a stable association of Psb30 (Ycf12) with photosystem II core complex in the cyanobacterium *Synechocystis* sp. PCC 6803. Photosynth Res 98: 323–335

Jakovljevic V, Leonardy S, Hoppert M and Sogaard-Andersen L (2008) PilB and PilT are ATPases acting antagonistically in type IV pilus function in *Myxococcus xanthus*. J Bacteriol 190: 2411–2421

Jordan P, Fromme P, Witt HT, Klukas O, Saenger W and Krauss N (2001) Three-dimensional structure of cyanobacterial photosystem I at 2.5 A resolution. Nature 411: 909–917

Jurcisek JA, Bookwalter JE, Baker BD, Fernandez S, Novotny LA, Munson RS Jr and Bakaletz LO (2007) The PilA protein of non-typeable *Haemophilus influenzae* plays a role in biofilm formation, adherence to epithelial cells and colonization of the mammalian upper respiratory tract. Mol Microbiol 65: 1288–1299

Kaneko T, Sato S, Kotani H, Tanaka A, Asamizu E, Nakamura Y, Miyajima N, Hirosawa M, Sugiura M, Sasamoto S, Kimura T, Hosouchi T, Matsuno A, Muraki A, Nakazaki N, Naruo K, Okumura S, Shimpo S, Takeuchi C, Wada T, Watanabe A, Yamada M, Yasuda M and Tabata S (1996) Sequence analysis of the genome of the unicellular cyanobacterium *Synechocystis* sp. strain PCC6803. II. Sequence determination of the entire genome and assignment of potential protein-coding regions. DNA Res 3: 109–136

Karradt A, Sobanski J, Mattow J, Lockau W and Baier K (2008) NblA, a key protein of phycobilisome degradation, interacts with ClpC, a HSP100 chaperone partner of a cyanobacterial Clp protease. J Biol Chem 283: 32394–32403
Katayama M, Tsinoremas NF, Kondo T and Golden SS (1999) *CpmA*, a gene involved in an output pathway of the cyanobacterial circadian system. J Bacteriol 181: 3516–3524
Kawano M, Reynolds AA, Miranda-Rios J and Storz G (2005) Detection of 5′- and 3′-UTR-derived small RNAs and *cis*-encoded antisense RNAs in *Escherichia coli*. Nucleic Acids Res 33: 1040–1050
Kennan RM, Dhungyel OP, Whittington RJ, Egerton JR and Rood JI (2001) The type IV fimbrial subunit gene (*fimA*) of *Dichelobacter nodosus* is essential for virulence, protease secretion, and natural competence. J Bacteriol 183: 4451–4458
Khudiakov I and Gromov BV (1973) Temperate cyanophage A-4 (L) of the blue-green alga *Anabaena variabilis*. Mikrobiologiia 42: 904–907
Koksharova OA and Wolk CP (2002a) Genetic tools for cyanobacteria. Appl Microbiol Biotechnol 58: 123–137
Koksharova OA and Wolk CP (2002b) A novel gene that bears a DnaJ motif influences cyanobacterial cell division. J Bacteriol 184: 5524–5528
Kondo T, Tsinoremas NF, Golden SS, Johnson CH, Kutsuna S and Ishiura M (1994) Circadian clock mutants of cyanobacteria. Science 266: 1233–1236
Kunert A, Hagemann M and Erdmann N (2000) Construction of promoter probe vectors for *Synechocystis* sp. PCC 6803 using the light-emitting reporter systems Gfp and LuxAB. J Microbiol Methods 41: 185–194
Labarre J, Chauvat F and Thuriaux P (1989) Insertional mutagenesis by random cloning of antibiotic resistance genes into the genome of the cyanobacterium *Synechocystis* strain PCC 6803. J Bacteriol 171: 3449–3457
Lanka E and Wilkins BM (1995) DNA processing reactions in bacterial conjugation. Annu Rev Biochem 64: 141–169
Lee J, Lee HJ, Shin MK and Ryu WS (2004) Versatile PCR-mediated insertion or deletion mutagenesis. Biotechniques 36: 398–400
Leganes F, Fernandez-Pinas F and Wolk CP (1994) Two mutations that block heterocyst differentiation have different effects on akinete differentiation in *Nostoc ellipsosporum*. Mol Microbiol 12: 679–684
Liu Y, Tsinoremas NF, Johnson CH, Lebedeva NV, Golden SS, Ishiura M and Kondo T (1995) Circadian orchestration of gene expression in cyanobacteria. Genes Dev 9: 1469–1478
Lluisma AO, Karmacharya N, Zarka A, Ben-Dov E, Zaritsky A and Boussiba S (2001) Suitability of *Anabaena* PCC7120 expressing mosquitocidal toxin genes from *Bacillus thuringiensis* subsp. *israelensis* for biotechnological application. Appl Microbiol Biotechnol 57: 161–166
Loll B, Kern J, Saenger W, Zouni A and Biesiadka J (2005) Towards complete cofactor arrangement in the 3.0 A resolution structure of photosystem II. Nature 438: 1040–1044
Lorenz MG and Wackernagel W (1994) Bacterial gene transfer by natural genetic transformation in the environment. Microbiol Rev 58: 563–602
Mackey SR, Ditty JL, Clerico EM and Golden SS (2007) Detection of rhythmic bioluminescence from luciferase reporters in cyanobacteria. Methods Mol Biol 362: 115–129
Marraccini P, Bulteau S, Cassier-Chauvat C, Mermet-Bouvier P and Chauvat F (1993) A conjugative plasmid vector for promoter analysis in several cyanobacteria of the genera *Synechococcus* and *Synechocystis*. Plant Mol Biol 23: 905–909
Matsuoka M, Takahama K and Ogawa T (2001) Gene replacement in cyanobacteria mediated by a dominant streptomycin-sensitive *rps12* gene that allows selection of mutants free from drug resistance markers. Microbiology 147: 2077–2087
Mattick JS (2002) Type IV pili and twitching motility. Annu Rev Microbiol 56: 289–314
Mattick JS, Whitchurch CB and Alm RA (1996) The molecular genetics of type-4 fimbriae in *Pseudomonas aeruginosa*—a review. Gene 179: 147–155
McDaniel L, Houchin LA, Williamson SJ and Paul JH (2002) Lysogeny in marine *Synechococcus*. Nature 415: 496

Mermet-Bouvier P, Cassier-Chauvat C and Marraccini P and Chauvat F (1993) Transfer and replication of RSF1010-derived plasmids in several cyanobacterial of the genera *Synechocystis* and *Synechococcus*. Curr Microbiol 27: 323–327

Merz AJ, So M and Sheetz MP (2000) Pilus retraction powers bacterial twitching motility. Nature 407: 98–102

Meyer BJ, Bartman AE and Schottel JL (1996) Isolation of a mRNA instability sequence that is *cis*-dominant to the *ompA* stability determinant in *Escherichia coli*. Gene 179: 263–270

Mitschke J, Georg J, Scholz I, Sharma CM, Dienst D, Bantscheff J, Voß B, Steglich C, Wilde A, Vogel J, Hess WR (2011) An experimentally anchored map of transcriptional start sites in the model cyanobacterium Synechocystis sp. PCC6803. Proc Natl Acad Sci U S A [Epub ahead of print]

Nakasugi K and Neilan BA (2006) Gene transfer in cyanobacteria. In: Pandalai S (ed) Recent Developments in Nucleic Acids Research, Vol 2, pp 83–114. Transworld Research Network, Kerala.

Nakasugi K, Svenson CJ and Neilan BA (2006) The competence gene, *comF*, from *Synechocystis* sp. strain PCC 6803 is involved in natural transformation, phototactic motility and piliation. Microbiology 152: 3623–3631

Nakasugi K, Alexova R, Svenson CJ and Neilan BA (2007) Functional analysis of PilT from the toxic cyanobacterium *Microcystis aeruginosa* PCC 7806. J Bacteriol 189: 1689–1697

Nowaczyk MM, Hebeler R, Schlodder E, Meyer HE, Warscheid B and Rögner M (2006) Psb27, a cyanobacterial lipoprotein, is involved in the repair cycle of photosystem II. Plant Cell 18: 3121–3131

Nudleman E and Kaiser D (2004) Pulling together with type IV pili. J Mol Microbiol Biotechnol 7: 52–62

Okamoto S and Ohmori M (2002) The cyanobacterial PilT protein responsible for cell motility and transformation hydrolyzes ATP. Plant Cell Physiol 43: 1127–1136

Ozaki H, Ikeuchi M, Ogawa T, Fukuzawa H and Sonoike K (2007) Large-scale analysis of chlorophyll fluorescence kinetics in *Synechocystis* sp. PCC 6803: identification of the factors involved in the modulation of photosystem stoichiometry. Plant Cell Physiol 48: 451–458

Peter E, Salinas A, Wallner T, Jeske D, Dienst D, Wilde A, Grimm B (2009) Differential requirement of two homologous proteins encoded by sll1214 and sll1874 for the reaction of Mg protoporphyrin monomethylester oxidative cyclase under aerobic and micro-oxic growth conditions. Biochim Biophys Acta 1787: 1458–1467

Provvedi R and Dubnau D (1999) ComEA is a DNA receptor for transformation of competent *Bacillus subtilis*. Mol Microbiol 31: 271–280

Qi Q, Hao M, Ng WO, Slater SC, Baszis SR, Weiss JD and Valentin HE (2005) Application of the *Synechococcus nirA* promoter to establish an inducible expression system for engineering the *Synechocystis* tocopherol pathway. Appl Environ Microbiol 71: 5678–5684

Rauhut R and Klug G (1999) mRNA degradation in bacteria. FEMS Microbiol Rev 23: 353–370

Rawlings DE and Tietze E (2001) Comparative biology of IncQ and IncQ-like plasmids. Microbiol Mol Biol Rev 65: 481–496, table of contents

Rodriguez-Ezpeleta N, Brinkmann H, Burey SC, Roure B, Burger G, Loffelhardt W, Bohnert HJ, Philippe H and Lang BF (2005) Monophyly of primary photosynthetic eukaryotes: green plants, red algae, and glaucophytes. Curr Biol 15: 1325–1330

Sakr S, Jeanjean R, Zhang CC and Arcondeguy T (2006) Inhibition of cell division suppresses heterocyst development in *Anabaena* sp. strain PCC 7120. J Bacteriol 188: 1396–1404

Schwartz SH, Black TA, Jager K, Panoff JM and Wolk CP (1998) Regulation of an osmoticum-responsive gene in *Anabaena* sp. strain PCC 7120. J Bacteriol 180: 6332–6337

Shen G and Vermaas WF (1994) Chlorophyll in a *Synechocystis* sp. PCC 6803 mutant without photosystem I and photosystem II core complexes. Evidence for peripheral antenna chlorophylls in cyanobacteria. J Biol Chem 269: 13904–13910

Shestakov SV and Khyen NT (1970) Evidence for genetic transformation in blue-green alga *Anacystis nidulans*. Mol Gen Genet 107: 372–375

Shestakov SV, Anbudurai PR, Stanbekova GE, Gadzhiev A, Lind LK and Pakrasi HB (1994) Molecular cloning and characterization of the *ctpA* gene encoding a carboxyl-terminal processing protease. Analysis of a spontaneous photosystem II-deficient mutant strain of the cyanobacterium *Synechocystis* sp. PCC 6803. J Biol Chem 269: 19354–19359

Shibata M, Ohkawa H, Kaneko T, Fukuzawa H, Tabata S, Kaplan A and Ogawa T (2001) Distinct constitutive and low-CO_2-induced CO_2 uptake systems in cyanobacteria: genes involved and their phylogenetic relationship with homologous genes in other organisms. Proc Natl Acad Sci U S A 98: 11789–11794

Singh S, Kate BN and Banerjee UC (2005) Bioactive compounds from cyanobacteria and microalgae: an overview. Crit Rev Biotechnol 25: 73–95

Smart L, Anderson S and McIntosh L (1991) Targeted genetic inactivation of the photosystem I reaction center in the cyanobacterium *Synechocystis* sp. PCC 6803. EMBO J 10: 3289–3296

Smart LB, Bowlby NR, Anderson SL, Sithole I and McIntosh L (1994) Genetic manipulation of the cyanobacterium *Synechocystis* sp. PCC 6803. Plant Physiol 104: 349–354

Sobotka R, Duhring U, Komenda J, Peter E, Gardian Z, Tichy M, Grimm B and Wilde A (2008) Importance of the cyanobacterial Gun4 protein for chlorophyll metabolism and assembly of photosynthetic complexes. J Biol Chem 283: 25794–25802

Sparling PF (1966) Genetic transformation of *Neisseria gonorrhoeae* to streptomycin resistance. J Bacteriol 92: 1364–1371

Steglich C, Futschik ME, Lindell D, Voss B, Chisholm SW and Hess WR (2008) The challenge of regulation in a minimal photoautotroph: non-coding RNAs in *Prochlorococcus*. PLoS Genet 4: e1000173

Stone BJ and Kwaik YA (1999) Natural competence for DNA transformation by *Legionella pneumophila* and its association with expression of type IV pili. J Bacteriol 181: 1395–1402

Takahama K, Matsuoka M, Nagahama K and Ogawa T (2004) High-frequency gene replacement in cyanobacteria using a heterologous *rps12* gene. Plant Cell Physiol 45: 333–339

Tamagnini P, Leitao E, Oliveira P, Ferreira D, Pinto F, Harris DJ, Heidorn T and Lindblad P (2007) Cyanobacterial hydrogenases: diversity, regulation and applications. FEMS Microbiol Rev 31: 692–720

Taroncher-Oldenburg G and Stephanopoulos G (2000) Targeted, PCR-based gene disruption in cyanobacteria: inactivation of the polyhydroxyalkanoic acid synthase genes in *Synechocystis* sp. PCC6803. Appl Microbiol Biotechnol 54: 677–680

Thiel T and Poo H (1989) Transformation of a filamentous cyanobacterium by electroporation. J Bacteriol 171: 5743–5746

Timms AR, Steingrimsdottir H, Lehmann AR and Bridges BA (1992) Mutant sequences in the *rpsL* gene of *Escherichia coli* B/r: mechanistic implications for spontaneous and ultraviolet light mutagenesis. Mol Gen Genet 232: 89–96

Tous C, Vega-Palas MA and Vioque A (2001) Conditional expression of RNase P in the cyanobacterium *Synechocystis* sp. PCC6803 allows detection of precursor RNAs. Insight in the *in vivo* maturation pathway of transfer and other stable RNAs. J Biol Chem 276: 29059–29066

Tsinoremas NF, Kutach AK, Strayer CA and Golden SS (1994) Efficient gene transfer in *Synechococcus* sp. strains PCC 7942 and PCC 6301 by interspecies conjugation and chromosomal recombination. J Bacteriol 176: 6764–6768

Waterbury JB and Valois FW (1993) Resistance to co-occurring phages enables marine *Synechococcus* communities to coexist with cyanophages abundant in seawater. Appl Environ Microbiol 59: 3393–3399

Williams J (1988) Construction of specific mutations in photosystem II photosynthetic reaction center by genetic engineering methods in *Synechocystis* 6803. Methods Enzymol 167: 766–778

Williams JG and Szalay AA (1983) Stable integration of foreign DNA into the chromosome of the cyanobacterium *Synechococcus* R2. Gene 24: 37–51

Wolfgang M, Lauer P, Park HS, Brossay L, Hebert J and Koomey M (1998) PilT mutations lead to simultaneous defects in competence for natural transformation and twitching motility in piliated *Neisseria gonorrhoeae*. Mol Microbiol 29: 321–330

Wolk CP, Vonshak A, Kehoe P and Elhai J (1984) Construction of shuttle vectors capable of conjugative transfer from *Escherichia coli* to nitrogen-fixing filamentous cyanobacteria. Proc Natl Acad Sci U S A 81: 1561–1565

Wolk CP, Cai Y, Cardemil L, Flores E, Hohn B, Murry M, Schmetterer G, Schrautemeier B and Wilson R (1988) Isolation and complementation of mutants of *Anabaena* sp. strain PCC 7120 unable to grow aerobically on dinitrogen. J Bacteriol 170: 1239–1244

Wolk CP, Cai Y and Panoff JM (1991) Use of a transposon with luciferase as a reporter to identify environmentally responsive genes in a cyanobacterium. Proc Natl Acad Sci U S A 88: 5355–5359

Wu X, Liu D, Lee MH and Golden JW (2004) *PatS* minigenes inhibit heterocyst development of *Anabaena* sp. strain PCC 7120. J Bacteriol 186: 6422–6429

Yang Y, Yin C, Li W and Xu X (2008) Alpha-tocopherol is essential for acquired chill-light tolerance in the cyanobacterium *Synechocystis* sp. strain PCC 6803. J Bacteriol 190: 1554–1560

Yoon HS and Golden JW (1998) Heterocyst pattern formation controlled by a diffusible peptide. Science 282: 935–938

Yoshida T, Nagasaki K, Takashima Y, Shirai Y, Tomaru Y, Takao Y, Sakamoto S, Hiroishi S and Ogata H (2008) Ma-LMM01 infecting toxic *Microcystis aeruginosa* illuminates diverse cyanophage genome strategies. J Bacteriol 190: 1762–1772

Yoshihara S and Ikeuchi M (2004) Phototactic motility in the unicellular cyanobacterium *Synechocystis* sp. PCC 6803. Photochem Photobiol Sci 3: 512–518

Yoshihara S, Geng X, Okamoto S, Yura K, Murata T, Go M, Ohmori M and Ikeuchi M (2001) Mutational analysis of genes involved in pilus structure, motility and transformation competency in the unicellular motile cyanobacterium *Synechocystis* sp. PCC 6803. Plant Cell Physiol 42: 63–73

Yoshihara S, Geng X and Ikeuchi M (2002) *PilG* Gene cluster and split *pilL* genes involved in pilus biogenesis, motility and genetic transformation in the cyanobacterium *Synechocystis* sp. PCC 6803. Plant Cell Physiol 43: 513–521

Yura K, Toh H and Go M (1999) Putative mechanism of natural transformation as deduced from genome data. DNA Res 6: 75–82

Zhang L, McSpadden B, Pakrasi HB and Whitmarsh J (1992) Copper-mediated regulation of cytochrome c553 and plastocyanin in the cyanobacterium *Synechocystis* 6803. J Biol Chem 267: 19054–19059

Zhu J, Jager K, Black T, Zarka K, Koksharova O and Wolk CP (2001) HcwA, an autolysin, is required for heterocyst maturation in *Anabaena* sp. strain PCC 7120. J Bacteriol 183: 6841–6851

Zinchenko VV, Piven IV, Melnik VA and Shestakov SV (1999) Vectors for the complementation analysis of cyanobacterial mutants. Russ J Genet 35: 228–232

Index

A

AAA+ type ATPase, 687
ABC transporter family, 425, 449
Abderitic philosophers, 3
Abiogenic O_2, 31
A-branch PSI, 317, 319
Acaryochloris marina, 164, 173, 177, 265, 659, 664
  MBIC 11017, 211, 473
Acceptor A_0, 318
Acceptor A_1, 318
Acceptor side PSI, 318
Acceptor side PSII, 299, 364
Acetate algae, 34
Acidianus ambivalens, 476
Adaptation, 36, 669
Adaptive behaviour, 124
Adaptive response, 125
Adenylate Kinase, 17
Adenylate System, 15
ADP, 18, 239
Aequorea victoria, 696
Aerobic respiration, 40, 89
Agmenellum quadruplicatum PR-6, 686
Akilia island, 268
Akinete fossils, 279
Akinetes, 279
Anoxygenic phototrophs, 276
Allochromatium vinosum, 147
Allophycocyanin (APC), 36, 226, 315, 397, 415
Allosterically regulated enzymes, 18
Alternative nitrogenases, 139
Amicyanins, 616
Anabaena
  7120, 48
  azotica, 142, 151
  cylindrica, vi
Anabaena sp., 51, 141, 547
  7119, 519, 520
  7120, 519
  PCC 7119, 520
  PCC 7120, 457, 458, 634, 641, 643,
  variabilis 29413/7937, 519
Anabaena variabilis, 119, 140, 149, 151, 523, 547, 581, 658, 664, 672, 674–676
  Anabaena variabilis ATCC 29413, 472, 634
Anabolism, 23, 33
Anacystis nidulans, 43, 46, 51, 127, 131, 132, 135, 202, 515, 531
  Anacystis nidulans (*Synechococcus* sp. PCC6301), 46
  Anacystis nidulans R2, 686, 690
Anaerobic carbonate respiration, 34
Anaerobic CO_2-cycle, 41
Anaerobic RET, 41
Anaplerotic reactions, 33
Anastomoses, 44
anfH, 140
anfHDGK, 139
Anoxic ecological niches, 32
Anoxygenic
  organisms, 593
  photosynthesis, 266, 274
  photosynthetic bacteria, 616
  phototrophs, 272
  procyanobacteria, 279
Anoxyphototrophs, 44
Antenna complexes, 37
Antenna system, 273
  PSI, 305.311, 320
  PSII, 304
Antimalarial drugs, 192
APC core, 227
Aphanizomenon flos-aquae, 523

G. A. Peschek et al. (eds.), *Bioenergetic Processes of Cyanobacteria*,
DOI 10.1007/978-94-007-0388-9,

Apicomplexa, 30, 43, 92
Aquifex aeolicus, 249
Arabidopsis thaliana, 233, 460, 529
Archaea, 41, 274
Archaeoglobus fulgidus, 474
Archaeon, 8
Archean origin, 271
Aristotle, 3
Arthrospira, 620
Arthrospira maxima, 516, 665
Arthrospira PCC9438, 245, 246
Arthrospira sp., 92
  maxima sp. CS-328, 472, 527, 533
  Ascaris suum, 474
Assimilation, 19, 23
Astasia longa, 92
Atmospheric hydrogen, 279
Atmospheric oxygen, 267, 269, 270
ATP, 17, 46, 55, 111, 198, 239
  binding, 250
  cycle, 18, 287, 474
ATPase, 26
ATP-binding cassette transporter FutABC, 618
Atx1, 617
Autopoietically, 128
Autotrophic CO_2 fixation, 29
7-azatryptophan, 153
Azolla, 142
Azomonas macrocytogenes, 139
Azospirillum brasilense Cd, 139
Azotobacter chroococcum, 139
Azotobacter vinelandii, 29, 139
Azurins, 616

B

b_6f complex, 38, 49, 197
ba_3 type oxidase, 371, 377
Bacillus PS3, 241, 243, 245
Bacillus spp., 671
Bacillus subtilis, 688
Bacillus (*B.*) *licheniformis*, 505
Bacillus (*B.*) *subtilis*, 474, 487
Bacteria, 274
Bacterial-type desaturase (CrtI), 224
Bacteriochlorophyll-based photosynthesis, 274
Bacterioferritins (BfrA and BfrB), 618
Bacteriopheophytin, 273
Bacteriorhodopsin, 232, 273
Baeocytes, 279
Banded iron formation, 269
Barstar, 556
B-branch PSI, 317, 319
bc_1 complex, 38, 96, 197, 589
BChl α, 276
bd-quinol oxidase, 54
bd-type, 96
Berkner-Marshall point, 613
Bertalanffy, Ludwig von, 19, 29
b_H/c_i couple, 594
Bicarbonate transporters, 449, 455
Bidirectional hydrogenase, 146–148
Bilin lyases, 221
Binding energy, 559
Binuclear center (BNC), 366, 372, 374, 378
  Binuclear copper center (CuA), 201
  Binuclear reaction center, 51
Bioavailability, 611
Bioenergetic processes, 23
Bioenergetics, 3
Biofertilizers, 151
Biofilms, 148
Biological Big-Bang, 13
Biological combustion, 29
Biological dichotomies, 19
Biological energy conversion, 17
Biomarkers, 271
Biomass, 30
Biosphere, 159
Biosynthesis of ATP, 30
Biotechnological applications, 685
Black smokers, 30
Blue-green algae, 265
BLUF domain, 413
bo cytochrome, 201
bo_3-type, 96
Boltzmann, 21
bo-type quinol oxidase, 658
Boyer, P.D., 242
Bradyrhizobium japonicum, 594
Brassica rapa, 547, 548
Bronze Age, 608
Buck Reef Chert, 267
Burkholderia pseudomallei, 164
  Transporter BCH-1, BicA, SbtA, 449

C

C550, 520
caa_3 oxidase, 372
Calothrix sp, 92
Calvin cycle, 23, 34, 40, 141, 279
Calvin-Benson cycle, 222
Cambrian Period, 267
Campylobacter (*C.*) *jejuni*, 461, 474
Car triplet (^{3}Car), 307, 323
Carbon concentration mechanism, 222
Carbon isotopes, 268
Carbon nutrition, 21

Carbon storage, 127
Carbonate rocks, 267
Carboniferous period, 13
Carboxysome, 449
Car_{D1}, 346
Car_{D2}, 346
Carotenoid, 37, 595
Carotenoid biosynthesis, 218
Carotenoids (Cars), 301, 306, 323
ß-carotene monoketolase, 413
β-carotene, 226
β-carotene hydroxylase, 218, 221
β-carotene ketolases, 221
  chlorophyll-based photosynthesis, 274
ζ-carotene desaturase, 218
Catabolism, 19, 23
Catalase-peroxidase (KatG), 36, 160, 162
Catalases, 160
cbb_3 oxidase, 378
Cc_6, 616, 618, 621, 623
CCMP1986, 518
Cell membranes, 229
Chance, 24
Channel I, II of PSII, 300
Channels H_2O, O_2, H^+, 359, 377
Chaos Game Representation (CGR), 101
Charge separation, 40, 199, 578
Charge transfer state, 414
Chemiosmosis, 17
Chemiosmotic electron transport phosphorylation, 21
Chemolitho(auto)trophs, 25
Chemotrophic organism, 33
Chemotrophy, 21
Chert, 267
Chl *a*, 232
Chl *a*/Chl *a*' dimer, 317
Chl *a*/Chl *a*' heterodimer, 224
Chl triplet (^{3}Chl), 323
Chl Z_{D1}, 292, 297, 304, 346
Chl Z_{D2}, 292, 297, 304, 346
Chlamydomonas reinhardtii, 49, 245, 319, 545, 547, 552, 573, 576, 582, 584, 585, 587, 595
Chlamydomonas, 589
Chl_{D1}, Chl_{D2}, 301, 346
ChlLNB, 218
Chlorella pyrenoidosa, 430
Chlorobi group, 276
Chlorobiaceae, 273
Chlorobium tepidum, 274, 275
Chlorobium, 198
Chloroflexi, 276
Chloroflexus aurantiacus, 274, 275
Chloroflexus, 278
Chloromitochondrion, 89
Chlorophyll
  Chlorophyll (Chl) *d*, 211
  Chlorophyll *a*, 594
  Chlorophyll a, vi
  Chlorophyll b, 278
  Chlorophyll b, vi
  Chlorophyll, 37, 191
Chloroplast, 89, 92, 189, 338, 340, 461
Chloroplast ATPase, 191
Chlostridium
  paradoxum, 245
  pasteurianum, 139, 143
  perfringens, 634
Chromatium spp. (*Chlorobium* spp.), 41
Chromatophores, 44, 58, 662, 673
Chroococcidiopsis thermalis, 150
α-chymotrypsin, 554
C_2H_2-reduction, 145, 152
C_i transport, 48
Citric acid cycle, 474
Class I *c*-type cytochrome, 618
Clean Energy, 151
ClpC, 695
[4Fe-4S] Cluster, 138
CN-insensitive respiration, 229
$^{12}CO_2$, 268
Coacervate, 14
ComF, 688
Common evolutionary origin, 273
Comparative genomics, 213
Complementation studies, 693
Complex I, 197, 365, 445
Complex II, 365
Complex III, 38
Conjugation, 686
Continuous cultures, 109
Convergent evolution, 34, 38, 97
Convergent evolutionary path, 32
Conversion Hypothesis, 9, 42, 95
Cooperative catalysis, 250
Coordination chemistry, 610
Copper, 608, 614
Copper dimer, 201
Copper sulphides, 561
Copper-transporting P-type ATPases, 617
Cosmological Big-Bang, 13
Coulombic energy, 559
COX, 32, 43, 51, 52, 55, 134, 197, 202, 342, 345, 365, 375, 562, 580, 607, 658, 664
  coxA, 669
  coxBAC, 658
  coxBAC, 671

CP43 (PsbC), 289, 294, 304, 307, 345
CP43, 232, 234
CP47 (PsbB), 289, 294, 304, 307, 345
CP47, 232
CpcC, 221
CpcD2, 221
CpcJ, 221, 226
CpeC, 221
CpeG, 221
c-ring, 251
Crocosphaera watsonii, 142, 674
Crocosphaera watsonii sp. WH8501, 407
Crocosphaera, 470, 673
Croococcales, 470
CRR1, 460
crtB, 218
crtI, 218
crtO, 218
crtW, 218
Crystal structure
  PSI of cyanobacteria, 291, 309
  Plants, 309, 315
  PSII, 289, 293, 302, 306, 308, 343, 428
  COX, 368
  Mntc, 425
  OPC, 409, 415
  QFR, 485, 492, 499
  SQOR, 478, 502
CtaA, 617, 618
ctaCDE, 658
CtpA protease, 429, 433
c-type heme, 38
Cu_A center, 52, 662, 664
Cu_A, 52, 366, 372
Cu_B center, 202
Cu_B, 51, 52, 366, 373
CupA, 454
CupB, 455
CupS, 454
Cyanobacteria, 43
Cyanobacterial
  ancestor, 278
  biomarkers, 267
  fermentation, 148
Cyanophora paradoxa, 43
Cyanothece, 470, 673
Cyanothece sp.
  ATCC 51142, 162
  ATTC51142, 405, 470
  CCY0110, 164, 169, 674
  PCC7424., 169
  CCY0110, 470
  PCC7424, 470
  PCC7425, 408, 470
  PCC7822, 408, 470
  PCC8801, 408, 470
  PCC8802, 470
Cyclic electron flow, 448, 457
Cyclic electron transport, 273, 580
Cyclic photophosphorylation, 579
Cyclophilin, 176
1-Cys Peroxiredoxin, 177
2-Cys Peroxiredoxin, 177
2-Cys Prx, 178
Cyt b_6f, 191, 228, 591, 607
Cyt b_6f complex, 37, 96, 199, 216, 286, 515, 516, 562, 564, 574, 576, 573, 589, 665
Cyt b_6f structure, 573
Cyt b_6, 576, 581
Cyt bc_1, 49
Cyt *bd* oxidase, 591
Cyt *bd* quinol oxidase, 228, 229
Cyt *c* oxidase, 228
Cyt c550, 515, 517, 534, 535
Cyt c552, 515
Cyt c554, 515
Cyt c_6, 49, 52, 224, 228, 299, 310, 314, 515–518, 520, 521, 523, 526, 527, 529, 536
Cyt cM, 516, 517, 520, 531, 533, 534
Cyt *f*, 576, 578, 581, 590
Cyt oxidase, 516
Cytochrome 190, 552, 371
Cytochrome a_3-Cu_B binuclear center, 202
Cytochrome *b* proteins, 500
Cytochrome b/Fe–S complex, 276
Cytochrome *b*559, 296, 300, 433
Cytochrome ba_3 oxidase, 562
Cytochrome *bc* complex, 145
Cytochrome bc_1 complex, 564
Cytochrome bc_1, 199
Cytochrome *bd* oxidase, 580
Cytochrome *bd* quinol oxidase, 51, 658
Cytochrome *bf* complex, 545
Cytochrome *bf*, 545
Cytochrome *c* oxidase, *see* COX
Cytochrome *c* peroxidase, 546
Cytochrome *c*, 203, 370, 546
Cytochrome c_2, 564
Cytochrome c_{553}, 132
Cytochrome c_6, 38, 52, 96, 199, 203, 204, 561, 562, 574, 607, 657, 665, 676
Cytochrome c_M, 55, 669
Cytochrome *f*, 37, 38, 55, 541, 545, 547–549, 553, 554, 557, 561, 562, 564, 565
Cytochrome oxidase, 132

Cytoplasm, 113, 427, 474, 580
Cytoplasmic membrane, 44, 48, 131, 458, 474, 657, 669
Cytosol, 286, 449

D

D1/D2/CP47/Cyt b559 complex, 288
D1/D2/Cyt b559 complex, 288
D1/D2 heterodimer, 289, 292, 310, 344, 428
D1-protein, 234
Darwinian evolution, 5
D-channel, 664
5-Deazariboflavin, 551
Debye-Hückel theory, 552
Dehydroascorbate reductase, 160
Dehydrogenase complex, 456
Dehydrogenase module, 47
DELSEED sequence, 253
Demokritos, 3
Desulfotomaculum sp., 33
Desulfovibrio, 536
Desulfovibrio (*D.*) *gigas*, 631
Desulfovibrio sp., 33, 143
Desulfuricants, 41
Diazotrophs, 56, 671
2, 5-Dibromo-3-methyl-6-isopropyl-*p*-benzoquinone, 585
Diffusional interaction, 544
Digalactosyldiacylglacerol (DGDG), 308
Di-heme cytochrome c peroxidases (DiHCCPs), 173
Diiron centre, 631, 644
Diiron-containing domain, 641
Di-manganese catalase, 26, 31
Di-manganese cluster, 171
Dimerization domain PSII, 298, 308
Dinitrogenase reductase, 138
Dinitrogen fixing organisms, 137
Dissimilation, 23
Dissimilatory, 19
Distance matrices, 102
DK, 138
Double membranes, 95
D pathway, 376
DrgA, 451
Dryopteris crassirhizoma, 547
Dual-function biomembrane, 54
Dual-function PET-RET, 44

E

E.coli plasmid pBR322, 688
Echinenone, 226
Ecological sulfur cycle, 41
Elastic Power Transmission, 249
Electrochemical proton potential, 474, 475
Electron transfer, 341, 370, 503, 542, 544, 662
Electron transport, 24, 199
  chain, 39, 145, 293, 299, 317, 574
  protagonists, 35
Electronic coupling, 543
Electrostatic interactions, 559
Electrostatic potentials, 559
Embden-Meyerhof-Parnas pathway, 25
Emerson Enhancement Effect, vi
Emerson, Robert, 194
Endergonic, 24
ENDOR, 356
Endosymbiont event, 8
Endosymbiont hypothesis, 7, 43, 89
  generalized, 30
  unified, 30
Endosymbiosis, 92, 104, 685
Energetic net efficiency, 25
Energized membrane, 24
Energy
  barrier, 558
  charge, 16
  conversion, 22, 211
  transfer, 227
Enterobacteria, 632
Enteromorpha prolifera, 547
Entropy, 21
Environmental (geological) constraints, 41
Environmental reducing power, 33
EPR, 302, 356
Ergotrophic combinations, 25
Ergotrophic Hexagon, 4, 33
Escherichia coli, 147, 201, 533, 645, 646
  complex I, 456
  QFR, SQR, 474
  QfrA, 477
E state COX, 373
Ethane-formation, 139
Etiolation, 25
ETP, 21, 33, 34
Euglena gracilis, 92
Eukarya, 274
Eukaryotic plant cell, 5
Evolution of photosynthesis, 276, 340
Evolutionary transition, 26
EXAFS, 302, 356, 426
Excisase, 141
Excitation energy transfer (EET), 320, 397
Exergonic, 24
External phosphate, 113
Extracellular sheaths, 267

F

F_1 part, 240
F_1-ATPase, 254
$F_{420}H_2$ cofactor, 632
F_A, 315, 319
F_B, 315, 319
FAD prosthetic group, 476
binding domain, 485
FdxH, 138, 141, 142
FeCys4 centre, 631
Fe-Mo cofactor, 671
Fe-nitrogenase, 139, 140
Fe-protein, 138
Fermentation, 22
Fermi's Golden Rule, 542, 574
Ferredoxin, 23, 47, 138, 141, 144, 316, 576, 579
Ferredoxin:plastoquinone oxidoreductase (FQR), 461, 579
Ferredoxin-$NADP^+$ reductase (FNR), 39
Ferredoxin-NADPH-Oxidoreductase, 254
Ferredoxin-type FdxH, 138
FeS cluster, 48, 311, 445, 461, 476, 487
Filamentous microfossils, 270
Firmicutes, 276
Fix genes, 138
Flavin mononucleotide (FMN), 631
Flavodiiron proteins (FDP), 631
Flavodoxin, 138, 287, 461
Flavodoxin-like domain, 641
Flavohemoglobin family, 632
Flavorubredoxin (FlRd), 632
Flow-equilibrium, 19, 33
Flow-force relationship, 117, 119
FlRd, 646
flrd gene, 646
FNR, 48, 143, 149, 451, 577, 579
F_o part, 240
Foerster Resonance Energy Transfer (FRET), 244
F_oF_1-ATPase, 26, 239
Formaldehyde, 34
Formation of NH_3, 138
Fossil record, 90
Four helix bundle, 500
F_R state COX, 373
Franck, J., 194
Franck-Condon factor, 543, 544
Free-living proto-chloromitochondria, 35
Fremyella diplosiphon, 227
Frequency Chaos Game Representation (FCGR), 101
matrices, 100
FtsZ, 696
F-type ATPases, 59
Fumarate hydratase, 222
FurA, 674
Fusion model, 277
Fusion organism, 277
FutA2, 618
FutB, 618
FutC, 618
F_x, 311, 319

G

Galderia sulfuraria, 314
Gallus gallus, 477
Gamow's tunnelling equation, 543
Generalized endosymbiont hypothesis, 43
Generatio spontanea, 5
Genetic manipulation of cyanobacteria, 685
Genome signature, 90, 98, 100
Geobacter sulfurreducens, 529
Giardia intestinalis, 645
Gibbs (free) energy, 20, 337, 357, 365, 426
Gibbs-Helmholtz equation, 4, 20
GlcP, 222
Gleobacter violaceus PCC 7421, 401, 423, 471
Gleobacteria, 471
Global carbon cycle, 271
Gloeobacter, 664
Gloeobacter violaceus, 36, 131, 164, 216, 227, 534, 581, 580, 659, 667
Gloeobacter violaceus PCC 7421, 211, 520
Gloeothece, 673
Glutaredoxin, 176
Glutathione peroxidase, 160, 172, 179
Glutathione, 160, 176
Glycogen, 148
Glycogen formation, 127
Glycogen pool, 111
Glycolate oxidase, 54
Glycolysis, 22, 222
Gram-positive bacteria, 474
Gram-negative bacteria, 141
Grand Canyon, 271
Granick hypothesis, 275
Great oxidation event (GOE), 269, 272
Green fluorescent protein (GFP), 696
Green sulfur bacteria, 49, 234, 277
Greenhouse effect, 42
Grotthus type mechanism, 377
Gun4, 695
Gunflint Formation, 271

H

3'-hydroxyechinenone (hECN), 408, 410, 414
H+/ ATP stoichiometry, 111
H+/ATP ratio, 246
H^+/Na^+ antiporters, 55
H_2-dependent O_2-uptake, 145
H_2O_2-containing atmosphere, 31
H_2-production, 152
H_2-uptake, 145
Haematin, 192
Haemozoin, 192
Haldane, J. B. S., 11
Haloarcula marismortui, 164
Halobacterial (retinal-based) photosynthesis, 26
H-bond (network), 317, 319, 352, 416
Heidegger, M., 3, 14
Heliobacillus mobilis, 247, 275
Heliobacter pylori, 461, 474
Heliobacteria, 273, 277, 593
Heliobacteriaceae, 49
Heliobacterium gestii, 140
Heme a, 368
Heme a_3, 368
Heme b_H, 592
Heme b_L, 578, 591
Heme catalases, 36
Heme $c_{i,}$ 592, 593
Heme group(s), 474
Heme-copper oxidase, 49, 58, 658
Herakleitos, 3
Heterocystous cyanobacteria, 674
Heterocysts, 139, 279, 674
Heterotrophy, 33
HetR, 141, 280, 674, 675
H/D exchange effect, 353
H–H cycle, 29
High-energy/energy-rich compound, 22
Hill, R., 195
His-tag, 452, 458
Histohaematins, 190
Homo sapiens sapiens, 30
Homologous recombination, 690
Horse-heart cyt c, 516
HoxE, 146
HoxEF, 147
HoxEFUYH, 147
HoxFU, 146
HoxFUYH, 146
HoxH, 147
HoxUYHWhypAB, 147
HoxW, 147
HoxYH, 146
Hudson Bay, 271
hupL, 145
hupL, 146
hupS, 145
hupSL, 145, 146
hupW, 146
Hydrogen bacteria, 25
Hydrogen fermentation, 42
Hydrogen peroxide, 192
Hydrogen relief valve, 42, 47
Hydrogenase, 46, 47, 137, 144
Hydrogenosomes, 94, 99
Hydrolysis of ATP, 250
Hydroxyl radicals, 192, 341, 423
Hyp genes, 147

I

Ilyobacter tatraricus, 245
IncP, 689
IncQ vectors, 689
Inorganic phosphate, 18, 239
Intracytoplasmic membranes (ICM), 44, 46, 48, 131, 133, 657
Invaginations, 94
Ionic strength, 203
Iron Age, 608
Iron, 608, 614
Iron–sulfur clusters, 99
IsiA, 315, 400
IsiAB, 696
Isoenergetic, 19
Isotope fractionation, 268
Iron starvation, 400

K

K-channel, 664
Keilin, 193
Kinetic models, 621
Kinetics, 550
Klebsiella pneumoniae, 138
Kluyver, 5
Kok cycle, 351
K pathway, 377
"Knallgas" reaction, 145

L

lacI-lacO, 694
Lactic acid fermentation, 23
Lactobacillus plantarum, 31, 169
Lateral gene transfer, 19, 104
Leibniz, G. W., 3
Leucothrix sp., 92
Leukippos, 3
LexA, 147

LHCI, 315
LHCII complexes, 191
Light compensation point, 228
Light harvesting complexes (LHCs), 286, 396
Light intensity, 253
Light-harvesting systems, 272
Light induced charge separation, 345
  "transfer to the trap limited", 349
  "trapping limited", 349
Light-inhibition of respiration, 54
Light-limited growth, 44
Linker polypeptides, 226
Linker proteins, 221
  Linker protein L_{CM}, 397
Lipids, 307, 322
Lithotrophic nutrition, 21
Lithotrophy, 33
LOV domain, 415
Low potential chain, 578
Low potential cyt c550, 517
LUCA (last universal common ancestor), 20
Luciferase, 696
Lumen, 289, 343, 353
Lycopene
  biosynthesis, 224
  cyclase, 218
  synthesis, 221
Lyngbya, 673
Lyngbya sp.
  8106, 520
  PCC8106, 472
  PCC8106, 674
Lyngbya wollei, 528, 536

M

M. chthonoplates 7420, 520
MacMunn, C.A., 189
Maintenance energy, 43, 44
Manganese catalase, 36, 160, 161, 168
Manganese cluster, 577
Marcus, R.A., 17, 544
Marine cyanobacteria, 147
Mastigocladus laminosus, 37, 49, 573, 576, 585, 587, 596, 597
Mat-building cyanobacteria, 268
Mayer, R.J., 337
Mehler reaction, 228
Membrane biogenesis, 233
Membrane principle, 17
Memory phenomena, 129
Menaquinol, 478, 501
Menaquinone, 48, 221
Mereschkowsky, C., 7
Mesosomes, 94
Metabolic hydrogen, 42
Metabolism, 19
Metallo-β-lactamase, 632
Methanobacterium spp., 41
Methanococcus maripaludis, 140
Methanogenesis, 271
Methanogenic archaea, 271, 632
Methanogens, 34
Methanosarcina acetivorans, 139
Methanosarcina barkeri 27, 139
Methanothermobacter marburgensis, 643
Methylotrophs, 23, 33, 34
Mg-ADP release, 250
Microbial carbon and sulfur cycles, 271
Microbial S reduction, 270
Microcoleus chthonplastes, 403
  7420, 520
Microcoleus chthonplastes sp.PCC7420, 472
Microcystis aeruginosa, 169, 178, 534, 686
  NIES-843, 470
Microcystis, 470
Microfossils, 266, 267, 279
Microplasmodesmata, 140
Midpoint potential, 296, 299, 301, 347, 500, 501
Miller, 11
MIT9215, 518
Mitchell equation, 28
Mitchell, Peter, 17, 28
Mitochondria, 43, 89, 92, 102, 189, 338, 340, 366, 378, 475
  Matrix, 338, 475
  Membrane, 474
Mitosomes, 99
Mixed-valence copper, 52
Mn_4O_xCa cluster, 31, 302, 354, 357, 424, 428, 431
  Ligands, 302
Mn accumulation, 426
Mn chaperon, 429
Mn cluster, 37
"Mn only" model, 360
Mn R/S system, 427
Mn sensor, 427
Mn transporter (MntABC), 425
Mob genes, 688
Mo-bearing sulfide minerals, 270
MoFe-cofactor, 138
MoFe-protein, 138, 139
Molecular oxygen, 13, 159
Mo-nitrogenase, 138, 145
Monod equation, 127
Monodehydroascorbate reductase, 160

Monogalactosyldiacylglycerol (MGDG), 308
Monophyletic clade, 280
Monophyletic nature of mitochondria, 93
Monophyletic origin of life, 19
Moorella thermoacetica, 637, 643
MQ-4, 224
MQ biosynthetic pathway, 221
Mutation rates, 98
Mycobacterium tuberculosis, 164, 168
Myxothiazol, 585
2-Methylhopanes, 271

N

N_2-fixation, 657, 672
NaCl, 132
NAD(P)H, 48, 160
  dehydrogenase (NDH-1), 142, 197, 228
  dehydrogenase activity, 456, 461
  flavin reductases, 633
NAD(P)H:ferredoxin oxidoreductase, 143, 148
NAD(P)H-PQ reductase, 197
NAD^+-reducing hydrogenase, 146
NADH, 48
  dehydrogenases, 47
  oxidoreductase, 637
NADH:FlRd oxidoreductase, 633, 646
NADH:rubredoxin oxidoreductase NRO, 631, 632
$NADP^+$ reductase (FNR), 386, 576
Natromonas pharaonis, 474
ncRNAs, 691
NDH-1, 47, 48, 96, 580
NDH-1 complexes, 457, 461
NDH-1 M, 197
NDH-1 subunits, 447, 453
NDH-1L, NDH-1MS, NDH-1S, 197, 452, 453, 455, 460
NDH-2, 47, 48, 96
NDH subunits, 446, 450, 452
Negentropy, 21
Neisseria gonorrhoeae, 687
nifD, 141, 275
nifDK, 140, 671
NiFe active site, 144
nifH, 138, 140
Ni-hydrogenases, 144
nirA, 695
Nitrate respiration, 40, 42
Nitric oxide, 631
Nitrogen metabolism, 222
Nitrogenase, 47, 56, 137, 222, 671, 674
  protein-1, 138
  protein-2, 138
  reductase, 138
  substrates, 138
Nitrogen fixing organisms, 137
Nodularia spumigema CCY9414, 519, 520
Nodularia spumigena, 401, 676
  CCY9414, 518
Nonadiabatic electron transfer (NET), 352
Non-coding nucleotide Sequences, 98
Non-equilibrium thermodynamics, 116
Nonheme iron (NHFe), 305, 345
Non-heterocystous filamentous cyanobacterium, 672
Non-heterocystous species, 142
Non-oxygenic photosynthesis, 612
Nonphotochemical quenching (NPQ), 396
Nostacales, 51, 58, 178, 279, 472, 665
Nostoc, 549, 553, 620
  8009, 48
  MAC, 51
  muscorum, 152
  PCC 7120, 582, 583
  PCC7119, 667
  PCC7120, 169, 658, 674, 675
Nostoc sp.
  PCC 7119, 561, 563, 622
  PCC 7120, 160, 576, 597, 691, 695–697
  strain ATCC 29133, 697
Nostoc punctiformae, 142, 160, 169, 405, 582, 676
  sp. ATCC29133, 472, 485
  29133/73102, 519
  PCC73102, 162
n-side, 28
NtcA, 141, 146, 674, 675
NTF2 domain, 408
NTG-treatment, 153
Nuclear genomes, 102
2-*n*-nonyl-4-hydroxyquinoline-*N*-oxide (NQNO), 593

O

O_2 back pressure effect, 362
O_2 reduction, 341, 365
O_2-cycle, 30
Obligate anaerobes, 32
Obligate photoautotroph, 222
OEC, 30, 32, 37, 234, *see also* WOC
Oligomycin-sensitivity-conferring protein (OSCP), 247
Oligotrophic lakes, 109
OmcF, 529
Ontological nothingness, 14

Oparin, A. 11
Open reading frames (ORFs), 403, 408
Orange carotenoid protein (OCP), 396, 399
 structure, 408
 photoconversion, 410, 417
 red form, 411
 orange form, 410
 his-tagged, 411
Organellar genomes, 102
Organelles, 89
Organotrophy, 33
Origin of life, 9
Origin of phototrophy, 271
oriT region, 688
Oscillatoria, 472, 533
Oscillatoria sp., 92
Oscillatoriales, 56, 472, 534, 662, 672
Osmotically inert structures, 110
O stateCOX, 373
Outer membrane, 199
Oxic biosphere, 40
Oxidative pentose phosphate pathway, 148
Oxidative phosphorylation, 26, 28, 94, 579
Oxidative water splitting, 349
Oxidized iron formations, 269
Oxo-ferryl, 373
Oxoglutarate, 148
Oxygen evolution, 30, 223, 342
Oxygenic photosynthesis, 29, 30, 89, 93, 131, 266, 270, 613
Oxygenic photosynthetic apparatus (PSA), 272
Oxygenic respiration, 211
Oxygen-radical damage, 31
Oxyphototrophs, 47
Ozone layer, 13, 340
"Oxo-radical" model, 360
(2S, 2′S)-oscillol 2, 2′-di(α-lfucoside), 226

P

$^{1}P^{*}_{RC}$, 346
$^{1}P680^{*}$, 299, 301, 343
$^{3}P680$, 301
$^{3}P_{RC}$, 346
$P^{+\bullet}_{RC}$, 346
P/O value, 58
P680, 301, 346
$P680^{+\bullet}$, 301, 306343, 346
P700, 317
$P700^{+}$, 448
PacS, 617, 618
 Paenibacillus macerans, 474
Paracoccus (P.) denitrificans, 7, 49, 51, 367, 371, 374, 376, 474, 659
Parmenides, 3
PAS superfamily, 414
Pasteur point, 613
Pasteur, Louis, 5
Paulinella chromatophora, 636
PBRC(s), 289, 292, 345
PBS, *see* phycobilisome
PC, 49, 52, 96, 227, 616, 618, 621, 623
PC/Cyt c_6, 312
Pcb protein, 286
P-cluster, 138
pD1 protein, 429, 437
P_{D1}/P_{D2}, 301, 346
Pentameric enzyme, 146
Pentose phosphate pathway, 222
Periplasm, 427
Peroxidase-catalase superfamily, 171
Peroxidases, 160
Peroxiredoxins (Prx), 36, 160, 172, 176
Peroxiredoxin Q (Prx Q), 179
PET, 36, 46, 54, 55, 669
petA, 216
petA gene, 576
petB, 216
petC, 216, 576
 petC1, 582, 583
 petC2, 582, 583
 petC3, 582, 583
 petC4, 583
petD, 216, 576
petE, 38, 52, 218, 623, 667, 671, 695
petE gene, 516
petG, 216, 581, 584
petJ, 38, 52, 218, 623, 667, 671, 694
petJ gene, 516
petL, 573, 581, 584
petM, 573, 581, 584
petN, 216, 581, 584
petO, 576, 584
petP, 573, 576, 584
Petroselinum crispum, 547
PEWY Sequence, 591
PFOs, 143
Phanerozoic Eon, 270
Phaseolus vulgaris, 547
$Pheo_{D1}$, 299, 346
Pheophytin, 37, 299
Pheophytin-quinone type II RCs, 273
PH-gradient, 112
Phormidium, 673
Phormidium laminosum, 546–549, 553, 557, 563, 622

Phormidium luminosum, 204
Phosphate flow, 116
Phosphate group transfer potentials, 17
Phosphate pulses, 120
Phosphate, 109
Phosphatedeficient *Anacystis nidulans*, 114
Phosphoglycerol (PG), 308, 324
Phosphorylation, 21, 46
Photoactivation, 430
Photoactive pigment P680, 288, 301
Photoactive yellow protein (PYP), 414
Photoautotrophic conditions, 265
Photoautotrophs, 337
  facultative, 25
  obligate, 25
Photochemical redox reactions, 31
Photodamage/Photoinhibition, 292, 301, 304, 323, 397
Photohetrotrophs, 337
Photoperiod, 672
Photophosphorylation, 26, 28, 11, 579
Photoprotection, 323, 418
Photorespiration, 54
Photosynthesis, 195, 211, 621
Photosynthetic
  ancestor, 273
  bacteria, 234
  electron transfer, 193
  electron transport, *see* PET
  electron transport chain, 657
  reaction centers, 32
Photosystem I, *see* PSI
Photosystem II, *see* PSII
Phototrophic archaeon, 46
Phototrophy, 21
Phototropin(s), 414, 416
Phycobilins, 272
Phycobiliproteins, 221, 226, 232, 272
Phycobilisome (PBS), 36, 226, 265, 286, 397
Phycochromaceae, 265
Phycocyanin (PC), 36, 226, 397
Phycoerythrin (PE), 36, 232, 397
Phycoerythrocyanin (PEC), 287, 397
Phylloquinol (Vitamin K), 478
Phylloquinone, 22, 318, 478
Phylogenetic relationships, 98
Phylogenetic trees, 99
Phytoene desaturase, 218
Picea sitchensis, 636
Piericidin A, 47
pilA, 687
pilA1, 687
pilB1, 687
Pilbarra Craton, 270
pilM, 687
pilM-O, 687
pilN, 687
pilO, 687
Plant-type photosynthesis, 159
  Plasma membrane (PM), 199, 202, 426, 434, 457
Plastocyanin (PetE), 38, 199, 204, 224, 228, 310, 515–518, 541, 545–549, 553, 554, 557, 561, 562, 564, 565, 574, 607, 657, 665, 676
Plastoquinol, 56, 657, 664
Plastoquinol oxidase, 201
Plastoquinone (PQ), 37, 48, 96, 199, 574
Plastoquinone pool, 145
Plectonema boryanum, 142, 523, 673
Plexaura homomalla, 162
  P_M state, 373
PMF, 226, 246
P_{nirA}, 694
Poisson-Boltzmann method, 559
Polyphosphate pool, 110
Populus nigra var. *italica*, 546
Porphyra, 177
P_{petE}, 694
P_{petJ}, 694
PQ, 577
PQ pool, 230, 286, 300, 397, 448
PQH_2/PQ exchange, 300, 308, 363
P_R state, 373
P_{RC}, 346
Prebiotic phase, 276
Precambriam rocks, 279
Primary producers, 265, 266
Primary production, 30
Primordial cyanobacterium, 31
Primordial RC, 277
  erimordial RC, 277cte
Prochlorales, 473
Prochlorococcus, 178, 275, 469, 474, 485, 535, 664, 669
Prochlorococcus marinus, 169, 276, 518, 639
  sp. CCMP1375, 473
  sp.MIT9303, 473
  sp.MIT9313, 473
  subsp. Pastoris str. CCMP1986, 516
Prochlorophytes, 30, 397, 657
Prochlorothrix, 549
Prochlorothrix hollandica, 547, 549
Procyanobacteria, 278, 279
Production of H_2-gas, 138
Prokaryotic cell, 5

Promethean fire of evolution, 31
Propiogenium modestum, 245
Proteobacteria, 89, 272
Proteomics, 453
Proteorhodopsin, 232
Protocell, 14
Protochloromitochondrion, 7
Protochlorophyll, 276
Protochlorophyllide oxidoreductase, 218
Protoeukaryotes, 90
Proto-Mitochondria, 94
Proton channel, 240
Proton electrochemical gradient, 28
Proton loading site (PLS), 374
Proton Motive Force (PMF), 239
Proton pump, 232, 365
Proton pumping, 662
Proton transfer (PT), 341
Proton transfer pathways, 664
Protontranslocating ATPases, 55
Protozoa, 30
PS I, 40, 132, 151, 197, 199, 223, 224, 265, 272, 285, 309, 396, 515, 545, 564, 607
  trimer, 309
  PSI/NDH-1 supercomplex, 459
  PSI-ferredoxin complex, 316
  PSI/NDH-1 supercomplex, 459
PS II, 37, 159, 199, 223, 225, 228, 265, 272, 277, 285, 287, 342, 343, 396, 574, 577, *see* Photosystem II
  assembly, 434
  complex, 427
  core antenna, 287, 294, 304
  core, 286
PSA, 275
PsaA, 216, 224, 292, 310, 315
PsaB, 216, 224, 233, 292, 310, 315
PsaC, 216, 224, 312, 315, 319
PsaD, 216, 224, 312, 315, 319
PsaE, 151, 216, 224, 315, 319
PsaF, 216, 224, 310, 314
PsaI, 310, 312, 313
PsaJ, 310, 314
PsaK, 310
PsaL, 216, 224, 310, 312, 313, 314
Psalteriomonas lanterna, 43, 92
PsaM, 216, 224, 310, 312, 313
PsaX, 310, 314, 315
PsaZ, 224
Psb27, 429
PsbA, 225
PsbB, 216, 225
PsbC, 216, 225
PsbD, 216, 225
PsbE, 216, 294, 296, 307, 308
PsbF, 216, 294, 296, 307, 308
PsbH, 216, 294, 288, 297
PsbI, 297
PsbJ, 294, 296, 297, 307
PsbK, 216, 294, 297, 307
PsbL, 216, 298, 307
PsbM, 216, 298, 307
PsbN, 216, 276, 297
PsbO, 216, 225, 234, 288, 298, 343, 428
PsbP, 216, 428
PsbQ, 428
PsbS, 396
PsbT, 216, 288, 298, 307
PsbU, 216, 225, 234, 288, 298, 345, 428
PsbV (Cyt c_{550}), 225, 234, 288, 298, 345, 428, 517
PsbV2, 517, 534, 535
PsbX, 294, 308
PsbY, 294, 296, 308
PsbZ, 288, 294
Pseudoanabaena (*Limnothrix*), 673
Pseudomonas
  aeruginosa, 687
  ocalaticus, 34
  stutzeri, 40
PSI core antenna, 307, 311, 320
PSI reduction, 625
p-side, 28
P-type ATPases, 55
Purple bacteria, 234
pVZ vectors, 689
pVZ321 vector, 690
Pyrite, 269
Pyruvate phosphoroclastic reaction, 143
Pyruvate:ferredoxin oxidoreductase, 143, 149, 632
α-proteobacteria, 90, 97, 102, 276
β-Proteobacteria, 276
γ-Proteobacteria, 276

Q

$Q_{A,}$ 286, 299, 346, 429
$Q_{B,}$ 286, 299, 308, 363, 429
$Q_{C,}$ 300, 308, 363
Q-cycle, 38
qE, 396, 414
qE_{cya}, 398, 408, 411, 413
Q_i pocket, 578
Q_i quinone-binding site, 592
QOX, 52, 664
qoxA, 669
Quantum requirement, 195

Quantum yield, 194
Quinol oxidase, 51, 581, 658
Quinol pool, 198
Quinol:fumarate reductases (QFR), 474, 478
Quinone/quinol binding sites, 576, 585

R

R. sphaeroides, 659
Radiation damage, 31, 303, 359, *see also* X-ray damage
Radical ion pairs PSII, 347
Ralstonia eutropha, 146, 151
Rat liver mitochondria, 241
RC 1, 277
RC 2, 276, 277
RC, 234
rDNA, 213
Reaction center (RC) domain, 306
Reaction coordinate WOC, 358
Reactive oxygen species (ROS), 35, 159, 342, 357, 374, 429
"Red" Chls, 323
Redox
  complex, 546
  couple O_2/H_2O, 340
  disproportionation, 32
  potentials, 23, 614, *see also* mid point potential
  reactions, 22
  regulation, 253, 254, 581
Regulatory mechanisms, 250
Repair cycle PSII, 306, 308
Respiration, 11, 29, 365, 399, 621
  aerobic, 475
  anaerobic, 475
Respiration-early hypothesis, 40, 93
Respiratory chain/transport, 32, 193, 365, 445, 448, 469, 657
Respiratory complex, 201, 366, 378
Respiratory electron transport (RET), 36, 43, 44, 46, 54, 55
Respiratory protection, 56
Reverse electron transport, 24
Reverse evolution, 40
Rhizobium, 145
Rhizoselenia, 142
Rhodobacter (*Rb.*), 51, 564, 197
  capsulatus, 140, 274, 423
  sphaeroides, 51, 371, 375, 564
Rhodopseudomonas palustris, 139, 275
Rhodospirillum rubrum, 140
Richelia intracellularis, 142
Rickettsiae, 8
Rickettsiales, 43, 102
Rieske 2Fe–2S protein, 573, 576, 578, 581, 588
Rieske iron-sulfur protein, 37
Rieske isoforms, 582
RNase, 556
rnpA, 695
rnpB, 695
Rotary catalysis, 242
Rotary molecular machine, 241
Rotenone, 47
Rotor, 241
RP-4, 689
rps12, 692, 693
rps12-R43, 693
RRL, 135
rRNA genes, 273
RSF1010, 689
R state, 373
Rubisco, 449
Rubredoxin:oxygen oxidoreductase, 641
Ruminants, 32
Ruminococcus spp., 41, 42
16S r-DNA phylogeny, 137

S

S12 ribosomal protein subunit, 693
sacB, 691
Saccharomyces cerevesiae, 245, 252
Saline pratensis, 546
Salmonella, 632
Salt adapted cells, 135
Saprospira sp., 92
Sartre, J.-P., 14
Schiff base, 411
Schizomycetes, 265
SDH (succinate dehydrogenase), 48, 96
Second-stalk-complex, 247
Self-replicating plasmid, 694
Self-shadowing, 153
Sequence diversification, 98
S_i states, 351, 356
S_{-i} states, 351, 432
Singlet oxygen, 342, 349, 429
slr0197, 688
slr0388, 688
Small integral subunits PSII, 294, 297
Spinach chloroplast, 241
  thylakoids, 252
Spirulina platensis, 254
Spirulina sp., 92
Stanier, R. Y., 29
State transition(s), 397

Stator, 241
Stator stalk, 248
Stigmatellin, 585
Stigonematales, 279
Stored phosphate, 119
Stress, 669
    conditions, 55
    defence, 43
Stroma, 286
Stromatolites, 268, 271
Substrate-level phosphorylation (SLP), 21, 33, 34
Subunit b, 248
Subunit(s)
    I COX, 367
    II COX, 369
    III COX, 367, 369
    IV COX, 370
    IV, 581
    γ, 243, 247
    δ, 248, 252
    ε, 243
    ε, 251, 252
    D1 (PsbA), D2 (PsbD), 292
    L, M, 292
Succinate dehydrogenase (SDH), 228, 580
Succinate:quinone oxidoreductase (SQOR), 469, 474
    Subunit A, 474, 478
    Subunit B, 474, 487, 492
    Subunit C, 474, 497
    Subunit D, 474
    Subunit E, 492
Succinate:quinone reductase (SQR), 474
    SQR flavoproteins, 478
Sulfate, 269
Sulfide, 613
Sulfolobus acidocaldarius, 476
Sulfoquinovosyl diacylglycerol (SQDG), 221, 308, 596
Sulfur-sulfur cycle, 29
Sulphenic acid, 176
Superoxide dismutases, 36
Superoxide, 192
Sus scrofa, 477
Symbiosis, 266
Symploca, 673
Synechococcus, 30, 474, 518, 522, 529, 534, 657
    6301, 48
Synechococcus el. BP-1, 169
Synechococcus PCC
    6301, 43, 48, 55
    7002, 164, 180, 529, 669
    7002, 582, 585, 587, 591, 594, 595
    7942, 151, 160, 164
Synechococcus elongatus (*Anacystis nidulans*), 245, 407, 427, 457, 664, 686
    PCC 6301, 164
    PCC 7942, 162, 176, 178, 179, 686, 696
    sp. PCC 7942, 407, 471
    sp. PCC 6301, 407, 470, 477
Synechococcus sp.
    7002, 516
    BG 043511, 145
    BL107, 471, 497
    PCC 7002, 450, 471, 497
    PCC 7942, 149
    PCC 9605, 407
    RS 9917, 526
    WH5701, 403, 471
    WH8102, 471, 458
Synechococcus sp.A, 407
Synechococcus sp.B, 407
Synechococcus sp.JA-2-3-B'a(2-13), 471, 497
Synechococcus sp. PCC
    7002 a, 597
    7002, 686
    7335, 520
    7942, 547
    7942, 516
Synechococcus sp. strain PCC 7942, 695
Synechococcus vulcanus, 658
Synechocystis, 620
Synechocystis 6803, 48
Synechocystis PCC 6803, 51, 52, 55, 132, 135, 147, 149, 164, 168, 176, 177, 179, 248, 254, 531, 533, 582, 583, 591, 658, 664, 667, 669, 671, 675, 685
Synechocystis sp., 274
    PCC 6714, 434
    PCC 6803 *petC1* strain, 581
    PCC 6803, 373, 397, 423, 426, 429, 433, 447, 450, 471
    PCC 6803, 216, 515, 516, 547, 563, 564, 585, 597, 639, 643, 646, 686, 690, 695
Syntrophic community, 29, 41, 42

T

Tamers of O_2, 36
TCA cycle, 222
Teleological, 19
Temporal segregation, 56, 673
Terrestrial biosphere, 29
Tetranuclear Mn4OxCa cluster, 31
Tfp, 687
Thermoplasma acidophilum, 474

Thermosynechococcus, 216, 667
Thermosynechococcus (*T.*) *elongatus*, 39, 248, 249, 252, 254, 288, 298, 343, 347, 407, 452, 517, 534, 535, 582, 664, 695
Thermosynechococcus elongatus BP-1, 39, 686
Thermosynechococcus elongatus BP-1 The. elo, 471
Thermosynechococcus (*T.*) *vulcanus*, 288, 349, 359, 517, 534
Thermotoga maritima, 643
Thermus (*T.*) *thermophilus*, 31, 169, 202, 371, 377, 445, 562, 671
Thiocapsa roseopersicina, 147
Thiol peroxidases, 161
Thiol-specific peroxidases, 172
Thioredoxin, 146, 176, 254
Thiothrix sp., 92
Thylakoid
- centres, 94
- lumen, 291, 344, 545, 617
- membrane, 93, 199, 131, 202, 213, 229, 338, 426, 434, 457, 580, 669

Thylakosomes, 30, 43, 92
Time-resolved fluorescence, 227
Tn5, 696
Transduction, 686
Transformation, 686, 688
Transmembrane helices, 289, 292, 294, 297, 310, 312, 343, 357, 368, 376
Transposon mutagenesis, 696
Treshold value, 115
Tricarboxylic acid cycle, 148
Trichodesmium erythraeum, 178, 275, 520, 534, 672
Trichodesmium, 56, 142, 662, 672
Trichomonas vaginalis, 632, 633, 637, 645
Trichromatic vision, 189
Tridecylstigmatellin, 585
Triple dichotomy, 33, 44, 47
TRO, 44
Trp-Tyr-Met adduct, 167
Trychodesmium erythraeum sp. IMS101, 472
Tryparedoxin, 176
Type 1 RCs, 276
Type 2 RCs, 276
Type II Peroxiredoxin, 178
Type IV pili (Tfp), 687
Type-1 blue copper protein, 618
Typical (monofunctional) catalase, 160, 162
Tyr radical, 356, 374

U
Ubiquinone, 48, 96, 199, 476
Ulva pertusa, 547
Unified/generalized endosymbiont hypothesis, 43
Unitary principle of biochemistry, 5
Universal dogma of life, 5
Uptake experiments, 114
Uptake hydrogenase, 47, 138, 141, 142, 144
Uptake rate, 117
Urey, H.C., 11

V
van Niel, C.B., 195
Vanadium peroxidases, 161
Vanadium-containing peroxidases, 173
Vegetative cells, 141
Vibrio harveyi, 696
Violaxanthin, 396
Virchov, R., 7
Vitreoscilla sp., 92
vnfDGKEN cluster, 142
vnfH, 140
vnfHDGK, 139
V-nitrogenase, 139, 140
V-type ATPases, 246

W
Wächtershäuser, G., 11
Warburg effect, 54
Warburg, O., 193
Warrawoona sediments, 267
Water oxidation, 234
Water oxidizing complex (WOC), 30, 37, 225, 302, 346, 350, 363, 434, 430
Water-plastoquinone:oxido-reductase, 348
Weber-Fechner's law, 125
Willstatter, R., 191
WOC, 30, 37, 234, 277, 354
Wollinella (*W.*) *succinogenes*, 474
Wollinella (*W.*) *succinogenes* sp.DSM1740, 474
"Working" potential, 347

X
XANES, 356
XisA, 141
XisC, 146
X ray damage, 303, *see also* radiation damage
X ray diffraction crystallography (XRDC), 343, 354
Xanthophyll cycle, 396

Y

Y_Z^{OX}, 353, 362
Ycf12, 294
Ycf37, 695
Y_D, 364
Yeast mitochondria, 241
Y_z, 346, 350, 352

Z

Zeaxanthin, 396, 413
Zinc β-lactamases, 641
Z scheme, 193

MIX
Papier aus verantwortungsvollen Quellen
Paper from responsible sources
FSC® C105338

If you have any concerns about our products,
you can contact us on
ProductSafety@springernature.com

In case Publisher is established outside the EU,
the EU authorized representative is:
Springer Nature Customer Service Center GmbH
Europaplatz 3, 69115 Heidelberg, Germany

Printed by Libri Plureos GmbH
in Hamburg, Germany